An Introduction to Dynamical Systems

Continuous and Discrete, Second Edition

Pure and Applied
UNDERGRADUATE TEXTS · 19

An Introduction to Dynamical Systems

Continuous and Discrete, Second Edition

R. Clark Robinson

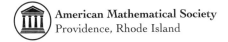

American Mathematical Society
Providence, Rhode Island

EDITORIAL COMMITTEE

Paul J. Sally, Jr. (Chair) Joseph Silverman
Francis Su Susan Tolman

2010 *Mathematics Subject Classification.* Primary 34Cxx, 37Cxx, 37Dxx, 37Exx, 37Nxx, 70Kxx.

This book was previously published by: Pearson Education, Inc.

For additional information and updates on this book, visit
www.ams.org/bookpages/amstext-19

Library of Congress Cataloging-in-Publication Data
Robinson, R. Clark (Rex Clark), 1943–
    An introduction to dynamical systems : continuous and discrete / R. Clark Robinson. – Second edition.
        pages cm. – (Pure and applied undergraduate texts ; volume 19)
    Includes bibliographical references and index.
    ISBN 978-0-8218-9135-3 (alk. paper)
  1. Differentiable dynamical systems. 2. Nonlinear theories. Chaotic behavior in systems. I. Title.

QA614.8.R65 2012
515′.39–dc23
                                                                                                2012025520

Copying and reprinting. Individual readers of this publication, and nonprofit libraries acting for them, are permitted to make fair use of the material, such as to copy a chapter for use in teaching or research. Permission is granted to quote brief passages from this publication in reviews, provided the customary acknowledgment of the source is given.

Republication, systematic copying, or multiple reproduction of any material in this publication is permitted only under license from the American Mathematical Society. Requests for such permission should be addressed to the Acquisitions Department, American Mathematical Society, 201 Charles Street, Providence, Rhode Island 02904-2294 USA. Requests can also be made by e-mail to reprint-permission@ams.org.

© 2012 by the American Mathematical Society. All rights reserved.
The American Mathematical Society retains all rights
except those granted to the United States Government.
Printed in the United States of America.

∞ The paper used in this book is acid-free and falls within the guidelines
established to ensure permanence and durability.
Visit the AMS home page at http://www.ams.org/
            10 9 8 7 6 5 4 3 2 1   17 16 15 14 13 12

How many are your works, O Lord!
In wisdom you made them all;
the earth is full of your creatures.
–Psalm 104:24

Contents

Preface	xiii
Historical Prologue	xvii

Part 1. Systems of Nonlinear Differential Equations

Chapter 1. Geometric Approach to Differential Equations	3
Chapter 2. Linear Systems	11
2.1. Fundamental Set of Solutions	13
Exercises 2.1	19
2.2. Constant Coefficients: Solutions and Phase Portraits	21
Exercises 2.2	48
2.3. Nonhomogeneous Systems: Time-dependent Forcing	49
Exercises 2.3	52
2.4. Applications	52
Exercises 2.4	56
2.5. Theory and Proofs	59
Chapter 3. The Flow: Solutions of Nonlinear Equations	75
3.1. Solutions of Nonlinear Equations	75
Exercises 3.1	83
3.2. Numerical Solutions of Differential Equations	84
Exercises 3.2	96
3.3. Theory and Proofs	97
Chapter 4. Phase Portraits with Emphasis on Fixed Points	109
4.1. Limit Sets	109

Exercises 4.1	114
4.2. Stability of Fixed Points	114
Exercises 4.2	119
4.3. Scalar Equations	119
Exercises 4.3	124
4.4. Two Dimensions and Nullclines	126
Exercises 4.4	133
4.5. Linearized Stability of Fixed Points	134
Exercises 4.5	143
4.6. Competitive Populations	145
Exercises 4.6	150
4.7. Applications	152
Exercises 4.7	158
4.8. Theory and Proofs	159
Chapter 5. Phase Portraits Using Scalar Functions	169
5.1. Predator–Prey Systems	169
Exercises 5.1	172
5.2. Undamped Forces	173
Exercises 5.2	182
5.3. Lyapunov Functions for Damped Systems	183
Exercises 5.3	190
5.4. Bounding Functions	191
Exercises 5.4	195
5.5. Gradient Systems	195
Exercises 5.5	199
5.6. Applications	199
Exercises 5.6	210
5.7. Theory and Proofs	210
Chapter 6. Periodic Orbits	213
6.1. Introduction to Periodic Orbits	214
Exercises 6.1	218
6.2. Poincaré–Bendixson Theorem	219
Exercises 6.2	226
6.3. Self-Excited Oscillator	229
Exercises 6.3	232
6.4. Andronov–Hopf Bifurcation	232
Exercises 6.4	240
6.5. Homoclinic Bifurcation	242

Exercises 6.5	246
6.6. Rate of Change of Volume	247
Exercises 6.6	249
6.7. Poincaré Map	251
Exercises 6.7	261
6.8. Applications	262
Exercises 6.8	271
6.9. Theory and Proofs	272
Chapter 7. Chaotic Attractors	**285**
7.1. Attractors	285
Exercises 7.1	289
7.2. Chaotic Attractors	291
Exercise 7.2	296
7.3. Lorenz System	297
Exercises 7.3	312
7.4. Rössler Attractor	313
Exercises 7.4	316
7.5. Forced Oscillator	317
Exercises 7.5	319
7.6. Lyapunov Exponents	320
Exercises 7.6	328
7.7. Test for Chaotic Attractors	329
Exercises 7.7	331
7.8. Applications	331
7.9. Theory and Proofs	336

Part 2. Iteration of Functions

Chapter 8. Iteration of Functions as Dynamics	**343**
8.1. One-Dimensional Maps	343
8.2. Functions with Several Variables	349
Chapter 9. Periodic Points of One-Dimensional Maps	**353**
9.1. Periodic Points	353
Exercises 9.1	362
9.2. Iteration Using the Graph	362
Exercises 9.2	366
9.3. Stability of Periodic Points	367
Exercises 9.3	382
9.4. Critical Points and Basins	386

Exercises 9.4	390
9.5. Bifurcation of Periodic Points	391
Exercises 9.5	404
9.6. Conjugacy	406
Exercises 9.6	411
9.7. Applications	412
Exercises 9.7	416
9.8. Theory and Proofs	417
Chapter 10. Itineraries for One-Dimensional Maps	**423**
10.1. Periodic Points from Transition Graphs	424
Exercises 10.1	435
10.2. Topological Transitivity	437
Exercises 10.2	441
10.3. Sequences of Symbols	442
Exercises 10.3	451
10.4. Sensitive Dependence on Initial Conditions	451
Exercises 10.4	454
10.5. Cantor Sets	455
Exercises 10.5	463
10.6. Piecewise Expanding Maps and Subshifts	464
Exercises 10.6	473
10.7. Applications	475
Exercises 10.7	478
10.8. Theory and Proofs	479
Chapter 11. Invariant Sets for One-Dimensional Maps	**487**
11.1. Limit Sets	487
Exercises 11.1	490
11.2. Chaotic Attractors	490
Exercises 11.2	505
11.3. Lyapunov Exponents	507
Exercises 11.3	513
11.4. Invariant Measures	514
Exercises 11.4	533
11.5. Applications	534
11.6. Theory and Proofs	537
Chapter 12. Periodic Points of Higher Dimensional Maps	**541**
12.1. Dynamics of Linear Maps	541

Exercises 12.1	555
12.2. Classification of Periodic Points	555
Exercises 12.2	566
12.3. Stable Manifolds	567
Exercises 12.3	575
12.4. Hyperbolic Toral Automorphisms	575
Exercises 12.4	580
12.5. Applications	580
Exercises 12.5	594
12.6. Theory and Proofs	595
Chapter 13. Invariant Sets for Higher Dimensional Maps	597
13.1. Geometric Horseshoe	598
Exercises 13.1	611
13.2. Symbolic Dynamics	612
Exercises 13.2	632
13.3. Homoclinic Points and Horseshoes	636
Exercises 13.3	639
13.4. Attractors	639
Exercises 13.4	649
13.5. Lyapunov Exponents	650
Exercises 13.5	661
13.6. Applications	662
13.7. Theory and Proofs	664
Chapter 14. Fractals	669
14.1. Box Dimension	670
Exercises 14.1	679
14.2. Dimension of Orbits	680
Exercises 14.2	684
14.3. Iterated-Function Systems	684
Exercises 14.3	696
14.4. Theory and Proofs	697
Appendix A. Background and Terminology	705
A.1. Calculus Background and Notation	705
A.2. Analysis and Topology Terminology	707
A.3. Matrix Algebra	713
Appendix B. Generic Properties	717
Bibliography	721
Index	727

Preface

Preface to the Second Edition

In the second edition of this book, much of the material has been rewritten to clarify the presentation. It also has provided the opportunity for correcting many minor typographical errors or mistakes. Also, the definition of a chaotic attractor has been changed to include the requirement that the chaotic attractor is transitive. This is the usual definition and it eliminates some attractors that should not be called chaotic. Several new applications are included for systems of differential equations in Part 1. I would encourage readers to email me with suggestions and further corrections that are needed.

R. Clark Robinson
March 2012

Preface to the First Edition

This book is intended for an advanced undergraduate course in dynamical systems or nonlinear ordinary differential equations. There are portions that could be beneficially used for introductory master level courses. The goal is a treatment that gives examples and methods of calculation, at the same time introducing the mathematical concepts involved. Depending on the selection of material covered, an instructor could teach a course from this book that is either strictly an introduction into the concepts, that covers both the concepts on applications, or that is a more theoretically mathematical introduction to dynamical systems. Further elaboration of the variety of uses is presented in the subsequent discussion of the organization of the book.

The assumption is that the student has taken courses on calculus covering both single variable and multivariables, a course on linear algebra, and an introductory course on differential equations. From the multivariable calculus, the material on partial derivatives is used extensively, and in a few places multiple integrals and surface integrals are used. (See Appendix A.1.) Eigenvalues and eigenvectors are the main concepts used from linear algebra, but further topics are listed in Appendix A.3. The material from the standard introductory course on differential equations is used only in Part 1; we assume that students can solve first-order equations by separation of variables, and that they know the form of solutions from second-order scalar equations. Students who have taken an introductory course on differential equations are usually familiar with linear systems with constant coefficients (at least the real-eigenvalue case), but this material is repeated in Chapter 2, where we also introduce the reader to the phase portrait. At Northwestern, some students have taken the course covering part one on differential equations without this introductory course on differential equations; they have been able to understand the new material when they have been willing to do the extra work in a few areas that is required to fill in the missing background. Finally, we have not assumed that the student has had a course on real analysis or advanced calculus. However, it is convenient to use some of the terminology from such a course, so we include an appendix with terminology on continuity and topology. (See Appendix A.)

Organization

This book presents an introduction to the concepts of dynamical systems. It is divided into two parts, which can be treated in either order: The first part treats various aspects of systems of nonlinear ordinary differential equations, and the second part treats those aspects dealing with iteration of a function. Each separate part can be used for a one-quarter course, a one-semester course, a two-quarter course, or possibly even a year course. At Northwestern University, we have courses that spend one quarter on the first part and two quarters on the second part. In a one-quarter course on differential equations, it is difficult to cover the material on chaotic attractors, even skipping many of the applications and proofs at the end of the chapters. A one-semester course on differential equations could also cover selected topics on iteration of functions from Chapters 9–11. In the course on discrete dynamical systems using Part 2, we cover most of the material on iteration of one-dimensional functions (Chapters 9–11) in one quarter. The material on iteration of functions in higher dimensions (Chapters 12–13) certainly depends on the one-dimensional material, but a one-semester course could mix in some of the higher dimensional examples with the treatment of Chapters 9–11. Finally, Chapter 14 on fractals could be treated after Chapter 12. Fractal dimensions could be integrated into the material on chaotic attractors at the end of a course on differential equations. The material on fractal dimensions or iterative function systems could be treated with a course on iteration of one-dimensional functions.

The main concepts are presented in the first sections of each chapter. These sections are followed by a section that presents some applications and then by a section that contains proofs of the more difficult results and more theoretical material. The division of material between these types of sections is somewhat arbitrary. The theorems proved at the end of the chapter are restated with their original theorem number. The material on competitive populations and predator–prey systems is contained in one of the beginning sections of the chapters in which these topics are covered, rather than in the applications at the end of the chapters, because these topics serve to develop the main techniques presented. Also, some proofs are contained in the main sections when they are more computational and serve to make the concepts clearer. Longer and more technical proofs and further theoretical discussion are presented separately at the end of the chapter.

A course that covers the material from the primary sections, without covering the sections at the end of the chapter on applications and more theoretical material, results in a course on the concepts of dynamical systems with some motivation from applications.

The applications provide motivation and illustrate the usefulness of the concepts. None of the material from the sections on applications is necessary for treating the main sections of later chapters. Treating more of this material would result in a more applied emphasis.

Separating the harder proofs allows the instructor to determine the level of theory of the course taught using this book as the text. A more theoretic course could consider most of the proofs at the end of the chapters.

Computer Programs

This book does not explicitly cover aspects of computer programming. However, a few selected problems require computer simulations to produce phase portraits of differential equations or to iterate functions. Sample Maple worksheets, which the students can modify to help with some of the more computational problems, will be available on the webpage:

http://www.math.northwestern.edu/~clark/dyn-sys.

(Other material on corrections and updates of the book will also be available at this website.) There are several books available that treat dynamical systems in the context of Maple or Mathematica: two such books are [**Kul02**] by M. Kulenović and [**Lyn01**] by S. Lynch. The book [**Pol04**] by J. Polking and D. Arnold discusses using Matlab to solve differential equations using packages available at http://math.rice.edu/~dfield. The book [**Nus98**] by H. Nusse and J. Yorke comes with its own specialized dynamical systems package.

Acknowledgments

I would like to acknowledge some of the other books I have used to teach this material, since they have influenced my understanding of the material, especially with regard to effective ways to present material. I will not attempt to list more advanced books which have also affected my understanding. For the material on differential equations, I have used the following books: F. Brauer and J. Nohel [**Bra69**], M. Hirsch and S. Smale [**Hir74**], M. Braun [**Bra73**], I. Percival and D. Richards [**Per82**], D.W. Jordan and P. Smith [**Jor87**], J. Hale and H. Koçak [**Hal91**], and S. Strogatz [**Str94**]. For the material on iteration of functions, I have used the following books: the two books by R. Devaney [**Dev89**] and [**Dev92**], D. Gulick [**Gul92**], and K. Alligood, T. Sauer, and J. Yorke [**All97**].

I would also like to thank three professors under whom I studied while a graduate student: Charles Pugh, Morris Hirsch, and Stephen Smale. These people introduced me to the subject of dynamical systems and taught me many of the ideas and methods that I have used throughout my career. Many of my colleagues at Northwestern have also deeply influenced me in different ways: these people include John Franks, Donald Saari, and Robert Williams.

I thank the following reviewers for their comments and useful suggestions for improvement of the manuscript:

John Alongi, Pomona College

Pau Atela, Smith College

Peter Bates, Brigham Young University

Philip Bayly, Washington University

Michael Brin, University of Maryland

Roman Grigoriev, Georgia Technological Institute

Palle Jorgensen, University of Iowa

Randall Pyke, Ryerson University

Joel Robbin, University of Wisconsin

Bjorn Sandstede, Ohio State University

Douglas Shafer, University of North Carolina at Charlotte

Milena Stanislavova, University of Kansas

Franz Tanner, Michigan Technological University

Howard Weiss, Pennsylvania State University

I also thank Miguel Lerma for help in solving various LaTeX and graphics problems, Marian Gidea for help with Adobe Illustrator, and Kamlesh Parwani for help with some Maple worksheets.

I gratefully acknowledge the photograph by Julio Ottino and Paul Swanson used on the cover of the first edition of the book depicting mixing of fluids. This photo had previously appeared in the article [**Ott89b**]. A brief discussion of his research is given in Section 11.5.4.

Most of all, I am grateful to my wife Peggie, who endured, with patience, understanding, and prayer, the ups and downs of writing this book from inception to conclusion.

R. Clark Robinson
clark@math.northwestern.edu
October 2003

Historical Prologue

The theory of differential equations has a long history, beginning with Isaac Newton. From the early Greeks through Copernicus, Kepler, and Galileo, the motions of planets had been described directly in terms of their properties or characteristics, for example, that they moved on approximately elliptical paths (or in combinations of circular motions of different periods and amplitudes). Instead of this approach, Newton described the laws that determine the motion in terms of the forces acting on the planets. The effect of these forces can be expressed by differential equations. The basic law he discovered was that the motion is determined by the gravitational attraction between the bodies, which is proportional to the product of the two masses of the bodies and one over the square of the distance between the bodies. The motion of one planet around a sun obeying these laws can then be shown to lie on an ellipse. The attraction of the other planets could then explain the deviation of the motion of the planet from the elliptic orbit. This program was continued by Euler, Lagrange, Laplace, Legendre, Poisson, Hamilton, Jacobi, Liouville, and others.

By the end of the nineteenth century, researchers realized that many nonlinear equations did not have explicit solutions. Even the case of three masses moving under the laws of Newtonian attraction could exhibit very complicated behavior and an explicit solution was not possible (e.g., the motion of the sun, earth, and moon cannot be given explicitly in terms of known functions). Short term solutions could be given by power series, but these were not useful in determining long-term behavior. Poincaré, working from 1880 to 1910, shifted the focus from finding explicit solutions to discovering geometric properties of solutions. He introduced many of the ideas in specific examples, which we now group together under the heading of chaotic dynamical systems. In particular, he realized that a deterministic system (in which the outside forces are not varying and are not random) can exhibit behavior that is apparently random (i.e., it is chaotic).

In 1898, Hadamard produced a specific example of geodesics on a surface of constant negative curvature which had this property of chaos. G. D. Birkhoff continued the work of Poincaré and found many different types of long-term limiting behavior, including the α- and ω-limit sets introduced in Sections 4.1 and 11.1. His work resulted in the book [**Bir27**] from which the term "dynamical systems" comes.

During the first half of the twentieth century, much work was carried out on nonlinear oscillators, that is, equations modeling a collection of springs (or other physical forces such as electrical forces) for which the restoring force depends non-linearly on the displacement from equilibrium. The stability of fixed points was studied by several people including Lyapunov. (See Sections 4.5 and 5.3.) The existence of a periodic orbit for certain self-excited systems was discovered by Van der Pol. (See Section 6.3.) Andronov and Pontryagin showed that a system of differential equations was structurally stable near an attracting fixed point, [**And37**] (i.e., the solutions for a small perturbation of the differential equation could be matched with the solutions for the original equations). Other people carried out research on nonlinear differential equations, including Bendixson, Cartwright, Bogoliubov, Krylov, Littlewood, Levinson, and Lefschetz. The types of solutions that could be analyzed were the ones which settled down to either (1) an equilibrium state (no motion), (2) periodic motion (such as the first approximations of the motion of the planets), or (3) quasiperiodic solutions which are combinations of several periodic terms with incommensurate frequencies. See Section 2.2.4. By 1950, Cartwright, Littlewood, and Levinson showed that a certain forced nonlinear oscillator had infinitely many different periods; that is, there were infinitely many different initial conditions for the same system of equations, each of which resulted in periodic motion in which the period was a multiple of the forcing frequency, but different initial conditions had different periods. This example contained a type of complexity not previously seen.

In the 1960s, Stephen Smale returned to using the topological and geometric perspective initiated by Poincaré to understand the properties of differential equations. He wrote a very influential survey article [**Sma67**] in 1967. In particular, Smale's "horseshoe" put the results of Cartwright, Littlewood, and Levinson in a general framework and extended their results to show that they were what was later called chaotic. A group of mathematicians worked in the United States and Europe to flesh out his ideas. At the same time, there was a group of mathematicians in Moscow lead by Anosov and Sinai investigating similar ideas. (Anosov generalized the work of Hadamard to geodesics on negatively curved manifolds with variable curvature.) The word "chaos" itself was introduced by T.Y. Li and J. Yorke in 1975 to designate systems that have aperiodic behavior more complicated than equilibrium, periodic, or quasiperiodic motion. (See [**Li,75**].) A related concept introduced by Ruelle and Takens was a *strange attractor*. It emphasized more the complicated geometry or topology of the attractor in phase space, than the complicated nature of the motion itself. See [**Rue71**]. The theoretical work by these mathematicians supplied many of the ideas and approaches that were later used in more applied situations in physics, celestial mechanics, chemistry, biology, and other fields.

The application of these ideas to physical systems really never stopped. One of these applications, which has been studied since earliest times, is the description and determination of the motion of the planets and stars. The study of the mathematical model for such motion is called *celestial mechanics*, and involves a finite number of bodies moving under the effects of gravitational attraction given by the Newtonian laws. Birkhoff, Siegel, Kolmogorov, Arnold, Moser, Herman, and many others investigated the ideas of stability and found complicated behavior for systems arising in celestial mechanics and other such physical systems, which could be described by what are called *Hamiltonian differential equations*. (These equations preserve energy and can be expressed in terms of partial derivatives of the energy function.) K. Sitnikov in [**Sit60**] introduced a situation in which three masses interacting by Newtonian attraction can exhibit chaotic oscillations. Later, Alekseev showed that this could be understood in terms of a "Smale horseshoe", [**Ale68a**], [**Ale68b**], and [**Ale69**]. The book by Moser, [**Mos73**], made this result available to many researchers and did much to further the applications of horseshoes to other physical situations. In the 1971 paper [**Rue71**] introducing strange attractors, Ruelle and Takens indicated how the ideas in nonlinear dynamics could be used to explain how turbulence developed in fluid flow. Further connections were made to physics, including the periodic doubling route to chaos discovered by Feigenbaum, [**Fei78**], and independently by P. Coullet and C. Tresser, [**Cou78**].

Relating to a completely different physical situation, starting with the work of Belousov and Zhabotinsky in the 1950s, certain mathematical models of chemical reactions that exhibit chaotic behavior were discovered. They discovered some systems of differential equations that not only did not tend to an equilibrium, but also did not even exhibit predictable oscillations. Eventually, this bizarre situation was understood in terms of chaos and strange attractors.

In the early 1920s, A.J. Lotka and V. Volterra independently showed how differential equations could be used to model the interaction of two populations of species, [**Lot25**] and [**Vol31**]. In the early 1970s, May showed how chaotic outcomes could arise in population dynamics. In the monograph [**May75**], he showed how simple nonlinear models could provide "mathematical metaphors for broad classes of phenomena." Starting in the 1970s, applications of nonlinear dynamics to mathematical models in biology have become widespread. The undergraduate books by Murray [**Mur89**] and Taubes [**Tau01**] afford good introductions to biological situations in which both oscillatory and chaotic differential equations arise. The books by Kaplan and Glass [**Kap95**] and Strogatz [**Str94**] include a large number of other applications.

Another phenomenon that has had a great impact on the study of nonlinear differential equations is the use of computers to find numerical solutions. There has certainly been much work done on deriving the most efficient algorithms for carrying out this study. Although we do discuss some of the simplest of these, our focus is more on the use of computer simulations to find the properties of solutions. E. Lorenz made an important contribution in 1963 when he used a computer to study nonlinear equations motivated by the turbulence of motion of the atmosphere. He discovered that a small change in initial conditions leads to very different outcomes in a relatively short time; this property is called *sensitive dependence on initial conditions* or, in more common language, the *butterfly effect*. Lorenz used the latter term because he interpreted the phenomenon to mean that a butterfly flapping its wings in Australia today could affect the weather in the United States a month later. We describe more of his work in Chapter 7. It was not until the 1970s that Lorenz's work became known to the more theoretical mathematical community. Since that time, much effort has gone into showing that Lorenz's basic ideas about these equations were correct. Recently, Warwick Tucker has shown, using a computer-assisted proof, that this system not only has sensitive dependence on initial conditions, but also has what is called a "chaotic attractor". (See Chapter 7.) About the same time as Lorenz, Ueda discovered that a periodically forced Van der Pol system (or other nonlinear oscillator) has what is now called a chaotic attractor. Systems of this type are also discussed in Chapter 7. (For a later publication by Ueda, see also [**Ued92**].)

Starting about 1970 and still continuing, there have been many other numerical studies of nonlinear equations using computers. Some of these studies were introduced as simple examples of certain phenomena. (See the discussion of the Rössler Attractor given in Section 7.4.) Others were models for specific situations in science, engineering, or other fields in which nonlinear differential equations are used for modeling. The book [**Enn97**] by Enns and McGuire presents many computer programs for investigation of nonlinear functions and differential equations that arise in physics and other scientific disciplines.

In sum, the last 40 years of the twentieth century saw the growing importance of nonlinearity in describing physical situations. Many of the ideas initiated by Poincaré a century ago are now much better understood in terms of the mathematics involved and the way in which they can be applied. One of the main contributions of the modern theory of dynamical systems to these applied fields has been the idea that erratic and complicated behavior can result from simple situations. Just because the outcome is chaotic, the basic environment does not need to contain stochastic or random perturbations. The simple forces themselves can cause chaotic outcomes.

There are three books of a nontechnical nature that discuss the history of the development of "chaos theory": the best seller *Chaos: Making a New Science* by James Gleick [**Gle87**], *Does God Play Dice?, The Mathematics of Chaos* by Ian Stewart [**Ste89**], and *Celestial Encounters* by Florin Diacu and Philip Holmes [**Dia96**]. Stewart's book puts a greater emphasis on the role of mathematicians in the development of the subject, while Gleick's book stresses the work of researchers making the connections with applications. Thus, the perspective of Stewart's book is closer to the one of this book, but Gleick's book is accessible to a broader audience and is more popular. The book by Diacu and Holmes has a good treatment of Poincaré's contribution and the developments in celestial mechanics up to today.

Part 1

Systems of Nonlinear Differential Equations

Chapter 1

Geometric Approach to Differential Equations

In a basic elementary differential equations course, the emphasis is on linear differential equations. One example treated is the linear second-order differential equation

(1.1) $$m\ddot{x} + b\dot{x} + kx = 0,$$

where we write $\dot{x}$ for $\dfrac{dx}{dt}$ and $\ddot{x}$ for $\dfrac{d^2x}{dt^2}$, and $m, k > 0$ and $b \geq 0$. This equation is a model for a linear spring with friction and is also called a *damped harmonic oscillator*.

To determine the future motion, we need to know both the current position x and the current velocity $\dot{x}$. Since these two quantities determine the motion, it is natural to use them for the coordinates; writing $v = \dot{x}$, we can rewrite this equation as a linear system of first-order differential equations (involving only first derivatives), so $m\dot{v} = m\ddot{x} = -kx - bv$:

(1.2) $$\dot{x} = v,$$
$$\dot{v} = -\frac{k}{m}x - \frac{b}{m}v.$$

In matrix notation, this becomes

$$\begin{pmatrix} \dot{x} \\ \dot{v} \end{pmatrix} = \begin{pmatrix} 0 & 1 \\ -\frac{k}{m} & -\frac{b}{m} \end{pmatrix} \begin{pmatrix} x \\ v \end{pmatrix}.$$

For a linear differential equation such as (1.1), the usual solution method is to seek solutions of the form $x(t) = e^{\lambda t}$, for which $\dot{x} = \lambda e^{\lambda t}$ and $\ddot{x} = \lambda^2 e^{\lambda t}$. We need

$$m\lambda^2 e^{\lambda t} + b\lambda e^{\lambda t} + k e^{\lambda t} = 0,$$

or

(1.3)
$$m\lambda^2 + b\lambda + k = 0,$$

which is called the *characteristic equation*.

When $b = 0$, the solution of the characteristic equation is $\lambda^2 = -k/m$ or $\lambda = \pm i\omega$, where $\omega = \sqrt{k/m}$. Since $e^{i\omega t} = \cos(\omega t) + i\sin(\omega t)$, the real and imaginary parts, $\cos(\omega t)$ and $\sin(\omega t)$, are each solutions. Linear combinations are also solutions, so the general solution is

$$x(t) = A\cos(\omega t) + B\sin(\omega t),$$
$$v(t) = -\omega A\sin(\omega t) + \omega B\cos(\omega t),$$

where A and B are arbitrary constants. These solutions are both periodic, with the same period $2\pi/\omega$. See Figure 1.

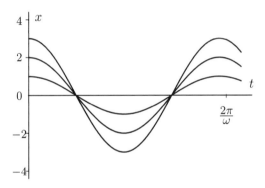

Figure 1. Solutions for the linear harmonic oscillator: x as a function of t

Another way to understand the solutions when $b = 0$ is to find the energy that is preserved by this system. Multiplying the equation $\dot{v} + \omega^2 x = 0$ by $v = \dot{x}$, we get

$$v\dot{v} + \omega^2 x\dot{x} = 0.$$

The left-hand side of this equation is the derivative with respect to t of

$$E(x, v) = \frac{1}{2}v^2 + \frac{\omega^2}{2}x^2,$$

so this function $E(x, v)$ is a constant along a solution of the equation. This *integral of motion* shows clearly that the solutions move on ellipses in the (x, v)-plane given by level sets of E. There is a fixed point or equilibrium at $(x, v) = (0, 0)$ and other solutions travel on periodic orbits in the shape of ellipses around the origin. For this linear equation, all the orbits have the same shape and period which is independent of size: We say that the local and global behavior is the same. See Figure 2.

For future reference, a point $\mathbf{x}^*$ is called a *fixed point* of a system of differential equations $\dot{\mathbf{x}} = \mathbf{F}(\mathbf{x})$ provided $\mathbf{F}(\mathbf{x}^*) = \mathbf{0}$. The solution starting at a fixed point has zero velocity, so it just stays there. Therefore, if $\mathbf{x}(t)$ is a solution with $\mathbf{x}(0) = \mathbf{x}^*$, then $\mathbf{x}(t) = \mathbf{x}^*$ for all t. Traditionally, such a point was called an *equilibrium point* because the forces are in balance and the point does not move.

A point $\mathbf{x}^*$ is called *periodic* for a system of differential equations $\dot{\mathbf{x}} = \mathbf{F}(\mathbf{x})$, provided that there is some $T > 0$ such that the solution $\mathbf{x}(t)$ with initial condition $\mathbf{x}(0) = \mathbf{x}^*$ has $\mathbf{x}(T) = \mathbf{x}^*$ but $\mathbf{x}(t) \neq \mathbf{x}^*$ for $0 < t < T$. This value T is called the *period* or *least period*. It follows that $\mathbf{x}(t+T) = \mathbf{x}(t)$ (i.e., it repeats itself after T units of time). The set of points $\{\mathbf{x}(t) : 0 \leq t \leq T\}$ is called the *periodic orbit*.

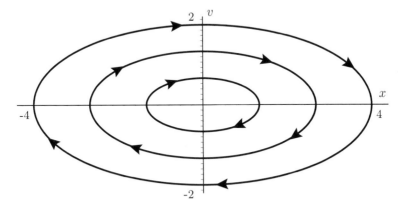

Figure 2. Solutions for the linear harmonic oscillator in (x, v)-space.

The set of curves in the (x, v)-plane determined by the solutions of the system of differential equations is an example of a phase portrait, which we use throughout this book. These portraits are a graphical (or geometric) way of understanding the solutions of the differential equations. Especially for nonlinear equations, for which we are often unable to obtain analytic solutions, the use of graphical representations of solutions is important. Besides determining the phase portrait for nonlinear equations by an energy function such as the preceding, we sometimes use geometric ideas such as the nullclines introduced in Section 4.4. Other times, we use numerical methods to draw the phase portraits.

Next, we consider the case for $b > 0$. The solutions of the characteristic equation (1.3) is
$$\lambda = \frac{-b \pm \sqrt{b^2 - 4km}}{2m} = -c \pm i\mu,$$
where
$$c = \frac{b}{2m} \quad \text{and} \quad \mu = \sqrt{\frac{k}{m} - c^2}.$$
(These roots are also the eigenvalues of the matrix of the system of differential equations.) The general solution of the system of equations is
$$x(t) = e^{-ct}\left[A\cos(\mu t) + B\sin(\mu t)\right],$$
$$v(t) = e^{-ct}\left[-(A\mu + Bc)\sin(\mu t) + (B\mu - Ac)\cos(\mu t)\right],$$
where A and B are arbitrary constants.

Another way to understand the properties of the solutions in this case is to use the "energy function",
$$E(x, v) = \frac{1}{2}v^2 + \frac{\omega^2}{2}x^2,$$

for $\omega = \sqrt{k/m}$, which is preserved when $b = 0$. For the system with $b > 0$ and so $c > 0$,

$$\frac{d}{dt} E(x, v) = v \dot{v} + \omega^2 x \dot{x}$$
$$= v(-2cv - \omega^2 x) + \omega^2 xv$$
$$= -2cv^2 \leq 0.$$

This shows that the energy is nonincreasing. A simple argument, which we give in the section on Lyapunov functions, shows that all of the solutions go to the fixed point at the origin. The use of this real-valued function $E(x, v)$ is a way to show that the origin is *attracting* without using the explicit representation of the solutions.

The system of equations (1.2) are linear, and most of the equations we consider are nonlinear. A simple nonlinear example is given by the pendulum

$$mL\ddot{\theta} = -mg\sin(\theta).$$

Setting $x = \theta$ and $v = \dot{\theta}$, we get the system

(1.4)
$$\dot{x} = v,$$
$$\dot{v} = -\frac{g}{L}\sin(x).$$

It is difficult to get explicit solutions for this nonlinear system of equations. However, the "energy method" just discussed can be used to find important properties of the solutions. By a derivation similar to the foregoing, we see that

$$E(x, v) = \frac{v^2}{2} + 1 - \frac{g}{L}\cos(x)$$

is constant along solutions. Thus, as in the linear case, the solution moves on a level set of E, so the level sets determine the path of the solution. The level sets are given in Figure 3 without justification, but we shall return to this example in Section 5.2 and give more details. See Example 5.3.

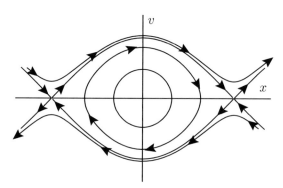

Figure 3. Level sets of energy for the pendulum

There are fixed points at $(x, v) = (0, 0), (\pm\pi, 0)$. The solutions near the origin are periodic, but those farther away have x either monotonically increasing or decreasing. We can use the level curves and trajectories in the phase space (the (x, v)-space) to get information about the solutions.

Even this simple nonlinear equation illustrates some of the differences between linear and nonlinear equations. For the linear harmonic oscillator with $b = 0$, the local behavior near the fixed point determines the behavior for all scales; for the pendulum, there are periodic orbits near the origin and nonperiodic orbits farther away. Second, for a linear system with periodic orbits, all the periods are the same; the period varies for the pendulum. The plots of the three solutions which are time periodic are given in Figure 4. Notice that the period changes with the amplitude. Finally, in Section 2.2 we give an algorithm for solving systems of linear differential equations with constant coefficients. On the other hand, there is no simple way to obtain the solutions of the pendulum equation; the energy method gives geometric information about the solutions, but not explicit solutions.

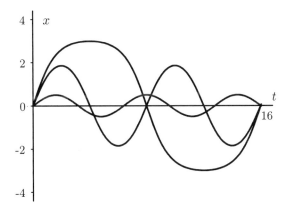

Figure 4. Time plots of solutions with different amplitudes for the pendulum showing the variation of the period

Besides linear systems and the pendulum equation, we consider a few other types of systems of nonlinear differential equation including two species of populations, which either compete or are in a predator–prey relationship. See Sections 4.6 and 5.1. Finally, the Van der Pol oscillator has a unique periodic orbit with other solutions converging to this periodic orbit. See Figure 5 and Section 6.3. Basically, nonlinear differential equations with two variables cannot be much more complicated than these examples. See the discussion of the Poincaré–Bendixson Theorem 6.1.

In three and more dimensions, there can be more complicated systems with apparently "chaotic" behavior. Such motion is neither periodic nor quasiperiodic, appears to be random, but is determined by explicit nonlinear equations. An example is the Lorenz system of differential equations

(1.5)
$$\dot{x} = -10\,x + 10\,y,$$
$$\dot{y} = 28\,x - y - xz,$$
$$\dot{z} = -\frac{8}{3}z + xy.$$

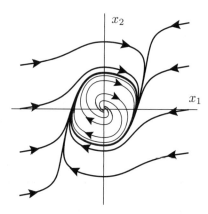

Figure 5. Phase portrait of the Van der Pol oscillator

Trajectories originating from nearby initial conditions move apart. Such systems are said to have *sensitive dependence on initial conditions.* See Figure 6 where two solutions are plotted, each starting near the origin. As these solutions are followed for a longer time, they fill up a set that is a fuzzy surface within three dimensions. (The set is like a whole set of sheets stacked very close together.) Even though the equations are deterministic, the outcome is apparently random or chaotic. This type of equation is discussed in Chapter 7. The possibility of chaos for nonlinear equations is an even more fundamental difference from linear equations.

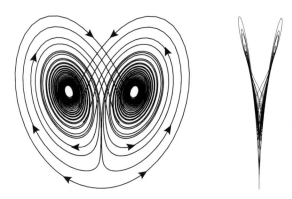

Figure 6. Two views of the Lorenz attractor

Finally, we highlight the differences that can occur for time plots of solutions of differential equations. Figure 7(a) shows an orbit that tends to a constant value (i.e., to a fixed point). Figure 7(b) shows a periodic orbit that repeats itself after the time increases by a fixed amount. Figure 7(c) has what is called a quasiperiodic orbit; it is generated by adding together functions of two different periods so it never exactly repeats. (See Section 2.2.4 for more details.) Figure 7(d) contains a chaotic orbit; Section 7.2 contains a precise definition, but notice that there is no obvious regularity to the length of time of the different types of oscillation.

1. Geometric Approach to Differential Equations

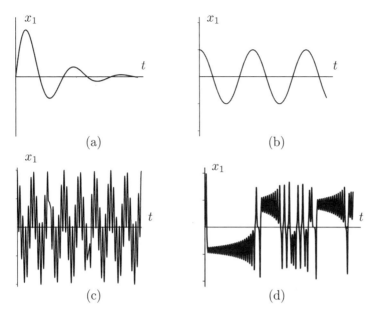

Figure 7. Plots of one position variable as a function of t: (a) orbits tending to a fixed point, (b) a periodic orbit, (c) a quasiperiodic orbit, and (d) a chaotic orbit.

Chapter 2

Linear Systems

This book is oriented toward the study of nonlinear systems of differential equations. However, aspects of the theory of linear systems are needed at various places to analyze nonlinear equations. Therefore, this chapter begins our study by presenting the solution method and properties for a general n-dimensional linear system of differential equations. We quickly present some of the material considered in a beginning differential equation course for systems with two variables while generalizing the situation to n variables and introducing the phase portrait and other new aspects of the qualitative approach.

Remember that throughout the book we write $\dot{x}$ for $\dfrac{dx}{dt}$ and $\ddot{x}$ for $\dfrac{d^2x}{dt^2}$.

In Chapter 1 we mentioned that a linear second-order differential equation of the form

$$m\ddot{x} + kx + b\dot{x} = 0$$

can be written as a linear system of first-order differential equations; setting $x_1 = x$ and $x_2 = \dot{x}$,

$$\dot{x}_1 = x_2,$$
$$\dot{x}_2 = -\frac{k}{m}x_1 - \frac{b}{m}x_2,$$

or in matrix notation

$$\dot{\mathbf{x}} = \begin{pmatrix} 0 & 1 \\ -\frac{k}{m} & -\frac{b}{m} \end{pmatrix} \mathbf{x},$$

where $\mathbf{x} = \begin{pmatrix} x_1 \\ x_2 \end{pmatrix}$ is a vector.

In this chapter, we study general systems of linear differential equations of this type with n variables with constant coefficients,

$$\dot{x}_1 = a_{1,1}\, x_1 + a_{1,2}\, x_2 + \cdots + a_{1,n}\, x_n,$$
$$\dot{x}_2 = a_{2,1}\, x_1 + a_{2,2}\, x_2 + \cdots + a_{2,n}\, x_n,$$
$$\vdots \qquad \vdots$$
$$\dot{x}_n = a_{n,1}\, x_1 + a_{n,2}\, x_2 + \cdots + a_{n,n}\, x_n,$$

where all the $a_{i,j}$ are constant real numbers. This system can be rewritten using vector and matrix notation. Let $\mathbf{A}$ be the $n \times n$ matrix with given constant real entries $a_{i,j}$, and $\mathbf{x}$ is the (column) vector in $\mathbb{R}^n$ of the variables,

$$\mathbf{x} = (x_1, \ldots, x_n)^\mathsf{T} = \begin{pmatrix} x_1 \\ \vdots \\ x_n \end{pmatrix}.$$

Here, $(x_1, \ldots, x_n)^\mathsf{T}$ is the transpose of the row vector which yields a column vector. The system of linear differential equations can be written as

(2.1) $$\dot{\mathbf{x}} = \mathbf{A}\mathbf{x}.$$

At times, we study linear systems that depend on time, so the coefficients $a_{i,j}(t)$ and entries in the matrix $\mathbf{A}(t)$ are functions of time t. This results in a time-dependent linear system

(2.2) $$\dot{\mathbf{x}} = \mathbf{A}(t)\, \mathbf{x}.$$

This type of equation could arise if the spring constants could somehow be controlled externally to give an equation of the form

$$\ddot{x} + k(t)\, x = 0,$$

where $k(t)$ is a known function of t. In this book, these equations mainly occur by "linearizing" along a solution of a nonlinear equation. See the First Variation Equation, Theorem 3.4.

As discussed in Chapter 1, equation (1.1) has the two linearly independent solutions $e^{-ct}\cos(\mu t)$ and $e^{-ct}\sin(\mu t)$, where $c = b/2m$ and $\mu^2 = k/m - c^2$. All solutions are linear combinations of these two solutions. In Section 2.1, we show that for linear systems with n-variables, we need to find n independent solutions. We also check that we can take linear combinations of these solutions and get all other solutions, so they form a "fundamental set of solutions." In Section 2.2, we see how to find explicit solutions for linear systems in which the coefficients do not depend on time, i.e., *linear systems with constant coefficients*. We also introduce the phase portrait of a system of differential equations. Section 2.2.4 contains examples with four variables in which the solutions have sines and cosines with two different frequencies and not just one, i.e., the solution is *quasiperiodic*. The motion of such a quasiperiodic solution is complicated but does not appear random. Therefore, when we define *chaotic sets* for nonlinear systems in Section 7.2, we require that there is an orbit that winds throughout the set and is neither periodic nor quasiperiodic. Finally, Section 2.3 considers linear equations in which there is a "forcing vector" besides the linear term. Nonhomogeneous systems are used several times later in the book: (1) to show that the linear terms dominate the higher order terms near a fixed point (Section 4.8), (2) to give an example of a Poincaré map (Section 6.7), and (3) to give an example of a calculation of a Lyapunov exponent (Section 7.6).

2.1. Fundamental Set of Solutions

In the next section, we give methods of finding solutions for linear equations with constant coefficients. This section focuses on showing that we need n independent solutions for a system with n variables. We also show that a linear combination of solutions is again a solution. Finally, we show how the exponential of a matrix can be used to find an analytical, but not very computable, solution of any system with constant coefficients. This exponential is used in the next section to motivate the method of finding solutions of systems with repeated eigenvalues.

We give the definitions for the time-dependent case, but there the definitions of the time-independent case given by equation (2.1) are essentially the same.

Solution

A *solution* of equation (2.2) is a function $\mathbf{x}(t)$ of time on some open interval of time where $\mathbf{A}(t)$ is defined such that

$$\frac{d}{dt}\mathbf{x}(t) = \mathbf{A}(t)\,\mathbf{x}(t).$$

A solution *satisfies the initial condition* $\mathbf{x}_0$, provided $\mathbf{x}(0) = \mathbf{x}_0$.

Existence of solutions

In Section 2.2, we construct solutions for constant coefficient systems of linear equations. Together with the discussion of fundamental matrix solutions, this discussion shows that, given an initial condition $\mathbf{x}_0$, there is a solution $\mathbf{x}(t)$ satisfying the differential equation with $\mathbf{x}(0) = \mathbf{x}_0$. For time-dependent linear systems, there is no such constructive proof of existence; however, the result certainly follows from Theorem 3.2 for nonlinear systems of differential equations. There are also easier proofs of existence that apply to just such time-dependent linear systems, but we do not discuss them.

Uniqueness of solutions

The uniqueness of solutions follows from Theorem 3.2 for systems of nonlinear differential equations. However, we can give a much more elementary proof in this case, so we state the result here and give its proof in Section 2.5.

Theorem 2.1. *Let $\mathbf{A}(t)$ be an $n \times n$ matrix whose entries depend continuously on t. Given $\mathbf{x}_0$ in $\mathbb{R}$, there is at most one solution $\mathbf{x}(t)$ of $\dot{\mathbf{x}} = \mathbf{A}(t)\mathbf{x}$ with $\mathbf{x}(0) = \mathbf{x}_0$.*

Linear combinations of solutions

For $j = 1, \ldots, k$, assume that $\mathbf{x}^j(t)$ are solutions of equation (2.2) (defined on a common time interval) and c_j are real or complex scalars. Using the properties of differentiation and matrix multiplication,

$$\frac{d}{dt}\bigl(c_1\mathbf{x}^1(t)+\cdots+c_k\mathbf{x}^k(t)\bigr) = c_1\dot{\mathbf{x}}^1(t)+\cdots+c_k\dot{\mathbf{x}}^k(t)$$
$$= c_1\mathbf{A}(t)\mathbf{x}^1(t)+\cdots+c_k\mathbf{A}(t)\mathbf{x}^k(t)$$
$$= \mathbf{A}(t)\bigl(c_1\mathbf{x}^1(t)+\cdots+c_k\mathbf{x}^k(t)\bigr),$$

so the linear combination $c_1\mathbf{x}^1(t)+\cdots+c_k\mathbf{x}^k(t)$ is also a solution. Thus, a linear combination of solutions is a solution.

Matrix solutions

To explain why it is enough to find n solutions for a linear system with n variables, we introduce matrix solutions and then the Wronskian. If $\mathbf{x}^j(t)$ for $1 \le j \le n$ are n-solutions of equations (2.1) or (2.2), we can combine them into a single matrix by putting each solution as a column,

$$\mathbf{M}(t) = \bigl(\mathbf{x}^1(t),\ldots,\mathbf{x}^n(t)\bigr).$$

By properties of differentiation and matrix multiplication,

$$\frac{d}{dt}\mathbf{M}(t) = \bigl(\dot{\mathbf{x}}^1(t),\ldots,\dot{\mathbf{x}}^n(t)\bigr)$$
$$= \bigl(\mathbf{A}(t)\mathbf{x}^1(t),\ldots,\mathbf{A}(t)\mathbf{x}^n(t)\bigr)$$
$$= \mathbf{A}(t)\bigl(\mathbf{x}^1(t),\ldots,\mathbf{x}^n(t)\bigr)$$
$$= \mathbf{A}(t)\mathbf{M}(t).$$

Thus, $\mathbf{M}(t)$ is a matrix that satisfies the differential equation.

Any $n\times k$ matrix $\mathbf{M}(t)$ that satisfies

$$\frac{d}{dt}\mathbf{M}(t) = \mathbf{A}(t)\mathbf{M}(t)$$

is called a *matrix solution*. If $\mathbf{M}(t)$ is a matrix solution with columns $\{\mathbf{x}^j(t)\}_{j=1}^{k}$ and we write $\mathbf{c}$ for the constant vector with entries c_j, then

$$\frac{d}{dt}\mathbf{M}(t)\mathbf{c} = \mathbf{A}(t)\mathbf{M}(t)\mathbf{c},$$

and so $\mathbf{M}(t)\mathbf{c} = c_1\mathbf{x}^1(t)+\cdots+c_k\mathbf{x}^k(t)$ is the vector solution that satisfies

$$\mathbf{M}(0)\mathbf{c} = c_1\mathbf{x}^1(0)+\cdots+c_k\mathbf{x}^k(0)$$

when $t=0$. In particular, if we let $\mathbf{c} = \mathbf{u}^j$ be the vector with a 1 in the jth coordinate and 0 in the other coordinates, then we see that $\mathbf{M}(t)\mathbf{u}^j = \mathbf{x}^j(t)$, which is the jth column of $\mathbf{M}(t)$, is a vector solution. The next theorem summarizes the preceding discussion.

Theorem 2.2. *Assume that $\mathbf{x}^j(t)$ are solutions of equation (2.2) for $j=1,\ldots,k$.*

  a. *If c_j are real or complex scalars for $j=1,\ldots,k$, then $\sum_{j=1}^{k} c_j\mathbf{x}^j(t)$ is a (vector) solution.*

  b. *The matrix $\mathbf{M}(t) = \bigl(\mathbf{x}^1(t),\ldots,\mathbf{x}^k(t)\bigr)$ is a matrix solution.*

  c. *If $\mathbf{c} = (c_1,\ldots,c_k)^{\mathsf{T}}$ is a constant vector and $\mathbf{M}(t)$ is an $n\times k$ matrix solution, then $\mathbf{M}(t)\mathbf{c}$ is a (vector) solution.*

d. *If $\mathbf{M}(t)$ is a matrix solution of equation (2.2), then the columns of $\mathbf{M}(t)$ are vector solutions.*

Fundamental set of solutions

A set of vectors $\left\{\mathbf{v}^j\right\}_{j=1}^n$ is *linearly independent* provided that, whenever a linear combination of these vectors gives the zero vector,

$$c_1\mathbf{v}^1 + \cdots + c_n\mathbf{v}^n = \mathbf{0},$$

then all the $c_j = 0$. If $\left\{\mathbf{v}^j\right\}_{j=1}^n$ is a set of n vectors in $\mathbb{R}^n$, then they are linearly independent if and only if

$$\det\left(\mathbf{v}^1, \ldots, \mathbf{v}^n\right) \neq 0.$$

If $\left\{\mathbf{v}^j\right\}_{j=1}^n$ is a set of n linearly independent vectors in $\mathbb{R}^n$, then for any $\mathbf{x}^0$ in $\mathbb{R}^n$ there exist $y_1, \ldots, y_n$ such that

$$\mathbf{x}^0 = y_1\mathbf{v}^1 + \cdots + y_n\mathbf{v}^n.$$

For this reason, a linearly independent set of vectors $\left\{\mathbf{v}^j\right\}_{j=1}^n$ is called a *basis of* $\mathbb{R}^n$.

A set of solutions $\{\mathbf{x}^1(t), \ldots, \mathbf{x}^n(t)\}$ is *linearly independent* provided that, whenever

$$c_1\mathbf{x}^1(t) + \cdots + c_n\mathbf{x}^n(t) = \mathbf{0} \quad \text{for all } t,$$

then all the $c_j = 0$. When a set of n solutions $\{\mathbf{x}^1(t), \ldots, \mathbf{x}^n(t)\}$ is linearly independent, then it is called a *fundamental set of solutions*, and the corresponding matrix solution $\mathbf{M}(t)$ is called a *fundamental matrix solution*.

Just as for vectors, we can relate the condition of being linearly independent to a determinant. If $\mathbf{M}(t)$ is an $n \times n$ matrix solution of equations (2.1) or (2.2), then the determinant

$$W(t) = \det\left(\mathbf{M}(t)\right)$$

is called the *Wronskian* of the system of vector solutions given by the columns of $\mathbf{M}(t)$.

If $\mathbf{M}(t)$ is an $n \times n$ matrix solution with $W(0) = \det(\mathbf{M}(0)) \neq 0$, then for any $\mathbf{x}_0$, we can solve the equation $\mathbf{M}(0)\mathbf{c} = \mathbf{x}_0$ for $\mathbf{c}$ to give a solution $\mathbf{x}(t) = \mathbf{M}(t)\mathbf{c}$ with $\mathbf{x}(0) = \mathbf{x}_0$. Notice that, by linear algebra, we need at least n columns in order to be able to solve $\mathbf{M}(0)\mathbf{c} = \mathbf{x}_0$ for all initial conditions $\mathbf{x}_0$.

The *Liouville formula*, (2.3), given in the next theorem, relates $W(t)$ and $W(t_0)$ at different times. This formula implies that, if $W(t_0) \neq 0$ for any time t_0, then $W(t) \neq 0$ for all times, the solutions are linearly independent, and we can solve for any initial conditions. The proof of the theorem is given at the end of this chapter. An alternative proof is given later in the book using the Divergence Theorem from calculus.

Theorem 2.3 (Liouville Formula). *Let $\mathbf{M}(t)$ be a fundamental matrix solution for a linear system as given in equation (2.1) or equation (2.2), and let $W(t) = \det(\mathbf{M}(t))$*

be the Wronskian. Then,

$$\frac{d}{dt} W(t) = \text{tr}(\mathbf{A}(t)) \, W(t) \quad \text{and}$$

(2.3)
$$W(t) = W(t_0) \exp\left(\int_{t_0}^{t} \text{tr}(\mathbf{A}(s)) \, ds \right),$$

where $\exp(z) = e^z$ is the exponential function. In particular, if $W(t_0) \neq 0$ for any time t_0, then $W(t) \neq 0$ for all times t.

For a constant coefficient equation, if $\mathbf{M}(t)$ is a fundamental matrix solution, the formula becomes

$$\det(\mathbf{M}(t)) = \det(\mathbf{M}(0)) \, e^{\text{tr}(\mathbf{A}) \, t}.$$

Example 2.1. The Euler differential equation with nonconstant coefficients,

$$t^2 \ddot{y} - 2 t \dot{y} + 2 y = 0,$$

can be used to give an example of the previous theorems. This second-order scalar equation has solutions of the form t^r, where r satisfies

$$0 = r(r-1) - 2r + 2 = r^2 - 3r + 2 = (r-1)(r-2).$$

Thus, there are two solutions $y^1(t) = t$ and $y^2(t) = t^2$. This second-order differential equation corresponds to the first-order system of differential equations

$$\dot{x}_1 = x_2,$$
$$\dot{x}_2 = \frac{2}{t^2} x_1 + \frac{2}{t} x_2.$$

The preceding two scalar solutions correspond to the vector solutions

$$\mathbf{x}^1(t) = (t, 1)^{\mathsf{T}} \quad \text{and} \quad \mathbf{x}^2(t) = (t^2, 2t)^{\mathsf{T}},$$

which have Wronskian

$$W(t) = \det \begin{pmatrix} t & t^2 \\ 1 & 2t \end{pmatrix} = t^2.$$

On the other hand, the trace of the matrix of coefficients is $2/t$, so for $t, t_0 > 0$,

$$W(t) = W(t_0) \exp\left(\int_{t_0}^{t} \frac{2}{s} \, ds \right)$$
$$= t_0^2 \exp\left(2 \ln(t) - 2 \ln(t_0)\right)$$
$$= t_0^2 \left(e^{\ln(t^2)} / e^{\ln(t_0^2)} \right) = t_0^2 \left(t^2 / t_0^2 \right)$$
$$= t^2.$$

Thus, the result from the Liouville formula (2.3) agrees with the calculation using the solutions.

Matrix Exponential

For the constant coefficient case, equation (2.1), there is a general way of using the exponential of a matrix to get a matrix solution that equals the identity when t equals 0. This exponential is usually not easy to compute, but is a very useful conceptual solution and notation. Also, we shall use it to derive the form of the solution when the eigenvalues are repeated.

For a scalar equation $\dot{x} = a\,x$, the solution is $x(t) = x_0\, e^{at}$ for an arbitrary constant x_0. For equation (2.1) with constant coefficients, we consider $e^{\mathbf{A}t}$ as a candidate for a matrix solution. This expression involves taking the exponential of a matrix. What should this mean? The exponential is a series in powers, so we write $\mathbf{A}^n$ for the nth power of the matrix. Using the power series for the exponential, but substituting a matrix for the variable, we define

$$e^{\mathbf{A}t} = \mathbf{I} + t\,\mathbf{A} + \frac{t^2}{2!}\mathbf{A}^2 + \cdots + \frac{t^n}{n!}\mathbf{A}^n + \cdots = \sum_{n=0}^{\infty} \frac{t^n}{n!}\mathbf{A}^n.$$

We will not worry about the convergence of this series of matrices, but in fact, it does converge because of the factor of $n!$ in the denominator. Differentiating with respect to t term by term (which can be justified),

$$\frac{d}{dt}e^{\mathbf{A}t} = 0 + \mathbf{A} + \frac{t}{1!}\mathbf{A}^2 + \frac{t^2}{2!}\mathbf{A}^3 + \cdots + \frac{t^{n-1}}{(n-1)!}\mathbf{A}^n + \cdots$$

$$= \mathbf{A}\left(\mathbf{I} + t\,\mathbf{A} + \frac{t^2}{2!}\mathbf{A}^2 + \cdots + \frac{t^{n-1}}{(n-1)!}\mathbf{A}^{n-1} + \cdots\right)$$

$$= \mathbf{A}\left(e^{\mathbf{A}t}\right).$$

Since $e^{\mathbf{A}0} = \mathbf{I}$, $e^{\mathbf{A}t}$ is a fundamental matrix solution that equals the identity when t equals 0. If $\mathbf{v}$ is any vector, then $\mathbf{x}(t) = e^{\mathbf{A}t}\mathbf{v}$ is a solution, with $\mathbf{x}(0) = \mathbf{v}$.

We give a few special cases where it is possible to calculate $e^{\mathbf{A}t}$ directly.

Example 2.2. Consider a diagonal matrix,

$$\mathbf{A} = \begin{pmatrix} a & 0 \\ 0 & b \end{pmatrix}.$$

Then

$$\mathbf{A}^n = \begin{pmatrix} a^n & 0 \\ 0 & b^n \end{pmatrix}$$

and

$$e^{\mathbf{A}t} = \begin{pmatrix} 1 & 0 \\ 0 & 1 \end{pmatrix} + \begin{pmatrix} at & 0 \\ 0 & bt \end{pmatrix} + \frac{1}{2!}\begin{pmatrix} a^2t^2 & 0 \\ 0 & b^2t^2 \end{pmatrix} + \cdots + \frac{1}{n!}\begin{pmatrix} a^n t^n & 0 \\ 0 & b^n t^n \end{pmatrix} + \cdots$$

$$= \begin{pmatrix} e^{at} & 0 \\ 0 & e^{bt} \end{pmatrix}.$$

For any $n \times n$ diagonal matrix, the exponential of the matrix is also the diagonal matrix whose entries are the exponential of the entries on the diagonal. See Exercise 8 in this section.

Example 2.3. Another simple example where the exponential can be calculated explicitly is
$$\mathbf{B} = \begin{pmatrix} 0 & -\omega \\ \omega & 0 \end{pmatrix}.$$
Then,
$$\mathbf{B}^2 = \begin{pmatrix} -\omega^2 & 0 \\ 0 & -\omega^2 \end{pmatrix}, \qquad \mathbf{B}^3 = \begin{pmatrix} 0 & \omega^3 \\ -\omega^3 & 0 \end{pmatrix},$$
$$\mathbf{B}^4 = \begin{pmatrix} \omega^4 & 0 \\ 0 & \omega^4 \end{pmatrix}, \qquad \mathbf{B}^5 = \begin{pmatrix} 0 & -\omega^5 \\ \omega^5 & 0 \end{pmatrix},$$
$$\mathbf{B}^{2n} = \begin{pmatrix} (-1)^n \omega^{2n} & 0 \\ 0 & (-1)^n \omega^{2n} \end{pmatrix},$$
$$\mathbf{B}^{2n+1} = \begin{pmatrix} 0 & (-1)^{n+1} \omega^{2n+1} \\ (-1)^n \omega^{2n+1} & 0 \end{pmatrix}.$$
Writing out the series and combining the even and odd terms, we get
$$e^{\mathbf{B}t} = \begin{pmatrix} 1 & 0 \\ 0 & 1 \end{pmatrix} + \frac{1}{2!}\begin{pmatrix} -\omega^2 t^2 & 0 \\ 0 & -\omega^2 t^2 \end{pmatrix} + \frac{1}{4!}\begin{pmatrix} \omega^4 t^4 & 0 \\ 0 & \omega^4 t^4 \end{pmatrix} + \cdots$$
$$+ \frac{1}{(2n)!}\begin{pmatrix} (-1)^n \omega^{2n} t^{2n} & 0 \\ 0 & (-1)^n \omega^{2n} t^{2n} \end{pmatrix} + \cdots$$
$$+ \begin{pmatrix} 0 & -\omega t \\ \omega t & 0 \end{pmatrix} + \frac{1}{3!}\begin{pmatrix} 0 & \omega^3 t^3 \\ -\omega^3 t^3 & 0 \end{pmatrix} + \frac{1}{5!}\begin{pmatrix} 0 & -\omega^5 t^5 \\ \omega^5 t^5 & 0 \end{pmatrix} + \cdots$$
$$+ \frac{1}{(2n+1)!}\begin{pmatrix} 0 & (-1)^{n+1} \omega^{2n+1} t^{2n+1} \\ (-1)^n \omega^{2n+1} t^{2n+1} & 0 \end{pmatrix} + \cdots$$
$$= \begin{pmatrix} \cos(\omega t) & 0 \\ 0 & \cos(\omega t) \end{pmatrix} + \begin{pmatrix} 0 & -\sin(\omega t) \\ \sin(\omega t) & 0 \end{pmatrix}$$
$$= \begin{pmatrix} \cos(\omega t) & -\sin(\omega t) \\ \sin(\omega t) & \cos(\omega t) \end{pmatrix}.$$

Example 2.4. Let $\mathbf{A} = a\mathbf{I}$ be a 2×2 diagonal matrix with equal entries and $\mathbf{B}$ be as in the preceding example. Since $\mathbf{AB} = \mathbf{BA}$, Theorem 2.13 at the end of this chapter shows that $e^{(\mathbf{A}+\mathbf{B})t} = e^{\mathbf{A}t} e^{\mathbf{B}t}$. (This formula is not true for matrices which do not commute.) Therefore,
$$\exp\begin{pmatrix} at & -\omega t \\ \omega t & at \end{pmatrix} = \exp\begin{pmatrix} at & 0 \\ 0 & at \end{pmatrix} \exp\begin{pmatrix} 0 & -\omega t \\ \omega t & 0 \end{pmatrix}$$
$$= \begin{pmatrix} e^{at} & 0 \\ 0 & e^{at} \end{pmatrix} \begin{pmatrix} \cos(\omega t) & -\sin(\omega t) \\ \sin(\omega t) & \cos(\omega t) \end{pmatrix}$$
$$= e^{at} \begin{pmatrix} \cos(\omega t) & -\sin(\omega t) \\ \sin(\omega t) & \cos(\omega t) \end{pmatrix}.$$

Example 2.5. Consider the matrix $\mathbf{N} = \begin{pmatrix} 0 & 1 \\ 0 & 0 \end{pmatrix}$. Then, $\mathbf{N}^2 = \mathbf{0}$ so $\mathbf{N}^k = \mathbf{0}$ for $k \geq 2$. Therefore,
$$e^{\mathbf{N}t} = \mathbf{I} + \mathbf{N}t + \mathbf{0} + \cdots = \begin{pmatrix} 1 & t \\ 0 & 1 \end{pmatrix}.$$

Example 2.6. Consider the matrix $\mathbf{A} = a\,\mathbf{I} + \mathbf{N} = \begin{pmatrix} a & 1 \\ 0 & a \end{pmatrix}$ with $\mathbf{N}$ as the previous example. Then,

$$e^{\mathbf{A}t} = e^{a\mathbf{I}t} e^{\mathbf{N}t} = e^{at} \begin{pmatrix} 1 & t \\ 0 & 1 \end{pmatrix}$$
$$= \begin{pmatrix} e^{at} & t\,e^{at} \\ 0 & e^{at} \end{pmatrix}.$$

It is only possible to calculate the exponential directly for a matrix in very special "normal form", such as the examples just considered. Therefore, for linear equations with constant coefficients, we give another method to find a fundamental set of solutions in the next section. For such a a fundamental set of solutions, $\{\mathbf{x}^{(j)}(t)\}_{j=1}^{n}$, let $\mathbf{M}(t)$ be the corresponding fundamental matrix solution, and define

$$\widetilde{\mathbf{M}}(t) = \mathbf{M}(t)\mathbf{M}(0)^{-1}.$$

By multiplying by the inverse of $\mathbf{M}(0)$, the initial condition when $t = 0$ becomes the identity, $\widetilde{\mathbf{M}}(0) = \mathbf{M}(0)\mathbf{M}(0)^{-1} = \mathbf{I}$. Because the inverse is on the right, this is still a matrix solution, since

$$\widetilde{\mathbf{M}}'(t) = \mathbf{M}'(t)\,\mathbf{M}(0)^{-1}$$
$$= [\mathbf{A}\mathbf{M}(t)]\,\mathbf{M}(0)^{-1}$$
$$= \mathbf{A}\widetilde{\mathbf{M}}(t).$$

By uniqueness of solutions, $\widetilde{\mathbf{M}}(t) = e^{\mathbf{A}t}$. Thus, it is possible to calculate $e^{\mathbf{A}t}$ by finding any fundamental set of solutions and constructing $\widetilde{\mathbf{M}}(t)$ as in the preceding discussion.

Exercises 2.1

1. Show that the following functions are linearly independent:

   $$e^{t} \begin{pmatrix} 1 \\ 1 \\ 0 \end{pmatrix}, \quad e^{2t} \begin{pmatrix} 0 \\ 2 \\ 1 \end{pmatrix}, \quad \text{and} \quad e^{-t} \begin{pmatrix} 1 \\ 0 \\ 1 \end{pmatrix}.$$

2. Consider the system of equations

   $$\dot{x}_1 = x_2,$$
   $$\dot{x}_2 = -q(t)\,x_1 - p(t)\,x_2,$$

   where $q(t)$ and $p(t)$ are continuous functions on all of $\mathbb{R}$. Find an expression for the Wronskian of a fundamental set of solutions.

3. Assume that $\mathbf{M}(t)$ is a fundamental matrix solution of $\dot{\mathbf{x}} = \mathbf{A}(t)\mathbf{x}$ and $\mathbf{C}$ is a constant matrix with $\det(\mathbf{C}) \neq 0$. Show that $\mathbf{M}(t)\mathbf{C}$ is a fundamental matrix solution.

4. Assume that $\mathbf{M}(t)$ and $\mathbf{N}(t)$ are two fundamental matrix solutions of $\dot{\mathbf{x}} = \mathbf{A}(t)\mathbf{x}$. Let $\mathbf{C} = \mathbf{M}(0)^{-1}\mathbf{N}(0)$. Show that $\mathbf{N}(t) = \mathbf{M}(t)\,\mathbf{C}$.

5. (Hale and Koçak) Alternative proof of Theorem 2.13.
   a. Prove that, if $\mathbf{AB} = \mathbf{BA}$, then $\mathbf{B}e^{t\mathbf{A}} = e^{t\mathbf{A}}\mathbf{B}$. Hint: Show that both $\mathbf{B}e^{t\mathbf{A}}$ and $e^{t\mathbf{A}}\mathbf{B}$ are matrix solutions of $\dot{\mathbf{x}} = \mathbf{A}\mathbf{x}$ and use uniqueness.
   b. Prove Theorem 2.13 using part (a). Hint: Using part (a), verify that both $e^{t(\mathbf{A}+\mathbf{B})}$ and $e^{t\mathbf{A}}e^{t\mathbf{B}}$ are matrix solutions of $\dot{\mathbf{x}} = (\mathbf{A}+\mathbf{B})\mathbf{x}$.

6. Let $\mathbf{A} = \begin{pmatrix} 0 & 1 \\ 0 & 0 \end{pmatrix}$ and $\mathbf{B} = \begin{pmatrix} 0 & 0 \\ -1 & 0 \end{pmatrix}$.
   a. Show that $\mathbf{AB} \neq \mathbf{BA}$.
   b. Show that $e^{\mathbf{A}} = \begin{pmatrix} 1 & 1 \\ 0 & 1 \end{pmatrix}$ and $e^{\mathbf{B}} = \begin{pmatrix} 1 & 0 \\ -1 & 1 \end{pmatrix}$. Hint: Use Example 2.5 with $t = 1$ for $e^{\mathbf{A}}$ and a similar calculation for $e^{\mathbf{B}}$.
   c. Show that $e^{\mathbf{A}}e^{\mathbf{B}} \neq e^{\mathbf{A}+\mathbf{B}}$. Hint: $\mathbf{A}+\mathbf{B}$ is of the form considered in Example 2.3.

7. Let $\mathbf{A}$ be an $n \times n$ matrix, and let $\mathbf{P}$ be a nonsingular $n \times n$ matrix. Show that $e^{\mathbf{P}^{-1}\mathbf{A}\mathbf{P}} = \mathbf{P}^{-1}e^{\mathbf{A}}\mathbf{P}$. Hint: Write out the power series for $e^{\mathbf{P}^{-1}\mathbf{A}\mathbf{P}}$.

8. Calculate the exponential $e^{\mathbf{D}t}$ for the following diagonal matrices:

   a. $\mathbf{D} = \begin{pmatrix} a_1 & 0 & 0 \\ 0 & a_2 & 0 \\ 0 & 0 & a_3 \end{pmatrix}$,
   b. $\mathbf{D} = \begin{pmatrix} a_1 & 0 & \cdots & 0 \\ 0 & a_2 & \cdots & 0 \\ \vdots & \vdots & \ddots & \vdots \\ 0 & 0 & \cdots & a_n \end{pmatrix}$.

9. Calculate the exponential $e^{\mathbf{N}t}$ for the following matrices:

   a. $\mathbf{N} = \begin{pmatrix} 0 & 1 & 0 \\ 0 & 0 & 1 \\ 0 & 0 & 0 \end{pmatrix}$, b. $\mathbf{N} = \begin{pmatrix} 0 & 1 & 0 & 0 \\ 0 & 0 & 1 & 0 \\ 0 & 0 & 0 & 1 \\ 0 & 0 & 0 & 0 \end{pmatrix}$, c. $\mathbf{N} = \begin{pmatrix} 0 & 1 & 0 & \cdots & 0 & 0 \\ 0 & 0 & 1 & \cdots & 0 & 0 \\ 0 & 0 & 0 & \cdots & 0 & 0 \\ \vdots & \vdots & \vdots & \ddots & \vdots & \vdots \\ 0 & 0 & 0 & \cdots & 0 & 1 \\ 0 & 0 & 0 & \cdots & 0 & 0 \end{pmatrix}$.

   Hint: (a) $\mathbf{N}^3 = \mathbf{0}$. (b) $\mathbf{N}^4 = \mathbf{0}$. (c) $\mathbf{N}^n = \mathbf{0}$ if $\mathbf{N}$ is $n \times n$.

10. Calculate the exponential $e^{\mathbf{A}t}$ for the following matrices:

    a. $\mathbf{A} = \begin{pmatrix} a & 1 & 0 \\ 0 & a & 1 \\ 0 & 0 & a \end{pmatrix}$, b. $\mathbf{A} = \begin{pmatrix} a & 1 & 0 & 0 \\ 0 & a & 1 & 0 \\ 0 & 0 & a & 1 \\ 0 & 0 & 0 & a \end{pmatrix}$, c. $\mathbf{A} = \begin{pmatrix} a & 1 & 0 & \cdots & 0 & 0 \\ 0 & a & 1 & \cdots & 0 & 0 \\ 0 & 0 & a & \cdots & 0 & 0 \\ \vdots & \vdots & \vdots & \ddots & \vdots & \vdots \\ 0 & 0 & 0 & \cdots & a & 1 \\ 0 & 0 & 0 & \cdots & 0 & a \end{pmatrix}$.

    Hint: Write $\mathbf{A} = \mathbf{D} + \mathbf{N}$ where $\mathbf{D}$ and $\mathbf{N}$ are of the form of the preceding two problems.

11. Assume that $\mathbf{A}$ is an $n \times n$ matrix such that $\mathbf{A}^T = -\mathbf{A}$ (antisymmetric). Prove that $e^{\mathbf{A}}$ is orthogonal, i.e., $\left(e^{\mathbf{A}}\right)^T e^{\mathbf{A}} = \mathbf{I}_n$. Hint: Exponentiate the two sides of the equation $\mathbf{O}_n = \mathbf{A} + \mathbf{A}^T$ where $\mathbf{O}_n$ is the zero matrix. Use the fact that $\mathbf{A}^T\mathbf{A} = (-\mathbf{A})\mathbf{A} = \mathbf{A}(-\mathbf{A}) = \mathbf{A}\mathbf{A}^T$ and $e^{(\mathbf{A}^T)} = \left(e^{\mathbf{A}}\right)^T$.

2.2. Constant Coefficients: Solutions and Phase Portraits

We motivate the form of the solutions of linear systems of equations by starting with a linear scalar equation. The differential equation $\dot{x} = ax$ has a solution $x(t) = x_0 e^{at}$ for arbitrary x_0. If we combine two such scalar equations into a single system, $\dot{x}_1 = a_1 x_1$ and $\dot{x}_2 = a_2 x_2$, then it has a solution of the form $x_1(t) = c_1 e^{a_1 t}$ and $x_2(t) = c_2 e^{a_2 t}$, where c_1 and c_2 are arbitrary constants. Combining these into a vector solution,
$$\mathbf{x}(t) = c_1 e^{a_1 t} \mathbf{u}^1 + c_2 e^{a_2 t} \mathbf{u}^2,$$
where $\mathbf{u}^j$ is the unit vector with a 1 in the jth place and 0 in the other entry. Notice that this vector solution $e^{a_j t} \mathbf{u}^j$ is always a scalar multiple of the vector $\mathbf{u}^j$, and it moves along the line determined by $\mathbf{u}^j$.

With this as motivation, we try for a solution of the constant coefficient equation (2.1) that is of the form
$$\mathbf{x}(t) = e^{\lambda t} \mathbf{v}.$$
For it to be a solution, we need
$$\mathbf{A} e^{\lambda t} \mathbf{v} = \frac{d}{dt}\left(e^{\lambda t} \mathbf{v}\right) = \lambda e^{\lambda t} \mathbf{v}.$$
Since $e^{\lambda t} \neq 0$, this equality is equivalent to
$$\mathbf{A}\mathbf{v} = \lambda \mathbf{v} \quad \text{or}$$
$$(\mathbf{A} - \lambda \mathbf{I})\mathbf{v} = \mathbf{0}.$$
Here, $\mathbf{I}$ is the *identity matrix* with 1s down the diagonal and 0s in the entries off the diagonal. Thus, $\mathbf{x}(t) = e^{\lambda t} \mathbf{v}$ is a nonzero solution if and only if λ is an eigenvalue with eigenvector $\mathbf{v}$. The way to find such solutions is (i) to solve for eigenvalues λ_j as roots of the characteristic equation
$$\det(\mathbf{A} - \lambda \mathbf{I}) = 0,$$
and then (ii) for each root, to solve for a corresponding eigenvector from the equation
$$(\mathbf{A} - \lambda_j \mathbf{I})\mathbf{v} = \mathbf{0}.$$
If there are n real distinct eigenvalues, $\lambda_1, \ldots, \lambda_n$, with corresponding eigenvectors, $\mathbf{v}^1, \ldots, \mathbf{v}^n$, then the eigenvectors are linearly independent and we get n solutions that form a fundamental set of solutions. In this case, the general solution is of the form
$$c_1 e^{\lambda_1 t} \mathbf{v}^1 + \cdots + c_n e^{\lambda_n t} \mathbf{v}^n.$$

For a two-dimensional linear system
$$\begin{pmatrix} \dot{x}_1 \\ \dot{x}_2 \end{pmatrix} = \begin{pmatrix} a & b \\ c & d \end{pmatrix} \begin{pmatrix} x_1 \\ x_2 \end{pmatrix},$$
the characteristic equation is
$$0 = \det\begin{pmatrix} a - \lambda & b \\ c & d - \lambda \end{pmatrix}$$
$$= (a - \lambda)(d - \lambda) - bc = \lambda^2 - (a + d)\lambda + (ad - bc)$$
$$= \lambda^2 - \tau \lambda + \Delta,$$

where $\tau = a+d = \text{tr}(A)$ is the trace of the matrix and $\Delta = ad - bc = \det(A)$ is the determinant. Therefore, the characteristic equation is especially easy to determine in this case.

We proceed to solve some specific examples for the fundamental set of solutions. We also draw the representative solutions in the phase space.

Example 2.7 (Saddle). Consider

$$\dot{\mathbf{x}} = \begin{pmatrix} 1 & 3 \\ 3 & 1 \end{pmatrix} \mathbf{x}.$$

This has characteristic equation $\lambda^2 - 2\lambda - 8 = 0$, with roots $\lambda = -2, 4$.

For the eigenvalue $\lambda_1 = -2$, $\mathbf{A} - \lambda_1 \mathbf{I} = \mathbf{A} + 2\mathbf{I} = \begin{pmatrix} 3 & 3 \\ 3 & 3 \end{pmatrix}$. This can be row reduced to $\begin{pmatrix} 1 & 1 \\ 0 & 0 \end{pmatrix}$, and so there is an eigenvector $\mathbf{v}^1 = \begin{pmatrix} 1 \\ -1 \end{pmatrix}$. Thus, the first solution is given by

$$\mathbf{x}^1(t) = e^{-2t} \begin{pmatrix} 1 \\ -1 \end{pmatrix}.$$

The components of this solution go to zero at the rate e^{-2t}. The plot of the first component of this solution versus time, $(t, x_1^1(t))^\mathsf{T}$, is shown in Figure 1a. We call this plot of one component of the solution versus t, a *time plot of the solution*.

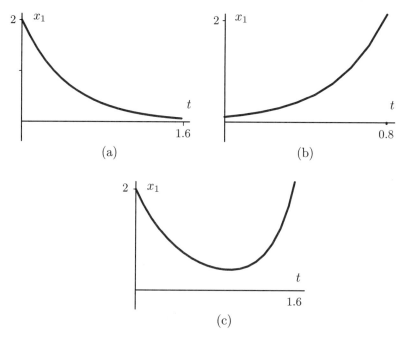

Figure 1. Time plot of solutions for Example 2.7 with a saddle: (a) $(t, x_1^1(t))$ for $(x_1^1(0), x_2^1(0)) = (2, -2)$, (b) $(t, x_1^2(t))$ for $(x_1^2(0), x_2^2(0)) = (0.1, 0.1)$, and (c) $(t, x_1^3(t))$ for $(x_1^3(0), x_2^3(0)) = (2, -1.995)$.

2.2. Constant Coefficients

Similarly, the second eigenvalue $\lambda_2 = 4$ has an eigenvector $\mathbf{v}^2 = \begin{pmatrix} 1 \\ 1 \end{pmatrix}$. The second solution is given by

$$\mathbf{x}^2(t) = e^{4t} \begin{pmatrix} 1 \\ 1 \end{pmatrix}.$$

The components of the solution grow at a rate of e^{4t}. See Figure 1b for a time plot of this second solution, $(t, x_1^2(t))$.

The matrix solution from these two solutions is given by

$$\begin{pmatrix} e^{-2t} & e^{4t} \\ -e^{-2t} & e^{4t} \end{pmatrix}.$$

The Wronskian of the system is

$$W(t) = \det \begin{pmatrix} e^{-2t} & e^{4t} \\ -e^{-2t} & e^{4t} \end{pmatrix} = 2\,e^{2t}.$$

Evaluating it at t equal to zero gives the determinant of the matrix with eigenvectors as the columns:

$$W(0) = \det(\mathbf{v}^1, \mathbf{v}^2) = \det \begin{pmatrix} 1 & 1 \\ -1 & 1 \end{pmatrix} = 2 \neq 0.$$

These two eigenvectors are independent, as must be the case from linear algebra, since the two eigenvalues are different. Thus, the general solution is of the form

$$c_1 e^{-2t} \begin{pmatrix} 1 \\ -1 \end{pmatrix} + c_2 e^{4t} \begin{pmatrix} 1 \\ 1 \end{pmatrix}.$$

The plot of the variable $x_1(t)$ as a function of t for different values of c_1 and c_2 is given in Figure 1.

In addition to giving the time plot of the solution, it is often very instructive to plot the solution in the (x_1, x_2)-plane as curves, without indicating the actual times, but just the direction of motion by means of arrows. This plot of the solutions in the **x**-space is called the *phase portrait*. When there are just two variables as in this example, the **x**-space is also called the *phase plane*; in higher dimensions, it is called the *phase space*.

Taking $c_1 = 1$ and $c_2 = 0$, we get the solution $e^{-2t} \begin{pmatrix} 1 \\ -1 \end{pmatrix}$, which moves along the straight line of scalar multiples of the vector $\begin{pmatrix} 1 \\ -1 \end{pmatrix}$. Notice that e^{-2t} goes to zero as t goes to infinity, but never equals zero, so the trajectory $e^{-2t} \begin{pmatrix} 1 \\ -1 \end{pmatrix}$ approaches the origin as t goes to infinity, but it never actually reaches the origin. This is evident in the time plot, Figure 1a. We can draw this trajectory in the (x_1, x_2)-plane with an arrow to mark the direction to move along the curve. See the straight line going toward the origin from the lower right in Figure 2. If we set $c_1 = -1$ and $c_2 = 0$, we get the solution $e^{-2t} \begin{pmatrix} -1 \\ 1 \end{pmatrix}$, which also moves along a straight line from the upper left.

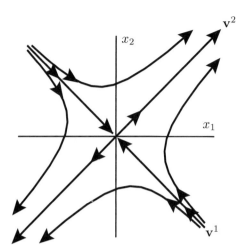

Figure 2. Phase portrait for Example 2.7 with a saddle

If we set $c_1 = 0$ and $c_2 = 1$, we get the solution $e^{4t}\begin{pmatrix}1\\1\end{pmatrix}$, which also moves along a straight line of scalar multiples of its eigenvector, but it grows in size as t increases: This solution goes to infinity as t goes to infinity, and goes to the origin as t goes to minus infinity. This trajectory is along the half-line in the upper right of Figure 2. If we set $c_1 = 0$ and $c_2 = -1$, we get the solution $e^{4t}\begin{pmatrix}-1\\-1\end{pmatrix}$, which moves away from the origin along the half-line in the lower left of Figure 2.

For a solution that contains components along both eigenvectors, $c_1 \neq 0$ and $c_2 \neq 0$,

$$c_1 e^{-2t}\begin{pmatrix}1\\-1\end{pmatrix} + c_2 e^{4t}\begin{pmatrix}1\\1\end{pmatrix} = e^{4t}\left[c_1 e^{-6t}\begin{pmatrix}1\\-1\end{pmatrix} + c_2 \begin{pmatrix}1\\1\end{pmatrix}\right],$$

the term inside the square bracket approaches $c_2 \begin{pmatrix}1\\1\end{pmatrix}$ as t goes to infinity, so the trajectory has the line generated by the eigenvector $\begin{pmatrix}1\\1\end{pmatrix}$ as an asymptote as t goes to infinity. Similarly, the trajectory has the line generated by the eigenvector $\begin{pmatrix}1\\-1\end{pmatrix}$ as an asymptote as t goes to minus infinity. For example, if both c_1 and c_2 are positive, the trajectory is like the one moving on the hyperbola at the right of Figure 2. For

$$\mathbf{x}^3(t) = 1.9975\,\mathbf{x}^1(t) + 0.0025\,\mathbf{x}^2(t),$$

with $\mathbf{x}^3(0) = (2, -1.995)$, the plot of $(t, x_1^3(t))$ is given in Figure 1c.

If $c_1 < 0$ and $c_2 > 0$, the trajectory is like the one at the top of Figure 2.

The origin for such a linear system with one positive real eigenvalue and one negative real eigenvalue is called a *saddle*. Sometimes we abuse terminology by calling the linear system a saddle.

2.2. Constant Coefficients

Table 1 summarizes the procedure for drawing the phase portrait near a saddle linear system in two dimensions.

Phase portrait for a saddle linear system

(1) Draw all the trajectories that move along straight lines in or out of the origin by finding the eigenvectors. Mark each of these half-lines with the direction that the solution is moving as t increases.

(2) In each of the four regions between the straight line solutions, draw in representative solutions, which are linear combinations of the two primary solutions.

When using a computer program, such as Maple, Mathematica, or Matlab, to draw the phase portrait, certain steps are helpful. Assume that $\lambda_1 < 0 < \lambda_2$ are the eigenvalues, with corresponding eigenvectors $\mathbf{v}^1$ and $\mathbf{v}^2$.

1. Take two initial conditions that are small multiples of $\pm \mathbf{v}^1$ and follow them forward in time (e.g., $\pm 0.01\,\mathbf{v}^1$).
2. Take two other initial conditions $\pm c\mathbf{v}^2$ that are large scalar multiples of $\pm \mathbf{v}^2$, so the points are about on the edge of the region displayed and follow the solutions forward in time. (Or, take two initial conditions that are small multiples of $\pm \mathbf{v}^2$ and follow them backward in time.)
3. Take at least four other initial conditions that are slightly off the scalar multiples of $\pm \mathbf{v}^2$ (e.g., $\pm c\mathbf{v}^2 \pm 0.1\,\mathbf{v}^1$).

Table 1

Assume we have a system in $\mathbb{R}^2$ with two distinct negative real eigenvalues, $0 > \lambda_1 > \lambda_2$, with corresponding eigenvectors $\mathbf{v}^1$ and $\mathbf{v}^2$. Then, the general solution is of the form

$$c_1 e^{\lambda_1 t} \mathbf{v}^1 + c_2 e^{\lambda_2 t} \mathbf{v}^2.$$

The solutions with either (i) $c_1 \neq 0$ and $c_2 = 0$ or (ii) $c_1 = 0$ and $c_2 \neq 0$ move toward the origins along straight lines. If both $c_1, c_2 \neq 0$, then the solution can be written as

$$e^{\lambda_1 t}\left(c_1 \mathbf{v}^1 + c_2 e^{(\lambda_2 - \lambda_1)t} \mathbf{v}^2\right).$$

Because $\lambda_2 < \lambda_1$, $\lambda_2 - \lambda_1 < 0$ and the second term in the sum goes to zero as t goes to infinity. This shows that any solution with $c_1 \neq 0$ approaches the origin asymptotic to the line generated by the eigenvector $\mathbf{v}^1$ (i.e., asymptotic to the line generated by the eigenvector corresponding the eigenvalue that is less negative). As t goes to minus infinity, the term involving $e^{\lambda_2 t}$ goes to zero faster and the solution becomes more parallel to the line generated by the vector $\mathbf{v}^2$. Figure 3 is a sketch of the phase portrait, which exhibits the four trajectories that move along straight lines toward the origins, as well as other solutions that have both $c_1, c_2 \neq 0$. The time plots of the solutions all go to zero at an exponential rate; however, the second solution goes at a faster rate.

The origin for a system of this type (or the linear system itself) with all eigenvalues real, negative, and distinct is called a *stable node*.

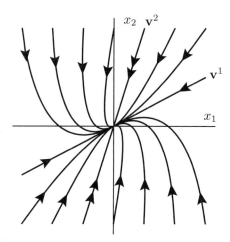

Figure 3. Phase portrait for Example 2.9 with a stable node

Remark 2.8. The word *node* means a point at which subsidiary parts originate. It is used in both biology for the point on the stem where a leaf is inserted, and in electrical circuits for the points where the different branches of the circuit come together. In the preceding phase plane, the solutions branch out from the origin; hence, the name.

The adjective *stable* is used because all the solutions tend to the origin as t goes to infinity.

Example 2.9 (Stable Node). The system

$$\dot{\mathbf{x}} = \begin{pmatrix} -4 & -2 \\ 3 & -11 \end{pmatrix} \mathbf{x}$$

is an example having a stable node. The matrix has eigenvalues $\lambda_1 = -5$ and $\lambda_2 = -10$ with corresponding eigenvectors $\mathbf{v}^1 = \begin{pmatrix} 2 \\ 1 \end{pmatrix}$ and $\mathbf{v}^2 = \begin{pmatrix} 1 \\ 3 \end{pmatrix}$. The general solution is

$$c_1 e^{-5t} \begin{pmatrix} 2 \\ 1 \end{pmatrix} + c_2 e^{-10t} \begin{pmatrix} 1 \\ 3 \end{pmatrix},$$

with phase portrait as shown in Figure 3. The Wronskian of the solutions at t equal zero is given by

$$W(0) = \det \begin{pmatrix} 2 & 1 \\ 1 & 3 \end{pmatrix} = 5 \neq 0;$$

thus, the two eigenvectors are independent, so the two solutions are independent.

Table 2 summarizes the procedure for drawing the phase portrait near a stable node with distinct eigenvalues in two dimensions.

2.2. Constant Coefficients

Phase portrait for a stable node

We assume that the eigenvalues are distinct, $0 > \lambda_1 > \lambda_2$.

(1) Draw all the trajectories that move along straight lines toward the origin. Mark each of these half-lines with the direction that the solution is moving as t increases. Mark each of the trajectories approaching the origin at the faster rate (corresponding to the more negative eigenvalue) with a double arrow.

(2) In each of the four regions between the straight line solutions, draw in representative solutions, which are linear combinations of the two primary solutions. Make sure that, as the trajectories approach the origin (as t goes to infinity), they are tangent to the line generated by the eigenvector corresponding to the eigenvalue which is less negative. Make sure that, as the trajectories go off toward infinity (as t goes to minus infinity), they become parallel to the eigenvector corresponding to the eigenvalue which is more negative.

When using a computer program to draw the phase portrait, certain steps are helpful. Assume $0 > \lambda_1 > \lambda_2$ are the eigenvalues with corresponding eigenvectors $\mathbf{v}^1$ and $\mathbf{v}^2$.

1. Take two initial conditions $\pm c_1 \mathbf{v}^1$ that are large multiples of $\pm \mathbf{v}^1$ so the points are about on the edge of the region displayed and follow the solutions forward in time.
2. Take two other initial conditions $\pm c_2 \mathbf{v}^2$ that are large scalar multiples of $\pm \mathbf{v}^2$, $\pm c_2 \mathbf{v}^2$, so the points are about on the edge of the region displayed and follow the solutions forward in time.
3. Take at least one initial condition in each of the four regions separated by $\pm \mathbf{v}^1 \pm \mathbf{v}^2$ that are on the edge of the region displayed and follow the solutions forward in time.

Table 2

Example 2.10 (Unstable Node). If both eigenvalues of a system on $\mathbb{R}^2$ are positive and distinct, then the origin for the linear system is called an *unstable node*. The phase portrait is similar to the preceding example, with the direction of the motion reversed. See Figure 4.

Drawing the phase portrait for an unstable node is similar to a stable node with obvious changes: the arrows are reversed and the trajectory converges to the origin as t goes to minus infinity.

Example 2.11 (Zero Eigenvalue). The differential equation

$$\dot{\mathbf{x}} = \begin{pmatrix} -1 & -3 \\ -1 & -3 \end{pmatrix} \mathbf{x}$$

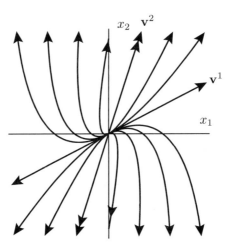

Figure 4. Phase portrait for Example 2.10 with an unstable node

has eigenvalues 0 and -4. The vector $\begin{pmatrix} 3 \\ -1 \end{pmatrix}$ is an eigenvector for 0 and $\begin{pmatrix} 1 \\ 1 \end{pmatrix}$ is an eigenvector for -4. Since $e^{0\,t} = 1$, the general solution is

$$\mathbf{x}(t) = c_1 \begin{pmatrix} 3 \\ -1 \end{pmatrix} + c_2 e^{-4t} \begin{pmatrix} 1 \\ 1 \end{pmatrix}.$$

Notice that all points that are multiples of $\begin{pmatrix} 3 \\ -1 \end{pmatrix}$ are fixed points. (These are the initial conditions obtained by taking $c_2 = 0$.) Since e^{-4t} goes to zero as t goes to infinity, all other solutions tend to this line of fixed points. See Figure 5.

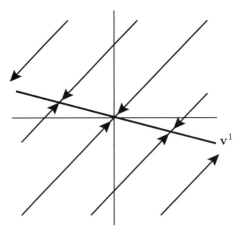

Figure 5. Phase portrait for Example 2.11 with a zero eigenvalue

The origin is always a fixed point for a linear equation, $\dot{\mathbf{x}} = \mathbf{A}\mathbf{x}$, since $\mathbf{A}\mathbf{0} = \mathbf{0}$. The only way that there are other fixed points is for 0 to be an eigenvalue; if

2.2. Constant Coefficients

$\dot{\mathbf{x}} = \mathbf{A}\mathbf{x}_0 = \mathbf{0}$ for some nonzero point $\mathbf{x}_0$, then $\mathbf{x}_0$ is an eigenvector for the eigenvalue 0.

2.2.1. Complex Eigenvalues. To be able to consider complex eigenvalues, we need to understand the exponential of a complex number. By comparing power series expansions, we can see that

$$e^{i\beta t} = \cos(\beta t) + i\sin(\beta t) \quad \text{and}$$
$$e^{(\alpha+i\beta)t} = e^{\alpha t}e^{i\beta t} = e^{\alpha t}\left(\cos(\beta t) + i\sin(\beta t)\right).$$

The next theorem shows how to use these formulas to find two real solutions for a complex eigenvalue.

Theorem 2.4. *Let $\mathbf{A}$ be an $n \times n$ matrix with constant real entries.*

a. *Assume that $\mathbf{z}(t) = \mathbf{x}(t) + i\mathbf{y}(t)$ is a complex solution of $\dot{\mathbf{z}} = \mathbf{A}\mathbf{z}$, where $\mathbf{x}(t)$ and $\mathbf{y}(t)$ are real. Then, $\mathbf{x}(t)$ and $\mathbf{y}(t)$ are each real solutions of the equation.*

b. *In particular, if $\lambda = \alpha + i\beta$ is a complex eigenvalue with a complex eigenvector $\mathbf{v} = \mathbf{u} + i\mathbf{w}$, where α and β real numbers and $\mathbf{u}$ and $\mathbf{v}$ real vectors, then*

$$e^{\alpha t}\left(\cos(\beta t)\mathbf{u} - \sin(\beta t)\mathbf{w}\right) \quad \text{and}$$
$$e^{\alpha t}\left(\sin(\beta t)\mathbf{u} + \cos(\beta t)\mathbf{w}\right)$$

are two real solutions of $\dot{\mathbf{x}} = \mathbf{A}\mathbf{x}$.

Proof. Part (a) follows directly from the rules of differentiation and matrix multiplication:

$$\dot{\mathbf{x}}(t) + i\dot{\mathbf{y}}(t) = \dot{\mathbf{z}}(t)$$
$$= \mathbf{A}\mathbf{z}(t)$$
$$= \mathbf{A}(\mathbf{x}(t) + i\mathbf{y}(t))$$
$$= \mathbf{A}\mathbf{x}(t) + i\mathbf{A}\mathbf{y}(t).$$

By equating the real and imaginary parts, we get $\dot{\mathbf{x}}(t) = \mathbf{A}\mathbf{x}(t)$ and $\dot{\mathbf{y}}(t) = \mathbf{A}\mathbf{y}(t)$, which gives the first part of the theorem.

(b) The complex solution

$$e^{(\alpha+i\beta)t}(\mathbf{u} + i\mathbf{w}) = e^{\alpha t}\left(\cos(\beta t) + i\sin(\beta t)\right)(\mathbf{u} + i\mathbf{w})$$
$$= e^{\alpha t}\left(\cos(\beta t)\mathbf{u} - \sin(\beta t)\mathbf{w}\right) + i\,e^{\alpha t}\left(\sin(\beta t)\mathbf{u} + \cos(\beta t)\mathbf{w}\right)$$

can be written as the sum of a real function plus a purely imaginary function. By part (a), the real and the imaginary parts are each real solutions, giving the result claimed. □

Example 2.12 (Elliptic Center). Consider the differential equation

$$\dot{\mathbf{x}} = \begin{pmatrix} 0 & 4 \\ -1 & 0 \end{pmatrix}\mathbf{x}.$$

The characteristic equation is $\lambda^2 + 4 = 0$, and the eigenvalues are $\lambda = \pm 2i$. Using the eigenvalue $2i$, we have

$$\mathbf{A} - (2i)\mathbf{I} = \begin{pmatrix} -2i & 4 \\ -1 & -2i \end{pmatrix}.$$

The two rows are (complex) multiples of each other, so the eigenvector satisfies the equation
$$-2i\,v_1 + 4\,v_2 = 0,$$
and an eigenvector is $\mathbf{v} = \begin{pmatrix} 2 \\ i \end{pmatrix}$. Using the eigenvector, we get the two real solutions

$$\mathbf{x}^1(t) = \cos(2t)\begin{pmatrix} 2 \\ 0 \end{pmatrix} - \sin(2t)\begin{pmatrix} 0 \\ 1 \end{pmatrix} \quad \text{and}$$

$$\mathbf{x}^2(t) = \sin(2t)\begin{pmatrix} 2 \\ 0 \end{pmatrix} + \cos(2t)\begin{pmatrix} 0 \\ 1 \end{pmatrix},$$

which have initial conditions $\begin{pmatrix} 2 \\ 0 \end{pmatrix}$ and $\begin{pmatrix} 0 \\ 1 \end{pmatrix}$ at $t = 0$, respectively. Notice that

$$\det\begin{pmatrix} 2 & 0 \\ 0 & 1 \end{pmatrix} = 2 \neq 0,$$

so the solutions are linearly independent. Both of these solutions are periodic with period $T = 2\pi/2 = \pi$, so the solution comes back to the same point after a time of π. The solution moves on ellipses with axes twice as long in the x_1-direction as in the x_2-direction. When $x_1 = 0$ and $x_2 > 0$, $\dot{x}_1 = 4\,x_2 > 0$, so the solution goes around in a clockwise direction. Such an example, with purely imaginary eigenvalues, is called an *elliptic center*. See Figure 6. The plot of $(t, x_1^1(t))$ for solution $\mathbf{x}^1(t)$ is given in Figure 7. Notice that the component $x_1^1(t)$ is a periodic function of t and that the period does not depend on the amplitude.

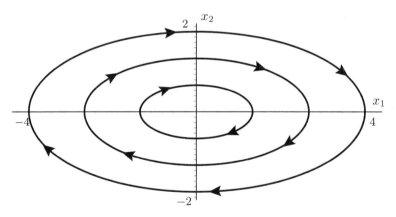

Figure 6. Phase portrait for Example 2.12 with an elliptic center

Example 2.13 (Stable Focus). This example has complex eigenvalues with negative real parts. Consider
$$\dot{\mathbf{x}} = \begin{pmatrix} -4 & 5 \\ -5 & 2 \end{pmatrix}\mathbf{x}.$$
The characteristic equation is $\lambda^2 + 2\lambda + 17 = 0$, and the eigenvalues are $\lambda = -1 \pm 4\,i$. Using the eigenvalue $-1 + 4\,i$, we have
$$\mathbf{A} - (-1 + 4\,i)\mathbf{I} = \begin{pmatrix} -3 - 4i & 5 \\ -5 & 3 - 4i \end{pmatrix}.$$

2.2. Constant Coefficients

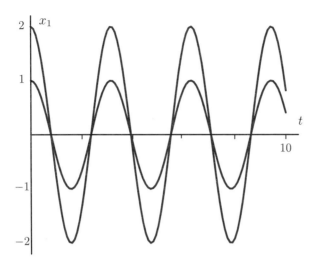

Figure 7. Plot of x_1 as a function of t, initial conditions $(1,0)$ and $(2,0)$: Example 2.12 with an elliptic center

Multiplying the first row by $-3+4i$, which is the complex conjugate of $-3-4i$, yields $(25, -15 + 20i)$, which is a multiple of the second row. Therefore, the eigenvector satisfies the equation
$$5\,v_1 + (-3 + 4i)\,v_2 = 0,$$
and an eigenvector is
$$\mathbf{v} = \begin{pmatrix} 3 - 4i \\ 5 \end{pmatrix} = \begin{pmatrix} 3 \\ 5 \end{pmatrix} + i \begin{pmatrix} -4 \\ 0 \end{pmatrix}.$$

A complex solution of the system is given by
$$e^{-t}\left(\cos(4t) + i\,\sin(4t)\right)\left[\begin{pmatrix} 3 \\ 5 \end{pmatrix} + i\begin{pmatrix} -4 \\ 0 \end{pmatrix}\right]$$
$$= e^{-t}\left[\cos(4t)\begin{pmatrix} 3 \\ 5 \end{pmatrix} - \sin(4t)\begin{pmatrix} -4 \\ 0 \end{pmatrix}\right]$$
$$+ i\,e^{-t}\left[\sin(4t)\begin{pmatrix} 3 \\ 5 \end{pmatrix} + \cos(4t)\begin{pmatrix} -4 \\ 0 \end{pmatrix}\right].$$

Taking the real and imaginary parts, the two real solutions are
$$\mathbf{x}^1(t) = e^{-t}\left(\cos(4t)\begin{pmatrix} 3 \\ 5 \end{pmatrix} - \sin(4t)\begin{pmatrix} -4 \\ 0 \end{pmatrix}\right) \quad \text{and}$$
$$\mathbf{x}^2(t) = e^{-t}\left(\sin(4t)\begin{pmatrix} 3 \\ 5 \end{pmatrix} + \cos(4t)\begin{pmatrix} -4 \\ 0 \end{pmatrix}\right).$$

The initial conditions of the two solutions are
$$\mathbf{x}^1(0) = \begin{pmatrix} 3 \\ 5 \end{pmatrix} \quad \text{and} \quad \mathbf{x}^2(0) = \begin{pmatrix} -4 \\ 0 \end{pmatrix}$$
and
$$\det\begin{pmatrix} 3 & -4 \\ 5 & 0 \end{pmatrix} = 20 \neq 0,$$

so the solutions are independent. The sine and cosine terms have period $2\pi/4 = \pi/2$; but the exponential factor decreases for this example as t increases and contracts by $e^{-\pi/2}$ every revolution around the origin. The solutions tend asymptotically toward the origin as t goes to infinity. When $x_1 = 0$ and $x_2 > 0$, $\dot{x}_1 = 5x_2 > 0$, so the solution go around in a clockwise direction. This example, with a negative real part and nonzero imaginary part of the eigenvalue, is called a *stable focus*. It is stable because solutions tend to the origin as t goes to infinity, and it is a focus because the solutions spiral. See Figure 8. The plot of $(t, x_1^1(t))$ for solution $\mathbf{x}^1(t)$ is given in Figure 9. Notice that the component $x_1^1(t)$ oscillates as a function of t as it goes to zero.

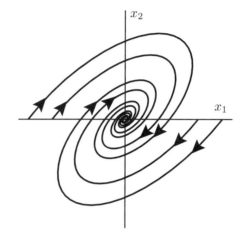

Figure 8. Phase portrait for Example 2.13 with a stable focus

Table 3 summarizes the procedure for drawing the phase portrait for a linear system with complex eigenvalues in two dimensions.

Example 2.14. For an example in $\mathbb{R}^3$, consider
$$\dot{\mathbf{x}} = \begin{pmatrix} 0 & 0 & 1 \\ 1 & 1 & -1 \\ -1 & 4 & -2 \end{pmatrix} \mathbf{x}.$$

The characteristic equation is
$$0 = \lambda^3 + \lambda^2 + 3\lambda - 5 = (\lambda - 1)(\lambda^2 + 2\lambda + 5),$$
and the eigenvalues are $\lambda = 1, -1 \pm 2i$.

Using the eigenvalue 1, we have the row reduction
$$\mathbf{A} - \mathbf{I} = \begin{pmatrix} -1 & 0 & 1 \\ 1 & 0 & -1 \\ -1 & 4 & -3 \end{pmatrix} \sim \begin{pmatrix} 1 & 0 & -1 \\ 0 & 4 & -4 \\ 0 & 0 & 0 \end{pmatrix}$$
$$\sim \begin{pmatrix} 1 & 0 & -1 \\ 0 & 1 & -1 \\ 0 & 0 & 0 \end{pmatrix},$$

2.2. Constant Coefficients

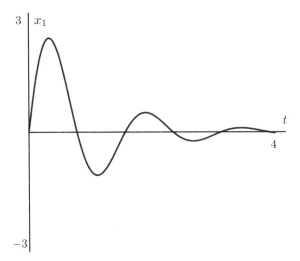

Figure 9. Plot of x_1 versus t, initial condition $(x_1(0), x_2(0)) = (0, 3)$: Example 2.13 with a stable focus

---

Phase portrait for a pair of complex eigenvalues

Assume the eigenvalues are $\lambda = \alpha \pm i\beta$ with $\beta \neq 0$.

(1) If $\alpha = 0$, then the origin is an *elliptic center*, with all the solutions periodic. The direction of motion can be either clockwise or counterclockwise.

(2) If $\alpha < 0$, then the origin is a *stable focus*, which spirals either clockwise or counterclockwise.

(3) If $\alpha > 0$, then the solutions spiral outward and the origin is an *unstable focus*, which spirals either clockwise or counterclockwise.

(4) In any of the three case, the direction the solution goes around the origin can be determined by checking whether $\dot{x}_1$ is positive or negative when $x_1 = 0$. If it is positive, then the direction is clockwise, and if it is negative, then the direction is counterclockwise.

Table 3

---

and an eigenvector is $(1, 1, 1)^\mathsf{T}$.

Using the eigenvalue $-1 + 2i$, we have

$$\mathbf{A} - (-1 + 2i)\mathbf{I} = \begin{pmatrix} 1 - 2i & 0 & 1 \\ 1 & 2 - 2i & -1 \\ -1 & 4 & -1 - 2i \end{pmatrix}.$$

Interchanging the first and third rows and multiplying the new third row by the complex conjugate of $1 - 2i$ (i.e., by $1 + 2i$) to make the first entry of the third row

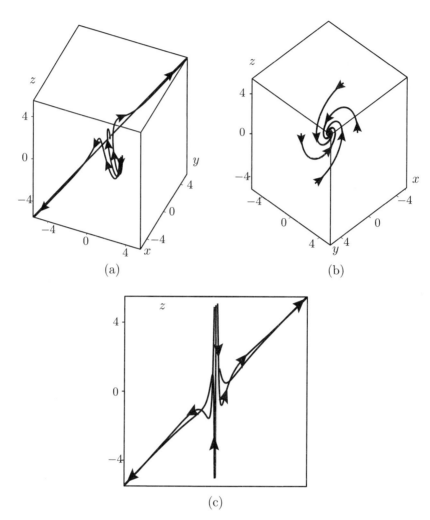

Figure 10. Three views of the phase portrait for Example 2.14: (a) a three-dimensional view of all the directions, (b) down the unstable direction onto the stable directions, and (c) one stable and the unstable direction. The initial conditions include points on the line spanned by $(1, 1, 1)$, on the plane spanned by $(1, -1, -5)$ and $(2, -2, 0)$, and points slightly off the plane.

real, we get
$$\mathbf{A} - (-4 + 2\,i)\mathbf{I} \sim \begin{pmatrix} -1 & 4 & -1 - 2i \\ 1 & 2 - 2i & -1 \\ 5 & 0 & 1 + 2i \end{pmatrix}.$$

Performing the row operations to make the entries of the first column, except for the top one, equal to zero, we get

$$\begin{pmatrix} -1 & 4 & -1 - 2i \\ 0 & 6 - 2i & -2 - 2i \\ 0 & 20 & -4 - 8i \end{pmatrix} \sim \begin{pmatrix} 1 & -4 & 1 + 2i \\ 0 & 3 - i & -1 - i \\ 0 & 5 & -1 - 2i \end{pmatrix},$$

2.2. Constant Coefficients

where in the second step we multiply the first row by -1, the second row by $1/2$, and the third row by $1/4$. To make the first entry in the second row real, we multiply the second row by the complex conjugate of the first entry, $1 - 3i$, resulting in

$$\begin{pmatrix} 1 & -4 & 1+2i \\ 0 & 10 & -2-4i \\ 0 & 5 & -1-2i \end{pmatrix}.$$

Performing further row operations, we get the following sequence of matrices:

$$\begin{pmatrix} 1 & -4 & 1+2i \\ 0 & 5 & -1-2i \\ 0 & 0 & 0 \end{pmatrix} \sim \begin{pmatrix} 1 & 1 & 0 \\ 0 & 5 & -1-2i \\ 0 & 0 & 0 \end{pmatrix}.$$

One eigenvector is

$$\begin{pmatrix} 1+2i \\ -1-2i \\ -5 \end{pmatrix} = \begin{pmatrix} 1 \\ -1 \\ -5 \end{pmatrix} + i \begin{pmatrix} 2 \\ -2 \\ 0 \end{pmatrix}.$$

Combining, three independent solutions are

$$e^t \begin{pmatrix} 1 \\ 1 \\ 1 \end{pmatrix}, \quad e^{-t}\cos(2t)\begin{pmatrix} 1 \\ -1 \\ -5 \end{pmatrix} - e^{-t}\sin(2t)\begin{pmatrix} 2 \\ -2 \\ 0 \end{pmatrix} \quad \text{and}$$

$$e^{-t}\sin(2t)\begin{pmatrix} 1 \\ -1 \\ -5 \end{pmatrix} + e^{-t}\cos(2t)\begin{pmatrix} 2 \\ -2 \\ 0 \end{pmatrix}.$$

The initial conditions of these three solutions are

$$\begin{pmatrix} 1 \\ 1 \\ 1 \end{pmatrix}, \quad \begin{pmatrix} 1 \\ -1 \\ -5 \end{pmatrix}, \text{ and } \begin{pmatrix} 2 \\ -2 \\ 0 \end{pmatrix}.$$

Since

$$\det \begin{pmatrix} 1 & 1 & 2 \\ 1 & -1 & -2 \\ 1 & -5 & 0 \end{pmatrix} = -20 \neq 0,$$

the solutions are independent.

See Figure 10 for three different views of the phase portrait.

2.2.2. Repeated Real Eigenvalues.

Example 2.15 (Enough eigenvectors for a repeated eigenvalue). As a first example with repeated eigenvalues, consider

$$\dot{\mathbf{x}} = \begin{pmatrix} -2 & -2 & -4 \\ 0 & 0 & 4 \\ 0 & 2 & 2 \end{pmatrix} \mathbf{x}.$$

The characteristic equation is $0 = \lambda^3 - 12\lambda - 16 = (\lambda - 4)(\lambda + 2)^2$. Therefore, -2 is a repeated eigenvalue with multiplicity two. The matrix

$$\mathbf{A} + 2\mathbf{I} = \begin{pmatrix} 0 & -2 & -4 \\ 0 & 2 & 4 \\ 0 & 2 & 4 \end{pmatrix}$$

is row reducible to
$$\begin{pmatrix} 0 & 1 & 2 \\ 0 & 0 & 0 \\ 0 & 0 & 0 \end{pmatrix}.$$
This matrix has rank one and, therefore, $3 - 1 = 2$ independent eigenvectors: $v_2 = -2v_3$ with v_1 and v_3 as free variables, or eigenvectors
$$\begin{pmatrix} 1 \\ 0 \\ 0 \end{pmatrix} \quad \text{and} \quad \begin{pmatrix} 0 \\ -2 \\ 1 \end{pmatrix}.$$
For this example, $\lambda = -2$ has as many independent eigenvectors as the multiplicity from the characteristic equation.

The eigenvalue $\lambda = 4$ has an eigenvector $\begin{pmatrix} 1 \\ -1 \\ -1 \end{pmatrix}$. Therefore, the general solution is
$$\mathbf{x}(t) = c_1 e^{-2t} \begin{pmatrix} 1 \\ 0 \\ 0 \end{pmatrix} + c_2 e^{-2t} \begin{pmatrix} 0 \\ -2 \\ 1 \end{pmatrix} + c_3 e^{4t} \begin{pmatrix} 1 \\ -1 \\ -1 \end{pmatrix}.$$
Notice that, in this case, the solution looks very much like the case for distinct eigenvalues.

Example 2.16 (Not enough eigenvectors). As an example with not enough eigenvectors, consider
$$\dot{\mathbf{x}} = \begin{pmatrix} 0 & -1 & 1 \\ 2 & -3 & 1 \\ 1 & -1 & -1 \end{pmatrix} \mathbf{x}.$$
The characteristic equation is $0 = (\lambda + 1)^2(\lambda + 2)$, so $\lambda = -1$ is a repeated eigenvalue.

For the eigenvalue $\lambda = -2$, the matrix is
$$\mathbf{A} + 2\mathbf{I} = \begin{pmatrix} 2 & -1 & 1 \\ 2 & -1 & 1 \\ 1 & -1 & 1 \end{pmatrix} \sim \begin{pmatrix} 1 & -1 & 1 \\ 0 & 1 & -1 \\ 0 & 0 & 0 \end{pmatrix},$$
and -2 has an eigenvector $(0, 1, 1)^\mathsf{T}$.

For the eigenvalue $\lambda = -1$, the matrix
$$\mathbf{A} - (-1)\mathbf{I} = \begin{pmatrix} 1 & -1 & 1 \\ 2 & -2 & 1 \\ 1 & -1 & 0 \end{pmatrix} \sim \begin{pmatrix} 1 & -1 & 0 \\ 0 & 0 & 1 \\ 0 & 0 & 0 \end{pmatrix},$$
which has rank two. Therefore, -1 has only one independent eigenvector $(1, 1, 0)^\mathsf{T}$. So far we have only found two independent solutions
$$e^{-2t} \begin{pmatrix} 0 \\ 1 \\ 1 \end{pmatrix} \quad \text{and} \quad e^{-t} \begin{pmatrix} 1 \\ 1 \\ 0 \end{pmatrix}.$$

2.2. Constant Coefficients

We will next discuss how to find a third independent solution in a general situation, and then we will return to the preceding example.

Let λ be a multiple eigenvalue for $\mathbf{A}$ with an eigenvector $\mathbf{v}$, $\mathbf{A}\mathbf{v} = \lambda\mathbf{v}$. As we discussed earlier, $e^{\mathbf{A}t}\mathbf{w}$ is a solution for any vector $\mathbf{w}$ using the matrix exponential. We want to rewrite this solution somewhat for special choices of $\mathbf{w}$. If two matrices $\mathbf{A}$ and $\mathbf{B}$ commute ($\mathbf{AB} = \mathbf{BA}$), then

$$e^{(\mathbf{A}+\mathbf{B})t} = e^{\mathbf{A}t}e^{\mathbf{B}t}.$$

This can be shown by multiplying the series and rearranging terms. See Theorem 2.13 at the end of the chapter. In our situation, $(\mathbf{A} - \lambda\mathbf{I})(\lambda\mathbf{I}) = (\lambda\mathbf{I})(\mathbf{A} - \lambda\mathbf{I})$, since $\lambda\mathbf{I}$ is a scalar multiple of the identity, so

$$e^{\mathbf{A}t}\mathbf{w} = e^{((\lambda\mathbf{I})+(\mathbf{A}-\lambda\mathbf{I}))t}\mathbf{w} = e^{(\lambda\mathbf{I})t}e^{(\mathbf{A}-\lambda\mathbf{I})t}\mathbf{w} = e^{\lambda t}\mathbf{I}e^{(\mathbf{A}-\lambda\mathbf{I})t}\mathbf{w}$$
$$= e^{\lambda t}\left(\mathbf{I}\mathbf{w} + t(\mathbf{A}-\lambda\mathbf{I})\mathbf{w} + \frac{t^2}{2!}(\mathbf{A}-\lambda\mathbf{I})^2\mathbf{w} + \cdots\right).$$

If we have $\mathbf{w}$ such that $(\mathbf{A}-\lambda\mathbf{I})\mathbf{w} = \mathbf{v}$, where $\mathbf{v}$ is the eigenvector for the eigenvalue λ, then

$$(\mathbf{A}-\lambda\mathbf{I})^2\mathbf{w} = (\mathbf{A}-\lambda\mathbf{I})\mathbf{v} = \mathbf{0}, \quad \text{so}$$
$$(\mathbf{A}-\lambda\mathbf{I})^n\mathbf{w} = (\mathbf{A}-\lambda\mathbf{I})^{n-1}\mathbf{v} = \mathbf{0} \quad \text{for } n \geq 2.$$

Therefore, the infinite series for $e^{\mathbf{A}t}\mathbf{w}$ is actually finite, and we get a second solution,

$$\mathbf{x}^2(t) = e^{\lambda t}(\mathbf{w} + t\mathbf{v}).$$

Note the similarity to the second solution $te^{\lambda t}$ of a second-order scalar equation for a repeated root of the characteristic equation.

In fact, a direct check shows that $\mathbf{x}^2(t) = e^{\lambda t}(\mathbf{w} + t\mathbf{v})$ is a solution. We use the fact that $\lambda\mathbf{v} = \mathbf{A}\mathbf{v}$ and $\lambda\mathbf{w} + \mathbf{v} = \mathbf{A}\mathbf{w}$ to obtain the following:

$$\dot{\mathbf{x}}^2(t) = \lambda e^{\lambda t}(\mathbf{w} + t\mathbf{v}) + e^{\lambda t}\mathbf{v}$$
$$= e^{\lambda t}(\lambda\mathbf{w} + \mathbf{v}) + e^{\lambda t}t\lambda\mathbf{v}$$
$$= e^{\lambda t}\mathbf{A}\mathbf{w} + e^{\lambda t}t\mathbf{A}\mathbf{v}$$
$$= \mathbf{A}e^{\lambda t}(\mathbf{w} + t\mathbf{v})$$
$$= \mathbf{A}\mathbf{x}^2(t).$$

Therefore, for an eigenvalue λ, which has multiplicity two but only one independent eigenvector $\mathbf{v}$ for λ, we solve $(\mathbf{A}-\lambda\mathbf{I})\mathbf{v} = \mathbf{0}$ for an eigenvector $\mathbf{v}$. Then, we solve the equation $(\mathbf{A}-\lambda\mathbf{I})\mathbf{w} = \mathbf{v}$ for $\mathbf{w}$. Such a vector $\mathbf{w}$ is called a *generalized eigenvector* for the eigenvalue λ. Two solutions of the linear differential equation are $e^{\lambda t}\mathbf{v}$ and $e^{\lambda t}(\mathbf{w} + t\mathbf{v})$. Note that, if the multiplicity is greater than two, then more complicated situations can arise, but the general idea is the same.

Returning to Example 2.16. For the repeated eigenvalue $\lambda = -1$,

$$\mathbf{A} + \mathbf{I} = \begin{pmatrix} 1 & -1 & 1 \\ 2 & -2 & 1 \\ 1 & -1 & 0 \end{pmatrix}.$$

and an eigenvector is $\mathbf{v} = (1,1,0)^\mathsf{T}$. We also need to solve the nonhomogeneous equation $(\mathbf{A} + \mathbf{I})\mathbf{w} = \mathbf{v}$. Separating the matrix $\mathbf{A} + \mathbf{I}$ from the vector $\mathbf{v}$ by a vertical line, the augmented matrix is

$$\left(\begin{array}{ccc|c} 1 & -1 & 1 & 1 \\ 2 & -2 & 1 & 1 \\ 1 & -1 & 0 & 0 \end{array}\right) \sim \left(\begin{array}{ccc|c} 1 & -1 & 0 & 0 \\ 0 & 0 & -1 & -1 \\ 0 & 0 & -1 & -1 \end{array}\right) \sim \left(\begin{array}{ccc|c} 1 & -1 & 0 & 0 \\ 0 & 0 & 1 & 1 \\ 0 & 0 & 0 & 0 \end{array}\right).$$

Therefore, a generalized eigenvector is $\mathbf{w} = (0,0,1)^\mathsf{T}$ and another solution of the differential equation is

$$\mathbf{x}^3(t) = e^{-t} \left[\begin{pmatrix} 0 \\ 0 \\ 1 \end{pmatrix} + t \begin{pmatrix} 1 \\ 1 \\ 0 \end{pmatrix} \right] = e^{-t} \begin{pmatrix} t \\ t \\ 1 \end{pmatrix}.$$

We have now found three solutions and the general solution is

$$c_1 e^{-2t} \begin{pmatrix} 0 \\ 1 \\ 1 \end{pmatrix} + c_2 e^{-t} \begin{pmatrix} 1 \\ 1 \\ 0 \end{pmatrix} + c_3 e^{-t} \begin{pmatrix} t \\ t \\ 1 \end{pmatrix}.$$

The determinant of the matrix formed by putting the initial conditions of the three solutions at $t = 0$ in as columns is nonzero,

$$\det \begin{pmatrix} 0 & 1 & 0 \\ 1 & 1 & 0 \\ 1 & 0 & 1 \end{pmatrix} = -1 \neq 0,$$

so the three solutions are independent. ∎

Example 2.17 (Degenerate Stable Node). A two-dimensional example with a repeated eigenvalue is given by

$$\dot{\mathbf{x}} = \begin{pmatrix} -2 & 1 \\ -1 & 0 \end{pmatrix} \mathbf{x}.$$

This has characteristic equation $0 = \lambda^2 + 2\lambda + 1 = (\lambda + 1)^2$, and has repeated eigenvalue -1. The matrix

$$\mathbf{A} + \mathbf{I} = \begin{pmatrix} -1 & 1 \\ -1 & 1 \end{pmatrix}$$

has rank one, and only one independent eigenvector $\begin{pmatrix} 1 \\ 1 \end{pmatrix}$. To solve the equation

$$(\mathbf{A} + \mathbf{I})\mathbf{w} = \mathbf{v},$$

we consider the augmented matrix

$$\left(\begin{array}{cc|c} -1 & 1 & 1 \\ -1 & 1 & 1 \end{array}\right) \sim \left(\begin{array}{cc|c} -1 & 1 & 1 \\ 0 & 0 & 0 \end{array}\right)$$

or $-w_1 + w_2 = 1$. Thus, a generalized eigenvector is $w_1 = 0$ and $w_2 = 1$, $\mathbf{w} = (0,1)^\mathsf{T}$. The second solution is

$$\mathbf{x}^2(t) = e^{-t}\left[\begin{pmatrix} 0 \\ 1 \end{pmatrix} + t\begin{pmatrix} 1 \\ 1 \end{pmatrix}\right] = e^{-t}\begin{pmatrix} t \\ 1+t \end{pmatrix},$$

2.2. Constant Coefficients

and the general solution is

$$c_1 e^{-t} \begin{pmatrix} 1 \\ 1 \end{pmatrix} + c_2 e^{-t} \begin{pmatrix} t \\ 1+t \end{pmatrix}.$$

The first component of the second solution is $x_1^2(t) = te^{-t}$. This term goes to zero as t goes to infinity, because the exponential goes to zero faster than t goes to infinity. Alternatively, apply l'Hôpital's rule to t/e^t and see that it goes to zero. In the same way, the second component $x_2^2(t) = e^{-t} + te^{-t}$ goes to zero as t goes to infinity. Combining, $\mathbf{x}^2(t)$ goes to the origin as t goes to infinity. Moreover,

$$\mathbf{x}^2(t) = t e^{-t} \left[\frac{1}{t} \begin{pmatrix} 0 \\ 1 \end{pmatrix} + \begin{pmatrix} 1 \\ 1 \end{pmatrix} \right]$$

approaches

$$t e^{-t} \begin{pmatrix} 1 \\ 1 \end{pmatrix}$$

as t goes to infinity, so the solution comes in to the origin in a direction asymptotic to the line generated by the eigenvector.

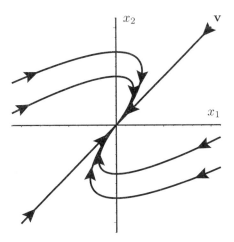

Figure 11. Phase portrait for Example 2.17 with a degenerate stable node

Clearly $\mathbf{x}^1(t)$ tends to the origin as t goes to infinity, so any solution, which is a linear combination of the two independent solutions, goes to the origin as t goes to infinity. However, in this case, there is only one solution that moves along a straight line. All other solutions approach the origin in a direction asymptotic to the line generated by the eigenvector. This system is called a *degenerate stable node*. See Figure 11. The plot of $(t, x_1(t))$ for $c_1 = 2$ and $c_2 = -2$ is given in Figure 12. This solution has initial conditions $x_1(0) = 2$ and $x_2(0) = 0$.

Table 4 summarizes the procedure for drawing the phase portrait for a stable linear system with real equal eigenvalues in two dimensions. Table 5 summarizes the process for drawing the phase portrait for any linear system in two dimensions.

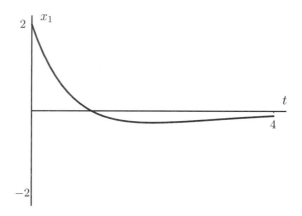

Figure 12. Plot of x_1 versus t for $c_1 = 2$ and $c_2 = -2$ for Example 2.17 with a degenerate stable node

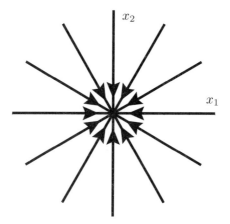

Figure 13. Phase portrait with a stable star

Example 2.18 (Multiplicity Three). For the situation with greater multiplicity in higher dimensions, there are many cases; we give just one example as an illustration:
$$\dot{\mathbf{x}} = \begin{pmatrix} -2 & 1 & 0 \\ 0 & -2 & 1 \\ 0 & 0 & -2 \end{pmatrix} \mathbf{x}.$$
This has characteristic equation $0 = (\lambda + 2)^3$, and has eigenvalue $\lambda = -2$ with multiplicity three. The vector $\mathbf{v} = (1, 0, 0)^\mathsf{T}$ is an eigenvector, $(\mathbf{A} + 2\mathbf{I})\mathbf{v} = \mathbf{0}$; the generalized eigenvector $\mathbf{w} = (0, 1, 0)^\mathsf{T}$ satisfies
$$(\mathbf{A} + 2\mathbf{I})\mathbf{w} = \mathbf{v} \quad \text{and} \quad (\mathbf{A} + 2\mathbf{I})^2 \mathbf{w} = (\mathbf{A} + 2\mathbf{I})\mathbf{v} = \mathbf{0};$$
finally, the generalized eigenvector $\mathbf{z} = (0, 0, 1)^\mathsf{T}$ satisfies
$$(\mathbf{A} + 2\mathbf{I})\mathbf{z} = \mathbf{w}, \quad (\mathbf{A} + 2\mathbf{I})^2 \mathbf{z} = (\mathbf{A} + 2\mathbf{I})\mathbf{w} = \mathbf{v} \quad \text{and}$$
$$(\mathbf{A} + 2\mathbf{I})^3 \mathbf{z} = (\mathbf{A} + 2\mathbf{I})^2 \mathbf{w} = \mathbf{0}.$$

2.2. Constant Coefficients

Phase portrait for two equal real eigenvalues

We consider the stable case; the case for unstable systems is similar, with obvious changes between t going to infinity and minus infinity. First, assume that there are two independent eigenvectors (and the matrix is diagonal).

(1) If there are two independent eigenvectors, then all solutions go straight in toward the origin. The origin of this system is called a *stable star*. See Figure 13.

Next, assume that there is only one independent eigenvector $\mathbf{v}$, and a second generalized eigenvector $\mathbf{w}$, where $(\mathbf{A} - \lambda \mathbf{I})\mathbf{w} = \mathbf{v}$.

(1) Draw the two trajectories, which move along straight lines toward the origin along the line generated by $\mathbf{v}$. Mark each of these half-lines with the direction that the solution is moving as t increases.

(2) Next, draw the trajectory, which has initial condition $\mathbf{w}$ and then comes in toward the origin along the half-line generated by positive multiples of the vector $\mathbf{v}$ (i.e., the trajectory $e^{\lambda t}\mathbf{w} + t e^{\lambda t}\mathbf{v}$ is nearly equal to the curve $t e^{\lambda t}\mathbf{v}$).

(3) Draw the trajectory with initial condition $-\mathbf{w}$, which should be just the reflection through the origin of the previous trajectory.

Table 4

A third solution with initial condition $\mathbf{w}$ is

$$e^{\mathbf{A}t}\mathbf{z} = e^{-2t}\left(\mathbf{I}\mathbf{z} + t(\mathbf{A} + 2\mathbf{I})\mathbf{z} + \frac{t^2}{2}(\mathbf{A} + 2\mathbf{I})^2 \mathbf{z} + \frac{t^3}{3!}(\mathbf{A} + 2\mathbf{I})^3\mathbf{z} + \cdots\right)$$

$$= e^{-2t}\left(\mathbf{z} + t\mathbf{w} + \frac{t^2}{2}\mathbf{v} + \mathbf{0}\right).$$

Therefore, the three independent solutions are

$$e^{-2t}\begin{pmatrix}1\\0\\0\end{pmatrix}, \quad e^{-2t}\left(\begin{pmatrix}0\\1\\0\end{pmatrix} + t\begin{pmatrix}1\\0\\0\end{pmatrix}\right), \quad \text{and} \quad e^{-2t}\left(\begin{pmatrix}0\\0\\1\end{pmatrix} + t\begin{pmatrix}0\\1\\0\end{pmatrix} + \frac{t^2}{2}\begin{pmatrix}1\\0\\0\end{pmatrix}\right).$$

2.2.3. Second-order Scalar Equations. Many of you have had a first course is differential equations in which the solution for second-order scalar linear differential equations was discussed. This section shows how those solutions are related to the solutions of systems of linear differential equations that we have presented. This section can be skipped without loss of continuity.

Consider

(2.4) $$y'' + ay' + by = 0,$$

where a and b are constants. This equation is called *second-order* since it involves derivatives up to order two. Assume that $y(t)$ is a solution (2.4), set $x_1(t) = y(t)$,

> **Phase portrait for a linear system in two dimensions**
>
> (1) From the characteristic equation $\lambda^2 - \tau\lambda + \Delta = 0$, where τ is the trace and Δ is the determinant of the coefficient matrix, determine the eigenvalues.
> (2) Classify the origin as a stable node, unstable node, stable focus, unstable focus, center, repeated real eigenvalue, or zero eigenvalue case. In the case of a repeated real eigenvalue, classify the origin as a star system if it is diagonal and degenerate node otherwise.
> (3) Proceed to draw the phase portrait in each case as listed previously.
>
> When using a computer program, such as Maple, Mathematica, or Matlab, to draw the phase portrait, certain steps are helpful.
>
> (1) Since the size of the region in the phase space plotted does not affect the appearance of the phase portrait for a linear system, pick a region of any size centered about the origin (e.g., $-5 \leq x_1 \leq 5$ and $-5 \leq x_2 \leq 5$). (The region plotted is sometimes called the *window* plotted.)
> (2) If you know what type of linear system it is, pick initial conditions that reveal the important behavior for that type of equation. Otherwise, experiment with initial conditions to determine the type of behavior of the linear system being drawn.
> (3) For a system with real eigenvalues, try to take initial conditions near, but on either side of, each of the half-lines that are scalar multiples of the eigenvectors. For unstable directions, either follow the solutions for negative time or start with initial conditions very near the origin. For stable directions, either follow the solutions for positive time or start with initial conditions near the origin and follow the trajectory for negative time.
> (4) For systems that oscillate (have complex eigenvalues), take enough initial conditions to reveal the phase portrait.
>
> **Table 5**

$x_2(t) = y'(t)$, and consider the vector $\mathbf{x}(t) = (x_1(t), x_2(t))^\mathsf{T} = (y(t), y'(t))^\mathsf{T}$. Then

$$(2.5) \qquad \mathbf{x}'(t) = \begin{bmatrix} y'(t) \\ y''(t) \end{bmatrix} = \begin{bmatrix} x_2(t) \\ -ax_2(t) - bx_1(t) \end{bmatrix} = \begin{bmatrix} 0 & 1 \\ -b & -a \end{bmatrix} \begin{bmatrix} x_1(t) \\ x_2(t) \end{bmatrix},$$

since $y''(t) = -ay'(t) - by(t) = -ax_2(t) - bx_1(t)$. We have shown that if $y(t)$ is a solution of the equation (2.4) then $\mathbf{x}(t) = (x_1(t), x_2(t))^\mathsf{T} = (y(t), y'(t))^\mathsf{T}$ is a solution of $\dot{\mathbf{x}} = \mathbf{A}\mathbf{x}$, where

$$(2.6) \qquad \mathbf{A} = \begin{bmatrix} 0 & 1 \\ -b & -a \end{bmatrix}.$$

Notice that the characteristic equation of (2.6) is $\lambda^2 + a\lambda + b = 0$, which is simply related to the original second-order equation (2.4). For the linear system $\mathbf{x}' = \mathbf{A}\mathbf{x}$ with $\mathbf{A}$ given by (2.6), we have to specify initial conditions of both $x_1(t_0)$ and

2.2. Constant Coefficients

$x_2(t_0)$. Therefore, for the second-order scalar equation $y'' + ay' + by = 0$, we have to specify initial conditions of both $y(t_0) = x_1(t_0)$ and $y'(t_0) = x_2(t_0)$.

Starting with a solution $\mathbf{x}(t) = (x_1(t), x_2(t))^\mathsf{T}$ of $\mathbf{x}' = \mathbf{A}\mathbf{x}$ with $\mathbf{A}$ given in (2.6), then the first coordinate $y(t) = x_1(t)$ satisfies

$$\ddot{y} = \dot{x}_2 = -b\,x_1 - a\,x_2 = -b\,y - a\,\dot{y},$$

and $y(t)$ is a solution of $y'' + ay' + by = 0$.

Since second-order scalar equations can be solved by converting them to a first order linear system, we do not need to give a separate method of solving them. However, the following result gives the more direct solution method that is discussed in elementary courses on differential equations.

Theorem 2.5. **a.** *The characteristic equation of* (2.6) *is* $\lambda^2 + a\lambda + b = 0$.

  b. *A function* $y = e^{rt}$ *is a solution of* $y'' + ay' + by = 0$ *if and only if* r *is a root of* $\lambda^2 + a\lambda + b = 0$.

  c. *If* r *is a double root of* $\lambda^2 + a\lambda + b = 0$, *then two independent solutions are* e^{rt} *and* $t\,e^{rt}$.

  d. *If* $r = \alpha \pm \beta i$ *are complex roots of* $\lambda^2 + a\lambda + b = 0$, *then two independent real solutions are* $e^{\alpha t}\cos(\beta t)$ *and* $e^{\alpha t}\sin(\beta t)$.

Proof. The proof of part (**a**) is direct and left to the reader.

(**b**) Let $y = e^{rt}$. Then

$$y'' + a\,y' + b\,y = r^2 e^{rt} + a r e^{rt} + b e^{rt} = e^{rt}\,(r^2 + ar + b).$$

Since $e^{rt} \neq 0$, e^{rt} is a solution if and only if r is a root of $\lambda^2 + a\lambda + b = 0$.

(**c**) This part is usually checked directly. However, if r is a double root, then the equation is $0 = (\lambda - r)^2 = \lambda^2 - 2r\lambda + r^2$ and the equation is $y'' - 2ry' + r^2 y = 0$. The matrix of the linear system is

$$\mathbf{A} = \begin{pmatrix} 0 & 1 \\ -r^2 & 2r \end{pmatrix} \quad \text{and} \quad \mathbf{A} - r\mathbf{I} = \begin{pmatrix} -r & 1 \\ -r^2 & 2r \end{pmatrix}.$$

An eigenvector for $\lambda = r$ is $(1, r)^\mathsf{T}$ and a generalized eigenvector is $(0, 1)^\mathsf{T}$. Thus, the two independent solutions of the linear system are

$$e^{rt}\begin{pmatrix} 1 \\ r \end{pmatrix} \quad \text{and} \quad e^{rt}\left[\begin{pmatrix} 0 \\ 1 \end{pmatrix} + t\begin{pmatrix} 1 \\ r \end{pmatrix}\right].$$

The first components of the two solutions are e^{rt} and $t\,e^{rt}$.

(**d**) The characteristic equation for the complex roots $\alpha \pm \beta i$ is $0 = \lambda^2 - 2\alpha\lambda + \alpha^2 + \beta^2$, and the matrix is

$$\mathbf{A} = \begin{pmatrix} 0 & 1 \\ -\alpha^2 - \beta^2 & 2\alpha \end{pmatrix} \quad \text{and} \quad \mathbf{A} - (\alpha + \beta i)\mathbf{I} = \begin{pmatrix} -\alpha - \beta i & 1 \\ -\alpha^2 - \beta^2 & \alpha - \beta i \end{pmatrix}.$$

A complex eigenvector for $\alpha + \beta i$ is $(1, \alpha + i\beta)^\mathsf{T}$. The first component of the complex solution is

$$e^{t\alpha}\left[\cos(\beta t) + i\sin(\beta t)\right],$$

which has real and imaginary parts as given in the statement of the theorem. □

2.2.4. Quasiperiodic Systems. This subsection considers linear systems in higher dimensions to introduce the idea of quasiperiodicity, a function with two or more frequencies which have no common period. The first example uses two harmonic oscillators that are not coupled (i.e., the equations of one oscillator do not contain the variables from the other oscillator). The second example again has two frequencies, but in the original variables, the equations of motion involve both position variables (i.e., the harmonic oscillators are coupled).

Example 2.19 (Two uncoupled harmonic oscillators). Consider two uncoupled oscillators given by
$$\ddot{x}_1 = -\omega_1^2 x_1 \quad \text{and} \quad \ddot{x}_2 = -\omega_2^2 x_2.$$
We can complete this to a first-order system of equations by letting $\dot{x}_1 = x_3$ and $\dot{x}_2 = x_4$. Then,
$$\dot{x}_1 = x_3,$$
$$\dot{x}_2 = x_4,$$
$$\dot{x}_3 = -\omega_1^2 x_1,$$
$$\dot{x}_4 = -\omega_2^2 x_2.$$
A direct calculation shows that this system has eigenvalues $\pm i\,\omega_1$ and $\pm i\,\omega_2$. The x_1 and x_2-components of a general solution are
$$\begin{pmatrix} x_1(t) \\ x_2(t) \end{pmatrix} = c_1 \cos(\omega_1 t) \begin{pmatrix} 1 \\ 0 \end{pmatrix} + c_2 \sin(\omega_1 t) \begin{pmatrix} 1 \\ 0 \end{pmatrix}$$
$$+ c_3 \cos(\omega_2 t) \begin{pmatrix} 0 \\ 1 \end{pmatrix} + c_4 \sin(\omega_2 t) \begin{pmatrix} 0 \\ 1 \end{pmatrix}$$
$$= R_1 \cos(\omega_1(t - \delta_1)) \begin{pmatrix} 1 \\ 0 \end{pmatrix} + R_2 \cos(\omega_2(t - \delta_2)) \begin{pmatrix} 0 \\ 1 \end{pmatrix},$$
where
$$c_1 = R_1 \cos(\omega_1 \delta_1), \quad c_2 = R_1 \sin(\omega_1 \delta_1),$$
$$c_3 = R_2 \cos(\omega_2 \delta_2), \quad c_4 = R_2 \sin(\omega_2 \delta_2).$$
Thus, the four arbitrary constants which give the general solution can be taken to be R_1, R_2, δ_1, and δ_2, rather than $c_1, \ldots, c_4$. If both R_1 and R_2 are nonzero, then the solution can be periodic with period T, if and only if there exist integers k and m such that
$$T\omega_1 = k\,2\pi \quad \text{and} \quad T\omega_2 = m\,2\pi,$$
$$T = \frac{k\,2\pi}{\omega_1} = \frac{m\,2\pi}{\omega_2}, \quad \text{or}$$
$$\frac{\omega_2}{\omega_1} = \frac{m}{k}.$$
Therefore, to be periodic, the ratio of the frequencies needs to be a rational number. When this ratio is irrational, the solution cannot be periodic but is generated by two frequencies. Such a solution is called *quasiperiodic*. For the case in which $\omega_1 = 1$ and $\omega_2 = \sqrt{2}$, a plot of a solution in the (x_1, x_2)-space is given in Figure 14 for $R_1 = R_2 = 1$ and $\delta_1 = \delta_2 = 0$. Notice that the solution tends to fill up

2.2. Constant Coefficients

the whole square $-1 \leq x_1, x_2 \leq 1$. The plot of $x_1 + x_2$ as a function of t for these constants is given in Figure 15.

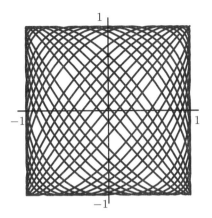

Figure 14. Quasiperiodic solution for Example 2.19: plot of $(x_1(t), x_2(t))$

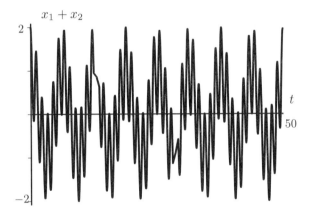

Figure 15. Quasiperiodic solution for Example 2.19: plot of $x_1 + x_2$ versus t

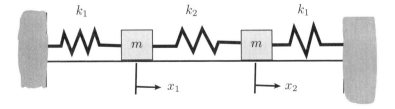

Figure 16. Two coupled oscillators

Example 2.20 (Coupled harmonic oscillators). Consider two equal masses m attached by springs between two fixed walls and supported by a floor so gravity is not a factor. See Figure 16. We assume that there is no friction, but only forces induced by the springs. Let the first body be the one of the left. There is a linear spring attached to the first body and the left wall with spring constant k_1. The second body is attached to the right wall by a spring with the same spring constant k_1. Between the two bodies, there is a third spring with spring constant k_2. Let x_1 be the displacement of the first body from the equilibrium position and x_2 be the displacement of the second body from its equilibrium position. If the first body is moved to the right (a positive value of x_1), there is a restoring force of $-k_1 x_1$ because the first spring is stretched and a restoring force of $-k_2 x_1$ because the middle spring is compressed. If the second body is now moved to the right, the middle spring is stretched and there is a force of $k_2 x_2$ exerted on the first body. Altogether, the forces on the first body are $-k_1 x_1 - k_2 x_1 + k_2 x_2 = -(k_1 + k_2)x_1 + k_2 x_2$. This force has to equal the mass times the acceleration of the first body, so the equation of motion is

$$m\ddot{x}_1 = -(k_1 + k_2)x_1 + k_2 x_2.$$

There are similar forces on the second body, and so it has a similar equation of motion with the indices 1 and 2 interchanged. The combined system of equations is

$$m\ddot{x}_1 = -(k_1 + k_2)\,x_1 + k_2\,x_2,$$
$$m\ddot{x}_2 = k_2\,x_1 - (k_1 + k_2)\,x_2.$$

This system can be made to be first order by setting $\dot{x}_1 = x_3$ and $\dot{x}_2 = x_4$. The system of equations then becomes

$$\begin{pmatrix}\dot{x}_1\\\dot{x}_2\\\dot{x}_3\\\dot{x}_4\end{pmatrix} = \begin{pmatrix} 0 & 0 & 1 & 0 \\ 0 & 0 & 0 & 1 \\ -\dfrac{k_1+k_2}{m} & \dfrac{k_2}{m} & 0 & 0 \\ \dfrac{k_2}{m} & -\dfrac{k_1+k_2}{m} & 0 & 0 \end{pmatrix} \begin{pmatrix}x_1\\x_2\\x_3\\x_4\end{pmatrix}.$$

Let

$$\mathbf{K} = \begin{pmatrix} -\dfrac{k_1+k_2}{m} & \dfrac{k_2}{m} \\ \dfrac{k_2}{m} & -\dfrac{k_1+k_2}{m} \end{pmatrix}$$

be the 2×2 matrix in the bottom left. Let λ be an eigenvalue for the total matrix $\mathbf{A}$ with eigenvector $\begin{pmatrix}\mathbf{v}\\\mathbf{w}\end{pmatrix}$, where $\mathbf{v}$ and $\mathbf{w}$ are each vectors with two components each. Then,

$$\lambda \begin{pmatrix}\mathbf{v}\\\mathbf{w}\end{pmatrix} = \mathbf{A} \begin{pmatrix}\mathbf{v}\\\mathbf{w}\end{pmatrix} = \begin{pmatrix}\mathbf{0} & \mathbf{I}\\\mathbf{K} & \mathbf{0}\end{pmatrix}\begin{pmatrix}\mathbf{v}\\\mathbf{w}\end{pmatrix} = \begin{pmatrix}\mathbf{w}\\\mathbf{Kv}\end{pmatrix};$$

setting the components equal, we get

$$\mathbf{w} = \lambda \mathbf{v} \quad \text{and}$$
$$\mathbf{Kv} = \lambda \mathbf{w} = \lambda^2 \mathbf{v}.$$

2.2. Constant Coefficients

Therefore, the square of the eigenvalues of **A** are actually the eigenvalues of **K**, and the characteristic equation of **A** is

$$\lambda^4 + \frac{2(k_1 + k_2)}{m}\lambda^2 + \frac{(k_1 + k_2)^2 - k_2^2}{m^2} = 0.$$

This equation has solutions

$$\lambda^2 = \frac{-(k_1 + k_2) \pm k_2}{m} = -\frac{k_1}{m}, \; -\left(\frac{k_1 + 2k_2}{m}\right), \quad \text{or}$$

$$\lambda = \pm i\,\omega_1, \pm i\,\omega_2$$

where $\omega_1 = \sqrt{k_1/m}$ and $\omega_2 = \sqrt{(k_1 + 2k_2)/m}$. The eigenvectors of **K** for the eigenvalues $-\frac{k_1}{m}$ and $-\left(\frac{k_1 + 2k_2}{m}\right)$ are, respectively,

$$\begin{pmatrix} 1 \\ 1 \end{pmatrix} \quad \text{and} \quad \begin{pmatrix} 1 \\ -1 \end{pmatrix}.$$

These vectors give the first two components of the eigenvectors of the total matrix **A**. Therefore, the first two components of the general solution are given by

$$\begin{pmatrix} x_1(t) \\ x_2(t) \end{pmatrix} = R_1 \cos(\omega_1(t - \delta_1)) \begin{pmatrix} 1 \\ 1 \end{pmatrix} + R_2 \cos(\omega_2(t - \delta_2)) \begin{pmatrix} 1 \\ -1 \end{pmatrix},$$

where R_1, R_2, δ_1, and δ_2 are arbitrary constants. Notice that the first solution in the sum corresponds to the two bodies moving in the same direction and the middle spring remaining unstretched and uncompressed. The resulting frequency is determined by the two end springs with spring constants k_1. The second solution corresponds to the bodies moving symmetrically with respect to the center position, pulsating toward and away from each other. In this latter case, all three springs are stretched and compressed, so the frequency involves both k_1 and k_2.

If both R_1 and R_2 are nonzero, then the solution is periodic if and only if the ratio ω_2/ω_1 is rational. If the ratio is irrational, then the solution never repeats but is generated by the two frequencies ω_1 and ω_2. In this latter case, the solution is quasiperiodic, just as in the case of the uncoupled oscillators of Example 2.19.

A general quasiperiodic function can involve more than two frequencies. A finite set of frequencies $\omega_1, \ldots, \omega_n$ is called *rationally dependent* provided that there are integers $k_1, \ldots, k_n$, not all of which are zero, such that

$$k_1\omega_1 + \cdots + k_n\omega_n = 0.$$

A finite set of frequencies $\omega_1, \ldots, \omega_n$ is called *rationally independent* provided that it is not rationally dependent; that is, if

$$k_1\omega_1 + \cdots + k_n\omega_n = 0$$

for integers $k_1, \ldots, k_n$, then all the k_j must equal zero. If the frequencies are rationally dependent, then one of them can be written as a rational linear combination of the other ones; for example, if $k_n \neq 0$, then

$$\omega_n = -\frac{k_1}{k_n}\omega_1 - \cdots - \frac{k_{n-1}}{k_n}\omega_{n-1}.$$

If the frequencies are rationally independent, then none of the frequencies can be written as a rational combination of the other frequencies. A *quasiperiodic function* $h(t)$ is a function generated by a finite number of such rationally independent frequencies $\omega_1, \ldots, \omega_n$ and can be written as $h(t) = g(\omega_1 t, \ldots, \omega_n t)$ where g is periodic of period 1 in each of its arguments. For example, $h(t)$ could be a linear combination of $\cos(\omega_j t)$ and $\sin(\omega_j t)$ for $j = 1, \ldots, n$.

Exercises 2.2

1. For each of the following linear systems of differential equations, (i) find the general real solution, (ii) show that the solutions are linearly independent, and (iii) draw the phase portrait.

   a.
   $$\dot{\mathbf{x}} = \begin{pmatrix} 6 & -3 \\ 2 & 1 \end{pmatrix} \mathbf{x},$$

   b.
   $$\dot{\mathbf{x}} = \begin{pmatrix} -2 & 1 \\ -4 & 3 \end{pmatrix} \mathbf{x},$$

   c.
   $$\dot{\mathbf{x}} = \begin{pmatrix} -4 & 1 \\ -1 & -2 \end{pmatrix} \mathbf{x},$$

   d.
   $$\dot{\mathbf{x}} = \begin{pmatrix} 0 & -1 \\ 1 & 0 \end{pmatrix} \mathbf{x},$$

   e.
   $$\dot{\mathbf{x}} = \begin{pmatrix} 0 & 1 \\ -1 & 0 \end{pmatrix} \mathbf{x},$$

   f.
   $$\dot{\mathbf{x}} = \begin{pmatrix} 5 & -5 \\ 2 & -1 \end{pmatrix} \mathbf{x},$$

   g.
   $$\dot{\mathbf{x}} = \begin{pmatrix} 1 & -2 \\ 3 & -4 \end{pmatrix} \mathbf{x}.$$

2. Consider the system of linear differential equations $\dot{\mathbf{x}} = \mathbf{A}\mathbf{x}$, where the matrix
   $$\mathbf{A} = \begin{pmatrix} 6 & 6 & 6 \\ 5 & 11 & -1 \\ 1 & -5 & 7 \end{pmatrix},$$
   which has eigenvalues 0, 12, and 12. Find the general solution.

3. Consider the system of linear differential equations $\dot{\mathbf{x}} = \mathbf{A}\mathbf{x}$, where the matrix
   $$\mathbf{A} = \begin{pmatrix} 0 & 3 & 1 \\ 4 & 1 & -1 \\ 2 & 7 & -5 \end{pmatrix},$$
   which has eigenvalues 4, and $-4 \pm 2i$. Find the general solution.

4. The motion of a *damped harmonic oscillator* is determined by $m\ddot{y}+b\dot{y}+ky = 0$, where $m > 0$ is the mass, $b \geq 0$ is the damping constant or friction coefficient, and $k > 0$ is the spring constant.
   a. Rewrite the differential equation as a first-order system of linear equations. (See Chapter 1.)
   b. Classify the type of linear system depending on the size of $b \geq 0$.
5. Find the Jordan canonical form for each of the matrices in Exercise 1.
6. Assume that $\lambda_1 \neq \lambda_2$ are two real eigenvalues of $\mathbf{A}$ with corresponding eigenvectors $\mathbf{v}^1$ and $\mathbf{v}^2$. Prove that $\mathbf{v}^1$ and $\mathbf{v}^2$ are linearly independent. Hint: Assume that $\mathbf{0} = C_1 \mathbf{v}^1 + C_2 \mathbf{v}^2$, and also consider the equation $\mathbf{0} = \mathbf{A}\mathbf{0} = C_1 \mathbf{A} \mathbf{v}^1 + C_2 \mathbf{A} \mathbf{v}^2 = C_1 \lambda_1 \mathbf{v}^1 + C_2 \lambda_2 \mathbf{v}^2$.
7. Assume that $\mathbf{A}$ is a 2×2 matrix with two real distinct negative eigenvalues λ_1 and λ_2.
   a. Let $\mathbf{v}^1$ and $\mathbf{v}^2$ be the corresponding eigenvectors and $\mathbf{P} = (\mathbf{v}^1\ \mathbf{v}^2)$ the matrix with eigenvectors as columns. Prove that $\mathbf{AP} = \mathbf{PD}$, where $\mathbf{D} = \text{diag}(\lambda_1, \lambda_2)$ is the diagonal matrix having the eigenvalues as entries (i.e., $\mathbf{A} = \mathbf{PDP}^{-1}$).
   b. Let $\alpha = \min\{-\lambda_1, -\lambda_2\} > 0$. Prove that there is a positive constant K such that
   $$\|e^{t\mathbf{A}}\mathbf{x}\| \leq K\, e^{-t\alpha}\, \|\mathbf{x}\|$$
   for all $t \geq 0$ and all $\mathbf{x}$ in $\mathbb{R}^2$.

2.3. Nonhomogeneous Systems: Time-dependent Forcing

The last topic of this chapter, before considering applications, is linear systems with a time-dependent forcing term, or a *time-dependent linear differential equation*. The general *nonhomogeneous linear system of differential equations* we consider is

(2.7) $$\dot{\mathbf{x}} = \mathbf{A}\mathbf{x} + \mathbf{g}(t).$$

Given such an equation, we associate the corresponding homogeneous linear system of differential equations,

(2.8) $$\dot{\mathbf{x}} = \mathbf{A}\mathbf{x}.$$

The next theorem indicates the relationship between the solutions of equations (2.7) and (2.8).

Theorem 2.6. **a.** *Let $\mathbf{x}^1(t)$ and $\mathbf{x}^2(t)$ be two solutions of the nonhomogeneous linear differential equation (2.7). Then, $\mathbf{x}^1(t) - \mathbf{x}^2(t)$ is a solution of the homogeneous linear differential equation (2.8).*

b. *Let $\mathbf{x}^p(t)$ be any solution of the nonhomogeneous linear differential equation (2.7) and $\mathbf{x}^h(t)$ be a solution of the homogeneous linear differential equation (2.8). Then, $\mathbf{x}^p(t) + \mathbf{x}^h(t)$ is a solution of the nonhomogeneous linear differential equation (2.7).*

c. *Let $\mathbf{x}^p(t)$ be a solution of the nonhomogeneous linear differential equation (2.7) and $\mathbf{M}(t)$ be a fundamental matrix solution of the homogeneous linear differential equation (2.8). Then, any solution of the nonhomogeneous linear differential equation (2.7) can be written as $\mathbf{x}^p(t) + \mathbf{M}(t)\mathbf{c}$ for some vector $\mathbf{c}$.*

The preceding theorem says that it is enough to find one particular solution of the nonhomogeneous differential equation (2.7) and add to it the general solution of the homogeneous differential equation (2.8). Just as in the case of second-order scalar equations, sometimes it is possible to guess a solution. (This method is often called the method of undetermined coefficients.) A more general method is that of variation of parameters. This method is more cumbersome, but it always works. We state this method for nonhomogeneous linear systems with constant coefficients in the next theorem, using the exponential of a matrix. The more general form uses a fundamental matrix solution, but we do not state it that way because it looks messier. However, in an example, the matrix exponential is usually calculated using any fundamental matrix solution.

Theorem 2.7 (Variation of parameters). *The solution $\mathbf{x}(t)$ of the nonhomogeneous linear differential equation with initial condition $\mathbf{x}(0) = \mathbf{x}_0$ can be written as*

$$\mathbf{x}(t) = e^{\mathbf{A}t}\mathbf{x}_0 + e^{\mathbf{A}t} \int_0^t e^{-\mathbf{A}s}\mathbf{g}(s)\,ds.$$

If $\mathbf{M}(t)$ is a fundamental matrix solution, then

$$\mathbf{x}(t) = \mathbf{M}(t)\mathbf{M}(0)^{-1}\mathbf{x}_0 + \mathbf{M}(t)\int_0^t \mathbf{M}(s)^{-1}\mathbf{g}(s)\,ds.$$

The proof is at the end of the chapter. The case for a fundamental matrix solution follows from the exponential case since $e^{\mathbf{A}t} = \mathbf{M}(t)\mathbf{M}(0)^{-1}$ and $e^{-\mathbf{A}s} = \mathbf{M}(0)\mathbf{M}(s)^{-1}$. Notice that for the fundamental matrix solution found by the solution method, $\mathbf{M}(s)^{-1}$ does not equal $\mathbf{M}(-s)$ unless $\mathbf{M}(0) = \mathbf{I}$.

Note that the first term is the solution of the homogeneous equation. In the integral, the integrand can be thought of as pulling back the effects of the nonhomogeneous term to $t=0$ by means of the fundamental matrix of the homogeneous equation. The integral adds up these perturbations and then the fundamental matrix transfers them to time t by means of the fundamental matrix.

These calculations are usually very cumbersome. The following simple example is one which can be carried out.

Example 2.21. Consider

$$\begin{pmatrix} \dot{x}_1 \\ \dot{x}_2 \end{pmatrix} = \begin{pmatrix} 0 & \omega_0 \\ -\omega_0 & 0 \end{pmatrix}\begin{pmatrix} x_1 \\ x_2 \end{pmatrix} + B\begin{pmatrix} 0 \\ \sin(\omega t) \end{pmatrix},$$

with $\omega \neq \omega_0$. The solution of the homogeneous equation is

$$e^{\mathbf{A}t} = \begin{pmatrix} \cos(\omega_0 t) & \sin(\omega_0 t) \\ -\sin(\omega_0 t) & \cos(\omega_0 t) \end{pmatrix} \quad \text{and}$$

$$e^{-\mathbf{A}s} = \begin{pmatrix} \cos(\omega_0 s) & -\sin(\omega_0 s) \\ \sin(\omega_0 s) & \cos(\omega_0 s) \end{pmatrix}.$$

2.3. Nonhomogeneous

The integral term given in the theorem is

$$\begin{aligned}
\mathbf{x}^p(t) &= Be^{\mathbf{A}t} \int_0^t e^{-\mathbf{A}s} \begin{pmatrix} 0 \\ \sin(\omega s) \end{pmatrix} ds \\
&= Be^{\mathbf{A}t} \int_0^t \begin{pmatrix} -\sin(\omega_0 s)\sin(\omega s) \\ \cos(\omega_0 s)\sin(\omega s) \end{pmatrix} ds \\
&= \frac{B}{2} e^{\mathbf{A}t} \int_0^t \begin{pmatrix} \cos((\omega_0+\omega)s) - \cos((\omega_0-\omega)s) \\ \sin((\omega_0+\omega)s) + \sin((-\omega_0+\omega)s) \end{pmatrix} ds \\
&= \frac{B}{2} e^{\mathbf{A}t} \begin{pmatrix} \frac{1}{\omega_0+\omega}\sin((\omega_0+\omega)t) - \frac{1}{\omega_0-\omega}\sin((\omega_0-\omega)t) \\ \frac{-1}{\omega_0+\omega}\cos((\omega_0+\omega)t) - \frac{1}{\omega-\omega_0}\cos((\omega-\omega_0)t) \end{pmatrix} \\
&\quad + \frac{B}{2} e^{\mathbf{A}t} \begin{pmatrix} 0 \\ \frac{1}{\omega_0+\omega} + \frac{1}{\omega-\omega_0} \end{pmatrix}.
\end{aligned}$$

Using some algebra and identities from trigonometry, we get

$$\mathbf{x}^p(t) = \frac{B}{\omega_0^2-\omega^2} \begin{pmatrix} \omega_0 \sin(\omega t) \\ \omega \cos(\omega t) \end{pmatrix} + \frac{B}{2} e^{\mathbf{A}t} \begin{pmatrix} 0 \\ \frac{1}{\omega+\omega_0} + \frac{1}{\omega-\omega_0} \end{pmatrix}.$$

The general solution is

$$\begin{aligned}
\mathbf{x}(t) &= e^{\mathbf{A}t} \mathbf{x}_0 + \mathbf{x}^p(t) \\
&= e^{\mathbf{A}t} \left(\mathbf{x}_0 + \frac{B}{2} \begin{pmatrix} 0 \\ \frac{1}{\omega_0+\omega} + \frac{1}{\omega-\omega_0} \end{pmatrix} \right) \\
&\quad + \frac{B}{\omega_0^2-\omega^2} \begin{pmatrix} \omega_0 \sin(\omega t) \\ \omega \cos(\omega t) \end{pmatrix}.
\end{aligned}$$

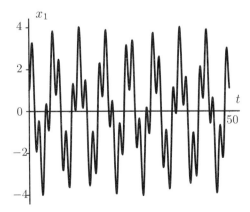

Figure 17. Plot of quasiperiodic solution for Example 2.21

The first term has period $\dfrac{2\pi}{\omega_0}$ and the second term has period $\dfrac{2\pi}{\omega}$. For this sum to be periodic it is necessary that there exists a common time T such that $T\omega_0 = k\,2\pi$ and $T\omega = m\,2\pi$, so $\dfrac{\omega}{\omega_0}$ must be rational. Therefore, the particular solution is periodic if and only if the ratio of the frequencies is a rational number. When the ratio is irrational, the solution has two frequencies with no common period and the frequencies are called rationally independent. Such combinations of rationally independent frequencies are called *quasiperiodic* just as we discussed for coupled and uncoupled harmonic oscillators. See Figure 17.

Exercises 2.3

1. Find the general solution of the differential equation
$$\dot{\mathbf{x}} = \begin{pmatrix} 0 & 1 \\ -1 & 0 \end{pmatrix} \mathbf{x} + \begin{pmatrix} 1 \\ 1 \end{pmatrix}.$$

2. Find the general solution of the differential equation
$$\dot{\mathbf{x}} = \begin{pmatrix} -2 & 1 \\ 1 & -2 \end{pmatrix} \mathbf{x} + \begin{pmatrix} \cos(t) \\ 0 \end{pmatrix}.$$

3. Find the general solution the nonhomogeneous linear (scalar) equation
$$\dot{x} = -x + \sin(t).$$
Note that the variation of parameters formula is equivalent to what is usually called "solving the nonhomogeneous linear (scalar) equation by means of an integrating factor."

4. Assume that μ is not an eigenvalue of the constant $n \times n$ matrix $\mathbf{A}$ and $\mathbf{b}$ is any constant n-vector. Show that the nonhomogeneous linear system of equations
$$\dot{\mathbf{x}} = \mathbf{A}\mathbf{x} + e^{\mu t}\mathbf{b}$$
has a solution of the form $\phi(t) = e^{\mu t}\mathbf{a}$ for some n-vector $\mathbf{a}$.

5. Find the general solution the nonhomogeneous linear equation
$$\dot{\mathbf{x}} = \begin{bmatrix} 1 & 0 \\ 0 & 2 \end{bmatrix} \mathbf{x} + \begin{bmatrix} e^{-t} \\ 0 \end{bmatrix}$$
with $\mathbf{x}(0) = (1, 3)^T$.

2.4. Applications

2.4.1. Model for Malignant Tumors.
Kaplan and Glass [**Kap95**] present a model for the metastasis of malignant tumors, based on the research of Liotta and DeLisi. In the experiments of Liotta and DeLisi, tumor cells were added to the blood stream of laboratory mice. The blood carried these cells to the capillaries of the lungs and then some entered the tissue of the lungs. By radioactively marking the cells, the levels of cancer cells could be measured over time. The rate of decay did not follow a simple exponential decay model which would be the case if there

2.4. Applications

were a complete transfer. Therefore, they use two variables: The first variable x_1 measures the number of cancer cells in the capillaries, and the second variable x_2 measures the number of cancer cells in the lung tissue. The rate of transfer of cancer cells from the capillaries to the lung tissue was hypothesized as a linear function of x_1, $\beta_2 x_1$. The rate of loss of cancer cells from the capillaries due to their being dislodged and carried away by the blood was given as $-\beta_1 x_1$. Finally, the loss of cancer cells in the lung tissue was given as $-\beta_3 x_2$. Thus, the system of differential equations is linear and is given as

$$\dot{x}_1 = -(\beta_1 + \beta_2)\, x_1,$$
$$\dot{x}_2 = \beta_2\, x_1 - \beta_3\, x_2.$$

This description is a *compartmental model*, with the number of cancer cells divided between those in the capillaries and those in the lung tissue, and with a rate of transfer between these two "compartments". The cells are assumed to be homogeneous in each compartment so the reaction rate depends only on the amount in each compartment and not on any distribution within the compartment. This assumption implies that the concentration instantaneously becomes uniformly distributed within each compartment as the amounts change.

The matrix of this system is

$$\begin{pmatrix} -(\beta_1 + \beta_2) & 0 \\ \beta_2 & -\beta_3 \end{pmatrix},$$

which has eigenvalues $-(\beta_1 + \beta_2)$ and $-\beta_3$, and eigenvectors

$$\begin{pmatrix} \beta_3 - (\beta_1 + \beta_2) \\ \beta_2 \end{pmatrix} \quad \text{and} \quad \begin{pmatrix} 0 \\ 1 \end{pmatrix}.$$

If the initial conditions are $x_1(0) = N$ and $x_2(0) = 0$, then the solution is

$$\begin{pmatrix} x_1(t) \\ x_2(t) \end{pmatrix} = \left(\frac{N}{\beta_3 - (\beta_1 + \beta_2)} \right) \begin{pmatrix} \beta_3 - (\beta_1 + \beta_2) \\ \beta_2 \end{pmatrix} e^{-(\beta_1 + \beta_2)t}$$
$$- \left(\frac{\beta_2 N}{\beta_3 - (\beta_1 + \beta_2)} \right) \begin{pmatrix} 0 \\ 1 \end{pmatrix} e^{-\beta_2 t}.$$

In the experiment, what could be measured was the total amount of radioactivity in both the capillaries and lung tissue, or

$$x_1(t) + x_2(t) = \left(\frac{(\beta_3 - \beta_1)\, N}{\beta_3 - (\beta_1 + \beta_2)} \right) e^{-(\beta_1 + \beta_2)t}$$
$$- \left(\frac{\beta_2 N}{\beta_3 - (\beta_1 + \beta_2)} \right) e^{-\beta_2 t}.$$

This sum has two rates of decay, rather than just one as the result of the two compartments. By matching the data of the laboratory experiments on mice, they found that the best fit was with $\beta_1 = 0.32$, $\beta_2 = 0.072$, and $\beta_3 = 0.02$ measured in units per hour.

Using this type of model, it is possible to fit the parameters with different treatments for the cancer and decide which treatments are most effective.

2.4.2. Detection of Diabetes.

M. Braun presents a model for the detection of diabetes in the book [**Bra78**] based on the work in [**Ack69**]. The two variables g and h are the deviations of the levels of glucose and hormonal concentration from base level after several hours of fasting. When the patient enters the hospital, the blood glucose level is increased to a level $g(0)$ and then the body's response is measured for positive time. The initial hormonal level is taken as $h(0) = 0$. Then the response is measured starting at $t = 0$ when the glucose is administered. If the responses are assumed linear (or the system is linearized near the equilibrium as in equation (4.1) in Section 4.5), the resulting system of differential equations is

$$\dot{g} = -m_1 g - m_2 h,$$
$$\dot{h} = m_4 g - m_3 h,$$

where the m_j are parameters. If $\tau = m_1 + m_3$ and $\Delta = m_1 m_3 + m_2 m_4$, then the characteristic equation of the homogeneous part is $\lambda^2 + \tau\lambda + \Delta$. We assume that $\tau^2 - 4\Delta = -\omega^2 < 0$, so the eigenvalues are $-\tau/2 \pm i\omega$.

All solutions have a factor of $e^{-\tau t/2}$ and either $\cos(\omega t)$ or $\sin(\omega t)$. Just as for scalar second-order scalar solutions, the g-component of an arbitrary solution can be written as

$$g(t) = A\,e^{-\tau t/2}\,\cos(\omega(t-\delta)),$$

for some constants A and δ. For a given patient, the m_j constants, as well as the amplitude A and phase shift δ, are unknown. Rather than trying to determine the m_j, it is enough to determine the quantities τ and ω, in addition to A and δ.

By measuring the level of glucose when the patient arrives for a base level, and then again at later times, it is possible to determine g_j at times t_j. Since there are four unknown constants, we need at least four readings at times t_j to solve equations

$$g_j = A\,e^{-\tau t_j/2}\,\cos(\omega(t_j - \delta))$$

for the constants. Rather than using just four readings, it is better to take more and then use least squares to minimize the quantity

$$E = \sum_{j=1}^{n} [g_j - A\,e^{-\tau t_j/2}\,\cos(\omega(t_j - \delta))]^2.$$

(See [**Lay01**] for a discussion of least squares.) When this was carried out in a medical study as reported in [**Ack69**], it was found that a slight error in the reading leads to large errors in the constant τ. However, the constant ω was much more reliable; thus, ω was a better quantity to determine whether a person had diabetes. ω is essentially the period of oscillation of the levels of hormones and glucose in the blood. A person without diabetes had a period of $T_0 = 2\pi/\omega$ less than four hours, while a person with diabetes had T_0 greater than four hours.

2.4.3. Model for Inflation and Unemployment.

A model for inflation and unemployment is given in [**Chi84**] based on a Phillips relation as applied by M. Friedman. The variables are the expected rate of inflation π and the rate of unemployment U. There are two other auxiliary variables which are the actual rate of inflation p and the rate of growth of wages w. There are two quantities that are fixed externally, exogenous variables or parameters: The rate of monetary expansion is $m > 0$ and the increase in productivity is $T > 0$. The model assumes the following:

$$w = \alpha - \beta U,$$
$$p = h\pi - \beta U + \alpha - T \quad \text{with } 0 < h \leq 1,$$
$$\frac{d\pi}{dt} = j(p - \pi) = -j(1-h)\pi - j\beta U + j(\alpha - T),$$
$$\frac{dU}{dt} = k(p - m) = kh\pi - k\beta U + k(\alpha - T - m),$$

where α, β, h, j, and k are parameters of the modeling differential equation, so

$$\frac{d}{dt}\begin{bmatrix}\pi \\ U\end{bmatrix} = \begin{bmatrix}-j(1-h) & -j\beta \\ kh & -k\beta\end{bmatrix}\begin{bmatrix}\pi \\ U\end{bmatrix} + \begin{bmatrix}j(\alpha - T) \\ k(\alpha - T - m)\end{bmatrix}.$$

All the parameters are assumed to be positive.

The equilibrium is obtained by finding the values where the time derivatives are zero, $\frac{d\pi}{dt} = 0 = \frac{dU}{dt}$. This yields a system of two linear equations that can be solved to give $\pi^* = m$ and $U^* = \frac{1}{\beta}[\alpha - T - m(1-h)]$. At the equilibrium, both the expected rate of inflation π^* and the actual rate of inflation p^* equal the rate of monetary expansion m, and the unemployment is $U^* = \frac{1}{\beta}[\alpha - T - m(1-h)]$.

The variables $x_1 = \pi - \pi^*$ and $x_2 = U - U^*$ giving the displacement from equilibrium satisfy $\mathbf{x}' = \mathbf{A}\mathbf{x}$, where $\mathbf{A}$ is the coefficient matrix given above. A direct calculations shows that the trace and determinant are as follows:

$$\text{tr}(\mathbf{A}) = -j(1-h) - k\beta < 0,$$
$$\det(\mathbf{A}) = jk\beta > 0.$$

Since the eigenvalues r_1 and r_2 satisfy $r_1 + r_2 = \text{tr}(\mathbf{A}) < 0$ and $r_1 r_2 = \det(\mathbf{A}) > 0$, the real part of both eigenvalues must be negative. Therefore, any solutions of the linear equations in the $\mathbf{x}$-variables goes to $\mathbf{0}$ and $(\pi(t), U(t))^\mathsf{T}$ goes to $(\pi^*, U^*)^\mathsf{T}$ as t goes to infinity.

2.4.4. Input-Ouput Economic Model. A model for adjustment of output in a Leontief input-output model is given in [Chi84] where the rate of adjustment is given by the excess demand.

In the static Leontief input-output model with n sectors, the vector $\mathbf{x}$ is the production vector of the sectors, the final demand vector is $\mathbf{d}(t)$, which we allow to depend on time, the intermediate demand need for production is $\mathbf{Cx}$, where $\mathbf{C}$ is a given $n \times n$ matrix. The excess demand is $\mathbf{Cx} + \mathbf{d}(t) - \mathbf{x}$. If we assume that the rate of adjust of production is equal to the excess demand, then we get the system of differential equations

$$\mathbf{x}' = (\mathbf{C} - \mathbf{I})\mathbf{x} + \mathbf{d}(t).$$

As an example of an external final demand that grows exponentially with time, consider a two sector economy with final demand $\mathbf{d}(t) = e^{rt}(1,1)^\mathsf{T}$ with $r > 0$

a given parameter. If we guess at a solution $\mathbf{x}_p(t) = e^{rt}(b_1, b_2)^\mathsf{T}$ with b_1 and b_2 undetermined coefficients, then we need

$$re^{rt}\begin{bmatrix} b_1 \\ b_2 \end{bmatrix} = (\mathbf{C} - \mathbf{I})\, e^{rt}\begin{bmatrix} b_1 \\ b_2 \end{bmatrix} + e^{rt}\begin{bmatrix} 1 \\ 1 \end{bmatrix},$$

$$((r+1)\mathbf{I} - \mathbf{C})\begin{bmatrix} b_1 \\ b_2 \end{bmatrix} = \begin{bmatrix} 1 \\ 1 \end{bmatrix} \quad \text{or}$$

$$\begin{bmatrix} b_1 \\ b_2 \end{bmatrix} = \frac{1}{\Delta}\begin{bmatrix} r+1-c_{22} & -c_{12} \\ -c_{21} & r+1-c_{11} \end{bmatrix}\begin{bmatrix} 1 \\ 1 \end{bmatrix}$$

$$= \frac{1}{\Delta}\begin{bmatrix} r+1-c_{22}-c_{12} \\ r+1-c_{11}-c_{21} \end{bmatrix}$$

where $\Delta = \det((r+1)\mathbf{I} - \mathbf{C}) = (r+1-c_{11})(r+1-c_{22}) - c_{12}c_{21}$. Therefore, we have solved for the *undetermined coefficients* b_1 and b_2 in terms of the known parameters. For the particular solution, the output of the two sectors will grow at the same rate as the growth of the final demand.

The eigenvalues of the homogeneous system satisfy

$$\lambda^2 - [\operatorname{tr}(\mathbf{C}) - 2]\,\lambda + \det(\mathbf{C} - \mathbf{I}) = 0.$$

If $\operatorname{tr}(\mathbf{C}) - 2 < 0$ and $\det(\mathbf{C} - \mathbf{I}) = \det(\mathbf{C}) - \operatorname{tr}(\mathbf{C}) + 1 > 0$, then the eigenvalues both have negative real parts; if these inequalities hold, then any solution of the nonhomogeneous solution converges to the particular solution $\mathbf{x}_p(t)$ as t goes to infinity.

Exercises 2.4

1. (From Kaplan and Glass [**Kap95**]) An intravenous administration of a drug can be described by two-compartment model, with compartment 1 representing the blood plasma and compartment 2 representing body tissue. The dynamics of evolution of the system are given by the system of differential equations

   $$\dot{C}_1 = -(K_1 + K_2)\,C_1 + K_3\,C_2,$$
   $$\dot{C}_2 = K_1\,C_1 - K_3\,C_2,$$

   with K_1, K_2, and K_3 all positive.
   a. Draw a schematic diagram that shows the compartments and the flows into and out of them.
   b. Solve this system of differential equations for the initial conditions $C_1(0) = N$ and $C_2(0) = 0$ for the special case with $K_1 = 0.5$ and $K_2 = K_3 = 1$. What happens in the limit as t goes to infinity?
   c. Sketch the phase plane for the case when $K_1 = 0.5$ and $K_2 = K_3 = 1$.

2. Consider the mixing of concentrations of salt in three different tanks which are connected as shown in Figure 18. We assume that the volume V of each tank is the same. We assume that the substance is uniformly distributed within a given tank (due to rapid diffusion or mixing), so its concentration in the jth tank at a given time is given by a single number $C_j(t)$. Thus, $C_j(t)V$ is the amount of salt in this tank at time t. Let K_1 be the constant rate of flow from

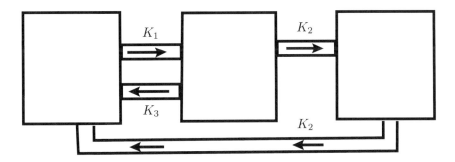

Figure 18. Flow between tanks for the closed system

the first tank to the second in gallons per minute. Let K_2 be the constant rate of flow from the second tank to the third tank and from the third tank to the first tank. Finally, let K_3 be the constant rate of flow back from the second tank to the first tank. We assume that $K_1 = K_2 + K_3$, so the amount of flow into and out of each tank balances. The system of differential equations for the amount of material in each tank is

$$\frac{d}{dt} V C_1 = -K_1 C_1 + (K_1 - K_2) C_2 + K_2 C_3,$$

$$\frac{d}{dt} V C_2 = K_1 C_1 - K_1 C_2,$$

$$\frac{d}{dt} V C_1 = K_2 C_2 - K_2 C_3.$$

Dividing by V, setting $k = K_2/V$, $bk = K_3/V$, and $K_1/V = k(1+b)$, we get the equations $K_1/V = k(1+b)$, we get the equations

$$\begin{pmatrix} \dot{C}_1 \\ \dot{C}_2 \\ \dot{C}_3 \end{pmatrix} = k \begin{pmatrix} -(1+b) & b & 1 \\ (1+b) & -(1+b) & 0 \\ 0 & 1 & -1 \end{pmatrix} \begin{pmatrix} C_1 \\ C_2 \\ C_3 \end{pmatrix}.$$

a. Show that the characteristic equation is
$$0 = -\lambda[\lambda^2 + (3+2b)k\lambda + (3+3b)k^2].$$

b. Show that the eigenvalues are roots 0 and $\dfrac{-(3+2b)k \pm k\sqrt{4b^2 - 3}}{2}$.

c. Show that the eigenvector for $\lambda = 0$ is $(1, 1, 1)^\mathsf{T}$. How does this relate to the fact that $\dfrac{d}{dt}(C_1 + C_2 + C_3) = 0$?

d. Show that the other two eigenvalues have negative relate part. Interpret the properties of a solution which starts with with $C_1(0) + C_2(0) + C_3(0) > 0$.

3. Consider the mixing of concentrations of a salt in two different tanks of equal volume V, which are connected as shown in Figure 19. The rate of flow into tank one from the outside is kV; the rate flowing out the system from tank two is also kV; the rate of flow from tank two to tank one is bkV; and the rate of flowing from tank one to tank two is $(1+b)kV$. We assume that the concentration of the salt in the fluid coming into the first tank from the outside

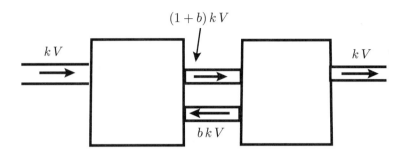

Figure 19. Flow between tanks with input and output

is $C_0(t)$, a given function of time. The system of differential equations for the change of concentrations in the two tanks is

$$\dot{C}_1 = -k(1+b)C_1 + kbC_2 + kC_0(t),$$
$$\dot{C}_2 = kC_1 - k(1+b)C_2,$$

or

$$\begin{pmatrix} \dot{C}_1 \\ \dot{C}_2 \end{pmatrix} = k \begin{pmatrix} -(1+b) & b \\ (1+b) & -(1+b) \end{pmatrix} \begin{pmatrix} C_1 \\ C_2 \end{pmatrix} + k \begin{pmatrix} C_0(t) \\ 0 \end{pmatrix}.$$

   a. Show that the eigenvalues of the matrix are $\lambda = -k(1+b) \pm k\sqrt{b+b^2}$. Also show that they are real and negative.
   b. Explain why any solution converges to a unique particular solution, depending on the rate of inflow of the salt.
   c. If $C_0(t) = 2+\sin(t)$, what are the limiting concentrations in the two tanks?
4. Consider an LRC electric circuit with $R > 0$, $L > 0$, and $C > 0$. Sketch the phase portrait for the three cases (a) $R^2 > 4L/C$, (b) $R^2 = 4L/C$, and (c) $R^2 < 4L/C$. What happens when $R = 0$ but $L > 0$ and $C > 0$.
5. Consider a Cournot model of a duopoly with profits * $\pi_1 = (6 - q_1 - q_2)q_1$ and $\pi_2 = (6 - q_1 - q_2)q_2$. The levels of output that maximize the profits of each firm fixing the output of the other firm satisfy $q_1 = 3 - q_2/2$ and $q_2 = 3 - q_1/2$. (These are found by setting the partial derivatives or marginal profits equal to zero.) Let these amounts be $x_1 = 3 - q_2/2$ and $x_2 = 3 - q_1/2$. Assume that the dynamics are proportional to to displacement from equilibrium,

$$\frac{dq_1}{dt} = k_1(x_1 - q_1) = 3k_1 - k_1 q_1 - \frac{k_1}{2}q_2,$$
$$\frac{dq_2}{dt} = k_2(x_2 - q_2) = 3k_2 - \frac{k_2}{2}q_1 - k_2 q_2,$$

with $k_1, k_2 > 0$.
   a. Find the equilibrium of the dynamic model.
   (Note that is also the solution of the static problem of finding the simultaneous maximum.)
   b. Show that the dynamic equilibrium is stable.
6. A simple model of dynamics for a Keynesian IS-LM continuous model for national income is given in Chapter 10 of [**Sho02**]. Let y be real income and

r the nominal interest rate. The assumptions are that (i) $m_0 + ky - ur$ is the demand for money with $m_0, k, u > 0$, and (ii) consumers' expenditure is $a + b(1-T)y - hr$ where a is the autonomous expenditure, b is the marginal propensity to consume, T is the marginal tax rate, and $h > 0$ the coefficient of investment in response to r. Let $A = 1 - b(1-T)$ with $0 < A < 1$. Assuming that income responds according linearly to the excess demand in that market with constant and that the interest rate responds linearly to the excess demand in the money market with proportionality constants $\gamma > 0$ and $\beta > 0$, then the system of differential equations are given by

$$y' = -\gamma A y - \gamma h r + \gamma a,$$
$$r' = \beta k y - \beta u r - \beta m_0.$$

a. Find the equilibrium for the system of differential equations.
b. Determine whether the equilibrium is stable or unstable.

2.5. Theory and Proofs

Fundamental set of solutions

The linear systems we study mainly have constant coefficients

(2.9)
$$\dot{\mathbf{x}} = \mathbf{A}\mathbf{x},$$

where the entries in the matrix $\mathbf{A}$ are constants. However, there are times when we need to consider the case in which the entries of $\mathbf{A}$ depend on time, $\mathbf{A}(t)$. (We always assume that the entries of $\mathbf{A}(t)$ are bounded for all time.) The differential equation formed using the matrix $\mathbf{A}(t)$,

(2.10)
$$\dot{\mathbf{x}} = \mathbf{A}(t)\mathbf{x},$$

is called a *time-dependent linear differential equation*. A linear combination of solutions of such an equation is still a solution of the equation. The method of solution given in Section 2.2 does not apply to such systems, but a fundamental matrix solution still exists even if it is very difficult to find.

Uniqueness of solutions

As noted in Section 2.1, uniqueness of solutions follows from the general treatment of nonlinear equations given in Theorem 3.2. However, we can give a much more elementary proof in the linear case, as we do it here.

We need a lemma covering the derivative of the inverse of a matrix.

Lemma 2.8. *Let $\mathbf{M}(t)$ be a square matrix whose entries depend differentiably on t. Also assume that $\mathbf{M}(t)$ is invertible (i.e., $\det(\mathbf{M}(t)) \neq 0$). Then, the derivative with respect to t of the inverse of $\mathbf{M}(t)$ is given as follows:*

$$\frac{d}{dt}\mathbf{M}(t)^{-1} = -\mathbf{M}(t)^{-1}\left(\frac{d}{dt}\mathbf{M}(t)\right)\mathbf{M}(t)^{-1}.$$

In particular, if $\mathbf{M}(t)$ is the fundamental matrix solution of $\dot{\mathbf{x}} = \mathbf{A}(t)\mathbf{x}$, then

$$\frac{d}{dt}\mathbf{M}(t)^{-1} = -\mathbf{M}(t)^{-1}\mathbf{A}(t).$$

Proof. The product of $\mathbf{M}(t)$ with its inverse is the identity matrix, so

$$\mathbf{0} = \frac{d}{dt}\mathbf{I}$$
$$= \frac{d}{dt}\left(\mathbf{M}(t)^{-1}\mathbf{M}(t)\right)$$
$$= \left(\frac{d}{dt}\mathbf{M}(t)^{-1}\right)\mathbf{M}(t) + \mathbf{M}(t)^{-1}\left(\frac{d}{dt}\mathbf{M}(t)\right)$$

and

$$\left(\frac{d}{dt}\mathbf{M}(t)^{-1}\right)\mathbf{M}(t) = -\mathbf{M}(t)^{-1}\left(\frac{d}{dt}\mathbf{M}(t)\right),$$
$$\left(\frac{d}{dt}\mathbf{M}(t)^{-1}\right) = -\mathbf{M}(t)^{-1}\left(\frac{d}{dt}\mathbf{M}(t)\right)\mathbf{M}(t)^{-1}.$$

The second statement about the fundamental matrix solution follows from the first by substituting $\mathbf{A}(t)\mathbf{M}(t)$ for $\frac{d}{dt}\mathbf{M}(t)$:

$$\frac{d}{dt}\mathbf{M}(t)^{-1} = -\mathbf{M}(t)^{-1}\left(\mathbf{A}(t)\mathbf{M}(t)\right)\mathbf{M}(t)^{-1}$$
$$= -\mathbf{M}(t)^{-1}\mathbf{A}(t).$$

□

Restatement of Theorem 2.1. *Let $\mathbf{A}(t)$ be a $n \times n$ matrix whose entries depend continuously on t. Given $\mathbf{x}_0$ in $\mathbb{R}^n$, there is at most one solution $\mathbf{x}(t; \mathbf{x}_0)$ of $\dot{\mathbf{x}} = \mathbf{A}(t)\mathbf{x}$ with $\mathbf{x}(0; \mathbf{x}_0) = \mathbf{x}_0$.*

Proof. Let $\mathbf{x}(t)$ be any solution with this initial condition, and let $\mathbf{M}(t)$ be a fundamental matrix solution. We want to understand how $\mathbf{x}(t)$ differs from $\mathbf{M}(t)\mathbf{M}(0)^{-1}\mathbf{x}_0$, so we introduce the quantity $\mathbf{y}(t) = \mathbf{M}(t)^{-1}\mathbf{x}(t)$. Then,

$$\dot{\mathbf{y}}(t) = \left(\frac{d}{dt}\mathbf{M}(t)^{-1}\right)\mathbf{x}(t) + \mathbf{M}(t)^{-1}\dot{\mathbf{x}}(t)$$
$$= -\mathbf{M}(t)^{-1}\mathbf{A}(t)\mathbf{x}(t) + \mathbf{M}(t)^{-1}\mathbf{A}(t)\mathbf{x}(t)$$
$$= \mathbf{0}.$$

The derivative of $\mathbf{y}(t)$ is zero, so it must be a constant, $\mathbf{y}(t) = \mathbf{y}(0) = \mathbf{M}(0)^{-1}\mathbf{x}_0$, and $\mathbf{x}(t) = \mathbf{M}(t)\mathbf{y}(t) = \mathbf{M}(t)\mathbf{M}(0)^{-1}\mathbf{x}_0$. This proves that any solution with the given initial conditions must equal $\mathbf{M}(t)\mathbf{M}(0)^{-1}\mathbf{x}_0$. (In essence, this is just the proof of the variation of parameters formula for solutions to nonhomogeneous equations that we give later in Section 2.5.) □

Existence for all time

Solutions of linear equations exist for all time. This is not always the case for nonlinear equations. The proof uses the idea of the norm of a matrix. We introduce this first.

Given a matrix $\mathbf{A}$, there is a number C such that

$$\|\mathbf{A}\mathbf{v}\| \leq C\|\mathbf{v}\|$$

2.5. Theory and Proofs

for all vectors $\mathbf{v}$. The smallest such number is written $\|\mathbf{A}\|$, so

$$\|\mathbf{A}\mathbf{v}\| \leq \|\mathbf{A}\| \cdot \|\mathbf{v}\|.$$

This number is called the *norm* of the matrix. (See Appendix A.3.) Usually, we do not need to calculate it explicitly, but just know that it exists. However, from the definition, it follows that

$$\begin{aligned}\|\mathbf{A}\| &= \max_{\|\mathbf{v}\|\neq 0} \frac{\|\mathbf{A}\mathbf{v}\|}{\|\mathbf{v}\|} \\ &= \max_{\|\mathbf{w}\|=1} \|\mathbf{A}\mathbf{w}\|.\end{aligned}$$

This form of the norm does not provide an easy way to calculate it. We can consider the square of the length of the image of a unit vector by the matrix

$$\|\mathbf{A}\mathbf{w}\|^2 = \mathbf{w}^\mathsf{T} \mathbf{A}^\mathsf{T} \mathbf{A} \mathbf{w}.$$

The product $\mathbf{A}^\mathsf{T}\mathbf{A}$ is always a symmetric matrix, so it has real eigenvalues with eigenvectors that are perpendicular. It follows that the maximum of $\|\mathbf{A}\mathbf{w}\|^2$ for unit vectors $\mathbf{w}$ is the largest eigenvalue of $\mathbf{A}^\mathsf{T}\mathbf{A}$. Therefore, the norm of $\mathbf{A}$ is the square root of the largest eigenvalue of $\mathbf{A}^\mathsf{T}\mathbf{A}$. This largest eigenvalue is (at least theoretically) computable.

Now, we return to the result dealing with differential equations.

Theorem 2.9. *Assume* $\mathbf{A}(t)$ *is a real* $n \times n$ *matrix with bounded entries. Then solutions of* $\dot{\mathbf{x}} = \mathbf{A}(t)\mathbf{x}$ *exist for all time.*

The proof of this theorem is more advanced than most other topics considered in this chapter, and can easily be skipped.

Proof. We consider the square of the length of a solution $\mathbf{x}(t)$:

$$\begin{aligned}\frac{d}{dt}\|\mathbf{x}(t)\|^2 &= \frac{d}{dt}\left(\mathbf{x}(t)\cdot\mathbf{x}(t)\right) \\ &= 2\,\mathbf{x}(t)\cdot\dot{\mathbf{x}}(t) \\ &= 2\,\mathbf{x}(t)\cdot\mathbf{A}(t)\mathbf{x}(t) \\ &\leq 2\,\|\mathbf{x}(t)\|\cdot\|\mathbf{A}(t)\mathbf{x}(t)\| \\ &\leq 2\,\|\mathbf{A}(t)\|\cdot\|\mathbf{x}(t)\|^2.\end{aligned}$$

Because of the assumption on the entries of $\mathbf{A}(t)$, there is a constant C such that $\|\mathbf{A}(t)\| \leq C$ for all t. (In the case of constant coefficients, we can take $C = \|\mathbf{A}\|$.) Therefore,

$$\frac{d}{dt}\|\mathbf{x}(t)\|^2 \leq 2C\|\mathbf{x}(t)\|^2$$

for some constant C. In this case, by dividing by $\|\mathbf{x}(t)\|^2$ and integrating, we see that

$$\|\mathbf{x}(t)\|^2 \leq \|\mathbf{x}(0)\|^2 e^{2Ct} \quad \text{or}$$
$$\|\mathbf{x}(t)\| \leq \|\mathbf{x}(0)\| e^{Ct}.$$

(This is a special case of Gronwall's inequality given in Lemma 3.8, which covers the case when $\mathbf{x}(t) = \mathbf{0}$ for some t.) This quantity does not go to infinity in finite time, so if the solution is defined for $0 \leq t < t_+ < \infty$, then

$$\|\mathbf{A}(t)\mathbf{x}(t)\| \leq C \|\mathbf{x}(0)\| e^{Ct_+} = K$$

and

$$\|\mathbf{x}(t_2) - \mathbf{x}(t_1)\| = \left\| \int_{t_1}^{t_2} \mathbf{A}(t)\mathbf{x}(t)\, dt \right\| \leq \int_{t_1}^{t_2} \|\mathbf{A}(t)\mathbf{x}(t)\|\, dt$$
$$\leq \int_{t_1}^{t_2} K\, dt = K\, |t_2 - t_1|.$$

Therefore, $\mathbf{x}(t)$ is Cauchy with respect to t and must converge to a limit as t converges to t_+, the solution is defined at time t_+, and the time interval on which the solution is defined is closed. It is open, because we can always find a solution on a short time interval. This contradicts the assumption that $t_+ < \infty$, and the solution can be extended for all time. $\square$

Vector space of solutions

Theorem 2.10. *Consider the differential equation* $\dot{\mathbf{x}} = \mathbf{A}(t)\mathbf{x}$ *where* $\mathbf{A}(t)$ *is a real* $n \times n$ *matrix with bounded entries. Let*

$$\mathscr{S} = \{\, \mathbf{x}(\cdot) : \mathbf{x}(t) \text{ is a solutions of the differential equation}\,\}.$$

Then $\mathscr{S}$ *is a vector space of dimension* n.

Proof. Each solution in $\mathscr{S}$ exists for all time so we can take a linear combinations of solutions and get a solution by Theorem 2.2. Therefore, $\mathscr{S}$ is a vector space.

If $\mathbf{u}^1, \ldots, \mathbf{u}^n$ is the standard basis of $\mathbb{R}^n$ (or any other basis, in fact), then there are solutions $\mathbf{x}^j(t)$ with initial conditions $\mathbf{u}^j$ for $j = 1, \ldots, n$. For any solution $\mathbf{x}(t)$ in $\mathscr{S}$, its initial condition $\mathbf{x}_0$ can be written as $\mathbf{x}_0 = a_1 \mathbf{u}^1 + \cdots + a_n \mathbf{u}^n$. Then, $a_1 \mathbf{x}^1(t) + \cdots + a_n \mathbf{x}^n(t)$ is also a solution with initial condition $\mathbf{x}_0$. By uniqueness, $\mathbf{x}(t) = a_1 \mathbf{x}^1(t) + \cdots + a_n \mathbf{x}^n(t)$ is a linear combination of the solutions $\mathbf{x}^j(t)$. This shows that these solutions span $\mathscr{S}$.

If $a_1 \mathbf{x}^1(t) + \cdots + a_n \mathbf{x}^n(t) = \mathbf{0}$, then at time $t = 0$, $a_1 \mathbf{u}^1 + \cdots + a_n \mathbf{u}^n = \mathbf{0}$. The basis vectors $\{\mathbf{u}^j\}$ are independent, so all the $a_j = 0$. This shows that the $\{\mathbf{x}^j(t)\}$ are linearly independent. Thus, they are a basis and $\mathscr{S}$ has dimension n. $\square$

Change of volume

We next give a proof of what is called the Liouville formula for the determinant of a fundamental set of solutions in terms of an integral. This formula is used later in the discussion of Lyapunov exponents for nonlinear systems. We give a proof at that time using the divergence theorem. The proof given here is direct, but is somewhat messy in n dimensions.

2.5. Theory and Proofs

Restatement of Theorem 2.3. *Let $\mathbf{M}(t)$ be a fundamental matrix solution for a linear system of differential equations $\dot{\mathbf{x}} = \mathbf{A}(t)\mathbf{x}$, and let $W(t) = \det(\mathbf{M}(t))$ be the Wronskian. Then,*

$$\frac{d}{dt} W(t) = \operatorname{tr}(\mathbf{A}(t))\, W(t) \qquad \text{and}$$

$$W(t) = W(t_0) \exp\left(\int_{t_0}^{t} \operatorname{tr}(\mathbf{A}(s))\, ds \right),$$

where $\exp(z) = e^z$ is the exponential function and $\operatorname{tr}(\mathbf{A}(s))$ is the trace of the matrix $\mathbf{A}(s)$. In particular, if $W(t_0) \neq 0$ for any time t_0, then $W(t) \neq 0$ for all times t.

For a constant coefficient equation, $e^{\mathbf{A}t}$ is a fundamental matrix solution with

$$\det\left(e^{\mathbf{A}t}\right) = e^{\operatorname{tr}(\mathbf{A})\, t}.$$

Proof. The main property of determinants used is that we can add a scalar multiple of one column to another column and not change the determinant. Therefore, if a column is repeated the determinant is zero. Also, the determinant is linear in the columns, so the product rule applies.

Considering the derivative at time $t = t_0$ and using $\mathbf{u}^j$ for the standard basis of $\mathbb{R}^n$,

$$\frac{d}{dt} \det\left(\mathbf{M}(t)\mathbf{M}(t_0)^{-1}\right)\Big|_{t=t_0}$$

$$= \frac{d}{dt} \det\left(\mathbf{M}(t)\mathbf{M}(t_0)^{-1}\mathbf{u}^1, \ldots, \mathbf{M}(t)\mathbf{M}(t_0)^{-1}\mathbf{u}^n\right)\Big|_{t=t_0}$$

$$= \sum_j \det\Big(\mathbf{M}(t_0)\mathbf{M}(t_0)^{-1}\mathbf{u}^1, \ldots, \mathbf{M}(t_0)\mathbf{M}(t_0)^{-1}\mathbf{u}^{j-1},$$

$$\mathbf{M}'(t_0)\mathbf{M}(t_0)^{-1}\mathbf{u}^j, \mathbf{M}(t_0)\mathbf{M}(t_0)^{-1}\mathbf{u}^{j+1}, \ldots, \mathbf{M}(t_0)\mathbf{M}(t_0)^{-1}\mathbf{u}^n\Big)$$

$$= \sum_j \det\left(\mathbf{u}^1, \ldots, \mathbf{u}^{j-1}, \mathbf{A}(t_0)\mathbf{M}(t_0)\mathbf{M}(t_0)^{-1}\mathbf{u}^j, \mathbf{u}^{j+1}, \ldots, \mathbf{u}^n\right)$$

$$= \sum_j \det\left(\mathbf{u}^1, \ldots, \mathbf{u}^{j-1}, \mathbf{A}(t_0)\mathbf{u}^j, \mathbf{u}^{j+1}, \ldots, \mathbf{u}^n\right)$$

$$= \sum_j \det\left(\mathbf{u}^1, \ldots, \mathbf{u}^{j-1}, \sum_i a_{i,j}(t_0)\mathbf{u}^i, \mathbf{u}^{j+1}, \ldots, \mathbf{u}^n\right)$$

$$= \sum_j a_{j,j}(t_0)$$

$$= \operatorname{tr}(\mathbf{A}(t_0)).$$

The next to the last equality holds because $\det\left(\mathbf{u}^1, \ldots, \mathbf{u}^{j-1}, \mathbf{u}^i, \mathbf{u}^{j+1}, \ldots, \mathbf{u}^n\right) = 0$ unless $i = j$. Since $\det\left(\mathbf{M}(t)\mathbf{M}(t_0)^{-1}\right) = \det\left(\mathbf{M}(t)\right)\det\left(\mathbf{M}(t_0)\right)^{-1}$, we get

$$\frac{d}{dt}\det\left(\mathbf{M}(t)\right)\Big|_{t=t_0} = \operatorname{tr}(\mathbf{A}(t))\det\left(\mathbf{M}(t_0)\right).$$

This proves the differential equation given in the theorem. This scalar differential equation has a solution for $\det\left(\mathbf{M}(t)\right)$ as given in the theorem.

For the second part of the theorem, $\det(e^{\mathbf{A} \, 0}) = \det(\mathbf{I}) = 1$. Also, $\operatorname{tr}(\mathbf{A})$ is a constant, so $\int_0^t \operatorname{tr}(\mathbf{A}) \, ds = \operatorname{tr}(\mathbf{A})t$ and

$$\det(e^{\mathbf{A}t}) = \det(e^{\mathbf{A} \, 0}) \exp\left(\int_0^t \operatorname{tr}(\mathbf{A}) \, ds\right) = e^{\operatorname{tr}(\mathbf{A})t}.$$

$\square$

Convergence of the exponential

To show that the series for the matrix exponential converges, we first give a lemma describing how large the entries of a power of a matrix can be. We use the notation $(A^k)_{ij}$ for the (i,j) entry of the power A^k.

Lemma 2.11. *Let $\mathbf{A}$ be a constant $n \times n$ matrix with $|\mathbf{A}_{ij}| \leq C$ for all $1 \leq i, j \leq n$. Then $\left|(\mathbf{A}^m)_{ij}\right| \leq n^{m-1} C^m$.*

Proof. We prove the lemma by induction on m. It is true for $m = 1$, by assumption. For clarity, we check the case for $m = 2$:

$$\left|(\mathbf{A}^2)_{ij}\right| = \left|\sum_{k=1}^n \mathbf{A}_{ik} \mathbf{A}_{kj}\right| \leq \sum_{k=1}^n |\mathbf{A}_{ik}| \cdot |\mathbf{A}_{kj}| \leq \sum_{k=1}^n C \cdot C = n C^2.$$

Next, assume the lemma is true for m. Then,

$$\left|(\mathbf{A}^{m+1})_{ij}\right| = \left|\sum_{k=1}^n \mathbf{A}_{ik} (\mathbf{A}^m)_{kj}\right| \leq \sum_{k=1}^n |\mathbf{A}_{ik}| \cdot \left|(\mathbf{A}^m)_{kj}\right|$$

$$\leq \sum_{k=1}^n C \cdot n^{m-1} C^m = n^m \, C^{m+1}.$$

Thus, the lemma is true for $m+1$. By induction, we are done. $\square$

Theorem 2.12. *Let $\mathbf{A}$ be a constant $n \times n$ matrix. For each pair (i,j), the series for $(e^{t\mathbf{A}})_{ij}$ converges.*

Proof. Let $C > 0$ be such that $|\mathbf{A}_{ij}| \leq C$ for all $1 \leq i, j \leq n$. For a fixed (i,j)-entry, the sum of the absolute values of this entry for the series of $e^{t\mathbf{A}}$ satisfies

$$\sum_{m=0}^\infty \frac{|t|^m}{m!} |(\mathbf{A}^m)_{ij}| \leq \sum_{m=0}^\infty \frac{|t|^m}{m!} \frac{(nC)^m}{n} = \frac{1}{n} e^{|t|nC}.$$

Thus, the series for this entry converges absolutely, and so converges. $\square$

One of the important features of the exponential of real numbers is the fact that $e^{a+b} = e^a e^b$. In general, the corresponding formula is not true for matrices. As an example, consider

$$\mathbf{A} = \begin{pmatrix} 0 & 1 \\ 0 & 0 \end{pmatrix} \quad \text{and} \quad \mathbf{B} = \begin{pmatrix} 0 & 0 \\ -1 & 0 \end{pmatrix}.$$

2.5. Theory and Proofs

Then,

$$e^{t\mathbf{A}} = \begin{pmatrix} 1 & 0 \\ 0 & 1 \end{pmatrix} + t \begin{pmatrix} 0 & 1 \\ 0 & 0 \end{pmatrix} = \begin{pmatrix} 1 & t \\ 0 & 1 \end{pmatrix},$$

$$e^{t\mathbf{B}} = \begin{pmatrix} 1 & 0 \\ 0 & 1 \end{pmatrix} + t \begin{pmatrix} 0 & 0 \\ -1 & 0 \end{pmatrix} = \begin{pmatrix} 1 & 0 \\ -t & 1 \end{pmatrix},$$

$$e^{t\mathbf{A}} e^{t\mathbf{B}} = \begin{pmatrix} 1 - t^2 & t \\ -t & 1 \end{pmatrix},$$

$$e^{t(\mathbf{A}+\mathbf{B})} = \begin{pmatrix} \cos(t) & \sin(t) \\ -\sin(t) & \cos(t) \end{pmatrix}.$$

Thus, $e^{t(\mathbf{A}+\mathbf{B})} \neq e^{t\mathbf{A}} e^{t\mathbf{B}}$.

Two matrices are said to *commute* provided that $\mathbf{A}\mathbf{B} = \mathbf{B}\mathbf{A}$. In order to prove that $e^{t(\mathbf{A}+\mathbf{B})} = e^{t\mathbf{A}} e^{t\mathbf{B}}$ it is necessary for the matrices to commute, so similar powers of $\mathbf{A}$ and $\mathbf{B}$ can be combined.

Theorem 2.13. *Let $\mathbf{A}$ and $\mathbf{B}$ be two $n \times n$ matrices that commute. Then,*

$$e^{\mathbf{A}+\mathbf{B}} = e^{\mathbf{A}} e^{\mathbf{B}}.$$

Proof. We need to check that the series for $e^{\mathbf{A}+\mathbf{B}}$ is the same as the product of the series for $e^{\mathbf{A}}$ and $e^{\mathbf{B}}$. Because the matrices commute,

$$(\mathbf{A}+\mathbf{B})^2 = \mathbf{A}^2 + \mathbf{A}\mathbf{B} + \mathbf{B}\mathbf{A} + \mathbf{B}^2$$
$$= \mathbf{A}^2 + 2\mathbf{A}\mathbf{B} + \mathbf{B}^2.$$

Similarly, for higher powers,

$$(\mathbf{A}+\mathbf{B})^n = \mathbf{A}^n + \frac{n!}{(n-1)!\,1!}\mathbf{A}^{n-1}\mathbf{B} + \cdots + \frac{n!}{(n-k)!\,k!}\mathbf{A}^{n-k}\mathbf{B}^k + \cdots + \mathbf{B}^n.$$

Writing out the power series for the exponential of the sum, we get

$$e^{\mathbf{A}+\mathbf{B}} = \mathbf{I} + (\mathbf{A}+\mathbf{B}) + \frac{1}{2!}(\mathbf{A}+\mathbf{B})^2$$
$$+ \frac{1}{3!}(\mathbf{A}+\mathbf{B})^3 + \cdots + \frac{1}{n!}(\mathbf{A}+\mathbf{B})^n + \cdots$$
$$= \mathbf{I} + (\mathbf{A}+\mathbf{B}) + \frac{1}{2!}\left(\mathbf{A}^2 + 2\mathbf{A}\mathbf{B} + \mathbf{B}^2\right)$$
$$+ \frac{1}{3!}\left(\mathbf{A}^3 + \frac{3!}{2!\,1!}\mathbf{A}^2\mathbf{B} + \frac{3!}{1!\,2!}\mathbf{A}\mathbf{B}^2 + \mathbf{B}^3\right) + \cdots$$
$$+ \frac{1}{n!}\left(\mathbf{A}^n + \frac{n!}{(n-1)!\,1!}\mathbf{A}^{n-1}\mathbf{B}\right.$$
$$\left. + \frac{n!}{(n-2)!\,2!}\mathbf{A}^{n-2}\mathbf{B}^2 + \cdots + \mathbf{B}^n\right) + \cdots$$

$$e^{\mathbf{A}+\mathbf{B}} = \mathbf{I} + (\mathbf{A}+\mathbf{B}) + \left(\frac{1}{2!}\mathbf{A}^2 + \mathbf{AB} + \frac{1}{2!}\mathbf{B}^2\right)$$
$$+ \left(\frac{1}{3!}\mathbf{A}^3 + \frac{1}{2!}\mathbf{A}^2\mathbf{B} + \mathbf{A}\frac{1}{2!}\mathbf{B}^2 + \frac{1}{3!}\mathbf{B}^3\right) + \cdots$$
$$+ \left(\frac{1}{n!}\mathbf{A}^n + \frac{1}{(n-1)!}\mathbf{A}^{n-1}\mathbf{B}\right.$$
$$\left.+ \frac{1}{(n-2)!}\mathbf{A}^{n-2}\frac{1}{2!}\mathbf{B}^2 + \cdots + \frac{1}{n!}\mathbf{B}^n\right) + \cdots$$
$$= \left(\mathbf{I} + \mathbf{A} + \frac{1}{2!}\mathbf{A}^2 + \cdots\right)\left(\mathbf{I} + \mathbf{B} + \frac{1}{2!}\mathbf{B}^2 + \cdots\right)$$
$$= e^{\mathbf{A}} e^{\mathbf{B}}.$$

□

Constant coefficients

We first give results related to the independence of the eigenvectors.

Theorem 2.14. *Let $\mathbf{A}$ be a real matrix.*

a. *Assume $\lambda_1, \ldots, \lambda_k$ are distinct eigenvalues with eigenvectors $\mathbf{v}^1, \ldots, \mathbf{v}^k$. Then, the eigenvectors $\mathbf{v}^1, \ldots, \mathbf{v}^k$ are linearly independent.*

b. *Assume $\lambda = \alpha + i\beta$ is a complex eigenvector ($\beta \neq 0$) with complex eigenvector $\mathbf{v} = \mathbf{u} + i\mathbf{w}$. Then, (i) both $\mathbf{u} \neq \mathbf{0}$ and $\mathbf{w} \neq \mathbf{0}$, and (ii) $\mathbf{u}$ and $\mathbf{w}$ are linearly independent.*

Proof. (a) Assume the result is false and the vectors are linearly dependent. Assume that $\mathbf{v}^1, \ldots, \mathbf{v}^m$ is the smallest set of linearly dependent vectors. Then,

(2.11) $$\mathbf{0} = c_1 \mathbf{v}^1 + \cdots + c_m \mathbf{v}^m.$$

Since it is the smallest set, $c_m \neq 0$. Acting on equation (2.11) by $\mathbf{A}$, we get

(2.12) $$\mathbf{0} = c_1 \mathbf{A}\mathbf{v}^1 + \cdots + c_m \mathbf{A}\mathbf{v}^m$$
$$= c_1 \lambda_1 \mathbf{v}^1 + \cdots + c_m \lambda_m \mathbf{v}^m.$$

Multiplying equation (2.11) by λ_m and subtracting it from equation (2.12), we get

$$\mathbf{0} = c_1(\lambda_1 - \lambda_m)\mathbf{v}^1 + \cdots + c_{m-1}(\lambda_{m-1} - \lambda_m)\mathbf{v}^{m-1}.$$

Because the eigenvalues are distinct, we get all the coefficients nonzero. This contradicts the fact that this is the smallest set of linearly dependent vectors. This proves part (**a**).

(**b.i**)

$$\mathbf{A}\mathbf{u} + i\mathbf{A}\mathbf{w} = \mathbf{A}(\mathbf{u} + i\mathbf{w})$$
$$= (\alpha + i\beta)(\mathbf{u} + i\mathbf{w})$$
$$= (\alpha\mathbf{u} - \beta\mathbf{w}) + i(\beta\mathbf{u} + \alpha\mathbf{w}).$$

If $\mathbf{w} = \mathbf{0}$, then the left side of the equality is real, and the right side has nonzero imaginary part, $i\beta\mathbf{u}$. This is a contradiction. Similarly, $\mathbf{u} = \mathbf{0}$ leads to a contradiction.

2.5. Theory and Proofs

(b.ii) Assume that

(2.13)
$$\mathbf{0} = c_1\,\mathbf{u} + c_2\,\mathbf{w}.$$

Acting on this equation by $\mathbf{A}$ gives

(2.14)
$$\begin{aligned}\mathbf{0} &= c_1\,\mathbf{A}\mathbf{u} + c_2\,\mathbf{A}\mathbf{w} \\ &= c_1\,(\alpha\mathbf{u} - \beta\mathbf{w}) + c_2\,(\beta\mathbf{u} + \alpha\mathbf{w}) \\ &= (c_1\alpha + c_2\beta)\,\mathbf{u} + (-c_1\beta + c_2\alpha)\,\mathbf{w}.\end{aligned}$$

Multiplying equation (2.13) by $c_1\alpha + c_2\beta$ and equation (2.14) by c_1 and subtracting, we get

$$\begin{aligned}\mathbf{0} &= c_2\,(c_1\alpha + c_2\beta)\,\mathbf{w} - c_1\,(-c_1\beta + c_2\alpha)\,\mathbf{w} \\ &= (c_2^2 + c_1^2)\,\beta\mathbf{w}.\end{aligned}$$

Since $\mathbf{w} \neq \mathbf{0}$ and $\beta \neq 0$, this implies that $c_2^2 + c_1^2 = 0$ and $c_1 = c_2 = 0$. Thus, we have shown that $\mathbf{u}$ and $\mathbf{w}$ are linearly independent. $\square$

Section 2.2 gives the solution method for various types of eigenvalues. From the theory of matrices, it follows that we have considered all the cases except the case of a repeated complex eigenvalue. This follows by showing that there is a basis that puts the matrix in a special form called the *Jordan canonical form* of the matrix.

If the matrix $\mathbf{A}$ is real and symmetric, then all the eigenvalues are real and there exists a basis of eigenvectors. Thus, in this case, there are real numbers λ_1, ..., λ_n (possibly repeated) and eigenvectors $\mathbf{v}_1$, ..., $\mathbf{v}_n$ such that $\mathbf{A}\mathbf{v}^j = \lambda_j\,\mathbf{v}^j$. Letting $\mathbf{V} = (\mathbf{v}_1, \ldots, \mathbf{v}_n)$ and $\Lambda = \mathrm{diag}(\lambda_1, \ldots, \lambda_n)$, $\mathbf{A}\mathbf{V} = \mathbf{V}\Lambda$ and $\mathbf{V}^{-1}\mathbf{A}\mathbf{V} = \Lambda$ is linearly conjugate to a diagonal matrix. Thus, in terms of the basis of vectors $\{\mathbf{v}_1, \ldots, \mathbf{v}_n\}$, the matrix is diagonal and the solutions are given by

$$\mathbf{x}(t) = \sum_{j=1}^{n} c_j e^{\lambda_j t}\mathbf{v}^j.$$

In any case where the matrix is linearly conjugate to a diagonal matrix (i.e., has a basis of eigenvectors), the solutions are of that same form, even though the matrix might not be symmetric.

If the eigenvalue $\lambda_j = \alpha_j + i\beta_j$ is complex, then its eigenvector $\mathbf{v}^j = \mathbf{u}^j + i\,\mathbf{w}^j$ must be complex with both $\mathbf{u}^j, \mathbf{w}^j \neq \mathbf{0}$. Since

$$\mathbf{A}(\mathbf{u}^j + i\,\mathbf{w}^j) = (\alpha_j\mathbf{u}^j - \beta_j\mathbf{w}^j) + i\,(\beta_j\mathbf{u}^j + \alpha_j\mathbf{w}^j),$$

equating the real and imaginary parts yields

$$\begin{aligned}\mathbf{A}\mathbf{u}^j &= \alpha_j\mathbf{u}^j - \beta_j\mathbf{w}^j \qquad \text{and} \\ \mathbf{A}\mathbf{w}^j &= \beta_j\mathbf{u}^j + \alpha_j\mathbf{w}^j.\end{aligned}$$

Using the vectors $\mathbf{u}^j$ and $\mathbf{w}^j$ as part of a basis yields a sub-block of the matrix of the form

$$\mathbf{B}_j = \begin{pmatrix} \alpha_j & \beta_j \\ -\beta_j & \alpha_j \end{pmatrix}.$$

Thus, if $\mathbf{A}$ has a basis of complex eigenvectors, then there is a real basis $\{\mathbf{z}^1, \ldots, \mathbf{z}^n\}$ in terms of which
$$\mathbf{A} = \mathrm{diag}(\mathbf{A}_1, \ldots, \mathbf{A}_q),$$
where each $\mathbf{A}_k$ is either: (i) a 1×1 block with real entry λ_k, or (ii) of the form $\mathbf{B}_j$ previously given. The 1×1 blocks give solutions of the form
$$e^{\lambda_k t}\mathbf{v}^k,$$
and the 2×2 blocks give solutions of the form
$$e^{\alpha_k t}(\cos(\beta_k t)\mathbf{u}^k - \sin(\beta_k t)\mathbf{w}^k) \quad \text{and}$$
$$e^{\alpha_k t}(\sin(\beta_k t)\mathbf{u}^k + \cos(\beta_k t)\mathbf{w}^k).$$

Next, we turn to the case of repeated eigenvalues in which there are fewer eigenvectors than the multiplicity of the eigenvalue. The Cayley–Hamilton theorem states that, if
$$p(x) = (-1)^n x^n + a_{n-1} x^{n-1} + \cdots + a_0$$
is the characteristic polynomial for $\mathbf{A}$, then
$$\mathbf{0} = (-1)^n \mathbf{A}^n + a_{n-1} \mathbf{A}^{n-1} + \cdots + a_0 \mathbf{I}.$$
In particular, if $\lambda_1, \ldots, \lambda_q$ are the distinct eigenvalues of $\mathbf{A}$ with algebraic multiplicities $m_1, \ldots, m_q$,
$$p(x) = (x - \lambda_1)^m_1 \cdots (x - \lambda_q)^m_q,$$
then
$$S_k = \{\mathbf{v} : (\mathbf{A} - \lambda_k \mathbf{I})^{m_k} \mathbf{v} = \mathbf{0}\}$$
is a vector subspace of dimension m_k. Thus, the geometric multiplicity, which is the dimension of S_k, is the same as the algebraic multiplicity, which is the multiplicity in the characteristic equation. Vectors in S_k are called *generalized eigenvectors*.

Take a real λ_k of multiplicity $m_k > 1$. Assume $\mathbf{v}^{(r)}$ is a vector with
$$(\mathbf{A} - \lambda_k \mathbf{I})^r \mathbf{v}^{(r)} = \mathbf{0}, \quad \text{but}$$
$$(\mathbf{A} - \lambda_k \mathbf{I})^{r-1} \mathbf{v}^{(r)} \neq \mathbf{0}.$$
If there is not a basis of eigenvectors of S_k, then the general theory says that there is such an r with $1 < r \leq k$ for which this is true. Setting
$$\mathbf{v}^{(r-j)} = (\mathbf{A} - \lambda_k \mathbf{I})^j \mathbf{v}^{(r)},$$
we get
$$(\mathbf{A} - \lambda_k \mathbf{I})\mathbf{v}^{(r)} = \mathbf{v}^{(r-1)},$$
$$(\mathbf{A} - \lambda_k \mathbf{I})\mathbf{v}^{(r-1)} = \mathbf{v}^{(r-2)},$$
$$\vdots \quad \vdots$$
$$(\mathbf{A} - \lambda_k \mathbf{I})\mathbf{v}^{(2)} = \mathbf{v}^{(1)},$$
$$(\mathbf{A} - \lambda_k \mathbf{I})\mathbf{v}^{(1)} = \mathbf{0}.$$

2.5. Theory and Proofs

In terms of this partial basis, there is an $r \times r$ sub-block of the form

$$\mathbf{C}_\ell = \begin{pmatrix} \lambda_k & 1 & 0 & \cdots & 0 & 0 \\ 0 & \lambda_k & 1 & \cdots & 0 & 0 \\ 0 & 0 & \lambda_k & \cdots & 0 & 0 \\ \vdots & \vdots & \vdots & & \vdots & \vdots \\ 0 & 0 & 0 & \cdots & \lambda_k & 1 \\ 0 & 0 & 0 & \cdots & 0 & \lambda_k \end{pmatrix}.$$

The general theory says that there are enough blocks of this form together with eigenvectors to span the total subspace S_k. Therefore, in the case of a repeated real eigenvector, the matrix $\mathbf{A}$ on S_k can be represented by blocks of the form $\mathbf{C}_\ell$ plus a diagonal matrix.

Summarizing the above discussion, we get the following theorem.

Theorem 2.15. a. *Assume that $\mathbf{A}$ has a real eigenvalue r with multiplicity m and a generalized eigenvector $\mathbf{w}$. Then there is the solution of $\mathbf{x}' = \mathbf{A}\mathbf{x}$ with $\mathbf{x}(0) = \mathbf{w}$ is given by*

$$\mathbf{x}(t) = e^{t\mathbf{A}}\mathbf{w} = e^{rt}\left[\mathbf{w} + t\,(\mathbf{A} - r\mathbf{I})\mathbf{w} + \cdots + \frac{t^{m-1}}{(m-1)!}(\mathbf{A} - r\mathbf{I})^{m-1}\mathbf{w}\right].$$

b. *Assume $\mathbf{A}$ has a real eigenvalue r with algebraic multiplicity 2 but geometric multiplicity 1. Assume that $\mathbf{v}$ is an eigenvector and $\mathbf{w}$ is a generalized eigenvector solving $(\mathbf{A} - r\mathbf{I})\mathbf{w} = \mathbf{v}$. Then $\mathbf{x}' = \mathbf{A}\mathbf{x}$ has two independent solutions $\mathbf{x}_1(t) = e^{t\mathbf{A}}\mathbf{v} = e^{rt}\mathbf{v}$ and $\mathbf{x}_2(t) = e^{t\mathbf{A}}\mathbf{w} = e^{rt}\mathbf{w} + te^{rt}\mathbf{v}$. Notice that $\mathbf{x}_1(0) = \mathbf{v}$ and $\mathbf{x}_2(0) = \mathbf{w}$.*

Next, we assume that $\lambda_k = \alpha_k + i\beta_k$ is a complex eigenvalue of multiplicity $m_k > 1$. If there are not as many eigenvectors as the dimension of the multiplicity, then there are blocks of the from

$$\mathbf{D}_\ell = \begin{pmatrix} \mathbf{B}_k & \mathbf{I} & \cdots & 0 & 0 \\ 0 & \mathbf{B}_k & \cdots & 0 & 0 \\ \vdots & \vdots & & \vdots & \vdots \\ 0 & 0 & \cdots & \mathbf{B}_k & \mathbf{I} \\ 0 & 0 & \cdots & 0 & \mathbf{B}_k \end{pmatrix},$$

where $\mathbf{B}_k$ is the 2×2 block with entries α_k and $\pm\beta_k$ just given. If the multiplicity is two, then the exponential of this matrix is given by

$$e^{\mathbf{D}_\ell t} = \begin{pmatrix} e^{\mathbf{B}_k t} & te^{\mathbf{B}_k t} \\ \mathbf{0} & e^{\mathbf{B}_k t} \end{pmatrix}.$$

Therefore, if the basis of vectors are $\mathbf{u}^1 + i\mathbf{w}^1$ and $\mathbf{u}^2 + i\mathbf{w}^2$, then there are solutions of the form

$$e^{\alpha_k t}(\cos(\beta_k t)\mathbf{u}^1 - \sin(\beta_k t)\mathbf{w}^1),$$
$$e^{\alpha_k t}(\sin(\beta_k t)\mathbf{u}^1 + \cos(\beta_k t)\mathbf{w}^1),$$
$$e^{\alpha_k t}(\cos(\beta_k t)\mathbf{u}^2 - \sin(\beta_k t)\mathbf{w}^2) + te^{\alpha_k t}(\cos(\beta_k t)\mathbf{u}^1 - \sin(\beta_k t)\mathbf{w}^1),$$
$$e^{\alpha_k t}(\sin(\beta_k t)\mathbf{u}^2 + \cos(\beta_k t)\mathbf{w}^2) + te^{\alpha_k t}(\sin(\beta_k t)\mathbf{u}^1 + \cos(\beta_k t)\mathbf{w}^1).$$

Again, there are enough blocks of this form to give the total subspaces for $\alpha_k \pm i\beta_k$.

The total matrix $\mathbf{A}$ can be decomposed into blocks of these four forms. This shows the form of the solutions given in the next theorem.

Theorem 2.16. *Given a real $n \times n$ constant matrix $\mathbf{A}$ and $\mathbf{x}_0 \in \mathbb{R}^n$, there is a solution $\phi(t; \mathbf{x}_0)$ of the differential equation $\dot{\mathbf{x}} = \mathbf{A}\mathbf{x}$ with $\phi(0; \mathbf{x}_0) = \mathbf{x}_0$. Moreover, each coordinate function of $\phi(t; \mathbf{x}_0)$ is a linear combination of functions of the form*

$$t^k e^{\alpha t} \cos(\beta t) \quad \text{and} \quad t^k e^{\alpha t} \sin(\beta t),$$

where $\alpha + i\beta$ is an eigenvalue of $\mathbf{A}$ and k is less than or equal to the algebraic multiplicity of the eigenvalue. In particular, we allow $k = 0$ so the solutions do not have the term t^k, and $\beta = 0$ so the solutions do not have any terms involving $\cos(\beta t)$ or $\sin(\beta t)$.

Quasiperiodic equations

For the solutions of Example 2.19, we could take a type of polar coordinates defined by

$$x_j = \rho_j \sin(2\pi \tau_j) \quad \text{and} \quad \dot{x}_j = \omega_j \rho_j \cos(2\pi \tau_j), \quad \text{or}$$

$$\tan(2\pi \tau_j) = \frac{\omega_j x_j}{\dot{x}_j} \quad \text{and} \quad \rho_j^2 = x_j^2 + \frac{\dot{x}_j^2}{\omega_j^2}.$$

The variables τ_1 and τ_2 are taken modulo 1, so $2\pi\tau_1$ and $2\pi\tau_2$ are taken modulo 2π (i.e., the variable τ_j is obtained by subtracting an integer so the variable τ_j satisfies $0 \leq \tau_j < 1$). We write the variables as "$\tau_j \ (\bmod \ 1)$" and "$2\pi\tau_j \ (\bmod \ 2\pi)$". Then,

$$\rho_j \dot{\rho}_j = x_j \dot{x}_j + \frac{\dot{x}_j \ddot{x}_j}{\omega_j^2} = x_j \dot{x}_j + \frac{\dot{x}_j (-\omega_j^2 x_j)}{\omega_j^2}$$

$$= x_j \dot{x}_j - \dot{x}_j x_j = 0 \quad \text{and}$$

$$2\pi \sec^2(2\pi \tau_j) \, \dot{\tau}_j = \frac{\omega_j \dot{x}_j \dot{x}_j - \omega_j x_j \ddot{x}_j}{\dot{x}_j^2}$$

$$= \frac{\omega_j \dot{x}_j^2 - \omega_j x_j (-\omega_j^2 x_j)}{\dot{x}_j^2}$$

$$= \frac{\omega_j \left(\dot{x}_j^2 + \omega_j^2 x_j^2 \right)}{\dot{x}_j^2}$$

$$= \frac{\omega_j^3 \rho_j^2}{\dot{x}_j^2}$$

$$= \omega_j \, \sec^2(2\pi \tau_j)$$

and so,

$$\dot{\tau}_j = \frac{\omega_j}{2\pi}.$$

2.5. Theory and Proofs

Thus, the differential equations for the variables are

$$\dot{\tau}_1 = \frac{\omega_j}{2\pi} = \alpha_1 \pmod{1},$$
$$\dot{\tau}_2 = \frac{\omega_j}{2\pi} = \alpha_2 \pmod{1},$$
$$\dot{\rho}_1 = 0,$$
$$\dot{\rho}_2 = 0.$$

The solution of this system is clearly

$$\tau_1(t) = \tau_1(0) + \alpha_1 t \pmod{1},$$
$$\tau_2(t) = \tau_2(0) + \alpha_2 t \pmod{1},$$
$$\rho_1(t) = \rho_1(0),$$
$$\rho_2(t) = \rho_2(0).$$

The next theorem states that the trajectories are dense in the phase portrait if $\omega_1/\omega_2 = \alpha_1/\alpha_2$ is irrational.

Theorem 2.17. *Assume that $\omega_1/\omega_2 = \alpha_1/\alpha_2$ is irrational. Then, for any initial condition $(\tau_1(0), \tau_2(0))$, the solution $(\tau_1(t), \tau_2(t))$ is dense in the phase portrait*

$$\{(\tau_1, \tau_2) : \text{where each } \tau_j \text{ is taken } (\bmod 1)\}.$$

Proof. For time $T = 1/\alpha_2$,

$$\tau_2(T) - \tau_2(0) = 0 \pmod{1} \quad \text{and}$$
$$\tau_1(T) - \tau_1(0) = \frac{\alpha_1}{\alpha_2} \pmod{1}.$$

Taking times equal to multiples of T,

$$\tau_2(nT) - \tau_2(0) = 0 \pmod{1} \quad \text{and}$$
$$\tau_1(nT) - \tau_1(0) = n\frac{\alpha_1}{\alpha_2} \pmod{1}.$$

Since the ratio α_1/α_2 is irrational, $n\alpha_1/\alpha_2$ cannot be an integer. These distinct points must get close in the interval $[0, 1]$. Given any $\epsilon > 0$, there must be three integers $n > m > 0$ and k such that

$$|\tau_1(nT) - \tau_1(mT) - k| < \epsilon.$$

Then

$$|\tau_1((n-m)T) - \tau_1(0) - k| = |\tau_1(nT) - \tau_1(mT) - k| < \epsilon.$$

Letting $q = n - m$, and taking multiples of qT,

$$|\tau_1((j+1)qT) - \tau_1(jqT) - k| < \epsilon$$

for all j. These points must fill up the interval $[0, 1]$ to within ϵ of each point. Because $\epsilon > 0$ is arbitrary, the forward orbit $\{\tau(nT)\}$ is dense in the interval $[0, 1]$. Taking the flow of these points for times $0 \leq t \leq T$, we see that the orbit $(\tau_1(t), \tau_2(t))$ is dense in the phase portrait

$$\{(\tau_1, \tau_2) : \text{where each } \tau_j \text{ is taken } (\bmod 1)\},$$

as claimed. $\square$

Nonhomogeneous systems

Restatement of Theorem 2.6. a. *Let $\mathbf{x}^1(t)$ and $\mathbf{x}^2(t)$ be two solutions of the nonhomogeneous linear differential equation (2.7). Then, $\mathbf{x}^1(t) - \mathbf{x}^2(t)$ is a solution of the homogeneous differential equation (2.8).*

b. *Let $\mathbf{x}^{(p)}(t)$ be a solution of the nonhomogeneous linear differential equation (2.7) and $\mathbf{x}^{(h)}(t)$ a solution of the homogeneous linear of differential equation (2.8). Then, $\mathbf{x}^{(p)}(t) + \mathbf{x}^{(h)}(t)$ is a solution of the nonhomogeneous linear differential equation (2.7).*

c. *Let $\mathbf{x}^{(p)}(t)$ be a solution of nonhomogeneous linear differential equation (2.7) and $\mathbf{M}(t)$ a fundamental matrix solution of homogeneous linear differential equation (2.8). Then, any solution of the nonhomogeneous linear differential equation (2.7) can be written as $\mathbf{x}^{(p)}(t) + \mathbf{M}(t)\mathbf{c}$ for some vector $\mathbf{c}$.*

Proof. (a) Taking $\mathbf{x}^1(t)$ and $\mathbf{x}^2(t)$ as in the statement of the theorem,

$$\frac{d}{dt}\left(\mathbf{x}^1(t) - \mathbf{x}^2(t)\right) = \mathbf{A}\mathbf{x}^1(t) + \mathbf{g}(t) - \left(\mathbf{A}\mathbf{x}^2(t) + \mathbf{g}(t)\right)$$
$$= \mathbf{A}\left(\mathbf{x}^1(t) - \mathbf{x}^2(t)\right),$$

which shows that $\mathbf{x}^1(t) - \mathbf{x}^2(t)$ is a solution of the homogeneous equation (2.8) as claimed.

(b) Let $\mathbf{x}^{(p)}(t)$ be a solution of the nonhomogeneous equation (2.7) and $\mathbf{x}^{(h)}(t)$ be a solution of homogeneous equation (2.8). Then,

$$\frac{d}{dt}\left(\mathbf{x}^{(p)}(t) + \mathbf{x}^{(h)}(t)\right) = \left(\mathbf{A}\mathbf{x}^{(p)}(t) + \mathbf{g}(t)\right) + \mathbf{A}\mathbf{x}^{(h)}(t)$$
$$= \mathbf{A}\left(\mathbf{x}^{(p)}(t) + \mathbf{x}^{(h)}(t)\right) + \mathbf{g}(t),$$

which shows that $\mathbf{x}^{(p)}(t) + \mathbf{x}^{(h)}(t)$ is a solution of nonhomogeneous equation (2.7), as claimed.

(c) Let $\mathbf{x}^{(p)}(t)$ be a solution of the nonhomogeneous equation (2.7) and $\mathbf{M}(t)$ a fundamental matrix solution of the homogeneous equation (2.8). Let $\mathbf{x}(t)$ be an arbitrary solution of the nonhomogeneous equation (2.7). Then, $\mathbf{x}(t) - \mathbf{x}^{(p)}(t)$ is a solution of (2.8) by part (a). But any solution of equation (2.8) can be written as $\mathbf{M}(t)\mathbf{c}$ for some vector $\mathbf{c}$. Therefore,

$$\mathbf{x}(t) - \mathbf{x}^{(p)}(t) = \mathbf{M}(t)\mathbf{c} \quad \text{and}$$
$$\mathbf{x}(t) = \mathbf{x}^{(p)}(t) + \mathbf{M}(t)\mathbf{c}$$

for some vector $\mathbf{c}$, as claimed. $\square$

Restatement of Theorem 2.7. *The solution $\mathbf{x}(t)$ of the nonhomogeneous linear differential equation with initial condition $\mathbf{x}(0) = \mathbf{x}_0$ can be written as*

$$\mathbf{x}(t) = e^{\mathbf{A}t}\left(\mathbf{x}_0 + \int_0^t e^{-\mathbf{A}s}\mathbf{g}(s)\,ds\right).$$

Proof. We derive the form of the solution $\mathbf{x}(t)$ of the nonhomogeneous equation (2.7). A solution of the homogeneous equation (2.8) can be written as $e^{\mathbf{A}t}\mathbf{c}$. For

2.5. Theory and Proofs

a solution of the nonhomogeneous equation (2.7), we investigate the possibility of writing it in this form, where **c** varies with t, by considering

$$\mathbf{x}(t) = e^{\mathbf{A}t}\mathbf{y}(t).$$

The extent to which $\mathbf{y}(t)$ varies measures how much the solution varies from a solution of the homogeneous system. Solving for $\mathbf{y}(t) = e^{-\mathbf{A}t}\mathbf{x}(t)$, we get

$$\begin{aligned}\dot{\mathbf{y}}(t) &= -\mathbf{A}e^{-\mathbf{A}t}\mathbf{x}(t) + e^{-\mathbf{A}t}\dot{\mathbf{x}}(t) \\ &= -\mathbf{A}e^{-\mathbf{A}t}\mathbf{x}(t) + e^{-\mathbf{A}t}\mathbf{A}\mathbf{x}(t) + e^{-\mathbf{A}t}\mathbf{g}(t) \\ &= e^{-\mathbf{A}t}\mathbf{g}(t).\end{aligned}$$

Integrating from 0 to t gives

$$\mathbf{y}(t) = \mathbf{y}(0) + \int_0^t e^{-\mathbf{A}s}\mathbf{g}(s)\,ds, \qquad \text{or}$$

$$\mathbf{x}(t) = e^{\mathbf{A}t}\mathbf{y}(0) + e^{\mathbf{A}t}\int_0^t e^{-\mathbf{A}s}\mathbf{g}(s)\,ds.$$

The first term gives the general solution of the homogeneous equation, and the integral gives one particular solution of the nonhomogeneous equation. Since $\mathbf{x}(0) = \mathbf{y}(0)$, we can set $\mathbf{x}_0 = \mathbf{y}(0)$. Rearranging terms gives the form as stated in the theorem. □

Chapter 3

The Flow: Solutions of Nonlinear Equations

Before we start the specific material on the phase portraits for nonlinear differential equations in the next chapter, we present the basic results on existence and uniqueness of solutions in Section 3.1. Not only does this answer the question about existence, but it emphasizes the dependence of the solution on the initial condition and how the solution varies as the initial condition varies. These considerations form the foundation for the analysis of the nonlinear equations treated in the rest of the book.

In Section 3.2, we consider the standard numerical methods for solving differential equations. Although they are usually presented in a beginning differential equations course for scalar equations, we emphasize the vector case. The reason some understanding of numerical methods is important is that we use various programs to help draw the phase space for the nonlinear equations. Such programs do not solve the differential equation, but implement some form of numerical integration of the type we discuss or a more complicated type. Since it is important to have some idea of what the computer program is doing, we introduce at least the simplest type of numerical methods for systems of equations. However, this material on numerical methods can be skipped without any loss of continuity.

3.1. Solutions of Nonlinear Equations

For linear equations with constant coefficients, we gave a constructive proof that solutions exist by showing the form of the solutions. For nonlinear equations, there is no general explicit solution from which we can deduce the existence and uniqueness of solutions. In this section, we discuss the condition which ensures that the solution exists and is unique, and some properties that follow from the uniqueness of solutions. We first consider scalar differential equations, and then differential equations with more variables.

We will state a general theorem. However, first we present a solution method utilizing integrals that can be used for many scalar equations. We start with an example, which is discussed further in Section 4.3.

Example 3.1 (Logistic Equation). The differential equation

$$\dot{x} = rx\left(1 - \frac{x}{K}\right),$$

with positive parameters r and K, is called the *logistic equation*. It can be solved explicitly by separation of variables and integration by partial fractions as follows. First, we separate all of the x variables to the left side and apply partial fractions:

$$\frac{\dot{x}}{x\left(1 - \frac{x}{K}\right)} = r,$$

$$\frac{\dot{x}}{x} + \frac{\dot{x}}{K - x} = r.$$

Then, we integrate with respect to t (on the left-hand side, the term $\dot{x}\,dt$ changes to an integral with respect to x):

$$\int \frac{1}{x}\,dx + \int \frac{1}{K-x}\,dx = \int r\,dt,$$

$$\ln(|x|) - \ln(|K-x|) = rt + C_1,$$

$$\frac{|x|}{|K-x|} = C\,e^{rt}, \qquad \text{where } C = e^{C_1}.$$

Next, we solve for x by performing some algebra, while assuming that $0 < x < K$, so we can drop the absolute value signs:

$$x = CK\,e^{rt} - C\,e^{rt}x,$$

$$(1 + C\,e^{rt})\,x = CK\,e^{rt},$$

$$x(t) = \frac{CK\,e^{rt}}{1 + C\,e^{rt}}.$$

Using the initial condition x_0 when $t = 0$, we have

$$x_0 = \frac{KC}{1+C} \qquad \text{or}$$

$$C = \frac{x_0}{K - x_0}.$$

Substituting into the solution and doing some algebra yields

$$x(t) = \frac{x_0 K}{x_0 + (K - x_0)e^{-rt}}.$$

Notice that for $t = 0$, this solution yields x_0. Also, once this form is derived, we can check that it works for all values of x_0 and not just $0 < x_0 < K$.

3.1. Solutions

The solution method in the preceding example applies to any equation of the form $\dot{x} = f(x)$, reducing its solution to simply solving the integral

$$F(x) = \int \frac{1}{f(x)} \, dx = \int dt = t + C.$$

This method is called *separation of variables*. In practice, the particular integral could be difficult or impossible to evaluate by known functions to determine $F(x)$. Even after the integral is evaluated, the result is an implicit solution, $F(x) = t + C$, and it is often difficult to solve it for x as an explicit function of t. Therefore, even for scalar equations, there might not be a practical method to find an explicit solution.

We now state the general result about the existence and uniqueness of solutions in one dimension.

Theorem 3.1 (Existence and uniqueness for scalar differential equations). *Consider the scalar differential equation $\dot{x} = f(x)$, where $f(x)$ is a function from an open interval (a, b) to $\mathbb{R}$, such that both $f(x)$ and $f'(x)$ are continuous.*

a. *For an initial condition $a < x_0 < b$, there exists a solution $x(t)$ to $\dot{x} = f(x)$ defined for some time interval $-\tau < t < \tau$ such that $x(0) = x_0$. Moreover, the solution is unique in the sense, that if $x(t)$ and $y(t)$ are two such solutions with $x(0) = x_0 = y(0)$, then they must be equal on the largest interval of time about $t = 0$ where both solutions are defined. Let $\phi(t; x_0) = x(t)$ be this unique solution with $\phi(0; x_0) = x_0$.*

b. *The solution $\phi(t; x_0)$ depends continuously on the initial condition x_0. Moreover, let $T > 0$ be a time for which $\phi(t; x_0)$ is defined for $-T \leq t \leq T$. Let $\epsilon > 0$ be any bound on the distance between solutions. Then, there exists a $\delta > 0$ which measures the distance between allowable initial conditions, such that if $|y_0 - x_0| < \delta$, then $\phi(t; y_0)$ is defined for $-T \leq t \leq T$ and*

$$|\phi(t; y_0) - \phi(t; x_0)| < \epsilon \qquad \text{for } -T \leq t \leq T.$$

c. *In fact, the solution $\phi(t; x_0)$ depends differentiably on the initial condition, x_0.*

Definition 3.2. The function $\phi(t; x_0)$, which gives the solution as a function of the initial condition x_0, is called the *flow* of the differential equation. This notation is used when the solutions exist and are unique (e.g., the components of the vector field defining the system of differential equations has continuous partial derivatives).

The proof of the multidimensional version of the preceding theorem is given in Section 3.3 at the end of the chapter. It uses what is called the *Picard iteration scheme*. Starting with any curve $y^0(t)$ (e.g., $y^0(t) \equiv x_0$), by induction, the next curve is constructed from the previous one by the integral equation

$$y^n(t) = x_0 + \int_0^t f(y^{n-1}(s)) \, ds.$$

The proof shows that this sequence of curves converges to a solution on some time interval $[-\tau, \tau]$.

It is possible to show that solutions exist provided $f(x)$ is continuous. However, without some additional condition, there are examples with nonunique solutions. The next example does not have a derivative at one point and the solutions are nonunique.

Example 3.3 (Nonunique solutions). Consider the differential equation
$$\dot{x} = \sqrt[3]{x} \qquad \text{with } x_0 = 0.$$
One solution is $x_1(t) \equiv 0$ for all t. On the other hand, separation of variables yields a different solution:
$$\int x^{-\frac{1}{3}}\, dx = \int dt,$$
$$\frac{3}{2} x^{\frac{2}{3}} = t - t_0,$$
$$x(t) = \left(\frac{2(t-t_0)}{3} \right)^{\frac{3}{2}}.$$

In the preceding calculation, we took the constant of integration to be $-t_0$. This solution can be extended to be equal to 0 for $t < t_0$. Thus for any t_0, there is a solution
$$z(t; t_0) = \begin{cases} 0 & \text{for } t \leq t_0, \\ \left(\dfrac{2(t-t_0)}{3} \right)^{\frac{3}{2}} & \text{for } t \geq t_0. \end{cases}$$

These solutions are zero up to the time t_0 and then become positive for $t > t_0$. Since t_0 can be made arbitrarily small, there can be branching as close as we like to $t = 0$ (even at 0). In addition, there is the solution $x_1(t)$ which is identically equal to zero. Therefore, there are many solutions with the same initial condition. See Figure 1. Notice that $f'(x) = \frac{1}{3} x^{-2/3}$ and $f'(0)$ is undefined (or equal to ∞). Therefore, Theorem 3.1 does not apply to guarantee unique solutions.

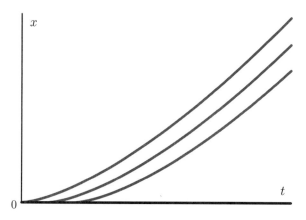

Figure 1. Nonunique solutions for Example 3.3

3.1. Solutions

Now, we return to consider differential equations whose right-hand sides are differentiable and so have unique solutions. Since any two solutions with initial condition x_0 agree on their common interval of definition, we can continue a solution to some maximal interval of definition $t_{x_0}^- < t < t_{x_0}^+$. Sometimes $t_{x_0}^+$ is infinity and other times it can be a finite time. We illustrate these different time intervals with the logistic equation.

For the logistic equation, the unique solution is given by the explicit form of the solution,

$$\phi(t; x_0) = x(t) = \frac{x_0 K}{x_0 + (K - x_0)e^{-rt}}.$$

Notice that for $0 \leq x_0 \leq K$, the denominator is never zero and the solution is defined for all t. On the other hand, if $x_0 > K > 0$, then $K - x_0 < 0$ and the denominator becomes zero when

$$e^{rt} = \frac{x_0 - K}{x_0} < 1$$

and

$$t_{x_0}^- = \frac{1}{r}\Big(\ln(x_0 - K) - \ln(x_0)\Big) < 0,$$

while $t_{x_0}^+ = \infty$. Finally, if $x_0 < 0$, then $t_{x_0}^- = -\infty$, $K-x_0/-x_0 > 1$, and

$$t_{x_0}^+ = \frac{1}{r}\Big(\ln(K - x_0) - \ln(|x_0|)\Big) > 0.$$

Thus, all initial conditions outside of $[0, K]$ go to infinity in either finite forward time or finite backward time.

Linear differential equations $\dot{x} = ax + b$ have solutions that are defined for all time, as can be seen by looking at the explicit solution. See also Theorem 2.9.

In general, if $t_{x_0}^+ < \infty$, then it can be shown that $|\phi(t; x_0)|$ goes to infinity (or perhaps $\phi(t; x_0)$ goes to the boundary of the points where $f(x)$ is defined).

3.1.1. Solutions in Multiple Dimensions. The results on the existence and uniqueness of differential equations in higher dimensions are similar to those for differential equations in one dimension. We already know how to find solutions for constant coefficient linear systems of differential equations, but there is no general method for finding explicit solutions to systems of nonlinear differential equations.

Before stating the general results, we give an example of the type of system of nonlinear differential equation we consider later in the book.

Example 3.4 (Competitive System). Consider the system of equations

$$\dot{x}_1 = x_1(K - x_1 - ax_2),$$
$$\dot{x}_2 = x_2(L - bx_1 - x_2).$$

There is no explicit solution to even these quadratic equations. In Section 4.6, we discuss the properties of the solutions by using phase plane analysis.

In general, we consider a differential equation in $\mathbb{R}^n$ given by

(3.1)
$$\dot{x}_1 = F_1(x_1, \ldots, x_n),$$
$$\dot{x}_2 = F_2(x_1, \ldots, x_n),$$
$$\vdots = \vdots$$
$$\dot{x}_n = F_n(x_1, \ldots, x_n),$$

or, in vector notation,

$$\dot{\mathbf{x}} = \mathbf{F}(\mathbf{x}),$$

where $\mathbf{x}$ is a point in $\mathbb{R}^n$ and $\mathbf{F}(\mathbf{x})$ is a vector field in $\mathbb{R}^n$ (i.e., $\mathbf{F}$ takes a point in $\mathbb{R}^n$ and gives a vector $\mathbf{F}(\mathbf{x})$ associated to that point).

The statement of the existence and uniqueness result is very similar to that for scaler differential equations.

Theorem 3.2. *Consider the vector differential equation* $\dot{\mathbf{x}} = \mathbf{F}(\mathbf{x})$, *where* $\mathbf{F}$ *is a function from* $\mathbb{R}^n$ *to* $\mathbb{R}^n$, *such that both* $\mathbf{F}(\mathbf{x})$ *and* $\dfrac{\partial F_i}{\partial x_j}(\mathbf{x})$ *are continuous. Let* $\mathbf{x}_0$ *be an initial condition in* $\mathbb{R}^n$. *(Actually* $\mathbf{F}$ *only needs to be defined at all points near* $\mathbf{x}_0$.*)*

a. *Then there exists a solution* $\mathbf{x}(t)$ *to* $\dot{\mathbf{x}} = \mathbf{F}(\mathbf{x})$, *defined for some time interval* $-\tau < t < \tau$, *such that* $\mathbf{x}(0) = \mathbf{x}_0$. *Moreover, the solution is unique in the sense, that if* $\mathbf{x}(t)$ *and* $\mathbf{y}(t)$ *are two such solutions with* $\mathbf{x}(0) = \mathbf{y}(0) = \mathbf{x}_0$, *then they must be equal on the largest interval of time about* $t = 0$ *where both solutions are defined. Let* $\phi(t; \mathbf{x}_0)$ *be this unique solution with* $\phi(0; \mathbf{x}_0) = \mathbf{x}_0$.

b. *The solution* $\phi(t; \mathbf{x}_0)$ *depends continuously on the initial condition* $\mathbf{x}_0$. *Moreover, let* $T > 0$ *be a time for which* $\phi(t; \mathbf{x}_0)$ *is defined for* $-T \leq t \leq T$. *Let* $\epsilon > 0$ *be any bound on the distance between solutions. Then, there exists a* $\delta > 0$, *such that if* $\|\mathbf{y}_0 - \mathbf{x}_0\| < \delta$, *then* $\phi(t; \mathbf{y}_0)$ *is defined for* $-T \leq t \leq T$ *and*

$$\|\phi(t; \mathbf{y}_0) - \phi(t; \mathbf{x}_0)\| < \epsilon \qquad \text{for } -T \leq t \leq T.$$

c. *In fact, the solution* $\phi(t; \mathbf{x}_0)$ *depends differentiably on the initial condition,* $\mathbf{x}_0$.

Definition 3.5. The function $\phi(t; \mathbf{x}_0)$, which gives the solution as a function of the time t and initial condition $\mathbf{x}_0$, is called the *flow of the differential equation*. For each fixed $\mathbf{x}_0$, the function $\phi(t; \mathbf{x}_0)$ is a parametrized curve in the higher dimensional space $\mathbb{R}^n$. The set of points on this curve is called the *orbit* or *trajectory with initial condition* $\mathbf{x}_0$.

Notice that, for a linear system $\dot{\mathbf{x}} = \mathbf{A}\mathbf{x}$, the flow $\phi(t; \mathbf{x}_0) = e^{\mathbf{A}t}\mathbf{x}_0$.

A consequence of uniqueness relates to what is called the group property of the flow. Let $\mathbf{x}_1 = \phi(t_1; \mathbf{x}_0)$. Then, both

$$\phi(t + t_1; \mathbf{x}_0) \qquad \text{and} \qquad \phi(t; \mathbf{x}_1)$$

are solutions that equal $\mathbf{x}_1$ when $t = 0$, so

$$\phi(t + t_1; \mathbf{x}_0) = \phi(t; \mathbf{x}_1) = \phi(t; \phi(t_1; \mathbf{x}_0)).$$

3.1. Solutions

See Figure 2. This says that the solution $\phi(t+t_1;\mathbf{x}_0)$ at time t_1+t is the same as the one found by the following procedure: go for time t_1, stop and get a new initial condition $\mathbf{x}_1$, and then go futher by time t, $\phi(t;\mathbf{x}_1) = \phi(t;\phi(t_1;\mathbf{x}_0))$.

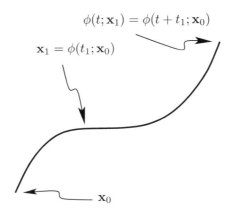

Figure 2. Group property for the flow

The group property has implications for solutions which return to the initial condition at a later time. If
$$\phi(T;\mathbf{x}_0) = \mathbf{x}_0$$
for some positive $T > 0$, but not all t, then, by the group property
$$\phi(t+T;\mathbf{x}_0) = \phi(t;\phi(T;\mathbf{x}_0)) = \phi(t;\mathbf{x}_0)$$
(i.e., it repeats the same points in the phase space). Thus, the orbit is *periodic*. The smallest value T that works is called the *period* or *least period*.

A point $\mathbf{x}^*$ is called a *fixed point*, provided that $\mathbf{F}(\mathbf{x}^*) = \mathbf{0}$. The solution starting at a fixed point has zero velocity, so it just stays there and $\phi(t;\mathbf{x}^*) = \mathbf{x}^*$ for all t. This justifies the name of a fixed point. Traditionally, such a point was called an *equilibrium point*, because the forces were in equilibrium and the mass did not move. The origin is always a fixed point for a linear system, $e^{\mathbf{A}t}\mathbf{0} \equiv \mathbf{0}$. This is the only fixed point of a linear system, unless 0 is an eigenvalue.

In the phase space analysis that we begin to consider in Chapter 4, an important property of the flow is given in the next theorem: different trajectories cannot intersect or cross each other.

Theorem 3.3. *Assume that the differential equation* $\dot{\mathbf{x}} = \mathbf{F}(\mathbf{x})$ *satisfies the assumptions required to give unique solutions for differential equations. Assume that there are two initial conditions* $\mathbf{x}_0$ *and* $\mathbf{y}_0$ *such that the trajectories intersect for some times along the two trajectories,* $\phi(t_0;\mathbf{x}_0) = \phi(t_1;\mathbf{y}_0)$. *Then, by the group property,* $\mathbf{y}_0 = \phi(t_0 - t_1;\mathbf{x}_0)$, *so* $\mathbf{x}_0$ *and* $\mathbf{y}_0$ *must be on the same trajectory.*

In two dimensions, the fact that orbits cannot cross each other is a strong restriction, as we will see in Section 6.2 on the Poincaré–Bendixson theorem: A bounded orbit either has to repeatedly come near a fixed point or it approaches a periodic orbit. In three dimensions and higher, a two-dimensional plot of the trajectories can look like they cross each other because one trajectory is in front and the other is behind. Still the orbits themselves do not intersect without actually being the same trajectory.

We let $t^-_{\mathbf{x}_0} < t < t^+_{\mathbf{x}_0}$ be the largest interval of definition. Again, if $t^+_{\mathbf{x}_0} < \infty$, then as t converges to $t^+_{\mathbf{x}_0}$, either: (i) $\|\phi(t;\mathbf{x}_0)\|$ goes to infinity, or (ii) the solution $\phi(t;\mathbf{x}_0)$ goes to the boundary of points where $\mathbf{F}(\mathbf{x})$ is defined.

A further result concerning the properties of solutions concerns the dependence on initial conditions. Not only is the dependence continuous, it is also differentiable. The next result explains how the derivatives of the flow vary along a solution curve. This formula for the derivative with respect to time of the linearization of the flow is called the *first variation equation*. We use the first variation equation later when we consider stability near a periodic orbit and Lyapunov exponents, which measure the growth of infinitesimal displacements.

Before giving the statement of the theorem, we need some notation. Let $\mathbf{u}^j$ be the unit vector with a 1 in the jth coordinate and zeroes elsewhere. In three dimensions, $\mathbf{i}$, $\mathbf{j}$, and $\mathbf{k}$ are often used for what we are labeling $\mathbf{u}^1$, $\mathbf{u}^2$, and $\mathbf{u}^3$. Let

$$\frac{\partial \phi_i}{\partial x_j}(t;\mathbf{x}_0) = \frac{d}{ds}\phi_i(t;\mathbf{x}_0 + s\,\mathbf{u}^j)\Big|_{s=0}$$

be the derivative of the ith coordinate with respect to a change in the jth variable. We denote by

$$D\mathbf{F}_{(\mathbf{x})} = \left(\frac{\partial F_i}{\partial x_j}(\mathbf{x})\right)$$

the *matrix of partial derivatives* or *derivative* of the vector field. Similarly, we denote by

$$D_{\mathbf{x}}\phi_{(t;\mathbf{x}_0)} = \left(\frac{\partial \phi_i}{\partial x_j}(t;\mathbf{x}_0)\right)$$

the $n \times n$ matrix of partial derivatives of $\phi(t;\mathbf{x}_0)$ with respect to initial conditions.

Theorem 3.4 (First Variation Equation). *Consider the differential equation* $\dot{\mathbf{x}} = \mathbf{F}(\mathbf{x})$. *Assume that both* $\mathbf{F}(\mathbf{x})$ *and* $\frac{\partial F_i}{\partial x_j}(\mathbf{x})$ *are continuous for* $\mathbf{x}$ *on* $\mathbb{R}^n$. *Then, the solution* $\phi(t;\mathbf{x}_0)$ *depends differentiably on the initial condition* $\mathbf{x}_0$, *so the matrix of partial derivatives with respect to initial conditions exists. Also, this matrix of partial derivatives satisfies the linear differential equation*

(3.2) $$\frac{d}{dt}D_{\mathbf{x}}\phi_{(t;\mathbf{x}_0)} = D\mathbf{F}_{(\phi(t;\mathbf{x}_0))}\,D_{\mathbf{x}}\phi_{(t;\mathbf{x}_0)}.$$

Moreover, if $\mathbf{x}_s$ *is a curve of initial conditions and*

$$\mathbf{v}(t) = \frac{\partial}{\partial s}\left(\phi(t;\mathbf{x}_s)\right)\Big|_{s=0},$$

then

(3.3) $$\frac{d}{dt}\mathbf{v}(t) = D\mathbf{F}_{(\phi(t;\mathbf{x}_0))}\,\mathbf{v}(t).$$

Equation (3.3) is called the *first variation equation*.

The proof that the solution is differentiable takes some work. However, if we assume it is differentiable and we can interchange the order of differentiation, the derivation of the first variation equation is straightforward. We present this derivation in the section at the end of this chapter.

If we know the solution curve, then the quantity $D\mathbf{F}_{(\phi(t;\mathbf{x}_0))}$ in equations (3.2) and (3.3) can be considered as a known time-dependent matrix. Therefore, these equations are linear time-dependent equations. If the solution is known only numerically, then it is possible to solve the variation equation at the same time as the position of the solution; that is, it is possible to solve numerically the equations

$$\dot{\mathbf{x}} = \mathbf{F}(\mathbf{x}),$$
$$\dot{\mathbf{v}} = D\mathbf{F}_{(\phi(t;\mathbf{x}_0))}\,\mathbf{v},$$

simultaneously. This idea of solving the equations numerically is especially relevant when we use the first variation equation in the discussion of the Lyapunov exponents in Chapter 7.

Exercises 3.1

1. Show that the differential equation $\dot{x} = x^{1/5}$, with initial condition $x(0) = 0$, has nonunique solutions. Why does the theorem on uniqueness of solutions not apply?

2. Consider the differential equation $\dot{x} = x^2$, with initial condition $x(0) = x_0 \neq 0$.
   a. Verify that the theorem on existence and uniqueness applies.
   b. Solve for an explicit solution.
   c. What is the maximal interval of definition if $x_0 > 0$?

3. Consider the differential equation $\dot{x} = x^2 - 1$, with initial condition $x(0) = x_0$.
   a. Verify that the theorem on existence and uniqueness applies.
   b. Solve for an explicit solution.
   c. What is the maximal interval of definition?

4. Consider the differential equation $\dot{x} = x^2 + 1$, with initial condition $x(0) = x_0$.
   a. Verify that the theorem on existence and uniqueness applies.
   b. Solve for an explicit solution.
   c. What is the maximal interval of definition?

5. Let $\phi(t; x_0)$ be the flow of the differential equation $\dot{x} = 1 + x^{2n}$ for $n \geq 2$ and let $\psi(t; x_0)$ be the flow of the differential equation $\dot{x} = 1 + x^2$
   a. Verify that $\phi(t; x_0) \geq \psi(t; x_0)$ for $x_0 \geq 1$ and $t \geq 0$. What can you say about the comparison of solutions for $x_0 < 1$?
   b. Show that the trajectories $\phi(t; x_0)$ go to infinity in finite time.

6. Let $\phi(t; x_0)$ be the flow of the differential equation $\dot{x} = p(x)$, where $p(x)$ is a polynomial of degree greater than or equal to one. Assume that $p(x) > 0$ for all $x > a$. Show that $\phi(t; x_0)$ goes to infinity in finite time for any $x_0 > a$.

7. Consider the differential equation $\dot{x} = x^2 - 1$, with initial condition $x_0 \neq \pm 1$.
   a. Solve for an explicit solution.

b. What is the maximal interval of definition if $x_0 > 1$? If $-1 < x_0 < 1$?

8. Consider the differential equation $\dot{x} = x + 1$, with $x(0) = 1$. Find the first six terms determined by the Picard iteration scheme, starting with $y^0(t) \equiv 1$. See Section 3.1 or 3.3 for the integral defining the sequence of curves.

9. Consider the linear system of differential equations
$$\dot{x}_1 = x_2,$$
$$\dot{x}_2 = -x_1,$$
with initial conditions $x_1(0) = 0$ and $x_2(0) = 1$. Find the first six terms by the Picard iteration scheme given by the integral in Section 3.3.

10. Let C_0, C_1, and C_2 be positive constants. Assume that $v(t)$ is a continuous nonnegative real-valued function on $\mathbb{R}$ such that
$$v(t) \leq C_0 + C_1 |t| + C_2 \left| \int_0^t v(s)\, ds \right|.$$
Prove that
$$v(t) \leq C_0 \, e^{C_2 |t|} + \frac{C_1}{C_2} \left[e^{C_2 |t|} - 1 \right].$$
Hint: Use
$$U(t) = C_0 + C_1 t + C_2 \int_0^t v(s)\, ds.$$

11. Let C_1 and C_2 be positive constants. Assume that $\mathbf{F} : \mathbb{R}^n \to \mathbb{R}^n$ is a C^1 function such that $\|\mathbf{F}(\mathbf{x})\| \leq C_1 + C_2 \|\mathbf{x}\|$ for all $\mathbf{x} \in \mathbb{R}^n$. Prove that the solutions of $\dot{\mathbf{x}} = \mathbf{F}(\mathbf{x})$ exist for all t. Hint: Use the previous exercise.

12. Consider the differential equation depending on a parameter
$$\dot{\mathbf{x}} = \mathbf{F}(\mathbf{x}; \mu),$$
where $\mathbf{x}$ is in $\mathbb{R}^n$, μ is in $\mathbb{R}^p$, and $\mathbf{F}$ is a C^1 function jointly in $\mathbf{x}$ and μ. Let $\phi^t(\mathbf{x}; \mu)$ be the solution with $\phi^0(\mathbf{x}; \mu) = \mathbf{x}$. Prove that $\phi^t(\mathbf{x}; \mu)$ depends continuously on the parameter μ.

3.2. Numerical Solutions of Differential Equations

For a scalar differential equation, it is possible to reduce the solution to an integral. In principle, this gives a solution in at least implicit form for any such problem. However, the integrals can be difficult to evaluate, and it may be difficult or impossible to solve for an explicit solution in which $x(t; x_0)$ is given as a function of t and the initial condition. Moreover, in higher dimensions, it is not even always or usually possible to reduce the problem to one of evaluating integrals and it is not possible to find an explicit solution. In fact, when computer programs are used to plot solutions of differential equation, what is usually plotted is a numerical solution and not an explicit solution of the differential equation.

For these reasons, numerical solutions are useful for analyzing differential equations. They should be used in connection with geometric properties of the differential equations, but they can give good insight into the behavior of the solutions. We start with scalar differential equations and then discuss the same numerical techniques for equations in higher dimensions, or vector differential equations. We

3.2. Numerical Solutions

discuss three techniques: the Euler method, the Heun or improved Euler method, and the (fourth-order) Runge–Kutta method; these methods correspond to the three numerical methods of evaluating integrals of Riemann sums using the left end point, the trapezoid method, and Simpson's method.

This material is not necessary for any of the rest of the book. It is included so that the reader will have some idea what the computer is doing when it plots solutions of differential equations, either for scalar equations or systems of equations. The reader who wants more details on the theory of numerical methods can consult [**Bur78**].

We consider an equation of the form
$$\dot{x} = f(t, x)$$
that can depend on time. Besides being more general, it allows us to indicate the time at which various evaluations of f are taken.

Euler method

Assume that we are given an initial time t_0 and an initial position $x(t_0) = x_0$. We are also given a step size $h = \Delta t$. The solution at time $t_1 = t_0 + h$ is given by the integral
$$\begin{aligned} x(t_1) &= x_0 + \int_{t_0}^{t_1} \dot{x}(t)\, dt \\ &\approx x_0 + h\, \dot{x}(t_0) \\ &= x_0 + h\, f(t_0, x_0), \end{aligned}$$
where we have used the value at the left-hand end of the interval to approximate the integral. This approximation defines the *Euler method* for the approximation at the next step:
$$x_1 = x_0 + h\, f(t_0, x_0).$$
At the nth step, if we are given t_n and position x_n at the nth step, the $(n+1)^{\text{th}}$ step is given by

(3.4)
$$\begin{aligned} k_n &= f(t_n, x_n), \\ x_{n+1} &= x_n + h\, k_n, \\ t_{n+1} &= t_n + h. \end{aligned}$$

We illustrate this method with a simple example.

Example 3.6 (**Euler Method**). Consider

(3.5) $$\dot{x} = x(1 - x^2) \qquad \text{with } x_0 = 2 \text{ at } t_0 = 0,$$

and step size $h = 0.1$ for the Euler method. By using separation of variables and partial fractions, the exact solution can be found to be
$$x(t) = \sqrt{\frac{4\, e^{2t}}{4\, e^{2t} - 3}}.$$

Table 1 contains the calculations for 10 steps up to $t = 1$. The reader should check a few of these entries. Also, the entries were calculated to a greater accuracy, but only three digits to the right of the decimal point are displayed in Table 1.

n	t_n	x_n	k_n	Exact	Error
0	0.0	2.000	-6.000	2.000	0.0×10^0
1	0.1	1.400	-1.344	1.610	2.1×10^{-1}
2	0.2	1.266	-0.762	1.418	1.5×10^{-1}
3	0.3	1.189	-0.493	1.304	1.1×10^{-1}
4	0.4	1.140	-0.342	1.228	8.8×10^{-2}
5	0.5	1.106	-0.247	1.175	6.9×10^{-2}
6	0.6	1.081	-0.183	1.137	5.5×10^{-2}
7	0.7	1.063	-0.138	1.108	4.5×10^{-2}
8	0.8	1.049	-0.106	1.086	3.6×10^{-2}
9	0.9	1.039	-0.082	1.068	3.0×10^{-2}
10	1.0	1.030	-0.064	1.055	2.5×10^{-2}

Table 1. Euler method for equation (3.5)

Heun or improved Euler method

The *improved Euler method* or *Heun method* corresponds to the trapezoid rule,

$$x(t_1) = x_0 + \int_{t_0}^{t_1} f(t, x(t))\, dt$$
$$\approx x_0 + h\left(\frac{f(t_0, x_0) + f(t_1, z)}{2}\right),$$

where z should be a position corresponding to time t_1. Therefore, we need a preliminary position z to use in the formula. Assume that we have calculated x_n as an approximation of $x(t_n)$. We start by using the Euler method to find the approximation z_n of the solution at time $t_{n+1} = t_n + h$, and then we use this point to evaluate the second "slope" $f(t_{n+1}, z_n)$. Thus, we carry out the following steps:

(3.6)
$$k_{1,n} = f(t_n, x_n),$$
$$z_n = x_n + h\, k_{1,n},$$
$$k_{2,n} = f(t_n + h, z_n),$$
$$k_{\text{avg},n} = \tfrac{1}{2}\left(k_{1,n} + k_{2,n}\right),$$
$$x_{n+1} = x_n + h\, k_{\text{avg},n},$$
$$t_{n+1} = t_n + h.$$

We illustrate this method with the same simple example we used for the Euler method.

Example 3.7 (Heun method). Again, consider

$$\dot{x} = x(1 - x^2) \quad \text{with } x_0 = 2 \text{ at } t_0 = 0,$$

and step size $h = 0.1$. The exact solution is the same as for the preceding example. Table 2 contains the calculations for 10 steps up to $t = 1$. The reader should check a few of these entries.

3.2. Numerical Solutions

n	t_n	x_n	$k_{1,n}$	z_n	$k_{2,n}$	$k_{\text{avg},n}$	Exact	Error
0	0.0	2.000	-6.000	1.400	-1.344	-3.672	2.000	0.0
1	0.1	1.633	-2.720	1.361	-1.159	-1.940	1.610	-2.3×10^{-2}
2	0.2	1.439	-1.540	1.285	-0.836	-1.188	1.418	-2.1×10^{-2}
3	0.3	1.320	-0.980	1.222	-0.603	-0.791	1.304	-1.6×10^{-2}
4	0.4	1.241	-0.670	1.174	-0.444	-0.557	1.228	-1.3×10^{-2}
5	0.5	1.185	-0.480	1.137	-0.334	-0.407	1.175	-1.0×10^{-2}
6	0.6	1.145	-0.355	1.109	-0.255	-0.305	1.137	-8.0×10^{-3}
7	0.7	1.114	-0.269	1.087	-0.198	-0.233	1.108	-6.4×10^{-3}
8	0.8	1.091	-0.207	1.070	-0.155	-0.181	1.086	-5.2×10^{-3}
9	0.9	1.073	-0.161	1.056	-0.123	-0.142	1.068	-4.2×10^{-3}
10	1.0	1.058	-0.127	1.046	-0.098	-0.113	1.055	-3.4×10^{-3}

Table 2. Heun method for equation (3.5)

Runge–Kutta method

The *Runge–Kutta method* corresponds to Simpson's method for numerical integration of integrals in which we use $g(t)$ for $f(t, x(t))$ at some point $x(t)$:

$$x(t_{n+1}) = x_n + \int_{t_n}^{t_n+h} g(t)\,dt$$
$$\approx x_n + h\left(\frac{g(t_n) + 4g(t_n + \frac{h}{2}) + g(t_n + h)}{6}\right).$$

To implement this method for differential equations, we split $4g(t_n + \frac{h}{2})$ two pieces, $2g(t_n + \frac{h}{2}) + 2g(t_n + \frac{h}{2})$, and we evaluate these two at different intermediate points, $z_{1,n}$ and $z_{2,n}$. Since these two points correspond to $t_n + \frac{h}{2}$ and not $t_n + h$, we use a step size of $h/2$ to find the two points $z_{1,n}$ and $z_{2,n}$. Finally, the term $g(t_n + h)$ corresponds to $t_n + h$, so we find a third intermediate point $z_{3,n}$ that we get from x_n by taking a full step size of h. This is implemented as follows:

$$(3.7) \quad \begin{aligned} k_{1,n} &= f(t_n, x_n), & z_{1,n} &= x_n + \frac{h}{2} k_{1,n}, \\ k_{2,n} &= f\left(t_n + h/2, z_{1,n}\right), & z_{2,n} &= x_n + \frac{h}{2} k_{2,n}, \\ k_{3,n} &= f\left(t_n + h/2, z_{2,n}\right), & z_{3,n} &= x_n + h\,k_{3,n}, \\ k_{4,n} &= f\left(t_n + h, z_{3,n}\right), & k_{\text{avg},n} &= \tfrac{1}{6}\left(k_{1,n} + 2k_{2,n} + 2k_{3,n} + k_{4,n}\right) \end{aligned}$$

and then

$$x_{n+1} = x_n + h\,k_{\text{avg},n},$$
$$t_{n+1} = t_n + h.$$

Thus, the final result uses slopes evaluated at $t = t_n$, $t_n + \frac{h}{2}$, $t_n + \frac{h}{2}$, and $t_n + h$ with weights 1, 2, 2, and 1.

We use the same example as in the proceding cases to compare the various methods.

Example 3.8 (Runge–Kutta Method). Consider

$$\dot{x} = x(1-x^2) \quad \text{with } x_0 = 2 \text{ at } t_0 = 0,$$

and step size $h = 0.1$. The exact solution is the same as for the last two examples. Table 3 contains the calculations for 10 steps up to $t = 1$. The reader should check a few of these entries.

t_n	x_n	$k_{1,n}$	$z_{1,n}$	$k_{2,n}$	$z_{2,n}$	$k_{3,n}$	$z_{3,n}$	$k_{4,n}$	$k_{\text{avg},n}$
0.0	2.000000	-6.000	1.700	-3.213	1.839	-4.384	1.562	-2.247	-3.906
0.1	1.609335	-2.559	1.481	-1.770	1.521	-1.997	1.410	-1.391	-1.913
0.2	1.417949	-1.433	1.346	-1.094	1.363	-1.170	1.301	-0.901	-1.143
0.3	1.303580	-0.912	1.258	-0.733	1.267	-0.767	1.227	-0.620	-0.755
0.4	1.228069	-0.624	1.197	-0.518	1.202	-0.535	1.175	-0.446	-0.529
0.5	1.175142	-0.448	1.153	-0.379	1.156	-0.389	1.136	-0.331	-0.385
0.6	1.136555	-0.332	1.120	-0.285	1.122	-0.291	1.107	-0.251	-0.289
0.7	1.107644	-0.251	1.095	-0.218	1.097	-0.222	1.085	-0.193	-0.220
0.8	1.085548	-0.194	1.076	-0.169	1.077	-0.172	1.068	-0.151	-0.171
0.9	1.068409	-0.151	1.061	-0.133	1.062	-0.135	1.055	-0.119	-0.134
1.0	1.054965	-0.119	1.049	-0.105	1.050	-0.107	1.044	-0.095	-0.106

Table 3. Runge–Kutta method for equation (3.5)

It is interesting to compare the accuracy of the solutions by the different numerical methods: at time 1, the Euler method has an error of 2.5×10^{-2}, the Heun method has an error of 3.4×10^{-3}, and the Runge–Kutta method has an error of 7.6×10^{-6}. See Table 4. Thus the extra evaluations in the Runge–Kutta method did indeed result in much greater accuracy.

t_n	x_n	Exact	Error
0.0	2.000000	2.000000	0.0
0.1	1.609335	1.609657	3.2×10^{-4}
0.2	1.417949	1.418105	1.6×10^{-4}
0.3	1.303580	1.303668	8.7×10^{-5}
0.4	1.228069	1.228124	5.4×10^{-5}
0.5	1.175142	1.175178	3.6×10^{-5}
0.6	1.136555	1.136581	2.5×10^{-5}
0.7	1.107644	1.107662	1.8×10^{-5}
0.8	1.085548	1.085561	1.3×10^{-5}
0.9	1.068409	1.068419	1.0×10^{-5}
1.0	1.054965	1.054973	7.6×10^{-6}

Table 4. Runge–Kutta method: Comparison with exact solution

3.2. Numerical Solutions

When working a problem, displaying the calculations in a chart helps to organize the work and avoid using the wrong number in the next steps. These calculations are easier to do with a spreadsheet, programming language such as C, or a general purpose package such as Maple, Mathematica, or Matlab.

Errors: local and global

To understand the error in one step of the Euler method, we use the Taylor expansion of the solution as a function of t (assuming that it is differentiable). At time $t_{n+1} = t_n + h$, the solution is given by

$$x(t_{n+1}) = x_n + h\,\dot{x}(t_n) + \frac{h^2}{2}\ddot{x}(t'_n)$$
$$= x_n + h\,f(t_n, x_n) + O(h^2),$$

where t'_n is a time between t_n and t_{n+1} and the term labeled by $O(h^2)$ is an error term that goes to zero like h^2 (i.e., there is a constant $C > 0$ such that $|O(h^2)| \leq C\,h^2$). In fact, if $|\ddot{x}(t)| \leq 2\,C$, then this value of C gives a bound. We provide a few more details in Section 3.3 at the end of the chapter. This error in one step is called the *local error*, and we have discussed why it is $O(h^2)$ for the Euler method.

In in Section 3.3, we show that the local error for the Heun method is $O(h^3)$. The analysis of the local error of the Runge–Kutta method is more complicated and we do not give it, but it can be shown that the local error for one step is $O(h^5)$.

When a time T is fixed and the interval from 0 to T is divided into N steps, so $h = T/N$, then the accumulated error $|x_N - x(T)|$ is called the *global error*. The errors from the local errors accumulate. If the local error of one step is bounded by $h\,E(h)$, since there are $N = T/h$ steps, a naive estimate is that the global error is about $N\,h\,E(h) = (T/h)\,h\,E(h) = T\,E(h)$. In in Section 3.3 at the end of the chapter, we show that the global error is estimated to be

$$|x_N - \phi(T; x_0)| = \frac{E(h)\,[e^{TL} - 1]}{L},$$

where L is a bound on $\frac{df}{dx}(x,t)$. This estimate shows that, for the length of time T fixed, the global error goes to zero with h. However, this bound grows exponentially with the length of the time interval, and so it is not very useful on long intervals and is useless for infinite time. The reason for the growth of the bound on the error is as follows: once a local error is made, then the flow from this new point can diverge exponentially as time increases, as we discussed in the preceding section in connection with the continuity of solutions with respect to initial conditions. See Figure 3.

For the Euler method, the local error is bounded by $M\,h^2$, so $E(h) = M\,h$ and the global error is linear in h. Since the global error is linear in h, the Euler method is called a *first-order method*.

For the Heun method, the local error for one step is $h\,E(h) = O(h^3)$ and the global error is $E(h) = O(h^2)$. Since the global error goes to zero quadratically in h, the Heun method is called a *second-order method*.

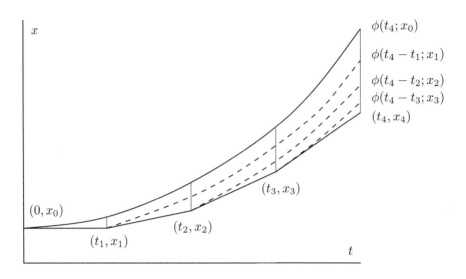

Figure 3. Propagation of local errors

For the Runge–Kutta method, the local error for one step is $h\,E(h) = O(h^5)$ and the global error is $E(h) = O(h^4)$. Since the global error goes to zero like h^4, the Runge–Kutta method is called a *fourth-order method*. Even though the Runge–Kutta method requires more evaluations for one step than the Euler or Heun methods, it is more accurate and also takes into account the bending of solutions. For this reason, the fourth-order Runge–Kutta method is often used.

Graphical solutions

Euler's method can be thought of as starting at each point (t_n, x_n) and moving along the direction field $(1, f(t_n, x_n))$ to the new point

$$(t_{n+1}, x_{n+1}) = (t_n, x_n) + h\,(1, f(t_n, x_n))$$
$$= (t_n + h, x_n + h\,f(t_n, x_n)).$$

The true solution moves in the (t, x) space, so it is tangent at every point to the *direction field* $(1, f(t, x))$. A graphical way to understand the properties of the solutions is to draw the direction fields $(1, f(t, x))$ at a number of points in the (t, x)-plane and then draw some curves in a manner that looks tangent to the direction fields.

Example 3.9. Consider the differential equation

$$\dot{x} = x(1 - x^2) = f(x).$$

Since the right-hand side does not depend on t, the direction field will be the same at any points with the same of value x. We calculate the value of $f(x)$ at 21 points between $x = -2$ and $x = 2$. We plot those direction fields for $0 \leq t \leq 10$. See Figure 4, which was actually plotted using Maple rather than by hand (but it could have been roughly plotted by hand.)

In Figure 5, curves are added with initial conditions $x(0) = -2, -1\, -0.25$, 0.25, 1, and 2. Again, approximations of these curves could be drawn manually.

3.2. Numerical Solutions

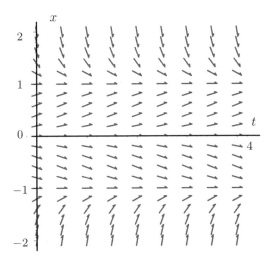

Figure 4. Direction field for Example 3.9

From the plot, it seems that solutions with $x_0 > 0$ tend to $x = 1$ as t increases, and solutions with $x_0 < 0$ tend to $x = -1$.

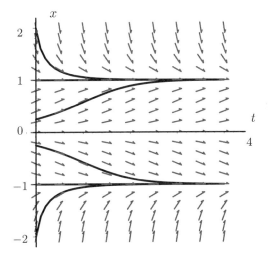

Figure 5. Direction field and solutions for Example 3.9

3.2.1. Numerical Methods in Multiple Dimensions. The form of the numerical methods in higher dimensions for $\dot{\mathbf{x}} = \mathbf{F}(t, \mathbf{x})$ is the same as the methods in one dimension, but everything is considered a vector.

For the Euler method,

(3.8)
$$\mathbf{x}_{n+1} = \mathbf{x}_n + h\,\mathbf{F}(t_n, \mathbf{x}_n) \quad \text{and}$$
$$t_{n+1} = t_n + h.$$

For the Heun method,
$$\begin{aligned}
\mathbf{k}_{1,n} &= \mathbf{F}(t_n, \mathbf{x}_n), \\
\mathbf{z}_n &= \mathbf{x}_n + h\,\mathbf{k}_{1,n}, \\
\mathbf{k}_{2,n} &= \mathbf{F}(t_n + h, \mathbf{z}_n), \\
\mathbf{k}_{\text{avg},n} &= \tfrac{1}{2}\left(\mathbf{k}_{1,n} + \mathbf{k}_{2,n}\right), \\
\mathbf{x}_{n+1} &= \mathbf{x}_n + h\,\mathbf{k}_{\text{avg},n} \\
t_{n+1} &= t_n + h.
\end{aligned} \qquad (3.9)$$

For the Runge–Kutta method,
$$\begin{aligned}
\mathbf{k}_{1,n} &= \mathbf{F}(t_n, \mathbf{x}_n), \\
\mathbf{z}_{1,n} &= \mathbf{x}_n + \frac{h}{2}\,\mathbf{k}_{1,n}, \\
\mathbf{k}_{2,n} &= \mathbf{F}\!\left(t_n + \tfrac{1}{2}h, \mathbf{z}_{1,n}\right), \\
\mathbf{z}_{2,n} &= \mathbf{x}_n + \frac{h}{2}\,\mathbf{k}_{2,n}, \\
\mathbf{k}_{3,n} &= \mathbf{F}\!\left(t_n + \tfrac{1}{2}h, \mathbf{z}_{2,n}\right), \\
\mathbf{z}_{3,n} &= \mathbf{x}_n + h\,\mathbf{k}_{3,n}, \\
\mathbf{k}_{4,n} &= \mathbf{F}\!\left(t_n + h, \mathbf{z}_{3,n}\right), \\
\mathbf{k}_{\text{avg},n} &= \tfrac{1}{6}\left(\mathbf{k}_{1,n} + 2\,\mathbf{k}_{2,n} + 2\,\mathbf{k}_{3,n} + \mathbf{k}_{4,n}\right), \\
\mathbf{x}_{n+1} &= \mathbf{x}_n + h\,\mathbf{k}_{\text{avg},n}, \\
t_{n+1} &= t_n + h.
\end{aligned} \qquad (3.10)$$

Example 3.10. We consider a simple linear example,
$$\begin{aligned}
\dot{x}_1 &= x_2, \\
\dot{x}_2 &= -4\,x_1,
\end{aligned}$$
with $\mathbf{x}_0 = \begin{pmatrix} 0 \\ 1 \end{pmatrix}$. Let
$$\mathbf{F}(\mathbf{x}) = \begin{pmatrix} x_2 \\ -4\,x_1 \end{pmatrix}$$
be the vector field defined by the right-hand side of the system of differential equations. We use step size $h = 0.1$ for each of the methods.

We start with the Euler method. Evaluating $\mathbf{F}$ at $\mathbf{x}_0$, we have
$$\mathbf{F}\begin{pmatrix} 0 \\ 1 \end{pmatrix} = \begin{pmatrix} 1 \\ 0 \end{pmatrix} \qquad \text{and}$$
$$\mathbf{x}_1 = \begin{pmatrix} 0 \\ 1 \end{pmatrix} + 0.1 \begin{pmatrix} 1 \\ 0 \end{pmatrix} = \begin{pmatrix} 0.1 \\ 1 \end{pmatrix}.$$
Repeating for the next step yields
$$\mathbf{F}\begin{pmatrix} 0.1 \\ 1 \end{pmatrix} = \begin{pmatrix} 1 \\ -0.4 \end{pmatrix} \qquad \text{and}$$
$$\mathbf{x}_2 = \begin{pmatrix} 0.1 \\ 1 \end{pmatrix} + 0.1 \begin{pmatrix} 1 \\ -0.4 \end{pmatrix} = \begin{pmatrix} 0.2 \\ 0.96 \end{pmatrix}.$$

3.2. Numerical Solutions

By comparison, using the Heun method, we have

$$\mathbf{k}_{1,0} = \mathbf{F}\begin{pmatrix} 0 \\ 1 \end{pmatrix} = \begin{pmatrix} 1 \\ 0 \end{pmatrix},$$

$$\mathbf{z}_0 = \mathbf{x}_0 + 0.1\,\mathbf{k}_{1,0} = \begin{pmatrix} 0 \\ 1 \end{pmatrix} + 0.1\begin{pmatrix} 1 \\ 0 \end{pmatrix} = \begin{pmatrix} 0.1 \\ 1 \end{pmatrix},$$

$$\mathbf{k}_{2,0} = \mathbf{F}\begin{pmatrix} 0.1 \\ 1 \end{pmatrix} = \begin{pmatrix} 1 \\ -0.4 \end{pmatrix},$$

$$\mathbf{k}_{\text{avg},0} = \tfrac{1}{2}\left[\begin{pmatrix} 1 \\ 0 \end{pmatrix} + \begin{pmatrix} 1 \\ -0.4 \end{pmatrix}\right] = \begin{pmatrix} 1 \\ -0.2 \end{pmatrix},$$

$$\mathbf{x}_1 = \begin{pmatrix} 0 \\ 1 \end{pmatrix} + 0.1\begin{pmatrix} 1 \\ -0.2 \end{pmatrix} = \begin{pmatrix} 0.1 \\ 0.98 \end{pmatrix}.$$

Repeating gives

$$\mathbf{k}_{1,1} = \mathbf{F}\begin{pmatrix} 0.1 \\ 0.98 \end{pmatrix} = \begin{pmatrix} 0.98 \\ -0.4 \end{pmatrix},$$

$$\mathbf{z}_0 = \begin{pmatrix} 0.1 \\ 0.98 \end{pmatrix} + 0.1\begin{pmatrix} 0.98 \\ -0.4 \end{pmatrix} = \begin{pmatrix} 0.198 \\ 0.94 \end{pmatrix},$$

$$\mathbf{k}_{2,1} = \mathbf{F}\begin{pmatrix} 0.198 \\ 0.94 \end{pmatrix} = \begin{pmatrix} 0.94 \\ -0.792 \end{pmatrix},$$

$$\mathbf{k}_{\text{avg},1} = \tfrac{1}{2}\left[\begin{pmatrix} 0.98 \\ -0.4 \end{pmatrix} + \begin{pmatrix} 0.94 \\ -0.792 \end{pmatrix}\right] = \begin{pmatrix} 0.96 \\ -0.596 \end{pmatrix},$$

$$\mathbf{x}_2 = \begin{pmatrix} 0.1 \\ 0.98 \end{pmatrix} + 0.1\begin{pmatrix} 0.96 \\ -0.596 \end{pmatrix} = \begin{pmatrix} 0.196 \\ 0.9204 \end{pmatrix}.$$

We next give one step for the Runge–Kutta method for $h = 0.1$.

$$\mathbf{k}_{1,0} = \mathbf{F}\begin{pmatrix} 0 \\ 1 \end{pmatrix} = \begin{pmatrix} 1 \\ 0 \end{pmatrix},$$

$$\mathbf{z}_{1,0} = \mathbf{x}_0 + \frac{h}{2}\mathbf{k}_{1,0} = \begin{pmatrix} 0 \\ 1 \end{pmatrix} + 0.05\begin{pmatrix} 1 \\ 0 \end{pmatrix} = \begin{pmatrix} 0.05 \\ 1 \end{pmatrix},$$

$$\mathbf{k}_{2,0} = \mathbf{F}(\mathbf{z}_{1,0}) = \mathbf{F}\begin{pmatrix} 0.05 \\ 1 \end{pmatrix} = \begin{pmatrix} 1 \\ -0.2 \end{pmatrix},$$

$$\mathbf{z}_{2,0} = \mathbf{x}_0 + \frac{h}{2}\mathbf{k}_{2,0} = \begin{pmatrix} 0 \\ 1 \end{pmatrix} + 0.05\begin{pmatrix} 1 \\ -0.2 \end{pmatrix} = \begin{pmatrix} 0.05 \\ 0.99 \end{pmatrix},$$

$$\mathbf{k}_{3,0} = \mathbf{F}(\mathbf{z}_{2,0}) = \mathbf{F}\begin{pmatrix} 0.05 \\ 0.99 \end{pmatrix} = \begin{pmatrix} 0.99 \\ -0.2 \end{pmatrix},$$

$$\mathbf{z}_{3,0} = \mathbf{x}_0 + h\,\mathbf{k}_{3,0} = \begin{pmatrix} 0 \\ 1 \end{pmatrix} + 0.1\begin{pmatrix} 0.99 \\ -0.2 \end{pmatrix} = \begin{pmatrix} 0.099 \\ 0.98 \end{pmatrix},$$

$$\mathbf{k}_{4,0} = \mathbf{F}(\mathbf{z}_{3,0}) = \mathbf{F}\begin{pmatrix} 0.099 \\ 0.98 \end{pmatrix} = \begin{pmatrix} 0.98 \\ -0.396 \end{pmatrix},$$

$$\mathbf{k}_{\text{avg},0} = \tfrac{1}{6}\left[\begin{pmatrix}1\\0\end{pmatrix} + 2\begin{pmatrix}1\\-0.2\end{pmatrix} + 2\begin{pmatrix}0.99\\-0.2\end{pmatrix} + \begin{pmatrix}0.98\\-0.396\end{pmatrix}\right] = \begin{pmatrix}0.9933\\-0.1993\end{pmatrix},$$
$$\mathbf{x}_1 = \begin{pmatrix}0\\1\end{pmatrix} + 0.1\begin{pmatrix}0.9933\\-0.1993\end{pmatrix} = \begin{pmatrix}0.09933\\0.98007\end{pmatrix}.$$

Because the Euler method takes into consideration the vector field at only the starting point, it ignores the manner in which the trajectory is bending. For this reason, this method has some consistent geometric errors in the approximate solution found. If we consider the simple harmonic oscillator with no friction,

$$\dot{x} = y,$$
$$\dot{y} = -4x,$$

then the solution found by the Euler method spirals out, even though the true solution moves on an ellipse. The reason for this is that the method moves along the tangent line to the ellipse and goes to a position outside the original ellipse. The next step again moves along the tangent line through the new position and again moves outside this ellipse. Continuing, the approximate solution spirals outward. Figure 6 shows these curves in the (x, y)-plane.

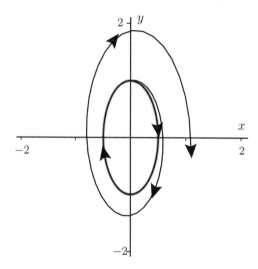

Figure 6. Comparison of a solution from Euler method and a true solution for an elliptic center

Because the Heun and Runge–Kutta methods average the vector field at points farther along, they take into consideration the curvature. For this reason, we use the fourth-order Runge–Kutta method to construct the phase portraits of the systems drawn in this book. It is a good compromise between a good approximation and a relatively efficient method.

For more complex systems, researchers use more complicated numerical methods. One improvement is to use a method that has a variable step size. The idea is to adjust the step size as necessary so that reasonable progress is made, while being careful in regions of the phase space where it is necessary. Near fixed points,

3.2. Numerical Solutions

the step size is increased so the progress does not slow down too much. In areas where the solution is moving more rapidly, the step size is decreased.

The second adjustment made is variable order. In parts of the phase space where the solution is bending a lot, the order is increased. Where the solutions are relatively straight and well behaved, the order is decreased.

Graphical solutions

Just as the direction fields plotted in the (t, x)-plane give information about the solutions for scalar equations, so plotting the vector field in the **x**-plane gives information about systems with more variables.

Example 3.11. Consider the system of equations

$$\dot{x}_1 = x_2,$$
$$\dot{x}_2 = -4\, x_1.$$

We can plot the vector field defined by the right-hand side of the system, which is

$$\mathbf{F}(\mathbf{x}) = \begin{pmatrix} x_2 \\ -4\, x_1 \end{pmatrix}.$$

In Figure 7, we plot this vector field at various positions: At various points **x**, we put a vector $\mathbf{F}(\mathbf{x})$ with base point at **x**. Thus, we get a collection of vectors (arrows) on the plane, with the magnitude and direction of the vector changing from point to point. Then, in Figure 8, we add solution curves with $(x_1(0), x_2(0)) = (0.5, 0)$, $(1, 0)$, and $(1.5, 0)$.

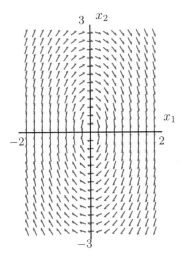

Figure 7. Vector field for Example 3.11

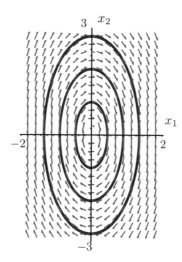

Figure 8. Vector field and solution curves for Example 3.11

Exercises 3.2

1. Consider the differential equation $\dot{x} = x^2$, with initial condition $x(0) = 1$.
   a. Using step size $h = 0.1$, estimate $x(0.3)$ using the following methods: Euler method, improved Euler method, and Runge–Kutta method.
   b. Sketch the direction field on the (t, x)-plane. Give an approximation of the trajectory $\phi(t; 1)$ for $0 \leq t \leq 1$ by drawing a curve that is everywhere tangent to the direction field.

   Remark: You do not need to find the exact solution, since you are not asked to compare the numerical methods with the exact solution.

2. Consider the differential equation $\dot{x} = x - x^2$.
   a. Sketch the direction field on the (t, x)-plane.
   b. Add to the sketch curves that are everywhere tangent to the direction field which approximate the trajectories $\phi(t; -0.25)$, $\phi(t; 0)$, $\phi(t; 0.25)$, $\phi(t; 1)$, and $\phi(t; 2)$.

3. Consider the equations
$$\begin{pmatrix} \dot{x} \\ \dot{y} \end{pmatrix} = \begin{pmatrix} y \\ -2x \end{pmatrix},$$
with initial condition $\mathbf{x}(0) = \begin{pmatrix} 1 \\ 0 \end{pmatrix}$.
   a. Using step size $h = 0.1$, estimate $x(0.2)$ using the following methods: Euler method, improved Euler method, and Runge–Kutta method.
   b. Sketch the direction field on the (x, y)-plane. Give an approximation of the trajectory $\phi(t; (1, 0))$ for $0 \leq t \leq 1$ by drawing a curve that is everywhere tangent to the direction field.

   Remark: You do not need to find the exact solution, since you are not asked to compare the numerical methods with the exact solution.

4. Consider the equations
$$\begin{pmatrix} \dot{x} \\ \dot{y} \end{pmatrix} = \begin{pmatrix} y \\ 4x \end{pmatrix},$$
with initial condition $\mathbf{x}(0) = \begin{pmatrix} 1 \\ 1 \end{pmatrix}$.

   a. Using step size $h = 0.1$, estimate $x(0.2)$ using the following methods: Euler method, improved Euler method, and Runge–Kutta method.

   b. Sketch the direction field on the (x, y)-plane. Give an approximation of the trajectory $\phi(t; (1, 1))$ for $0 \leq t \leq 1$ by drawing a curve that is everywhere tangent to the direction field.

   Remark: You do not need to find the exact solution, since you are not asked to compare the numerical methods with the exact solution.

5. Consider the equation
$$\dot{x} = 2x + e^{-t},$$
with initial conditions $x(0) = 1/3$. The exact solution is $x(t) = \frac{2}{3} e^{2t} - \frac{1}{3} e^{-t}$.

   a. Using the Euler method and $h = 0.1$, estimate $x(0.5)$. Compare the value with the exact solution.

   b. Using the global estimate in Theorem 3.10, how small must the value of h be to ensure that the error is less than 0.1 at $T = 0.5$.

3.3. Theory and Proofs

Properties of solutions

We give the proofs only for differential equations that do not depend on time. If we start with an equation that depends on time explicitly, $\dot{\mathbf{x}} = \mathbf{F}(t, \mathbf{x})$, it can be made into one that is independent of time by adding another variable $\dot{\tau} = 1$, and considering
$$\dot{\tau} = 1,$$
$$\dot{\mathbf{x}} = \mathbf{F}(\tau, \mathbf{x}).$$

We also use the notation of a system of differential equations that has several dependent variables. This certainly includes the case of scalar equations that have a single dependent real variable.

In the proofs of the existence and uniqueness of solutions, we need to use the fact that $\mathbf{F}(\mathbf{x})$ has continuous partial derivatives to get an estimate on how large $\|\mathbf{F}(\mathbf{x}) - \mathbf{F}(\mathbf{x}')\|$ is in terms of $\|\mathbf{x} - \mathbf{x}'\|$. This property is called the *Lipschitz* condition. We do not prove the most general result, but one that is enough for our use.

Theorem 3.5. *Assume that $\mathbf{F}(\mathbf{x})$ has continuous partial derivatives in a region of $\mathbb{R}^n$ which contains a closed ball $\overline{\mathbf{B}}(\mathbf{x}_0; r) = \{\mathbf{x} : \|\mathbf{x} - \mathbf{x}_0\| \leq r\}$. Then, there is a constant L such that*
$$\|\mathbf{F}(\mathbf{x}) - \mathbf{F}(\mathbf{x}')\| \leq L \|\mathbf{x} - \mathbf{x}'\|$$
for any two points $\mathbf{x}$ and $\mathbf{x}'$ in $\bar{B}(\mathbf{x}_0; r)$.

Proof. We want to reduce it to the mean value theorem for a function of one variable. The line segment from $\mathbf{x}$ to $\mathbf{x}'$ can be parametrized by
$$(1-t)\mathbf{x} + t\mathbf{x}' = \mathbf{x} + t\,\mathbf{v},$$
where $\mathbf{v} = \mathbf{x}' - \mathbf{x}$ and $0 \leq t \leq 1$. We define the function $g(t)$ to be the values of $\mathbf{F}$ along this line segment:
$$\begin{aligned}\mathbf{g}(t) &= \mathbf{F}\left((1-t)\mathbf{x} + t\mathbf{x}'\right) \\ &= \mathbf{F}(\mathbf{x} + t\,\mathbf{v}).\end{aligned}$$
Then, $\mathbf{g}(0) = \mathbf{F}(\mathbf{x})$ and $\mathbf{g}(1) = \mathbf{F}(\mathbf{x}')$. By the chain rule (applied to each component of $\mathbf{g}$),
$$\begin{aligned}\mathbf{g}'(t) &= \sum_{j=1}^{n} \frac{\partial \mathbf{g}}{\partial x_j}(\mathbf{x} + t\,\mathbf{v})\,v_j \\ &= D\mathbf{g}_{(\mathbf{x}+t\,\mathbf{v})}\mathbf{v}.\end{aligned}$$
Therefore,
$$\begin{aligned}\mathbf{F}(\mathbf{x}') - \mathbf{F}(\mathbf{x}) &= \mathbf{g}(1) - \mathbf{g}(0) \\ &= \int_0^1 \mathbf{g}'(t)\,dt \\ &= \int_0^1 D\mathbf{g}_{(\mathbf{x}+t\,\mathbf{v})}\mathbf{v}\,dt.\end{aligned}$$
In the last step we take absolute values. For the integrand,
$$\|D\mathbf{g}_{(\mathbf{x}+t\,\mathbf{v})}\mathbf{v}\| \leq \|D\mathbf{g}_{(\mathbf{x}+t\,\mathbf{v})}\| \cdot \|\mathbf{v}\|,$$
where we use the norm of the matrix of partial derivatives, $\|D\mathbf{g}_{(\mathbf{x}+t\,\mathbf{v})}\|$. Since the partial derivatives are continuous, there is a single constant $L > 0$ such that $\|D\mathbf{g}_{(\mathbf{y})}\| \leq L$ for all points $\mathbf{x}$ in the closed ball $\bar{B}(\mathbf{x}_0; r)$ and so, in particular, at the points $\mathbf{x} + t\,\mathbf{v}$. Therefore,
$$\begin{aligned}\|\mathbf{F}(\mathbf{x}') - \mathbf{F}(\mathbf{x})\| &\leq \int_0^1 \|D\mathbf{g}_{(\mathbf{x}+t\,\mathbf{v})}\mathbf{v}\|\,dt \\ &\leq \int_0^1 L\|\mathbf{v}\|\,dt \\ &\leq L\|\mathbf{v}\| \\ &= L\|\mathbf{x} - \mathbf{x}\|.\end{aligned}$$
This proves the theorem. $\square$

Now we turn to Theorem 3.2 itself, considering it in smaller steps.

Theorem 3.6. *Consider the differential equation $\dot{\mathbf{x}} = \mathbf{F}(\mathbf{x})$. Assume that both $\mathbf{F}(\mathbf{x})$ and $\dfrac{\partial F_i}{\partial x_j}(\mathbf{x})$ are continuous for $\mathbf{x}$ in some open set U in $\mathbb{R}^n$, and that $\mathbf{x}_0$ is a point in U. Then there exists a solution $\mathbf{x}(t)$ defined for some time interval $-\tau < t < \tau$ such that $\mathbf{x}(0) = \mathbf{x}_0$.*

3.3. Theory and Proofs

Proof. The proof constructs a sequence of curves which converge to a solution. If $\mathbf{x}(t) = \mathbf{x}(t; \mathbf{x}_0)$ is a solution, then

$$\mathbf{x}(t) = \mathbf{x}_0 + \int_0^t \mathbf{F}(\mathbf{x}(s))\, ds.$$

We use this integral equality to define an iteration scheme. To start, take a trial solution and obtain a new curve that better approximates the solution. We start with a trial solution $\mathbf{y}(t)$ and define the next trial solution $\mathbf{z}(t) = \mathscr{G}(\mathbf{y})(t)$ by

$$\mathbf{z}(t) = \mathscr{G}(\mathbf{y})(t) = \mathbf{x}_0 + \int_0^t \mathbf{F}(\mathbf{y}(s))\, ds.$$

In particular, we could start with $\mathbf{y}^0(t) \equiv \mathbf{x}_0$ and define $\mathbf{y}^i(t)$ by induction as

$$\mathbf{y}^i(t) = \mathscr{G}(\mathbf{y}^{i-1})(t).$$

This process is called the *Picard iteration scheme*.

We need to check whether this sequence of trial solutions has a number of properties in order to determine that it converges to a true solution. First, we pick some constants and a length of time interval used in the argument. Let $r > 0$ be such that the closed ball $\bar{B}(\mathbf{x}_0; r)$ is contained in the domain of the differential equation. Let K and L be positive constants such that $\|\mathbf{F}(\mathbf{x})\| \leq K$ for $\mathbf{x}$ in $\bar{B}(\mathbf{x}_0; r)$ and $\mathbf{F}$ has a Lipschitz constant L as given in the previous theorem. We take a length of time interval

$$\tau = \min\left\{\frac{r}{K}, \frac{1}{L}\right\}$$

and consider $-\tau \leq t \leq \tau$. The interval is shorter than r/K to insure that the curve does not leave the ball $\bar{B}(\mathbf{x}_0; r)$ in the time interval. It is shorter than $1/L$ to insure that the sequence of curves converges to a limit curve.

By induction, we show that the curves $\mathbf{y}^i(t)$ are continuous, $\mathbf{y}^i(0) = \mathbf{x}_0$, and take values in $\bar{B}(\mathbf{x}_0; r)$. The first two properties follow from the definitions and the fact that the integral of a continuous function is differentiable with respect to the upper limit of integration t, and so continuous. The distance from the initial condition is estimated by the following:

$$\|\mathbf{y}^i(t) - \mathbf{x}_0\| = \left\|\int_0^t \mathbf{F}(\mathbf{y}^{i-1}(s))\, ds\right\| \leq \left|\int_0^t \|\mathbf{F}(\mathbf{y}^{i-1}(s))\|\, ds\right|$$

$$\leq \left|\int_0^t K\, ds\right| = K|t| \leq K\tau \leq r.$$

This shows that all the values are in the closed ball.

We want to show that the curves converge. We let C_0 be a bound on how much $\mathbf{y}^1(t)$ and $\mathbf{y}^0(t)$ differ:

$$C_0 = \max\left\{\|\mathbf{y}^1(t) - \mathbf{y}^0(t)\| : -\tau \leq t \leq \tau\right\}.$$

Let L be the Lipschitz constant for $\mathbf{F}$ as mentioned above. Let $L\tau = \lambda$, which is less than one by the choice of τ. We show by induction that $\|\mathbf{y}^{j+1}(t) - \mathbf{y}^j(t)\| \leq C_0 \lambda^j$.

This is true for $j = 0$ by the definition of C_0. Assume that it is true for $j-1$. Then,

$$\begin{aligned}\|\mathbf{y}^{j+1}(t) - \mathbf{y}^j(t)\| &= \left\|\int_0^t \mathbf{F}(\mathbf{y}^j(s)) - \mathbf{F}(\mathbf{y}^{j-1}(s))\, ds\right\| \\ &\leq \left|\int_0^t \|\mathbf{F}(\mathbf{y}^j(s)) - \mathbf{F}(\mathbf{y}^{j-1}(s))\|\, ds\right| \\ &\leq \left|\int_0^t L\|\mathbf{y}^j(s) - \mathbf{y}^{j-1}(s)\|\, ds\right| \\ &\leq \left|\int_0^t L C_0 \lambda^{j-1}\, ds\right| \quad \text{by the induction hypothesis} \\ &= L C_0 \lambda^{j-1}|t| \\ &\leq C_0 \lambda^{j-1} \tau L \\ &= C_0 \lambda^j.\end{aligned}$$

This proves the induction step and the estimate on successive curves.

To prove the convergence, we need to compare curves whose indices differ by more than one:

$$\|\mathbf{y}^{j+k}(t) - \mathbf{y}^j(t)\| \leq \sum_{i=0}^{k-1} \|\mathbf{y}^{j+i+1}(t) - \mathbf{y}^{j+i}(t)\| \leq \sum_{i=0}^{k-1} C_0 \lambda^{j+i}$$

$$\leq C_0 \lambda^j \sum_{i=0}^{\infty} \lambda^i = C_0 \frac{\lambda^j}{1-\lambda}.$$

By taking j large, we can insure that the difference is less than a predetermined size. This shows that the sequence of curves has what is called the Cauchy property. It is proved in a standard course in analysis, that for any fixed value of t, the sequence will have to converge. See [**Lay01**], [**Mar93**], or [**Lew93**]. Therefore, we can define a limiting curve

$$\mathbf{y}^\infty(t) = \lim_{j \to \infty} \mathbf{y}^j(t).$$

Also, because the convergence is uniform in t (the bounds insuring the convergence do not depend on t), the limiting sequence is continuous. Finally, the limiting curve satisfies the differential equation because

$$\begin{aligned}\mathbf{y}^\infty(t) &= \lim_{j \to \infty} \mathbf{y}^{j+1}(t) = \lim_{j \to \infty} \mathscr{G}(\mathbf{y}^j)(t) \\ &= \lim_{j \to \infty} \mathbf{x}_0 + \int_0^t \mathbf{F}(\mathbf{y}^j(s))\, ds = \mathbf{x}_0 + \int_0^t \mathbf{F}(\mathbf{y}^\infty(s))\, ds.\end{aligned}$$

Differentiating this integral equation gives the differential equation. □

We prove the uniqueness at the same time we prove that the solutions depend continuously on initial condition. We use what is known as the Gronwall inequality in the proof. We give the statement of this result as a lemma in the middle of the proof.

Theorem 3.7. *Consider the differential equation* $\dot{\mathbf{x}} = \mathbf{F}(\mathbf{x})$. *Assume that both* $\mathbf{F}(\mathbf{x})$ *and* $\dfrac{\partial F_i}{\partial x_j}(\mathbf{x})$ *are continuous for* $\mathbf{x}$ *in some open set U in* $\mathbb{R}^n$, *and that* $\mathbf{x}_0$ *is*

3.3. Theory and Proofs

a point in U. The solution $\mathbf{x}(t)$ with initial condition $\mathbf{x}(0) = \mathbf{x}_0$ is unique in the sense, that if $\mathbf{x}(t)$ and $\mathbf{y}(t)$ are two such solutions, then they must be equal on the largest interval of time about $t = 0$ where both solutions are defined. Moreover, the solution depends continuously on $\mathbf{x}_0$.

Proof. Assume that $\mathbf{x}(t)$ and $\mathbf{y}(t)$ are two solutions with initial conditions $\mathbf{x}(t) = \mathbf{x}_0$ and $\mathbf{y}(t) = \mathbf{y}_0$. Then,

$$\|\mathbf{x}(t) - \mathbf{y}(t)\| = \left\|\mathbf{x}_0 - \mathbf{y}_0 + \int_0^t \mathbf{F}(\mathbf{x}(s)) - \mathbf{F}(\mathbf{y}(s))\, ds\right\|$$

$$\leq \|\mathbf{x}_0 - \mathbf{y}_0\| + \left|\int_0^t \|\mathbf{F}(\mathbf{x}(s)) - \mathbf{F}(\mathbf{y}(s))\|\, ds\right|$$

$$\leq \|\mathbf{x}_0 - \mathbf{y}_0\| + \left|\int_0^t L\|\mathbf{x}(s) - \mathbf{y}(s)\|\, ds\right|.$$

If we let $v(t) = \|\mathbf{x}(t) - \mathbf{y}(t)\|$, we have

$$v(t) \leq v(0) + \left|\int_0^t Lv(s)\, ds\right|.$$

The next lemma, called Gronwall's inequality, derives a bound for $v(t)$.

Lemma 3.8 (Gronwall's Inequality). *Let $v(t)$ be a continuous nonnegative real-valued function on an interval $-\tau \leq t \leq \tau$, $L, C \geq 0$, such that*

$$v(t) \leq C + \left|\int_0^t Lv(s)\, ds\right|.$$

Then,

$$v(t) \leq Ce^{L|t|}.$$

In particular, if $C = 0$, then $v(t) \equiv 0$.

Proof. We give the proof for $t \geq 0$. The case for $t < 0$ can be derived by putting t as the lower limit in the integral.

We cannot differentiate an inequality and get an inequality. However, if we define

$$U(t) = C + \int_0^t Lv(s)\, ds,$$

then $v(t) \leq U(t)$ and we can differentiate $U(t)$, yielding

$$U'(t) = Lv(t) \leq LU(t).$$

We first consider the case when $C > 0$, which ensures that $U(t) > 0$, so we can divide by it:

$$\frac{U'(t)}{U(t)} \leq L,$$

$$\ln\left(\frac{U(t)}{U(0)}\right) \leq Lt, \quad \text{and}$$

$$U(t) \leq U(0)e^{Lt} = Ce^{Lt}.$$

Therefore, $v(t) \leq U(t) \leq Ce^{Lt}$, as claimed.

If $C = 0$, then we can take a decreasing sequence C_n going to 0. By the previous case, we have
$$v(t) \leq C_n e^{Lt}$$
for all n, so $v(t) \leq 0$. Since $v(t)$ is nonnegative, $v(t) \equiv 0$. □

Returning to the proof of the theorem and applying Gronwall's inequality yields
$$\|\mathbf{x}(t) - \mathbf{y}(t)\| \leq \|\mathbf{x}_0 - \mathbf{y}_0\| e^{L|t|}.$$
This inequality shows that the solution depends continuously on the initial condition. If $\mathbf{x}_0 = \mathbf{y}_0$, it follows that $\|\mathbf{x}(t) - \mathbf{y}(t)\| = 0$ for the time interval, which gives uniqueness. □

Explanation of First Variation Equation Theorem 3.4. The proof that the solution is differentiable with respect to initial conditions takes some work. However, if we assume that it is differentiable and we can interchange the order of differentiation, then the derivation of the first variation equation is straightforward. Let $\mathbf{x}_0 + s\,\mathbf{v}$ be a straight line through $\mathbf{x}_0$ at $s = 0$:
$$\mathbf{v}(t) = \left.\frac{\partial}{\partial s}\phi(t;\mathbf{x}_0 + s\,\mathbf{v})\right|_{s=0}.$$
Then,
$$\begin{aligned}
\frac{d}{dt}\mathbf{v}(t) &= \frac{d}{dt}\left(\left.\frac{\partial}{\partial s}\phi(t;\mathbf{x}_0 + s\,\mathbf{v})\right|_{s=0}\right) \\
&= \left.\frac{\partial}{\partial s}\left(\frac{d}{dt}\phi(t;\mathbf{x}_0 + s\,\mathbf{v})\right)\right|_{s=0} \\
&= \left.\frac{\partial}{\partial s}\mathbf{F}\left(\phi(t;\mathbf{x}_0 + s\,\mathbf{v})\right)\right|_{s=0} \\
&= \left.\sum_k \frac{\partial F_i}{\partial x_k}(\phi(t;\mathbf{x}_0))\,\frac{\partial}{\partial s}\left(\phi_k(t;\mathbf{x}_0 + s\,\mathbf{v})\right)\right|_{s=0} \\
&= \left.D\mathbf{F}_{(\phi(t;\mathbf{x}_0))}\,\frac{\partial}{\partial s}\phi(t;\mathbf{x}_0 + s\,\mathbf{v})\right|_{s=0} \\
&= D\mathbf{F}_{(\phi(t;\mathbf{x}_0))}\,\mathbf{v}(t).
\end{aligned}$$
The fourth equality is the chain rule for a function of several variables. The fifth equality rewrites the chain rule in terms of a matrix and vectors, where the matrix of partial derivatives $\left(\frac{\partial F_i}{\partial x_k}(\mathbf{x})\right)$ is denoted by $D\mathbf{F}_{(\mathbf{x})}$ and $\mathbf{v}(t)$ is as previously defined.

The differential equation for $D_\mathbf{x}\phi_{(t;\mathbf{x}_0)}$ follows from the preceding formula for vectors by taking the standard unit vectors for the vector $\mathbf{v}$. Let $\mathbf{u}^j$ be the unit vector with a 1 in the jth coordinate and other coordinates equal to zero, This results in
$$\mathbf{v}^j(t) = \left.\frac{\partial}{\partial s}\phi(t;\mathbf{x}_0 + s\,\mathbf{u}^j)\right|_{s=0} = \frac{\partial}{\partial x_j}\phi(t;\mathbf{x}_0).$$

3.3. Theory and Proofs

By the preceding formula applied to these $\mathbf{v}^j(t)$, we have

$$\frac{d}{dt}\frac{\partial}{\partial x_j}\phi(t;\mathbf{x}_0) = D\mathbf{F}_{(\phi(t;\mathbf{x}_0))}\frac{\partial}{\partial x_j}\phi(t;\mathbf{x}_0).$$

Since $\dfrac{\partial}{\partial x_j}\phi(t;\mathbf{x}_0)$ is the jth column of $D_{\mathbf{x}}\phi_{(\phi(t;\mathbf{x}_0))}$, putting these in as the columns of one matrix results in the matrix differential equation

$$\frac{d}{dt}D_{\mathbf{x}}\phi_{(\phi(t;\mathbf{x}_0))} = D\mathbf{F}_{(\phi(t;\mathbf{x}_0))}\, D_{\mathbf{x}}\phi_{(\phi(t;\mathbf{x}_0))},$$

which is the equation desired. Notice that, at $t = 0$, $\phi(0;\mathbf{x}_0) = \mathbf{x}_0$ so $D_{\mathbf{x}}\phi_{(0;\mathbf{x}_0)} = \mathbf{I}$. Thus, $D_{\mathbf{x}}\phi_{(\phi(t;\mathbf{x}_0))}$ is the fundamental matrix solution for the time-dependent linear system of differential equations

$$\frac{d}{dt}\mathbf{v} = D\mathbf{F}_{(\phi(t;\mathbf{x}_0))}\,\mathbf{v},$$

with $D_{\mathbf{x}}\phi_{(0;\mathbf{x}_0)} = \mathbf{I}$. $\square$

Numerical solutions: local error

In Section 3.2, we discussed the local errors for the Euler, Heun, and Runge–Kutta methods. In this section we give a more explicit bound for the Euler method and derive the bound for the Heun method. The derivation of the local error for the fourth-order Runge–Kutta method is more complicated and we skip it. For a more detailed treatment of numerical methods, see the book by Burden and Faires [Bur78].

In order to specify a solution of a time-dependent differential equation of the form $\dot{x} = f(t,x)$, it is necessary to not only specify the initial condition x_0, but also the time $t = \tau$ at which the initial condition is attained. Thus, we write $\phi(t;\tau,a)$ for the solution for which $\phi(\tau;\tau,x_0) = x_0$ and

$$\frac{d}{dt}\phi(t;\tau,x_0) = f(t,\phi(t;\tau,x_0))$$

for all t.

Theorem 3.9. *Consider a differential equation $\dot{x} = f(t,x)$, where x is a scalar variable.*

a. *Assume that the solution stays in a region where*

$$\left|\frac{\partial f}{\partial t}(t,x) + \frac{\partial f}{\partial x}(t,x)\,f(t,x)\right| \leq 2M.$$

Then, the local error for the Euler method is bounded by Mh^2:

$$|x_{n+1} - \phi(h+t_n;,t_n,x_n)| \leq M\,h^2.$$

b. *The local error for the Heun method is $O(h^3)$.*

Proof. (a) We start at t_0 and x_0 and assume that we have obtained $t_n = t_0 + nh$ and x_n by the Euler process. We check the estimate of $|\phi(h+t_n;t_n,x_n) - x_{n+1}|$. We let $x(t) = \phi(t;t_n,x_n)$.

The Taylor expansion of the solution $x(t) = \phi(t; t_n, x_n)$ as a function of t about time t_n and evaluated at $t = t_{n+1} = t_n + h$ is given by

$$\phi(t_{n+1}; t_n, x_n) = x_n + h \left. \frac{d}{dt} \phi(t; t_n, x_n) \right|_{t=t_n} + \frac{h^2}{2} \left(\left. \frac{d^2}{dt^2} \phi(t; t_n, x_n) \right|_{t=\tau_n} \right),$$

where τ_n is a time between t_n and $t_{n+1} = t_n + h$. Since $\left. \frac{d}{dt} \phi(t; t_n, x_n) \right|_{t=t_n} = f(t_n, x_n)$, $x_{n+1} = x_n + h \left. \frac{d}{dt} \phi(t; t_n, x_n) \right|_{t=t_n}$, and

$$|\phi(h; t_n, x_n) - x_{n+1}| = \frac{h^2}{2} \left| \left. \frac{d^2}{dt^2} \phi(t; t_n, x_n) \right|_{t=\tau_n} \right|.$$

Thus, we need to know only that there is a bound on the second time derivative $\frac{d^2}{dt^2} \phi(t; t_n, x_n)$. Since $\frac{d}{dt} \phi(t; t_n, x_n) = f(t, \phi(t; t_n, x_n))$, by the chain rule,

$$\frac{d^2}{dt^2} \phi(t; t_n, x_n) = \frac{\partial f}{\partial t}(t, \phi(t; t_n, x_n)) + \frac{\partial f}{\partial x}(t, \phi(t; t_n, x_n)) \frac{d}{dt} \phi(t; t_n, x_n)$$
$$= \frac{\partial f}{\partial t}(t, \phi(t; t_n, x_n)) + \frac{\partial f}{\partial x}(t, \phi(t; t_n, x_n)) f(t, \phi(t; t_n, x_n)).$$

Thus, if there is a bound

$$\left| \frac{\partial f}{\partial t}(t, x) + \frac{\partial f}{\partial x}(t, x) f(t, x) \right| \leq 2M$$

in the region in which the solution remains, then the local error is bounded by $M h^2$:

$$|\phi(h; t_n, x_n) - x_{n+1}| \leq M h^2.$$

(b) Next, we turn to estimating the local error for the Heun method. Let

$$g(t, x) = \tfrac{1}{2} \big[f(t, x) + f(t + h, x + h f(t, x)) \big].$$

Then, for the Heun method,

$$x_{n+1} = x_n + h \, g(t_n, x_n).$$

Again, we estimate $|\phi(t_{n+1}; t_n, x_n) - x_{n+1}|$. We use Taylor expansions to represent both $\phi(t_{n+1}; t_n, x_n)$ and $g(t_n, x_n)$. First, there is a τ_n between t_n and $t_{n+1} = t_n + h$ such that

$$\phi(t_{n+1}; t_n, x_n) = x_n + h \left. \frac{d}{dt} \phi(t; t_n, x_n) \right|_{t=t_n} + \frac{h^2}{2} \left. \frac{d^2}{dt^2} \phi(t; t_n, x_n) \right|_{t=t_n}$$
$$+ \frac{h^3}{3!} \left. \frac{d^3}{dt^3} \phi(t; t_n, x_n) \right|_{t=\tau_n} + O(h^4)$$
$$= x_n + h \, f(t_n, x_n) + \frac{h^2}{2} \left. \frac{d}{dt} f(t, \phi(t; t_n, x_n)) \right|_{t=t_n} + O(h^3)$$
$$= x_n + h \, f(t_n, x_n)$$
$$+ \frac{h^2}{2} \left[\frac{\partial f}{\partial t}(t_n, x_n) + \frac{\partial f}{\partial x}(x_n, t_n) f(t_n, x_n) \right] + O(h^3).$$

3.3. Theory and Proofs

Notice, that we have just represented the term
$$\frac{h^3}{3!}\left(\frac{d^3}{dt^3}\phi(t;t_n,x_n)\Big|_{t=\tau_n}\right)$$
by $O(h^3)$. Then, turning to $g(t_n, x_n)$, we have
$$g(t_n, x_n) = \tfrac{1}{2}\left[f(t_n, x_n) + f(t_n + h, x_n + hf(t_n, x_n))\right]$$
$$= \tfrac{1}{2} f(t_n, x_n) + \tfrac{1}{2} f(t_n, x_n) + \tfrac{1}{2} \frac{\partial f}{\partial t}(t_n, x_n)\, h$$
$$+ \tfrac{1}{2} \frac{\partial f}{\partial x}(t_n, x_n)\, h\, f(t_n, x_n) + O(h^2)$$
$$= f(t_n, x_n) + \frac{h}{2} \frac{\partial f}{\partial t}(t_n, x_n)$$
$$+ \frac{h}{2} \frac{\partial f}{\partial x}(t_n, x_n)\, f(t_n, x_n) + O(h^2),$$

where the $O(h^2)$ contains the terms involving second partial derivatives of f. Combining, all the terms except the $O(h^3)$ cancel,

$$\phi(t_{n+1}; t_n, x_n) - x_{n+1}$$
$$= \phi(h; t_n, x_n) - [x_n + h\, g(t_n, x_n)]$$
$$= x_n + h\, f(t_n, x_n)$$
$$+ \frac{h^2}{2}\left[\frac{\partial f}{\partial t}(t_n, x_n) + \frac{\partial f}{\partial x}(t_n, x_n)\, f(t_n, x_n)\right] + O(h^3)$$
$$- x_n - h\, f(t_n, x_n) - \frac{h^2}{2} \frac{\partial f}{\partial t}(t_n, x_n)$$
$$- \frac{h^2}{2} \frac{\partial f}{\partial x}(t_n, x_n)\, f(t_n, x_n) + O(h^3)$$
$$= O(h^3).$$

This completes the derivation of the local error for the Heun method. $\square$

Global error

Theorem 3.10. *Assume that the local error of a numerical method is $hE(h)$. Assume that the time interval is fixed at $T = t_N - t_0$ and we take N steps so $h = T/N$. Assume that for $t_0 \leq t \leq t_N = t_0 + Nh$, the solution stays in a region where*
$$\left|\frac{\partial f}{\partial x}(t, x)\right| \leq L.$$
Then, there is a bound on the global error given by
$$|x_N - \phi(t_N; t_0, x_0)| \leq \frac{E(h)\left[e^{L(t_N - t_0)} - 1\right]}{L}.$$

a. *For the Euler method, the local error is bounded by $hE(h) = Mh^2$, where*
$$\left|\frac{\partial f}{\partial t}(t, x) + \frac{\partial f}{\partial x}(t, x)\, f(t, x)\right| \leq 2M,$$

so $E(h) = Mh$ and we get

$$|x_N - \phi(t_N; t_0, x_0)| \leq \frac{M h \left[e^{L(t_N - t_0)} - 1\right]}{L}.$$

This term is $O(h)$ when the time $T = t_N - t_0$ is fixed.

b. For the Heun method, the local error is bounded by $hE(h) = Mh^3$, so $E(h) = Mh^2$, and we get

$$|x_N - \phi(t_N; t_0, x_0)| \leq \frac{M h^2 \left[e^{L(t_N - t_0)} - 1\right]}{L}.$$

c. For the Runge-Kutta method, $h E(h) = M h^5$, so $E(h) = M h^4$, and we get

$$|x_N - \phi(t_N; t_0, x_0)| \leq \frac{M h^4 \left[e^{L(t_N - t_0)} - 1\right]}{L}.$$

Proof. We write the difference between the final numerical term and the actual flow as the sum of terms that differ in only one step of the numerical method, namely

$$\begin{aligned}
x_N &- \phi(t_N; t_0, x_0) \\
&= x_N - \phi(t_N; t_{N-1}, x_{N-1}) \\
&\quad + \phi(t_N; t_{N-1}, x_{N-1}) - \phi(t_N; t_{N-2}, x_{N-2}) \\
&\quad + \cdots + \phi(t_N; t_1, x_1) - \phi(t_N; t_0, x_0) \\
&= x_N - \phi(t_N; t_{N-1}, x_{N-1}) \\
&\quad + \phi(t_N; t_{N-1}, x_{N-1}) - \phi(t_N; t_{N-1}, \phi(t_{N-1}; t_{N-2}, x_{N-2})) \\
&\quad + \phi(t_N; t_{N-2}, x_{N-2}) - \phi(t_N; t_{N-2}, \phi(t_{N-2}; t_{N-3}, x_{N-3})) \\
&\quad + \cdots + \phi(t_N; t_1, x_1) - \phi(t_N; t_1, \phi(t_1; t_0, x_0))
\end{aligned}$$

and

$$\begin{aligned}
|x_N &- \phi(t_N; t_0, x_0)| \\
&\leq |x_N - \phi(t_{N-1} + h; t_{N-1}, x_{N-1})| \\
&\quad + |\phi(t_N; t_{N-1}, x_{N-1}) - \phi(t_N; t_{N-1}, \phi(t_{N-1}; t_{N-2}, x_{N-2}))| \\
&\quad + |\phi(t_N; t_{N-2}, x_{N-2}) - \phi(t_N; t_{N-2}, \phi(t_{N-2}; t_{N-3}, x_{N-3}))| \\
&\quad + \cdots + |\phi(t_N; t_1, x_1) - \phi(t_N; t_1, \phi(t_1; t_0, x_0))| \\
&\leq |x_N - \phi(t_N; t_{N-1}, x_{N-1})| \\
&\quad + e^{hL} |x_{N-1} - \phi(t_{N-1}; t_{N-2}, x_{N-2})| \\
&\quad + e^{2hL} |x_{N-2} - \phi(t_{N-2}; t_{N-3}, x_{N-3})| \\
&\quad + \cdots + e^{(N-1)hL} |x_1 - \phi(t_1; t_0, x_0)|,
\end{aligned}$$

where, in the last inequality, we use the propagation of errors by the flow derived when we discussed continuity of solutions for times of 0, $h = t_N - t_{N-1}$, $2h = t_N - t_{N-2}$, ..., $(N-1)h = t_N - t_1$. Substituting the local error and using $T = Nh$,

3.3. Theory and Proofs

we get

$$|x_N - \phi(t_N; t_0, x_0)| \leq E(h)\, h \left[1 + e^{hL} + e^{2hL} + \cdots + e^{(N-1)hL}\right]$$

$$= E(h)\, h \left[\frac{e^{NhL} - 1}{e^{hL} - 1}\right]$$

$$= E(h)\, [e^{TL} - 1]\left[\frac{h}{e^{hL} - 1}\right].$$

However,

$$\frac{e^{hL} - 1}{h} = L + \frac{1}{2}L^2 h + \cdots$$

$$\geq L,$$

so

$$\frac{h}{e^{hL} - 1} \leq \frac{1}{L}$$

and

$$|x_N - \phi(t_N; t_0, x_0)| \leq \frac{E(h)\,[e^{TL} - 1]}{L}.$$

The errors for the three different method considered follow directly. □

All of these methods have the bounds mention in the earlier section on an interval of fixed length. However, as the time interval grows in length, T goes to infinity, the bound grows exponentially and the bound is not useful.

Chapter 4

Phase Portraits with Emphasis on Fixed Points

This chapter starts the analysis of nonlinear differential equations by means of phase portraits. This process works especially well in two dimensions where the phase space is the phase plane. This chapter emphasizes the examples in which the fixed points are the most important feature of the phase portrait. Chapter 5 uses energy functions (or other test functions) to determine the phase portrait. Chapter 6 considers phase portraits with periodic orbits, and Chapter 7 emphasizes chaotic differential equations.

This chapter begins by we introducing the limit set, which is the set of points that an orbit approaches as time goes to plus or minus infinity. The limit set is used to give definitions of different types of stability of fixed points in Section 4.2.

We begin the consideration of the phase space by considering one-dimensional equations, where the situation is the simplest possible, and in which the material is more closely connected to material covered in calculus and an elementary differential equation course.

With this background, we turn to the discussion of the phase plane analysis in Section 4.4, where we separate the phase plane into regions in which $\dot{x}_1$ and $\dot{x}_2$ have one sign. In Section 4.5, the linearized equations are used to determine the behavior near fixed points. These ideas are applied in Section 4.6 to equations arising from the interaction of two populations of different species, which compete with each other. Further applications are given in Section 4.7.

4.1. Limit Sets

When determining a phase portrait, we are are interested in the long-term behavior of trajectories as time goes to plus or minus infinity: The ω-limit set relates to time going to plus infinity and the α-limit set relates to time going to minus infinity.

This concept is used in the next section for one of the definitions of stability of a fixed point.

We assume through out this section that we have a system of differential equations $\dot{\mathbf{x}} = \mathbf{F}(\mathbf{x})$ with continuous partial derivatives of the components of $\mathbf{F}$, so solutions exist and are unique for a given initial condition. We use the notation of the flow $\phi(t; \mathbf{x}_0)$ of $\dot{\mathbf{x}} = \mathbf{F}(\mathbf{x})$; that is,

$$\frac{d}{dt}\phi(t; \mathbf{x}_0) = \mathbf{F}(\phi(t; \mathbf{x}_0)) \quad \text{and} \quad \phi(0; \mathbf{x}_0) = \mathbf{x}_0.$$

Definition 4.1. A point $\mathbf{q}$ is an ω-*limit point* of the trajectory of $\mathbf{x}_0$, provided that $\phi(t; \mathbf{x}_0)$ keeps coming near $\mathbf{q}$ as t goes to infinity, i.e., there is a sequence of times t_j, with t_j going to infinity as j goes to infinity, such that $\phi(t_j; \mathbf{x}_0)$ converges to $\mathbf{q}$. Certainly, if $\|\phi(t; \mathbf{x}_0) - \mathbf{x}^*\|$ goes to zero as t goes to infinity, then $\mathbf{x}^*$ is the only ω-limit point of $\mathbf{x}_0$. There can be more than one point that is an ω-limit point of $\mathbf{x}_0$. The set of all ω-limit points of $\mathbf{x}_0$ is denoted by $\omega(\mathbf{x}_0)$ and is called the ω-*limit set* of $\mathbf{x}_0$.

Similarly, a point $\mathbf{q}$ is an α-limit point of $\mathbf{x}_0$, provided that $\phi(t; \mathbf{x}_0)$ keeps coming near $\mathbf{q}$ as t goes to minus infinity. In particular, if $\|\phi(t; \mathbf{x}_0) - \mathbf{x}^*\|$ goes to zero as t goes to minus infinity, then $\mathbf{x}^*$ is the only α-limit point of $\mathbf{x}_0$. The set of all α-limit points of $\mathbf{x}_0$ is denoted by $\alpha(\mathbf{x}_0)$ and is called the α-*limit set* of $\mathbf{x}_0$.

Note that α is the first letter of the Greek alphabet and ω is the last letter. Thus, the limit with time going to plus infinity is denoted using ω and the limit with time going to minus infinity is denoted using α

If $\mathbf{x}_0$ is a fixed point, then $\omega(\mathbf{x}_0) = \alpha(\mathbf{x}_0) = \{\mathbf{x}_0\}$. For a linear elliptic center, such as in Example 2.12, the ω-limit set of a point is the whole periodic orbit on which it lies, since the orbit repeatedly goes through each point on the ellipse. This is true for any periodic point: If $\mathbf{x}_0$ is on a periodic orbit, then $\omega(\mathbf{x}_0) = \alpha(\mathbf{x}_0) = \{\phi(t; \mathbf{x}_0)\}_{t \in \mathbb{R}}$ and is the set of points on the periodic orbit.

The following example illustrates the fact that a point off a periodic orbit can have its ω-limit equal to the periodic orbit.

Example 4.2. Consider the system of equations

$$\dot{x} = y + x(1 - x^2 - y^2),$$
$$\dot{y} = -x + y(1 - x^2 - y^2).$$

The orbits and limit sets of this example can easily be determined by using polar coordinates. The polar coordinate r satisfies $r^2 = x^2 + y^2$, so by differentiating with respect to t and using the differential equations we get

$$r\dot{r} = x\dot{x} + y\dot{y}$$
$$= xy + x^2(1 - r^2) - xy + y^2(1 - r^2)$$
$$= r^2(1 - r^2), \quad \text{or}$$
$$\dot{r} = r(1 - r^2).$$

4.1. Limit Sets

Similarly, the angle variable θ satisfies $\tan(\theta) = y/x$, so the derivative with respect to t yields

$$\sec^2(\theta)\,\dot\theta = x^{-2}\left[-x^2 + xy\,(1-r^2) - y^2 - xy\,(1-r^2)\right]$$
$$= -\frac{r^2}{x^2}, \quad \text{so}$$
$$\dot\theta = -1.$$

Thus, the solution goes clockwise around the origin at unit angular speed.

The α and ω-limit sets of various initial conditions are as follows.

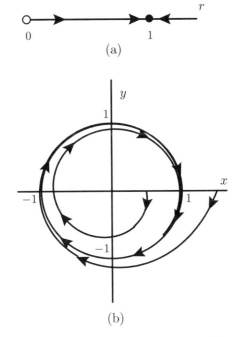

Figure 1. Limit cycle for Example 4.2: (a) r-variable, (b) Phase portrait in (x, y)-variables

The origin is a fixed point so $\alpha(\mathbf{0}) = \omega(\mathbf{0}) = \{\mathbf{0}\}$.

If $r_0 > 1$, then $r(t)$ is decreasing and tends to 1 as t goes to infinity. See Figure 1a. Thus, a trajectory $\phi(t; \mathbf{p}_0)$ starting outside the unit circle tends inward toward the unit circle. It does not converge to any one point in the circle, but given any point $\mathbf{z}$ in the circle, $\phi(t; \mathbf{p}_0)$ keeps coming back near $\mathbf{z}$ every 2π units of time. Therefore, there is a sequence of times t_n going to infinity such that $\phi(t_n; \mathbf{p}_0)$ converges to $\mathbf{z}$. Obviously, if we take a different point $\mathbf{z}$ we get different times t_n. Letting $\Gamma = \{r = 1\}$, the ω-limit set of $\mathbf{p}_0$ with $\|\mathbf{p}_0\| > 1$ is Γ, $\omega(\mathbf{p}_0) = \Gamma$. See Figure 1b.

Similarly, if $0 < r_0 < 1$, then $r(t)$ is increasing and tends to 1 as t goes to infinity. Thus, if $\|\mathbf{p}_0\| < 1$, then $\omega(\mathbf{p}_0) = \Gamma$. In fact, if $\mathbf{p}_0$ is any point different from $(0,0)$, then $\omega(\mathbf{p}_0) = \Gamma$.

If $\|\mathbf{p}_0\| < 1$, then the α-limit set equals the origin, $\alpha(\mathbf{p}_0) = \{\mathbf{0}\}$. If $\|\mathbf{p}_0\| = 1$, then $\alpha(\mathbf{p}_0) = \Gamma$. If $\|\mathbf{p}_0\| > 1$, the backward orbit goes off to infinity, so $\alpha(\mathbf{p}_0) = \emptyset$.

Example 4.3. Example 2.19 considers a system of two uncoupled harmonic oscillators,

$$\dot{x}_1 = x_3,$$
$$\dot{x}_2 = x_4,$$
$$\dot{x}_3 = -\omega_1^2 x_1,$$
$$\dot{x}_4 = -\omega_2^2 x_2.$$

If we introduce the coordinates which are like two sets of polar coordinates given by

$$x_j = \rho_j \sin(2\pi \tau_j) \quad \text{and}$$
$$\dot{x}_j = \omega_j \rho_j \cos(2\pi \tau_j),$$

then

$$\dot{\rho}_j = 0,$$
$$\dot{\tau}_j = \frac{\omega_j}{2\pi}.$$

Assume that ω_1/ω_2 is irrational. Take an initial condition $(x_{1,0}, x_{2,0}, x_{3,0}, x_{4,0})$ with $r_{1,0}^2 = \omega_1^2 x_{1,0}^2 + x_{3,0}^2$ and $r_{2,0}^2 = \omega_2^2 x_{2,0}^2 + x_{4,0}^2$. Then, Theorem 2.17 states that the trajectory $\phi(t; \mathbf{x}_0)$ is dense in set

$$\left\{ (x_1, x_2, x_3, x_4) : \omega_1^2 x_1^2 + x_3^2 = r_{1,0}^2 \text{ and } \omega_1^2 x_2^2 + x_4^2 = r_{2,0}^2 \right\}.$$

In fact, a similar argument shows that the limit sets are this same subset of phase space:

$$\alpha(\mathbf{x}_0) = \omega(\mathbf{x}_0) = \left\{ (x_1, x_2, x_3, x_4) : \omega_1^2 x_1^2 + x_3^2 = r_{1,0}^2 \text{ and } \omega_1^2 x_2^2 + x_4^2 = r_{2,0}^2 \right\}.$$

Notice that the trajectory does not pass through each point in the limit set, but keeps coming back closer and closer to each such point.

In Section 5.4, we give some more complicated limit sets. We next present some of the properties of limit sets. These properties are implicit in the examples we consider, and are used explicitly in a proof of Theorem 5.2.

One of these properties involves the invariance of a set, so we give a precise definition first.

Definition 4.4. Let $\phi(t; \mathbf{x})$ be the flow of a system of differential equations. A subset of the phase space $\mathbf{S}$ is called *positively invariant* provided that $\phi(t; \mathbf{x}_0)$ is in $\mathbf{S}$ for all $\mathbf{x}_0$ in $\mathbf{S}$ and all $t \geq 0$. A subset of the phase space $\mathbf{S}$ is called *negatively invariant* provided that $\phi(t; \mathbf{x}_0)$ is in $\mathbf{S}$ for all $\mathbf{x}_0$ in $\mathbf{S}$ and all $t \leq 0$. Finally, a subset of the phase space $\mathbf{S}$ is called *invariant* provided that $\phi(t; \mathbf{x}_0)$ is in $\mathbf{S}$ for all $\mathbf{x}_0$ in $\mathbf{S}$ and all real t.

4.1. Limit Sets

Theorem 4.1. *Assume that $\phi(t; \mathbf{x}_0)$ is a trajectory. Then, the following properties of the ω-limit set are true:*

- **a.** *The limit set depends only on the trajectory and not on the particular point, so $\omega(\mathbf{x}_0) = \omega(\phi(t; \mathbf{x}_0))$ for any real time t.*
- **b.** *The $\omega(\mathbf{x}_0)$ is invariant: if $\mathbf{z}_0 \in \omega(\mathbf{x}_0)$, then the orbit $\phi(t; \mathbf{z}_0)$ is in $\omega(\mathbf{x}_0)$ for all positive and negative t.*
- **c.** *The $\omega(\mathbf{x}_0)$ is closed (i.e., $\omega(\mathbf{x}_0)$ contains all its limit points).*
- **d.** *If $\mathbf{y}_0$ is a point in $\omega(\mathbf{x}_0)$, then $\omega(\mathbf{y}_0) \subset \omega(\mathbf{x}_0)$.*

In addition, assume that the trajectory $\phi(t; \mathbf{x}_0)$ stays bounded for $t \geq 0$ (i.e., there is a constant $C > 0$ such that $\|\phi(t; \mathbf{x}_0)\| \leq C$ for $t \geq 0$). Then, properties (e) and (f) are true.

- **e.** *The $\omega(\mathbf{x}_0)$ is nonempty.*
- **f.** *The $\omega(\mathbf{x}_0)$ is connected; it is not made up of more than one piece.*

Similar properties hold for the α-limit set if $\phi(t; \mathbf{x}_0)$ stays bounded for $t \leq 0$.

The detailed proof is given in Section 4.8, but we give the intuitive justification here.

Property (**a**) follows easily from the definition of the ω-limit set. We implicitly use the fact that the ω-limit set depends only on the orbit and not the particular point in all our discussion.

Property (**b**) follows from the continuity of the flows. For Example 4.2, the ω-limit set is a periodic orbit, which is invariant. For Example 4.3, the ω-limit set is $\{\rho_1 = r_{1,0} \; \rho_2 = r_{2,0}\}$, which is invariant.

Property (**c**) follows from combining the limits of points converging to $\mathbf{y}^j$ in $\omega(\mathbf{x}_0)$, and $\mathbf{y}^j$ converging to $\mathbf{y}^\infty$. For Example 4.2, the ω-limit set is a periodic orbit, which is closed. For Example 4.3, the ω-limit set is $\{\rho_1 = r_{1,0} \; \rho_2 = r_{2,0}\}$, which is closed.

Property (**d**) follows properties (b) and (c).

Property (**e**) is true because the trajectory has to repeatedly return to some part of the space. Thus, the only way that one of the limit sets to be empty is for the orbit to go off to infinity.

Property (**f**) follows from the fact that the trajectory itself is connected. Again, all our examples have connected limit sets.

The ω-limit set can also be represented as the intersection over all positive times $T > 0$ of the closure of the trajectory for times larger than T:

$$\omega(\mathbf{x}_0) = \bigcap_{T \geq 0} \text{closure}\{ \phi(t; \mathbf{x}_0) : t \geq T \}.$$

We do not prove this property because we do not use it. Some other books take this as the definition of the ω-limit set.

Exercises 4.1

1. What are all the possible α- and ω-limit sets of points for the linear system of equations $\dot{\mathbf{x}} = \mathbf{A}\mathbf{x}$ given by the following matrices $\mathbf{A}$.

   (a) $\mathbf{A} = \begin{pmatrix} -4 & -2 \\ 3 & -11 \end{pmatrix}$,

   (b) $\mathbf{A} = \begin{pmatrix} 1 & 3 \\ 3 & 1 \end{pmatrix}$,

   (c) $\mathbf{A} = \begin{pmatrix} 4 & 5 \\ -5 & -2 \end{pmatrix}$.

2. Consider the system of equations
$$\dot{x} = y + x\left(x^2 + y^2 - 1\right),$$
$$\dot{y} = -x + y\left(x^2 + y^2 - 1\right),$$
which, in polar coordinates, is given by
$$\dot{r} = r\left(r^2 - 1\right),$$
$$\dot{\theta} = -1.$$
What are all the possible α- and ω-limit sets of points in the plane?

3. Consider the system of equations
$$\dot{x} = y + x\left(1 - x^2 - y^2\right)\left(x^2 + y^2 - 4\right),$$
$$\dot{y} = -x + y\left(1 - x^2 - y^2\right)\left(x^2 + y^2 - 4\right),$$
which, in polar coordinates, has the form
$$\dot{r} = r(1 - r^2)(r^2 - 4),$$
$$\dot{\theta} = -1.$$
What are all the possible α- and ω-limit sets of points in the plane?

4.2. Stability of Fixed Points

For a fixed point, we want to identify how many solutions tend toward it. If all nearby initial conditions converge, then the fixed point is asymptotically stable; if some converge and others go away then the fixed point is a saddle point. We give definitions of these concepts that are in in any dimension and for nonlinear systems as well as linear systems.

In discussing the stability of fixed points, we use the set of points whose α- and ω-limit sets equal the fixed point. A notation for these sets are given in the following definition.

Definition 4.5. For a fixed point $\mathbf{x}^*$, the *stable manifold* $W^s(\mathbf{x}^*)$ is the set of all points which tend to the fixed point as t goes to plus infinity:
$$W^s(\mathbf{x}^*) = \{\,\mathbf{p}_0 : \phi(t;\mathbf{p}_0)\text{ tends to }\mathbf{x}^*\text{ as }t\to\infty\,\} = \{\,\mathbf{p}_0 : \omega(\mathbf{p}_0) = \{\mathbf{x}^*\}\,\}.$$

4.2. Stability of Fixed Points

In this context, if the orbit converges to a single point $\mathbf{x}^*$ as t goes to infinity, then the ω-limit set equals this single point, $\omega(\mathbf{p}_0) = \{\mathbf{x}^*\}$. If the stable manifold contains all points within some small distance of $\mathbf{x}^*$, then $W^s(\mathbf{x}^*)$ is called the *basin of attraction of* $\mathbf{x}^*$.

In the same way, the *unstable manifold of the fixed point* $W^u(\mathbf{x}^*)$ is the set of all the points which tend to the fixed point as t goes to minus infinity:

$$W^u(\mathbf{x}^*) = \{\, \mathbf{p}_0 : \phi(t; \mathbf{p}_0) \text{ tends to } \mathbf{x}^* \text{ as } t \to -\infty \,\} = \{\, \mathbf{p}_0 : \alpha(\mathbf{p}_0) = \{\mathbf{x}^*\} \,\}.$$

Here, if an orbit converges to a single point $\mathbf{x}^*$ as t goes to minus infinity, then the α-limit set equals this single point, $\alpha(\mathbf{p}_0) = \{\mathbf{x}^*\}$. In Section 4.5, we discuss more about the properties of these stable and unstable manifolds.

We now proceed to give several different ways of designating that a fixed point is stable. Notice that, for linear systems, trajectories for an elliptic center stay approximately the same distance from the origin as they move on their elliptic orbit; at least the trajectories do not move away from the origin. This property is captured in the concept of Lyapunov stability, which we define first. For a linear system for which the real parts of all the eigenvalues are negative, the trajectories not only stay near the origin but they also tend toward the origin. This idea is captured in the definitions of asymptotic stability given below.

Definition 4.6. A fixed point $\mathbf{x}^*$ is said to be *Lyapunov stable* or *L-stable*, provided that any solution $\phi(t; \mathbf{x}_0)$ stays near $\mathbf{x}^*$ for all $t \geq 0$ if the initial condition $\mathbf{x}_0$ starts near enough to $\mathbf{x}^*$. More precisely, a fixed point $\mathbf{x}^*$ is called *L*-stable, provided that for any $\epsilon > 0$, there is a $\delta > 0$ such that, if $\|\mathbf{x}_0 - \mathbf{x}^*\| < \delta$, then $\|\phi(t; \mathbf{x}_0) - \mathbf{x}^*\| < \epsilon$ for all $t \geq 0$.

Notice that a linear elliptic center is *L*-stable, but it is necessary to take δ smaller than ϵ, since a solution moves on an ellipse and not a circle. A linear stable focus is also *L*-stable, and it is also often necessary to take δ smaller than ϵ.

Definition 4.7. A fixed point is called *unstable*, provided that it is not *L*-stable (i.e., there exists an $\epsilon_1 > 0$ such that for any $\delta > 0$ there is some point $\mathbf{x}_\delta$ with $\|\mathbf{x}_\delta - \mathbf{x}^*\| < \delta$ and a time $t_1 > 0$ depending on the point $\mathbf{x}_\delta$ with $\|\phi(t_1; \mathbf{x}_\delta) - \mathbf{x}^*\| > \epsilon_1$). Thus, trajectories that start as near to $\mathbf{x}^*$ as we would like to specify move at least a distance of ϵ_1 away from $\mathbf{x}^*$.

Definition 4.8. A fixed point $\mathbf{x}^*$ is called ω-*attracting* provided that there exists a $\delta_1 > 0$ such that $\omega(\mathbf{x}_0) = \{\mathbf{x}^*\}$ for all $\|\mathbf{x}_0 - \mathbf{x}^*\| < \delta_1$ (i.e., $\|\phi(t; \mathbf{x}_0) - \mathbf{x}^*\|$ goes to zero as t goes to infinity for all $\|\mathbf{x}_0 - \mathbf{x}^*\| < \delta_1$). Thus, a fixed point is ω-attracting stable, provided that the stable manifold contains all points in a neighborhood of the fixed point (i.e., all points sufficiently close). Warning: The concept which we refer to as ω-attracting does not have a standardized name.

A fixed point $\mathbf{x}^*$ is called *asymptotically stable*, provided that it is both *L*-stable and ω-attracting. An asymptotically stable fixed point is also called a fixed point *sink*. We also use the word *attracting* to mean asymptotically stable, but in other contexts it means only ω-attracting; hopefully, the context will make clear which concept is intended. An asymptotically stable fixed point whose stable manifold is the whole phase space is called *globally asymptotically stable*.

Definition 4.9. A fixed point is called *repelling* or a *fixed point source*, provided that it is asymptotically stable backward in time (i.e., (i) for any closeness $\epsilon > 0$, there is a $\delta > 0$ such that, if $\|\mathbf{x}_0 - \mathbf{x}^*\| < \delta$, then $\|\phi(t; \mathbf{x}_0) - \mathbf{x}^*\| < \epsilon$ for all $t \leq 0$ and (ii) there exists a $\delta_1 > 0$ such that $\alpha(\mathbf{x}_0) = \{\mathbf{x}^*\}$ for all $\|\mathbf{x}_0 - \mathbf{x}^*\| < \delta_1$).

For nonlinear systems, the following two examples show that there are cases in which all nearby solutions eventually tend to the fixed point, but it is not L-stable; for this reason, it is necessary to add the assumption that the fixed point is L-stable as well as ω-attracting in the definition of asymptotic stability.

Example 4.10. The system of equations
$$\dot{x} = x - y - x(x^2 + y^2) + \frac{xy}{\sqrt{x^2 + y^2}},$$
$$\dot{y} = x + y - y(x^2 + y^2) - \frac{x^2}{\sqrt{x^2 + y^2}},$$
in polar coordinates, takes the form
$$\dot{r} = r(1 - r^2),$$
$$\dot{\theta} = 2\sin^2(\theta/2).$$

The circle $r = 1$ is invariant and attracts nearby trajectories. On the circle, $\dot{\theta} \geq 0$ and is zero only at $\theta = 0$. Therefore, all trajectories starting near the circle have $(x, y) = (1, 0)$ as their limit; however, points with $r = 1$ and $\theta > 0$ small must go all the way around the circle before they tend to the fixed point. Therefore, $(1, 0)$ is ω-attracting, but not asymptotically stable. See Figure 2.

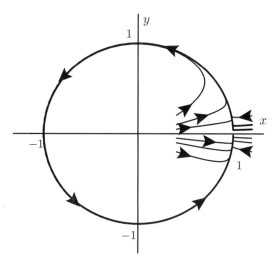

Figure 2. A fixed point that is ω-attracting but not asymptotically stable, Example 4.10

The next example is not as artificial as the previous one, but is harder to analyze. For that reason, we show only the phase portrait and give a reference for a complete analysis.

4.2. Stability of Fixed Points

Example 4.11. In 1957, Vinograd gave the following system of nonlinear differential equations, with the origin an isolated fixed point which is ω-attracting but not Lyapunov stable:

$$\dot{x} = \frac{x^2(y-x) + y^5}{(x^2+y^2)[1+(x^2+y^2)^2]},$$
$$\dot{y} = \frac{y^2(y-2x)}{(x^2+y^2)[1+(x^2+y^2)^2]}.$$

The phase portrait is given in Figure 3. The origin is ω-attracting and all the trajectories tend to the origin. However, there are trajectories that come arbitrarily close to the origin and then go out near $x = 0.5$ and $y = 0.5$ before eventually limiting on the origin: Therefore, the origin is not L-stable. An analysis of this system of equations is given in [**Hah67**] (page 191).

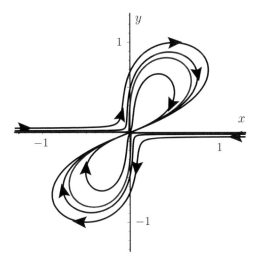

Figure 3. A fixed point that is ω-attracting but not asymptotically stable, Example 4.11

For linear systems, the explicit form of solutions implies that the real parts of the eigenvalues determine the stability type of the origin as summarized in the next theorem.

Theorem 4.2. *Consider the linear differential equation* $\dot{\mathbf{x}} = \mathbf{A}\mathbf{x}$.

a. *If all of the eigenvalues λ of $\mathbf{A}$ have negative real parts, then the origin is asymptotically stable. In particular, stable nodes, degenerate stable nodes, and stable foci are all asymptotically stable.*

b. *If one of the eigenvalues λ_1 has a positive real part, then the origin is unstable. In particular, saddles, unstable nodes, unstable degenerate nodes, and unstable foci are all unstable. A saddle has some directions that are attracting and others that are expanding, but it still satisfies the condition to be unstable.*

In two dimensions, the determinant and the trace determine the type of linear system. It is convenient to have these results summarized so we can quickly tell the stability type from these quantities, which are easy to compute.

Theorem 4.3. *Let* $\mathbf{A}$ *be a* 2×2 *matrix with determinant* Δ *and trace* τ.

a. *The sum of the eigenvalues is* τ *and their product is* Δ.

b. *If* $\Delta < 0$, *then the origin is a saddle, and therefore unstable.*

c. *If* $\Delta > 0$ *and* $\tau > 0$, *then the origin is unstable.* (i) *If* $\tau^2 - 4\Delta > 0$, *then it is an unstable node.* (ii) *If* $\tau^2 - 4\Delta = 0$, *then it is a degenerate unstable node.* (iii) *If* $\tau^2 - 4\Delta < 0$, *then it is an unstable focus.*

d. *If* $\Delta > 0$ *and* $\tau < 0$, *then the origin is asymptotically stable.* (i) *If* $\tau^2 - 4\Delta > 0$, *then it is a stable node.* (ii) *If* $\tau^2 - 4\Delta = 0$, *then it is a degenerate stable node.* (iii) *If* $\tau^2 - 4\Delta < 0$, *then it is a stable focus.*

e. *If* $\Delta > 0$ *and* $\tau = 0$, *then the eigenvalues are purely imaginary,* $\pm i\beta$, *and the origin is L-stable but not asymptotically stable.*

f. *If* $\Delta = 0$, *then one or more of the eigenvalues is zero.* (i) *If* $\tau > 0$, *then the second eigenvalue is positive and the origin is unstable.* (ii) *If* $\tau = 0$, *then both eigenvalues are zero.* (iii) *If* $\tau < 0$, *then the second eigenvalue is negative and the origin is L-stable but not asymptotically stable.*

Proof. (a) The characteristic equation is given by $\lambda^2 - \tau\lambda + \Delta = 0$, with roots

$$\lambda_\pm = \frac{\tau \pm \sqrt{\tau^2 - 4\Delta}}{2}.$$

Then,

$$(\lambda - \lambda_+)(\lambda - \lambda_-) = \lambda^2 - (\lambda_+ + \lambda_-)\lambda + \lambda_+\lambda_-.$$

Equating the coefficients of λ and the constants in the preceding equation with those in $\lambda^2 - \tau\lambda + \Delta = 0$, we get

$$\tau = \lambda_+ + \lambda_- \quad \text{and}$$
$$\Delta = \lambda_+ \lambda_-.$$

(b) If $\Delta < 0$, then

$$\tau^2 - 4\Delta > \tau^2 > 0,$$
$$\sqrt{\tau^2 - 4\Delta} > |\tau|,$$
$$\lambda_+ = \frac{\tau + \sqrt{\tau^2 - 4\Delta}}{2} > 0,$$
$$\lambda_- = \frac{\tau - \sqrt{\tau^2 - 4\Delta}}{2} < 0.$$

Thus, the origin is a saddle point with real eigenvalues.

(c) Assume that $\Delta > 0$ and $\tau > 0$. (i) If $\tau^2 - 4\Delta > 0$, then the eigenvalues are real. Since $\tau^2 - 4\Delta < \tau^2$, $\sqrt{\tau^2 - 4\Delta} < \tau$ and

$$\lambda_\pm = \frac{\tau \pm \sqrt{\tau^2 - 4\Delta}}{2} > 0.$$

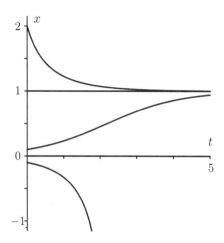

Figure 4. Logistic equation: plot of t versus x for $k = 1$ and $r = 1$

Figure 5. Logistic equation: phase portrait for $k = 1$ and $r = 1$

In the section on existence and uniqueness, we gave the explicit solution of the logistic equation,
$$\phi(t; x_0) = \frac{x_0 K}{x_0 + (K - x_0)e^{-rt}}.$$
The solution is well defined as long as the denominator is nonzero. The denominator is zero when
$$e^{rt} = \frac{x_0 - K}{x_0}.$$
Therefore, for the solution to be undefined, the right-hand side of the last equation must be positive, which occurs for $x_0 > K$ and $x_0 < 0$. For $x_0 > K$, the solution becomes undefined for a negative time,
$$t_{x_0}^- = \frac{1}{r} \ln\left(\frac{x_0 - K}{x_0}\right) < 0,$$
but is defined for all positive time, so $t_{x_0}^+ = \infty$. For $x_0 < 0$, the solution becomes undefined for a positive time,
$$t_{x_0}^+ = \frac{1}{r} \ln\left(\frac{x_0 - K}{x_0}\right) = \frac{1}{r} \ln\left(\frac{|x_0| + K}{|x_0|}\right) > 0,$$
but $t_{x_0}^- = -\infty$. Finally, for $0 \leq x_0 \leq K$, $t_{x_0}^- = -\infty$ and $t_{x_0}^+ = \infty$.

In particular, for any $x_0 > 0$, the solution is well defined for all $t \geq 0$, and as t goes to infinity, the solution converges to
$$x_\infty = \frac{x_0 K}{x_0 + 0} = K.$$

For $0 \le x_0 < K$, the solution is also well defined for all $t \le 0$, the denominator becomes arbitrarily large, and the solution converges to 0 as t goes to minus infinity. For $x_0 > K$, the denominator goes to 0 as t converges down to $t_{x_0}^-$, so the solution goes to infinity.

For $x_0 < 0$, the solution is defined for $-\infty < t < t_{x_0}^+$. The denominator is positive and the numerator is negative, so the solution goes to minus infinity as t converges up to $t_{x_0}^+$.

The explicit solution yields the same behavior as we previously deduced using only the form of the equations.

In the discussion of the logistic equation, we argued that solutions with positive initial conditions converge to the fixed point K. This convergence is a consequence of the next theorem about the limiting behavior of a bounded solution (and the uniqueness of solutions given in Section 3.1).

Theorem 4.4. *Consider a differential equation $\dot{x} = f(x)$ on $\mathbb{R}$, for which $f(x)$ has a continuous derivative. Assume that $x(t) = \phi(t; x_0)$ is the solution, with initial condition x_0. Assume that the maximum interval containing 0 for which it can be defined is (t^-, t^+).*

a. *Further assume that the solution $\phi(t; x_0)$ is bounded for $0 \le t < t^+$, i.e., there is a constant $C > 0$ such that $|\phi(t; x_0)| \le C$ for $0 \le t < t^+$. Then as t converges to t^+, $\phi(t; x_0)$ must converge either to a fixed point or to a point where $f(x)$ is undefined.*

b. *Similarly, if the solution $\phi(t; x_0)$ is bounded for $t^- < t \le 0$, then, as t converges to t^-, $\phi(t; x_0)$ must converge either to a fixed point or to a point where $f(x)$ is undefined.*

c. *Assume that $f(x)$ is defined for all x in $\mathbb{R}$. (i) If $f(x_0) > 0$, assume that there is a fixed point $x^* > x_0$, and in fact, let x^* be the smallest fixed point larger than x_0. (ii) If $f(x_0) < 0$, assume that there is a fixed point $x^* < x_0$, and in fact, let x^* be the largest fixed point less than x_0. Then, $t^+ = \infty$ and $\phi(t; x_0)$ converges to x^* as t goes to infinity.*

The idea of the proof is that, by the uniqueness theorem, the trajectory cannot cross any fixed point x^* where $f(x^*) = 0$. If $\phi(t_1; x_0) = x^*$ for some time t_1, then $x_0 = \phi(-t_1; x^*) = x^*$ and $x_0 = x^*$. Therefore, a nonequilibrium solution must stay in some interval in which $f(x)$ has one sign, either positive or negative. Assume that it is positive. Then, $\phi(t; x_0) \le C$ and is increasing. Because the solution is bounded, $\phi(t; x_0)$ must converge to a limiting value x_∞ by the properties of the real numbers. (See Lemma 4.11 for the general argument.) Because the solution $\phi(t; x_0)$ approaches x_∞ but does not cross it, it must slow down and so $f(x_\infty) = 0$. Therefore, $\phi(t; x_0)$ must converge to a fixed point. The same argument applies when $f(x) < 0$ and $\phi(t; x_0)$ must decrease down to a fixed point.

The theorem applies to the logistic equation to show that, for $0 < x_0 < K$, $\phi(t; x_0)$ converges to K as t goes to infinity, and converges to 0 as t goes to minus infinity.

4.3. Scalar Equations

The discussion given for the logistic equation shows how the stability type of a fixed point can be determined by the derivative of the function defining the differential equation.

Theorem 4.5. *Consider a fixed point x^* for the differential equation $\dot{x} = f(x)$, where f and f' are continuous.*

   a. *If $f'(x^*) < 0$, then x^* is an attracting fixed point.*
   b. *If $f'(x^*) > 0$, then x^* is a repelling fixed point.*
   c. *If $f'(x^*) = 0$, then the derivative does not determine the stability type.*

Example 4.13. We apply the criteria of Theorem 4.5 for the stability of a fixed point to the differential equation
$$\dot{x} = x^3 - x = f(x).$$
The fixed points are found by setting $0 = f(x) = x^3 - x = x\,(x^2 - 1)$, or $x = 0, \pm 1$. The derivative is $f'(x) = 3\,x^2 - 1$, and $f'(0) = -1$ and $f'(\pm 1) = 2$. Therefore, the fixed point $x = 0$ has $f'(0) < 0$ and is attracting. The fixed points $x = \pm 1$ have $f'(\pm 1) > 0$ and are repelling. See Figure 6.

Figure 6. Phase portrait for Example 4.13

The next examples show that, when $f'(x^*) = 0$, the fixed point can be attracting, repelling, or neither. Therefore, the test is inconclusive.

Example 4.14 (Semistable Fixed Point). The differential equation
$$\dot{x} = x^2$$
has a fixed point at 0, where $f'(0) = 2x|_{x=0} = 0$, so the derivative test does not apply. The derivative $\dot{x} = x^2$ is positive for both $x < 0$ and $x > 0$, so is attracting from below and repelling from above. Therefore the fixed point is semistable. This fixed point is unstable, but not repelling. See Figure 7.

Figure 7. Semistable Fixed Point

Example 4.15 (Weakly Stable Fixed Point). The differential equation
$$\dot{x} = -x^3$$
again has a fixed point at 0, where $f'(0) = -3x^2|_{x=0} = 0$. The derivative $\dot{x} = -x^3$ is positive for $x < 0$ and negative for $x > 0$, so 0 is an attracting fixed point. The attraction is not caused by the linear term but by the cubic term, therefore, the fixed point is called *weakly attracting*. See Figure 8.

Figure 8. Weakly Attracting Fixed Point

Example 4.16 (Weakly Repelling Fixed Point). The differential equation
$$\dot{x} = x^3$$
again has a fixed point at 0, where $f'(0) = 3x^2|_{x=0} = 0$. The derivative $\dot{x} = x^3$ is negative for $x < 0$ and positive for $x > 0$, so 0 is a repelling fixed point. The repulsion is not caused by the linear term but by the cubic term. Therefore, the fixed point is called weakly repelling. See Figure 9.

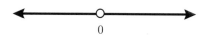

Figure 9. Weakly Repelling Fixed Point

As a preceding example with $f'(x^*) = 0$ at a fixed point, we consider an example with many fixed points.

Example 4.17 (Nonisolated Fixed Point). Consider the differential equation
$$\dot{x} = x^2 \sin(1/x).$$
The fixed points are $x = 0$, and $1/x_n = n\pi$ or $x_n = 1/(n\pi)$. These fixed points converge to the fixed point at 0. The intervals $(-1/(n\pi), 1/(n\pi))$ are invariant and arbitrarily small. Therefore, the fixed point 0 is L-stable. However, the points $1/(n\pi)$ are arbitrarily close to 0, but their solutions do not converge to 0. Therefore, 0 is not asymptotically stable.

Exercises 4.3

1. Consider the differential equation
$$\dot{x} = x^2 - 9.$$
   a. Find the stability type of each fixed point.
   b. Sketch the phase portrait on the line.
   c. Sketch the graph of $x(t) = \phi(t; x_0)$ in the (t, x)-plane for several representative initial conditions x_0.

2. Consider the differential equation
$$\dot{x} = x(x-1)(2-x) = -x^3 + 3x^2 - 2x.$$
   a. Find the stability type of each fixed point.
   b. Draw the phase portrait on the real line.

Exercises 4.3

   c. Sketch the graph of $x(t) = \phi(t; x_0)$ in the (t, x)-space for several representative initial conditions x_0. Describe in words which initial conditions converge to which fixed points.
3. Consider the differential equation
$$\dot{x} = rx\left(1 - \frac{x}{K}\right) - H$$
   for $x \geq 0$, which models a population of fish that is governed by the logistic equation and a number of fish H caught and removed every unit of time. This last factor is often called *harvesting*. In this equation, the harvesting is called a *constant rate* or *constant yield* because the amount taken out does not depend on the population $x(t)$ present at the given time. Here the parameters r, K, and H are all positive.
   a. Assume that $H < rK/4$. Draw the phase portrait on the real line.
   b. If $H < rK/4$, what levels of initial fish population lead to a fish population which does not die out? (If the population of fish becomes negative, then the fish population has died out and the fish no longer exist in the system.)
   c. Assume that $H > rK/4$. What happens to the population of fish as t increases?
4. Consider the differential equation
$$\dot{x} = rx\left(1 - \frac{x}{K}\right) - Ex$$
   for $x \geq 0$, which models a population of fish that is governed by the logistic equation and the number of fish Ex caught and removed every unit of time, which depends on the population of fish present at that time. This type of harvesting is called a *constant effort*, because the amount taken out is proportional to the population, which is thought to be realistic if the same effort is made. Here the parameters r, K, and E are all positive, and $E < r$.
   a. Draw the phase portrait on the real line.
   b. What is the effect of the harvesting on the steady state of fish present in the long-term behavior?
5. Consider equations of the form $\dot{x} = f(x)$ that have exactly 7 fixed points (as usual $f(x)$ is a C^1 function).
   a. Give an example of such an equation.
   b. What are the possible values for the number of stable fixed points? Sketch functions $f(x)$ which would provide each number you think is possible. For the numbers you consider impossible, explain why briefly. (Don't forget that there are 7 fixed points in total.)
6. Consider the differential equation $\dot{\theta} = \sin(\theta)$, where θ is an angle variable that is taken modulo 2π. Draw the phase portrait on the line and on the circle.
7. Compare the rates of convergence of solutions of the weak stable fixed point for $\dot{x} = -x^3$ and those for the linear equation $\dot{x} = -x$. The qualitative phase portraits are the same, but the rates of convergence are different. (You can either use the analytic form of the solutions, or compare the plots of the solutions as functions of t using a computer simulation.
8. The Solow-Swan model for economic growth is given by $\dot{K} = sAK^\beta L^{1-\beta} - \delta K$ and $\dot{L} = nL$, where K is the capital, L is the labor force, $AK^\beta L^{1-\beta}$ is the

production function with $0 < \beta < 1$ and $A > 0$, $0 < s < 1$ is the rate of reinvestment income, $\delta > 0$ is the rate of depreciation of capital, and $n > 0$ is the rate of growth of the labor force.

a. If the variable $k = K/L$ is the capital per capita (of labor), show that
$$\dot{k} = k^a \left[sA - (\delta + n) k^{1-a} \right].$$

b. Find the fixed points of the scalar differential equation.

c. Argue that for any initial condition $k_0 > 0$, $\phi(t; k_0)$ coverges to the fixed point $k^* > 0$.

4.4. Two Dimensions and Nullclines

In Section 3.2, we discussed the idea of plotting the vector field as a means of determining the geometric properties of the flow. In this section, we refine this geometric approach to draw the phase portraits for nonlinear planar differential equations. These methods apply even when we cannot obtain explicit analytic solutions. We relate this geometric approach to phase portraits drawn using a computer program.

We start by reviewing the relationship between a vector field and a system of differential equations in the plane. For the system of differential equations given by
$$\dot{x} = f(x, y),$$
$$\dot{y} = g(x, y),$$
the *vector field for the system of equations* is given by
$$\mathbf{F}(x, y) = \begin{pmatrix} f(x, y) \\ g(x, y) \end{pmatrix}.$$

The idea is that with every point (x, y) in the plane, there is associated a vector $\mathbf{F}(x, y)$, with base point at (x, y). Thus, there is a collection of vectors (arrows) on the plane, with the magnitude and direction of the vector changing from point to point. A trajectory of the system goes along a curve such that the velocity vector of the curve at each point is the vector field at that same point.

The basic geometric approach to determining the phase portrait is to find the fixed points, and then determine the signs of $\dot{x}$ and $\dot{y}$ in different regions of the plane. Since these time derivatives can change sign only at the points where they equal zero (or are undefined), we find the curves where $\dot{x} = 0$ and $\dot{y} = 0$. These two curves are called *nullclines*, since they are curves on which one component of the vector field for the system of differential equations is zero (i.e., is either vertical or horizontal). Other books call these curves *isoclines*, because the vector field has the same direction along one of these curves. This information is used to draw the *phase portrait* for the nonlinear system of differential equations.

Example 4.18. Consider the system of differential equations
$$\dot{x} = -x,$$
$$\dot{y} = -2y + 2x^3.$$

4.4. Two Dimensions

The vector field is formed by taking the right-hand side of the system as the components,
$$\mathbf{F}(\mathbf{x}) = \begin{pmatrix} -x \\ -2y + 2x^3 \end{pmatrix}.$$
The only fixed point is at the origin, as can be seen by solving the two equations
$$0 = -x \quad \text{and}$$
$$0 = -2y + 2x^3.$$
The nullclines where $\dot{x} = 0$ and $\dot{y} = 0$ are given by $x = 0$ and $y = x^3$ respectively. These two nullclines separate each other into four parts:

(N1) $y = x^3$ with $x > 0$,
(N2) $x = 0$ with $y > x^3$,
(N3) $y = x^3$ with $x < 0$, and
(N4) $x = 0$ with $y < x^3$.

They also divide the plane into four regions:

(R1) $x > 0$ and $y > x^3$,
(R2) $x < 0$ and $y > x^3$,
(R3) $x < 0$ and $y < x^3$, and
(R4) $x > 0$ and $y < x^3$.

See Figure 10 for the regions.

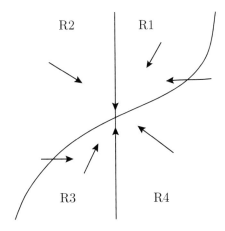

Figure 10. Nullclines for Example 4.18

We next want to find the general directions of the vector field in the regions and whether the vector field points up, down, right, or left on the nullclines. If we start on the curve $y = x^3$ and increase y, then $\dot{y}$ becomes negative. Therefore, $\dot{y} < 0$ on the curve (N2) and in regions (R1) and (R2). (Each of these derivatives must have one sign in one of the regions.) In the same way, decreasing y from the curve

$y = x^3$, then $\dot{y}$ becomes positive; therefore $\dot{y} > 0$ on the curve (N4) and in regions (R3) and (R4). Similarly, if we start on $x = 0$ and increase x, then $\dot{x}$ becomes negative; therefore, $\dot{x} < 0$ on curve (N1) and regions (R1) and (R4). Finally, $\dot{x} > 0$ on curve (N3) and in regions (R2) and (R3). We can summarize the signs of $\dot{x}$ and $\dot{y}$ on the various pieces of the nullclines and in the various regions in the following list.

(N1) $y = x^3$ with $x > 0$: horizontal with $\dot{x} < 0$.
(N2) $x = 0$ with $y > x^3$: vertical with $\dot{y} < 0$.
(N3) $y = x^3$ with $x < 0$: horizontal with $\dot{x} > 0$.
(N4) $x = 0$ with $y < x^3$: vertical with $\dot{y} > 0$.
(R1) $x > 0$ and $y > x^3$: $\dot{x} < 0$ and $\dot{y} < 0$.
(R2) $x < 0$ and $y > x^3$: $\dot{x} > 0$ and $\dot{y} < 0$.
(R3) $x < 0$ and $y < x^3$: $\dot{x} > 0$ and $\dot{y} > 0$.
(R4) $x > 0$ and $y < x^3$: $\dot{x} < 0$ and $\dot{y} > 0$.

See Figure 10.

On the nullcline $x = 0$, the vector field is everywhere tangent to the line (since $\dot{x} = 0$). Therefore, any trajectory $\phi(t; (0, y_0))$ starting on the line stays on the line (i.e., the line $x = 0$ is invariant). For $y > 0$, $\dot{y} < 0$, and for $y < 0$, $\dot{y} > 0$, so the trajectories move toward the origin.

For region (R1), one boundary (N2) is invariant (where $x = 0$), and trajectories are entering along (N1); therefore, region (R1) is positively invariant. A trajectory starting in region (R1) stays in the region for $t \geq 0$, where both $\dot{x} < 0$ and $\dot{y} < 0$. Therefore, x variable and the y variable are monotonically decreasing, and so the trajectory must converge to the fixed point at the origin.

Similarly, region (R3) is positively invariant, and a trajectory in this region is monotonically increasing in both variables and also converges to the fixed point at the origin.

A trajectory starting in region (R4) either stays in region (R4) for all positive time or it enters region (R1). If it stays in region (R4), then the x variable is decreasing and the y variable is increasing, and the solution must converge to the fixed point at the origin. If the solution enters region (R1), then it converges to the fixed point by the argument previously given. In either case, a trajectory must converge to the fixed point at the origin.

A trajectory starting in region (R2) must either stay in region (R2) for all positive time or enter region (R3). By an argument similar to that for region (R4), any such trajectory must converge to the fixed point at the origin.

By considering all four regions, we have shown that the basin of attraction for $\mathbf{0}$ (or its stable manifold) is the whole plane,

$$W^s(\mathbf{0}) = \{ (x, y) : \omega(x, y) = \mathbf{0} \} = \mathbb{R}^2,$$

and the origin is globally asymptotically attracting.

4.4. Two Dimensions

Important features of the phase portrait

(1) Location of the fixed points.
(2) Regions where $\dot{x}$ and $\dot{y}$ have a constant sign.
(3) Behavior near the fixed point. (Emphasis of the next section.) This includes whether the fixed point is stable or unstable, and how trajectories may pass near the fixed point and then leave.
(4) Location of periodic orbits. This is much harder and is the focus of Chapter 6.

Phase portraits using a computer program

There are general purpose programs such as Maple, Mathematica, or Matlab, as well as programs more specifically written to draw phase portraits. These programs can usually draw the vector field at a grid of points in a selected region. These are also called *direction fields*. Also, by picking good representative initial conditions, a fairly complete phase portrait can be drawn. The phase portraits in this book are drawn using Maple and the Runge-Kutta numerical method. For example, see Figures 12, 15, and 16.

A trajectory is a curve in the phase plane that is tangent to the direction field at each point. It is often possible to find the basic properties of the trajectories by using these direction fields.

(1) If the fixed points are known, then pick a window in which to draw the phase portrait, which includes some or all the fixed points. (If there is a finite number of fixed points, then include all the fixed points. If there are infinitely many fixed points as is the case for the system $\dot{x} = y$ and $\dot{y} = -\sin(x)$, then include a representative sample of the fixed points.)

(2) Experiment with initial conditions to determine the behavior of trajectories near the fixed points.

(3) In Section 4.5, we discuss the stable and unstable manifolds for saddle fixed points. When these are present, try to include initial conditions near the unstable manifold and follow these trajectories to find out where they go. Also, include initial conditions near the stable manifold very close to the fixed point and follow them backward, or find initial conditions away from the fixed point which closely follow the stable manifold near the fixed point. See Section 4.5.

(4) Try to show the behavior of representative trajectories in all parts of the window in the phase plane.

Example 4.19. Consider the system of differential equations

$$\dot{x} = x - y,$$
$$\dot{y} = 1 - x y^2.$$

The sets where $\dot{x} = 0$ and $\dot{y} = 0$ are given by $y = x$ and $x = \dfrac{1}{y^2}$ respectively. See Figure 11. The fixed points are the intersections of these two curves. Algebraically, the fixed points are found by solving the pair of equations

$$0 = x - y,$$
$$0 = 1 - x y^2.$$

The only solution of this system of equations is $(x, y) = (1, 1)$, and so this is the only fixed point.

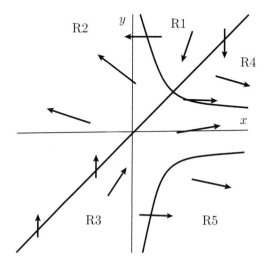

Figure 11. Nullclines for Example 4.19

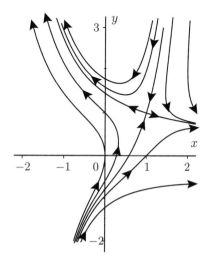

Figure 12. Computer plot of solution for Example 4.19

4.4. Two Dimensions

These nullclines divide themselves into five curves and divide the plane into five regions:

(N1) $y = x$ with $x > 1$.
(N2) $y = x$ with $x < 1$.
(N3) $x = 1/y^2$ with $y > 1$.
(N4) $x = 1/y^2$ with $0 < y < 1$.
(N5) $x = 1/y^2$ with $y < 0$.
(R1) $0 < y^{-2} < x < y$.
(R2) $x < y$ and $x < y^{-2}$ (including the part of $x < y$ where $x \leq 0$).
(R3) $y < x < y^{-2}$.
(R4) $0 < y < x$ and $y^{-2} < x$.
(R5) $y < 0$ and $y^{-2} < x$.

To determine the direction arrows, first note that they are vertical on the nullcline $\dot{x} = 0$ ($y = x$), and horizontal on the nullcline $\dot{y} = 0$ ($x = 1/y^2$). Moving up off $y = x$, causes $\dot{x} = x - y$ to become negative. Therefore, $\dot{x}$ is negative in regions (R1) and (R2) and on curve (N3). Similarly, $\dot{x} = x - y$ becomes positive as x is increased (or as y decreased) off the line $y = x$ and the point enters regions (R3), (R4), and (R5), and on curves (N4) and (N5). Turning to the other component, $\dot{y} = 1 - x y^2$ becomes negative as y is increased off (N3) and (N4). Thus, $\dot{y}$ is negative in regions (R1) and (R4) and on the curve (N1). Similarly, it is negative in region (R5). On the other hand, as $|y|$ is decreased off any part of the nullcline made up of (N3), (N4), and (N5), $\dot{y}$ becomes positive. Thus, $\dot{y}$ is positive in regions (R2) and (R3) and on (N1). Note that, in the argument we are using, the sign of $\dot{y}$ can change only on the nullcline. This information is summarized in the following list.

(N1) $y = x$ with $x > 1$: vertical and $\dot{y} < 0$.
(N2) $y = x$ with $x < 1$: vertical and $\dot{y} > 0$.
(N3) $x = 1/y^2$ with $y > 1$: horizontal and $\dot{x} < 0$.
(N4) $x = 1/y^2$ with $0 < y < 1$: horizontal and $\dot{x} > 0$.
(N5) $x = 1/y^2$ with $y < 0$: horizontal and $\dot{x} > 0$.
(R1) $0 < y^{-2} < x < y$: $\dot{x} < 0$ and $\dot{y} < 0$.
(R2) $x < y$ and $x < y^{-2}$ (including $x \leq 0$): $\dot{x} < 0$ and $\dot{y} > 0$.
(R3) $y < x < y^{-2}$: $\dot{x} > 0$ and $\dot{y} > 0$.
(R4) $0 < y < x$ and $y^{-2} < x$: where $\dot{x} > 0$ and $\dot{y} < 0$.
(R5) $y < 0$ and $y^{-2} < x$: $\dot{x} > 0$ and $\dot{y} < 0$.

Solutions are leaving region (R1) on its boundary. However, any solution that stays in the region for all time has $x(t)$ and $y(t)$ both decreasing. By arguments like those used for one-dimensional equations, they must approach points where $\dot{x} = 0$ and $\dot{y} = 0$ (i.e., the trajectory must approach a fixed point, which must be $(1, 1)$). Because some solutions leave on the left side and some leave on the right side, there must be at least one such trajectory that is asymptotic to the fixed point.

Region (R3) is similar to region (R1) except that $\dot{x} > 0$ and $y > 0$, so there must again be at least one trajectory that approaches the fixed point.

Regions (R2), (R4), and (R5) are positively invariant because points on the boundary either enter the region or remain on the boundary. Therefore, trajectories starting in region (R2) stay in region (R2), the x variable is monotonically decreasing, and the y variable is monotonically decreasing. Since there are no fixed points in these regions, $\|(x(t), y(t))\|$ goes to ∞ as t goes to infinity. Similarly, trajectories starting in region (R4) and (R5) have $\|(x(t), y(t))\|$ going to ∞ as t goes to ∞.

Because trajectories that leave regions (R1) and (R3) must enter regions (R2), (R4), or (R5), this describes the behavior of all the trajectories as t goes to ∞.

Considering t going to $-\infty$, trajectories in regions (R2) and (R4) either leave or go to the fixed point $(1, 1)$ as t goes to $-\infty$, Trajectories starting in region (R5) must leave and enter region $(R3)$ as t decreases. On the other hand, trajectories in regions (R1) and (R3) must have $\|(x(t), y(t))\|$ going to ∞ as t goes to $-\infty$.

Using linearization at the fixed point, which is discussed in the next section, we see that the fixed point is a saddle point. There is only one orbit in region (R1) and one orbit in region (R3) going to $(1,1)$ as t goes to ∞. Similarly, there is only one orbit in region (R2) and one orbit in region (R4) that goes to $(1,1)$ as t goes to $-\infty$. See Figure 11 for the sketch of the nullclines.

Figure 12 shows a computer plot of some trajectories. To draw the unstable manifolds of the fixed point $(1,1)$, initial conditions where taken slightly off the fixed point in directions of expansion, $(1.01, 1.003)$ and $(0.99, 0.997)$. To draw the stable manifold of the fixed point $(1,1)$, initial conditions where taken slightly off the fixed point in directions of contraction, $(1.01, 1.033)$ and $(0.99, 0.967)$, and trajectories were followed backward in time. Other representative initial conditions were taken to display the trajectories moving past the fixed point.

Example 4.20. Consider the example
$$\dot{x} = -x + y,$$
$$\dot{y} = xy - 1.$$

The nullclines are
$$\{\dot{x} = 0\} = \{y = x\} \quad \text{and}$$
$$\{\dot{y} = 0\} = \{y = 1/x\}.$$

Solving the nonlinear equations shows that the fixed points are $(1, 1)$ and $(-1, -1)$. Around the fixed point at $(-1, -1)$, trajectories move in a clockwise fashion, entering and leaving each of the four regions nearby defined by the nullclines. See Figure 13. Thus, by itself, the method of nullclines and directions of flow does not determine whether the fixed point is stable or unstable. In the next section, we will use the method of linearization at the fixed point to show that it is asymptotically stable.

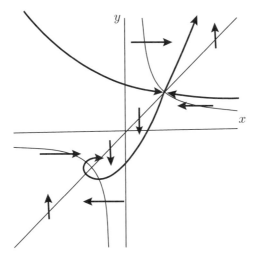

Figure 13. Nullclines for Example 4.20

Exercises 4.4

1. Consider the system of differential equations
$$\dot{x} = -2x + y - x^3,$$
$$\dot{y} = -y + x^2.$$
   a. Determine the fixed points.
   b. Determine the nullclines and the signs of $\dot{x}$ and $\dot{y}$ in the various regions of the plane.
   c. Using the information from parts (a) and (b), sketch by hand a rough phase portrait. Explain and justify your sketch.

2. Consider the system of differential equations
$$\dot{x} = y - x^2,$$
$$\dot{y} = x - y.$$
   a. Determine the fixed points.
   b. Determine the nullclines and the signs of $\dot{x}$ and $\dot{y}$ in the various regions of the plane.
   c. Using the information from parts (a) and (b), sketch by hand a rough phase portrait. Explain and justify your sketch.

3. Consider the system of differential equations
$$\dot{x} = y - x^3,$$
$$\dot{y} = -y + x^2.$$
   a. Determine the fixed points.
   b. Determine the nullclines and the signs of $\dot{x}$ and $\dot{y}$ in the various regions of the plane.

c. Using the information from parts (a) and (b), sketch by hand a rough phase portrait. Explain and justify your sketch.
d. Using the computer, sketch the direction field and trajectories for enough initial conditions to show the important features of the phase portrait.

4. Consider a competitive market of two commodities whose prices are p_1 and p_2. Let $E_i(p_1, p_2)$ be the excess demand (demand minus supply) of the ith commodity. For $i = 1, 2$, assume that the price adjustment is given by $p_i' = k_i E_i(p_1, p_2)$ for constants $k_i > 0$. Based on economic assumptions, Shone on pages 353-357 of [**Sho02**] gives the following information about the nullclines and vector field in the (p_1, p_2)-space: Both nullclines $E_i = 0$ have positive slope in the (p_1, p_2)-space and cross at one point $\mathbf{p}^* = (p_1^*, p_2^*)$ with both $p_i^* > 0$. The nullcline $E_1 = 0$ crosses from below $E_2 = 0$ to above at $\mathbf{p}^*$ as p_1 increases. The excess demand E_1 is: (i) positive to the left of $E_1 = 0$ and (ii) negative to the right of $E_1 = 0$. The excess demand E_2 is: (i) positive below $E_2 = 0$ and (ii) negative above $E_2 = 0$.
   a. Draw a qualitative picture of the nullclines and give the signs of p_1' and p_2' in the different regions of the first quadrant of the (p_1, p_2)-space.
   b. Using the type of argument given for two competitive populations, argue that for any initial conditions $(p_{1,0}, p_{2,0})$ for prices with $p_{1,0}, , p_{2,0} > 0$ have solutions that converge to $\mathbf{p}^*$.

4.5. Linearized Stability of Fixed Points

Often, the knowledge of the locations of the nullclines and of the signs of $\dot{x}$ and $\dot{y}$ in different regions is not enough to completely understand the properties of the trajectories. This is especially true when solutions cycle through a sequence of these regions as illustrated in Example 4.20. Another important technique is to linearize the vector field near the fixed points. The linear terms often dominate near the fixed point, and in that case, they determine the behavior of the nonlinear equations near the fixed point; in particular, the linearized system usually determines the stability of the fixed point.

In Example 4.18, the linearized system at the origin is obtained by dropping the $2x^3$ term from the $\dot{y}$ equation,

$$\dot{x} = -x$$
$$\dot{y} = -2y.$$

The origin is a stable node for this linear system. For an attracting fixed point, the linear terms dominate the higher order terms near the fixed point, so the solutions of the system of nonlinear equations behave similarly to those of the linear system. In particular, the nonlinear system of differential equations is asymptotically stable. See Theorem 4.6.

Example 4.21. Consider the system of differential equations

$$\dot{x} = 4y,$$
$$\dot{y} = x + 4x^3.$$

4.5. Linearized Stability

The linearized system at the origin is obtained by dropping the $4x^3$ term:
$$\dot{x} = 4y,$$
$$\dot{y} = x.$$

The origin is a saddle with eigenvalues ± 2. The eigenvector for -2 is $\begin{pmatrix} 2 \\ -1 \end{pmatrix}$, and the eigenvector for 2 is $\begin{pmatrix} 2 \\ 1 \end{pmatrix}$.

The linear system has two trajectories that tend toward the origin from opposite sides along straight lines:
$$\pm e^{-2t} \begin{pmatrix} 2 \\ -1 \end{pmatrix}.$$

Theorem 4.7 shows that the nonlinear equations have two nonlinear trajectories that tend to the fixed point from opposite sides and are tangent to the vectors:
$$\pm \begin{pmatrix} 2 \\ -1 \end{pmatrix}.$$

The points of these two trajectories make up what is called the *stable manifold of the fixed point* $W^s(\mathbf{0})$ (i.e., the set of all points $\mathbf{p}_0$ such that the solution $\phi(t; \mathbf{p}_0)$ tends to the fixed point as t goes to infinity and $\omega(\mathbf{p}_0) = \{\mathbf{0}\}$). See Figure 14.

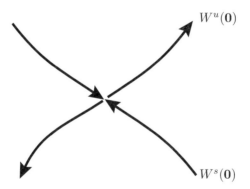

Figure 14. Stable and Unstable Manifolds

In the same way, the linear equation has two trajectories that tend to the fixed point as t goes to $-\infty$,
$$\pm e^{2t} \begin{pmatrix} 2 \\ 1 \end{pmatrix}.$$

Again, the nonlinear equations have two nonlinear trajectories that tend to the fixed point from opposite sides as t goes to $-\infty$ and are tangent at the origin to the vectors
$$\pm \begin{pmatrix} 2 \\ 1 \end{pmatrix}.$$

The points of these two trajectories make up what is called the *unstable manifold of the fixed point* $W^u(\mathbf{0})$ (i.e., the set of all points $\mathbf{p}_0$ such that the solution $\phi(t; \mathbf{p}_0)$ tends to the fixed point as t goes to $-\infty$ and $\alpha(\mathbf{p}_0) = \{\mathbf{0}\}$).

Example 4.22. As another example, consider the system of differential equations
$$\dot{x} = -x + y,$$
$$\dot{y} = 2 - 2xy^2.$$
The nullclines are $y = x$ and $xy^2 = 1$; there is a unique fixed point $x = 1$ and $y = 1$. Using the signs of $\dot{x}$ and $\dot{y}$ is not enough to determine the behavior of the solutions near the fixed point. However, using the Taylor expansion about the fixed point for the two differential equations
$$\dot{x} = f(x, y) = -x + y,$$
$$\dot{y} = g(x, y) = 2 - 2xy^2,$$
letting $u = x - 1$ and $v = y - 1$, we have
$$\dot{u} = \dot{x} = f(1,1) + u\frac{\partial f}{\partial x}(1,1) + v\frac{\partial f}{\partial y}(1,1) + \cdots = -u + v,$$
$$\dot{v} = \dot{y} = g(1,1) + u\frac{\partial g}{\partial x}(1,1) + v\frac{\partial g}{\partial y}(1,1) + \cdots = -2u - 4v + \cdots.$$
The coefficient matrix for the linear terms has eigenvalues -2 and -3, with eigenvectors $\begin{pmatrix} 1 \\ -1 \end{pmatrix}$ and $\begin{pmatrix} 1 \\ -2 \end{pmatrix}$, respectively. The linearized system has a stable node at the origin. The linear terms dominate near the fixed point, so the nonlinear equations also have an attracting fixed point. Also, most solutions approach the fixed point with an asymptote of the line $y - 1 = -(x - 1)$ (i.e., with an asymptotic displacement in the direction of the vector $\begin{pmatrix} 1 \\ -1 \end{pmatrix}$). See Figure 15. The stable manifold of the fixed point $(1, 1)$ $W^s((1, 1))$ certainly contains a disk about the fixed point. (The figure indicates that it contains at least all the points with $0 \leq x \leq 2$ and $0 \leq y \leq 2$.) Therefore, $W^s((1, 1))$ is called the *basin of attraction* of $(1, 1)$.

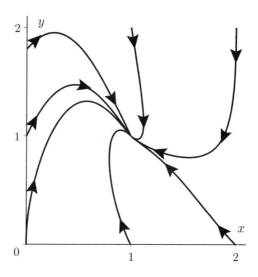

Figure 15. Nonlinear Sink, Example 4.22

4.5. Linearized Stability

For a general equation

$$\dot{x} = f(x, y),$$
$$\dot{y} = g(x, y),$$

the linearized system at a fixed point (x^*, y^*) is given by

$$(4.1) \quad \begin{pmatrix} \dot{u} \\ \dot{v} \end{pmatrix} = \begin{pmatrix} \frac{\partial f}{\partial x} & \frac{\partial f}{\partial y} \\ \frac{\partial g}{\partial x} & \frac{\partial g}{\partial y} \end{pmatrix} \begin{pmatrix} u \\ v \end{pmatrix},$$

where all the partial derivatives in the matrix are evaluated at (x^*, y^*). When comparing the solutions of the linearized system with the nonlinear system, the coordinates (u, v) for the linearized system must be compared with $(x, y) = (u + x^*, v + y^*)$ for the nonlinear system. If

$$\mathbf{F}(x, y) = \begin{pmatrix} f(x, y) \\ g(x, y) \end{pmatrix},$$

then, as a shorthand notation, we write

$$D\mathbf{F}_{(x^*, y^*)} = \begin{pmatrix} \frac{\partial f}{\partial x}(x^*, y^*) & \frac{\partial f}{\partial y}(x^*, y^*) \\ \frac{\partial g}{\partial x}(x^*, y^*) & \frac{\partial g}{\partial y}(x^*, y^*) \end{pmatrix},$$

for the *matrix of partial derivatives* or *derivative*.

For n variables, if

$$\mathbf{F}(x_1, \ldots, x_n) = \begin{pmatrix} F_1(x_1, \ldots, x_n) \\ \vdots \\ F_n(x_1, \ldots, x_n) \end{pmatrix}$$

and $\mathbf{x}^* = (x_1^*, \ldots, x_n^*)$ is a fixed point, then we write

$$D\mathbf{F}_{(\mathbf{x}^*)} = \left(\frac{\partial F_i}{\partial x_j}(\mathbf{x}^*) \right)$$

for the $n \times n$ matrix of partial derivatives or derivative. The linearized system is

$$\dot{\mathbf{u}} = D\mathbf{F}_{(\mathbf{x}^*)} \mathbf{u}.$$

Definition 4.23. If $\mathbf{x}^*$ is a fixed point of $\dot{\mathbf{x}} = \mathbf{F}(\mathbf{x})$, then we refer to the eigenvalues of the matrix of partial derivatives $D\mathbf{F}_{(\mathbf{x}^*)}$ as the *eigenvalues of the fixed point* or the *eigenvalues of* $\mathbf{x}^*$.

A fixed point $\mathbf{x}^*$ is called *hyperbolic* provided that the real parts of all the eigenvalues of the matrix $D\mathbf{F}_{(\mathbf{x}^*)}$ are nonzero. A hyperbolic fixed point is called a *saddle fixed point* provided that it has at least one eigenvalue with a positive real part and at least one other eigenvalue with a negative real part.

If a fixed point is hyperbolic, then the stability type of the fixed point for the nonlinear system is the same as that for the linearized system. The following result states this more precisely.

Theorem 4.6. *Consider a differential equation* $\dot{\mathbf{x}} = \mathbf{F}(\mathbf{x})$ *in n variables, with a hyperbolic fixed point* $\mathbf{x}^*$. *Assume that* $\mathbf{F}$, $\dfrac{\partial F_i}{\partial x_j}(\mathbf{x})$, *and* $\dfrac{\partial^2 F_i}{\partial x_j \partial x_k}(\mathbf{x})$ *are all continuous. Then, the stability type of the fixed point for the nonlinear system is the same as that for the linearized system at that fixed point. In particular, we have the following.*

a. *If the real parts of all the eigenvalues of* $D\mathbf{F}_{(\mathbf{x}^*)}$ *are negative, then the fixed point is asymptotically stable for the nonlinear equation (i.e., if the origin is asymptotically stable for the linearized system, then* $\mathbf{x}^*$ *is asymptotically stable for the nonlinear equation). In this case, the basin of attraction* $W^s(\mathbf{x}^*)$ *contains all the points within a small distance of the fixed point.*

b. *If at least one eigenvalue of* $D\mathbf{F}_{(\mathbf{x}^*)}$ *has a positive real part, then the fixed point* $\mathbf{x}^*$ *is unstable for the nonlinear equation. (The linearized system can be a saddle, unstable node, unstable focus, etc.)*

c. *If one of the eigenvalues of* $D\mathbf{F}_{(\mathbf{x}^*)}$ *has a zero real part, then the situation is more delicate. In particular, for* $n = 2$, *if the fixed point is an elliptic center (eigenvalues* $\pm i\beta$*) or one eigenvalue is 0 of multiplicity one, then the linearized system does not determine the stability type of the fixed point.*

Remark 4.24. The solutions of the nonlinear and linear equations are similar enough that most trajectories in Example 4.22 must approach the asymptotically stable fixed point in a direction that is asymptotic to plus or minus the eigenvector for the weaker eigenvalue of -2 (i.e., tangent to the line $y - 1 = x - 1$).

We next consider the behavior near a nonlinear saddle fixed point.

Definition 4.25. The *stable manifold of the fixed point* $W^s(\mathbf{x}^*)$ is the set of all the points that tend to the fixed point as t goes to plus infinity,

$$W^s(\mathbf{x}^*) = \{\, \mathbf{p}_0 : \phi(t; \mathbf{p}_0) \text{ tends to } \mathbf{x}^* \text{ as } t \to \infty \,\} = \{\, \mathbf{p}_0 : \omega(\mathbf{p}_0) = \{\mathbf{x}^*\} \,\}.$$

Notice that the orbit $\phi(t; \mathbf{p}_0)$ converges to the single fixed point $\mathbf{x}^*$ as t goes to infinity if and only if $\omega(\mathbf{p}_0) = \{\mathbf{x}^*\}$.

We also consider all those points in the stable manifold whose whole forward orbit stays near the fixed point. For a small $r > 0$, the *local stable manifold of size r of the fixed point* $\mathbf{x}^*$ is the set of points in $W^s(\mathbf{x}^*)$ whose whole forward orbit stays within a distance r of $\mathbf{x}^*$,

$$W_r^s(\mathbf{x}^*) = \{\, \mathbf{p}_0 \in W^s(\mathbf{x}^*) : \|\phi(t; \mathbf{p}_0) - \mathbf{x}^*\| < r \text{ for all } t \geq 0 \,\}.$$

Example 4.10 shows that there can be points on the stable manifold of a fixed point that start near the fixed point, go away from the fixed point, and finally return and converge to the fixed point; such points are not on the local stable manifold.

Definition 4.26. The *unstable manifold of the fixed point* $W^u(\mathbf{x}^*)$ is the set of all the points that tend to the fixed point as t goes to minus infinity,

$$W^u(\mathbf{x}^*) = \{\, \mathbf{p}_0 : \phi(t; \mathbf{p}_0) \text{ tends to } \mathbf{x}^* \text{ as } t \to -\infty \,\} = \{\, \mathbf{p}_0 : \alpha(\mathbf{p}_0) = \{\mathbf{x}^*\} \,\}.$$

4.5. Linearized Stability

For a small $r > 0$, the *local unstable manifold of size r of the fixed point* $\mathbf{x}^*$ is the set of points in $W^u(\mathbf{x}^*)$ whose whole backward orbit stays within a distance r of $\mathbf{x}^*$,

$$W_r^u(\mathbf{x}^*) = \{\, \mathbf{p}_0 \in W^s(\mathbf{x}^*) : \|\phi(t; \mathbf{p}_0) - \mathbf{x}^*\| < r \text{ for all } t \leq 0 \,\}.$$

In two dimensions, the stable and unstable manifolds of a saddle fixed point are curves, each of which separates the phase plane into the two sides of the curve. For this reason, in two dimensions, the stable and unstable manifolds of a saddle fixed point are often called *separatrices*. For a sink, the stable manifold is a whole region in the plane and is called the *basin of attraction* of the fixed point. (The set is open as defined in Appendix A.2.) The unstable manifold of a sink is only the fixed point itself. For a source, the stable manifold is just the fixed point and the unstable manifold is a whole region in the plane.

The Lorenz system considered in Chapter 7 has a a two-dimensional stable manifold in $\mathbb{R}^3$. The word *manifold* is a term in mathematics which includes curves, surfaces, and higher dimensional objects. The word manifold is used to refer to $W^s(\mathbf{x}^*)$ and $W^u(\mathbf{x}^*)$ because these sets can have a variety of dimensions. A more thorough discussion of manifolds is given in Appendix A.2.

The next theorem states that the local stable and unstable manifolds can be represented as graphs. This graph is expressed in terms of subspaces associated with the eigenvalues at a fixed point, which we define next.

The *stable subspace* at a fixed point is the linear subspace spanned by the set of all the generalized eigenvectors of the linearized equations at the fixed point associated with eigenvalues having negative real parts,

$$\mathbb{E}^s = \text{span}\{\, \mathbf{v} : \mathbf{v} \text{ is a generalized eigenvector for } D\mathbf{F}_{(\mathbf{x}^*)}$$
$$\text{whose associated eigenvalue has negative real part}\,\}.$$

In this subspace, we consider all vectors whose length is less than r,

$$\mathbb{E}^s(r) = \{\, \mathbf{v} \in \mathbb{E}^s : \|\mathbf{v}\| \leq r \,\}.$$

Similarly, the *unstable subspace* at a fixed point is the linear subspace spanned by the set of all the generalized eigenvectors of the linearized equations at the fixed point associated with eigenvalues having positive real parts,

$$\mathbb{E}^u = \text{span}\{\, \mathbf{v} : \mathbf{v} \text{ is a generalized eigenvector for } D\mathbf{F}_{(\mathbf{x}^*)}$$
$$\text{whose associated eigenvalue has positive real part}\,\} \text{ and}$$
$$\mathbb{E}^u(r) = \{\, \mathbf{v} \in \mathbb{E}^u : \|\mathbf{v}\| \leq r \,\}.$$

If the fixed point is hyperbolic, then $\mathbb{R}^n = \mathbb{E}^s + \mathbb{E}^u$. Thus, we can identify $(\mathbf{y}, \mathbf{z}) \in \mathbb{E}^s \times \mathbb{E}^u$ with $\mathbf{y} + \mathbf{z} \in \mathbb{R}^n$.

The next theorem states that the local stable (respectively, unstable manifold) can be expressed as a graph from $\mathbb{E}^s(r)$ into $\mathbb{E}^u(r)$ (respectively, from $\mathbb{E}^u(r)$ into $\mathbb{E}^s(r)$). In two dimensions, a local stable manifold of a saddle fixed point is a curve segment through the fixed point; it can be represented as a graph over an interval of length $2r$ in the line generated by the stable eigenvector. In higher dimensions, it is a graph over a ball in the subspace of all of the contracting directions for the linearized equations.

Theorem 4.7. *Assume that a differential equation* $\dot{\mathbf{x}} = \mathbf{F}(\mathbf{x})$ *has a saddle fixed point at* $\mathbf{x}^*$.

a. *In two dimensions, the eigenvalues for* $\mathbf{x}^*$ *are real, and one is positive and one is negative,* $\lambda_s < 0 < \lambda_u$. *Let* $\mathbf{v}^s$ *be an eigenvector for* λ_s *and* $\mathbf{v}^u$ *be an eigenvector for* λ_u. *Then, for sufficiently small* $r > 0$, *the local stable manifold of size* r *for the fixed point* $\mathbf{x}^*$, $W_r^s(\mathbf{x}^*)$, *is a curve passing through* $\mathbf{x}^*$ *that is tangent to* $\mathbf{v}^s$ *at* $\mathbf{x}^*$. *Similarly, the local unstable manifold of size* r, $W^u(\mathbf{x}^*)$, *is a curve passing through* $\mathbf{x}^*$ *that is tangent to* $\mathbf{v}^u$ *at* $\mathbf{x}^*$.

b. *In n dimensions, let* $\mathbb{E}^s$ *and* $\mathbb{E}^u$ *be the stable and unstable subspaces previously defined. Then, for sufficiently small* $r > 0$, *the local stable manifold of size* r *for the fixed point* $\mathbf{x}^*$, $W_r^s(\mathbf{x}^*)$, *is a "surface", which is a graph of a function* σ^s *from* $\mathbb{E}^s(r)$ *into* $\mathbb{E}^u(r)$ *that is tangent to the subspace* $\mathbb{E}^s$ *at* $\mathbf{x}^*$,

$$W_r^s(\mathbf{x}^*) = \{\, \mathbf{x}^* + (\mathbf{y}, \sigma^s(\mathbf{y})) : \mathbf{y} \in \mathbb{E}^s(r) \,\}.$$

Here, the pair $(\mathbf{y}, \sigma^s(\mathbf{y}))$ *with* $\mathbf{y} \in \mathbb{E}^s(r)$ *and* $\sigma^s(\mathbf{y}) \in \mathbb{E}^u(r)$ *is identified with* $\mathbf{y} + \sigma^s(\mathbf{y}) \in \mathbb{R}^n$.

In the same way, for sufficiently small $r > 0$, *the local unstable manifold of size* r *for the fixed point* $\mathbf{x}^*$, $W_r^u(\mathbf{x}^*)$, *is a "surface", which is a graph of a function* σ^u *from* $\mathbb{E}^u(r)$ *into* $\mathbb{E}^s(r)$ *that is tangent to the subspace* $\mathbb{E}^u$ *at* $\mathbf{x}^*$,

$$W_r^u(\mathbf{x}^*) = \{\, \mathbf{x}^* + (\sigma^u(\mathbf{z}), \mathbf{z}) : \mathbf{z} \in \mathbb{E}^u(r) \,\}.$$

c. *In any dimension, the (global) stable manifold is the set of points whose forward orbit eventually gets into the local stable manifold of size* r,

$$W^s(\mathbf{x}^*) = \{\, \mathbf{p} : \phi(t; \mathbf{p}) \in W_r^s(\mathbf{x}^*) \text{ for some } t \geq 0 \,\} = \bigcup_{t \leq 0} \phi(t; W_r^s(\mathbf{x}^*)).$$

In the same way, the (global) unstable manifold is the set of points whose backward orbit eventually gets into the local unstable manifold of size r,

$$W^u(\mathbf{x}^*) = \{\, \mathbf{p} : \phi(t; \mathbf{p}) \in W_r^u(\mathbf{x}^*) \text{ for some } t \leq 0 \,\} = \bigcup_{t \geq 0} \phi(t; W_r^u(\mathbf{x}^*)).$$

See [**Rob99**] for a proof.

The preceding theorem says that stable and unstable manifolds exist, but does not say how to determine them. In general, there is no good way to get an analytical formula for these curves. It is possible to represent them as a power series near the fixed point, but this is difficult to calculate and not that useful. To find a numerical approximation using a computer, an initial condition is taken a short distance away from the fixed point in the direction of the eigenvector, $\mathbf{x}_0 = \mathbf{x}^* + \epsilon \mathbf{v}^u$; then, the solution $\{\phi(t; \mathbf{x}_0) : t \geq 0\}$ gives an approximation for the unstable manifold. To find an approximation for the stable manifold, an initial condition $\mathbf{x}_0 = \mathbf{x}^* + \epsilon \mathbf{v}^s$ is used, and the solution is calculated for negative time.

The theorem also deals with only the stable and unstable manifolds (curves) near the fixed point. Away from the fixed point, the stable manifold cannot cross itself, but can wind around in the phase space and spiral in toward another fixed point or periodic orbit. It can also connect with the unstable manifold of this or another fixed point, as occurs for the undamped pendulum discussed in Example 5.3.

4.5. Linearized Stability

Example 4.27. Consider the phase portrait for the system given in Example 4.20,
$$\dot{x} = -x + y,$$
$$\dot{y} = xy - 1.$$

First, we find the fixed points: $\dot{x} = 0$ implies $x = y$, and $\dot{y} = 0$ implies that $1 = xy = x^2$ or $x = \pm 1$. Thus, there are two fixed points: $(1, 1)$ and $(-1, -1)$.

The linearization is
$$D\mathbf{F}_{(x,y)} = \begin{pmatrix} -1 & 1 \\ y & x \end{pmatrix}.$$

At $(1, 1)$,
$$D\mathbf{F}_{(1,1)} = \begin{pmatrix} -1 & 1 \\ 1 & 1 \end{pmatrix},$$
which has characteristic equation $\lambda^2 - 2 = 0$ and eigenvalues $\lambda = \pm\sqrt{2}$, so $(1, 1)$ is a saddle. The eigenvectors are $\begin{pmatrix} 1 \\ 1 \pm \sqrt{2} \end{pmatrix}$.

At $(-1, -1)$,
$$D\mathbf{F}_{(-1,-1)} = \begin{pmatrix} -1 & 1 \\ -1 & -1 \end{pmatrix},$$
which has characteristic equation $\lambda^2 + 2\lambda + 2 = 0$ and eigenvalues $\lambda = -1 \pm i$, so $(1, -1)$ is a stable focus.

The nullclines are $\{\dot{x} = 0\} = \{y = x\}$ and $\{\dot{y} = 0\} = \{y = 1/x\}$. See Figure 13. The orbit coming out of the saddle below goes down to the stable focus and spirals into it. The saddle has a trajectory from either side going asymptotically into it. These trajectories extend out to infinity. All initial conditions starting below the stable manifold of the saddle end up by spiraling into the stable focus. Initial conditions above the stable manifold have trajectories that eventually go off to infinity.

Much of this phase portrait can be determined by analyzing the stability of the fixed points and finding the nullclines. However, computers are very helpful in understanding the behavior. In drawing a phase portrait using a computer program, the first question is to determine the *window* in which to display the phase portrait. Certainly, it should include the fixed points and should be big enough to show how solutions approach the fixed points. Often, it is best not to make the window too big so that the important features are apparent.

Once the window is chosen, various initial conditions must be picked to display important behavior of the phase portrait. This takes some experimentation, especially if not much is known about the equations. By varying the initial conditions, it is possible to find solutions that go near the stable manifold of a saddle and then progress onward to another part of the phase portrait. See Figure 16 for the choices made for the present example.

Example 4.28 (Linear Center). Consider the system of nonlinear differential equations
$$\dot{x} = -y + ax(x^2 + y^2),$$
$$\dot{y} = x + ay(x^2 + y^2).$$

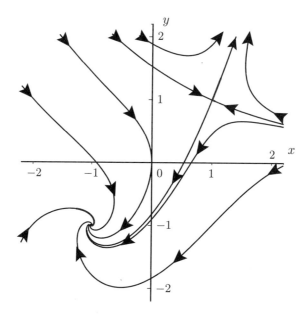

Figure 16. Phase plane for Example 4.27

The eigenvalues of the matrix of partial derivatives at the origin are $\pm i$, no matter what the value is of the parameter a. Therefore, the origin is a center for the linearized differential equations. We show that by varying the parameter a, the fixed point of the nonlinear system of differential equations can be attracting, repelling, or L-stable.

As we have rotation, it is convenient to change to polar coordinates. We have $r^2 = x^2 + y^2$, so

$$r\dot{r} = x\dot{x} + y\dot{y} = x(-y + axr^2) + y(x + ayr^2) = ar^2(x^2 + y^2) = ar^4 \qquad \text{or}$$
$$\dot{r} = a\,r^3.$$

Similarly, $\tan(\theta) = y/x$, and

$$\sec^2(\theta)\,\dot{\theta} = \frac{x\dot{y} - y\dot{x}}{x^2}$$
$$= \frac{x^2 + axyr^2 + y^2 - axyr^2}{x^2}$$
$$= \frac{r^2}{x^2}, \qquad \text{or}$$
$$\dot{\theta} = 1.$$

If $a < 0$, then $\dot{r} < 0$ for $r > 0$, and the origin is attracting. Thus, the origin is not linearly attracting, but it attracts due to the cubic terms in the differential equations. Similarly, if $a > 0$, then $\dot{r} > 0$ for $r > 0$, and the origin is repelling due to the cubic terms. Finally, the equations with $a = 0$ give a linear center. This shows that, when the linear terms give a center, they are not enough to determine the stability type of the fixed point.

Exercises 4.5

1. Consider the system of differential equations

   $$\dot{x} = -x - y + 4,$$
   $$\dot{y} = 3 - xy.$$

   a. Find the fixed points.
   b. Determine the type of the linearized system at each fixed point (saddle, stable focus, etc.).
   c. Determine the nullclines and the signs of $\dot{x}$ and $\dot{y}$ on the nullclines and in the various regions determined by them.
   d. Draw the phase portrait by hand for the system using the information from parts (a) through (c). Explain and justify your sketch.
   e. Use a computer program to draw the phase portrait for representative initial conditions. Be sure to include enough initial conditions to reveal the important features of the system.

2. Consider the system of differential equations

   $$\dot{x} = y - x^3,$$
   $$\dot{y} = xy - 1.$$

   a. Find the (two) fixed points.
   b. Determine the type of the linearized system at each fixed point (saddle, stable focus, etc.).
   c. Determine the nullclines and the signs of $\dot{x}$ and $\dot{y}$ on the nullclines and in the various regions determined by them.
   d. Draw the phase portraits for the system using the information from parts (a) through (c). Explain and justify your sketch.
   e. Using the computer, trajectories for enough initial conditions to show the important features of the phase portrait.

3. Consider the damped pendulum equation $\ddot{\theta} + c\dot{\theta} + \sin(\theta) = 0$, where $c > 0$. Find and classify the fixed points of the related system of first-order equations.

4. Consider the system of differential equations

   $$\dot{x} = -2x + y,$$
   $$\dot{y} = 4 - 2xy.$$

   a. Find the fixed points.
   b. Determine the type of the linearized system at each fixed point (saddle, stable focus, etc.).
   c. Determine the nullclines and the signs of $\dot{x}$ and $\dot{y}$ on the nullclines and in the various regions determined by them.
   d. Draw the phase portrait for the system using the information from parts (a) through (c). Explain and justify your sketch.

5. Consider the system of differential equations
$$\dot{x} = -\frac{x}{2} + y,$$
$$\dot{y} = 1 - y^2.$$
   a. Determine the fixed points and linear type of each fixed point (saddle, stable focus, etc.).
   b. Determine the nullclines.
   c. Draw the phase portrait for the system using the information from parts (a) and (b). Explain and justify your sketch.

6. Consider the system of differential equations
$$\dot{x} = xy,$$
$$\dot{y} = 1 - y - (1 + y)x.$$
   a. Determine the fixed points and linear type of each fixed point (saddle, stable focus, etc.).
   b. Determine the nullclines.
   c. Draw the phase portrait for the system using the information from parts (a) and (b). Explain and justify your sketch.

7. Consider the system of differential equations
$$\dot{x} = (x-1)(y-1),$$
$$\dot{y} = 3 - xy,$$
which has fixed points at $(1, 3)$ and $(3, 1)$.
   a. Determine the type of the linearized equations at each fixed point (saddle, stable node, etc.).
   b. Determine the nullclines and the signs of $\dot{x}$ and $\dot{y}$ in the regions determined by the nullclines.
   c. Draw the phase portrait for the system using the information from parts (a) and (b). Explain your sketch of the phase portrait.

8. Consider the system of differential equations
$$\dot{x} = x^2 - y - 1,$$
$$\dot{y} = y - y(x - 2).$$
   a. Show that there are three fixed points and classify them.
   b. Show that there are no closed orbits (look at what the vector field does along the three lines which go through a pair of fixed points).
   c. Sketch the phase portrait of the system.

9. Consider the system of differential equations
$$\dot{x} = x - y + x^2 - xy,$$
$$\dot{y} = -y + x^2.$$
   a. Show that there are three fixed points and classify them.
   b. Sketch the phase portrait of the system using the nullclines and the information of part (a).

10. Find the fixed points and classify them for the system of equations
$$\dot{x} = v,$$
$$\dot{v} = -x + \omega x^3,$$
$$\dot{\omega} = -\omega.$$

11. Find the fixed points and classify them for the Lorenz system of equations
$$\dot{x} = -10x + 10y,$$
$$\dot{y} = 28x - y - xz,$$
$$\dot{z} = -\frac{8}{3}z + xy.$$

12. Prove that a fixed point source of a system of differential equations in the plane is unstable. Hint: Compare with Theorem 4.12 in Section 4.8.

4.6. Competitive Populations

In this section, we discuss two populations which interact and have either positive or negative effects on the growth of the other population. Just as in the case of one population, $\dot{x}/x$ and $\dot{y}/y$ are the rates of growth per unit population. A general set of equations would have these rates of growth to be functions of both populations,

$$\begin{aligned}\frac{\dot{x}}{x} &= f(x,y), \\ \frac{\dot{y}}{y} &= g(x,y),\end{aligned} \quad \text{or} \quad \begin{aligned}\dot{x} &= x\,f(x,y), \\ \dot{y} &= y\,g(x,y).\end{aligned}$$

One classical model takes these dependencies to be linear functions of x and y, yielding the following equations which are called the *Lotka–Volterra equations*:

$$\dot{x} = x\,(B_1 + A_{1,1}\,x + A_{1,2}\,y),$$
$$\dot{y} = y\,(B_2 + A_{2,1}\,x + A_{2,2}\,y).$$

We consider two cases. In the first case, each population has a negative effect on the growth rate of the other population, and $A_{1,2}, A_{2,1} < 0$. This case is called *competitive species*. (We often also assume that the population has a negative effect on its own growth rate, so $A_{1,1}, A_{2,2} < 0$.)

In the second case, one of the populations has a positive effect on the growth rate of the other (it is the prey), and the second population has a negative effect on the growth rate of the first (it is the predator): $A_{2,1} > 0$ and $A_{1,2} < 0$. This case is called a *predator–prey system*, and is discussed in the next chapter, because the solution method is more similar to those discussed in the next chapter than those of the present chapter. See Section 5.1.

Competitive Populations

We first consider competitive systems in which each population has a negative effect on the growth rate of the other and the dependency is linear:

$$\frac{\dot{x}}{x} = K - x - ay,$$
$$\frac{\dot{y}}{y} = L - bx - y,$$

or
$$\dot{x} = x(K - x - ay),$$
$$\dot{y} = y(L - bx - y).$$

We consider only cases in which all the parameters are positive in the form given here. The constants K and L measure the growth rates of x and y, respectively, when the populations are small; we assume that these growth rates are positive. Each population has a negative effect on its own growth rate and the growth rate of the other population. We have assumed (or scaled the variables so) that the coefficient of x in the first equation and of y in the second equation is -1.

The nullclines are $x = 0$ or $K - x - ay = 0$ for the $\dot{x} = 0$ equation, and $y = 0$ or $L - bx - y = 0$ for $\dot{y} = 0$. For a fixed point, if $x = 0$, then either $y = 0$ or $y = L$. There is also a fixed point with $y = 0$ and $x = K$. Finally, there is the fixed point (x^*, y^*), which is the solution of the two equations
$$0 = K - x^* - ay^*,$$
$$0 = L - bx^* - y^*.$$

We are interested in nonnegative populations, so this preceding fixed point is of interest only if both $x^* > 0$ and $y^* > 0$. This occurs if the lines cross in the first quadrant. Since the x and y intercepts of the lines are K, K/a, L/b, and L, there is a fixed point inside the first quadrant if either
$$\frac{L}{b} < K \quad \text{and} \quad L > \frac{K}{a}, \quad \text{or}$$
$$\frac{L}{b} > K \quad \text{and} \quad L < \frac{K}{a}.$$

Thus, there are four fixed points, $(0,0)$, $(K,0)$, $(0,L)$, and (x^*, y^*), that satisfy the two preceding equations.

The matrix of partial derivatives is
$$\begin{pmatrix} (K - x - ay) - x & -ax \\ -by & (L - bx - y) - y \end{pmatrix}.$$

I. The first case with an interior fixed point has
$$L < bK \quad \text{and} \quad aL > K.$$

Then, the nullclines cross in the first quadrant, so $x^* > 0$ and $y^* > 0$. The matrix of partial derivatives at the fixed point $(K, 0)$ is
$$\begin{pmatrix} -K & -aK \\ 0 & L - bK \end{pmatrix}.$$

Since $L - bK < 0$, $(K, 0)$ is a stable node. Similarly, the matrix of partial derivatives at $(0, L)$ is
$$\begin{pmatrix} K - aL & 0 \\ -bL & -L \end{pmatrix},$$
and $(0, L)$ is a stable node. Finally, at (x^*, y^*),
$$\begin{pmatrix} -x^* & -ax^* \\ -by^* & -y^* \end{pmatrix}$$

4.6. Competitive Populations

has determinant $x^*y^*(1-ab)$. Since $L < bK < b(aL)$, $1 - ab < 0$ and the fixed point is a saddle.

The nullclines create four regions in the first quadrant. See Figure 17. The bottom region abutting on $(K, 0)$, to the right of $\dot{y} = 0$ and below $\dot{x} = 0$ has solutions moving down and to the right. Since they cannot leave the region, they must go to the attracting fixed point at $(K, 0)$.

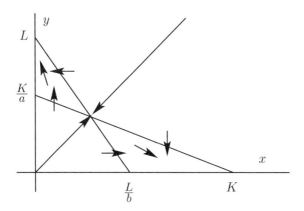

Figure 17. Competitive Exclusion

Similarly, the points in the region abutting on $(0, L)$ above $\dot{x} = 0$ and below $\dot{y} = 0$ are moving up and to the left and they converge to the fixed point at $(0, L)$.

For any points in the region abutting on $(0,0)$ below $\dot{x} = 0$ and $\dot{y} = 0$, both $\dot{x}$ and $\dot{y}$ are positive. The solutions either leave and enter one of the two regions just analyzed, or they must stay in the region forever and tend toward the fixed point (x^*, y^*) (i.e., they must be on the stable manifold of (x^*, y^*)). Similarly, points above and to the right of both $\dot{x} = 0$ and $\dot{y} = 0$ must either be on the stable manifold of (x^*, y^*), or enter the region where trajectories go to $(K, 0)$ or the region where they go to $(0, L)$. Thus, the stable manifold $W^s(x^*, y^*)$ separates the first quadrant into the points that are in the basin of $(K, 0)$ and those in the basin of $(0, L)$. The points that are on $W^s(x^*, y^*)$ tend to the saddle point (x^*, y^*). Thus, we have competitive exclusion, with most initial conditions leading to one of the populations dying out as t goes to infinity.

II. The second case with an interior fixed point has

$$L > bK \quad \text{and} \quad aL < K.$$

Again, the nullclines cross in the first quadrant, so $x^* > 0$ and $y^* > 0$. See Figure 18. Since $L > bK > baL$, $1 - ab > 0$, the determinant of the matrix of partial derivatives at (x^*, y^*) is positive and the trace is negative. Thus, (x^*, y^*) is asymptotically stable. In fact, both eigenvalues are real. Each of $(K, 0)$ and $(0, L)$ is a saddle, which attracts orbits along the respective axis and repels into the interior of the first quadrant. The two regions between the two lines $\dot{x} = 0$ and $\dot{y} = 0$ are invariant, with trajectories tending monotonically toward (x^*, y^*). Solutions in the other two regions either tend directly toward (x^*, y^*) or enter one of the other two regions. Therefore, eventually, all trajectories tend toward (x^*, y^*) (i.e., the basin of attraction of (x^*, y^*) is the entire open first quadrant). The stable manifold of $(K, 0)$ is just the positive x-axis. Similarly, $W^s(0, L)$ is the positive y-axis. Thus, we have competitive coexistence of all initial conditions which contain both species.

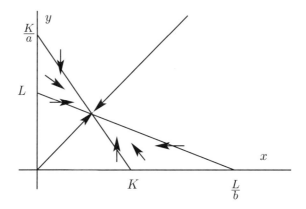

Figure 18. Competitive Coexistence

4.6.1. Three Competitive Populations. In this section, we do not consider the general case of three competitive populations. Instead, we consider one example of the Lotka–Volterra system for which the ω-limit set of most points is not a single fixed point but a cycle of fixed points. Based on what can happen with two populations, a conjecture might be that the ω-limit set of any trajectory is a single fixed point. However, the next example from the book [**Hof88**] by Hofbauer and Sigmund shows that this conjecture is false.

Consider the competitive system of three populations given by

$$\dot{x}_1 = x_1 (1 - x_1 - \alpha x_2 - \beta x_3),$$
$$\dot{x}_2 = x_2 (1 - \beta x_1 - x_2 - \alpha x_3),$$
$$\dot{x}_3 = x_3 (1 - \alpha x_1 - \beta x_2 - x_3),$$

with $0 < \beta < 1 < \alpha$ and $\alpha + \beta > 2$. We consider only points with nonnegative coordinates,

$$\{ (x_1, x_2, x_3) : x_1 \geq 0, \ x_2 \geq 0, \ x_3 \geq 0 \ \}.$$

The fixed points are $\mathbf{u}_1 = (1, 0, 0)$, $\mathbf{u}_2 = (0, 1, 0)$, $\mathbf{u}_3 = (0, 0, 1)$, and $\mathbf{x}^*$, where $x_1^* = x_2^* = x_3^* = 1/(1 + \alpha + \beta)$. The matrix of partial derivatives at $\mathbf{x}^*$ is

$$D\mathbf{F}_{(\mathbf{x}^*)} = \begin{pmatrix} -1 & -\alpha & -\beta \\ -\beta & -1 & -\alpha \\ -\alpha & -\beta & -1 \end{pmatrix}.$$

Because of the symmetry in the problem, it is possible to show that the eigenvalues are -1 and

$$\lambda_2 = \bar{\lambda}_3 = \frac{1}{1 + \alpha + \beta} \left(-1 - \alpha \, e^{i 2\pi/3} - \beta \, e^{i 4\pi/3} \right),$$

4.6. Competitive Populations

and the real parts of λ_2 and λ_3 are

$$\frac{1}{1+\alpha+\beta}\left(-1+\frac{\alpha+\beta}{2}\right) = \frac{\alpha+\beta-1}{1+\alpha+\beta} > 0.$$

The eigenvector for -1 is $(1,1,1)$. Thus, $\mathbf{x}^*$ has a one-dimensional stable manifold and a two-dimensional unstable manifold. The diagonal Δ are the points $\mathbf{x}$ for which $x_1 = x_2 = x_3$. Since $\dot{x}_1 = \dot{x}_2 = \dot{x}_3$ at points of Δ, it is invariant and the stable manifold of $\mathbf{x}^*$ consists of the points on the diagonal with positive entries,

$$W^s(\mathbf{x}) = \{(x_1, x_2, x_3) : x_1 = x_2 = x_3 > 0\}.$$

The matrix of partial derivatives at $\mathbf{u}_1$ is

$$D\mathbf{F}_{(\mathbf{u}_1)} = \begin{pmatrix} -1 & -\alpha & -\beta \\ 0 & 1-\beta & 0 \\ 0 & 0 & 1-\alpha \end{pmatrix}.$$

The the eigenvalues are $-1 < 0$, $1-\alpha < 0$, and $1-\beta > 0$. Thus, this fixed point has a two-dimensional stable manifold and a one-dimensional unstable manifold. The other two fixed points $\mathbf{u}_2$ and $\mathbf{u}_3$ have similar eigenvalues and the same dimensional stable and unstable manifolds.

We start by showing that most orbits tend to the union of the coordinate planes, $\{x_1 = 0\}$, $\{x_2 = 0\}$, and $\{x_3 = 0\}$. Let

$$P = x_1 x_2 x_3 \quad \text{and}$$
$$S = x_1 + x_2 + x_3.$$

Then,

$$\begin{aligned}\dot{S} &= x_1 + x_2 + x_3 - [x_1^2 + x_2^2 + x_3^2 + (\alpha+\beta)(x_1 x_2 + x_2 x_3 + x_3 x_1)] \\ &\leq x_1 + x_2 + x_3 - [x_1^2 + x_2^2 + x_3^2 + 2(x_1 x_2 + x_2 x_3 + x_3 x_1)] \\ &= S - S^2 \\ &= S(1-S).\end{aligned}$$

Therefore, any orbit must be asymptotic to the set for which $S = 1$ (unless $S = 0$) and has to enter and remain in the set where $S \leq 2$. This shows that all orbits are bounded. Turning to P,

$$\begin{aligned}\dot{P} &= \dot{x}_1 x_2 x_3 + x_1 \dot{x}_2 x_3 + x_1 x_2 \dot{x}_3 \\ &= P[3 - (1+\alpha+\beta)S].\end{aligned}$$

To see that solutions go to the set for which $P(\mathbf{x}) = 0$, a computation shows that

$$\frac{d}{dt}(PS^{-3}) = PS^{-4}\left(1 - \frac{\alpha+\beta}{2}\right)[(x_1-x_2)^2 + (x_2-x_3)^2 + (x_3-x_1)^2]$$
$$\leq 0.$$

The preceding derivative is strictly negative if $P(\mathbf{x}) > 0$ and $\mathbf{x}$ is not on the diagonal Δ. Since the orbits are bounded, we show in Section 5.3 that, in such a situation, PS^{-3} is a strict Lyapunov function and the trajectories must go to the minimum of PS^{-3}, which has $P(\mathbf{x})S^{-3}(\mathbf{x}) = 0$ (i.e., to the coordinate planes where at least one coordinate variable is zero).

On the coordinate plane $\{x_3 = 0\}$, the nullcline $\dot{x}_2 = 0$ is strictly above the nullcline $\dot{x}_1 = 0$, and all orbits having initial conditions with $x_1 > 0$ and $x_2 > 0$ tend to the fixed point $\mathbf{u}_2$. Thus,

$$W^s(\mathbf{u}_2) \supset \{(x_1, x_2, x_3) : x_3 = 0,\ x_1 \geq 0,\ x_2 > 0\} \subset \{x_3 = 0\}.$$

We saw from the eigenvalues that the fixed point $\mathbf{u}_1$ has a one-dimensional unstable manifold, and it has a one-dimensional unstable manifold in the coordinate plane $\{x_3 = 0\}$. Therefore, $W^u(\mathbf{u}_1)$ is contained in this plane and

$$W^u(\mathbf{u}_1) \subset W^s(\mathbf{u}_2) \subset \{x_3 = 0\}.$$

The coordinate planes $\{x_1 = 0\}$ and $\{x_2 = 0\}$ are similar, with

$$W^u(\mathbf{u}_2) \subset W^s(\mathbf{u}_3) \subset \{x_1 = 0\} \quad \text{and}$$
$$W^u(\mathbf{u}_3) \subset W^s(\mathbf{u}_1) \subset \{x_2 = 0\}.$$

Thus, the three fixed points $\mathbf{u}_1$, $\mathbf{u}_2$, and $\mathbf{u}_3$ form a cycle of fixed points connected by stable and unstable manifolds.

Take any initial condition (x_1^0, x_2^0, x_3^0) with all three coordinates positive. Then, its ω-limit set $\omega(x_1^0, x_2^0, x_3^0)$ must be in the set

$$P^{-1}(0) \cap S^{-1}(1) = \{(x_1, x_2, x_3) : P(x_1, x_2, x_3) = 0 \text{ and } S(x_1, x_2, x_3) = 1\}.$$

Since the stable manifolds of the three fixed points do not enter the positive octant where all coordinates are positive, $\omega(x_1^0, x_2^0, x_3^0)$ cannot be just a single fixed point. It must be positively invariant. Indeed, the orbit cycles near each fixed point in succession and then repeats the process. Therefore, the ω-limit set is the whole triangular set

$$\omega(x_1^0, x_2^0, x_3^0) = P^{-1}(0) \cap S^{-1}(1).$$

Thus, the ω-limit set is not a single fixed point but a cycle of unstable manifolds running from one fixed point to the next.

Exercises 4.6

1. Consider the system of differential equations for a competitive pair of populations

$$\dot{x} = x(K - x - ay),$$
$$\dot{y} = y(L - bx - y).$$

   Assume that $K > L/b$ and $K/a > L$.

   a. Find the fixed points for $x \geq 0$ and $y \geq 0$, and determine the type of the linearized equations at each fixed point (e.g., stable node, stable focus, saddle, unstable node, etc.).

   b. Sketch the phase portrait for $x \geq 0$ and $y \geq 0$, including the nullclines.

   c. Do most solutions with $x_0 > 0$ and $y_0 > 0$ tend to a fixed point? Which fixed point?

2. Consider the system of differential equations for a competitive pair of populations

$$\dot{x} = x(2 - x - y),$$
$$\dot{y} = y(3 - x - y).$$

   a. Find the fixed points for this system and classify them.
   b. Use the nullclines and the stability type of the fixed points to sketch the phase portrait. Indicate the basin of attraction of any attracting fixed points.

3. Consider the system of differential equations for a competitive pair of populations

$$\dot{x} = x(K - ay),$$
$$\dot{y} = y(L - bx).$$

   Note that there is no limit to the growth of each population separately.
   a. Find the fixed points for $x \geq 0$ and $y \geq 0$, and determine the type of the linearized equations at each fixed point (e.g., stable node, stable focus, saddle, unstable node, etc.).
   b. Sketch the phase portrait for $x \geq 0$ and $y \geq 0$, including the nullclines.

4. Consider a competitive market of two commodities whose prices (against a numeraire) are p_1 and p_2. Let $E_i(p_1, p_2)$ be the excess demand (demand minus supply) of the ith commodity. For $i = 1, 2$, assume that the price adjustment is given by $\dot{p}_i = k_i E_i(p_1, p_2)$ for constants $k_i > 0$. Based on economic assumptions, [Sho02] (pages 353-357) gives the following information about the nullclines and vector field in the (p_1, p_2)-space: Both nullclines $E_i = 0$ have positive slope in the (p_1, p_2)-space and cross at one point $\mathbf{p}^* = (p_1^*, p_2^*)$ with both $p_i^* > 0$. The nullcline $E_1 = 0$ crosses from below $E_2 = 0$ to above at $\mathbf{p}^*$ as p_1 increases. The partial derivatives satisfy $\frac{\partial E_1}{\partial p_1} < 0$ and $\frac{\partial E_2}{\partial p_2} < 0$.
   a. Draw a qualitative picture of the nullclines and give the signs of $\dot{p}_1$ and $\dot{p}_2$ in the different regions of the first quadrant of the (p_1, p_2)-space.
   b. Argue that for any initial conditions $(p_{1,0}, p_{2,0})$ for prices with $p_{1,0}, , p_{2,0} > 0$ have solutions that converge to $\mathbf{p}^*$.

5. Consider the situation in which both species have a positive effect on the growth of the other, given by the system of differential equations

$$\dot{x} = x(K + ay - cx),$$
$$\dot{y} = y(L + bx - ey),$$

   where all the parameters are positive. $K, a, c, L, b, e > 0$. Also assume that $c/a > b/e$.
   a. Show that there is a fixed point (x^*, y^*) with $x^* > 0$ and $y^* > 0$.
   b. Show that the fixed point (x^*, y^*) from part (a) is asymptotically stable. Hint: For the matrix of partial derivatives, show that the trace $\tau < 0$ and the determinant $\Delta = (ce - ab)x^*y^* > 0$.
   c. Draw the phase portrait for $x \geq 0$ and $y \geq 0$, including the nullclines. Discuss the basin of attraction of (x^*, y^*) and the stable and unstable manifolds of other fixed points.

4.7. Applications

4.7.1. Chemostats.

A simple chemostat

A chemostat is a laboratory apparatus used to grow micro-organisms under controlled experimental conditions. In this apparatus, a constant supply of nutrients is pumped into a vessel containing the micro-organism and the same amount of fluid containing the cells, nutrients, and byproducts flows out. Because the flow of fluid into and out of the vessel is the same, the volume inside the chemostat remains constant.

In the first model of a simple chemostat, the concentration of the micro-organism in the effluent tends to a constant value (an equilibrium), as would be expected in the chemical reaction. This source of an effluent with a constant concentration of the micro-organism could be used in a further experiment, or the equilibrium concentration could be used to study the rate of cell growth or the rate of metabolic product formation.

Sections 6.2.1 (Brusselator) and 6.8.1 (Field–Noyes model for the Belousov–Zhabotinsky chemical reaction) discuss models of chemical processes that tend to periodic levels of concentrations in the effluent. These examples need at least three chemicals, even though the example in Section 6.2 contains only two variables. In fact, people have found situations in which the concentrations are chaotic.

Let S be the concentration of the nutrient and x the concentration of the micro-organism. The concentration of the nutrient pumped in is C, if the nutrient were not consumed by the micro-organism, the rate of change of S would be proportional to $C-S$, $\dot S = (C-S)D$. The nutrient is consumed by the micro-organism at a rate that is a function of the concentration of the nutrient $r(S)$ times the concentration of the micro-organism x. A model commonly used is

$$r(S) = \frac{mS}{a+S},$$

with $m > 0$ and $a > 0$, which was introduced by Monod in 1950: it has the property that $r(0) = 0$ and $r(S)$ approaches a limit value of m as S approaches infinity. Using this model, if C is the concentration of the nutrient pumped into the vessel, the differential equations become

$$\dot S = (C-S)\,D - \frac{\beta\, m\, S\, x}{a+S},$$
$$\dot x = \frac{m\, S\, x}{a+S} - D\, x,$$

where D is a constant determined by the rate of flow of fluid in and out of the vessel and so can be adjusted. We assume that all of the parameters are positive and $m > D$. This last condition implies that the rate $r(S)$ for large S is greater than the amount removed by the flow of fluid out of the vessel. The initial conditions

4.7. Applications

satisfy $x(0) > 0$ and $S(0) \geq 0$. The nullclines are

$$\{\dot{x} = 0\} = \left\{ S = \frac{aD}{m-D} \right\} \quad \text{and}$$

$$\{\dot{S} = 0\} = \left\{ x = \frac{D(C-S)(S+a)}{\beta m S} \right\}.$$

The second nullcline has $x = 0$ when $S = C$. Also, the slope of the nullcline is

$$\frac{dx}{dS} = \frac{d}{dS}\left(-\frac{D}{\beta m}S + \frac{D(C-a)}{\beta m} + \frac{DaC}{\beta m S}\right)$$
$$= -\frac{D}{\beta m} - \frac{DaC}{\beta m S^2}$$
$$< 0,$$

and x goes to ∞ as S goes to 0. Therefore, there is a fixed point in the interior of the first quadrant if and only if

$$\lambda = \frac{aD}{m-D} < C.$$

Theorem 4.8. *For the preceding system of differential equations, assume that $m > D$ and $\lambda < C$. Then, the interior fixed point (S^*, x^*) is given by*

$$S^* = \lambda \quad \text{and} \quad x^* = \frac{C - \lambda}{\beta},$$

and it is asymptotically stable. The only other fixed point is given by $S = C$ and $x = 0$, and it is a saddle point. In fact, all initial conditions with $x_0 > 0$ and $S_0 > 0$ tend to the interior fixed point S^* and x^*.

Proof. A little algebra shows that the unique fixed point in the interior has the form given in the theorem. The matrix of partial derivatives at this fixed point is

$$\begin{pmatrix} -D - \dfrac{\beta(m-D)x^*}{a+S^*} & -\dfrac{\beta m S^*}{a+S^*} \\ \dfrac{(m-D)x^*}{a+S^*} & 0 \end{pmatrix}.$$

The trace is negative and the determinant is positive. By Theorem 4.3, this interior fixed point is asymptotically stable.

The matrix of partial derivatives at the second fixed point $(C, 0)$ is

$$\begin{pmatrix} -D & \dfrac{\beta m C}{a+C} \\ 0 & \dfrac{mC}{a+C} - D \end{pmatrix}.$$

Since $\lambda < C$, the eigenvalues are

$$-D < 0 \quad \text{and} \quad \frac{mC}{a+C} - D > 0,$$

so this fixed point is an unstable saddle point. $\square$

The next result gives the results when there are no interior fixed points. We leave the details to the reader.

Theorem 4.9. *For the preceding system of differential equations, assume that $m > D$ and $\lambda > C$. Then, there is no interior fixed point in the first quadrant and the fixed point $(C, 0)$ is asymptotically stable.*

For references, see the books on population biology [**Bra01**] and [**Wal83**].

Two micro-organisms competing for one nutrient

We next consider two micro-organisms in a vessel, with concentrations x and y, that are competing for the same nutrient supplied as the simple chemostat, but are not directly affecting each other. Using D for the rate of flow through the vessel, the system of equations becomes

$$\dot{S} = (C - S)\,D - \frac{\beta_1\,m_1\,S\,x}{a_1 + S} - \frac{\beta_2\,m_2\,S\,y}{a_2 + S},$$

$$\dot{x} = \frac{m_1\,S\,x}{a_1 + S} - D\,x,$$

$$\dot{y} = \frac{m_2\,S\,y}{a_2 + S} - D\,y.$$

We are interested only in values with $S \geq 0$, $x \geq 0$, and $y \geq 0$.

Rather than considering these equations with all three variables, we use an argument to reduce the problem to two variables. We define the new variable

$$z = C - S - \beta_1\,x - \beta_2\,y,$$

whose derivative is

$$\dot{z} = -(C - S)\,D + \frac{\beta_1\,m_1\,S\,x}{a_1 + S} + \frac{\beta_2\,m_2\,S\,y}{a_2 + S}$$
$$- \beta_1 \frac{m_1\,S\,x}{a_1 + S} + \beta_1\,D\,x - \beta_2 \frac{m_2\,S\,y}{a_2 + S} + \beta_2\,D\,y$$
$$= -D\,z.$$

Therefore, the solution $z(t)$ decays to zero,

$$z(t) = z(0)\,e^{-Dt}.$$

Since we are interested only in the long-term behavior in the equations, we can take $z = 0$ and use it to solve for S in terms of x and y. Note that, since we consider only $S \geq 0$, we need to restrict our consideration to

$$\beta_1\,x + \beta_2\,y = C - S \leq C.$$

On this set with $z = 0$, the differential equations for x and y become

$$\dot{x} = x\left(\frac{m_1\,(C - \beta_1\,x - \beta_2\,y)}{a_1 + C - \beta_1\,x - \beta_2\,y} - D\right),$$

$$\dot{y} = y\left(\frac{m_2\,(C - \beta_1\,x - \beta_2\,y)}{a_2 + C - \beta_1\,x - \beta_2\,y} - D\right).$$

Since $\beta_1\,x + \beta_2\,y \leq C$, the denominators of the expressions are positive.

4.7. Applications

We want to consider only parameters for which each organism would approach a nonzero limit concentration if it were present alone. So, based on the simple chemostat, we assume

$$m_1 > D,$$
$$m_2 > D,$$
$$\lambda_1 = \frac{a_1 D}{m_1 - D} < C, \quad \text{and}$$
$$\lambda_2 = \frac{a_2 D}{m_2 - D} < C.$$

By renumbering the organisms, we can assume that

$$\lambda_1 < \lambda_2.$$

We analyze the phase portrait by considering the nullclines for the differential equations in only the x and y variables. With a little bit of algebra, we find that the two nullclines are the following:

$$C - \lambda_1 = \beta_1 x + \beta_2 y \quad \text{and}$$
$$C - \lambda_2 = \beta_1 x + \beta_2 y.$$

These two lines are parallel, and since $\lambda_1 < \lambda_2$, the first line is outside the second. Therefore, the nullclines and direction arrows are as in Figure 19. The fixed points are $(x^*, 0)$ and $(0, y^*)$, where

$$x^* = \frac{C - \lambda_1}{\beta_1} \quad \text{and} \quad y^* = \frac{C - \lambda_2}{\beta_1}.$$

By phase plane analysis similar to that in Section 4.6 for competitive systems, any trajectory $\phi(t;(x_0,y_0))$ with $x_0 > 0$ and $y_0 > 0$ tends to the fixed point at $(x^*, 0)$ as t goes to infinity.

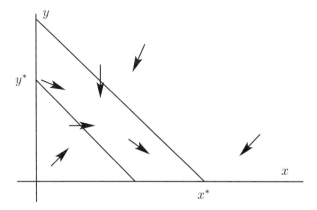

Figure 19. Nullclines for the competitive chemostat

The discussion of this competitive chemostat is based on the treatment in [**Wal83**]. For further details, see that reference. That book also considers other related situations such as a food chain, in which the organism x lives off the first nutrient S, and the second organism y lives off the first organism x.

4.7.2. Epidemic Model. We present a model of epidemics, in which S is the susceptible population, I is the infected population, and R is the recovered population. Because of the variables used for different types of populations considered, the model is called an *SIR model of epidemics*. The model we present includes births in the susceptible group at a constant rate μK. (If the total population were a constant K, then this would be the births occurring with a rate of μ.) We also assume that there is a death rate of $-\mu$ from each class of population. There is an infection rate of people in the susceptible group who become infected which is proportional to the contacts between the two groups βSI. There is a recovery of γI from the infected group into the recovered group. Finally, the disease is fatal to some of the infected group, which results in the removal rate of $-\alpha I$ from the infected population. Putting this together, the system of differential equations becomes

$$\dot{S} = \mu K - \beta S I - \mu S,$$
$$\dot{I} = \beta S I - \gamma I - \mu I - \alpha I,$$
$$\dot{R} = \gamma I - \mu R.$$

Notice that the change in total population is

$$\dot{S} + \dot{I} + \dot{R} = \mu(K - S - I - R) - \alpha I.$$

Thus, without the term $-\alpha I$, a total population of K would be preserved.

Because the first two equations do not involve R, we can solve them separately, and then use the solution to determine the solution for R. Setting $\delta = \alpha + \gamma + \mu$, the system of equations becomes

$$\dot{S} = \mu K - \beta S I - \mu S,$$
$$\dot{I} = \beta S I - \delta I.$$

The nullclines for the equations are

$$I = \frac{\mu K}{\beta S} - \frac{\mu}{\beta} \quad \text{for } \{\dot{S} = 0\} \quad \text{and}$$
$$S = \frac{\delta}{\beta} \quad \text{or } I = 0 \quad \text{for } \{\dot{I} = 0\}.$$

Thus, the fixed points are

$$S = K, I = 0 \quad \text{and} \quad S^* = \frac{\delta}{\beta}, \; I^* = \frac{\mu K}{\delta} - \frac{\mu}{\beta}.$$

The value $I^* < 0$ when $\beta K/\delta < 1$, and $I^* > 0$ when $\beta K/\delta > 1$.

To check the stability of the fixed points, we consider the matrix of partial derivatives

$$\begin{pmatrix} -\beta I - \mu & -\beta S \\ \beta I & \beta S - \delta \end{pmatrix}.$$

If $\beta K/\delta < 1$, then the only fixed point with nonnegative values is at $S = K$ and $I = 0$. The matrix of partial derivatives is

$$\begin{pmatrix} -\mu & -\beta K \\ 0 & \beta K - \delta \end{pmatrix}.$$

4.7. Applications

This has eigenvalues $-\mu$ and $\beta K - \delta$, both of which are negative. It is not hard to see that this fixed point attracts all trajectories with $S_0 \geq 0$ and $I_0 \geq 0$. See Figure 20. Thus, there is no epidemic, but the disease dies out.

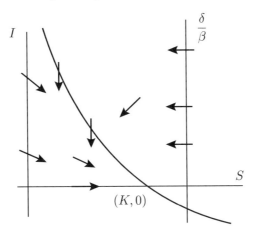

Figure 20. Nullclines for the epidemic model with $\beta K / \delta < 1$

Next, assume that $\beta K / \delta > 1$. The fixed point $S = K$ and $I = 0$ now becomes a saddle point because $\beta K - \delta > 0$. The matrix of partial derivatives at (S^*, I^*) is

$$\begin{pmatrix} -\beta I^* - \mu & -\beta S^* \\ \beta I^* & 0 \end{pmatrix}.$$

This matrix has a trace of $-\beta I^* - \mu$, which is negative, and has a determinant of $\beta^2 I^* S^* > 0$. Thus, this fixed point is asymptotically stable, and is either a stable focus or a stable node. By considering the nullclines and direction arrows, it follows that any solution for initial conditions with $S_0 > 0$ and $I_0 > 0$ goes around the fixed point (S^*, I^*). In fact, such a solution tends to the fixed point (S^*, I^*). Thus, the epidemic reaches a steady state. See Figure 21

This system of equations is one of the models considered in [**Bra01**] (Section 7.4). For more discussion of the interpretation and other models, see this reference.

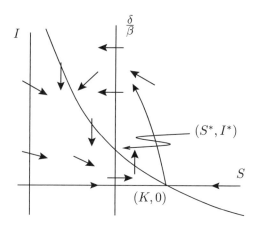

Figure 21. Nullclines for the epidemic model with $\beta K/\delta > 1$

Exercises 4.7

1. Consider the SIR model for a population with vaccination given by
$$\dot{S} = (1-p)\,\mu\,K - \beta\,S\,I - \mu\,S,$$
$$\dot{I} = \beta\,S\,I - \gamma\,I - \mu\,I,$$
$$\dot{R} = p\,\mu\,K + \gamma\,I - \mu\,R.$$

   Here the population R contains: (i) those who have recovered and are immune and (ii) those who have been vaccinated and are immune. The parameter p, which satisfies $0 < p < 1$, is the proportion of those born who are vaccinated immediately and are transferred to the immune state. Also, $\mu > 0$, $\beta > 0$, and $\gamma > 0$. The parameter μ is the birth rate and death rate, β is the rate of infection, and γ is the rate of recovery from infected to recovered.

   a. Show that $\frac{d}{dt}(S+I+R) = \mu\,(K - S - I - R)$ and so decays to a total population $S + I + R = K$. Also, why can we consider the system of differential equations for (S, I) first and then find $R(t)$? The rest of the problem considers only the system of differential equations for (S, I).

   b. Find the two fixed points for the system of differential equations for (S, I), $(\bar{S}, 0)$ and (S^*, I^*) with $I^* \neq 0$. For what parameter values is $I^* > 0$?

   c. Find the matrix of partial derivatives.

   d. What is the matrix of partial derivatives at the fixed point $(\bar{S}, 0)$? For what parameter values is this fixed point asymptotically stable?

   e. What is the matrix of partial derivatives at the fixed point (S^*, I^*)? What is the trace and determinant? For what parameter values is this fixed point asymptotically stable?

2. For this epidemic model, let x and y be the healthy and sick populations, respectively. Assume that the system of differential equations is given by

$$\dot{x} = -kxy,$$
$$\dot{y} = kxy - ly,$$

where k and l are both positive.
   a. Find and classify all the fixed points.
   b. Sketch the nullclines and the direction of the vector field in each region bounded by the nullclines.
   c. Indicate the set of points in the first quadrant at which the epidemic is getting worse (i.e. where $dy/dt > 0$).
   d. Find an expression for dy/dx by taking the quotient of the equations. Then separate variables and find a function of x and y that is conserved by the system.
   e. Sketch the phase portrait.

4.8. Theory and Proofs

Limit sets

Restatement of Theorem 4.1. *Assume that $\phi(t; \mathbf{x}_0)$ is a trajectory. Then, the following properties of the ω-limit set are true:*

   a. *The limit set depends only on the trajectory and not on the particular point, so $\omega(\mathbf{x}_0) = \omega(\phi(t; \mathbf{x}_0))$ for any real time t.*
   b. *The $\omega(\mathbf{x}_0)$ is invariant: if $\mathbf{z}_0 \in \omega(\mathbf{x}_0)$, then the orbit $\phi(t; \mathbf{z}_0)$ is in $\omega(\mathbf{x}_0)$ for all positive and negative t.*
   c. *The $\omega(\mathbf{x}_0)$ contains all its limit points (i.e., $\omega(\mathbf{x}_0)$ is closed).*
   d. *If $\mathbf{y}_0$ is a point in $\omega(\mathbf{x}_0)$, then $\omega(\mathbf{y}_0) \subset \omega(\mathbf{x}_0)$.*

In addition, assume that the trajectory $\phi(t; \mathbf{x}_0)$ stays bounded for $t \geq 0$ (i.e., there is a constant $C > 0$ such that $\|\phi(t; \mathbf{x}_0)\| \leq C$ for $t \geq 0$). Then, properties (e) and (f) are true.

   e. *The $\omega(\mathbf{x}_0)$ is nonempty.*
   f. *The $\omega(\mathbf{x}_0)$ is connected; it is not made up of more than one piece.*

Similar properties hold for the α-limit set if $\phi(t; \mathbf{x}_0)$ stays bounded for $t \leq 0$.

Proof. (a) Assume that $\mathbf{z}_0 \in \omega(\mathbf{x}_0)$, so there exists a sequence of times t_n going to infinity such that $\phi(t_n; \mathbf{x}_0)$ converges to $\mathbf{z}_0$. Then $\phi(t_n - t; \phi(t; \mathbf{x}_0)) = \phi(t_n; \mathbf{x}_0)$ converges to $\mathbf{z}$, so $\mathbf{z}$ is in $\omega(\phi(t; \mathbf{x}_0))$, $\omega(\mathbf{x}_0) \subset \omega(\phi(t; \mathbf{x}_0))$. Reversing the roles of $\mathbf{x}_0$ and $\phi(t; \mathbf{x}_0)$ gives the other inclusion, proving (a).

(b) Assume that $\mathbf{z}_0 \in \omega(\mathbf{x}_0)$, so there exists a sequence of times t_n going to infinity such that $\phi(t_n; \mathbf{x}_0)$ converges to $\mathbf{z}_0$. By continuity of the flow, $\phi(t+t_n; \mathbf{x}_0) = \phi(t; \phi(t_n; \mathbf{x}_0))$ converges to $\phi(t; \mathbf{z}_0)$. This proves $\phi(t; \mathbf{z}_0)$ is in $\omega(\mathbf{x}_0)$.

(c) Assume that $\mathbf{z}_j$ is a sequence of points in $\omega(\mathbf{x}_0)$ converging to a point $\mathbf{z}_\infty$ in $\mathbb{R}^n$. In fact, we can take the sequence so that

$$\|\mathbf{z}_j - \mathbf{z}_\infty\| < 1/j.$$

Each of the $\mathbf{z}_j$ is an ω-limit point, so there exist sequences of times $t_{j,n}$ such that

$$\|\phi(t_{j,n}; \mathbf{x}_0) - \mathbf{z}_j\| < 1/n.$$

(We have taken the times so that the points are within $1/n$.) Then

$$\|\phi(t_{n,n}; \mathbf{x}_0) - \mathbf{z}_\infty\| \leq \|\phi(t_{n,n}; \mathbf{x}_0) - \mathbf{z}_n\| + \|\mathbf{z}_n - \mathbf{z}_\infty\|$$
$$\leq \frac{1}{n} + \frac{1}{n} = \frac{2}{n}.$$

Since $2/n$ is arbitrarily small, the points $\phi(t_{n,n}; \mathbf{x}_0)$ converge to $\mathbf{z}_\infty$, and $\mathbf{z}_\infty$ is in the ω-limit set of $\mathbf{x}_0$.

(d) Assume that $\mathbf{z}_0$ is a point in $\omega(\mathbf{x}_0)$. Assume that $\mathbf{y}$ is an ω-limit point of $\mathbf{z}_0$, so there exist times t_n going to infinity such that $\phi(t_n, \mathbf{z}_0)$ converges to $\mathbf{y}$. By (b), the points $\phi(t_n; \mathbf{z}_0)$ are all contained in $\omega(\mathbf{x}_0)$. Because the limit set is closed by (c) and $\phi(t_n; \mathbf{z}_0)$ converges to $\mathbf{y}$, the limit point $\mathbf{y}$ must be in $\omega(\mathbf{x}_0)$. This shows that any limit point of $\mathbf{z}_0$ is contained in $\omega(\mathbf{x}_0)$, or $\omega(\mathbf{z}_0) \subset \omega(\mathbf{x}_0)$.

(e) The sequence of points $\phi(n; \mathbf{x}_0)$ is bounded, so $|\phi(n; \mathbf{x}_0)| \leq C$ for some $C > 0$. It follows from compactness of the set

$$\{\, \mathbf{x} \in \mathbb{R}^n : \|\mathbf{x}\| \leq C \,\}$$

that there must be a subsequence $\phi(n_k; \mathbf{x}_0)$ which converges to a point $\mathbf{z}$ in $\mathbb{R}^n$. This point $\mathbf{z}$ must be an ω-limit point of $\mathbf{z}_0$. See a book on real analysis such as [**Lay01**], [**Lew93**], or [**Mar93**] for discussion of compactness and convergence of a bounded sequence in $\mathbb{R}^n$.

(f) Assume that $\omega(\mathbf{x}_0)$ is not connected and there are two pieces. Then there are two disjoint open sets $\mathbf{U}_1$ and $\mathbf{U}_2$ such that $\omega(\mathbf{x}_0) \subset \mathbf{U}_1 \cup \mathbf{U}_2$ and $\omega(\mathbf{x}_0) \cap \mathbf{U}_j \neq \emptyset$ for $j = 1, 2$. (The intuition is that $\mathbf{U}_1$ and $\mathbf{U}_2$ are a positive distance apart, but this is not necessarily the case. We just assume both are open so that a curve going from one set to the other must pass through points outside of both sets.) Then, the trajectory must come close to points in both $\mathbf{U}_1$ and $\mathbf{U}_2$ for arbitrarily large times, so there must be times t_n going to infinity such that $\phi(t_n; \mathbf{x}_0)$ is in neither $\mathbf{U}_1$ nor $\mathbf{U}_2$. Since the trajectory is bounded, there must be a subsequence t_{n_k} such that $\phi(t_{n_k}; \mathbf{x}_0)$ converges to a point $\mathbf{z}$ that is in neither $\mathbf{U}_1$ nor $\mathbf{U}_2$. But this contradicts the fact that $\omega(\mathbf{x}_0) \subset \mathbf{U}_1 \cup \mathbf{U}_2$. □

Stability of fixed points

In this section we use the norm of a matrix introduced in Section 2.5 to prove that, if all the eigenvalues of $\mathbf{A}$ have negative real parts, then the origin is attracting at an exponential rate determined by the eigenvalues. We start with an example, which illustrates the fact that the Euclidean norm is not always strictly decreasing.

4.8. Theory and Proofs

Example 4.29. Consider the system of linear differential equations with constant coefficients given by
$$\dot{x} = -x - 9y,$$
$$\dot{y} = x - y.$$
The eigenvalues are $-1 \pm 3i$, and the general solution is given by
$$\begin{pmatrix} x(t) \\ y(t) \end{pmatrix} = e^{-t} \begin{pmatrix} \cos(3t) & -3\sin(3t) \\ \frac{1}{3}\sin(3t) & \cos(3t) \end{pmatrix} \begin{pmatrix} x_0 \\ y_0 \end{pmatrix}.$$
The Euclidean norm is not strictly decreasing: its derivative is given by
$$\frac{d}{dt}(x^2 + y^2)^{1/2} = \frac{1}{2}(x^2 + y^2)^{-1/2}(2x\dot{x} + 2y\dot{y})$$
$$= (x^2 + y^2)^{-1/2}(-x^2 - y^2 - 8xy).$$
Along the line $y = -x$, the derivative is positive and the norm is increasing, while along the line $y = x$ the derivative is negative and the norm is decreasing. Therefore, the Euclidean norm does not go to zero monotonically. However, the length of the solution does go to zero exponentially:
$$\left\| \begin{pmatrix} x(t) \\ y(t) \end{pmatrix} \right\| \leq e^{-t} \left\| \begin{pmatrix} \cos(3t) & -3\sin(3t) \\ \frac{1}{3}\sin(3t) & \cos(3t) \end{pmatrix} \right\| \cdot \left\| \begin{pmatrix} x_0 \\ y_0 \end{pmatrix} \right\|$$
$$\leq 3 e^{-t} \left\| \begin{pmatrix} x_0 \\ y_0 \end{pmatrix} \right\|.$$
Here, we have used the fact that the length of a matrix times a vector is less than or equal to the norm of the matrix time the length of the vector. The norm of a matrix is briefly discussed in Appendix A.3. The exact constant in the preceding inequality is not very important. The entries of the matrix are periodic, so the norm has a maximum. In fact, the maximum can be taken to be three. Thus, the Euclidean norm goes to zero at a rate determined by the real parts of the eigenvalues $-1 \pm 3i$, with a coefficient determined by the eigenvectors.

The next theorem gives the general result.

Theorem 4.10. *Let $\mathbf{A}$ be a real $n \times n$ matrix, and consider the differential equation $\dot{\mathbf{x}} = \mathbf{A}\mathbf{x}$. The following conditions are equivalent:*

(i) There exist constants $a > 0$ and $C \geq 1$ such that, for an initial condition $\mathbf{x}_0$ in $\mathbb{R}^n$, the solution satisfies
$$\|e^{\mathbf{A}t}\mathbf{x}_0\| \leq Ce^{-at}\|\mathbf{x}_0\|$$
for all $t \geq 0$.

(ii) The real parts of all the eigenvalues of $\mathbf{A}$ are negative.

Proof. First, we show that (i) implies (ii). Suppose that (ii) is not true and one of the eigenvalues $\lambda = \alpha + i\beta$ has a nonnegative real part $\alpha \geq 0$. (If the eigenvalue is real, then $\beta = 0$.) The solution method shows there is a solution of the form
$$e^{\alpha t}\bigl(\sin(\beta t)\mathbf{u} + \cos(\beta t)\mathbf{w}\bigr).$$
Since $\alpha \geq 0$, this solution does not go to zero, which contradicts (i). This shows that if (ii) is false, then (i) is false, or if (i) is true, then (ii) is true.

Next, we show the converse, that (ii) implies (i). Let $\mathbf{v}^1, \ldots, \mathbf{v}^n$ be a basis of generalized eigenvectors, each of length one (unit vectors). We let $\lambda_j = \alpha_j + i\beta_j$ be the eigenvector associated with $\mathbf{v}^j$. Then, the solutions $e^{\mathbf{A}t}\mathbf{v}^j$ involve terms dependent on t, such as $t^k e^{\alpha_j t}\cos(\beta_j t)$ and $t^k e^{\alpha_j t}\sin(\beta_j t)$ for some k less than the multiplicity of the eigenvalue for $\mathbf{v}^j$, so $0 \le k \le n$. Assume $\operatorname{Re}(\lambda) < -a < 0$ for all the eigenvalues λ of $\mathbf{A}$. Let $\epsilon > 0$ be such that $\operatorname{Re}(\lambda) < -a - \epsilon$ for all the eigenvalues λ of $\mathbf{A}$. By the form of the solutions, there is $C'_j \ge 1$ such that

$$\begin{aligned}\|e^{\mathbf{A}t}\mathbf{v}^j\| &\le (1+t^n)e^{\alpha_j t}C'_j\|\mathbf{v}^j\| \\ &\le (1+t^n)e^{-at-\epsilon t}C'_j\|\mathbf{v}^j\| \\ &\le e^{-at}(t^n e^{-\epsilon t})C'_j\|\mathbf{v}^j\| \\ &\le e^{-at}C_j\|\mathbf{v}^j\|\end{aligned}$$

where $(t^n e^{-\epsilon t})C'_j \le C_j$ for all $t \ge 0$. Any initial condition $\mathbf{x}_0$ can be written as a combination of the $\mathbf{v}^j$, $\mathbf{x}_0 = \sum_{j=1}^n y_j \mathbf{v}^j$, so the solution is

$$e^{\mathbf{A}t}\mathbf{x}_0 = \sum_{j=1}^n y_j e^{\mathbf{A}t}\mathbf{v}^j$$

and

$$\|e^{\mathbf{A}t}\mathbf{x}_0\| \le e^{-at}\sum_{j=1}^n |y_j|C_j\|\mathbf{v}^j\|.$$

Letting $C' = \max\{C_j\}$,

$$\|e^{\mathbf{A}t}\mathbf{x}_0\| \le e^{-at}C'\sum_{j=1}^n |y_j|\cdot\|\mathbf{v}^j\| \le e^{-at}C'\sum_{j=1}^n |y_j|$$

$$\le e^{-at}C'n\max\{|y_j|:1\le j\le n\} \le e^{-at}C'n\sqrt{\sum_{j=1}^n |y_j|^2}.$$

Let $\mathbf{V}$ be the matrix whose columns are the vectors $\mathbf{v}^j$ and let $\mathbf{y}$ be the vector with entries y_j. Then,

$$\mathbf{x}_0 = \sum_{j=1}^n y_j \mathbf{v}^j = \mathbf{V}\mathbf{y},$$

$$\mathbf{y} = \mathbf{V}^{-1}\mathbf{x}_0,$$

$$\|\mathbf{y}\| = \sqrt{\sum_{j=1}^n |y_j|^2} \le \|\mathbf{V}^{-1}\|\cdot\|\mathbf{x}_0\|.$$

Combining with the preceding inequality, we have

$$\begin{aligned}\|e^{\mathbf{A}t}\mathbf{x}_0\| &\le e^{-at}C'n\|\mathbf{V}^{-1}\|\cdot\|\mathbf{x}_0\| \\ &= Ce^{-at}\|\mathbf{x}_0\|,\end{aligned}$$

where $C = C'n\|\mathbf{V}^{-1}\|$. $\square$

One-dimensional differential equations

Restatement of Theorem 4.4. *Consider a differential equation $\dot{x} = f(x)$ on $\mathbb{R}$, for which $f(x)$ has a continuous derivative. Assume that $x(t) = \phi(t; x_0)$ is the solution, with initial condition x_0. Assume that the maximum interval containing 0 for which it can be defined is (t^-, t^+).*

a. *Further assume that the solution $\phi(t; x_0)$ is bounded for $0 \le t < t^+$ (i.e., there is a constant $C > 0$ such that $|\phi(t; x_0)| \le C$ for $0 \le t < t^+$). Then, $\phi(t; x_0)$ must converge either to a fixed point or to a point where $f(x)$ is undefined as t converges to t^+.*

b. *Similarly, if the solution $\phi(t; x_0)$ is bounded for $t^- < t \le 0$, then $\phi(t; x_0)$ must converge either to a fixed point or to a point where $f(x)$ is undefined as t converges to t^-.*

c. *Assume that $f(x)$ is defined for all x in $\mathbb{R}$. (i) If $f(x_0) > 0$, assume that there is a fixed point $x^* > x_0$, and in fact, let x^* be the smallest fixed point larger than x_0. (ii) If $f'(x_0) < 0$, assume that there is a fixed point $x^* < x_0$, and in fact, let x^* be the largest fixed point less than x_0. Then, $t^+ = \infty$ and $\phi(t; x_0)$ converges to x^* as t goes to infinity.*

Proof. This theorem states that a bounded solution for a differential equation on the real line must converge to either a fixed point or to a point where $f(x)$ is undefined as t converges to either t^+ or t^-.

We give a proof which treats both parts (a) and (c). The proof of part (b) is similar. By the uniqueness of solutions, the solution cannot cross points where $f(x) = 0$. Thus, the solution $\phi(t; x_0)$ must stay in a region where $f(x)$ has one sign. Assume that $f(\phi(t; x_0)) > 0$. Thus, $\phi(t; x_0)$ is increasing and is bounded above. By properties of the real numbers, it must converge to a limiting value as t converges to t^+, the supremum of times for which the flow is defined. (The value $t^+ = \infty$ if $f(x)$ is defined everywhere on the line.) Call this limiting value x_∞. Because $\phi(t; x_0)$ converges to x_∞ and $f(\phi(t; x_0)) > 0$, $f(x_\infty) \ge 0$ or is undefined. If either $f(x_\infty) = 0$ or $f(x_\infty)$ is undefined, we are done. Otherwise, if $f(x_\infty) > 0$, then $\phi(t; x_\infty)$ goes from $x_- = \phi(-\delta; x_\infty) < x_\infty$ to $x_+ = \phi(\delta; x_\infty) > x_\infty$ as t goes from $-\delta$ to δ. But there is some time t_1 such that $x_- < \phi(t_1; x_0) < x_\infty$. Then, $x_\infty < x_+ = \phi(2\delta; x_-) < \phi(2\delta + t_1; x_0)$, so the solution starting at x_0 goes past x_∞. This contradicts the fact that $f(x_\infty) > 0$ and we are done. □

A very similar proof shows the next lemma.

Lemma 4.11. *Assume that, for all $t \ge 0$, $g(t)$ is defined, $g(t)$ is bounded, and $g'(t) = \frac{d}{dt} g(t) > 0$.*

a. *Then, $g(t)$ approaches a limiting value as t goes to infinity.*

b. *If $g'(t)$ is also uniformly continuous, then $g'(t)$ approaches 0 as t goes to infinity.*

Proof. (a) By the properties of the real numbers, the monotone values, $g(t)$, must approach a limiting value g_∞ as t goes to infinity.

(b) Since it is increasing, $g(t) < g_\infty$ for all $t \geq 0$. The $\liminf_{t\to\infty} g'(t) \geq 0$ because $g'(t) > 0$ for all t. If $\limsup_{t\to\infty} g'(t) = L > 0$, then there is a sequence of times t_n such that $g'(t_n) \geq L/2$. By uniform continuity, there is a $\delta > 0$ such that $g'(t) \geq L/4$ for $t_n \leq t \leq t_n + \delta$, so $g(t_n + \delta) \geq g(t_n) + \delta L/4$. For sufficiently large n, $g(t_n) \geq g_\infty - \delta L/8$, so

$$g(t_n + \delta) \geq g(t_n) + \delta L/4$$
$$\geq g_\infty - \delta L/8 + \delta L/4$$
$$= g_\infty + \delta L/8$$
$$> g_\infty.$$

This contradicts the fact that $g(t) < g_\infty$ for all $t \geq 0$. □

Restatement of Theorem 4.5. *Consider a fixed point x^* for the differential equation $\dot{x} = f(x)$ on $\mathbb{R}$, where f and f' are continuous.*

   a. *If $f'(x^*) < 0$, then x^* is an attracting fixed point.*
   b. *If $f'(x^*) > 0$, then x^* is a repelling fixed point.*
   c. *If $f'(x^*) = 0$, then the derivative does not determine the stability type.*

Proof. (a) If $f'(x^*) < 0$, then there is an interval about x^*, $(x^* - \delta, x^* + \delta)$, such that $f'(x) < 0$ for all x in this interval. Then, for any x with $x^* - \delta < x < x^* + \delta$, $f(x) = f(x^*) + f'(z_x)x = f'(z_x)x$ for some z_x between x^* and x. Therefore, $f(x) > 0$ for $x^* - \delta < x < x^*$ and $f(x) > 0$ for $x^* < x < x^* + \delta$.

Now take an initial condition x_0 with $x^* - \delta < x_0 < x^*$. The solution $\phi(t; x_0)$ is increasing because $f(x) > 0$. By the uniqueness given in Section 3.1, the solution cannot cross the point x^*. Therefore, the solution must stay in the interval $(x^* - \delta, x^*)$, where $f'(x) > 0$. The solution must be increasing and must converge to x^*. This shows that x^* is attracting from the left. A similar argument shows it is attracting from the right and so it is an attracting fixed point.

(b) The proof for $f'(x^*) > 0$ is similar to the preceding proof and is left to the reader.

(c) The examples of Section 4.3 show that, if $f'(x^*) = 0$, then the fixed point can be attracting, repelling, or attracting from one side or can have other possibilities. Therefore, the derivative does not determine the stability type. □

Linearized stability of fixed points

In this subsection we prove that a fixed point whose eigenvalues all have negative real parts is asymptotically stable.

Restatement of Theorem 4.6a. *Let $\mathbf{x}^*$ be a fixed point for $\dot{\mathbf{x}} = \mathbf{F}(\mathbf{x})$, where $\mathbf{F}$ is C^2. (The theorem is actually true for $\mathbf{F}$ is C^1 but the proof uses C^2.) Assume all the eigenvalues of $D\mathbf{F}_{(\mathbf{x}^*)}$ have negative real part. Then $\mathbf{x}^*$ is asymptotically stable.*

Proof. We consider a Taylor expansion of $\mathbf{F}$ about $\mathbf{x}^*$. Let $\mathbf{A} = D\mathbf{F}_{(\mathbf{x}^*)}$ and $\mathbf{y} = \mathbf{x} - \mathbf{x}^*$, the displacement away from the fixed point. Since $\mathbf{F}(\mathbf{x}^*) = \mathbf{0}$,

$$\dot{\mathbf{y}} = \mathbf{A}\mathbf{y} + \mathbf{g}(\mathbf{y}),$$

where $\|\mathbf{g}(\mathbf{y})\| \leq M\|\mathbf{y}\|^2$ for $\|\mathbf{y}\| \leq \delta_0$. Along a solution $\mathbf{y}(t)$, we can think of
$$\mathbf{G}(t) = \mathbf{g}(\mathbf{y}(t))$$
as known. So we get the time-dependent linear equation,
$$\dot{\mathbf{y}} = \mathbf{A}\mathbf{y} + \mathbf{G}(t),$$
which has a solution
$$\mathbf{y}(t) = e^{\mathbf{A}t}\mathbf{y}_0 + \int_0^t e^{\mathbf{A}(t-s)}\mathbf{G}(s)\,ds.$$
For the matrix $e^{\mathbf{A}t}$ there is a scalar $K \geq 1$ such that $\|e^{\mathbf{A}t}\| \leq Ke^{-\alpha t}$, so
$$\|e^{\mathbf{A}t}\mathbf{y}\| \leq \|e^{\mathbf{A}t}\| \cdot \|\mathbf{y}\| \leq K\,e^{-\alpha t}\|\mathbf{y}\|.$$
If $\|\mathbf{y}\|$ is sufficiently small, then $M\|\mathbf{y}\|^2 \leq m\|\mathbf{y}\|$, where m is small. So,
$$\|\mathbf{y}(t)\| \leq K\,e^{-\alpha t}\|\mathbf{y}_0\| + \int_0^t K\,e^{-\alpha(t-s)}\,M\|\mathbf{y}(s)\|^2\,ds$$
$$\leq K\,e^{-\alpha t}\|\mathbf{y}_0\| + \int_0^t K\,e^{-\alpha(t-s)}\,m\|\mathbf{y}(s)\|\,ds,$$
and multiplying by $e^{\alpha t}$, we obtain
$$e^{\alpha t}\|\mathbf{y}(t)\| \leq K\,\|\mathbf{y}_0\| + \int_0^t mK\,e^{\alpha s}\|\mathbf{y}(s)\|\,ds.$$
Considering $\phi(t) = e^{\alpha t}\|\mathbf{y}(t)\|$, this is like a differential inequality $\phi'(t) \leq mK\phi(t)$, with $\phi(0) = K\|\mathbf{y}_0\|$, so it is reasonable that $\phi(t) \leq \phi(0)e^{mKt} = K\|\mathbf{y}_0\|e^{mKt}$. Indeed, by Gronwall's inequality, this is true, so
$$e^{\alpha t}\|\mathbf{y}(t)\| \leq K\|\mathbf{y}_0\|e^{mKt} \quad \text{and}$$
$$\|\mathbf{y}(t)\| \leq K\|\mathbf{y}_0\|e^{(mK-\alpha)t}.$$
If $\|\mathbf{y}_0\| \leq \delta_0/K$ is sufficiently small and m is sufficiently small so that $mK - \alpha < 0$, then
$$\|\mathbf{y}(t)\| \leq K\|\mathbf{y}_0\|e^{(mK-\alpha)t} \leq K\,\|\mathbf{y}_0\| \leq \delta_0,$$
and $\|\mathbf{y}(t)\|$ goes to zero exponentially. Also, it is clearly L-stable, because for any $\epsilon > 0$, if
$$\|\mathbf{y}_0\| \leq \delta = \min\{\epsilon/K, \delta_0/K\},$$
then the solution stays less than both ϵ and δ_0 for all $t \geq 0$. $\square$

We have said that a saddle point has an unstable manifold; we have not proved this since it is not easy to do. See [**Rob99**] for a proof. We can prove that a saddle point is unstable, which we do in the next theorem.

Theorem 4.12. *Assume that $\mathbf{x}^*$ is a saddle fixed point for a system of differential equations in the plane. Then, $\mathbf{x}^*$ is unstable.*

Remark 4.30. A similar result is true for an unstable fixed point. We leave this result as an exercise. Also, the result is true in any dimension. We restrict to two variables just to make the notation and details simpler.

Proof. By changing the variable from $\mathbf{x}$ to $\mathbf{u} = \mathbf{x} - \mathbf{x}^*$, we can assume that the fixed point is at the origin. A saddle fixed point in two dimensions has real eigenvalues. We can also use the coordinates from the eigenvectors to put the equations in the following form:

$$\dot{x} = a\,x + R(x, y),$$
$$\dot{y} = -b\,y + S(x, y),$$

where $a, b > 0$. Let $m = \min\{a, b\}$. We also assume that $R(x, y)$ and $S(x, y)$ have zero partial derivatives at $(0, 0)$: they contain quadratic terms and higher, so we can find constants $C > 0$ and $\delta > 0$ such that

$$|R(x, y)| + |S(x, y)| \leq C(x^2 + y^2)$$

for $\|(x, y)\| \leq \delta$.

To show that trajectories move away from the fixed point, we use the test function

$$L(x, y) = \frac{x^2 - y^2}{2}.$$

If $L(x, y) \geq 0$ (i.e., $|y| \leq |x|$) and $\|(x, y)\| \leq \delta$, we have

$$\begin{aligned}
\dot{L} &= x\,\dot{x} - y\,\dot{y} \\
&= x\,(a\,x + R(x, y)) - (-b\,y + S(x, y)) \\
&= a\,x^2 + b\,y^2 + x\,R(x, y) - y\,S(x, y) \\
&\geq m\,(x^2 + y^2) - |x|\,|R(x, y)| - |y|\,|S(x, y)| \\
&\geq m\,(x^2 + y^2) - |x|\,(|R(x, y)| + |S(x, y)|) \\
&\geq m\,(x^2 + y^2) - |x|\,C\,(x^2 + y^2) \\
&= (m - C\,|x|)\,(x^2 + y^2) \\
&\geq \frac{m}{2}(x^2 + y^2) \\
&> 0
\end{aligned}$$

for $|x| \leq m/(2C)$. Since this derivative is positive, a trajectory $\phi(t; (x_0, y_0))$ stays in the region where $L(x, y) \geq 0$ until either $|x(t)| = m/(2C)$ or $\|\phi(t; (x_0, y_0))\| = \delta$. Thus, we can find initial conditions arbitrarily near the fixed point which move to a distance away of at least $\min\{\delta, m/(2C)\}$. This proves that the fixed point is unstable. □

Conjugacy of phase portraits

To a large extent, the linearized equations at a fixed point determine the features of the phase portrait near the fixed point. In particular, two differential equations are conjugate provided there is a change of coordinates which takes the trajectories of one system into the trajectories of the other, preserving time. Sometimes the change of coordinates preserves the sense of direction along the trajectories, but not the time: Two systems related in this way are called topologically equivalent. In this section, we make these ideas more precise; in particular, we need to be more specific about how many derivatives the change of coordinates has. The

4.8. Theory and Proofs

results in this section have proofs beyond the scope of this text. We merely state the theorems and provide references for the proofs.

Definition 4.31. For an integer $r \geq 1$, a system of nonlinear differential equations $\dot{\mathbf{x}} = \mathbf{F}(\mathbf{x})$ is said to be C^r provided $\mathbf{F}$ is continuous and all partial derivatives up to order r exist and are continuous. If partial derivatives of all orders exist (i.e., $\mathbf{F}$ is C^r for all r), then it is called C^∞.

Definition 4.32. Let $\mathbf{U}$ and $\mathbf{V}$ be two open sets in $\mathbb{R}^n$. A *homeomorphism* $\mathbf{h}$ from $\mathbf{U}$ to $\mathbf{V}$ is a continuous map onto $\mathbf{V}$, which has a continuous inverse $\mathbf{k}$ from $\mathbf{V}$ to $\mathbf{U}$ (i.e., $\mathbf{k} \circ \mathbf{h}$ is the identity on $\mathbf{U}$ and $\mathbf{k} \circ \mathbf{h}$ is the identity on $\mathbf{V}$). It follows that $\mathbf{h}$ is one to one (i.e., if $\mathbf{h}(\mathbf{x}_1) = \mathbf{h}(\mathbf{x}_2)$, then $\mathbf{x}_1 = \mathbf{x}_2$).

For an integer $r \geq 1$, a C^r *diffeomorphism* $\mathbf{h}$ from $\mathbf{U}$ to $\mathbf{V}$ is a homeomorphism from $\mathbf{U}$ to $\mathbf{V}$ such that $\mathbf{h}$ and its inverse $\mathbf{k}$ each have all partial derivatives up to order r and they are continuous. It follows that $\det(D\mathbf{h}_{(\mathbf{x})}) \neq 0$ at all points $\mathbf{x}$.

A homeomorphism (respectively, diffeomorphism) can be considered a continuous (respectively, differentiable) change of coordinates.

Definition 4.33. Two flows $\phi(t;\cdot)$ and $\psi(t;\cdot)$ are called *topologically conjugate* on open sets $\mathbf{U}$ and $\mathbf{V}$ provided there is a homeomorphism $\mathbf{h}$ from $\mathbf{U}$ to $\mathbf{V}$ such that
$$\mathbf{h} \circ \phi(t;\mathbf{x}_0) = \psi(t;\mathbf{h}(\mathbf{x}_0)).$$
Such a map $\mathbf{h}$ is a continuous change of coordinates.

These two flows are called *topologically equivalent* provided that there is a homeomorphism $\mathbf{h}$ from $\mathbf{U}$ to $\mathbf{V}$ which takes the trajectories of $\phi(t;\cdot)$ into the trajectories of $\psi(t;\cdot)$, preserving the sense of direction of time, but not necessarily the exact same time.

For an integer $r \geq 1$, these flows are called C^r *conjugate* on open sets $\mathbf{U}$ and $\mathbf{V}$ provided there is a C^r diffeomorphism $\mathbf{h}$ from $\mathbf{U}$ to $\mathbf{V}$ such that
$$\mathbf{h} \circ \phi(t;\mathbf{x}_0) = \psi(t;\mathbf{h}(\mathbf{x}_0)).$$
Such a map $\mathbf{h}$ is a differentiable change of coordinates.

These two flows are called C^r *equivalent* on open sets $\mathbf{U}$ and $\mathbf{V}$ provided that there is a C^r diffeomorphism $\mathbf{h}$ from $\mathbf{U}$ to $\mathbf{V}$, which takes the trajectories of $\phi(t;\cdot)$ into the trajectories of $\psi(t;\cdot)$, preserving the sense of direction of time.

As we state shortly, there is a general theorem called the Grobman–Hartman theorem that states that a nonlinear system with a hyperbolic fixed point is topologically conjugate to its linearized flow. Unfortunately, this topological conjugacy does not tell us much about the features of the phase portrait. Any two linear systems in the same dimension for which the origin is asymptotically stable are topologically conjugate on all of $\mathbb{R}^n$. See [**Rob99**]. In particular, a linear system for which the origin is a stable focus is topologically conjugate to one for which the origin is a stable node. Thus, a continuous change of coordinates does not preserve the property that trajectories spiral in toward a fixed point or do not spiral. A differentiable change of coordinates does preserve such features. Unfortunately, the theorems that imply the existence of a differentiable change of coordinates require more assumptions.

We now state these results.

Theorem 4.13 (Grobman–Hartman). *Let $\dot{\mathbf{x}} = \mathbf{F}(\mathbf{x})$ be a C^1 nonlinear system of differential equations that has a hyperbolic fixed point $\mathbf{x}^*$ and flow $\phi(t;\cdot)$. Let $\mathbf{A} = D\mathbf{F}_{(\mathbf{x}^*)}$ be the matrix of partial derivatives, and $\psi(t;\mathbf{x}_0) = e^{\mathbf{A}t}\mathbf{x}_0$ the flow of the linearized system of equations, with fixed point at the origin. Then, there are open sets $\mathbf{U}$ containing $\mathbf{x}^*$ and $\mathbf{V}$ containing the origin, such that the flow $\phi(t;\cdot)$ on $\mathbf{U}$ is topologically conjugate to the flow $\psi(t;\cdot)$ on $\mathbf{V}$.*

See [**Rob99**] for a proof, since the proof of this theorem is more involved than we want to pursue.

To get a differentiable change of coordinates, we need to treat special cases or add more assumptions. A result with more derivatives was shown by Sternberg, but requires a "nonresonance" requirement on the eigenvalues.

Theorem 4.14 (Sternberg). *Let $\dot{\mathbf{x}} = \mathbf{F}(\mathbf{x})$ be a C^∞ system of differential equations with a hyperbolic fixed point $\mathbf{x}^*$ and flow $\phi(t;\cdot)$. Let $\lambda_1,\ldots,\lambda_n$ be the eigenvalues at the fixed point and assume that each λ_j has algebraic multiplicity one. Moreover, assume that*
$$m_1\lambda_1 + m_2\lambda_2 + \cdots + m_n\lambda_n \neq \lambda_k,$$
for any k and any nonnegative integers m_j with $\sum_j m_j \geq 2$. (This condition is called nonresonance of the eigenvalues.) Then, there exist open sets $\mathbf{U}$ containing $\mathbf{x}^*$ and $\mathbf{V}$ containing the origin, such that the flow $\phi(t;\cdot)$ on $\mathbf{U}$ is C^∞ conjugate to the flow for the linearized system on $\mathbf{V}$.

The proof of this theorem is much harder than the one for the Grobman–Hartman theorem. See [**Har82**] for a proof. It also requires the nonresonance of the eigenvalues. The following two theorems give results about a C^1 conjugacy without the nonresonance of the eigenvalues.

Theorem 4.15 (Hartman). *Let $\dot{\mathbf{x}} = \mathbf{F}(\mathbf{x})$ be a C^2 nonlinear system of differential equations with an asymptotically stable fixed point $\mathbf{x}^*$ and flow $\phi(t;\cdot)$. Then, there exist open sets $\mathbf{U}$ containing $\mathbf{x}^*$ and $\mathbf{V}$ containing the origin, such that the flow $\phi(t;\cdot)$ on $\mathbf{U}$ is C^1 conjugate to the flow for the linearized system on $\mathbf{V}$.*

Theorem 4.16 (Belickii). *Let $\dot{\mathbf{x}} = \mathbf{F}(\mathbf{x})$ be a C^2 system of differential equations with a hyperbolic fixed point $\mathbf{x}^*$ and flow $\phi(t;\cdot)$. Assume that the eigenvalues satisfy a nonresonance assumption that says that $\lambda_k \neq \lambda_i + \lambda_j$ for any three eigenvalues. Then, there exist open sets $\mathbf{U}$ containing $\mathbf{x}^*$ and $\mathbf{V}$ containing the origin, such that the flow $\phi(t;\cdot)$ on $\mathbf{U}$ is C^1 conjugate to the flow for the linearized system on $\mathbf{V}$.*

See [**Bel72**].

Chapter 5

Phase Portraits Using Scalar Functions

This chapter continues our consideration of phase portraits for systems of nonlinear differential equations. In the previous chapter, the systems considered could be analyzed using nullclines and the linearization at the fixed points. Those considered in this chapter are analyzed using a real-valued function, an "energy function". The first section treats the predator–prey model for two populations; this continues our treatment of Lotka–Volterra equations, but for the case in which one population has a positive effect on the growth of the other. In Section 5.2, we consider nonlinear oscillators such as the pendulum, for which we can use energy to analyze the phase plane. When friction is added to these nonlinear oscillators, the energy decreases. The energy in these situations is an example of what is called a Lyapunov function. The Lyapunov function can be used in certain classes of problems to determine the set of initial conditions which have trajectories tending to a fixed point. Section 5.4 uses a real-valued function (test function) to determine the limit sets of some more complicated examples. These examples are used in the next few chapters to illustrate concepts which are introduced. Finally, for a system given as the gradient of a function G, the function G is strictly decreasing along trajectories not at a fixed point, so any trajectory that is bounded tends to a fixed point.

5.1. Predator–Prey Systems

In Section 4.6, we introduced the Lotka–Volterra equations, but looked mainly at the case of two competitive populations. In this section, we consider two populations for which the first population has a positive effect on the second population, but the second population has a negative effect on the first. This situation is called a *predator–prey system*. The simplest system of equations of this type is given by

$$\dot{x} = x(a - by),$$
$$\dot{y} = y(-c + ex),$$

with all parameters positive, $a, b, c, e > 0$. Notice that x is the prey population and y is the predator population. We have not included any limit to growth or any possible negative effect either population has on its own growth rate. Without any of the predators around, the prey grows at a linear positive rate of a. As the predator population increases, the growth rate of x decreases linearly in y. Without any prey, the predator population has a negative growth rate of $-c$. As the prey population increases, the growth rate of y increases linearly in x.

These equations have fixed points at $(0,0)$ and $(c/e, a/b)$. Analyzing the signs of $\dot{x}$ and $\dot{y}$ in the regions separated by the nullclines leads to the possibility of cycles, or periodic orbits. The matrix of partial derivatives at $(c/e, a/b)$ is

$$\begin{pmatrix} a - by & -bx \\ ey & ex - c \end{pmatrix}_{x=c/e, y=a/b} = \begin{pmatrix} 0 & -\dfrac{bc}{e} \\ \dfrac{ae}{b} & 0 \end{pmatrix}.$$

This matrix has characteristic equation $\lambda^2 + ac = 0$, and the eigenvalues $\pm i\sqrt{ac}$ are purely imaginary. Thus, $(c/e, a/b)$ is an elliptic center.

For this system, there is a way to show that the orbits close when they go once around the fixed point. Notice that

$$\frac{\dot{y}}{\dot{x}} = \frac{y(-c + ex)}{x(a - by)},$$

so we can separate variables

$$\left(\frac{a}{y} - b\right)\dot{y} = \left(-\frac{c}{x} + e\right)\dot{x}.$$

This equality can be integrated with respect to t to yield

$$K + a\ln(y) - by = -c\ln(x) + ex \qquad \text{or}$$

$$K = \bigl(ex - c\ln(x)\bigr) + \bigl(by - a\ln(y)\bigr),$$

where K is an arbitrary constant. Thus, K is a conserved quantity that can be considered as a function of x and y on the phase plane. This function is the sum of a function of x and a function of y,

$$f(x) = ex - c\ln(x) \qquad \text{and} \qquad g(y) = by - a\ln(y).$$

To understand the function f, note that

$$f'(x) = e - \frac{c}{x} \qquad \text{and} \qquad f''(x) = \frac{c}{x^2} > 0.$$

The graph of f is concave up and has a unique minimum at c/e. See Figure 1. Similarly, the graph of $g(y)$ is concave up and has a unique minimum at a/b. When a trajectory starting on the line $y = a/b$ goes around the fixed point and returns to this line at time t_1, the value of $f(x)$ must be the same as when it left, so $x(t_1) = x(0)$ and $y(t_1) = y(0) = a/b$. Therefore, all these solutions are periodic, and the fixed point is not only a linear center, but is also a *nonlinear center* that is surrounded by periodic orbits.

Since the function $f(x) + g(y)$ is a conserved quantity along trajectories, each level set of this function corresponds to a trajectory of the system of differential equations. Thus, to analyze the phase portrait more carefully and more completely,

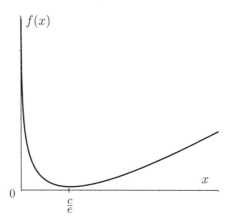

Figure 1. Graph of $f(x)$

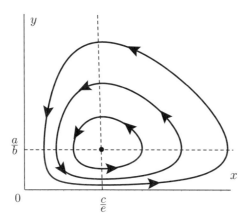

Figure 2. Predator–prey system

we need to analyze the level curves in the first quadrant of the function $f(x)+g(y)$. The minimum value is $K_0 = f(c/e) + g(a/b)$. The level curve $f(x) + g(y) = K_0$ is just the fixed point $(c/e, a/b)$. If $K < K_0$, then level curve $f(x)+g(y) = K$ is empty. Finally, we consider $K > K_0$ and let $K' = K - K_0 > 0$. Therefore, for a given x, we need to find the y that satisfy

$$g(y) - g(a/b) = K' - \left[f(x) - f(c/e)\right].$$

For x with $f(x) - f(c/e) > K'$, the right-hand side is negative and there are no solutions for y. There are two values of x, x_1 and x_2, for which $f(x) - f(c/e) = K'$. For these x_j, the right-hand side is zero and the only solution is $y = a/b$. For $x_1 < x < x_2$, the right-hand side is positive and there are two values of y which satisfy the equations. One of these values of y will be less than a/b and the other will be greater. Therefore, the level curve for $K > f(c/e) + g(a/b)$ is a closed curve that surrounds the fixed point. Compare with the trajectories drawn in Figure 2.

Exercises 5.1

1. Consider the predator–prey system
$$\dot{x} = x(1-y),$$
$$\dot{y} = y\left(-1 + x^{\frac{1}{2}}\right).$$
   a. Find the nullclines and fixed point with $x^* > 0$ and $x^* > 0$.
   b. Show that there is a function $L(x,y)$ that is constant along trajectories and has a minimum at (x^*, y^*). Conclude that all trajectories in the first quadrant other than the fixed point are on closed curves that surround the fixed point.

2. Consider the SIR epidemic model with no births or deaths given by
$$\dot{S} = -\beta SI,$$
$$\dot{I} = \beta SI - \gamma I,$$
   for $S \geq 0$ and $I \geq 0$. (In terms of the equations given in Section 4.7, we are taking $\mu = 0$ and $\alpha = 0$, because there are no deaths or fatalities, and $\mu K = 0$ because there are no births.) We do not include the equation $\dot{R} = \gamma I$ because it can be solved once the solution $I(t)$ is known.
   a. What are the nullclines and fixed points?
   b. Show that there is a real-valued function $G(S,I)$ that is constant along trajectories. Hint: Consider $\dot{I}/\dot{S}$.
   c. Draw the phase portrait for $S \geq 0$ and $I \geq 0$ using the function G. Hint: Write a level curve of G in the form of I as a function of S.

3. Consider the predator–prey system with terms added corresponding to limited growth,
$$\dot{x} = x(a - by - mx),$$
$$\dot{y} = y(-c + ex - ny),$$
   where all the constants are positive, $a > 0$, $b > 0$, $c > 0$, $e > 0$, $m > 0$, and $n > 0$. Also assume that $a/m > c/e$.
   a. Show that there is a fixed point (x^*, y^*) with $x^* > 0$ and $y^* > 0$.
   b. Show that the fixed point (x^*, y^*) is asymptotically stable.
   c. Draw a sketch of the phase portrait for $x \geq 0$ and $y \geq 0$. (You may use the fact that there are no periodic orbits, which is true but not trivial to show. See Example 6.24.)

4. Consider the system of differential equations given by
$$\dot{x} = xy^2,$$
$$\dot{y} = -4yx^2.$$
   Show that there is a real-valued function $G(x,y)$ that is constant along trajectories. Hint: Consider $\dot{y}/\dot{x}$.

5.2. Undamped Forces

In this section, we consider a particle of mass m moving with position given by only one scalar variable, x. To describe its complete state, we need to also specify its velocity, $\dot{x}$. We assume the motion is determined by forces $F(x)$ that depend only on the position and not on the velocity (i.e., there is no damping or friction). Since the mass times the acceleration equals the forces acting on the particle, the differential equation which determines the motion is given by $m\ddot{x} = F(x)$. This equation can be written as a system of differential equations with only first derivatives by setting $y = \dot{x}$,

$$\dot{x} = y,$$
$$\dot{y} = \frac{1}{m} F(x).$$

Such a system is said to have *one degree of freedom*, even though the system of equations has the two variables, position and velocity.

We want to solve for a quantity (energy) that is conserved along trajectories by "integrating" the differential equation. We can multiply both sides of the equation $m\ddot{x} = F(x)$ by $\dot{x}$, obtaining

$$m\ddot{x}\dot{x} = F(x)\dot{x},$$
$$m\dot{x}\ddot{x} - F(x)\dot{x} = 0,$$
$$\frac{d}{dt}\left[\frac{1}{2} m\dot{x}^2 - \int_{x_0}^{x} F(s)\,ds\right] = 0, \quad \text{or}$$
$$\frac{1}{2} m\dot{x}^2 - \int_{x_0}^{x} F(s)\,ds = E,$$

where E is a constant along trajectories. The *potential energy* of the force field is defined to be the integral of the negative of the force field,

$$V(x) = -\int_{x_0}^{x} F(s)\,ds.$$

Therefore,

$$E(x,\dot{x}) = \frac{1}{2} m\dot{x}^2 + V(x)$$

is a constant along solutions. The quantity $E(x,\dot{x})$ is called the *energy* or *total energy*, and is the sum of the potential energy $V(x)$ and the *kinetic energy* $\frac{1}{2}m\dot{x}^2$. Because there is a function that is constant along trajectories, these systems are often called *conservative systems*. We usually consider E a function of the phase plane variables x and y.

In physics, the potential energy is interpreted as the work (or energy) needed to move the particle from the position x_0 to x. The term $\frac{1}{2}my^2$ is the kinetic energy determined by the velocity of the particle. For simplicity, in what follows, we usually take the mass coefficient to be equal to one, or incorporate it into F.

An equation of the form $m\ddot{x} = F(x)$ is often called an *oscillator* because, frequently, the resulting motion is periodic with the particle going back and forth. See Section 5.6.1 for further discussion of the connection with oscillators.

For a real-valued function $L(\mathbf{x})$, we often use the following notation for the level set for which L has the value C:
$$L^{-1}(C) = \{\,\mathbf{x} : L(\mathbf{x}) = C\,\}.$$
If $\mathbf{I}$ is any set of real values (e.g., an interval), then we also denote the set of points taking values in $\mathbf{I}$ by
$$L^{-1}(\mathbf{I}) = \{\,\mathbf{x} : L(\mathbf{x}) \in \mathbf{I}\,\}.$$

Example 5.1 (Double-Well Potential). The first example we consider is called the Duffing equation, whose potential has two minima. Consider the system
$$\dot{x} = y,$$
$$\dot{y} = x - x^3.$$
The fixed points are $x = \pm 1, 0$ for $y = 0$. The matrix of partial derivatives is
$$D\mathbf{F}_{(x,y)} = \begin{pmatrix} 0 & 1 \\ 1 - 3x^2 & 0 \end{pmatrix},$$
whose evaluation at the fixed points equals
$$D\mathbf{F}_{(0,0)} = \begin{pmatrix} 0 & 1 \\ 1 & 0 \end{pmatrix} \quad \text{and} \quad D\mathbf{F}_{(\pm 1,0)} = \begin{pmatrix} 0 & 1 \\ -2 & 0 \end{pmatrix}.$$
Therefore, the fixed point at $(0,0)$ is a saddle and the fixed points at $(\pm 1, 0)$ are centers.

The potential energy is
$$V(x) = \int_0^x -s + s^3\,ds = -\frac{x^2}{2} + \frac{x^4}{4},$$
which has a local maximum at $x = 0$ and minima at $x = \pm 1$. Since the potential function has two minima, it is called a *double-well potential*. The values of the potential energy at the critical points are $V(0) = 0$ and $V(\pm 1) = -1/4$. See the graph of V in Figure 3. The total energy is
$$E(x,y) = -\frac{x^2}{2} + \frac{x^4}{4} + \frac{1}{2}y^2.$$

We consider values of the energy $C_2 > V(0) = 0$, C_1 with $V(\pm 1) = -1/4 < C_1 < 0 = V(0)$, $V_0 = -1/4$, and finally $V_1 = 0 = V(0)$. See Figure 3.

We take $C_2 = 2 > 0$ so the points come out easily; therefore, we consider the level set
$$E^{-1}(2) = \left\{\,(x,y) : E(x,y) = \frac{y^2}{2} + V(x) = 2\,\right\}.$$
Since $y^2/2 \geq 0$, the only allowable x have $V(x) \leq 2$. Solving for the x with $V(x) = 2$, we have
$$2 = -\frac{x^2}{2} + \frac{x^4}{4},$$
$$0 = x^4 - 2x^2 + 8,$$
$$0 = (x^2 - 4)(x^2 + 1),$$

5.2. Undamped Forces   175

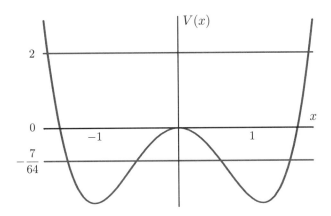

Figure 3. Double-well potential function $V(x)$ for Example 5.1 (not drawn to scale)

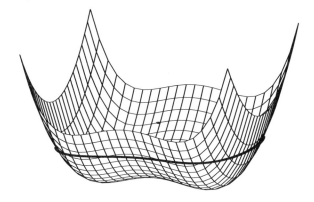

Figure 4. Level curve $E^{-1}(2)$ on three-dimensional graph of energy function for Example 5.1

and x must be ± 2. For $|x| > 2$, $V(x) > 2$ and there are no solutions of $E(x,y) = 2$. For $x = \pm 2$, $V(\pm 2) = 2$, and the only allowable value of y must solve $y^2/2 = 2 - V(\pm 2) = 0$, so y must be zero. For $-2 < x < 2$, $V(x) < 2$, so there are two values of y for which $E(x,y) = 2$, one the negative of the other, $y = \pm[4 - 2V(x)]^{1/2}$. Therefore, the level set $E^{-1}(2) = \{(x,y) : E(x,y) = 2\}$ is a closed curve around the three fixed points and it extends from $x = -2$ to $x = 2$. See Figure 4 for a plot of the level curve on the three-dimensional graph of the energy function $E(x,y)$. See Figure 5 for the level curve in the (x,y)-plane. A similar analysis of $E^{-1}(C_2)$ holds for any $C_2 > 0 = V(0)$ with just a change of the endpoints in the x values.

Next, for a value of C_1 with $V(\pm 1) = -1/4 < C_1 < 0 = V(0)$, we take $C_1 = -7/64 = V(\pm 1/2) = V(\pm\sqrt{7}/2)$. Again, we need $V(x) \leq -7/64$, so

$$-\frac{7}{64} \leq -\frac{x^2}{2} + \frac{x^4}{4},$$
$$0 \geq x^4 - 2x^2 + \frac{7}{16}.$$

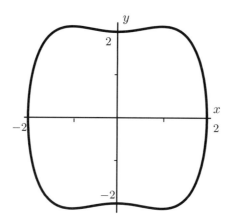

Figure 5. Level curve $E^{-1}(2)$ for Example 5.1

The right-hand side is zero when

$$x^2 = \frac{2 \pm \sqrt{4 - \frac{7}{4}}}{2} = \frac{4 \pm \sqrt{16-7}}{4}.$$

$$= \frac{4 \pm 3}{4} = \frac{7}{4} \text{ and } \frac{1}{4}.$$

Therefore, the allowable x are $-\sqrt{7}/2 \leq x \leq -1/2$ or $1/2 \leq x \leq \sqrt{7}/2$. For the range $-\sqrt{7}/2 \leq x \leq -1/2$, $E^{-1}\left(-7/64\right)$ is a closed curve around $(-1, 0)$, and for the range $1/2 \leq x \leq \sqrt{7}/2$, $E^{-1}(-7/64)$ is a closed curve around $(1, 0)$. See Figure 6 for the level curve $E^{-1}\left(-7/64\right)$ in the (x, y)-plane. Similarly, for any $-1/4 < C_1 < 0$, $E^{-1}(C_1)$ is the union of two closed curves, one around $(-1, 0)$ and the other around $(1, 0)$.

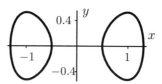

Figure 6. Level curve $E^{-1}\left(-\frac{7}{64}\right)$ for Example 5.1

The level set $E^{-1}\left(-1/4\right)$ is just the two points $(-1, 0)$ and $(1, 0)$.

Finally, we consider the level $E^{-1}(0)$. The level set has values of x with $V(x) \leq 0$,

$$0 \geq \frac{x^4}{4} - \frac{x^2}{2},$$
$$0 \geq x^4 - 2x^2 = x^2(x^2 - 2).$$

Since $x^2 \geq 0$, we need $x^2 \leq 2$, or $-\sqrt{2} \leq x \leq \sqrt{2}$. For $x = 0, \pm\sqrt{2}$, the only value of y is 0. For $-\sqrt{2} < x < 0$ or $0 < x < \sqrt{2}$, there are two values of y. Thus, the level set is a "figure eight", which crosses nontangentially at $x = 0$. See

Figure 7. For a point $\mathbf{p}_0 = (x_0, y_0)$ on $E^{-1}(0)$ with $x_0 \neq 0$, the trajectory $\phi(t; \mathbf{p}_0)$ approaches $(0,0)$ as t goes to both $\pm\infty$. Such an orbit, which is asymptotic to the same fixed point as t goes to both $\pm\infty$, is called a *homoclinic orbit*. Homoclinic roughly means self-inclined; the orbits are called this because they are inclined to the same point both forward and backward in time.

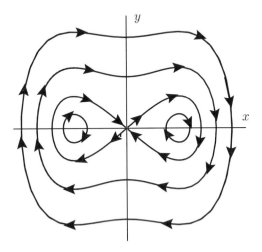

Figure 7. Phase plane for Example 5.1

Thus, the two fixed points $(\pm 1, 0)$ are surrounded by periodic orbits. There are also periodic orbits that surround all three fixed points. The two types of periodic orbits are separated by the level set $E^{-1}(0)$ which is made up of two homoclinic orbits.

This example indicate the steps needed to analyze a system without damping forces. See Table 1 on the next page.

Remark 5.2. Computer programs, such as Maple, Mathematica, or Matlab, can also help with the preceding analysis by using a command to plot the contour curves of the real-valued function E.

Example 5.3 (Pendulum). Consider the system of equations for a pendulum
$$\dot{x} = y,$$
$$\dot{y} = -\sin(x).$$
The potential function is
$$V(x) = \int_0^x \sin(s)\, ds = 1 - \cos(x).$$
The potential function has a minimum at $x = 0$ and multiples of 2π, $V(0) = 0 = V(n\, 2\pi)$. It has maxima at $x = \pm\pi$ and other odd integer multiples of π, $V(\pi) = V(\pi + n\, 2\pi) = 2$. See Figure 8. A direct check shows that the fixed points $(n\, 2\pi, 0)$ are centers, and $(\pi + n\, 2\pi, 0)$ are saddles. Therefore, we need to understand the level sets of the energy for C_1 with $0 < C_1 < 2$, for 2, and for C_2 with $2 < C_2$.

> **Steps to analyze a system with undamped forces**
> 1. Plot the potential function v.
>   a. Determine the local maxima and local minima of V.
>   b. List of all the critical values $V_0 < V_1 < \cdots < V_k$ that are the values attained by $V(x_j)$ at critical points x_j, where $V'(x_j) = 0$. (These include the local maximum, local minimum, and possibly inflection points.)
>   c. Sketch the graph of V.
> 2. Take values C_j with $C_0 < V_0$, $V_{j-1} < C_j < V_j$ for $1 \leq j \leq k-1$, and $C_{k+1} > V_k$.
>   a. For each of these values, determine the allowable range of x-values, $\{x : V(x) \leq C_j\}$, using the graph of V.
>   b. Using the range of allowable x-values for C_j, determine the level set $E^{-1}(C_j)$ in the phase space. Note that if a component of $\{x : V(x) \leq C_j\}$ is a closed interval for one of these values of the energy which is not a critical value, then the corresponding component of $E^{-1}(C_j)$ is a periodic orbit.
> 3. Repeat the last step for the level sets $E^{-1}(V_j)$, $1 \leq j \leq k$.
>   a. If V_0 is an absolute minimum, then the level set $E^{-1}(V_0)$ is just the fixed points with this energy.
>   b. Some of these values correspond to local maxima of V, and so the level set $E^{-1}(V_j)$ contains the stable and unstable manifolds of the saddle fixed point.
>   (i) If a component of $\{x : V(x) \leq V_j\}$ is a closed interval that does not contain any other critical points of V, then the corresponding component of the energy level set $E^{-1}(V_j)$ closes back up on itself, resulting in homoclinic orbits. This is the same as case 2b.
>   (ii) If the potential function has the same value at two local maxima x_1 and x_2 of V, $V(x_1) = V(x_2) = V_j$, and if $V(x) < V_j$ for $x_1 < x < x_2$, then the energy level set $E^{-1}(V_j)$ connects the two saddle fixed points $\mathbf{p}_1 = (x_1, 0)$ and $\mathbf{p}_2 = (x_2, 0)$ with one curve with positive y and one with negative y. Thus, there are two orbits each of whose ω-limit set is one fixed point and whose α-limit set is the other fixed point: such orbits are called *heteroclinic orbits*. See Example 5.3 and Figure 9.
>
> **Table 1**

For a constant $0 < C_1 < 2$, the level set $E^{-1}(C_1)$ in $-\pi < x < \pi$ is a level curve surrounding $(0,0)$. This repeats itself in every strip $-\pi + n\,2\pi < x < \pi + n\,2\pi$. Thus $(0,0)$ is surrounded by periodic orbits.

5.2. Undamped Forces

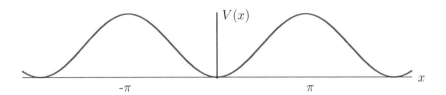

Figure 8. Pendulum: potential energy

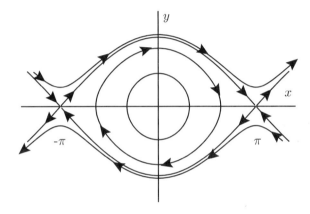

Figure 9. Pendulum: phase plane

The level set $E^{-1}(2)$ in $-\pi \leq x \leq \pi$ is the union of two curves that come together at $(-\pi, 0)$ and $(\pi, 0)$. The trajectory on the top curve is asymptotic to $(\pi, 0)$ as t goes to ∞ and is asymptotic to $(-\pi, 0)$ as t goes to $-\infty$. The bottom orbit is asymptotic to the opposite fixed points as t goes to plus and minus infinity. This picture repeats itself in every strip $-\pi + n2\pi < x < \pi + n2\pi$. Such orbits are called *heteroclinic orbits*, because they tend to different fixed points at plus and minus infinity.

The potential function $V(x) \leq 2$ for all x. Therefore, for $C_2 > 2$, there are two values of y for each x in $E^{-1}(C_2)$. Thus, $E^{-1}(C_2)$ is the union of two curves, one with $y > 0$ and one with $y < 0$. The top trajectory has x increasing to ∞ as t goes to ∞ and the bottom trajectory has x decreasing to $-\infty$ as t goes to ∞. If we consider x as a periodic angular variable, then each of these two trajectories is periodic and each just goes around and around in a single direction; these two different trajectories go around in opposite directions. They are called *rotary solutions*. See Figure 9. As noted in Chapter 1, for the pendulum, there are periodic orbits near the origin and nonperiodic rotary orbits farther away.

Also as noted in Chapter 1, the period of the orbits surrounding the origin depends on the amplitude. The plots of the three solutions, which are time periodic, are plotted in Figure 10. This is also true for Example 5.1.

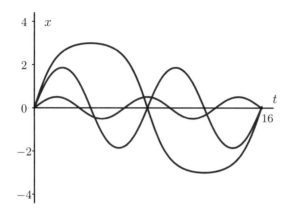

Figure 10. Time plots of solutions with different amplitudes for the pendulum showing the variation of the period

Notice that, for any system of the form

$$\dot{x} = y,$$
$$\dot{y} = F(x),$$

the matrix of partial derivatives at a fixed point $(x^*, 0)$ is

$$\begin{pmatrix} 0 & 1 \\ F'(x^*) & 0 \end{pmatrix} = \begin{pmatrix} 0 & 1 \\ -V''(x^*) & 0 \end{pmatrix},$$

with a characteristic equation $\lambda^2 + V''(x^*) = 0$ and eigenvalues $\lambda = \pm\sqrt{-V''(x^*)}$. Thus, if the point x^* is a local minimum of $V(x)$ with $V''(x^*) > 0$, then the fixed point has purely imaginary eigenvalues and is a center for the linearized system. In fact, the analysis given in this section shows that such a fixed point is surrounded by periodic orbits, and so is a *nonlinear center*. If the point x^* is a local maximum of $V(x)$ with $V''(x^*) < 0$, then the fixed point has one positive and one negative real eigenvalue and is a saddle for both the linearized system and the nonlinear system. Compare with the preceding examples.

Example 5.4. As a final example, consider the system

$$\dot{x} = y,$$
$$\dot{y} = -x - x^2.$$

The potential energy is

$$V(x) = \frac{x^2}{2} + \frac{x^3}{3}$$

and the energy function is

$$E(x,y) = V(x) + \frac{y^2}{2}.$$

The fixed points, which are critical points of the potential function, have $x = 0, -1$. The potential function has $V(0) = 0$ and $V(-1) = 1/6$. Since $V(0) = 0$ is a local minimum of the potential function, it corresponds to a center for the differential

5.2. Undamped Forces

equation; since $V(-1) = 1/6$ is a local maximum, it corresponds to a saddle for the differential equation. See the graph of the potential function in Figure 11.

We need to consider level sets of the energy function for $C_0 < 0 = V(0)$, for C_1 with $V(0) = 0 < C_1 < 1/6 = V(-1)$, and for $C_2 > 1/6$.

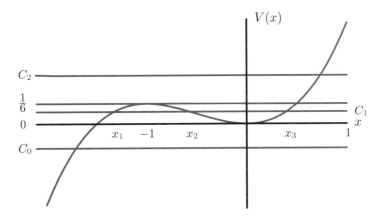

Figure 11. Potential energy function, V, for Example 5.4

First, consider C_1 with $0 < C_1 < 1/6$. There are three points x_1, x_2, and x_3 with $x_1 < -1 < x_2 < 0 < x_3$, such that $V(x) = C_1$. The set

$$\{\, x : V(x) \leq C_1 \,\}$$

has two parts: an interval $(-\infty, x_1]$ from minus infinity up to a value x_1 and the closed interval $[x_2, x_3]$. The corresponding level set $E^{-1}(C_1)$ has only the value of $y = 0$ for $x = x_1, x_2, x_3$, and two values of y for $x < x_1$ and $x_2 < x < x_3$. The part with $x_2 \leq x \leq x_3$ closes up for $E^{-1}(C_1)$ to give a single closed curve surrounding $(0, 0)$. The part with $x \leq x_1$ gives a curve that opens up to the left in the plane to give a trajectory which comes and goes to infinity. See Figure 12.

For $C_2 > 1/6$ and $C_0 < 0$, the sets

$$\{\, x : V(x) \leq C_0 \,\} \quad \text{and} \quad \{\, x : V(x) \leq C_2 \,\}$$

form intervals from minus infinity up to a value $x_j = V^{-1}(C_j)$. The level set $E^{-1}(C_j)$ for $j = 1, 2$ is a single curve that crosses the axis $y = 0$ once at $x = x_j$ and then runs off to $x = -\infty$ as t goes to both $\pm\infty$. See Figure 12.

For the value of $1/6$, the line of constant height $1/6$ is tangent to the potential function at the local maximum at $x = -1$. The potential function equals $1/6$ at $x = -1$ and $1/2$:

$$0 = V(x) - \frac{1}{6} = \frac{1}{6}\left(2x^3 + 3x^2 - 1\right)$$
$$= \frac{1}{6}(x+1)^2(2x-1).$$

For $x = 0.5, -1$, the only y is $y = 0$ on $E^{-1}(1/6)$. For $x < -1$ or $-1 < x < 0.5$, there are two values of y on $E^{-1}(1/6)$. Thus, this level curve $E^{-1}(1/6)$ has a homoclinic orbit to the right of the fixed point $x = -1$; to the left of the fixed

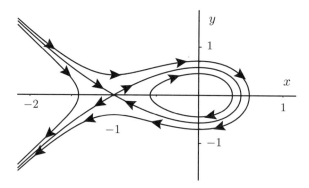

Figure 12. Phase portrait for Example 5.4

point, there is an unbounded orbit on $W^s(-1,0)$ coming in from the left and an unbounded orbit on $W^u(-1,0)$ going off to the left. See Figure 12. Figure 16 in the next section shows the level sets $E^{-1}(0)$ and $E^{-1}(1/6)$.

Exercises 5.2

1. Consider the system of differential equations,

$$\dot{x} = y,$$
$$\dot{y} = -x + x^3.$$

   a. Find the fixed points.
   b. Find the potential function $V(x)$ and draw its graph. Classify the type of each fixed point.
   c. Draw the phase portrait for the system of differential equations.

2. Consider the system of differential equations,

$$\dot{x} = y,$$
$$\dot{y} = -x + x^2.$$

   a. Find the fixed points.
   b. Find the potential function $V(x)$ and draw its graph. Classify the type of each fixed point.
   c. Draw the phase portrait for the system of differential equations.

3. Consider the system of differential equations for a pendulum with constant torque,

$$\dot{x} = y,$$
$$\dot{y} = -\sin(x) + L,$$

   for $0 < L < 1$.

a. Find the fixed points. Hint: You will not be able to give exact numerical values of x at the fixed points, but merely give an expression for them. In which intervals of $[0, \pi/2]$, $[\pi/2, \pi]$, $[\pi, 3\pi/2]$, or $[3\pi/2, 2\pi]$ do they lie?

b. Find the potential function $V(x)$ and draw its graph. Classify the type of each fixed point. Hint: The critical points (with zero derivative) correspond to the fixed points of the differential equation.

c. Draw the phase portrait for the system of differential equations.

d. How do the answers change if $L = 1$ or $L > 1$? How about $0 > L > -1$, $L = -1$, or $L < -1$?

4. Consider the system of differential equations,

$$\dot{x} = y,$$
$$\dot{y} = -x(x-1)(x+2)(x^2-0) = -x^5 - x^4 + 11x^3 + 9x^2 - 18x.$$

a. Find the fixed points.

b. Find the potential function $V(x)$ and draw its graph. Classify the type of each fixed point. Hint: $V(-2) > V(1)$.

c. Draw the phase portrait for the system of differential equations.

5. Consider the system of differential equations,

$$\dot{x} = y,$$
$$\dot{y} = x^3 + x^2 - 2x.$$

a. Find the potential function $V(x)$ and draw its graph.

b. Find the fixed points of the system of differential equations and classify the type of each.

c. Draw the phase portrait for the system of differential equations. Pay special attention to the location of the stable and unstable manifolds of the saddle fixed points.

6. Consider the system of differential equations,

$$\dot{x} = y,$$
$$\dot{y} = -V'(x),$$

where the potential function is given in Figure 13. Draw the phase portrait for the system of differential equations. Pay special attention to the location of the stable and unstable manifolds of the saddle fixed points.

7. Give the ω-limit sets and α-limit sets for Example 5.1.

5.3. Lyapunov Functions for Damped Systems

The preceding section considered physical systems that preserve energy. By adding a damping term (or friction), the energy is no longer preserved but decreases and the equations are called *dissipative*.

A *damping* term is a force caused by the motion of the particle; thus, the damping $G(x, y)$ depends on the velocity y and possibly on the position x. Since the damping slows the particle down, a positive value of y results in a negative

Figure 13. For Exercise 6 in Section 5.2

value of damping $G(x, y)$. The simplest type of damping depends linearly on the velocity and is independent of position, yielding the system of differential equations

(5.1)
$$\dot{x} = y,$$
$$\dot{y} = F(x) - by.$$

We start by adding a linear damping term to the pendulum equations considered in the preceding section. We use an argument about the basin of attraction of the origin $W^s(\mathbf{0})$ based on consideration of energy. Then, we abstract the ideas from this example to define Lyapunov functions.

Example 5.5 (Pendulum with Damping). For the pendulum with linear damping, we have the equations

$$\dot{x} = y,$$
$$\dot{y} = -\sin(x) - by,$$

where $b > 0$. The term $-by$ is the damping. (If $b = 0$, we are back in the situation with no damping.) The fixed points are again $y = 0$ and $x = n\pi$ for any integer n. The linearization is

$$\begin{pmatrix} 0 & 1 \\ -\cos(x) & -b \end{pmatrix}.$$

A direct check shows that the eigenvalues at $x = \pm\pi$ are $\lambda = \dfrac{-b \pm \sqrt{b^2 + 4}}{2}$, and these fixed points are still saddles. At $x = 0$ or even multiples of π, $-\cos(0) = -1$, and the characteristic equation is $\lambda^2 + b\lambda + 1$, which has eigenvalues

$$\lambda = \frac{-b \pm \sqrt{b^2 - 4}}{2} = -\frac{b}{2} \pm i\sqrt{1 - (b^2/4)}.$$

These eigenvalues are complex for $0 < b < 2$, and real for $b \geq 2$. In any case, the fixed point is attracting. The linearization does not tell much about the size of the basin of attraction, but it does say that nearby orbits tend to the fixed point.

To find out more about the basin of attraction, consider the energy obtained for the system with no damping,

$$E(x, y) = \frac{y^2}{2} + (1 - \cos(x)).$$

5.3. Lyapunov Functions

The time derivative along a solution is given as follows:

$$\dot{E} = \frac{d}{dt}\left[\frac{y^2}{2} + (1 - \cos(x))\right] = y\dot{y} + \sin(x)\dot{x}$$
$$= y[-\sin(x) - by] + \sin(x)y$$
$$= -by^2 \leq 0.$$

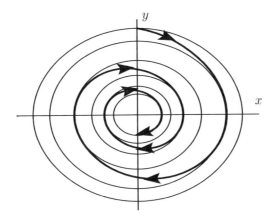

Figure 14. A Trajectory of damped pendulum crossing the level sets of $E(x, y)$, Example 5.5

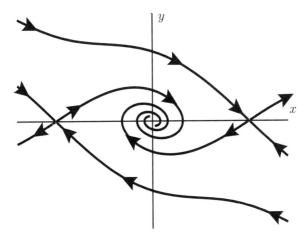

Figure 15. Phase Plane for Example 5.5

The time derivative of E is strictly negative, except when $y = 0$, or off the line $\mathbf{Z} = \{(x, 0)\}$. Thus, the energy is strictly decreasing except when the velocity y is zero. If a trajectory has $y = 0$ but is not at a fixed point, then $\dot{y} \neq 0$ and the trajectory moves off the line $y = 0$ and E starts decreasing again. Thus, a trajectory that starts with $E(x_0, y_0) < 2$ and $-\pi < x < \pi$, will have decreasing energy along the whole trajectory. Whenever it crosses $y = 0$, it will have $\dot{E} = 0$ for an instant, but will then start decreasing again. The energy must decrease toward the minimum at $(0,0)$; thus, it will tend asymptotically toward this fixed point. See Figures 14 and 15. Therefore, all these initial conditions are in the basin of attraction of $(0,0)$.

Definition 5.6. Assume $\mathbf{x}^*$ is a fixed point for the differential equation $\dot{\mathbf{x}} = \mathbf{F}(\mathbf{x})$. A real-valued function L is called a *weak Lyapunov function* for the differential equation provided there is a region $\mathbf{U}$ about $\mathbf{x}^*$ on which L is defined and (i) $L(\mathbf{x}) > L(\mathbf{x}^*)$ for all $\mathbf{x}$ in $\mathbf{U}$ but distinct from $\mathbf{x}^*$, and (ii) $\dot{L}(\mathbf{x}) \leq 0$ for all $\mathbf{x}$ in $\mathbf{U}$. Notice that we allow the time derivative to be zero. In particular, the energy function for a conservative differential equation is a weak Lyapunov function near a local minimum of the energy function. The function L is called a *Lyapunov function* or *strict Lyapunov function* on a region $\mathbf{U}$ about $\mathbf{x}^*$ provided it is a weak Lyapunov function which satisfies $\dot{L}(\mathbf{x}) < 0$ for all $\mathbf{x}$ in $\mathbf{U}$ but distinct from $\mathbf{x}^*$.

To use a Lyapunov function to analyze the phase portrait of a system of equations, somehow we must discover or invent the function L. However, we do not need to be able to calculate the solutions in order to calculate $\dot{L}$. In particular, we do not have explicit representations of the solutions of the damped pendulum problem.

The following result gives the first result about stability of the fixed points with a weak Lyapunov function or Lyapunov function.

Theorem 5.1. *Let $\mathbf{x}^*$ be a fixed point of the differential equation $\dot{\mathbf{x}} = \mathbf{F}(\mathbf{x})$.*

    a. *Assume that L is a weak Lyapunov function in a neighborhood $\mathbf{U}$ of $\mathbf{x}^*$. Then, $\mathbf{x}^*$ is L-stable.*

    b. *Assume that L is a (strict) Lyapunov function in a neighborhood $\mathbf{U}$ of $\mathbf{x}^*$. Then, $\mathbf{x}^*$ is attracting. Further assume that $L_0 > L(\mathbf{x}^*)$ is a value for which*
$$\mathbf{U}_{L_0} = \{\mathbf{x} \in \mathbf{U} : L(\mathbf{x}) \leq L_0\}$$
is contained inside $\mathbf{U}$, is bounded, and is bounded away from the boundary of $\mathbf{U}$ (i.e., contained in the interior of $\mathbf{U}$). Then, the set $\mathbf{U}_{L_0}$ is contained in the basin of attraction of $\mathbf{x}^*$, $\mathbf{U}_{L_0} \subset W^s(\mathbf{x}^*)$.

Proof. (a) Let $\mathbf{x}(t) = \phi(t; \mathbf{x}_0)$ be a trajectory. Because the time derivative is less than or equal to zero,
$$L(\mathbf{x}(t)) - L(\mathbf{x}_0) = \int_0^t \dot{L}(\mathbf{x}(s))\,ds \leq 0,$$
provided $\mathbf{x}(s)$ is in $\mathbf{U}$ for $0 \leq s \leq t$. Thus, the solution cannot cross the level curves to a higher value of L. Fix an $\epsilon > 0$. By taking a sufficiently small value of L_0, the set
$$\mathbf{U}_{L_0} = \{\mathbf{x} \in \mathbf{U} : L(\mathbf{x}) \leq L_0\} \subset \{\mathbf{x} : \|\mathbf{x} - \mathbf{x}^*\| < \epsilon\},$$
and it must be positively invariant. On the other hand, there is a $\delta > 0$ such that
$$\{\mathbf{x} : \|\mathbf{x} - \mathbf{x}^*\| < \delta\} \subset \mathbf{U}_{L_0}.$$
Therefore, any initial condition $\mathbf{x}_0$ with $\|\mathbf{x}_0 - \mathbf{x}^*\| < \delta$ must have $\mathbf{x}_0$ in $\mathbf{U}_{L_0}$, $\phi(t; \mathbf{x}_0)$ in $\mathbf{U}_{L_0}$ for $t \geq 0$, and so $\|\phi(t; \mathbf{x}_0) - \mathbf{x}^*\| < \epsilon$ for all $t \geq 0$. Since ϵ is arbitrarily small, the fixed point is L-stable.

5.3. Lyapunov Functions

(b) Now assume that $\dot{L} < 0$. Take $\mathbf{x}_0$ in $\mathbf{U}_{L_0}$, but not equal to $\mathbf{x}^*$. If $\phi(t; \mathbf{x}_0)$ did not converge to $\mathbf{x}^*$, then $\dot{L}(\phi(t; \mathbf{x}_0)) \leq -K < 0$ for $t \geq 0$. But then $L(\phi(t; \mathbf{x}_0)) \leq L(\mathbf{x}_0) - Kt$ goes to $-\infty$. This is impossible since it must stay greater than $L(\mathbf{x}^*)$. This contradiction shows that $\phi(t; \mathbf{x}_0)$ must converge to $\mathbf{x}^*$; that is, $\mathbf{x}^*$ must be attracting. $\square$

Example 5.7. Consider
$$\dot{x} = -y - x^3,$$
$$\dot{y} = x - y^3.$$
Let $L(x, y) = \frac{1}{2}(x^2 + y^2)$, half of the distance from the origin squared. Then
$$\dot{L} = \frac{\partial L}{\partial x}\dot{x} + \frac{\partial L}{\partial y}\dot{y} = x\dot{x} + y\dot{y} = x(-y - x^3) + y(x - y^3) = -x^4 - y^4,$$
which is negative except at $(0, 0)$. Thus, the square of the distance to the origin is strictly decreasing as t increases, and the trajectory must eventually limit on the fixed point $(0, 0)$. Therefore, $(0, 0)$ is asymptotically stable, and the basin of attraction is the whole plane.

Unfortunately, Theorem 5.1 shows only that the pendulum with damping is L-stable and not that it is asymptotically stable. To understand the key aspects that make the damped pendulum asymptotically stable, we need to consider more carefully the points where $\dot{L} = 0$.

Theorem 5.2. *Let $\mathbf{x}^*$, L, and $\mathbf{U}$ be as in the preceding theorem, where L is assumed to be a weak Lyapunov function. Define*
$$\mathbf{Z}_\mathbf{U} = \{\mathbf{x} \in \mathbf{U} : \dot{L}(\mathbf{x}) = 0\}.$$
Assume that $\mathbf{U}$ is sufficiently small such that, for any $\mathbf{x}_1 \in \mathbf{Z}_\mathbf{U} \setminus \{\mathbf{x}^*\}$, the trajectory $\phi(t; \mathbf{x}_1)$ moves off $\mathbf{Z}_\mathbf{U}$ into $\mathbf{U} \setminus \mathbf{Z}_\mathbf{U}$ for small positive t (i.e., $\{\mathbf{x}^*\}$ is the largest positively invariant set in $\mathbf{Z}_\mathbf{U}$). Then, $\mathbf{x}^*$ is asymptotically stable.

Further assume that $L_0 > L(\mathbf{x}^*)$ is a value for which
$$\mathbf{U}_{L_0} = \{\mathbf{x} \in \mathbf{U} : L(\mathbf{x}) \leq L_0\}$$
is contained inside $\mathbf{U}$, is bounded, and is bounded away from the boundary of $\mathbf{U}$. Then, the set $\mathbf{U}_{L_0}$ is contained in the basin of attraction of $\mathbf{x}^*$, $\mathbf{U}_{L_0} \subset W^s(\mathbf{x}^*)$.

The proof of this theorem is much the same as the analysis for the pendulum with damping and is given in Example 5.5. A general proof is given in Section 5.7.

Sometimes we can use this theorem to prove that a fixed point is asymptotically stable even though its linearized equation is a center (has eigenvalues with zero real part). When we want to emphasize the fact that the linearized system at a fixed point is a center, we call such an attracting fixed point *weakly attracting* rather than attracting. When we want to emphasize that the linearized system has an attracting fixed point as well as the nonlinear system, we call the fixed point *strongly attracting*. In particular, we use this terminology when discussing the Andronov–Hopf bifurcation in Section 6.4.

If we reverse the signs of the time derivatives, then we can get a result about the unstable manifold of the fixed point.

Theorem 5.3. *Let $\mathbf{x}^*$, L, and $\mathbf{U}$ be as in the preceding theorem, but now we assume that*
$$\dot{L}(\mathbf{x}) \geq 0$$
for all $\mathbf{x}$ in $\mathbf{U}$. We still assume that $L(\mathbf{x}) > L(\mathbf{x}^*)$ for $\mathbf{x}$ in $\mathbf{U}$ but distinct from $\mathbf{x}^*$. Define
$$\mathbf{Z_U} = \{\mathbf{x} \in \mathbf{U} : \dot{L}(\mathbf{x}) = 0\}.$$
Assume that $\mathbf{U}$ is sufficiently small such that, for any $\mathbf{x}_1 \in \mathbf{Z_U} \setminus \{\mathbf{x}^*\}$, the trajectory $\phi(t; \mathbf{x}_1)$ moves off $\mathbf{Z_U}$ into $\mathbf{U} \setminus \mathbf{Z_U}$ for small negative t (i.e., $\{\mathbf{x}^*\}$ is the largest negatively invariant set in $\mathbf{Z_U}$). Then, the unstable manifold of $\mathbf{x}^*$ contains a whole neighborhood of $\mathbf{x}^*$; that is, $\mathbf{x}^*$ is weakly repelling.

Further, assume that $L_0 > L(\mathbf{x}^*)$ is a value for which
$$\mathbf{U}_{L_0} = \{\mathbf{x} \in \mathbf{U} : L(\mathbf{x}) \leq L_0\}$$
is contained inside $\mathbf{U}$, is bounded, and is bounded away from the boundary of $\mathbf{U}$. Then, the set $\mathbf{U}_{L_0}$ is contained in the unstable manifold of $\mathbf{x}^*$, $\mathbf{U}_{L_0} \subset W^u(\mathbf{x}^*)$.

Example 5.8. Consider the basin of attraction for $(0, 0)$ for the system formed by adding a damping term to Example 5.4:
$$\dot{x} = y,$$
$$\dot{y} = -(x + x^2) - 0.3\, y.$$

The natural energy is obtained by considering the system without the term $-0.3y$ in the $\dot{y}$ equation:
$$L(x, y) = \frac{1}{2}y^2 + \frac{1}{2}x^2 + \frac{1}{3}x^3.$$
Let $V(x) = x^2/2 + x^3/3$ be the potential energy term. The fixed points have $x = 0, -1$ with $y = 0$.

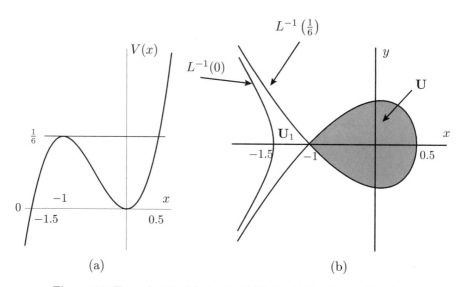

Figure 16. Example 5.8: (a) Graph of $V(x)$. (b) Level sets $L(x, y) = 0$, $L(x, y) = \frac{1}{6}$, and $\mathbf{U}$.

5.3. Lyapunov Functions

The value of the potential function $V(0) = 0$, so we must consider values of x for which $V(x) \geq 0$, i.e., $x \geq -1.5$. See Figure 16(a).

As before, $\dot{L} = L_x \dot{x} + L_y \dot{y} = (x + x^2)y + y(-x - x^2 - 0.3y) = -0.3\,y^2 \leq 0$. The set where $\dot{L}$ is zero is
$$\mathbf{Z_U} = \{(x,0) \in \mathbf{U}\}.$$
To make the fixed point at $(0,0)$ be the maximal invariant set in the set $\mathbf{Z_U}$, we need to exclude the other fixed point at $(-1,0)$ from $\mathbf{U}$. Since $L(-1,0) = V(-1) = \frac{1}{6}$, we take
$$\mathbf{U} = \left\{ (x,y) : 0 \leq L(x,y) < \frac{1}{6},\ x > -1 \right\}.$$
See Figure 16(b). If we take any $L_0 < \frac{1}{6}$, $\mathbf{U}_{L_0}$ is contained in the basin of attraction for $\mathbf{0}$. Since $L_0 < \frac{1}{6}$ is arbitrary, all of $\mathbf{U}$ is contained in the basin of attraction of $\mathbf{0}$. The set $\mathbf{U}$ is equal to the set of points inside the saddle connection for the saddle fixed point for the equations without damping of Example 5.4. Thus, all these points are in the basin of attraction. See Figure 17 for a plot of trajectories and the basin of attraction.

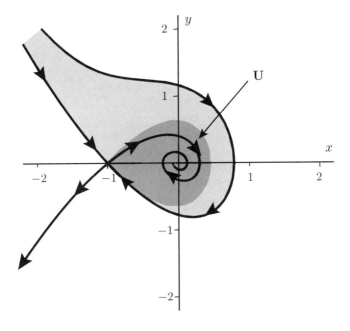

Figure 17. Basin of attraction for Example 5.8

Remark 5.9. Only special types of equations have Lyapunov functions. There is no general way of finding such a function. We have given a method of finding one for a system from an undamped force with "damping" added. If the equation under consideration is formed by adding terms to an equation that is known to have a Lyapunov function L, then sometimes a slight variation of L will be a Lyapunov function for the new equation. The book by LaSalle and Lefschetz [**LaS61**] has many such examples.

Exercises 5.3

1. Consider the system

$$\dot{x} = y,$$
$$\dot{y} = -x + x^3 - y.$$

   a. Use a Lyapunov function to find a (reasonably large) region contained in the basin of attraction for **0**.
   b. Discuss why solutions go to the fixed point.
   c. Draw the phase portrait by hand.
   d. Use a computer program to draw the phase portrait for representative initial conditions. Be sure to include enough initial conditions to reveal the important features of the system, including the stable and unstable manifolds of any saddle fixed points.

2. Consider the system

$$\dot{x} = y,$$
$$\dot{y} = x - x^3 - y.$$

   a. Use a Lyapunov function to find a (reasonably large) region contained in the basin of attraction for the fixed point sinks.
   b. Discuss why solutions go to the fixed point.
   c. Draw the phase portrait by hand.
   d. Use a computer program to draw the phase portrait for representative initial conditions. Be sure to include enough initial conditions to reveal the important features of the system, including the stable and unstable manifolds of any saddle fixed points.

3. Consider the system of differential equations

$$\dot{x} = -x^3 + xy^2,$$
$$\dot{y} = -2x^2y - y^3.$$

   Let $L(x, y) = x^2 + y^2$.
   a. Show that L is a Lyapunov function.
   b. What does $\dot{L}$ tell about the solutions of the system?

4. Consider the Lotka–Volterra equation for a *food chain* of three species given by

$$\dot{x}_1 = x_1(r_1 - a_{11}x_1 - a_{12}x_2) = x_1 w_1,$$
$$\dot{x}_2 = x_2(r_2 + a_{21}x_1 - a_{22}x_2 - a_{23}x_3) = x_2 w_2,$$
$$\dot{x}_3 = x_3(r_3 + a_{32}x_2 - a_{33}x_3) = x_3 w_3,$$

   where all the $r_i > 0$ and $a_{ij} > 0$. The quantities w_i are defined to be the quantities in the parentheses on the same line. (Notice that, (i) x_1 and x_2 have a negative effect on the growth rate of x_1; (ii) x_1 has a positive effect and x_2 and x_3 have a negative effect on the growth rate of x_2; and (iii) x_2 has a positive effect and x_3 have a negative effect on the growth rate of x_3.

That is why it is called a food chain.) Assume that there is an equilibrium $(x_1, x_2, x_3) = (p_1, p_2, p_3)$ with all the $p_i > 0$, so

$$r_1 = a_{11}p_1 + a_{12}p_2,$$
$$r_2 = -a_{21}p_1 + a_{22}p_2 + a_{23}p_3,$$
$$r_3 = -a_{32}p_2 + a_{33}p_3.$$

a. Show that the quantities w_i can be rewritten as

$$w_1 = -a_{11}(x_1 - p_1) - a_{12}(x_2 - p_2),$$
$$w_2 = a_{21}(x_1 - p_1) - a_{22}(x_2 - p_2) - a_{23}(x_3 - p_3),$$
$$w_3 = a_{32}(x_2 - p_2) - a_{33}(x_3 - p_3).$$

b. Define the function

$$L(x_1, x_2, x_3) = \sum_{i=1}^{3} c_i(x_i - p_i \ln(x_i)),$$

where the $c_i > 0$ are chosen so that

$$\frac{c_2}{c_1} = \frac{a_{12}}{a_{21}} \quad \text{and} \quad \frac{c_3}{c_2} = \frac{a_{23}}{a_{32}}.$$

Show that

$$\dot{L} = -\sum_{i=1}^{3} c_i a_{ii}(x_i - p_i)^2.$$

c. Show that the basin of attraction of the fixed point (p_1, p_2, p_3) includes the whole first octant

$$\{(x_1, x_2, x_3) : x_1 > 0, \ x_2 > 0, \ x_3 > 0\}.$$

5. Assume that L is a weak Lyapunov function for the system $\dot{\mathbf{x}} = \mathbf{F}(\mathbf{x})$ in $\mathbb{R}^n$. Assume that the forward orbit from $\mathbf{x}_0$ is bounded. Prove that $\dot{L}(\mathbf{y}) = 0$ at all points $\mathbf{y}$ in $\omega(\mathbf{x}_0)$.

5.4. Bounding Functions

In this section, for a given differential equation $\dot{\mathbf{x}} = \mathbf{F}(\mathbf{x})$, we try to pick a real-valued function $L(\mathbf{x})$ to determine the positively invariant sets and limit sets. The time derivative $\dot{L}$ is usually both positive and negative in different parts of the phase space so it is not a Lyapunov function, but we are still able to use it to find regions that are positively invariant. We call any such real-valued function a *test function*. We call L a bounding function if some value C can be used to locate a positively invariant set surrounded by $L^{-1}(C)$.

Definition 5.10. A test function L from $\mathbb{R}^n$ to $\mathbb{R}$ is called a *bounding function* for $\dot{\mathbf{x}} = \mathbf{F}(\mathbf{x})$ *and a subset* $\mathbf{U}$ *of* $\mathbb{R}^n$ provided that there is is a connected piece $\mathbf{B}$ of $L^{-1}(C)$ for some value C such that $\mathbf{U}$ is the set of all points inside $\mathbf{B}$ with $\mathbf{B}$ the boundary of $\mathbf{U}$ and they satisfy the following conditions.

(i) The sets $\mathbf{U}$ and $\mathbf{B}$ are bounded (do not go off to infinity).
(ii) The gradient $\nabla L(\mathbf{x}) \neq \mathbf{0}$ at all points $\mathbf{x}$ of $\mathbf{B}$.
(iii) If $\mathbf{x}$ is a point of $\mathbf{B}$, then $\phi(\mathbf{x}; t)$ is in $\mathbf{U} \setminus \mathbf{B}$ for all small enough $t > 0$.

If conditions (i)–(iii) are satisfied, then the set $\mathbf{U} \cup \mathbf{B}$ is then positively invariant and the level set $\mathbf{B}$ bounds the forward trajectories starting inside. Note that condition (ii) implies that $\mathbf{B}$ is a smooth level curve in dimension two and a smooth level surface in dimension three.

If the time derivative $\dot{L}$ is nonzero (and so of one sign) on the level set $\mathbf{B}$, then the trajectories cross this level set in one direction. So, (iii′) if $\dot{L}(\mathbf{x}) < 0$ and $\nabla L(\mathbf{x})$ points to the outside of $\mathbf{U}$, then condition (iii) is satisfied.

Example 5.11. This example has a limit set that contains both a fixed point and orbits that have both their ω-limit and α-limit at the fixed point. Consider

$$\dot{x} = y,$$
$$\dot{y} = x - 2x^3 + y(x^2 - x^4 - y^2).$$

If the term containing y were not there in the $\dot{y}$ equation, the energy function would be

$$L(x,y) = \frac{-x^2 + x^4}{2} + \frac{y^2}{2}.$$

The time derivative of this L for the whole original equations is given by

$$\dot{L} = (-x + 2x^3)y + y\bigl(x - 2x^3 + y(x^2 - x^4 - y^2)\bigr)$$
$$= -2y^2 L \begin{cases} \geq 0 & \text{when } L < 0, \\ = 0 & \text{when } L = 0, \\ \leq 0 & \text{when } L > 0. \end{cases}$$

Because its derivative is both positive and negative, it is not a Lyapunov function.

If $L(x_0, y_0) = 0$, then the orbit stays on the level set $L^{-1}(0)$, which contains the fixed point at the origin. This level set is bounded and one-dimensional, so all the trajectories must converge to the fixed point at the origin as t goes to plus or minus infinity; that is, for these points, $\omega(x_0, y_0) = \alpha(x_0, y_0) = \mathbf{0}$ and

$$(x_0, y_0) \in W^s(\mathbf{0}) \cap W^u(\mathbf{0}).$$

Let

$$\Gamma = \{(x,y) : L(x,y) = 0\}.$$

Since the stable and unstable manifolds of the origin are curves and contain Γ, they must equal Γ,

$$\Gamma = W^s(\mathbf{0}) = W^u(\mathbf{0}).$$

Any $C > 0$ and $\mathbf{B} = L^{-1}(C)$ satisfies conditions (i)–(iii) and so L is a bounding function for the region inside $L^{-1}(C)$. (Note that $\dot{L}(\mathbf{x})$ is only ≤ 0, but it still satisfies (iii).)

If $C = L(x_0, y_0) > 0$, then the trajectory has to stay inside $L^{-1}(C)$ and so is bounded. The value $L(\phi(x_0, y_0))$ decreases along the orbit with its time derivative equals to zero only when the orbit crosses the x-axis (where $y = 0$). This orbit $\phi(x_0, y_0)$ continues to cross the level set $L^{-1}(C')$ for $0 < C' < C$, going in toward the level set $L^{-1}(0) = \Gamma$. Thus, these orbits spiral down and accumulate on Γ; for any point $\mathbf{q} \in \Gamma$, there is a sequence of times t_n going to infinity such that $\phi(t_n; (x_0, y_0))$ is close to $\mathbf{q}$, so $\mathbf{q} \in \omega(x_0, y_0)$. Since this is true for each such point

5.4. Bounding Functions

$\mathbf{q} \in \Gamma$, $\omega(x_0, y_0) = \Gamma$. See Figure 18. For negative time, the orbit goes off to infinity and $\alpha(x_0, y_0) = \emptyset$.

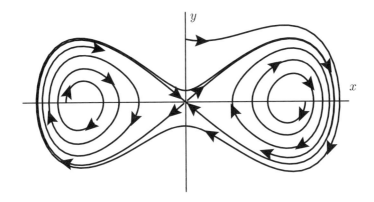

Figure 18. Phase portrait for Example 5.11

For a point (x_1, y_1) with $-1/8 < L(x_1, y_1) < 0$ and $x_1 > 0$, the orbit spirals out and accumulates on the right half of Γ,

$$\omega(x_1, y_1) = \Gamma \cap \{(x, y) : x \geq 0\}.$$

See Figure 18. Going backward in time, the orbit approaches the fixed point $(1/\sqrt{2}, 0)$, so the α-limit is $\alpha(x_1, y_1) = \{(1/\sqrt{2}, 0)\}$.

Similarly, for a point (x_2, y_2) with $-1/8 < L(x_2, y_2) < 0$ and $x_2 < 0$, the orbit spirals out and accumulates on the left half of Γ,

$$\omega(x_2, y_2) = \Gamma \cap \{(x, y) : x \leq 0\}.$$

Also $\alpha(x_2, y_2) = \{(-1/\sqrt{2}, 0)\}$.

For any of these orbits, whenever the limit set is nonempty, it contains one of the fixed points.

The next example illustrates that it is not necessary for both sides of the unstable manifold to return, as in the preceding example.

Example 5.12. Consider

$$\dot{x} = y,$$
$$\dot{y} = x - x^2 + y(3x^2 - 2x^3 - 3y^2),$$

and $L(x, y) = -\frac{x^2}{2} + \frac{x^3}{3} + \frac{y^2}{2}$. In this case, only the right side of the unstable manifold of the saddle point at the origin is in the stable manifold:

$$\Gamma^+ = W^u(\mathbf{0}) \cap W^s(\mathbf{0}) = \{(x, y) : L(x, y) = 0, \ x \geq 0\}.$$

If $-1/6 < L(x_0, y_0) < 0$ and $x_0 > 0$, then $\omega(x_0, y_0) = \Gamma^+$.

In each of the preceding two examples, the divergence of the vector field at the saddle point (origin) is zero, which is not the usual way this type of behavior occurs. The generic properties of a planar differential equation that cause the homoclinic connection to attract nearby orbits are the following: (i) There is a saddle fixed point **p** with at least one branch Γ^+ of the unstable manifold $W^u(\mathbf{p})$ contained in the stable manifold $W^s(\mathbf{p})$, $\Gamma^+ \subset W^u(\mathbf{p}) \cap W^s(\mathbf{p})$. (ii) Assume that the divergence of the vector field at **p** is negative. Then orbits starting at $\mathbf{q}_0$ in the region bounded by Γ^+ have

$$\omega(\mathbf{q}_0) = \{\mathbf{p}\} \cup \Gamma^+.$$

Example 5.13. This is an example of a system of differential equations on $\mathbb{R}^3$ with the limit set equal to an ellipsoid together with a line segment. Consider the system of differential equations

$$\dot{x} = -\beta y + 2xz,$$
$$\dot{y} = \beta x + 2yz,$$
$$\dot{z} = 1 - x^2 - y^2 - z^2 + z(x^2 + y^2)\left(1 - \frac{1}{3}(x^2 + y^2) - z^2\right),$$

which in cylindrical coordinates is

$$\dot{\theta} = \beta,$$
$$\dot{r} = 2rz,$$
$$\dot{z} = 1 - r^2 - z^2 + r^2 z\left(1 - \frac{1}{3}r^2 - z^2\right).$$

This example is motivated by an equation considered in [**Guc83**] used as a bifurcation from a fixed point, with a pair of purely imaginary eigenvalues and zero as the other eigenvalue.

We first consider the system in the (r, z)-space for $r \geq 0$. The fixed points are $(r, z) = (0, \pm 1)$ and $(1, 0)$. (The fixed point $(1, 0)$ corresponds to a periodic orbit in $\mathbb{R}^3$.) We use the test function

$$L(r, z) = r\left(1 - \frac{1}{3}r^2 - z^2\right).$$

Notice that $\dot{z} = 1 - r^2 - z^2 + rzL$. The time derivative of L is

$$\dot{L} = \dot{r}\left(1 - \frac{1}{3}r^2 - z^2\right) + r\left(-\frac{2}{3}r\dot{r} - 2z\dot{z}\right)$$
$$= 2zL - \frac{4}{3}r^3 z - 2rz\left(1 - r^2 - z^2 + rzL\right)$$
$$= 2zL - 2zr\left(1 - \frac{1}{3}r^2 - z^2\right) - 2r^2 z^2 L$$
$$= -2r^2 z^2 L = \begin{cases} > 0 & \text{when } L < 0, \ r > 0, \text{ and } z \neq 0, \\ = 0 & \text{when } L = 0, \\ < 0 & \text{when } L > 0, \ r > 0, \text{ and } z \neq 0. \end{cases}$$

The level set

$$L^{-1}(0) = \{(0, z)\} \cup \left\{(r, z) : 1 = \frac{1}{3}r^2 + z^2, \ r \geq 0\right\},$$

is invariant, which is the z-axis together with a semi-ellipse.

Points inside the ellipse with $r > 0$ and $z \neq 0$ have $L > 0$ and $\dot{L} < 0$. The only point here that stays on the set $\{z = 0\}$ is the fixed point $(1, 0)$. Therefore, if we start at a point inside the ellipse $\mathbf{p}_0 = (r_0, z_0) \neq (1, 0)$ and $r_0 > 0$, then the orbit $\phi(t; \mathbf{p}_0)$ converges to the set $L^{-1}(0)$, the semi-ellipse together with the z-axis between $z = -1$ and 1. Therefore,

$$\omega(\mathbf{p}_0) \subset \{(0, z) : -1 \leq z \leq 1\} \cup \left\{(r, z) : 1 = \frac{1}{3}r^2 + z^2,\ r \geq 0\right\} \subset L^{-1}(0).$$

Turning to the flow on $\mathbb{R}^3$, the only fixed points are $(x, y, z) = (0, 0, \pm 1)$. There is a periodic orbit for $z = 0$, $r = 1$, and θ arbitrary. The semi-ellipse in the plane corresponds to an ellipsoid in $\mathbb{R}^3$,

$$L^{-1}(0) = \{(0, 0, z)\} \cup \left\{(x, y, z) : 1 = \frac{1}{3}(x^2 + y^2) + z^2\right\}.$$

If we start at a point $\mathbf{q}_0$ inside the ellipsoid, but not on the z-axis and not on the periodic orbit with $z = 0$ and $r = 1$, then by the analysis in $\mathbb{R}^2$,

$$\omega(\mathbf{q}_0) \subset \{(0, 0, z) : -1 \leq z \leq 1\} \cup \left\{(x, y, z) : 1 = \frac{1}{3}(x^2 + y^2) + z^2\right\}.$$

Since the orbit goes past the fixed points $(0, 0, \pm 1)$ slowly, it seems likely that the ω-limit set contains the whole ellipsoid together with the line segment in the z-axis.

We give further examples of bounding functions when considering periodic orbits in Section 6.2 and chaotic attractors in Chapter 7.

Exercises 5.4

1. Consider the system of differential equations
$$\dot{x} = y,$$
$$\dot{y} = -(x + x^2) - y\,(3\,y^2 + 3\,x^2 + 2\,x^3 - 1).$$
Give the ω-limit sets and α-limit sets. Hint: Use the test function
$$L(x, y) = \frac{1}{2}y^2 + \frac{1}{2}x^2 + \frac{1}{3}x^3.$$

2. Suppose that $\mathbf{A}$ is a closed and bounded set that is invariant by a flow $\phi(t; \mathbf{x})$ and there are no nonempty invariant closed proper subsets of $\mathbf{A}$.
   a. Prove that every trajectory is dense in $\mathbf{A}$.
   b. Prove that $\omega(\mathbf{x}_0) = \alpha(\mathbf{x}_0) = \mathbf{A}$ for each $\mathbf{x}_0$ in $\mathbf{A}$.

5.5. Gradient Systems

In this section we consider a type of equations for which all the α- and ω-limit sets are fixed points and there can be no periodic orbits. Let $G(\mathbf{x})$ be a real-valued function on $\mathbb{R}^n$, and let

$$\nabla G(\mathbf{x}) = \begin{pmatrix} \dfrac{\partial G}{\partial x_1}(\mathbf{x}) \\ \vdots \\ \dfrac{\partial G}{\partial x_n}(\mathbf{x}) \end{pmatrix}$$

be its *gradient of G* at the point **x** (written as a column vector). The associated system of differential equations

$$\dot{x}_1 = -\frac{\partial G}{\partial x_1}(\mathbf{x}),$$
$$\vdots = \vdots$$
$$\dot{x}_n = -\frac{\partial G}{\partial x_n}(\mathbf{x}),$$

or in vector notation

$$\dot{\mathbf{x}} = -\nabla G(\mathbf{x}),$$

is called a *gradient system of differential equations*.

Any point $\mathbf{x}^*$ where $\nabla G(\mathbf{x}^*) = \mathbf{0}$ is called a *critical point* of the function and is a fixed point of the gradient system of differential equations.

Theorem 5.4. *Consider a gradient system of differential equations* $\dot{\mathbf{x}} = s - \nabla G(\mathbf{x})$, *where G is a real-valued function. Then, G is a Lyapunov function (at least near a local minimum) for the gradient system of differential equations formed from G. More precisely, $\dot{G}(\mathbf{x}) \leq 0$ at all points and $\dot{G}(\mathbf{x}) = 0$ if and only if **x** is a fixed point. Away from the fixed points, G is strictly decreasing along trajectories.*

Proof.
$$\dot{G} = \sum_{i=1}^{n} \frac{\partial G}{\partial x_i} \dot{x}_i = -\sum_{i=1}^{n} \left(\frac{\partial G}{\partial x_i}\right)^2,$$

or in vector notation

$$\dot{G} = \nabla G(\mathbf{x}) \cdot \dot{\mathbf{x}} = \nabla G(\mathbf{x}) \cdot (-\nabla G(\mathbf{x}))$$
$$= -\|\nabla G(\mathbf{x})\|^2 \leq 0.$$

This time derivative is zero only at points where $\nabla G(\mathbf{x}) = \mathbf{0}$; that is, at fixed points of the differential equation. Away from the fixed points, $\dot{G} < 0$ and G is strictly decreasing. □

Calculus books usually mention that the gradient is perpendicular to any vector tangent to a level surface, so the vectors tangent to the level surface are just those with $\nabla G(\mathbf{x}) \cdot \mathbf{v} = 0$.

Theorem 5.5. *Let $\dot{\mathbf{x}} = -\nabla G(\mathbf{x})$ be a gradient system of differential equations. Then, the vector field $\nabla G(\mathbf{x})$ is perpendicular to the level sets of G at any point that is not a critical point (i.e., it is perpendicular to any vector tangent to the level set).*

Proof. Any tangent vector **v** that is tangent to the level set, can be given as the tangent vector to a curve lying in the level set. Therefore, there is a curve $\mathbf{x}(s)$

5.5. Gradient Systems

with $G(\mathbf{x}(s)) \equiv G_0$, a constant. Letting $\mathbf{v} = \frac{d}{ds}\mathbf{x}(s)\big|_{s=0}$,

$$\begin{aligned} 0 &= \frac{d}{ds}G_0\big|_{s=0} \\ &= \frac{d}{ds}G(\mathbf{x}(s))\big|_{s=0} \\ &= \nabla G(\mathbf{x}(0)) \cdot \mathbf{v}. \end{aligned}$$

Therefore, $\nabla G(\mathbf{x}(0))$ is perpendicular to $\mathbf{v}$. $\square$

Example 5.14. Consider the real-valued function associated to the two-welled potential,

$$G(x,y) = \frac{-x^2 + x^4 + y^2}{2}.$$

It has critical points at $(\pm 1, 0)$ and $(0,0)$. The points $(\pm 1, 0)$ are local minima and $(0,0)$ is a saddle. The gradient system is

$$\dot{x} = x - x^3,$$
$$\dot{y} = -y.$$

The solutions $\phi(t;(0,y_0)) = (0, y_0 e^{-t})$ all tend directly to the saddle point at the origin. Thus, the stable manifold of the origin is the y-axis. Any solution $\phi(t;(x_0,y_0))$ with $x_0 > 0$ cannot cross the stable manifold of the origin, and so must stay in the right half-plane where $x > 0$. Therefore, all these points must converge to the fixed point $(1,0)$. Similarly, if $x_0 < 0$, then $\phi(t;(x_0,y_0))$ converges to the attracting fixed point at $(-1,0)$. See Figure 19 for the level sets of G and a few representative trajectories.

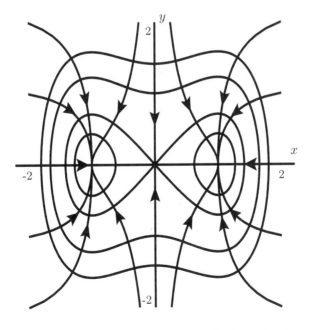

Figure 19. Level sets and trajectories for Example 5.14

There is a close connection between the type of critical point of the function G and the type of fixed point for the associated gradient system of differential equations.

Theorem 5.6. *Let $\mathbf{x}^*$ be a critical point of the real-valued function G and, therefore, a fixed point of the associated gradient system of differential equations. Then, all the eigenvalues of the fixed point of the system of differential equations are real. (i) If the critical point is a local minimum of G, then it is an asymptotically stable fixed point (sink) for the associated gradient system of differential equations. (ii) If the critical point is a saddle of G, then it is a saddle fixed point for the associated gradient system of differential equations, and so it is unstable. (iii) If the critical point is a local maximum of G, then it is a source for the associated gradient system of differential equations, and so it is unstable.*

Proof. The vector field giving the differential equation is

$$\mathbf{F}(\mathbf{x}) = -\nabla G(\mathbf{x}) = \left(-\frac{\partial G}{\partial x_i}(\mathbf{x})\right).$$

Therefore, the linearization at fixed points is

$$D\mathbf{F}_{(\mathbf{x}^*)} = \left(-\frac{\partial^2 G}{\partial x_i \partial x_j}(\mathbf{x}^*)\right),$$

which is the *negative of the Hessian of G*. This matrix is symmetric, so it has real eigenvalues.

At a minimum of G, all the eigenvalues of the Hessian are positive, so all the eigenvalues of $D\mathbf{F}_{(\mathbf{x})}$ are negative; therefore, it is a sink and is asymptotically stable. At a maximum of G, all the eigenvalues of the Hessian are negative, so all the eigenvalues of $D\mathbf{F}_{\mathbf{x}}$ are positive; therefore, it is a source. At a saddle of G, there are some positive and some negative eigenvalues of the Hessian and also of $D\mathbf{F}_{\mathbf{x}}$, so it is a saddle fixed point. □

Theorem 5.7. *Let $\dot{\mathbf{x}} = -\nabla G(\mathbf{x})$ be a gradient system of differential equations. Let $\mathbf{z}$ be an α-limit point or an ω-limit point. Then, $\mathbf{z}$ is a fixed point for the gradient system of differential equations.*

Proof. This proof is essentially the same as the proof of Theorem 4.4. Assume that $\mathbf{z}$ is in $\omega(\mathbf{x}_0)$. Then, $G(\phi(t; \mathbf{x}_0))$ is decreasing. There is a sequence of times t_j going to infinity such that $\phi(t_j; \mathbf{x}_0)$ approaches $\mathbf{z}$. We can take the times so that $t_1 < t_2 < \cdots < t_j < t_{j+1} < \cdots$, so

$$G(\phi(t_1; \mathbf{x}_0)) > G(\phi(t_2; \mathbf{x}_0)) > \cdots > G(\phi(t_j; \mathbf{x}_0)) > G(\phi(t_{j+1}; \mathbf{x}_0)) > \cdots.$$

By continuity of G, $G(\phi(t_j; \mathbf{x}_0))$ approaches $G(\mathbf{z})$. The value $G(\phi(t; \mathbf{x}_0))$ can never get below $G(\mathbf{z})$ because it is strictly decreasing. Therefore, $\dot{G}(\phi(t; \mathbf{x}_0))$ must go to zero. Again, by continuity, $\dot{G}(\phi(t_j; \mathbf{x}_0))$ converges to $\dot{G}(\mathbf{z})$ so this must be zero; that is, $\mathbf{z}$ must be a fixed point. □

The theorem says that the ω-limit points must be fixed points. In fact, most points go to sinks, since the stable manifolds of saddles are lower dimensional. Since sinks are local minima of G, most points have ω-limits at minima of G. Thus, going along the trajectories of the gradient system of differential equations is one way to find a local minimum of G.

Exercises 5.5

1. Consider the real-valued function
$$G(x_1, x_2) = -x_1^2 + x_2^2.$$
Plot the phase portrait for the gradient system $\dot{\mathbf{x}} = -\nabla G_{(\mathbf{x})}$. Include in the sketch some representative level sets of G.

2. Consider the real-valued function
$$G(x_1, x_2) = \frac{x_1^2 - x_1^4 + x_2^2}{2}.$$
Plot the phase portrait for the gradient system $\dot{\mathbf{x}} = -\nabla G_{(\mathbf{x})}$. Include in the sketch some representative level sets of G.

3. Consider the real-valued function
$$G(x_1, x_2) = \frac{3\,x_1^2 - 2\,x_1^3 + 3\,x_2^2}{6}.$$
Plot the phase portrait for the gradient system $\dot{\mathbf{x}} = -\nabla G_{(\mathbf{x})}$. Include in the sketch some representative level sets of G.

4. Show that any differential equation $\dot{x} = f(x)$ on the line is a gradient differential equation. What is the function $G(x)$ such that $f(x) = -G'(x)$?

5.6. Applications

5.6.1. Nonlinear Oscillators. In Section 5.2, we discussed undamped forces and showed how they preserved an energy function. However, we did not discuss the relationship between the different equations and physical oscillators. In this section, we discuss such connections.

We consider an equation of the form
$$\ddot{x} = f(x)$$
or the corresponding system
(5.2)
$$\dot{x} = y,$$
$$\dot{y} = f(x).$$

The function $f(x)$ is the force acting on the particle. (We have taken the mass equal to one or incorporated it into the function f for simplicity.) If this is some type of spring or restoring force, then often the force for $-x$ is equal to $-f(x)$ (i.e., $f(-x) = -f(x)$ and f is an odd function). The harmonic oscillator or linear oscillator has $f(x) = -\omega^2 x$, which is a linear odd restoring force. The pendulum given in Example 5.3 has $f(x) = -\sin(x)$, which is a nonlinear odd restoring force. Example 5.1 has $f(x) = x - x^3$ which is an odd function, but is repelling for small displacements and is only attracting for displacements with $|x| > 1$. In Example

5.4, the force $f(x) = -x - x^2$ is not an odd function and the effect of positive and negative displacements from $x = 0$ is very different.

If $\mathbf{x}(t) = \phi(t; \mathbf{x}_0)$ is periodic, then each of the coordinates $x_j(t)$ is periodic and the motion is an *oscillation*. Any nonlinear system of the form of equation (5.2) that has a periodic orbit for some initial condition can be called an *undamped nonlinear oscillator* or just an *oscillator*. With this general definition, Examples 5.1, 5.3, and 5.4 are undamped nonlinear oscillators. However, since the force for Example 5.4 is not odd and the orbits do not surround the origin, this system is not what is usually called an oscillator.

Duffing equation

The simplest odd function contains both linear and cubic terms,
$$f(x) = a\,x + b\,x^3.$$
The differential equation with this cubic force is called the *Duffing equation*. For small values of $|x|$, the linear term is largest and dominates, while for large values of $|x|$, the cubic term is bigger. The potential function is given by
$$V(x) = -a\,\frac{x^2}{2} - b\,\frac{x^4}{4}.$$

Comparing with the examples in Section 5.2, notice that for $a < 0$, the origin is a center. If both $a < 0$ and $b < 0$, then the restoring force becomes stronger for larger values of $|x|$. This is called a *hard spring*. In this case, the potential function has a single critical point at $x = 0$, and all the orbits are periodic.

If $a < 0$ but $b > 0$, then the restoring force becomes weaker for larger values of $|x|$. In fact for sufficiently large values of $|x|$, the force is a repelling force rather than an attracting one. For this reason, it is called a *soft spring*. The potential function has three critical points: $x = 0$ and $\pm\sqrt{|a/b|}$. The point $x = 0$ is a local minimum for V, so it is a center for the system of differential equations. The points $x = \pm\sqrt{|a/b|}$ are maxima for V, so they are saddles for the system of differential equations. In this system, the origin is surrounded by periodic orbits, but other orbits go off to infinity. This system has some similarity to the pendulum equation.

If $a > 0$, then the force is a repelling for small displacements. The potential has a local maximum at $x = 0$, and the origin is a saddle. If $b < 0$, then for larger values of $|x|$, the force is a restoring force; it is what we called the two-well potential. Thus, there are two center fixed points at $(x, y) = (\pm\sqrt{|a/b|}, 0)$. See Example 5.1. There are some periodic orbits that surround one of the fixed points $(\pm\sqrt{|a/b|}, 0)$, and other periodic orbits that surround all three fixed points.

If both $a > 0$, and $b > 0$, then the origin is the only critical point, and the origin is the only fixed point and is a saddle. This system is repelling for all displacements.

Six-twelve potential

Interatomic interactions are often described by a model using what is called the *six-twelve potential*, given by
$$V(x) = \frac{a}{12\,x^{12}} - \frac{b}{6\,x^6},$$

5.6. Applications

for the force
$$f(x) = -\frac{a}{x^{13}} + \frac{b}{x^7},$$
where $a > 0$ and $b > 0$. The potential has a unique critical point at $(a/b)^{1/6}$, which is a minimum. As x goes to zero, $V(x)$ goes to plus infinity. As x goes to infinity, $V(x)$ goes zero. Thus, there is a center for the system of differential equations that is surrounded by periodic orbits. Points with $E(x,y) = V(x) + y^2/2 > 0$ have motion that is unbounded. We leave the details of the phase portrait as an exercise.

5.6.2. Interacting Populations. We have considered competitive and predator-prey models for two interacting populations. In this section, we consider interactions with more populations, where the equations are still of the type called *Lotka–Volterra equations*,

$$(5.3) \qquad \dot{x}_i = x_i \left(r_i + \sum_{j=1}^{n} a_{ij} x_j \right) \qquad \text{for } 1 \leq i \leq n.$$

A general reference for more results on these systems is [**Hof88**]. The book by Hirsch and Smale [**Hir74**] considers systems where the part inside the parentheses can be a nonlinear function. In this section, we let $\mathbb{R}^n_+ = \{ \mathbf{x} : x_i > 0 \text{ for all } i \}$.

We next show that if there is no fixed point in $\mathbb{R}^n_+$, then the ω-limit set of a point has to be contained in the set with at least one of the populations equal to zero. This result is a general type of competitive exclusion.

Theorem 5.8 (Exclusion property). *Assume that a Lotka–Volterra system (5.3) has no fixed point in $\mathbb{R}^n_+$. Then for any $\mathbf{x}_0 \in \mathbb{R}^n_+$, $\omega(\mathbf{x}_0) \cap \mathbb{R}^n_+ = \emptyset$ and $\alpha(\mathbf{x}_0) \cap \mathbb{R}^n_+ = \emptyset$.*

Proof. The sets $\{ \mathbf{r} + \mathbf{A}\mathbf{x} : \mathbf{x} \in \mathbb{R}^n_+ \}$ and $\{\mathbf{0}\}$ are each convex and disjoint since there are no fixed points. A standard theorem in the theory of convex sets says that there is a $\mathbf{c} \in \mathbb{R}^n \setminus \{\mathbf{0}\}$ such that $\mathbf{c} \cdot (\mathbf{r} + \mathbf{A}\mathbf{x}) < \mathbf{c} \cdot \mathbf{0} = 0$ for all $\mathbf{x} \in \mathbb{R}^n_+$.

We consider the Lyapunov function
$$L(\mathbf{x}) = \sum_i c_i \ln(x_i).$$
The time derivative is as follows:
$$\dot{L} = \sum_i c_i \frac{\dot{x}_i}{x_i} = \sum_i c_i \left(r_i + (\mathbf{A}\mathbf{x})_i \right) = \mathbf{c} \cdot (\mathbf{r} + \mathbf{A}\mathbf{x}) < 0.$$
Since this is strictly negative on all of $\mathbb{R}^n_+$, $\omega(\mathbf{x}_0) \cap \mathbb{R}^n_+ = \emptyset$ and $\alpha(\mathbf{x}_0) \cap \mathbb{R}^n_+ = \emptyset$ for any $\mathbf{x}_0 \in \mathbb{R}^n_+$. $\square$

The following theorem is a generalization of the fact that the time average of a periodic orbit for a predator-prey system is equal to the fixed point.

Theorem 5.9. *Consider a Lotka–Volterra system (5.3) where there is an orbit $\mathbf{x}(t) = \phi(t; \mathbf{x}_0)$ with $\mathbf{x}_0 \in \mathbb{R}^n_+$ such that*
$$0 < a \leq x_i(t) \leq A < \infty \qquad \text{for } 1 \leq i \leq n.$$

Denote the time average of the orbit by

$$\mathbf{z}(T) = \frac{1}{T} \int_0^T \mathbf{x}(t)\, dt.$$

Then there has to be a fixed point of (5.3) $\mathbf{p} \in \mathbb{R}_+^n$ *in the ω-limit of* $\mathbf{z}(T)$. *Further, if there is a unique fixed point* $\mathbf{p} \in \mathbb{R}_+^n$, *then* $\mathbf{z}(T)$ *converges to* $\mathbf{p}$ *as* T *goes to infinity.*

Proof. We integrate $\dfrac{\dot{x}_i(t)}{x_i}$ from 0 to T.

(*) $\quad \dfrac{\ln(x_i(T)) - \ln(x_i(0))}{T} = \dfrac{1}{T}\int_0^T \dfrac{\dot{x}_i(t)}{x_i}\,dt$

$$= \frac{1}{T}\int_0^T r_i\,dt + \sum_j a_{ij} \frac{1}{T}\int_0^T x_j(t)\,dt$$

$$= r_i + \sum_j a_{ij}\, z_j(T).$$

Since $a \leq x_i(T) \leq A$, the limit of the left side of (*) is zero. Since $a \leq z_i(T) \leq A$, there has to be a subsequence of times T_n such that $\mathbf{z}(T_n)$ converges to a point $\mathbf{p}$ with $a \leq p_i \leq A$ and

$$\mathbf{0} = \mathbf{r} + \mathbf{A}\mathbf{p}.$$

So $\mathbf{p}$ is a fixed point in $\mathbb{R}_+^n$. If the fixed point is unique, then it is isolated. Thus $\mathbf{z}$ cannot have any other ω-limit points and the limit must exist. □

5.6.3. Replicator equations. In this chapter, we have considered the population models given by the Lotka–Volterra equations. They are examples of equations of the form $\dot{x}_i = x_i\, f_i(\mathbf{x})$, called *Volterra equations*. For the Lotka–Volterra equations, the functions f_i are linear. In this subsection, we consider replicator equations where the functions f_i contain quadratic terms.

Evolutionary game theory developed by applying some concepts from biology to the situation in game theory where a population competes against itself. The replicator differential equations adds continuous adjustment toward Nash equilibria into evolutionary game theory. Thus, the replicator differential equations introduce dynamics into game theory and a type of stability for some Nash equilibria. We content ourselves by introducing the replicator differential equations, giving a few basic results, and presenting some examples. The book [**Gin00**] by Gintis discusses the connection of the replicator differential equations with evolutionary game theory. The book [**Hof88**] by Hofbauer and Sigmund gives a much more complete treatment of the replicator differential equations. In writing this section, we have also consulted the notes by Schecter [**Sch06**] that are used for a course taught using [**Gin00**].

In the game theoretic context of the replicator equation, individuals within a large population can choose from a finite number of choices of actions $\{s_1, \ldots, s_n\}$. A square $n \times n$ matrix $\mathbf{A}$ is specified that gives the payoff of playing one action against another: a_{ij} is the payoff for the individual playing action s_i against another

5.6. Applications

playing s_j. This is called a *symmetric game* because the playoff for the second player playing s_j against s_i is a_{ji}. The matrix $\mathbf{A}$ is not necessarily symmetric.

Let
$$\mathbf{S} = \left\{ \mathbf{p} = (p_1, \ldots, p_n)^T : p_i \geq 0 \text{ for all } i, \sum_i p_1 = 1 \right\}$$
be the simplex of probability vectors. An element $\mathbf{p} \in \mathbf{S}$ is called a *state* of the population and corresponds to individuals within a large population playing the action s_j with probability p_j. The payoff of an individual who takes action s_i against a population with state $\mathbf{p}$ is assumed to be
$$\sum_j a_{ij} p_j = (\mathbf{e}^i)^T \mathbf{A}\mathbf{p} = \mathbf{e}^i \cdot \mathbf{A}\mathbf{p},$$
where $\mathbf{e}^i$ is the standard unit vector with a one in the ith-place. In the same way, the payoff of a state $\mathbf{q}$ playing against a state $\mathbf{p}$ is assumed to be
$$\sum_{ij} q_i a_{ij} p_j = \mathbf{q}^T \mathbf{A}\mathbf{p} = \mathbf{q} \cdot \mathbf{A}\mathbf{p}.$$

For the *replicator system of differential equations*, the relative rate of growth $\dot{p}_i/p_i$ is set equal to the difference between the payoff of playing action s_i and the payoff of playing $\mathbf{p}$,

(5.4) $$\dot{p}_i = p_i \left(\mathbf{e}^i \cdot \mathbf{A}\mathbf{p} - \mathbf{p} \cdot \mathbf{A}\mathbf{p} \right) \quad \text{for } 1 \leq i \leq n.$$

Thus, if $\mathbf{e}^i \cdot \mathbf{A}\mathbf{p} > \mathbf{p} \cdot \mathbf{A}\mathbf{p}$, then the proportion of the population playing action s_i increases, and if $\mathbf{e}^i \cdot \mathbf{A}\mathbf{p} < \mathbf{p} \cdot \mathbf{A}\mathbf{p}$, then the proportion of the population playing action s_i decreases.

In the analysis of the dynamics, we use the "faces" of $\mathbf{S}$ where certain p_i are zero. For a nonempty subset I of $\{1, \ldots, n\}$, let
$$\mathbf{S}_I = \{ \mathbf{p} \in \mathbf{S} : p_i > 0 \text{ if } i \in I, \ p_j = 0 \text{ if } j \notin I \},$$
$$\hat{\mathbf{S}}_I = \{ \mathbf{p} \in \mathbf{S} : p_i > 0 \text{ for all } i \in I \} \quad \text{and}$$
$$I(\mathbf{p}) = \{ i : p_i > 0 \}.$$
Note that
$$\mathbf{S} = \bigcup \{ \mathbf{S}_I : I \subset \{1, \ldots, n\} \} \quad \text{and}$$
$$\text{cl}(\mathbf{S}_I) = \{ \mathbf{p} \in \mathbf{S} : p_j = 0 \text{ if } j \notin I \}.$$
We have the following straightforward result.

Theorem 5.10. *The replicator system of equations (5.4) has the following properties.*

a. $\mathbf{S}$ *is invariant.*

b. *For any nonempty subset I of $\{1, \ldots, n\}$, $\text{cl}(\mathbf{S}_I)$, $\mathbf{S}_I$ and $\hat{\mathbf{S}}_I$ are invariant.*

Proof. (a)
$$\frac{d}{dt} \left(\sum_i p_i \right) = \sum_i p_i \mathbf{e}^i \cdot \mathbf{A}\mathbf{p} - \sum_i p_i \mathbf{p} \cdot \mathbf{A}\mathbf{p}$$
$$= \mathbf{p} \cdot \mathbf{A}\mathbf{p} - \left(\sum_i p_i \right) (\mathbf{p} \cdot \mathbf{A}\mathbf{p})$$
$$= \left(1 - \sum_i p_i \right) (\mathbf{p} \cdot \mathbf{A}\mathbf{p}).$$

It follows that the set of the vectors $\mathbf{S}' = \{\, \mathbf{p} \in \mathbb{R}^n : \sum_i p_i = 1 \,\}$ is invariant. The derivative $\dot{p}_i = 0$ on the hyper-plane $\{p_i = 0\}$, so it is invariant and a solution starting with $p_i \geq 0$ cannot cross into the region with $p_i < 0$. Thus, a solution starting in $\mathbf{S}$ must stay in both $\mathbf{S}'$ and the region $\{\, \mathbf{p} : p_i \geq 0 \text{ for } 1 \leq i \leq n \,\}$, i.e., $\mathbf{S}$ is invariant.

(b) A solution starting in $\text{cl}(\mathbf{S}_I)$ starts and remains in the subset of $\mathbf{S}$ with $p_i = 0$ for $i \notin I$, so in $\text{cl}(\mathbf{S}_I)$, i.e., $\text{cl}(\mathbf{S}_I)$ is invariant.

Similarly, the set $\text{cl}(\mathbf{S}_I) \setminus \mathbf{S}_I = \{\, \mathbf{p} \in \text{cl}(\mathbf{S}_I) : p_i = 0 \text{ for some } i \in I \,\}$ is invariant. It follows that $\mathbf{S}_I = \text{cl}(\mathbf{S}_I) \setminus (\text{cl}(\mathbf{S}_I) \setminus \mathbf{S}_I)$ is invariant.

Since $\hat{\mathbf{S}}_I = \bigcup_{J \supset I} \mathbf{S}_J$, $\hat{\mathbf{S}}_I$ is also invariant. □

Theorem 5.11. *A point $\hat{\mathbf{p}}$ is a fixed point of (5.4) if and only if $\mathbf{e}^i \cdot \mathbf{A}\hat{\mathbf{p}} = \hat{\mathbf{p}} \cdot \mathbf{A}\hat{\mathbf{p}}$ whenever $\hat{p}_i > 0$. So, $\mathbf{e}^i \cdot \mathbf{A}\hat{\mathbf{p}}$ has the same value for all i with $\hat{p}_i > 0$.*

The proof follows directly from the replicator system of differential equations (5.4).

Theorem 5.12. *If the replicator system of differential equations (5.4) does not have any fixed points in the interior $\text{int}(\mathbf{S}) = \mathbf{S}_{\{1,\ldots,n\}}$, then $\omega(\mathbf{p}) \subset \partial(\mathbf{S}) = \mathbf{S} \setminus \text{int}(\mathbf{S})$ and $\alpha(\mathbf{p}) \subset \partial(\mathbf{S})$.*

Remark 5.15. Theorem 5.8 gives a similar result for Lotka–Volterra equations in arbitrary dimensions. If there is no fixed point in $\mathbb{R}^n_+ = \{\, \mathbf{x} : x_i > 0 \text{ for all } i \,\}$, then $\omega(\mathbf{p}) \cap \mathbb{R}^n_+ = \emptyset$ for any $\mathbf{p}$.

Proof. We consider the case for the ω-limit set, and the proof for the α-limit set is similar. Also, because the boundary is invariant, if $\mathbf{p} \in \partial(\mathbf{S})$, then $\omega(\mathbf{p}) \subset \partial((\mathbf{S}))$. Therefore, we can assume that $\mathbf{p} \in \text{int}(\mathbf{S})$. We next construct a strictly decreasing Lyapunov function on $\text{int}(\mathbf{S})$.

For the proof, we use the following two convex sets: $\Delta = \{\, \mathbf{y} \in \mathbb{R}^n : y_1 = \cdots = y_n \,\}$ and $\mathbf{W} = \mathbf{A}(\text{int}(\mathbf{S}))$. The set Δ is a subspace and so is convex; $\mathbf{W}$ is the linear image of a convex set, so is convex. It follows from Theorem 5.11 that $\mathbf{p} \in \text{int}(\mathbf{S})$ is a fixed point if and only if $\mathbf{A}\mathbf{p} \in \Delta$. Because there are no fixed points in the interior, $\mathbf{W} \cap \Delta = \emptyset$. Since these two convex sets are disjoint, a standard theorem in the theory of convex sets says there is a $\mathbf{c} \in \mathbb{R}^n$ such that $\mathbf{c} \cdot \mathbf{z} < \mathbf{c} \cdot \mathbf{y}$ for all $\mathbf{z} \in \mathbf{W}$ and all $\mathbf{y} \in \Delta$. But $\mathbf{c} \cdot \mathbf{y} = \sum_i c_i y_i = y_1 \sum_i c_i$. For $\mathbf{c} \cdot \mathbf{z} < y_1 \sum_i c_i$ to hold for both large positive and negative y_1, we need $\sum_i c_i = 0$. Since each $\mathbf{z} \in \mathbf{W}$ equals $\mathbf{A}\mathbf{x}$ for $\mathbf{x} \in \text{int}(\mathbf{S})$, $\mathbf{c} \cdot \mathbf{A}\mathbf{x} < 0$ for all $\mathbf{x} \in \text{int}(\mathbf{S})$.

Using this $\mathbf{c}$, define $V(\mathbf{x}) = \sum_i c_i \ln(x_i)$ on $\text{int}(\mathbf{S})$. Then

$$\dot{V}(\mathbf{x}) = \sum_i c_i (\mathbf{A}\mathbf{x})_i - \mathbf{x} \cdot \mathbf{A}\mathbf{x} \sum_i c_i = \mathbf{c} \cdot \mathbf{A}\mathbf{x} < 0.$$

By the argument of Theorem 5.1(b), if there were an $\mathbf{x} \in \omega(\mathbf{p}) \cap \text{int}(\mathbf{S})$, then $\dot{V}(\mathbf{x}) = 0$. Thus, $\omega(\mathbf{p}) \cap \text{int}(\mathbf{S}) = \emptyset$ and $\omega(\mathbf{p}) \subset \partial(\mathbf{S})$. □

Example 5.16. When $\mathbf{A}$ is 2×2, the differential equation is determined by a scalar equation by setting $x = p_1$ and $p_2 = 1 - x$. The differential equation for

5.6. Applications

$\begin{pmatrix} -1 & 2 \\ 0 & 1 \end{pmatrix}$ is

$$\begin{aligned} \dot{x} &= x\left[-x + 2(1-x) + x^2 - 2x(1-x) - (1-x)^2\right] \\ &= x\left[-x(1-x) + 2(1-x)^2 - (1-x)^2\right] \\ &= x(1-x)\left[-x + 1 - x\right] \\ &= x(1-x)(1-2x). \end{aligned}$$

The fixed points are $x = 0$, $1/2$, 1. The two end points 0 and 1 are repelling, and $1/2$ is asymptotically stable.

Example 5.17. The differential equation for $\begin{pmatrix} 1 & 0 \\ 0 & 2 \end{pmatrix}$ is

$$\begin{aligned} \dot{x} &= x\left[x - x^2 - 2(1-x)^2\right] \\ &= x(1-x)(3x - 2). \end{aligned}$$

The fixed points are $x = 0$, $2/3$, 1. The two end points 0 and 1 are asymptotically stable and $2/3$ is repelling.

Example 5.18. The 3×3 matrix $\begin{pmatrix} 0 & -1 & 1 \\ 1 & 0 & -1 \\ -1 & 1 & 0 \end{pmatrix}$ has $a_{ji} = -a_{ij}$ so $\mathbf{p} \cdot \mathbf{Ap} = -\mathbf{p} \cdot \mathbf{Ap} = 0$, and the differential equations are $\dot{p}_i = p_i\left(\mathbf{e}^i \cdot \mathbf{Ap}\right)$. Therefore,

$$\begin{aligned} \dot{p}_1 &= p_1\left(-p_2 + p_3\right), \\ \dot{p}_2 &= p_2\left(p_1 - p_3\right), \\ \dot{p}_3 &= p_3\left(-p_1 + p_2\right). \end{aligned}$$

Letting $p_3 = 1 - p_1 - p_2$, we get a system with two variables,

$$\begin{aligned} \dot{p}_1 &= p_1(1 - p_1 - 2p_2), \\ \dot{p}_2 &= p_2(2p_1 + p_2 - 1). \end{aligned}$$

The fixed points are $(0,0)$, $(1,0)$, $(0,1)$, and $(1/3, 1/3)$. A direct check shows that $(0,0)$, $(1,0)$, and $(0,1)$, are saddles and $(1/3, 1/3)$ is a center.

For the original equations, $\frac{d}{dt}(-p_1 p_2 p_3) = 0$, so $L = -p_1 p_2 p_3$ is constant along solutions so is an integral of motion. For equations with two variables, the integral becomes $L(p_1, p_2) = -p_1 p_2(1 - p_1 - p_2)$, which has a minimum at $(1/3, 1/3)$ and is a weak Lyapunov function for this fixed point. Therefore $(1/3, 1/3)$ is Lyapunov stable but not asymptotically stable.

Example 5.19. (Hofbauer and Sigmund) The 3×3 matrix $\begin{pmatrix} 0 & 6 & -4 \\ -3 & 0 & 5 \\ -1 & 3 & 0 \end{pmatrix}$ has the system of differential equations

$$\begin{aligned} \dot{p}_1 &= p_1\left(6p_2 - 4p_3 - 3p_1 p_2 + 5p_1 p_3 - 8p_2 p_3\right), \\ \dot{p}_2 &= p_2\left(-3p_1 + 5p_3 - 3p_1 p_2 + 5p_1 p_3 - 8p_2 p_3\right), \\ \dot{p}_3 &= p_3\left(-p_1 + 3p_2 - 3p_1 p_2 + 5p_1 p_3 - 8p_2 p_3\right). \end{aligned}$$

Setting $p_3 = 1 - p_1 - p_2$,
$$\dot{p}_1 = p_1\left(-4 + 9p_1 + 2p_2 - 5p_1^2 + 8p_2^2\right),$$
$$\dot{p}_2 = p_2\left(5 - 3p_1 - 13p_2 - 5p_1^2 + 8p_2^2\right).$$

This system has fixed points at $(0,0)$, $(1,0)$, $(0,1)$, $(0, 5/8)$, $(4/5, 0)$, and $(1/3, 1/3)$. The matrix of partial derivatives is

$$DF_{(p_1, p_2)} = \begin{pmatrix} -4 + 18p_1 + 2p_2 - 15p_1^2 + 8p_2^2 & 2p_1 + 16p_1 p_2 \\ -3p_2 - 10p_1 p_2 & 5 - 3p_1 - 26p_2 - 5p_1^2 + 24p_2^2 \end{pmatrix}.$$

The matrix at the various fixed points and their stability type are as follows:

$DF_{(0,0)} = \begin{pmatrix} -4 & 0 \\ 0 & 5 \end{pmatrix},$ $\qquad (0,0)$ is a saddle,

$DF_{(1,0)} = \begin{pmatrix} -1 & 2 \\ 0 & -3 \end{pmatrix},$ $\qquad (1,0)$ is asymptotically stable,

$DF_{(0,1)} = \begin{pmatrix} 6 & 0 \\ -3 & 3 \end{pmatrix},$ $\qquad (1,0)$ is repelling,

$DF_{(\frac{4}{5}, 0)} = \begin{pmatrix} \frac{4}{5} & \frac{8}{5} \\ 0 & -\frac{3}{5} \end{pmatrix},$ $\qquad \left(\frac{4}{5}, 0\right)$ is a saddle,

$DF_{(0, \frac{5}{8})} = \begin{pmatrix} \frac{3}{8} & 0 \\ -\frac{15}{8} & -\frac{15}{8} \end{pmatrix},$ $\qquad \left(0, \frac{5}{8}\right)$ is a saddle,

$DF_{(\frac{1}{3}, \frac{1}{3})} = \begin{pmatrix} \frac{17}{9} & \frac{22}{9} \\ -\frac{19}{9} & -\frac{23}{9} \end{pmatrix},$ $\qquad \left(\frac{1}{3}, \frac{1}{3}\right)$ is asymptotically stable.

The characteristic equation at $(1/3, 1/3)$ is $\lambda^2 + \frac{2}{3}\lambda + \frac{1}{3} = 0$ with eigenvalues $-\frac{1}{3} \pm \frac{\sqrt{2}i}{3}$. Thus, $(1/3, 1/3)$ is asymptotically stable. See Figure 20 for the phase portrait.

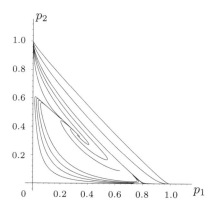

Figure 20. Example 5.19

5.6.4. Monetary policy. A model of N. Obst for inflation and monetary policy is given in [**Chi84**]. The amount of the national product is Q, the "price" of the national product is P, and the value of the national product is PQ. The money supply and demand are given by M_s and M_d, and the variable $\mu = M_d/M_s$ is the ratio. It is assumed that $M_d = aPQ$ where $a > 0$ is a constant, so $\mu = aPQ/M_s$. The following rates of change are given:

$$\frac{1}{P}\frac{dP}{dt} = p \quad \text{is the rate of inflation,}$$

$$\frac{1}{Q}\frac{dQ}{dt} = q \quad \text{is the rate of growth of national product, and}$$

$$\frac{1}{M_s}\frac{dM_s}{dt} = m \quad \text{is the rate of monetary expansion.}$$

We assume q is a constant, and p and m are variables. The rate of change of inflation is assumed to be given by

$$\frac{dp}{dt} = h\left(\frac{M_s - M_d}{M_s}\right) = h(1 - \mu),$$

where $h > 0$ is a constant.

Obst argued that the policy that sets m should not be a function of p but a function of $-\frac{dp}{dt}$, so of μ. For simplicity of discussion, we assume that $m = m_1\mu + m_0$ with $m_1 > 0$. Differentiating $\ln(\mu)$ with respect to t, we get the following:

$$\ln(\mu) = \ln(a) + \ln(P) + \ln(Q) - \ln(M_s),$$

$$\frac{1}{\mu}\frac{d\mu}{dt} = 0 + \frac{1}{P}\frac{dP}{dt} + \frac{1}{Q}\frac{dQ}{dt} - \frac{1}{M_s}\frac{dM_s}{dt} = p + q - (m_1\mu + m_0).$$

Combining, we have the system of nonlinear equations

$$\frac{dp}{dt} = h(1 - \mu),$$

$$\frac{d\mu}{dt} = (p + q - m_1\mu - m_0)\mu.$$

At the equilibrium, $\mu^* = 1$ and $p^* = m_1 + m_0 - q$. To avoid deflation, the monetary policy should be made with $m_1 + m_0 \geq q$, i.e., the monetary growth needs to be large enough to sustain the growth of the national product. The system can be linearized at (p^*, μ^*) by forming the matrix of partial derivatives,

$$\begin{bmatrix} 0 & -h \\ \mu & (p + q - m_1\mu - m_0) - m_1\mu \end{bmatrix}_{(p^*, \mu^*)} = \begin{bmatrix} 0 & -h \\ 1 & -m_1 \end{bmatrix}.$$

The determinant is $h > 0$, and the trace is $-m_1 < 0$. Therefore, the eigenvalues at the equilibrium, $-m_1/2 \pm \sqrt{m_1^2/4 - h}$, have negative real parts. For $4h > m_1^2$, they are complex, $-m_1/2 \pm i\sqrt{h - m_1^2/4}$. Since the real parts of the eigenvalues are nonzero, the linear terms dominate the behavior of solutions near the equilibrium (just like for critical points of a real-valued function) and solutions near the equilibrium spiral in toward the equilibrium. See Figure 21.

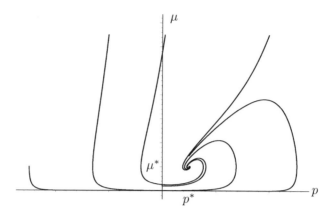

Figure 21. Monetary policy

By itself, the linearization does not tell what happens to solutions far from the equilibrium, but another method shows that all solutions with $\mu_0 > 0$ do converge to the equilibrium. We form a real-valued Lyapunov function (by judicious guessing and integrating),

$$L(p, \mu) = \frac{p^2}{2} + (q - m_1 - m_0)p + h\mu - h\ln(\mu).$$

This function is defined on the upper half-plane, $\mu > 0$, and has a minimum at (p^*, μ^*). The time derivative of L along solution of the differential equation is given as follows:

$$\frac{d}{dt}L = p\frac{dp}{dt} + (q - m_1 - m_0)\frac{dp}{dt} + h\frac{d\mu}{dt} - \frac{h}{\mu}\frac{d\mu}{dt}$$
$$= (p + q - m_1 - m_0)\,h(1 - \mu) + h(\mu - 1)(p + q - m_1\mu + m_1 - m_0)$$
$$= h(\mu - 1)\,(m_1 - m_1\mu)$$
$$= -hm_1(\mu - 1)^2 \leq 0.$$

This function decreases along solutions, and it can be shown that it must go to the minimum value so the solution goes to the equilibrium.

5.6.5. Neural Networks. A neural network is a model for the human brain, in which there are many neurons that separately decay to some steady state but are coupled together. Hopfield [**Hop82**] discussed some models and constructed a Lyapunov function for his model. Let N be the number of neurons, with the state of the ith neuron given by the real variable x_i. The constants $c_i > 0$ measure the decay of the ith neuron without any stimulation. There is a matrix $\mathbf{T} = (t_{i,j})$, called the *connection matrix*, which determines whether the jth neuron affects the ith neuron or not, and in what manner. If $t_{i,j} > 0$, then the action is excitatory, if $t_{i,j} < 0$, then it is inhibitory, and if $t_{i,j} = 0$, then there is no direct interaction. Typically, there are many neurons which do not interact with each other, so there are many 0's in the connection matrix. A function $f_j(x_j)$, called the *neuron response function*, measures the effective action of the jth neuron on other neurons, with the net effect of the jth neuron on the ith neuron given by $t_{i,j}f_j(x_j)$. We assume that

5.6. Applications

each of the functions f_j is differentiable, $f_j'(x_i) > 0$, and bounded above and below. Finally, there are constant inputs b_i. Combining these effects, we assume that the differential equation for the ith neuron is given by

$$\dot{x}_i = -c_i\, x_i + \sum_{j=1}^{N} t_{i,j}\, f_j(x_j) + b_i,$$

or the system of differential equations

(5.5) $$\dot{\mathbf{x}} = -\mathbf{C}\,\mathbf{x} + \mathbf{T}\,\mathbf{f}(\mathbf{x}) + \mathbf{b},$$

where $\mathbf{C}$ is the diagonal matrix with entries c_i along the diagonal, the function $\mathbf{f}$ has the form, $\mathbf{f}(\mathbf{x}) = (f_1(x_1), \ldots, f_n(x_n))^\mathsf{T}$, and $\mathbf{b} = (b_1, \ldots, b_n)^\mathsf{T}$. For simplicity in the following, we assume the $\mathbf{T}$ is invertible and $\mathbf{B} = \mathbf{T}^{-1}\mathbf{b}$, so

$$\dot{\mathbf{x}} = -\mathbf{C}\,\mathbf{x} + \mathbf{T}\,\mathbf{f}(\mathbf{x}) + \mathbf{T}\,\mathbf{B}.$$

For the case when $\mathbf{T}$ is symmetric, Hopfield [**Hop82**] was able to construct a Lyapunov function which forces the trajectories to tend to fixed points. (It is possible to have more than one fixed point with just these assumptions.) The assumption that $\mathbf{T}$ is symmetric is not very realistic, and recently A. Williams [**Wil02**] was able to generalize to the case for which $\mathbf{T}$ is not symmetric with other appropriate assumptions. Let

$$F(\mathbf{x}) = \sum_i c_i \int_0^{f_i(x_i)} f_i^{-1}(s)\, ds.$$

Then, the Lyapunov function is given by

(5.6) $$V(\mathbf{x}) = F(\mathbf{x}) - \frac{1}{2} [\mathbf{f}(\mathbf{x}) + \mathbf{B}]^\mathsf{T}\, \mathbf{T}\, [\mathbf{f}(\mathbf{x}) + \mathbf{B}].$$

We give this result in the next theorem.

Theorem 5.13. *Consider the system of differential equations given by equation (5.5) with $\mathbf{C}$ a diagonal matrix. Assume that each of the functions f_i is differentiable, bounded above and below, and that $f_i'(x_i) > 0$. Further assume that the connection matrix $\mathbf{T}$ is invertible and symmetric. Then, the function $V(\mathbf{x})$ given by equation (5.6) is a Lyapunov function, with $\dot{V}(\mathbf{x}) < 0$ except at the fixed points. Thus, all solutions tend to a fixed points as time goes to infinity.*

Proof. Taking the time derivative of V, we get

$$\dot{V}(\mathbf{x}) = \sum_i c_i\, f_i^{-1}(f_i(x_i))\, f_i'(x_i)\, \dot{x}_i - \sum_i [\mathbf{f}(\mathbf{x}) + \mathbf{B}]^\mathsf{T}\, \mathbf{T}\, \frac{\partial \mathbf{f}}{\partial x_i}(\mathbf{x})\, \dot{x}_i$$

$$= [\mathbf{C}\,\mathbf{x}]^\mathsf{T}\, D\mathbf{f}_{(\mathbf{x})}\, \dot{\mathbf{x}} - [\mathbf{T}\,\mathbf{f}(\mathbf{x}) + \mathbf{T}\,\mathbf{B}]^\mathsf{T}\, D\mathbf{f}_{(\mathbf{x})}\, \dot{\mathbf{x}}$$

$$= [\mathbf{C}\,\mathbf{x} - \mathbf{T}\,\mathbf{f}(\mathbf{x}) - \mathbf{T}\,\mathbf{B}]^\mathsf{T}\, D\mathbf{f}_{(\mathbf{x})}\, \dot{\mathbf{x}}$$

$$= -\dot{\mathbf{x}}^\mathsf{T}\, D\mathbf{f}_{(\mathbf{x})}\, \dot{\mathbf{x}}.$$

Since $D\mathbf{f}_{(\mathbf{x})}$ is a diagonal matrix with positive entries $f_i'(x_i)$, $\dot{V}(\mathbf{x}) \leq 0$ and $\dot{V}(\mathbf{x})$ is zero if and only if $\dot{\mathbf{x}}$ is zero. This is what we need to prove. □

Remark 5.20. Notice that, if $\mathbf{T}$ is not symmetric, then the derivative of the second term gives the symmetric part of $\mathbf{T}$ and the substitution for $\dot{\mathbf{x}}$ cannot be made.

The neural network gives an output for any input **b**. The next theorem says that distinct inputs yield distinct outputs.

Theorem 5.14. *Consider the system of differential equations given by equation (5.5) with* **C**, **T**, *and* **f** *fixed. Assume that each of the coordinate functions f_i is differentiable, bounded above and below, and that $f_i'(x_i) > 0$. Assume that* $\mathbf{b}^1$ *and* $\mathbf{b}^2$ *are two inputs with* $\mathbf{b}^1 \neq \mathbf{b}^2$. *Then, any steady state for* $\mathbf{b}^1$ *is distinct from any steady state for* $\mathbf{b}^2$.

Proof. Assume that $\bar{\mathbf{x}}^j$ is a steady state for $\mathbf{b}^j$ for $j = 1, 2$. Then,
$$0 = -\mathbf{C}\,\bar{\mathbf{x}}^1 + \mathbf{T}\,\mathbf{f}(\bar{\mathbf{x}}^1) + \mathbf{b}^1 \quad \text{and}$$
$$0 = -\mathbf{C}\,\bar{\mathbf{x}}^2 + \mathbf{T}\,\mathbf{f}(\bar{\mathbf{x}}^2) + \mathbf{b}^2,$$
so
$$\mathbf{C}[\bar{\mathbf{x}}^1 - \bar{\mathbf{x}}^2] + \mathbf{T}\,[-\mathbf{f}(\bar{\mathbf{x}}^1) + \mathbf{f}(\bar{\mathbf{x}}^2)] = \mathbf{b}^1 - \mathbf{b}^2.$$
Notice that, if $\bar{\mathbf{x}}^1 = \bar{\mathbf{x}}^2$, then the left side of the last equation is zero, so $\mathbf{b}^1 = \mathbf{b}^2$. Thus, if $\mathbf{b}^1 \neq \mathbf{b}^2$, then $\bar{\mathbf{x}}^1 \neq \bar{\mathbf{x}}^2$. □

Exercises 5.6

1. Consider the system of differential equations with the six-twelve potential
$$\dot{x} = y,$$
$$\dot{y} = -x - \frac{a}{x^{13}} + \frac{b}{x^7},$$
where $a > 0$ and $b > 0$.
   a. Find the potential function $V(x)$ and draw its graph. Classify the type of the fixed point.
   b. Draw the phase portrait for the system of differential equations.

2. Consider the replicator equations for the matrix $\mathbf{A} = \begin{pmatrix} 0 & 3 \\ 4 & 1 \end{pmatrix}$.
   a. Determine the scalar replicator equations.
   b. Determine the fixed points and their stability.

5.7. Theory and Proofs

Lyapunov functions

Restatement of Theorem 5.2. *Let* $\mathbf{x}^*$ *be a fixed point, and let* L *be a weak Lyapunov function in a neighborhood* **U** *of* $\mathbf{x}^*$. *Define*
$$\mathbf{Z_U} = \{\mathbf{x} \in \mathbf{U} : \dot{L}(\mathbf{x}) = 0\}.$$
Assume that **U** *is sufficiently small such that, for any* $\mathbf{x}_1 \in \mathbf{Z_U} \setminus \{\mathbf{x}^*\}$, *the trajectory* $\phi(t; \mathbf{x}_1)$ *moves off* $\mathbf{Z_U}$ *into* $\mathbf{U} \setminus \mathbf{Z_U}$ *for small positive t (i.e., $\{\mathbf{x}^*\}$ is the largest positively invariant set in* $\mathbf{Z_U}$). *Then,* $\mathbf{x}^*$ *is asymptotically stable.*

Further assume that $L_0 > L(\mathbf{x}^*)$ is a value for which
$$\mathbf{U}_{L_0} = \{\mathbf{x} \in \mathbf{U} : L(\mathbf{x}) \leq L_0\}$$

5.7. Theory and Proofs

is contained inside $\mathbf{U}$, is bounded, and is bounded away from the boundary of $\mathbf{U}$. Then, the set $\mathbf{U}_{L_0}$ is contained in the basin of attraction of $\mathbf{x}^*$, $\mathbf{U}_{L_0} \subset W^s(\mathbf{x}^*)$.

Proof. Let $L^* = L(\mathbf{x}^*)$, and $L_0 > L^*$ be such that
$$\mathbf{U}_{L_0} = \{\mathbf{x} \in \mathbf{U} : L(\mathbf{x}) \leq L_0\}$$
is contained inside $\mathbf{U}$, is bounded, and is bounded away from the boundary of $\mathbf{U}$. We show that $\mathbf{U}_{L_0}$ is contained in $W^s(\mathbf{x}^*)$.

Let $\mathbf{x}_0$ be a point in $\mathbf{U}_{L_0} \setminus \{\mathbf{x}^*\}$. The set $\mathbf{U}_{L_0}$ is positively invariant, so the whole forward orbit $\phi(t; \mathbf{x}_0)$ must be contained in $\mathbf{U}_{L_0}$. Since $\dot{L} \leq 0$, the value $L(\phi(t; \mathbf{x}_0))$ is decreasing as t increases and is bounded below by L^*. Therefore, it must have a limit value L_∞ as t goes to infinity. Since $L(\phi(t; \mathbf{x}_0))$ is decreasing, $L(\phi(t; \mathbf{x}_0)) \geq L_\infty$ for all $t \geq 0$.

We complete the proof using two different ways of expressing the argument. The first is closer to the argument given for the damped pendulum and the second uses the properties of the ω-limit set.

First Argument: If $L_\infty = L^*$, then we are done, since
$$L^{-1}(L^*) \cap \mathbf{U} = \{\mathbf{x}^*\}.$$

Assume that $L_\infty > L^*$. In order for $L(\phi(t; \mathbf{x}_0))$ to have L_∞ as a limit as t goes to infinity, $\dot{L}(\phi(t; \mathbf{x}_0))$ must go to zero. The orbit $\phi(t; \mathbf{x}_0)$ is bounded, so it must keep coming near some point $\mathbf{z}$. (This actually uses the fact that $\mathbf{U}_{L_0}$ is compact.) Therefore, there is a sequence of times t_n going to infinity such that $\phi(t_n; \mathbf{x}_0)$ approaches $\mathbf{z}$ as n goes to infinity. Therefore, $L(\phi(t_n; \mathbf{x}_0))$ limits on $L(\mathbf{z})$ and also L_∞, so $L(\mathbf{z}) = L_\infty$. Because $\dot{L}(\phi(t; \mathbf{x}_0))$ goes to zero, $\dot{L}(\mathbf{z}) = 0$ and $\mathbf{z}$ must be a point in
$$\mathbf{Z}_\mathbf{U} \cap L^{-1}(L_\infty) \subset \mathbf{U}_{L_0}.$$
Since $\mathbf{z} \neq \mathbf{x}^*$, for small $\hat{t} > 0$, $\phi(\hat{t}; \mathbf{z})$ must leave $\mathbf{Z}_\mathbf{U}$ and $L(\phi(\hat{t}; \mathbf{z}))$ must be a value less than L_∞. Then, $\phi(\hat{t} + t_n; \mathbf{x}_0) = \phi(\hat{t}; \phi(t_n; \mathbf{x}_0))$ converges to $\phi(\hat{t}; \mathbf{z})$ and $L(\phi(\hat{t} + t_n; \mathbf{x}_0))$ converges to $L(\phi(\hat{t}; \mathbf{z}))$ which is less than L_∞. This contradicts the fact that $L(\phi(t; \mathbf{x}_0)) \geq L_\infty$ for all $t \geq 0$. Therefore, $L_\infty = L^*$, and the trajectory converges to $\mathbf{x}^*$.

Second Argument: Because the trajectory is contained inside $\mathbf{U}_{L_0}$, the ω-limit set $\omega(\mathbf{x}_0)$ is nonempty. For any point $\mathbf{z}$ in $\omega(\mathbf{x}_0)$, there is a sequence of times t_n going to infinity such that $\phi(t_n; \mathbf{x}_0)$ approaches $\mathbf{z}$. Therefore, the limit of $L(\phi(t_n; \mathbf{x}_0))$ is $L(\mathbf{z})$ and also L_∞. Thus, $L(\mathbf{z}) = L_\infty$ for any point in $\omega(\mathbf{x}_0)$.

Take any point $\mathbf{z}$ in $\omega(\mathbf{x}_0)$. The limit set is invariant, so $\phi(t; \mathbf{z})$ must remain in $\omega(\mathbf{x}_0)$ for all $t \geq 0$,
$$L(\phi(t; \mathbf{z})) \equiv L_\infty,$$
$$\frac{d}{dt} L(\phi(t; \mathbf{z})) \equiv 0,$$
and $\phi(t; \mathbf{z})$ must stay in $\mathbf{Z}_\mathbf{U}$ for all $t \geq 0$. Since $\mathbf{x}^*$ is the only point in $\mathbf{Z}_\mathbf{U}$ whose whole forward orbit stays inside $\mathbf{Z}_\mathbf{U}$, the only point in $\omega(\mathbf{x}_0)$ must be $\mathbf{x}^*$. This shows that $\omega(\mathbf{x}_0) = \{\mathbf{x}^*\}$, and the trajectory $\phi(t; \mathbf{x}_0)$ must converge to $\mathbf{x}^*$. $\square$

Chapter 6

Periodic Orbits

Although several systems already considered have contained periodic orbits, they have not been the focus of the study until this chapter. In the first section, we give the basic definitions and some simple examples that have periodic orbits. These examples are ones that are simple in polar equations, so we can use one-dimensional analysis of the radial component to show that there is a periodic orbit. Section 6.2 presents a general result about differential equations in the plane, called the Poincaré–Bendixson theorem. This theorem states that an orbit, which stays away from fixed points and does not go to infinity, must approach a periodic orbit. Although, this result holds only in two dimensions, it is an important way to show that certain systems must have periodic orbits. Most of the applications given of this theorem have a repelling fixed point and use a test function to show that there is a positively invariant bounded region: There must be at least one periodic orbit in this bounded region. Section 6.3 considers a certain class of nonlinear self-excited oscillators that have unique attracting periodic orbits. These equations again have a repelling fixed point at the origin, but have an attracting force only outside a vertical strip. Orbits with large amplitudes must cross the strip where energy is added to the system, but spend more time in the region where energy is taken out of the system. A calculation is needed to show that the net effect on large amplitude orbits is to lose energy when a complete trip around the fixed point at the origin is made.

The next two sections consider mechanisms of creating periodic orbits by varying parameters. The first "bifurcation" is the creation of a periodic orbit coming out of a fixed point, called the *Andronov–Hopf bifurcation*. There are many systems arising from applications for which the Andronov–Hopf bifurcation applies, of which we give a couple in this chapter; in Chapter 7, we note that it also occurs for the Lorenz system. The second "bifurcation" is the creation of a periodic orbit from an orbit with both α- and ω- limit sets at a fixed point; this is called a *homoclinic bifurcation*. This bifurcation occurs less often than the Andronov–Hopf bifurcation, but it is another mechanism for the creation of periodic orbits; it, too, arises for the Lorenz system considered in Chapter 7.

Section 6.6 shows the relationship between the divergence of the vector field of a differential equation and the way the flow changes the area or volume of a region in the phase space. This result is used to show that certain systems of differential equations in the plane do not have periodic orbits contained entirely in a region of the phase plane. In Chapter 7, these results are used in the discussion of "chaotic attractors." In Section 6.7, we return to the discussion of stable periodic orbits and the connection with a function obtained by going around once near the periodic orbit. This function is what is called the Poincaré map. It is introduced in Section 6.1 in terms of simple examples, and it plays the central role in the analysis of the Lienard equations in Section 6.3.

6.1. Introduction to Periodic Orbits

Let $\phi(t; \mathbf{x}_0)$ be the flow of the differential equation $\dot{\mathbf{x}} = \mathbf{F}(\mathbf{x})$. In Section 3.1, we defined a *periodic point of period* T to be a point $\mathbf{x}_0$ such that $\phi(T; \mathbf{x}_0) = \mathbf{x}_0$, but $\phi(t; \mathbf{x}_0) \neq \mathbf{x}_0$ for $0 < t < T$. If $\mathbf{x}_0$ is periodic, with period T, then the set of all the points $\{\phi(t; \mathbf{x}_0) : 0 \leq t \leq T\}$ is called a *periodic orbit* or *closed orbit*. It is called a periodic orbit because the flow is periodic in time, $\phi(t + T; \mathbf{x}_0) = \phi(t; \mathbf{x}_0)$ for all t. It is called a closed orbit because the orbit "closes" up on itself after time T and the set of points on the whole orbit $\{\phi(t; \mathbf{x}_0) : -\infty < t < \infty\}$ is a closed set. (The definition of a closed set is given in Appendix A.2.)

Definition 6.1. Just as a fixed point can be L-stable or asymptotically stable, so a periodic orbit can have different types of stability.

A periodic orbit $\gamma = \{\phi(t; \mathbf{x}_0) : 0 \leq t \leq T\}$ is called *orbitally L-stable* provided that the following condition holds: Given any $\epsilon > 0$, there is a $\delta > 0$ such that, if $\mathbf{x}_0$ is an initial condition within a distance δ of γ, then $\phi(t; \mathbf{x}_0)$ is within a distance ϵ of γ for all $t \geq 0$. Thus, an orbit is orbitally L-stable if all solutions which start near the orbit stay near it for all positive time.

The periodic orbit is called *orbitally ω-attracting* provided there is a $\delta_1 > 0$ such that any initial condition $\mathbf{x}_0$ within δ_1 of γ has the distance between $\phi(t; \mathbf{x}_0)$ and γ go to zero as t goes to infinity (i.e., $\omega(\mathbf{x}_0) = \gamma$).

The periodic orbit is called *orbitally asymptotically stable* provided it is orbitally L-stable and orbitally ω-attracting. Notice, that the definition does not require that the orbit approach any particular solution in the closed orbit, but merely the set of all points in the orbit. This last fact is the reason the adverb "orbitally" is used in the term. An orbitally asymptotically stable periodic orbit is also called a *periodic sink* or an *attracting periodic orbit*.

An orbit is called *orbitally unstable* provided it is not orbitally L-stable; in other words, there is a distance $\epsilon_0 > 0$, initial conditions $\mathbf{x}_j$ arbitrarily close to γ, and times $t_j > 0$, such that the distance from $\phi(t_j; \mathbf{x}_j)$ to γ is greater than $\epsilon_0 > 0$.

Finally, an orbit is called repelling if all orbits move away: A periodic orbit is called *repelling* provided that, as t goes to minus infinity, it satisfies conditions similar to those for orbitally L-stable and orbitally ω attracting, that is, given $\epsilon > 0$, there is a $\delta > 0$ such that, if $\mathbf{x}_0$ is within δ of γ, then $\phi(t; \mathbf{x}_0)$ is within ϵ of γ for

6.1. Introduction to Periodic Orbits

all $t \leq 0$, and the distance between $\phi(t; \mathbf{x}_0)$ and γ goes to zero as t goes to minus infinity (i.e., $\alpha(\mathbf{x}_0) = \gamma$). A repelling periodic orbit is also called a *periodic source*.

Notice that the periodic orbits for the pendulum are orbitally L-stable but not orbitally asymptotically stable (because they occur in a whole band of periodic orbits of varying periods).

Periodic orbits in the plane can either be contained in a band of periodic orbits, like the pendulum equation, or they can be isolated in the sense that nearby orbits are not periodic. The latter case is called a limit cycle: A *limit cycle* is an isolated periodic orbit for a system of differential equations in the plane. On each side of a limit cycle, the other trajectories can be either spiraling in toward the periodic orbit or spiraling away (i.e., on each side, either the α- or ω-limit set equals the periodic orbit). If trajectories on both sides are spiraling in, then the periodic orbit is attracting or orbitally asymptotically stable in the sense previously defined. If trajectories on both sides are spiraling away, then the periodic orbit is repelling in the sense previously defined. Finally, the case of attracting from one side and repelling from the other side is orbitally unstable and is called *orbitally semistable*.

The periodic orbit for the system of equations given earlier in Example 4.2 is the simplest example of an attracting limit cycle. It is analyzed using polar coordinates.

In the rest of this section, we consider one example in the plane, which is easy to analyze using polar coordinates, and we mention the example of the Van der Pol system of differential equations, which we consider in Section 6.3. In the rest of the chapter, we consider more complicated situations which have periodic orbits.

Example 6.2. As in Example 4.2, the following system of equations can be analyzed using polar coordinates. In rectangular coordinates, the example is given by

$$\dot{x} = y + x(1 - x^2 - y^2)(4 - x^2 - y^2),$$
$$\dot{y} = -x + y(1 - x^2 - y^2)(4 - x^2 - y^2),$$

which, in polar coordinates, has the form

$$\dot{r} = r(1 - r^2)(4 - r^2),$$
$$\dot{\theta} = -1.$$

We do not find explicit solutions, but just consider the signs of $\dot{r}$:

$$\begin{cases} \dot{r} > 0 & \text{if } 0 < r < 1, \\ \dot{r} < 0 & \text{if } 1 < r < 2, \\ \dot{r} > 0 & \text{if } 2 < r. \end{cases}$$

Therefore, if r_0 is an initial condition with $0 < r_0 < 2$, then $r(t; r_0)$ converges to 1 as t goes to infinity. If $2 < r_0$, then $r(t; r_0)$ goes to infinity as t goes to infinity, and converges to 2 as t goes to minus infinity. Considering the differential equations in Cartesian coordinates, there is an attracting limit cycle at $r = 1$ and a repelling limit cycle at $r = 2$. See Figure 1

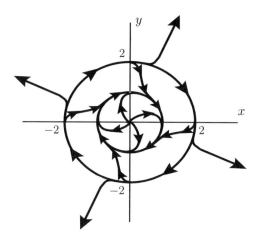

Figure 1. Phase portrait for Example 6.2

Example 6.3 (Van der Pol). A more complicated example is the Van der Pol equation given by $\ddot{x} - \mu(1-x^2)\dot{x} + x = 0$, which corresponds to the system of differential equations

$$\dot{x} = v,$$
$$\dot{v} = -x + \mu(1-x^2)v.$$

For $x^2 < 1$, $\mu(1-x^2) > 0$ and the effect is a force in the direction of the velocity, or an "antifriction". Using polar coordinates,

$$r\dot{r} = x\dot{x} + v\dot{v} = xv - vx + \mu(1-x^2)v^2 \quad \text{or}$$
$$\dot{r} = \frac{v^2(1-x^2)}{r}.$$

Therefore,

$$\begin{cases} \dot{r} > 0 & \text{for } -1 < x < 1, \\ \dot{r} < 0 & \text{for } |x| > 1. \end{cases}$$

This shows that no periodic orbit can be contained entirely in the strip $-1 < x < 1$. We show in Section 6.3, that there is a periodic orbit that passes through the region $|x| > 1$ as well as the strip $|x| \leq 1$. See Figure 7.

Several times in this chapter and later in the book, we use the differential equation to define a function which takes a point on a line in the phase plane and assigns to it the point on the line obtained by following the trajectory until it returns again to the same line. This function is called the *first return map*, or *Poincaré map*. We discuss this construction more completely in Section 6.7, but give an introduction to the idea at this time. For the Van der Pol equation, we take an initial condition on the positive y-axis, $\{(0,y) : y > 0\}$, and follow the trajectory $\phi(t;(0,y))$ until it returns again to the positive y-axis at a later time $T(y)$. The point of this first return can be defined as the Poincaré map, $\phi(T(y),(0,y)) = (0,P(y))$. Although this construction is discussed more completely in Section 6.3, the following example illustrates this analysis for the simple equations given in Example 4.2.

6.1. Introduction to Periodic Orbits

Example 6.4 (Poincaré map for Example 4.2). The differential equations are
$$\dot{x} = y + x\,(1 - x^2 - y^2),$$
$$\dot{y} = -x + y\,(1 - x^2 - y^2),$$
or in polar coordinates
$$\dot{r} = r\,(1 - r^2),$$
$$\dot{\theta} = -1.$$

We discuss the first return of trajectories from the half-line $\{(x,0) : x > 0\}$ to itself. In polar coordinates, this amounts to following solutions from $\theta = 0$ to $\theta = -2\pi$.

After separation of variables, a functional relation between r and t is obtained by integrals:

$$\int_0^t 2\,dt = 2\int_{r_0}^{r(t)} \frac{1}{r(1-r^2)}\,dr = 2\int_{r_0}^{r(t)} \frac{1}{r}\,dr + \int_{r_0}^{r(t)} \frac{1}{1-r}\,dr - \int_{r_0}^{r(t)} \frac{1}{1+r}\,dr,$$

$$2t = \ln\left(\frac{r(t)^2}{1-r(t)^2}\right) - \ln\left(\frac{r_0^2}{1-r_0^2}\right).$$

Solving for $r(t)$, we have

$$r(t)^2 = \frac{r_0^2 e^{2t}}{1 - r_0^2 + r_0^2 e^{2t}}$$
$$= \frac{1}{1 + e^{-2t}(r_0^{-2} - 1)},$$
$$r(t) = r_0[r_0^2 + e^{-2t}(1 - r_0^2)]^{-1/2}$$
$$= [1 + e^{-2t}(r_0^{-2} - 1)]^{-1/2}.$$

The solution for θ is
$$\theta(t) = \theta_0 - t.$$

Thus, it takes a length of time of 2π to go once around the origin from $\theta_0 = 0$ to $\theta(t) = -2\pi$. So, evaluating $r(t)$ at 2π gives the radius after one revolution in terms of the original radius as

$$r_1 = r(2\pi) = [1 + e^{-4\pi}(r_0^{-2} - 1)]^{-1/2}.$$

We can think of this as taking a point r_0 on the half-line $\{\theta = 0, r_0 > 0\}$ and following its trajectory until it returns to this same half-line at the point $r_1 = P(r_0) = r(2\pi)$. This first return map to this half-line is called the *Poincaré map*. The half-line used to construct the Poincaré map is called the *transversal* or *cross section* to the flow. In the current example,

$$P(r_0) = r(2\pi)$$
$$= r_0[r_0^2 + e^{-4\pi}(1 - r_0^2)]^{-1/2}$$
$$= r_0[1 - (1 - e^{-4\pi})(1 - r_0^2)]^{-1/2}$$
$$= [1 + e^{-4\pi}(r_0^{-2} - 1)]^{-1/2}.$$

In Section 6.7, we present a complete discussion of the properties of the graph of P. However, it should not be too difficult to see that the graph of P given in

Figure 2 is basically the correct shape. The points $r = 0$ and $r = 1$ return to themselves, $P(0) = 0$ and $P(1) = 1$, as can be seen from the first equation for $P(r_0)$. For other initial values, $P(r_0) \neq r_0$. We show in Section 6.7, using the Poincaré map, that the periodic orbit $r = 1$ is orbitally asymptotically stable and attracts all orbits other than the fixed point at the origin.

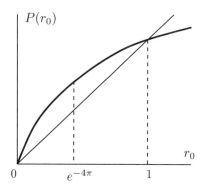

Figure 2. Poincaré map for Example 6.4

Limit Cycles for Polynomial Systems

D. Hilbert conjectured in 1900 that there was a bound on the number of limit cycles in terms of the degree of a system of polynomial differential equations in the plane. This conjecture has not been proved or disproved yet. Yu. Ilyahenko has proved that any system of polynomial differential equations in the plane can have only a finite number of limit cycles. Song Ling Shi in 1980 gave an example of a quadratic system with four limit cycles. The following example of Chin Chu is a simpler example with two limit cycles.

Example 6.5. Consider the system of differential equations
$$\dot{x} = -y - y^2,$$
$$\dot{y} = \frac{1}{2}x - \frac{1}{5}y + xy - \frac{6}{5}y^2.$$

The fixed points are $(0, 0)$ and $(-2, -1)$. Figure 3 shows an attracting limit cycle around $(-2, -1)$ and a repelling limit cycle around $(0, 0)$. Chin Chu proved that these limit cycles exist.

Exercises 6.1

1. Show that the system
$$\dot{x} = y,$$
$$\dot{y} = -x + y(1 - x^2 - y^2),$$
has a periodic orbit. Hint: Find the equation for $\dot{r}$ and then find a value of r for which $\dot{r} = 0$.

6.2. Poincaré–Bendixson Theorem

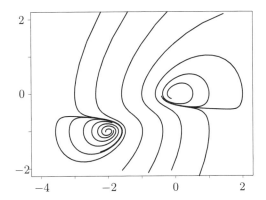

Figure 3. Two limit cycles for Example 6.5.

6.2. Poincaré–Bendixson Theorem

In Section 6.1, we considered some simple equations for which the limit cycles could be found by using polar coordinates. In this section, we will give a slightly more complicated example for which the limit cycle is not a circle and it cannot be determined as easily: There is an annular region for which the trajectories are entering into the region on both boundaries, and a limit cycle is forced to exist somewhere inside. Indeed, a general theorem, called the Poincaré–Bendixson theorem, implies the existence of a periodic orbit in this and similar examples.

Example 6.6. Consider the equations

$$\dot{x} = y,$$
$$\dot{y} = -x + y(4 - x^2 - 4y^2),$$

These equations contain the terms $y(4 - x^2 - 4y^2)$ in the $\dot{y}$ equation, which has the same sign as y for small $\|(x, y)\|$ ("antifriction") and which has the opposite sign as y for large $\|(x, y)\|$ ("friction"); thus, this factor $y(4 - x^2 - 4y^2)$ acts as a generalized nonlinear "friction" in the system. We could use the polar radius to measure the behavior of the system, but instead, we use the corresponding test or bounding function $L(x, y) = \frac{1}{2}(x^2 + y^2)$. The time derivative is

$$\dot{L} = y^2(4 - x^2 - 4y^2) = \begin{cases} \geq 0 & \text{if} \quad 2L(x,y) = x^2 + y^2 \leq 1, \\ \leq 0 & \text{if} \quad 2L(x,y) = x^2 + y^2 \geq 4. \end{cases}$$

These inequalities imply that the annulus

$$\mathscr{A} = \left\{ (x, y) : \frac{1}{2} \leq L(x, y) \leq 2 \right\}$$

is positively invariant. The only fixed point is at the origin, which is not in $\mathscr{A}$. The next theorem then implies that there is a periodic orbit in $\mathscr{A}$. The idea is that this system has a term that is damping on the the outer boundary $L^{-1}(2)$, and an antidamping term on the inner boundary $L^{-1}(1/2)$, so that the orbits are trapped in the annulus where there are no fixed points. See Figure 4.

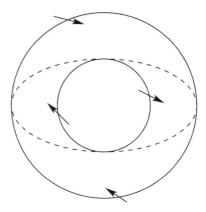

Figure 4. Positively invariant region for Example 6.6

Theorem 6.1 (Poincaré–Bendixson Theorem). *Consider a differential equation $\dot{\mathbf{x}} = \mathbf{F}(\mathbf{x})$ on $\mathbb{R}^2$.*

  a. *Assume that $\mathbf{F}$ is defined on all of $\mathbb{R}^2$ and that a forward orbit $\{\phi(t; \mathbf{q}) : t \geq 0\}$ is bounded. Then, $\omega(\mathbf{q})$ either: (i) contains a fixed point or (ii) is a periodic orbit.*

  b. *Assume that $\mathscr{A}$ is a closed (includes its boundary) and bounded subset of $\mathbb{R}^2$ that is positively invariant for the differential equation. We assume that $\mathbf{F}(\mathbf{x})$ is defined at all points of $\mathscr{A}$ and has no fixed point in $\mathscr{A}$. Then, given any $\mathbf{x}_0$ in $\mathscr{A}$, the orbit $\phi(t; \mathbf{x}_0)$ is either: (i) periodic or (ii) tends toward a periodic orbit as t goes to ∞, and $\omega(\mathbf{x}_0)$ equals this periodic orbit.*

Remark 6.7. In order for a connected region $\mathscr{A}$ in the plane to be positively invariant, but not contain any fixed points, it must be "annular" with just one hole. Thus, it has two boundaries, which are each closed curves (but are not necessarily circles).

Remark 6.8. To get the region $\mathscr{A}$ to be positively invariant, it is enough for the vector field of the system to be coming into the region on both boundaries.

Remark 6.9. With appropriate modifications of Theorem 6.1, the region $\mathscr{A}$ could just as easily be negatively invariant, with the trajectories going out on both boundaries.

Idea of the Proof. We indicate the key ideas of the proof here. In Section 6.9, we include a complete proof. The proof definitely uses the continuity of the flow with respect to initial conditions.

 The trajectory $\phi(t; \mathbf{x}_0)$ stays inside $\mathscr{A}$, so the trajectory needs to repeatedly come near some point $\mathbf{z}$, and there is a sequence of times t_n which go to infinity such that $\phi(t_n; \mathbf{x}_0)$ converges to $\mathbf{z}$. This is the idea called *compactness* in mathematics. Just as an increasing sequence of points that are bounded in the line must converge, so a bounded sequence in $\mathbb{R}^2$ needs to repeatedly come near some point. The point $\mathbf{z}$ is not fixed, so nearby trajectories all go in roughly the same direction. Let $\mathbf{S}$ be a short line segment through $\mathbf{z}$ such that the other trajectories are crossing in the same direction. Then, for sufficiently large n, we can adjust the times t_n so that each $\phi(t_n; \mathbf{x}_0)$ is in $\mathbf{S}$. Taking the piece of the trajectory

$$\{\phi(t; \mathbf{x}_0) : t_n \leq t \leq t_{n+1}\},$$

together with the part $\mathbf{S}'$ of $\mathbf{S}$ between $\phi(t_n; \mathbf{x}_0)$ and $\phi(t_{n+1}; \mathbf{x}_0)$, we get a closed curve which separates the plane into two pieces. See Figure 5. The trajectory going out from $\phi(t_{n+1}; \mathbf{x}_0)$ enters either the region inside or the outside of Γ. All the trajectories starting in $\mathbf{S}'$ enter the same region as that starting at $\phi(t_{n+1}; \mathbf{x}_0)$. Therefore, $\phi(t; \mathbf{x}_0)$ can never reenter the other region for $t > t_{n+1}$. This means that the intersections of the trajectory with $\mathbf{S}$ occur in a monotone manner. They must converge to $\mathbf{z}$ from one side. A further argument shows that, if $\mathbf{z}$ were not periodic, then these nearby trajectories could not return. Therefore, $\mathbf{z}$ must be periodic. $\square$

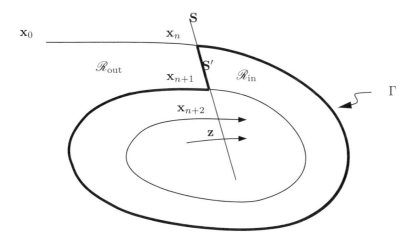

Figure 5. Transversal at $\mathbf{z}$

The proof actually shows more about the convergence to a periodic orbit. If one orbit accumulates on a limit cycle from one side, then the Poincaré map is monotone and attracting on that side. It follows that all initial conditions which are nearby the limit cycle on that side also have to have the limit cycle as their ω-limit, and the limit cycle is orbitally asymptotically stable from that one side. Thus, if there are points on both sides of a periodic orbit which have the periodic orbit as their ω-limit set, then the periodic orbit is orbitally asymptotically stable (from both sides). We summarize this reasoning in the next corollary.

Corollary 6.2. *Consider a differential equation $\dot{\mathbf{x}} = \mathbf{F}(\mathbf{x})$ on $\mathbb{R}^2$ with an isolated periodic orbit γ.*

a. *Assume that a point $\mathbf{p}$ not on γ has $\omega(\mathbf{p}, \phi) = \gamma$. Then, all points $\mathbf{q}$ near enough to γ on the same side of γ as $\mathbf{p}$ also have $\omega(\mathbf{q}, \phi) = \gamma$. In particular, γ is orbitally asymptotically stable from that one side.*

b. *Assume that $\omega(\mathbf{p}_1, \phi) = \gamma = \omega(\mathbf{p}_2, \phi)$ for points $\mathbf{p}_1$ and $\mathbf{p}_2$ that are on different sides of γ. Then, γ is orbitally asymptotically stable (from both sides).*

Example 6.10. As a somewhat more complicated example, consider
$$\dot{x} = y,$$
$$\dot{y} = -x^3 + y(4 - x^2 - 4y^2).$$
We use the corresponding "energy" function
$$L(x,y) = \frac{1}{2}y^2 + \frac{1}{4}x^4$$
as a bounding function. The time derivative is
$$\dot{L} = y\dot{y} + x^3\dot{x} = -yx^3 + y^2(4 - x^2 - 4y^2) + x^3y$$
$$= y^2(4 - x^2 - 4y^2).$$
It is zero on the ellipse $x^2 + 4y^2 = 4$, $\dot{L} \geq 0$ inside the ellipse, and $\dot{L} \leq 0$ outside the ellipse. Letting $W(x,y) = x^2 + 4y^2$ be the function determined by $\dot{L}$, we can find the maximum and minimum of L on $W(x,y) = 4$ using Lagrange multipliers. Solving
$$x^3 = \lambda 2x,$$
$$y = \lambda 8y,$$
$$4 = x^2 + 4y^2,$$
we get the critical points $(\pm 2, 0)$, $(0, \pm 1)$, and $(\pm\frac{1}{2}, \pm\frac{\sqrt{15}}{4})$; the values of L at these points are 4, $1/2$, and $31/64$, respectively. Thus, the largest invariant annulus given by level curves of L is
$$\mathscr{A} = \left\{(x,y) : \frac{31}{64} \leq L(x,y) \leq 4\right\}.$$
The trajectories are coming into the annulus on the outer and the inner boundaries, so $\mathscr{A}$ is positively invariant and contains a periodic orbit.

What makes this example work is the fact that the equations have a term $y(4 - x^2 - y^2)$, which has a damping effect on a large level curve of L like $L^{-1}(4)$, and has an antidamping effect on a small level curve like $L^{-1}(31/64)$. The region between is trapped and contains no fixed points, so it must contain a periodic orbit.

We cannot completely analyze the phase portrait just using the Poincaré–Bendixson theorem. If, however, we assume that there is a unique periodic orbit for this system (which is, in fact, true), then we can state more about the α- and ω-limit sets of points. Let **p** be an initial condition in the region bounded by the periodic orbit Γ. Since the fixed point at the origin is repelling, $\omega(\mathbf{p})$ cannot contain only fixed points. The orbit is bounded because it is inside Γ. Therefore, the only possibility is that $\omega(\mathbf{p}) = \Gamma$. On the other hand, if **p** is outside Γ, the orbit cannot escape to infinity because L is decreasing for values greater than 4. The orbit cannot approach **0**, so the only possibility is that its ω-limit set is a periodic orbit. If there is only one periodic orbit, it must be Γ. In any case, any initial condition other than the origin has an ω-limit set equal to a periodic orbit.

Turning to α-limit sets, notice that there are trajectories on both the inside and outside that approach Γ; so, no trajectory can be moving away from Γ as time increases or can be approaching it as time goes to minus infinity. Therefore, no point can have Γ as its α-limit set. Therefore, if **p** is inside Γ, its α-limit set must

6.2. Poincaré–Bendixson Theorem

equal $\{\mathbf{0}\}$. On the other hand, if $\mathbf{p}$ is outside Γ, there is no other fixed point or periodic orbit, so the backward trajectory must leave the annulus and enter the region where $L(\mathbf{x}) > 4$. Once in this region, $\dot{L} < 0$, so the value of L goes to infinity as t goes to minus infinity, and $\alpha(\mathbf{p}) = \emptyset$.

Example 5.11 in the section on limit sets has an ω-limit set equal to a fixed point, together with two orbits that are on both the stable and unstable manifold of the fixed point. This example illustrates a more general form of the Poincaré–Bendixson theorem given in the following result.

Theorem 6.3. *Consider a differential equation $\dot{\mathbf{x}} = \mathbf{F}(\mathbf{x})$ on $\mathbb{R}^2$. Assume that $\mathscr{A}$ is a closed (includes its boundary) and bounded subset of $\mathbb{R}^2$, which is positively invariant for the differential equation. We assume that $\mathbf{F}(\mathbf{x})$ is defined at all points of $\mathscr{A}$ and has a finite number of fixed points in $\mathscr{A}$. Then, given any $\mathbf{x}_0$ in $\mathscr{A}$, $\omega(\mathbf{x}_0)$ is of one of the following three types:*

i. *The ω-limit set $\omega(\mathbf{x}_0)$ is a periodic orbit,*

ii. *The ω-limit set $\omega(\mathbf{x}_0)$ is a single fixed point.*

iii. *The ω-limit set $\omega(\mathbf{x}_0)$ is a finite number of fixed points $\mathbf{q}_1, \ldots, \mathbf{q}_m$, together with a (finite or denumerable) collection of orbits $\{\gamma_j\}$, such that for each orbit γ_j, $\alpha(\gamma_j)$ is a single fixed point $\mathbf{q}_{i(j,\alpha)}$ and $\omega(\gamma_j)$ is a single fixed point $\mathbf{q}_{i(j,\omega)}$:*

$$\omega(\mathbf{x}_0) = \{\mathbf{q}_1, \ldots, \mathbf{q}_m\} \cup \bigcup_j \gamma_j, \quad \text{with}$$

$$\alpha(\gamma_j) = \mathbf{q}_{i(j,\alpha)} \quad \text{and} \quad \omega(\gamma_j) = \mathbf{q}_{i(j,\omega)}.$$

Moreover, given any two fixed points $\mathbf{q}_{i_1} \neq \mathbf{q}_{i_2}$, there must be a directed choice of connecting orbits $\gamma_{j_1}, \ldots, \gamma_{j_k}$ with $\alpha(\gamma_{j_1}) = \mathbf{q}_{i_1}$, $\omega(\gamma_{j_p}) = \alpha(\gamma_{j_{p+1}})$ for $1 \leq k-1$, and $\omega(\gamma_{j_k}) = \mathbf{q}_{i_2}$.

For a proof, see [**Hal69**].

Another consequence of the Poincaré–Bendixson theorem is that any periodic orbit of a differential equation in the plane has to surround a fixed point.

Theorem 6.4. *Consider a differential equation $\dot{\mathbf{x}} = \mathbf{F}(\mathbf{x})$ on $\mathbb{R}^2$. Let γ be a periodic orbit which encloses the open set $\mathbf{U}$. Then, $\mathbf{F}(\mathbf{x})$ has a fixed point in $\mathbf{U}$ or there are points in $\mathbf{U}$ at which $\mathbf{F}(\mathbf{x})$ is undefined.*

Proof. We argue by contradiction. Assume that γ is a periodic orbit which encloses the open set $\mathbf{U}$, $\mathbf{U}$ contains no fixed point, and $\mathbf{F}(\mathbf{x})$ is defined at all points of $\mathbf{U}$. We can take such a γ of minimal area, so there can be no other periodic orbits inside γ. Take an initial condition $\mathbf{x}_0$ in $\mathbf{U}$. We argue that the limit sets $\omega(\mathbf{x}_0)$ and $\alpha(\mathbf{x}_0)$ cannot be both equal to γ as follows. Assume $\omega(\mathbf{x}_0) = \gamma$. By the proof of the Poincaré–Bendixson theorem, the Poincaré map is monotone on the inside of γ and all nearby points $\mathbf{y}_0$ inside γ have $\omega(\mathbf{y}_0) = \gamma$ and $\alpha(\mathbf{y}_0)$ cannot equal γ. Then, $\alpha(\mathbf{x}_0)$ is nonempty and must be either a periodic orbit or a fixed point in $\mathbf{U}$: Either conclusion contradicts the assumptions, since $\mathbf{U}$ contains no fixed points and the minimality of the areas implies there can be no periodic orbit inside $\mathbf{U}$. This contradiction proves the theorem. $\square$

6.2.1. Chemical Reaction Model.
The Brusselator, which is a system of equations that model a hypothetical chemical reaction, has the equations

$$\dot{x} = a - bx + x^2 y - x,$$
$$\dot{y} = bx - x^2 y.$$

We show, using the Poincaré–Bendixson theorem, that if $b > 1+a^2$ and $a > 0$, then the system has a periodic orbit. We include this example in the main sections of the chapters rather in the applications, because the method of finding the positively invariant region is an interesting one that is different from the other examples considered so far.

For another system of differential equations that model chemical reactions, which have periodic limiting behavior, see the oregonator system of differential equations in Section 6.8.1.

To find the fixed points, adding the $\dot{y} = 0$ equation to the $\dot{x} = 0$ equation, we get $0 = a - x$, or $x = a$. Substituting $x = a$ into the $\dot{y} = 0$ equation yields $0 = ba - a^2 y$, or $y = b/a$. Thus, the only fixed point is $(x^*, y^*) = (a, b/a)$.

The matrix of partial derivatives at the fixed point is

$$\begin{pmatrix} 2xy - b - 1 & x^2 \\ b - 2xy & -x^2 \end{pmatrix}_{x=a, y=b/a} = \begin{pmatrix} b - 1 & a^2 \\ -b & -a^2 \end{pmatrix}.$$

This matrix has its determinant equal to a^2, which is positive, and has trace $b - 1 - a^2$, which we assumed is positive. Therefore, the fixed point is either an unstable node or an unstable focus. If we remove a small elliptical disk D of the correct shape about the fixed point, then the trajectories will leave this elliptical disk and enter the region outside. Thus, the boundary of D is the inside boundary of the annular region $\mathscr{A}$ we are seeking to find.

To get the outside boundary of the region $\mathscr{A}$, we take a quadrilateral with sides (1) $x = 0$, (2) $y = 0$, (3) $y = A - x$, for a value of A to be determined, and (4) $y = B + x$, for a value of B to be determined. Let the region $\mathscr{A}$ be between the boundary of D and the boundary formed by these four sides. See Figure 6

Along side (1), $\dot{x} = a > 0$, so trajectories are entering into the region $\mathscr{A}$.

Along side (2), $\dot{y} = bx > 0$ (except at $x = 0$), so trajectories are entering $\mathscr{A}$ along this boundary.

Side (3) is in the zero level set of the function $L_3(x, y) = y + x - A$, with $\mathscr{A}$ on the side where L_3 is negative. The time derivative is

$$\dot{L}_3 = \dot{x} + \dot{y}$$
$$= a - x,$$

which is negative if $x > a$. Therefore, we need to take A and B so that the top vertex of $\mathscr{A}$, where the sides (3) and (4) come together, has $x \geq a$. In fact, in the following, we take them so that $x = a$ at this top vertex.

Side (4) is in the zero level set of the function $L_4(x, y) = y - x - B$, with $\mathscr{A}$ on the side where L_4 is negative. The derivative is

$$\dot{L}_4 = \dot{y} - \dot{x} = 2bx - 2x^2 y - a + x.$$

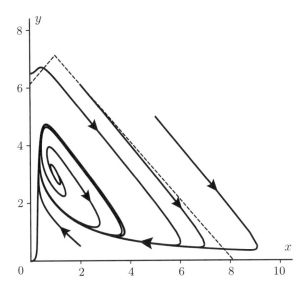

Figure 6. Positively invariant region and trajectories for Brusselator

This is negative if
$$y > \frac{(2b+1)x - a}{2x^2}.$$
Let $h(x)$ equal the right-hand side of the preceding inequality. Then, (i) $h(x)$ goes to minus infinity as x goes to zero, (ii) $h(x)$ goes to zero as x goes to plus infinity, (iii) $h(x)$ equals zero at $a/(2b+1)$, and (iv) $h(x)$ has a unique maximum at $x = 2a/(2b+1)$. The maximum of $h(x)$ is as indicated because
$$h'(x) = \frac{2a - (2b+1)x}{2x^3},$$
where $h'(x)$ is zero at $x = 2a/(2b+1)$, $h'(x)$ is positive for $x < 2a/(2b+1)$, and $h'(x)$ is negative for $x > 2a/(2b+1)$. The value at the maximum is
$$h\left(\frac{2a}{2b+1}\right) = \frac{(2b+1)^2}{8a}.$$
If we take $B = (2b+1)^2/(8a)$, then certainly $y = B + x$ is greater than $h(x)$ along side (4), since $y \geq B = \max\{h(x) : x > 0\}$. Having chosen B, set $A = B + 2a$, so the two edges intersect when $x = a$ and $y = B + a = A - a$.

With these choices for the boundary of $\mathscr{A}$, all the points on the boundary have trajectories that are entering the region $\mathscr{A}$. Since the only fixed point is not in the region $\mathscr{A}$, there must be a periodic orbit in the region $\mathscr{A}$ by the Poincaré–Bendixson theorem.

Exercises 6.2

1. Consider the system of differential equations
$$\dot{x} = y,$$
$$\dot{y} = -4x + y(1 - x^2 - y^2).$$
Show that the system has a periodic orbit. Hint: Use a bounding function.

2. Consider the system of equations
$$\dot{x} = x - 2y - x(x^2 + 3y^2),$$
$$\dot{y} = 2x + y - y(x^2 + 3y^2).$$
   a. Classify the fixed point at the origin.
   b. Rewrite the system of differential equations in polar coordinates.
   c. Find the maximum radius r_1 for the circle on which all the solutions are crossing outward across it.
   d. Find the minimum radius r_2 for the circle on which all the solutions are crossing inward across it.
   e. Prove that there is a periodic orbit somewhere in the annulus $r_1 \leq r \leq r_2$.
   f. Use a computer program to plot the periodic solution.

3. Consider the system of equations
$$\dot{x} = 3x + 2y - x(x^2 + y^2),$$
$$\dot{y} = -x + y - y(x^2 + y^2).$$
   a. Classify the fixed point at the origin.
   b. Show that $(0, 0)$ is the only fixed point.
   c. Calculate $r\dot{r}$ in terms of x and y.
   d. Show that $\dot{r}$ is positive for small r and negative for large r. Hint: To show the quadratic terms are positive definite (positive for all $(x, y) \neq (0, 0)$), either (i) complete the square or (ii) use the test for a minimum of a function.
   e. Prove that the system has a periodic orbit.

4. Assume that the system of differential equations in the plane $\dot{\mathbf{x}} = \mathbf{F}(\mathbf{x})$ has an annulus
$$\{\mathbf{x} : r_1 \leq \|\mathbf{x}\| \leq r_2\},$$
where the trajectories are coming in on the boundary and there are no fixed points in the annulus.
   a. Assume that there is exactly one periodic orbit in the annulus. Prove that it is orbitally asymptotically stable. Hint: Use Corollary 6.2, as well as the Poincaré–Bendixson theorem.
   b. Assume that there are exactly two periodic orbits in the annulus. Prove that one of the periodic orbits is orbitally asymptotically stable.
   c. Prove that there is an orbitally asymptotically stable periodic orbit in the annulus if there are exactly n periodic orbits, where $n = 3$ or 4. (The result is true for any positive integer n.)

5. Consider the system of equations

$$\dot{x} = 3x - 2y - x(x^2 + 4y^2),$$
$$\dot{y} = 4x - y - y(x^2 + 4y^2).$$

   a. Show that the linearization of this system at the origin is an unstable spiral.
   b. Show that $(0,0)$ is the only fixed point. Hint: One way to do this is to consider the equations which say that (x_*, y_*) is a fixed point. Express them as a matrix equation. This matrix equation says something that contradicts what you discovered in (a).
   c. Let $L(x,y) = \frac{x^2}{2} + \frac{y^2}{2}$ and let $M(x,y) = \frac{x^2}{2} + \frac{(x-y)^2}{2}$. Show that $\dot{M} > 0$ if (x,y) is close to $(0,0)$ and $\dot{L} < 0$ if (x,y) is sufficiently far from $(0,0)$.
   d. Show that there is a trapping region of the form $\{(x,y) : a^2 \leq M(x,y)$ and $L(x,y) \leq A^2\}$, where a is small and A is large.
   e. Show that there is a closed orbit in the set defined in (d).

6. In [**Mur89**], Murray presents a system of differential equations based on a trimolecular reaction given by

$$\dot{u} = a - u + u^2 v,$$
$$\dot{v} = b - u^2 v,$$

   with $a > 0$ and $b > 0$. In this exercise, we consider only the case in which $a = 1/8$ and $b = 1/2$. (Notice that these equations are similar to the Brusselator given in the text.)
   a. Find the fixed point.
   b. Show that the fixed point is a source, with the real parts of both eigenvalues positive.
   c. Show that the following region is positively invariant:

$$\mathscr{R} = \{(u,v) : u \geq \frac{1}{16},\ 0 \leq v \leq 128,\ u + v \leq 130\}.$$

   d. Explain why there is a periodic orbit in the region $\mathscr{R}$.

7. In [**Bra01**], Brauer and Castillo–Chávez give the system,

$$\dot{x} = x\left(1 - \frac{x}{30} - \frac{y}{x+10}\right),$$
$$\dot{y} = y\left(\frac{x}{x+10} - \frac{1}{3}\right),$$

   as an example of a more general type of predator–prey system.
   a. Show that the only fixed points are $(0,0)$, $(30,0)$, and $(5, 12.5)$. Also show that $(0,0)$ and $(30,0)$ are saddle fixed points and that $(5, 12.5)$ is a repelling fixed point (a source).
   b. Show that the region

$$\mathscr{R} = \{(x,y) : 0 \leq x,\ 0 \leq y,\ x + y \leq 50\}$$

   is positively invariant. Hint: Show that the line $x + y = 50$ lies above the curve where $\dot{x} + \dot{y} = 0$, so $\dot{x} + \dot{y} < 0$ on $x + y = 50$.

c. Use Theorem 6.3 to show that, for any point $\mathbf{p}_0 = (x_0, y_0)$ in $\mathscr{R}$ with $x_0 > 0$, $y_0 > 0$, and $\mathbf{p}_0 \neq (5, 12.5)$, $\omega(\mathbf{p}_0)$ must be a periodic orbit. Hints: (i) If $\omega(\mathbf{p}_0)$ contained either $(0,0)$ or $(30,0)$ then it must contain both. (ii) Thus, $\omega(\mathbf{p}_0)$ would contain an orbit γ with $\alpha(\gamma) = (30,0)$ and $\omega(\gamma) = (0,0)$, i.e., $\gamma \subset W^u(30,0) \cap W^s(0,0)$. Since there are no such orbits, this is impossible and $\omega(\mathbf{p}_0)$ must be a periodic orbit.

8. A model by Selkov for the biochemical process of glycolysis is given by
$$\dot{x} = -x + ay + x^2 y,$$
$$\dot{y} = b - ay - x^2 y,$$
where a and b are positive constants.

   a. Show that the only fixed point is $(x^*, y^*) = \left(b, \dfrac{b}{a+b^2}\right)$.

   b. Show that the fixed point (x^*, y^*) is unstable for $a = 0.1$ and $b = 0.5$.

   c. To get an invariant region in the next part, we use one boundary of the form $x + y$ equal to a constant. We want the line to go through the point $(b, b/a)$, where $b = x^*$ and b/a is the y-intercept of $\dot{y} = 0$. Therefore, we consider $x + y = b + b/a$. We also consider the nullcline $\dot{x} = 0$. Show that there is a unique point (x_1, y_1), such that
$$b + \frac{b}{a} = x + y,$$
$$y = \frac{x}{a + x^2}.$$
   Hint: Plot the two curves.

   d. Show that the region
$$\mathscr{R} = \{(x,y) : 0 \leq x \leq x_1,\ 0 \leq y \leq \frac{b}{a},\ x + y \leq b + \frac{b}{a}\}$$
   is positively invariant, where x_1 is the value found in the previous part.

   e. For $a = 0.1$, $b = 0.5$, and $(x_0, y_0) \neq (x^*, y^*)$ in $\mathscr{R}$, show that $\omega((x_0, y_0))$ is a periodic orbit.

9. Consider the system of differential equations
$$\dot{x} = y,$$
$$\dot{y} = -x - 2x^3 + y(4 - x^2 - 4y^2).$$
   Show that the system has a periodic orbit. Hint: Find an "energy function" $L(x,y)$ by disregarding the terms involving y in the $\dot{y}$ equation. Find the maximum and minimum of L on the curve $\dot{L} = 0$. Use those values to find a region that is positively invariant without any fixed point.

10. A special case of the Rowenzweig model for a predator-prey system had the following system of differential equations
$$\dot{x} = x(6 - x) - y x^{\frac{1}{2}},$$
$$\dot{y} = y\left(x^{\frac{1}{2}} - 1\right).$$

    a. Find all the fixed points.

b. Show that the fixed point (x^*, y^*) with $x^*, y^* > 0$ is repelling and the other (two) fixed points are saddles.

c. Assume that every orbit in this first quadrant is bounded for $t \geq 0$, prove that there must be a limit cycle. *Note:* This model gives a more stable periodic limit cycle for a predator-prey system, than the periodic orbits for earlier models considered.

6.3. Self-Excited Oscillator

In this section, we present a class of differential equations, called Lienard equations, which have increasing "amplitude" for small oscillations and decreasing "amplitude" for large oscillations. The result is a unique periodic orbit or limit cycle. Since the amplitudes increase for small oscillations without external forcing, they are called *self-excited oscillators*. These equations were originally introduced to model a vacuum tube that settled down to periodic output. Later, a similar type of phenomenon was found to occur in many other situations, and the equations have become a nontrivial model of a situation with an attracting limit cycle.

The simplest example of this type of system is given by

(6.1) $$\ddot{x} = -x - (x^2 - 1)\,\dot{x},$$

which is called the *Van der Pol equation*. When written as a system of differential equations, the equation becomes

(6.2) $$\dot{x} = v,$$
$$\dot{v} = -x - (x^2 - 1)\,v,$$

which is called the *Van der Pol system*. This system can be generalized to the *Lienard system*,

(6.3) $$\dot{x} = v,$$
$$\dot{v} = -g(x) - f(x)\,v,$$

where assumptions on $g(x)$ and $f(x)$ make them similar to the corresponding factors in the Van der Pol system. In particular, $g(x) > 0$ for $x > 0$, so we have a restoring force. The function $f(x)$ is negative for small $|x|$ and then becomes positive for large $|x|$. We give the exact requirements in what follows.

Before we change coordinates, note that if we use

$$L(x, v) = G(x) + \frac{v^2}{2}, \quad \text{where} \quad G(x) = \int_0^x g(s)\,ds,$$

then $\dot{L} = v^2(1 - x^2)$ is both positive and negative. It is greater or equal to zero on a small level curve $L^{-1}(C)$ contained strictly in the strip

$$\{(x, v) : -1 \leq x \leq 1\}.$$

However, there is no large level curve on which $\dot{L}$ is less than or equal to zero. Each such level curve passes through the strip where $\dot{L}$ is positive. Therefore, we cannot use just the type of argument we gave in the preceding section in the applications of the Poincaré–Bendixson theorem. Instead, the proof discusses the total amount that a function like L changes when a trajectory makes a whole passage around the origin.

Before we proceed further with the discussion, we rewrite the Lienard equations in a form that is easier to analyze. Let $y = v + F(x)$, where

$$F(x) = \int_0^x f(s)\, ds.$$

Then,

$$\dot{x} = y - F(x),$$
$$\dot{y} = \dot{v} + F'(x)\,\dot{x} = -g(x) - f(x)\,v + f(x)\,v = -g(x).$$

Thus, we get the equivalent set of equations

(6.4)
$$\dot{x} = y - F(x),$$
$$\dot{y} = -g(x).$$

For the Van der Pol system, $g(x) = x$ and $F(x) = x^3/3 - x$. The phase portrait of Van der Pol system in the form of the equations (6.4) is given in Figure 7.

The assumptions we make on the equations are the following.

(1) The functions $F(x)$ and $g(x)$ are continuously differentiable.

(2) The function g is an odd function, $g(-x) = -g(x)$, with $g(x) > 0$ for $x > 0$. Thus, g is a restoring force.

(3) The function F is an odd function, $F(-x) = -F(x)$. There is a single positive value a for which $F(a) = 0$, $F(x) < 0$ for $0 < x < a$, $F(x) > 0$ for $x > a$, and $F(x)$ is nondecreasing for $x \geq a$.

In Section 6.8.2, we show how the second form of the Lienard equations arises in a nonlinear electric RLC circuit.

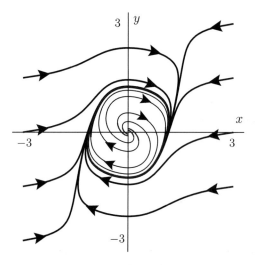

Figure 7. Phase portrait for Van der Pol system in the form of the equations (6.4)

6.3. Self-Excited Oscillator

Theorem 6.5. *With the preceding assumptions (1)–(3) on $g(x)$ and $F(x)$, the Lienard system of equations (6.4) has a unique attracting periodic orbit.*

The proof of this theorem is given in Section 6.9. For the rest of this section, we sketch the ideas of the proof.

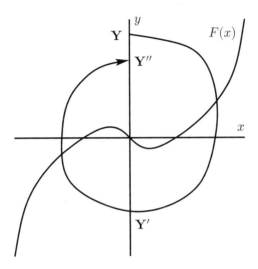

Figure 8. Trajectory path for Lienard equations

A trajectory starting at $\mathbf{Y} = (0, y)$, with $y > 0$, has $\dot{x} > 0$ and $\dot{y} < 0$, so it goes to the right and hits the curve $y = F(x)$. Then, it enters the region where $\dot{x} < 0$ and $\dot{y} < 0$. It is possible to show that the trajectory must next hit the negative y-axis at a point $\mathbf{Y}'$. See Figure 8. Subsequently, the trajectory enters the region with $x < 0$ and $y < F(x)$, where $\dot{x} < 0$ and $\dot{y} > 0$, so it intersects the part of the curve $y = F(x)$ with both x and y negative. Next the trajectory returns to the positive y axis at a point $\mathbf{Y}''$. This map P from the positive y-axis to itself, which takes the point $\mathbf{Y}$ to the point $\mathbf{Y}''$, where the trajectory intersects the positive y-axis the next time, is called the *first return map* or *Poincaré map*. The half-line $\{(y, 0) : y > 0\}$ is the *transversal* for the Poincaré map. The orbit is periodic if and only if $\mathbf{Y}'' = \mathbf{Y}$. But, $(x(t), y(t))$ is a solution if and only if $(-x(t), -y(t))$ is a solution. Therefore, if $\mathbf{Y}' = -\mathbf{Y}$ then $\mathbf{Y}'' = -\mathbf{Y}' = \mathbf{Y}$ and the trajectory is periodic. On the other hand, we want to show that it is not periodic if $\mathbf{Y}' \neq -\mathbf{Y}$. First, assume that $\|\mathbf{Y}'\| < \|\mathbf{Y}\|$. The trajectory starting at $-\mathbf{Y}$ goes to the point $-\mathbf{Y}'$ on the positive axis. The trajectory starting at $\mathbf{Y}'$ is inside that trajectory so $\|\mathbf{Y}''\| < \|\mathbf{Y}'\| < \|\mathbf{Y}\|$ and it is not periodic. Similarly, if $\|\mathbf{Y}'\| > \|\mathbf{Y}\|$, then $\mathbf{Y}'' > \|\mathbf{Y}'\| > \mathbf{Y}$ and it is not periodic. Therefore, the trajectory is periodic if and only if $\mathbf{Y}' = -\mathbf{Y}$.

The idea of the proof is to use a real-valued function to determine whether the orbit is moving closer or farther from the origin. We use the potential function

$$G(x) = \int_0^x g(s)\, ds.$$

Since g is an odd function, with $g(x) > 0$ for $x > 0$, $G(x) > 0$ for all $x \neq 0$. Let
$$L(x, y) = G(x) + \frac{y^2}{2},$$
which is a measurement of the distance from the origin. The orbit is periodic if and only if $L(\mathbf{Y}) = L(\mathbf{Y'})$.

A direct calculation shows that $\dot{L} = -g(x)F(x)$, which is positive for $-a < x < a$ and negative for $|x| > a$. The quantity L increases for x between $\pm a$. Therefore, the origin is repelling and there can be no periodic orbits contained completely in the region $-a \leq x \leq a$.

As the size of the orbit gets bigger, a comparison of integrals shows that the change in L, $L(\mathbf{Y'}) - L(\mathbf{Y})$ decreases as $\mathbf{Y}$ moves farther up the y-axis. A further argument shows that $L(\mathbf{Y'}) - L(\mathbf{Y})$ goes to minus infinity as the point goes to infinity along the axis. Therefore, there is one and only one positive y for which $\mathbf{Y} = (0, y)$ is periodic. See Section 6.9 for details.

Exercises 6.3

1. Prove that the system
$$\dot{x} = y,$$
$$\dot{y} = -x - x^3 - \mu(x^4 - 1)y,$$
has a unique periodic orbit when $\mu > 0$.

2. The Raleigh differential equation is given by
$$\ddot{z} + \epsilon \left(\frac{1}{3}\dot{z}^3 - \dot{z}\right) + z = 0.$$
   a. Show that the substitution $x = \dot{z}$ transforms the Raleigh differential equation into the Van der Pol equation (6.1). Hint: What is $\ddot{x}$?
   b. Also, show that the system of differential equations for the variables $x = \dot{z}$ and $y = -z$ is of the form given in equation (6.4).
   c. Conclude that the Raleigh differential equation has a unique limit cycle.

6.4. Andronov–Hopf Bifurcation

In this section, we discuss the spawning of a periodic orbit, as a parameter varies and a fixed point with complex eigenvalue makes a transition from a stable focus to an unstable focus. Such a change in the phase portrait is called a *bifurcation*, and the value of the parameter at which the change takes place is called the *bifurcation value*. The word bifurcation itself is a French word that is used to describe a fork in a road or other such splitting apart.

We start with a model problem and then state a general theorem.

Example 6.11 (Model Problem). Consider the differential equations
$$\dot{x} = \mu x - \omega y + K x (x^2 + y^2),$$
$$\dot{y} = \omega x + \mu y + K y (x^2 + y^2),$$

6.4. Andronov–Hopf Bifurcation

with $\omega > 0$. We consider μ as the parameter that is varying. The eigenvalues of the fixed point at the origin are $\lambda_\mu = \mu \pm i\omega$. For $\mu = 0$, the linear equations have a center, and as μ varies from negative to positive, the fixed point changes from a stable focus to an unstable focus. In polar coordinates, the equations become

$$\dot{r} = \mu r + K r^3,$$
$$\dot{\theta} = \omega.$$

The $\dot{\theta}$ equation implies that θ is increasing and is periodic when t increases by $2\pi/\omega$ units of time.

We first consider the case when $K < 0$. For $\mu < 0$, all solutions with $r(0) > 0$ tend to 0 at an exponential rate. For $\mu = 0$, $\dot{r} = Kr^3 < 0$ for $r > 0$, so still all solutions with $r(0) > 0$ tend to 0, but now at a slower rate. Thus, for $\mu = 0$, the fixed point is weakly attracting. For $\mu > 0$, 0 is a repelling fixed point in the r-space, and there is a fixed point at $r = \sqrt{|\mu/K|}$, that is attracting. See Figure 9.

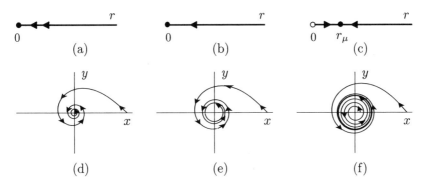

Figure 9. $K < 0$, Supercritical bifurcation. The dynamics in the r-space for (a) $\mu < 0$, (b) $\mu = 0$, and (c) $\mu > 0$. The phase plane in (x, y)-variables for (d) $\mu < 0$, (e) $\mu = 0$, and (f) $\mu > 0$.

In the Cartesian (x, y)-space, the origin changes from a stable focus for $\mu < 0$, to a weakly attracting focus for $\mu = 0$, to an unstable focus surrounded by an attracting limit cycle for $\mu > 0$. Notice that, for this system, when the fixed point is weakly attracting at the bifurcation value $\mu = 0$, the limit cycle appears when the fixed point is unstable, $\text{Re}(\lambda_\mu) = \mu > 0$, and the limit cycle is attracting. This is called a *supercritical bifurcation*, because the limit cycle appears after the real part of the eigenvalue has become positive.

Next, we want to consider the case when $K > 0$. The change in the eigenvalues is the same as before. Now the fixed point is weakly repelling for the nonlinear equations at the bifurcation value $\mu = 0$. For $\mu > 0$, there is no positive fixed point in the r equation, and in the (x, y)-plane, the origin is repelling. There are no periodic orbits. Now, the second fixed point in the r equation appears for $\mu < 0$ at $r = \sqrt{|\mu/K|}$, and it is unstable. Thus, the limit cycle appears before the bifurcation, when the real part of the eigenvalue at the fixed point is negative. See Figure 10. This case is called the *subcritical bifurcation*.

Notice that, in both cases the limit cycle has radius $\sqrt{|\mu/K|}$ for the values of the parameter having a limit cycle.

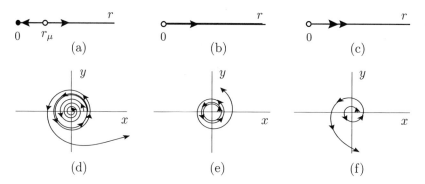

Figure 10. $K > 0$, Subcritical bifurcation. The r variables for (a) $\mu < 0$, (b) $\mu = 0$, and (c) $\mu > 0$. The phase plane in (x, y)-variables for (d) $\mu < 0$, (e) $\mu = 0$, and (f) $\mu > 0$.

The general result concerns the change of a fixed point from a stable focus to an unstable focus and the resulting periodic orbit for parameter values either before of after the bifurcation parameter value. The key assumptions of the theorem, which indicate when it can be applied, are as follows:

(1) There is a system of differential equations

$$\dot{\mathbf{x}} = \mathbf{F}_\mu(\mathbf{x})$$

in the plane depending on a parameter μ. For a range of the parameters, there is a fixed point $\mathbf{x}_\mu$ that varies smoothly. (In Example 6.11, $\mathbf{x}_\mu = \mathbf{0}$ for all μ.) The fixed point has complex eigenvalues, $\lambda_\mu = \alpha_\mu \pm i\beta_\mu$. The bifurcation value μ_0 is determined so that the eigenvalues are purely imaginary (i.e., the real parts of the eigenvalues $\alpha_{\mu_0} = 0$, and the imaginary parts $\pm\beta_{\mu_0} \neq 0$).

(2) The real part of each eigenvalue is changing from negative to positive or positive to negative as the parameter μ varies. For simplicity of statement of the theorem, we assumed that α_μ is increasing with μ,

$$\frac{d}{d\mu}\alpha_\mu\Big|_{\mu_0} > 0,$$

but the results could be rewritten if this derivative were negative. Thus, the fixed point is a stable focus for $\mu < \mu_0$, but μ near μ_0, and it is an unstable focus for $\mu > \mu_0$, but μ near μ_0.

(3) The system is weakly attracting or weakly repelling for $\mu = \mu_0$. We show three ways to check this nonlinear stability of the fixed point at the bifurcation value: (i) We check most of our examples by using a test function and apply Theorem 5.2. (ii) We give an expression for a constant K_{μ_0} that can be calculated by converting the system to polar coordinates and averaging some of the coefficients. This constant plays the same role as K in the model problem. (iii) There is a constant M that can be calculated using partial derivatives in Cartesian coordinates, that is similar to K_μ.

6.4. Andronov–Hopf Bifurcation

Theorem 6.6 (Andronov–Hopf bifurcation). *Let $\dot{\mathbf{x}} = \mathbf{F}_\mu(\mathbf{x})$ be a system of differential equations in the plane, depending on a parameter μ. Assume that there is a fixed point $\mathbf{x}_\mu$ for the parameter value μ, with eigenvalues $\lambda_\mu = \alpha_\mu \pm i\beta_\mu$. Assume that $\beta_\mu \neq 0$, $\alpha_{\mu_0} = 0$, and*

$$\frac{d}{d\mu}(\alpha_\mu)\Big|_{\mu=\mu_0} > 0.$$

a. *(Supercritical bifurcation) If $\mathbf{x}_{\mu_0}$ is weakly attracting for $\mu = \mu_0$, then there is a small attracting periodic orbit for $|\mu - \mu_0|$ small and $\mu > \mu_0$. The period of the orbit is equal to $2\pi/\beta_{\mu_0}$ plus terms which go to zero as μ goes to μ_0. See Figure 9.*

b. *(Subcritical bifurcation) If $\mathbf{x}_{\mu_0}$ is weakly repelling for $\mu = \mu_0$, then there is a small repelling periodic orbit for $|\mu - \mu_0|$ small and $\mu < \mu_0$. The period of the orbit is equal to $2\pi/\beta_{\mu_0}$ plus terms that go to zero as μ goes to μ_0. See Figure 10.*

Remark 6.12. If the rest of the assumptions of the theorem are true, but

$$\frac{d}{d\mu}\alpha_\mu\Big|_{\mu_0} < 0,$$

then the periodic orbit appears for the parameter values on the opposite side of the bifurcation value. In particular, for the supercritical bifurcation case, the stable periodic orbit appears for $\mu < \mu_0$ (when the fixed point is unstable), and for the subcritical bifurcation case, the unstable periodic orbit appears for $\mu > \mu_0$ (when the fixed point is stable).

A proof of part of this theorem is contained in Section 6.9.

The next theorem indicates two ways to check the weak stability or weak instability of the fixed point at the bifurcation value; the first is given in terms of polar coordinates and the second is given in terms of partial derivatives in Cartesian coordinates. A third way is to use a test function as we demonstrate in some of the examples.

Theorem 6.7 (Stability criterion for Andronov–Hopf bifurcation). *Assume that the system of differential equations in the plane satisfies Theorem 6.6. Then, there is a constant K_μ that determines the stability of the fixed point at $\mu = \mu_0$ and whether the bifurcation is supercritical or subcritical. (i) If $K_{\mu_0} < 0$, then the fixed point is weakly attracting at the lowest order terms possible when $\mu = \mu_0$, and there is a supercritical bifurcation of an attracting periodic orbit of approximate radius $\sqrt{|\alpha_\mu/K_{\mu_0}|}$ for $|\mu - \mu_0|$ small and $\mu > \mu_0$. (ii) If $K_{\mu_0} > 0$, then the fixed point is repelling at the lowest order terms possible when $\mu = \mu_0$, and there is a subcritical bifurcation of a repelling periodic orbit of approximate radius $\sqrt{|\alpha_\mu/K_{\mu_0}|}$ for $|\mu - \mu_0|$ small and $\mu < \mu_0$. (iii) In fact, there is a change of coordinates to (R, Θ) such that the equations become*

$$\dot{R} = \alpha_\mu R + K_\mu R^3 + \text{ higher order terms},$$
$$\dot{\Theta} = \beta_\mu + \text{ higher order terms}.$$

(These are polar coordinates, but are not given in terms of x_1 and x_2.)

The following two parts give two formulas for the constant K_{μ_0}.

a. Let $\mathbf{x}_\mu = (x_{\mu,1}, x_{\mu,2})$. Assume the system in polar coordinates $x_1 - x_{\mu,1} = r\cos(\theta)$ and $x_2 - x_{\mu,2} = r\sin(\theta)$ is

$$\dot{r} = \alpha_\mu r + C_3(\theta,\mu)\, r^2 + C_4(\theta,\mu)\, r^3 + \text{higher order terms},$$
$$\dot{\theta} = \beta_\mu + D_3(\theta,\mu)\, r + D_4(\theta,\mu)\, r^2 + \text{higher order terms},$$

where C_j^k and D_j^k are homogeneous of degree j in $\sin(\theta)$ and $\cos(\theta)$. Then, the constant K_{μ_0} is given as follows:

$$K_{\mu_0} = \frac{1}{2\pi} \int_0^{2\pi} \left[C_4(\theta,\mu_0) - \frac{1}{\beta_0} C_3(\theta,\mu_0)\, D_3(\theta,\mu_0) \right] d\theta.$$

b. Assume that the system at $\mu = \mu_0$ is given by $\dot{x}_1 = f(x_1, x_2)$ and $\dot{x}_2 = g(x_1, x_2)$ (i.e., the first coordinate function at μ_0 is $F_{\mu_0,1} = f$, and the second coordinate function is $F_{\mu_0,2} = g$). We denote the partial derivatives by subscripts, with a subscript x representing the partial derivative with respect to the first variable and a subscript y representing the partial derivative with respect to the second variable:

$$\frac{\partial f}{\partial x_1}(\mathbf{x}_{\mu_0}) = f_x, \qquad \frac{\partial^3 f}{\partial x_1^2 \partial x_2}(\mathbf{x}_{\mu_0}) = f_{xxy}, \qquad \frac{\partial^2 g}{\partial x_2^2}(\mathbf{x}_{\mu_0}) = g_{yy},$$

etc. Assume that

$$f_x = 0, \qquad f_y \neq 0,$$
$$g_x \neq 0, \qquad g_y = 0,$$

and $f_y g_x < 0$. This ensures that the eigenvalues are $\pm i \sqrt{|f_y g_x|}$. Then,

$$K_{\mu_0} = \frac{1}{16} \left(\left|\frac{f_y}{g_x}\right|^{1/2} f_{xxx} + \left|\frac{g_x}{f_y}\right|^{1/2} f_{xyy} + \left|\frac{f_y}{g_x}\right|^{1/2} g_{xxy} + \left|\frac{g_x}{f_y}\right|^{1/2} g_{yyy} \right)$$
$$+ \frac{1}{16\, g_x} \left|\frac{g_x}{f_y}\right|^{1/2} \left(f_{xy} f_{xx} + \left|\frac{g_x}{f_y}\right| f_{xy} f_{yy} + \left|\frac{g_x}{f_y}\right| f_{yy} g_{yy} \right)$$
$$+ \frac{1}{16\, f_y} \left|\frac{f_y}{g_x}\right|^{1/2} \left(g_{xy} g_{yy} + \left|\frac{f_y}{g_x}\right| g_{xy} g_{xx} + \left|\frac{f_y}{g_x}\right| f_{xx} g_{xx} \right).$$

We do not prove this theorem, but only discuss its meaning. Part (a) gives a formula in terms of "averaging" terms given in polar coordinates. See [**Cho77**], [**Cho82**], or [**Rob99**] for the derivation of the formula given in part (a). The average of $C_4(\theta,\mu_0)$ in the formula seems intuitively correct, but the weighted average of $C_3(\theta,\mu_0)$ is not obvious.

Part (b) gives a formula in terms of usual partial derivatives and is easier to calculate, but is less clear and much messier. The reader can check that, if only one term of the type given by the partial derivatives in the first four terms appears, then the result is correct by using a test function. See Example 6.13. We have written it in a form to appear symmetric in the first and second variables. We have assumed that $f_x = 0 = g_y$, but not that $g_x = -f_y$ as our references do. If $f_x \neq 0$ or $g_y \neq 0$, but the eigenvalues are still purely imaginary, then a preliminary change of coordinates is necessary (using the eigenvectors) to make these two terms equal zero. Note that, in the second line (of terms five through seven), the coefficient outside the parenthesis has the same sign as g_x, while in the third line (of terms eight through ten), the coefficient outside the parenthesis has the same sign as f_y. In [**Mar76**], this constant is related to a Poincaré map. The Poincaré map P is formed by following the trajectories as the angular variable increases by 2π. At the bifurcation value, $P'(0) = 1$ and $P''(0) = 0$. This reference shows that the third derivative of the Poincaré map, $P'''(0)$, equals a scalar multiple of this constant. The derivation is similar to the proof of Part (a) given in the references cited. It is derived in [**Guc83**] using complex variables: This calculation looks much different.

In some equations, it is possible to verify that the fixed point is weakly attracting or repelling at the bifurcation value by means of a test function, rather than calculating the constant K_{μ_0}. This calculation shows whether the bifurcation is supercritical or subcritical and whether the periodic orbit is attracting or repelling. The following example illustrates this possibility.

Example 6.13. Consider the system of equations given by

$$\dot{x} = y,$$
$$\dot{y} = -x - x^3 + \mu y + k x^2 y,$$

with parameter μ. The only fixed point is at the origin. The linearization at the origin is

$$\begin{pmatrix} 0 & 1 \\ -1 & \mu \end{pmatrix}.$$

The characteristic equation is $\lambda^2 - \mu \lambda + 1 = 0$, with eigenvalues

$$\lambda = \frac{\mu}{2} \pm i \sqrt{1 - \frac{\mu^2}{4}}.$$

Thus, the eigenvalues are complex (for $-2 < \mu < 2$) and the real part of the eigenvalues is $\alpha_\mu = \mu/2$. The real part of the eigenvalues is zero for $\mu = 0$, so this is the bifurcation value. The sign of the real part of the eigenvalue changes as μ passes through 0,

$$\frac{d}{d\mu}\alpha_\mu = \frac{1}{2} > 0,$$

so the stability of the fixed point changes as μ passes through 0.

We need to check the weak stability of the fixed point at the bifurcation value $\mu = 0$. (Notice that the fixed point has eigenvalues $\pm i$, and is a linear center for $\mu = 0$.) The constant in Theorem 6.7(b) is

$$K_0 = \frac{k}{8},$$

because $f_y = 1$, $g_x = -1$, $g_{xxy} = 2k$, and the other partial derivatives in the formula are zero. Applying the theorem, we see that the origin is weakly attracting for $k < 0$ and weakly repelling for $k > 0$. We will obtain this same result using a test function in what follows.

For $\mu = 0$, the system of equations becomes

$$\dot{x} = y,$$
$$\dot{y} = -x - x^3 + k x^2 y.$$

Because the term x^2y is like a damping term, we use the test function
$$L(x,y) = \frac{x^2}{2} + \frac{x^4}{4} + \frac{y^2}{2}$$
to measure the stability of the origin. The time derivative
$$\begin{aligned}\dot{L} &= (x + x^3)\,\dot{x} + y\,\dot{y} \\ &= (x + x^3)\,y + y\,(-x - x^3 + k\,x^2\,y) \\ &= k\,x^2\,y^2.\end{aligned}$$
We first consider the case in which $k < 0$. Then, this time derivative is less than or equal to zero. It equals zero only on
$$\mathbf{Z} = \{\,(x,y) : x = 0 \text{ or } y = 0\,\}.$$
The only point whose entire orbit is totally within the set $\mathbf{Z}$ is the origin. Thus, this system satisfies the hypothesis of Theorem 5.2. The function L is positive on the whole plane except at the origin, so we can use the whole plane for $\mathbf{U}$, and the basin of attraction of the origin is the whole plane. Thus, the origin is weakly attracting for $\mu = 0$. Notice that, for $\mu < 0$, we have
$$\dot{L} = y^2\,(\mu + k\,x^2),$$
which is still less than or equal to zero. Now, the set $\mathbf{Z}$ is only the set $y = 0$, but the argument still implies that the origin is attracting. Thus, for any $\mu \leq 0$, the origin is asymptotically attracting, with a basin of attraction consisting of the whole plane. The bifurcation theorem applies and there is a supercritical bifurcation to a stable periodic orbit for $\mu > 0$. Notice that, for $\mu = 0$, the system is attracting across some small level curve $L^{-1}(C)$. This will still be true on this one level curve for small $\mu > 0$. However, now the fixed point is repelling, so the ω-limit set of a point different than the fixed point must be a periodic orbit and not the fixed point. Theorem 6.6 says that there is a unique periodic orbit near the fixed point and that all points near the fixed point have this periodic orbit as their ω-limit sets.

Now consider $k > 0$. The time derivative of L is still $\dot{L} = y^2(\mu + k\,x^2)$, but this is now greater than or equal to zero for $\mu \geq 0$. Thus, the origin is weakly repelling for $\mu = 0$ and strongly repelling for $\mu > 0$. Therefore, there are no periodic orbits for any $\mu \geq 0$. However, by the theorem, there is a repelling periodic orbit for $\mu < 0$. In fact, for $\mu = 0$, the system comes in toward the origin across a small level curve $L^{-1}(C)$ backward in time. This will still be true for small $\mu < 0$. However, the fixed point at the origin is now attracting, so a trajectory starting inside the level curve $L^{-1}(C)$ must have an α-limit set that is different from the fixed point, and so there must be a repelling periodic orbit.

There can be other periodic orbits for the system besides the one caused by the Andronov–Hopf bifurcation. The following example illustrates this possibility in the case of a subcritical Andronov–Hopf bifurcation, where there is another attracting periodic orbit whose radius is bounded away from zero.

Example 6.14. Consider the equations given in polar coordinates by
$$\begin{aligned}\dot{r} &= \mu\,r + r^3 - r^5, \\ \dot{\theta} &= \omega.\end{aligned}$$

6.4. Andronov–Hopf Bifurcation

This is a subcritical bifurcation because the coefficient of r^3 is positive. The nonzero fixed points in the $\dot{r}$ equation are given by

$$r^2 = \frac{1 \pm \sqrt{1+4\mu}}{2}.$$

There is always one positive root

$$r_1^2 = \frac{1 + \sqrt{1+4\mu}}{2} \approx 1 + \mu.$$

This is not the orbit bifurcating from the fixed point. The second fixed point is real when

$$r_2^2 = \frac{1 - \sqrt{1+4\mu}}{2} \approx -\mu$$

is positive, or $\mu < 0$. Notice that $r_2 \approx \sqrt{-\mu}$ and $r_1 \approx \sqrt{1+\mu} \approx 1$. The r equation has fixed points at $0 < r_2 < r_1$ for $\mu < 0$, and at $0 < r_1$ for $\mu > 0$. The fixed point at r_1 is always stable. The fixed point at r_2 is unstable when it exists. The fixed point 0 is stable for $\mu < 0$ and unstable for $\mu > 0$.

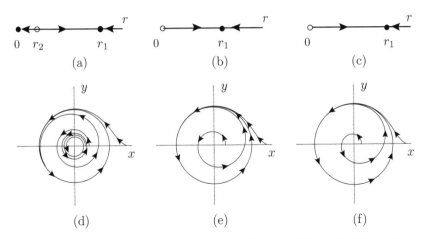

Figure 11. Example 6.14. r variables for: (a) $\mu < 0$, (b) $\mu = 0$, (c) $\mu > 0$. Phase plane in (x, y) variables for (d) $\mu < 0$, (e) $\mu = 0$, (f) $\mu > 0$. In each case, the outer periodic orbit is attracting. There is an inner repelling periodic orbit for $\mu < 0$.

In terms of the Cartesian coordinates, there is always an attracting periodic orbit of radius approximately $\sqrt{1+\mu}$. For $\mu < 0$, the fixed point at the origin is stable, and there is another repelling periodic orbit of radius approximately $\sqrt{-\mu}$. See Figure 11.

Example 6.15 (Vertical Hopf). Consider an oscillator with friction

$$\dot{x} = y,$$
$$\dot{y} = -x + \mu y.$$

This has an attracting fixed point for $\mu < 0$ and a repelling fixed point for $\mu > 0$. In any case, there are no periodic orbits for $\mu \neq 0$. For $\mu = 0$, the whole plane is filled with periodic orbits. These orbits are not isolated, so they are not called limit cycles. They all appear for the same parameter value. See Figure 12.

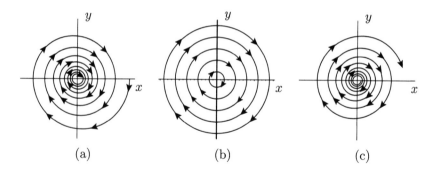

Figure 12. Phase plane in (x, y)-variables for vertical Hopf of Example 6.15

A small change of the preceding example is given by the pendulum equation with friction,

$$\dot{x} = y,$$
$$\dot{y} = -\sin(x) + \mu y.$$

Again, for $\mu = 0$, there is a whole band of periodic orbits. For $\mu \neq 0$, there are no periodic orbits. Thus, this system has neither a subcritical nor a supercritical bifurcation.

Exercises 6.4

1. Consider the system of differential equations

$$\dot{x} = y,$$
$$\dot{y} = -x - 2x^3 - 3x^5 + y(2\mu + x^2 + y^2).$$

   a. What are the eigenvalues at the fixed point at the origin for different values of μ?
   b. For $\mu = 0$, is the origin weakly attracting or repelling?
   c. Show that there is an Andronov–Hopf bifurcation for $\mu = 0$. Is it a subcritical or supercritical bifurcation, and is the periodic orbit attracting or repelling? Explain your answer.

2. Consider the system

$$\dot{x} = y + \mu x,$$
$$\dot{y} = -x + \mu y - x^2 y.$$

   a. Show that the origin is weakly attracting or repelling for $\mu = 0$ by using a Lyapunov function.
   b. Show that there is an Andronov–Hopf bifurcation as μ varies.
   c. Is the bifurcation subcritical or supercritical? Is the periodic orbit attracting or repelling?

Exercises 6.4

3. Consider the variation of a predator–prey system

$$\dot{x} = x\,(1 + 2x - x^2 - y),$$
$$\dot{y} = y\,(x - a),$$

with $a > 0$. The x population is the prey; by itself, its rate of growth increases for small populations and then decreases for $x > 1$. The predator is given by y, and it dies out when no prey is present. The parameter is given by a. You are asked to show that this system has an Andronov–Hopf bifurcation.

   a. Find the fixed point (x_a, y_a) with both x_a and y_a positive.
   b. Find the linearization and eigenvalues at the fixed point (x_a, y_a). For what parameter value, are the eigenvalues purely imaginary? If α_a is the real part of the eigenvalues, show that

$$\frac{d}{da}(\alpha_a) \neq 0$$

   for the bifurcation value.
   c. Using Theorem 6.7, check whether the fixed point is weakly attracting or repelling at the bifurcation value.
   d. Does the periodic orbit appear for a larger than the bifurcation value or less? Is the periodic orbit attracting or repelling?

4. Consider the system of Van der Pol differential equations with parameter given by

$$\dot{x} = y + \mu x - x^3,$$
$$\dot{y} = -x.$$

   a. Check whether the origin is weakly attracting or repelling for $\mu = 0$, by using a test function.
   b. Show that there is an Andronov–Hopf bifurcation as μ varies.
   c. Is the bifurcation subcritical or supercritical? Is the periodic orbit attracting or repelling?

5. Mas-Colell gave a Walrasian price and quantity adjustment that undergoes an Andronov-Hopf bifurcation. Let Y denote the quantity produced and p the price. Assume that the marginal wage cost is $\psi(Y) = 1 + 0.25\,Y$ and the demand function is $D(p) = -0.025\,p^3 + 0.75\,p^2 - 6p + 48.525$. The model assumes that the price and quantity adjustment are determined by the system of differential equations

$$\dot{p} = 0.75\,[D(p) - Y],$$
$$\dot{Y} = b\,[p - \psi(Y)].$$

   a. Show that $p^* = 11$ and $Y^* = 40$ is a fixed point.
   b. Show that the eigenvalues have zero real part for $b^* = 4.275$. Assuming the fixed point is weakly attracting for b^*, show that this is an attracting limit cycle for b near b^* and $b < b^*$.

6. Consider the Holling-Tanner predator-prey system given by
$$\dot{x} = x\left(1 - \frac{x}{6} - \frac{5y}{3(1+x)}\right),$$
$$\dot{y} = \mu y\left(1 - \frac{y}{x}\right).$$

   a. Find the fixed point (x^*, y^*) with $x^* > 0$ and $y^* > 0$.
   b. Find the value of $\mu^* > 0$ for which the eigenvalues for the fixed point (x^*, y^*) have zero real part.
   c. Assuming the fixed point is weakly attracting for μ^*, show that this is an attracting limit cycle for μ near μ^* and $\mu < \mu^*$.

6.5. Homoclinic Bifurcation

Example 5.11 gives a system of differential equations for which the ω-limit set of certain points was equal to a homoclinic loop. In this section, we add a parameter to this system to see how a periodic orbit can be born or die in a homoclinic loop as a parameter varies. After considering this example, which is not typical, but which illustrates the relationship between periodic orbits and homoclinic loops, we present a more general theorem. We consider this bifurcation mainly because it occurs for the Lorenz system, which is discussed in Chapter 7.

Example 6.16. Consider the system of differential equations
$$\dot{x} = y,$$
$$\dot{y} = x - x^2 + y\left(6x^2 - 4x^3 - 6y^2 + 12\mu\right),$$
where μ is a parameter. If the term containing y in the $\dot{y}$ equation were not there, there would be an energy function
$$L(x, y) = -\frac{x^2}{2} + \frac{x^3}{3} + \frac{y^2}{2}.$$
For the complete equations, we use L as a bounding function. See Figure 13 for a plot of some level curves for L. Its time derivative is
$$\dot{L} = y^2\left(6x^2 - 4x^3 - 6y^2 + 12\mu\right) = -12y^2(L - \mu) \begin{cases} \geq 0 & \text{when } L < \mu, \\ = 0 & \text{when } L = \mu, \\ \leq 0 & \text{when } L > \mu. \end{cases}$$

First we consider the situation when the parameter $\mu = 0$. Then, there is a homoclinic loop
$$\Gamma_0 = \{(x, y) : L(x, y) = 0, \ x \geq 0\}.$$
For (x_0, y_0) with $L(x_0, y_0) < 0$ and $x_0 > 0$, $\omega(x_0, y_0) = \Gamma_0$, as we discussed before. See Figure 14(b).

Next, consider the situation when the parameter $-1/6 < \mu < 0$. Then, the level curve
$$\Gamma_\mu = \{(x, y) : L(x, y) = \mu, \ x > 0\}$$
is invariant, and is a closed curve inside the loop Γ_0. Since it is invariant and contains no fixed points, it must be a periodic orbit. If (x_0, y_0) is a point with $x_0 > 0$ and $-1/6 < L(x_0, y_0) < \mu$, then, again, $\dot{L} \geq 0$ and $\omega(x_0, y_0) = \Gamma_\mu$. Similarly,

6.5. Homoclinic Bifurcation

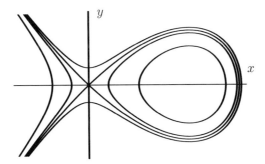

Figure 13. The level curves $L^{-1}(-0.5)$, $L^{-1}(-0.1)$, $L^{-1}(0)$, and $L^{-1}(0.1)$ for the test function for Example 6.16

if $\mu < L(x_0, y_0) \leq 0$ and $x_0 > 0$, then $\dot{L} \leq 0$, the orbit is trapped inside $L^{-1}(0)$, and so $\omega(x_0, y_0) = \Gamma_\mu$. In particular, the branch of the unstable manifold of the saddle point at the origin with $x > 0$ must spiral down onto the periodic orbit. The periodic orbit Γ_μ is an attracting limit cycle. See Figure 14(a). If (x_0, y_0), with $x_0 > 0$, is on the stable manifold of the origin, the quantity $L(\phi(t; (x_0, y_0)))$ increases as t becomes more negative, $\phi(t; (x_0, y_0))$ must go outside the unstable manifold of the origin, and $\phi(t; (x_0, y_0))$ goes off to infinity. (In Figure 14(a), we follow this trajectory until it intersects the positive y axis.) In particular, there is no longer a homoclinic loop,

$$W^u(\mathbf{0}) \cap W^s(\mathbf{0}) = \{\mathbf{0}\},$$

and the stable and unstable manifolds intersect only at the fixed point. See Figure 14(a). As μ goes to zero, the limit cycle Γ_μ passes closer and closer to the fixed point at the origin; it takes longer and longer to go past the fixed point, and so the period goes to infinity.

Next, we consider $\mu > 0$. The invariant curve $L^{-1}(\mu)$ is unbounded with no closed loops. In fact, we argue that there are no periodic orbits. On the unstable manifold of the origin, the value of L starts out near 0 for a point near the origin, which is less than μ. The derivative $\dot{L}$ is positive and the value of L increases. Therefore, the orbit cannot come back to the origin, but goes outside the stable manifold of the origin. Thus, the homoclinic loop is destroyed. In fact, if (x_0, y_0) is any point inside Γ_0 but not the fixed point $(1, 0)$ (i.e., $x_0 > 0$ and $0 > L(x_0, y_0) > -1/6$), then (x_0, y_0) is either on the stable manifold of the origin or eventually $\phi(t; (x_0, y_0))$ goes off to infinity as t increases. The time derivative $\dot{L}$ is positive on the loop $\Gamma_0 = L^{-1}(0)$. Therefore, no trajectory can enter the inside of this loop as time increases, and any point starting outside the loop must remain outside the loop. In fact, it is not hard to argue that points not on the stable manifold of the origin must go off to infinity. Therefore, there are no periodic orbits. See Figure 14(c) for the full phase portrait of this case.

Remember that the *divergence of a vector field* $\mathbf{F}(\mathbf{x})$ is given by

$$\nabla \cdot \mathbf{F}_{(\mathbf{x})} = \sum_{i=1}^{n} \frac{\partial F_i}{\partial x_i}(\mathbf{x}).$$

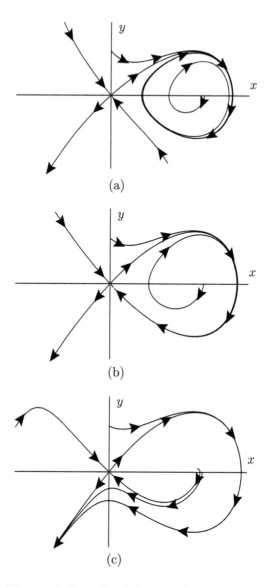

Figure 14. Homoclinic bifurcation for Example 6.16

For a system of differential equations $\dot{\mathbf{x}} = \mathbf{F}(\mathbf{x})$, we often call the divergence of the vector field $\mathbf{F}$ the *divergence of the system of differential equations*. The main difference between the preceding example and the generic situation is that the divergence of the vector field in the example is 0 at the fixed point when $\mu = 0$. The typical manner in which such a homoclinic bifurcation occurs is for the divergence to be nonzero at the fixed point, as assumed in the next theorem.

Theorem 6.8 (Homoclinic bifurcation). *Assume that the system of differential equations* $\dot{\mathbf{x}} = \mathbf{F}_\mu(\mathbf{x})$, *with parameter μ, has a saddle fixed point $\mathbf{p}_\mu$ and satisfies the following conditions:*

6.5. Homoclinic Bifurcation

(i) For $\mu = \mu_0$, one branch $\Gamma_{\mu_0}^+$ of the unstable manifold $W^u(\mathbf{p}_{\mu_0}, \mathbf{F}_{\mu_0})$ is contained in the stable manifold $W^s(\mathbf{p}_{\mu_0}, \mathbf{F}_{\mu_0})$.

(ii) For $\mu = \mu_0$, assume that the divergence of $\mathbf{F}_{\mu_0}$ at $\mathbf{p}_{\mu_0}$ is negative. Then orbits starting at $\mathbf{q}_0$, inside $\Gamma_{\mu_0}^+$ but near $\Gamma_{\mu_0}^+$, have

$$\omega(\mathbf{q}_0, \mathbf{F}_{\mu_0}) = \{\mathbf{p}_{\mu_0}\} \cup \Gamma_{\mu_0}^+.$$

(iii) Further assume that the homoclinic loop $\Gamma_{\mu_0}^+$ is broken for $\mu \neq \mu_0$, with the unstable manifold of the fixed point $\mathbf{p}_\mu$ inside the stable manifold for $\mu < \mu_0$ and outside the stable manifold for $\mu > \mu_0$.

Then, for μ near μ_0 and $\mu < \mu_0$, there is an attracting periodic orbit Γ_μ whose period goes to infinity as μ goes to 0, and is such that Γ_μ converges to $\Gamma_{\mu_0}^+$.

If we have the same situation except that the divergence of $\mathbf{F}_{\mu_0}$ at $\mathbf{p}_{\mu_0}$ is positive, then there is an unstable periodic orbit for μ near μ_0 and $\mu > \mu_0$.

The creation of a periodic orbit from a homoclinic loop as given in the preceding theorem is called a *homoclinic bifurcation*.

We explain the reason that the homoclinic loop is attracting when the divergence is negative at the fixed point. By differentiable linearization at a saddle fixed point as given in Theorem 4.14 or 4.16, there is a differentiable change of variables, so the equations are

$$\dot{x} = a\,x,$$
$$\dot{y} = -b\,y,$$

where $a - b < 0$, or $a < b$. (We are writing these equations as if the coefficients do not depend on the parameter μ.) Then, a trajectory starting at the cross section $y = y_0$ has initial conditions (x, y_0) and solution $x(t) = x\,e^{at}$ and $y(t) = y_0\,e^{-bt}$. If we follow the solutions until $x = x_1$, then the time $t = \tau$ satisfies

$$x_1 = x\,e^{a\tau},$$
$$e^{-\tau} = x_1^{-1/a}\,x^{1/a}.$$

The y value for $t = \tau$ is

$$y(\tau) = y_0\,(e^{-\tau})^b$$
$$= y_0\,x_1^{-b/a}\,x^{b/a}.$$

Thus, the Poincaré map past the fixed point is

$$y = P_{\mu,1}(x) = y_0\,x_1^{-b/a}\,x^{b/a},$$

which has an exponent on x that is greater than one. The derivative is $P'_{\mu,1}(x) = y_0\,x_1^{-b/a}\,x^{-1+b/a}$, so $P'_{\mu,1}(0) = 0$ and $|P'_{\mu,1}(x)| \ll 1$ for small x. The Poincaré map $x = P_{\mu,2}(y)$ from $x = x_1$ back to $y = y_0$ is a differentiable map. Since the time to travel between these two transversals is bounded, $P'_{\mu,2}(y) \neq 0$ and is finite. The point $y = 0$ is on the unstable manifold, so $P_{\mu,2}(0)$ is the intersection of the unstable manifold with the transversal at $y = y_0$. This point changes with μ, so we let

$$P_{\mu,2}(0) = \delta_\mu,$$

which has the same sign as μ. The composition of $P_{\mu,1}$ and $P_{\mu,2}$ is the Poincaré map all the way around the loop from $y = y_0$ to itself,

$$P_\mu(x) = P_{\mu,2}(P_{\mu,1}(x)) = P_{\mu,2}(x^{b/a}).$$

By the chain rule,

$$P'_\mu(x) = P'_{\mu,2}(x^{b/a})P'_{\mu,1}(x),$$
$$P'_\mu(0) = P'_{\mu,2}(0)P'_{\mu,1}(0) = 0, \quad \text{and}$$
$$|P'_\mu(x)| < 1$$

for all small x. Using the mean value theorem, for μ small and x small, we have

$$|P_\mu(x) - P_\mu(0)| = |P'_\mu(x')| \cdot |x - 0|.$$

For $\mu = 0$ and small $x > 0$,

$$0 = P_0(0) < P_0(\hat{x}) < \hat{x}.$$

For $\mu > 0$ and sufficiently small,

$$0 < \delta_\mu = P_\mu(0) < P_\mu(\hat{x}) < \hat{x},$$

and P_μ contracts the interval $[0, \hat{x}]$ into itself. Therefore, there is a point x_μ with $P_\mu(x_\mu) = x_\mu$ and $P'_\mu(x_\mu) < 1$. (See Theorem 9.3 and Lemma 10.1 for more details on this type of argument to get a fixed point.) This results in an attracting periodic orbit for $\mu > 0$, as we wanted to show.

If $a > b$, the same argument applies, but $P'_{\mu,1}(0) = \infty$, and

$$P'_\mu(x) = P'_{\mu,2}(x^{b/a})P'_{\mu,1}(x) > 1$$

for small x. Thus, for $\mu < 0$,

$$P_\mu(0) = \delta_\mu < 0 < \hat{x} < P_\mu(\hat{x}),$$

and P_μ stretches the interval $[0, \hat{x}]$ across itself. Therefore, there is a point x_μ with $P_\mu(x_\mu) = x_\mu$ and $P'_\mu(x_\mu) > 1$. This gives a repelling periodic orbit for $\mu < 0$, as we wanted to show.

Exercises 6.5

1. Consider the system of differential equations

    $$\dot{x} = y,$$
    $$\dot{y} = x - 2x^3 + y(\mu + x^2 - x^4 - y^2),$$

    where μ is a parameter.
    a. Using the test function

    $$L(x, y) = \frac{-x^2 + x^4 + y^2}{2},$$

    show that there are (i) two homoclinic orbits for $\mu = 0$, (ii) one attracting limit cycle for $\mu > 0$, and (iii) two attracting limit cycles for $-\frac{1}{4} < \mu < 0$.
    b. What are the ω-limit sets for various points for $\mu < 0$, $\mu = 0$, and $\mu > 0$.
    c. Draw the phase portrait for $\mu > 0$, $\mu = 0$, $-\frac{1}{4} < \mu < 0$, and $\mu < -\frac{1}{4}$

6.6. Rate of Change of Volume

In this section, we consider how the flow changes the area or volume of a region in its domain. In two dimensions, this can be used to rule out the existence of a periodic orbit in some part of the phase plane. This result is used in Section 7.1 to show that any invariant set for the Lorenz equations must have zero volume. We also use the Liouville formula from this section to calculate the derivative of the Poincaré map in the next section.

The next theorem considers the area of a region in two dimensions or volume in dimensions three and larger. By volume in dimensions greater than three, we mean the integral of $dV = dx_1 \cdots dx_n$ over the region. Although we state the general result, we only use the theorem in dimensions two and three, so this higher dimensional volume need not concern the reader.

Theorem 6.9. *Let $\dot{\mathbf{x}} = \mathbf{F}(\mathbf{x})$ be a system of differential equations in $\mathbb{R}^n$, with flow $\phi(t; \mathbf{x})$. Let $\mathscr{D}$ be a region in $\mathbb{R}^n$, with (i) finite n-volume and (ii) smooth boundary $\mathrm{bd}\,\mathscr{D}$. Let $\mathscr{D}(t)$ be the region formed by flowing along for time t,*

$$\mathscr{D}(t) = \{\phi(t; \mathbf{x}_0) : \mathbf{x}_0 \in \mathscr{D}\}.$$

a. *Let $V(t)$ be the n-volume of $\mathscr{D}(t)$. Then*

$$\frac{d}{dt} V(t) = \int_{\mathscr{D}(t)} \nabla \cdot \mathbf{F}_{(\mathbf{x})} \, dV,$$

where $\nabla \cdot \mathbf{F}_{(\mathbf{x})}$ is the divergence of $\mathbf{F}$ at the point $\mathbf{x}$ and $dV = dx_1 \cdots dx_n$ is the element of n volume. If $n = 2$, then $V(t)$ should be called the area and dV should be replaced by the element of area $dA = dx_1 dx_2$.

b. *If the divergence of $\mathbf{F}$ is a constant (independent of $\mathbf{x}$), then*

$$V(t) = V(0)\, e^{t \nabla \cdot \mathbf{F}}.$$

Thus, if the vector field is divergence free, with $\nabla \cdot \mathbf{F} \equiv 0$, then the volume is preserved; if $\nabla \cdot \mathbf{F} < 0$, then the volume decays exponentially; if $\nabla \cdot \mathbf{F} > 0$, then the volume grows exponentially.

In the next chapter, we use the version of the preceding theorem with constant divergence to discuss the Lorenz equations. This system of differential equations in three dimensions has a constant negative divergence. It also has a positively invariant region $\mathscr{D}$ that is taken inside itself for positive time. The volume $V(t)$ of $\mathscr{D}(t) = \phi(t; \mathscr{D})$ has to decrease as t increases. Thus, this set converges to an invariant attracting set that must have zero volume. (It is less than any positive number by the preceding theorem.)

We will also use the manner in which the determinant of the derivative of the flow varies along a solution curve. This formula is called the *Liouville formula*. Since the determinant of $D_\mathbf{x} \phi_{(t; \mathbf{x}_0)}$ is the volume of an infinitesimal n-cube, this formula is related to the preceding theorem.

Theorem 6.10 (Liouville formula). *Consider a system of differential equations*
$$\dot{\mathbf{x}} = \mathbf{F}(\mathbf{x})$$
in $\mathbb{R}^n$ with both $\mathbf{F}(\mathbf{x})$ and $\dfrac{\partial F_i}{\partial x_j}(\mathbf{x})$ continuous. Then, the determinant of the matrix of partial derivatives of the flow can be written as the exponential of the integral of the divergence of the vector field, $\nabla \cdot \mathbf{F}_{(\mathbf{x})}$, along the trajectory,
$$\det\left(D_{\mathbf{x}}\phi_{(t;\mathbf{x}_0)}\right) = \exp\left(\int_0^t (\nabla \cdot \mathbf{F})_{(\phi(s;\mathbf{x}_0))}\,ds\right).$$

The equation for the derivative with respect to time of the determinant of $D_{\mathbf{x}}\phi_{(t;\mathbf{x}_0)}$ follows by combining the first variation formula given in Theorem 3.4 with the formula for the determinant of a fundamental matrix solution of a time-dependent linear differential equation given in Theorem 2.3. We give another proof in Section 6.9 based on the change of volume given in the preceding theorem.

We can also use these theorems to show that it is impossible to have a periodic orbit completely contained in a region of the plane.

Theorem 6.11. *Suppose $\dot{\mathbf{x}} = \mathbf{F}(\mathbf{x})$ is a differential equation in the plane. Let γ be a periodic orbit which surrounds the region $\mathscr{D}$, and the differential equation is defined at all points of $\mathscr{D}$.*

a. (**Bendixson Criterion**) *Then,*
$$\iint_{\mathscr{D}} (\nabla \cdot \mathbf{F})_{(\mathbf{x})}\,dA = 0.$$
In particular, if $\mathscr{R}$ is a region with no holes and divergence has one sign in $\mathscr{R}$, then there can be no periodic orbit that can be completely contained in $\mathscr{R}$.

b. (**Dulac Criterion**) *Let $g(x, y)$ be any real-valued function defined and differentiable at least on the points of $\mathscr{D}$. Then,*
$$\iint_{\mathscr{D}} \nabla \cdot (g\,\mathbf{F})_{(\mathbf{x})}\,dA = 0.$$
In particular, if $\mathscr{R}$ is a region with no holes and the divergence of $g(x, y)\,\mathbf{F}(x, y)$ has one sign in $\mathscr{R}$, then there can be no periodic orbit that can be completely contained in $\mathscr{R}$.

Part (b) follows from Part (a) because $\dot{\mathbf{x}} = g(x, y)\,\mathbf{F}(x, y)$ has the same trajectories as $\dot{\mathbf{x}} = \mathbf{F}(x, y)$; they just travel at different speeds.

This theorem can be applied to some examples to show there are no periodic orbits.

Example 6.17. Consider the Van der Pol equation given in the form
$$\dot{x} = y,$$
$$\dot{y} = -x + \mu\,(1 - x^2)\,y,$$
This has divergence equal to $\mu\,(1 - x^2)$. Therefore, for $\mu \neq 0$, there cannot be any periodic orbit contained completely in the strip $-1 \leq x \leq 1$. This same fact can be shown using the test function $L(x, y) = (x^2 + y^2)/2$, as we did for the Lienard equation in Section 6.3.

Example 6.18. Consider the equations

$$\dot{x} = x(a - by - fx),$$
$$\dot{y} = y(-c + ex - hy),$$

where all the parameters a, b, c, e, f, and h are positive. Notice that this is just the predator–prey system, with a negative impact of each population on itself. We assume that $a/f > c/e$, so there is a fixed point (x^*, y^*) in the first quadrant. The first quadrant is invariant, because $\dot{x} = 0$ along $x = 0$ and $\dot{y} = 0$ along $y = 0$. The divergence does not have one sign in the first quadrant:

$$(\nabla \cdot \mathbf{F})_{(x,y)} = a - by - 2fx - c + ex - 2hy.$$

However, if we let $g(x, y) = 1/xy$, then

$$\nabla \cdot (g\mathbf{F})_{(x,y)} = -\frac{f}{y} - \frac{h}{x},$$

which is strictly negative in the first quadrant. Therefore, there cannot be any periodic orbits inside the first quadrant. Note that $g(x, y) = 1/xy$ is not defined when either x or y equals zero, but it would be well defined inside any periodic orbit contained entirely in the first quadrant.

In the next section, we give another argument that this example does not have any periodic orbits by using the Poincaré map. See Example 6.24.

Exercises 6.6

1. Use Theorem 6.11 to show that the following system of differential equations does not have a periodic orbit in the first quadrant:

$$\dot{x} = x(a - by),$$
$$\dot{y} = y(-a - bx).$$

Both a and b are positive parameters.

2. Consider the predator–prey system of differential equations

$$\dot{x} = x(a - x - y),$$
$$\dot{y} = y(-3a + x),$$

where a is a positive parameter.
   a. Find the fixed points and determine the stability type of each fixed point (saddle, attracting, repelling, etc.).
   b. Use Theorem 6.11 to show that the system does not have a periodic orbit in the first quadrant,
   c. What is the basin of attraction of any attracting fixed point?
   d. Sketch the phase portrait of the system using the information from parts (a–c) and the nullclines. Explain the location of the stable and unstable manifolds of any saddle fixed points.

3. Show that the following system of differential equations does not have a periodic orbit:
$$\dot{x} = -x + y^2,$$
$$\dot{y} = x^2 - y^3.$$

4. Consider the forced damped pendulum given by
$$\dot{\tau} = 1 \quad (\bmod\ 2\pi),$$
$$\dot{x} = y \quad (\bmod\ 2\pi),$$
$$\dot{y} = -\sin(x) - y + \cos(\tau).$$
   a. What is the divergence of the system of equations?
   b. Show that the region $\mathscr{R} = \{(\tau, x, y) : |y| \leq 3\}$ is a positively invariant region. Hint: What is the sign of $\dot{y}$ when $y = \pm 3$?
   c. Let $\mathscr{R}(t) = \phi(t; \mathscr{R})$. What is the volume of $\mathscr{R}(t)$ in terms of the volume of $\mathscr{R}$? What happens to the volume of $\mathscr{R}(t)$ as t goes to infinity?

5. Consider the system of differential equations given by
$$\dot{x} = y,$$
$$\dot{y} = x - 2x^3 - y,$$
with flow $\phi(t; \mathbf{x})$,
   a. What is the divergence of the system?
   b. If $\mathscr{D}$ is a region in the plane with positive area, what is the area of $\mathscr{D}(t)$ in terms of the area of $\mathscr{D}$?

6. Consider the system of differential equations
$$\dot{x} = y,$$
$$\dot{y} = -x - y + x^2 + y^2.$$
Using the scalar function $g(x,y) = e^{-2x}$ in the Dulac Criterion, show that the system does not have a periodic orbit.

7. Rikitake gave a model to explain the reversal of the magnetic poles of the earth. The system of differential equations is
$$\dot{x} = -\mu x + y z,$$
$$\dot{y} = -\mu y + (z - a) x,$$
$$\dot{z} = 1 - x y,$$
with $\mu > 0$ and $a > 0$.
   a. Show that the total system decreases volume as t increases.
   b. Show that there are two fixed points (x^*, y^*, z^*), which satisfy $y^* = 1/x^*$, and $z^* = \mu (x^*)^2$. $(x^*)^2 - (x^*)^{-2} = a/\mu$,
   c. Show that the characteristic equation of the linearization at the fixed points is
$$\lambda^3 + 2\mu\lambda^2 + \frac{a}{\mu}\lambda + 2a = 0.$$
   Show that $\lambda_1 = -2\mu$ is one eigenvalue, and the other two eigenvalues are $\pm i \sqrt{(a/\mu)}$.

6.7. Poincaré Map

The self-excited oscillator given by the Lienard system of differential equations was shown to have a unique attracting periodic orbit by following the trajectories from the y-axis until the first time they return to the y-axis. This first return map is an example of a Poincaré map.

We start this section by calculating the Poincaré map for the simple example of the first section of this chapter, for which the solutions can be calculated using polar coordinates. After this example, we turn to the general definition and examples in which the Poincaré map has more variables. We give a condition for the periodic orbit to be asymptotically stable in terms of the eigenvalues of the derivative of the Poincaré map. We also relate the eigenvalues of the derivative of the Poincaré map to the eigenvalues of the derivative of the flow. In the case of two dimensions, there is a condition in terms of the divergence which can be used to determine whether a periodic orbit is attracting or repelling. We apply this to the Van der Pol equation to give another verification that its periodic orbit is attracting.

Example 6.19 (Return to the Poincaré map for Example 6.4). The differential equations are

$$\dot{x} = y + x(1 - x^2 - y^2),$$
$$\dot{y} = -x + y(1 - x^2 - y^2),$$

or

$$\dot{r} = r(1 - r^2),$$
$$\dot{\theta} = -1,$$

in polar coordinates. In Example 6.4, we showed that the first return of trajectories from the half-line $\{(x, 0) : x > 0\}$ to itself, the Poincaré map, is given by

$$P(r) = r[r^2 + e^{-4\pi}(1 - r^2)]^{-1/2}$$
$$= r[1 - (1 - e^{-4\pi})(1 - r^2)]^{-1/2}$$
$$= [1 + e^{-4\pi}(r^{-2} - 1)]^{-1/2}.$$

The derivative of the first form of the Poincaré map, is

$$P'(r) = \frac{e^{-4\pi}}{[r^2 + e^{-4\pi}(1 - r^2)]^{3/2}} = \frac{e^{-4\pi} P(r)^3}{r^3} > 0.$$

For the rest of the discussion of this example, we use the Poincaré map to show that the periodic orbit $r = 1$ is orbitally asymptotically stable, and attracts all orbits other than the fixed point at the origin. This can be seen from the differential equations directly, but we want to derive this from the properties of the Poincaré map in order to introduce the relationship of the Poincaré map and the stability of the periodic orbit.

(a) The points $r = 0$ and $r = 1$ return to themselves, $P(0) = 0$ and $P(1) = 1$, as can be seen from the first equation for $P(r)$.

(b) Next, we consider $r > 1$. The denominator in the second representation of $P(r)$ has

$$1 - (1 - e^{-4\pi})(1 - r^2) > 1, \quad \text{so}$$
$$P(r) < r.$$

Similarly, the denominator in the third representation of $P(r)$ has

$$1 + e^{-4\pi}(r^{-2} - 1) < 1, \quad \text{so}$$
$$1 < P(r).$$

Combining, we have $1 < P(r) < r$, and the orbit returns nearer the periodic orbit $r = 1$. See the graph of $P(r)$ in Figure 15. Note that the graph of $P(r)$ is below the diagonal (r, r) for $r > 1$.

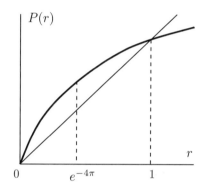

Figure 15. Poincaré map for Example 6.4

If $r_0 > 1$, then $1 < r_1 = P(r_0) < r_0$. Continuing by induction, we see that $1 < r_n = P(r_{n-1}) < r_{n-1}$, so

$$1 < r_n = P(r_{n-1}) < r_{n-1} = P(r_{n-2}) < \cdots < r_2 = P(r_1) < r_1 = P(r_0) < r_0.$$

This decreasing sequence of points must converge to a fixed point of P, so r_n converges to 1.

In fact, the mean value theorem can be used to show this convergence more explicitly. For $r > 1$, $P(r)/r < 1$, so

$$P'(r) = \frac{e^{-4\pi} P(r)^3}{r^3} < e^{-4\pi} < 1.$$

For the preceding sequence of points r_j, there are points r'_j between r_j and 1 such that

$$P(r_j) - P(1) = P'(r'_j)(r_j - 1),$$

6.7. Poincaré Map

so

$$r_n - 1 = P(r_{n-1}) - P(1) = P'(r'_{n-1})(r_{n-1} - 1)$$
$$< e^{-4\pi}(r_{n-1} - 1) = e^{-4\pi}(P(r_{n-2}) - P(1))$$
$$= e^{-4\pi} P'(r'_{n-2})(r_{n-2} - 1) < e^{-4\pi} e^{-4\pi}(r_{n-2} - 1) = e^{-2 \cdot 4\pi}(r_{n-2} - 1)$$
$$\vdots$$
$$< e^{-n4\pi}(r_0 - 1).$$

Since $e^{-4\pi} < 1$, $e^{-n4\pi}$ goes to zero as n goes to infinity, and r_n must converge to 1. In fact, this gives an explicit bound on the rate at which the sequence of return positions r_n converges to 1.

(**c**) Finally, we consider $0 < r < 1$. The quantity $0 < (1 - e^{-4\pi})(1 - r^2) < 1$, so the denominator

$$0 < [1 - (1 - e^{-4\pi})(1 - r^2)]^{1/2} < 1 \qquad \text{and}$$
$$0 < r < P(r).$$

Using the second equation for $P(r)$, we have

$$1 < 1 + e^{-4\pi}(r^{-2} - 1), \qquad \text{so}$$
$$P(r) = [1 + e^{-4\pi}(r^{-2} - 1)]^{-1/2} < 1.$$

(We can also see that $P(r) < 1$ because two different trajectories cannot cross and $r = 1$ returns to $P(1) = 1$.) Combining, we get

$$0 < r < P(r) < 1,$$

and the orbit returns nearer the periodic orbit $r = 1$. Taking an r_0 with $0 < r_0 < 1$, we have

$$0 < r_0 < r_1 < \cdots < r_{n-1} < r_n < 1.$$

This increasing sequence of points must converge to a fixed point of P, so r_n converges to 1.

We can also apply the mean value theorem as in the earlier case. For $0 < r < 1$, $P(r) < 1$, so

$$P'(r) \leq e^{-4\pi} r^{-3}.$$

Therefore, $0 < P'(r) < 1$ for $r > e^{-4\pi/3}$. On the other hand, $P'(0) = e^{2\pi} > 1$, and so $P'(r) > 1$ for small $r > 0$. These estimates on the derivative indicate that the graph of the Poincaré map $P(r)$ indeed looks like the graph in Figure 15. Once the sequence gets in the range where $P'(r) < 1$, the argument using the mean value theorem can be used again to get an explicit bound on the rate of convergence to 1.

The derivative of the Poincaré map at the periodic orbit is $P'(r) = e^{-4\pi} < 1$. Because $P'(1)$ is a single number, it can be thought of as the eigenvalue of the 1×1 matrix whose entry is $P'(1)$.

Before giving the general conditions, we discuss a specific example of a non-homogeneous linear differential equation. The Poincaré map for such an equation requires the evaluation of the fundamental matrix solution at the period of the forcing. For the example, the derivative of the Poincaré map can be given fairly explicitly in terms of the fundamental matrix of the homogeneous linear equations. This allows us to discuss the conditions on the eigenvalues of the Poincaré map.

Example 6.20. Consider the nonhomogeneous linear system
$$\dot{x}_1 = x_2,$$
$$\dot{x}_2 = -x_1 - x_2 + \cos(\omega t).$$

The forcing term $\cos(\omega t)$ has period $2\pi/\omega$ in t, $\cos\left(\omega(t + 2\pi/\omega)\right) = \cos(\omega t + 2\pi) = \cos(\omega t)$. In order to change this into a system that does not depend on time explicitly, we introduce the variable τ whose time derivative is identically one and is taken as a periodic variable with period $2\pi/\omega$ (the period of the forcing term):
$$\dot{\tau} = 1 \quad (\text{mod } 2\pi/\omega),$$
$$\dot{x}_1 = x_2,$$
$$\dot{x}_2 = -x_1 - x_2 + \cos(\omega \tau).$$

Here we write "$(\text{mod } 2\pi/\omega)$" as an abbreviation for modulo $2\pi/\omega$, which means that we subtract the integer multiples of $2\pi/\omega$ and leave only the remaining part, which is a fraction of $2\pi/\omega$. Since the variable τ is periodic, we can consider it an "angle variable". We put the differential equation for τ first so the results look more like the general case considered later in the section. Taking initial conditions $(0, x_{1,0}, x_{2,0})$, we know by variation of parameters that the solution is given by

$$\begin{pmatrix} \tau(t) \\ x_1(t) \\ x_2(t) \end{pmatrix} = \left(e^{\mathbf{A}t} \begin{pmatrix} x_{1,0} \\ x_{2,0} \end{pmatrix} + \int_0^t e^{\mathbf{A}(t-s)} \begin{pmatrix} 0 \\ \cos(\omega s) \end{pmatrix} ds \right),$$

where
$$A = \begin{pmatrix} 0 & 1 \\ -1 & -1 \end{pmatrix}$$
is the matrix of the linear system. The variable τ comes back to itself at time $2\pi/\omega$. We can take the *transversal* to be
$$\Sigma = \{\tau = 0\}.$$

Writing only the **x**-coordinates on Σ, the Poincaré map from Σ to itself is given by
$$\mathbf{P}(\mathbf{x}_0) = e^{\mathbf{A}2\pi/\omega}\mathbf{x}_0 + \mathbf{v},$$
where
$$\mathbf{v} = \int_0^{\frac{2\pi}{\omega}} e^{\mathbf{A}((2\pi/\omega)-s)} \begin{pmatrix} 0 \\ \cos(\omega s) \end{pmatrix} ds$$
is a constant vector. The derivative of the Poincaré map is $DP_{(\mathbf{x})} = e^{\mathbf{A}2\pi/\omega}$. The eigenvalues of $\mathbf{A}$ are $\lambda^{\pm} = -\frac{1}{2} \pm i\frac{\sqrt{3}}{2}$, and those of $e^{\mathbf{A}2\pi/\omega}$ are $e^{2\pi\lambda^{\pm}/\omega} = e^{-\pi/\omega}e^{\pm i\sqrt{3}\pi/\omega}$, which have absolute value equal to $e^{-\pi/\omega}$. The point $\mathbf{x}^*$ is fixed by the Poincaré map, provided that
$$e^{\mathbf{A}2\pi/\omega}\mathbf{x}^* + \mathbf{v} = \mathbf{x}^* \qquad \text{or}$$
$$\mathbf{x}^* = \left(\mathbf{I} - e^{\mathbf{A}2\pi/\omega}\right)^{-1}\mathbf{v}.$$

6.7. Poincaré Map

Thus, there is one fixed point for $\mathbf{P}$ and one periodic orbit for the system of differential equations. For any other initial condition,

$$\mathbf{P}(\mathbf{x}_0) - \mathbf{x}^* = \mathbf{P}(\mathbf{x}_0) - \mathbf{P}(\mathbf{x}^*) = e^{\mathbf{A}2\pi/\omega}\mathbf{x}_0 + \mathbf{v} - \left(e^{\mathbf{A}2\pi/\omega}\mathbf{x}^* + \mathbf{v}\right)$$
$$= e^{\mathbf{A}2\pi/\omega}\left(\mathbf{x}_0 - \mathbf{x}^*\right) = e^{-\pi/\omega}e^{(\mathbf{A}+\frac{1}{2}\mathbf{I})2\pi/\omega}\left(\mathbf{x}_0 - \mathbf{x}^*\right).$$

Continuing by induction, the nth iterate $\mathbf{x}_n = P(\mathbf{x}_{n-1}) = \mathbf{P}^n(\mathbf{x}_0)$ satisfies

$$\mathbf{P}^n(\mathbf{x}_0) - \mathbf{x}^* = \mathbf{P}^n(\mathbf{x}_0) - \mathbf{P}^n(\mathbf{x}^*) = e^{-n\pi/\omega}e^{(\mathbf{A}+\frac{1}{2}\mathbf{I})n2\pi/\omega}\left(\mathbf{x}_0 - \mathbf{x}^*\right).$$

The eigenvalues of $e^{(\mathbf{A}+\frac{1}{2}\mathbf{I})n2\pi/\omega}$ are $e^{\pm in\sqrt{3}\pi/\omega}$ and the points

$$e^{(\mathbf{A}+\frac{1}{2}\mathbf{I})n2\pi/\omega}\left(\mathbf{x}_0 - \mathbf{x}^*\right)$$

move on an ellipse. The term $e^{-n\pi/\omega}$ goes to zero as n increases. Therefore, for any initial condition, $\mathbf{P}^n(\mathbf{x}_0)$ converges to the fixed point $\mathbf{x}^*$ of $\mathbf{P}$, which is on the periodic orbit of the flow. The fact that makes this work is that $e^{-\pi/\omega} < 1$. But $e^{-\pi/\omega}$ equals the absolute value of each eigenvalue of the matrix $e^{\mathbf{A}2\pi/\omega}$. The matrix $e^{\mathbf{A}2\pi/\omega}$, in turn, is the matrix of partial derivatives of the Poincaré map.

In terms of the whole flow in three dimensions, the time $2\pi/\omega$ map of the flow is

$$\phi\left(\frac{2\pi}{\omega}; (\tau_0, \mathbf{x}_0)\right) = \begin{pmatrix} \tau_0 + \frac{2\pi}{\omega} \\ \mathbf{P}(\mathbf{x}_0) \end{pmatrix}$$
$$= \begin{pmatrix} \tau_0 + \frac{2\pi}{\omega} \\ e^{\mathbf{A}2\pi/\omega}\mathbf{x}_0 + \int_0^{\frac{2\pi}{\omega}} e^{\mathbf{A}((2\pi/\omega)-s)}\begin{pmatrix} 0 \\ \cos(\omega s + \tau_0) \end{pmatrix} ds \end{pmatrix}.$$

The derivative with respect to the initial conditions in all the variables $(\tau_0, \mathbf{x}_0)$ is

$$D_{(\tau_0, \mathbf{x}_0)}\phi(2\pi/\omega; (\tau_0, \mathbf{x}_0)) = \begin{pmatrix} 1 & \mathbf{0} \\ * & D\mathbf{P}_{(\mathbf{x}_0)} \end{pmatrix},$$

where the entry designated by the star is some possibly nonzero term. Therefore, the eigenvalues of the derivative with respect to initial conditions of the flow after one period are 1 and the eigenvalues of $D\mathbf{P}_{(\mathbf{x}_0)}$.

We now state a general definition of a Poincaré map.

Definition 6.21. Consider a system of differential equations $\dot{\mathbf{x}} = \mathbf{F}(\mathbf{x})$, and a point $\mathbf{x}^*$ for which one of the coordinate functions $F_k(\mathbf{x}^*) \neq 0$. The hyperplane through $\mathbf{x}^*$ formed by setting the kth coordinate equal to a constant,

$$\Sigma = \{\mathbf{x} : x_k = x_k^*\},$$

is called a *transversal*, because trajectories are crossing it near $\mathbf{x}^*$. Assume that $\phi(\tau^*; \mathbf{x}^*)$ is again in Σ for some $\tau^* > 0$. We also assume that there are no other intersections of $\phi(t; \mathbf{x}^*)$ with Σ near $\mathbf{x}^*$. For $\mathbf{x}$ near $\mathbf{x}^*$, there is a nearby time $\tau(\mathbf{x})$ such that $\phi(\tau(\mathbf{x}); \mathbf{x})$ is in Σ. Then,

$$\mathbf{P}(\mathbf{x}) = \phi(\tau(\mathbf{x}); \mathbf{x})$$

is the *Poincaré map*.

We use the notation $P^n(\mathbf{x}_0) = P(P^{n-1}(\mathbf{x}_0)) = P \circ P \circ \cdots \circ P(\mathbf{x}_0)$ for the nth *iterate*. Thus, P^n is the composition of P with itself n times. If $\mathbf{x}_j = P(\mathbf{x}_{j-1})$ for $j = 1, \ldots, n$, then $\mathbf{x}_n = P^n(\mathbf{x}_0)$.

We next state the result that compares the eigenvalues of the Poincaré map with the eigenvalues of the period map of the flow.

Theorem 6.12. *Assume that $\mathbf{x}^*$ is on a periodic orbit with period T. Let Σ be a hyperplane through $\mathbf{x}^*$, formed by setting one of the variables equal to a constant; that is, for some $1 \leq k \leq n$ with $F_k(\mathbf{x}^*) \neq 0$, let*

$$\Sigma = \{\mathbf{x} : x_k = x_k^*\}.$$

Let $\mathbf{P}$ be the Poincaré map from a neighborhood of $\mathbf{x}^*$ in Σ back to Σ. Then, the n eigenvalues of $D_{\mathbf{x}}\phi_{(T;\mathbf{x}^*)}$ consist of the $n-1$ eigenvalues of $D\mathbf{P}_{(\mathbf{x}^*)}$, together with 1, the latter resulting from the periodicity of the orbit.

Idea of the proof. We showed earlier that

$$D_{\mathbf{x}}\phi_{(T;\mathbf{x}^*)}\mathbf{F}(\mathbf{x}^*) = \mathbf{F}(\phi(T;\mathbf{x}^*)) = \mathbf{F}(\mathbf{x}^*).$$

Therefore, 1 is an eigenvalue for the eigenvector $\mathbf{F}(\mathbf{x}^*)$.

The Poincaré map is formed by taking the time $\tau(\mathbf{x})$ such that $\phi(\tau(\mathbf{x});\mathbf{x})$ is back in Σ:

$$\mathbf{P}(\mathbf{x}) = \phi(\tau(\mathbf{x});\mathbf{x}).$$

Therefore, if we take a vector $\mathbf{v}$ lying in Σ, we have

$$D\mathbf{P}_{(\mathbf{x}^*)}\mathbf{v} = D_{\mathbf{x}}\phi_{(\tau(\mathbf{x}^*);\mathbf{x}^*)}\mathbf{v} + \left(\left.\frac{\partial \phi}{\partial t}\right|_{(\tau(\mathbf{x}^*);\mathbf{x}^*)}\right)\left(\left.\frac{\partial \tau}{\partial \mathbf{x}}\right|_{\mathbf{x}^*}\right)\mathbf{v}$$
$$= D_{\mathbf{x}}\phi_{(T;\mathbf{x}^*)}\mathbf{v} + \mathbf{F}(\mathbf{x}^*)\left(D\tau_{(\mathbf{x}^*)}\right)\mathbf{v}.$$

In the second term,

$$\left(D\tau_{(\mathbf{x}^*)}\right)\mathbf{v} = \left(\left.\frac{\partial \tau}{\partial \mathbf{x}}\right|_{\mathbf{x}^*}\right)\mathbf{v}$$

is a scalar multiplied by the vector field $\mathbf{F}(\mathbf{x}^*)$ and represents the change in time to return from Σ back to Σ. Thus, if we take a vector $\mathbf{v}$ lying in Σ, we get

$$D_{\mathbf{x}}\phi_{(T;\mathbf{x}^*)}\mathbf{v} = D\mathbf{P}_{(\mathbf{x}^*)}\mathbf{v} - \mathbf{F}(\mathbf{x}^*)\left(D\tau_{(\mathbf{x}^*)}\right)\mathbf{v}.$$

Therefore, using a basis of $\mathbf{F}(\mathbf{x}^*)$ and $n-1$ vectors along Σ, we have

$$(6.5) \qquad D_{\mathbf{x}}\phi_{(T;\mathbf{x}^*)} = \begin{pmatrix} 1 & -D\tau_{(\mathbf{x}^*)} \\ 0 & D\mathbf{P}_{(\mathbf{x}^*)} \end{pmatrix},$$

and the eigenvalues are related as stated in the theorem. $\square$

Definition 6.22. At a periodic orbit, the $(n-1)$ eigenvalues of the Poincaré map $D\mathbf{P}_{(\mathbf{x}^*)}$ are called the *characteristic multipliers*. The eigenvalues of $D_{\mathbf{x}}\phi_{(T;\mathbf{x}^*)}$ consist of the $(n-1)$ characteristic multipliers together with the eigenvalue 1, which is always an eigenvalue.

If all the characteristic multipliers have absolute value not equal to one, then the periodic orbit is called *hyperbolic*.

Now, we can state the main theorems.

6.7. Poincaré Map

Theorem 6.13. *Assume that $\mathbf{x}^*$ is on a periodic orbit with period T.*

a. *If all the characteristic multipliers of the periodic orbit have absolute values less than one, then the orbit is attracting (i.e., orbitally asymptotically stable).*

b. *If at least one of the characteristic multipliers of the periodic orbit has an absolute value greater than one, then the periodic orbit is not orbitally Lyapunov stable (i.e., it is an unstable periodic orbit).*

It is only in very unusual cases that we can calculate the Poincaré map explicitly: nonhomogeneous linear system or simple system where the solutions for the variables can be calculated separately (in polar coordinates). In other cases, we need to use a more indirect means to determine something about the Poincaré map. In the plane, the characteristic multiplier of a periodic orbit can be determined from an integral of the divergence.

Theorem 6.14. *Consider a differential equation in the plane, and assume that $\mathbf{x}^*$ is on a periodic orbit, with period T. Let $(\nabla \cdot \mathbf{F})_{(\mathbf{x})}$ be the divergence of the vector field. Let y be the scalar variable along the transversal Σ, and y^* its value at $\mathbf{x}^*$. Then, the derivative of the Poincaré map is given by*

$$P'(y^*) = \exp\left(\int_0^T (\nabla \cdot \mathbf{F})_{(\phi(t;\mathbf{x}^*))} \, dt\right).$$

Proof. In two dimensions, using equation (6.5), we have

$$\det(D_{\mathbf{x}}\phi_{(T;\mathbf{x}^*)}) = P'(y^*),$$

where we use prime for the derivative in one variable. By the Liouville formula, Theorem 6.10, we have

$$\det(D_{\mathbf{x}}\phi_{(T;\mathbf{x}^*)}) = \exp\left(\int_0^T (\nabla \cdot \mathbf{F})_{(\phi(t;\mathbf{x}^*))} \, dt\right) \det(D_{\mathbf{x}}\phi_{(0;\mathbf{x}^*)})$$

$$= \exp\left(\int_0^T (\nabla \cdot \mathbf{F})_{(\phi(t;\mathbf{x}^*))} \, dt\right),$$

because $D_{\mathbf{x}}\phi_{(0;\mathbf{x}^*)} = \mathbf{I}$. Combining, we have the result. $\square$

We give a few nonlinear examples that use the preceding theorem.

Example 6.23. Consider the time-dependent differential equation

$$\dot{x} = (a + b\cos(t))\, x - x^3,$$

where both a and b are positive. We can rewrite this as a system which does not explicitly depend on t:

$$\dot{\theta} = 1 \pmod{2\pi},$$
$$\dot{x} = (a - b\cos(\theta))\, x - x^3.$$

The flow starting at $\theta = 0$ can be given as

$$\phi(t;(0,x_0)) = (t, \psi(t;x_0)),$$

where $\theta(t) = t$. We can write the Poincaré map from $\theta = 0$ to $\theta = 2\pi$ as
$$\phi(2\pi; (0, x_0)) = (2\pi, P(x_0)) \quad \text{or}$$
$$\psi(2\pi; x_0) = P(x_0).$$

Since $\dot{x} = 0$ for $x = 0$, $\psi(t; 0) \equiv 0$, $P(0) = 0$, 0 is a fixed point of the Poincaré map, and $(0,0)$ is on a periodic orbit for the system of differential equations.

For $x > \sqrt{a+b}$,
$$\dot{x} < (a+b)x - x^3 = x(a+b-x^2) < 0,$$
so $0 < \psi(t;x) < x$ and $0 < P(x) < x$. Similarly, for $x < -\sqrt{a+b}$, $0 > P(x) > x$. Therefore, any other fixed points for P must lie between $\pm\sqrt{a+b}$.

The matrix of partial derivatives is
$$\begin{pmatrix} 0 & 0 \\ -b\sin(\theta)x & a+b\cos(\theta) - 3x^2 \end{pmatrix}.$$

Note that the equation for $\dot{\theta}$ is a constant, so it has zero derivatives. Therefore, the divergence is $a + b\cos(\theta) - 3x^2$.

Along $\phi(t; (0,0)) = (t, 0)$, the divergence is $a + b\cos(t)$. Using Theorem 6.14, we have
$$P'(0) = e^{\int_0^{2\pi} a+b\cos(t)\,dt} = e^{a2\pi} > 1.$$

Therefore, 0 is a repelling fixed point of P, and an unstable periodic orbit for the flow.

Next we consider a periodic orbit with $x_0 \neq 0$, so $P(x_0) = x_0$. Then, $\psi(t; x_0) \neq 0$ for any t, by uniqueness of the solutions. We can write the divergence as
$$a + b\cos(t) - 3x^2 = 3\left(a + b\cos(t) - x^2\right) - 2a - 2b\cos(t)$$
$$= \frac{3\dot{x}}{x} - 2a - 2b\cos(t).$$

Then,
$$P'(x_0) = e^{\int_0^{2\pi} 3\dot{x}/x - 2a - 2b\cos(t)\,dt} = e^{3\ln|P(x_0)| - 3\ln|x_0|} e^{-a4\pi} = e^{-a4\pi} < 1.$$

Now, $P'(0) > 1$, so $P(x_1) > x_1$ for small positive x_1. But, $P(x) < x$ for $x > \sqrt{a+b}$. Therefore, there has to be at least one x_0 between 0 and $\sqrt{a+b}$ for which $P(x_0) = x_0$. At each of these points, $P'(x_0) < 1$ and the graph of P is crossing from above the graph of x to below this graph as x increases. Therefore, there can be only one such point, and so, only one positive fixed point of P; this fix point corresponds to an attracting periodic orbit for the system of differential equations.

For $x < 0$, a similar argument shows that there is a unique fixed point of P, so there is a unique periodic orbit for the system of differential equations.

The periodicity of the equation means that we cannot explicitly find the solutions, but the basic behavior is similar to that of the equation with the periodic term dropped.

We describe an application of the preceding theorem to show that a predator–prey system, with limit to growth, has no periodic orbit.

Example 6.24. Consider the system
$$\dot{x} = x\,(a - b\,y - f\,x),$$
$$\dot{y} = y\,(-c + e\,x - h\,y),$$
where all the parameters a, b, c, e, f, and h are positive. Notice, that these are just the predator–prey equations, with a negative impact of each population on itself. We assume that $a/f > c/e$, so there is a fixed point (x^*, y^*) in the first quadrant. For these values, the fixed point satisfies the equations
$$0 = a - b\,y^* - f\,x^*,$$
$$0 = -c + e\,x^* - h\,y^*.$$
The matrix of partial derivatives at (x^*, y^*) is
$$\begin{pmatrix} -fx^* & -bx^* \\ ey^* & -hy^* \end{pmatrix}.$$
This matrix has a negative trace, so the fixed point is attracting.

In the preceding section, we proved that this equation does not have any periodic orbits by applying the Dulac criterion. We now show this same fact using an argument involving the Poincaré map. To show that there is no periodic orbit, we use Theorem 6.14 to determine the derivative of the Poincaré map at a periodic orbit, if it exists. The divergence is given by
$$\nabla \cdot \mathbf{F}_{(x,y)} = (a - b\,y - f\,x) - f\,x + (-c + e\,x - h\,y) - h\,y$$
$$= \frac{\dot{x}}{x} + \frac{\dot{y}}{y} - f\,x - h\,y.$$
To apply the theorem, we consider the Poincaré map P from
$$\Sigma = \{\,(x, y^*) : x > x^*\,\}$$
to itself. If the system has a periodic orbit and $P(x^1) = x^1$, then the integral giving the derivative of P can be calculated as
$$\int_0^T (\nabla \cdot \mathbf{F})_{(\phi(t;(x^1,y^*)))}\,dt = \int_0^T \frac{\dot{x}}{x} + \frac{\dot{y}}{y} - f\,x - h\,y\,dt$$
$$= \ln(x(T)) - \ln(x(0)) + \ln(y(T)) - \ln(y(0)) - \int_0^T f\,x + h\,y\,dt$$
$$= -\int_0^T f\,x + h\,y\,dt$$
$$< 0,$$
since $x(T) = x(0) = x^1$ and $y(T) = y(0) = y^*$ on a periodic orbit. Thus,
$$P'(x) = \exp\left(\int_0^T (\nabla \cdot \mathbf{F})_{(\phi(t;(x^1,y^*)))}\,dt\right) < 1.$$
The Poincaré map has $P(0) = 0$ and $0 < P'(x) < 1$, so the graph of $P(x)$ is increasing, but below the diagonal. Therefore, there are no periodic orbits. If (x_0, y_0) is an initial condition not at the fixed point (x^*, y^*) and with $x_0 > 0$ and $y_0 > 0$, then the backward orbit cannot limit on any periodic orbit or fixed points and so must be unbounded.

Hale and Koçak, [**Hal91**], apply Theorem 6.14 to the Van der Pol system to give another proof that the periodic orbit is stable.

Theorem 6.15. *Any periodic orbit for the Van der Pol system is stable. Therefore, there can be only one periodic orbit.*

Proof. We write the Van der Pol equations in the form
$$\dot{x} = y,$$
$$\dot{y} = -x + (1 - x^2) y.$$

The divergence of **F** is given as follows:
$$\nabla \cdot \mathbf{F} = \frac{\partial}{\partial x}(y) + \frac{\partial}{\partial y}(-x + (1 - x^2) y) = 1 - x^2.$$

Consider the real-valued function $L(x, y) = \frac{1}{2}(x^2 + y^2)$. Then,
$$\dot{L} = y^2(1 - x^2).$$

If L has a minimum along a periodic orbit $\phi(t; \mathbf{x}^*)$ at a time t_1, then $\dot{L}(\phi(t_1; \mathbf{x}^*)) = 0$, and $x = \pm 1$ or $y = 0$. If $x \neq \pm 1$, $1 - x^2$ and so $\dot{L}$ do not change sign, so it cannot be a minimum. Therefore, the minimum along a periodic orbit occurs where $x = \pm 1$. The minimum has $y \neq 0$ because the points $(\pm 1, 0)$ are not on a periodic orbit. Therefore,
$$L(\phi(t; \mathbf{x}^*)) \geq L(\phi(t_1; \mathbf{x}^*)) > L(1, 0) = \frac{1}{2},$$

and $2L - 1 > 0$ along the periodic orbit. We want to relate the divergence and the time derivative of L, $\dot{L} = y^2(1 - x^2)$. Using that $y^2 = 2L - x^2 = 2L - 1 + (1 - x^2)$,
$$\dot{L} = (1 - x^2)\left(2L - 1 + (1 - x^2)\right)$$
$$= (\nabla \cdot \mathbf{F})(2L - 1) + (1 - x^2)^2.$$

Since $2L - 1 > 0$, we can solve for $\nabla \cdot \mathbf{F}$:
$$\nabla \cdot \mathbf{F} = \frac{\dot{L}}{2L - 1} - \frac{(1 - x^2)^2}{2L - 1}.$$

By Theorem 6.14,
$$P'(\mathbf{x}^*) = \exp\left(\int_0^T (\nabla \cdot \mathbf{F})_{\phi(t;\mathbf{x}^*)}\, dt\right) = \exp\left(\int_0^T \frac{\dot{L}}{2L - 1}\, dt - \int_0^T \frac{(1 - x^2)^2}{2L - 1}\, dt\right).$$

The first integral under the exponential is
$$\frac{1}{2}\left(\ln(2L(\phi(T; \mathbf{x}^*)) - 1) - \ln(2L(\phi(0; \mathbf{x}^*)) - 1)\right),$$

which is zero, since $\phi(T; \mathbf{x}^*) = \mathbf{x}^* = \phi(0; \mathbf{x}^*)$. Therefore,
$$P'(\mathbf{x}^*) = \exp\left(-\int_0^T \frac{(1 - x^2)^2}{2L - 1}\, dt\right) < 1.$$

This shows that any periodic orbit is attracting. Since two attracting periodic orbits must be separated by a repelling one, there can be only one periodic orbit.

Notice that we were able to determine the sign of the integral of the divergence without actually calculating it. The divergence can be positive and negative, but we split it up into a term that is strictly negative and one that integrates to zero around a periodic orbit. □

Exercises 6.7

1. Consider the 2π-periodically forced scalar differential equation
$$\dot{x} = -x + \sin(t).$$
   a. Find the solution of the equation by the variation of parameters formula. (In one dimension, the variation of parameters formula is usually called solving the nonhomogeneous linear equation by using an integration factor.)
   b. Find the Poincaré map P from $t = 0$ to $t = 2\pi$.
   c. Find all the fixed points of the Poincaré map and determine their stability.

2. Consider the 2π-periodically forced scalar differential equation
$$\dot{x} = (1 + \cos(t))\,x - x^2.$$
   a. Show that $x = 0$ lies on an unstable periodic orbit.
   b. Show that there are no periodic orbits with $x < 0$. Hint: Show that $\dot{x} < 0$.
   c. Let P be the Poincaré map from $t = 0$ to $t = 2\pi$. Show that $P(x) < x$ for $x \geq 1$. Show that there are no periodic orbits with $x > 2$ and that there is at least one periodic orbit with $0 < x_0 < 2$.
   d. Using Theorem 6.14, show that any periodic orbit with $0 < x_0$ is orbitally stable. Hint: Use a method similar to that of Example 6.23.
   e. Conclude that there are two periodic orbits, one stable and one unstable.

3. Consider the forced equation
$$\dot{\tau} = 1 \pmod{2\pi},$$
$$\dot{x} = x(1-x)[1 + \cos(\tau)].$$
   Notice that there are two periodic orbits: $(\tau, x) = (\tau, 0)$ for $0 \leq \tau \leq 2\pi$ and $(\tau, x) = (\tau, 1)$ for $0 \leq \tau \leq 2\pi$.
   a. What is the divergence of the system of equations?
   b. Find the derivative of the Poincaré map at the two periodic points $(\tau, x) = (0, 0)$ and $(\tau, x) = (0, 1)$. Hint: Use the formula for the derivative of the Poincaré map
$$P'(x) = \exp\left(\int_0^{2\pi} \nabla \cdot \mathbf{F}_{\phi(t;(0,x))}\, dt\right),$$
   where $\nabla \cdot \mathbf{F}$ is the divergence of the differential equations and the exponential $\exp(u) = e^u$.

4. Consider the system of differential equations
$$\dot{x} = y,$$
$$\dot{y} = -x - 2x^3 + y(\mu - y^2 - x^2 - x^4)$$
for $\mu > 0$.

   a. Show that
   $$\gamma = \left\{ \begin{pmatrix} x \\ y \end{pmatrix} : \mu = y^2 + x^2 + x^4 \right\}$$
   is a periodic orbit. Hint: Use a test function.

   b. Show that the divergence of the system is $-2y^2$ along the periodic orbit γ. Hint: Use that $\mu - y^2 - x^2 - x^4 = 0$ along γ.

   c. Let $\Sigma = \left\{ \begin{pmatrix} x \\ 0 \end{pmatrix} : x > 0 \right\}$, and P be the Poincaré map from Σ to itself. Show that the periodic orbit γ is orbitally asymptotically stable (attracting) by showing that the derivative of the Poincaré map satisfies $0 < P'(x_0) < 1$ where $\begin{pmatrix} x_0 \\ 0 \end{pmatrix} \in \gamma$. Hint: Use Theorem 6.14.

5. Consider the system of differential equations
$$\dot{x} = -y + x(1 - x^2 - y^2),$$
$$\dot{y} = x + yx(1 - x^2 - y^2),$$
$$\dot{z} = -z.$$

   a. Show that $x^2 + y^2 = 1$ is a periodic orbit.
   b. What are the characteristic multipliers for the periodic orbit $x^2 + y^2 = 1$?

6. Assume that $\mathbf{x}_0$ is on a periodic orbit γ for a flow and that Σ is a transversal at $\mathbf{x}_0$, with Poincaré map $\mathbf{P}$ from an open set $\mathbf{U}$ in Σ back to Σ. Then, $\mathbf{x}_0$ is a fixed point for $\mathbf{P}$. This fixed point is called *L-stable* for $\mathbf{P}$ provided that, for every $\epsilon > 0$, there is a $\delta > 0$ such that, if $\mathbf{x}$ in $\mathbf{U}$ has $\|\mathbf{x} - \mathbf{x}_0\| < \delta$, then $\|\mathbf{P}^j(\mathbf{x}) - \mathbf{P}^j(\mathbf{x}_0)\| < \epsilon$ for all $j \geq 0$. This fixed point is called *asymptotically stable* provided that it is L-stable and that there is a $\delta_0 > 0$ such that, for any $\mathbf{x}$ in $\mathbf{U}$ with $\|\mathbf{x} - \mathbf{x}_0\| < \delta_0$,
$$\lim_{j \to \infty} \|\mathbf{P}^j(\mathbf{x}) - \mathbf{P}^j(\mathbf{x}_0)\| = 0.$$

   a. Prove that, if $\mathbf{x}_0$ is L-stable for the Poincaré map $\mathbf{P}$, then the periodic orbit γ is orbitally L-stable for the flow. .
   b. Prove that, if $\mathbf{x}_0$ is asymptotically stable for the Poincaré map $\mathbf{P}$, then the periodic orbit γ is orbitally asymptotically stable for the flow.

6.8. Applications

6.8.1. Chemical Oscillation. In the early 1950s, Boris Belousov discovered a chemical reaction that did not go to an equilibrium, but continued to oscillate as time continued. Later, Zhabotinsky confirmed this discovery and brought this result to the attention of the wider scientific community. Field and Noyes made a kinetic model of the reactions in a system of differential equations they called the "Oregonator," named after the location where the research was carried out. For

6.8. Applications

further discussion of the chemical reactions and the modeling by the differential equations see [**Mur89**], [**Str94**], or [**Enn97**]. In this subsection, we sketch the reason why this system of differential equations with three variables has a periodic orbit. This presentation is based on the treatment in [**Smi95**].

The *Oregonator system of differential equations* can be written in the form

(6.6)
$$\epsilon \dot{x} = y - xy + x(1 - qx),$$
$$\dot{y} = -y - xy + 2fz,$$
$$\dot{z} = \delta(x - z),$$

where the parameters ϵ, q, f, and δ are positive. We restrict this discussion to the case where $0 < q < 1$.

A positively invariant region is given by the cube

$$\mathscr{R} = \left\{ (x, y, z) : 1 \leq x \leq q^{-1}, \ \frac{2fq}{1+q} \leq y \leq \frac{f}{q}, \ 1 \leq z \leq q^{-1} \right\}.$$

(The reader can check the signs of $\dot{x}$, $\dot{y}$, or $\dot{z}$ on each of the faces of the boundary.)

The fixed points are the solutions of the three equations

$$z = x, \qquad y = \frac{2fx}{1+x}, \qquad \text{and} \qquad 0 = (1-x)\frac{2fx}{1+x} + x - qx^2.$$

Multiplying the third equation by $-(1+x)/x$, we obtain

$$0 = qx^2 + (2f + q - 1)x - (2f + 1).$$

This last equation has one positive solution,

$$x^* = \frac{1}{2q}\left((1 - q - 2f) + \left[(1 - q - 2f)^2 + 4q(2f + 1)\right]^{1/2}\right).$$

Let $z^* = x^*$ and $y^* = \dfrac{2fx^*}{1 + x^*}$. The other fixed points are $(0, 0, 0)$ and one with negative coordinates.

The matrix of partial derivatives is

$$\begin{pmatrix} \epsilon^{-1}(1 - y - 2qx) & \epsilon^{-1}(1 - x) & 0 \\ -y & -(1 + x) & 2f \\ \delta & 0 & -\delta \end{pmatrix}.$$

A calculation shows that the determinant at (x^*, y^*, z^*) is

$$\det(D\mathbf{F}_{(x^*, y^*, z^*)}) = -\frac{\delta}{\epsilon}(2y^* + qx^* + q(x^*)^2),$$

which is negative. Therefore, either all three eigenvalues have negative real parts, or one has a negative real part and two have positive real parts. It is shown in [**Mur89**] that both cases can occur for different parameter values.

For the case in which two of the eigenvalues of (x^*, y^*, z^*) have positive real parts, [**Smi95**] argues that there must be a nontrivial periodic orbit. First, the system can be considered "competitive" by considering the signs of the off-diagonal terms of the matrix of partial derivatives. The $(1, 2)$ and $(2, 1)$ terms are negative for the points in $\mathscr{R}$. The sign of the $(2, 3)$ and $(3, 1)$ terms are positive, but by using $-z$, which multiplies the third row and third column by -1, these terms become negative. The other two terms, $(1,3)$ and $(3,2)$, are zero. Thus, the system for $(x, y, -z)$ is competitive by his definition.

Smith in [**Smi95**] has a theorem that shows that a limit set of a competitive system in $\mathbb{R}^3$ is topologically equivalent to a flow on a compact invariant set of a two-dimensional system. The system may not have a derivative, but only be of Lipschitz type, but this is good enough to permit application of the Poincaré–Bendixson theorem. See Theorem 3.4 in [**Smi95**].

The results of these arguments in the next theorem.

Theorem 6.16. *Suppose that the parameter values for system* (6.6) *are chosen so that the fixed point* (x^*, y^*, z^*) *has two eigenvalues with positive real parts, so the stable manifold* $W^s(x^*, y^*, z^*)$ *is one-dimensional. If* **q** *is an initial condition in the previously defined region* $\mathscr{R}$, *but not on* $W^s(x^*, y^*, z^*)$, *then* $\omega(\mathbf{q})$ *is a periodic orbit.*

Remark 6.25. Smith mentions that much of the work on competitive systems given in his book started with the work of Morris Hirsch in the 1980s.

6.8.2. Nonlinear Electric Circuit. In this subsection, we show how the Van der Pol equation can arise from an electric circuit in a single loop with one resistor (R), one inductor (L), and one capacitor (C). Such a circuit is called a RLC circuit. The part of the circuit containing one element is called a branch. The points where the branches connect are called nodes. In this simplest example, there are three branches and three node. See Figure 16. We let i_R, i_L, and i_C be the current in the resistor, inductor, and capacitor, respectively. Similarly let v_R, v_L, and v_C be the voltage drop across the three branches of the circuit. If we think of water flowing through pipes, then the current is like the rate of flow of water, and the voltage is like water pressure. Kirchhoff's current law states that the total current flowing into a node must equal the current flowing out of that node. In the circuit being discussed, this means that $|i_R| = |i_L| = |i_C|$ with the correct choice of signs. We orient the branches in the direction given in Figure 16, so

$$x = i_R = i_L = i_C.$$

Kirchhoff's voltage law states that the sum of the voltage drops around any loop is zero. For the present example, this just means that

$$v_R + v_L + v_C = 0.$$

Next, we need to describe the properties of the elements and the laws that determine how the variables change. A resistor is determined by a relationship between the current i_R and voltage v_R. For the Van der Pol system, rather than a linear resistor, we assume that the voltage in the resistor is given by

$$v_R = -i_R + i_R^3 = -x + x^3.$$

Thus, for small currents the voltage drop is negative (antiresistance) and for large currents it is positive (resistance).

A capacitor is characterized by giving the time derivative of the voltage, $\frac{dv_C}{dt}$, in terms of the current i_C,

$$C \frac{dv_C}{dt} = i_C,$$

6.8. Applications

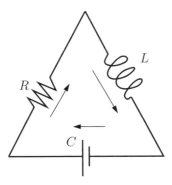

Figure 16. RLC electric circuit

where the constant $C > 0$ is called the capacitance. Classically, a capacitor was constructed by two parallel metal plates separated by some insulating material. To make the equation look like the form of the Van der Pol equation, we let $y = -v_C$ and the equations becomes

$$-C\frac{dy}{dt} = x.$$

An inductor is characterized by giving the time derivative of the current, $\frac{di_L}{dt}$, in terms of the voltage v_C: Faraday's law says that

$$L\frac{di_L}{dt} = v_L = -v_C - v_R = y - (-x + x^3),$$

where the constant $L > 0$ is called the inductance. Classically, an inductor was constructed by making a coil of wire. Then, the magnetic field induced by the change of current in the coil creates a voltage drop across the coil.

Summarizing, system of differential equations is

$$L\frac{dx}{dt} = y - (-x + x^3),$$
$$C\frac{dy}{dt} = -x.$$

This system is just the Lienard system in the form given in equation 6.4 with $g(x) = x/C$ and $F(x) = -x/L + x^3/L$. These functions clearly satisfy the assumptions needed on the functions g and F from that context.

By changing the voltage drop for the resistor to include a parameter in the coefficient of i_R,

$$v_R = -\mu\, i_R + i_R^3 = -\mu x + x^3,$$

this system has an Andronov–Hopf bifurcation at $\mu = 0$, as the reader can check.

6.8.3. Predator–Prey System with an Andronov–Hopf Bifurcation.

In [**Wal83**] and [**Bra01**], a variation of the predator–prey system is given that exhibits an Andronov–Hopf bifurcation; The so called *Rosenzweig-MacArthur model* is given by

$$\dot{x} = x\left(1 - x - \frac{2y}{1+2x}\right),$$
$$\dot{y} = y\left(\frac{2x}{1+2x} - \mu\right).$$

(We have changed the labeling of the variables to make them conform to the notation used in other examples. Also we have picked specific values for some of the parameters given in [**Wal83**].) The x population is the prey; by itself, it has a positive growth rate for small populations and has a carrying capacity of $x = 1$, with negative growth rates for $x > 1$. The predator is given by y, and it dies out when no prey is present. The interaction between the two species involves the product xy, but the benefit to the predator decreases with increasing population of the prey: The net growth rate for the predator approaches $1 - \mu$ as x approaches infinity. The parameter is given by μ, which we assume is positive, but less than one, so the growth rate for the predator is positive for x sufficiently large.

To find the fixed points with x and y positive, the $\dot{y} = 0$ equation yields

$$2x^* = \mu(1 + 2x^*),$$
$$(2 - 2\mu)x^* = \mu, \quad \text{and so}$$
$$x^* = \frac{\mu}{2(1-\mu)}.$$

For future use, notice that

$$1 + 2x^* = \frac{2x^*}{\mu} = \frac{1}{1-\mu} \quad \text{and}$$
$$1 - x^* = \frac{2 - 2\mu - \mu}{2(1-\mu)} = \frac{2 - 3\mu}{2(1-\mu)}.$$

The value of y at the fixed point is

$$y^* = \frac{(1-x^*)(1+2x^*)}{2} = \frac{(1-x^*)}{2(1-\mu)}$$
$$= \frac{2-3\mu}{2(1-\mu)} \cdot \frac{1}{2(1-\mu)} = \frac{2-3\mu}{4(1-\mu)^2}.$$

The linearization at this fixed point is

$$\begin{pmatrix} 1 - 2x^* - \frac{2y^*}{(1+2x^*)^2} & -\frac{2x^*}{1+2x^*} \\ \frac{2y^*}{(1+2x^*)^2} & 0 \end{pmatrix}.$$

The determinant is

$$\Delta = \frac{4x^*y^*}{(1+2x^*)^3} > 0.$$

6.8. Applications

The trace is

$$\tau = 1 - 2x^* - \frac{2y^*}{(1+2x^*)^2}$$
$$= \frac{2-2\mu}{2(1-\mu)} - \frac{2\mu}{2(1-\mu)} - \frac{2-3\mu}{2(1-\mu)^2} \cdot (1-\mu)^2$$
$$= \frac{2-2\mu - 2\mu - (2-5\mu+3\mu^2)}{2(1-\mu)}$$
$$= \frac{\mu(1-3\mu)}{2(1-\mu)}.$$

The eigenvalues are purely imaginary, $\pm i\sqrt{\Delta}$, when $\tau = 0$, or $\mu_0 = 1/3$. For $\mu_0 = 1/3$,

$$x^* = \frac{1}{4}, \quad y^* = \frac{9}{16}, \quad \text{and} \quad 1+2x^* = \frac{3}{2}.$$

The real part of the eigenvalue is $\alpha_\mu = \tau/2$, so

$$\alpha_\mu = \frac{\mu(1-3\mu)}{4(1-\mu)} = \frac{3\mu(1-\mu) + 2(1-\mu) - 2}{4(1-\mu)}$$
$$= \frac{3}{4}\mu + \frac{1}{2} - \frac{1}{2(1-\mu)}$$

and

$$\left.\frac{d\alpha_\mu}{d\mu}\right|_{\mu=1/3} = \frac{3}{4} - \left.\frac{1}{2(1-\mu)^2}\right|_{\mu=1/3}$$
$$= \frac{3}{4} - \frac{9}{8}$$
$$= -\frac{3}{8} < 0.$$

Therefore, the real part of the eigenvalues is decreasing with μ.

To check the stability at $\mu = 1/3$, we apply Theorem 6.7. In terms of the notation of that theorem,

$$f_x = 1 - 2x - \left.\frac{2y^*}{(1+2x^*)^2}\right|_{\mu=1/3} = 0, \quad f_y = -\left.\frac{2x^*}{1+2x^*}\right|_{\mu=1/3} = -\frac{1}{3},$$
$$g_x = \frac{2y^*}{(1+2x^*)^2} = \frac{2(3/16)}{9/4} = \frac{1}{6}, \quad g_y = 1 - \mu - \left.\frac{1}{1+2x^*}\right|_{\mu=1/3} = 0,$$

$$f_{xx} = -2 - \left.\frac{2y^*(-2)(2)}{(1+2x^*)^3}\right|_{\mu=1/3} = -2 + \frac{8(3/16)}{27/9} = \frac{-18+4}{9} = \frac{-14}{9},$$
$$f_{xxx} = \left.\frac{8x^*(-3)(2)}{(1+2x^*)^4}\right|_{\mu=1/3} = -\frac{48(3/16)}{(81/81)} = -\frac{16}{9},$$

$$f_{xy} = -\frac{2}{(1+2x^*)^2}\bigg|_{\mu=1/3} = -\frac{2}{(9/4)} = -\frac{8}{9},$$

$$f_{xxy} = \frac{8}{(1+2x^*)^3}\bigg|_{\mu=1/3} = \frac{8}{(27/8)} = \frac{64}{27},$$

$$f_{yy} = 0 = f_{xyy},$$

$$g_{xx} = \frac{2y^*(-2)(2)}{(1+2x^*)^3}\bigg|_{\mu=1/3} = -\frac{8(9/16)}{(27/8)} = -\frac{4}{3},$$

$$g_{xxx} = -\frac{8y^*(-3)(2)}{(1+2x^*)^4}\bigg|_{\mu=1/3} = \frac{48(9/16)}{(81/16)} = \frac{48}{9} = \frac{16}{3},$$

$$g_{xy} = \frac{2}{(1+2x^*)^2}\bigg|_{\mu=1/3} = \frac{2}{(9/4)} = \frac{8}{9},$$

$$g_{xxy} = \frac{2(-2)(2)}{(1+2x^*)^3}\bigg|_{\mu=1/3} = -\frac{8}{(27/8)} = -\frac{64}{27},$$

$$g_{yy} = 0 = g_{yyy} = g_{xyy},$$

and other partial derivatives are zero. Therefore,

$$16\,K_{1/3} = \left(\left|\frac{f_y}{g_x}\right|^{1/2} f_{xxx} + \left|\frac{g_x}{f_y}\right|^{1/2} f_{xyy} + \left|\frac{f_y}{g_x}\right|^{1/2} g_{xxy} + \left|\frac{g_x}{f_y}\right|^{1/2} g_{yyy}\right)$$

$$+ \frac{1}{g_x}\left|\frac{g_x}{f_y}\right|^{1/2}\left(f_{xy}f_{xx} + \left|\frac{g_x}{f_y}\right| f_{xy}f_{yy} + \left|\frac{g_x}{f_y}\right| f_{yy}g_{yy}\right)$$

$$+ \frac{1}{f_y}\left|\frac{f_y}{g_x}\right|^{1/2}\left(g_{xy}g_{yy} + \left|\frac{f_y}{g_x}\right| g_{xy}g_{xx} + \left|\frac{f_y}{g_x}\right| f_{xx}g_{xx}\right).$$

$$= \sqrt{2}\left(\frac{-16}{9}\right) + 0 + \sqrt{2}\left(-\frac{64}{27}\right) + 0$$

$$+ \frac{6}{\sqrt{2}}\left(\left(-\frac{8}{9}\right)\left(-\frac{14}{9}\right) + 0 + 0\right)$$

$$- \frac{6}{\sqrt{2}}\left(0 + 2\left(\frac{8}{9}\right)\left(-\frac{4}{3}\right) + 2\left(-\frac{14}{9}\right)\left(-\frac{4}{3}\right)\right)$$

$$= \frac{\sqrt{2}}{27}(-48 - 64 + 112 + 192 - 336)$$

$$= -\frac{144\sqrt{2}}{27} < 0.$$

Therefore, the fixed point is weakly attracting at $\mu = 1/3$, and the bifurcation is supercritical with a stable periodic orbit appearing for $\mu < 1/3$ (since the derivative of the real part of the eigenvalue is negative).

We now give an alternative way to check that the fixed point is weakly attracting when $\mu = 1/3$, using the Dulac criterion from Theorem 6.11. We can use this criterion to show that there are no periodic orbits in the first quadrant for $\mu \geq 1/3$. We multiply by the function

$$g(x,y) = \left(\frac{1+2x}{2x}\right) y^{\alpha-1},$$

6.8. Applications

where $\alpha = \dfrac{1}{2(1-\mu)}$. The only justification for this function is that it eliminates the xs from some terms in the $\dot{x}$ equation, and the power on y makes the final outcome (at the end of the following calculation) negative. With this choice, we obtain

$$\nabla \cdot (g\mathbf{F}) = \frac{\partial}{\partial x}\left[\frac{y^{\alpha-1}(1-x)(1+2x)}{2} - y^\alpha\right] + \frac{\partial}{\partial y}\left[y^\alpha\left(\frac{2x - \mu(1+2x)}{2x}\right)\right]$$

$$= \frac{y^{\alpha-1}}{2}\left[-1(1+2x) + 2(1-x)\right] + \frac{\alpha y^{\alpha-1}}{2x}\left[2x - \mu - 2\mu x\right]$$

$$= \frac{y^{\alpha-1}}{2x}\left[x(-4x+1) + \frac{2(1-\mu)}{2(1-\mu)}x - \frac{\mu}{2(1-\mu)}\right]$$

$$= \frac{y^{\alpha-1}}{2x}\left[-4x^2 + 2x - \frac{\mu}{2(1-\mu)}\right].$$

The quantity inside the brackets has a maximum at $x = 1/4$ and a maximum value of $\dfrac{1}{4} - \dfrac{\mu}{2(1-\mu)}$. Then, $\nabla \cdot (g\mathbf{F}) \leq 0$ in the first quadrant, provided that

$$\frac{1}{4} \leq \frac{\mu}{2(1-\mu)}, \quad 1 - \mu \leq 2\mu, \quad \text{or} \quad \frac{1}{3} \leq \mu.$$

Thus for $\mu \geq 1/3$, the integral of $\nabla \cdot (g\mathbf{F})$ inside a closed curve in the positive quadrant must be strictly negative. Therefore, by Theorem 6.11, there can be no periodic orbits.

There is a saddle fixed point at $(1,0)$ whose unstable manifold enters the first quadrant and then reaches the line $x = x^*$ before it gets to infinity. Then, $\dot{y}$ becomes negative, so it is trapped. This forces all the points below this to have nonempty ω-limit sets, which must be (x^*, y^*) since it cannot be a periodic orbit. In fact, with more argument, it is possible to show that any initial (x_0, y_0) must have a trajectory which intersects $x = x^*$ and must then become trapped. Therefore, for $\mu \geq 1/3$ and any $x_0 > 0$ and $y_0 > 0$, $\omega(x_0, y_0) = (x^*, y^*)$, and the basin of attraction includes the whole first quadrant.

6.8.4. No Limit Cycles for Lotka–Volterra Systems. We saw in Section 5.1 that for certain predator-prey systems in two dimensions any initial condition in the first quadrant lies on a periodic orbit or fixed point. In particular, these systems do not have any limit cycles. In this section, we will show that more general Lotka–Volterra systems with two variables do not have limit cycles.

Theorem 6.17. *Consider a Lotka–Volterra system in two variables,*

$$(6.7) \qquad \dot{x}_i = x_i\left[r_i + \sum_{j=1}^{2} a_{ij} x_j\right] \quad \text{for } i = 1, 2,$$

where $\mathbf{A} = (a_{ij})$ *is invertible. Then there are no limit cycles.*

Proof. Let $\hat{\mathbf{x}}$ solve $\mathbf{r} + \mathbf{A}\hat{\mathbf{x}} = \mathbf{0}$. Then the equation can be rewritten as

$$\dot{x}_i = x_i \sum_{j=1}^{2} a_{ij}(x_j - \hat{x}_j) = F_i(x_1, x_2).$$

We calculate the divergence of $\mathbf{F}$ at $\hat{\mathbf{x}}$:
$$\frac{\partial F_i}{\partial x_i} = x_i\, a_{ii} + \sum_j a_{ij}\, (x_j - \hat{x}_j),$$
$$\delta = \operatorname{div}(\mathbf{F})(\hat{\mathbf{x}}) = \sum_i a_{ii}\, \hat{x}_i.$$

Choose exponents $\mathbf{c} = (c_1, c_2)^T$ for a rescaling factor that solve
$$\mathbf{A}^T \mathbf{c} + (a_{11}, a_{22})^T = \mathbf{0}.$$

Set
$$\hat{\mathbf{F}}(\mathbf{x}) = x_1^{c_1-1} x_2^{c_2-1}\, \mathbf{F}(\mathbf{x}), \qquad \text{so}$$
$$\hat{F}_i(\mathbf{x}) = x_i\, x_1^{c_1-1} x_2^{c_2-1} \sum_j a_{ij}\, (x_j - \hat{x}_j).$$

Any limit cycle has to be in a single quadrant. Since the orbits of $\hat{\mathbf{F}}$ are just reparametrizations of orbits of $\mathbf{F}$, if $\hat{\mathbf{F}}$ has no limit cycles then $\mathbf{F}$ has no limit cycles. We use the divergence to show that $\hat{\mathbf{F}}$ does not have any limit cycles.

$$\frac{\partial \hat{F}_i}{\partial x_i} = c_i\, x_1^{c_1-1} x_2^{c_2-1} \sum_j a_{ij}\, (x_j - \hat{x}_j) + a_{ii}\, x_i\, x_1^{c_1-1} x_2^{c_2-1},$$

$$\operatorname{div}(\hat{\mathbf{F}}) = x_1^{c_1-1} x_2^{c_2-1} \left[\sum_i \sum_j c_i\, a_{ij}\, (x_j - \hat{x}_j) + \sum_i a_{ii}\, x_i \right]$$

$$= x_1^{c_1-1} x_2^{c_2-1} \left[\sum_j \sum_i c_i\, a_{ij}\, (x_j - \hat{x}_j) + \sum_j a_{jj}\, (x_j - \hat{x}_j) + \sum_j a_{jj}\, \hat{x}_j \right]$$

$$= x_1^{c_1-1} x_2^{c_2-1} \left[\sum_j \left(\sum_i a_{ij}\, c_i - a_{jj} \right) (x_j - \hat{x}_j) + \delta \right]$$

$$= x_1^{c_1-1} x_2^{c_2-1}\, \delta.$$

In this calculation, we used the definition of δ and the fact that $\sum_i a_{ij}\, c_i - a_{jj} = 0$ by the choice of $\mathbf{c}$. The $\operatorname{div}(\hat{\mathbf{F}})$ has the same sign in any single quadrant. If δ is nonzero, there can be no periodic orbits by the Bendixson Criterion, Theorem 6.11. If $\delta = 0$, then $\hat{\mathbf{F}}$ preserves area, and it cannot have any orbits that are attracting or repelling from even one side. □

Exercises 6.8

1. A simplified model for oscillation in a chemical reaction given by Lengyel is

$$\dot{x} = a - x - \frac{4xy}{1+x^2},$$

$$\dot{y} = bx\left(1 - \frac{y}{1+x^2}\right),$$

with $a > 0$ and $b > 0$.
   a. Plot the nullclines and find the equilibrium (x^*, y^*) with $x^* > 0$ and $y^* > 0$.
   b. Show that the determinant Δ and trace τ at the equilibrium (x^*, y^*) are

$$\Delta = \frac{5bx^*}{1+(x^*)^2} > 0,$$

$$\tau = \frac{3(x^*)^2 - 5 - bx^*}{1+(x^*)^2}.$$

   Find an inequality on b which ensures that the fixed point is repelling.
   c. Show that a box $x_1 \leq x \leq x_2$ and $y_1 \leq y \leq y_2$ is positively invariant provided that $0 < x_1 < x_2$, $0 < y_1 < 1 < y_2$, $y_2 = 1 + x_2^2$, $4x_2 y_1 = (a - x_2)(1+x_2^2)$, and $y_2 < \frac{(a-x_1)(1+x_1^2)}{4x_1}$.
   d. Show that, if the parameters satisfy the conditions given in part (b), which ensure that the fixed point is repelling, then there is a limit cycle.

2. (Fitzhugh-Nagumo model for neurons) The variable v is related to the voltage and represents the extent of excitation of the cell. The variable is scaled so $v = 0$ is the resting value, $v = a$ is the value at which the neuron fires, and $v = 1$ is the value above which the amplification turns to damping. It is assumed that $0 < a < 1$. The variable w denotes the strength of the blocking mechanism. Finally, J is the extent of external stimulation. The Fitzhugh-Nagumo system of differential equations are

$$\dot{v} = -v(v-a)(v-1) - w + J,$$
$$\dot{w} = \epsilon(v - bw),$$

where $0 < a < 1$, $0 < b < 1$, and $0 < \epsilon \ll 1$. For simplicity, we take $\epsilon = 1$. See [Bra01] for more details and references.
   a. Take $J = 0$. Show that $(0,0)$ is the only fixed point. (Hint: $4/b > 4 > (1-a)^2$.) Show that $(0,0)$ is asymptotically stable.
   b. Let (v_J, w_J) the the fixed point for a value of J. Using the fact that $J > 0$ lifts the graph of $w = -v(v-a)(v-1) + J$, explain why the value of v_J increases as J increases.
   c. Take $a = 1/2$ and $b = 3/8$. Find the value of $v^* = v_{J^*}$ for which the fixed point (v_{J^*}, w_{J^*}) is a linear center. Hint: Find the value of v and not of J.
   d. Keep $a = 1/2$ and $b = 3/8$. Using Theorem 6.6, show that there is a supercritical Andronov-Hopf bifurcation of an attracting periodic orbit.

Remark: The interpretation is that when the stimulation exceeds this threshold, then the neuron becomes excited.

6.9. Theory and Proofs

Poincaré–Bendixson theorem

Restatement of Theorem 6.1. *Consider a system of differential equations $\dot{\mathbf{x}} = \mathbf{F}(\mathbf{x})$ on $\mathbb{R}^2$.*

  a. *Assume that $\mathbf{F}$ is defined on all of $\mathbb{R}^2$. Assume that a forward orbit $\{\phi(t;\mathbf{q}) : t \geq 0\}$ is bounded. Then, $\omega(\mathbf{q})$ either: (i) contains a fixed point or (ii) is a periodic orbit.*

  b. *Assume that $\mathscr{A}$ is a closed (includes its boundary) and bounded subset of $\mathbb{R}^2$ that is positively invariant for the differential equation. We assume that $\mathbf{F}(\mathbf{x})$ is defined at all points of $\mathscr{A}$ and has no fixed point in $\mathscr{A}$. Then, given any $\mathbf{x}_0$ in $\mathscr{A}$, the orbit $\phi(t;\mathbf{x}_0)$ is either:(i) periodic or (ii) tends toward a periodic orbit as t goes to ∞, and $\omega(\mathbf{x}_0)$ equals this periodic orbit.*

Proof. The proof is based on the one in [Har82].

As mentioned in the main section, the proof definitely uses the continuity of the flow with respect to initial conditions.

Part (b) follows from part (a) because the ω-limit set of a point $\mathbf{x}_0$ in $\mathscr{A}$ must be contained in $\mathscr{A}$ (since it is positively invariant) and cannot contain a fixed point (since there are none in $\mathscr{A}$), and so must be a periodic orbit.

Turning to part (a), the limit set $\omega(\mathbf{x}_0)$ must be nonempty because the orbit is bounded; let $\mathbf{q}$ be a point in $\omega(\mathbf{x}_0)$. The limit set $\omega(\mathbf{x}_0)$ is bounded because the forward orbit of $\mathbf{x}_0$ is bounded. We need to show that $\mathbf{q}$ is a periodic point, so first we look at where its orbit goes. The limit set $\omega(\mathbf{q})$ is contained in $\omega(\mathbf{x}_0)$ by Theorem 4.1 and is nonempty because the forward orbit of $\mathbf{q}$ is contained in $\omega(\mathbf{x}_0)$, which is bounded. Let $\mathbf{z}$ be a point in $\omega(\mathbf{q})$. Since $\omega(\mathbf{x}_0)$ does not contain any fixed points, $\mathbf{z}$ is not a fixed point.

We can take a line segment $\mathbf{S}$ through $\mathbf{z}$ such that the orbits through points of $\mathbf{S}$ are crossing $\mathbf{S}$. They all must cross in the same direction, since none of them are tangent. The forward orbits of both $\mathbf{x}_0$ and $\mathbf{q}$ repeatedly come near $\mathbf{z}$, so must repeatedly intersect $\mathbf{S}$. There is a sequence of times t_n going to infinity such that $\mathbf{x}_n = \phi(t_n, \mathbf{x}_0)$ is in $\mathbf{S}$. As discussed in the sketch of the proof, taking the piece of the trajectory
$$\{\phi(t;\mathbf{x}_0) : t_n \leq t \leq t_{n+1}\},$$
together with the part $\mathbf{S}'$ of $\mathbf{S}$ between $\mathbf{x}_n = \phi(t_n;\mathbf{x}_0)$ and $\mathbf{x}_{n+1} = \phi(t_{n+1};\mathbf{x}_0)$, we get a closed curve Γ which separates the plane into two pieces,
$$\mathbb{R}^2 \setminus \Gamma = \mathscr{R}_{\text{in}} \cup \mathscr{R}_{\text{out}}.$$
See Figure 17. The trajectory going out from $\phi(t_{n+1};\mathbf{x}_0)$ enters either the region $\mathscr{R}_{\text{in}}$ on the inside of Γ or the region $\mathscr{R}_{\text{out}}$ on the outside of Γ. We assume that it enters $\mathscr{R}_{\text{in}}$ to be specific in the discussion, but the other case is similar. All the trajectories starting in $\mathbf{S}'$ also enter the same region $\mathscr{R}_{\text{in}}$ as that starting at $\phi(t_{n+1};\mathbf{x}_0)$. Therefore, $\phi(t;\mathbf{x}_0)$ can never reenter $\mathscr{R}_{\text{out}}$ for $t > t_{n+1}$. This means

6.9. Theory and Proofs

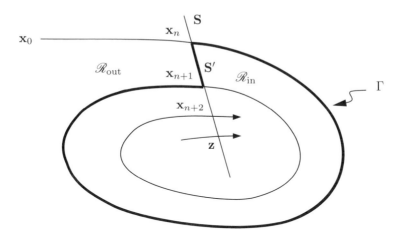

Figure 17. Curve Γ and regions $\mathscr{R}_{\text{in}}$ and $\mathscr{R}_{\text{out}}$

that the intersections of the trajectory with **S** occur in a monotone manner. They must converge to **z** from one side, and the orbit of $\mathbf{x}_0$ can accumulate only on one point in **S**; therefore, $\omega(\mathbf{x}_0) \cap \mathbf{S} = \{\,\mathbf{z}\,\}$ is one point.

By the same argument applied to **q**, there are increasing times s_n going to infinity such that $\phi(s_n; \mathbf{q})$ accumulates on **z**. Since the whole orbit of **q** is in $\omega(\mathbf{x}_0)$, these points $\phi(s_n; \mathbf{q})$ must be in $\omega(\mathbf{x}_0) \cap \mathbf{S} = \{\,\mathbf{z}\,\}$. Therefore, all the points $\phi(s_n; \mathbf{q}) = \mathbf{z}$ are the same, $\phi(s_n; \mathbf{q}) = \phi(s_{n+1}; \mathbf{q})$, and **q** is periodic.

We have left to show that $\omega(\mathbf{x}_0) = \mathcal{O}(\mathbf{q})$. Assume that $\omega(\mathbf{x}_0) \setminus \mathcal{O}(\mathbf{q}) \neq \emptyset$. By Theorem 4.1, the set $\omega(\mathbf{x}_0)$ is connected. Therefore, there would have to exist points $\mathbf{y}_j \in \omega(\mathbf{x}_0) \setminus \mathcal{O}(\mathbf{q})$ that accumulate on a point $\mathbf{y}^*$ in $\mathcal{O}(\mathbf{q})$. Taking a transversal **S** through $\mathbf{y}^*$, we can adjust the points $\mathbf{y}_j$ so they lie on **S**. But, we showed above that $\omega(\mathbf{x}_0) \cap \mathbf{S}$ has to be a single point. This contradiction shows that $\omega(\mathbf{x}_0) \setminus \mathcal{O}(\mathbf{q}) = \emptyset$ and $\omega(\mathbf{x}_0) = \mathcal{O}(\mathbf{q})$, i.e., $\omega(\mathbf{x}_0)$ is a single periodic orbit. $\square$

Self–excited oscillator

Restatement of Theorem 6.5. *With the assumptions (1)–(3) on $g(x)$ and $F(x)$ given in Section 6.3, the Lienard system of equations (6.4) has a unique attracting periodic orbit.*

Proof. A trajectory starting at $\mathbf{Y} = (0, y)$, with $y > 0$, has $\dot{x} > 0$ and $\dot{y} < 0$, so it goes to the right and hits the curve $y = F(x)$. Then, it enters the region where $\dot{x} < 0$ and $\dot{y} < 0$. It is possible to show that the trajectory must next hit the negative y-axis at a point $\mathbf{Y}'$. Subsequently, the trajectory enters the region with $x < 0$ and $y < F(x)$, where $\dot{x} < 0$ and $\dot{y} > 0$, so it intersects the part of the curve $y = F(x)$, with both x and y negative. Next, the trajectory returns to the positive y axis at a point $\mathbf{Y}''$. This first return map P from the positive y-axis to itself, which takes the point $\mathbf{Y}$ to the point $\mathbf{Y}''$, is called the *first return map* or *Poincaré map*. The orbit is periodic if and only if $\mathbf{Y}'' = \mathbf{Y}$. But, $(x(t), y(t))$ is a solution if and only if $(-x(t), -y(t))$ is a solution. Therefore, if $\mathbf{Y}' = -\mathbf{Y}$ then $\mathbf{Y}'' = -\mathbf{Y}' = \mathbf{Y}$ and the trajectory is periodic. On the other hand, we want to show that it is not periodic if $\mathbf{Y}' \neq -\mathbf{Y}$. First, assume that $\|\mathbf{Y}'\| < \|\mathbf{Y}\|$. The trajectory starting at $-\mathbf{Y}$ goes to the point $-\mathbf{Y}'$ on the positive axis. The trajectory starting at $\mathbf{Y}'$ is inside that trajectory so $\|\mathbf{Y}''\| < \|\mathbf{Y}'\| < \|\mathbf{Y}\|$ and it is not periodic. Similarly, if $\|\mathbf{Y}'\| > \|\mathbf{Y}\|$, then $\|\mathbf{Y}''\| > \|\mathbf{Y}'\| > \|\mathbf{Y}\|$ and it is not periodic. Therefore, the trajectory is periodic if and only if $\mathbf{Y}' = -\mathbf{Y}$.

To measure the distance from the origin, we use the potential function
$$G(x) = \int_0^x g(s)\,ds.$$
Since g is an odd function, with $g(x) > 0$ for $x > 0$, $G(x) > 0$ for all $x \neq 0$. Let
$$L(x,y) = G(x) + \frac{y^2}{2},$$
which is a measurement of the distance form the origin. The orbit is periodic if and only if $L(\mathbf{Y}) = L(\mathbf{Y}')$.

We let
$$L_{\mathbf{PQ}} = L(\mathbf{Q}) - L(\mathbf{P})$$
be the change of L from $\mathbf{P}$ to $\mathbf{Q}$. Also, let
$$K(\mathbf{Y}) = L_{\mathbf{YY}'} = L(\mathbf{Y}') - L(\mathbf{Y}).$$
Then, the orbit is periodic if and only if $K(\mathbf{Y}) = 0$.

We can measure K in terms of an integral along the trajectory,
$$K(\mathbf{Y}) = \int_0^{t(\mathbf{Y})} \dot{L}\,dt,$$
where $t(\mathbf{Y})$ is the time to go from $\mathbf{Y}$ to $\mathbf{Y}'$. But,
$$\dot{L} = g(x)\,\dot{x} + y\,\dot{y} = g(x)\,(y - F(x)) - y\,g(x) = -g(x)\,F(x) = F(x)\,\dot{y}.$$
Thus,
$$K(\mathbf{Y}) = \int_0^{t(\mathbf{Y})} F(x)\,\dot{y}\,dt = \int_{\mathbf{Y}}^{\mathbf{Y}'} F(x)\,dy = -\int_{\mathbf{Y}'}^{\mathbf{Y}} F(x)\,dy.$$

Now, let $\mathbf{Y}_0$ be the point such that the trajectory becomes tangent to $x = a$ and returns to the y-axis. Then, if $0 < \mathbf{Y} < \mathbf{Y}_0$, $K(\mathbf{Y}) = L_{\mathbf{YY}'} > 0$, so it cannot be periodic. (We saw this earlier by considering the $\dot{r}$ equation for the Van der Pol equation.) For $\mathbf{Y}$ above $\mathbf{Y}_0$, let $\mathbf{B}$ be the first point where the trajectory hits $x = a$ and let $\mathbf{B}'$ be the second point. See Figure 18. We give three lemmas that finish the proof of the theorem. □

Lemma 6.18. *For $\mathbf{Y}$ above $\mathbf{Y}_0$, $L_{\mathbf{YB}} > 0$ and $L_{\mathbf{B'Y'}} > 0$, and $L_{\mathbf{YB}} + L_{\mathbf{B'Y'}}$ is monotonically decreasing as $\mathbf{Y}$ increases.*

Proof. The quantity
$$L_{\mathbf{YB}} = \int_{\mathbf{Y}}^{\mathbf{B}} -g(x)\,F(x)\left[\frac{\dot{x}}{y - F(x)}\right]dt = \int_{\mathbf{Y}}^{\mathbf{B}} (-F(x))\left[\frac{g(x)}{y - F(x)}\right]dx > 0,$$

6.9. Theory and Proofs

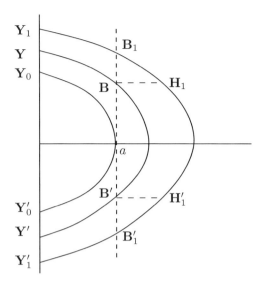

Figure 18. Map for Lienard equation

since both terms in the integrand are positive. The value of y on the trajectory starting at $\mathbf{Y}_1$ above $\mathbf{Y}$ is always larger on the trajectory from $\mathbf{Y}_1$ to $\mathbf{B}_1$, so $[y - F(x)]^{-1}$ and the integrand is smaller, and $L_{\mathbf{Y}_1\mathbf{B}_1} < L_{\mathbf{Y}\mathbf{B}}$. Similarly, $L_{\mathbf{B}'_1\mathbf{Y}'_1} < L_{\mathbf{B}'\mathbf{Y}'}$. Adding, we get the result. □

Lemma 6.19. *For $\mathbf{Y}$ above $\mathbf{Y}_0$, $L_{\mathbf{BB}'}$ is monotonically decreasing,*

$$L_{\mathbf{BB}'} > L_{\mathbf{B}_1\mathbf{B}'_1},$$

if $\mathbf{Y}_1$ is above $\mathbf{Y}$, which is above $\mathbf{Y}_0$.

Proof.

$$L_{\mathbf{B}_1\mathbf{B}'_1} = -L_{\mathbf{B}'_1\mathbf{B}_1} = -\int_{\mathbf{B}'_1}^{\mathbf{B}_1} F(x)\, dy.$$

Let $\mathbf{H}_1$ and $\mathbf{H}'_1$ be the points on the trajectory starting at $\mathbf{Y}_1$ with the same y values as $\mathbf{B}$ and $\mathbf{B}'$ respectively. See Figure 18. Then $F(x) > 0$ on this path, so if we shorten the path to the one from $\mathbf{H}'_1$ to $\mathbf{H}_1$, the quantity gets less negative, and

$$L_{\mathbf{B}_1\mathbf{B}'_1} < -\int_{\mathbf{H}'_1}^{\mathbf{H}_1} F(x)\, dy < -\int_{\mathbf{B}'}^{\mathbf{B}} F(x)\, dy = L_{\mathbf{BB}'},$$

because $-F(x)$ is less negative on the integral from $\mathbf{B}'$ to $\mathbf{B}$ than from $\mathbf{H}'_1$ to $\mathbf{H}_1$. □

Therefore, the graph of $K(0, y)$ is positive for $0 < y \leq y_0$, and then is monotonically decreasing. Thus, we have shown that K has at most one zero and there is at most one limit cycle. The following lemma shows that $K(0, y)$ goes to $-\infty$ as y goes to ∞, so there is exactly one zero and therefore, exactly one limit cycle.

Lemma 6.20. $K(0, y)$ *goes to $-\infty$ as y goes to ∞.*

Proof. Fix a value of $x = x^* > a$. Let **P** and **P**$'$ be the two points where the trajectory starting at **Y** intersects the line $x = x^*$. For $x > x^*$, $F(x) > F(x^*) > 0$. (This uses the fact that F is monotonically increasing for $x > a$.) Then,

$$L_{\mathbf{BB}'} = -\int_{\mathbf{B}'}^{\mathbf{B}} F(x)\,dy \leq -\int_{\mathbf{P}'}^{\mathbf{P}} F(x)\,dy < -\int_{\mathbf{P}'}^{\mathbf{P}} F(x^*)\,dy = -F(x^*)\,\|\mathbf{PP}'\|.$$

As **Y** goes to infinity, $\|\mathbf{PP}'\|$ must go to infinity, so $L_{\mathbf{BB}'} < -F(x^*)\,\|\mathbf{PP}'\|$ goes to $-\infty$. $\square$

This completes the proof of Theorem 6.5.

Andronov–Hopf bifurcation

This section presents a partial proof of Theorem 6.6. The proof given is a combination of those found in [**Car81**] and [**Cho82**], and closely follows that given in [**Rob99**]. This proof is more involved than most given in this book. However, the outline is straightforward if somewhat tedious. Using polar coordinates, we calculate the manner in which the radius changes after making one revolution around the fixed point. Using the radius of the trajectory at $\theta = 0$ as a variable, we show that we can solve for the parameter value μ (or α in the following proof) for which the trajectory closes up to give a periodic orbit. This fact is shown using the implicit function theorem, as explained in the proof. We do not actually show which parameter values have the periodic orbit nor do we check their stability type: We leave these aspects of the proof to the references given. Thus, the result proved is actually the next theorem, where we assume the bifurcation parameter value is $\mu_0 = 0$.

Theorem 6.21. *Consider a one-parameter family of systems of differential equations*

(6.8) $$\dot{\mathbf{x}} = \mathbf{F}_\mu(\mathbf{x}),$$

with $\mathbf{x}$ in $\mathbb{R}^2$. Assume that the family has a family of fixed points $\mathbf{x}_\mu$. Assume that the eigenvalues at the fixed point are $\alpha(\mu) \pm i\,\beta(\mu)$, with $\alpha(\mu_0) = 0$, $\beta_0 = \beta(\mu_0) \neq 0$, and

$$\frac{d\alpha}{d\mu}(\mu_0) \neq 0,$$

so the eigenvalues are crossing the imaginary axis at $\mu = 0$. Then, there exists an $\epsilon_0 > 0$ such that for $0 \leq \epsilon \leq \epsilon_0$, there are (i) differentiable functions $\mu(\epsilon)$ and $T(\epsilon)$, with $T(0) = 2\pi/\beta_0$, $\mu(0) = \mu_0$, and $\mu'(0) = 0$, and (ii) a $T(\epsilon)$-periodic solution $\mathbf{x}^*(t, \epsilon)$ of (6.8) for the parameter value $\mu = \mu(\epsilon)$, with initial conditions in polar coordinates given by $r^*(0, \epsilon) = \epsilon$ and $\theta(0, \epsilon) = 0$.

Proof. By using the change of variables $\mathbf{y} = \mathbf{x} - \mathbf{x}_\mu$, we can assume that the fixed points are all at the origin. Because,

$$\frac{d\alpha}{d\mu}(\mu_0) \neq 0,$$

$\alpha(\mu)$ is a monotone function of μ, and given an α, we can find a unique μ that corresponds to it. Therefore, we can consider that α as the independent parameter of the system of differential equations and can solve for μ in terms of α. The

6.9. Theory and Proofs

eigenvalues are now functions of α, $\alpha \pm i\beta(\alpha)$. Note that α is the real part of each of the eigenvalues.

By taking a linear change of basis, using the eigenvectors for the complex eigenvalues, we get

$$D(\mathbf{F}_\alpha)_{\mathbf{0}} = \begin{pmatrix} \alpha & -\beta(\alpha) \\ \beta(\alpha) & \alpha \end{pmatrix} \equiv \mathbf{A}(\alpha).$$

The differential equations become

$$(6.9) \qquad \dot{\mathbf{x}} = \mathbf{A}(\alpha)\mathbf{x} + \begin{pmatrix} B_2^1(x_1, x_2, \alpha) \\ B_2^2(x_1, x_2, \alpha) \end{pmatrix} + \begin{pmatrix} B_3^1(x_1, x_2, \alpha) \\ B_3^2(x_1, x_2, \alpha) \end{pmatrix} + O(r^4),$$

where $B_j^k(x_1, x_2, \alpha)$ is a homogeneous polynomial of degree j in x_1 and x_2, and $O(r^4)$ contains the higher order terms. More specifically, we write $O(r^j)$ to mean that there is a constant $C > 0$ such that

$$\|O(r^j)\| \leq C\, r^j.$$

Since we want to determine the effect of going once around the fixed points, we use polar coordinates. The following lemma gives the equations in polar coordinates in terms of those in rectangular coordinates.

Lemma 6.22. *The differential equations* (6.9) *in polar coordinates are given by*

$$(6.10) \qquad \begin{aligned} \dot{r} &= \alpha r + r^2\, C_3(\theta, \alpha) + r^3\, C_4(\theta, \alpha) + O(r^4), \\ \dot{\theta} &= \beta(\alpha) + r\, D_3(\theta, \alpha) + r^2\, D_4(\theta, \alpha) + O(r^3), \end{aligned}$$

where $C_j(\theta, \alpha)$ and $D_j(\theta, \alpha)$ are homogeneous polynomials of degree j in terms of $\sin(\theta)$ and $\cos(\theta)$. In terms of the B_j^k,

$$\begin{aligned} C_{j+1}(\theta, \alpha) &= \cos(\theta)B_j^1(\cos(\theta), \sin(\theta), \alpha) + \sin(\theta)B_j^2(\cos(\theta), \sin(\theta), \alpha), \\ D_{j+1}(\theta, \alpha) &= -\sin(\theta)B_j^1(\cos(\theta), \sin(\theta), \alpha) + \cos(\theta)B_j^2(\cos(\theta), \sin(\theta), \alpha). \end{aligned}$$

Moreover,

$$\int_0^{2\pi} C_3(\theta, \alpha)\, d\theta = 0.$$

Proof. The proof is a straightforward calculation of the change of variables into polar coordinates. Taking the time derivatives of the equations $x_1 = r\cos(\theta)$ and

$x_2 = r \sin(\theta)$, we get

$$\begin{pmatrix} \dot{x}_1 \\ \dot{x}_2 \end{pmatrix} = \begin{pmatrix} \cos(\theta) & -r\sin(\theta) \\ \sin(\theta) & r\cos(\theta) \end{pmatrix} \begin{pmatrix} \dot{r} \\ \dot{\theta} \end{pmatrix},$$

$$\begin{pmatrix} \dot{r} \\ \dot{\theta} \end{pmatrix} = \frac{1}{r} \begin{pmatrix} r\cos(\theta) & r\sin(\theta) \\ -\sin(\theta) & \cos(\theta) \end{pmatrix} \begin{pmatrix} \dot{x}_1 \\ \dot{x}_2 \end{pmatrix}$$

$$= \begin{pmatrix} \cos(\theta)\dot{x}_1 + \sin(\theta)\dot{x}_2 \\ -r^{-1}\sin(\theta)\dot{x}_1 + r^{-1}\cos(\theta)\dot{x}_2 \end{pmatrix}$$

$$= \begin{pmatrix} \alpha r \\ \beta(\alpha) \end{pmatrix} + \begin{pmatrix} \cos(\theta) B_2^1 + \sin(\theta) B_2^2 \\ -r^{-1}\sin(\theta) B_2^1 + r^{-1}\cos(\theta) B_2^2 \end{pmatrix}$$

$$+ \begin{pmatrix} \cos(\theta) B_3^1 + \sin(\theta) B_3^2 \\ -r^{-1}\sin(\theta) B_3^1 + r^{-1}\cos(\theta) B_3^2 \end{pmatrix} + \begin{pmatrix} O(r^4) \\ O(r^3) \end{pmatrix},$$

where the B_j^k are functions of $r\cos(\theta)$, $r\sin(\theta)$, and α, $B_j^k(r\cos(\theta), r\sin(\theta), \alpha)$. Since the B_j^k are homogeneous of degrees j, we can factor out r from the B_j^k, and get the results stated in the lemma.

The term $C_3(\theta, \alpha)$ is a homogeneous cubic polynomial in $\sin(\theta)$ and $\cos(\theta)$ and, therefore, has an integral equal to zero. $\square$

Since $\dot\theta \neq 0$, instead of considering $\dot{r}$, we use $\frac{dr}{d\theta}$ and find out how r changes when θ increase by 2π. Using equation (6.10), we get

$$\frac{dr}{d\theta} = \frac{\dot{r}}{\dot{\theta}} = \frac{\alpha r + r^2 C_3(\theta, \alpha) + r^3 C_4(\theta, \alpha) + O(r^4)}{\beta(\alpha) + r D_3(\theta, \alpha) + r^2 D_4(\theta, \alpha) + O(r^3)}$$

$$= \frac{\alpha}{\beta} r + r^2 \left[\frac{1}{\beta} C_3(\theta, \alpha) - \frac{\alpha}{\beta^2} D_3(\theta, \alpha) \right]$$

$$+ r^3 \left[\frac{1}{\beta} C_4(\theta, \alpha) - \frac{1}{\beta^2} C_3(\theta, \alpha) D_3(\theta, \alpha) \right.$$

$$\left. - \frac{\alpha}{\beta^2} D_4(\theta, \alpha) + \frac{\alpha}{\beta^3} D_3(\theta, \alpha)^2 \right] + O(r^4).$$

The idea is to find a periodic orbit whose radius is ϵ for some α. So, we switch parameters from α to the radius ϵ of the periodic orbit. Let $r(\theta, \epsilon, \alpha)$ be the solution for the parameter value α, with $r(0, \epsilon, \alpha) = \epsilon$. We want to solve $r(2\pi, \epsilon, \alpha) = \epsilon$, which gives a periodic orbit. The solution with $\epsilon = 0$ has $r(\theta, 0, \alpha) = 0$, since the origin is a fixed point. Therefore, ϵ is a factor of $r(\theta, \epsilon, \alpha)$, and we can define

$$g(\epsilon, \alpha) = \frac{r(2\pi, \epsilon, \alpha) - \epsilon}{\epsilon}.$$

If $g(\epsilon, \alpha) = 0$ for $\epsilon > 0$, then this gives a periodic orbit as desired.

To show that we can solve for α as a function of ϵ, $\alpha(\epsilon)$, such that $g(\epsilon, \alpha(\epsilon)) = 0$, we use the implicit function theorem. We show that $g(0, 0) = 0$ and $\frac{\partial g}{\partial \alpha}(0, 0) \neq 0$.

6.9. Theory and Proofs

The method of implicit differentiation, then says that

$$\frac{\partial g}{\partial \epsilon}(0,0) + \frac{\partial g}{\partial \alpha}(0,0)\frac{d\alpha}{d\epsilon}(0) = 0,$$

$$\frac{d\alpha}{d\epsilon}(0) = -\frac{\frac{\partial g}{\partial \epsilon}(0,0)}{\frac{\partial g}{\partial \alpha}(0,0)}.$$

The implicit function theorem says, that when $\frac{\partial g}{\partial \alpha}(0,0) \neq 0$ (and implicit differentiation works), it is possible to solve for α as a differentiable function of ϵ, which makes this equation remain equal to zero.

We show the preceding requirements on g and its partial derivatives by scaling the variable r by ϵ, $r = \epsilon \rho$. We show that $r(2\pi, \epsilon, 0) - \epsilon = O(\epsilon^3)$, so $g(\epsilon, 0) = O(\epsilon^2)$. Let $\rho(\theta, \epsilon, \alpha) = r(\theta, \epsilon, \alpha)/\epsilon$, so it is the solution for ρ, with $\rho(0, \epsilon, \alpha) = r(0, \epsilon, \alpha)/\epsilon = \epsilon/\epsilon = 1$. The differential equation for ρ is given by

$$\frac{d\rho}{d\theta} = \frac{\alpha}{\beta}\rho + \epsilon\rho^2\left[\frac{1}{\beta}C_3(\theta,\alpha) - \frac{\alpha}{\beta^2}D_3(\theta,\alpha)\right]$$

$$+ \epsilon^2\rho^3\left[\frac{1}{\beta}C_4(\theta,\alpha) - \frac{1}{\beta^2}C_3(\theta,\alpha)D_3(\theta,\alpha)\right.$$

$$\left. - \frac{\alpha}{\beta^2}D_4(\theta,\alpha) + \frac{\alpha}{\beta^3}D_3(\theta,\alpha)^2\right] + O(\epsilon^3\rho^4).$$

Taking the linear term to the left-hand side, and using the integrating factor $e^{-\alpha\theta/\beta}$, the derivative of $e^{-\alpha\theta/\beta}\rho$ is

$$\frac{d}{d\theta}\left(e^{-\alpha\theta/\beta}\rho\right) = \epsilon e^{-\alpha\theta/\beta}\rho^2\left[\frac{1}{\beta}C_3 - \frac{\alpha}{\beta^2}D_3\right]$$

$$+ \epsilon^2 e^{-\alpha\theta/\beta}\rho^3\left[\frac{1}{\beta}C_4 - \frac{1}{\beta^2}C_3 D_3 - \frac{\alpha}{\beta^2}D_4 + \frac{\alpha}{\beta^3}D_3^2\right] + O(\epsilon^3\rho^4).$$

Integrating form 0 to 2π, we obtain

$$e^{-2\pi\alpha/\beta}\rho(2\pi,\epsilon,\alpha) - 1$$

$$= \epsilon\int_0^{2\pi} e^{-\alpha\theta/\beta}\rho(\theta,\epsilon,\alpha)^2\left[\frac{1}{\beta}C_3(\theta,\alpha) - \frac{\alpha}{\beta^2}D_3(\theta,\alpha)\right]d\theta$$

$$+ \epsilon^2\int_0^{2\pi} e^{-\alpha\theta/\beta}\rho(\theta,\epsilon,\alpha)^3\left[\frac{1}{\beta}C_4(\theta,\alpha)\right.$$

$$\left. - \frac{1}{\beta^2}C_3(\theta,\alpha)D_3(\theta,\alpha) - \frac{\alpha}{\beta^2}D_4(\theta,\alpha) + \frac{\alpha}{\beta^3}D_3(\theta,\alpha)^2\right]d\theta + O(\epsilon^3\rho^4)$$

$$= \epsilon\, h(\epsilon,\alpha).$$

This last line defines $h(\epsilon, \alpha)$. By the definitions of g and h, we have

$$g(\epsilon,\alpha) = \rho(2\pi,\epsilon,\alpha) - 1$$

$$= \left[e^{2\pi\alpha/\beta} - 1\right] + \epsilon\, e^{2\pi\alpha/\beta} h(\epsilon,\alpha).$$

Now we can calculate the quantities for g mentioned above. The value of g at $(\epsilon, \alpha) = (0, 0)$ is

$$g(0,0) = e^0 - 1 = 0.$$

The partial derivative with respect to α is

$$\frac{\partial g}{\partial \alpha}(0,0) = \left[\left(\frac{2\pi}{\beta}\right) e^{2\pi\alpha/\beta} - \left(\frac{2\pi\alpha}{\beta^2}\right)\left(\frac{\partial \beta}{\partial \alpha}\right) e^{2\pi\alpha/\beta} \right.$$
$$\left. + \epsilon \frac{\partial}{\partial \alpha}\left(e^{2\pi\alpha/\beta} h(\epsilon, \alpha)\right)\right]\bigg|_{\epsilon=0,\alpha=0}$$
$$= \frac{2\pi}{\beta(0)} \neq 0.$$

This verifies the assumptions previously stated, for the implicit function theorem, so we can solve for α as a function $\alpha(\epsilon)$ of ϵ such that $g(\epsilon, \alpha(\epsilon)) \equiv 0$. These give points that are periodic with $1 = \rho(0, \epsilon, \alpha(\epsilon)) = \rho(2\pi, \epsilon, \alpha(\epsilon))$, so $\epsilon = r(0, \epsilon, \alpha(\epsilon)) = r(2\pi, \epsilon, \alpha(\epsilon))$.

We previously noted that implicit differentiation gives

$$\frac{d\alpha}{d\epsilon}(0) = -\frac{\frac{\partial g}{\partial \epsilon}(0,0)}{\frac{\partial g}{\partial \alpha}(0,0)}.$$

Therefore, to determine $\frac{d\alpha}{d\epsilon}(0)$, we need to calculate $\frac{\partial g}{\partial \epsilon}(0,0)$.

By the previous expression for $g(\epsilon, \alpha)$, we have

$$g(\epsilon, 0) = \epsilon h(\epsilon, 0)$$
$$= \epsilon \int_0^{2\pi} \rho(\theta, \epsilon, 0)^2 \left(\frac{1}{\beta}\right) C_3(\theta, 0)\, d\theta$$
$$+ \epsilon^2 \int_0^{2\pi} \rho(\theta, \epsilon, 0)^3 \left[\frac{1}{\beta} C_4(\theta, 0) - \frac{1}{\beta^2} C_3(\theta, 0) D_3(\theta, 0)\right] d\theta$$
$$+ O(\epsilon^3 \rho^4).$$

Substituting $\rho(\theta, \epsilon, 0) = 1$, we obtain

$$\frac{\partial g}{\partial \epsilon}(0,0) = \frac{1}{\beta}\int_0^{2\pi} C_3(\theta, 0)\, d\theta = 0.$$

The integral is zero because $C_3(\theta, 0)$ is a homogeneous polynomial of degree three in $\sin(\theta)$ and $\cos(\theta)$. It follows that

$$\frac{\partial \alpha}{\partial \epsilon}(0) = -\frac{\frac{\partial g}{\partial \epsilon}(0,0)}{\frac{\partial g}{\partial \alpha}(0,0)} = 0.$$

This completes the proof of the theorem. □

Although we leave the details to the reference [**Rob99**], we indicate the outline of the manner in which we can determine the parameters for which the periodic orbit appears and the stability type of the periodic orbits.

By applying implicit differentiation again to the equation

$$0 = \frac{\partial g}{\partial \epsilon}(\epsilon, \alpha(\epsilon)) + \frac{\partial g}{\partial \alpha}((\epsilon, \alpha(\epsilon))\alpha'(\epsilon),$$

6.9. Theory and Proofs

we get
$$0 = \frac{\partial^2 g}{\partial \epsilon^2}(0,0) + 2\frac{\partial^2 g}{\partial \epsilon \partial \alpha}(0,0)\,\alpha'(0) + \frac{\partial^2 g}{\partial \alpha^2}(0,0)\,\alpha'(0)^2 + \frac{\partial g}{\partial \alpha}(0,0)\,\alpha''(0).$$

Using the fact that $\alpha'(0) = 0$, we can solve for $\alpha''(0)$:
$$\alpha''(0) = -\frac{\frac{\partial^2 g}{\partial \epsilon^2}(0,0)}{\frac{\partial g}{\partial \alpha}(0,0)}$$
$$= -\left(\frac{\beta_0}{2\pi}\right)\frac{\partial^2 g}{\partial \epsilon^2}(0,0).$$

As stated in Section 6.4, we define
$$K_\alpha = \frac{1}{2\pi}\int_0^{2\pi} C_4(\theta,\alpha) - \frac{1}{\beta(\alpha)}C_3(\theta,\alpha)\,D_3(\theta,\alpha)\,d\theta.$$

It is shown in [**Rob99**] that
$$\frac{\partial^2 g}{\partial \epsilon^2}(0,0) = \frac{4\pi}{\beta_0}\,K_0,$$

so
$$\alpha''(0) = -2\,K_0.$$

This shows that the periodic orbit (i) has $\alpha(\epsilon) - K_0\,\epsilon^2 + O(\epsilon^3) > 0$, when $K_0 < 0$ and the fixed point is weakly attracting at $\alpha = 0$, and (ii) has $\alpha(\epsilon) < 0$, when $K_0 > 0$ and the fixed point is weakly repelling at $\alpha = 0$.

As a check of the stability of the periodic orbit, the Poincaré map is
$$P(\epsilon,\alpha) = r(2\pi,\epsilon,\alpha).$$

A calculation in [**Rob99**] shows that
$$\frac{\partial P}{\partial \epsilon}(\epsilon) = 1 + \left(\frac{4\pi}{\beta_0}\right)K_0\,\epsilon^2 + O(\epsilon^3).$$

Therefore, the periodic orbit is attracting when $K_0 < 0$ and is repelling when $K_0 > 0$. See the reference for details.

Change of volume by the flow

We first prove the change of volume formula and then use it to prove the Liouville formula.

Restatement of Theorem 6.9. *Let $\dot{\mathbf{x}} = \mathbf{F}(\mathbf{x})$ be a system of differential equations in $\mathbb{R}^n$, with flow $\phi(t;\mathbf{x})$. Let $\mathscr{D}$ be a region in $\mathbb{R}^n$, with (i) finite n-volume and (ii) smooth boundary* bd $\mathscr{D}$. *Let $\mathscr{D}(t)$ be the region formed by flowing along for time t, $\mathscr{D}(t) = \{\phi(t;\mathbf{x}_0) : \mathbf{x}_0 \in \mathscr{D}\}$.*

a. *Let $V(t)$ be the n-volume of $\mathscr{D}(t)$. Then*
$$\frac{d}{dt}V(t) = \int_{\mathscr{D}(t)} \nabla \cdot \mathbf{F}_{(\mathbf{x})}\,dV,$$

where $\nabla \cdot \mathbf{F}_{(\mathbf{x})}$ is the divergence of $\mathbf{F}$ at the point $\mathbf{x}$ and $dV = dx_1 \cdots dx_n$ is the element of n volume. If $n = 2$, then $V(t)$ should be called the area and dV should be replaced by the element of area $dA = dx_1 dx_2$.

b. *If the divergence of $\mathbf{F}$ is a constant (independent of $\mathbf{x}$), then*
$$V(t) = V(0)\, e^{\nabla \cdot \mathbf{F}\, t}.$$
Thus, if the vector field is divergence free, with $\nabla \cdot \mathbf{F} \equiv 0$, then the volume is preserved; if $\nabla \cdot \mathbf{F} < 0$, then the volume decays exponentially; if $\nabla \cdot \mathbf{F} > 0$, then the volume grows exponentially.

Proof. We present the proof using the notation for dimension three, but the proof in dimension two or higher dimensions is the same.

The time derivative of $V(t)$ is equal to the flux of the vector field $\mathbf{F}$ across the boundary $\partial \mathscr{D}(t)$ of $\mathscr{D}(t)$,
$$\dot{V}(t) = \int_{\partial \mathscr{D}(t)} (\mathbf{F} \cdot \mathbf{n})\, dA,$$
where $\mathbf{n}$ is the outward normal. Notice that it is only the component $\mathbf{F}$ in the direction of $\mathbf{n}$ that contributes to the change of volume. If this quantity is positive, then the volume is increasing, and if it is negative, the volume is decreasing. Applying the divergence theorem, we see that the right-hand side is the integral of the divergence, so
$$\dot{V}(t) = \int_{\mathscr{D}(t)} \nabla \cdot \mathbf{F}_{(\mathbf{x})}\, dV.$$
This proves part (a).

If the divergence is a constant, then the integral gives the volume multiplied by this constant,
$$\dot{V}(t) = (\nabla \cdot \mathbf{F}) V(t).$$
This scalar equation has a solution
$$V(t) = V(0)\, e^{(\nabla \cdot \mathbf{F})\, t}.$$
The remaining statements in part (b) follow directly from this formula for $V(t)$, because the divergence is constant. $\square$

We indicated earlier how Liouville's formula can be proved by differentiating the determinant for a time-dependent linear differential equation. It can also be proved using the preceding formula derived from the divergence theorem, as we indicate next.

Restatement of Theorem 6.10. *Consider a system of differential equations $\dot{\mathbf{x}} = \mathbf{F}(\mathbf{x})$ in $\mathbb{R}^n$. Assume that both $\mathbf{F}(\mathbf{x})$ and $\frac{\partial F_i}{\partial x_j}(\mathbf{x})$ are continuous for $\mathbf{x}$ in some open set U in $\mathbb{R}^n$. Then the determinant of the matrix of partial derivatives of the flow can be written as the exponential of the integral of the divergence of the vector field, $\nabla \cdot \mathbf{F}_{(\mathbf{x})}$, along the trajectory,*
$$\det\left(D_{\mathbf{x}} \phi_{(t; \mathbf{x}_0)}\right) = \exp\left(\int_0^t (\nabla \cdot \mathbf{F})_{(\phi(s; \mathbf{x}_0))}\, ds\right).$$

Proof. We use the general change of variables formula in multidimensions given by
$$\int_{\mathbf{G}(\mathscr{D})} h(\mathbf{y})\, dV = \int_{\mathscr{D}} h(\mathbf{G}(\mathbf{x}))\, |\det(D\mathbf{G}_{(\mathbf{x})})|\, dV.$$

6.9. Theory and Proofs

Because $\phi(-t; \phi(t; \mathbf{x}_0)) = \mathbf{x}_0$,
$$D_\mathbf{x}\phi_{(t;\phi(t;\mathbf{x}_0))}\, D_\mathbf{x}\phi_{(t;\mathbf{x}_0)} = \mathbf{I},$$
and $D_\mathbf{x}\phi_{(t;\mathbf{x}_0)}$ is invertible for all t. At $t=0$, it is the identity, so it has a positive determinant. Therefore, $\mathbf{G}(\mathbf{x}_0) = \phi(t; \mathbf{x}_0)$ has a positive determinant of the matrix of partial derivatives for all t,
$$\det\left(D_\mathbf{x}\phi_{(t;\mathbf{x}_0)}\right) > 0.$$
This fact allows us to drop the absolute value signs for our application of the change of variables formula. By the previous theorem, we have
$$\dot{V}(t) = \int_{\mathbf{x}\in\mathscr{D}(t)} \nabla\cdot \mathbf{F}_{(\mathbf{x})}\, dV = \int_{\mathbf{x}_0\in\mathscr{D}(0)} \nabla\cdot \mathbf{F}_{(\phi(t;\mathbf{x}_0))} \det\left(D_\mathbf{x}\phi_{(t;\mathbf{x}_0)}\right) dV.$$
But, we also have
$$\dot{V}(t) = \frac{d}{dt}\int_{\mathscr{D}(t)} dV = \frac{d}{dt}\int_{\mathscr{D}(0)} \det\left(D_\mathbf{x}\phi_{(t;\mathbf{x}_0)}\right) dV = \int_{\mathscr{D}(0)} \frac{d}{dt}\det\left(D_\mathbf{x}\phi_{(t;\mathbf{x}_0)}\right) dV.$$
Setting these equal to each other, we get
$$\int_\mathscr{D} \nabla\cdot \mathbf{F}_{(\phi(t;\mathbf{x}_0))} \det\left(D_\mathbf{x}\phi_{(t;\mathbf{x}_0)}\right) dV = \int_\mathscr{D} \frac{d}{dt}\det\left(D_\mathbf{x}\phi_{(t;\mathbf{x}_0)}\right) dV$$
for any region $\mathscr{D}$. Applying this formula to the solid balls $\mathscr{B}_r = \mathbf{B}(\mathbf{x}_0, r)$ of radius r about a point $\mathbf{x}_0$, and dividing by the volume of $\mathscr{B}_r$, we get
$$\lim_{r\to 0} \frac{1}{\mathrm{vol}(\mathscr{B}_r)} \int_{\mathscr{B}_r} \frac{d}{dt}\det\left(D_\mathbf{x}\phi_{(t;\mathbf{x}_0)}\right) dV$$
$$= \lim_{r\to 0} \frac{1}{\mathrm{vol}(\mathscr{B}_r)} \int_{\mathscr{B}_r} \nabla\cdot \mathbf{F}_{(\phi(t;\mathbf{x}_0))} \det\left(D_\mathbf{x}\phi_{(t;\mathbf{x}_0)}\right) dV.$$
Because each side of the equality is the average of a continuous function over a shrinking region, in the limit,
$$\frac{d}{dt}\det\left(D_\mathbf{x}\phi_{(t;\mathbf{x}_0)}\right) = \nabla\cdot \mathbf{F}_{(\phi(t;\mathbf{x}_0))} \det\left(D_\mathbf{x}\phi_{(t;\mathbf{x}_0)}\right),$$
which is the differential equation given in the theorem.

If we let $J(t) = \det\left(D_\mathbf{x}\phi_{(t;\mathbf{x}_0)}\right)$, and $T(t) = \nabla\cdot \mathbf{F}_{(\phi(t;\mathbf{x}_0))}$, we get the scalar differential equation
$$\frac{d}{dt}J(t) = T(t)\, J(t),$$
which has the solution
$$J(t)) = \exp\left(\int_0^t T(s)\, ds\right),$$
since $J(0) = \det(I) = 1$. Substituting in the quantities for $J(t)$ and $T(s)$, we get the integral form given in the theorem. $\square$

Chapter 7

Chaotic Attractors

The Poincaré–Bendixson theorem implies that a system of differential equations in the plane cannot have behavior that should be called chaotic: The most complicated behavior for a limit set of these systems involves a limit cycle, or stable and unstable manifolds connecting fixed points. Therefore, the lowest dimension in which chaotic behavior can occur is three, or two dimensions for periodically forced equations.

As mentioned in the Historical Prologue, Lorenz discovered a system of differential equations with three variables that had very complicated dynamics on a limit set. At about this same time, Ueda discovered a periodically forced nonlinear oscillator which also had complicated dynamics on its limit set. This type of limit sets came to be called a *chaotic attractor*. See [**Ued73**] or [**Ued92**] for historical background.

In Sections 7.1 and 7.2, we present the main concepts involved in chaotic behavior, including attractors, sensitive dependence on initial conditions, and chaotic attractors. We then discuss a number of systems which have chaotic attractors, or at least numerical studies indicate that they do: We consider the Lorenz system in Section 7.3, the Rössler system in Section 7.4, and periodically forced system of equations related to those considered by Ueda in Section 7.5. Section 7.6 introduces a numerical measurement related to sensitive dependence on initial conditions, called the Lyapunov exponents. Then, Section 7.7 discusses a more practical manner of testing to see whether a system has a chaotic attractor.

In general, this chapter is more qualitative and geometric and less quantitative than the earlier material.

7.1. Attractors

Previous chapters have discussed attracting fixed points and attracting periodic orbits. In this section, we extend these concepts by considering more complicated attracting sets. For an attracting set or attractor $\mathbf{A}$, we want at least that all points $\mathbf{q}$ near to $\mathbf{A}$ have $\omega(\mathbf{q}) \subset \mathbf{A}$. However, just like the definition of an asymptotically stable fixed point, we require a stronger condition that a set of nearby points collapses down onto the attracting set as the time for the flow goes to infinity.

If a system of differential equations has a strict Lyapunov function L for a fixed point $\mathbf{p}$ with $L_0 = L(\mathbf{p})$, the set $\mathbf{U}$ equal to a component of $L^{-1}([L_0, L_0 + \delta])$ around $\mathbf{p}$ is a positively invariant region, for small $\delta > 0$. In fact for $t > 0$,

$$\phi^t(\mathbf{U}) \subset \{\, \mathbf{x} \in \mathbf{U} : L(\mathbf{x}) < L_0 + \delta \,\} \quad \text{and}$$

$$\bigcap_{t>0} \phi^t(\mathbf{U}) = \{\, \mathbf{p} \,\}.$$

In this chapter, we consider situations where a more complicated set $\mathbf{A}$ is attracting. Because we want $\mathbf{A}$ to have a property similar to L-stability, we give the following definition in terms of the intersection of the forward orbit of a whole neighborhood rather than just using ω-limit sets of points. We require that $\mathbf{A}$ has a neighborhood $\mathbf{U}$ such that $\phi^t(\mathbf{U})$ squeezes down onto $\mathbf{A}$ as t goes to infinity. For such a set, every point $\mathbf{q}$ in $\mathbf{U}$ will have its ω-limit set as a subset of $\mathbf{A}$.

We specify these conditions using the standard terminology from topology. Also see Appendix A.2 for more about these concepts.

Definition 7.1. For a point $\mathbf{p}$ in $\mathbb{R}^n$, the *open ball* or radius r about $\mathbf{p}$ is the set

$$\mathbf{B}(\mathbf{p}, r) = \{\, \mathbf{q} \in \mathbb{R}^n : \|\mathbf{q} - \mathbf{p}\| < r \,\}.$$

The *boundary* of a set $\mathbf{U}$ in $\mathbb{R}^n$, $\mathrm{bd}(\mathbf{U})$, is the set of all the points that are arbitrarily close to both points in $\mathbf{U}$ and the complement of $\mathbf{U}$,

$$\mathrm{bd}(\mathbf{U}) = \{\, \mathbf{p} \in \mathbb{R}^n : \mathbf{B}(\mathbf{p}, r) \cap \mathbf{U} \neq \emptyset \text{ and } \mathbf{B}(\mathbf{p}, r) \cap (\mathbb{R}^n \smallsetminus \mathbf{U}) \neq \emptyset \text{ for all } r > 0 \,\}.$$

A set $\mathbf{U}$ in $\mathbb{R}^n$ is *open* provided that no boundary point is in $\mathbf{U}$, $\mathbf{U} \cap \mathrm{bd}(\mathbf{U}) = \emptyset$. This is equivalent to saying that for every point $\mathbf{p}$ in $\mathbf{U}$, there is an $r > 0$ such that $\mathbf{B}(\mathbf{p}, r) \subset \mathbf{U}$.

A set $\mathbf{U}$ in $\mathbb{R}^n$ is *closed* provided that it contains its boundary, $\mathrm{bd}(\mathbf{U}) \subset \mathbf{U}$. This is equivalent to saying that the complement of $\mathbf{U}$, $\mathbb{R}^n \smallsetminus \mathbf{U}$, is open.

The *interior* of a set $\mathbf{U}$ is the set minus its boundary, $\mathrm{int}(\mathbf{U}) = \mathbf{U} \smallsetminus \mathrm{bd}(\mathbf{U})$. The interior is the largest open set contained inside $\mathbf{U}$.

The *closure* of a set $\mathbf{U}$ is the set union with its boundary, $\mathrm{cl}(\mathbf{U}) = \mathbf{U} \cup \mathrm{bd}(\mathbf{U})$. The closure is the smallest closed set containing $\mathbf{U}$.

A set $\mathbf{U}$ is *bounded* provided that there is some (big) radius $r > 0$ such that $\mathbf{U} \subset \mathbf{B}(\mathbf{0}, r)$.

A subset of $\mathbb{R}^n$ is *compact* provided that it is closed and bounded.

Definition 7.2. A *trapping region* for a flow $\phi(t; \mathbf{x})$ of a system of differential equations is a closed and bounded set $\mathbf{U}$ for which $\phi(t; \mathbf{U}) \subset \mathrm{int}(\mathbf{U})$ for all $t > 0$. This condition means that the set is positively invariant and the boundary of $\mathbf{U}$ is moved a positive distance into the set for a $t > 0$.

We usually defined a trapping region using a test function $L : \mathbb{R}^n \to \mathbb{R}$,

$$\mathbf{U} = L^{-1}((-\infty, C]) = \{\, \mathbf{x} : L(\mathbf{x}) \leq C \,\}.$$

7.1. Attractors

Since L is continuous, a standard theorem in analysis states that $\mathbf{U}$ is closed and $L^{-1}((-\infty, C))$ is open. If the gradient of L is nonzero on the set $L^{-1}(C)$, then the boundary is $\mathrm{bd}(\mathbf{U}) = L^{-1}(C)$ and $\mathrm{int}(\mathbf{U}) = L^{-1}((-\infty, C))$. In this case, the condition for the trapping region can be stated as

$$\phi(t; \mathbf{U}) \subset L^{-1}((-\infty, C)) = \{\, \mathbf{x} : L(\mathbf{x}) < C \,\} \qquad \text{for } t > 0.$$

Definition 7.3. A set $\mathbf{A}$ is called an *attracting set* for the trapping region $\mathbf{U}$ provided that

$$\mathbf{A} = \bigcap_{t \geq 0} \phi^t(\mathbf{U}).$$

Since $\mathbf{A}$ is invariant and any invariant set contained in $\mathbf{U}$ must be a subset of $\mathbf{A}$, it is the largest invariant set contained in $\mathbf{U}$.

An *attractor* is an attracting set $\mathbf{A}$ which has no proper subset $\mathbf{A}'$ that is also an attracting set (i.e., if $\emptyset \neq \mathbf{A}' \subset \mathbf{A}$ is an attracting set, then $\mathbf{A}' = \mathbf{A}$). In particular, an attracting set $\mathbf{A}$ for which there is a point $\mathbf{x}_0$ (in the ambient phase space) such that $\omega(\mathbf{x}_0) = \mathbf{A}$ is an attractor.

An invariant set $\mathbf{S}$ is called *topologically transitive*, or just *transitive*, provided that there is a point $\mathbf{x}_0$ in $\mathbf{S}$ such that the orbit of $\mathbf{x}_0$ comes arbitrarily close to every point in $\mathbf{S}$ (i.e., the orbit is dense in $\mathbf{S}$). We usually verify this condition by showing that $\omega(\mathbf{x}_0) = \mathbf{S}$.

Thus, a *transitive attractor* is an attracting set $\mathbf{A}$ for which there is a point $\mathbf{x}_0$ in $\mathbf{A}$ with a dense orbit. (This condition automatically shows there is no proper subset that is an attracting set.)

Notice that an asymptotically stable fixed point or an orbitally asymptotically stable periodic orbit is a transitive attractor by the preceding definition.

Theorem 7.1. *The following properties hold for an attracting set $\mathbf{A}$ for a trapping region $\mathbf{U}$.*

a. *The set $\mathbf{A}$ is closed and bounded, compact.*

b. *The set $\mathbf{A}$ is invariant for both positive and negative time.*

c. *If $\mathbf{x}_0$ is in $\mathbf{U}$, then $\omega(\mathbf{x}_0) \subset \mathbf{A}$.*

d. *If $\mathbf{x}_0$ is a hyperbolic fixed point in $\mathbf{A}$, then $W^u(\mathbf{x}_0)$ is contained in $\mathbf{A}$.*

e. *If γ is a hyperbolic periodic orbit in $\mathbf{A}$, then $W^u(\gamma)$ is contained in $\mathbf{A}$.*

The proof is given in Section 7.9.

Example 7.4. Consider the system of equations given in Example 5.11:

$$\begin{aligned} \dot{x} &= y, \\ \dot{y} &= x - 2x^3 + y(x^2 - x^4 - y^2). \end{aligned}$$

The time derivative of the test function
$$L(x,y) = \frac{-x^2 + x^4}{2} + \frac{y^2}{2} \quad \text{is}$$
$$\dot{L} = -2y^2 L \begin{cases} \geq 0 & \text{when } L < 0, \\ = 0 & \text{when } L = 0, \\ \leq 0 & \text{when } L > 0. \end{cases}$$

On the set $\Gamma = L^{-1}(0) = \{(x,y) : L(x,y) = 0\}$, $\dot{L} = 0$, so Γ is invariant. This set Γ is a "figure eight" and equals the origin together with its stable and unstable manifolds. See Figure 18. At the other two fixed points, $L(\pm 2^{-1/2}, 0) = -1/8$. We use the values $\pm 1/16$, which satisfy $-1/8 < -1/16 < 0$ and $0 < 1/16$, and consider the closed set
$$\mathbf{U}_1 = \left\{ (x,y) : -\frac{1}{16} \leq L(x,y) \leq \frac{1}{16} \right\}.$$
This set $\mathbf{U}_1$ is a thickened set around the Γ. For $t > 0$, $\phi^t(\mathbf{U}_1)$ is a thinner neighborhood about Γ. In fact, for any $r > 0$, there is a time $t_r > 0$, such that $\phi^{t_r}(\mathbf{U}_1) \subset L^{-1}(-r,r)$. Since $r > 0$ is arbitrary,
$$\Gamma = \bigcap_{t>0} \phi^t(\mathbf{U}_1),$$
and Γ is an attracting set. We have argued previously that there is a point (x_0, y_0) outside Γ with $\omega(x_0, y_0) = \Gamma$, so Γ is an attractor. Since every point inside Γ has its ω-limit equal to $\{\mathbf{0}\}$, it is not a transitive attractor.

There are other attracting sets for this system of equations. For example we could fill the holes of $\mathbf{U}_1$ to get a different trapping region:
$$\mathbf{U}_2 = \left\{ (x,y) : L(x,y) \leq \frac{1}{16} \right\}.$$
For any $t > 0$, $\phi^t(\mathbf{U}_2)$ contains the set
$$\mathbf{A}_2 = \{(x,y) : L(x,y) \leq 0\}.$$
This set is Γ together with the regions inside Γ. Again, as t goes to infinity, $\phi^t(\mathbf{U}_2)$ collapses down onto $\mathbf{A}_2$, so $\mathbf{A}_2$ is an attracting set. Since Γ is a proper subset of $\mathbf{A}_2$ that is also an attracting set, $\mathbf{A}_2$ is not an attractor.

There are two other attracting sets, which fill in only one hole of Γ,
$$\mathbf{A}_3 = \Gamma \cup \{(x,y) \in \mathbf{A}_2 : x \geq 0\} \quad \text{and}$$
$$\mathbf{A}_4 = \Gamma \cup \{(x,y) \in \mathbf{A}_2 : x \leq 0\}.$$
Again, neither of these attracting sets is an attractor.

Since the ω-limit set of any point is a subset of $\mathbf{A}_2$, any attracting set is a subset of $\mathbf{A}_2$ and contains Γ. If any point from one of the holes of Γ is in the attracting set then the whole side inside Γ must be included. Therefore, the only attracting sets are Γ, $\mathbf{A}_2$, $\mathbf{A}_3$, and $\mathbf{A}_4$.

Remark 7.5. The idea of the definition of an attracting set is that all points in a neighborhood of $\mathbf{A}$ must have their ω-limit set inside $\mathbf{A}$. C. Conley showed that an attracting set with no proper sub-attracting set has a type of recurrence. Every point in such a set $\mathbf{A}$ is chain recurrent and there is an ϵ-chain between any two points in **A**. What this means is, that given two points **p** and **q** in **A** and any small size of jumps $\epsilon > 0$, there are a sequence of points, $\mathbf{x}_0 = \mathbf{p}, \mathbf{x}_1, \ldots, \mathbf{x}_n = \mathbf{q}$, and a sequence of times, $t_1 \geq 1, t_2 \geq 1, \ldots, t_n \geq 1$, such that $\|\phi(t_i; \mathbf{x}_{i-1}) - \mathbf{x}_i\| < \epsilon$ for $1 \leq i \leq n$. Thus, there is an orbit with errors from **p** to **q**. There is also such an ϵ-chain from **p** back to **p** again. See Section 13.4 or [**Rob99**] for more details.

Remark 7.6. Some authors completely drop the assumption on the transitivity or ω-limit from the definition of an attractor. Although there is no single way to give the requirement, we feel that the set should be minimal in some sense even though it can only be verified by computer simulation is some examples.

Remark 7.7. J. Milnor has a definition of an attractor that does not require that it attract all points in a neighborhood of the attractor. We introduce some notation to give his definition. The *basin of attraction* of a closed invariant set **A** is the set of points with ω-limit set in **A**,

$$\mathscr{B}(\mathbf{A}; \phi) = \{\, \mathbf{x}_0 : \omega(\mathbf{x}_0; \phi) \subset \mathbf{A} \,\}.$$

The *measure* of a set $\mathscr{B}(\mathbf{A}; \phi)$ is the generalized "area" for subsets of $\mathbb{R}^2$ and generalized "volume" for sets in $\mathbb{R}^3$ or higher dimensions.

A closed invariant set **A** for a flow ϕ is called a *Milnor attractor* provided that (i) the basin of attraction of **A** has positive measure, and (ii) there is no smaller closed invariant set $\mathbf{A}' \subset \mathbf{A}$ such that $\mathscr{B}(\mathbf{A}; \phi) \setminus \mathscr{B}(\mathbf{A}'; \phi)$ has measure zero.

The idea is that it is possible to select a point in a set of positive measure. If **A** is an attractor in Milnor's sense, then it is possible to select a point $\mathbf{x}_0$ such that $\omega(\mathbf{x}_0) \subset \mathbf{A}$. Since the set of initial conditions with the ω-limit in a smaller set has measure zero, $\omega(\mathbf{x}_0)$ is likely to be the whole attractor.

If we use Milnor's definition, then a fixed point that is attracting from one side is called an attractor; for example, 0 is a Milnor attractor for

$$\dot{x} = x^2.$$

Other Milnor attractors, which are not attractors in our sense, are the point $(1, 0)$ for Example 4.10 and the origin for Example 4.11. We prefer not to call these sets attractors, but reserve the word for invariant sets that attract whole neighborhoods using a trapping region.

Exercises 7.1

1. Consider the system of differential equations given in polar coordinates by
$$\dot{r} = r(1 - r^2)(r^2 - 4),$$
$$\dot{\theta} = 1.$$

   a. Draw the phase portrait for $\dot{r}$.
   b. Draw the phase portrait in the (x, y)-plane.
   c. What are the attracting sets and attractors for this system? Explain your answer, including what properties these sets have to make them attracting sets and attractors. Hint: Consider all attracting points and intervals for $\dot{r}$.

2. Let $V(x) = x^6/6 - 5x^4/4 + 2x^2$ and $L(x,y) = V(x) + y^2/2$. Notice that $V'(x) = x^5 - 5x^3 + 4x$ and $0 = V'(0) = V'(\pm 1) = V'(\pm 2)$. Consider the system of differential equations

$$\dot{x} = y,$$
$$\dot{y} = -V'(x) - y\left[L(x,y) - L(1,0)\right].$$

   a. Show that $\dot{L} = -y^2\left[L(x,y) - L(1,0)\right]$.
   b. Draw the phase portrait.
   c. What are the attracting sets and attractors for this system of differential equations.

3. Consider the forced damped pendulum given by

$$\dot{\tau} = 1 \pmod{2\pi},$$
$$\dot{x} = y \pmod{2\pi},$$
$$\dot{y} = -\sin(x) - y + \cos(\tau).$$

   Here, the x variable is considered an angle variable and is taken modulo 2π. The forcing variable τ is also taken modulo 2π.
   a. What is the divergence of the system of equations?
   b. If V_0 is the volume of a region $\mathscr{D}$, what is the volume of the region $\mathscr{D}(t) = \phi(t; \mathscr{D})$? (The region $\mathscr{D}(t)$ is the region formed by following the trajectories of points starting in $\mathscr{D}$ at time 0 and following them to time t.)
   c. Show that the region $\mathscr{R} = \{(\tau, x, y) : |y| \leq 3\}$ is a trapping region and $\bigcap_{t \geq 0} \phi(t; \mathscr{R})$ is an attracting set. Hint: Check $\dot{y}$ on $y = 3$ and $y = -3$.

4. Consider the system of differential equations

$$\dot{x} = y,$$
$$\dot{y} = x - 2x^3 - y,$$

   with flow $\phi(t; \mathbf{x})$. Let $V(x) = \frac{1}{2}(x^4 - x^2)$ and $L(x,y) = V(x) + \frac{1}{2}y^2$.
   a. Show that $\mathscr{D} = L^{-1}((-\infty, 1])$ is a trapping region and $\mathbf{A} = \bigcap_{t \geq 0} \phi(t; \mathscr{D})$ is an attracting set. Why are $W^u(\mathbf{0})$ and $\left(\pm\sqrt{1/2}, 0\right)$ contained in $\mathbf{A}$?
   b. What is the divergence? What is the area of $\phi(t; \mathscr{D})$ in terms of the area of $\mathscr{D}$? What is the area of $\mathbf{A}$?

5. Consider the system of differential equations

$$\dot{x} = -x + yz,$$
$$\dot{y} = -y + xz,$$
$$\dot{z} = -xy + z - z^3,$$

   with flow $\phi(t; \mathbf{x})$.
   a. Using the test function $L(x,y,z) = \frac{1}{2}(x^2 + y^2) + z^2$, show that $\mathscr{R} = \{(x,y,z) : L(x,y,z) \leq 4\}$ is a trapping region. Hint: Find the maximum of $\dot{L}$ on $L^{-1}(4)$.
   b. Explain why the attracting set for the trapping region $\mathscr{R}$ has zero volume.

7.2. Chaotic Attractors

In the Historical Prologue, we discussed the butterfly effect, in which a small change in initial conditions leads to a large change in the location at a later time. This property is formally called *sensitive dependence on initial conditions*. Such a system of differential equations would amplify small changes. E. Lorenz discovered this property for the system of equations (7.1). We define a chaotic system as one with sensitive dependence on initial conditions.

We give a mathematical definition of a chaotic attractor in terms of sensitive dependence. Section 13.4 defines these same concepts for maps, and some more aspects of the subject are developed in that section. In Section 7.7, we give an alternative test in terms of quantities which are more easily verified for computer simulations or from experimental data.

If two solutions are bounded, then the distance between the orbits cannot be arbitrarily large but merely bigger than some predetermined constant, r. This constant $r > 0$ should be large enough so the difference in the two orbits is clearly observable, but we do not put a restriction on its value in the definition. Also, we do not want the distance to grow merely because the orbits are being traversed at different speeds, as illustrated in Example 7.11. Therefore, we have to allow a reparametrization of one of the orbits in the following definition. A *reparametrization* is a strictly increasing function $\tau : \mathbb{R} \to \mathbb{R}$ with $\lim_{t \to \infty} \tau(t) = \infty$ and $\lim_{t \to -\infty} \tau(t) = -\infty$.

Definition 7.8. The flow $\phi(t; \mathbf{x})$ for a system of differential equations has *sensitive dependence on initial conditions at* $\mathbf{x}_0$ provided that there is an $r > 0$ such that, for any $\delta > 0$, there is some $\mathbf{y}_0$ with $\|\mathbf{y}_0 - \mathbf{x}_0\| < \delta$ for which the orbits of $\mathbf{x}_0$ and $\mathbf{y}_0$ move apart by a distance of at least r. More specifically, for any reparametrization $\tau : \mathbb{R} \to \mathbb{R}$ of the orbit of $\mathbf{y}_0$, there is some time $t_1 > 0$ such that

$$\|\phi(\tau(t_1); \mathbf{y}_0) - \phi(t_1; \mathbf{x}_0)\| \geq r.$$

Definition 7.9. The system is said to have *sensitive dependence on initial conditions on a set* **S** provided that it has sensitive dependence on initial conditions for any initial condition $\mathbf{x}_0$ in **S**. More precisely, there is one $r > 0$ (that works for all points) such that, for any $\mathbf{x}_0$ in **S** and any $\delta > 0$, there is some $\mathbf{y}_0$ in the phase space such that the orbits of $\mathbf{x}_0$ and $\mathbf{y}_0$ move apart by a distance of at least r in the sense given in the previous definition. In this definition, the second point $\mathbf{y}_0$ is allowed to be outside of the set **S** in the ambient phase space.

If **S** is an invariant set, we say that the system has *sensitive dependence on initial conditions when restricted to* **S**, provided that there is an $r > 0$ such that, for any $\mathbf{x}_0$ in **S** and any $\delta > 0$, there is some $\mathbf{y}_0$ in **S** such that the orbits of $\mathbf{x}_0$ and $\mathbf{y}_0$ move apart by a distance of at least r in the sense given in the previous definition. In this definition, the second point $\mathbf{y}_0$ must be selected within the invariant set **S**.

Example 7.15 gives an example that shows the difference in these two ways of considering sensitive dependence on initial conditions on an invariant set.

Example 7.10. For a simple example with sensitive dependence, consider the system in polar coordinates given by

$$\dot{r} = r(-1 + r^2),$$
$$\dot{\theta} = 1.$$

The set $r = 1$ is an unstable periodic orbit and is invariant. Any initial condition (x_0, y_0) on the periodic orbit has sensitive dependence, because an initial condition (x_1, y_1) with $r_1 \neq 1$ eventually has $|r(t) - 1| > 1/2$, and $\|\phi(t;(x_1,y_1)) - \phi(t;(x_0,y_0))\| > 1/2$. This system does not have sensitive dependence when restricted to the invariant set $r = 1$, since points on the periodic orbit stay the same distance apart as they flow along.

Example 7.11. This example illustrates why it is necessary to allow reparametrizations in the definition of sensitive dependence.

Consider the system of differential equations given by

$$\dot{\theta} = 1 + \frac{1}{2}\sin(\psi) \pmod{2\pi},$$
$$\dot{\psi} = 0 \pmod{2\pi},$$
$$\dot{y} = -y.$$

The set
$$\mathbf{U} = \{(\theta, \psi, y) : 0 \leq \theta \leq 2\pi,\ 0 \leq \psi \leq 2\pi,\ |y| \leq 1\,\}$$
is clearly a trapping region,
$$\phi^t(\mathbf{U}) = \{(\theta, \psi, y) : 0 \leq \theta \leq 2\pi,\ 0 \leq \psi \leq 2\pi,\ |y| \leq e^{-t}\,\}.$$
Its attracting set is
$$\mathbf{A} = \{(\theta, \psi, 0) : 0 \leq \theta \leq 2\pi,\ 0 \leq \psi \leq 2\pi\,\}.$$
Since $\dot{\theta} > 0$, any invariant set must contain arbitrary θ. Also, $\psi(t)$ is constant. The values of ψ also cannot be split up by a trapping region. Therefore, $\mathbf{A}$ has no proper subsets that are attracting sets and it is an attractor. However, since $\psi(t)$ is constant, there is no point whose ω-limit equals all of $\mathbf{A}$. Therefore, $\mathbf{A}$ is not a transitive attractor.

Next, we explain why this example does not have sensitive dependence on initial conditions when restricted to $\mathbf{A}$. Consider a point $(\theta_0, \psi, 0)$ in $\mathbf{A}$ and a second point $(\theta_1, \psi, 0)$ in $\mathbf{A}$ that is nearby. Let $\phi(t;(\theta_j,\psi_j,0)) = (\theta_j(t), \psi_j, 0)$. If $\sin(\psi_0) \neq \sin(\psi_1)$, then

$$|\theta_1(t) - \theta_0(t)| = |\theta_1 - \theta_0 + t/2\,(\sin(\psi_1) - \sin(\psi_0))|.$$

This will grow until it is as far apart as possible modulo 2π, which is a distance π. However, we can reparametrize the second orbit so that $\theta_1(\tau(t)) = \theta_1 + t + t/2\sin(\psi_0)$. Then,

$$|\theta_1(\tau(t)) - \theta_0(t)| = |\theta_1 - \theta_0|,$$

and the distance between the two orbits with reparametrization is a constant. Thus, the flow does not have sensitive dependence on initial conditions when restricted to the attractor.

7.2. Chaotic Attractors

Definition 7.12. A *chaotic attractor* is a transitive attractor $\mathbf{A}$ for which the flow has sensitive dependence on initial conditions when restricted to $\mathbf{A}$.

Remark 7.13. For a chaotic attractor, we require that the system have sensitive dependence when restricted to the attractor and not just sensitive dependence on initial conditions at all points in the attractor. Example 7.15 shows that these two conditions are not the same: Our stronger assumption eliminates the type of nonchaotic dynamic behavior this example exhibits.

Remark 7.14. As indicated in the Historical Prologue, T.Y. Li and J. Yorke introduced the word "chaos" in [**Li,75**]. Although their paper contains many mathematical properties of the system studied, which ones were necessary to call the system chaotic was not precisely stated.

Devaney was the first to give a precise mathematical definition in [**Dev89**]. His definition concerned a chaotic set (or a chaotic system, after it was restricted to an invariant set). The three properties he required were that the system have sensitive dependence on initial conditions, the periodic points be dense in the set, and there exist a point with a dense orbit in the set. Having a dense orbit is equivalent to being able to take the $\mathbf{x}_0$ in $\mathbf{A}$ with $\omega(\mathbf{x}_0) = \mathbf{A}$, i.e., it is a transitive attractor. See Appendix B for a discussion of the reason why a "generic" system has the periodic orbits dense in any attractor.

Martelli in [**Mar99**] states the following condition, which combines transitivity and the condition that the system has sensitive dependence on initial conditions when restricted to the attractor $\mathbf{A}$.

> There is an initial condition $\mathbf{x}_0$ in the attractor $\mathbf{A}$ such that $\omega(\mathbf{x}_0) = \mathbf{A}$ and the forward orbit of $\mathbf{x}_0$ is unstable in $\mathbf{A}$ (i.e., there is an $r > 0$ such that for any $\delta > 0$, there is a point $\mathbf{y}_0$ in $\mathbf{A}$ with $\|\mathbf{y}_0 - \mathbf{x}_0\| < \delta$ and a $t_1 > 0$ such that $\|\phi(t_1; \mathbf{x}_{t_0}) - \phi(t_1; \mathbf{y}_0)\| \geq r$).

Ruelle and Takens introduced the related concept in 1971 of a *strange attractor*. See [**Rue71**] or [**Guc83**]. This concept emphasizes the fact that the attractor has a complicated appearance, geometry, or topology. In Section 14.1, we introduce the box dimension which is a measurement of some aspects of the complicated geometry of a chaotic attractor.

Our or Devaney's definition of chaos does not directly require that orbits exhibit random behavior as a function of time. However, transitivity together with sensitive dependence implies that is is true.

Example 7.15 (Nonchaotic attractor). We consider the same system of differential equations considered in Examples 5.11 and 7.4,

$$\dot{x} = y,$$
$$\dot{y} = x - 2x^3 - y\left(-x^2 + x^4 + y^2\right).$$

In Example 7.4, we showed that $\Gamma = L^{-1}(0)$ is an attractor by using the test function

$$L(\tau, x, y) = \frac{-x^2 + x^4 + y^2}{2}.$$

The attractor Γ equals the union of the stable and unstable manifolds of the fixed point at $\mathbf{0}$.

If we take the initial point $\mathbf{p}_0$ in Γ and take the nearby point $\mathbf{q}_0$ outside Γ, then $\phi(t;\mathbf{p}_0)$ tends toward the origin $\mathbf{0}$ while $\phi(t;\mathbf{q}_0)$ repeatedly passes near $(\pm 1, 0)$, even with a reparametrization. These orbits are a distance approximately 1 apart, and the system does have sensitive dependence on initial conditions at all points on Γ within the ambient space.

However, if we take nearby points $\mathbf{p}_0$ and $\mathbf{q}_0$ both in Γ, then $\phi^t(\mathbf{p}_0)$ and $\phi^t(\mathbf{q}_0)$ both tend to the fixed point $\mathbf{0}$ as t goes to infinity and so they get closer together: therefore, the system does not have sensitive dependence on initial conditions when restricted to Γ. Thus, this systems is not a chaotic attractor.

There are several aspects of the dynamics of this system that make it reasonable not to call it chaotic. For any point $\mathbf{p}_0$ in Γ, the ω-limit set $\omega(\mathbf{p}_0)$ is just the fixed point $\mathbf{0}$ and not all of Γ. So, any point $\mathbf{p}_1$ with $\omega(\mathbf{p}_1) = \Gamma$ is not in Γ but $\mathbf{p}_1$ is outside Γ. The trajectory for such a $\mathbf{p}_1$ spends periods of time near the fixed point, then makes an excursion near one branch of the homoclinic connection, and then spends even a longer period of time near the fixed point before making an excursion near the other branch of the homoclinic connection. See Figure 1 for the plot of x as a function of t for a point starting slightly outside of $\mathbf{A}$. This trajectory in phase space and the plot of one coordinate versus time do not seem "chaotic" in the normal sense of the term; they seem almost periodic, with bursts of transition of a more or less fixed duration. The length of time spent near the fixed point $\mathbf{0}$ increases each time the trajectory passes by $\mathbf{0}$.

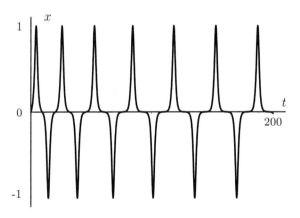

Figure 1. Plot of x as a function of t for initial conditions $(0, 0.05, 0)$ for Example 7.15

Example 7.16 (Nonchaotic attractor with sensitive dependence). We modify Example 7.11 to make it have sensitive dependence on initial conditions when restricted to the attractor. Consider the system of differential equations given by

$$\dot{\tau} = 1 \pmod{2\pi},$$
$$\dot{\theta} = 1 + \frac{1}{2}\sin(\psi) \pmod{2\pi},$$
$$\dot{\psi} = 0 \pmod{2\pi},$$
$$\dot{y} = -y.$$

This system still has an attractor with $\{y = 0\}$. For two points $(\tau_j, \theta_j, \psi_j, 0)$ in the attractor, in order for the τ variables to stay close, we cannot reparametrize the trajectories. If $\sin(\psi_0) \neq \sin(\psi_1)$, then $|\theta_0(t) - \theta_1(t)|$ will become large until it reaches a value of π. Thus, the system has sensitive dependence on initial conditions when restricted to the attractor. However, the attractor is not transitive since ψ is a constant. Therefore, this is not a chaotic attractor.

Example 7.17 (Quasiperiodic attractor). This example illustrates the fact that a quasiperiodic systems does not have sensitive dependence on initial conditions when restricted to its attractor. Consider the system of differential equations

$$\dot{x}_1 = y_1 + x_1 \left(1 - x_1^2 - y_1^2\right),$$
$$\dot{y}_1 = -x_1 + y_1 \left(1 - x_1^2 - y_1^2\right),$$
$$\dot{x}_2 = \sqrt{2}\, y_2 + 3\, x_2 \left(1 - x_2^2 - y_2^2\right),$$
$$\dot{y}_2 = -\sqrt{2}\, x_2 + 3\, y_2 \left(1 - x_2^2 - y_2^2\right).$$

If we introduce polar coordinates $r_1^2 = x_1^2 + y_1^2$, $\tan(\theta_1) = y_1/x_1$, $r_2^2 = x_2^2 + y_2^2$, and $\tan(\theta_2) = y_2/x_2$, then the differential equations become

$$\dot{r}_1 = r_1 \left(1 - r_1^2\right),$$
$$\dot{\theta}_1 = 1 \pmod{2\pi},$$
$$\dot{r}_2 = 3\, r_2 \left(1 - r_2^2\right),$$
$$\dot{\theta}_2 = \sqrt{2} \pmod{2\pi}.$$

The set $r_1 = 1 = r_2$ is easily determined to be an attracting set. Since the motion on this set is quasiperiodic, there is a dense orbit and the set is an attractor. Two trajectories, with initial conditions $(\theta_{1,0}^1, \theta_{2,0}^1)$ and $(\theta_{1,0}^2, \theta_{2,0}^2)$ in the angles, satisfy

$$(\theta_1^1(t), \theta_2^1(t)) = (\theta_{1,0}^1 + t, \theta_{2,0}^1 + t\sqrt{2}) \quad \text{and}$$
$$(\theta_1^2(t), \theta_2^2(t)) = (\theta_{1,0}^2 + t, \theta_{2,0}^2 + t\sqrt{2}).$$

Therefore, the distance between the trajectories stays a constant, and there is not sensitive dependence on initial conditions,

$$\|(\theta_1^1(t), \theta_2^1(t)) - (\theta_1^2(t), \theta_2^2(t))\|$$
$$= \|(\theta_{1,0}^1 + t, \theta_{2,0}^1 + t\sqrt{2}) - (\theta_{1,0}^2 + t, \theta_{2,0}^2 + t\sqrt{2})\|$$
$$= \|(\theta_{1,0}^1 - \theta_{1,0}^2, \theta_{2,0}^1 - \theta_{2,0}^2)\|.$$

Since the system does not have sensitive dependence on initial conditions, the set is not a chaotic attractor. Since quasiperiodic motion is not "chaotic" in appearance, this classification makes sense for this example.

The system considered in Example 7.11 is another example of an attractor that does not have sensitive dependence and also is not transitive. Therefore, there are a couple of reason it does not qualify as a chaotic attractor.

So far, we have not given very complicated examples with sensitive dependence and no examples with sensitive dependence when restricted to an invariant set, so no chaotic attractors. In Section 7.3, we indicate why the Lorenz system has a chaotic attractor, both (1) by a computer-generated plot of trajectories and (2) by reducing the Poincaré map to a one-dimensional function. In Sections 7.4 and 7.5, computer-generated plots indicate that both the Rössler system of differential equations and a periodically forced system possess a chaotic attractor.

Exercise 7.2

1. Consider the system of differential equations given in polar coordinates by
$$\dot{r} = r(1-r^2)(r^2-4),$$
$$\dot{\theta} = 1.$$

   a. Does the system have sensitive dependence on initial conditions at points in either of the sets $\{r=1\}$ or $\{r=2\}$. Explain your answer.
   b. Does the system have sensitive dependence on initial conditions when restricted to either of the sets $\{r=1\}$ or $\{r=2\}$. Explain your answer.
   c. Is either of the sets $\{r=1\}$ or $\{r=2\}$ a chaotic attractor?

2. Consider the system of differential equations given in exercise 2 in section 7.1. Let $C_0 = L(1,0)$.
   a. Does the system have sensitive dependence on initial conditions at points in the set $L^{-1}(C_0)$? Explain your answer.
   b. Does the system have sensitive dependence on initial conditions when restricted to the set $L^{-1}(C_0)$? Explain your answer.
   c. Is the set $L^{-1}(C_0)$ a chaotic attractor?

3. Consider the system of differential equations given by
$$\dot{r} = 1 + \frac{1}{2}\sin(\theta) \pmod{2\pi},$$
$$\dot{\theta} = 0 \pmod{2\pi},$$
$$\dot{x} = -x.$$

   a. Show that the system has sensitive dependence on initial conditions when restricted to $\mathbf{A} = \{(r,\theta,0) : 0 \leq r \leq 2\pi,\ 0 \leq \theta \leq 2\pi\}$.
   Hint: Note that $\dot{r} > 0$ at all points but the value changes with θ.
   b. Discuss why $\mathbf{A}$ is not a chaotic attractor. Does it seems chaotic?

4. Let $V(x) = -2x^6 + 15x^4 - 24x^2$, for which $V'(x) = -12x^5 + 60x^3 - 48x$, $V'(0) = V'(\pm 1) = V'(\pm 2) = 0$, $V(0) = 0$, $V(\pm 1) = -11$, and $V(\pm 2) = 16$. Also, $V(x)$ goes to $-\infty$ as x goes to $\pm\infty$. Let $L(x,y) = V(x) + y^2/2$.
   a. Plot the potential function $V(x)$ and sketch the phase portrait for the system of differential equations
$$\dot{x} = y,$$
$$\dot{y} = -V'(x).$$

   Notice that $L(\pm 2, 0) > L(0,0)$.

b. Sketch the phase portrait for the system of differential equations

$$\dot{x} = y$$
$$\dot{y} = -V'(x) - y L(x,y).$$

Pay special attention to the stable and unstable manifolds of saddle fixed points.

c. Let $\mathbf{A} = \{(x,y) \in L^{-1}(0) : -2 < x < 2\}$. Does the system of part (b) have sensitive dependence on initial conditions at points of the set $\mathbf{A}$? Explain why. Does it have have sensitive dependence on initial conditions when restricted to $\mathbf{A}$? Explain why.

d. What are the attracting sets and attractors for the system of part (b)?

7.3. Lorenz System

In this section we discuss the Lorenz system of differential equations, which is given by

(7.1)
$$\dot{x} = -\sigma x + \sigma y,$$
$$\dot{y} = rx - y - xz,$$
$$\dot{z} = -bz + xy.$$

The quantities σ, r, and b are three positive parameters. The values of the parameters that Lorenz considered are $\sigma = 10$, $b = 8/3$, and $r = 28$. We will fix $\sigma = 10$ and $b = 8/3$, but consider the dynamics for various values of r, including Lorenz's original value of $r = 28$.

In Section 7.8.1, we discuss some of the history of E. Lorenz's development of this system and how it models convection rolls in the atmosphere. In this section, we discuss various properties of this system, including why it has a chaotic attractor.

This system gives a good example of a chaotic attractor, as defined in Section 7.2. It has only linear and quadratic terms in three variables and is deterministic, because the equations and parameters are fixed and there are no external "stochastic" inputs. Nevertheless, this simple deterministic system of equations has complicated dynamics, with apparently random behavior. The chaotic behavior reveals itself in at least two ways. First, the plot of an individual trajectory can seem random. See Figure 2. Second, Figure 3 shows the results of numerical integration that indicate that it has sensitive dependence on initial conditions restricted to its invariant set, because nearby initial conditions lead to solutions that are eventually on opposite sides of the invariant set. The initial conditions $(1.4, 1.4, 28)$ and $(1.5, 1.5, 28)$ are followed to time $t = 1.5$ in Figure 3a and $t = 2.3$ in Figure 3b. Notice that the solution with smaller initial conditions switches sides on the second time around, while the other solution stays on the same side. After time 2.3, the first solution has crossed back over to the right side, and the second solution is now on the left side. The trajectories of these two solutions have very little to do with each other, even though they have relatively similar initial conditions. The fact that systems with so little nonlinearity can behave so differently from linear systems demonstrates that even a small amount of nonlinearity is enough to cause chaotic outcomes.

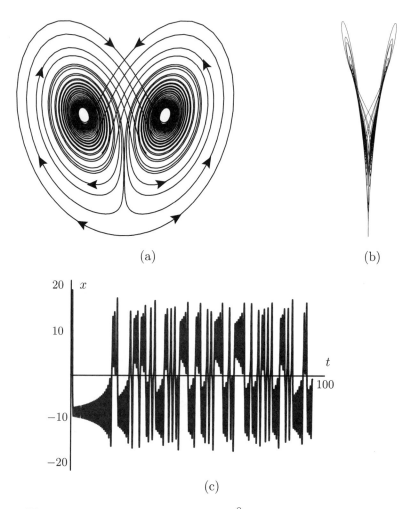

Figure 2. Lorenz attractor for $\sigma = 10$, $b = 8/3$, and $r = 28$: Two views of the phase portrait and the plot of x as a function of t

Warwick Tucker [**Tuc99**] has proved that the Lorenz system has sensitive dependence on initial conditions and a chaotic attractor for the parameter values $\sigma = 10$, $b = 8/3$, and $r = 28$. His proof is computer assisted in the sense that it used the computer to numerically integrate the differential equations with estimates to verify the conditions.

7.3.1. Fixed Points. The fixed points of the the Lorenz system (7.1) satisfy $y = x$, $x(r - 1 - z) = 0$, and $bz = x^2$. The origin is always a fixed point, and if $r > 1$, then there are two other fixed points,

$$\mathbf{P}^+ = \left(\sqrt{b(r-1)}, \sqrt{b(r-1)}, r-1\right) \quad \text{and}$$
$$\mathbf{P}^- = \left(-\sqrt{b(r-1)}, -\sqrt{b(r-1)}, r-1\right).$$

7.3. Lorenz System

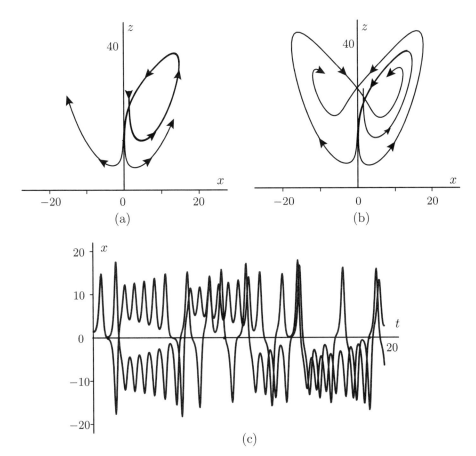

Figure 3. Sensitive dependence for the Lorenz system of equations: Initial conditions $(1.4, 1.4, 28)$ and $(1.5, 1.5, 28)$ followed up to $t = 1.5$ in (a) and $t = 2.3$ in the (b); figure (c) gives the plot of t versus x up to time 20

The matrix of partial derivatives of the system of differential equations is

$$D\mathbf{F}_{(x,y,x)} = \begin{pmatrix} -\sigma & \sigma & 0 \\ r-z & -1 & -x \\ y & x & -b \end{pmatrix}.$$

At the origin, this reduces to

$$D\mathbf{F}_{(\mathbf{0})} = \begin{pmatrix} -\sigma & \sigma & 0 \\ r & -1 & 0 \\ 0 & 0 & -b \end{pmatrix}.$$

The characteristic equation for the origin is

$$(\lambda + b)\left(\lambda^2 + (\sigma + 1)\lambda + \sigma(1 - r)\right) = 0.$$

The eigenvalues are

$$\lambda_s = -b < 0,$$
$$\lambda_{ss} = \frac{-(\sigma+1) - \sqrt{(\sigma+1)^2 + 4\sigma(r-1)}}{2} < 0, \quad \text{and}$$
$$\lambda_u = \frac{-(\sigma+1) + \sqrt{(\sigma+1)^2 + 4\sigma(r-1)}}{2}.$$

The first two eigenvalues are always negative, and λ_s is labeled with s for "weak stable" and λ_{ss} is labeled with ss for strong stable. For $r > 1$ and $\sigma > b - 1$,

$$\lambda_{ss} < \lambda_s < 0.$$

The last eigenvalue λ_u is negative for $0 < r < 1$ and positive for $r > 1$. We label it with a u for "unstable" since it is positive for most of the parameter values. Thus, the origin is attracting for $0 < r < 1$; for $r > 1$, the origin is a saddle with all real eigenvalues, two negative (attracting) and one positive (repelling).

For $r > 1$, the two other fixed points $\mathbf{P}^+$ and $\mathbf{P}^-$ have the same eigenvalues, so we consider only $\mathbf{P}^+$. At the fixed point, $r - z = 1$ and $y = x$, so the matrix of partial derivatives is

$$\begin{pmatrix} -\sigma & \sigma & 0 \\ 1 & -1 & -x \\ x & x & -b \end{pmatrix},$$

and has characteristic equation

$$0 = p_r(\lambda) = \lambda^3 + \lambda^2(\sigma + b + 1) + \lambda(\sigma b + b + x^2) + 2\sigma x^2$$
$$= \lambda^3 + \lambda^2(\sigma + b + 1) + \lambda b(r + \sigma) + 2b\sigma(r - 1).$$

All the coefficients are positive, so it has a negative real root λ_1: One of the eigenvalues is always a negative real number. (This is true because the characteristic polynomial $p_r(\lambda)$ is positive at $\lambda = 0$ and goes to $-\infty$ as λ goes to $-\infty$: Since $p_r(\lambda)$ varies continuously as λ varies from 0 to $-\infty$, it must be zero for some negative value of λ.)

We now discuss the manner in which the other two eigenvalues of $\mathbf{P}^\pm$ vary as r varies and $\sigma = 10$ and $b = 8/3$ stay fixed. The eigenvalues are all real for r near 1. Plotting the graph of $p_r(\lambda)$ as r increases indicates that for about $r \geq 1.3456$, there is only one real root and so another pair of complex eigenvalues. See Figure 4.

For small $r > 1.3456$, the real part of the complex eigenvalues is negative. It is possible to solve for the parameter values, which results in purely imaginary eigenvalues by substituting $\lambda = i\omega$ in $p_r(\lambda) = 0$ and equating the real and imaginary parts of $p_r(i\omega)$ to zero. For $\lambda = i\omega$, $\lambda^2 = -\omega^2$ and $\lambda^3 = -i\omega^3$, so

$$0 = -i\omega^3 - \omega^2(\sigma + b + 1) + i\omega b(r + \sigma) + 2b\sigma(r - 1)$$
$$= i\omega(-\omega^2 + b(r + \sigma)) - \omega^2(\sigma + b + 1) + 2b\sigma(r - 1).$$

7.3. Lorenz System

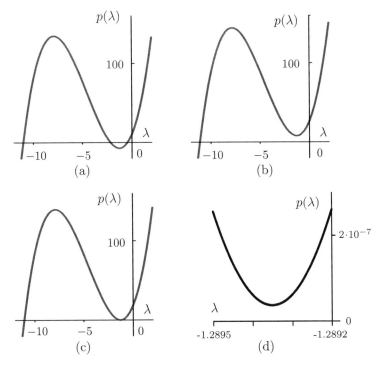

Figure 4. Plot of the characteristic polynomial for (a) $r = 1.2$, (b) $r = 1.5$, (c) $r = 1.3456171795$, and (d) $r = 1.3456171795$.

So, equating the real and imaginary parts to zero, we get
$$\omega^2 = b(r + \sigma) = \frac{2b\sigma(r-1)}{\sigma + b + 1},$$
$$(r + \sigma)(\sigma + b + 1) = 2\sigma(r - 1),$$
$$\sigma(\sigma + b + 3) = r(\sigma - b - 1), \quad \text{or}$$
$$r_1 = \frac{\sigma(\sigma + b + 3)}{\sigma - b - 1}.$$

For $\sigma = 10$ and $b = 8/3$, this gives
$$r_1 = \frac{470}{19} \approx 24.74.$$

At this parameter value r_1, there are a pair of purely imaginary eigenvalues and another negative real eigenvalue. For values of r near r_1, there is a "center manifold", a two-dimensional surface, toward which the system attracts. It has been verified that there is a subcritical Andronov–Hopf bifurcation of an periodic orbit within the center manifold. The orbit appears for values of $r < r_1$ and is unstable within the center manifold. Since there is still one contracting direction, the periodic orbit is a saddle in the whole phase space. The fixed point goes from stable to unstable as r increases past r_1, and there is a saddle periodic orbit for $r < r_1$.

Besides the Andronov–Hopf bifurcation, there is a homoclinic bifurcation as r varies. For small values of r, the unstable manifold of the origin stays on the same

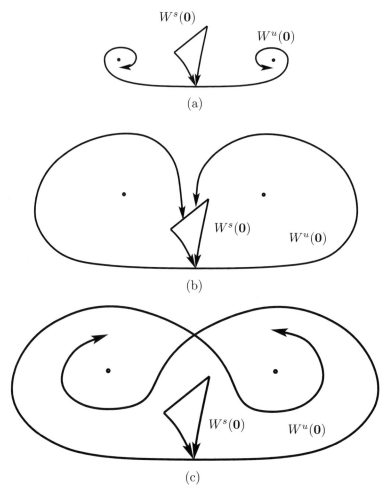

Figure 5. Unstable manifold of the origin for (a) $r < r_0$, (b) $r = r_0$, and (c) $r > r_0$

side of the stable manifold of the origin and spirals into the fixed points $\mathbf{P}^{\pm}$. See Figure 5(a). For $r = r_0 \approx 13.926$, the unstable manifold is homoclinic and goes directly back into the origin. See Figure 5(b). For $r > r_0$, the unstable manifold crosses over to the other side of the stable manifold. See Figure 5(c). Thus, there is a homoclinic bifurcation at $r = r_0$. For $r = r_0$, the strong contraction $e^{\lambda_{ss}} \ll 1$ causes the solution to contract toward a surface containing the homoclinic connections. The sum of the eigenvalues $\lambda_u + \lambda_s > 0$ so the homoclinic bifurcation spawns an unstable periodic orbit in this surface for $r > r_0$. Numerical studies indicate that this periodic orbit continues up to $r = r_1$, when it dies in the subcritical Andronov–Hopf bifurcation at the fixed points $\mathbf{P}^{\pm}$. See Sparrow [**Spa82**] for more details.

7.3.2. Attracting Sets.

Example 7.18 (Fixed point sink for Lorenz system). Assume that $0 < r < 1$. For these parameter values, the Lorenz systems has only one fixed point. By using a Lyapunov function, it is possible to show that the origin is globally attracting for these values. Let

$$L_1(x,y,z) = \frac{x^2}{2\sigma} + \frac{y^2}{2} + \frac{z^2}{2}.$$

This function is positive and measures the square of the distance from the origin with a scaling in the x coordinate. Then,

$$\begin{aligned}\dot{L}_1 &= \frac{1}{\sigma}x\dot{x} + y\dot{y} + z\dot{z} \\ &= x(y-x) + y(rx - y - xz) + z(-bz + xy) \\ &= -x^2 + (r+1)xy - y^2 - bz^2 \\ &= -\left(x - \frac{r+1}{2}y\right)^2 - \left(1 - (\frac{r+1}{2})^2\right)y^2 - bz^2,\end{aligned}$$

where we completed the square in the last equality. If $r < 1$, then $r + 1 < 2$, $(r+1)/2 < 1$, and the coefficient of y^2 is negative. If $\dot{L}_1 = 0$, then $y = z = 0$ and $0 = x - \frac{r+1}{2}y$; it follows that $x = \frac{r+1}{2}y = 0$. Therefore, $\dot{L}_1 = 0$ only when $(x,y,z) = (0,0,0)$, and $\dot{L}_1$ is strictly negative on all of $\mathbb{R}^3 \setminus \{\mathbf{0}\}$. This shows that L_1 is a global Lyapunov function and that the basin of attraction of the origin is all of $\mathbb{R}^3$.

Example 7.19 (Attracting set for Lorenz system). We now consider the Lorenz system for $r > 1$, where the origin is no longer attracting. For this range of r, we use the test function

$$L_2(x,y,z) = \frac{x^2 + y^2 + (z - r - \sigma)^2}{2}$$

to show that all the solutions enter a large sphere, but we cannot use it to determine the nature of their ω-limit sets. The time derivative is

$$\begin{aligned}\dot{L}_2 &= x(-\sigma x + \sigma y) + y(rx - y - xz) + (z - r - \sigma)(-bz + xy) \\ &= -\sigma x^2 + \sigma xy + rxy - y^2 - xyz - bz^2 + xyz + rbz - rxy + \sigma bz - \sigma xy \\ &= -\sigma x^2 - y^2 - b\left(z^2 - (r+\sigma)z\right) \\ &= -\sigma x^2 - y^2 - b\left(z - \frac{r+\sigma}{2}\right)^2 + \frac{b(r+\sigma)^2}{4}.\end{aligned}$$

The sum of the first three terms is negative, and $\dot{L}_2$ is negative for

$$\sigma x^2 + y^2 + b\left(z - \frac{r+\sigma}{2}\right)^2 > \frac{b(r+\sigma)^2}{4}.$$

The level set formed by replacing the inequality by an equality is an ellipsoid with the ends of the axes at the following six points,

$$\left(\frac{\pm\sqrt{b}(r+\sigma)}{2\sqrt{\sigma}}, 0, \frac{r+\sigma}{2}\right), \qquad \left(0, \frac{\pm\sqrt{b}(r+\sigma)}{2}, \frac{r+\sigma}{2}\right),$$

$$(0,0,0), \qquad\qquad (0,0,r+\sigma).$$

The values of L_2 at these points are

$$\frac{(r+\sigma)^2}{8}\left(\frac{b}{\sigma}+1\right), \quad \frac{(r+\sigma)^2}{8}(b+1), \quad \frac{(r+\sigma)^2}{2}, \quad \text{and} \quad 0.$$

For $\sigma > 1$ and $b < 3$, the largest of these is $\dfrac{(r+\sigma)^2}{2}$. Let C_2 be any number slightly greater than this value,

$$C_2 > \frac{(r+\sigma)^2}{2}.$$

Let

$$\mathbf{U}_2 = L_2^{-1}((-\infty, C_2]) = \{(x,y,z) : L_2(x,y,z) \leq C_2\}.$$

The interior of $\mathbf{U}_2$ is the set with a strict inequality,

$$\operatorname{int}(\mathbf{U}_2) = \{(x,y,z) : L_2(x,y,z) < C_2\}.$$

If $\phi(t;\mathbf{x})$ is the flow for the Lorenz system, then the calculation of the time derivative of L_2 shows that $\phi(t;\mathbf{U}_2) \subset \operatorname{int}(\mathbf{U}_2)$ for $t > 0$. Thus, $\mathbf{U}_2$ is a trapping region, and

$$\mathbf{A} = \bigcap_{t \geq 0} \phi(t;\mathbf{U}_2)$$

is an *attracting set*.

For $r = 28$, the fixed points $\mathbf{P}^{\pm}$ are saddles and computer simulation shows that they are not part of the attractor. Therefore, to get a trapping region for an attractor, we need to remove tubes around the stable manifolds $W^s(\mathbf{P}^+) \cup W^s(\mathbf{P}^-)$ from the open set $\mathbf{U}_2$. At the end of this section, we discuss how the resulting attracting set can be shown to be a chaotic attractor.

In fact, the trajectories for the Lorenz system tend to a set of zero volume. To see this, we need to consider how volume changes as we let a region be transported by the flow. The next theorem shows how the volume of any region goes to zero as it is flowed along by the solutions of the differential equation.

Theorem 7.2. *Consider the Lorenz system of differential equations.*

  a. *Let $\mathscr{D}$ be a region in $\mathbb{R}^3$ with smooth boundary $\partial \mathscr{D}$. Let $\mathscr{D}(t)$ be the region formed by flowing along for time t,*

$$\mathscr{D}(t) = \{\phi(t;\mathbf{x}_0) : \mathbf{x}_0 \in \mathscr{D}\}.$$

  Finally, let $V(t)$ be the volume of $\mathscr{D}(t)$. Then

$$V(t) = V(0)\, e^{-(\sigma+1+b)t}.$$

  Thus, the volume decreases exponentially fast.

  b. *If $\mathbf{A}$ is an invariant set with finite volume, it follows that it must have zero volume. Thus, the invariant set for the Lorenz equations has zero volume.*

7.3. Lorenz System

Proof. By Theorem 6.9, we have
$$\dot{V}(t) = \int_{\mathscr{D}(t)} \nabla \cdot \mathbf{F} \, dV.$$
For the Lorenz equations, $\nabla \cdot \mathbf{F} = -\sigma - 1 - b$ is a negative constant. Therefore, for the Lorenz equations,
$$\dot{V}(t) = -(\sigma + 1 + b)V(t),$$
and the volume is decreasing. This equation has a solution
$$V(t) = V(0)\, e^{-(\sigma+1+b)t},$$
as claimed in the theorem. The rest of the results of the theorem follow directly.

The invariant set $\mathbf{A}$ is contained in the regions $\mathscr{D}(t)$ of finite volume. Therefore,
$$\mathrm{vol}(A) \leq \mathrm{vol}(\mathscr{D}(t)) = \mathrm{vol}(\mathscr{D}(0))\, e^{-(\sigma+1+b)t},$$
and it must be zero. $\square$

Table 1 summarizes the information about the bifurcations and eigenvalues of the fixed points for the Lorenz system of differential equations for $\sigma = 10$ and $b = 8/3$, treating r as the parameter which is varying.

	Bifurcations for the Lorenz system
$1 < r$	The origin is unstable, with two negative eigenvalues (two contracting directions) and one positive eigenvalue (one expanding direction).
$1 < r < r_1$	The fixed points $\mathbf{P}^\pm$ are stable. For r larger than about 1.35, there is one negative real eigenvalue and a pair of complex eigenvalues with negative real parts.
$r = r_0$	There is a homoclinic bifurcation of a periodic orbit which continues up to the subcritical Andronov–Hopf bifurcation at $r = r_1$.
$r = r_1$	There is a subcritical Andronov–Hopf bifurcation. The periodic orbit can be continued numerically back to the homoclinic bifurcation at $r = r_0$.
$r_1 < r$	The fixed points $\mathbf{P}^\pm$ are unstable. There is one negative real eigenvalue and a pair of complex eigenvalues with positive real parts.
$r = 28$	A chaotic attractor is observed.

Table 1. $r_0 \approx 13.926$, and $r_1 \approx 24.74$

When the Lorenz system became more widely analyzed in the 1970s, mathematicians recognized that computer simulation for $r = 28$ indicates its attracting set possesses certain properties that would imply that it is transitive and has sensitive dependence on initial conditions and so a chaotic attractor. (1) There is a strong contraction toward an attracting set that is almost a surface. (2) Another direction "within the surface-like object" is stretched or expanded. (3) This expansion can occur and and still have the orbits bounded because the attracting set is cut into two pieces as orbits flow on different sides of the stable manifold of the origin. (4) Finally, the two sheets of the surface are piled back on top of each other. See Figure 2. Because conditions (1) and (2) cannot be verified point-by-point but are global, no one has been able to prove the existence of a chaotic attractor by purely human reasoning and calculation. The mathematical theory does provide a basis for a geometric model of the dynamics introduced by J. Guckenheimer in [**Guc76**], which does possess a chaotic attractor. The geometric model is based on assuming certain properties of the Poincaré map for the Lorenz system as discussed in Section 7.3.3. These properties of Poincaré map in turn imply that there is a chaotic attractor. As mentioned earlier, Warwick Tucker [**Tuc99**] has given a computer-assisted proof that the actual Lorenz system for $r = 28$, $\sigma = 10$, and $b = 8/3$ has a chaotic attractor; hence, the name *Lorenz attractor* has been justified by a computer-assisted proof.

7.3.3. Poincaré Map. Most of the analysis of the Lorenz system has used the Poincaré map, which we discuss in this section.

By looking at successive maximum z_n of the z-coordinate along an orbit, Lorenz developed an argument that the orbits within the attracting set were diverging. He noticed that, if a plot is made of successive pairs of maxima (z_n, z_{n+1}), the result has very little thickness and can be modeled by the graph of a real-valued function $h(z_n)$. See Figure 6. The function $h(z_n)$ appears to have a derivative with absolute value greater than one on each side. So, if z_n and z'_n are on the same side, then $|h(z_n) - h(z'_n)| = |h'(\zeta)| \cdot |z_n - z'_n| > |z_n - z'_n|$, where ζ is some point between z_n and z'_n. This argument can be used to show that the system has sensitive dependence on initial conditions, provided that the reduction to the one-dimensional map h can be justified.

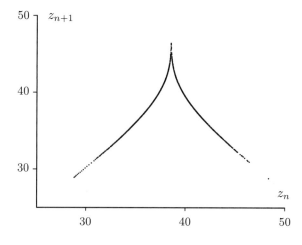

Figure 6. Plot of z_n versus z_{n+1}

Instead of using Lorenz's approach, we connect the argument back to Poincaré maps. It is difficult to prove the actual results about the Poincaré map, but Warwick

7.3. Lorenz System

Tucker [**Tuc99**] has published a computer-assisted proof of these facts. He uses the computer to calculate the Poincaré map with estimates that verify that his analysis is valid for the actual equations. Rather than try to develop these ideas, we connect the observed behavior to the *geometric model of the Lorenz equations* that was introduced by Guckenheimer. The geometric model was further analyzed by Guckenheimer and Williams. See [**Guc76**] and [**Guc80**].

The idea of the geometric model is to take the Poincaré map from $z = z_0$ to itself, only considering intersections between the two fixed points $\mathbf{P}^+$ and $\mathbf{P}^-$. A numerical plot of the return points of one orbit is shown in Figure 7.

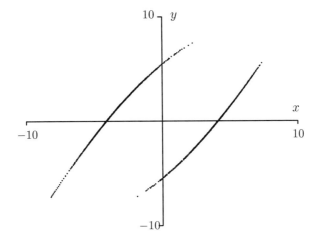

Figure 7. Plot of the orbit of one point by the two-dimensional Poincaré map P

The Lorenz system near the origin is nearly linear, so we approximate the dynamics by

$$\dot{u} = au,$$
$$\dot{v} = -cv,$$
$$\dot{z} = -bz,$$

where $a = \lambda_u \approx 11.83$, $c = -\lambda_{ss} \approx 22.83$, and $b = -\lambda_s = 8/3$. The variable u is in the expanding direction and v is in the strong contracting direction. For these values of a, b, and c, we have $0 < b < a < c$, so $0 < b/a < 1$ and $c/a > 1$. We consider the flow from $\{z = z_0\}$ to $\{u = \pm u_1\}$. We restrict the transversal Σ by

$$\Sigma = \{(u, v, z_0) : |u|, |v| \leq \alpha\},$$

and define

$$\Sigma_1 = \{(\pm u_1, v, z) \, |v|, |z| \leq \beta\}.$$

We take α equal to the value of u for the first time $W^u(\mathbf{0})$ crosses $z = z_0$ between $\mathbf{P}^+$ and $\mathbf{P}^-$. Thus, α is small enough that the cross section is between the two fixed points $\mathbf{P}^+$ and $\mathbf{P}^-$. The points on $\Sigma \cap W^s(\mathbf{0})$ tend directly toward the origin and never reach Σ_1, so we must look at

$$\Sigma' = \Sigma \setminus \{(u, v, z_0) : u \neq 0\} = \Sigma^+ \cup \Sigma^-,$$

where Σ^+ has positive u and Σ^- has negative u.

The solution of the linearized equations is given by $u(t) = e^{at}u(0)$, $v(t) = e^{-ct}v(0)$, and $z(t) = e^{-bt}z(0)$. We take an initial condition (u, v, z_0) in Σ' ($u \neq 0$) and follow it until it reaches Σ_1. We use $u > 0$, or a point in Σ^+, to simplify the notation. The time τ to reach Σ_1 satisfies

$$e^{a\tau}u = u_1,$$

$$e^{\tau} = \left(\frac{u_1}{u}\right)^{1/a}.$$

Substituting in this time,

$$v(\tau) = (e^{\tau})^{-c} v$$

$$= \left(\frac{u_1}{u}\right)^{-c/a} v$$

$$= \frac{u^{c/a} v}{u_1^{c/a}},$$

$$z(\tau) = (e^{\tau})^{-b} z_0$$

$$= \frac{z_0 u^{b/a}}{u_1^{b/a}}.$$

Thus of the linear system, the Poincaré map $P_1 : \Sigma' \to \Sigma_1$ is given by

$$P_1(u, v) = (z(\tau), v(\tau))$$
$$= \left(z_0 u_1^{-b/a} |u|^{b/a}, u_1^{-c/a} |u|^{c/a} v\right).$$

The image of Σ' is given in two pieces, $P_1(\Sigma^+) = \Sigma_1^+$ and $P_1(\Sigma^-) = \Sigma_1^-$. See Figure 8.

The map P_2 from Σ_1 back to Σ takes a finite length of time, so it is differentiable. For the actual equations, this is difficult to control and has only been done by a computer-assisted proof. The geometric model makes the simplifying assumption that the flow takes $(u_1, z(\tau), v(\tau))$ in Σ_1^+ back to $(u, v, z) = (-\alpha + z(\tau), v(\tau), z_0)$. What is most important is that P_2 takes horizontal lines (u_1, v, z_1) back into lines in Σ with the same value of u; that is, the new u value depends only on z and not on v. The Poincaré map for u negative is similar, but the u coordinate is $\alpha - u_1^{-b/a} z_0 |u|^{b/a}$. Therefore, we get the Poincaré map, $P = P_2 \circ P_1$, of the geometric model

$$P(u, v) = (f(u), g(u, v)) = \begin{cases} \left(-\alpha + z_0 u_1^{-b/a} u^{b/a}, u_1^{-c/a} u^{c/a} v\right) & \text{for } u > 0, \\ \left(\alpha - z_0 u_1^{-b/a} |u|^{b/a}, u_1^{-c/a} |u|^{c/a} v\right) & \text{for } u < 0. \end{cases}$$

Notice that the first coordinate function depends only on u and is independent of v. The second coordinate function is linear in v. The image of Σ' by P is as in Figure 9. Compare it with the numerically computed map in Figure 7, where the wedges are very very thin.

The derivative of the first coordinate function $f(u)$ is

$$f'(u) = z_0 u_1^{-b/a} |u|^{-1+b/a} \quad \text{for } u \text{ either positive or negative}.$$

7.3. Lorenz System

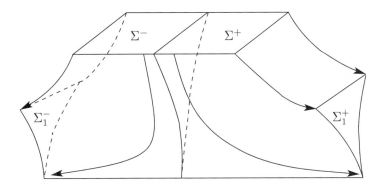

Figure 8. Flow of Σ past the origin

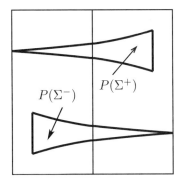

Figure 9. Image of Σ' by Poincaré map

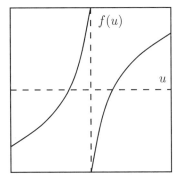

Figure 10. Graph of f

Since $b/a < 1$, it has a negative exponent, and $f'(u)$ is greater than one if u is small enough. This gives the expansion in the u-direction of the Poincaré map. The partial derivative of the second coordinate function is

$$\frac{\partial g}{\partial v} = u_1^{-c/a} |u|^{c/a} < 1 \quad \text{for } |u| < u_1.$$

Therefore, the Poincaré map is a (strong) contraction in this direction,
$$|g(u_0, v_1) - g(u_0, v_2)| < |v_1 - v_2|.$$
The contraction in the v-direction and the fact that the first coordinate depends only on u, allows the reduction to the one-dimensional map, which takes u to $f(u)$. The map f has a graph as shown in Figure 10 and replaces the one-dimensional map $h(z)$ (approximating the plot of (z_n, z_{n+1}) given in Figure 6) used by Lorenz. The fact that the derivative of the one-dimensional Poincaré map is greater than one implies that the flow has stretching in a direction within the attractor; this stretching causes the system to have sensitive dependence on initial conditions as is discussed in the next section. The cutting by the stable manifold of the origin is reflected in the discontinuity of the one-dimensional Poincaré map; it allows the attractor to be inside a bounded trapping region even though there is stretching. Thus, a combination of cutting and stretching results in a chaotic attractor.

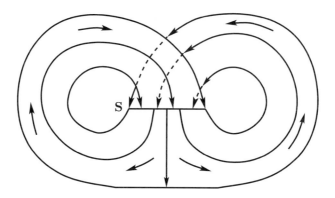

Figure 11. Flow on the branched manifold

The fact that there is a consistent contracting direction over the attracting set is the most important assumption made about the geometric model for the Lorenz system. This condition allows the reduction to a one-dimensional Poincaré map. This conditions is difficulty to verify for the actual system for $r = 28$, and has only been done with computer assistance.

7.3.4. One-dimensional Maps. We saw that the Lorenz system of differential equations has a trapping region for an attracting set. We verify this directly for the one-dimensional maps we are considering. We also verify that they have a chaotic attractor, and so the geometric model for the Lorenz system has a chaotic attractor.

We start with stating the definition of a chaotic attractor in this context.

Definition 7.20. A closed interval $[p, q]$ is a *trapping interval* for $f : \mathbb{R} \to \mathbb{R}$ provided that
$$\operatorname{cl}(f([p, q])) \subset (p, q).$$
We add the closure to the earlier definition, because the image does not have to be closed since the map can have a discontinuity.

7.3. Lorenz System

We denote the composition of f with itself k times by $f^k(x)$. For a trapping interval $[p, q]$, $\bigcap_{k \geq 0} \text{cl}(f^k([p, q]))$ is called an *attracting set* for $[p, q]$. An *attractor* is an attracting set $\mathbf{A}$ for which there is a point x_0 with $\omega(x_0) = \mathbf{A}$. An attractor $\mathbf{A}$ is *transitive* provided that there is a point x_0 in $\mathbf{A}$ for which $\omega(x_0) = \mathbf{A}$. An attractor is called *chaotic* provided that it is transitive and has sensitive dependence on initial conditions when restricted to the attractor.

The one-dimensional Poincaré map has one point of discontinuity and derivative greater than one at all points in the interval. The following theorem of R. Williams gives sufficient conditions for such maps to be transitive and have sensitive dependence on initial conditions.

Theorem 7.3. *Assume that $a < c < b$ and that f is a function defined at all points of the intervals $[a, c)$ and $(c, b]$, taking values in $[a, b]$ and the following three conditions hold. (i) The map f is differentiable at all points of $[a, c) \cup (c, b]$, with a derivative that satisfies $f'(x) \geq \gamma > \sqrt{2}$ at all these points. (ii) The point c is the single discontinuity, and*

$$\lim_{x<c,\, x \to c} f(x) = b \quad \text{and} \quad \lim_{x>c,\, x \to c} f(x) = a.$$

(iii) The iterates of the end points stay on the same side of the discontinuity c for two iterates,

$$a \leq f(a) \leq f^2(a) < c < f^2(b) \leq f(b) \leq b.$$

Then, the following two results are true:

   a. *The map f has sensitive dependence on initial conditions at all points of $[a, b]$.*

   b. *There is a point x^* in $[a, b]$ such that $\omega(x^*) = [a, b]$.*

   c. *If $a < f(a)$ and $f(b) < b$, then the interval $[a, b]$ is a chaotic attractor for f.*

The proof of this general result in given in Chapter 11; see Theorem 11.5. In this chapter, rather than prove this theorem about nonlinear maps, we simplify to a map that is linear on each of the two sides. For a constant $1 < \gamma \leq 2$, let

$$f_\gamma(x) = \begin{cases} \gamma x + 1 & \text{for } x < 0, \\ \gamma x - 1 & \text{for } x \geq 0. \end{cases}$$

The only discontinuity is $x = 0$, and the interval corresponding to the one in Theorem 7.3 is $[-1, 1]$. The images $f_\gamma([-1, 0)) = [1 - \gamma, 1)$ and $f_\gamma([0, 1]) = [-1, -1 + \gamma]$ together cover $[-1, 1)$, $f[-1, 1] = [-1, 1)$. (The discontinuity causes the image not to be closed.) For $1 < \gamma < 2$, $f_\gamma([-1, 0)) = [1 - \gamma, 1) \neq [-1, 1)$ and $f_\gamma([0, 1]) = [-1, -1 + \gamma] \neq [-1, 1)$. Each subinterval separately does not go all the way across the interval but together cover $(-1, 1)$, just as is the case for the Lorenz map f. However, for $\gamma = 2$, f_2 takes both $[-1, 0)$ and $[0, 1]$ all the way across, $f_2([-1, 0)) = [-1, 1)$ and $f_2([0, 1]) = [-1, 1]$.

The following theorem gives the results for the map f_γ.

Theorem 7.4. *Consider the map $f_\gamma(x)$ for $\sqrt{2} < \gamma \leq 2$ defined previously.*

   a. *The map f_γ has sensitive dependence on initial conditions when restricted to $[-1, 1]$.*

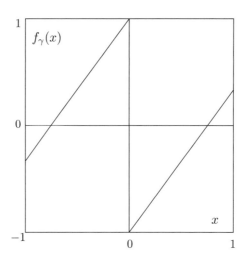

Figure 12. Graph of $f_\gamma(x)$

b. There is a point z^* in $[-1, 1]$ such that $\omega(z^*, f_\gamma) = [-1, 1]$.

c. For $\sqrt{2} < \gamma < 2$, the interval $[-1, 1]$ is a chaotic attractor for f_γ.

The proof is given in Section 7.9.

Exercises 7.3

1. (Elliptical trapping region for the Lorenz system) This exercise considers an alternative bounding function given by
$$L_3(x, y, z) = \frac{r\,x^2 + \sigma\,y^2 + \sigma\,(z - 2r)^2}{2}.$$
   a. Calculate $\dot{L}_3$.
   b. Find a value $C_3 > 0$ such that, if $L_3(x, y, z) > C_3$, then $\dot{L}_3 < 0$.
   c. Conclude that all solutions enter and remain in the region
$$\mathbf{U}_3 = \{(x, y, z) : L_3(x, y, z) \leq C_3\}$$
   (i.e., that $\mathbf{U}_3$ is a trapping region).

2. Consider the tripling map $T(x) = 3x \pmod{1}$ on $[0, 1]$ defined by
$$T(x) = \begin{cases} 3x & \text{for } 0 \leq x < \frac{1}{3}, \\ 3x - 1 & \text{for } \frac{1}{3} \leq x < \frac{2}{3}, \\ 3x - 2 & \text{for } \frac{2}{3} \leq x < 1, \\ 0 & \text{for } 1 = x. \end{cases}$$
Show that this map has sensitive dependence on initial conditions. Hint: If $y_n = x_n + \delta 3^n$ is on a different side of a discontinuity from x_n, then $y_{n+1} = x_{n+1} + 3(\delta 3^n) - 1$. Try using $r = 1/4$.

3. Consider the extension of f_2 to all of $\mathbb{R}$ defined by
$$F(x) = \begin{cases} \frac{1}{2}x - \frac{1}{2} & \text{for } x < -1, \\ 2x + 1 & \text{for } -1 \leq x < 0, \\ 2x - 1 & \text{for } 0 \leq x < 1, \\ \frac{1}{2}x + \frac{1}{2} & \text{for } x > 1. \end{cases}$$

   a. Show that $[-2, 2]$ is a trapping interval.
   b. Show that $[-1, 1]$ is the attracting set for the trapping interval $[-2, 2]$.
   Hint: In this case, you must calculate $F^k([-2, 2])$ for $k \geq 1$.
   c. Explain why $[-1, 1]$ is a chaotic attractor for F.

7.4. Rössler Attractor

The chaos for the Lorenz system is caused by stretching and cutting. Rather than cutting, a more common way to pile the two sets of orbits on top of each other is by folding a region on top of itself. Stretching, folding, and placing back on top of itself can create an attracting set that is chaotic and transitive, a chaotic attractor. Rössler developed the system of differential equations given in this section as a simple example of a system of differential equations that has this property, and computer simulation indicates that it has a chaotic attractor. However, verifying that a system has a chaotic attractor with folding rather than cutting is even more delicate. The Smale horseshoe map discussed in Section 13.1 was one of the first cases with folding that was thoroughly analyzed, but it is not an attractor. The Hénon map, considered in Section 13.4, also has stretching and folding. Much work has been done to show rigorously that it has a chaotic attractor.

The Rössler system of differential equations is given by
$$\dot{x} = -y - z,$$
$$\dot{y} = x + ay,$$
$$\dot{z} = b + z(x - c).$$

Computer simulation indicates that this systems has a chaotic attractor for $a = b = 0.2$ and $c = 5.7$.

If we set $z = 0$, the (x, y) equations become
$$\dot{x} = -y,$$
$$\dot{y} = x + ay.$$

The eigenvalues of the origin in this restricted system are
$$\lambda^{\pm} = \frac{a \pm i\sqrt{4 - a^2}}{2},$$

which have positive real parts, and a solution spirals outward. In the total three-dimensional system, as long as z stays near $z = b/(c - x)$ where $\dot{z} = 0$, the solution continues to spiral outward. When $x - c$ becomes positive, z increases. For these larger z, x decreases and becomes less than c, so z decreases. To understand the entire attracting set, we take a line segment $x_1 \leq x \leq x_2$, $y = 0$, and $z = 0$, and follow it once around the z-axis until it returns to $y = 0$. Since the orbits on the outside of the band of orbits lift up and come into the inside of the band while those on the inside merely spiral out slightly, the result is a fold when the system makes a revolution around the z-axis. All the orbits come back with $z \approx 0$; the inside and the outside orbits come back with the smallest values of x; and the middle orbits return with the largest values of x. See Figure 13.

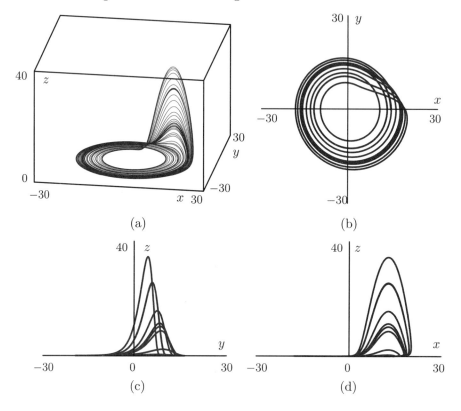

Figure 13. Rössler attractor for $a = 0.1$, $b = 0.1$, and $c = 14$

To completely analyze this system theoretically is difficult or impossible. However, its two-dimensional Poincaré map is very thin and it is like a one-dimensional map. (Because of the folding, this reduction is not really valid but gives some indication of why the system appears to have a chaotic attractor.) The one-dimensional map is nonlinear but is similar to the following linear map which we call the *shed map*:

$$g_a = \begin{cases} 2 - a + ax & \text{if } x \leq c, \\ a - ax & \text{if } x \geq c. \end{cases}$$

where $c = 1 - 1/a$, so $g_a(c) = 1$ for both definitions. We take $\sqrt{2} < a \leq 2$. This map has slope $\pm a$ everywhere, so its absolute value is greater than $\sqrt{2}$. Also, $0 < g_a(0) = 2 - a < 1 - 1/a = c$.

For $a = 2$, we get the map called the tent map. See Section 10.3. By a proof very similar to that for the Lorenz one-dimensional map, we show that this shed map has sensitive dependence and is transitive.

7.4. Rössler Attractor

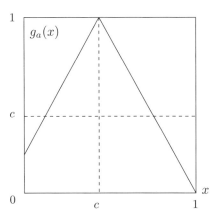

Figure 14. Shed map

Theorem 7.5. *Consider the shed map g_a for $\sqrt{2} < a \leq 2$.*

a. *The shed map g_a has sensitive dependence on initial conditions when restricted to $[0, 1]$.*

b. *The shed map has a point z^* in $[0, 1]$ such that $\omega(z^*) = [0, 1]$.*

c. *For $\sqrt{2} < a < 2$, the interval $[0, 1]$ is a chaotic attractor.*

The proof is similar to that for the one-dimensional Lorenz map and is given in Section 7.9.

7.4.1. Cantor Sets and Attractors. The Poincaré map for the Rössler attractor stretches within the attractor, contracts in toward the attractor, and then folds the trapping region over into itself. The result is that one strip is replaced by two thinner strips. If the process is repeated, then the two strips are stretched and folded to become four strips. These strips are connected within the attractor at the fold points, but we concentrate on the part of the attractor away from the fold points and look in the contracting direction only. Thus, we idealize the situation in what follows, and seek to understand a much-simplified model situation.

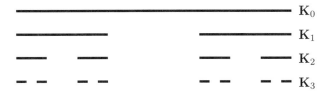

Figure 15. The sets $\mathbf{K}_0, \ldots \mathbf{K}_3$ for the Cantor set

The situation in the Lorenz system is similar, except that the sheet is cut rather than folded. In Figure 7, the sheets are so thin that the gaps between the sheets are not visible. However, this system also has the structure of a "Cantor set" in the direction coming out of the attractor.

The Cantor-like structure of the attractor in the contracting direction is part of the topological nature of these chaotic attractors, which motivated Ruelle and Takens to call them *strange attractors*.

We introduce Cantor sets in Section 10.5 in terms of ternary expansion of numbers. The middle-third Cantor set **K** consists of all points in the interval $[0, 1]$ that can be expressed using only 0's and 2's in their ternary expansion. It is the intersection of sets $\mathbf{K}_n$, which are shown for $0 \leq n \leq 3$ in Figure 15. The Cantor set has empty interior, is closed and bounded, and is uncountable.

Exercises 7.4

1. Prove that the tent map T given in this chapter has a point z^* with $\omega(z^*) = [0, 1]$. Hint: Compare with Theorem 7.4(b).

2. Consider the saw toothed map
$$S(x) = \begin{cases} \frac{1}{2}x & \text{for } x < 0, \\ 3x & \text{for } 0 \leq x \leq \frac{1}{3}, \\ 2 - 3x & \text{for } \frac{1}{3} \leq x \leq \frac{2}{3}, \\ 3x - 2 & \text{for } \frac{2}{3} \leq x \leq 1, \\ \frac{1}{2}x + \frac{1}{2} & \text{for } x > 1. \end{cases}$$

   a. Show that S restricted to $[0, 1]$ has sensitive dependence on initial conditions.

   b. Let $J = [-1, 2]$. Calculate $J_1 = S(J)$, $J_2 = S(J_1)$, ..., $J_n = S(J_{n-1})$. Argue why J is a trapping region for the attracting set $[0, 1]$.

   c. Explain why $[0, 1]$ is a chaotic attractor for S.

3. Consider the tent map with slope ± 4,
$$T(x) = T_4(x) = \begin{cases} 4x & \text{for } 0 \leq x \leq 0.5, \\ 4 - 4x & \text{for } 0.5 \leq x \leq 1. \end{cases}$$

   Let
   $$K_n(4) = \{\, x_0 : T^j(x_0) \in [0, 1] \text{ for } 0 \leq j \leq n \,\}$$
   and
   $$K(4) = \bigcap_{n \geq 0} K_n(4) = \{\, x_0 : T^j(x_0) \in [0, 1] \text{ for } 0 \leq j < \infty \,\}.$$

   a. Describe the sets $K_1(4)$ and $K_2(4)$.

   b. How many intervals does $K_n(4)$ contain, what is the length of each interval, and what is the total length of $K_n(4)$? Why does $K(4)$ have "length" or measure zero?

   c. Using an expansion of numbers base 4, which expansions are in the Cantor set $K(4)$?

   d. Give two numbers which are in $K(4)$ but are not end points of any of the sets $K_n(4)$.

   e. Show that the set $K(4)$ is nondenumerable.

4. Consider the tent map with slope $\pm\beta$, where $\beta > 2$,

$$T(x) = T_\beta(x) = \begin{cases} \beta x & \text{for } 0 \leq x \leq 0.5, \\ \beta - \beta x & \text{for } 0.5 \leq x \leq 1. \end{cases}$$

Let

$$K_n(\beta) = \{\, x_0 : T^j(x_0) \in [0,1] \text{ for } 0 \leq j \leq n \,\}$$

and

$$K(\beta) = \bigcap_{n \geq 0} K_n(\beta) = \{\, x_0 : T^j(x_0) \in [0,1] \text{ for } 0 \leq j < \infty \,\}.$$

  a. Describe the sets $K_1(\beta)$ and $K_2(\beta)$. Let $\alpha = 1 - 2\beta$. Why is it reasonable to call the Cantor set $K(\beta)$ the *middle α Cantor set*?
  b. How many intervals does $K_n(\beta)$ contain, what is the length of each interval, and what is the total length of $K_n(\beta)$? Why does $K(\beta)$ have "length" or measure zero?

7.5. Forced Oscillator

Ueda observed attractors for periodically forced nonlinear oscillators. See [**Ued73**] and [**Ued92**]. In order to get an attracting set, we include damping in the system.

Example 7.21. We add forcing to the double-welled Duffing equation considered in Example 5.1. Thus, the system of equations we consider is $\ddot{x} + \delta\dot{x} - x + x^3 = F\cos(\omega t)$, or

(7.2)
$$\begin{aligned} \dot{\tau} &= 1 \quad (\bmod {}^{2\pi}\!/_\omega), \\ \dot{x} &= y, \\ \dot{y} &= x - x^3 - \delta y + F\cos(\omega \tau). \end{aligned}$$

The forcing term has period ${}^{2\pi}\!/_\omega$ in τ, so we consider the Poincaré map from $\tau = 0$ to $\tau = {}^{2\pi}\!/_\omega$. The divergence is $-\delta$, which is negative for $\delta > 0$. Therefore, the system decreases volume in the total space, and any invariant sets have zero volume, or zero area in the two-dimensional cross section $\tau = 0$.

For $F = 0$ and $\delta > 0$, the system has two fixed point sinks at $(x, y) = (\pm 1, 0)$, which attract all the orbits except the stable manifolds of the saddle point at the origin. In terms of the potential function, the orbits tend to one of the two minima, or the bottom of one of the two "wells". For small F, the orbits continue to stay near the bottom of one of the two wells (e.g., $F = 0.18$, $\delta = 0.25$, and $\omega = 1$). See Figure 16 for a time plot of x as a function of t.

For a sufficiently large value of F, the system is shaken hard enough that a trajectory in one well is pulled over to the other side. Therefore, orbits can move back and forth between the two sides. For $F = 0.40$, the plot of $x(t)$ is erratic. See Figure 17 for a time plot of x as a function of t. The plot of $(x(t), y(t))$ crosses itself since the actual equations take place in three dimensions; therefore, these plots do not reveal much. The Poincaré map shows only where the solution is at certain times. It is a stroboscopic image, plotting the location every $2\pi/\omega$ units of time when the strobe goes off. The numerical plot of the Poincaré map for $F = 0.4$,

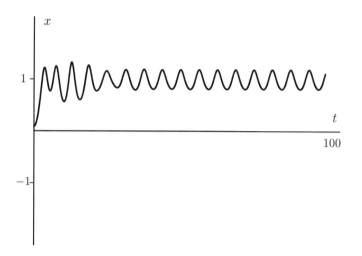

Figure 16. Plot x as a function of t for system (7.2), with $F = 0.18$, $\delta = 0.25$, and $\omega = 1$

$\delta = 0.25$, and $\omega = 1$ indicates that the system has a chaotic attractor. See Figure 18.

The stretching within the attractor is caused by what remains of the saddle periodic orbit $x = 0 = y$ for the unforced system. The periodic forcing breaks the homoclinic orbits and causes the folding within the resulting attractor. The chaotic attractor is observed via numerical calculation, but it is extremely difficult to give a rigorous mathematical proof.

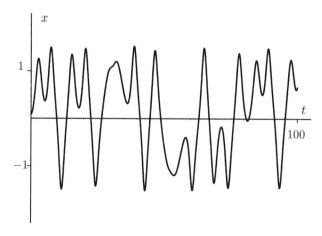

Figure 17. Plot x as a function of t for system (7.2), with $F = 0.4$, $\delta = 0.25$, and $\omega = 1$

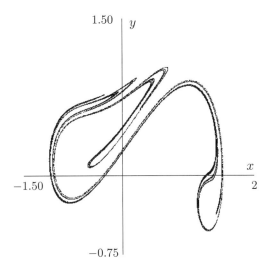

Figure 18. Plot of one orbit of Poincaré map for system (7.2), with $F = 0.4$, $\delta = 0.25$, and $\omega = 1$

Example 7.22. In this example, we add forcing to Example 7.15 and consider the system of equations

(7.3)
$$\dot{\tau} = 1 \quad (\text{mod } 2\pi),$$
$$\dot{x} = y,$$
$$\dot{y} = x - 2x^3 + y(x^2 - x^4 - y^2) + F\cos(\tau).$$

For these equations, an arbitrarily small forcing term makes the stable and the unstable manifolds of the saddle periodic orbit (which was formerly at the origin) become transverse rather than coincide. When this happens, the numerical result is a chaotic attractor with a dense orbit. See Figure 19. This indicates that a "generic" perturbation of the system of Example 7.15 results in a system with a chaotic attractor.

This system has stretching and folding like the Rössler system. The stretching comes from orbits coming near the fixed point and being stretched apart by the expansion in the unstable direction. The folding is caused by the different locations of orbits when the periodic forcing changes its direction.

Exercises 7.5

1. Consider the forced Duffing system of differential equations
$$\dot{\tau} = 1 \quad (\text{mod } {}^{2\pi}/\omega),$$
$$\dot{x} = y,$$
$$\dot{y} = x - x^3 - \delta y + F\cos(\omega \tau).$$

   Using the divergence, show that any attractor for $\delta > 0$ must have zero volume, or zero area for the Poincaré map.

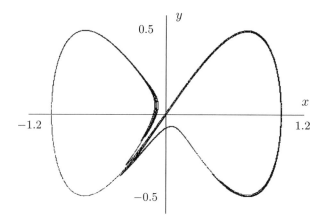

Figure 19. Plot of one orbit of Poincaré map for System (7.3), with $F = 0.01$

2. Show that the attractor for the system of equations
$$\dot{\tau} = 1 \quad (\bmod\ 2\pi),$$
$$\dot{x} = y,$$
$$\dot{y} = x - 2x^3 + y(x^2 - x^4 - y^2) + F\cos(\tau)$$
must have zero volume, or zero area for the Poincaré map.

3. Consider the forced system of equations
$$\dot{\tau} = 1 \quad (\bmod\ 2\pi),$$
$$\dot{x} = y,$$
$$\dot{y} = x^3 - \delta y + F\cos(\omega\tau).$$

Using a computer, show numerically that the system has a chaotic attractor for $\delta = 0.1$ and $F = 12$.

7.6. Lyapunov Exponents

We have discussed sensitive dependence on initial conditions and how it relates to chaos. It is well defined mathematically, can be verified in a few model cases (see Theorems 7.4 and 7.5), and it can be observed in numerical studies. In order to quantify the rate at which nearby points move apart, we introduce the related concept of Lyapunov exponents. They are more computable with computer simulations than sensitive dependence and arises from considering infinitesimal displacements. They measure the average divergence rate per unit time of infinitesimal displacements and generalize the concept of eigenvalues of a fixed point and characteristic multiplier of a periodic orbit.

By using the linear approximation for fixed t, we get
$$\phi(t; \mathbf{y}_0) - \phi(t; \mathbf{x}_0) \approx D_{\mathbf{x}}\phi_{(t;\mathbf{x}_0)}(\mathbf{y}_0 - \mathbf{x}_0).$$

7.6. Lyapunov Exponents

for small displacements $\mathbf{y}_0 - \mathbf{x}_0$. If $\|\mathbf{y}_0 - \mathbf{x}_0\|$ is "infinitesimally small", then the "infinitesimal displacement" at time t is $\mathbf{v}(t) = D_{\mathbf{x}}\phi_{(t;\mathbf{x}_0)}(\mathbf{y}_0 - \mathbf{x}_0)$. To make the "infinitesimal displacements" rigorous, we consider tangent vectors to curves of initial conditions, $\mathbf{x}_s$ of initial conditions. Letting

$$\mathbf{v}_0 = \left.\frac{\partial \mathbf{x}_s}{\partial s}\right|_{s=0} \quad \text{and}$$

$$\mathbf{v}(t) = \left.\frac{\partial}{\partial s}\phi(t; \mathbf{x}_s)\right|_{s=0} = \left.D_{\mathbf{x}}\phi_{(t;\mathbf{x}_0)}\frac{\partial \mathbf{x}_s}{\partial s}\right|_{s=0} = D_{\mathbf{x}}\phi_{(t;\mathbf{x}_0)}\mathbf{v}_0,$$

then $\mathbf{v}(t)$ satisfies the first variation equation, Theorem 3.4,

$$(7.4) \qquad \frac{d}{dt}\mathbf{v}(t) = D\mathbf{F}_{(\phi(t;\mathbf{x}_0))}\mathbf{v}(t).$$

The growth rate of $\|\mathbf{v}(t)\|$ is a number ℓ such that

$$\|\mathbf{v}(t)\| \approx C\,e^{\ell t}.$$

Taking logarithms, we want

$$\ln\|\mathbf{v}(t)\| \approx \ln(C) + \ell\,t,$$

$$\frac{\ln\|\mathbf{v}(t)\|}{t} \approx \frac{\ln(C)}{t} + \ell,$$

$$\ell = \lim_{t\to\infty}\frac{\ln\|\mathbf{v}(t)\|}{t}.$$

We use this last equation to define the Lyapunov exponent for initial condition $\mathbf{x}_0$ and initial infinitesimal displacement $\mathbf{v}_0$. There can be more than one Lyapunov exponent for a given initial condition $\mathbf{x}_0$; in fact, there are usually n Lyapunov exponents counting multiplicity.

Definition 7.23. Let $\mathbf{v}(t)$ be the solution of the first variation equation (7.4), starting at $\mathbf{x}_0$ with $\mathbf{v}(0) = \mathbf{v}_0$. The *Lyapunov exponent* for initial condition $\mathbf{x}_0$ and initial infinitesimal displacement $\mathbf{v}_0$ is defined to be

$$\ell(\mathbf{x}_0; \mathbf{v}_0) = \lim_{t\to\infty}\frac{\ln\|\mathbf{v}(t)\|}{t},$$

whenever this limit exists.

For most initial conditions $\mathbf{x}_0$ for which the forward orbit is bounded, the Lyapunov exponents exist for all vectors $\mathbf{v}$. For a system in n dimensions and an initial condition $\mathbf{x}_0$, there are at most n distinct values for $\ell(\mathbf{x}_0; \mathbf{v})$ as $\mathbf{v}$ varies. If we count multiplicities, then there are exactly n values, $\ell_1(\mathbf{x}_0) = \ell(\mathbf{x}_0; \mathbf{v}_1)$, $\ell_2(\mathbf{x}_0) = \ell(\mathbf{x}_0; \mathbf{v}_2)$, ..., $\ell_n(\mathbf{x}_0) = \ell(\mathbf{x}_0; \mathbf{v}_n)$. We can order these so that

$$\ell_1(\mathbf{x}_0) \geq \ell_2(\mathbf{x}_0) \geq \cdots \geq \ell_n(\mathbf{x}_0).$$

The Oseledec multiplicative ergodic theorem 13.17, which we discuss in term of iteration of functions in Section 13.5, is the mathematical basis of the Lyapunov exponents.

First, we consider the Lyapunov exponents at a fixed point.

Theorem 7.6. *Assume that $\mathbf{x}_0$ is a fixed point of the differential equation. Then, the Lyapunov exponents at the fixed point are the real parts of the eigenvalues of the fixed point.*

Proof. We do not give a formal proof, but consider the different possibilities. First assume that $\mathbf{v}^j$ is an eigenvector for the real eigenvalue λ_j. Then, the solution is

$$\mathbf{v}(t) = e^{\lambda_j t} \mathbf{v}^j.$$

Taking the limit of the quantity involved, we get

$$\begin{aligned}\ell(\mathbf{x}_0; \mathbf{v}^j) &= \lim_{t \to \infty} \frac{\ln \|\mathbf{v}(t)\|}{t} \\ &= \lim_{t \to \infty} \frac{\ln(e^{\lambda_j t}) + \ln \|\mathbf{v}^j\|}{t} \\ &= \lambda_j + \lim_{t \to \infty} \frac{\ln \|\mathbf{v}^j\|}{t} \\ &= \lambda_j.\end{aligned}$$

The third equality holds because $\ln \|\mathbf{v}^j\|$ is constant and the denominator t goes to infinity.

If $\lambda_j = \alpha_j + i\beta_j$ is a complex eigenvalue, then there is a solution of the form

$$\mathbf{v}(t) = e^{\alpha_j t} \Big(\cos(\beta_j t) \mathbf{u}^j + \sin(\beta_j t) \mathbf{w}^j \Big).$$

Taking the limit, we have

$$\begin{aligned}\ell(\mathbf{x}_0; \mathbf{u}^j) &= \lim_{t \to \infty} \frac{\ln \|\mathbf{v}(t)\|}{t} \\ &= \lim_{t \to \infty} \frac{\ln(e^{\alpha_j t})}{t} + \frac{\ln \|\cos(\beta_j t)\mathbf{u}^j + \sin(\beta_j t)\mathbf{w}^j\|}{t} \\ &= \alpha_j + \lim_{t \to \infty} \frac{\ln \|\cos(\beta_j t)\mathbf{u}^j + \sin(\beta_j t)\mathbf{w}^j\|}{t} \\ &= \alpha_j.\end{aligned}$$

The last equality holds because the numerator in the limit oscillates but is bounded so the limit is zero.

Finally, in the case in which there is a repeated real eigenvalue, there is a solution of the form

$$\mathbf{w}(t) = e^{\lambda_j t} \mathbf{w}^j + t e^{\lambda_j t} \mathbf{v}^j.$$

Taking the limit, we get

$$\begin{aligned}\ell(\mathbf{x}_0; \mathbf{w}^j) &= \lim_{t \to \infty} \frac{\ln \|\mathbf{w}(t)\|}{t} \\ &= \lim_{t \to \infty} \frac{\ln(e^{\lambda_j t})}{t} + \frac{\ln \|\mathbf{w}^j + t\mathbf{v}^j\|}{t} \\ &= \lambda_j + \lim_{t \to \infty} \frac{\ln \|\mathbf{w}^j + t\mathbf{v}^j\|}{t} \\ &= \lambda_j.\end{aligned}$$

The last equality holds because the numerator of the last term grows as $\ln(t)$ and the denominator grows as t, so the limit is 0.

If we have the sum of two solutions of the preceding form, the rate of growth is always the larger value. For example, if $\mathbf{v}(t) = C_1 e^{\lambda_1 t} \mathbf{v}^1 + C_2 e^{\lambda_2 t} \mathbf{v}^2$, with both

7.6. Lyapunov Exponents

$C_1, C_2 \neq 0$ and $\lambda_1 > \lambda_2$, then

$$\ell(\mathbf{x}_0; C_1\mathbf{v}^1 + C_2\mathbf{v}^2) = \lim_{t\to\infty} \frac{\ln\|\mathbf{v}(t)\|}{t}$$
$$= \lim_{t\to\infty} \frac{\ln(e^{\lambda_1 t})}{t} + \frac{\ln\|C_1\mathbf{v}^1 + C_2 e^{(\lambda_2-\lambda_1)t}\mathbf{v}^2\|}{t}$$
$$= \lambda_1 + \lim_{t\to\infty} \frac{\ln\|C_1\mathbf{v}^1 + C_2 e^{(\lambda_2-\lambda_1)t}\mathbf{v}^2\|}{t}$$
$$= \lambda_1.$$

The last equality holds because $\lambda_2 - \lambda_1$ is negative, so the term $e^{(\lambda_2-\lambda_1)t}$ goes to zero, the numerator is bounded, and the limit is zero. $\square$

When we consider solutions which are not fixed points, the vector in the direction of the orbit usually corresponds to a Lyapunov exponent zero.

Theorem 7.7. *Let $\mathbf{x}_0$ be an initial condition such that $\phi(t; \mathbf{x}_0)$ is bounded and $\omega(\mathbf{x}_0)$ does not contain any fixed points. Then,*

$$\ell(\mathbf{x}_0; \mathbf{F}(\mathbf{x}_0)) = 0.$$

This says that there is no growth or decay in the direction of the vector field, $\mathbf{v} = \mathbf{F}(\mathbf{x}_0)$.

Proof. The idea is that

$$\mathbf{F}(\phi(t;\mathbf{x}_0)) = \frac{\partial}{\partial s}\phi(s;\phi(t;\mathbf{x}_0))\Big|_{s=0} = \frac{\partial}{\partial s}\phi(s+t;\mathbf{x}_0)\Big|_{s=0}$$
$$= \frac{\partial}{\partial s}\phi(t;\phi(s;\mathbf{x}_0))\Big|_{s=0} = D_{\mathbf{x}}\phi_{(t;\mathbf{x}_0)}\frac{\partial}{\partial s}\phi(s;\mathbf{x}_0)\Big|_{s=0}$$
$$= D_{\mathbf{x}}\phi_{(t;\mathbf{x}_0)}\mathbf{F}(\mathbf{x}_0).$$

This says that the linearization of the flow at time t takes the vector field at the initial condition to the vector field at the position of the flow at time t. If we let $\mathbf{x}_s = \phi(s;\mathbf{x}_0)$, then

$$\mathbf{v}(t) = \frac{\partial}{\partial s}\phi(t;\mathbf{x}_s)\Big|_{s=0} = \mathbf{F}(\phi(t;\mathbf{x}_0))$$

is the solution of the first variation equation; that is,

$$\frac{d}{dt}\mathbf{F}(\phi(t;\mathbf{x}_0)) = D\mathbf{F}_{(\phi(t;\mathbf{x}_0))}\mathbf{F}(\phi(t;\mathbf{x}_0)).$$

If the orbit is bounded and does not have a fixed point in its ω-limit set, then the quantity $\|\mathbf{F}(\phi(t;\mathbf{x}_0))\|$ is bounded and bounded away from 0. Therefore,

$$\ell(\mathbf{x}_0;\mathbf{F}(\mathbf{x}_0)) = \lim_{t\to\infty}\frac{\ln\|D_{\mathbf{x}}\phi_{(t;\mathbf{x}_0)}\mathbf{F}(\mathbf{x}_0)\|}{t} = \lim_{t\to\infty}\frac{\ln\|\mathbf{F}(\phi(t;\mathbf{x}_0))\|}{t} = 0.$$

$\square$

Definition 7.24. Let $\mathbf{x}_0$ be in initial condition such that $\phi(t;\mathbf{x}_0)$ is bounded and the Lyapunov exponent in the direction $\mathbf{F}(\mathbf{x}_0)$ is zero. If the other $n-1$ Lyapunov exponents exist, then they are called the *principal Lyapunov exponents*.

Theorem 7.8. *Let $\mathbf{x}_0$ be an initial condition on a periodic orbit of period T. Then, the $(n-1)$ principal Lyapunov exponents are given by the $(\ln|\lambda_j|)/T$, where λ_j are the characteristic multipliers of the periodic orbit and the eigenvalues of the Poincaré map.*

We calculate the Lyapunov exponents for a periodic orbit in a simple example.

Example 7.25. Consider
$$\dot{x} = -y + x(1 - x^2 - y^2),$$
$$\dot{y} = x - y(1 - x^2 - y^2).$$

This has a periodic orbit $(x(t), y(t)) = (\cos t, \sin t)$ of radius one. Along this orbit the matrix of partial derivatives is given by

$$\mathbf{A}(t) = \begin{pmatrix} 1 - r^2 - 2x^2 & -1 - 2yx \\ 1 - 2xy & 1 - r^2 - 2y^2 \end{pmatrix}$$
$$= \begin{pmatrix} -2\cos^2 t & -1 - 2\cos t \sin(t) \\ 1 - 2\cos t \sin t & -2\sin^2 t \end{pmatrix}.$$

In this form, the solutions of the equation $\dot{\mathbf{v}}(t) = \mathbf{A}(t)\mathbf{v}(t)$ are not obvious. However, a direct check shows that

$$\mathbf{v}(t) = e^{-2t}\begin{pmatrix} \cos t \\ \sin t \end{pmatrix} \quad \text{and} \quad \begin{pmatrix} -\sin(t) \\ \cos t \end{pmatrix}$$

satisfy $\frac{d}{dt}\mathbf{v}(t) = \mathbf{A}(t)\mathbf{v}(t)$ (i.e., they are solutions of the time-dependent linear differential equation). (The second expression is the vector along the orbit.) Since $(x(0), y(0)) = (1, 0)$,

$$e^{-2t}\begin{pmatrix} \cos t \\ \sin t \end{pmatrix}\bigg|_{t=0} = \begin{pmatrix} 1 \\ 0 \end{pmatrix} \quad \text{and}$$
$$\begin{pmatrix} -\sin(t) \\ \cos t \end{pmatrix}\bigg|_{t=0} = \begin{pmatrix} 0 \\ 1 \end{pmatrix},$$

one Lyapunov exponent is

$$\ell((1,0);(1,0)) = \lim_{t \to \infty} \frac{1}{t} \ln \left\| e^{-2t}\begin{pmatrix} \cos t \\ \sin t \end{pmatrix} \right\|$$
$$= \lim_{t \to \infty} \frac{\ln(e^{-2t})}{t}$$
$$= -2,$$

and the other Lyapunov exponent is

$$\ell((1,0);(0,1)) = \lim_{t \to \infty} \frac{1}{t} \ln \left\| \begin{pmatrix} -\sin(t) \\ \cos t \end{pmatrix} \right\|$$
$$= \lim_{t \to \infty} \frac{\ln(1)}{t}$$
$$= 0.$$

7.6. Lyapunov Exponents

The Lyapunov exponents are easier to calculate using polar coordinates, where
$$\dot{r} = r(1 - r^2),$$
$$\dot{\theta} = 1.$$
At $r = 1$, the partial derivative in the r variable is
$$\left.\frac{\partial}{\partial r}r(1-r^2)\right|_{r=1} = \left.1 - 3r^2\right|_{r=1} = -2.$$
Therefore, the linearized equation of the component in the r direction is
$$\frac{d}{dt}v_r = -2v_r,$$
so, $v_r(t) = e^{-2t}v_r(0)$. Again, the growth rate is -2. The angular component of a vector v_θ satisfies the equation,
$$\frac{d}{dt}v_\theta = 0,$$
and the growth rate in the angular direction is 0. Thus, the two Lyapunov exponents are -2 and 0.

The following theorem states that two orbits that are exponentially asymptotic have the same Lyapunov exponents.

Theorem 7.9. a. *Assume that $\phi(t; \mathbf{x}_0)$ and $\phi(t; \mathbf{y}_0)$ are two orbits for the same differential equation, which are bounded. and whose separation converges exponentially to zero (i.e., there are constants $-a < 0$ and $C \geq 1$ such that*
$$\|\phi(t; \mathbf{x}_0) - \phi(t; \mathbf{y}_0)\| \leq C e^{-at}$$
for $t \geq 0$). Then, the Lyapunov exponents for $\mathbf{x}_0$ and $\mathbf{y}_0$ are the same. So, if the limits defining the Lyapunov exponents exist for one of the points, they exist for the other point. The vectors which give the various Lyapunov exponents can be different at the two points.

b. *In particular, if an orbit is asymptotic to (i) a fixed point with all the eigenvalues having nonzero real parts or (ii) a periodic orbit with all the characteristic multipliers of absolute value not equal to zero, then the Lyapunov exponents are determined by those at the fixed point or periodic orbit.*

The Liouville formula given in Theorem 6.10 implies that the sum of the Lyapunov exponents gives the growth rate of the volume. This leads to the next theorem, which gives the connection between the sum of the Lyapunov exponents and the divergence.

Theorem 7.10. *Consider the system of differential equations $\dot{\mathbf{x}} = \mathbf{F}(\mathbf{x})$ in $\mathbb{R}^n$. Assume that $\mathbf{x}_0$ is a point such that the Lyapunov exponents $\ell_1(\mathbf{x}_0), \ldots \ell_n(\mathbf{x}_0)$ exist,*

a. *Then, the sum of the Lyapunov exponents is the limit of the average of the divergence along the trajectory,*
$$\sum_{j=1}^n \ell_j(\mathbf{x}_0) = \lim_{T \to \infty} \frac{1}{T} \int_0^T \nabla \cdot \mathbf{F}_{\phi(t;\mathbf{x}_0)} \, dt.$$

b. *In particular, if the system has constant divergence δ, then the sum of the Lyapunov exponents at any point must equal δ.*

c. In three dimensions, assume that the divergence is a constant δ and that $\mathbf{x}_0$ is a point for which the positive orbit is bounded and $\omega(\mathbf{x}_0)$ does not contain any fixed points. If $\ell_1(\mathbf{x}_0)$ is a nonzero Lyapunov exponent at $\mathbf{x}_0$, then the other two Lyapunov exponents are 0 and $\delta - \ell_1$.

Applying the preceding theorem to the Lorenz system, the divergence is $-\sigma - 1 - b$, which is negative, so the sum of the Lyapunov exponents must be negative. For the parameter values $\sigma = 10$ and $b = 8/3$, the sum of the exponents is -13.67. Numerical calculations have found $\ell_1 = 0.90$ to be positive (an expanding direction), $\ell_2 = 0$ along the orbit, so $\ell_3 = -13.67 - \ell_1 = -14.57$ must be negative (a contracting direction). Since $\ell_1 + \ell_3$ is equal to the divergence, to determine all the Lyapunov exponents, it is necessary only to calculate one of these values numerically.

7.6.1. Numerical Calculation. For most points $\mathbf{x}_0$ for which the forward orbit is bounded, the Lyapunov exponents exist. Take such a point. Then for most $\mathbf{v}_0$, $\ell(\mathbf{x}_0, \mathbf{v}_0)$ is the largest Lyapunov exponent, because the vector $D\phi_{(t,\mathbf{x}_0)}\mathbf{v}_0$ tends toward the direction of greatest expansion as t increases. (Look at the flow near a linear saddle. Unless the initial condition $\mathbf{v}_0$ is on the stable manifold, $e^{\mathbf{A}t}\mathbf{v}_0$ tends toward the unstable manifold as t goes to infinity.) Thus, the largest Lyapunov exponent can be calculated numerically by integrating the differential equation and the first variation equation at the same time,

$$\dot{\mathbf{x}} = \mathbf{F}(\mathbf{x}),$$
$$\dot{\mathbf{v}} = D\mathbf{F}_{(\mathbf{x})}\mathbf{v}.$$

The growth rate of the resulting vector $\mathbf{v}(t)$ gives the largest Lyapunov exponent,

$$\ell_1(\mathbf{x}_0) = \ell(\mathbf{x}_0, \mathbf{v}_0) = \lim_{t \to \infty} \frac{\|\mathbf{v}(t)\|}{t}.$$

It is more difficult to calculate the second or other Lyapunov exponents. We indicate the process to get all n exponents for a differential equation in $\mathbb{R}^n$.

First, we explain another way of understanding the smaller Lyapunov exponents. Let $J_k = D\phi_{(k;\mathbf{x}_0)}$ be the matrix of the derivative of the flow at time k. Let $\mathbf{U}$ be the set of all vectors of length one,

$$\mathbf{U} = \{\mathbf{x} : \|\mathbf{x}\| = 1\}.$$

The set $\mathbf{U}$ is often called the "sphere" in $\mathbb{R}^n$. The image of $\mathbf{U}$ by J_k $J_k(\mathbf{U})$ is an ellipsoid with axis of lengths $r_k^{(1)}, \ldots, r_k^{(n)}$. (If the matrix J_k were symmetric, it would be fairly clear that the image was an ellipse from topics usually covered in a first course in linear algebra. For nonsymmetric matrices, it is still true. It can be shown by considering $J_k^T J_k$ which is symmetric, where J_k^T is the transpose.) The vectors which go onto the axes of the ellipsoid change with k. However, for most points,

$$\ell_j(\mathbf{x}_0) = \lim_{k \to \infty} \frac{\ln(r_k^{(j)})}{k}.$$

Rather than considering the ellipsoids in all of $\mathbb{R}^n$, we consider only the ellipsoid determined by j longest directions within $J_k(\mathbf{U})$, which we label by $E_k^{(j)}$. The volume of $E_k^{(j)}$ is a constant involving π times $r_k^{(1)} \cdots r_k^{(j)}$, which is approximately

7.6. Lyapunov Exponents

$e^{k(\ell_1+\cdots+\ell_j)}$. Thus, the growth rate of the volume of the ellipsoids $E_k^{(j)}$ as k goes to infinity is the sum of the first j Lyapunov exponents, given by

$$\lim_{k\to\infty} \frac{\text{vol}(E_k^{(j)})}{k} = \ell_1 + \cdots + \ell_j.$$

Thus, we can determine the growth rate of the area of ellipses defined by two vectors; this gives the sum of the first two Lyapunov exponents.

If we take n vectors $\mathbf{v}_0^1 \ldots \mathbf{v}_0^n$, then each vector $D\phi_{(t,\mathbf{x}_0)}\mathbf{v}_0^j$ most likely tends toward the direction of greatest increase, so $\ell(\mathbf{x}_0, \mathbf{v}_0^j) = \ell_1(\mathbf{x}_0)$ are all the same value. However, the area of the parallelogram spanned by $D\phi_{(t,\mathbf{x}_0)}\mathbf{v}_0^1$ and $D\phi_{(t,\mathbf{x}_0)}\mathbf{v}_0^2$ grows at a rate equal to the sum of the two largest Lyapunov exponents, and the volume of the parallelepiped spanned by $\{ D\phi_{(t,\mathbf{x}_0)}\mathbf{v}_0^1, \ldots, D\phi_{(t,\mathbf{x}_0)}\mathbf{v}_0^j \}$ grows at a rate equal to the sum of the first j largest Lyapunov exponents. To calculate this growth, the set of vectors in converted into a set of perpendicular vectors by means of the Gram–Schmidt orthogonalization process. Thus, we take n initial vectors $\mathbf{v}_0^1 \ldots \mathbf{v}_0^n$, a point $\mathbf{x}_0$, and initial time $t_0 = 0$. We calculate

$$\mathbf{x}_1 = \phi(1; \mathbf{x}_0) \quad \text{and}$$
$$\mathbf{w}^j = D\phi_{(1;\mathbf{x}_0)}\mathbf{v}_0^j \quad \text{for } 1 \leq j \leq n$$

by solving the system of equations

$$\dot{\mathbf{x}} = \mathbf{F}(\mathbf{x}),$$
$$\dot{\mathbf{w}}^1 = D\mathbf{F}_{(\mathbf{x})} \mathbf{w}^1,$$
$$\vdots \quad \vdots$$
$$\dot{\mathbf{w}}^n = D\mathbf{F}_{(\mathbf{x})} \mathbf{w}^n,$$

for $0 = t_0 \leq t \leq t_0 + 1 = 1$, with initial conditions

$$\mathbf{x}(0) = \mathbf{x}_0,$$
$$\mathbf{w}^j(0) = \mathbf{v}_0^j \quad \text{for } 1 \leq j \leq n.$$

Then, we make this into a set of perpendicular vectors by means of the Gram–Schmidt orthogonalization process, setting

$$\mathbf{v}_1^1 = \mathbf{w}^1,$$
$$\mathbf{v}_1^2 = \mathbf{w}^2 - \left(\frac{\mathbf{w}^2 \cdot \mathbf{v}_1^1}{\|\mathbf{v}_1^1\|^2}\right)\mathbf{v}_1^1,$$
$$\mathbf{v}_1^3 = \mathbf{w}^3 - \left(\frac{\mathbf{w}^3 \cdot \mathbf{v}_1^1}{\|\mathbf{v}_1^1\|^2}\right)\mathbf{v}_1^1 - \left(\frac{\mathbf{w}^3 \cdot \mathbf{v}_1^2}{\|\mathbf{v}_1^2\|^2}\right)\mathbf{v}_1^2,$$
$$\vdots \quad \vdots$$
$$\mathbf{v}_1^n = \mathbf{w}^n - \left(\frac{\mathbf{w}^n \cdot \mathbf{v}_1^1}{\|\mathbf{v}_1^1\|^2}\right)\mathbf{v}_1^1 - \cdots - \left(\frac{\mathbf{w}^n \cdot \mathbf{v}_1^{n-1}}{\|\mathbf{v}_1^{n-1}\|^2}\right)\mathbf{v}_1^{n-1}.$$

The area of the parallelogram spanned by $\mathbf{v}_1^1$ and $\mathbf{v}_1^2$ is the same as the area for $\mathbf{w}^1$ and $\mathbf{w}^2$; for $3 \leq j \leq n$, the volume of the parallelepiped spanned by $\mathbf{v}_1^1 \ldots \mathbf{v}_1^j$ is the same as the volume for $\mathbf{w}^1 \ldots \mathbf{w}^j$. Therefore, we can use these vectors to calculate the growth rate of the volumes. We set $t_1 = t_0 + 1 = 1$.

We repeat the steps by induction. Assume that we have found the point $\mathbf{x}_k$, the n vectors $\mathbf{v}_k^1 \ldots \mathbf{v}_k^n$, and the time $t_k = k$. We calculate

$$\mathbf{x}_{k+1} = \phi(1; \mathbf{x}_k) \quad \text{and}$$
$$\mathbf{w}^j = D\phi_{(1;\mathbf{x}_k)} \mathbf{v}_k^j \quad \text{for } 1 \leq j \leq n$$

by solving the system of equations

$$\dot{\mathbf{x}} = \mathbf{F}(\mathbf{x}),$$
$$\dot{\mathbf{w}}^1 = D\mathbf{F}_{(\mathbf{x})} \mathbf{w}^1,$$
$$\vdots \quad \vdots$$
$$\dot{\mathbf{w}}^n = D\mathbf{F}_{(\mathbf{x})} \mathbf{w}^n,$$

for $0 = t_k \leq t \leq t_k + 1 = k + 1$, with initial conditions

$$\mathbf{x}(k) = \mathbf{x}_k,$$
$$\mathbf{w}^j(k) = \mathbf{v}_k^j \quad \text{for } 1 \leq j \leq n.$$

Then, we make this into a set of perpendicular vectors by setting

$$\mathbf{v}_{k+1}^1 = \mathbf{w}^1,$$
$$\mathbf{v}_{k+1}^2 = \mathbf{w}^2 - \left(\frac{\mathbf{w}^2 \cdot \mathbf{v}_{k+1}^1}{\|\mathbf{v}_{k+1}^1\|^2}\right) \mathbf{v}_{k+1}^1,$$
$$\mathbf{v}_{k+1}^3 = \mathbf{w}^3 - \left(\frac{\mathbf{w}^3 \cdot \mathbf{v}_{k+1}^1}{\|\mathbf{v}_{k+1}^1\|^2}\right) \mathbf{v}_{k+1}^1 - \left(\frac{\mathbf{w}^3 \cdot \mathbf{v}_{k+1}^2}{\|\mathbf{v}_{k+1}^2\|^2}\right) \mathbf{v}_{k+1}^2,$$
$$\vdots \quad \vdots$$
$$\mathbf{v}_{k+1}^n = \mathbf{w}^n - \left(\frac{\mathbf{w}^n \cdot \mathbf{v}_{k+1}^1}{\|\mathbf{v}_{k+1}^1\|^2}\right) \mathbf{v}_{k+1}^1 - \cdots - \left(\frac{\mathbf{w}^n \cdot \mathbf{v}_{k+1}^{n-1}}{\|\mathbf{v}_{k+1}^{n-1}\|^2}\right) \mathbf{v}_{k+1}^{n-1}.$$

We set $t_{k+1} = t_k + 1 = k + 1$. This completes the description of the process of constructing the vectors.

The growth rate of the length of the vectors $\mathbf{v}_k^1$ usually gives the largest Lyapunov exponent, and the growth rate of the length of vectors $\mathbf{v}_k^j$ give the jth largest Lyapunov exponent,

$$\ell_j(\mathbf{x}_0) = \lim_{k \to \infty} \frac{\|\mathbf{v}_k^j\|}{k} \quad \text{for } 1 \leq j \leq n.$$

Exercises 7.6

1. For a system of differential equations with constant coefficients, the sum of the Lyapunov exponents equals the divergence. What is the sum of the exponents

for the following double-well oscillator?
$$\dot{s} = 1 \pmod{2\pi/\omega},$$
$$\dot{x} = y,$$
$$\dot{y} = x - x^3 - \delta y + F\cos(\omega s).$$

The variable s is a modulo $2\pi/\omega$.

2. Consider the undamped system
$$\dot{x} = y,$$
$$\dot{y} = -\frac{\partial V}{\partial x}(x).$$

What are the Lyapunov exponents for any periodic orbit?

3. Consider the system of differential equations
$$\dot{x} = x + y - x(x^2 + y^2),$$
$$\dot{y} = -x + y - y(x^2 + y^2),$$
$$\dot{z} = -z.$$

   a. Show that the circle $x^2 + y^2 = 1$ and $z = 0$ is a periodic orbit.
   b. Determine the Lyapunov exponents of a point $(x_0, y_0, 0)$ with $x_0^2 + y_0^2 = 1$.
   c. What are the Lyapunov exponents of a point $(x_0, y_0, 0)$ with $x_0^2 + y_0^2 \neq 0$?

7.7. Test for Chaotic Attractors

We have introduced Lyapunov exponents as a more computable test for sensitive dependence on initial conditions. In this section, we return to the concept of a chaotic attractor and describe a test that is more easily verifiable for experiments or numerical work. We prefer to take the more mathematical version given earlier as our definition and take the conditions given here as a test. For "generic" systems this test probably implies our definition.

Theorem 7.7 shows that infinitesimal perturbations in the direction of the orbit usually have zero growth rate. However, the other $n-1$ directions can either expand or contract, and these $n-1$ Lyapunov exponents are called the *principal Lyapunov exponents*.

If one of the exponents is positive, then small displacements in that direction tend to grow exponentially in length (at least infinitesimally).

Test for a Chaotic Attractor

An experimental criterion for a set $\mathbf{A}$ to be a *chaotic attractor* for a flow $\phi(t; \mathbf{x})$ is that there are several initial conditions $\mathbf{x}_0, \ldots, \mathbf{x}_k$ with distinct orbits for which the following conditions hold for $0 \leq i \leq k$:

(1) The ω-limit set of each of these points equals to all of $\mathbf{A}$, $\omega(\mathbf{x}_i) = \mathbf{A}$.
(2) At least one of the Lyapunov exponents for each $\mathbf{x}_i$ is positive, $\ell_1(\mathbf{x}_i) > 0$.
(3) All the $n-1$ principal Lyapunov exponents $\ell_j(\mathbf{x}_i)$ are nonzero.

Remark 7.26. The fact that there are many distinct points with the same ω-limit set, indicates that the measure of the basin of attraction is positive and **A** is a Milnor attractor. There may be a trapping region for the ω-limit set, however, this is not easily verifiable experimentally or numerically.

If this set were a true attracting set, then the fact that $\omega(\mathbf{x}_i) = \mathbf{A}$ would imply that it is an attractor. We do not require that the points $\mathbf{x}_i$ are in **A**, so we do not necessarily have that the flow is topologically transitive on **A**. However, the fact that the ω-limits are all of **A** make it plausible that it is transitive.

In this criterion, a positive Lyapunov exponent replaces the assumption of sensitive dependence on initial conditions used in the more mathematical definition given in Section 7.2.

Most systems satisfying the preceding test also satisfy the earlier definition. Thus, we tacitly assume that the system is "generic" even though this is not really verifiable. For a "generic" system, the periodic points are dense in **A**, and the stable and unstable manifolds of fixed points and periodic orbits cross each other. We present a discussion of generic properties in Appendix B.

The system of differential equations with chaotic attractors that we have consider certainly satisfy this criterion: Lorenz system, Rössler system, and the periodically forced oscillator.

Example 7.16 is not transitive and it fails to satisfy all the assumptions of this Test for a Chaotic Attractor: This system has three Lyapunov exponents equal to zero and one negative. Example 7.15 satisfies the first assumption of this criterion, but numerical integration indicates that the Lyapunov exponents are zero for this system so it fails the test. We noted earlier that it is not transitive and does not have sensitive dependence on initial conditions when restricted to the attractor.

Example 7.27 (Quasiperiodic attractor revised). Consider the system of differential equations

$$\dot{x}_1 = y_1 + x_1(1 - x_1^2 - y_1^2),$$
$$\dot{y}_1 = -x_1 + y_1(1 - x_1^2 - y_1^2),$$
$$\dot{x}_2 = \sqrt{2}\,y_2 + 3x_2(1 - x_2^2 - y_2^2),$$
$$\dot{y}_2 = -\sqrt{2}\,x_2 + 3y_2(1 - x_2^2 - y_2^2),$$

considered in Example 7.17. If we introduce polar coordinates $r_1^2 = x_1^2 + y_1^2$, $\tan(\theta_1) = y_1/x_1$, $r_2^2 = x_2^2 + y_2^2$, and $\tan(\theta_2) = y_2/x_2$, then the differential equations become

$$\dot{r}_1 = r_1(1 - r_1^2),$$
$$\dot{\theta}_1 = 1 \pmod{2\pi},$$
$$\dot{r}_2 = 3r_2(1 - r_2^2),$$
$$\dot{\theta}_2 = \sqrt{2} \pmod{2\pi}.$$

The set $r_1 = 1 = r_2$ is easily determined to be an attracting set. Since the motion on this set is quasiperiodic, there is a dense orbit and the set is an attractor. Since the system is the product of two systems, the Lyapunov exponents are the exponents of the two subsystems. The exponents in the r_1 and θ_1 system are 0 and -2, while those in r_2 and θ_2 system are 0 and -6; therefore, the total system has Lyapunov exponents -2, -6, 0, and 0. Since 0 appears as a double Lyapunov exponent, one of the principal Lyapunov exponents is 0 and the set fails the test to be a chaotic attractor. We showed earlier that it failed to satisfy our definition of a chaotic attractor. Since quasiperiodic motion is not "chaotic" in appearance, this classification makes sense.

Exercises 7.7

1. Consider the system
$$\dot{x} = y,$$
$$\dot{y} = x - 2x^3 - y\left(-x^2 + x^4 + y^2\right).$$
We showed in Section 7.2 that the set
$$\Gamma = \{(\tau, x, y) : -x^2 + x^4 + y^2 = 0\}$$
is a nonchaotic attractor.
   a. Show that, for any point (x_0, y_0) in Γ, the Lyapunov exponents are nonzero with one positive exponent. (Note that numerical integration shows that both Lyapunov exponents are zero for (x_0, y_0) not in Γ with $\omega(x_0, y_0) = \Gamma$.)
   b. What is nongeneric about this example?

2. Consider the system of differential equations given by
$$\dot{\tau} = 1 + \frac{1}{2}\sin(\theta) \pmod{2\pi}$$
$$\dot{\theta} = 0 \pmod{2\pi}$$
$$\dot{x} = y$$
$$\dot{y} = x - 2x^3 - y\left(-x^2 + x^4 + y^2\right).$$
Let $L(\tau, \theta, x, y) = \dfrac{-x^2 + x^4 + y^2}{2}$ be a test function and $\mathbf{A} = L^{-1}(0)$.
   a. Show that the system has sensitive dependence on initial conditions when restricted to $\mathbf{A}$.
   b. Discuss why $\mathbf{A}$ is a chaotic attractor by our definition. Does it seems chaotic?
   c. Explain why $\mathbf{A}$ does not pass the test for a chaotic attractor given in this section. In particular, for any point $(\tau_0, \theta_0, x_0, y_0)$ in $\mathbf{A}$, explain why there are two Lyapunov exponents equal to zero and two nonzero Lyapunov exponents with one positive exponent.

7.8. Applications

7.8.1. Lorenz System as a Model. As mentioned in the Historical Prologue, Edward Lorenz found a system with sensitive dependence in 1963. In the late 1950s, he was using a computer to study the nonlinear effects on weather forecasting. He used the computer to study the effects of change in various parameters on the outcome. He started with a very simplified model that used only 12 measurements or coordinates. The printing was done with a line printer, which was very slow, taking 10 seconds to print each line. To speed up the calculation as he followed the solution, he printed out only a selection of data points, every so many calculations, and he printed out the numbers to only three digits accuracy, even though the computer calculated numbers to six decimal places. He wanted to look more closely at one computer run by displaying more frequent data points, so he input the initial conditions part way through the computer run using the truncated values obtained from the computer print out. The values of the new computer run very quickly diverged from the original computer run.

After further study, Lorenz concluded that the source of the deviation of the two computer runs resulted from sensitive dependence on initial conditions; the small change in the initial conditions was amplified by the system of equations so that the two orbits became vastly different.

Lorenz wanted to simplify the equations further. Barry Saltzman had been studying the equations for convection of a fluid that is heated from below and he had a model that involved just seven variables. After further simplification, Lorenz published his paper with the system of three differential equations, which is now called the Lorenz system of differential equations.

We now turn to a discussion of the manner in which the Lorenz system of differential equations arises as a model for two different physical situations: for a waterwheel and for convection rolls in the atmosphere, which Lorenz originally considered. Although Lorenz originally considered convection rolls in the atmosphere, we start with a model for a waterwheel with similar equations, which is easier to visualize.

Model for a waterwheel

In this model, water is added to chambers on a waterwheel, which is on an incline, and water leaks out the bottom of each chamber. With correct choices of the rates, the wheel makes an apparently random number of rotations in one direction before reversing itself. Willem Malkus and Lou Howard not only derived the equations, but also built a physical apparatus. See [**Str94**] for further discussion.

A wheel that can rotate is put on an incline. There are chambers at each angle around the wheel that hold water. See Figure 20. Let $m(\theta, t)$ be the quantity of the water at angle θ and time t. The quantity $Q(\theta)$ is the rate at which water is added at the angle θ. The quantity K is the leakage rate per unit volume, so the total quantity flowing out is $-Km(\theta, t)$. The wheel is rotating at an angular rate of $\omega(t)$, so by comparing the mass coming in and out at the angle θ, the rate of change of m with respect to t at the angle θ contains a term $-\omega \frac{\partial m}{\partial \theta}(\theta, t)$. Combining, we get

$$(7.5) \qquad \frac{\partial m}{\partial t}(\theta, t) = Q(\theta) - K\, m(\theta, t) - \omega \frac{\partial m}{\partial \theta}(\theta, t).$$

The equation for $\dot\omega$ is determined by balancing the torque on the wheel. Let I be the moment of inertia, R be the radius of the wheel, and α be the angle at which the wheel is tilted. Let ν be the rotational damping, so the equation for $I\dot\omega$ contains

7.8. Applications

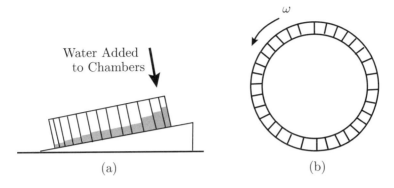

Figure 20. Apparatus for the waterwheel. (a) Side view. (b) Top view.

the term $-\nu\omega(t)$. The torque due to the mass at the angle θ is $g_\alpha \sin(\theta) m(\theta, t)$, where $g_\alpha = g\sin(\alpha)$, and the total torque is

$$g_\alpha \int_0^{2\pi} m(\theta, t) \sin(\theta)\, d\theta.$$

Combining the torque due to damping and the mass distribution, we have

(7.6) $$I\dot{\omega}(t) = -\nu\,\omega(t) + g_\alpha\, R \int_0^{2\pi} m(\theta, t) \sin(\theta)\, d\theta.$$

Equations (7.5) and (7.6) give the integro-differential equations of motion for the system.

The quantities m and Q are periodic in the angle θ; therefore, it is natural to express them in Fourier series, which are sums of quantities involving $\sin(n\theta)$ and $\cos(n\theta)$. Let

$$Q(\theta) = q_0 + \sum_{n=1}^{\infty} q_n \cos(n\theta) \quad \text{and}$$

$$m(\theta, t) = a_0(t) + \sum_{n=1}^{\infty} a_n(t)\cos(n\theta) + b_n(t)\sin(n\theta).$$

The quantity $Q(\theta)$ is given with known coefficients q_n; we assume that water is added symmetrically in θ, $Q(-\theta) = Q(\theta)$, so there are only the cosine terms and no sine terms. Because $\sin(0\,\theta) = 0$, there is no $b_0(t)$ term (i.e., $b_0(t) \equiv 0$). To evaluate the integrals, we use the following:

$$\int_0^{2\pi} \sin(\theta)\cos(n\theta)\, d\theta = 0 \quad \text{and}$$

$$\int_0^{2\pi} \sin(\theta)\sin(n\theta)\, d\theta = \begin{cases} 0 & \text{if } n \neq 1, \\ \pi & \text{if } n = 1. \end{cases}$$

Substituting the Fourier expansions and using the preceding integrals of sines and cosines, we get the following system of equations by equating the coefficients of

$\sin(n\theta)$ and of $\cos(n\theta)$:

(7.7)
$$\dot{a}_n(t) = -K\,a_n(t) - \omega\,n\,b_n(t) + q_n,$$
$$\dot{b}_n(t) = -K\,b_n(t) + \omega\,n\,a_n(t),$$
$$I\,\dot{\omega}(t) = -\nu\,\omega(t) + \pi\,g_\alpha\,R\,b_1(t).$$

The equation for $n = 0$ has $\dot{a}_0(t) = -Ka_0(t) + q_0$, so
$$a_0(t) = \big(a_0(0) - q_0/K\big)e^{-Kt} + q_0/K$$
goes to q_0/K as t goes to infinity. Thus, the average depth of water $b_0(t)$ is asymptotically equal to q_0/K (i.e., to the average rate of water flowing in divided by the leakage rate). The first interesting equations are for $n = 1$, which do not depend on the higher terms in the Fourier expansion:

(7.8)
$$\dot{a}_1(t) = -K\,a_1(t) - \omega\,b_1(t) + q_1,$$
$$\dot{b}_1(t) = -K\,b_1(t) + \omega\,a_1(t),$$
$$\dot{\omega}(t) = -\frac{\nu}{I}\omega(t) + \frac{\pi g_\alpha R}{I}b_1(t).$$

Notice that all the terms in this system of equations are linear except for two quadratic terms: $\omega a_1(t)$ in the $\dot{b}_1$ equation and $-\omega b_1$ in the $\dot{a}_1$ equation. People have worried about the effect of the higher Fourier coefficients on the type of dynamics. However, we do not pursue that discussion, but merely look at the lowest order terms.

This system of equations can be changed into a slight variation of the Lorenz system by proper scaling and substitution. Let
$$x = \omega, \quad y = B\,b_1, \quad z = r - B\,a_1, \quad \text{where}$$
$$B = \frac{\pi g_\alpha R}{\nu}, \quad r = \frac{\pi g_\alpha R q_1}{\nu K}, \quad \text{and} \quad \sigma = \frac{I}{\nu}.$$

A direct calculation shows that the system of equations becomes
$$\dot{x} = -\sigma x + \sigma y,$$
$$\dot{y} = r\,x - K\,y - x\,z,$$
$$\dot{z} = -K\,z + xy.$$

The difference between this system and the usual Lorenz system is the coefficients of y in the $\dot{y}$ equation and of z in the $\dot{x}$ equation are both $-K$, rather than -1 and $-b$ as in the Lorenz equations. This system has a nonzero fixed point provided that $r > K$, or
$$\frac{\pi g_\alpha R q_1}{\nu K^2} > 1.$$

Thus, the rate of inflow q_1 times the sine of the inclination of the waterwheel α must be larger than the rotational damping ν times the square of the leakage rate K.

The fact that water is being added to the chambers at the higher end of the waterwheel causes the wheel to "want" to turn to take them to the lower end. This causes the waterwheel to start to turn; so x, which corresponds to ω in the original variables, becomes nonzero. If the wheel turns fast enough, the chambers with

7.8. Applications

deeper water can be taken past the lower end before the water level drops very far. The chambers with deeper water are now rising which causes the waterwheel to slow down. If the rate of turning is just right, the waterwheel can reverse directions and start to turn in the opposite direction, so ω and x change sign.

If the parameters are just right (rate of adding water, angle of tipping the waterwheel, leakage rate, and damping torque), then we get some trajectories that progress through regions of space where x continues to change sign from positive to negative. Since x corresponds to ω in the original variables, this corresponds to motions in which the wheel alternates, going counterclockwise and clockwise. These are not just small oscillations; the wheel makes a number of rotations in one direction and then a number of rotations in the other direction. The number of rotations appears to be random or chaotic, even though the equations are deterministic, with fixed parameters.

Model for atmospheric convection

E. Lorenz used the system of differential equations (7.1) to model convection rolls in the atmosphere, occurring when the atmosphere is heated from below and cooled from above. He was interested in studying nonlinear effects on weather forecasting. For small differences in temperature, thermal conduction can carry the heat upward and the atmosphere will remain at rest. This equilibrium was originally studied by Lord Rayleigh in 1916. When the difference in temperature gets greater, the warmer air below rises and the cooler air above falls and convection develops. For relatively small differences, the air forms convection rolls, which take the warm air up and the cool air down. See Figure 21. The difference in temperature plays a role similar to the difference of the rate of adding water for the waterwheel of the preceding subsection; more specifically, the difference of rates of adding water for $\theta = 0$ and π (i.e. the coefficient q_1). Most of the discussion is similar. In particular, the chaotic orbits correspond to the atmosphere forming a convection roll in one direction for a length of time and then reversing and rolling in the other direction. The number of rolls in each direction before the reversal is apparently random. For further discussion, we refer the reader to the original paper, [**Lor63**].

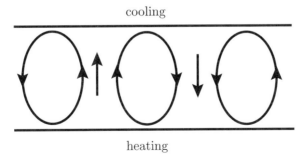

Figure 21. Convection rolls for fluid

7.9. Theory and Proofs

Attractors

Restatement of Theorem 7.1. *The following properties hold for an attracting set* $\mathbf{A}$ *for a trapping region* $\mathbf{U}$.

  a. *The set* $\mathbf{A}$ *is closed and bounded, compact.*
  b. *The set* $\mathbf{A}$ *is invariant for both positive and negative time.*
  c. *If* $\mathbf{x}_0$ *is in* $\mathbf{U}$, *then* $\omega(\mathbf{x}_0) \subset \mathbf{A}$.
  d. *If* $\mathbf{x}_0$ *is a hyperbolic fixed point in* $\mathbf{A}$, *then* $W^u(\mathbf{x}_0)$ *is contained in* $\mathbf{A}$.
  e. *If* γ *is a hyperbolic periodic orbit in* $\mathbf{A}$, *then* $W^u(\gamma)$ *is contained in* $\mathbf{A}$.

Proof. (a) (See Appendix A.2 for more discussion of open and closed sets, their interiors, and closures.) The set $\mathbf{U}$ is closed and bounded and so compact. For any two times $0 < t_1 < t_2$, $\phi^{t_2}(\mathbf{U}) \subset \phi^{t_1}(\mathbf{U}) \subset \mathbf{U}$, so the sets are "nested". A standard theorem in analysis states that the intersection of nested compact sets is a nonempty compact set, so $\mathbf{A}$ is a nonempty compact set.

(b) If $\mathbf{p}$ is in $\mathbf{A}$, then $\mathbf{p}$ is in $\phi(t; \mathbf{U})$ for all $t \geq 0$. Similarly, for any real number s, $\phi(s; \mathbf{p})$ is in $\phi(s; \phi(t; \mathbf{U})) = \phi(s+t; \mathbf{U})$ for all t with $s+t \geq 0$ or $t \geq -s$. But

$$\mathbf{A} = \bigcap_{t \geq -s} \phi(s+t; \mathbf{U}),$$

so $\phi(s; \mathbf{p})$ is in $\mathbf{A}$. This argument works for both positive and negative s, so $\mathbf{A}$ is invariant.

(c) Assume $\mathbf{x}_0$ be in $\mathbf{U}$. Then, $\phi(t; \mathbf{x}_0)$ is in $\phi(s; \mathbf{U})$ for any $t \geq s \geq 0$. Since the ω-limit is defined by taking t_n going to infinity, $\omega(\mathbf{x}_0)$ must be inside $\phi(s; \mathbf{U})$ for any $s \geq 0$. By taking the intersection over all such s, we see that $\omega(\mathbf{x}_0)$ must be inside $\mathbf{A}$.

(d) Since the fixed point $\mathbf{x}_0$ is in $\mathbf{A}$, it must be in the interior of $\mathbf{U}$. Therefore, the whole local unstable manifold of $\mathbf{x}_0$, $W_r^u(\mathbf{x}_0)$ for some $r > 0$, must be contained in $\mathbf{U}$. Since

$$\phi(t; \mathbf{U}) \supset \phi(t; W_r^u(\mathbf{x}_0)) \supset W_r^u(\mathbf{x}_0) \qquad \text{for all } t \geq 0,$$

$W_r^u(\mathbf{x}_0)$ must be contained in $\mathbf{A}$. But $\mathbf{A}$ is positively invariant, so

$$W^u(\mathbf{x}_0) = \bigcup_{t \geq 0} \phi(t; W_r^u(\mathbf{x}_0))$$

must be contained in $\mathbf{A}$.

The proof of part (e) is similar to the one for part (d). □

Poincaré map of Lorenz system

Let f_γ be the one-dimensional linear map associated with the Lorenz system given by

$$f_\gamma(x) = \begin{cases} \gamma x + 1 & \text{for } x < 0, \\ \gamma x - 1 & \text{for } x \geq 0. \end{cases}$$

7.9. Theory and Proofs

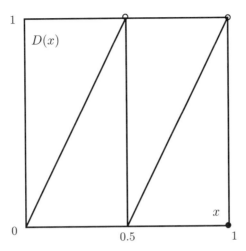

Figure 22. Doubling map

We show below that f_γ has sensitive dependence on initial conditions when restricted to $[-1, 1]$ for $\sqrt{2} < \gamma \leq 2$.

The map f_2 is closely related to the doubling map D defined on $[0, 1]$,

$$D(y) = 2y \pmod{1},$$
$$= \begin{cases} 2y & \text{for } 0 \leq y < 0.5, \\ 2y - 1 & \text{for } 0.5 \leq y < 1, \\ 0 & \text{for } y = 1. \end{cases}$$

The doubling map can be obtained from f_2 by making the linear change of coordinates $x = -1 + 2y$. Notice that $y = 0$ corresponds to $x = -1$, $y = 0.5$ corresponds to $x = 0$, and $y = 1$ corresponds to $x = 1$. If $x_1 = f_2(x_0)$, $2y_0 = x_0 + 1$, and $2y_1 = x_1 + 1$, then for $0 \leq x_0 < 1$,

$$2y_1 = x_1 + 1 = f_2(x_0) + 1 = 2x_0 - 1 + 1 = 2(2y_0 - 1), \quad \text{or}$$
$$y_1 = 2y_0 - 1 = 2y_0 \pmod{1} = D(y_0).$$

A similar calculation for $x_0 < 0$, also shows that $y_1 = D(y_0)$.

Thus, the map f_2 in the y coordinates gives the doubling map. (The definitions at $x = 1$ and $y = 1$ are slightly different.) Notice that the point of discontinuity for D is 0.5 rather than 0 for f_2. Again, the doubling map takes $[0, 0.5)$ and $[0.5, 1)$ all the way across $[0, 1)$,

$$D([0, 0.5)) = [0, 1) \quad \text{and} \quad D([0.5, 1)) = [0, 1).$$

while the Lorenz map f_γ for $\gamma < 2$ leaves little gaps at the ends, $f_\gamma(1) < 1$ and $f_\gamma(-1) > -1$.

Restatement of Theorem 7.4. *Consider the maps $f_\gamma(x)$ for $\sqrt{2} < \gamma \leq 2$ and $D(x)$.*

a. *Both the maps f_γ restricted to $[-1,1]$ and the doubling map D on $[0,1]$ have sensitive dependence on initial conditions.*

b. *There are points z_γ^* in $[-1,1]$ and z_D^* in $[0,1]$ such that $\omega(z_\gamma^*, f_\gamma) = [-1,1]$ and $\omega(z_D^*, D) = [0,1]$.*

c. *For $\sqrt{2} < \gamma < 2$, the interval $[-1,1]$ is a chaotic attractor for f_γ.*

Proof. (a) The proof for the doubling map using binary expansion of numbers is given in Chapter 10, Theorem 10.14. We given another proof here for f_γ that also implies the result for the doubling map.

For the map f_γ, we show that $r = 2/3$ works for sensitive dependence. In fact, the iterates of any two points gets a distance greater than $2/3$ apart. Let $y_0 = x_0 + \delta > x_0$. If x_0 and y_0 are on the same side of 0, then
$$y_1 = f_\gamma(y_0) = f_\gamma(x_0) + \gamma\delta = x_1 + \gamma\delta.$$
Continuing, let $y_n = f_\gamma(y_{n-1})$ and $x_n = f_\gamma(x_{n-1})$. If x_j and y_j are on the same side of 0 for $0 \leq j < n$, then
$$y_n = f_\gamma(y_{n-1}) = f_\gamma(x_{n-1}) + \gamma(y_{n-1} - x_{n-1}) = x_n + \gamma^n \delta.$$
Because $\gamma^n \delta$ become arbitrarily large, eventually $x_n < 0 < y_n$. Then,
$$x_{n+1} = 1 + \gamma x_n,$$
$$y_{n+1} = -1 + \gamma y_n = -1 + \gamma(x_n + a^n\delta), \quad \text{and}$$
$$x_{n+1} - y_{n+1} = 2 - \gamma(\gamma^n \delta) = 2 - \gamma^{n+1}\delta.$$
If $\gamma^n\delta = y_n - x_n < 2/3$, then
$$x_{n+1} - y_{n+1} > 2 - \gamma(2/3) \geq 2 - 2(2/3) = 2/3.$$
This shows that either $|y_n - x_n|$ or $|y_{n+1} - x_{n+1}|$ is greater than $2/3$.

(b) For the doubling map, it is possible to give an explicit binary expansion of a point whose ω-limit set is $[0,1]$. For the maps f_γ, we follow the method of Williams which does not give an explicit point whose ω-limit is $[-1,1]$ but is an existence proof.

In order to give the proof so that it applies to more situations, we let $c = 0$. We first state and prove the following lemma. We use that notation, that if $\mathbf{J}$ is a closed interval, let $\text{int}(\mathbf{J})$ be the open interval with the same end points as $\mathbf{J}$. Also, let $\lambda(\mathbf{J})$ be the length of an interval $\mathbf{J}$.

Lemma 7.11. *Let $\mathbf{J}_0 \subset [-1,1]$ be a closed interval of positive length.*

i. *There is a positive integer n such that $c \in \text{int}\left(f_\gamma^n(\mathbf{J}_0)\right)$ and $c \in \text{int}\left(f_\gamma^{n+1}(\mathbf{J}_0)\right)$.*

ii. *There is a positive integer n such that $f_\gamma^{n+4}(\mathbf{J}_0) = (-1,1)$.*

Proof. (i) If $c \in \text{int}(\mathbf{J})$, then let
$$\text{subinterval}(\mathbf{J} \setminus \{c\}) = \text{longer subinterval of } \mathbf{J} \setminus \{c\},$$
where we actually mean the closed subinterval with the same end points. Note that $\lambda(\text{subinterval}(\mathbf{J})) \geq \lambda(\mathbf{J})/2$.

7.9. Theory and Proofs

If $c \in \text{int}(\mathbf{J}_0)$ replace $\mathbf{J}_0$ by subinterval$(\mathbf{J}_0 \setminus \{c\})$. The length $\lambda(f_\gamma(\mathbf{J}_0)) = \gamma \lambda(\mathbf{J}_0) > \lambda(\mathbf{J}_0)$. The stretching by γ increases the length of intervals. Now let

$$\mathbf{J}_1 = \begin{cases} f_\gamma(\mathbf{J}_0) & \text{if } c \notin \text{int}\left(f_\gamma(\mathbf{J}_0)\right), \\ \text{subinterval}(f_\gamma(\mathbf{J}_0) \setminus \{c\}) & \text{if } c \in \text{int}\left(f_\gamma(\mathbf{J}_0)\right). \end{cases}$$

In the two cases,

$$\lambda(\mathbf{J}_1) \geq \begin{cases} \gamma \lambda(\mathbf{J}_0) & \text{if } c \notin \text{int}\left(f_\gamma(\mathbf{J}_0)\right), \\ \dfrac{\gamma}{2} \lambda(\mathbf{J}_0) & \text{if } c \in \text{int}\left(f_\gamma(\mathbf{J}_0)\right). \end{cases}$$

Continuing by induction, let

$$\mathbf{J}_k = \begin{cases} f_\gamma(\mathbf{J}_{k-1}) & \text{if } c \notin \text{int}\left(f_\gamma(\mathbf{J}_{k-1})\right), \\ \text{subinterval}(f_\gamma(\mathbf{J}_{k-1}) \setminus \{c\}) & \text{if } c \in \text{int}\left(f_\gamma(\mathbf{J}_{k-1})\right). \end{cases}$$

In the two cases,

$$\lambda(\mathbf{J}_k) \geq \begin{cases} \gamma \lambda(\mathbf{J}_{k-1}) & \text{if } c \notin \text{int}(f_\gamma(\mathbf{J}_{k-1})), \\ \dfrac{\gamma}{2} \lambda(\mathbf{J}_{k-1}) & \text{if } c \in \text{int}(f_\gamma(\mathbf{J}_{k-1})). \end{cases}$$

Now if $c \notin \text{int}(f_\gamma(\mathbf{J}_k))$ or $c \notin \text{int}(f_\gamma(\mathbf{J}_{k+1}))$, then

$$\lambda(\mathbf{J}_{k+2}) \geq \frac{\gamma^2}{2} \lambda(\mathbf{J}_k) > \lambda(\mathbf{J}_k).$$

Because the length of intervals continues to grow every two iterates, eventually there is a positive integer n such that $c \in \text{int}(f_\gamma(\mathbf{J}_n))$ and $c \in \text{int}(f_\gamma(\mathbf{J}_{n+1}))$. This proves part (i).

(ii) Let n be as in part (i). Because $c \in \text{int}(f_\gamma(\mathbf{J}_n))$, c is one of the end points of $\mathbf{J}_{n+1}$ and 1 or -1 is an end point of $f_\gamma(\mathbf{J}_{n+1})$. Because $c \in \text{int}(f_\gamma(\mathbf{J}_{n+1}))$, $f_\gamma(\mathbf{J}_{n+1}))$ contains either $(-1, c]$ or $[c, 1)$. We take the first case, but the second is similar:

$$f_\gamma^2(\mathbf{J}_{n+1})) \supset f_\gamma((-1, c)) = (1 - r, 1),$$
$$f_\gamma^{n+4}(\mathbf{J}_0) \supset f_\gamma^3(\mathbf{J}_{n+1})) \supset f_\gamma((1 - \gamma, c)) \cup f_\gamma((c, 1))$$
$$\supset (c, 1) \cup (-1, c] = (-1, 1).$$

This proves part (ii). $\square$

Because any interval, has a forward iterate which covers $(-1, 1)$, the Birkhoff transitivity Theorem 11.18 implies that there is a point whose ω limit is the whole interval $[-1, 1]$. This theorem is not constructive, but merely gives the existence of such points as stated in (b).

(c) Assume that $\gamma < 2$. We consider the interval $[-a_\gamma, a_\gamma]$ for $a_\gamma = 2/\gamma > 1$. Then $f_\gamma([-a_\gamma, a_\gamma]) = [-1, 1] \subset (-a_\gamma, a_\gamma)$, so $[-a_\gamma, a_\gamma]$ is a trapping region. Further iterates take the interval onto $[-1, 1]$, $\text{cl}(f_\gamma^2([-a_r, a_r]) = \text{cl}(f_\gamma([-1, 1]) = [-1, 1]$. Therefore, $[-1, 1]$ is an attracting set. It has a dense orbit by (b), so it is a transitive attractor. The map has sensitive dependence on initial conditions when restricted to $[-1, 1]$ by (a). Therefore it is a chaotic attractor. $\square$

Rössler attractor

Restatement of Theorem 7.5. *Consider the shed map g_a for $\sqrt{2} < a \leq 2$.*

   a. *The shed map g_a has sensitive dependence on initial conditions when restricted to $[0, 1]$.*

   b. *The shed map has a point z^* in $[0, 1]$ such that $\omega(z^*) = [0, 1]$.*

   c. *For $\sqrt{2} < a < 2$, the interval $[0, 1]$ is a chaotic attractor.*

Proof. (a) A proof like that of Theorem 7.4(a) does not work, since points on opposite sides of c get mapped to the same point. Thus, the folding does not immediately give sensitive dependence. However, for $0 < x_0 < 1$, if $\mathbf{J}_0^\delta = [x_0-\delta, x_0+\delta]$ is a small interval, then there is a positive integer such that $g_a^{n+3}(\mathbf{J}_0^\delta) = [0, 1]$. Thus, there are points close to x_0 that go to 0 and 1 by g_a^{n+3}. One of these points has to be at a distance at least $1/2$ from $g_a^{n+3}(x_0)$. Thus, g_a has sensitive dependence on initial conditions using $r = 1/2$.

(b) The first step in the proof is to modify the proof of Lemma 7.11 for a closed subinterval $\mathbf{J}_0 \subset [-1, 1]$ of positive length. Let $c = 1 - 1/a$. By a proof similar to the proof of Lemma 7.11(i), there must exist a positive integer n such that $c \in \text{int}\left(g_a^n(\mathbf{J}_0)\right)$ and $c \in \text{int}\left(g_a^{n+1}(\mathbf{J}_0)\right)$. Because $c \in \text{int}(g_a(\mathbf{J}_n))$, c is one of the end points of $\mathbf{J}_{n+1}$ and 1 is an end point of $g_a(\mathbf{J}_{n+1})$. Because $c \in \text{int}(g_a(\mathbf{J}_{n+1}))$, $g_a(\mathbf{J}_{n+1})) \supset [c, 1]$ and

$$g_a^{n+3}(\mathbf{J}_0) \supset g_a^2(\mathbf{J}_{n+1})) \supset g_a([c, 1])) = [0, 1].$$

This prove that g_a has a property similar to Lemma 7.11(ii).

Again, by the Birkhoff transitivity theorem, g_a is topologically transitive, proving part (b) of the theorem.

(c) The interval $\left[\frac{a-2}{a}, 1 + \frac{2-a}{2a^2}\right]$ has a negative left end point and a right end point greater than 1. It gets mapped to $g_a\left(\left[\frac{a-2}{a}, 1 + \frac{2-a}{2a^2}\right]\right) = \left[\frac{a-2}{2a}, 1\right] \subset \left(\frac{a-2}{a}, 1 + \frac{2-a}{2a^2}\right)$ so is a trapping interval. The next iterate is $g_a\left(\left[\frac{a-2}{2a}, 1\right]\right) = [0, 1]$. Further iterates all equal $[0, 1]$, so $[0, 1]$ is the attracting set. Combining with parts (a) and (b) gives that it is a chaotic attractor. □

Part 2

Iteration of Functions

Chapter 8

Iteration of Functions as Dynamics

A dynamical system is a rule which specifies the manner in which a system evolves as time progresses. Given an initial condition or "state" x_0, the rule determines exactly the state x_t at a future time t. For example, x_t could be a population at time t. We assume that the rule specifying the manner in which the system evolves is deterministic (i.e., if we know what the state is now, then we can, in principle, determine exactly what it will be in the future). The rule describing the manner in which the system evolves can be given by an ordinary differential equation or it can be determined by a function that gives the value of the next state one unit of "time" later. For short periods of time, the solution can be followed using the rule defining the system, either iterating the function or numerically integrating the differential equation. However, we are mainly interested in the long-term behavior of a system or the behavior of typical initial states; unless an explicit formula is known for the solution of the system, it is difficult or impossible to understand the long-term behavior by this method. We need to develop more geometric methods for understanding the long-term behavior.

In this part of the book, a function is the rule that determines the state at the next unit of "time", and repeated application or iteration of the function for states further in the future. Since, in calculus courses, functions are more often graphed than iterated, we start by showing some situations in which iteration of a function arises naturally. As the course progresses, iteration as a means of determining future states should seem more natural.

8.1. One-Dimensional Maps

We start this section by describing linear growth of money under compound interest, and linear growth of a population. For these situations, we can give a formula for a value at all later iterates. Next, we discuss a population growth with a crowding factor, or a limit to the growth. For this example, we cannot determine an analytical expression for the population at a much later time in terms of the initial population. The analysis of this type of function requires the development of the ideas discussed in the later chapters. We end the section on one-dimensional maps by mentioning a few other situations in which iteration of a function arises naturally.

Models for interest and population growth

Example 8.1. In this example, we explain how iteration of the function $f_\lambda(x) = \lambda x$ can model interest accumulated at discrete time intervals.

Assume that x_0 amount of money is deposited in a bank at 5 percent simple interest per year. One year later, $0.05\, x_0$ is paid in interest and the total amount of money in the account is $x_1 = x_0 + 0.05\, x_0 = 1.05\, x_0$. If all the money is left in the account, another year later, there would be $x_2 = 1.05\, x_1 = 1.05 \times 1.05\, x_0 = (1.05)^2 x_0$ in the account. Continuing, after n years, the amount of money would be $x_n = (1.05)^n x_0$, assuming none of the money is withdrawn and no additional money is invested.

In general, if the interest rate is I, then after one time period the amount of money in the account would be $x_1 = x_0 + I x_0 = (1+I) x_0 = \lambda x_0$ where $\lambda = 1 + I > 1$. If we define $f_\lambda(x) = \lambda x$, then $x_1 = f_\lambda(x_0)$. If all the money is left to accumulate interest, then after two time periods, there would be $x_2 = f_\lambda(x_1) = \lambda x_1 = \lambda^2 x_0$. We use the notation $f_\lambda^2(x_0)$ for the composition of f_λ with itself, $f_\lambda^2(x_0) = f_\lambda \circ f_\lambda(x_0) = f_\lambda(x_1) = x_2$. Continuing, the amount of money accumulated after n intervals of receiving interest is

$$x_n = \lambda\, x_{n-1} = f_\lambda(x_{n-1}) = f_\lambda^n(x_0),$$

where f_λ^n is the composition of f_λ with itself n times. In this example, there is a simple formula for the iterates,

$$x_n = \lambda^n x_0.$$

If $x_0 > 0$ and $\lambda > 1$, then the iterates $x_n = f_\lambda^n(x_0) = \lambda^n x_0$ go to infinity as n goes to infinity. In terms of money left in a bank account, this means that, if the money is left in the bank, then the amount on deposit becomes arbitrarily large as time passes.

If we take the same function, but with $0 < \lambda < 1$, then $f_\lambda(x)$ could model a situation in which $d = (1 - \lambda)$ proportion of the population dies off in each time period: $d x_0$ is the number of people who die, leaving $x_0 - d x_0 = \lambda x_0$ population left. In this case, after one time period, there is less population than there was initially. Continuing, the population decreases after each time period. The formula for the iterates are the same as for $\lambda > 1$, but now $x_n = \lambda^n x_0$ goes to zero as n goes to infinity since now $0 < \lambda < 1$ (i.e., the population is dying off at a rate determined by λ).

Example 8.2. In this example, we introduce iteration of a nonlinear function. We assume that the population at the $(n+1)^{\text{st}}$ time is given in terms of the population at the n^{st} time by some function $g(x)$,

$$x_{n+1} = g(x_n).$$

8.1. One-Dimensional Maps

A reason for using discrete time intervals rather than continuous time would be if the population reproduced at a certain time of year (e.g., in the spring) and the population was then measured once a year.

The ratio of the population at time $n+1$ and time n, x_{n+1}/x_n, gives the growth factor per unit population. In the previous example, this ratio is a constant. If this per capita growth factor changes with the population, then we can get more complicated examples. The simplest way this can depend on the population is linearly, as

$$\frac{x_{n+1}}{x_n} = a - b\,x_n \quad \text{or}$$
$$x_{n+1} = x_n\,(a - b\,x_n).$$

This function could model the way a population evolves in discrete time intervals in which there is a crowding factor where the rate of growth decreases as the population increases. Thus, the function

$$g_{a,b}(x) = x\,(a - b\,x)$$

determines the population at the $(n+1)^{\text{st}}$ step in terms of the population at the n^{st} step, by means of the rule

$$x_{n+1} = g_{a,b}(x_n) = x_n\,(a - b\,x_n).$$

For this function, the ratio of successive populations is $x_{n+1}/x_n = (a - bx_n)$. For $x_n > a/b$, this ratio is negative, which means the next population is actually negative (which may or may not have a physical meaning). Robert May discussed this function in terms of population models, [**May75**].

By rescaling the unit of measure using $x = ty$,

$$y_{n+1} = \frac{1}{t}x_{n+1} = \frac{1}{t}(ty_n)(a - bty_n) = y_n(a - bty_n).$$

By taking t so that $tb = a$, we obtain the standard form of what is called the *logistic function*,

$$g_a(y) = g_{a,a}(y) = ay(1 - y).$$

While the linear growth model has simple dynamics, the logistic function can exhibit very complicated dynamics under iteration, depending on the value of the parameter a.

If $a = 1.5$, then $g_{1.5}(1/3) = (3/2)(1/3)(2/3) = 1/3$ and $1/3$ is a fixed point or steady state. We will see later that any other $0 < x_0 < 1$ has iterates $g_{1.5}^n(x_0)$ which converge to this fixed point $1/3$. For example, if $x_0 = 0.25$, then

$$x_1 = g_{1.5}(0.25) = 0.28125,$$
$$x_2 = g_{1.5}(0.28125) = 0.3032227\ldots,$$
$$x_3 = g_{1.5}(0.3032227\ldots) = 0.3169180\ldots,$$
$$x_4 = g_{1.5}(0.3169180\ldots) = 0.3247215\ldots, \quad \text{and}$$
$$x_{21} = g_{1.5}^{21}(0.25) = 0.3333333\ldots.$$

Since the iterates of many initial conditions converge to the fixed point for this value of a, the fixed point $1/3$ is a carrying capacity of the population.

If $a_2 = 1+\sqrt{5} \approx 3.24$, then $g_{a_2}(1/2) = (1+\sqrt{5})/4$ and $g_{a_2}((1+\sqrt{5})/4) = 1/2$, so $\{\,1/2,\ (1+\sqrt{5})/4\,\}$ is a period-2 orbit. In fact, Theorem 9.8 proves that the iterates of most initial conditions x_0 with $0 < x_0 < 1$ converge to this period-2 orbit and not to a fixed point; $g_{a_2}^n(x_0)$ alternates between being close to $1/2$ and $(1+\sqrt{5})/4$ as n goes to infinity.

If $a = 3.839$, then there is a point of period three: $x_0 \approx 0.149888$, $x_1 \approx 0.489172$, $x_2 \approx 0.959300$, and $x_3 \approx 0.149888$ is the same value as x_0. These are not exact values but are rounded to six decimal places; still, there is a period three point approximately equal to this. Moreover, other initial conditions do not converge to this cyclical behavior. Therefore, there is no intrinsic carrying capacity of this iterative scheme, neither an attracting steady state nor an attracting cyclical state.

Finally, for $a = 4.0$ and for most initial conditions x_0, the iterates appears random even though the system is deterministic. As the course goes on, we will see why this is the case.

Example 8.2 indicates that iteration of a nonlinear function of one variable can result in very complicated behavior. The corresponding differential equation considered in the first part of the book has much simpler dynamical behavior. This example also illustrates that a very simple algebraic form of the function can result in chaotic behavior. This function is only slightly nonlinear (quadratic), but the dynamics can be very complicated. For this function, there is no simple formula which gives the value at all later iterates in terms of x_0: The only way to determine x_n for large n is to apply n times the function g_a.

There are other models for population growth which do not become negative for large values of x, but merely become very small. Two such models are

$$x_{n+1} = \alpha\, x_n\, e^{-\beta x_n}$$

for α and β positive, and

$$x_{n+1} = \frac{a\, x_n}{1 + b\, x_n}$$

for a and b positive. These two models are discussed in Section 9.7.2.

Iteration in economics

Much of economic theory involves finding equilibria, so it is a static theory. A dynamic theory of the manner in which economies adjust toward equilibria is not as well understood. However, there are several situations in which iteration of a function arises. One situation is capital formation, which considers the reinvestment and growth of production. The Cobb–Douglas production function is given by

$$Y = A K^\alpha L^{1-\alpha},$$

where K is the capital, L is the labor, and Y is the amount produced. The parameters are $A > 0$ and $0 < \alpha < 1$. It is assumed that σ is the fraction of the production that is available at the next unit of time, so $K_1 = \sigma Y$. The labor is assumed to grow by a factor of $1 + \lambda$, so $L_1 = (1 + \lambda)L$ gives the growth in the labor. Then, the ratio of capital to labor K/L changes from one period to the next

8.1. One-Dimensional Maps

by the equation

$$k_1 = \frac{K_1}{L_1} = \frac{\sigma A K^\alpha L}{L^\alpha (1+\lambda) L}$$
$$= \left(\frac{\sigma A}{1+\lambda}\right) \left(\frac{K}{L}\right)^\alpha = \frac{\sigma A k^\alpha}{1+\lambda}.$$

The iteration function becomes

(8.1) $$k_{n+1} = f(k_n) = \frac{\sigma A k_n^\alpha}{1+\lambda}.$$

Thus, the capital-to-labor ratio at one time period determines the ratio at the next time period. We return to this model in Sections 9.7.1 and 11.5.1.

Newton method for finding roots

Example 8.3. If $f(x)$ is a polynomial, or even a more general function, we are sometimes interested in determining its roots (i.e., the values x for which $f(x) = 0$). It is theoretically possible to factor any polynomial and find its real and complex roots. For a high degree polynomial, there is no algebraic expression for finding the roots. It is of practical importance to find an approximation of the roots when no exact expression for the roots exists. The Newton method (or Newton–Raphson method) is an iterative scheme to find an approximate root.

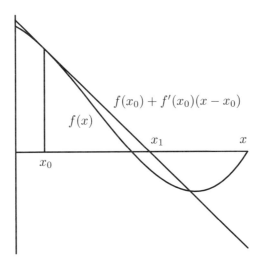

Figure 1. The Newton map from the plot of $f(x)$

Assume that x_0 is an initial guess of a root. The tangent line at x_0 to the graph of f is the best linear (affine) approximation of the function near x_0, and its equation is

$$y = f(x_0) + f'(x_0)(x - x_0).$$

This line crosses the x-axis at the point x_1 where y is zero, so
$$0 = f(x_0) + f'(x_0)(x_1 - x_0), \qquad \text{or}$$
$$x_1 = x_0 - \frac{f(x_0)}{f'(x_0)}.$$

This formula is valid as long as the tangent line is not horizontal, $f'(x_0) \neq 0$. See Figure 1. Starting again with x_1, we can get another point x_2 where the tangent line at x_1 crosses the x-axis:
$$x_2 = x_1 - \frac{f(x_1)}{f'(x_1)}.$$

If we let
$$N(x) = x - \frac{f(x)}{f'(x)},$$
then $x_1 = N(x_0)$ and $x_2 = N(x_1)$. This induced map N is called the *Newton map for the function* f. Repeating this process, if x_n is the nth approximation of the root, then the $(n+1)^{\text{st}}$ approximation is given by
$$x_{n+1} = N(x_n) = x_n - \frac{f(x_n)}{f'(x_n)}.$$

If the initial guess is good enough (or, in fact, for most initial guesses), the iterative scheme converges very rapidly to a root of the polynomial as n goes to infinity.

As an example, let $f(x) = x^4 - 2$. The Newton map for this f is given by
$$N(x) = x - \frac{x^4 - 2}{4x^3}.$$

If we take the initial guess of $x_0 = 1$, then
$$x_1 = N(1) = 1 - \frac{-1}{4} = 1.25,$$
$$x_2 = N(x_1) = 1.25 - \frac{1.25^4 - 2}{4 \cdot 1.25^3} \approx 1.1935000,$$
$$x_3 = N(x_2) \approx 1.1892302\ldots,$$
$$x_4 = N(x_3) \approx 1.1892071\ldots, \qquad \text{and}$$
$$x_5 = N(x_4) \approx 1.1892071\ldots.$$

After just four iterates, $|x_4 - x_3| < 3 \cdot 10^{-5}$. The real roots are $x^{\pm} = \sqrt[4]{2} = 1.18920711\ldots$, so the preceding process did, in fact, give the correct root to 7 decimal places, $x_4 \approx 1.1892071$.

For higher degree polynomials, the derived Newton map can have periodic points or points with chaotic iterates: For such initial conditions, the iterates do not converge to a root of the polynomial. Still, most initial conditions do converge to a root of the polynomial, and certainly, if we start near a root, then the iterative process does converge rapidly to the nearby root. We discuss this map further in Sections 9.3.1 and 12.5.2.

Poincaré maps for differential equations in the plane

Example 8.4. In Part 1, we used the Poincaré map to analyze several systems of differential equations. In Section 6.3 for the Lienard system (or Van der Pol system), the return map from the positive y-axis to itself is a map of a single variable. Figure 2 shows a trajectory leaving the y-axis at $\mathbf{Y}_0 = (0, y_0)$, moving into the first quadrant, and eventually returning at $\mathbf{Y}_1 = (0, y_1)$ with $y_1 > 0$. The assignment of y_1 from y_0 is a function $y_1 = P(y_0)$ from the positive axis to itself. We used this map P to show that the system has a unique periodic orbit. In the process of that proof, we developed some ideas of the dynamics of iteration of functions. This second part of the book treats the dynamics of iteration of functions more systematically than is done in part one.

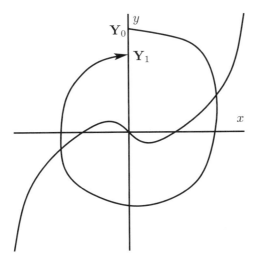

Figure 2. Trajectory path for Lienard equations

8.2. Functions with Several Variables

Population growth with several stages

Example 8.5. In many situations, a reasonable model of the population growth of a species must consider several different stages of life. The flour beetle first exists in a larva stage, then a pupa stage, and finally in the adult stage. Costantino, Desharnais, Cushing, and Dennis [**Cos95**] argue that its population should be modeled by iteration because of the discrete nature of its generations. At the nth time step, let L_n be the population of larva, P_n the population of pupa, and A_n the adult population. A simple linear model for the populations at the $(n+1)^{\text{st}}$ step in terms of those at n^{th} step is given by

$$L_{n+1} = b\, A_n,$$
$$P_{n+1} = (1 - \mu_L)\, L_n,$$
$$A_{n+1} = (1 - \mu_P)\, P_n + (1 - \mu_A)\, A_n,$$

where b is the birth rate of larva in terms of the adult population, μ_L is the death rate of larva (which do not transform into pupa), μ_P is the death rate of pupa (which do not transform into adults), and μ_A is the death rate of adults. In practice, very few pupae die before becoming adults, so we take $\mu_P = 0$.

For the flour beetle, there are also effects on the population caused by cannibalism. When overpopulation is present, the adult beetles will eat pupae and unhatched eggs (future larvae), and the larvae will also eat eggs. These factors affect the population of larvae and adults at the next time step. A model that incorporates such factors is given by

$$L_{n+1} = b\,A_n e^{-C_{LA} A_n} e^{-C_{LL} L_n},$$
$$P_{n+1} = (1 - \mu_L)\,L_n,$$
$$A_{n+1} = P_n\, e^{-C_{PA} A_n} + (1 - \mu_A)\,A_n,$$

where the quantities $\exp(-C_{LA} A_n)$ and $\exp(-C_{LL} L_n)$ are the probabilities that an egg is not eaten by the adult population A_n and larvae population L_n. The quantity $\exp(-C_{PA} A_n)$ is the survival probability of the pupae into adulthood.

For parameter values $b = 4.88$, $C_{LL} = 0$, $C_{LA} = 0.01$, $C_{PA} = 0.005$, $\mu_L = 0.2$, and $\mu_A = 0.01$, a numerical simulation converges to the survival equilibrium with $L \approx 21.79775$, $P \approx 15.97775$, and $A \approx 530.87362$. Thus, there is a stable fixed point which attracts initial positive populations. If the value μ_A is changed to 0.96 and C_{PA} to 0.5, then the fixed point becomes unstable and the resulting numerical iteration does not tend to a fixed point or to a low period periodic point, but exhibits complex dynamics, "chaos". See Section 12.5.3 for further discussion of this model.

Hénon map

Example 8.6. This example is more artificial than the previous ones. However, it introduces an important function with two variables that has been extensively studied. As in the previous example, assume that there are two different age groups, with a population measured by x and y. Assume that the population of y at the $(n+1)^{\text{st}}$ stage is equal to the population of x at the nth stage, $y_{n+1} = x_n$. The population of x at the $(n+1)^{\text{st}}$ stage is given in terms of both x and y at the nth stage by means of the equation $x_{n+1} = a - b\,y_n - x_n^2$, where a and b are parameters which are fixed during iteration. Combining these two into a single function from $\mathbb{R}^2$ to $\mathbb{R}^2$, we get the *Hénon map*

$$\begin{pmatrix} x_{n+1} \\ y_{n+1} \end{pmatrix} = F_{a,b} \begin{pmatrix} x_n \\ y_n \end{pmatrix} = \begin{pmatrix} a - b\,y_n - x_n^2 \\ x_n \end{pmatrix}.$$

M. Hénon introduced this function as an explicit map which could be studied numerically on the computer and which exhibited many interesting dynamical properties. For $a = 1.4$ and $b = -0.3$, iteration of a single initial condition like $(x_0, y_0) = (0, 0)$ converges toward a very complicated set of points called the *Hénon attractor*, given in Figure 3. The figure shows the plot of points on a single orbit, $\{(x_n, y_n) : 0 \leq n \leq N\}$ for large N.

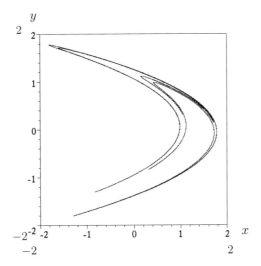

Figure 3. Hénon attractor

Markov chains

This example is linear. Assume that there are two locations for renting cars. Let p_{ij} be the probability of returning a car that was rented from the ith lot to the jth lot. Since we assume that every car is returned to one of the two lots,

$$p_{11} + p_{12} = 1 \quad \text{and}$$
$$p_{21} + p_{22} = 1.$$

Assume that we start with $x_1^{(0)}$ proportion of the cars at the first lot and $x_2^{(0)}$ at the second lot. If all the cars are rented out and then all are returned, then we will have $x_1^{(1)} = p_{11} x_1^{(0)} + p_{21} x_2^{(0)}$ at the first lot and $x_2^{(1)} = p_{12} x_1^{(0)} + p_{22} x_2^{(0)}$ at the second lot at time one. We can write this transition by matrix multiplication:

$$\left(x_1^{(1)}, x_2^{(1)}\right) = \left(x_1^{(0)}, x_2^{(0)}\right) \begin{pmatrix} p_{11} & p_{12} \\ p_{21} & p_{22} \end{pmatrix}.$$

This equation gives the distribution of cars at time one in terms of the distribution at time zero. We assume the proportion of cars from one location returned to another location at the next stage stays constant over time. Thus, this process can be repeated to give the distribution of cars between the two lots at time $n+1$ in terms of the distribution at time n:

$$\left(x_1^{(n+1)}, x_2^{(n+1)}\right) = \left(x_1^{(n)}, x_2^{(n)}\right) \begin{pmatrix} p_{11} & p_{12} \\ p_{21} & p_{22} \end{pmatrix}.$$

Such a process of transition from one state to another is called a *Markov chain*. The long-term steady state can be found as a row eigenvector for the eigenvalue of 1.

As an example, take

$$\begin{pmatrix} p_{11} & p_{12} \\ p_{21} & p_{22} \end{pmatrix} = \begin{pmatrix} 0.8 & 0.2 \\ 0.3 & 0.7 \end{pmatrix}.$$

Then the steady state is

$$\left(x_1^{(\infty)}, x_2^{(\infty)}\right) = (0.6, 0.4).$$

The reader can check that this steady state is taken to itself. (Notice that it is a row eigenvector for the eigenvalue 1.) Section 12.5.1 presents further discussion of Markov chains.

Time-dependent systems of differential equations

As mentioned previously, in Part 1 we discussed the Poincaré map in several situations for systems of differential equations. For the Lorenz system, the return map from $z = 28$ to itself determines a map of two variables. We also argued that the important aspects of this map are reflected by a map of one variable, but that is particular to this system of differential equations.

Example 8.7. In another context, in Section 7.5, we briefly discussed a forced Duffing equation, which results in a system of differential equations

$$\dot{\tau} = 1 \quad \left(\text{mod } \tfrac{2\pi}{\omega}\right),$$
$$\dot{x} = y,$$
$$\dot{y} = x - x^3 - \delta y + F\cos(\omega \tau),$$

where τ is the variable representing the time-dependent terms. Taking an initial condition $(\tau, x, y) = (0, x_0, y_0)$ and following it until $\tau = {2\pi}/{\omega}$, results in a map from (x_0, y_0) to two new (x_1, y_1). Thus, the Poincaré map is a map of two variables. Figure 4 shows the plot of a single orbit of the Poincaré map for $F = 0.4$, $\delta = 0.25$, and $\omega = 1$.

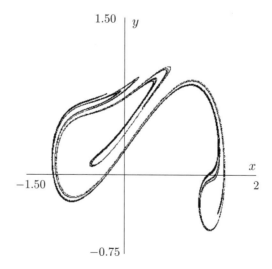

Figure 4. Plot of Poincaré map for system (7.2), with $F = 0.4$, $\delta = 0.25$, and $\omega = 1$

Chapter 9

Periodic Points of One-Dimensional Maps

In the preceding chapter, we discussed several situations in which iteration of functions occurs in a natural way. This chapter begins our presentation of the methods of analyzing the iteration of a function. The calculations and geometry are simpler for maps from the real line $\mathbb{R}$ to itself than for higher dimensional space, so we start with this type of function in Chapters 9–11. This chapter mostly considers periodic points of such maps and the sets of all points which tend to periodic orbits under iteration (i.e., their basins of attraction). The next two chapters consider more complicated or chaotic dynamics of one-dimensional maps. Chapters 12 and 13 consider the dynamics of maps on higher dimensional spaces.

9.1. Periodic Points

In this section, we start our study of iterates of points and periodic points. For certain functions, we can easily find points of low period by solving some equations algebraically. For certain other functions, we use properties of the graph of the function and the graph of the iterates of the function to find periodic points.

Before giving some definitions, we consider a simple example of iteration.

Example 9.1. Consider the cosine function, $f(x) = \cos(x)$, where x is measured in radians. If we take $x_0 = 1$, then

$$x_1 = \cos(x_0) = 0.5403023\ldots$$
$$x_2 = \cos(x_1) = 0.8575532\ldots$$
$$x_3 = \cos(x_2) = 0.6542898\ldots$$
$$x_4 = \cos(x_3) = 0.7934804\ldots$$
$$x_5 = \cos(x_4) = 0.7013688\ldots$$
$$x_6 = \cos(x_5) = 0.7639597\ldots$$

$$x_7 = \cos(x_6) = 0.7221024\ldots$$
$$x_8 = \cos(x_7) = 0.7504178\ldots$$
$$x_9 = \cos(x_8) = 0.7314040\ldots$$

give the first nine iterates of the initial point. Eventually, this iteration converges toward $x_\infty = 0.7390851332\cdots$. For this limiting value, $\cos(x_\infty) = x_\infty$, and x_∞ is taken to itself under iteration. See Figure 1 for a plot of the values x_n plotted versus n. Again, there is no explicit formula that gives the value of x_n in terms of x_0; it is necessary to repeatedly apply the cosine function in order to find a sequence of points converging to x_∞.

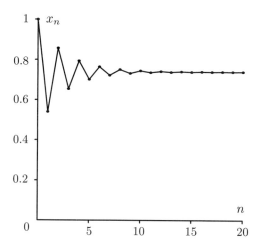

Figure 1. Plot of $x_n = \cos(x_{n-1})$ versus n

A *fixed point* for the cosine function is a value x^* for which $x^* = \cos(x^*)$. For such a value, the point given by the pair $(x^*, \cos(x^*)) = (x^*, x^*)$ must lie on the graph of $y = x$. Thus, by looking at the graph of $y = \cos(x)$, and considering the points where the graph intersects the graph of $y = x$, we find all the points x^* where $x^* = \cos(x^*)$. There is certainly one such point between 0 and $\pi/2$. See Figure 2. This point is exactly the point x_∞ found by iteration above.

Although we rarely want to calculate the formula for the composition of a function with itself, it is convenient to have a notation for these iterates. We write $f^0(x) = x$ for the identity (that is, 0th iterate), $f^1(x)$ for $f(x)$, and
$$f^2(x) = f(f(x))$$
for the composition of f with f. Continuing by induction, we write
$$f^n(x) = f(f^{n-1}(x))$$
for the composition of f with itself n times. The map f^n is called the nth *iterate* of f. It is important to remember that $f^2(x)$ does not stand for the square of the

9.1. Periodic Points

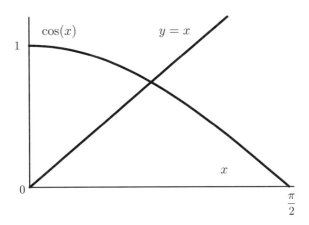

Figure 2. Graph of $\cos(x)$ and $y = x$

formula for the function, $f^2(x) \neq [f(x)]^2$, but it is the composition of f with itself. For example, if $f(x) = x^3$, then

$$f^2(x) = f(x^3) = x^9 \neq x^6 = [x^3]^2 = [f(x)]^2.$$

Using this notation, for the initial condition x_0, $x_1 = f(x_0)$, $x_2 = f(x_1) = f^2(x_0)$, and $x_n = f(x_{n-1}) = f^n(x_0)$.

The *forward orbit of x_0 by f* is the set of all iterates of x_0 and is denoted

$$\mathcal{O}_f^+(x_0) = \{\, f^j(x_0) : j \geq 0 \,\}.$$

The plus sign in the notation is used because we are considering only positive iterates. Later, when considering invertible maps, we consider the whole orbit with both positive and negative iterates.

Example 9.2. Let $f(x) = -x^3$. The fixed points satisfy

$$x = f(x) = -x^3,$$
$$0 = -x^3 - x$$
$$= -x(x^2 + 1).$$

Since $x^2 + 1 \neq 0$, the only fixed point is $x = 0$.

For this map, we can calculate the points which are fixed by $f^2(x)$ by calculating $f^2(x)$ and solving $0 = f^2(x) - x$:

$$f^2(x) = f(-x^3) = -(-x^3)^3 = x^9,$$
$$0 = f^2(x) - x = x^9 - x$$
$$= x(x^8 - 1) = x(x^4 - 1)(x^4 + 1)$$
$$= x(x^2 - 1)(x^2 + 1)(x^4 + 1)$$
$$= x(x - 1)(x + 1)(x^2 + 1)(x^4 + 1).$$

The real roots of this equation are $x = 0$ and ± 1, since $(x^2 + 1)(x^4 + 1) \neq 0$. The point $x = 0$ is a fixed point, so the only new points fixed by f^2 that are not fixed by f are ± 1. Since $f(1) = -1$ and $f(-1) = 1$, these points are the period-2 points and $\mathcal{O}_f^+(1) = \mathcal{O}_f^+(-1) = \{1, -1\}$.

Definition 9.3. Let f be a function from $\mathbb{R}$ to itself. A point p is a *fixed point for f* provided that $f(p) = p$. The set of all fixed points or period-1 points of f is denoted by
$$\mathrm{Fix}(f) = \{x_0 : f(x_0) = x_0\}.$$

A point p is a *period-n point for f* provided that $f^n(p) = p$ but $f^j(p) \neq p$ for $0 < j < n$. The positive integer n is called the *period for p* if p is a period-n point for f (i.e., $f^n(p) = p$ but $f^j(p) \neq p$ for $0 < j < n$). Sometimes, to emphasize the fact that n is the least integer such that $f^n(p) = p$, this integer n is called the *least period*. The set of all period-n points of f is denoted by
$$\mathrm{Per}(n, f) = \{x_0 : f^n(x_0) = x_0 \text{ and } f^j(x_0) \neq x_0 \text{ for } 0 < j < n\}.$$
If p is a period-n point, then $\mathcal{O}_f^+(p) = \{f^j(p) : 0 \leq j < n\}$ is called a *period-n orbit*.

Remark 9.4. Notice that, if p is a period-n point and $p_j = f^j(p)$, then
$$f^n(p_j) = f^n(f^j(p)) = f^{n+j}(p) = f^j(f^n(p)) = f^j(p) = p_j,$$
and p_j is also a period-n point for any j.

Definition 9.5. A point p is an *eventually period-n point* provided that there is a positive integer k such that $f^k(p)$ is a period-n point (i.e., $f^n(f^k(p)) = f^k(p)$).

The next example for which we find periodic points uses the notation $z = y$ (mod 1), or z *equals y modulo 1*. This notation means that we drop the integer part of y leaving only the fractional part z with $0 \leq z < 1$. For example,
$$1.4 \pmod{1} = 0.4 \quad \text{and}$$
$$2.0 \pmod{1} = 0.$$
In general, "$y \pmod{a} = z$" means we subtract the integer multiples of a, leaving only the remaining fraction part of a, $y = k\,a + z$ with $0 \leq z < a$.

Example 9.6. We let D be the function which doubles a number in $[0, 1]$ and then takes the integer part,
$$D(x) = 2\,x \pmod{1}.$$
This map is called the *doubling map*. Thus, $D(x) = 2\,x - k$ where k is an integer depending on x so that $0 \leq D(x) < 1$. For example,
$$D\left(\frac{2}{3}\right) = \frac{4}{3} \pmod{1} = \frac{1}{3} \quad \text{and}$$
$$D\left(\frac{1}{2}\right) = 1 \pmod{1} = 0.$$
For $0 \leq x < 0.5$, $0 \leq 2x < 1$ and $D(x) = 2x$; for $0.5 \leq x < 1$, $1 \leq 2x < 2$, so $D(x) = 2x - 1$; finally, $D(1) = 2 - 2 = 0$. Therefore,
$$D(x) = \begin{cases} 2\,x & \text{for } 0 \leq x < 0.5, \\ 2\,x - 1 & \text{for } 0.5 \leq x < 1, \\ 0 & \text{for } x = 1. \end{cases}$$

9.1. Periodic Points

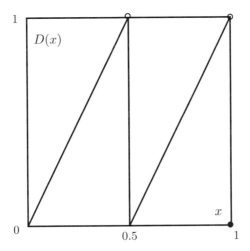

Figure 3. Doubling map

See Figure 3 for its graph.

It is necessary to solve for the fixed points in the intervals $[0, 0.5)$, $[0.5, 1)$, and $\{1\}$ separately. For x in the interval $[0, 0.5)$, the fixed point satisfies $D(x) = 2x = x$, so $x = 0$ is a fixed point. In the second interval $[0.5, 1)$, the fixed point would have to satisfy $D(x) = 2x - 1 = x$, or $x = 1$, which is not in the interval $[0.5, 1)$; so, there is no fixed point in $[0.5, 1)$. Finally, $D(1) = 0$ is not fixed. Thus, there is a single fixed point at $x = 0$.

Also,
$$D^2(x) = 2^2 x \quad (\text{mod } 1) \quad \text{and}$$
$$D^n(x) = 2^n x \quad (\text{mod } 1).$$

Thus, $D^n(x) = x$ is the same as $0 = D^n(x) - x = 2^n x - x - k$, where k is a nonnegative integer that depends on x and n; the period-n points must satisfy $(2^n - 1) x = k$ or
$$x = \frac{k}{2^n - 1}.$$

Some period-4 points are $\frac{1}{2^4 - 1} = \frac{1}{15}$, $\frac{3}{2^4 - 1} = \frac{3}{15} = \frac{1}{5}$, $\frac{2 \cdot 3}{2^4 - 1} = \frac{2}{5}$, $\frac{3 \cdot 3}{2^4 - 1} = \frac{3}{5}$, and $\frac{4 \cdot 3}{2^4 - 1} = \frac{4}{5}$. Note that the last four points are on the same orbit: $D\left(\frac{1}{5}\right) = \frac{2}{5}$, $D\left(\frac{2}{5}\right) = \frac{4}{5}$, $D\left(\frac{4}{5}\right) = \frac{3}{5}$, and $D\left(\frac{3}{5}\right) = \frac{1}{5}$.

A point x is an eventually period-n point if
$$D^{n+j}(x) - D^j(x) = 0,$$
$$(2^{n+j} - 2^j) x = k, \quad \text{or}$$
$$x = \frac{k}{2^j (2^n - 1)}.$$

In particular, the point $\frac{1}{28} = \frac{1}{2^2(2^3-1)}$ is an eventually period-3 point. Because of the formulas given, any periodic or eventually periodic point must be rational.

In fact, the set of eventually periodic points is exactly the set of rational points in $[0, 1]$. We start by considering a rational number in $(0, 1)$ of the form $x = p/q$, where $q \geq 3$ is odd. First, we show that $D(j/q) \neq 0$ for $0 < j < q$. The iterate $D(j/q) = 2j/q - k$, where $k = 0, 1$. If $0 = D(j/q) = 2j/q - k$, then $k \neq 0$ since $j > 0$. Thus, $k = 1$ and $2j = q$, so 2 divides q, which is a contradiction. Thus, $D(j/q) \neq 0$ for $0 < j < q$, and the map D takes the set of points

$$\mathbf{Q}_q = \left\{ \frac{1}{q}, \frac{2}{q}, \ldots, \frac{q-1}{q} \right\}$$

to itself. There are only $q - 1$ points in $\mathbf{Q}_q$, so the iterates must repeat and each point in $\mathbf{Q}_q$ is eventually periodic. (A little more work shows all these points are periodic period-n.)

The point 0 is fixed, and $D(1) = 0$, so 1 is eventually fixed. Any rational point in $(0, 1)$ can be written as $x = \frac{p}{2^j q}$, where q is odd and p is odd if $j > 1$. If $q = 1$, then $D^j\left(\frac{p}{2^j}\right) = p \pmod 1 = 0$ is a fixed point, and $\frac{p}{2^j}$ is eventually fixed. If $q \geq 3$, then $D^j\left(\frac{p}{2^j q}\right) = \frac{p}{q} \pmod 1 = \frac{p}{q}$, which is eventually periodic as we showed above. Thus, all rational points are eventually periodic.

The next theorem states which iterates of a period-n point return to its starting point.

Theorem 9.1. **a.** *If p is a period-n point, then $f^{kn}(p) = p$ for any positive integer k. Moreover, if j is not a multiple of n, then $f^j(p) \neq p$.*

b. *If $f^m(p) = p$, then m is a multiple of the period of p and the period divides m.*

Proof. (a) If $f^n(p) = p$, then $f^{2n}(p) = f^n(f^n(p)) = f^n(p) = p$ and p is fixed by f^{2n}. Then, by induction,

$$f^{kn}(p) = f^n(f^{(k-1)n}(p)) = f^n(p) = p.$$

If j is not a multiple of n, then it is possible to write $j = kn + i$ with $0 < i < n$, so

$$f^j(p) = f^{kn+i}(p) = f^i(f^{kn}(p)) = f^i(p) \neq p,$$

because n is the least period.

Part **(b)** follows directly from part (a). □

While we used the expression of the doubling map to find its periodic points, in the next example, we use the graph of the tent map to determine the existence of periodic points. Therefore, we explicitly define the graph and diagonal.

Definition 9.7. The *graph of a function* f is the set of points

$$\{(x, f(x))\}.$$

The *diagonal*, denoted by Δ, is the graph of the identity function that takes x to x:

$$\Delta = \{(x, x)\}.$$

9.1. Periodic Points

A point p is fixed for a function f if and only if $(p, f(p))$ is on the diagonal Δ. If the graph of f^k intersects the diagonal at a point p, then $f^k(p) = p$ and p is periodic with a period that divides k. Therefore, finding periodic points reduces to finding points where the graphs of iterates of f intersect the diagonal.

Example 9.8. We let T be the *tent map* defined by

$$T(x) = \begin{cases} 2x & \text{for } x \leq 0.5, \\ 2(1-x) & \text{for } x \geq 0.5. \end{cases}$$

See Figure 4(a) for the graph of T and the diagonal. The period-n points are among the intersections of the graph of T^n and the diagonal Δ. Thus, we need to determine the form of the graph of T^n.

Before determining the graph of T^n and the period-n points, notice that

$$T\left(\frac{2}{7}\right) = \frac{4}{7},$$
$$T\left(\frac{4}{7}\right) = 2 - \frac{8}{7} = \frac{6}{7}, \quad \text{and}$$
$$T\left(\frac{6}{7}\right) = 2 - \frac{12}{7} = \frac{2}{7},$$

which is a period-3 orbit.

We first determine $T^2(x)$ by using the formula for T. For $0 \leq x \leq 1/2$, $T^2(x) = T(2x)$. Then, for $0 \leq 2x \leq 1/2$ or $0 \leq x \leq 1/4$, $T^2(x) = 2(2x) = 2^2 x$. For $1/2 \leq 2x \leq 1$ or $1/4 \leq x \leq 1/2$, $T^2(x) = T(2x) = 2(1-2x) = 2^2(1/2 - x)$.

For $1/2 \leq x \leq 1$, $T^2(x) = T(2(1-x))$. Then, for $1/2 \leq 2(1-x) \leq 1$ or $1/2 \leq x \leq 3/4$, $T^2(x) = T(2(1-x)) = 2[1 - 2(1-x)] = 2^2(x - 1/2)$. For $0 \leq 2(1-x) \leq 1/2$ or $3/4 \leq x \leq 1$, $T^2(x) = T(2(1-x)) = 2[2(1-x)] = 2^2(1-x)$.

Summarizing, we have

$$T^2(x) = \begin{cases} 2^2 x & \text{for } 0 \leq x \leq 1/4, \\ 2^2 \left(1/2 - x\right) & \text{for } 1/4 \leq x \leq 1/2, \\ 2^2 \left(x - 1/2\right) & \text{for } 1/2 \leq x \leq 3/4, \\ 2^2 (1-x) & \text{for } 3/4 \leq x \leq 1. \end{cases}$$

The graph of T^2 goes from 0 up to 1, back to 0, up to 1, and back down to 0. See Figure 4(b).

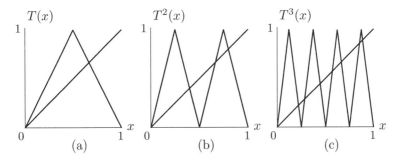

Figure 4. Graph of the tent map T, T^2, and T^3

To understand the graph of T^2 in terms of the graph of T, notice that T maps the subinterval $[0, 1/2]$ across the whole interval $[0, 1]$: As x runs from 0 to $1/2$, the value $2x$ goes across $[0, 1]$. Therefore, $T^2(x) = T(2x)$ reproduces the graph of T above the subinterval $[0, 1/2]$. In particular, there is a point ($x = 1/4$) that is mapped to $1/2$ by T, and then T maps $1/2$ to 1, so T^2 maps this point ($x = 1/4$) to 1. Thus, T^2 maps the interval $[0, 1/4]$ once across $[0, 1]$. By the time x increases to $1/2$, $1/2$ gets mapped to 1 by T and then to 0 by the second iterate of T, so T^2 takes $1/2$ to 0: Thus, T^2 maps the interval $[1/4, 1/2]$ again across $[0, 1]$.

Similarly, T maps the subinterval $[1/2, 1]$ across the whole interval $[0, 1]$, but going from 1 back to 0. Therefore, $T^2(x) = T(2(1 - x))$ reproduces the graph of T in a backward fashion over the subinterval $[1/2, 1]$. Since the backward graph of T looks like the forward graph, T has a second "tent" over the interval $[1/2, 1]$. In particular, on the subinterval $[1/2, 1]$, there is another point ($x = 3/4$) that is mapped to $1/2$ by T, and then T maps $1/2$ to 1, so T^2 maps this point ($x = 3/4$) to 1. Thus, T maps 1 to zero, which goes to zero, so $T^2(1) = 0$. Thus, T^2 maps $[1/2, 3/4]$ to $[0, 1]$ and also $[3/4, 1]$ to $[0, 1]$.

Combining the results on $[0, 1/2]$ and $[1/2, 1]$, we see that T^2 has two similar "tents" on the interval $[0, 1]$. The graph of T^2 intersects the diagonal twice for every tent (i.e., 2^2 times).

Now consider the third iterate, $T^3(x) = T^2(T(x))$. The first iterate T maps the subinterval $[0, 1/2]$ across the whole interval $[0, 1]$, so $T^3(x) = T^2(T(x)) = T^2(2x)$ reproduces the graph of T^2 above the subinterval $[0, 1/2]$. Then, T maps the subinterval $[1/2, 1]$ across the whole interval $[0, 1]$, but going from 1 back to 0; so, $T^3(x) = T^2(T(x)) = T^2(2(1 - x))$ reproduces the graph of T^2 in a backward fashion over the subinterval $[1/2, 1]$. Thus, T^3 has twice as many "tents", or 2^2 tents over $[0, 1]$. The graph of T^3 intersects the diagonal twice for every tent, or 2^3 times. See Figure 4(c).

By induction, $T^n(x)$ has 2^{n-1} "tents" and its graph intersects the diagonal 2^n times. Thus, the number of points in $\text{Fix}(f^n)$ is 2^n.

In Table 1, we make a chart that lists the period n, the number of fixed points of T^n, the number of those which have lower period, the number of least period n, and the number of orbits of period n. We have to use Theorem 9.1 to eliminate points with lower periods.

There are two fixed points of T; there are no lower periods, so column three is 0; $\#\text{Per}(1, T)$ in column four is $2 - 0 = 2$. Each fixed point is on a separate orbit so the number of fixed points in column five is also 2.

For $n = 2$, there are $2^2 = 4$ points fixed by T^2, $\#\text{Fix}(T^2) = 4$. The fixed points are also fixed by T^2 so there is a 2 in column three. The entry in column four is $\#\text{Per}(2, T) = 4 - 2 = 2$. There are two points on each orbit, so there is only $1 = 2/2$ orbit of period-2 points.

We leave it to the reader to check the rows corresponding to n equal to 3, 4, and 5.

For period 6, there are $2^6 = 64$ points fixed by T^6, $\#\text{Fix}(T^6) = 64$. Note, to find those points that are fixed by T^6 which have lower period, we need to subtract the number of period-1, period-2, and period-3 points since 1, 2, and 3 divide 6.

9.1. Periodic Points

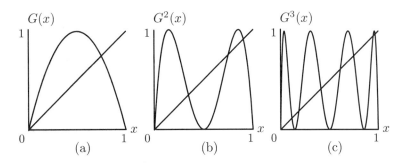

Figure 5. Graph of the logistic map G, G^2, and G^3

Thus, the number of points of period-6, $\#\operatorname{Per}(6,T)$ is $64 - (2+2+6) = 54$. Each period-6 orbit contains 6 points, so there are $54/6 = 9$ separate period-6 orbits.

n	Number of Points in $\operatorname{Fix}(f^n)$	Number of Points in $\operatorname{Fix}(f^n)$ of Lower Period	Number of Points in $\operatorname{Per}(n,f)$	Number of Period-n Orbits
1	2	0	2	2
2	$2^2 = 4$	2	2	1
3	$2^3 = 8$	2	6	2
4	$2^4 = 16$	$2+2 = 4$	12	3
5	$2^5 = 32$	2	30	6
6	$2^6 = 64$	$2+2+6 = 10$	54	9

Table 1. Number of periodic points for f either the tent map T or the logistic map G

Remark 9.9. In a calculation as in the previous example, the number of orbits of period-n is found by dividing the number of points of period-n by n. If this does not divide an integral number of times, there has been some mistake in the counting and the earlier work must be corrected.

Example 9.10. This example is similar to the preceding example in many ways, even though the formula for the function is very different. Let
$$G(x) = 4\,x(1-x).$$
This is a special case of the logistic map $g_a(x) = a\,x(1-x)$, which we consider in the next section. The function G takes the subinterval $[0, 1/2]$ all the way across the interval $[0, 1]$ in an increasing direction. It takes the subinterval $[1/2, 1]$ all the way across the interval $[0, 1]$ in a decreasing direction. By the graphical discussion similar to that for the tent map, G^n has 2^{n-1} hills, which reach all the way from 0 to 1 and back to 0. Thus, the graph of G^n intersects the diagonal 2^n times. The rest of the count of period-n points and period-n orbits is just the same as for the tent map of the preceding example. See Figure 5 and Table 1.

Exercises 9.1

1. Use a calculator, spreadsheet, or computer (with mathematical program) to calculate the first five iterates with the initial value $x_0 = 1$, for each of the following functions: (Use radians for the trigonometric functions.)
   a. $S(x) = \sin(x)$,
   b. $f(x) = 2\sin(x)$,
   c. $T(x) = \tan(x)$,
   d. $E(x) = e^x$,
   e. $F(x) = \frac{1}{4}e^x$.

2. Continue Table 1 for the number of periodic points for the tent map, for all $n \leq 10$.

3. Let f be the tripling map $f(x) = 3x \pmod{1}$. Make a table like Table 1 for the number of period-n points for f, for all $n \leq 6$.

4. Let f be the tripling map $f(x) = 3x \pmod{1}$. Determine the complete orbit of the points $1/8$ and $1/72$. Indicate whether each of these points is periodic, eventually periodic, or neither.

5. Let $f(x)$ be a function from the line $\mathbb{R}$ to itself. Suppose that, for every $k \geq 1$, f^k has $(3^k - 2^k)$ fixed points (e.g., there are $(3-2) = 1$ points fixed by $f = f^1$ and $(9-4) = 5$ points fixed by f^2). Make a table for $1 \leq k \leq 4$ showing the following: (i) k, (ii) the number of fixed points of f^k, (iii) the number of these points fixed by f^k that have lower period, (iv) number of points of period k, and (v) number of orbits of period k.

6. Consider the tent map T. Show that a point x in $[0,1]$ is eventually periodic if and only if x is a rational number. Hint: Show that $T^n(x) = $ integer $\pm 2^n x$.

7. Consider the doubling map D. Show that any rational point of the form p/q where q is odd and greater than one is a periodic point (and not just an eventually periodic point).

8. Prove that the tent map T and the quadratic map G have at least one period-k orbit for each positive integer k.

9. Consider the tent map defined by
$$T(x) = \begin{cases} 2x & \text{for } x \leq 0.5, \\ 2(1-x) & \text{for } x \geq 0.5. \end{cases}$$
   Find periodic points with the following properties:
   a. A period-3 point $x_0 < 0.5$ such that $T(x_0) < 0.5$, and $T^2(x_0) > 0.5$.
   b. A period-5 point $x_0 < 0.5$ such that $T^j(x_0) < 0.5$ for $0 \leq j < 4$ and $T^4(x_0) > 0.5$.

9.2. Iteration Using the Graph

In the preceding section, we used the graphs of the function and its iterates to determine the number of periodic points. In this section, we introduce a graphical method for finding these iterates that uses the graph of the function and the

9.2. Iteration

diagonal to determine the iterates. It is often easier to understand the long-term dynamics of iteration using this approach than can be obtained using the formula for the function being iterated. By long-term dynamics, we mean something like the fact that, for the cosine function, the sequence x_n given in Example 9.1 converges to x_∞.

Example 9.11 (Introduction of the graphical method of iteration). Example 8.2 presented the *logistic function* as a model for population growth. Here we use it to introduce the graphical method of iteration. For a parameter a, let

$$g_a(x) = ax(1-x) \quad \text{and}$$
$$g(x) = g_{1.5}(x) = 1.5\,x(1-x)$$

for the parameter value $a = 1.5$. The fixed points of g are those x for which $x = g(x)$. We can solve for the fixed points algebraically as follows:

$$x = 1.5\,x - 1.5\,x^2,$$
$$0 = 1.5\,x^2 - 0.5\,x$$
$$= 0.5\,x(3\,x - 1),$$

so $x = 0$ or $1/3$. These are just the points where the graph of g intersects the diagonal. Also, note that $g(1) = 0$ and $g\left(2/3\right) = 1/3$, so 1 and $2/3$ are eventually fixed points.

To understand the behavior of other points under iteration, we plot both the graph of g and the diagonal $\Delta = \{(x,x)\}$ on the same figure. Assume that we start with an initial condition x_0, with $0 < x_0 < 1/3$. On this figure, we draw a vertical line segment, starting at the initial condition $(x_0, 0)$ on the first axis, up to the graph of g at the point $(x_0, g(x_0)) = (x_0, x_1)$. Then, we draw the horizontal line segment from this point over to the diagonal at the point $(g(x_0), g(x_0)) = (x_1, x_1)$. The first coordinate of the preceding pair of points equals the first iterate of x_0, $x_1 = g(x_0)$. These line segments form the first "step" in our graphical iteration process. See Figure 6.

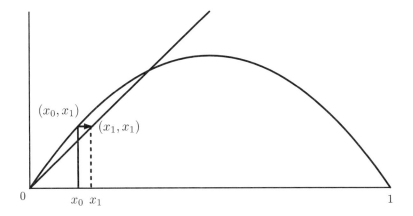

Figure 6. Graphical iteration of $g_{1.5}$ for $x_0 = 0.1$

Starting where we left off, we draw the vertical line segment from (x_1, x_1) to $(x_1, g(x_1))$ on the graph of g, followed by the horizontal line over to $(g(x_1), g(x_1)) = (x_2, x_2)$ on the diagonal, which is over the next iterate $x_2 = g(x_1)$. The line segments from (x_1, x_1) to $(x_1, g(x_1))$ to (x_2, x_2) form the second "step" in our graphical iteration process.

Continuing this process, we draw the vertical line segment from (x_2, x_2) to $(x_2, g(x_2))$ on the graph of g, followed by a horizontal line segment to (x_3, x_3) on the diagonal. At the nth stage, we draw the vertical line segment from (x_{n-1}, x_{n-1}) to $(x_{n-1}, g(x_{n-1})) = (x_{n-1}, x_n)$ on the graph of g, followed by a horizontal line segment from (x_{n-1}, x_n) to (x_n, x_n) on the diagonal. See Figure 7.

This method of exhibiting the iterates is called the *graphical method of iteration*. Together, these line segments look like stair steps, so the method is also called the *stair step method*. In other contexts, these stair steps double back on themselves and become very intertwined, so some authors call this the *cobweb method*.

For this example and for $0 < x < 1/3$, the graph of $(x, g(x))$ is above the diagonal, but $g(x) < 1/3$. Therefore, the point $(x_0, g(x_0))$ is above the diagonal but below the horizontal line $\{(x, 1/3)\}$; $(x_1, x_1) = (g(x_0), g(x_0))$ moves to the right from (x_0, x_1) and $x_0 < x_1 < 1/3$. By induction, we can repeat this to get

$$0 < x_0 < x_1 < x_2 < \cdots < x_n < 1/3.$$

We can see in Figure 7 that these points converge to the fixed point $1/3$: Theorem 9.2 at the end of this section states explicitly the conditions which imply this convergence to a fixed point. For example if $x_0 = 0.1$, then

$$\begin{aligned} x_0 = 0.1 &< x_1 = 0.13500 \\ &< x_2 = 0.17516\ldots \\ &< x_3 = 0.21672\ldots \\ &< \cdots \\ &< x_{10} = 0.33129\ldots \\ &< 1/3. \end{aligned}$$

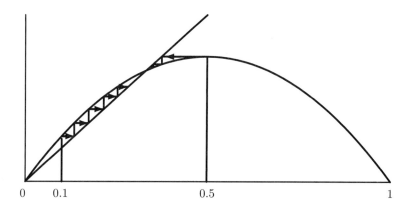

Figure 7. Graphical iteration of $g_{1.5}$ for $x_0 = 0.1$ and $x_0 = 0.5$

9.2. Iteration

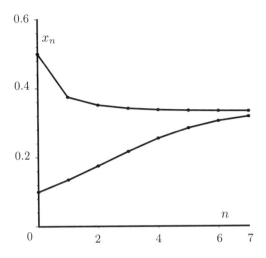

Figure 8. Plot of x_n versus n for the iteration by $g_{1.5}$ for $x_0 = 0.1$ and $x_0 = 0.5$

The graph from $1/3$ to $1/2$ is increasing and below the diagonal. Therefore, if we start with a point x_0 with $1/3 < x_0 \leq 1/2$, then

$$\frac{1}{2} \geq x_0 > x_1 > x_2 > \cdots > x_n > \frac{1}{3}.$$

Again, such x_n converge to the fixed point at $1/3$. See Figure 7 for the iterates of 0.1 and $1/2$ on the graph of g and Figure 8 for the plots of $g^{(n}(0.1)$ and $g^n(0.5)$ versus n.

Finally, for $1/2 < x_0 < 1$, $0 < x_1 = g(x_0) < 1/2$, so

$$x_{n+1} = g^{n+1}(x_0) = g^n(g(x_0)) = g^n(x_1)$$

converges to $1/3$. Thus for any point x_0 with $0 < x_0 < 1$, $g^n(x_0)$ converges to the fixed point $1/3$. We say that all these points are in the *basin of attraction* of $1/3$.

We have introduced the graphical method of iteration a particular function. The next definition gives the procedure for an arbitrary function.

Definition 9.12. *Graphical method of iteration* (or *stair step method of iteration*, or *cobweb method of iteration*) of the function f calculates the iterates of an initial condition x_0 by f as follows. Plot the graph of the function f and the diagonal on the same figure. Draw the vertical line segment from $(x_0, 0)$ on the first axis up or down to $(x_0, f(x_0)) = (x_0, x_1)$ on the graph. Then, draw the horizontal line segment from the point (x_0, x_1) on the graph of f to (x_1, x_1) on the diagonal. This gives the first iterate over the point $x_1 = f(x_0)$. Continuing, at the nth stage, draw the vertical line segment from (x_{n-1}, x_{n-1}) on the diagonal up or down to $(x_{n-1}, f(x_{n-1})) = (x_{n-1}, x_n)$ on the graph. Then, draw the horizontal line segment from (x_{n-1}, x_n) on the graph of f to (x_n, x_n) on the diagonal. This gives the nth iterate over the point $x_n = f(x_{n-1})$.

Definition 9.13. Let f be a function and p a fixed point. Then, the *basin of attraction* of p is the set of all initial conditions x_0 such that $f^n(x_0)$ converges to p

as n goes to infinity (i.e., the distance $|f^n(x_0)-p|$ goes to zero as n goes to infinity). We denote the basin of attraction by $\mathscr{B}(p;f)$, so

$$\mathscr{B}(p;f) = \{\, x_0 : |f^n(x_0) - p| \text{ goes to 0 as } n \text{ goes to } \infty \,\}.$$

For a period-k point p, the definition needs to be only slightly changed. The *basin of attraction* of a period-k point p is the set of all initial conditions x_0 such that $|f^n(x_0) - f^n(p)|$ goes to zero as n goes to infinity:

$$\mathscr{B}(p;f) = \{\, x_0 : |f^n(x_0) - f^n(p)| \text{ goes to zero as } n \text{ goes to } \infty \,\}.$$

The *basin of the period-k orbit* $\mathcal{O}_f^+(p,f)$ is the union of the basins for points on the orbit:

$$\mathscr{B}(\mathcal{O}_f^+(p);f) = \bigcup_{i=0}^{k-1} \mathscr{B}(f^i(p);f).$$

In the next theorem, we identify the features of Example 9.11 that allowed us to find the basin of attraction. The details of the proof are given in Section 9.8, at the end of the chapter.

Theorem 9.2. *Let f be a continuous function on $\mathbb{R}$ and let x^* be a fixed point.*

  a. *If $y_1 < x^*$ and $x < f(x) < x^*$ for $y_1 < x < x^*$, then the basin of attraction of x^* includes the interval (y_1, x^*), $(y_1, x^*) \subset \mathscr{B}(x^*;f)$.*

  b. *If $x^* < y_2$ and $x^* < f(x) < x$ for $x^* < x < y_2$, then the basin of attraction of x^* includes the interval (x^*, y_2), $(x^*, y_2) \subset \mathscr{B}(x^*;f)$.*

Exercises 9.2

1. Let $f(x) = (x + x^3)/2$.
    a. Find the fixed points.
    b. Use the graphical analysis to determine the dynamics for all points on $\mathbb{R}$. In other words, for initial conditions x_0 in different intervals, describe where the iterates $f^n(x_0)$ tend.

2. Let $f(x) = \frac{3}{2}(x - x^3)$.
    a. Find the fixed points.
    b. Use the graphical analysis to determine the dynamics for all points on $\mathbb{R}$. Hint: Split up the line by the fixed points and describe where the iterates $f^n(x_0)$ tend for points x_0 in each interval.

3. Let $f(x) = \frac{5}{4}(x - x^3)$.
    a. Find the fixed points.
    b. Use the graphical analysis to determine the dynamics for all points on $\mathbb{R}$. Hint: Split up the line by the fixed points and consider points in each interval separately.

4. For each of the following functions, (i) find the fixed points and (ii) use the graphical analysis to determine the dynamics for all points on $\mathbb{R}$:
    a. $S(x) = \sin(x)$.
    b. $f(x) = 2\sin(x)$.

c. $T(x) = \tan(x)$.
   d. $E(x) = e^x$.
   e. $F(x) = \frac{1}{4} e^x$.

5. Assume that f is a continuous function with $f([0,1]) = [0,1]$.
   a. Show that f must have at least one fixed point in $[0,1]$. Hint: There are points x_0 and x_1 with $f(x_0) = 0$ and $f(x_1) = x$.
   b. Show that f^2 must have at least two fixed points. Hint: Either f or f^2 must have points $y_0 < y_1$, which get mapped to 0 and 1 respectively.
   c. Assume in addition that $f(0) \neq 0$ and $f(1) \neq 1$. Show that f^2 must have at least three fixed points.

9.3. Stability of Periodic Points

In the preceding section, we introduced the graphical method of iteration. We showed that the fixed point $1/3$ for the logistic function $g_{1.5}(x)$ attracted an interval of initial conditions about the fixed point. In this section, we give terminology used to refer to such a fixed point whose basin of attraction contains a surrounding interval. We also present a criterion, in terms of the derivative of the function, for the fixed point to be attracting.

Many of the theorems involve assumptions about the derivatives of the function. So, before stating the definitions about stability, we introduce terminology relating to the existence of continuous derivatives.

Definition 9.14. A function f from $\mathbb{R}$ to $\mathbb{R}$ is said to be *continuously differentiable* or C^1 provided that f is continuous and $f'(x)$ is a continuous function of x. The function is said to be *continuously differentiable of order r* or C^r, for some integer $r \geq 1$, provided that f is continuous and the first r derivatives of f are continuous functions of x. If the function is C^r for all positive integers r, then it is called C^∞.

Example 9.15. This example contains a function which has a fixed point that we do not want to call attracting, even though the iterates of all points do converge to this fixed point. Let

$$f(x) = x + \frac{1}{2}x(1-x) \pmod{1}.$$

We think of the interval $[0,1)$ as the circle by wrapping $[0,1]$ around and identifying both 0 and 1 as one point, and the distance from 0 to a point x is the minimum of $|x|$ and $|1-x|$. Thus, points which get close to 1 are thought to get close to 0. The only fixed point is $x = 0$. For $0 < x_0 < 1$, $0 < x_0 < f(x_0) < 1$, so all points are moving to the right. Since there are no other fixed points, $\lim_{j \to \infty} f^j(x_0) = 1 \pmod{1} = 0$. Thus, the iterates of all points converge to 0.

However, if we take a point such as $x_0 = 0.01$, then iterates become about distance $1/2$ away from 0 before they return back close to 0. We do not want to call this behavior attracting, so we add a condition in the definition to disallow calling such examples attracting.

Definition 9.16. A period-n point p_0 is called *Lyapunov stable* or *L-stable* provided that, for all $r > 0$, there is a $\delta > 0$ such that, for $|x_0 - p_0| < \delta$,
$$|f^k(x_0) - f^k(p_0)| < r \qquad \text{for all } k \geq 0.$$

A period-n point p_0 is called *attracting* provided that it is Lyapunov stable and there is a $\delta_1 > 0$ such that, for $|x_0 - p_0| < \delta_1$,
$$\lim_{j \to \infty} |f^j(x_0) - f^j(p_0)| = 0.$$

Such a point is also called a *period-n sink* or an *asymptotically stable period-n point*. (Some authors call these stable period-n points, but we do not do so.)

In the definition of attracting, the condition of the limit is the most important. Example 9.15 gives a map on the circle which satisfies this limit condition but is not L-stable. For a function on $\mathbb{R}$, without taking numbers modulo any number, the condition of being L-stable follows from the limit condition (although this fact is not obvious). We include the L-stability condition to make the definition similar to the one necessary in higher dimensions and for systems of differential equations.

Definition 9.17. A period-n point p_0 is called *unstable* provided it is not L-stable; that is, there is an $r_0 > 0$ such that, for all $\delta > 0$, there is a point x_δ with $|x_\delta - p_0| < \delta$ and a $k \geq 0$ such that
$$|f^k(x_\delta) - f^k(p_0)| \geq r_0.$$

This means that, arbitrarily near p_0, there are points that move away from the orbit of p_0 by a distance of at least r_0.

For Example 9.15, 0 can be shown to be an unstable fixed point using $r_0 = 1/4$.

Definition 9.18. A period-n point p_0 is called a *repelling* provided that there is an $r_1 > 0$ such that, if $x \neq p_0$ and $|x - p_0| < r_1$, then there exists a $k \geq 0$ (which depends on x) such that
$$|f^k(x) - f^k(p_0)| \geq r_1.$$

Thus for a repelling periodic point, all points nearby move away. A repelling periodic point is also called a *source*.

Definition 9.19. A period-n point p_0 is called *semistable* provided it is attracting from one side and repelling on the other side.

The next theorem relates the absolute value of the derivative at a fixed point with the stability type of the fixed point, attracting or repelling. In Example 9.11, $g'_{1.5}(x) = \left(\frac{3}{2}\right)(1 - 2x)$, and the derivative of the function at the fixed point $1/3$ is
$$g'_{1.5}\left(\frac{1}{3}\right) = \left(\frac{3}{2}\right)\left(1 - \frac{2}{3}\right) = \frac{1}{2} < 1.$$

The fact that this number is less than one (in absolute value) is what made the basin of attraction contain an interval about the fixed point. Also, $g'_{1.5}(0) = 3/2 > 1$, and 0 is repelling.

Theorem 9.3. *Let f be a continuously differentiable function from $\mathbb{R}$ to itself.*

9.3. Stability of Periodic Points

a. *Assume that p_0 is a fixed point for $f(x)$. The following three statements give the relationship between the absolute value of the derivative and the stability of p_0:*
   (i) *If $|f'(p_0)| < 1$, then p_0 is an attracting fixed point.*
   (ii) *If $|f'(p_0)| > 1$, then p_0 is a repelling fixed point.*
   (iii) *If $|f'(p_0)| = 1$, then p_0 can be attracting, repelling, semistable, or none of these.*

b. *Assume that p_0 is a period-n point, and $p_j = f^j(p_0)$. By the chain rule,*
$$|(f^n)'(p_0)| = |f'(p_{n-1})| \cdots |f'(p_1)| \cdot |f'(p_0)|,$$
where $p_j = f^j(p_0)$. The following three statements give the relationship between the size of this derivative and the stability of p_0:
   (i) *If $|(f^n)'(p_0)| < 1$, then p_0 is an attracting period-n point.*
   (ii) *If $|(f^n)'(p_0)| > 1$, then p_0 is a repelling period-n point.*
   (iii) *If $|(f^n)'(p_0)| = 1$, then the periodic orbit can be attracting, repelling, semistable, or none of these.*

We leave a rigorous proof using the mean value theorem to Section 9.8 at the end of this chapter.

The case for a fixed point for which $0 < f'(p_0) < 1$ can be understood by looking at the graphical method of iteration. In this case, the graph of f is crossing from above the diagonal to below the diagonal; from Theorem 9.2, there is an interval about p_0 that is in the basin of attraction. This is the case for the logistic map for $a = 1.5$. See Figure 7. In the case in which $-1 < f'(p_0) < 0$, the iterate of a nearby point switches sides of p_0 with every iterate, but comes back to the same side but closer on the second iterate. See Figures 9 and 10 for $g_{2.8}(x) = 2.8\,x\,(1-x)$. For a period-n point,
$$|f^n(x) - p_0| \approx |(f^n)'(p_0)| \cdot |x - p_0| < |x - p_0|$$
if $|(f^n)'(p_0)| < 1$.

If $f'(p_0) > 1$ for a fixed point p_0, the graph of f is crossing from below the diagonal to above the diagonal; using the stair step method of iteration, it follows that p_0 is a repelling fixed point. See Figure 11. Finally, Figures 12 and 13, for $g_{3.3}(x) = 3.3\,x\,(1-x)$, illustrate a fixed point $p_{3.3}$ for which $g'_{3.3}(p_{3.3}) < -1$; the orbit shown moves away from the fixed point and converges to a period-2 orbit.

Definition 9.20. A period-n point p_0 is called *superattracting* for a map f if p_0 is also a critical point, $(f^n)'(p_0) = 0$.

Note that, by Taylor series expansion at a superattracting period-n point p_0,
$$|f^n(x) - p_0| \approx \tfrac{1}{2} |(f^n)''(p_0)| \cdot |x - p_0|^2.$$
For $\epsilon > 0$ small and $|x - p_0| < \dfrac{2\,\epsilon}{|(f^n)''(p_0)|}$,
$$|f^n(x) - p_0| < \epsilon\,|x - p_0| \ll |x - p_0|.$$
Thus, the attraction is very fast, hence the name.

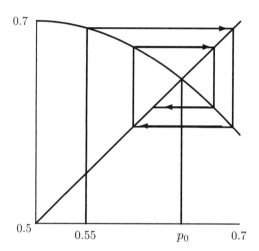

Figure 9. Graphical iteration of $x_0 = 0.55$ for $g_{2.8}(x) = 2.8x(1-x)$

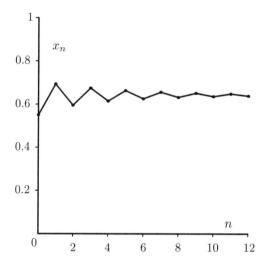

Figure 10. Plot of iteration of $x_0 = 0.55$ for $g_{2.8}(x) = 2.8x(1-x)$

Example 9.21. Consider $f(x) = -x^3$. We showed in Example 9.2 that f has a fixed point at 0 and a period-2 orbit at $\{-1, 1\}$. The derivative is $f'(x) = -3x^2$. Therefore, $f'(0) = 0$ and 0 is a superattracting fixed point.

The derivative at the period-2 point is
$$|(f^2)'(1)| = |f'(-1)| \cdot |f'(1)| = 3 \cdot 3 = 9 > 1,$$
so this period-2 orbit is repelling. Notice that we did not need to calculate the derivative of f^2, but took the product of the derivative of f.

Example 9.22. Consider the function
$$f(x) = x^3 - 1.25\,x.$$

9.3. Stability of Periodic Points

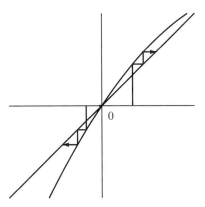

Figure 11. Graphical iteration of $x_0 = -0.5$ and $x_0 = 0.1$ for $g_{1.5}(x) = 1.5x(1-x)$

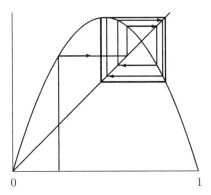

Figure 12. Graphical iteration of $x_0 = 0.25$ for $g_{3.3}(x) = 3.3\,x(1-x)$

The fixed points are 0 and ± 1.5, as can be found directly. Since $f'(x) = 3\,x^2 - 1.25$, $|f'(0)| = 1.25 > 1$ and $|f'(\pm 1.5)| = |3(2.25) - 1.25| = 5.5 > 1$, all the fixed points are repelling.

Because the function is odd, $f(-x) = -f(x)$, it seems reasonable to search for a period-2 point which satisfies $x_2 = f(x_1) = -x_1$ and $x_1 = f(x_2) = -x_2$. Thus, we solve $-x = f(x)$ or $0 = f(x) + x$:

$$0 = f(x) + x = x^3 - \frac{5}{4}x + x = x\left(x^2 - \frac{1}{4}\right).$$

The solution $x = 0$ is a fixed point, but $x = \pm 0.5$ gives a period-2 orbit, $f(0.5) = -0.5$ and $f(-0.5) = 0.5$. Then,

$$\begin{aligned}(f^2)'(0.5) &= f'(-0.5)f'(0.5) \\ &= \bigl(3(0.25) - 1.25\bigr)^2 \\ &= 0.25 < 1,\end{aligned}$$

so ± 0.5 is attracting. See Figure 14.

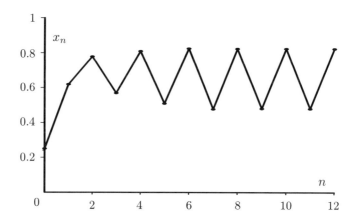

Figure 13. Plot of iteration of $x_0 = 0.25$ for $g_{3.3}(x) = 3.3\,x(1-x)$

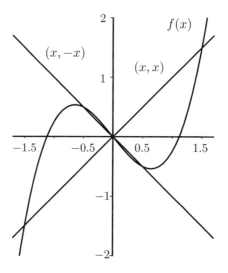

Figure 14. Plot of the graph for Example 9.22

The critical points of the function satisfy $3x^2 - 5/4 = 0$, so they are $x_c^\pm = \pm\sqrt{5/12} \approx 0.655$, and $|x_c^\pm| > |\pm 0.5|$. The critical points are outside the period-2 orbit.

We still use the usual diagonal Δ to calculate orbits using the graphical iteration method, but the comparison of the graph of $f(x)$ and the line $\Delta^- = \{\,(x,-x)\,\}$ can be used to show that the intervals $(x_c^-, 0)$ and $(0, x_c^+)$ are in the basin of attraction of the period-2 orbit. Notice that the graph of $f(x)$ is below Δ^- for $0 < x < 0.5$, so $-0.5 < f(x) < -x < 0$ and $0 < |x| < |f(x)| < 0.5$. Similarly, the graph $f(x)$ is above Δ^- for $-0.5 < x < 0$, so $0 < -x < f(x) < 0.5$ and $0 < |x| < |f(x)| < 0.5$.

9.3. Stability of Periodic Points

Combining, for $-0.5 < x < 0.5$ but $x \neq 0$,

$$0 < |x| < |f(x)| < 0.5 \quad \text{and}$$
$$0 < |x| < |f(x)| < |f^2(x)| < 0.5.$$

Thus, for $0 < x_0 < 0.5$,

$$0 < x_0 < f^2(x_0) < f^4(x_0) < \cdots < 0.5,$$

and $f^{2j}(x_0)$ converges to 0.5. By continuity, the whole orbit converges to the orbit of 0.5 and x_0 is in the basin of attraction of the period-2 orbit. A similar argument applies to $-0.5 < x_0 < 0$.

In the intervals $[x_c^-, -0.5) \cup (0.5, x_c^+]$, $0.5 < |f(x)| < |x| \leq |x_c^\pm|$, so the iterates converge down to the periodic orbit and are in the basin of attraction. Combining, we get

$$[x_c^-, 0) \cup (0, x_c^+] \subset \mathscr{B}(\mathcal{O}(0.5); f).$$

Next, on the interval $(-1.5, x_c^-)$, the graph of f is always above the usual diagonal Δ and below the value $f(x_c^-) < x_c^+$, so iterates of points keep increasing until they land in $[x_c^-, x_c^+]$. If they do not land exactly on 0, further iteration takes them to the period-2 orbit. Similar analysis applies to $(x_c^+, 1.5)$, except that the graph is below the diagonal and above the value $f(x_c^+) > x_c^+$. Letting $\mathscr{B}(0; f)$ be the set of all eventually fixed points at 0,

$$\mathscr{B}(\mathcal{O}(0.5); f) \supset (-1.5, 1.5) \smallsetminus \mathscr{B}(0; f).$$

Since points outside $[-1.5, 1.5]$ go to $\pm\infty$ under iteration, this is exactly the basin of attraction of the period-2 orbit.

Another consequence of Theorem 9.3 is that the derivative $(f^n)'(x_i)$ must be equal at all points x_i on the same period-n orbit. We leave to the exercises the use of this fact to determine the two separate period-3 orbits for $G(x) = 4\,x\,(1-x)$, using the graph of G^3 given in Figure 15.

9.3.1. Newton Map. We introduced the Newton method for finding the root of a polynomial or function of one variable in Section 8.1. If $f(x)$ is a polynomial, or even a more general function, we are sometimes interested in determining the roots (i.e., the values of x where $f(x) = 0$). It is theoretically possible to factor a polynomial and find the real and complex roots. For a high degree polynomial, there is no algebraic expression for finding the roots. It is of practical importance to find an approximation of the roots when no exact expression for the roots exists. The Newton method (or Newton–Raphson method) is an iterative scheme to find an approximate root.

Assume that x_0 is an initial guess of a root. The tangent line to the graph of f at x_0 is given by

$$y = f(x_0) + f'(x_0)(x - x_0).$$

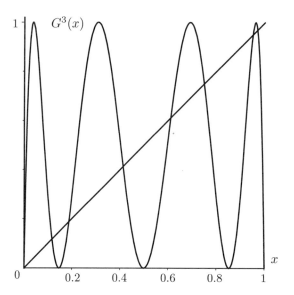

Figure 15. Plot of the graph of G^3

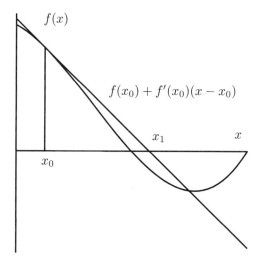

Figure 16. The Newton map from the plot of $f(x)$

The tangent line is the best linear (affine) approximation of the function near x_0. This line crosses the x-axis at the point x_1 where y is zero; that is,

$$0 = f(x_0) + f'(x_0)(x_1 - x_0) \quad \text{or}$$
$$x_1 = x_0 - \frac{f(x_0)}{f'(x_0)}.$$

This formula is valid as long as the tangent line is not horizontal, $f'(x_0) \neq 0$. See Figure 16. Starting again with x_1, we can get another point x_2 where the tangent

9.3. Stability of Periodic Points

line at x_1 crosses the x-axis:
$$x_2 = x_1 - \frac{f(x_1)}{f'(x_1)}.$$

If we let

(9.1) $$N_f(x) = x - \frac{f(x)}{f'(x)},$$

then $x_1 = N_f(x_0)$ and $x_2 = N_f(x_1)$. This induced map N_f is called the *Newton map for the function* f. Repeating this process, if x_n is the nth approximation of the root, then the $(n+1)^{\text{st}}$ approximation is given by

$$x_{n+1} = N_f(x_n) = x_n - \frac{f(x_n)}{f'(x_n)}.$$

The next theorem shows that if the initial guess is good enough (or, in fact, for most initial guesses), the iterative scheme converges to a root of the polynomial as n goes to infinity.

Theorem 9.4. *Assume $f(x)$ is a function with two continuous derivatives. Assume that x^* is a simple zero of f, so that $f(x^*) = 0$ but $f'(x^*) \neq 0$. Then, x^* is a superattracting fixed point of N_f (i.e., $N_f(x^*) = x^*$ and $N_f'(x^*) = 0$).*

Proof. If $f(x^*) = 0$, then

$$N_f(x^*) = x^* - \frac{f(x^*)}{f'(x^*)} = x^* - \frac{0}{f'(x^*)} = x^*,$$

so x^* is a fixed point of N_f. Also,

$$N_f'(x^*) = 1 - \frac{f'(x^*)^2 - f(x^*)f''(x^*)}{f'(x^*)^2}$$
$$= 1 - \frac{f'(x^*)^2}{f'(x^*)^2}$$
$$= 0.$$

Thus, x^* is a superattracting fixed point and has a basin of attraction which contains an open interval about x^*. □

Example 9.23. Let $f(x) = x^2 - 7$. The roots are $\pm\sqrt{7}$. The Newton map for this f is given by

$$N_f(x) = x - \frac{x^2 - 7}{2x}.$$

The point $\bar{x} = \sqrt{7}$ is a superattracting fixed point for the Newton map by Theorem 9.4. If we take the initial guess of $x_0 = 1$, then

$$x_1 = N_f(1) = 1 - \frac{(-6)}{2} = 4.00000,$$
$$x_2 = N_f(x_1) = 4 - \frac{9}{8} = 2.87500,$$
$$x_3 = N_f(x_2) = 2.87500 - \frac{2.87500^2 - 7}{2 \times 2.87500} \approx 2.65489,$$

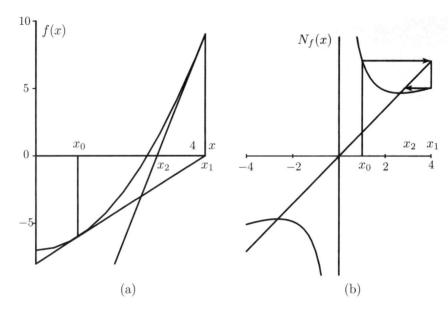

Figure 17. For $f(x) = x^2 - 7$: (a) the tangent lines to the graph of $f(x)$, (b) the Newton map

$$x_4 = N_f(x_3) \approx 2.65489 - \frac{2.65489^2 - 7}{2 \times 2.65489} \approx 2.64577,$$

$$x_5 = N_f(x_4) \approx 2.64577 - \frac{2.64577^2 - 7}{2 \times 2.64577} \approx 2.64575.$$

Even with a bad guess of $x_0 = 1$, after just five iterates, $|x_5 - x_4| < 2 \times 10^{-5}$. Since $\sqrt{7} \approx 2.6457513$ and the preceding process did, in fact, give the correct root to five decimal places, $x_5 \approx 2.64575$. See Figure 17.

For higher degree polynomials, the derived Newton map can have periodic points or points with chaotic iterates: for such initial conditions, the iterates do not converge to a root of the polynomial. See Section 10.7.1. Still, most initial conditions do converge to a root of the polynomial, and certainly if we start near a root, then the iterative process does converge rapidly to the nearby root.

See the exercises for the stability of a zero with higher multiplicity.

9.3.2. Fixed and Period-2 Points for the Logistic Family. We have already considered the logistic family of maps $g_a(x) = a\,x\,(1 - x)$ several times. In this section, we discuss the stability of the fixed points and the basin of attraction of the fixed point in the range of parameters for which it is attracting. Also, we find the period-2 points and discuss their stability.

Theorem 9.5. *The fixed points of the logistic family of maps $g_a(x) = a\,x\,(1 - x)$ are 0 and $p_a = \frac{a-1}{a}$. For $a > 1$, $0 < p_a < 1$. The stability of the fixed points is as follows.*

The fixed point 0 is attracting for $0 < a < 1$ and repelling for $a > 1$.

9.3. Stability of Periodic Points

The fixed point p_a is repelling and negative for $0 < a < 1$, is attracting for $1 < a < 3$, and is repelling with $g'_a(p_a) < -1$ for $a > 3$.

Proof. The fixed points of g_a satisfy
$$x = g_a(x) = a\,x - a\,x^2,$$
$$0 = x\,(a\,x - a + 1),$$

which have solutions as given in the statement of the theorem.

(a) To check the stability, we compute $g'_a(x) = a - 2\,a\,x$. At the fixed point 0, $g'_a(0) = a$, so 0 is attracting for $0 < a < 1$ and repelling for $a > 1$. We ignore the parameter values $a \leq 0$.

(b) For the fixed point p_a,
$$g'_a(p_a) = a - 2\,a\left(\frac{a-1}{a}\right) = a - 2\,a + 2 = 2 - a.$$

For $0 < a < 1$, p_a is repelling (and p_a is negative). For $1 < a < 3$, $|2 - a| < 1$ and the fixed point is attracting. For $a > 3$, $g'_a(p_a) = 2 - a < -1$ and the fixed point is repelling. □

Theorem 9.6. *The logistic family of maps $g_a(x) = a\,x\,(1-x)$ has a single period-2 orbit that occurs only for $a > 3$; it is given by*
$$q_a^\pm = \frac{1}{2} + \frac{1}{2a} \pm \frac{1}{2a}\sqrt{(a-3)(a+1)}.$$

The period-2 orbit is attracting for $3 < a < 1 + \sqrt{6}$ and repelling for $a > 1 + \sqrt{6}$.

Proof. The points fixed by g_a^2 satisfy
$$0 = g_a^2(x) - x$$
$$= g_a(ax(1-x)) - x$$
$$= a[ax(1-x)][1 - ax(1-x)] - x$$
$$= -x\,(ax - a + 1)[a^2 x^2 - (a^2 + a)x + (a+1)].$$

The first two factors are zero at the fixed points 0 and p_a. Therefore, the period-2 points are determined by setting the last factor equal to zero. The solutions of setting this last factor equal to zero are
$$q_a^\pm = \frac{1}{2} + \frac{1}{2a} \pm \frac{1}{2a}\sqrt{(a-3)(a+1)}.$$

These roots are real for $a \geq 3$. Since the iterate of a period-2 point also has period 2, $g_a(q_a^+) = q_a^-$ and $g_a(q_a^-) = q_a^+$. For $a = 3$, these are the same point as the fixed point p_3, $p_3 = 2/3 = q_3^\pm$. Therefore, the period-2 orbit exists for $a > 3$.

To check the stability, we calculate

$$(g_a^2)'(q_a^\pm) = g_a'(q_a^-) \cdot g_a'(q_a^+)$$
$$= (a - 2\,a\,q_a^-)(a - 2\,a\,q_a^+)$$
$$= \left(-1 + \sqrt{(a-3)(a+1)}\right)\left(-1 - \sqrt{(a-3)(a+1)}\right)$$
$$= 1 - (a^2 - 2a - 3)$$
$$= -a^2 + 2\,a + 4.$$

The derivative equals 1 when $0 = a^2 - 2\,a - 3 = (a-3)(a+1)$, or at $a = 3$; it equals -1 when $0 = a^2 - 2\,a - 5$, or $a = 1 + \sqrt{6}$; it is between 1 and -1 for $3 < a < 1 + \sqrt{6}$. Thus, the periodic orbit is attracting for $3 < a < 1 + \sqrt{6}$, and is repelling for $a > 1 + \sqrt{6}$. □

Basins of attraction

The next proposition specifies the basin of attraction for p_a for the parameter values for which it is attracting.

Proposition 9.7. *Consider the logistic family of maps $g_a(x)$ for $1 < a \leq 3$. Then, the basin of attraction of the fixed point p_a is the whole open interval $(0,1)$, $\mathscr{B}(p_a; g_a) = (0,1)$. In particular, there are no other periodic points than the fixed points at 0 and p_a.*

Proof. In Example 9.11, we proved this result for $a = 1.5$. The argument given there applies for all a with $1 < a \leq 2$. For all this range of parameters, the fixed point p_a is less than the critical point 0.5, where the derivative is zero.

Therefore, we consider the case for which $2 < a \leq 3$ and $0.5 < p_a$. (Note that for $a = 2.5$, $p_{2.5} = 0.6$.) We first find the points that maps to 0.5. We solve

$$0.5 = a\,x - a\,x^2$$
$$0 = a\,x^2 - a\,x + 0.5, \quad \text{yielding}$$

$$x_+ = \frac{1 + \sqrt{1 - \frac{2}{a}}}{2a} \quad \text{and} \quad x_- = \frac{1 - \sqrt{1 - \frac{2}{a}}}{2}.$$

Since $g_a(x_+) = 0.5 < p_a = g_a(p_a)$, $0.5 < p_a < x_+$. We want to show the graph of g_a^2 on $[0.5, x_+]$ looks like Figure 19(b). We use the fact that $g_a'(x) < 0$ on $(0.5,]$ so g_a is decreasing. The iterate of 0.5, $p_a < g_a(0.5) = a/4$. The following calculation shows that $g_a(0.5) = a/4 < x_+ = \frac{1}{2}\left(1 + \sqrt{1 - 2/a}\right)$, so inside the interval $[0.5, x_+]$,

$$a \stackrel{?}{<} 2 + 2\sqrt{1 - 2/a},$$
$$a - 2 \stackrel{?}{<} 2\sqrt{1 - 2/a},$$
$$a^2 - 4\,a + 4 \stackrel{?}{<} 4 - \frac{8}{a},$$
$$a^3 - 4\,a^2 + 8 \stackrel{?}{<} 0,$$

9.3. Stability of Periodic Points

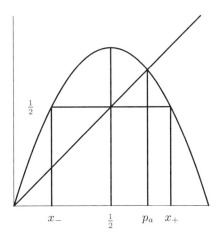

Figure 18. Points $x_\pm$

$$(a-2)(a^2 - 2a - 4) \stackrel{?}{<} 0,$$

$$a^2 - 2a - 4 \stackrel{?}{<} 0.$$

The left-hand side of the last line is zero at $a = 1 \pm \sqrt{5}$, and it is negative for $1 - \sqrt{5} < a < 1 + \sqrt{5}$. Since $1 - \sqrt{5} < 2$ and $3 < 1 + \sqrt{5}$, the left-hand side is negative for $2 < a \leq 3$. Therefore, for $2 < a \leq 3$, both end points of $[0.5, x_+]$ map into $[0.5, x_+]$. Since $g'_a(x), 0$ on $(0.5.x_+)$ and $g_a((0.5, x_+)) = (0.5, ^a/4)$, g_a^2 is monotone decreasing on $[0.5, x_+]$, and $g_a([0.5, x_+]) \subset ([0.5, x_+]$. Because the interval $[0.5, x_+]$ is mapped into itself by g_a, we call it positively invariant.

Moreover, since $g(x) \leq g_a(0.5) < x_+$ for all x in $[0, 1]$, it follows that

$$g_a^2(0.5) > g_a(x_+) = 0.5 \quad \text{and} \quad g_a^2(x_+) = g_a(0.5) < x_+.$$

The function g_a^2 is monotone increasing on $[0.5, x_+]$ and has only 0 and p_a as fixed points, so its graph lies above the diagonal and $g_a^2(x) > p_a$ on $[0.5, p_a)$. Similarly, the graph of g_a^2 lies below the diagonal and $g_a^2(x) < p_a$ on $(p_a, x_+]$. See Figure 19(b). Applying Theorem 9.2 to g_a^2, we get that $[0.5, x_+] \subset \mathscr{B}(p_a; g_a^2)$. See Figure 20.

Since $g_a([x_-, 0.5]) = g_a([0.5, x_+])$ and $g_a^2([x_-, 0.5]) = g_a^2([0.5, x_+])$,

$$[x_-, 0.5] \subset \mathscr{B}(p_a; g_a^2).$$

Next, for any point $0 < x_0 < 0.5$, each iterate by g_a^2 is larger, so

$$0 < x_0 < x_2 < x_4 < \cdots < x_{2m}$$

until $x_{2n} > 0.5$. Since $0.5 < x_{2n} \leq g_a(0.5) < x_+$, $x_{2n} \in [0.5, x_+] \subset \mathscr{B}(p_a; g_a^2)$ and x_0 must be in $\mathscr{B}(p_a; g_a^2)$. This shows that the interval $(0, x_+)$ is in $\mathscr{B}(p_a; g_a^2)$.

Then, for $x_+ < x_0 < 1$, $g_a^2(x_0) = x_2$ must be in $(0, g_a(0.5)] \subset (0, x_+]$, so by the previous cases, x_2 and x_0 must be in $\mathscr{B}(p_a; g_a^2)$. We have shown that all of $(0, 1)$ is in $\mathscr{B}(p_a; g_a^2)$.

Finally, we need to worry about the odd iterates. Since p_a is attracting, there is a small interval about p_a, $\mathbf{J} = (p_a - \delta, p_a + \delta)$ that is in the basin of attraction of p_a

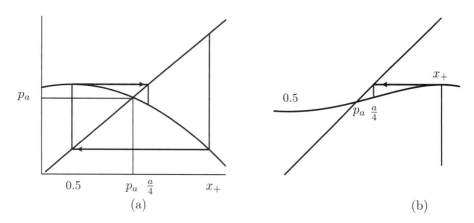

Figure 19. Logistic map for $2 < a \leq 3$: (a) Iteration of x_+. (b) Graph of g_a^2 between 0.5 and x^+.

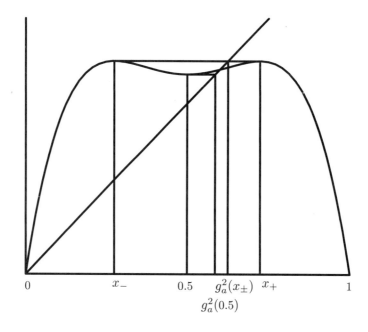

Figure 20. For $2 < a \leq 3$, the graph of g_a^2 and iterates $g_a^2(0.5)$ and $g_a^2(x_-) = g_a^2(x_+)$.

for g_a. For any x_0 in $(0,1)$, there is an even iterate $x_{2k} \in \mathbf{J}$. Then, $g_a^{2k+j}(x_0) \in \mathbf{J}$ and must converge to p_a. Therefore, all iterates converge to p_a, and x_0 must be in the basin of attraction of p_a for g_a and not just g_a^2. This completes the proof. □

Example 9.24. We are interested in the basin of attraction for the period-2 orbit. Consider the case for $a = 3.3$. The periodic-2 orbit is made up of the two points $q^-3.3 \approx 0.4794$ and $q^+3.3 \approx 0.8236$. Table 2 gives the orbits for the three initial conditions 0.25, 0.5, and 0.95. Notice that all three converge to the period-2 orbit. These orbits do not actually land on the period-2 orbit, but merely become equal to the number of decimal places displayed. Figure 12 shows the corresponding stair

9.3. Stability of Periodic Points

step plot for $x_0 = 0.25$. Figure 21 shows the plot of x_j versus j for $x_0 = 0.5, 0.95$, and $f(0.25) = 0.61875$.

j	$g_{3.3}^j(0.25)$	$g_{3.3}^j(0.5)$	$g_{3.3}^j(0.95)$
0	0.2500	0.5000	0.9500
1	0.6188	0.8250	0.1568
2	0.7785	0.4764	0.4362
3	0.5691	0.8232	0.8116
4	0.8093	0.4804	0.5047
5	0.5094	0.8237	0.8249
6	0.8247	0.4792	0.4766
7	0.4771	0.8236	0.8232
8	0.8233	0.4795	0.4803
9	0.4802	0.8236	0.8237
10	0.8237	0.4794	0.4792
11	0.4792	0.8236	0.8236
12	0.8236	0.4794	0.4795

Table 2. Three orbits for $g_{3.3}$

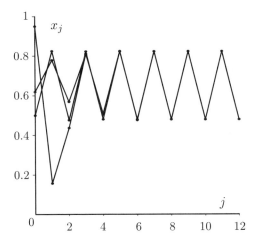

Figure 21. Plot of x_j versus j for $g_{3.3}(x) = 3.3\,x(1-x)$ for $x_0 = 0.5, 0.95$, and $f(0.25) = 0.61875$

To determine the basin of attraction for the period-2 orbit takes more work. The only points that are attracted to the fixed point p_a are the points in $(0, 1)$ that land exactly on the fixed point p_a for some iterate (i.e., they are eventually fixed points), so they are not in the basin of the period-2 orbit; these points could be called the basin of attraction of p_a. We write $g_a^{-j}(p_a)$ for the set of all points whose j^{th} iterate land on p_a; that is,

$$g_a^{-j}(p_a) = \{\, x : g_a^j(x) = p_a \,\}.$$

Thus, the set of eventually fixed points in $(0,1)$ is the set

$$\mathscr{B}(p_a; g_a) = \bigcup_{j=0}^{\infty} g_a^{-j}(p_a),$$

which is a countable set. The remaining set of points in $(0,1)$ equals the basin of the period-2 orbit as the next theorem states.

Theorem 9.8. *Take $3 < a < 1 + \sqrt{6}$ for the logistic family of maps $g_a(x)$. Then, the basin of attraction of the period-2 orbit is the set of all points in $(0,1)$ which are not eventually fixed points, so*

$$\mathscr{B}(\mathcal{O}(q_a^{\pm}); g_a) = (0,1) \smallsetminus \bigcup_{j=0}^{\infty} g_a^{-j}(p_a).$$

In particular, the only periodic points are the two fixed points and the period-2 orbit.

For values of the parameter $a > 1 + \sqrt{6}$, there are other periodic points. However, the family can have at most one attracting periodic orbit for any one-parameter value, as we see in Section 9.4. Immediately after $1 + \sqrt{6}$, there is an attracting period-4 orbit, then an attracting period-8 orbit, etc. Experimentally, using computer simulation, we find that there is an attracting periodic orbit whose period is a power of 2 for $3 < a < a_\infty \approx 3.569946$. These periods go to infinity as a approaches a_∞ via a cascade of period doubling, as we discuss in Section 9.5.1. It is impossible to determine all these period-2^k points algebraically by means of computing the iterates of the function. Therefore, we restrict our algebraic consideration to finding the fixed points and period-2 points, and exhibit all other periodic points for this family by means of computer simulation.

Applying the Graphical Method of Iteration

We summarize the way we can apply the graphical method of iteration to examples where fixed points and period-2 points are the only periodic attracting orbits.

1. Around any attracting fixed point with positive derivative, try to find the maximal interval where Theorem 9.2 applies.

2. Besides breaking up the line with the fixed points, it is sometimes necessary to use critical points and points that map to them to find a positively invariant interval. See Example 9.22 and the proof of Theorem 9.7.

3. Once an interval about an attractive periodic point that is in in the basin is found, any interval that maps into this set must also be included in the basin.

Exercises 9.3

1. Consider the function $f(x) = \frac{5}{4}x - x^3$.
    a. Find the fixed points and classify each of them as attracting, repelling, or neither.

Exercises 9.3

b. Plot the graph of f and the diagonal together. Pay special attention to the intervals where f is above and below the diagonal. Also, note the critical points $\pm\sqrt{5/12}$ and the intervals on which f is increasing or decreasing.

c. Describe in words the orbits of representative points in different subintervals of $-\sqrt{5/12} \leq x \leq \sqrt{5/12}$ using the graphical method of iteration as a tool.

2. For each of the following functions, show that 0 is a fixed point, and the derivative of the function is 1 at $x = 0$. Describe the dynamics of points near 0. Is 0 attracting, repelling, semi-attracting, or none of these?

   a. $f(x) = x - x^2$.
   b. $g(x) = x + x^3$.
   c. $h(x) = x - x^3$.
   d. $k(x) = \tan(x)$.

3. Determine the stability of all the fixed points of the following functions:

   a. $S(x) = \sin(x)$.
   b. $f(x) = 2\sin(x)$.
   c. $T(x) = \tan(x)$.
   d. $E(x) = e^x$.
   e. $F(x) = \frac{1}{4}e^x$.

4. Consider the function $Q(x) = x^2 - c$, with $c = 1.11$.

   a. Find the fixed points and determine whether they are attracting or repelling.
   b. Find the points of period two and determine whether they are attracting or repelling. Hint: The points of period two satisfy
   $0 = Q^2(x) - x = x^4 - 2cx^2 - x + c^2 - c = (x^2 - x - c)(x^2 + x + 1 - c)$.

5. Find the period-2 orbit for $f(x) = 2x^2 - 5x$. Is this orbit attracting, repelling, or neither?

6. Consider the function $f(x) = x^3 - \frac{3}{2}x$.

   a. Find the fixed points and period-2 points, and classify each of them as attracting, repelling, or neither.
   Hint: $f^2(x) - x = \frac{1}{8}x(2x^2 - 5)(2x^2 - 1)(2x^4 - 3x^2 + 2)$. They also solve $f(x) = -x$ or $f(x) + x = 0$.
   b. Plot the graph of f and the diagonal together. Pay special attention to the intervals where f is above and below the diagonal. Also, note the critical points and the intervals on which f is increasing or decreasing.
   c. Plot the graph of f^2 and the diagonal together. Note that a point a, which is mapped by f to a critical point of f, is a critical point of f^2, $(f^2)'(a) = f'(f(a))f'(a) = 0$. What are the values $f^2(a)$ for these points?
   d. Use the graphical method of iteration to determine the basin of attraction of the period-2 orbit. Hint: Note that some points can land exactly on 0 and are eventually fixed points. Therefore, describe the basin as an interval minus certain points.

7. For $a = 3.839$, the logistic map $g_a(x) = ax(1 - x)$ has a period-3 cycle at approximately the points $x_0 = 0.959$, $x_1 = g_a(x_0) = 0.150$, and $x_2 = g_a^2(x_0) =$

0.489. Assuming that these numerical values are exact, determine the stability of this cycle.

8. Consider the function $f(x) = r\,x\,e^{r(1-x/K)}$, with $r > 0$ and $K > 0$. Find the fixed points and determine their stability type.

9. Consider the function $f(x) = -\frac{3}{2}x^2 + \frac{5}{2}x + 1$. Notice that $f(0) = 1$, $f(1) = 2$, and $f(2) = 0$ is a period-3 orbit. Are these period-3 points attracting or repelling?

10. Consider the tent map of slope r given by
$$T_r(x) = \begin{cases} r\,x & \text{if } x \leq \frac{1}{2}, \\ r\,(1-x) & \text{if } x \geq \frac{1}{2}. \end{cases}$$
In this exercise, we restrict to the values $1 < r \leq 2$.
    a. Find the fixed points and determine their stability.
    b. Find an expression for $T_r^2(x)$ on $[0,1]$. Hint: Divide the interval into four subintervals on which the slope is constant.
    c. Find the points of period two and determine their stability. How large does r have to be for there to be a period-2 orbit?

11. Consider the logistic map $G(x) = 4\,x\,(1-x)$. Let $q_0 = 0 < q_1 < q_2 < q_3 \cdots < q_7$ be the eight points left fixed by G^3. Determine which are the two fixed points and which other points are grouped together into period-3 orbits. Hint: Remember that the fixed points are 0 and $3/4$. Also, $(G^3)'(x)$ must have the same value along an orbit, in particular, it must have the same sign. Finally, use the plot of the graph of G^3 given in Figure 15.

12. The graph of f^2 is shown in Figure 22 for a certain function f, including the five points $q_1 < q_2 < q_3 < q_4 < q_5$, which are fixed by f^2. Why must at least three of these five points be fixed by f, and which are they? Which pair of points might be a period-2 orbit? Hint: If p is a fixed point, then $(f^2)'(p) = f'(p)^2 \geq 0$. If $\{q_1, q_2\}$ is a period-2 point, then $(f^2)'(q_1) = f'(q_2)\,f'(q_1) = (f^2)'(q_2)$ and the slopes must be equal.

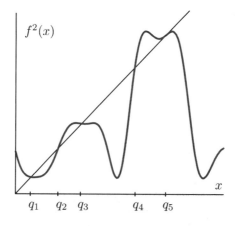

Figure 22. Graph of f^2 for Exercise 12

13. Consider the map $f(x) = \frac{3}{2}x - \frac{1}{2}x^3$.
    a. Find the fixed points and classify each of them as attracting, repelling, or neither.
    b. Use the graphical method of iteration to determine the dynamic behavior for all the points between the fixed points.
    c. What happens to points greater than all the fixed points and those less than all the fixed points?

14. Let $p(x) = (x - x_0)^k g(x)$, with $k > 1$ and $g(x_0) \neq 0$. Assume that $p(x)$ and $g(x)$ have at least two continuous derivatives. Show that the derivative of the Newton map for p at x_0 is $(k-1)/k$, so x_0 is an attracting fixed point for N_p.

15. Consider $f(x) = x^3 - 5x$ and its Newton Map $N_f(x)$.
    a. Check that the points $x = 0$ and $\pm\sqrt{5}$ are the only fixed points of N_f and that they are all attracting.
    b. Check that N_f has a period-2 orbit ± 1, $N_f(1) = -1$ and $N_f(-1) = 1$. Show that it is repelling.
    c. Show that the interval $\left(\sqrt{5/3}, \infty\right)$ is in the basin of attraction of $\sqrt{5}$. Similarly, show that the interval $\left(-\infty, -\sqrt{5/3}\right)$ is in the basin of attraction of $-\sqrt{5}$. Note that almost all of the points in the two intervals $\left(-\sqrt{5/3}, -1\right)$ and $\left(1, \sqrt{5/3}\right)$ are in one of these two basins, but we do not ask you to show this.
    d. Show that the basin of attraction of 0 contains $(-1, 1)$. Hint: Plot $N(x)$ and $-x$ on the same plot. The graphs of $N(x)$ and $-x$ intersect at 0 and ± 1. Show that $0 > N(x) > -x$ for $0 < x < 1$, so $|N(x)| < |x|$. Similarly, show that $0 < N(x) < -x$ for $-1 < x < 0$, so $|N(x)| < |x|$. Now, using the plot of $N(x)$ and x, calculate the iteration of points in $(-1, 1)$. The iterates switch sides but the preceding estimates show that the points converge to 0.
    e. What happens to points in the intervals $\left(-\sqrt{5/3}, -1\right)$ and $\left(1, \sqrt{5/3}\right)$? Notice that some points from each side are in the basins of $\pm\sqrt{5}$.

16. Consider the function $f(x) = \cos(x) - x$, using the variable x in radians. Calculate enough iterates of the Newton map starting at $x_0 = 0$ (using a hand calculator or computer) to estimate the point x^* with $\cos(x^*) = x^*$ to four decimal places.

17. Assume that $f(x)$ has $f(a) = a$, $f'(a) = 1$, and $f''(a) = 0$.
    a. If $f'''(a) < 0$, explain why a is an attracting fixed point.
    b. If $f'''(a) > 0$, explain why a is a repelling fixed point.

18. Assume that $\mathbf{F}$ is a map from $\mathbb{R}^n$ to $\mathbb{R}^n$ that is a contraction (i.e., there is a constant r with $0 < r < 1$ such that $|\mathbf{F}(\mathbf{x}) - \mathbf{F}(\mathbf{y})| \leq r |\mathbf{x} - \mathbf{y}|$ for all $\mathbf{x}$ and $\mathbf{y}$ in $\mathbb{R}^n$). Take any point $\mathbf{x}_0$ in $\mathbb{R}^n$ and define $\mathbf{x}_j = \mathbf{F}^j(\mathbf{x}_0)$.
    a. Show that
    $$|\mathbf{x}_{k+1} - \mathbf{x}_k| \leq r |\mathbf{x}_k - \mathbf{x}_{k-1}|,$$
    and by induction, that
    $$|\mathbf{x}_{k+1} - \mathbf{x}_k| \leq r^k |\mathbf{x}_1 - \mathbf{x}_0|.$$

b. Show that
   $$|\mathbf{x}_{k+j} - \mathbf{x}_k| \leq \frac{r^k}{1-r}|\mathbf{x}_1 - \mathbf{x}_0|.$$
   c. Part (b) shows that the sequence of points $\mathbf{x}_k$ in $\mathbb{R}^n$ is what is called a Cauchy sequence and must converge to a limit point $\mathbf{x}^*$. Since both $\mathbf{x}_k$ and $\mathbf{x}_{k+1} = \mathbf{F}(\mathbf{x}_k)$ converge to $\mathbf{x}^*$, show that $\mathbf{x}^*$ must be a fixed point.
   d. Show that there can be at most one fixed point.
19. Let
$$f(x) = \frac{2}{3}x^3 - \frac{1}{2}x.$$
   a. Find the fixed points and determine their stability type as attracting or repelling.
   b. Find the critical points, where $f'(x) = 0$. Call the critical points $x_c^{\pm}$.
   c. For $x_c^- \leq x \leq x_c^+$, show that $-1/2 \leq f'(x) \leq 0$. By the Mean Value Theorem, conclude that
   $$|f(x) - f(0)| \leq \frac{1}{2}|x - 0| \qquad \text{for } x_c^- \leq x \leq x_c^+.$$
   d. Determine the basin of attraction of 0.

9.4. Critical Points and Basins

In Section 9.3.2, we saw that the logistic family g_a has an attracting fixed point p_a for $1 < a < 3$, and an attracting period-2 orbit for $3 < a < 1 + \sqrt{3}$. For larger parameter values, g_a has attracting periodic orbits of periods 4, 8, and then all powers 2^k. (See the discussion in Section 9.5.1.) In this section, we explain why there are no parameter values for which the logistic family has two attracting periodic orbits for the same parameter value. The reason is based on what is called the Schwarzian derivative. In 1978, D. Singer [**Sin78**] realized that a quantity that had previously been used in complex analysis could be applied in this situation to obtain the result that had previously been observed experimentally.

The logistic map g_a has derivative $g_a'(x) = a - 2ax$, which is zero only at the point $x = 0.5$ (i.e., g_a has a single *critical point* $x = 0.5$). The main theorem of this section shows that any attracting periodic orbit for g_a must have this critical point in its basin. Since there is only one critical point, and it can be in the basin of only one orbit, there can be only one attracting periodic orbit.

This restriction on attracting periodic orbits of g_a is proved by considering a measurement on the nonlinearity defined in terms of the derivatives of g_a, called the Schwarzian derivative $S_{g_a}(x)$. If $S_{g_a}(x)$ is always negative, then there is no iterate g_a^k for which a portion of its graph looks like that for $f(x) = \frac{3}{4}x + x^3$. This function $f(x)$ has three fixed points at 0 and ± 0.5, and the basin of 0 contains the whole interval $(-0.5, 0.5)$ between the other points, but there is no critical point in this interval. See Figure 23

The Schwarzian derivative $S_f(x)$ has the following two properties:

1. If $S_f(x) < 0$ for all x with $f'(x) \neq 0$, then the graph between three fixed points cannot be like that in Figure 23.

9.4. Critical Points and Basins

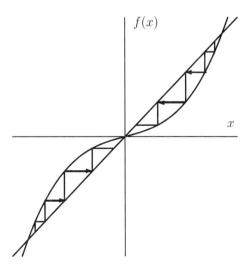

Figure 23. Impossible for negative Schwarzian derivative

2. If $S_f(x) < 0$ for all x with $f'(x) \neq 0$, then $S_{f^n}(x) < 0$ for all x with $(f^n)'(x) \neq 0$.

Property (2) implies that we can insure that $S_{f^n}(x) < 0$ without calculating the iterates of the map. Then, by property (1) applied to f^n, no iterate of the map can have a piece of its graph that looks like Figure 23 that does not have a critical point in the basin of attraction of an attracting periodic point.

To find such a quantity, we start with the second derivative, scaled by or relative to the first derivative,

$$N_f(x) = \frac{f''(x)}{f'(x)},$$

called the *nonlinearity* of f. (Note that, although we use the same notation, this map $N_f(x)$ has nothing to do with the Newton map.) If f is linear, then $f''(x) \equiv 0$ and the nonlinearity $N_f(x) \equiv 0$, as its name would suggest. Next, we consider the derivative of the nonlinearity, or the infinitesimal change in the nonlinearity:

$$N'_f(x) = \frac{f'''(x) f'(x) - f''(x)^2}{f'(x)^2}.$$

The proof of Proposition 9.11 shows that $N'_f(x)$ has property (1) listed above. However, $N'_f(x)$ does not have property (2). Therefore, we modify $N'_f(x)$ in such a way that it retains property (1), but now also has property (2):

$$S_f(x) = N'_f(x) - \frac{1}{2}[N_f(x)]^2$$
$$= \frac{f'''(x) f'(x) - \frac{3}{2} f''(x)^2}{f'(x)^2}.$$

An important fact is that, if $f''(x) = 0$, then $S_f(x) = N'_f(x)$. Also notice that $S_f(x)$ is not defined for points for which $f'(x) = 0$.

Definition 9.25. Let f be a C^3 function from $\mathbb{R}$ to $\mathbb{R}$. The ***Schwarzian derivative*** of f is defined by
$$S_f(x) = \frac{f'''(x) f'(x) - \frac{3}{2} f''(x)^2}{f'(x)^2}.$$

We say that f has *negative Schwarzian derivative* provided that $S_{f^n}(x) < 0$ whenever $(f^n)'(x) \neq 0$.

Lemma 9.9. *If f and g are C^3 functions from $\mathbb{R}$ to $\mathbb{R}$, then,*
$$S_{g \circ f}(x) = S_g(f(x)) \cdot |f'(x)|^2 + S_f(x).$$

The proof is a direct calculation. The term $-\frac{1}{2}[N_f(x)]^2$, which is added to the derivative of the nonlinearity, is chosen so this lemma is valid.

Proposition 9.10. *If f is C^3 and has negative Schwarzian derivative, then f^n has negative Schwarzian derivative. This gives property (2).*

Proof. First, we consider the second iterate. By Lemma 9.9,
$$S_{f^2}(x) = S_f(f(x)) \cdot |f'(x)|^2 + S_f(x),$$
which is less than zero because $|f'(x)|^2 > 0$, $S_f(f(x)) < 0$, and $S_f(x) < 0$.

Continuing by induction, we have
$$S_{f^n}(x) = S_{f^{n-1}}(f(x)) \cdot |f'(x)|^2 + S_f(x),$$
which is less than zero because $S_f(x) < 0$, $|f'(x)|^2 > 0$, and $S_{f^{n-1}}(f(x)) < 0$ by the induction hypothesis. □

The next result shows that the Schwarzian derivative (or $N_f'(x)$) satisfies property (1).

Proposition 9.11. *Assume that f is a C^3 function defined on $\mathbb{R}$ and has negative Schwarzian derivative.*

a. *If m is a point at which $g(x) = f'(x)$ has a local minimum, then $g(m) = f'(m) \leq 0$.*

b. *Assume that $a < p < b$ are three fixed points of f, $f(a) = a$, $f(p) = p$, and $f(b) = b$. Further assume that (i) $0 < f'(p) < 1$, (ii) $a < x < f(x) < p$ for $a < x < p$, and (iii) $p < f(x) < x < b$ for $p < x < b$. Then there exists a critical point x_c in (a,b) with $f'(x_c) = 0$. (This says that the Schwarzian derivative has property (1).)*

Proof. (a) At a point m at which $g(x) = f'(x)$ has a local minimum, $0 = g'(m) = f''(m)$ and $0 \leq g''(m) = f'''(m)$. If $f'(m) = 0$, then we are done. If $f'(m) \neq 0$, then
$$0 > S_f(m) = \frac{f'''(m) f'(m) - \frac{3}{2} f''(m)^2}{f'(m)^2} = \frac{f'''(m)}{f'(m)}.$$

Therefore, $f'''(m)$ and $f'(m)$ must both be nonzero and have different signs. Since $f'''(m) \geq 0$, it follows that $f'''(m) > 0$ and $f'(m) < 0$. This completes the proof of (a).

9.4. Critical Points and Basins

(b) By the assumptions on the graph of f, $f'(a) \geq 1$ and $f'(b) \geq 1$. Since $0 < f'(p) < 1$, the point m, at which $g(x) = f'(x)$ has a local minimum, must be in the open interval (a, b). By the first step, $f'(m) \leq 0$. If $f'(m) < 0$, then by the intermediate value theorem applied to $g(x) = f'(x)$, there must be a point x_c between m and p at which $f'(x_c) = 0$. □

The next theorem is the goal of the consideration of the Schwarzian derivative.

Theorem 9.12. *Assume that f is C^3 a function defined on $\mathbb{R}$, and has negative Schwarzian derivative. Assume that p is an attracting period-n point. Then, one of the two following cases is true about the basin of attraction for the orbit of p:*

1. *The basin $\mathscr{B}(\mathcal{O}_f^+(p); f)$ extends to either ∞ or $-\infty$.*
2. *The derivative $f'(x_c) = 0$ at some critical point x_c in the basin $\mathscr{B}(\mathcal{O}_f^+(p); f)$.*

The proof is given in Section 9.8 at the end of the chapter.

Example 9.26. We show that the logistic function has a negative Schwarzian derivative for any parameter value a. The derivatives

$$g'_a(x) = a - 2ax, \quad g''_a(x) = -2a, \quad \text{and} \quad g'''_a(x) = 0,$$

so

$$S_{g_a}(x) = \frac{(a - 2ax)0 - \frac{3}{2}(-2a)^2}{(a - 2ax)^2}$$
$$= \frac{-6}{(1 - 2x)^2} < 0.$$

The fact that the logistic family of functions has negative Schwarzian implies that all iterates of g_a have negative Schwarzian. Then, Theorem 9.12 implies that the basin of attraction for each attracting periodic orbit either contains a critical point or extends to infinity. Since all points outside $[0, 1]$ tend to minus infinity, no basin of attraction can extend to infinity, and any basin of attraction must contain a critical point. Because 0.5 is the only critical point, there can be at most one attracting periodic orbit for a given parameter value.

Example 9.27. Consider a scalar multiple of the arc tangent function:

$$A(x) = r \arctan(x).$$

Since

$$A'(x) = \frac{r}{1 + x^2}, \quad A''(x) = \frac{-r\, 2x}{(1 + x^2)^2}, \quad \text{and} \quad A'''(x) = \frac{r\,(6x^2 - 2)}{(1 + x^2)^3},$$

it follows that

$$S_A(x) = \frac{6x^2 - 2}{(1 + x^2)^2} - \frac{3}{2}\left(\frac{-2x}{1 + x^2}\right)^2 = \frac{-2}{(1 + x^2)^2} < 0.$$

For $0 < r < 1$, the origin is an attracting fixed point whose basin is the whole real line. There is no critical point, but this does not contradict the theorem because the basin extends to both $\pm\infty$.

For $r > 1$, there are three fixed points at $x = 0$ and $\pm x_0$, where $x_0 > 0$ solves

$$\arctan(x_0) = \frac{x_0}{r}.$$

The origin $x = 0$ is repelling and $\pm x_0$ are attracting. Again, there is no critical point, but this does not contradict the theorem because the basins of $\pm x_0$ extend to $\pm\infty$, respectively.

Exercises 9.4

1. Determine the Schwarzian derivative of the following functions:
   a. $f(x) = a x^2 + b x + c$,
   b. $g(x) = x^3 + \frac{1}{2} x$,
   c. $h(x) = x^n$ with $n \geq 3$.
   Is the Schwarzian derivative always negative?

2. Let $g(x) = \frac{2}{5} x^3 - \frac{7}{5} x$. The fixed points are 0 and $\pm\sqrt{6}$. There is a period-2 orbit of 1 and -1. The critical points are $\pm\sqrt{7/6}$.
   a. Calculate the Schwarzian derivative for g.
   b. The point $x_0 = \sqrt{7/6}$ belongs to the basin of attraction of what orbit? What about $-\sqrt{7/6}$? Hint: The map g has $g(-x) = -g(x)$ is odd, so $g^j\left(-\sqrt{7/6}\right) = -g^j\left(\sqrt{7/6}\right)$. Also, $g\left((-\sqrt{6}, \sqrt{6})\right) \subset (-\sqrt{6}, \sqrt{6})$.

3. Let $p(x)$ be a fourth degree polynomial such that its derivative $p'(x)$ has all real distinct roots; that is,

$$p'(x) = (x - a_1)(x - a_2)(x - a_3),$$

with $a_i \neq a_j$ for $i \neq j$.
   a. Show that

$$p''(x) = \frac{p'(x)}{x - a_1} + \frac{p'(x)}{x - a_2} + \frac{p'(x)}{x - a_3} \quad \text{and}$$

$$p'''(x) = \frac{2 p'(x)}{(x - a_1)(x - a_2)} + \frac{2 p'(x)}{(x - a_1)(x - a_3)} + \frac{2 p'(x)}{(x - a_2)(x - a_3)}.$$

   b. Show that the Schwarzian derivative is given by

$$S_p(x) = -\sum_{i=1}^{3} \left(\frac{1}{x - a_i}\right)^2 - \frac{1}{2}\left(\sum_{i=1}^{3} \frac{1}{x - a_i}\right)^2,$$

so it is always negative. Hint: Cancel the factor of $p'(x)$ from the form of $p'''(x)$ found in part (a). Also, since the final answer contains a term $-\frac{1}{2}\left[\frac{p''(x)}{p'(x)}\right]^2$ so expand $-1\left[\frac{p''(x)}{p'(x)}\right]^2$ and combine with $\frac{p'''(x)}{p'(x)}$.

4. Let $p(x)$ be a polynomial of degree $n+1$ such that its derivative $p'(x)$ has all real distinct roots; that is,

$$p'(x) = \prod_{i=1}^{n}(x - a_i) = (x - a_1) \cdots (x - a_n),$$

with $a_i \neq a_j$ for $i \neq j$.

  a. Show that

  $$p''(x) = \sum_{i=1}^{n} \frac{p'(x)}{x - a_i} \quad \text{and}$$

  $$p'''(x) = \sum_{i=1}^{n} \sum_{j \neq i} \frac{p'(x)}{(x-a_i)(x-a_j)}$$
  $$= \sum_{i=1}^{n} \sum_{j > i} \frac{2\, p'(x)}{(x-a_i)(x-a_j)}.$$

  See the previous exercise for the case when $n = 3$.

  b. Show that the Schwarzian derivative is given by

  $$S_p(x) = -\sum_{i}\left(\frac{1}{x - a_i}\right)^2 - \frac{1}{2}\left(\sum_{i}\frac{1}{x-a_i}\right)^2,$$

  so it is always negative.

5. Assume that the Schwarzian derivative of a function $f(x)$ is negative at all points of an interval (a, b). Show that, for each x in (a, b), either $f''(x) \neq 0$ or $f'''(x) \neq 0$.

6. Consider a third-degree polynomial

$$f(x) = a\,x^3 + b\,x^2 + c\,x + d,$$

with $a > 0$.

  a. Letting $z = x + b/(3a)$, show that

  $$a\,x^3 + b\,x^2 + c\,x + d = a\,z^3 + \left(c - \frac{b^2}{3a}\right)z + r = g(z)$$

  for some value of r.

  b. Show that $g'(z) = 0$ has two real roots if and only if $c < b^2/(3a)$, where $g(z)$ is defined in part (a).

  c. Show that $S_g(z) < 0$ for all z with $g'(z) \neq 0$ if $c < b^2/(3a)$.

  d. Show that the Schwarzian derivative of f is negative if f has two critical points.

9.5. Bifurcation of Periodic Points

In Section 9.3.2, we saw that the fixed point p_a for the logistic family became unstable at $a = 3$, and an attracting period-2 orbit is formed for $3 < a < 1 + \sqrt{3}$. The number and stability of the periodic orbits changes at $a = 3$, so the family of maps is said to undergo a *bifurcation* at $a = 3$, and $a = 3$ is called a *bifurcation value*. The word bifurcation itself is a French word that is used to describe a fork in a road or other such splitting apart or change. The particular bifurcation for the logistic family at $a = 3$ is called a *period doubling bifurcation* because the period of the attracting orbit goes from one to two. In this section, we discuss this bifurcation and the *tangential* or *saddle-node bifurcation*. In this latter bifurcation, two orbits of the same period are created (or destroyed) at the bifurcation value, where previously no periodic orbit existed. This bifurcation can occur for a fixed point x_0 and parameter value μ_0 for which $f(x_0; \mu_0) = x_0$ and $f_x(x_0; \mu_0) = \frac{\partial f}{\partial x}(x_0; \mu_0) = 1$ (i.e., the graph is tangent to the diagonal at x_0). Note in this section, we denote partial derivatives with respect to a variable by a subscript of that variable.

Example 9.28 (Tangential or Saddle-node Bifurcation). Let $f(x; \mu) = \mu + x - x^2$ for which $\frac{\partial f}{\partial x}(x; \mu) = f_x(x; \mu) = 1 - 2x$. The fixed points satisfy

$$x = \mu + x - x^2,$$
$$x^2 = \mu, \quad \text{so}$$
$$x_\mu^\pm = \pm\sqrt{\mu}.$$

For $\mu < 0$, there are no (real) fixed points; for $\mu > 0$, there are two fixed points. See Figure 24. The bifurcation can be understood graphically as follows: for $\mu < 0$, the graph of $f(x; \mu)$ is completely below the diagonal and does not intersect it; for $\mu = 0$, the graph of $f(x; 0)$ intersects the diagonal at the single point 0 at which it is tangent to the diagonal, $f_x(0; 0) = 1$. For $\mu > 0$, the graph of $f(x; \mu)$ intersects the diagonal in two places. Notice that, as μ increases, the graph is rising and this causes the change in the number of intersections of the graph with the diagonal.

For all $\mu > 0$,

$$f_x(x_\mu^+; \mu) = 1 - 2\sqrt{\mu} < 1 \quad \text{and} \quad f_x(x_\mu^-; \mu) = 1 + 2\sqrt{\mu} > 1,$$

so x_μ^- is always repelling. The derivative $f_x(x_\mu^+; \mu)$ equals -1 when

$$-1 = 1 - 2\sqrt{\mu}, \quad 2\sqrt{\mu} = 2, \quad \text{or} \quad \mu = 1.$$

Because the value of $f_x(x_\mu^+; \mu)$ varies continuously with μ, x_μ^+ is attracting for $0 < \mu < 1$ and becomes repelling for $\mu > 1$.

Notice that $x_\mu^+ = \sqrt{\mu}$ is not a differentiable function of μ, but the function of μ in terms of x given by $\mu = \left(x_\mu^\pm\right)^2$ is differentiable. In the theorem, we use this second function that expresses the μ-value in terms of the x value on the curve of fixed points.

In Figure 25, the set of all these fixed points is plotted in the (μ, x)-plane in what is called the *bifurcation diagram*. The branch of points x_μ^- is dotted because these points are repelling. The branch of points x_μ^+ is dotted for $\mu > 1$ for the same reason.

The preceding example is very representative of the general tangential bifurcation given in the next theorem. If the coefficient of x^2 has the opposite sign for the function $f(x; \mu)$, then the fixed points can occur before the bifurcation value rather than after. Also, rather than just use the coefficients of certain terms in the expression of the function, the theorem is expressed in terms of derivatives of the function with respect to x and the parameter μ. To express these derivatives, we

9.5. Bifurcation of Periodic Points   393

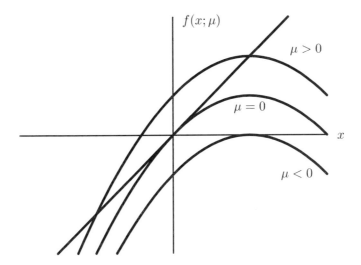

Figure 24. Creation of two fixed points for Example 9.28. $\mu < 0$, $\mu = 0$, and $\mu > 0$

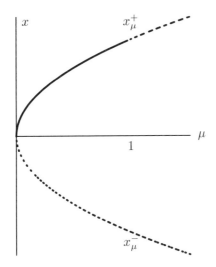

Figure 25. Bifurcation diagram for saddle-node bifurcation in Example 9.28

think of $f(x;\mu)$ as being a real-valued function of two variables (x,μ) (from $\mathbb{R}^2$). We let μ_0 be the bifurcation value of the parameter μ and x_0 the fixed point at the bifurcation, $f(x_0;\mu_0) = x_0$. At the bifurcation value, the graph of $f(x;\mu_0)$ is assumed to be tangent to the diagonal, or $f_x(x_0;\mu_0) = 1$. At the bifurcation value, we also assume that $f_{xx}(x_0;\mu_0) \neq 0$, so the graph of f_{μ_0} is on one side of the diagonal. The function in the previous example has a term μ that lifts the graph as the parameter varies. The corresponding assumption for the general function in the theorem is that $f_\mu(x_0,\mu_0) \neq 0$.

Theorem 9.13 (Tangential or saddle-node bifurcation). *Assume that $f(x;\mu)$ is a C^2 function from $\mathbb{R}^2$ to $\mathbb{R}$. Assume that there is a bifurcation value μ_0 that has a fixed point x_0, $f(x_0, \mu_0) = x_0$, such that the derivatives of f satisfy the following conditions:*

1. $f_x(x_0; \mu_0) = 1$.

2. *The second derivative $f_{xx}(x_0; \mu_0) \neq 0$, so the graph of $f(x; \mu_0)$ lies on one side of the diagonal for x near x_0.*

3. *The graph of $f(x; \mu)$ is moving up or down as the parameter μ varies, or more specifically, $f_\mu(x_0; \mu_0) \neq 0$.*

Then, there exist intervals $\mathbf{I}$ about x_0 and $\mathbf{J}$ about μ_0, and a differentiable function $\mu = m(x)$ from $\mathbf{I}$ into $\mathbf{J}$, such that the following conditions are satisfied:

(i) $f(x; m(x)) = x$ for all x in $\mathbf{I}$.

(ii) *The graph of the function $\mu = m(x)$ goes through (x_0, μ_0), $m(x_0) = \mu_0$.*

(iii) *The graph of $m(x)$ gives all the fixed points of $f(x; \mu)$ in $\mathbf{I} \times \mathbf{J}$.*

(iv) *The derivatives of the function $m(x)$ satisfy $m'(x_0) = 0$ and*

$$m''(x_0) = \frac{-f_{xx}(x_0; \mu_0)}{f_\mu(x_0; \mu_0)} \neq 0.$$

So $\mu = m(x) = \mu_0 + \frac{1}{2} m''(x_0)(x - x_0)^2 + O(|x - x_0|^3)$. If $m''(x_0) > 0$, then the fixed points appear for $\mu > \mu_0$; if $m''(x_0) < 0$, then the fixed points appear for $\mu < \mu_0$.

(v) *The fixed points on the graph of $m(x)$ are attracting on one side of x_0 and repelling on the other. In fact, $f_x(x, m(x)) = 1 + f_{xx}(x_0, \mu_0)(x - x_0) + O(|x - x_0|^2)$.*

Remark 9.29. In conclusions (iv) and (v), $O(|x - x_0|^k)$ is a term of order $(x - x_0)^k$ or higher; that is, there is a constant $C > 0$ such that $|O(|x - x_0|^k)| \leq C |x - x_0|^k$. For small displacements, these error terms are smaller than the terms given.

Idea of the proof: We show the calculation of the derivatives of the function $m(x)$ given in the conclusion of the theorem by implicit differentiation. The implicit function theorem says that, if implicit differentiation can solve for $m'(x) = \dfrac{\partial \mu}{\partial x}$, then, in fact, the function $\mu = m(x)$ exists, with the properties given in the statement of the theorem. See [**Wad00**] or [**Rob99**] for a precise statement of the implicit function theorem.

To find the fixed points, we need to solve

$$G(x, \mu) = f(x, \mu) - x = 0.$$

First, $G(x_0; \mu_0) = f(x_0; \mu_0) - x_0 = 0$. We proceed to calculate the partial derivatives of G at (x_0, μ_0):

$$G_x(x_0, \mu_0) = f_x(x_0, \mu_0) - 1 = 0,$$
$$G_\mu(x_0, \mu_0) = f_\mu(x_0, \mu_0) \neq 0.$$

9.5. Bifurcation of Periodic Points

The fact that $G(x_0, \mu_0) = 0$ and $G_\mu(x_0, \mu_0) \neq 0$ is what allows us to conclude that μ is a function of x. Let $\mu = m(x)$ be the function with $m(x_0) = \mu_0$ such that
$$0 = G(x, m(x))$$
for all x near x_0. Differentiating with respect to x gives
$$0 = G_x(x, m(x)) + G_\mu(x, m(x))\, m'(x).$$
Since $G_\mu(x_0, m(x_0)) = f_\mu(x_0, \mu_0) \neq 0$, we can solve for $m'(x_0)$. The result is
$$m'(x_0) = -\frac{G_x(x_0, \mu_0)}{G_\mu(x_0, \mu_0)} = 0,$$
as given in conclusion (iv). Differentiation again with respect to x yields
$$0 = G_{xx}(x, m(x)) + 2\, G_{x\mu}(x, m(x))\, m'(x)$$
$$\quad + G_{\mu\mu}(x, m(x))\, (m'(x))^2 + G_\mu(x, m(x))\, m''(x),$$
$$0 = G_{xx}(x_0, \mu_0) + 2\, G_{x\mu}(x_0, \mu_0) \cdot 0$$
$$\quad + G_{\mu\mu}(x_0, \mu_0) \cdot 0^2 + G_\mu(x_0, \mu_0)\, m''(x_0), \quad \text{and}$$
$$m''(x_0) = -\frac{G_{xx}(x_0, \mu_0)}{G_\mu(x_0, \mu_0)} = -\frac{f_{xx}(x_0, \mu_0)}{f_\mu(x_0, \mu_0)} \neq 0,$$
as given in conclusion (iv) of the theorem.

The stability of the fixed points can be determined using a Taylor series expansion of the function $f_x(x, \mu)$. We use the notation of $O(|x - x_0|^2)$ for terms that are at least quadratic in $|x - x_0|$, and similar notation for other terms. Since $f_x(x_0, \mu_0) = 1$, we obtain
$$f_x(x, \mu) = 1 + f_{xx}(x_0, \mu_0)\, (x - x_0) + f_{x\mu}(x_0, \mu_0)\, (\mu - \mu_0)$$
$$\quad + O(|x - x_0|^2) + O(|x - x_0| \cdot |\mu - \mu_0|) + O(|\mu - \mu_0|^2).$$
Using the fact that $m'(x_0) = 0$, we get $m(x) - \mu_0 = O(|x - x_0|^2)$ and
$$f_x(x, m(x)) - 1 = f_{xx}(x_0, \mu_0)\, (x - x_0) + O(|x - x_0|^2).$$
Because $f_{xx}(x_0, \mu_0) \neq 0$, the quantity $f_x(x, m(x)) - 1$ has opposite signs on the two sides of x_0, and one of the fixed points is attracting while the other is repelling. $\square$

We now turn to the period doubling bifurcation which occurs when $f(x_0; \mu_0) = x_0$ and $f_x(x_0; \mu_0) = -1$. We denote the application twice of $f(\cdot\, ; \mu)$ on a point x by $f(f(x; \mu); \mu) = f^2(x; \mu)$. We noted in the introduction to this section that the logistic family undergoes a period doubling bifurcation at $a = 3$. The next example is a more typical example of the derivatives at a period doubling bifurcation.

Example 9.30 (Period doubling bifurcation). Let
$$f(x; \mu) = -\mu\, x + a\, x^2 + b\, x^3,$$
with $a, b > 0$. Notice that $f(0; \mu) = 0$ for all parameter values μ. Also, $f_x(x; \mu) = -\mu + 2\, a\, x + 3\, b\, x^2$, $f_x(0; \mu) = -\mu$, $f_x(0; 1) = -1$, and the stability type of the fixed point changes from attracting to repelling as μ increases through 1. We want to find the period-2 orbit near 0 for μ near 1. The graph of $f^2(x; \mu(x))$ is given in

Figure 26 for $\mu < 1$, $\mu = 1$, and $\mu > 1$. Notice that there are two new intersections of the graph of $f^2(x;\mu)$ with the diagonal for $\mu > 1$.

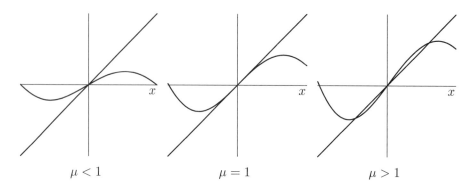

Figure 26. Graph of $f(x;\mu)$ for $\mu < 1$, $\mu = 1$, and $\mu > 1$ for $b + a^2 > 0$

To find the period-2 orbits, we need to find the solution of $0 = f^2(x;\mu) - x$. A direct calculation shows that

$$f^2(x;\mu) = \mu^2 x + (-a\mu + a\mu^2) x^2 + (-b\mu - 2a^2\mu - b\mu^3) x^3 + O(x^4),$$

where, again, $O(x^4)$ are terms of order x^4 or higher. Since we want to find the zeros of $f^2(x;\mu) - x$ distinct from the fixed point at 0, we divide by x and define

$$M(x,\mu) = \frac{f^2(x;\mu) - x}{x}$$
$$= \mu^2 - 1 + (-a\mu + a\mu^2) x + (-b\mu - 2a^2\mu - b\mu^3) x^2 + O(x^3).$$

It is not easy to solve explicitly $M(x,\mu) = 0$ for μ in terms of x. In Section 9.8, we use implicit differentiation to calculate the derivatives of $\mu = m(x)$ and show that

$$m(0) = 1, \quad m'(0) = 0, \quad \text{and} \quad m''(0) = 2\left(b + a^2\right).$$

Thus, $\mu = m(x) = 1 + (b + a^2) x^2 + O(x^3)$. In order for the sign of $m(x) - 1$ for the period-2 orbit to be determined by the quadratic terms, we need $b + a^2 \neq 0$. In particular, if $b + a^2 > 0$, then the period-2 orbit appears for $\mu > 1 = \mu_0$, and if $b + a^2 < 0$, then the period-2 orbit appears for $\mu < 1 = \mu_0$.

The stability type of the period-2 orbit is also determined by the sign of $b + a^2$. We show in Section 9.8 that

$$f^2_{xx}(x;m(x)) = 1 - 4\left(b + a^2\right) x^2 + O(x^3).$$

Thus, the period-2 orbit is attracting if $b + a^2 > 0$, and it is repelling if $b + a^2 < 0$. See Figure 27 for the bifurcation diagram.

The general result for period doubling is given in the next theorem. The constant β defined in the theorem plays the role of $b + a^2$ in the example. The constant α in the theorem measures the change of the derivative $f_x(x(\mu);\mu)$ as a function of μ; it is -1 for the previous example. Because $f_x(x_0;\mu_0) = -1 \neq 1$, the fixed point persists for nearby parameter values and there is a curve of fixed points $x(\mu)$ as

9.5. Bifurcation of Periodic Points

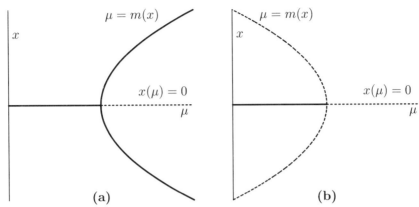

Figure 27. Bifurcation diagram for period doubling in Example 9.30: (a) for $b + a^2 > 0$, (b) for $b + a^2 < 0$.

a function of μ through x_0. The remark following the theorem further interprets these assumptions.

Theorem 9.14 (Period doubling bifurcation). *Assume that the $f(x, \mu)$ is a C^3 function from $\mathbb{R}^2$ to R. Assume that there is a bifurcation value μ_0 that has a fixed point x_0, $f(x_0, \mu_0) = x_0$, such that the derivatives of f satisfy the following conditions.*

1. $f_x(x_0; \mu_0) = -1$.
2. $\alpha = \left[f_{\mu x} + \frac{1}{2}(f_\mu)(f_{xx}) \right]\big|_{(x_0, \mu_0)} \neq 0$.
3. $\beta = \left(\frac{1}{3!} f_{xxx}(x_0, \mu_0) \right) + \left(\frac{1}{2!} f_{xx}(x_0, \mu_0) \right)^2 \neq 0$.

Then, there exist intervals $\mathbf{I}$ about x_0 and $\mathbf{J}$ about μ_0, and a differentiable function $\mu = m(x)$ from $\mathbf{I}$ into $\mathbf{J}$, such that the following conditions are satisfied:

(i) *The points $(x, m(x))$ have period-2: $f^2(x, m(x)) = x$, but $f(x, m(x)) \neq x$, for $x \neq x_0$.*

(ii) *The curve $\mu = m(x)$ goes through the point (x_0, μ_0), $m(x_0) = \mu_0$.*

(iii) *The graph of $m(x)$ gives all the period-2 points of f in $\mathbf{I} \times \mathbf{J}$.*

(iv) *The derivatives of $m(x)$ satisfy $m'(x_0) = 0$ and*
$$m''(x_0) = -\frac{\beta}{\alpha} \neq 0,$$
so $m(x) = \mu_0 - (\beta/2\alpha)(x - x_0)^2 + O(|x - x_0|^3)$. It follows that, if $-\beta/\alpha > 0$, then the period-2 orbit occurs for $\mu > \mu_0$ if $-\beta/\alpha > 0$, and it occurs for $\mu < \mu_0$ if $-\beta/\alpha < 0$.

(v) *The stability type of the period-2 orbit depends on the sign of β: if $\beta > 0$, then the period-2 orbit is attracting; if $\beta < 0$, then the period-2 orbit is repelling. In fact, $(f^2)_x(x, m(x)) = 1 - 4\beta(x - x_0)^2 + O(|x - x_0|^3)$.*

Remark 9.31. We interpret the conditions on the derivatives of f.

1. Because $f_x(x_0; \mu_0) \neq 1$, there is a differentiable curve of fixed points, $x(\mu) = f(x(\mu); \mu)$, with $x(\mu_0) = x_0$.

2. The proof of the theorem shows that the derivative of $f_x(x(\mu); \mu)$ with respect to μ at (x_0, μ_0) equals α,

$$\left.\frac{d}{d\mu} f_x(x(\mu); \mu)\right|_{\mu=\mu_0} = \left[f_{\mu x} + \tfrac{1}{2}(f_\mu)(f_{xx})\right]\Big|_{(x_0,\mu_0)} = \alpha.$$

We need to know this derivative is nonzero so that the stability of the fixed point changes at μ_0: if $\alpha > 0$, then the fixed point is attracting for $\mu > \mu_0$, and if $\alpha < 0$, then the fixed point is attracting for $\mu < \mu_0$.

3. The cubic term of the Taylor expansion of $f^2(x; \mu_0)$ equals β. If this term is nonzero, the graph of $f^2(x; \mu_0)$ has a nonzero cubic term in its tangency with the diagonal and crosses the diagonal.

The proof follows the idea of Example 9.30. See [**Rob99**] for details.

9.5.1. The Bifurcation Diagram for the Logistic Family. For the logistic family, there is a sequence of bifurcation values at which an attracting periodic orbit changes its stability type as the parameter varies. For $1 < a < 3$, the fixed point p_a is attracting and 0 is the only other periodic point. For $3 < a < 1 + \sqrt{6}$, both fixed points are repelling and the period-2 orbit $\{q_a^+, q_a^-\}$ is attracting. For a slightly above $1 + \sqrt{6}$, there is an attracting period-4 orbit. Figure 28 shows graphical iteration for $a = 3.56$ for which there is an attracting period-8 orbit. The first 200 iterates were not plotted and the next iterates start repeating every 8th iterate. There are only 8 vertical lines visible, which corresponds to a period 8 orbit. It is difficult to carry out the algebra to show that this happens, but it is seen numerically.

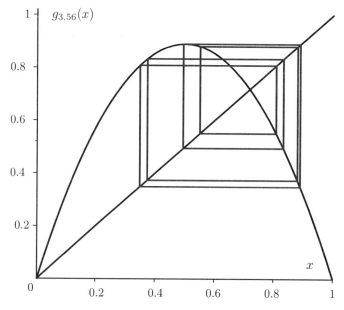

Figure 28. Graphical iteration for $a = 3.56$ of the 201 through 456 iterate of $x_0 = 0.5$

9.5. Bifurcation of Periodic Points

Continuing, as a increases, all the periodic orbits of period 2^k appear as sinks and then become repelling for a larger parameter value. In Section 10.1.1, we discuss the fact that it is necessary to add all the periodic orbits with periods equal to powers of two before adding any other periodic orbits. In Section 9.4, we prove, using the Schwarzian derivative, the fact that, for the logistic family, any periodic sink must have the critical point 0.5 in its basin, so the map can have just one attracting periodic orbit for each parameter value a. Let a_k be the parameter value at which the attracting periodic orbit changes from period 2^{k-1} to period 2^k. Table 3 contains a list of the first few values. This sequence of bifurcations is called the *period doubling cascade*. For the logistic family, the family runs through the whole sequence of attracting periodic points of period 2^k before

$$a_\infty \approx 3.569946.$$

Figure 29 shows the graphical iteration for $a = 3.572$: for this parameter value, it is more visible that the orbit does not converge to an attracting periodic orbit, since there are many different vertical lines in the stair step plot. (Note that, as in Figure 28, the first 200 iterates are dropped.)

2^k	a_k	F_k
2	3.000000...	
4	3.449490...	4.7514...
8	3.544090...	4.6562...
16	3.564407...	4.6683...
32	3.568759...	4.6686...
64	3.569692...	4.6692...
128	3.569891...	4.6694...
256	3.569934...	
∞	3.569946...	4.6692...

Table 3. Period doubling parameters for the logistic family.

It is possible to make a plot of the attracting periodic points. In Section 9.4, using the Schwarzian derivative, we show that the critical point 0.5 must be in the basin of attraction of any periodic sink. Start with $x_0 = 0.5$ and iterate it a number of times without plotting the points (e.g. one thousand iterates), then plot the next 200 points on a plot that includes the (a, x) coordinates. After doing this for one value of a, increase it by 0.01 and repeat the process. In fact, since the periodic orbit does not change much, it is more efficient to start the x value for the next parameter value at the final value of x for the previous parameter value. Figure 30 plots the bifurcation diagram for all the values of a between 1 and 4. For those parameter values with an attracting periodic point, the orbit plotted converges to this orbit and there are only a finite number of points above this value of a; for other parameter values (e.g., $a = 3.58$), there is no attracting periodic orbit, and the iterated point fills out a set which contains an interval (i.e., there is a more complicated set which is the "attractor"). Therefore, these plots are really bifurcation diagrams of the "attractor" and not of periodic points. Figure 31 shows only the shorter range of a between 2.9 and 3.6, where the period doubling takes

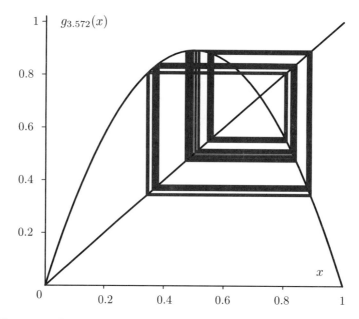

Figure 29. Graphical iteration for $a = 3.572$ of the 201 through 456 iterate of $x_0 = 0.5$

place. Finally, Figure 32 enlarges a portion of this period doubling range for a between 3.54 and 3.57; in this last figure, we restrict the values of x to lie between 0.47 and 0.57, so we get only a quarter of the orbit and do not show the other three quarters outside this interval.

M. Feigenbaum observed numerically that the rate at which the bifurcations occur is the same for several different families of functions, and indicated a reason why this should be true. See [**Fei78**]. P. Coullet and C. Tresser independently discovered similar facts about this bifurcation process. See [**Cou78**].

To see how to calculate the bifurcation rate, assume that

$$a_k \approx a_\infty - C\,F^{-k}.$$

Then,

$$a_k - a_{k-1} \approx -C\,F^{-k} + C\,F^{-(k-1)} = C\,(F-1)F^{-k}.$$

Defining F_k to be equal to the ratio of $a_k - a_{k-1}$ to $a_{k+1} - a_k$, then the limit of F_k is F:

$$F_k = \frac{a_k - a_{k-1}}{a_{k+1} - a_k}$$
$$\sim \frac{C\,(F-1)F^{-k}}{C\,(F-1)F^{-k-1}} = F.$$

In fact, we can use this sequence to define the quantity F as the limit of the F_k:

$$F = \lim_{k\to\infty} F_k = \lim_{k\to\infty} \frac{a_k - a_{k-1}}{a_{k+1} - a_k}.$$

9.5. Bifurcation of Periodic Points

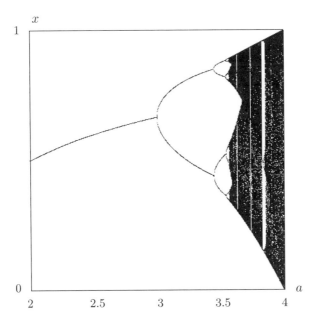

Figure 30. The bifurcation diagram for the family g_a: the horizontal direction is the parameter a between 2 and 4.0; the vertical direction is the space variable x between 0 and 1

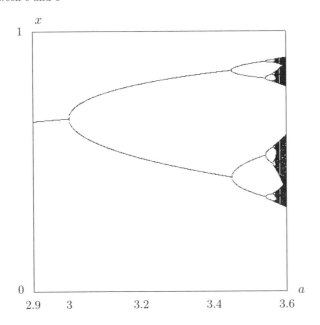

Figure 31. The bifurcation diagram for the family g_a: the horizontal direction is the parameter a between 2.9 and 3.6; the vertical direction is the space variable x between 0 and 1

See Table 3. M. Feigenbaum discovered that this limit is
$$4.669201609\ldots$$

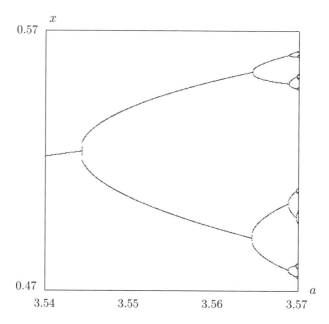

Figure 32. The bifurcation diagram for the family g_a: the horizontal direction is the parameter a between 3.54 and 3.57; the vertical direction is the space variable x between 0.47 and 0.57

not only for the logistic family, but also for the other families of functions that he considered, e.g. $f_r(x) = r\sin(x)$. Thus, this number is a *universal constant*, called the *Feigenbaum constant*.

Many people worked to show that Feigenbaum was correct that this number must be the same for any families satisfying certain reasonable conditions. O.E. Lanford [**Lan84**] gave the first complete proof for any one-parameter family of unimodal maps (one critical point) with negative Schwarzian derivative.

In a later range of parameters, for a between about 3.828 and 3.858, the bifurcation diagram reduces to three points (i.e., there is an attracting period-3 orbit in this parameter range). Figure 33 shows a more detailed view of the bifurcation diagram in this parameter range. The introduction of the period-3 orbit near a equal to 3.828 via a saddle node bifurcation can also be seen by looking at the graph of g_a^3 as the parameter a varies. Figure 34 shows how this periodic orbit is introduced in terms of the graph of g_a^3. For a equals 3.828, the only intersections of the graph of g_a^3 with the diagonal occur at the fixed points. For $a = 3.83$, there are six new intersections, three corresponding to a new period-3 sink and three corresponding to a new period-3 source. By the time a has reached 3.845, the period-3 sink has changed into a repelling periodic orbit. When the period 3 orbit first becomes repelling, there is a period doubling and a new period-6 sink is created. Thus, the bifurcation diagram in Figure 33 shows a new cascade of period sinks via period doubling bifurcations whose periods are of the form $3 \cdot 2^k$. The graphical representation of iteration of $g_{3.86}$, shown in Figure 35, indicates how an orbit wanders around in a complicated manner, so there is no attracting periodic orbit.

9.5. Bifurcation of Periodic Points

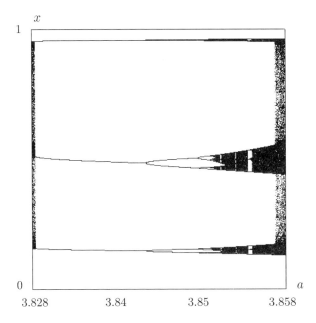

Figure 33. The bifurcation diagram for the family g_a: the horizontal direction is the parameter a between 3.828 and 3.858; the vertical direction is the space variable x between 0 and 1

Similar windows with other periods occur in the bifurcation diagram, but they are harder to see because they occur over a shorter range of parameters. After the introduction of one of these period-p sinks by a saddle-node bifurcation, there occurs a period doubling cascade of period-$p2^k$ sinks. The largest windows start with periods 3, 5, and 6.

Computer exercise

C. Ross and J. Sorensen, [**Ros00**], suggest that, rather than plotting the bifurcation diagram of the attractor, it is instructive to plot the bifurcation diagram of all the periodic points below a given level. This is possible to do using a computer program such as Maple or Mathematica, in a given range of iterates and parameters, say $1 \leq n \leq 10$ and $2.99 \leq a \leq 3.56$. Using the computer, it is possible to plot the graph of $g_a^n(x)$ for $1 \leq n \leq 10$ for various $2.5 \leq a \leq 3.56$ and to find the points where the graph crosses the diagonal. A point where the graph of g_a^n crosses the diagonal with derivative of absolute value less than one corresponds to an attracting periodic point and can be plotted with an "x"; a point where it crosses the diagonal with derivative of absolute value greater than one corresponds to a repelling periodic point and can be plotted with an "o" or small circle. Also, label the periods of the various points on the bifurcation diagram. Obviously, it is not possible to check all the values of the parameters, but it is necessary to increment a in small steps. The nearby x's of the same period can be connected with a solid curve and the nearby o's can be connected with a dotted curve. The result should be a bifurcation diagram that resembles Figures 25 and 27, but with many more curves. This project can be done in groups, with various people checking different parameter ranges.

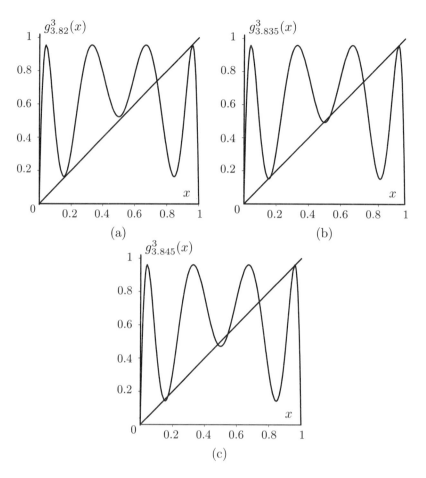

Figure 34. The graph of g_a^3 for (a) $a = 3.82$, (b) $a = 3.835$, and (c) $a = 3.845$.

Exercises 9.5

1. Let $E_\mu(x) = \mu e^x$. Show that the family E_μ undergoes a tangential bifurcation at $\mu = 1/e$. In particular, carry out the following steps:
    a. Plot the diagonal and the graph of $E_\mu(x)$ for $\mu < 1/e$, $\mu = 1/e$, and $\mu > 1/e$.
    b. Calculate the partial derivative of the assumptions of Theorem 9.13.
    c. Do the fixed points occur for $\mu < 1/e$ or $\mu > 1/e$?

2. Let $f_\mu(x) = \mu x - x^3$. What bifurcation occurs for $\mu = -1$? Check the derivatives of the relevant theorem, i.e., either Theorem 9.13 or 9.14.

3. Let $f_\mu(x) = \mu + x^2$.
    a. Find the bifurcation value and bifurcation point where the family undergoes a tangential bifurcation. Verify the assumptions of Theorem 9.13.

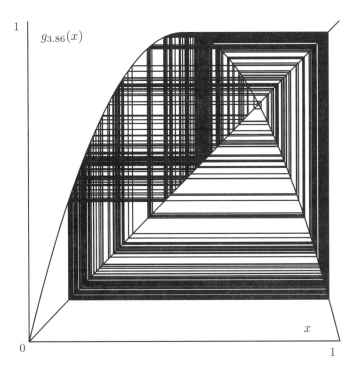

Figure 35. The plot of graphical iteration for g_a with $a = 3.86$

    b. Find the bifurcation value and bifurcation point where the family undergoes a period doubling bifurcation. Verify the assumptions of Theorem 9.14.

4. Let $g_a(x) = ax(1-x)$ be the logistic map. Use the computer to determine whether 0.5 is in the basin of attraction of an attracting periodic orbit or a more complicated set for $a = $ 3.5, 3.566, 3.58, 3.82, 3.83, and 3.86. Hint: Iterate the function a number times (e.g. 200) without plotting the points on a graphical iteration plot. Then, plot the next 256 iterates on the plot. If the values are printed out, as well as plotting the graph, then you can compare values to determine the period more accurately. If the graph is filled with many lines which either fill up space or do not repeat, then the orbit is not converging to a periodic sink but to a more complicated set.

5. Let $b_0 = 3$, and let $b_{k+1} = 1 + \sqrt{3 + b_k}$ for $k \geq 1$. Let $b_\infty = \lim_{k\to\infty} b_k$. It has been shown (but is difficult to verify) that $b_k \approx a_k$ and $b_\infty \approx a_\infty$, where a_k and a_∞ are the values for the logistic equation.
    a. What quadratic equation must b_∞ satisfy? What is the value of b_∞?
    b. Show that $\lim_{k\to\infty} c_k = 1 + \sqrt{17}$, where c_k are defined by

$$c_k = \frac{b_k - b_{k-1}}{b_{k+1} - b_k}.$$

9.6. Conjugacy

In Section 9.4, we discussed properties for the basin of attraction of a periodic sink for a map with a negative Schwarzian derivative. The preceding theorem in this section shows that two different maps are equivalent via a continuous change of coordinates on the basin of attractions of their fixed points. The argument is quite general, but we consider only a specific case. This idea of continuous change of coordinates is made precise in the definition of topological conjugacy given in this section.

In Section 9.1, we showed, by looking at each of their graphs, that the tent map T and the logistic map $G(x) = 4x(1-x)$ have the same number of points of all periods. In this section, we show that this follows because there is a continuous "change of coordinates," or a conjugacy, which transforms one of these functions into the other. We use this conjugacy in the next chapter to show that other dynamical properties, that are easier to verify for the tent map, are also true for the logistic map $G(x)$.

The concept of conjugacy arises in may subjects of mathematics. In linear algebra, the natural concept is a linear conjugacy. If $\mathbf{x}_1 = \mathbf{A}\mathbf{x}_0$ is a linear map, where $\mathbf{A}$ is a square n-by-n matrix and $\mathbf{y} = \mathbf{C}\mathbf{x}$ is a linear change of coordinates with an inverse linear map $\mathbf{x} = \mathbf{C}^{-1}\mathbf{y}$, then the "abstract linear map" is given in the new $\mathbf{y}$ coordinates by

$$\mathbf{y}_1 = \mathbf{C}\mathbf{x}_1 = \mathbf{C}\mathbf{A}\mathbf{x}_0 = \mathbf{C}\mathbf{A}\mathbf{C}^{-1}\mathbf{y}_0,$$

with a matrix $\mathbf{C}\mathbf{A}\mathbf{C}^{-1}$. The matrix $\mathbf{B} = \mathbf{C}\mathbf{A}\mathbf{C}^{-1}$ is called *similar* to $\mathbf{A}$.

In dynamical systems, we can sometimes find a linear, or a differentiable, change of coordinates, but usually the best we can obtain is a continuous change of coordinate. Therefore, the following definition considers continuous conjugacies as well as differentiable or linear ones. Usually, the maps we use are maps on $\mathbb{R}^n$, or an interval in $\mathbb{R}$. However, in the next chapter, we have one example in which one of the maps is defined on a "symbol space" rather than a Euclidean space. Therefore, we give the definition of conjugacy for functions defined on general spaces.

Definition 9.32. Consider two spaces $\mathbf{X}$ and $\mathbf{Y}$. A map h from $\mathbf{X}$ to $\mathbf{Y}$ is said to be *onto* $\mathbf{Y}$ provided that for each point y_0 in $\mathbf{Y}$ there is some x_0 in $\mathbf{X}$ with $h(x_0) = y_0$.

A map h from $\mathbf{X}$ to $\mathbf{Y}$ is said to be a *one-to-one* map provided that $h(x_1) \neq h(x_2)$ whenever $x_1 \neq x_2$.

If h is a one-to-one map from $\mathbf{X}$ onto $\mathbf{Y}$, then there is an *inverse* h^{-1} defined from $\mathbf{Y}$ to $\mathbf{X}$ by $h^{-1}(y) = x$ if and only if $h(x) = y$.

If there is a distance (or metric) defined on each of the spaces $\mathbf{X}$ and $\mathbf{Y}$, a map h from $\mathbf{X}$ to $\mathbf{Y}$ is called a *homeomorphism* provided that (i) h is continuous, (ii) h is onto $\mathbf{Y}$, (iii) h is a one-to-one map, and (iv) its inverse h^{-1} from $\mathbf{Y}$ to $\mathbf{X}$ is continuous. We think of h as a continuous change of coordinates.

Definition 9.33. Let f from $\mathbf{X}$ to $\mathbf{X}$ and g from $\mathbf{Y}$ to $\mathbf{Y}$ be two maps. A map h from $\mathbf{X}$ to $\mathbf{Y}$ is called a *topological conjugacy from f to g*, or just a *conjugacy*, provided that (i) h is a homeomorphism from $\mathbf{X}$ onto $\mathbf{Y}$, and (ii) $h \circ f(x) = g \circ h(x)$.

9.6. Conjugacy

We also say that f and g are *topologically conjugate*, or just *conjugate*. If h is merely a continuous function from $\mathbf{X}$ onto $\mathbf{Y}$ (but maybe not one to one) such that $h \circ f(x) = g \circ h(x)$, then h is called a *semiconjugacy*.

If both a conjugacy h and its inverse h^{-1} are C^r differentiable with $r \geq 1$, then we say that f and g are C^r *conjugate* or *differentiably conjugate*. If the conjugacy h is affine, $h(x) = ax + b$, then we say that f and g are *affinely conjugate*. If the conjugacy h is linear, $h(\mathbf{x}) = \mathbf{A}\mathbf{x}$, then we say that f and g are *linearly conjugate*.

We first note a simple consequence of the conjugacy equation, $h \circ f(x) = g \circ h(x)$. Since $g(y) = h \circ f \circ h^{-1}(y)$, it follows that

$$g^2(y) = h \circ f \circ h^{-1} \circ h \circ f \circ h^{-1}(y)$$
$$= h \circ f \circ f \circ h^{-1}(y)$$
$$= h \circ f^2 \circ h^{-1}(y),$$

and f^2 and g^2 are conjugate by h as well. Continuing by induction, we have

$$g^n(y) = g^{n-1} \circ g(y)$$
$$= h \circ f^{n-1} \circ h^{-1} \circ h \circ f \circ h^{-1}(y)$$
$$= h \circ f^{n-1} \circ f \circ h^{-1}(y)$$
$$= h \circ f^n \circ h^{-1}(y),$$

and f^n and g^n are also conjugate by h. Thus, a conjugacy of maps induces a conjugacy of iterates of the maps.

Let us see what this means in terms of the points on an orbit labeled by $x_n = f^n(x_0)$. Let $y_0 = h(x_0)$, so $x_0 = h^{-1}(y_0)$. Then,

$$y_n = g^n(y_0)$$
$$= h \circ f^n \circ h^{-1}(y_0)$$
$$= h \circ f^n(x_0)$$
$$= h(x_n),$$

and the conjugacy takes the nth iterate of x_0 by f to the nth iterate of y_0 by g.

We give some examples in which we can give explicit formulas for the conjugacy. Some of these conjugacies are linear or affine, and some are only continuous. In particular, Theorem 9.16 shows that the tent map and the logistic map G are topological conjugacy but not differentiably conjugate.

Theorem 9.15. *Consider the two families of maps*

$$f_b(x) = b\,x^2 - 1 \quad \text{and} \quad g_a(y) = ay(1-y).$$

The two families of maps are affinely conjugate by $h(x) = 0.5 - mx$ *for*

$$b = 0.25\,a^2 - 0.5\,a \quad \text{and} \quad m = {}^b\!/\!_a = 0.25\,a - 0.5.$$

In particular, $f_2(x) = 2\,x^2 - 1$ *is topologically conjugate to* $G(y) = g_4(y) = 4y(1-y)$ *by* $y = 0.5 - 0.5\,x$.

Proof. Both families of maps are quadratic, so it makes sense that they should be similar for the correct choices of the two parameters a and b. Near the critical point 0 of $f_b(x)$ there are two points $\pm\delta$ where f_b takes on the same value. Similarly, near the critical point 0.5 of g_a, g_a takes on the same values at the two points $0.5 \pm \delta$. Therefore, the conjugacy should take 0 to 0.5, $h(0) = 0.5$.

We start by seeking an affine conjugacy of the form $y = h_m(x) = 0.5 - mx$. (We put a minus sign in front of mx to make the parameter $m > 0$.) We calculate $g_a \circ h_m(x)$ and $h_m \circ f_b(x)$ to determine what value of m and which a and b work to give a conjugacy:

$$g_a \circ h_m(x) = g_a(0.5 - mx)$$
$$= a(0.5 - mx)(0.5 + mx)$$
$$= 0.25\,a - m^2\,a\,x^2;$$
$$h_m \circ f_b(x) = h_m(b\,x^2 - 1)$$
$$= 0.5 - m(b\,x^2 - 1)$$
$$= 0.5 + m - mb\,x^2.$$

In the expressions for $g_a \circ h_m(x)$ and $h_m \circ f_b(x)$, the constants and the coefficients of x^2 must be equal, so we get

$$0.25\,a = 0.5 + m,$$
$$m^2\,a = m\,b.$$

From the second of these equations, we get $m = b/a$. Substituting this in the first equation, we get

$$0.25\,a = \frac{b}{a} + 0.5, \qquad \text{or}$$
$$b = 0.25\,a^2 - 0.5\,a.$$

This last equation gives the relationship between the parameters of the two maps that are conjugate.

Notice that g_a has a fixed point at 0 and $g_a(1) = 0$. The map f_b has a fixed point at $-(1 + \sqrt{1+4b})/2b$ and $f_b\left((1 + \sqrt{1+4b})/2b\right) = -(1 + \sqrt{1+4b})/2b$. The conjugacy h matches not only the fixed points, but also the pre-fixed points $(1 + \sqrt{1+4b})/2b$ and 1.

Although we did not need to use the inverse of the affine conjugacy, notice that for $m \neq 0$, since $y = h_{m,B}(x) = mx + B$

$$mx = y - B,$$
$$x = h^{-1}(y) = \frac{y - B}{m}.$$

The inverse solves for the x that corresponds to the value of y. Since it is possible to solve this equation, h is one to one and has an inverse given by the preceding expression. □

We next state the result that G and the tent map are conjugate.

9.6. Conjugacy

Theorem 9.16. *The logistic map $G(u) = 4u(1-u)$ and the tent map $T(s)$ are topologically conjugate on $[0, 1]$ by the map*

$$u = h(s) = [\sin(\pi s/2)]^2.$$

We give two proofs. The first is a sequence of steps of change of variables. The combination, or compositions, of these steps gives the final conjugacy. The second proof is a direct check to show that $h(s)$ is a conjugacy between T and G.

Proof 1. We start by taking $\theta = h_1(s) = \pi s$. This takes the interval $[0, 1]$ and maps it into "angles" θ which run from 0 to π. The resulting map $\hat{T}(\theta) = h_1 \circ T \circ h_1^{-1}(\theta)$ is given by

$$\hat{T}(\theta) = \begin{cases} 2\theta & \text{if } 0 \leq \theta \leq \pi/2, \\ 2(\pi - \theta) & \text{if } \pi/2 \leq \theta \leq \pi. \end{cases}$$

This map can be visualized by taking a point on the upper semicircle, doubling the angle, and then folding it back on itself (i.e., the bottom semicircle is taken back up to the top semicircle, to the point just above it on the circle).

Next, we look at the map $x = h_2(\theta) = \cos(\theta)$, which assigns the point x in the interval $[-1, 1]$ that corresponds to the angle θ. The map $f(x) = h_2 \circ \hat{T} \circ h_2^{-1}(x)$ is given by

$$\begin{aligned} f(x) = h_2 \circ \hat{T} \circ h_2^{-1}(x) &= h_2 \circ \hat{T}\left(\cos^{-1}(x)\right) \\ &= \begin{cases} h_2\left(2\cos^{-1}(x)\right) & \text{if } 0 \leq x \leq 1, \\ h_2\left(2\pi - 2\cos^{-1}(x)\right) & \text{if } -1 \leq x \leq 0 \end{cases} \\ &= \cos\left(2\cos^{-1}(x)\right) \\ &= 2\cos^2\left(\cos^{-1}(x)\right) - 1 \\ &= 2x^2 - 1. \end{aligned}$$

By Theorem 9.15, $u = h_3(x) = 0.5 - 0.5\,x$ is a conjugacy from $f(x)$ to $G(u)$. Combining, we obtain

$$\begin{aligned} u = h(s) &= h_3 \circ h_2 \circ h_1(s) \\ &= h_3 \circ h_2(\pi s) \\ &= h_3(\cos(\pi s)) \\ &= 0.5 - 0.5\cos(\pi s) \\ &= \sin^2(\pi s/2) \end{aligned}$$

must be a conjugacy between $G(u)$ and $T(s)$. $\square$

Proof 2. We calculate $G \circ h(s)$ as follows:

$$\begin{aligned} G \circ h(s) &= 4\,h(s)\,[1 - h(s)] \\ &= 4\,[\sin(\pi s/2)]^2\,[\cos(\pi s/2)]^2 \\ &= [2\sin(\pi s/2)\cos(\pi s/2)]^2 \\ &= [\sin(\pi s)]^2. \end{aligned}$$

We now consider $h \circ T(s)$. For $0 \leq s \leq 0.5$,
$$h \circ T(s) = h(2s) = [\sin(\pi s)]^2.$$
For $0.5 \leq s \leq 1$,
$$h \circ T(s) = h(2 - 2s)$$
$$= [\sin(\pi - \pi s)]^2 = [-\sin(\pi s)]^2 = [\sin(\pi s)]^2.$$
Since $G \circ h(s) = h \circ T(s)$ for all $0 \leq s \leq 1$, it follows that $u = h(s)$ is a conjugacy from T to G. $\square$

Notice that the conjugacy $u = h(s) = [\sin(\pi s/2)]^2$ is differentiable, but $h'(0) = h'(1) = 0$. So, the inverse is not differentiable at 0 and 1. (Its derivative equals to infinity at these points.) It is interesting that h and its inverse are differentiable at 0.5, which is the critical point for G and the nondifferentiable point for T.

Proposition 9.17. a. *Assume that $f(x)$ and $g(y)$ are conjugate by $y = h(x)$. Then, the map h takes a periodic orbit of f to a periodic orbit of g with the same period.*

b. *Further assume that p_0 is a period-n point for f, $q_0 = h(p_0)$, and $h(x)$ is differentiable at p_0 with $h'(p_0)$ not equal to zero or infinity (so h^{-1} is differentiable at q_0 with $(h^{-1})'(q_0) \neq 0$). Then, $|(f^n)'(p_0)| = |(g^n)'(q_0)|$. So, one of the periodic orbits is attracting if and only if the other is attracting. Also, if one is repelling, then the other is repelling.*

Proof. (a) Since $g(y) = h \circ f \circ h^{-1}(y)$, we also have that $g^n(y) = h \circ f^n \circ h^{-1}(y)$. Thus, if $f^n(p_0) = p_0$ and $q_0 = h(p_0)$, then
$$g^n(q_0) = h \circ f^n \circ h^{-1}(q_0)$$
$$= h \circ f^n(p_0)$$
$$= h(p_0)$$
$$= q_0,$$
and q_0 is fixed by g^n. If q_0 had a lower period for g, then it would imply that p_0 had a lower period for f, so n must be the least period of q_0 for g.

(b) Taking the derivatives of the equation $h \circ f^n(x) = g^n \circ h(x)$ at p_0 yields
$$h'(p_0) \cdot (f^n)'(p_0) = (g^n)'(q_0) \cdot h'(p_0), \quad \text{so}$$
$$(f^n)'(p_0) = (g^n)'(q_0),$$
provided that $h'(p_0)$ does not equal zero or infinity. Therefore, $|(f^n)'(p_0)| = |(g^n)'(q_0)|$ and the stability of the periodic orbits for f and g must be the same. $\square$

Corollary 9.18. *All the Periodic orbits of $G(x) = 4x(1-x)$ are repelling.*

Proof. $G \circ h(s) = h \circ T(s)$ for $x = h(s) = \sin^2\left(\frac{\pi s}{2}\right)$ and $h'(s) \neq 0$ for $0 < s < 1$. All the periodic points of tent map have derivative $\pm 2^n$. Therefore, all nonzero periodic points of $G(x)$ have derivative $\pm 2^n$, and are repelling.

Note that $G'(0) = 4 > 1$ is repelling, but $T'(0) = 2 \neq 4 = G'(0)$ are different. This is not a contradiction, because $h'(0) = 0$. $\square$

Next, we turn to conjugacies on basins of attracting fixed points.

Proposition 9.19. *Let $f(x) = ax$ and $g(y) = by$, where $0 < a, b < 1$. Then, these two maps are conjugate by $y = h(x) = \text{sign}(x)\,|x|^\alpha$, where $b = a^\alpha$, or $\alpha = \ln(b)/\ln(a)$, and $\text{sign}(x)$ is 1 for $x > 0$ and -1 for $x < 0$.*

For $a \neq b$, either: (i) $h(x) = x^\alpha$ is not differentiable at 0 or (ii) $h'(0) = 0$ so $h^{-1}(y)$ is not differentiable at 0.

Proof. We try for a conjugacy of the form $y = h(x) = x^\alpha$. Then,
$$h \circ f(x) = h(ax) = a^\alpha x^\alpha \quad \text{and}$$
$$g \circ h(x) = g(x^\alpha) = bx^\alpha.$$

For these to be equal, all we need is that
$$b = a^\alpha \quad \text{or}$$
$$\alpha = \frac{\ln(b)}{\ln(a)},$$

as claimed in the statement of the proposition.

Since $\alpha \neq 1$ when $a \neq b$, either $h(x) = x^\alpha$ or $h^{-1}(y) = y^{1/\alpha}$ is not differentiable at 0. □

The next theorem states that two particular maps are topologically conjugate on basins of attractions of fixed points.

Theorem 9.20. *Let $f(x) = \frac{x+x^3}{2}$ and $g(y) = ay + by^3$, with $0 < a < 1$ and $b > 0$. Then, f on $[-1, 1]$ is topologically conjugate to g on $\left[-\sqrt{(1-a)b^{-1}}, \sqrt{(1-a)b^{-1}}\right]$. The conjugacy can be extended to be a conjugacy on the whole real line.*

Note that the map f has fixed points at $x = 0$ and ± 1, while the map g has fixed points at $y = 0$ and $\pm\sqrt{(1-a)b^{-1}}$. The proof is given in Section 9.8 at the end of the chapter.

Exercises 9.6

1. Let $f_a(x) = 1 - ax^2$ and $g_b(y) = y^2 - b$. Find a conjugacy $y = C_m(x)$ between the maps. What values of the parameter a correspond to what values of the parameter b? Hint: The point 0 is the critical point of both functions, $f'_a(0) = 0 = g'_b(0)$, so try a linear function of the form $y = C_m(x) = mx$. For what value of m is $C_m(f_a(x)) = g_b(C_m(x))$ satisfied?.

2. Let $f(x) = x^3$ and $g(y) = \frac{1}{4}y^3 + \frac{3}{2}y^2 + 3y$. Find an affine map $y = C(x) = ax+b$ which conjugates f and g. Verify that your $C(x)$ works. Hint: f has fixed points at -1, 0, and 1, and g has fixed points at -4, -2, and 0.

3. Let $f(x) = 4x(1-x)$. Determine the map $g(y)$ such that $y = h(x) = 3x - 3$ is a conjugacy from $f(x)$ to $g(y)$. Hint: $x = h^{-1}(y) = (y+3)/3$.

9.7. Applications

9.7.1. Capital Accumulation.
In Chapter 8, we used the Cobb–Douglas production function to obtain a function giving the capital formation on the ratio $k = K/L$ of capital to labor, shown in equation (8.1),

$$k_{n+1} = f(k_n) = \frac{\sigma A k_n^\alpha}{1+\lambda}.$$

The parameters are assumed to satisfy $0 < \alpha < 1$, $0 < \sigma < 1$, $A > 0$, and $0 < \lambda$. The fixed points are $k = 0$ and

$$k^* = \left(\frac{\sigma A}{1+\lambda}\right)^{\frac{1}{1-\alpha}} > 0.$$

The derivative is given by

$$f'(k) = \frac{\alpha \sigma A k^{\alpha-1}}{1+\lambda}.$$

Since $f'(k^*) = \alpha < 1$, this fixed point is attracting. Because $f'(k) > 0$, this function is monotonically increasing and satisfies

$$f(k) > k \quad \text{for } 0 < k < k^* \text{ and}$$
$$f(k) < k \quad \text{for } k^* < k.$$

Therefore, for any initial condition $k_0 > 0$, the iterates $f^n(k_0)$ converge to the fixed point k^*; that is, the basin of attraction of k^* is the set of all positive k, or

$$\mathscr{B}(k^*; f) = \{\, k : k > 0 \,\}.$$

9.7.2. Single Population Models.
The linear growth model for a population has unlimited growth if the growth constant is greater than one. The logistic model takes into account the fact that the rate of growth may decrease with increasing population. The logistic model has the following difficulty: If the population becomes sufficiently large, then the next iterate of the population becomes negative (i.e., it completely dies out). There are other models of the form $F(x) = xf(x)$ that take into account the fact that the growth of population may decrease without causing the population to become negative for large initial populations. We give just a quick overview of a few other models. Refer to the book [**Bra01**] by Brauer and Castillo–Chávez for more details. (We also used some of the material from the article [**Thu01**] by Thunberg, but it is more technical.)

Verhulst model. In 1945, Verhulst suggested the model

$$F(x) = \frac{rx}{A+x},$$

with both r and A positive. Notice that the ratio $F(x)/x = f(x) = r/(A+x)$ goes to zero as x goes to infinity, $F(x) > 0$ for $x > 0$, but $F(x)$ approaches $r > 0$ as x goes to infinity. We leave it to the reader to check that the Schwarzian derivative is identically equal to zero. Therefore, the theory about the basin of attraction given in this chapter does not apply to this function. We do not check further properties for this map.

9.7. Applications

Ricker model. In 1954, Ricker [**Ric54**] gave another model for population growth, which did not become negative, namely,

$$R(x) = a\,x\,e^{-bx},$$

for both a and b positive. Again, $R_{a,b}(x) > 0$ for $x > 0$, and both $R_{a,b}(x)$ and $R_{a,b}(x)/x$ go to zero as x goes to infinity. The most interesting range of parameters is for $a > 1$. We leave it as an exercise to check that the Ricker map has negative Schwarzian derivative and to find the fixed points.

Hassel model. Another model in the literature, given by Hassel [**Has74**] in 1974, is

$$H_{r,b}(x) = rx(1+x)^{-b},$$

for $r > 0$ and $b > 1$. Again, $H_{r,b}(x) > 0$ for $x > 0$, so the iterates of an initial condition $x_0 > 0$ never become negative. The derivative is given by

$$H'_{r,b}(x) = \frac{r}{(1+x)^b} - \frac{brx}{(1+x)^{b+1}}$$
$$= \frac{r(1+x-bx)}{(1+x)^{b+1}}.$$

We start with a lemma that states some of the basic properties.

Lemma 9.21. *Assume that $r > 0$ and $b > 1$.*

a. *Both $H_{r,b}(x)$ and $\dfrac{H_{r,b}(x)}{x}$ go to zero as x goes to infinity.*

b. *There is a unique critical point $x_c = (b-1)^{-1} > 0$ where $H'_{r,b}(x_c) = 0$. Moreover,*

$$H'_{r,b}(x) > 0 \quad \text{for } 0 < x < x_c \text{ and}$$
$$H'_{r,b}(x) < 0 \quad \text{for } x_c < x.$$

Proof. (a) The limit of $H_{r,b}(x)$ is zero, because the power in the denominator is larger than the numerator. (Note that we need $b > 1$ for $H_{r,b}(x)$ to go to zero as x goes to infinity.) The ratio $H_{r,b}(x)/x = r(1+x)^{-b}$ goes to zero as x goes to infinity.

(b) The critical point satisfies

$$H'_{r,b}(x_c) = \frac{r(1+x_c-bx_c)}{(1+x_c)^{b+1}} = 0,$$

so $1 = (b-1)x_c$, or $x_c = (b-1)^{-1}$. The signs of $H'_{r,b}(x)$ follow from the form of the derivative. □

Notice that $H_{r,b}(0) = 0$ and $H'_{r,b}(0) = r$. Therefore, 0 is attracting for $0 < r < 1$ and repelling for $r > 1$.

Proposition 9.22. *For $0 < r \leq 1$, $x = 0$ is a globally attracting fixed point.*

Proof. For $0 < r \leq 1$, $H_{r,b}(x) \leq x(1+x)^{-b} < x$ for $x > 0$. Therefore, by graphical iteration, the basin of attraction of 0 includes all positive numbers, $(0,\infty)$. □

From now on, we assume that $r > 1$.

Lemma 9.23. *Assume that $r > 1$. The only fixed point with $x > 0$ is $x^* = r^{1/b} - 1$.*

Proof. A nonzero fixed point x^* satisfies $x^* = rx^*(1 + x^*)^{-b}$, $(1 + x^*)^b = r$, or $x^* = r^{1/b} - 1$, which is unique. □

Lemma 9.24. *Assume that $r > 1$. If $b > 1$ is chosen so that $-1 < H'_{r,b}(x^*) < 1$, then the basin of attraction of x^* is the set of all $x > 0$. In particular, this is true for $1 < b < 2$.*

Proof. We split the proof into two cases: (a) $0 \leq H'_{r,b}(x^*) < 1$ and (b) $-1 < H'_{r,b}(x^*) < 0$.

(a) In the first case when $0 \leq H'_{r,b}(x^*) < 1$, it follows that $0 < x^* \leq x_c$, where $x_c = (b-1)^{-1}$ is the unique critical point. Also, $H'_{r,b}(0) = r > 1$. Therefore, for the parameter values indicated,

$$0 < x < H_{r,b}(x) < H_{r,b}(x^*) = x^* \quad \text{for } 0 < x < x^*,$$
$$x^* = H_{r,b}(x^*) < H_{r,b}(x) < x \quad \text{for } x^* < x \leq x_c.$$

By Theorem 9.2, the basin of attraction of x^* includes the whole interval $(0, x_c]$. The function is decreasing for x greater than x_c, so for a point $x > x_c$, the iterate $H_{r,b}(x) < H_{r,b}(x_c) \leq x_c$, and it falls into the basin of attraction.

(b) When $-1 < H'_{r,b}(x^*) < 0$, it follows that $x^* > x_c$. We must look at the images of the critical point. It can be shown that

$$x_c < H^2_{r,b}(x_c) < x^* < H_{r,b}(x_c),$$

so an analysis like the logistic map for $2 < a < 3$ shows that the basin of attraction includes all positive $x_0 > 0$.

(c) Now we assume that $1 < b < 2$. At the fixed point,

$$H'_{r,b}(x^*) = \frac{r}{(1+x^*)^b} - \frac{brx^*}{(1+x^*)^{b+1}}$$
$$= 1 - \frac{bx^*}{1+x^*}$$
$$= 1 - b(1 - r^{-1/b}).$$

For $1 < b < 2$,

$$1 > 1 - b(1 - r^{-1/b}) > 1 - b > 1 - 2 = -1,$$

so the fixed point is attracting. □

For $0 < r < 1$ or $1 < b < 2$, we know that the only periodic points are fixed points. The question arises as to what can happen for $b \geq 2$. Numerical iteration shows the following: for $b = 5$ and $r = 20$, there is period-2 sink, and for $b = 5$ and $r = 40$ there is a period-4 sink. Thus, we can get periodic orbits of higher period. Since there is only one critical point, the following lemma show that there can be only one periodic sink for all values of $b \geq 2$, and so for $b > 1$.

Lemma 9.25. *For $b \geq 2$, the Hassel family has a negative Schwarzian derivative.*

9.7. Applications

Proof. Since

$$H'_{r,b}(x) = r(1+x)^{-b} - brx(1+x)^{-b-1}$$
$$= r(1+x)^{-b-1}(1+x-bx),$$
$$H''_{r,b}(x) = -2br(1+x)^{-b-1} + b(b+1)rx(1+x)^{-b-2}$$
$$= r(1+x)^{-b-2}(-2b - 2bx + b^2x + bx)$$
$$= r(1+x)^{-b-2}(-2b + b^2x - bx),$$
$$H'''_{r,b}(x) = 3b(b+1)r(1+x)^{-b-2} - b(b+1)(b+2)rx(1+x)^{-b-3}$$
$$= r(1+x)^{-b-3}[3b(b+1)(1+x) - b(b+1)(b+2)x]$$
$$= r(1+x)^{-b-3}[3b(b+1) + b(b+1)(-b+1)x],$$

we get

$$S_{H_{r,b}}(x) = \frac{2(3b(b+1)(1+x) - b(b+1)(b+2)x)(1+x-bx)}{2(1+x)^2(1+x-bx)^2}$$
$$\frac{-3(-2b - 2bx + b^2x + bx)^2}{2(1+x)^2(1+x-bx)^2}$$
$$= -\left[\frac{b(b-1)}{2(1+x)^2(1+x-bx)^2}\right]\left[(b-1)(b-2)x^2 - 4(b-2)x + 6\right].$$

The first factor is positive whenever it is defined. The following paragraph shows that the second factor is positive for $b \geq 2$. Because of the minus sign in front, the Schwarzian derivative is negative as claimed in the lemma.

Using the quadratic formula to express the roots of the second factor of the Schwarzian derivative, the term under the square root is as follows:

$$16(b-2)^2 - 24(b-1)(b-2) = 8(b-2)\left[2(b-2) - 3(b-1)\right]$$
$$= 8(b-2)\left[-b-1\right].$$

Since this term is negative for $b \geq 2$, the roots are not real. Because it is positive for $x = 0$, it must be positive for all x. This completes the proof. $\square$

9.7.3. Blood Cell Population Model. A. Lasota [**Las77**] proposed a discrete model for the population of red blood cells that uses the function

$$f(x) = (1-a)x + b\,x^r e^{-sx}.$$

The term $-ax$ corresponds to the number of cells which die off in one time period. The term $b\,x^r e^{-sx}$ corresponds to the production of new cells. Based on data, the parameter values taken are

$$r = 8, \quad s = 16, \quad b = 1.1 \times 10^6,$$

with a left as a variable parameter, with $0 < a \leq 1$. For these parameters, denote the function by $f_a(x)$ to indicate its dependence on the parameter a.

The value $p_0 = 0$ is always an attracting fixed point for all parameter values, $f_a(0) = 0$ and $|f'_a(0)| = 1 - a < 1$.

For $0 < a \leq 1$, the map f_a has three fixed points, $p_0 = 0 < p_1 < p_2$. See Figure 36. Since the origin is attracting, the graph of f_a must be crossing from below the

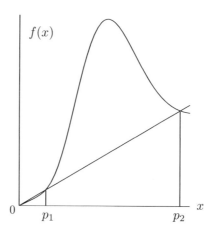

Figure 36. Graph of the blood cell model for $a = 0.2$

diagonal to above it at p_1, so p_1 is repelling for all these parameter values. For any initial condition $0 < x_0 < p_1$, $f_a^n(x_0)$ goes to zero as n goes to infinity, so the level of blood cells dies out. Thus, p_1 is a threshold level, and if the initial level of red blood cells is too low, then recovery is not possible without intervention.

The stability of p_2 depends on the parameter value of a. For $a_0 \approx 0.262727$, $f'_{a_0}(p_2) = -1$; for $0 < a < a_0$, p_2 is an attracting fixed point. In particular, this is true for $a = 0.2$. For $a = 0.2$ and $x_0 > p_1$, $f^n_{0.2}(x_0)$ converges to the positive attracting level of p_2 for the amount of red blood cells. Thus, if the rate of dying out of the blood cells is not too large, any sufficiently large level of initial blood cells converges to the constant level of p_2.

For $a > a_0$, the fixed point p_2 becomes unstable. It first spawns an attracting period-2 orbit by a period doubling bifurcation and then higher period attracting orbits. For $a = 0.81$, there are two period-3 orbits. In Section 10.1, we see that any map with a period-3 orbit has orbits of all periods. We return to this model in Section 11.5.3, where we give numerical results for $a = 0.81$, which indicate that there is a chaotic attractor (as defined in Section 11.2) rather than an attracting periodic orbit.

See [**Mar99**] for further discussion and references for this model of blood cell population.

Exercises 9.7

1. Consider the Ricker model for population growth given by $R(x) = a\,x\,e^{-bx}$ for both a and b positive.
    a. Find the fixed points and check their stability type for $a > 1$ and $b > 0$.
    b. For the range of parameters for which the second fixed point $x^* > 0$ is attracting, show that its basin of attraction is all $x > 0$.
    c. Show that the Schwarzian derivative is negative for all $a > 0$ and $b > 0$.

2. Consider the Hassel model with $b = 1$, given by $F(x) = \frac{rx}{1+ax}$, with $r > 1$ and $a > 0$.
   a. Show that there is a unique positive fixed point x^*.
   b. Show that the basin of attraction of x^* $\mathscr{B}(x^*)$ is the set of positive x, $\mathscr{B}(x^*) = \{x : x > 0\}$.

3. (Kaplan and Glass, [**Kap95**]) The following function has been considered as a mathematical model for a periodically stimulated biological oscillator by Bélair and Glass:
$$f(\phi) = \begin{cases} 6\phi - 12\phi^2 & \text{for } 0 \leq \phi \leq 0.5, \\ 12\phi^2 - 18\phi + 7 & \text{for } 0.5 \leq \phi \leq 1. \end{cases}$$
   a. Sketch the graph of $f(\phi)$. Be sure to show all the local maxima and minima and to compute the values of f at these extrema.
   b. Find all the fixed points and determine their stability type.
   c. Use the graph of f to find a period-2 orbit. What is its stability type.

4. Consider the function
$$f(x) = \frac{Rx}{1 + \left(\frac{x}{\theta}\right)^n},$$
where $R > 1$, $n \geq 2$, and $\theta > 1$ are parameters. Kaplan and Glass [**Kap95**] present this function as a possible model for the rate of division of cell nuclei.
   a. Sketch the graph of f for $x \geq 0$, for $R = 2$, $n = 3$, and $\theta = 5$. Include the critical point where $f'(x) = 0$.
   b. Determine the stability of the fixed point at $x = 0$.
   c. Find the second fixed point and determine its stability type for $R = 2$. Your answer will depend on the value of n.

5. Consider the function $f(x) = b\, x^r\, e^{-sx}$, for $b, r, s > 0$. How do the fixed points of this function compare with those of the blood model given in Section 9.7.3?

9.8. Theory and Proofs

Graphical analysis

Restatement of Theorem 9.2. *Let f be a continuous function on $\mathbb{R}$ and let x^* be a fixed point.*

   a. *If $y_1 < x^*$ and $x < f(x) < x^*$ for $y_1 < x < x^*$, then the basin of attraction of x^* includes the interval (y_1, x^*), $(y_1, x^*) \subset \mathscr{B}(x^*; f)$.*

   b. *If $x^* < y_2$ and $x^* < f(x) < x$ for $x^* < x < y_2$, then the basin of attraction of x^* includes the interval (x^*, y_2), $(x^*, y_2) \subset \mathscr{B}(x^*; f)$.*

Proof. We show only part (**a**), since the proof of part (**b**) is similar. Assume that x_0 is a point with $y_1 < x_0 < x^*$. Let $x_n = f^n(x_0)$. By the assumptions of the theorem, the graph of $f(x)$ is above the diagonal but below x^* for $y_1 < x < x^*$. As in Example 9.11, it follows that
$$y_1 < x_0 < x_1 < \cdots < x_{n-1} < x_n < x^*.$$
This sequence is increasing in the line and bounded above by x^*, so it must converge to some point $x_\infty \leq x^*$. By the continuity of the function f, $x_{n+1} = f(x_n)$ must converge to $f(x_\infty)$. Since it is a reindexed version of the original sequence, it also

converges to x_∞, and so $f(x_\infty) = x_\infty$. Because there are no fixed points in the interval (y_1, x^*), x_∞ must be the fixed point x^*, x_n converges to x^*, and x_0 is in $\mathscr{B}(x^*; f)$. Since x_0 is an arbitrary point in (y_1, x^*), $(y_1, x^*) \subset \mathscr{B}(x^*; f)$. □

Stability of periodic points

Restatement of Theorem 9.3. *Let f be a continuously differentiable function from $\mathbb{R}$ to itself.*

a. *Assume p_0 is a fixed point, $f(p_0) = p_0$. The following three statements give the relationship between the absolute value of the derivative and the stability of the fixed point:*

(i) *If $|f'(p_0)| < 1$, then p_0 is an attracting fixed point.*

(ii) *If $|f'(p_0)| > 1$, then p_0 is a repelling fixed point.*

(iii) *If $|f'(p_0)| = 1$, then p_0 can be attracting, repelling, semistable, or even none of these.*

b. *Assume that p_0 is a period-n point, and $p_j = f^j(p_0)$. By the chain rule,*

$$|(f^n)'(p_0)| = |f'(p_{n-1})| \cdots |f'(p_1)| \cdot |f'(p_0)|,$$

where $p_j = f^j(p_0)$. The following three statements give the relationship between the size of this derivative and the stability of the periodic point:

(i) *If $|(f^n)'(p_0)| < 1$, then p_0 is an attracting period-n point.*

(ii) *If $|(f^n)'(p_0)| > 1$, then p_0 is a repelling period-n point.*

(iii) *If $|(f^n)'(p_0)| = 1$, then the periodic orbit can be attracting, repelling, semistable, or even none of these.*

Proof. (a.i) Assume that $|f'(p_0)| < 1$. Let $\delta > 0$ be such that $|f'(x)| \leq r < 1$ for all x in $[p_0 - \delta, p_0 + \delta]$. By the mean value theorem, for any x in $[p_0 - \delta, p_0 + \delta]$, there is a y which depends on x that lies between x and p_0 such that

$$f(x) - f(p_0) = f'(y)(x - p_0).$$

Taking absolute values, we get

$$|f(x) - f(p_0)| = |f'(y)| \cdot |x - p_0|$$
$$\leq r |x - p_0|$$
$$< |x - p_0|.$$

Since $x_1 = f(x)$ is closer to p_0 than x is, it also lies in the interval $[p_0 - \delta, p_0 + \delta]$. Repeating the argument for x_1, we have

$$|f^2(x) - f(p_0)| = |f(x_1) - f(p_0)| = |f'(y)| \cdot |x_1 - p_0|$$
$$\leq r |x_1 - p_0| \leq r^2 |x - p_0|.$$

It follows that $f(x) = x_1$ is in the interval $[p_0 - \delta, p_0 + \delta]$. Continuing by induction, we see that all the points $x_n = f^n(x)$ are in the interval $[p_0 - \delta, p_0 + \delta]$, and

$$|f^r(x) - f(p_0)| \leq r^n |x - p_0|.$$

Since r^n goes to zero as n goes to infinity, this proves that p_0 is an attracting fixed point.

9.8. Theory and Proofs

(a.ii) If $|f'(p_0)| > 1$, then there is a $\delta > 0$ such that $|f'(x)| \geq r > 1$ for all x in the interval $[p_0 - \delta, p_0 + \delta]$. Take an x in $[p_0 - \delta, p_0 + \delta]$ with $x \neq p_0$. By the same argument we just used, as long as $x_j = f^j(x)$ stays in $[p_0 - \delta, p_0 + \delta]$ for $0 \leq j \leq n-1$, then
$$|f^n(x) - p_0| \geq r^n |x - p_0|,$$
so $x_n = f^n(x) \neq p_0$. This cannot last forever, so there is an $n_0 > 0$ such that
$$|f^{n_0}(x) - p_0| > \delta.$$
This proves that p_0 is a repelling fixed point.

(a.iii) We leave it to the reader to check, that for a neutral fixed point with $|f'(p_0)| = 1$, the fixed point can be either attracting, repelling, or neither.

(b.i) Assume that $f^n(p_0) = p_0$ and that $|(f^n)'(p_0)| < 1$. Part (a) proves that the point p_0 is an attracting fixed point for f^n. By the continuity of f^j for $0 < j < n$, the intermediate iterates also have to stay near the orbit. We leave the details to the reader, as well as parts **(b.ii)** and **(b.iii)**. $\square$

Critical Points and Basins

Restatement of Theorem 9.12. *Assume that f is a C^3 function defined on $\mathbb{R}$, and that $S_f(x) < 0$ whenever $f'(x) \neq 0$. Assume that p is an attracting period-n point. Then, one of the two following cases is true about the basin of attraction of the orbit of p:*

(1) *The basin $\mathscr{B}(\mathcal{O}_f^+(p))$ extends to either ∞ or $-\infty$.*

(2) *The derivative $f'(x_c) = 0$ at some critical point x_c in the basin $\mathscr{B}(\mathcal{O}_f^+(p))$.*

Proof. We consider only the case in which p is a fixed point. The other case is treated by considering f^n. We further divide up the proof into different cases:

Case a: If $f'(p) = 0$, then we are done.

Case b: Assume that $0 < f'(p) < 1$.

(i) If $p < f(x) < x$ for all $x > p$, then $\mathscr{B}(p)$ contains $[p, \infty)$ as needed.

(ii) If $x < f(x) < p$ for all $x < p$, then $\mathscr{B}(p)$ contains $(-\infty, p]$ as needed.

Therefore, we can assume that there exist $a < p < b$ such that $x < f(x) < p$ for $a < x < p$, $p < f(x) < x$ for $p < x < b$, and at least one of the following hold: (iii) $f(b) = p$, (iv) $f(a) = p$, or (v) $f(a) = a$ and $f(b) = b$.

If (iii) holds, then $\mathscr{B}(p)$ contains $[p, b]$ by Theorem 9.2. Since $f(p) = p = f(b)$, by the mean value theorem, there exists a point x_c between p and b such that $f'(x_c) = 0$. Since x_c is in $[p, b] \subset \mathscr{B}(p)$, we are done in this case.

If (iv) holds, then $[a, p] \subset \mathscr{B}(p)$ and there exists a $a < x_c < p$ with $f'(x_c) = 0$ by a proof similar to that for case (iii).

If (v) holds, then $f(a) = a$, $f(b) = b$, $f(p) = p$, and $0 < f'(p) < 1$. If $f'(x) \leq 0$ at some point x_1 in (a, b), then by the intermediate value theorem, $f'(x)$ would have to be zero at some point between x_1 and p. Therefore, we can assume $f'(x) > 0$ on (a, b). Then, (a, b) is in $\mathscr{B}(p)$ by Theorem 9.2. By Proposition 9.11, there is a point x_c in (a, b) at which $f'(x_c) = 0$. Since this point must be in $\mathscr{B}(p)$, we are done with the case, and all possible cases when $f'(p) > 0$.

Case c: The proof for the case when $-1 < f'(p) < 0$ uses f^2 in a way like (b) used f, but we do not give the details. □

Bifurcation of periodic points

Detailed calculations for Example 9.30:

Let $f(x;\mu) = -\mu x + a x^2 + b x^3$, with $a, b > 0$. Notice that $f_\mu(0) = 0$ for all parameter values μ. Also, $f_x(x:\mu) = -\mu + 2 a x + 3 b x^2$, $f_x(0;\mu) = -\mu$, $f'_x(0;1) = -1$, and the stability type of the fixed point changes from attracting to repelling as μ increases through 1.

A period-2 point satisfies $0 = f^2(x;\mu) - x$. By a direct calculation,
$$f^2(x;\mu) = \mu^2 x + (-a\mu + a\mu^2) x^2 + (-b\mu - 2 a^2 \mu - b\mu^3) x^3 + O(x^4).$$
Since we want to find the zeros of $f^2(x;\mu) - x$, other than the fixed point at 0, we divide by x and define
$$M(x,\mu) = \frac{f^2(x;\mu) - x}{x}$$
$$= \mu^2 - 1 + (-a\mu + a\mu^2) x + (-b\mu - 2 a^2 \mu - b\mu^3) x^2 + O(x^3).$$

It is not easy to solve explicitly $M(x,\mu) = 0$ for μ in terms of x, so we use implicit differentiation to calculate the derivatives of $\mu = m(x)$. Evaluating at $(0,1)$, $M(0,1) = 0$. The partial derivatives of M are as follows:
$$M_\mu(0,1) = f_\mu(0;1) = 2\mu\big|_{\mu=1} = 2 \neq 0,$$
$$M_x(0,1) = -a\mu + a\mu^2 + (-b\mu - 2 a^2 \mu - b\mu^3) 2x \big|_{x=0,\mu=1} = 0,$$
$$M_{xx}(0,1) = 2\left(-b\mu - 2 a^2 \mu - b\mu^3\right)\big|_{\mu=1} = -4\left(b + a^2\right).$$

Differentiating $0 = M(x,\mu)$ twice, we get
$$0 = M_x(x,m(x)) + M_\mu(x,m(x)) m'(x),$$
$$0 = M_{xx}(0,1) + 2 M_{x\mu}(0,1) m'(0) + M_{\mu\mu}(0,1) (m'(0))^2 + M_\mu(0,1) m''(0).$$

Evaluating the first equation at $(0,1)$ and using the derivatives of M already calculated, we get
$$0 = 0 + 2 m'(0) \quad \text{so} \quad m'(0) = 0;$$
from the second equation, we have
$$0 = -4\left(b + a^2\right) + 2 M_{x\mu}(0,1) \cdot 0 + M_{\mu\mu}(0,1) \cdot 0^2 + 2 m''(0),$$
$$m''(0) = 2\left(b + a^2\right).$$

Substituting into the expansion of μ,
$$\mu = m(x) = 1 + \left(b + a^2\right) x^2 + O(x^3).$$

We need $b + a^2 \neq 0$ for the sign of $m(x) - 1$ to be determined by the quadratic terms. In particular, the period-2 orbit appears for $\mu > 1 = \mu_0$ if $b + a^2 > 0$ and appears for $\mu < 1 = \mu_0$ if $b + a^2 < 0$.

The stability type of the period-2 orbit is also determined by the sign of $b+a^2$. We use a Taylor series expansion of $(f^2)_x(x, m(x))$ about $x = 0$ and $\mu = m(0) = 1$:

$$(f^2)_x(x; m(x)) = (f^2)_x(0; 1) + (f^2)_{xx}(0; 1)\, x$$
$$+ (f^2)_{\mu x}(0; 1)\,(m(x) - 1) + \tfrac{1}{2}\,(f^2)_{xxx}(0; 1)\, x^3 + \cdots.$$

Using the expansion of f^2 given earlier,

$$(f^2)_x(0; 1) = (f_x(0, 1))^2 = (-1)^2 = 1,$$
$$(f^2)_{xx}(0; 1) = 2\left(-a\,\mu + a\,\mu^2\right)\big|_{\mu=1} = 0,$$
$$(f^2)_{\mu x}(0; 1) = 2\,\mu\big|_{\mu=1} = 2,$$
$$\tfrac{1}{2}\,(f^2)_{xxx}(0; 1) = \left(\tfrac{1}{2}\right)6\left(-b\,\mu - 2\,a^2\,\mu - b\,\mu^3\right)\big|_{\mu=1} = -6\left(b + a^2\right).$$

Substituting $m(x) - 1 = (b + a^2)\, x^2 + O(x^3)$ yields

$$(f^2)_{\mu x}(0; 1)\,(m(x) - 1) = 2\,(b + a^2)\, x^2 + O(x^3).$$

Combining all the partial derivatives at $x = 0$ and $\mu = m(0) = 1$ results in

$$(f^2)_x(x; m(x)) = 1 + 0 \cdot x + 2\,(b + a^2)\, x^2 - 6\,(b + a^2)\, x^2 + O(x^3)$$
$$= 1 - 4\,(b + a^2)\, x^2 + O(x^3);$$

the period-2 orbit is attracting if $b + a^2 > 0$ and is repelling if $b + a^2 < 0$. $\square$

Conjugacy

Restatement of Theorem 9.20. *Let $f(x) = \tfrac{1}{2}(x + x^3)$ and $g(y) = ay + by^3$ with $0 < a < 1$ and $b > 0$. Then, f on $[-1, 1]$ is topologically conjugate to g on the interval between the fixed points, $\left[-\sqrt{(1-a)b^{-1}}, \sqrt{(1-a)b^{-1}}\right]$. The conjugacy can be extended to be a conjugacy on the whole real line.*

Proof. The map f has fixed points at $x = 0$ and ± 1, while the map g has fixed points at $y = 0$ and $\pm\sqrt{(1-a)b^{-1}}$.

Notice that $f'(x) = \tfrac{1}{2}(1 + 3\,x^2)$ and $g'(y) = a + 3by^2$ are both positive everywhere, which makes the proof easier, since f and g have inverses.

Also, $f'(0) = 0.5$ and $g'(0) = a$, so 0 is an attracting fixed point for both maps. The basins of attraction for these fixed points for f and g are

$$\mathscr{B}(0; f) = (-1, 1) \quad \text{and}$$
$$\mathscr{B}(0; g) = \left(-\sqrt{(1-a)b^{-1}}, \sqrt{(1-a)b^{-1}}\right).$$

Since every point x in $(0, 1)$ has exactly one preimage in $\left(f(\tfrac{1}{2}), \tfrac{1}{2}\right]$, $x_{-j} = f^{-j}(x)$ for some j; the corresponding closed interval $\left[f(\tfrac{1}{2}), \tfrac{1}{2}\right]$ is called a *fundamental domain*. Let $\bar{y} = \sqrt{(1-a)b^{-1}}$. Again, every point y in $(0, \bar{y})$ has exactly one iterate $y_{-j} = g^{-j}(y)$ in $(g(\bar{y}/2), \bar{y}/2]$.

Define the map $y = h_0(x)$ on $\left[f(\tfrac{1}{2}), \tfrac{1}{2}\right]$ to be affine, taking the interval $\left[f(\tfrac{1}{2}), \tfrac{1}{2}\right]$ to the interval $[g(\bar{y}/2), \bar{y}/2]$:

$$h_0(x) = \frac{\bar{y}}{2} - (0.5 - x)\,\frac{\bar{y}/2 - g(\bar{y}/2)}{0.5 - f(0.5)}.$$

The end points of the fundamental domain of f go to the end points of the fundamental domain of g, $h_0(0.5) = \bar{y}/2$, and $h_0\left(f(0.5)\right) = \bar{y}/2 - (\bar{y}/2 - g\left(\bar{y}/2\right)) = g\left(\bar{y}/2\right)$. Notice that $h_0\left(f(0.5)\right) = g\left(h_0(0,5)\right)$, which is the only point for which both compositions are defined.

For any point x in $(0,1)$, let $j(x)$ be such that $f^{-j(x)}(x) \in \left(f(1/2), 1/2\right]$, and let
$$h(x) = g^{j(x)} \circ h_0 \circ f^{-j(x)}(x).$$
Then, $j(f(x)) = j(x) + 1$, so
$$\begin{aligned} h\left(f(x)\right) &= g^{1+j(x)} \circ h_0 \circ f^{-1-j(x)}\left(f(x)\right) \\ &= g \circ g^{j(x)} \circ h_0 \circ f^{-j(x)}(x) \\ &= g(h(x)). \end{aligned}$$
This shows that $y = h(x)$ satisfies the conjugacy equation.

The definition of h from $(-1, 0)$ to $(-\bar{y}, 0)$ is similar, using the fundamental domains $\left[f(1/2), 1/2\right]$ and $[g(\bar{y}/2), \bar{y}/2]$. Also, define $h(0) = 0$, $h(1) = \bar{y}$, and $h(-1) = -\bar{y}$.

The map $h(x)$ is clearly continuous on $(-1, 0)$ and $(0, 1)$. As x converges to 0, $j(x)$ goes to infinity, so $h(x) = g^{j(x)} \circ h_0 \circ f^{-j(x)}(x)$ must converge to 0. (Notice that $h_0 \circ f^{-j(x)}(x)$ is in a closed interval in the basin of attraction for g.) This shows that h is continuous at 0.

The continuity at ± 1 is similar.

The extension of $h(x)$ to all of $\mathbb{R}$ is similar by using fundamental domains outside the fixed points. $\square$

Chapter 10

Itineraries for One-Dimensional Maps

In the proceding chapter, we found periodic points using the graphical method of iteration or algebra. We start this chapter by introducing the transition graph as a means of proving the existence of various periodic orbits. The next few sections use similar methods to show the existence of other more complicated dynamics on invariant sets. In the second section, we show the existence of an orbit for the tent map T and logistic map G that comes arbitrarily close to each point in the interval $[0, 1]$. The tool we use to find orbits with prescribed behavior goes under the name of *symbolic dynamics*, because we use strings of symbols to find orbits that move through a sequence of intervals in the specified order. This method also gives us another way to show the existence of periodic points for the two maps T and G. In fact, in the third section, we use symbolic dynamics to show that there is a periodic point arbitrarily close to every point in $[0, 1]$.

In the fourth section, we use the method of symbolic dynamics to show that several of these maps have a property called sensitive dependence on initial conditions. This property is key to what is usually called chaotic dynamics, which we discuss in the next chapter.

In the fifth section, symbolic dynamics are applied to invariant sets, which have many points but are filled with gaps; these sets are called Cantor sets. In the sixth section, the method is applied to maps for which only certain strings of symbols are allowed; hence, they are called subshifts.

All of these sections involve finding some structure for invariant sets, which are more complicated than periodic orbits. At one time in the development of the subject, it was popular to refer to the aspect of this subject as "finding order in chaos."

10.1. Periodic Points from Transition Graphs

We have found periodic points for a variety of maps. We now start considering the set $\mathscr{P}(f)$ of all periods that occur for a map f. For maps on the line, there is a general result about restrictions on $\mathscr{P}(f)$ called the Sharkovskii theorem [**Sha64**], which we discuss in Section 10.1.1. A special case of this theorem was proved by J. Yorke and T.Y. Li in the paper in which they introduced the term chaos, [**Li,75**].

Ironically, the more general paper by A. N. Sharkovskii appeared eleven years earlier than this specific case, but appeared in a journal that was not widely read in the United States. But the paper by Li and Yorke has still had a significant impact, because it did much to make the result of Sharkovskii known, as well as giving a big push to find chaotic behavior for deterministic systems in applications.

We also use this section to introduce some of the ideas of symbolic dynamics. We show the existence of orbits with a given itinerary, by writing down a sequence of intervals through which the orbit must pass. Each interval is given a symbol. For any string of allowable symbols (i.e., for any string of allowable intervals), there is an orbit which goes through these intervals in the prescribed order. In this section, we use this idea to get periodic orbits. In later sections, we use this method to get nonperiodic orbits whose orbit winds densely throughout an interval. Altogether, the ideas of symbolic dynamics are introduced in this section together with Sections 10.3 and 10.5.

To make the statements more convenient, we introduce the following notation.

Definition 10.1. For a closed bounded interval $\mathbf{I} = [a, b]$, the *interior of* $\mathbf{I}$ is the corresponding open interval (a, b), and we write $\text{int}(\mathbf{I}) = (a, b)$. (Later in the book we define the interior of more general sets.) We also write $\partial(\mathbf{I})$ for the two end points, $\partial(\mathbf{I}) = \{a, b\}$.

The first result is the basic lemma on the existence of a fixed point, on which we build the general result: Part (b) proves the existence of a fixed point and part (a) is used in the induction step to prove the existence of higher period points.

Lemma 10.1. *Let f be a continuous function from a closed interval $[a_1, a_2]$ to $\mathbb{R}$. Assume that $f([a_1, a_2]) \supset [b_1, b_2]$ for another closed interval $\mathbf{I} = [b_1, b_2]$. Then the following hold.*

    a. *There is a subinterval $\mathbf{J} = [x_1, x_2] \subset [a_1, a_2]$ such that $f(\mathbf{J}) = \mathbf{I}$, $f(\text{int}(\mathbf{J})) = \text{int}(\mathbf{I})$, and $f(\partial \mathbf{J}) = \partial \mathbf{I}$.*

    b. *If $f([a_1, a_2]) \supset [a_1, a_2]$, then f has a fixed point in $[a_1, a_2]$.*

Proof. (a) Since $f([a_1, a_2]) \supset [b_1, b_2]$, there are points $y_1, y_2 \in [a_1, a_2]$ with $y_1 < y_2$ such that $\{f(y_1), f(y_2)\} = \{b_1, b_2\}$. For specificity, assume that $f(y_1) = b_2$ and $f(y_2) = b_1$ as in Figure 1. Then, let x_1 be the greatest number $y_1 \leq x < y_2$ such that $f(x) = b_2$. Therefore, $f(x_1) = b_2$ and $f(x) < b_2$ for $x_1 < x \leq y_2$. Next, let x_2 be the smallest number $x_1 < x \leq y_2$ such that $f(x) = b_1$. Therefore, $f(x_2) = b_1$ and $b_1 < f(x) < b_2$ for $x_1 < x < x_2$. By the intermediate value theorem, $f([x_1, x_2])$ must realize every value between b_1 and b_2, $f([x_1, x_2]) = [b_1, b_2]$. The case with $f(y_1) = b_1$ and $f(y_2) = b_2$ is similar.

10.1. Transition Graphs

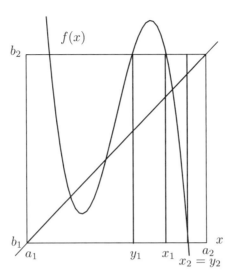

Figure 1. Subinterval for Lemma 10.1

(b) Assume that $f([a_1, a_2]) \supset [a_1, a_2]$ and $[x_1, x_2] \subset [a_1, a_2]$ is as in the proof of part (a). $f(x_2) = a_1 \leq x_2$. Setting $g(x) = f(x) - x$, $g(x_1) = f(x_1) - x_1 \geq 0$ and $g(x_2) = f(x_2) - x_2 \leq 0$ for either of the cases in the proof of part (a). Applying the intermediate value theorem to g, we see that there is a point x^* with $x_1 \leq x^* \leq x_2$ and $g(x^*) = 0$. For this x^*, $f(x^*) = x^*$. This proves (b). $\square$

Theorem 10.2. *Let f be a continuous function from $\mathbb{R}$ to $\mathbb{R}$ and $\mathbf{J}_0, \ldots, \mathbf{J}_n$ a finite string of closed bounded intervals such that $f(\mathbf{J}_j) \supset \mathbf{J}_{j+1}$ for $0 \leq j \leq n-1$. (The intervals are allowed to overlap and/or occur more than once.)*

a. *The set*
$$\mathscr{J} = \{\, x : f^j(x) \in \mathbf{J}_j \text{ for } 0 \leq j \leq n \,\} = \bigcap_{0 \leq j \leq n} f^{-j}(\mathbf{J}_j).$$
contains a nonempty closed interval and there is a point x^* such that $f^j(x^*)$ is in $\mathbf{J}_j$ for $0 \leq j \leq n$.

b. *If the first and last intervals are the same, $\mathbf{J}_0 = \mathbf{J}_n$, then the point x^* of part (a) can be chosen so that $f^n(x^*) = x^*$. Thus, the period of x^* must divide n.*

c. *Assume the following three conditions: (i) $\mathbf{J}_0 = \mathbf{J}_n$. (ii) The end points of $\mathbf{J}_0$ are not in $\mathscr{J}$, $\partial(\mathbf{J}_0) \cap \mathscr{J} = \emptyset$. (iii) $\mathrm{int}(\mathbf{J}_0) \cap \mathbf{J}_i = \emptyset$ for $0 < i < n$. Then, the point x^* has least period n for f.*

Remark 10.2. In the theorem (and repeatedly in the book), we write $f^{-j}(\mathbf{J}_j)$ for the set $\{\, x : f^j(x) \in \mathbf{J}_j \,\}$ (and not for the inverse of the function).

Remark 10.3. In Theorem 10.3, we given another condition to replace the assumption (iii) in part (c) of the theorem. The assumption given here is used in the more technical discussion at the end of the chapter and is taken from the paper by Burns and Hasselblatt [**Bur11**].

Proof. (a) Let $\mathbf{K}_n = \mathbf{J}_n$. By Lemma 10.1(b), there is a subinterval $\mathbf{K}_{n-1}$ contained in $\mathbf{J}_{n-1}$ such that $f(\mathbf{K}_{n-1}) = \mathbf{K}_n = \mathbf{J}_n$ and $f(\mathrm{int}(\mathbf{K}_{n-1})) = \mathrm{int}(\mathbf{J}_n)$. Continuing backward by induction, we find that there is a subinterval $\mathbf{K}_{n-j}$ contained in $\mathbf{J}_{n-j}$ such that $f(\mathbf{K}_{n-j}) = \mathbf{K}_{n-j+1}$ and $f(\mathrm{int}(\mathbf{K}_{n-j})) = \mathrm{int}(\mathbf{K}_{n-j+1})$ for $j = 1, \ldots, n$. The resulting interval $\mathbf{K}_0$ satisfies
$$f^j(\mathbf{K}_0) = \mathbf{K}_j \subset \mathbf{J}_j,$$
$$f^j(\mathrm{int}(\mathbf{K}_0)) = \mathrm{int}(\mathbf{K}_j) \subset \mathrm{int}(\mathbf{J}_j), \quad \text{and}$$
$$f^n(\mathbf{K}_0) = \mathbf{K}_n = \mathbf{J}_n.$$
By construction, the subinterval $\mathbf{K}_0$ is a nonempty closed interval and $\mathbf{K}_0 \subset \mathscr{J}$. Any point x^* in $\mathbf{K}_0$ has $f^j(x^*)$ in $\mathbf{J}_j$ for $0 \leq j \leq n$, so we have proved part (a).

(b) The interval $\mathbf{K}_0$ constructed in part (a) satisfies $f^n(\mathbf{K}_0) = \mathbf{J}_n = \mathbf{J}_0 \supset \mathbf{K}_0$. Applying Lemma 10.1(b) to f^n on $\mathbf{K}_0$, there is a point x^* in $\mathbf{K}_0$ with $f^n(x^*) = x^*$.

(c) We know that $f^n(x^*) = x^* \in \mathrm{int}(\mathbf{J}_0)$. Since $f^j(x^*) \in \mathbf{K}_j \subset \mathbf{J}_j$ for $0 < j < n$ and Since $\mathbf{J}_j \cap \mathrm{int}(\mathbf{J}_0) = \emptyset$ for $0 < j < n$, $f^j(x^*) \neq x^*$ for $0 < j < n$. This shows that x^* cannot have period less than n. □

The preceding theorem is used to find points of higher periods. We usually start with a finite collection of intervals and specify which interval the orbit should be in at each iterate (i.e., specify the *itinerary*). We use the finite collection of intervals to define a transition graph that specifies the allowable sequences of the choices of intervals. We give three examples before the general definitions and results.

Example 10.4. Let $G(x) = 4x(1-x)$ be the logistic map, with parameter 4. Let $\mathbf{I}_L = [0, 0.5]$ and $\mathbf{I}_R = [0.5, 1]$ be the left and right subintervals. Then,
$$G(\mathbf{I}_L) = [0, 1] = \mathbf{I}_L \cup \mathbf{I}_R \quad \text{and}$$
$$G(\mathbf{I}_R) = [0, 1] = \mathbf{I}_L \cup \mathbf{I}_R.$$
Since each interval covers both intervals, we make a graph with vertices labeled by L and R and directed edges from each symbol to itself and to the other symbol. See Figure 2. Theorem 10.3 shows that this transition graph implies that G has points of all periods.

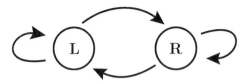

Figure 2. Transition graph for Example 10.4

Example 10.5. Let f have a graph as given in Figure 3, with the three subintervals $\mathbf{I}_1$, $\mathbf{I}_2$, and $\mathbf{I}_3$ as labeled. Then,
$$f(\mathbf{I}_1) \supset \mathbf{I}_3,$$
$$f(\mathbf{I}_2) \supset \mathbf{I}_2 \cup \mathbf{I}_1, \quad \text{and}$$
$$f(\mathbf{I}_3) \supset \mathbf{I}_1.$$

10.1. Transition Graphs

Thus, the transition graph has an arrow from 1 to 3, from 2 to 2, from 2 to 1, and from 3 to 1. See Figure 4. By Lemma 10.1, there is a fixed point in interval $\mathbf{I}_2$. By Theorem 10.3 that follows, there is a period-2 orbit that alternates between intervals $\mathbf{I}_1$ and $\mathbf{I}_3$.

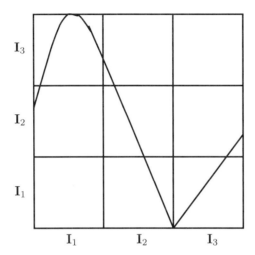

Figure 3. Examples 10.5: Graph of function

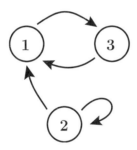

Figure 4. Examples 10.5: Transition graph

Example 10.6. Let $g(x) = (3.839)x(1-x)$ be the logistic map for $a = 3.839$. This map has two period-3 orbits: The orbit with values $x_0 \approx 0.1498$, $x_1 \approx 0.4891$, and $x_2 \approx 0.9593$ is attracting, while the one with values $y_0 \approx 0.1690$, $y_1 \approx 0.5392$, and $y_2 \approx 0.9538$ is repelling. Let $\mathbf{I}_L = [x_0, x_1]$ and $\mathbf{I}_R = [x_1, x_2]$ be intervals with end points on one of the period-3 orbits. By following the iterates of the end points, it follows that

$$g(\mathbf{I}_L) \supset [x_1, x_2] = \mathbf{I}_R \quad \text{and}$$
$$g(\mathbf{I}_R) \supset [x_0, x_2] = \mathbf{I}_R \cup \mathbf{I}_L.$$

We make a graph with vertices labeled by the symbols L and R. We draw an arrow connecting the symbol i to k, provided $g(\mathbf{I}_i) \supset \mathbf{I}_k$, Thus, there is an arrow from L to R, from R to R, and from R to L. See Figure 5. After stating Theorem 10.3, we show that this transition graph implies that g has points of all periods.

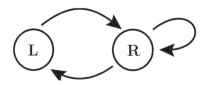

Figure 5. Transition graph for Example 10.6

The following definitions apply in situations such as the preceding examples. Since the number of intervals is an arbitrary number N, we use the numbers 1 to N as symbols rather than letters such as L, C, and R. (Other times, the symbols 0 to $N-1$ are used.)

Definition 10.7. A *partition of closed intervals* is a finite collection of closed bounded intervals $\mathbf{I}_1, \ldots, \mathbf{I}_N$ such that the associated open intervals are disjoint,

$$\text{int}(\mathbf{I}_i) \cap \text{int}(\mathbf{I}_j) = \emptyset \quad \text{if } i \neq j.$$

We say that a function f is *continuous on a partition* $\{\mathbf{I}_i\}_{i=1}^N$ provided that f is continuous on $\bigcup_{i=1}^N \mathbf{I}_i$.

Definition 10.8. Assume that f is a function from $\mathbb{R}$ to $\mathbb{R}$. Based on the nature of the function, we select a partition $\{\mathbf{I}_i\}_{i=1}^N$ of closed intervals such that f is continuous on the partition. The set $\mathbf{S} = \{1, \ldots, N\}$ of indices of the intervals is called the *set of symbols for the partition*, and an element of $\mathbf{S}$ is called a *symbol*. The *transition graph* for the partition and function is the directed graph with (i) vertices labeled by the symbols $\mathbf{S}$ and with (ii) a directed edge (an arrow) connecting the symbol i to k if and only if $f(\mathbf{I}_i) \supset \mathbf{I}_k$.

Remark 10.9. The directed graph with vertices and directed edges described in the preceding definition is a graph in the sense of graph theory, with vertices and directed edges connecting some of the vertices. It is not the kind of graph that is referred to in the graph of $y = x^2$.

Definition 10.10. Assume we have a transition graph with N vertices $\mathbf{S}$. A finite string of symbols $s_0 \ldots s_n$, where each s_j is one of symbols in $\{1, \ldots, N\}$, is also called a *word*. For a word $\mathbf{w} = s_0 \ldots s_n$, we write $\mathbf{w}^k$ for the word $\mathbf{w}$ repeated k times,

$$\mathbf{w}^k = s_0 \ldots s_n s_0 \ldots s_n \ldots s_0 \ldots s_n.$$

The infinitely repeated word is written $\mathbf{w}^\infty = (s_0 \ldots s_n)^\infty = s_0 \ldots s_n s_0 \ldots s_n \ldots$.

A word $s_0 \ldots s_n$ is called *allowable* provided that there is an edge in the transition graph from s_j to s_{j+1} for $0 \leq j \leq n-1$. From the definition of the transition graph, it follows that a string of symbols $s_0 \ldots s_n$ is allowable if and only if there is a path in the transition graph moving along directed edges from the vertices s_0 to s_n passing through all the s_j in the prescribed order. An infinite sequence $s_0 s_1 \ldots$

10.1. Transition Graphs

is *allowable* provided that there is an edge in the transition graph from s_j to s_{j+1} for all $0 \leq j$.

An allowable infinite sequence of symbols $\mathbf{s} = s_0 s_1 \ldots$ is called *periodic* provided that there exists an $n \geq 1$ such that $s_{n+k} = s_k$ for all $k \geq 0$, so $\mathbf{s} = (s_0 s_1 \ldots s_{n-1})^\infty$. An allowable word $s_0 \ldots s_n$ is called *periodic* provided that $s_n = s_0$, so $(s_0 s_1 \ldots s_{n-1})^\infty$ is an allowable periodic sequence.

An allowable periodic word $\mathbf{w} = s_0 s_1 \ldots s_n$ is called *reducible*, provided that $n = mp$ with $p > 1$ and $s_0 s_1 \ldots s_{n-1} s_n = (s_0 \ldots s_{m-1})^p s_n$. An allowable periodic word $\mathbf{s} = s_0 s_1 \ldots s_n$ that is not reducible is said to be *irreducible* and to have *least period* n. An irreducible periodic word $\mathbf{w} = s_0 s_1 \ldots s_n$ is called n-*periodic*.

An allowable periodic sequence $\mathbf{s} = (s_0 s_1 \ldots$ is said to have *least period* n and be n-*periodic* provided that $\mathbf{s} = (s_0 \ldots s_{n-1})^\infty$ and the word $s_0 \ldots s_{n-1} s_0$ is irreducible.

Definition 10.11. Assume that a function f is continuous on a partition of closed intervals $\{\mathbf{I}_i\}_{i=1}^N$. We use the subscript on the interval, $\mathbf{I}_s$, as a symbol for the given interval, so s is a choice from $\mathbf{S} = \{1, \ldots, N\}$. Given a string of such symbols, $s_0 \ldots s_n$ with each $s_j \in \mathbf{S}$, we seek to find a point x such that $f^j(x)$ is a point in the interval $\mathbf{I}_{s_j}$ for $0 \leq j \leq n$. In such a string of intervals we allow a given interval to be used more than once. We let

$$\mathbf{I}_{s_0 \ldots s_n} = \{\, x : f^j(x) \in \mathbf{I}_{s_j} \text{ for } 0 \leq j \leq n \,\} = \bigcap_{0 \leq j \leq n} f^{-j}(\mathbf{I}_{s_j}).$$

Most of the following theorem follows directly from Theorem 10.2. For part (c) we assume that we start with a partition for f, which allows us to conclude the point has least period n for f.

Theorem 10.3. *Let f be a function from $\mathbb{R}$ to $\mathbb{R}$ and it is continuous on a partition of closed intervals $\{\mathbf{I}_i\}_{i=1}^N$. Let $\mathbf{w} = s_0 \ldots s_n$ be an allowable word for the transition graph for the partition.*

    a. *The set $\mathbf{I}_{s_0 \ldots s_n} \neq \emptyset$, and there is a point x^* such that $f^j(x^*)$ is in $\mathbf{I}_{s_j}$ for $0 \leq j \leq n$.*

    b. *If the word $s_0 \ldots s_n$ is periodic with $s_n = s_0$, then the point x^* of part (a) can be chosen so that $f^n(x^*) = x^*$. Thus, the period of x^* must divide n.*

    c. *Assume that $\mathbf{w} = s_0 \ldots s_n$ is n-periodic and irreducible (i.e., n is the least period of the string). Assume that the end points of $\mathbf{I}_{s_0}$ are not in $\mathbf{I}_{s_0 \ldots s_n}$, or $\partial(\mathbf{I}_{s_0}) \cap \mathbf{I}_{s_0 \ldots s_n} = \emptyset$. Then, the point x^* has least period n for f.*

Remark 10.12. We usually verify the assumption in part (c) that the end points of $\mathbf{I}_{s_0}$ are not in $\mathbf{I}_{s_0 \ldots s_n}$ by using the fact that they are periodic with a period different than n.

Proof. Let $\mathbf{J}_i = \mathbf{I}_{s_i}$. Let $\mathbf{K}_i$ for $0 \leq i \leq n$ be the intervals constructed in the proof of Theorem 10.2. Parts (a) and (b) follow directly from that earlier theorem.

(c) We assumed that $\partial(\mathbf{I}_{s_0}) \cap \mathbf{I}_{s_0 \ldots s_n} = \emptyset$, so $f^n(x^*) = x^* \in \text{int}(\mathbf{I}_{s_0}) = \text{int}(\mathbf{I}_{s_n})$. Since $f^n(\text{int}(\mathbf{K}_0)) = \text{int}(\mathbf{I}_{s_n})$, $x^* \in \text{int}(\mathbf{K}_0)$. Then, $f^j(x^*) \in f^j(\text{int}(\mathbf{K}_0)) = \text{int}(\mathbf{K}_j) \subset \text{int}(\mathbf{I}_{s_j})$ must be in the open intervals corresponding to each $\mathbf{I}_{s_j}$. Since we started with a partition, the open intervals $\text{int}(\mathbf{I}_i)$ are disjoint. If $f^k(x^*) = x^*$ for $0 < k < n$, then the word $\mathbf{w} = s_0 \ldots s_n$ would have period $k < n$ which is impossible since it is irreducible. This shows that x^* cannot have period less than n and must have least period n. □

Remark 10.13. When we list the string of symbols for a periodic orbit, we usually write an infinite sequence of symbols, rather than the finite one used in the statement of the theorem. In the preceding theorem, a string such as $s_0 s_1 s_2 s_3 = RRLR$ corresponds to a period-3 point. The orbit continues to pass through this same string of intervals for higher iterates. Therefore, we can write an infinite sequence of symbols for a periodic point, which we call the *itinerary* of the point. For the preceding example, we write

$$s_0 s_1 \cdots = (RRL)^\infty = RRLRRLRRLRRL \cdots,$$

because $f^j(x_0)$ is in $\mathbf{I}_{s_j}$ for all $j \geq 0$. For a word $\mathbf{w}$, in order for an infinite sequence $\mathbf{w}^\infty$ to be allowable, it is necessary for the first symbol in the word $\mathbf{w}$ to be able to follow the last symbol in the word $\mathbf{w}$. Thus, for the string $\mathbf{w} = RRL$, for $\mathbf{w}^\infty$ to be an allowable sequence, it is necessary that the first symbol R can follow the last symbol L. In Section 10.3, an infinite sequence of symbols (an itinerary) is assigned to nonperiodic points, as well as to periodic points. Therefore, we start using this notation at this time.

We can apply the previous theorem to a function with a period-3 point to obtain the following result by T. Li and J. Yorke, [**Li,75**].

Proposition 10.4 (Li and Yorke). *Assume that f is a continuous function from $\mathbb{R}$ to itself. Assume assume that either (a) f has a period-3 point or (b) there is a point x_0 such that either (i) $f^3(x_0) \leq x_0 < f(x_0) < f^2(x_0)$ or (ii) $f^3(x_0) \geq x_0 > f(x_0) > f^2(x_0)$. Then, for each $n \in \mathbb{N}$, f has a period-n point.*

Proof. Case (b) includes case (a). Case (b.ii) is similar to (b.i) by reversing the orientation on the line, so we assume that condition (b.i) holds: $f(x_2) \leq x_0 < f(x_0) = x_1 < f(x_1) = x_2$. See Figure 6 for the graph of a function with period three. Take the partition of closed intervals $\mathbf{I}_L = [x_0, x_1]$ and $\mathbf{I}_R = [x_1, x_2]$. By following the iterates of the end points, it follows that

$$f(\mathbf{I}_L) \supset [x_1, x_2] = \mathbf{I}_R \quad \text{and} \quad f(\mathbf{I}_R) \supset [x_0, x_2] = \mathbf{I}_R \cup \mathbf{I}_L.$$

The transition graph for $\mathbf{I}_L = [x_0, x_1]$ and $\mathbf{I}_R = [x_1, x_2]$ is given in Figure 5. Table 1 indicates which sequence of symbols corresponds to a periodic point of the period indicated.

n	Symbol Sequence
1	R^∞
2	$(RL)^\infty$
3	$(RRL)^\infty$
n	$(R^{n-1}L)^\infty$

Table 1. Sequence of symbols for periods for a function with a period-3 point

10.1. Transition Graphs

Note that the points on the boundaries of the intervals all have period three, so the point given by Theorem 10.3(b) must be in the open intervals and it so must have the period indicated. Since all periods appear, the function must have points of all periods. □

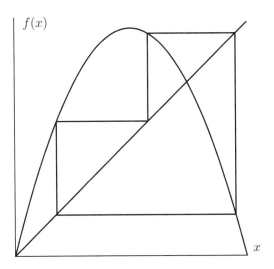

Figure 6. Plot of a period-3 orbit using the stair step plot

10.1.1. Sharkovskii Theorem. We first define the special ordering of the positive integers necessary to state the Sharkovskii theorem about the set of periods that can occur for a continuous function on $\mathbb{R}$.

We want to see which periods are forced by other periods. For a function f, let
$$\mathscr{P}(f) = \{\, k : f \text{ has a point of period } k \,\}.$$
Proposition 10.4 shows that if 3 is in $\mathscr{P}(f)$ for a continuous function on $\mathbb{R}$, then $\mathscr{P}(f)$ is the set of all positive integers.

Definition 10.14. We introduce an ordering of the positive integers, which are the possible periods, called the *Sharkovskii ordering*. When we write $m \triangleright n$, we want the existence of a period-m point to force the existence of a period-n point. So $3 \triangleright n$ for any positive integer n. The correct ordering is as follows. Let $\mathscr{J}$ be the set of all odd integers greater than one. These odd integers $\mathscr{J}$ are listed in the Sharkovskii ordering, with the ordering opposite the one they have in terms of how they appear on the line,
$$3 \triangleright 5 \triangleright 7 \triangleright 9 \triangleright 11 \triangleright \cdots.$$
Next, all the integers which are equal to 2 times an integer in $\mathscr{J}$ are added to the list, then 2^2 times an integer in $\mathscr{J}$ are added to the list, and then increasing powers of two times an integer in $\mathscr{J}$ are added to the list:
$$3 \triangleright 5 \triangleright 7 \triangleright \cdots \triangleright 2 \cdot 3 \triangleright 2 \cdot 5 \triangleright 2 \cdot 7 \triangleright \cdots$$
$$\triangleright 2^k \cdot 3 \triangleright 2^k \cdot 5 \triangleright 2^k \cdot 7 \triangleright \cdots \triangleright 2^{k+1} \cdot 3 \triangleright 2^{k+1} \cdot 5 \triangleright 2^{k+1} \cdot 7 \triangleright \cdots.$$

Finally, all the powers of 2 are added in decreasing powers:

$$3 \triangleright 5 \triangleright 7 \triangleright \cdots \triangleright 2^k \cdot 3 \triangleright 2^k \cdot 5 \triangleright 2^k \cdot 7 \triangleright \cdots$$
$$\cdots \triangleright 2^{k+1} \triangleright 2^k \triangleright 2^{k-1} \triangleright \cdots \triangleright 2^2 \triangleright 2 \triangleright 1.$$

We have now listed all the positive integers.

Theorem 10.5 (Sharkovskii). **a.** *Let f be a continuous function from $\mathbb{R}$ to itself. Assume that f has a period-n point and $n \triangleright k$. Then, f has a period-k point. Thus, if n is in $\mathscr{P}(f)$ and $n \triangleright k$, then k is in $\mathscr{P}(f)$.*

b. *For any $n \in \mathbb{N}$, there exists a continuous $f : \mathbb{R} \to \mathbb{R}$ such that $\mathscr{P}(f) = \{k : n \triangleright k\}$.*

Key steps in the proof involve the clever use of Theorem 10.3. We do not prove the general statement of part (a), but present some of the main ideas through the consideration of special cases. See [**Rob99**] gives a proof using Stephan cycles defined below. P. D. Straffin [**Str78**], B.-S. Du [**Du,02**], and K. Burns and B. Hasselblatt [**Bur11**] have published articles proving the Sharkovskii theorem that are accessible to anyone reading this book. We do give constructions to show that part (b) of the theorem is true and partial results toward the proof of part (a) in Section 10.8 at the end of the chapter.

Many of the proofs of this theorem (e.g., the one by Block, Guckenheimer, Misiurewicz, and Young [**Blo80**]) use the fact that, if there is a point of odd period n, then there is possibly another period-n point for which the order of its iterates in the real line is especially simple. This latter periodic orbit is called a Stefan cycle. Piecewise linear maps of the type given in Example 10.16 show the existence of a map with exactly the periods forced by an odd integer in the Sharkovskii order.

Definition 10.15. A *Stefan cycle* is a period-n orbit with n odd, for which the order of the iterates $x_j = f^j(x_0)$ on the line is one of the following:

$$x_{n-1} < \cdots < x_4 < x_2 < x_0 < x_1 < x_3 < \cdots < x_{n-2},$$

the exact reverse order

$$x_{n-2} < \cdots < x_3 < x_1 < x_0 < x_2 < x_4 < \cdots < x_{n-1},$$

or one of these two types of orbits starting at a different point on the orbit.

A periodic orbit of period 3 is always a Stefan cycle.

Example 10.16. Assume that f from $\mathbb{R}$ to itself is continuous and has a period-7 Stefan cycle $x_j = f^j(x_0)$ for $0 \leq j \leq 6$, with order on the line $x_6 < x_4 < x_2 < x_0 < x_1 < x_3 < x_5$. The associated partition of closed intervals for Stefan cycle is $\mathbf{I}_1 = [x_0, x_1]$, $\mathbf{I}_2 = [x_2, x_0]$, $\mathbf{I}_3 = [x_1, x_3]$, $\mathbf{I}_4 = [x_4, x_2]$, $\mathbf{I}_5 = [x_3, x_5]$, and $\mathbf{I}_6 = [x_6, x_4]$. See Figure 7. By considering the image of the end points of each $\mathbf{I}_j$,

10.1. Transition Graphs

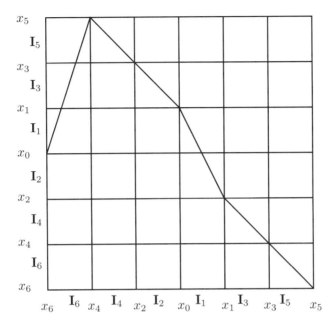

Figure 7. Example 10.16: Graph of an f with period-7 Stephan cycle

we get that

$$f(\mathbf{I}_1) \supset [x_2, x_1] = \mathbf{I}_1 \cup \mathbf{I}_2,$$
$$f(\mathbf{I}_2) \supset [x_1, x_3] = \mathbf{I}_3,$$
$$f(\mathbf{I}_3) \supset [x_4, x_2] = \mathbf{I}_4,$$
$$f(\mathbf{I}_4) \supset [x_3, x_5] = \mathbf{I}_5,$$
$$f(\mathbf{I}_5) \supset [x_6, x_4] = \mathbf{I}_6,$$
$$f(\mathbf{I}_6) \supset [x_5, x_0] = \mathbf{I}_1 \cup \mathbf{I}_3 \cup \mathbf{I}_5.$$

The graph of a piecewise linear map f with the preceding properties and its transition graph are shown in Figures 7 and 8. The Table 2 gives sequences of symbols for the various periods that are possible. The end points of $\mathbf{I}_1$ have period 7 so cannot have the symbol sequence given in the chart, so the point must have the period-n indicated. In particular, it shows that $\mathscr{P}(f)$ must contain all positive integers except possibly 3 and 5. In fact, if the function f is linear between the points x_i as is the case for the function in the figure, then $\mathscr{P}(f) = \{\, k : 7 \triangleright k \,\} = \{\, k : k \neq 3, 5 \,\}$. Other functions with a period-7 Stephan cycle could have other periods.

Example 10.17 (Doubling the periods). In this example, we show how to take a function f from $[0, 1]$ to itself and make a new function g defined on $[0, 1]$ such that

(10.1) $$\mathscr{P}(g) = \{1\} \cup 2\,\mathscr{P}(f)$$

(i.e., g has a fixed point, together with twice all the periods of f).

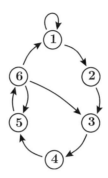

Figure 8. Example 10.16: Transition graph

n	Sequence of Symbol	Period Exists?
1	1^∞	Yes
2	$(56)^\infty$	Yes
3	None	No
4	$(3456)^\infty$	Yes
5	None	No
6	$(123456)^\infty$	Yes
n	$(1^{n-5}23456)^\infty$	Yes

Table 2. Strings of symbols giving various periods for a function with a period-7 Stefan cycle, Example 10.16.

Define g by

(10.2) $$g(x) = \begin{cases} \frac{2}{3} + \frac{1}{3} f(3x) & \text{if } 0 \leq x \leq \frac{1}{3}, \\ [2 + f(1)]\left(\frac{2}{3} - x\right) & \text{if } \frac{1}{3} \leq x \leq \frac{2}{3}, \\ x - \frac{2}{3} & \text{if } \frac{2}{3} \leq x \leq 1. \end{cases}$$

See Figure 9. Note that the graph of $\frac{1}{3}f(3x)$ reproduces the graph of f, but one-third as large. For the function g this graph is raised by the constant $\frac{2}{3}$.

The interval $[\frac{1}{3}, \frac{2}{3}]$ maps over itself by g, so there is a fixed point for g in $[\frac{1}{3}, \frac{2}{3}]$. No other interval maps to $[\frac{1}{3}, \frac{2}{3}]$, so there are no other periodic points in this interval.

The two intervals $[0, \frac{1}{3}]$ and $[\frac{2}{3}, 1]$ are interchanged by g,

$$g\big([0, \tfrac{1}{3}]\big) \subset [\tfrac{2}{3}, 1] \quad \text{and} \quad g\big([\tfrac{2}{3}, 1]\big) = [0, \tfrac{1}{3}],$$

so all periodic points in these two intervals must have periods that are a multiple of 2. On $[0, \frac{1}{3}]$,

$$g^2(x) = g\left(\frac{2}{3} + \frac{1}{3}f(3x)\right) = \frac{1}{3}f(3x),$$

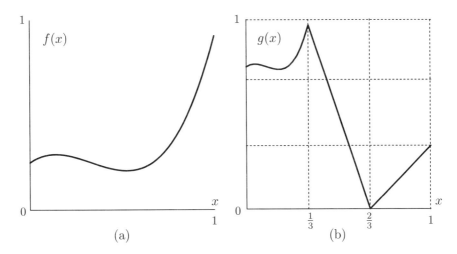

Figure 9. (a) Graph of a map $f(x)$ and (b) graph of $g(x)$ that has double the periods of $f(x)$

which simply reproduces a copy of f, but on a smaller scale. Thus, the periods of g^2 on $[0, 1/3]$ are the same as the periods of f. This is also true for g^2 on $[2/3, 1]$. Therefore, the periods of g for points starting in either $[0, 1/3]$ or $[2/3, 1]$ are just twice the periods of f. This completes the discussion about the periods of g.

For any $n \in \mathbb{N}$, we now know how to construct a continuous function $f: \mathbb{R} \to \mathbb{R}$ such that $\mathscr{P}(f) = \{k : n \triangleright k\}$. Using a function with only fixed points and a piecewise linear function with a period-n Stephan cycle, we can get examples for any odd n. Repeatedly applying the doubling of periods construction of the preceding example, we can get functions with $n = 2^k m$ for m odd. Exercise 5 in this section shows that there is a function whose periods contain all powers of two 2^k but no other periods.

For the proof of Theorem 10.5(a), the existence of a Stefan cycle of odd period implies the existence of all periods forces by the Sharkovskii ordering. We do not give a complete proof but prove some further results in Section 10.8 at the end of the chapter that illustrate the techniques used in the proof.

Exercises 10.1

1. Assume there is a period-9 Stefan cycle for a continuous map f on $\mathbb{R}$.
    a. Determine the transition graph. In particular, show that the interval $\mathbf{J}_8 = [p_8, p_6]$ maps over $\mathbf{J}_i$ for all odd i.
    b. Show that f must have all periods except, possibly, 3, 5, and 7.
2. Let f be a continuous function defined on the interval $[1, 6]$ with $f(1) = 5$, $f(2) = 6$, $f(3) = 4$, $f(4) = 1$, $f(5) = 2$, and $f(6) = 3$. Assume that the function is linear between these integers.
    a. Sketch the graph of f.
    b. Label the intervals between the integers and give the transition graph.

c. For which n is there a period-n orbit? Determine a symbol sequence in terms of the intervals that shows each period that exists.

3. Let f be a continuous function defined on the interval $[1,4]$, with $f(1) = 4$, $f(2) = 3$, $f(3) = 1$, and $f(4) = 2$. Assume that the function is linear between these integers.
   a. Sketch the graph of f.
   b. Label the intervals between the integers and give the transition graph.
   c. For which n is there a period-n orbit? Determine a symbol sequence in terms of the intervals that shows each period that exists.

4. Let $p(x) = (x^2 - 1)(x^2 - 4) = x^4 - 5x^2 + 4$. Let $N(x) = N_p(x)$ be the Newton map associated with p given by equation (9.1). Notice that $N(x)$ goes to $\pm\infty$ at $a_1 = -\sqrt{2.5}$, $a_2 = 0$, and $a_3 = \sqrt{2.5}$. The general shape of the graph of $N(x)$ is the same as that given in Figure 22, but this example is more symmetrical about 0.
   a. Sketch the graph of $N(x)$.
   b. Show that $x = -2, -1, 1$, and 2 are fixed point sinks of N.
   c. Let $\mathbf{I}_1$ be the closed interval in (a_1, a_2) that is mapped onto $[a_1, a_3]$ by N; let $\mathbf{I}_2$ be the closed interval in (a_2, a_3) that is mapped onto $[a_1, a_3]$ by N. Give the transition graph for $\mathbf{I}_1$ and $\mathbf{I}_2$.
   d. Show that the Newton map N has points of all periods. (Note that these are points for which the Newton iteration does not converge to a zero of the polynomial.)
   e. Show that the basin of attraction of the fixed point 2 for N contains the interval $(\sqrt{2.5}, \infty)$. (A similar argument should show that the basin of attraction of the fixed point -2 contains the interval $(-\infty, -\sqrt{2.5})$.)

5. Let f_0 be the function that is identically equal to $^1\!/_3$ on $[0, 1]$. As in Example 10.17, let f_1 be the function defined in terms of f_0 by equation (10.2). (The periods of f_1 and f_0 are related by equation (10.1).) By induction, let f_n be the function defined in terms of f_{n-1} by equation (10.2). Define f_∞ by $f_\infty(x) = \lim_{n\to\infty} f_n(x)$.
   a. Using equation (10.2) and induction, prove that each of the functions f_n has periods of points exactly equal to $\{1, 2, \ldots, 2^n\}$.
   b. Explain why f_∞ is continuous. In particular, why is $f_\infty(0) = 1$ and f_∞ continuous at $x = 0$? Hint: Draw the graphs of f_0, f_1, f_2, and f_3. Show by induction that for $k \geq n$, $f_k(x) = f_n(x)$ for $^1\!/_{3^n} \leq x \leq 1$ and $f_k(x) \geq \frac{2}{3}\left(1 + \frac{1}{3} + \cdots + \frac{1}{3^{n-1}}\right)$ for $0 \leq x \leq ^1\!/_{3^n}$.
   c. Prove that the periods of f_∞ are exactly $\{\, 2^i : 0 \leq i < \infty \,\}$. Further, explain why f_∞ has exactly one orbit of each period 2^n.

6. Assume that f is continuous and takes the closed interval $[a, b]$ into itself. Prove that f has a fixed point in $[a, b]$.

7. Suppose that $\mathbf{I}_1, \mathbf{I}_2, \mathbf{I}_3$, and $\mathbf{I}_4$ are disjoint closed intervals and $f(x)$ is a continuous map such that $f(\mathbf{I}_1) = \mathbf{I}_2$, $f(\mathbf{I}_2) = \mathbf{I}_3$, $f(\mathbf{I}_3) = \mathbf{I}_4$, and $f(\mathbf{I}_4) \supset \mathbf{I}_1 \cup \mathbf{I}_2 \cup \mathbf{I}_3$. Show that f has a period-3 point.

8. (Burns and Hasselblatt) Assume that a map f from $\mathbb{R}$ to itself has a period-9 orbit, $x_j = f^j(x)$ with $x_9 = x_0$, whose order on the line is

$$x_6 < x_4 < x_2 < x_7 < x_0 < x_1 < x_3 < x_8 < x_5.$$

(Note this orbit is not a Stefan cycle.) Denote six subintervals by

$$\mathbf{I}_1 = [x_0, x_1], \quad \mathbf{I}_2 = [x_7, x_1], \quad \mathbf{I}_3 = [x_4, x_1],$$
$$\mathbf{I}_4 = [x_4, x_1], \quad \mathbf{I}_5 = [x_0, x_5], \quad \mathbf{I}_6 = [x_6, x_4].$$

(Note that the interiors are not disjoint, but $\partial \mathbf{I}_6 \cap \mathbf{I}_j = \emptyset$ for $1 \leq j \leq 5$.)
  a. Show that these interval force at least the covering indicated by the transition graph in Figure 8. (There are many other coverings as well.)
  b. Using Theorem 10.3(c) and strings that start with the symbol 6, show that f has all periods forced by period 7.

10.2. Topological Transitivity

We next consider maps that have orbits that come close to every point in a whole interval. In this section we use binary expansions of a number to find such an orbit, while in the next section, we use symbolic dynamics. To make this concept precise, we define what we mean for a subset to be dense in the set which contains it. Since we are often considering a set of points which is taken to itself, we next define an invariant set. Then, if a map has an orbit that is dense in an invariant set, we call the map topologically transitive. We use binary expansions of numbers to show that the doubling map is topologically transitive on the unit interval.

Binary expansions

Any number x in $[0,1]$ can be represented as a *binary expansion*

$$x = \sum_{j=1}^{\infty} \frac{a_j}{2^j},$$

where each a_j is either 0 or 1. Just as with the decimal expansion, some numbers have two expansions. For example,

$$\sum_{j=2}^{\infty} \frac{1}{2^j} = \frac{1}{2^2} \left(\sum_{j=0}^{\infty} \frac{1}{2^j} \right) = \frac{1}{2^2} \left(\frac{1}{1-\frac{1}{2}} \right) = \frac{1}{2^2} \cdot \frac{2}{1} = \frac{1}{2}.$$

Notice that we used the formula for a *geometric series*. If $-1 < r < 1$, then

$$\sum_{j=0}^{\infty} r^j = \frac{1}{1-r}.$$

In general, a number whose expansion ends in an infinite string of 1's is the same as the number whose expansion changes the previous 0 to a 1 and replaces all the 1's with 0's,

$$\sum_{j=1}^{n-1} \frac{a_j}{2^j} + \sum_{j=n+1}^{\infty} \frac{1}{2^j} = \sum_{j=1}^{n-1} \frac{a_j}{2^j} + \frac{1}{2^n}.$$

If we start with a repeating binary expansion, we can find the rational number that it represents. For example,

$$\frac{1}{2} + \frac{1}{2^2} + \frac{0}{2^3} + \frac{1}{2^4} + \frac{1}{2^5} + \frac{0}{2^6} + \cdots = \left(\frac{1}{2} + \frac{1}{4}\right)\left(1 + \frac{1}{8} + \frac{1}{8^2} + \cdots\right)$$

$$= \frac{3}{4}\left(\frac{1}{1-\frac{1}{8}}\right) = \frac{3}{4} \cdot \frac{8}{7} = \frac{6}{7}.$$

If $x = 1$, then $x = \sum_{j=1}^{\infty} \frac{1}{2^j}$. If x is in $[0, 1)$, then we can find its binary expansion as follows: Multiply x by 2 and let a_1 be the integer part of $2x$ and f_1 the fractional part, $2x = a_1 + f_1$ with $a_1 = 0$ or 1 and $0 \leq f_1 < 1$; multiply f_1 by 2 and let a_2 be the integer part of $2f_1$ and f_2 the fractional part, $2f_1 = a_2 + f_2$; continue by induction to find a_j and f_j. Consider this process for the number $4/7$:

$$2 \cdot \frac{4}{7} = \frac{8}{7} = 1 + \frac{1}{7}, \qquad a_1 = 1 \text{ and } f_1 = \frac{1}{7},$$

$$2 \cdot \frac{1}{7} = \frac{2}{7} = 0 + \frac{2}{7}, \qquad a_2 = 0 \text{ and } f_2 = \frac{2}{7},$$

$$2 \cdot \frac{2}{7} = \frac{4}{7} = 0 + \frac{4}{7}, \qquad a_3 = 0 \text{ and } f_3 = \frac{4}{7},$$

$$2 \cdot \frac{4}{7} = \frac{8}{7} = 1 + \frac{1}{7}, \qquad a_4 = 1 \text{ and } f_4 = \frac{1}{7},$$

and the process is beginning to repeat, $a_{1+3j} = 1$ and $a_{2+3j} = a_{3j} = 0$. Indeed, this binary expansion does sum to $4/7$ as desired:

$$\frac{1}{2} + \frac{0}{2^2} + \frac{0}{2^3} + \frac{1}{2^4} + \cdots = \frac{1}{2}\left(1 + \frac{1}{8} + \frac{1}{8^2} + \cdots\right) = \frac{1}{2}\left(\frac{1}{1-\frac{1}{8}}\right)$$

$$= \frac{1}{2} \cdot \frac{8}{7} = \frac{4}{7},$$

Ternary expansions

In the same way, any number x in $[0, 1]$ can be represented as a *ternary expansion*

$$x = \sum_{j=1}^{\infty} \frac{a_j}{3^j},$$

where each a_j is either a 0, a 1, or a 2. A number whose ternary expansion ends in an infinite string of 2's is the same as the number whose expansion adds one to the previous entry and replaces all the 2's with 0's; that is,

$$\frac{a_n}{3^n} + \sum_{j=n+1}^{\infty} \frac{2}{3^j} = \frac{a_n + 1}{3^n}.$$

Dense sets

Definition 10.18. For a set $\mathbf{A} \subset \mathbf{S} \subset \mathbb{R}^n$, $\mathbf{A}$ is *dense* in $\mathbf{S}$ provided that arbitrarily close to each point $\mathbf{p}$ in $\mathbf{S}$, there is a point in $\mathbf{A}$. More precisely, for each $\mathbf{p}$ in $\mathbf{S}$ and each $\epsilon > 0$, there is a point $\mathbf{a}$ in $\mathbf{A}$ such that $\|\mathbf{a} - \mathbf{p}\| < \epsilon$.

10.2. Topological Transitivity

For sets in the line, this can be expressed by saying that for each p in $\mathbf{S}$ and each $\epsilon > 0$, the interval $(p - \epsilon, p + \epsilon)$ contains a point of $\mathbf{A}$,

$$(p - \epsilon, p + \epsilon) \cap \mathbf{A} \neq \emptyset.$$

Example 10.19. Let $\mathbb{Q}$ be the set of rational numbers, and $\mathbb{Q}^c = \mathbb{R} \setminus \mathbb{Q}$ be the set of irrational numbers. Then, both $\mathbb{Q} \cap [0,1]$ and $\mathbb{Q}^c \cap [0,1]$ are dense in $[0,1]$. This says that, arbitrarily close to any number, there is a rational number and an irrational number.

To see that the rational numbers are dense in $[0,1]$, take the decimal expansion of an arbitrary number in $[0,1]$, $x = \sum_{j=1}^{\infty} d_j/10^j$ with each d_j a choice from the set $\{0, 1, \ldots, 9\}$. For any finite n, the truncated expansion $x_n = \sum_{j=1}^{n} d_j/10^j$ is a rational number within 10^{-n} of x. This shows that the rational numbers are dense.

The irrational numbers can be shown to be dense, by showing that any number x can be approximated by $\sqrt{2}\,(p/q)$, where p and q are integers.

Example 10.20. Consider the set $\mathbf{S} = \left\{ \frac{p}{2^{100}} : 0 \leq p \leq 2^{100} \text{ is an integer} \right\}$. This set is within a distance of 2^{-101} of each point in $[0,1]$, but this distance is not arbitrarily small. In particular, $\mathbf{S}$ does not intersect the open interval $(0, 2^{-100})$. Therefore, $\mathbf{S}$ is not dense in $[0,1]$.

Invariant sets and topological transitivity

Since we eventually are going to consider an orbit that is dense in a set which is taken to itself by the map, we start by defining an invariant set. In the material through Chapter 11, the ambient space is line $\mathbb{R}$, but we give the definition for a general space $\mathbf{X}$ which can be $\mathbb{R}^n$ in later chapters.

Definition 10.21. Let f be a function from a set $\mathbf{X}$ to itself. A subset $\mathbf{A}$ of $\mathbf{X}$ is called *positively invariant* provided that, if x is in $\mathbf{A}$ then $f(x)$ is in $\mathbf{A}$ (i.e., $f(\mathbf{A}) \subset \mathbf{A}$). A subset $\mathbf{A}$ of $\mathbf{X}$ is called *invariant*, provided that (i) it is positively invariant and (ii) for every point b in $\mathbf{A}$, there is some a in $\mathbf{A}$ with $f(a) = b$ (i.e., $f(\mathbf{A}) = \mathbf{A}$).

Definition 10.22. If f is a map from $\mathbf{X}$ to itself, where $\mathbf{X}$ has a distance is defined (e.g., a subset of some $\mathbb{R}^n$), then f is *topologically transitive* on an invariant subset $\mathbf{A} \subset \mathbf{X}$, or just *transitive*, provided that there is a point x^* in $\mathbf{A}$ such that the orbit $\mathcal{O}_f^+(x^*)$ is dense in $\mathbf{A}$.

Doubling map

We show that the *doubling map* $D(x) = 2x \pmod{1}$ is topologically transitive on $[0,1]$ by using binary expansions. Consider a number $x = \sum_{j=1}^{\infty} a_j/2^j$, where each a_j is either a 0 or a 1. Using this expansion,

$$D\left(\sum_{j=1}^{\infty} \frac{a_j}{2^j}\right) = a_1 + \sum_{j=2}^{\infty} \frac{a_j}{2^{j-1}} \pmod{1} = \sum_{k=1}^{\infty} \frac{a_{k+1}}{2^k}.$$

The nth iterate simply shifts the expansion by n places:

$$D^n\left(\sum_{j=1}^{\infty}\frac{a_j}{2^j}\right) = \sum_{k=1}^{\infty}\frac{a_{k+n}}{2^k}.$$

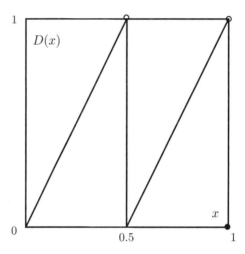

Figure 10. Graph of the doubling map

If a point x has an expansion that repeats every n places, $a_{j+n} = a_j$ for all j, then $D^n(x) = x$ and x is a period-n point.

The first n places in the expansion of a number also determine which subinterval of length 2^{-n} contains the point x; if $x = \sum_{j=1}^{\infty} a_j/2^j$ and $x_n = \sum_{j=1}^{n} a_j/2^j$, then $x_n \leq x \leq x_n + 2^{-n}$.

We can now show that D has a point with a dense orbit in the interval $[0, 1]$.

Theorem 10.6. *The doubling map D is topologically transitive on $[0, 1]$. In other words, there is a point x^* in $[0, 1]$ such that the orbit of x^* by the doubling map D is dense in $[0, 1]$ (i.e., the orbit of x^* comes arbitrarily close to every point in $[0, 1]$).*

Proof. We describe the point x^* by giving its binary expansion. Let $b_1^* = 0$ and $b_2^* = 1$; this lists both possible bits. Then we list all the strings of bits of length two, $b_3^* b_4^* = 00$, $b_5^* b_6^* = 10$, $b_7^* b_8^* = 01$, and $b_9^* b_{10}^* = 11$. At the third stage, we use b_{11}^* through b_{34}^* to list all the strings of bits of length three: 000, 100, 010, 110, 001, 101, 011, and 111. Continuing by induction, at the jth stage, we list all the strings of bits of length j. Let x^* be the point with this particular binary expansion: that is, $x^* = \sum_{j=1}^{n} b_j^*/2^j$.

We claim that the orbit of x^* comes arbitrarily close to every point in $[0, 1]$. Take any a point $x = \sum_{j=1}^{\infty} a_j/2^j$, where we have given its binary expansion. For any k, the first k bits $a_1 a_2 \ldots a_k$ appear as a string of bits for x^*; that is, $a_j = b_{j+m}^*$

for some fixed m and $1 \leq j \leq k$. Then, the binary expansion of $D^m(x^*)$ agrees with x in these first k bits, and

$$|D^m(x^*) - x| = \left| \sum_{j=k+1}^{\infty} \frac{b^*_{j+m} - a_j}{2^j} \right| \leq \sum_{j=k+1}^{\infty} \frac{1}{2^j}$$
$$= \frac{1}{2^{k+1}} \left(\frac{1}{1 - \frac{1}{2}} \right) = \frac{1}{2^k}.$$

Thus, the orbit of x^* comes within a distance less than 2^{-k} of x. Since 2^{-k} is arbitrarily small, we are done. □

Although we do not prove that the following example has a dense orbit until Chapter 11, we refer to it several times, so we state the result at this time.

Example 10.23. Let α be an irrational number, e.g., $\alpha = \sqrt{2}/2$. Let R_α be the map given by

$$R_\alpha(x) = x + \alpha \quad (\text{mod } 1).$$

Although we do not make this formal, we often identify the interval modulo one with the circle, and speak of this map R_α as an irrational rotation on the circle by α percentage of a full turn. Then,

$$R_\alpha^n(x) = x + n\alpha \quad (\text{mod } 1).$$

In Example 11.2, we show that orbit $\mathcal{O}^+_{R_\alpha}(0)$ is dense in $[0,1]$, so R_α is topologically transitive on $[0,1]$.

Exercises 10.2

1. Which of the following sets are dense in $[0,1]$?
   a. $\mathbf{S}_1$ is the set of all real numbers in $[0,1]$ of the form $\frac{p}{2^n}$, where p and n are arbitrary positive integers.
   b. $\mathbf{S}_2$ is the set of all real numbers in $[0,1]$ except those of the form $\frac{p}{2^n}$, where p and n are arbitrary positive integers (i.e., $\mathbf{S}_2 = [0,1] \setminus \mathbf{S}_1$).
   c. The set of numbers x in $[0,1]$ which have decimal expansions that use the digits 0, 2, 4, 6, and 8, but not 1, 3, 5, 7, or 9; that is,
   $$\mathbf{S}_3 = \left\{ x = \sum_{j=1}^{\infty} \frac{d_j}{10^j} : d_j \text{ is } 0, 2, 4, 6, \text{ or } 8 \right\}.$$

2. Express the number $5/24$ in terms of a ternary expansion.

3. Using the ternary expansion of numbers, show that the tripling map $f(x) = 3x \pmod 1$ is topologically transitive.

4. In terms of its binary expansion, describe an irrational number that does not have a dense orbit for the doubling map.

5. Let **S** be the set of all irrational numbers that have a dense orbit for the doubling map. Show that **S** is dense in $[0, 1]$. Hint: Consider numbers of the form $\sum_{j=1}^{k} \frac{a_j}{2^j} + \frac{1}{2^k} x^*$, where x^* is the point with a dense orbit for the doubling map.

10.3. Sequences of Symbols

In Section 10.1, we showed the existence of periodic points by means of specifying their itineraries, the sequence of intervals through which they pass. In this section, we extend these ideas to nonperiodic itineraries and orbits by using infinite sequences of symbols. In particular, we show that the tent and logistic maps are topologically transitive on $[0, 1]$. We start by giving a definition of the shift space.

Shift Space

Definition 10.24. We consider a *symbol space* **S** with N symbols. For this definition, we take the symbols as $\mathbf{S} = \{0, \ldots, N-1\}$. When there are just two symbols, we often use $\mathbf{S} = \{L, R\}$, as we do when considering the tent map in this section.

The *full shift space on N-symbols*, denoted by Σ_N^+, is the set of all infinite sequences of symbols $\mathbf{s} = s_0 s_1 s_2 \ldots$, where each s_j is an element of **S** (i.e., $s_j = 0$, ..., or $N-1$). The N indicates that there are N symbols; the plus sign indicates that the symbols are defined only for $j \geq 0$. The term "full" refers to the fact that any symbol is allowed for each s_j. If the context is clear, we just call Σ_N^+, the *shift space*.

The *shift map* σ from Σ_N^+, to itself is defined by dropping the first symbol s_0 in the sequence and shifting all the other symbols to the left one place,

$$\sigma(s_0 s_1 \ldots s_n \ldots) = s_1 s_2 \ldots s_n \ldots,$$

so that, $\sigma(\mathbf{s}) = \mathbf{s}'$, where $s'_j = s_{j+1}$.

Definition 10.25. For two symbols s and t in **S**, with $0 \leq s, t \leq N - 1$, let

$$\delta(s, t) = \begin{cases} 0 & \text{if } s = t, \\ 1 & \text{if } s \neq t. \end{cases}$$

For two elements **s** and **t** in Σ_N^+ (two sequences of symbols), we define the *distance* between them as

$$d(\mathbf{s}, \mathbf{t}) = \sum_{j=0}^{\infty} \frac{\delta(s_j, t_j)}{3^j}.$$

The distance function d is called a *metric* on Σ_N^+.

Proposition 10.7. *The metric d on Σ_N^+ just defined has the following properties that justify calling it a distance:*

i. $d(\mathbf{s}, \mathbf{t}) \geq 0$.
ii. $d(\mathbf{s}, \mathbf{t}) = 0$ *if and only if* $\mathbf{s} = \mathbf{t}$.
iii. $d(\mathbf{s}, \mathbf{t}) = d(\mathbf{t}, \mathbf{s})$.
iv. $d(\mathbf{r}, \mathbf{s}) + d(\mathbf{s}, \mathbf{t}) \geq d(\mathbf{r}, \mathbf{t})$.

10.3. Sequences of Symbols

The proof of this proposition and the proofs of the following two results are given in Section 10.8.

The next result shows that two sequences **s** and **t** are close in terms of the metric d provided they agree on the first finite number of entries.

Proposition 10.8. *Let Σ_N^+ be the shift space on N symbols, with distance d as previously defined. Given **t** fixed in Σ_N^+. Then,*

$$\left\{ \mathbf{s} \in \Sigma_N^+ : s_j = t_j \text{ for } 0 \leq j \leq k \right\} = \left\{ \mathbf{s} \in \Sigma_N^+ : d(\mathbf{s}, \mathbf{t}) \leq 3^{-k} 2^{-1} \right\}.$$

Proposition 10.9. *The shift map σ is continuous on Σ_N^+.*

Theorem 10.10. *The shift map σ on Σ_N^+ has the following two properties:*

a. *The periodic points are dense.*

b. *There exists $\mathbf{s}^*$ in Σ_N^+ such that $\mathcal{O}_\sigma^+(\mathbf{s}^*)$ is dense in Σ_N^+, so the shift map σ is topologically transitive on Σ_N^+.*

Proof. (a) Given any **t** in Σ_N^+ and $n \geq 1$, let $\mathbf{w} = t_0 \ldots t_{n-1}$ and

$$\mathbf{s} = \mathbf{w}^\infty = t_0 \ldots t_{n-1} t_0 \ldots t_{n-1} t_0 \ldots t_{n-1} \cdots .$$

Since $\sigma^n(\mathbf{w}^\infty) = \mathbf{w}^\infty$, it is a periodic point for σ. The distance $d(\mathbf{s}, \mathbf{t}) \leq 3^{-k+1} 2^{-1}$. Since n is arbitrary, the periodic points are dense.

(b) The construction of a point $\mathbf{s}^*$ with a dense orbit is similar to that given in Theorem 10.6 for the doubling map. Let $\mathbf{s}^*$ be the symbol sequence that includes all the symbol strings of length 1; then, all the symbol strings of length 2; and then, by induction, all the symbol strings of length n for each n. For $N = 2$,

$$\mathbf{s}^* = 0\ 1\ 00\ 01\ 10\ 11\ 000\ 001\ 010\ 011\ 100\ 101\ 110\ 111\ \ldots$$

Given any **t** in Σ_N^+ and any $k > 0$, there is an iterate m such that the first $k+1$ symbols of $\sigma^m(\mathbf{s}^*)$ are $t_0 \ldots t_k$. By Proposition 10.8, the distance $d(\mathbf{t}, \sigma^m(\mathbf{s}^*)) \leq 3^{-k} 2^{-1}$. Since k is arbitrary, this can be made arbitrarily small. This shows that the orbit of $\mathbf{s}^*$ is dense. □

Definition 10.26. Throughout the rest of this chapter and the next, we indicate the length of an interval **J** by $\lambda(\mathbf{J})$ where λ is the lower case Greek letter lambda.

Tent map

Consider the tent map with partition of closed intervals

$$\mathbf{I}_L = [0, 0.5] \quad \text{and} \quad \mathbf{I}_R = [0.5, 1].$$

Each of these intervals in the partition has length $1/2$, $\lambda(\mathbf{I}_L) = \lambda(\mathbf{I}_R) = 1/2$. Like the doubling map, both intervals $\mathbf{I}_L$ and $\mathbf{I}_R$ are mapped all the way across the union $[0, 1] = \mathbf{I}_L \cup \mathbf{I}_R$, $T([0, 0.5]) = [0, 1] = T([0.5, 1])$. See Figure 11. The difference from the doubling map is that the tent map is decreasing on the second half of the interval, $[0.5, 1]$. Therefore, rather than using a binary expansion of numbers, we specify a point by the sequence of intervals through which it passes, using the labels L and R on the two intervals in the partition as the symbols. These symbols play the role of 0 and 1 in the binary expansion for the doubling map.

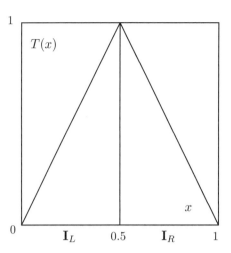

Figure 11. Intervals for the tent map T

For a point x, let

$$s_j = \begin{cases} L & \text{if } T^j(x) \in \mathbf{I}_L = [0, 0.5], \\ R & \text{if } T^j(x) \in \mathbf{I}_R = [0.5, 1], \end{cases}$$

and let

$$h(x) = \mathbf{s} = s_0 s_1 \ldots s_n \ldots$$

be the infinite sequence of symbols of L's or R's associated with x. This map h is called the *itinerary map*, and $h(x)$ gives the *itinerary of the point x* in terms of the partition. The only arbitrary choice of symbols occurs when $T^k(x) = 0.5$: to define h we set $s_k = R$, but $s_k = L$ is just as reasonable. However, if $T^k(x) = 0.5$, then $T^{k+1}(x) = 1$ and $T^i(x) = 0$ for $i \geq k + 2$, so $s_{k+1} = R$ and $s_i = L$ for $i \geq k + 2$. Thus, for any finite string of symbols $\mathbf{w}$, $\mathbf{w}LRL^\infty$ and $\mathbf{w}RRL^\infty$ could both be associated to the same point. (This ambiguity is related to the nonuniqueness of the decimal or binary expansions of a number.)

If $h(x) = \mathbf{s} = \{s_j\}_{j \geq 0}$, then $T^j(T(x)) = T^{j+1}(x)$, is in the interval $\mathbf{I}_{s_{j+1}}$, so the symbol sequence associated with $T(x)$ is the shift of the sequence $\mathbf{s}$ by σ, $\sigma(\mathbf{s}) = \sigma(h(x))$, or

$$h(T(x)) = \sigma(h(x)).$$

Thus, h satisfies the conjugacy equation for T and σ.

We next explain how we can start with a symbol sequence and associate a point in the interval. As we have done previously, let

$$\mathbf{I}_{s_0 \ldots s_n} = \{\, x : T^j(x) \in \mathbf{I}_{s_j} \text{ for } 0 \leq j \leq n \,\} = \bigcap_{j=0}^{n} T^{-j}(\mathbf{I}_{s_j})$$

be the interval associated with this finite string of these symbols. Since $T^j(T(x)) = T^{j+1}(x) \in \mathbf{I}_{s_{j+1}}$, $x \in \mathbf{I}_{s_0 \ldots s_n}$ if and only if $x \in \mathbf{I}_{s_0}$ and $T(x) \in \mathbf{I}_{s_1 \ldots s_n}$, so

$$\mathbf{I}_{s_0 \ldots s_n} = \mathbf{I}_{s_0} \cap T^{-1}(\mathbf{I}_{s_1 \ldots s_n}).$$

10.3. Sequences of Symbols

The subinterval $[0, 1/4] \subset \mathbf{I}_L$ is mapped onto $\mathbf{I}_L = [0, 1/2]$, so
$$[0, 1/4] = \mathbf{I}_L \cap T^{-1}(\mathbf{I}_L) = \mathbf{I}_{LL}.$$
Similarly, $[1/4, 1/2] \subset \mathbf{I}_L$ is mapped onto $\mathbf{I}_R = [1/2, 1]$, so
$$[1/4, 1/2] = \mathbf{I}_L \cap T^{-1}(\mathbf{I}_R) = \mathbf{I}_{LR}.$$
Notice that, T restricted to $\mathbf{I}_L = [0, 1/2]$ is increasing, so the order of $\mathbf{I}_L$ and $\mathbf{I}_R$ is reproduced with $\mathbf{I}_{LL}$ to the left of $\mathbf{I}_{LR}$. In the same way, $[1/2, 3/4] \subset \mathbf{I}_R$ is mapped onto $\mathbf{I}_R = [1/2, 1]$ and $[3/4, 1] \subset \mathbf{I}_R$ is mapped onto $\mathbf{I}_L = [0, 1/2]$, so
$$[1/2, 3/4] = \mathbf{I}_R \cap T^{-1}(\mathbf{I}_R) = \mathbf{I}_{RR} \quad \text{and} \quad [3/4, 1] = \mathbf{I}_R \cap T^{-1}(\mathbf{I}_L) = \mathbf{I}_{RL}.$$
Notice that, T restricted to $\mathbf{I}_R = [1/2, 1]$ is decreasing, so the order of $\mathbf{I}_L$ and $\mathbf{I}_R$ is reversed with $\mathbf{I}_{RR}$ to the left of $\mathbf{I}_{RL}$. Each of the intervals, with two symbols specified, has length 2^{-2}, $\lambda(\mathbf{I}_{s_0 s_1}) = 2^{-2}$. See Figure 12.

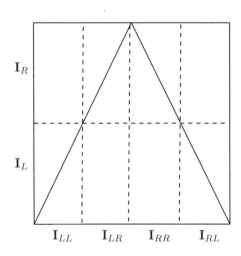

Figure 12. Intervals for strings of two symbols

To determine the order of intervals for which three symbols are specified, place the intervals with two symbols along the second axis of the plot. See Figure 13. The tent map T restricted to $\mathbf{I}_L = [0, 0.5]$ is increasing, so the order of $\mathbf{I}_{LL}$, $\mathbf{I}_{LR}$, $\mathbf{I}_{RR}$, and $\mathbf{I}_{RL}$ is reproduced with the order of intervals for three symbols whose first symbols is L:
$$\mathbf{I}_{LLL} = \mathbf{I}_L \cap T^{-1}(\mathbf{I}_{LL}), \qquad \mathbf{I}_{LLR} = \mathbf{I}_L \cap T^{-1}(\mathbf{I}_{LR}),$$
$$\mathbf{I}_{LRR} = \mathbf{I}_L \cap T^{-1}(\mathbf{I}_{RR}), \qquad \mathbf{I}_{LRL} = \mathbf{I}_L \cap T^{-1}(\mathbf{I}_{RL}).$$
On the other hand, T restricted to $\mathbf{I}_R = [0.5, 1]$ is decreasing, so the order of the intervals for three symbols whose first symbols is R is reversed:
$$\mathbf{I}_{RRL} = \mathbf{I}_L \cap T^{-1}(\mathbf{I}_{RL}), \qquad \mathbf{I}_{RRR} = \mathbf{I}_L \cap T^{-1}(\mathbf{I}_{RR}),$$
$$\mathbf{I}_{RLR} = \mathbf{I}_L \cap T^{-1}(\mathbf{I}_{LR}), \qquad \mathbf{I}_{RLL} = \mathbf{I}_L \cap T^{-1}((\mathbf{I}_{LL}).$$

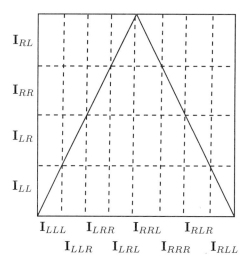

Figure 13. Intervals for strings of three symbols

Each of the intervals with three symbols specified has length 2^{-3}, $\lambda(\mathbf{I}_{s_0 s_1 s_2}) = 2^{-3}$. The eight intervals together cover all of $[0,1]$, or

$$[0,1] = \bigcup_{s_0, s_1, s_2} \mathbf{I}_{s_0 s_1 s_2}.$$

Continuing this process, each of the intervals obtained by specifying n symbols has length 2^{-n}, $\lambda(\mathbf{I}_{s_0 \ldots s_{n-1}}) = 2^{-n}$. The 2^n such intervals together cover all of $[0,1]$,

$$[0,1] = \bigcup_{s_0, \ldots, s_{n-1}} \mathbf{I}_{s_0 \ldots s_{n-1}}.$$

The order of these intervals could be determined.

Finally, we want to see that an infinite sequence of symbols determines a unique point. Just as a point can be specified by giving an infinite decimal expansion, an infinite sequence of symbols determines a point by the sequence of interval through which the point passes under iteration by the map. By the preceding construction, the intervals $\mathbf{I}_{s_0 \ldots s_n}$ are nested; that is,

$$\mathbf{I}_{s_0} \supset \mathbf{I}_{s_0 s_1} \supset \cdots \supset \mathbf{I}_{s_0 \ldots s_{n-1}} \supset \mathbf{I}_{s_0 \ldots s_n}.$$

Since these intervals are closed, the intersection is nonempty. (See Appendix A.2 for more discussion of the intersection of closed bounded sets.) In particular, if $\mathbf{I}_{s_0 \ldots s_n}$ has a left-end point of a_n and right-end point of b_n, then

$$a_0 \leq a_1 \leq \cdots \leq a_{n-1} \leq a_n < b_n \leq b_{n-1} \leq \cdots \leq b_1 \leq b_0.$$

The sequence of left-end points a_n is increasing and bounded above by b_0 (or any of the b_n), so it must converge to a point a_∞; the sequence of right b_n is decreasing and is bounded below by a_0, so it must converge to a point b_∞, and $a_\infty \leq b_\infty$. For each n and $j \geq n$, $a_n \leq a_j \leq b_j \leq b_n$, so $a_n \leq a_\infty \leq b_\infty \leq b_n$. The distance between the two end points of $\mathbf{I}_{s_0 \ldots s_n}$ is 2^{-n-1}, $b_\infty - a_\infty \leq b_n - a_n = 2^{-n-1}$ for all n, so $a_\infty = b_\infty$ and the intersection is a single point. Thus, we have shown that

10.3. Sequences of Symbols

the infinite sequence of symbols determines a unique point, which we denote by $k(\mathbf{s})$. This map k is essentially the inverse of the itinerary map h, and goes from the space of sequences, Σ_2^+, to the points in the interval.

The next theorem summarizes these results.

Theorem 10.11. *Let h be the itinerary map for the tent map with partition $\{\mathbf{I}_L, \mathbf{I}_R\}$. Let $\mathbf{s}$ in Σ_2^+ be any symbol sequence of L's and R's.*

a. *Then there is a unique point $x_\mathbf{s} = k(\mathbf{s})$ in $[0,1]$ such that $T^j(x_\mathbf{s})$ is in $\mathbf{I}_{s_j}$ for all j, or*

$$\{x_\mathbf{s}\} = \bigcap_{j=0}^{\infty} T^{-j}(\mathbf{I}_{s_j}) = \bigcap_{n=0}^{\infty} \mathbf{I}_{s_0 \ldots s_{n-1}}.$$

Moreover, the map k from Σ_2^+ to $[0,1]$ is continuous but not one to one, $k(\mathbf{w}LRL^\infty) = k(\mathbf{w}RRL^\infty)$ for any finite string $\mathbf{w}$.

b. *The map k is a semiconjugacy between the shift map σ and the tent map T; that is*

$$T \circ k(\mathbf{s}) = k \circ \sigma(\mathbf{s}).$$

c. *The maps k and h are essentially inverses of each other, $k \circ h(x_0) = x_0$. On the other hand, $h \circ k(\mathbf{s}) = \mathbf{s}$, except that, for $\mathbf{s} = \mathbf{w}LRL^\infty$ where $\mathbf{w}$ is a word of length j, $T^j(k(\mathbf{s})) = k(\sigma^j(\mathbf{s})) = k(LRL^\infty) = 0.5$ and $h \circ k(\mathbf{s}) = \mathbf{w}RRL^\infty$.*

Proof. (a) The proof that $k(\mathbf{s})$ is a single point is given in the discussion preceding the statement of the theorem. We prove the continuity using δ's and ϵ's. Given any $\epsilon > 0$, there is an m such that $2^{-(m+1)} < \epsilon$. Let $\delta = 1/(2 \cdot 3^m)$. If $d(\mathbf{s}, \mathbf{t}) \leq \delta$, then $s_j = t_j$ for $0 \leq j \leq m$. This implies that both $k(\mathbf{s})$ and $k(\mathbf{t})$ are in the same interval $\mathbf{I}_{s_0 \ldots s_m}$. The length $\lambda(I_{s_0 \ldots s_m}) = 2^{-(m+1)}$, so

$$|k(\mathbf{s}) - k(\mathbf{t})| \leq 2^{-(m+1)} < \epsilon.$$

This proves the continuity of k.

(b) Let $x_\mathbf{s} = k(\mathbf{s})$. The point $T^j(T(x_\mathbf{s}))$ is in the interval $\mathbf{I}_{s_{j+1}}$ for all $j \geq 0$, so $T(x_\mathbf{s})$ is in $\mathbf{I}_{s_1 s_2 \ldots}$ and $T(k(\mathbf{s})) = k(\sigma(\mathbf{s}))$.

(c) The maps are inverses by construction, except as noted. $\square$

Remark 10.27. This map k gives us another way to show the existence of periodic points for the map T. It also allow us show that the tent map is topologically transitive.

We have previously considered the intervals $\mathbf{I}_{s_0 \ldots s_{n-1}}$ for finite strings of symbols, for which there is a whole interval of points. The theorem says that, for an infinite sequence of symbols, there is a unique point which has this itinerary.

Note that an infinite binary expansion, which is used in the discussion of the doubling map, specifies a point by giving its location to greater and greater precision. By contrast, an infinite sequence of symbols from the itinerary specifies the point by giving the rough location for more and more iterates. (For the doubling map, the binary expansion and the itinerary are essentially linked together.)

Theorem 10.12. *Let k be the map from symbols in Σ_2^+ to points in $[0,1]$ for the tent map T given in the preceding theorem. Let h be the itinerary map.*

a. If $\mathbf{s} = (s_0 \ldots s_{n-1})^\infty$ is a symbol sequence that is periodic with period n (i.e., a period-n point for the shift map σ), then $k(\mathbf{s})$ is a period-n point for T. If $\mathbf{s}$ is a symbol sequence which is eventually periodic with period n, then $k(\mathbf{s})$ is an eventually period-n point for T.

b. If x_0 is a period-n point for T, then $h(x_0)$ is a period-n point for σ. If x_0 is an eventually period-n point for T, then $h(x_0)$ is an eventually period-n point for σ.

c. The periodic points for T are dense in $[0, 1]$.

d. The tent map T is topologically transitive on the interval $[0, 1]$. In other words, there is a point x^* in $[0, 1]$ such that the orbit of x^* by the tent map T is dense in $[0, 1]$.

Proof. (**a-b**) If the map k were one to one, then parts (a) and (b) would follow from Proposition 9.17. However, the only way in which k is not one to one and h is not uniquely defined involves points that go through 0.5 and these orbits are not periodic. Therefore, the parts (a) and (b) really follow from the earlier result. We leave the details to the reader.

(**c**) Let x_0 be an arbitrary point in $[0, 1]$. Given any $\epsilon > 0$, there is a n such that $2^{-n} < \epsilon$. We need to show that there is a periodic point within 2^{-n} of x_0. The point $x_0 \in \mathbf{I}_{s_0 \ldots s_{n-1}}$ for some finite string $\mathbf{w} = s_0 \ldots s_{n-1}$. Let $\mathbf{w}^\infty$ be the infinite sequence that repeats the string $\mathbf{w}$ over and over. Let $p = k(\mathbf{w}^\infty)$ be the point for the periodic sequence $\mathbf{w}^\infty$. Both p and $T^n(p)$ have the same symbol sequence, so $p = T^n(p)$ is periodic. Both x_0 and p are in $\mathbf{I}_\mathbf{w}$ with $\lambda(\mathbf{I}_\mathbf{w}) = 2^{-n}$, so the periodic point p satisfies $|x_0 - p| \leq 2^{-n} < \epsilon$.

(**d**) The proof is very similar to that for the doubling map. Let $\mathbf{s}^*$ be the symbol sequence constructed in the proof of Theorem 10.10(b). Let $x^* = k(\mathbf{s}^*)$ be the point with this symbol sequence. For any interval $\mathbf{I}_{t_0 \ldots t_{n-1}}$, this string of symbols $t_0 \ldots t_{n-1}$ appears somewhere in the sequence $\mathbf{s}^*$. Let m be the iterate, so that $\sigma^m(\mathbf{s})$ has $t_0 \ldots t_{n-1}$ as the first n symbols. Then,

$$T^m(x^*) \in \mathbf{I}_{t_0 \ldots t_{n-1}}.$$

Since the lengths of these intervals are arbitrarily small, $\lambda(\mathbf{I}_{t_0 \ldots t_{n-1}}) = 2^{-n}$, this proves that the orbit of x^* is dense in the interval $[0, 1]$. $\square$

Logistic map

Consider the logistic function for $a = 4$, $G(y) = g_4(y)$. Let $\mathbf{I}_L = [0, 0.5]$ and $\mathbf{I}_R = [0.5, 1]$ be the same intervals as those for the tent map. Then G takes each of these intervals onto $[0, 1]$, since $G(0.5) = 1$. See Figure 14. The results for the logistic map $G(y)$ are the same as those for the tent map. Let $\mathbf{I}_L$ and $\mathbf{I}_R$ be the same intervals as those for the tent map. See Figure 14. Given a symbol sequence $\mathbf{s}$, define the intervals

$$\mathbf{I}^G_{s_0 \ldots s_n} = \{ y : G^j(y) \in \mathbf{I}_{s_j} \text{ for } 0 \leq j \leq n \} = \bigcap_{j=0}^{n} G^{-j}(\mathbf{I}_{s_j}).$$

10.3. Sequences of Symbols

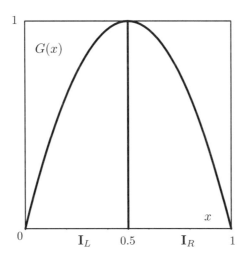

Figure 14. Intervals for the logistic map $G(y) = g_4(y)$

We showed in Proposition 9.16 that the map

$$y = h(x) = \sin^2\left(\frac{\pi x}{2}\right) = \frac{1 - \cos(\pi x)}{2}$$

is a conjugacy from $T(x)$ to $G(y)$. Because h is a conjugacy between T and G, with $h([0, 0.5]) = [0, 0.5]$ and $h([0.5, 1]) = [0.5, 1]$, we have

$$\mathbf{I}^G_{s_0 \ldots s_{n-1}} = \bigcap_{j=0}^{n-1} G^{-j}(\mathbf{I}_{s_j}) = \bigcap_{j=0}^{n-1} h \circ T^{-j} \circ h^{-1}(\mathbf{I}_{s_j})$$

$$= h\left(\bigcap_{j=0}^{n-1} T^{-j}(\mathbf{I}_{s_j})\right)$$

$$= h\left(\mathbf{I}^T_{s_0 \ldots s_{n-1}}\right),$$

where $\mathbf{I}^T_{s_0 \ldots s_{n-1}}$ is the interval for the tent map. There are bounds on the derivative of the conjugacy equation,

$$h'(x) = \frac{\pi}{2} \sin(\pi x) \quad \text{and}$$
$$|h'(x)| \le \frac{\pi}{2}.$$

Let x_0 and x_1 be the end points of the interval $\mathbf{I}^T_{s_0 \ldots s_{n-1}}$, so $|x_1 - x_0| = 2^{-n}$, and let $y_0 = h(x_0)$ and $y_1 = h(x_1)$ be the corresponding end points of the interval $\mathbf{I}^G_{s_0 \ldots s_{n-1}}$. By the mean value theorem, there is a point x_2 between x_0 and x_1 such that

$$y_1 - y_0 = h(x_1) - h(x_0)$$
$$= h'(x_2)(x_1 - x_0).$$

Therefore, the length of the interval is

$$\lambda\left(\mathbf{I}^G_{s_0\ldots s_{n-1}}\right) = |y_1 - y_0|$$
$$= |h'(x_2)|\,|x_1 - x_0|$$
$$\leq \frac{\pi}{2}|x_1 - x_0|$$
$$= \pi\, 2^{-n-1}.$$

Since the length of these intervals is going to zero as n goes to infinity, their intersection is a single point, which we define as $k(\mathbf{s})$,

$$\bigcap_{j=0}^{\infty} G^{-j}(\mathbf{I}_{s_j}) = \{\,k(\mathbf{s})\,\}.$$

We summarize the results in the next theorem.

Theorem 10.13. *Let h^G be the itinerary map for the logistic map G. Let $\mathbf{s}$ in Σ_2^+ be any symbol sequence of L's and R's.*

a. *The intervals $\mathbf{I}^G_{s_0\ldots s_{n-1}}$ have length at most $\pi 2^{-n-1}$, $\lambda(\mathbf{I}^G_{s_0\ldots s_{n-1}}) \leq \pi 2^{-n-1}$.*

b. *There is a unique point $x_\mathbf{s} = k^G(\mathbf{s})$ in $[0,1]$ such that $G^j(x_\mathbf{s})$ is in $\mathbf{I}_{s_j}$ for all j,*

$$\{\,x_\mathbf{s}\,\} = \bigcap_{j=0}^{\infty} G^{-j}(\mathbf{I}_{s_j}) = \bigcap_{n=0}^{\infty} \mathbf{I}^G_{s_0\ldots s_{n-1}}.$$

Moreover, the map k from Σ_2^+ to $[0,1]$ is continuous but not one to one, $k^G(\mathbf{w}LRL^\infty) = k^G(\mathbf{w}RRL^\infty)$ for any finite string $\mathbf{w}$.

c. *The map k^G is a semiconjugacy between the shift map σ and the logistic map G; that is,*

$$G \circ k^G(\mathbf{s}) = k^G \circ \sigma(\mathbf{s}).$$

d. *The maps k^G and h^G are essentially inverses of each other, $k^G \circ h^G(x) = x$. On the other hand, $h^G \circ k^G(\mathbf{s}) = \mathbf{s}$, except that, for $\mathbf{s} = \mathbf{w}LRL^\infty$ where $\mathbf{w}$ is a word of length j, $G^j(k^G(\mathbf{s})) = k^G(\sigma^j(\mathbf{s})) = k^G(LRL^\infty) = 0.5$ and $h^G \circ k^G(\mathbf{s}) = \mathbf{w}RRL^\infty$.*

e. *If $\mathbf{s}$ is a symbol sequence, which is periodic with period n (i.e., a period-n point for the shift map σ), then $k^G(\mathbf{s})$ is a period-n point for G. If $\mathbf{s}$ is a symbol sequence which is eventually periodic, with period n, then $k(\mathbf{s})$ is an eventually period-n point for G.*

f. *If y_0 is a period-n point for G, then $h^G(y_0)$ is a period-n point for σ. If y_0 is an eventually period-n point for G, then $h^G(y_0)$ is an eventually period-n point for σ.*

g. *The periodic points for G are dense in $[0,1]$.*

h. *The logistic map G is topologically transitive on $[0,1]$. In other words, there is a point $y^* = k^G(\mathbf{s}^*)$ in $[0,1]$ such that the orbit of y^* by G is dense in $[0,1]$.*

Exercises 10.3

1. Consider the intervals for the quadratic map G. Let x_0 be the point corresponding to the symbol string $LRRLLRRL$. Is x_0 less than or greater than $1/2$? Is $G^5(x_0)$ less than or greater than $1/2$?

2. Consider the tripling map $f(x) = 3x \pmod{1}$. Use the left, center, and right intervals given by $\mathbf{I}_L = [0, 1/3]$, $\mathbf{I}_C = [1/3, 2/3]$, and $\mathbf{I}_R = [2/3, 1]$. Give all the intervals for symbol strings of length less than or equal to three. What is the order of these intervals on the line?

3. Consider the "saw-toothed map" defined by
$$S(x) = \begin{cases} 3x & \text{if } 0 \leq x \leq \tfrac{1}{3}, \\ 2 - 3x & \text{if } \tfrac{1}{3} \leq x \leq \tfrac{2}{3}, \\ 3x - 2 & \text{if } \tfrac{2}{3} \leq x \leq 1. \end{cases}$$
Use the three symbols L, C, and R, with corresponding intervals $\mathbf{I}_L = [0, 1/3]$, $\mathbf{I}_C = [1/3, 2/3]$, and $\mathbf{I}_R = [2/3, 1]$.
   a. Give the order in the line of the nine intervals that correspond to strings of 2 symbols (e.g., $\mathbf{I}_{CR}$). (What you are asked to determine is which interval is the furthest to the left, which interval is the next to the furthest to the left, and straight through to the interval which is furthest to the right.)
   b. Consider the interval $\mathbf{I}_{CRL}$. It is a subset of which of the intervals involving two symbols, $\mathbf{I}_{t_0 t_1}$? This interval $\mathbf{I}_{t_0 t_1}$ is divided into three subintervals when giving intervals for strings of three symbols. Is $\mathbf{I}_{CRL}$ the furthest left of these three intervals, the center interval, or the furthest right of these three intervals.

4. Give all the sequences in Σ_2^+ that are period-3 points for σ. Which of these sequences are on the same orbit for σ?

10.4. Sensitive Dependence on Initial Conditions

In this section, we define sensitive dependence on initial conditions: the property of two orbits which start close to each other and move apart. This property is central to the notion of "chaotic dynamics". We start by considering the doubling map.

Example 10.28. Let $D(x) = 2x \pmod{1}$ be the doubling map on $[0, 1]$. This map demonstrates the property of sensitive dependence on initial conditions. Let $x_0 = 1/3$ and $y_0 = x_0 + 0.0001$. Table 3 shows the first 14 iterates of these points and the distances between their iterates. Notice that the first 12 iterates get farther and farther apart, and by the 12th iterate, they are more than a distance $1/3$ apart. After this number of iterates, they can get closer or farther apart on subsequent iterates, and their dynamic behavior seems independent of each other. Also, see Figures 15. When looking at the figures, remember that $x_0 = 1/3$ has period 2, so the iteration keeps repeating; the orbit for $y_0 = x_0 + 0.0001$ wanders off.

n	x_n	y_n	$\|y_n - x_n\|$
0	0.3333	0.3334	0.0001
1	0.6667	0.6669	0.0002
2	0.3333	0.3337	0.0004
3	0.6667	0.6675	0.0008
4	0.3333	0.3349	0.0016
5	0.6667	0.6699	0.0032
6	0.3333	0.3397	0.0064
7	0.6667	0.6795	0.0128
8	0.3333	0.3589	0.0256
9	0.6667	0.7179	0.0512
10	0.3333	0.4357	0.1024
11	0.6667	0.8715	0.2048
12	0.3333	0.7429	0.4096
13	0.6667	0.4859	0.1808
14	0.3333	0.9717	0.6384

Table 3. Sensitive dependence for the doubling map

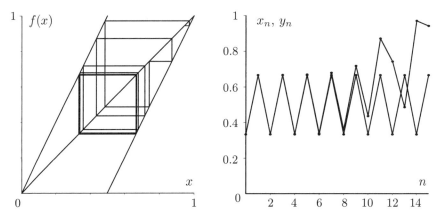

Figure 15. Iteration of doubling map showing sensitive dependence on initial conditions: $x_0 = 1/3$ and $y_0 = x_0 + 0.0001$. (**a**) Graphical iteration. (**b**) Iterate x_n and y_n as functions of n.

We now proceed to give the formal definition. The second definition concerning sensitive dependence when restricted to an invariant set is used in our discussion of attractors, especially in higher dimensions.

Definition 10.29. A map f on a space $\mathbf{X}$ has *sensitive dependence on initial conditions at points of* $\mathbf{A} \subset \mathbf{X}$ provided that there exists $r > 0$ such that for every x_0 in $\mathbf{A}$ and for any $\delta > 0$, there is a y_0 within δ of x_0, (i.e., $|y_0 - x_0| < \delta$) and an iterate $k > 0$ such that
$$|f^k(y_0) - f^k(x_0)| \geq r.$$

If the set $\mathbf{A}$ is invariant by f, then we say that the map has *sensitive dependence on initial conditions when restricted to* $\mathbf{A}$ provided that the new point y_0 can be

10.4. Sensitive Dependence

chosen within the set **A**; that is, there is an $r > 0$ such that, for any point x_0 in **A** and any $\delta > 0$, there is a y_0 in **A** with $|y_0 - x_0| < \delta$ and an iterate $k \geq 0$ such that
$$|f^k(y_0) - f^k(x_0)| \geq r.$$

We will show that D, T, and G each have sensitive dependence on initial conditions on $[0, 1]$.

The next definition of expansive differs from sensitive dependence in that all pairs of distinct points move apart.

Definition 10.30. A map f is *expansive* on an invariant subset **A** in **X**, provided that there exists $r > 0$ such that, for any y_0 and x_0 in **A** with $y_0 \neq x_0$, there is some iterate $k \geq 0$ with
$$|f^k(y_0) - f^k(x_0)| \geq r.$$

From the foregoing definitions, any expansive map on **A** has sensitive dependence on initial conditions when restricted to **A**, but the converse is not necessarily true.

Remark 10.31. A map with sensitive dependence on initial conditions has the "butterfly effect" mentioned in the historical introduction. A small change in initial conditions can be amplified to create large differences. E. Lorenz introduced this latter term because he interpreted it to mean that a butterfly flapping its wings in one part of the world could affect the weather in another part of the world a month later.

Theorem 10.14. *The doubling map D is expansive with expansive constant $r = 1/3$ and so has sensitive dependence on initial conditions at points in the whole domain $[0, 1]$.*

Proof. The doubling map is given by $D(x) = 2x - k$, where $k = 0$ for $0 \leq x < 1/2$, $k = 1$ for $1/2 \leq x < 1$, and $k = 2$ for $x = 1$.

Let $x_0 < y_0$ be two points in $[0, 1]$, $x_j = D^j(x_0)$, and $y_j = D^j(y_0)$. Let k_j and k'_j be determined by $x_{j+1} = 2x_j - k_j$ and $y_{j+1} = 2y_j - k'_j$. If $x_j < y_j$ are both in $[0, 1/2)$ or both in $[1/2, 1)$, then $k_j = k'_j$ and
$$y_{j+1} - x_{j+1} = 2y_j - k_j - (2x_j - k_j) = 2(y_j - x_j).$$
By induction, $y_j - x_j = 2^j(y_0 - x_0)$ for $0 \leq j \leq n$ until either (i) $y_n - x_n \geq 1/3$ or (ii) $0 < y_n - x_n < 1/3$ and x_n and y_n are not in the same subinterval so $k'_j = k_j + 1$. In case (i), we are done. In case (ii),
$$x_{n+1} - y_{n+1} = 2x_n - k_n - (2y_n - k_n - 1)$$
$$= 1 - 2(y_n - x_n)$$
$$> 1 - 2\left(\tfrac{1}{3}\right) = \tfrac{1}{3}.$$

Since the distance cannot double at each step and stay less than $1/3$ forever, eventually the distance must get greater than $1/3$. □

Theorem 10.15. *The tent map T and logistic map G both have sensitive dependence on initial conditions when restricted to the interval $[0, 1]$.*

Proof. A similar proof shows that either map has sensitive dependence on initial conditions, with constant $r = 1/2$. We write the proof for the logistic map, which in nonlinear.

Let x_0 be a point in $[0,1]$. Let $h^G(x_0) = \mathbf{s}$ be the symbol sequence for x_0.

Case (i): Assume that $\mathbf{s}$ does not end in repeated L's, $h^G(x_0) \neq s_0 \ldots s_n L^\infty$, so x_0 does not eventually go to the fixed point at 0. There is an arbitrarily large n for which $s_n = R$. Consider the sequence of symbols $\mathbf{t}$, with $t_j = s_j$ for $0 \leq j < n$ and $t_j = L$ for $j \geq n$. Let $y_0 = k^G(\mathbf{t})$ be the point with this symbol sequence. Both x_0 and y_0 are in $\mathbf{I}^G_{s_0 \ldots s_{n-1}}$, so $|x_0 - y_0| \leq \lambda(\mathbf{I}^G_{s_0 \ldots s_{n-1}}) \leq \pi/2^{n+1}$ by Theorem 10.13(a). We can make this arbitrarily small by taking n large with $s_n = R$. Taking the nth iterate, $h^G(G^n(y_0)) = L^\infty$ so $G^n(y_0) = 0$, while $G^n(x_0)$ is in $\mathbf{I}_R$, so $G^n(x_0) \geq 0.5$; therefore, $|G^n(x_0) - G^n(y_0)| \geq 0.5$. This shows that there is a point arbitrarily close to x_0 and an iterate such that the distance between $G^n(x_0)$ and $G^n(y_0)$ is greater than $1/2$; that is, G has sensitive dependence on initial conditions.

Case (ii): $\mathbf{s} = h(x_0) = s_0 \ldots s_{m-1} L^\infty$. Let $\mathbf{t} = s_0 \ldots s_{n-1} R^\infty$ for large $n \geq m$. Then, $|x_0 - y_0| \leq \lambda(\mathbf{I}_{s_0 \ldots s_{n-1}}) \leq \pi/2^{n+1}$ and $G^n(x_0) = k(L^\infty) = 0$, while $G^n(y_0) = k(R^\infty) \geq 1/2$. Again, $|G^n(x_0) - G^n(y_0)| \geq 0.5$. □

The proof that the logistic and tent maps have sensitive dependence used symbolic dynamics and did not use a bound on the derivative directly, as is done for the doubling map. However, a common feature of the doubling map and the tent map is that they have absolute value of the derivative bigger than one (i.e., they stretch intervals). In order to have the interval $[0,1]$ forward invariant by a map which stretches, the map must either fold like the tent map or cut like the doubling map. When we consider chaotic attractors in Section 11.2, we return to this idea of stretching and folding or stretching and cutting.

Exercises 10.4

1. Let $f(x) = 3x \pmod 1$ be the tripling map. Let $x_0 < y_0$ be two distinct points in $[0,1]$, $x_j = f^j(x_0)$, and $y_j = f^j(y_0)$. Consider the subintervals $\mathbf{I}_0 = [0, 1/3)$, $\mathbf{I}_1 = [1/3, 2/3)$, and $\mathbf{I}_2 = [2/3, 1)$.
   a. If $x_j < y_j$ and both are in the same subinterval $\mathbf{I}_k$, then show that $y_{j+1} - x_{j+1} = 3(y_j - x_j)$.
   b. If $x_j < y_j$ are in different adjacent intervals with $y_j - x_j \leq 1/4$, then show that $f(x_j) - f(y_j) = 1 - 3(y_j - x_j) \geq 1/4$.
   c. Show that f is expansive with expansive constant $1/4$.
   d. Find a pair of points whose distance is not tripled by the map.

2. Consider the "saw-toothed map" $S(x)$ defined in Exercise 10.3.3. Show that S has sensitive dependence on initial conditions.

3. Let p be a fixed point for f such that $|f'(p)| > 1$. Prove that f has sensitive dependence on initial conditions at p.

4. Let p be a period-n point for f such that $|(f^n)'(p)| < 1$. Prove that f does not have sensitive dependence on initial conditions at p.

5. Write out the details showing that the logistic map G has sensitive dependence on initial conditions.

6. Use a computer program to investigate the sensitive dependence for the quadratic map $G(x) = 4x(1-x)$. How many iterates does it take to get separation by 0.1 and 0.3 for the two different initial conditions x_0 and $x_0 + \delta$, for the choices $x_0 = 0.1$ and 0.48 and $\delta = 0.01$ and 0.001? (Thus, there are four pairs of points.)

10.5. Cantor Sets

So far we have used symbolic dynamics for maps that have an invariant interval, with no points leaving the interval. In this section, we consider the case in which some points leave the interval. The result is a very "thin" invariant set made up of infinitely many points, but containing no intervals. This set is called a Cantor set, named after a mathematician who worked around 1900 and introduced many of the ideas of rigorous set theory into mathematics.

Previously, we have used the tent map with slope 2 quite extensively. In this section, we consider tent maps with larger slopes to construct the Cantor set.

For $r > 0$, let T_r be the *tent map of slope r* given by

$$T_r(x) = \begin{cases} rx & \text{if } x \leq \tfrac{1}{2}, \\ r(1-x) & \text{if } x \geq \tfrac{1}{2}. \end{cases}$$

See Figure 16.

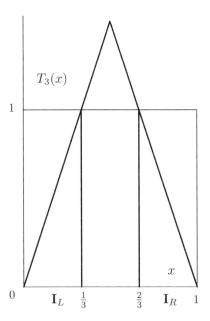

Figure 16. Graph of the tent map T_3

Lemma 10.16. Let $r > 2$. If $T_r^j(x_0)$ is bounded for all $j \geq 0$, then $T_r^j(x_0) \in [0,1]$ for all $j \geq 0$.

Proof. If $x_0 < 0$, then $T_r^j(x_0) = r^j x_0$, which goes to minus infinity as j goes to infinity.

If $x_0 > 1$, then $T_r(x_0) < 0$, and $T_r^{j+1}(x_0) = T_3^j(T_3(x_0)) = 3^j T_3(x_0)$ goes to minus infinity as j goes to infinity, by the previous case. $\square$

We consider all the points that stay in the interval $[0,1]$ for the first n iterates,

$$\mathbf{K}_n = \left\{ x : T_3^j(x) \in [0,1] \text{ for } 0 \leq j \leq n \right\}.$$

Note that $\mathbf{K}_0 = [0,1]$. From now on we consider the map T_3 of slope 3, but most of what we do is valid for any T_r with $r > 2$. The subintervals defined in the argument would have to be modified for other values of r.

If $1/3 < x_0 < 2/3$, then $T_3(x_0) > 1$, so $(1/3, 2/3)$ is not in $\mathbf{K}_1$. However, $T_3\left([0, 1/3]\right) = [0,1]$ and $T_3\left([2/3, 1]\right) = [0,1]$, so $\mathbf{K}_1$ is the union of two intervals

$$\mathbf{I}_L = \left[0, 1/3\right] \quad \text{and} \quad \mathbf{I}_R = \left[2/3, 1\right],$$

each of length $1/3$. The total length of $\mathbf{K}_1$ is $2\left(1/3\right) = 2/3$, $\lambda(\mathbf{K}_1) = 2/3$. Again, we use λ to denote the total length of a union of intervals.

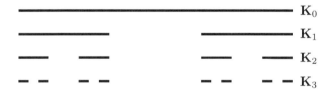

Figure 17. The sets $\mathbf{K}_0, \ldots \mathbf{K}_3$ for the Cantor set

A point x in $\mathbf{K}_2$ has $T_3(x)$, and $T_3^2(x)$ in $[0,1]$, so $T_3(x)$ is in $\mathbf{K}_1$. Also, $\mathbf{K}_2 \subset \mathbf{K}_1$, so it has a portion in $\mathbf{I}_L$ and a portion in $\mathbf{I}_R$. Thus, the set $\mathbf{K}_2$ is the set of the points in $\mathbf{I}_L \cup \mathbf{I}_R$ that map to $\mathbf{K}_1$, or

$$\mathbf{K}_2 = \left(\mathbf{I}_L \cap T_3^{-1}(\mathbf{K}_1)\right) \cup \left(\mathbf{I}_R \cap T_3^{-1}(\mathbf{K}_1)\right).$$

Since T_3 is monotonic with an expansion by a factor of 3 on each interval, $\mathbf{I}_L$ and $\mathbf{I}_R$, the sets $\mathbf{I}_L \cap T_3^{-1}(\mathbf{K}_1)$ and $\mathbf{I}_R \cap T_3^{-1}(\mathbf{K}_1)$ are each made up of two intervals of length $(1/3)^2$:

$$\mathbf{I}_L \cap T_3^{-1}(\mathbf{K}_1) = [0, 1/9] \cup [2/9, 1/3],$$
$$\mathbf{I}_R \cap T_3^{-1}(\mathbf{K}_1) = [2/3, 7/9] \cup [8/9, 1].$$

The part in $\mathbf{I}_L$ is obtained by shrinking $\mathbf{K}_1$ by a factor of $1/3$, and the part in $\mathbf{I}_R$ is obtained by shrinking $\mathbf{K}_1$ by a factor of $1/3$ and flipping it over. The total set

$$\mathbf{K}_2 = [0, 1/9] \cup [2/9, 1/3] \cup [2/3, 7/9] \cup [8/9, 1]$$

is the union of $4 = 2^2$ intervals, each of length $1/9 = (1/3)^2$; the total length of the intervals in $\mathbf{K}_2$ is $(2/3)^2$, $\lambda(\mathbf{K}_2) = (2/3)^2$.

10.5. Cantor Sets

Repeating by induction, we see that each of

$$\mathbf{I}_L \cap T_3^{-1}(\mathbf{K}_{n-1}) \quad \text{and}$$
$$\mathbf{I}_R \cap T_3^{-1}(\mathbf{K}_{n-1})$$

is the union of 2^{n-1} intervals, each of length $(1/3)(1/3)^{n-1} = (1/3)^n$, and

$$\mathbf{K}_n = \left(\mathbf{I}_L \cap T_3^{-1}(\mathbf{K}_{n-1})\right) \cup \left(\mathbf{I}_R \cap T_3^{-1}(\mathbf{K}_{n-1})\right)$$

is the union of $2^{n-1} + 2^{n-1} = 2^n$ intervals, each of length $(1/3)^n$; the total length of the intervals in $\mathbf{K}_n$ is $(2/3)^n$, $\lambda(\mathbf{K}_n) = (2/3)^n$.

Let

$$\mathbf{K} = \bigcap_{n \geq 0} \mathbf{K}_n = \left\{ x : T_3^j(x) \in [0,1] \text{ for } 0 \leq j < \infty \right\}.$$

The set $\mathbf{K}$ is called the *middle-third Cantor set*, since the middle third of each interval is removed at each stage. The set $\mathbf{K}$ is contained in each of the sets $\mathbf{K}_n$, so it must have length less than $(2/3)^n$ for any n. Therefore, the "length" or "Lebesgue measure" of $\mathbf{K}$ is zero, $\lambda(\mathbf{K}) = 0$. (See Section 11.4 for a more complete discussion of Lebesgue measure.)

Just as we did for the tent map T_2 in Section 10.3, we can label each of the intervals in $\mathbf{K}_n$ by means of the dynamics of T_3 rather than the order on the real line. For a finite string $s_0 s_1 \ldots s_{n-1}$, where each s_j equals R or L, let

$$\mathbf{I}_{s_0 s_1 \ldots s_{n-1}} = \left\{ x : T_3^j(x) \in \mathbf{I}_{s_j} \text{ for } 0 \leq j \leq n-1 \right\}.$$

In particular, $\mathbf{I}_{LL} = [0, 1/9]$, $\mathbf{I}_{LR} = [2/9, 1/3]$, $\mathbf{I}_{RR} = [2/3, 7/9]$, and $\mathbf{I}_{RL} = [8/9, 1]$. See Figure 18. In this case, these intervals are disjoint for different strings of the the same length, and $\mathbf{K}_n$ is the union of all the $\mathbf{I}_{s_0 s_1 \ldots s_{n-1}}$ over all the choices of the strings of symbols,

$$\mathbf{K}_n = \bigcup \left\{ \mathbf{I}_{s_0 s_1 \ldots s_{n-1}} : s_j \in \{L, R\} \text{ for } 0 \leq j < n \right\}.$$

We explore some of the properties of this set $\mathbf{K}$. At each stage, the set $\mathbf{K}_n$ has 2^{n+1} end points, which we label as $\mathbf{E}_n$. Thus,

$$\mathbf{E}_2 = \left\{ 0, \frac{1}{9}, \frac{2}{9}, \frac{1}{3}, \frac{2}{3}, \frac{7}{9}, \frac{8}{9}, 1 \right\}.$$

Let $\mathbf{E}$ be the union of all these sets; that is,

$$\mathbf{E} = \bigcup_{n \geq 0} \mathbf{E}_n,$$

which we call the *set of all the end points of the Cantor set* $\mathbf{K}$. In the construction, the part removed from $\mathbf{K}_m$ to form $\mathbf{K}_{m+1}$ comes from the interior of $\mathbf{K}_m$, so the points in $\mathbf{E}_n$ are never removed, so $\mathbf{E}_n$ and $\mathbf{E}$ are contained in $\mathbf{K}$. Since each of the sets $\mathbf{E}_n$ is finite, the union $\mathbf{E}$ is a *countable* set of points (i.e., there is a function from the positive integers onto the set $\mathbf{E}$).

We want to see that there are more points in the Cantor set than just the end points. One way to see this is to show that the Cantor set is uncountable, while we have shown that the set of end points is countable. An infinite set $\mathbf{S}$ is *uncountable*

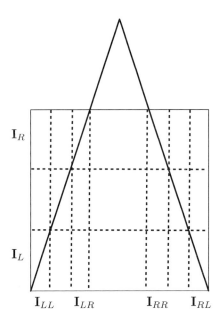

Figure 18. Intervals for strings of two symbols for T_3

if it is not countable (i.e., there is no function from the positive integers onto the set **S**). A standard result in analysis is that the unit interval $[0,1]$ is uncountable. See [**Mar93**] or [**Wad00**]. In the next theorem, we show that there is a function from the Cantor set onto the unit interval $[0,1]$, so the Cantor set is also uncountable. Thus, there are many more points in the entire Cantor set than just the end points.

Theorem 10.17. a. *The points in the set* **K** *are exactly those which can be given in a ternary expansion using only 0's and 2's, and such an expansion for points in* **K** *is unique. The set* **K** *is uncountable.*

b. *The set* **E** *is countable. The points in* **E** *have ternary expansions, which end in either repeated 0's (the expansion is finite), or repeated 2's.*

c. *The set*
$$\mathbf{K} \smallsetminus \mathbf{E} \neq \emptyset,$$
and points in this set have ternary expansions with only 0's and 2's, which do not end in repeated 0's or in repeated 2's.

Proof. (a) Using the ternary expansion, a point x in $[0,1]$ can be written as
$$x = \sum_{j=1}^{\infty} \frac{a_j}{3^j},$$
with each a_j equal to 0, 1, or 2. Notice that
$$\sum_{j\geq 1} \frac{2}{3^j} = \frac{2}{3} \sum_{j\geq 0} \frac{1}{3^j} = \frac{2}{3} \cdot \frac{1}{1-\frac{1}{3}} = \frac{2}{3} \cdot \frac{3}{2} = 1,$$
so we can write the number 1 with an expansion of the preceding form.

10.5. Cantor Sets

The sets $\mathbf{K}_1$ and $\mathbf{K}$ do not contain the open interval $(1/3, 2/3)$, whose points have $a_1 = 1$ and $1/3$ as part of their expansion. All the points in $[2/3, 1]$ start with $a_1 = 2$, while all those in $[0, 1/3)$ start with $a_1 = 0$. The point $1/3$ is in $\mathbf{K}_1$ and $\mathbf{K}$, but it can be written using only 0's and 2's in its expansion:

$$\sum_{j \geq 2} \frac{2}{3^j} = \frac{2}{3^2} \sum_{j \geq 0} \frac{1}{3^j} = \frac{2}{3^2} \cdot \frac{1}{1 - \frac{1}{3}} = \frac{2}{3^2} \cdot \frac{3}{2} = \frac{1}{3}.$$

Thus, all the points in K_1 and K can be represented with an expansion, with a_1 either 0 or 2. The left end points of the intervals in K_1 all have expansions which end in repeated 0's, while the right end points have expansions which end in repeated 2's.

At the second stage,

$$\mathbf{K}_2 = \mathbf{K}_1 \setminus \left((1/9, 2/9) \cup (7/9, 8/9) \right).$$

The interval $(1/9, 2/9)$, which contains points that start with the expansion $0/3 + 1/3^2$, is removed, and the interval $(7/9, 8/9)$, which contains points that start with the expansion $2/3 + 1/3^2$, is removed: both of these intervals have points with a 1 in their expansions. The points in

$$[0, 1/9] \cup [2/3, 7/9]$$

have $a_2 = 0$, while those in

$$[2/9, 1/3] \cup [8/9, 1]$$

have $a_2 = 2$. Again, the right end points can all be expressed using repeated 2's:

$$\frac{1}{9} = \sum_{j \geq 3} \frac{2}{3^j} \quad \text{and}$$

$$\frac{7}{9} = \frac{2}{3} + \sum_{j \geq 3} \frac{2}{3^j}.$$

Thus, all the points in K_2 and K can be represented with an expansion with a_2 either 0 or 2. The left end points of the intervals in K_2 all have expansions that end in repeated 0's, while the right end points have expansions which end in repeated 2's.

At the nth stage, all the points which need to have $1/3^n$ in their expansion are removed. The points at the right end points of $\mathbf{K}_n$ all have expansions that end in repeated 2's, and the left end points of $\mathbf{K}_n$ all have expansions that end in repeated 0's. Thus, any point in $\mathbf{K}$ can be given by a ternary expansion that uses only 0's and 2's and no 1's. By making these choices, for a point in $\mathbf{K}$, the ternary expansion is unique.

Now, form the map F from $\mathbf{K}$ to $[0, 1]$, defined by

$$F\left(\sum_{j=1}^{\infty} \frac{a_j}{3^j} \right) = \sum_{j=1}^{\infty} \frac{(a_j/2)}{2^j}.$$

This map realizes all possible binary expansions, so F is onto $[0, 1]$. Since $[0, 1]$ is uncountable, it follows that $\mathbf{K}$ is uncountable.

(b) We have already argued that the set of end points is countable. The proof of part (a) explains why they have the type of ternary expansions claimed in the theorem.

(c) Since $\mathbf{E}$ is countable, there are uncountably many points in $\mathbf{K} \smallsetminus \mathbf{E}$, and $\mathbf{K} \smallsetminus \mathbf{E} \neq \emptyset$. □

Example 10.32. To find a point in $\mathbf{K} \smallsetminus \mathbf{E}$, we take a ternary expansion that does not end in either repeated 0's or in repeated 2's. For example, if a_j is 0 and for odd j and a_j is 2 for even j, then

$$x = \frac{2}{3^2} + \frac{2}{3^4} + \cdots = \frac{2}{9} + \frac{2}{9^2} + \cdots$$
$$= \frac{2}{9}\left(1 + \frac{1}{9} + \frac{1}{9^2} + \cdots\right) = \frac{2}{9}\left(\frac{1}{1-\frac{1}{9}}\right)$$
$$= \frac{2}{9} \cdot \frac{9}{8} = \frac{1}{4}.$$

Therefore, $1/4$ is in $\mathbf{K}$ but not in $\mathbf{E}$.

Example 10.33. Let us check to see whether the point $9/13$ is in $\mathbf{K}$. To find its ternary expansion, we multiply by 3 and take the integer part, obtaining,

$$3 \cdot \frac{9}{13} = \frac{27}{13} = 2 + \frac{1}{13}, \qquad a_1 = 2,$$
$$3 \cdot \frac{1}{13} = 0 + \frac{3}{13}, \qquad a_2 = 0,$$
$$3 \cdot \frac{3}{13} = 0 + \frac{9}{13}, \qquad a_3 = 0,$$
$$3 \cdot \frac{9}{13} = \frac{27}{13} = 2 + \frac{1}{13}, \qquad a_4 = 2,$$

and the process is beginning to repeat. Therefore, $a_{1+3j} = 2$ and $a_{2+3j} = a_{3j} = 0$. Since the expansion uses only 0's and 2's, the point $9/13$ is in $\mathbf{K}$. Since the expansion does not end in repeated 0's or just repeated 2's, it is not in $\mathbf{E}$.

The two intervals, $\mathbf{I}_L = [0, 1/3]$ and $\mathbf{I}_R = [2/3, 1]$, can be used to define a map k^T from Σ_2^+ to $\mathbf{K}$,

$$\{k^T(\mathbf{s})\} = \bigcap_n \mathbf{I}_{s_0 \ldots s_n}.$$

Since $\mathbf{I}_{s_0 \ldots s_n} \cap \mathbf{I}_{t_0 \ldots t_n} = \emptyset$ if $s_0 \ldots s_n \neq t_0 \ldots t_n$, k^T is one to one. We get the next theorem using this function and symbolic dynamics as before.

Theorem 10.18. *Consider the tent map T_3. Use this to define the middle-third Cantor set $\mathbf{K}$.*

  a. *The Cantor set $\mathbf{K}$ is an invariant set for T_3.*

  b. *$k^T : \Sigma_2^+ \to \mathbf{K}$ is a one-to-one conjugacy from σ on Σ_2^+ to T_3 on $\mathbf{K}$.*

  c. *The periodic points for T_3 are dense in $\mathbf{K}$.*

  d. *The map T_3 has sensitive dependence on initial conditions when restricted to $\mathbf{K}$.*

  e. *The map T_3 is topologically transitive on $\mathbf{K}$.*

Length of preimages of intervals

When we turn to the logistic map, we need a comparison of the length of an interval and its image by a nonlinear map. This lemma is also used to give an estimate on the length of intervals defined by a symbol sequence. When the derivative is always greater than one, this will allow us to conclude that there is a unique point that realizes the symbol sequence.

Lemma 10.19. *Assume that $\beta > 0$ and f is a continuously differentiable map from a closed interval $[a, b]$ into $\mathbb{R}$ such that $|f'(x)| \geq \beta$ for all $a < x < b$. Assume that $\mathbf{J}_1$ is a closed subinterval of $[a, b]$ that is mapped onto the interval $\mathbf{J}_2 = f(\mathbf{J}_1)$. Then the lengths of $\mathbf{J}_1$ and $\mathbf{J}_2$ are related by*

$$\lambda(\mathbf{J}_1) \leq \frac{1}{\beta} \lambda(\mathbf{J}_2).$$

If $f(x)$ is linear on $[a, b]$ with $|f'(x)| = \beta$, then $\lambda(\mathbf{J}_1) = \frac{1}{\beta} \lambda(\mathbf{J}_2)$.

Proof. Let x_1 and x_2 be the end points of $\mathbf{J}_1$. By the mean value theorem,

$$f(x_1) - f(x_2) = f'(x_3)(x_1 - x_2),$$

where x_3 is a point between x_1 and x_2. So

$$\lambda(\mathbf{J}_2) = |f(x_1) - f(x_2)| = |f'(x_3)| \cdot |x_1 - x_2|$$
$$\geq \beta |x_1 - x_2| = \beta \lambda(\mathbf{J}_1),$$
$$\lambda(\mathbf{J}_1) \leq \frac{1}{\beta} \lambda(\mathbf{J}_2).$$

$\square$

Logistic family

The results for the tent map T_3 go over to the logistic family $g_a(x) = ax(1-x)$ for $a > 4$. Our proof uses the fact that the absolute value of the derivative is always greater than one on the Cantor set, which is true for $a > 2 + \sqrt{5} \approx 4.24$. Our argument about the Lebesgue measure of the Cantor set being zero uses $a > 2 + 2\sqrt{2} \approx 4.83$. Both conditions are true for $a = 5$, so we take $g(x) = g_5(x)$ to simplify the discussion.

The points that g maps to 1 can be found by solving $g(x) = 1$:

$$5x - 5x^2 = 1,$$
$$0 = 5x^2 - 5x + 1,$$
$$x_{\pm} = \frac{5 \pm \sqrt{25 - 20}}{10} = \frac{1}{2} \pm \frac{\sqrt{5}}{10}.$$

We let $\mathbf{I}_L^g = [0, x_-]$ and $\mathbf{I}_R^g = [x_+, 1]$. The derivative of g is $g'(x) = 5 - 10x$, and

$$g'(x_-) = 5 - 5 + \sqrt{5} = \sqrt{5} > 2,$$
$$g'(x_+) = 5 - 5 - \sqrt{5} = -\sqrt{5} < -2.$$

Since the absolute value of the derivative of g is even larger in $\mathbf{I}_L^g \cup \mathbf{I}_R^g$ than at the points $x_{\pm}$, $|g'(x)| \geq \sqrt{5}$ for all x in $\mathbf{I}_L^g \cup \mathbf{I}_R^g$.

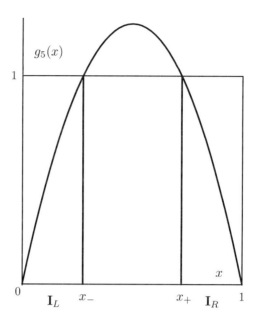

Figure 19. Graph of the logistic map g_5

Since $g(\mathbf{I}^g_{s_0}) = [0,1]$ for $s_0 = L$ or R, $\lambda(\mathbf{I}^g_{s_0}) < \lambda([0,1])/\sqrt{5} = 1/\sqrt{5} < 1/2$ by Lemma 10.19. Again, each of the intervals $\mathbf{I}^g_{s_0 s_1}$ maps onto the interval $\mathbf{I}^g_{s_1}$ by g. By Lemma 10.19, the length of any $\mathbf{I}^g_{s_0 s_1}$ satisfies

$$\lambda(\mathbf{I}^g_{s_0 s_1}) \leq \left(\frac{1}{\sqrt{5}}\right) \lambda(\mathbf{I}^g_{s_1}) \leq \left(\frac{1}{\sqrt{5}}\right)^2.$$

Continuing by induction, each of the intervals $\mathbf{I}^g_{s_0\ldots s_n}$ is mapped onto the interval $\mathbf{I}^g_{s_1\ldots s_n}$. By Lemma 10.19, the lengths of the intervals $\mathbf{I}^g_{s_0\ldots s_n}$ and $\mathbf{I}^g_{s_1\ldots s_n}$ satisfy

$$\lambda(\mathbf{I}^g_{s_0\ldots s_n}) \leq \left(\frac{1}{\sqrt{5}}\right) \lambda(\mathbf{I}^g_{s_1\ldots s_n}) \leq \left(\frac{1}{\sqrt{5}}\right)\left(\frac{1}{(\sqrt{5})}\right)^n = \left(\frac{1}{(\sqrt{5})}\right)^{n+1},$$

where the second inequality follows from the induction hypothesis. Let $\mathbf{K}^g_n$ be the union of the 2^n possible $\mathbf{I}^g_{s_0\ldots s_{n-1}}$ and $\Lambda_g = \bigcap_{n \geq 0} \mathbf{K}^g_n$. We can define a map $k^g : \Sigma^+_2 \to \Lambda_g$ by

$$\{k^g(\mathbf{s})\} = \bigcap_n \mathbf{I}^g_{s_0\ldots s_n}.$$

The total length $\lambda(\mathbf{K}^g_{n+1}) \leq (2/\sqrt{5})^{n+1}$ goes to zero as n goes to infinity, so Λ_g has Lebesgue measure zero. The set Λ_g has many of the properties of the middle-third Cantor set as summarized in the following theorem.

Theorem 10.20. *Consider the logistic map $g = g_a$, with $a \geq 2 + 2\sqrt{2}$, and its invariant set $\Lambda_g = \bigcap_{n \geq 0} g^{-n}([0,1])$.*

a. *The set Λ_g is uncountable and has Lebesgue measure equal to zero.*

b. *$k^g : \Sigma^+_2 \to \Lambda_g$ is a one-to-one conjugacy from σ on Σ^+_2 to g on Λ_g.*

c. *The periodic points for g are dense in Λ_g.*

d. *The map g has sensitive dependence on initial conditions when restricted to Λ_g.*
e. *The map g is topologically transitive on Λ_g.*

In Section 10.8 at the end of the chapter, we show that the sets **K** for T_3 and Λ_g for $g_5(x) = 5x(1-x)$ have the properties which characterize sets that are called Cantor sets.

Exercises 10.5

1. A sequence of intervals $\mathbf{J}_k$ is called nested, provided that $\mathbf{J}_k \supset \mathbf{J}_{k+1}$ for $k \geq 1$.
   a. Give a nested sequence of closed intervals $\mathbf{J}_k$ such that
   $$\bigcap_{k \geq 1} \mathbf{J}_k = \emptyset.$$
   b. Give a nested sequence of bounded intervals $\mathbf{J}_k$ such that
   $$\bigcap_{k \geq 1} \mathbf{J}_k = \emptyset.$$

2. Let
   $$T(x) = T_5(x) = \begin{cases} 5x & x \leq 0.5, \\ 5(1-x) & x \geq 0.5. \end{cases}$$
   a. Sketch the graph of T.
   b. What are the intervals that make up the set of points x such that x and $T(x)$ are both in $[0, 1]$?
   c. Describe the set of points x such that x, $T(x)$, and $T^2(x)$ are all in $[0, 1]$; that is, describe the set
   $$\mathbf{K}_2 = \{x : T^j(x) \in [0, 1] \text{ for } 0 \leq j \leq 2\}.$$
   It is made up of how many intervals of what length each? What is its total length?
   d. Without giving the intervals exactly, how many intervals are there, and what is length of each interval in the set
   $$\mathbf{K}_n = \{x : T^j(x) \in [0, 1] \text{ for all } 0 \leq j \leq n\}?$$
   e. Let
   $$\mathbf{K} = \{x : T^j(x) \in [0, 1] \text{ for all } j \geq 0\}.$$
   Explain which numbers in $[0, 1]$ belong to **K** in terms of the numbers expansion base 5, or $x = \sum_{j=1}^{\infty} a_j/5^j$.
   f. Give a number in **K** that is not an end point of one of the intervals in the finite process defining **K**.
   g. Are the numbers $23/25$ and $25/31$ in the set **K**?

3. Consider the logistic map $g_5(x) = 5x(1-x)$, with intervals $\mathbf{I}^{g_5}_{s_0 \ldots s_n}$, as defined in Section 10.5. Show that $\mathbf{I}^{g_5}_{LL}$ and $\mathbf{I}^{g_5}_{LR}$ have different lengths.

4. Consider the function $F(x) = 6x^3 - 5x$ on $[-1, 1]$. The points where $F(x) = 1$ are 1 and $-1/2 \pm \sqrt{3}/6$. The points where $F(x) = -1$ are -1 and $1/2 \pm \sqrt{3}/6$.

a. Sketch the graph of F.
b. Describe the set of points x such that both x and $F(x)$ are in $[-1, 1]$, or
$$\mathbf{K}_1 = \{ x : F^j(x) \in [-1,1] \text{ for } 0 \leq j \leq 1 \}$$
$$= [-1,1] \cap F^{-1}([-1,1]).$$
It is made up of how many intervals? What is the maximum length of these intervals?
c. Describe the set of points
$$\mathbf{K}_2 = \{ x : F^j(x) \in [-1,1] \text{ for } 0 \leq j \leq 2 \}$$
$$= \bigcap_{j=0}^{2} F^{-j}([-1,1]).$$
It is made up of how many intervals? What bound can you put on the length of each of the intervals in $\mathbf{K}_2$? Hint: Use a lower bound on the absolute value of the derivative on $\mathbf{K}_1$ (i.e., $|F'(x)| \geq \lambda$ for some $\lambda > 1$ and all x in $\mathbf{K}_1$).
d. Tell what bound you can put on the length of one of the intervals in
$$\mathbf{K}_n = \{ x : F^j(x) \in [-1,1] \text{ for } 0 \leq j \leq n \}$$
$$= \bigcap_{j=0}^{n} F^{-j}([-1,1]).$$
e. Explain why F has an invariant set that is like a Cantor set.

5. Consider the cotangent function $f(x) = \cot(x)$ on $[0, 2\pi]$. Explain why it has an invariant set that is like a Cantor set made up of points that stay in $[0, 2\pi]$ for all iterates.

6. Let
$$f(x) = \begin{cases} 5x + 4 & \text{for } x \leq -0.4, \\ -5x & \text{for } -0.4 \leq x \leq 0.4, \\ 5x - 4 & \text{for } 0.4 \leq x. \end{cases}$$
a. Sketch the graph of f. Notice that $f(-1) = -1$, $f(-0.4) = 2$, $f(0.4) = -2$, and $f(1) = 1$.
b. Consider the sets $\mathbf{K}_n = \{ x : f^j(x) \in [-1, 1] \text{ for } 0 \leq j \leq n \}$ How many intervals do $\mathbf{K}_1$, $\mathbf{K}_2$, and $\mathbf{K}_n$ contain? What is the length of each of these intervals in these sets? What is the total length of $\mathbf{K}_n$?

10.6. Piecewise Expanding Maps and Subshifts

In the preceding three sections, we used symbolic dynamics (itineraries) to find a dense orbit in an interval or invariant set and to show that a map has sensitive dependence on initial conditions. All the cases considered so far have allowed all possible transitions between intervals. On the other hand, in Section 10.1, we used transition graphs where only certain transitions were allowed between symbols. In this section, we combine transition graphs with symbolic dynamics to show that certain other maps are topologically transitive on an interval. The space

10.6. Expanding Maps

of sequences of symbols where there are restrictions on the transitions are called *subshifts of finite type*.

In the section on transition graphs for the Sharkovskii theorem, we got the existence of periodic orbits, but not uniqueness. In this section, we use the assumption that the map has a derivative with absolute value greater than one to get uniqueness of the point with as given symbol sequence.

Example 10.34. We define what is called the *shed map* by

$$f(x) = \begin{cases} \beta x + c & \text{if } 0 \leq x \leq c, \\ (1-x)\beta & \text{if } c \leq x \leq 1, \end{cases}$$

with $\beta > 1$ and $0 < c < 1/2$. See Figure 20. The absolute value of the derivative of this function is greater than one on two adjacent intervals, with a point of nondifferentiability at the point where the intervals meet. To insure that the map is continuous and takes on the same value 1 at $x = c$, we need

$$\beta c + c = 1 \quad \text{and}$$
$$(1-c)\beta = 1.$$

For these equations to hold, we need

$$c = \frac{1}{1+\beta},$$
$$1 = (1-c)\beta = \left(1 - \frac{1}{1+\beta}\right)\beta,$$
$$\beta + 1 = (\beta + 1 - 1)\beta,$$
$$0 = \beta^2 - \beta - 1, \quad \text{so}$$
$$\beta = \frac{1+\sqrt{5}}{2} \approx 1.618 > 1.$$

For these choices, f is continuous on $[0,1]$ and differentiable on the two open intervals $(0,c)$ and $(c,1)$ with $|f'(x)| = \beta > 1$.

Notice that

$$f([0,c]) = [c,1] \quad \text{and}$$
$$f([c,1]) = [0,1] = [0,c] \cup [c,1],$$

so the image of the left interval covers the right interval, and the image of the right interval covers both the left and the right intervals. Thus, the transition graph $\mathscr{G}$ allows either R or L to follow R, but only allows R to follow L. Any sequence $\mathbf{s} = s_0 s_1 \ldots$ that follows these rules is called an *allowable symbol sequence*. Let $\Sigma_{\mathscr{G}}^+$ be the set of all allowable symbol sequences.

The length of $\mathbf{I}_L$ is c and the length of $\mathbf{I}_R$ is $1-c$, so the length of each of these intervals is less than or equal to $1-c$. Let $\mathbf{s} = \{s_j\}_{j=0}^{\infty}$ be an allowable sequence based on the transition graph. The interval $\mathbf{I}_{s_0 \ldots s_n}$ maps to $\mathbf{I}_{s_1 \ldots s_n}$, so applying Lemma 10.19 and induction on n,

$$\lambda(\mathbf{I}_{s_0 \ldots s_n}) \leq \beta^{-1} \lambda(\mathbf{I}_{s_1 \ldots s_n}) \leq \beta^{-2} \lambda(\mathbf{I}_{s_2 \ldots s_n}) \leq \cdots \leq \beta^{-n} \lambda(\mathbf{I}_{s_n}) \leq \beta^{-n}(1-c).$$

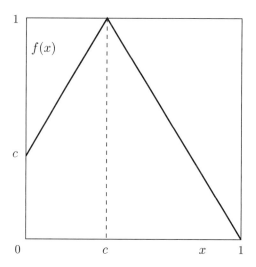

Figure 20. Shed map

Thus, for an allowable symbol sequence $\mathbf{s} = s_0 s_1 \ldots$, the intersection of these intervals $\mathbf{I}_{s_0\ldots s_n}$ for all positive n has length zero and contains a single point $k(\mathbf{s})$:

$$\bigcap_{n\geq 0} \mathbf{I}_{s_0\ldots s_n} = \{\, k(\mathbf{s})\,\} \neq \emptyset.$$

Any orbit of the shed map can be determined by a sequence of symbols that is allowable in terms of its transition graph because the image of each interval in the partition is the union of intervals in the partition. Because the union of the intervals in the partition is not positively invariant for maps like the tent map with slope greater than two, we express the following and subsequent definitions in a way that does not require this condition.

Remember that a function f is continuous on a partition of closed bounded intervals $\{\mathbf{I}_i\}_{i=1}^N$ provided that (i) $\text{int}(\mathbf{I}_i) \cap \text{int}(\mathbf{I}_j) = \emptyset$ for $i \neq j$ and (ii) f is continuous on $\bigcup_{i=1}^N \mathbf{I}_i$. (We do not assume that f is continuous everywhere.) If the function is one to one on $\mathbf{I}_i$, then $f^{-1}(\mathbf{I}_j) \cap \mathbf{I}_i$ is a single interval whenever it is nonempty.

Definition 10.35. Assume that f is a real-valued function defined on a domain $\mathscr{D} \subset \mathbb{R}$. A *Markov partition for f* is a partition of closed bounded intervals $\{\mathbf{I}_1, \ldots, \mathbf{I}_N\}$ such that (i) f is continuous on the partition, (ii) f is one to one on each $\mathbf{I}_i$, and (iii) if the image of the interior of one of the $\mathbf{I}_j$ by f intersects the interior of any $\mathbf{I}_i$, then the image $f(\mathbf{I}_j)$ goes all the way across $\mathbf{I}_i$; that is,

$$\text{if } f(\text{int}(\mathbf{I}_i)) \cap \text{int}(\mathbf{I}_j) \neq \emptyset, \quad \text{then } f(\mathbf{I}_i) \supset \mathbf{I}_j.$$

Condition (iii) is called the *Markov property for the partition*.

If a function is continuous on a partition and one to one on each interval, then it satisfies the Markov property if and only if, for each end point e of one of the intervals, $f(e)$ is not in any of the open intervals $\text{int}(\mathbf{I}_j)$.

10.6. Expanding Maps

A Markov partition for f determines a transition graph as defined in Section 10.1 with an edge from i to j if and only if $f(\mathbf{I}_i) \supset \mathbf{I}_j$. In this case, Theorem 10.22(c) proves that all the points that stay in the union of the intervals of the partition are coded by allowable sequences.

We have seen several examples of Markov partitions. (1) The tent map T_2 and the logistic map $G = g_4$ each has $\{\,[0, 0.5], [0.5, 1]\,\}$ as a Markov partition. (2) The tent map T_3 has Markov partition $\{\,[0, 1/3], [2/3, 1]\,\}$. However, $\{\,[0, 2/5], [3/5, 1]\,\}$ and $\{\,[0, 1/2], [1/2, 1]\,\}$ are two other Markov partitions for T_3. (3) The shed map given in Example 10.34 has $\{\,[0, c], [c, 1]\,\}$ as a Markov partition.

Definition 10.36. Given a Markov partition $\{\,\mathbf{I}_j\,\}_{j=1}^N$ for f, we say that f is *expanding on the Markov partition* with *expanding factor* $\beta > 1$ provided that f is differentiable on each open interval $\mathrm{int}(\mathbf{I}_j)$ and $|f'(x)| \geq \beta$ for all x in $\mathrm{int}(\mathbf{I}_j)$.

If a function is expanding on a Markov partition, then we can show that each allowable sequence of intervals has a unique point with that itinerary. See Theorem 10.22(c).

The shed map and functions in the homework for this section satisfy the stronger requirements of the following definition that imply they are are expanding on their Markov partition.

Definition 10.37. A map f from $\mathbb{R}$ to itself is called a *continuous piecewise expanding* on an interval $[a, b]$ provided that it is continuous on $[a, b]$ and there are a finite set of points

$$a = p_0 < p_1 < \cdots < p_N = b$$

and a constant $\beta > 1$ such that $|f'(x)| \geq \beta$ for $p_{j-1} < x < p_j$ and for $j = 1, \ldots, N$. The number β is called the *expanding factor* or *stretching factor*. In all of our examples, the derivative is a constant on each of the intervals (p_{j-1}, p_j), so $\beta = \min\{\,|f'(x_j)|\,\}$, where x_j is any point in (p_{j-1}, p_j).

Given the points $\{\,p_0, p_1, \ldots, p_N\,\}$, we denote the induced closed intervals by $\mathbf{I}_j = [p_{j-1}, p_j]$ $\{\,p_0, p_1, \ldots, p_N\,\}$, and $\mathrm{int}(\mathbf{I}_j) = (p_{j-1}, p_j)$ the corresponding open intervals on which f is differentiable. We call either the points $\{\,p_0, \ldots, p_N\,\}$ or the intervals $\{\,\mathbf{I}_j\,\}_{j=1}^N$ the *partition* for the expanding map.

Note that, for a continuous piecewise expanding map f, the partition $\{\,\mathbf{I}_j\,\}_{j=1}^N$ automatically satisfies conditions (i) and (ii) of a Markov partition. Therefore, we only need to check the images $f(p_i)$ of the points in the partition.

The fact that f is continuous means that the limits of f from the two sides of the p_j give the same value, and there are not any jumps at the p_j points. The tent maps T_2 and T_3 are examples of continuous piecewise expanding maps. The doubling map is piecewise expanding but not continuous. Much of what we say in this sections applies when f is allowed to have discontinuous jumps at the p_j's: We say a few words about this at the end of the section, but do not state a formal theorem that applies to these functions. The logistic map g_5 is not piecewise expanding because $g_5'(0.5) = 0$, but it is expanding on its natural Markov partition, $\{\,[0, x_-], [x_+, 1]\,\}$.

Example 10.38. Let $f(x)$ be defined by
$$f(x) = \begin{cases} \frac{2}{3} + \frac{4}{3}x & \text{for } 0 \leq x \leq \frac{1}{4}, \\ \frac{4}{3} - \frac{4}{3}x & \text{for } \frac{1}{4} \leq x \leq 1. \end{cases}$$
Then, $f(0) = {}^2\!/_3$, $f^2(0) = f({}^2\!/_3) = {}^4\!/_{3^2}$, $f^3(0) = f({}^4\!/_{3^2}) = {}^{20}\!/_{3^3}$, and $f^j(0) = {}^i\!/_{3^j}$ for some integer i that does not have 3 as a divisor. All of the points $f^j(0)$ would have to be end points of any Markov partition. Since this set of points does not repeat and so is infinite, there cannot be a finite set of dividing points for a Markov partition (i.e., f does not have a Markov partition of all of $[0,1]$). This continuous piecewise expanding map does not have a Markov partition of $[0,1]$ even though the slopes are rational.

Definition 10.39. Let $\mathscr{G}$ be a transition graph with N vertices **S**. A sequence **s** in Σ_N^+ is called *allowable* provided that there is an edge in $\mathscr{G}$ from s_j to s_{j+1} for all $j \geq 0$. Let $\Sigma_\mathscr{G}^+$ be the subset of Σ_N^+ of all allowable sequences. We refer to $\Sigma_\mathscr{G}^+$ as the *shift space*. We denote by $\sigma_\mathscr{G}$ the restriction of σ to $\Sigma_\mathscr{G}^+$. This shift map $\sigma_\mathscr{G}$ takes $\Sigma_\mathscr{G}^+$ to itself, so it is is called a *subshift*; it is called a *subshift of finite type* because the rules for allowable sequences can be expressed using a transition graph with a finite number of edges.

We next define the properties of the transition graph that are needed to imply topological transitivity.

Definition 10.40. A transition graph $\mathscr{G}$ is called *reducible* provided that there is some pair of vertices i and j such that there is no (finite) path in $\mathscr{G}$ from the ith vertex to the jth vertex. (If $i = j$, then it would mean that there is no path in the graph that leaves the vertex and returns again to the vertex.) A transition graph $\mathscr{G}$ is called *irreducible* provided that it is not reducible (i.e., for every pair of vertices i and j, including the case of $i = j$, there is some path in $\mathscr{G}$ from the ith vertex to the jth vertex).

The transition graph for the shed map in Example 10.34 is irreducible.

Example 10.41. The transition graph given in Figure 21(a) is reducible: there is no transition from vertices 3 or 4 back to vertices 1 or 2. A piecewise expanding map with this transition graph is the function given in Figure 21(b):
$$f(x) = \begin{cases} 2 - 2x & \text{for } 0 \leq x \leq 1, \\ 4x - 4 & \text{for } 1 \leq x \leq 2, \\ 8 - 2x & \text{for } 2 \leq x \leq 3, \\ -4 + 2x & \text{for } 3 \leq x \leq 4. \end{cases}$$

A function with a reducible transition graph could not possibly be topologically transitive on the union of all the intervals.

Theorem 10.21. *Assume that $\mathscr{G}$ is an irreducible transition graph with $N \geq 2$ vertices. Let $\Sigma_\mathscr{G}^+$ be the set of all allowable sequences for the transition graph $\mathscr{G}$ and $\sigma_\mathscr{G}$ be the restriction of the shift map σ to $\Sigma_\mathscr{G}^+$.*

  a. *Then $\sigma_\mathscr{G}$ is topologically transitive on $\Sigma_\mathscr{G}^+$.*

10.6. Expanding Maps

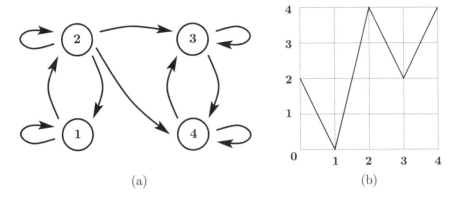

Figure 21. Reducible transition graph for Example 10.41

b. *If in addition $\mathscr{G}$ has some vertex that has more than one edge going out, then $\sigma_{\mathscr{G}}$ has sensitive dependence on initial conditions on $\Sigma_{\mathscr{G}}^+$.*

Proof. (a) We need to define a symbol sequence $\mathbf{s}^*$ which realizes all allowable string of all lengths. Let $\mathbf{w}_{ij}$ be a finite string of symbols such that $i\mathbf{w}_{ij}j$ is an allowable string, starting at i and ending at j. If j can follow i, then $\mathbf{w}_{ij}$ can be taken to be the empty string. Now, let $\mathbf{s}^*$ be the infinite sequence described as follows. Write down 1, followed by $\mathbf{w}_{12}$, followed by 2, continuing until $\mathbf{w}_{k-1\,k}$, followed by k. This gives all the strings of length one. Next, write down all allowable strings with two symbols, with a transition between them that makes the total sequence allowable. Continue, by induction, writing down all the allowable strings with n symbols, with a transition between them, to make the total sequence allowable. In this way $\mathbf{s}^*$ has all possible finite allowable strings. By the proof as before, $\mathbf{s}^*$ has a dense orbit in $\Sigma_{\mathscr{G}}$.

(b) There is a periodic orbit that goes through the vertex with more than one edge going out. This periodic orbit can be used in a way similar to the fixed point 0 in the proof of sensitive dependence for the tent map. We leave the details to the reader. $\square$

The next theorem summarizes the result for a function that is expanding on a Markov partition, which includes the case of a *continuous* piecewise expanding map. This result contains earlier results about the tent map, logistic map, and tent maps with invariant Cantor sets. It does not apply to maps such as the doubling map that are not continuous.

Theorem 10.22. *Assume that f is expanding on a Markov partition $\{\,\mathbf{I}_j\,\}_{j=1}^N$ with $N \geq 2$ and expanding factor $\beta > 1$. Note that f must be continuous on $\mathbf{X} = \bigcup_{j=1}^N \mathbf{I}_j$. Let $L = \max\{\,\lambda(\mathbf{I}_j) : 1 \leq j \leq N\,\}$ be the maximum of the length of the intervals. Let $\mathscr{G}$ be the transition graph induced by the partition, and $\Sigma_{\mathscr{G}}^+$ and $\sigma_{\mathscr{G}}$ be the associated subshift of finite type.*

a.
$$\Lambda_f = \bigcap_{n \geq 0} f^{-n}(\mathbf{X}) = \{\,x \in \mathbf{X} : f^j(x) \in \mathbf{X} \ \ \text{for all} \ j \geq 0\,\}$$

is a nonempty invariant set for f. Note that if $f(\mathbf{X}) = \mathbf{X}$ is invariant, then $\Lambda_f = \mathbf{X}$.

b. *Any finite allowable string $s_0 \ldots s_n$ of symbols corresponds to a nonempty closed interval $\mathbf{I}_{s_0 \ldots s_n}$ of length, at most, $L\beta^{-n}$.*

c. *An infinite allowable sequence $\mathbf{s}$ in $\Sigma_{\mathscr{G}}^+$ corresponds to a unique point $x_0 = k(\mathbf{s})$ in $\mathbb{R}$ such that $f^j(x_0)$ is in the interval $\mathbf{I}_{s_j}$ for all $j \geq 0$, $\{k(\mathbf{s})\} = \bigcap_{n \geq 0} \mathbf{I}_{s_0 \ldots s_n}$.*

Moreover, the map k from $\Sigma_{\mathscr{G}}^+$ to $\Lambda_f \subset \mathbf{X} \subset \mathbb{R}$ is continuous and onto Λ_f. It is a semiconjugacy between the shift map $\sigma_{\mathscr{G}}$ on $\Sigma_{\mathscr{G}}^+$ and f on Λ_f.

d. *If $\mathbf{s}$ is an allowable period-n sequence, then $k(\mathbf{s})$ is a period-n point for f. If $\mathbf{s}$ is an allowable eventually period-n sequence, then $k(\mathbf{s})$ is an eventually period-n point for f. If $k(\mathbf{s})$ is neither periodic nor eventually periodic, then $\mathbf{s}$ is neither periodic nor eventually periodic.*

e. *Assume that the transition graph $\mathscr{G}$ is irreducible and has some vertex that has more than one edge going out. (This latter assumption is automatic if the union of the intervals $\mathbf{X}$ is invariant.) Then the map f has sensitive dependence on initial conditions when restricted to Λ_f.*

f. *If the transition graph $\mathscr{G}$ is irreducible, then f is topologically transitive on Λ_f.*

Idea of Proof. Parts (**a**), (**b**), (**c**), and (**d**) follow as before. The analysis of Example 10.34 actually proves parts (**b**) and (**c**). Note that the points in Λ_f are exactly those points that stay in $\mathbf{X}$ for all iterates and so have an infinite itinerary defined. For a map like the tent map T_r with $r > 2$, Λ_f is a Cantor set.

Parts (**e**) and (**f**) follow from Theorem 10.21 since k is a semiconjugacy from $\Sigma_{\mathscr{G}}^+$ onto Λ_f: $x^* = k(\mathbf{s}^*)$ has a dense orbit in Λ_f. □

Remark 10.42. The point x^* passes through all intervals $\mathbf{I}_{s_0 \ldots s_n}$. For this discussion, assume that all transitions are allowable. For some string of 100 iterates, $f^j(x^*)$ is in interval $\mathbf{I}_1$. Later, it passes through the string of intervals $\mathbf{I}_1, \mathbf{I}_2, \ldots, \mathbf{I}_N$ a million times. Thus, the future behavior of the orbit is not at all predictable on the basis of the past and current location. This feature justifies calling this orbit *chaotic*.

Remark 10.43. Sometimes, we want to consider *discontinuous* expanding maps, such as the doubling map. In this case, all allowable sequences correspond to orbits except possibly, those sequences that would give the end points on one of the subintervals. For the doubling map, if we take $^1\!/_2$ in $[^1\!/_2, 1] = \mathbf{I}_1$, then $0 = D\left(^1\!/_2\right) = D^j\left(^1\!/_2\right)$ is in $\mathbf{I}_0$ for all $j \geq 1$. Thus an orbit that goes through the point $^1\!/_2$ ends in a symbol sequence 10^∞. On the other hand, the symbol sequence 01^∞ should correspond to a point p with $p \leq {}^1\!/_2$ and $D^j(^1\!/_2) = 1$ for all $j \geq 1$. Since such an orbit does not exist, there is no point that has a symbol sequence ending in 01^∞. (If we used $D(^1\!/_2) = 1$ and $D(1) = 1$, then we do have a point for the symbol sequence 01^∞, but none with the symbol sequence 10^∞.) The point is that although symbolic dynamics could be used to analyze the doubling map or other discontinuous maps which are piecewise expanding, the corresponding shift space is not of finite type and the preceding theorem *does not apply directly*.

10.6.1. Counting Periodic Points for Subshifts of Finite Type. In Section 10.1, we used transition graphs to determine periodic points. In this section, we introduce the transition matrix which makes it easier to count the number of periodic points for a subshift of finite type.

Definition 10.44. Assume that we start with a transition graph $\mathscr{G}$ with N vertices, labeled 1, ..., N. The associated *transition matrix* for the subshift $\Sigma_\mathscr{G}$ is the $N \times N$ matrix $\mathbf{T} = (t_{ij})$ of 0's and 1's, in which an entry $t_{ij} = 1$ whenever there is a transition from vertex i to vertex j, and $t_{ij} = 0$ whenever this transition is not allowed. Let $\Sigma_\mathbf{T}$ be the set of sequences of symbols $\mathbf{s} = \{s_k\}_{k=0}^\infty$, where each transition $s_k s_{k+1}$ is allowed, $t_{s_k s_{k+1}} = 1$. By the definitions, $\Sigma_\mathbf{T} = \Sigma_\mathscr{G}$. Let $\sigma_\mathbf{T}$ be the restriction of the shift map to $\Sigma_\mathbf{T}$. The space $\Sigma_\mathbf{T}$ with the map $\sigma_\mathbf{T}$ is called a *subshift of finite type for the matrix* $\mathbf{T}$. Another term for $\Sigma_\mathbf{T}$ is is a *topological Markov chain*.

If we start with the matrix, and not with the transition graph, then we need to specify the properties it must have to call it a transition matrix. A transition matrix $\mathbf{T} = (t_{ij})$ must satisfy the following properties: (i) It is a square $N \times N$ matrix, where each entry t_{ij} is either 0 or 1. (ii) For i fixed between 1 and N, the sum on the row $\sum_j t_{ij} \geq 1$. (This means that it is possible to go to some other symbol from the symbol i.) Notice, that if $\sum_i t_{ij} = 0$, then there is no way to get back to the symbol j.

A finite string of symbols $\mathbf{w} = s_k \ldots s_{k+m}$ is called an *allowable word* or an *allowable string* provided that $t_{s_j s_{j+1}} = 1$ for $j = k \ldots k+m-1$.

The trace of $\mathbf{T}$ is the sum of the entries down the diagonal, $\text{tr}(\mathbf{T}) = t_{11} + \cdots + t_{NN}$. For a transition matrix, the trace gives the number of symbols that can follow themselves (i.e., the symbols s for which s^∞ is an allowable sequence of symbols). These sequences correspond to fixed points of the shift map. Therefore, the trace of $\mathbf{T}$ gives the number of fixed points of the shift map $\sigma_\mathbf{T}$. The next theorem gives the corresponding result for higher powers.

Theorem 10.23. *Let $\mathbf{T}$ be a transition matrix, and let $\Sigma_\mathbf{T}$ be the corresponding shift space, with shift map $\sigma_\mathbf{T}$. Then, the number of fixed points of $\sigma_\mathbf{T}^k$ on $\Sigma_\mathbf{T}$, $N(k)$, equals the trace of $\mathbf{T}^k$, $\text{tr}(\mathbf{T}^k)$.*

This theorem follows from the next lemma about the number of words of length $k+1$ that start at symbol i and end at symbol j.

Lemma 10.24. *Assume that the ij-entry of $\mathbf{T}^k$ is p, $\left(\mathbf{T}^k\right)_{ij} = p$. Then, there are p allowable strings of symbols of length $k+1$, starting at i and ending at j (i.e., strings of the form $is_1 s_2 \ldots s_{k-1} j$).*

Proof. We prove the result by induction on k. Let $N(k; i, j)$ be the number of strings of symbols of length $k+1$ starting at i and ending at j.

The result is true for $k = 1$, since $t_{ij} = (\mathbf{T})_{ij}$ equals 0 or 1, depending on whether the transition from i to j is allowed or not (i.e., whether the string ij is allowed or not), $(\mathbf{T})_{ij} = N(1; i, j)$.

Assume that the result is true for $k - 1$, for all choices of i and j. Then, by matrix multiplication,

$$\left(\mathbf{T}^k\right)_{ij} = \left(\mathbf{T}^{k-1}\mathbf{T}\right)_{ij} = \sum_m \left(\mathbf{T}^{k-1}\right)_{im} t_{mj} = \sum_{t_{mj}=1} \left(\mathbf{T}^{k-1}\right)_{im}.$$

By the induction hypothesis, we see that $\left(\mathbf{T}^{k-1}\right)_{im}$ is the number of allowable strings, starting at i and ending at m, which is the same as the number of allowable strings of length $k+1$ that start at i and end in j and are equal to m in the next-to-last position. By adding up all of these for which it is possible to make a transition from m to j (i.e., for which $t_{mj} = 1$), we get $N(k; i, j)$, or

$$\left(\mathbf{T}^k\right)_{ij} = \sum_{t_{mj}=1} N(k-1; i, m) = N(k; i, j).$$

This completes the induction step, the proof of the lemma, and the proof of the theorem. □

Example 10.45. Consider the transition matrix

$$\mathbf{T} = \begin{pmatrix} 0 & 1 \\ 1 & 1 \end{pmatrix}.$$

Then, the sequence of $N(k)$ starts as

$$N(1) = \operatorname{tr}(\mathbf{T}) = 1 \quad \text{and} \quad N(2) = \operatorname{tr}\begin{pmatrix} 1 & 1 \\ 1 & 2 \end{pmatrix} = 3.$$

We leave it to the exercises to verify that

$$\operatorname{tr}(\mathbf{T}^k) = \operatorname{tr}(\mathbf{T}^{k-2}) + \operatorname{tr}(\mathbf{T}^{k-1}) \quad \text{or}$$
$$N(k) = N(k-2) + N(k-1).$$

Thus, $N(3) = 1 + 3 = 4$, $N(4) = 3 + 4 = 7$, and $N(5) = 4 + 7 = 11$. The recurrence relationship $N(k) = N(k-2) + N(k-1)$ for a sequence of integers is called the *Fibonacci recurrence relation*. The usual sequence starts with $N(1) = 1$ and $N(2) = 1$, but our sequence starts with $N(1) = 1$ and $N(2) = 3$.

Since $\#\operatorname{Fix}(\sigma_{\mathbf{T}}^k) = N(k)$, $\#\operatorname{Per}(k, \sigma_{\mathbf{T}})$ equals $\#\operatorname{Fix}(\sigma_{\mathbf{T}}^k)$ minus the points of lower period, and the number of orbits of period-k satisfies $\#\operatorname{Orb}(k, \sigma_{\mathbf{T}}) = \dfrac{\#\operatorname{Per}(k, \sigma_{\mathbf{T}})}{k}$, Table 4 gives the number of period-k points and orbits up to period seven.

There are certain types of transition matrices that we want to distinguish.

Definition 10.46. A transition matrix $\mathbf{T}$ is a *permutation matrix* provided that $\sum_j t_{ij} = 1$ for all i and $\sum_i t_{ij} = 1$ for all j (i.e., each symbol i goes to a unique symbol j and there is a unique symbol i which goes to each other symbol j).

A transition matrix $\mathbf{T}$ is called *reducible* provided that there is a pair (i, j) with $1 \leq i, j \leq n$ such that $(\mathbf{T}^k)_{ij} = 0$ for all $k \geq 1$. (This means that there is no way to get from the symbol i to the symbol j.) A transition matrix $\mathbf{T}$ is called *irreducible* provided that it is not reducible; that is, for each pair (i, j) with $1 \leq i, j \leq n$, there is an integer k that depends on the pair such that $(\mathbf{T}^k)_{ij} > 0$ (i.e., there is some allowable sequence from the symbol i to the symbol j). Notice

k	$\#\operatorname{Fix}(\sigma_{\mathbf{T}}^k)$	Lower Period	$\#\operatorname{Per}(k,\sigma_{\mathbf{T}})$	$\#\operatorname{Orb}(k,\sigma_{\mathbf{T}})$
1	1	0	1	1
2	3	1	2	1
3	4	1	3	1
4	7	3	4	1
5	11	1	10	2
6	18	6	12	2
7	29	1	28	4

Table 4. Number of Periodic Points for $\sigma_{\mathbf{T}}$ in Example 10.45

that, a transition matrix is irreducible if and only if its associated transition graph is irreducible.

Theorem 10.25. *For a transition matrix* $\mathbf{T}$, *the following two conditions are equivalent:*

i. *The matrix* $\mathbf{T}$ *is irreducible.*
ii. *The shift map* $\sigma_{\mathbf{T}}$ *has a dense forward orbit in* $\Sigma_{\mathbf{T}}^+$ *and* $\sum_i t_{ij} \geq 1$ *for each* j.

Proof. First, we assume condition (i) and prove that condition (ii) follows. For a fixed j, since it is possible to get from any symbol to j, we must have $\sum_i t_{ij} \geq 1$. The proof that $\sigma_{\mathbf{T}}$ has a dense forward orbit in $\Sigma_{\mathbf{T}}^+$ is essentially the same as that of 10.21(f) so we do not repeat it. Thus, we have shown that (i) implies (ii).

Now assume that condition (ii) is satisfied. Take an arbitrary pair (i,j). Let $\mathbf{a}$ be a sequence with $a_0 = i$. Since it is possible to reach j from another symbol, there is a sequence $\mathbf{b}$ with $b_m = j$, for some $m \geq 1$. The orbit of $\mathbf{s}^*$ comes close to both $\mathbf{a}$ and $\mathbf{b}$. Thus, there is some $k_1 \geq 0$ such that the first symbol of $\sigma^{k_1}(\mathbf{s})$ is i, so $s_{k_1}^* = i$. Because, the orbit comes near to $\mathbf{b}$, there is a $k_2 > k_1$ such that the first symbol of $\sigma^{k_2}(\mathbf{s})$ is j, so $s_{k_2}^* = j$. Thus, $s_{k_1}^* \ldots s_{k_2}^*$ is an allowable string that starts with i and ends with j. This shows that $(\mathbf{T}^{k_2-k_1})_{ij} \neq 0$. Since (i,j) is arbitrary, it follows that $\mathbf{T}$ is irreducible and condition (i) is satisfied. □

Exercises 10.6

1. Consider the continuous piecewise expanding map
$$f(x) = \begin{cases} \frac{1}{3} + 2x & \text{if } 0 \leq x \leq \frac{1}{3}, \\ 2 - 3x & \text{if } \frac{1}{3} \leq x \leq \frac{2}{3}, \\ 2x - \frac{4}{3} & \text{if } \frac{2}{3} \leq x \leq 1. \end{cases}$$

   a. Draw a graph of f (i.e., draw $f(x)$ versus x, not the transition graph).
   b. What is a Markov partition for f? What is the expanding factor for f on $[0,1]$?
   c. What is the transition graph for the Markov partition?

d. Why does f have a point with a dense orbit in $[0,1]$?
  e. How many periodic points does f have of each period, considering the periods up to and including 6?
2. Consider the discontinuous map
$$g(x) = \begin{cases} \frac{1}{3} + 2x & \text{if } 0 \leq x < \frac{1}{3}, \\ 3x - 1 & \text{if } \frac{1}{3} \leq x \leq \frac{2}{3}, \\ \frac{7}{3} - 2x & \text{if } \frac{2}{3} \leq x \leq 1. \end{cases}$$
  a. Draw the graph of g (i.e., draw $g(x)$ versus x, not the transition graph).
  b. What is a Markov partition for g? What is the expanding factor for g on $[0,1]$?
  c. What is the transition graph $\mathscr{G}$ for the Markov partition?
  d. Let $\Sigma_{\mathscr{G}}$ be the set of allowable symbol sequences in terms of the transition graph. Let h be the itinerary function for g. Since the function g is discontinuous, it is possible that an allowable symbol sequence $\mathbf{s}$ in $\Sigma_{\mathscr{G}}$ does not correspond to an orbit of g (i.e., $\mathbf{s}$ is not equal to $h(x)$ for any x in $[0,1]$). What is the set of allowable symbol sequences $\mathbf{s}$ in $\Sigma_{\mathscr{G}}$ for which $\mathbf{s} \neq h(x)$ for all x in $[0,1]$?
3. Consider the continuous piecewise expanding map
$$f(x) = \begin{cases} \sqrt{2}\,x & \text{for } 0 \leq x \leq 1/\sqrt{2}, \\ 2 - \sqrt{2}\,x & \text{for } 1/\sqrt{2} \leq x \leq 1. \end{cases}$$
  a. Find a Markov partition for the function on the interval $[0,1]$. Hint: Take enough iterates of the points 0, $1/\sqrt{2}$, and 1 until they start repeating.
  b. Given the transition graph for the Markov partition.
4. Show that the function
$$f(x) = \begin{cases} \frac{3}{5} + \frac{7}{5}x, & \text{for } 0 \leq x \leq \frac{2}{7}, \\ \frac{7}{5}(1 - x) & \text{for } \frac{2}{7} \leq x \leq 1 \end{cases}$$
does not have a Markov partition.
5. Consider the transition matrix
$$\mathbf{T} = \begin{pmatrix} 1 & 1 & 0 \\ 0 & 1 & 1 \\ 1 & 0 & 1 \end{pmatrix}.$$
How many periodic points does $\sigma_{\mathbf{T}}$ have for each period, considering the periods up to and including 4?
6. Let
$$\mathbf{T} = \begin{pmatrix} 0 & 1 \\ 1 & 1 \end{pmatrix}$$
and
$$\mathbf{T}^k = \begin{pmatrix} a_k & b_k \\ c_k & d_k \end{pmatrix}.$$
  a. For a vector (x_0, y_0), let $(x_k, y_k) = (x_0, y_0)\mathbf{T}^k$. Taking $y_{-1} = x_0$, show that $x_{k+1} = y_k$ and $y_{k+1} = y_k + y_{k-1}$. (Thus, the y_k satisfy the Fibonacci recurrence relation.)

b. Use the fact that $(1,0)\mathbf{T}^k = (a_k, b_k)$ and $(0,1)\mathbf{T}^k = (c_k, d_k)$ to show that $a_k = a_{k-1} + a_{k-2}$ and $d_k = d_{k-1} + d_{k-2}$ (i.e., they also satisfy the Fibonacci recurrence relation).

c. Show that $\operatorname{tr}(\mathbf{T}^k) = \operatorname{tr}(\mathbf{T}^{k-2}) + \operatorname{tr}(\mathbf{T}^{k-1})$.

7. Using symbolic dynamics, show that the shed map of Example 10.34 has sensitive dependence on initial conditions. Do not quote the theorem, but give an explicit argument.

10.7. Applications

10.7.1. Newton Map with Nonconvergent Orbits.

In this subsection, we return to considering the Newton map N_f for finding roots of polynomials. We show by the methods of Section 10.6, that this map can have many nonperiodic orbits that do not converge to the fixed points of N_f (i.e., do not converge to a zero of f). The invariant set given by the subshift of finite type has zero Lebesgue measure ("length"), so still most orbits converge to one of the fixed points. These Newton maps are not continuous at all points, since they have vertical asymptotes. However, we are still able to use symbolic dynamics to show the existence of points that do not converge to fixed points or periodic orbits.

Theorem 10.26. *Let f be a polynomial of degree $d \geq 4$ with all real roots that are distinct. Then, the associated Newton map $N_f(x)$ has many orbits that do not converge to fixed points and are not eventually periodic. In fact, the orbits can be specified by symbolic dynamics of subshift of finite type on $2d-4$ symbols.*

Proof. We have assumed that the roots are all real and distinct, so
$$f(x) = A(x - x_0)(x - x_1) \cdots (x - x_{d-1}),$$
with $x_0 < x_1 < \cdots < x_{d-1}$. For simplicity, assume that the coefficient of x^d, $A = 1$. By Rolle's theorem, between successive zeroes of $f(x)$, there is a zero of $f'(x)$: so there are
$$x_j < a_j < x_{j+1} \quad \text{with} \quad f'(a_j) = 0 \quad \text{for} \quad 0 \leq j \leq d-2.$$
Because $f'(x)$ has degree $d-1$, the a_j are exactly the zeros of $f'(x)$. Similarly, by Rolle's theorem applied to $f'(x)$, between successive zeroes of $f'(x)$ there is a zero of $f''(x)$: so there are
$$a_{j-1} < b_j < a_j \quad \text{with} \quad f''(b_j) = 0 \quad \text{for} \quad 1 \leq j \leq d-2.$$
Again, the zeros of $f''(x)$ are exactly b_j for $1 \leq j \leq d-2$.

Let $N(x) = N_f(x)$ be the Newton map associated with f. The map $N(x)$ and $N'(x)$ are
$$N(x) = x - \frac{f(x)}{f'(x)} \quad \text{and}$$
$$N'(x) = 1 - \frac{f'(x)^2 - f(x)f''(x)}{f'(x)^2} = \frac{f(x)f''(x)}{f'(x)^2}.$$

The Newton map has fixed points at the x_j and vertical asymptotes at the values a_j for $0 \leq j \leq d-2$. Since only $f'(x)$ changes sign as x crosses one of the a_j, the map N goes to ∞ on one side of a_j and to $-\infty$ on the other side. Since we are assuming the leading coefficient $A = 1$, $N(x)$ goes to ∞ for on the right side of a_{d-2}. Thus, $N(x)$ goes to $-\infty$ on the left side of a_{d-2}. This pattern has to repeat at each of the a_j.

The tangent line for $N(x)$ is horizontal at the points x_j for $0 \leq j \leq d-1$, and at b_j, for $1 \leq j \leq d-2$. In particular, the fixed points of N are superattracting, as we noted in the preceding chapter. Thus, there are two critical points x_j and b_j between the vertical asymptotes at a_{j-1} and a_j for $1 \leq j \leq d-2$. To make the argument simpler, we assume that $f(x)$ and $f''(x)$ do not have any common roots, so $x_j \neq b_j$. The order of these critical points x_j and b_j can vary. In Figure 22, we sketch the graph of $N(x)$ a polynomial of degree four for which $b_1 < x_1$ and $x_2 < b_2$.

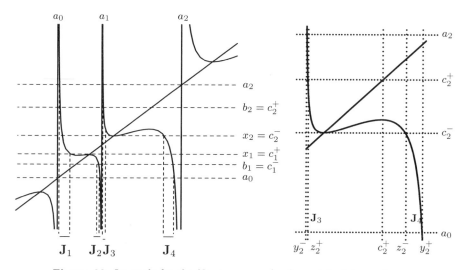

Figure 22. Intervals for the Newton map of polynomial of degree four. **a.** Plot from below a_0 to above a_2. **b.** Enlargement of plot between a_1 and a_2.

For each of the intervals $[a_j, a_{j+1}]$, we define two subintervals to use for the symbolic dynamics. We start by letting $y_j^\pm$, for $1 \leq j \leq d-2$, be chosen so that

$$N(y_j^+) = a_0,$$
$$N(y_j^-) = a_{d-2}, \quad \text{and}$$
$$a_{j-1} < y_j^- < y_j^+ < a_j.$$

If we were to use symbolic dynamics based on these $d-2$ intervals $[y_j^-, y_j^+]$ for $1 \leq j \leq d-2$, the result is the full shift on $d-2$ symbols. Many of the orbits are not fixed. Since these intervals contain the superattracting fixed points x_j and we want orbits which stay away from the fixed points, we remove open intervals about x_j which lie in their basin of attractions. By this process, we get $2d-4$ closed intervals without any fixed points for N. The resulting $2d-4$ intervals induce a subshift of finite type, which is not a permutation for $k \geq 4$.

We are assuming that $f(x)$ and $f''(x)$ do not have any common roots, so $x_j \neq b_j$. Let
$$c_j^- = \min\{x_j, b_j\} \quad \text{and}$$
$$c_j^+ = \max\{x_j, b_j\},$$
so $c_j^- < c_j^+$. Let z_j^+ and z_j^- be chosen such that
$$N(z_j^+) = c_j^+,$$
$$N(z_j^-) = c_j^-, \quad \text{and}$$
$$y_j^- < z_j^+ < c_j^- \leq x_j \leq c_j^+ < z_j^- < y_j^+.$$
See Figure 22. For $1 \leq j \leq d-2$, let
$$\mathbf{J}_{2j-1} = [y_j^-, z_j^+] \quad \text{and}$$
$$\mathbf{J}_{2j} = [z_j^-, y_j^+],$$
so x_j in not in $\mathbf{J}_{2j-1}$ nor in $\mathbf{J}_{2j}$. By the choices,
$$N(\mathbf{J}_{2j-1}) = [c_j^+, a_{d-2}] \supset [z_j^-, a_{d-2}] \supset \mathbf{J}_{2j} \cup \cdots \cup \mathbf{J}_{4d-4} \quad \text{and}$$
$$N(\mathbf{J}_{2j}) = [a_0, c_j^-] \supset [a_0, z_j^+] \supset \mathbf{J}_1 \cup \cdots \cup \mathbf{J}_{2j-1}.$$
For $d = 5$, the transition matrix is
$$\begin{pmatrix} 0 & 1 & 1 & 1 & 1 & 1 \\ 1 & 0 & 0 & 0 & 0 & 0 \\ 0 & 0 & 0 & 1 & 1 & 1 \\ 1 & 1 & 1 & 0 & 0 & 0 \\ 0 & 0 & 0 & 0 & 0 & 1 \\ 1 & 1 & 1 & 1 & 1 & 0 \end{pmatrix}.$$
This irreducible transition matrix defines a transitive subshift of finite type. By the construction, the intervals $\mathbf{J}_j$ do not contain any of the fixed points of N. Using these $2d-4$ intervals, we get symbolic dynamics of orbits whose itineraries pass through these intervals. Each point that corresponds to one of these sequences of intervals has an orbit that does not converge to a zero or $f(x)$. In fact, for the case considered, in which the zeros are distinct, it can be shown that each symbol sequence corresponds to exactly one point, but we do not give those details. For $d = 3$, all we get is one period-2 orbit, so we assume $d \geq 4$. $\square$

Remark 10.47. For more details about the complicated dynamics of Newton maps, see [**Saa84**] or [**Hur84**].

10.7.2. Complicated Dynamics for Population Growth Models. In Section 9.7.2, we discussed the fixed points for several population growth models. In this section, we indicate that these models can exhibit more complicated dynamics. The reader could also see the book by R. May [**May75**] or the more recent book [**Bra01**] by Brauer and Castillo–Chávez.

The population growth model given by Ricker is
$$R_{a,b}(x) = a\,x\,e^{-bx}$$

for both a and b positive. The most interesting range of parameters is for $a > 1$. For $a \approx 22.2$ and $b = 1$, the map exhibits a period-3 orbit for $x_0 \approx 0.05$. See Figure 23. In fact, by Proposition 10.4(b), all we need is that $R_{22.2,1}^3(0.05) \leq 0.05$ to imply that $R_{22.2,1}(x)$ must have periodic points of all periods. The dynamics is not as simple as that discussed in Section 9.7.2; in particular, the population does not always tend to a constant level. All of these periodic points could be unstable and the total set could be a Cantor set, so it does not follow that a randomly chosen point exhibits complicated dynamics. However, Figure 23 is the plot of a single orbit starting at $x_0 = 0.1$, and it seems to fill much of the interval from 0 up to 8.1, indicating that there is an invariant set with complicated dynamics. This occurs even though the absolute value of the derivative is not always greater than one: in this sense, this example is more like the logistic map G than the piecewise expanding maps. The next chapter, where we discuss chaotic attractors, describes much more carefully the types of complicated dynamics necessary to call an invariant set chaotic.

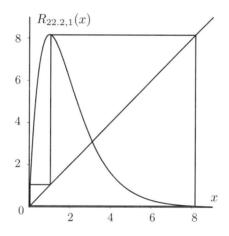

Figure 23. Period-3 orbit for $a \approx 22.2$, $b = 1$, and $x_0 \approx 0.05$.

Exercises 10.7

1. Let $p(x) = (x - x_0)(x - x_1)(x - x_2)(x - x_3)(x - x_4)$ be a polynomial of degree five, with all real roots that are distinct and with
$$x_0 < x_1 < x_2 < x_3 < x_5.$$
Describe the symbolic dynamics that results from picking intervals in a manner analogous to those used in the proof of Theorem 10.26 for four real roots.

2. Consider the polynomial $p(x) = x^2 + 1$, which has no real zeros.
    a. Determine the Newton map $N_p(x)$ and plot its graph.
    b. Let $D(x) = 2x \pmod 1$ be the doubling map and $C(x) = \cot(\pi x)$ be the cotangent function, which takes the interval $(0, 1)$ onto all of $\mathbb{R}$. Show that $C(D(x)) = N_p(C(x))$, so C is a conjugacy of D and N_p. Hint: Use the

10.8. Theory and Proofs

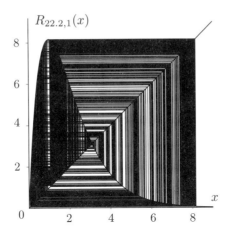

Figure 24. Stair step plot of Ricker model starting at $x_0 = 0.1$

fact that $\cot(\pi x + \pi) = \cot(\pi x)$ and

$$\cot(2\pi x) = \frac{\cos^2(\pi x) - \sin^2(\pi x)}{2\sin(\pi x)\cos(\pi x)}$$
$$= \frac{1}{2}\left(\cot(\pi x) - \frac{1}{\cot(\pi x)}\right).$$

c. Conclude that $N_p(x)$ has an orbit that is dense in the whole real line.

10.8. Theory and Proofs

Transition graphs

We prove a few results that give part of the proof of Theorem 10.5. K. Burns and B. Hasselblatt [**Bur11**] have published an article proving the Sharkovskii theorem that is accessible to anyone reading this book. Their proof does not use Stefan cycles, but ideas related to the proof of Proposition 10.28. Other recent proofs have been given by P. D. Straffin [**Str78**] and B.-S. Du [**Du,02**].

Proposition 10.27. *Let f be a continuous function on $\mathbb{R}$.*

a. *If f has a point of any period, then f has a fixed point.*

b. *If f has a periodic point that is not a fixed point, then f has a period-2 point.*

c. *If f has a point of period 2^m, then f has points of all periods 2^k, for $0 \le k \le m$.*

Proof. The proof is based on the paper of Barton and Burns, [**Bar00b**].

(a) Assume that p is a point with period greater than one, and let $x_1 < x_2 < \cdots < x_p$ be the points on the orbit $\mathcal{O}^+(p)$, labeled by their order on the line and not the order of the iterates. Let j be the largest index such that $f(x_j) > x_j$. Then, $f(x_{j+1}) \le x_j$, and $f([x_j, x_{j+1}]) \supset [x_j, x_{j+1}]$. By Lemma 10.1(a), there is a fixed point in $[x_j, x_{j+1}]$.

(b) Assume p is the smallest integer with $p > 1$ such that f has a period-p point. If $p = 2$ we are done, so we suppose that $p \geq 3$ and get a contradiction. Let $x_1 < x_2 < \cdots < x_p$ be the points on such a periodic orbit. Let $\mathbf{I}_i = [x_i, x_{i+1}]$, for $1 \leq i \leq p-1$, be the intervals between the adjacent points on the orbit. We consider the transition graph for these symbols. Since either $f(x_i) \neq x_{i+1}$ or $f(x_{i+1}) \neq x_i$, $f([x_i, x_{i+1}])$ must contain at least one of the other intervals $\mathbf{I}_k$ with $k \neq i$ (the two points x_i and x_{i+1} cannot just be interchanged). Therefore, for each vertex i in the transition graph, there has to be at least one directed edge going from i to some k with $k \neq i$.

Consider a path in the transition graph starting at $s_0 = 1$. It is always possible to continue this path, always picking the next vertex different than the present one. We get a sequence $s_0 s_1 \ldots$ with each $s_j \neq s_{j-1}$. Since there are only $p-1$ vertices or symbols, the path must repeat a vertex by the times there are p symbols, $s_k = s_\ell$ with $0 \leq k < k+2 \leq \ell \leq p-1$. The sequence $(s_k \ldots s_{\ell-1})^\infty$ is periodic with period q that divides $\ell - k$, so $2 \leq q \leq \ell \leq p-1$. This shows that there is a periodic path in By Theorem 10.3, there must be a point x^* with $f^q(x^*) = x^*$ and $2 \leq q \leq p-1$. The end points have period p, so x^* cannot be an end point and so must have period q, which is less than p and greater than 1. This contradicts the assumption that $p \geq 3$ was the minimum such period. Thus, there is a point of period 2.

(c) If $m = 1$ we are done. Assume that $m > 1$ and that x_m is the point of period 2^m. By part (b), f has a point of period 2. (This also follows from the following argument for $q = 1$.) Let $1 \leq q \leq m-1$. Take $g = f^{2^{q-1}}$. The point x_m has period $2^m/2^{q-1} = 2^{m-q+1} \geq 2^2$ for the map g. By part (b), there is a point x_q of period 2 for g, $g^2(x_q) = x_q$ but $g(x_q) \neq x_q$. Thus, $x_q = g^2(x_q) = f^{2^{q-1}+2^{q-1}}(x_q) = f^{2^q}(x_q)$ has a period for f which divides 2^q (i.e., period $1, 2, \ldots 2^q$). Since $f^{2^{q-1}}(x_q) = g(x_q) \neq x_q$, the period cannot be $1, 2, \ldots$, or 2^{q-1}, and so must be 2^q. We have shown that f must have points with periods 2^q for $1 \leq q \leq m-1$, as desired. By part (a), there is a fixed point, completing the proof. □

Proposition 10.28. *Assume that f is a continuous function on $\mathbb{R}$ and has a period-m point, where $m \geq 3$ is odd. Then, for every integer $n > m$, f also has a period-n point.*

Proof. The proof is based on the paper by Du, [**Du,02**].

Assume that
$$x_1 < x_2 < \cdots < x_m$$
is the period-m orbit in which the points are indexed by the order on the line and not the order of the iterates of the point. Let s be the largest index for which $f(x_s) > x_s$, and let $\mathbf{J} = [x_s, x_{s+1}]$. Since $f(x_{s+1}) < x_{s+1}$, $f(\mathbf{J}) \supset \mathbf{J}$. Let z be the fixed point in $\mathbf{J}$, which must exist by Lemma 10.1.

Claim 10.29. *There must exist a $t \neq s$ such that $f(x_t)$ and $f(x_{t+1})$ are on opposite sides of the interval $\mathbf{J}$.*

Proof. Assume the claim is false. Then, for $1 \leq i \leq s$, all the $f(x_i)$ must be on the same of $\mathbf{J}$. Since $f(x_s) \geq x_{s+1}$, it follows that $f(x_i)$ must be on the same side of

10.8. Theory and Proofs

J, $f(x_i) \geq x_{s+1}$ for all $i \leq s$. Similarly, $f(x_i) \leq x_s$ for all $i \geq s+1$. Thus, f takes $\{x_1, \ldots, x_s\}$ into $\{x_{s+1}, \ldots, x_m\}$ and $\{x_{s+1}, \ldots, x_m\}$ into $\{x_1, \ldots, x_s\}$. Since f is one to one on the orbit, f must interchange these two sets of points, the number of points on the orbit is the same on each side of **J**, and the total number of points m must be even. This contradicts the fact that the period m is odd and proves the claim. □

By the claim, there is some integer t with $t \neq s$ such that $f(x_t)$ and $f(x_{t+1})$ lie on opposite sides of **J**. Let $\mathbf{L} = [x_t, x_{t+1}]$. We assume that $x_t < x_s$, and leave it to the reader to check the case for $x_t > x_s$. Let q be the smallest integer such that $f^q(x_{s+1}) \leq x_t$. Because there are only m points on the orbit and $f^m(x_{s+1}) = x_{s+1}$, q must be less than or equal $m - 1$.

For any two distinct real numbers a and b, let $[a; b]$ denote the closed interval with end points a and b (i.e., we allow either $a < b$ or $a > b$). Let $\mathbf{K} = [z, x_{s+1}]$, where z is the fixed point in the interval **J**. By the choice of x_t, $f(\mathbf{L}) \supset \mathbf{J} \supset \mathbf{K}$. We use the following sequence of intervals, $\mathbf{I}_0, \ldots, \mathbf{I}_n$, for which $f(\mathbf{I}_i) \supset \mathbf{I}_{i+1}$:

$\mathbf{I}_0 = \mathbf{K} = [z, x_{s+1}]$, $\mathbf{I}_i = [z; f^i(x_{s+1})]$ for $1 \leq i \leq q - 1$,

$\mathbf{I}_q = \mathbf{L} = [x_t, x_{t+1}]$, $\mathbf{I}_i = \mathbf{J} = [x_s, x_{s+1}]$ for $q + 1 \leq i \leq n - 1$,

$\mathbf{I}_n = \mathbf{I}_0 = \mathbf{K}$.

By Theorem 10.3(a,b), there is a point $y \in \mathbf{I}_0 = \mathbf{K}$ such that $f^i(y)$ is in $\mathbf{I}_i$ for $0 \leq i \leq n-1$ and $f^n(y) = y$.

By the next claim, the orbit of y cannot intersect the original period-m orbit, so it cannot intersect the end points of $\mathbf{I}_q = \mathbf{L} = [x_t, x_{t+1}]$. Since $\operatorname{int}(\mathbf{I}_q) = \operatorname{int}(\mathbf{L})$ is disjoint from $\mathbf{I}_0 = \mathbf{K}$, $\mathbf{I}_i = [z; f^i(x_{s+1})]$ for $1 \leq i \leq q-1$, and $\mathbf{I}_i$ for $q+1 \leq i \leq n-1$, Theorem 10.3(c) under assumption (ii) implies that y has period-n. Thus, this final claim completes the proof of the proposition. □

Claim 10.30. *The orbit of y cannot intersect the original period-m orbit.*

Proof. If the orbit of y intersects the original period-m orbit, it has least period m and m divides n. Since $m \geq 3$ and $n > m$, $n - m \geq 3$.

The orbit of y is an element of **J** for $(n-1) - q$ times in a row. Since $q + 1 \leq m$, $n - (q+1) \geq n - m \geq 3$. But, the original x-orbit intersects **J** only twice at the end points. This shows that it is impossible for y to be on the original x-orbit and proves the claim. □

Properties of Cantor Sets

In order to state the properties that characterize a Cantor set, we need to introduce some additional terminology from analysis. This terminology is given in Appendix A.2 and some of it in Section 11.2.

For our purposes, the *boundary* of a set $\mathbf{S} \subset \mathbb{R}$, denoted by $\operatorname{bd}(\mathbf{S})$, is the set of all points p such that $(p - r, p + r)$ contains points in both $\mathbf{S}$ and $\mathbf{S}^c$ for all $r > 0$.

A set $\mathbf{S} \subset \mathbb{R}$ is *closed* provided that it contains all its boundary points, $\operatorname{bd}(\mathbf{S}) \subset \mathbf{S}$.

A closed set **S** is *nowhere dense* provided that, for any point p in **S**, there are points in $\mathbf{S}^c$ arbitrarily close to p; that is, for any point p in **S** and any $r > 0$, the interval $(p - r, p + r)$ contains points that are not in **S**).

A set **S** is called *perfect* provided that, for every point p in **S** and every $r > 0$, the open interval $(p - \delta, p + \delta)$ about p contains points of **S** distinct from p.

Any closed, nowhere dense, perfect set in $\mathbb{R}$ is called a *Cantor set*.

Theorem 10.31. *Let* **S** *be either the middle-third Cantor set* K *for* T_3 *or* Λ_g *for* $g(x) = ax(1 - x)$, *with* $a \geq 2 + \sqrt{5}$.

Then, **S** *has the following properties:*

a. *The set* **S** *is closed.*
b. *The set* **S** *is perfect.*
c. *The set* **S** *has empty interior, so it is nowhere dense. Also, the boundary of* **S** *is* **S**.

Proof. (a) The set **S** is the intersection of closed sets, so it is closed.

(b) Any interval of the form $(p - \delta, p + \delta)$ about a point in **S** contains an end point that is distinct from p. (There is always an end point of one of the intervals $\mathbf{I}_{s_0 \ldots s_n}$ that is distinct from p.) Therefore, **S** is perfect.

(c) The intervals at the nth stage are arbitrarily short, so $(p - \delta, p + \delta)$ contains points that are not in **S**. Thus, **S** is nowhere dense. □

Theory of symbolic dynamics

The metric d on Σ_N^+ is defined in Section 10.3.

Restatement of Proposition 10.7. *The metric d on Σ_N^+ has the following properties:*

i. $d(\mathbf{s}, \mathbf{t}) \geq 0$.
ii. $d(\mathbf{s}, \mathbf{t}) = 0$ *if and only if* $\mathbf{s} = \mathbf{t}$.
iii. $d(\mathbf{s}, \mathbf{t}) = d(\mathbf{t}, \mathbf{s})$.
iv. $d(\mathbf{r}, \mathbf{s}) + d(\mathbf{s}, \mathbf{t}) \geq d(\mathbf{r}, \mathbf{t})$.

Proof. (i) Each term in the sum defining d is nonnegative, so $d(\mathbf{s}, \mathbf{t}) \geq 0$.

(ii) Since $\delta(s_j, s_j) = 0$ for all j, $d(\mathbf{s}, \mathbf{s}) = 0$. Also, if $\mathbf{t} \neq \mathbf{s}$, then $s_k \neq t_k$ for some k and $d(\mathbf{s}, \mathbf{t}) \geq 1/3^k > 0$.

(iii) Since $\delta(s_j, t_j) = \delta(t_j, s_j)$ for all j, $d(\mathbf{s}, \mathbf{t}) = d(\mathbf{t}, \mathbf{s})$.

(iv) Since
$$\delta(r_j, s_j) + \delta(s_j, t_j) \geq \delta(r_j, t_j)$$
for all j,
$$d(\mathbf{s}, \mathbf{s}) + d(\mathbf{s}, \mathbf{t}) \geq d(\mathbf{r}, \mathbf{t}).$$
□

Restatement of Proposition 10.8. *Let Σ_N^+ be the shift space on N symbols, with distance d as previously defined. Fix a* $\mathbf{t}$ *in* Σ_N^+. *Then,*

$$\{ \mathbf{s} \in \Sigma_N^+ : s_j = t_j \text{ for } 0 \leq j \leq k \} = \{ \mathbf{s} \in \Sigma_N^+ : d(\mathbf{s}, \mathbf{t}) \leq 3^{-k} 2^{-1} \}.$$

10.8. Theory and Proofs

Proof. Take an element $\mathbf{s}$ in the set on the left-hand side of the equality of the statement. Then, $s_j = t_j$ for $0 \leq j \leq k$, and

$$d(\mathbf{s}, \mathbf{t}) = \sum_{j=0}^{\infty} \frac{\delta(s_j, t_j)}{3^j} \leq \sum_{j=k+1}^{\infty} \frac{1}{3^j} = 3^{-(k+1)} \sum_{j=0}^{\infty} \frac{1}{3^j}$$

$$= 3^{-(k+1)} \left(\frac{1}{1 - 3^{-1}} \right) = \frac{1}{3^{k+1} \left(\frac{2}{3} \right)} = \frac{1}{2 \cdot 3^k}.$$

This shows that $\mathbf{s}$ is an element of the set on the right-hand side of the equality of the statement. This, in turn, shows that the set on the left-hand side is a subset of the set on the right-hand side.

Now assume that $\mathbf{s}$ is not in the set on the left-hand side, so $s_i \neq t_i$ for some $0 \leq i \leq k$. Then,

$$d(\mathbf{s}, \mathbf{t}) = \sum_{j=0}^{\infty} \frac{\delta(s_j, t_j)}{3^j} \geq \frac{\delta(s_i, t_i)}{3^i}$$

$$= \frac{1}{3^i} > \frac{1}{2 \cdot 3^k}.$$

This shows that $\mathbf{s}$ is not an element of the set on the right-hand side of the equality of the statement. This, in turn, shows that the set on the left-hand side is not smaller than the one on the right-hand side, and they must be equal. $\square$

Restatement of Proposition 10.9. *The shift map σ is continuous on Σ_N^+.*

Proof. Let $\epsilon > 0$. Pick k such that $1/(2 \cdot 3^k) < \epsilon$, and let $\delta = 1/(2 \cdot 3^{k+1})$. If $d(\mathbf{s}, \mathbf{t}) < \delta$, then $s_j = t_j$ for $0 \leq j \leq k+1$. It follows that the $\sigma(\mathbf{s})$ and $\sigma(\mathbf{t})$ agree for $0 \leq j \leq k$, and so

$$d(\sigma(\mathbf{s}), \sigma(\mathbf{t})) \leq 1/(2 \cdot 3^k) < \epsilon.$$

This proves the continuity of σ. $\square$

Subshift for a logistic map

In Section 10.6, we discussed the manner in which subshifts of finite type occur for maps that are expanding everywhere. In this section, we sketch a situation in which one can arise for the logistic family. After removing part of the basin of attraction of a periodic sink, a power of the map is expanding on the remaining intervals. These features are much more like those a for typical nonlinear map which has a subsystem that is conjugate to a subshift of finite type. Another difference from the material on piecewise expanding maps is that the invariant set for this example is a Cantor set rather than a whole interval. The treatment given here is based on that given by Devaney in [**Dev89**].

As mentioned in Example 10.6, the logistic map has two period-3 orbits for $a = 3.839$:

$$\{x_0, x_1, x_2\} \approx \{0.1499, 0.4892, 0.9593\} \quad \text{and}$$
$$\{y_0, y_1, y_2\} \approx \{0.1690, 0.5392, 0.9538\}.$$

The first of these orbits is attracting and the second is repelling. Let $g(x) = g_{3.839}(x)$ be this map. There is a point $\hat{y}_i$ on the opposite side of x_i that is mapped by g^3 to $y_i = g^3(y_i)$, $g^3(\hat{y}_i) = y_i$. The order of these points is

$$\hat{y}_0 < x_0 < y_0 < \hat{y}_1 < x_1 < y_1 < y_2 < x_2 < \hat{y}_2.$$

Let $\mathbf{A}_0 = (\hat{y}_0, y_0)$, $\mathbf{A}_1 = (\hat{y}_1, y_1)$, and $\mathbf{A}_2 = (\hat{y}_2, y_2)$ be the intervals about the attracting periodic orbit. The map g takes $\mathbf{A}_0$ monotonically to $\mathbf{A}_1$, and takes $\mathbf{A}_2$ monotonically onto $\mathbf{A}_0$. The critical point 0.5 is in $\mathbf{A}_1$, but it is still mapped into $\mathbf{A}_2$. It follows by arguments like the ones in the proof of Proposition 9.7 that $\mathbf{A}_0 \subset \mathscr{B}(x_0; g^3)$, $\mathbf{A}_1 \subset \mathscr{B}(x_1; g^3)$, and $\mathbf{A}_2 \subset \mathscr{B}(x_2; g^3)$. Therefore, $\mathbf{A}_0 \cup \mathbf{A}_1 \cup \mathbf{A}_2 \subset \mathscr{B}(\mathcal{O}_g(x_0); g)$.

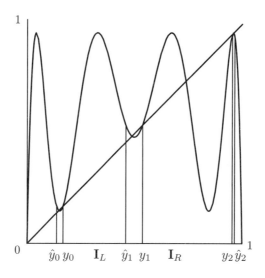

Figure 25. Plot of third iterate g^3 and the intervals $\mathbf{I}_L$ and $\mathbf{I}_L$

We can use the intervals

$$\mathbf{I}_L = [y_0, \hat{y}_1] \quad \text{and} \quad \mathbf{I}_R = [y_1, y_2]$$

for symbolic dynamics associated with g. See Figure 25. The set

$$\Lambda_g = \bigcap_{n \geq 0} g^{-n}(\mathbf{I}_L \cup \mathbf{I}_R)$$

contains most of the periodic orbits. The origin 0 is a repelling fixed point. The only periodic orbit in $\mathbf{A}_0 \cup \mathbf{A}_1 \cup \mathbf{A}_2$ is $\mathcal{O}_g^+(x_0)$, since this union is contained in the basin of attraction of this periodic sink. Points in $(0, \hat{y}_0)$ eventually are mapped into $\mathbf{A}_0 \cup \mathbf{I}_L$, and so can never return nor be periodic. Points in $(y_2, 1)$ get mapped to $(0, \hat{y}_0)$ and so eventually are mapped into $\mathbf{A}_0 \cup \mathbf{I}_L$; therefore, they can never return nor be periodic. It follows that all the periodic orbits for g have orbits entirely inside $\mathbf{I}_L \cup \mathbf{I}_R$, with the exception of the fixed point at 0 and the period-3 orbit $\mathcal{O}_g^+(x_0)$; that is, all periodic orbits are contained in $\Lambda_g \cup \mathcal{O}_g^+(x_0) \cup \{0\}$.

The image of the interval $\mathbf{I}_L$ equals $\mathbf{I}_R$ and the image of $\mathbf{I}_R$ covers both $\mathbf{I}_L$ and $\mathbf{I}_R$ (together with $\mathbf{A}_1$). See Figure 26. Therefore, for the symbolic dynamics, L can

10.8. Theory and Proofs

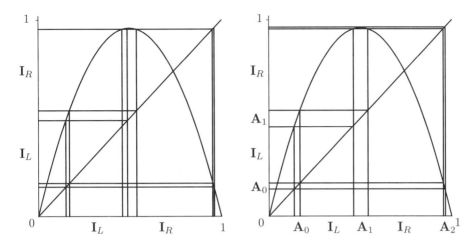

Figure 26. (a) The graph of g, the orbits $\{x_0, x_1, x_2\}$ and $\{y_0, y_1, y_2\}$, a vertical line at $\hat{y}_1$, and the intervals $\mathbf{I}_L$ and $\mathbf{I}_L$; (b) vertical lines at $\hat{y}_0$, y_0, $\hat{y}_1$, y_1, y_2 and $\hat{y}_2$, and the intervals $\mathbf{A}_0$, $\mathbf{A}_1$, and $\mathbf{A}_2$.

be followed only by R and L can be followed by either L or R (i.e., the transition graph is the same as in Figure 5).

By the usual construction, there is an itinerary map from Λ_g to the shift space allowed by this transition graph. To see that this set is a Cantor set and that the symbolic dynamics determine the iterates of the map, we need to see that the length of any of the intervals $\mathbf{I}_{s_0\ldots s_n}$ goes to zero as n goes to infinity.

The maps g and g^3 are not always expanding on $\mathbf{I}_L \cup \mathbf{I}_R$; indeed, g^3 is a contraction on three intervals $\mathbf{G}_1 \subset \mathbf{I}_L$, $\mathbf{G}_2 \subset \mathbf{I}_R$, and $\mathbf{G}_3 \subset \mathbf{I}_R$. See Figure 27. The interval $\mathbf{G}_3 = g^{-1}(\mathbf{A}_1) \cap \mathbf{I}_R$, so $g(\mathbf{G}_3) = \mathbf{A}_1$, $g^2(\mathbf{G}_3) \subset \mathbf{A}_2$, and $g^3(\mathbf{G}_3) \subset \mathbf{A}_0$.

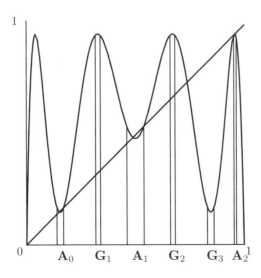

Figure 27. The intervals $\mathbf{G}_1$, $\mathbf{G}_2$, and $\mathbf{G}_3$

It follows that $\mathbf{G}_3 \cap \Lambda_g = \emptyset$. The intervals $\mathbf{G}_1$ and $\mathbf{G}_2$ are defined by $g^{-1}(\mathbf{G}_3) = \mathbf{G}_1 \cup \mathbf{G}_2$, so they are symmetrical with respect to 0.5, and $g(\mathbf{G}_1) = g(\mathbf{G}_2) = \mathbf{G}_3$, $g^2(\mathbf{G}_1) = g^2(\mathbf{G}_2) = g(\mathbf{G}_3) = \mathbf{A}_1$, and $g^3(\mathbf{G}_1) = g^3(\mathbf{G}_2) = g^2(\mathbf{G}_3) \subset \mathbf{A}_2$. Thus, $\mathbf{G}_1 \cap \Lambda_g = \mathbf{G}_2 \cap \Lambda_g = \emptyset$. Defining
$$\mathbf{S} = (\mathbf{I}_L \cup \mathbf{I}_R) \setminus (\mathbf{G}_1 \cup \mathbf{G}_2 \cup \mathbf{G}_3)$$
and combining the preceding results, we have $\Lambda_g \subset \mathbf{S}$. From the construction of the sets, it can be seen that
$$\Lambda_g = \bigcap_{n \geq 0} g^{-n} (\mathbf{I}_L \cup \mathbf{I}_R) = \bigcap_{k \geq 0} g^{-3k} (\mathbf{S}).$$
Thus, to see that a sequence of symbols corresponds to a unique point, it is enough to look at g^3 on $\mathbf{S}$.

We need to see that the derivative of g^3 on $\mathbf{S}$ is greater than one in absolute value. The interval $\mathbf{G}_2 \subset (0.661, 0.683)$, $(g^3)'(0.661) > 1$, and $(g^3)'(0.683) < -1$, so the derivative of g^3 on the boundaries of $\mathbf{G}_2$ and $\mathbf{G}_1$ is greater than one in absolute value. The derivative of g^3 at the end points of $\mathbf{G}_3$ is the same as the derivative at y_0 and $\hat{y}_0$, which are the end points of $\mathbf{A}_0$, and are greater than one, since y_0 is a repelling periodic point. The derivative of g^3 at all other points of $\mathbf{S}$ is larger than these values in absolute value, so there is some $\lambda > 1$ such that $|(g^3)'(x)| > 1$ at all the points x of $\mathbf{S}$. The fact that the absolute value of the derivative of g^3 is greater than one at all points of $\mathbf{S}$ is enough to show that the lengths of that intervals in
$$\bigcap_{k=0}^{n} g^{-3k} (\mathbf{S})$$
go to zero as n goes to infinity. Therefore, the lengths of the intervals in
$$\bigcap_{j=0}^{n} g^{-j} (\mathbf{I}_L \cup \mathbf{I}_R)$$
also go to zero as n goes to infinity. Thus, Λ_g is a Cantor set and the periodic points are dense in Λ_g.

We summarize the results in the next theorem.

Theorem 10.32. *Let $g = g_{3.839}$. Then g has a period-3 sink $\mathcal{O}_g^+(x_0)$ and a repelling fixed point at 0, and all the other periodic points are repelling and are in the invariant set*
$$\Lambda_g = \bigcap_{n \geq 0} g^{-n} (\mathbf{I}_L \cup \mathbf{I}_R).$$
Moreover, the map g restricted to Λ_g is topologically conjugate to the shift map for the subshift of finite type, where L can be followed only by R and L can be followed by either L or R (i.e., for the transition graph given in Figure 5). It follows that g restricted to Λ_g is topologically transitive and has sensitive dependence on initial conditions. The set Λ_g itself is a Cantor set (i.e., closed, perfect, and has empty interior).

Chapter 11

Invariant Sets for One-Dimensional Maps

In this chapter, we consider topics related to what are called *chaotic attractors*. We have already seen maps that are topologically transitive on a set. In the first section, we define the related ω-limit set that is the set of all points that an orbit approaches with iterates going to infinity. In the second section, chaotic attractors are defined in terms of ω-limit sets and sensitive dependence on initial conditions. Symbolic dynamics and other methods from the preceding chapter are used to show an attractor is transitive and has sensitive dependence on initial conditions.

The third section defines the Lyapunov exponent of an orbit, which measures the extent to which nearby orbits are separating. This concept is related to a more computational manner of detecting chaotic attractors.

The fourth section considers not merely what points are the limit of an orbit, but also the frequency with which the orbit returns near various points. This frequency of return is related to an invariant quantity that gives the "size" of sets, called a measure. A map that has an invariant measure that is spread over one or several intervals must have the complicated dynamics of a chaotic attractor.

11.1. Limit Sets

The tent map has a point x^* with a dense orbit in $[0,1]$. On the other hand, we have considered orbits in a basin of attraction that tend to a periodic orbit. In this section, we define the ω-limit set which can be used to describe both of these situations. Similar notions are given in Section 4.1 for systems of differential equations. In the next section, one of the defining properties of a chaotic attractor is given in terms of an ω-limit set.

Definition 11.1. Let f be a map on a space **X**, and let x_0 be an initial condition in **X**. A point q is an *ω-limit point of x_0* provided that there is a sequence of iterates k_j with $k_j < k_{j+1}$ such that $f^{k_j}(x_0)$ converges to q as k_j goes to infinity; that is,

the orbit of x_0 keeps coming back close to q. More precisely in $\mathbb{R}$, for any $N > 0$ and any $\epsilon > 0$, there exists an $n \geq N$ such that
$$|f^n(x_0) - q| < \epsilon.$$
The letter omega is used because it is the last letter in the Greek alphabet, and this limit is taken as the power of the iterate goes to infinity.

The *ω-limit set of x_0* is the set of all ω-limit points of x_0:
$$\omega(x_0; f) = \{\, q : q \text{ is an } \omega\text{-limit point of } x_0 \,\}.$$

Proposition 11.1. **a.** *If x_0 is a period-n point of f, then*
$$\omega(x_0; f) = \mathcal{O}_f^+(x_0) = \{\, f^j(x_0) : 0 \leq j < n \,\}.$$

b. *If x_0 is in the basin of attraction of a period-n point p, then $\omega(x_0; f) = \mathcal{O}_f^+(p)$.*

We leave the proofs to the reader.

Note that, if the orbit of x_0 goes through a point y once on the way toward a periodic sink, then y is not in the ω-limit set.

The next theorem is used to show that there are maps with ω-limit sets that are intervals.

Theorem 11.2. *Assume that $\mathcal{G}$ is an irreducible transition graph and let $\sigma_\mathcal{G}$ be the shift map restricted to the corresponding subshift of finite type $\Sigma_\mathcal{G}^+$. Then there is a sequence $\mathbf{s}^*$ such that $\omega(\mathbf{s}^*; \sigma_\mathcal{G}) = \Sigma_\mathcal{G}^+$.*

Proof. Let $\mathbf{s}^*$ be the sequence given in Theorem 10.21 for a subshift, or given in Theorem 10.10 for a full shift. The proof requires a little more argument about this sequence than the earlier proof. Given an allowable word $\mathbf{w}$, it appears as part of an arbitrary long string $\mathbf{w}\mathbf{w}'$, where $\mathbf{w}'$ is a long string that makes the concatenated string allowable. Therefore, $\mathbf{w}$ appears infinitely often in the symbol sequence $\mathbf{s}^*$. Using the proof of Theorem 10.21, the forward orbit of $\mathbf{s}^*$ repeatedly comes close to the string $\mathbf{w}$, so $\mathbf{w}$ is in the ω-limit set of $\mathbf{s}^*$. □

Theorem 11.3. *The tent map $T = T_2$ and the logistic map $G = g_4$ each have a point x_T^* and x_G^* such that $\omega(x_T^*; T) = [0, 1]$ and $\omega(x_G^*; G) = [0, 1]$.*

Proof. A direct proof shows that the point $x^* = k(\mathbf{s}^*)$ has this property where $\mathbf{s}^*$ is the sequence from Theorem 11.2 and $k : \Sigma_2^+ \to [0, 1]$ is given by Theorems 10.11 and 10.13. □

The tent map has other points which have ω-limit sets that are proper subsets of $[0, 1]$. In the exercises, we ask the reader to show that there is a point that has $\omega(x_0; T)$ equal to a Cantor subset of $[0, 1]$; this is an uncountable set, but not the whole interval.

Example 11.2. This example discusses a map that has a dense orbit in an interval, but has no periodic orbits and does not have sensitive dependence on initial conditions. The map is the rotation on the "circle" through an irrational part of one full rotation.

11.1. Limit Sets

We start by putting a distance on $[0, 1)$ by identifying 0 and 1 to make a circle: the distance from $1 - \delta$ to δ is $d(1 - \delta, \delta) = 2\delta$ for small δ. More specifically, for $0 \leq x \leq y < 1$, we define the distance
$$d(x, y) = \min\{\, y - x,\ 1 + x - y \,\}$$
$$= \min\{\, |y - x - k| : k \in \mathbb{Z} \,\}.$$

For $0 < \beta < 1$, let R_β be the "rotation" through a β fraction of a full turn on the interval $[0, 1]$ (which we are identifying with the circle in vocabulary if not in actuality); that is,
$$R_\beta(x) = x + \beta \quad (\bmod\ 1).$$
We write $R(x)$ for $R_\beta(x)$.

If we use the distance d formed by identifying 0 and 1, then
$$d(R(x), R(y)) = d(x + \beta, y + \beta) = d(x, y),$$
and R preserves the distance between points. Thus, R does not have sensitive dependence on initial conditions.

If x is eventually periodic, then $R^{n+j}(x) = R^j(x)$, so
$$x + (n + j)\beta = x + j\beta \pmod{1},$$
$$x + (n + j)\beta = x + j\beta + k \quad \text{for some integer } k,$$
$$\beta = \frac{k}{n}.$$
Thus, $R_\beta(x)$ has periodic or eventually periodic points if and only if β is rational.

Now assume that β is an irrational number (e.g., $\beta = \sqrt{2}/10$) so there are no periodic points. We show that $\omega(x_0, R_\beta) = [0, 1]$ for any point x_0. Pick any $\epsilon > 0$. The points $R^j(x_0)$ are all distinct, so there must exist n and $n + m$ such that
$$d(R^{n+m}(x_0), R^n(x_0)) < \epsilon.$$
Therefore, there is an integer k such that
$$\epsilon > d(R^{n+m}(x_0), R^n(x_0))$$
$$= |x_0 + (n + m)\beta - (x_0 + n\beta) - k|$$
$$= |x_0 + m\beta - x_0 - k|$$
$$= d(R^m(x_0), x_0).$$
This shows that x_0 comes back within ϵ of itself. Since R^{mj} preserves the distance,
$$d(R^{m(j+1)}(x_0), R^{mj}(x_0)) = d(R^m(x_0), x_0) < \epsilon$$
for any j. Assume that $R^m(x_0) > x_0$ for discussion purposes. Then,
$$x_0 < R^m(x_0) < R^{2m}(x_0) < R^{3m}(x_0) < \cdots$$
are evenly spaced and all within ϵ of the nearest points in the sequence. Eventually, $R^{km}(x_0)$ gets larger than $x_0 + 1$ modulo 1 and, so, returns to the beginning of the interval $[x_0, R^m(x_0)]$. This shows that $\{\, R^j(x_0) : 0 \leq j < km \,\}$ is within a distance ϵ of every point in the whole interval. Because the rotation preserves lengths, any string $\{\, R^j(x_0) : n \leq j < n + km \,\}$ of km successive iterates is within a distance ϵ of every point in the whole interval. For any point x in $[0, 1]$ and for any $\epsilon > 0$, the

forward orbit of x_0 keeps coming back to within ϵ of x, so x is in $\omega(x_0, R_\beta)$. The point x is arbitrary, so $\omega(x_0, R_\beta) = [0, 1]$ for any x_0 whenever β is irrational.

In Section 11.6 at the end of the chapter, we present a theorem that gives some basic properties of any ω-limit set.

Exercises 11.1

1. Show that, if the ω-limit set of a point x_0 contains an attracting periodic point p, then $\omega(x_0; f) = \mathcal{O}_f^+(p)$. Hint: From the treatment of attracting periodic orbits, show that there is a finite union of open intervals $\mathbf{U}$ that contain $\mathcal{O}_f^+(p)$ such that
$$\bigcap_j f^j(\mathbf{U}) = \mathcal{O}_f^+(p).$$
(In terms of Section 11.2, $\mathbf{U}$ is a trapping region for $\mathcal{O}_f^+(p)$.) Then, show why there has to be some iterate $f^k(x_0)$ that is contained in $\mathbf{U}$. Finally, explain why $\omega(x_0; f) = \omega(f^k(x_0); f) = \mathcal{O}_f^+(p)$.

2. Let $T(x)$ be the usual tent map with slope ± 2, let
$$\mathbf{I}_L = \left[\frac{1}{4}, \frac{7}{16}\right] \quad \text{and} \quad \mathbf{I}_R = \left[\frac{9}{16}, \frac{7}{8}\right],$$
and let
$$\Lambda = \bigcap_{j \geq 0} T^{-j}(\mathbf{I}_L \cup \mathbf{I}_R).$$
   a. Explain why there is a Markov partition for Λ. Also, what is the transition graph for this Markov partition?
   b. For an allowable sequence $\mathbf{s}$, why do the lengths of the intervals $\mathbf{I}_{s_0 \ldots s_n}$ go to zero as n goes to infinity?
   c. Explain why there is a (particular) point x^* such that $\omega(x^*; T) = \Lambda$.

11.2. Chaotic Attractors

A rough definition of an attractor $\mathbf{A}$ is that it is an invariant set such that all the points nearby have ω-limit sets in $\mathbf{A}$. However, rather than using ω-limit sets to define an attractor, we use the stronger condition that it has a trapping region, which we define in this section.

Once we have defined an attractor, we call it chaotic provided that it is transitive and has sensitive dependence on initial conditions when restricted to the attractor.

We give an example to motivate the formal definitions.

Example 11.3. The rooftop map, graphed in Figure 1, is defined by
$$f(x) = \begin{cases} \frac{1}{2}x & \text{if } x \leq 0, \\ 2x & \text{if } 0 \leq x \leq 0.5, \\ 2(1-x) & \text{if } 0.5 \leq x \leq 1, \\ \frac{1}{2}(1-x) & \text{if } 1 \leq x. \end{cases}$$

11.2. Chaotic Attractors

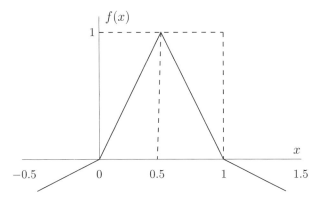

Figure 1. Rooftop map

The set $[0,1]$ is positively invariant, so for x_0 in $[0,1]$, $\omega(x_0; f) \subset [0,1]$. For x_0 not in $[0,1]$, graphical iteration shows that $\omega(x_0; f) = \{0\} \subset [0,1]$. These ω-limit sets are enough to show that $[0,1]$ is what we later defined as a Milnor attractor. However, we want a stronger condition that shows that nearby orbits go directly toward $[0,1]$ and do not make long excursions first. Note that for any closed interval $\mathbf{I}_L = [-L, 1+L]$ with $L > 0$, $f(\mathbf{I}_L) = [-\frac{1}{2}L, 1] \subset (-L, 1+L)$; any such interval $\mathbf{I}_L$ is called a "trapping region," since points that start in $\mathbf{I}_L$ can never get out and the set is mapped strictly inside itself. The intersection for forward iterates of the trapping region equals $[0,1]$, or

$$\bigcap_{j \geq 0} f^j(\mathbf{I}_L) = \bigcap_{j \geq 0} \left[-\frac{1}{2^j} L, 1 \right] = [0, 1].$$

Because the interval collapses onto $[0,1]$, it is called an attracting set. The map f restricted to the interval $[0,1]$ is just the tent map, so it has a point whose ω-limit set is all of $[0,1]$. Therefore, $[0,1]$ cannot be broken into a smaller attracting set, so it is called an attractor. Because the map f restricted to $[0,1]$ has sensitive dependence on initial conditions, $[0,1]$ is called a chaotic attractor.

For the rooftop map, the trapping region is a closed interval $\mathbf{I}_L = [-L, 1+L]$ for which $f(\mathbf{I}_L]) \subset (-L, 1+L)$. The fact that the image of the closed interval is contained in the corresponding open interval implies that the intersection of all forward iterates does not touch the end points of the original interval; it is well inside. In general, the corresponding open interval or intervals is the interior of the set, $(-L, 1+L) = \text{int}([-L, 1+L])$. For a function with discontinuities, the image of a closed bounded interval might not be a closed interval. Therefore, we have to add back in any possibly missing end points. For example, for the doubling map, $D([0.25, 0.5]) = [0.5, 1) \cup \{0\}$. The smallest closed set containing this image, called the closure, is $[0.5, 1] \cup \{0\}$. Although this interval is not mapped into itself, it illustrates the situation we might encounter in one of our examples of a function with jump discontinuities. The general definitions of interior, closure, and boundary are given in Appendix A.2. Be sure you understand what the following definitions mean for a set that is the union of intervals, which can be open intervals, closed intervals, or half open intervals.

Definition 11.4. The *boundary* of a union of intervals **S** is the set of all the end points of the intervals and is denoted by bd(**S**). In general for $\mathbf{S} \subset \mathbb{R}$,

$$\mathrm{bd}(\mathbf{S}) = \{\, p : (p-r, p+r) \text{ intersects both } \mathbf{S} \text{ \& } \mathbf{S}^c \text{ for all } r > 0 \,\}.$$

The set **S** is *closed* provided that it contains all its boundary points.

The *closure* of **S** is the set **S** with the boundary added, and is denoted by $\mathrm{cl}(\mathbf{S}) = \mathbf{S} \cup \mathrm{bd}(\mathbf{S})$. The closure of **S** is the smallest closed set that contains **S**.

The set **S** is *open* provided that it contains no boundary point.

The *interior* of **S** is the set **S** with the boundary removed, and is denoted by $\mathrm{int}(\mathbf{S}) = \mathbf{S} \smallsetminus \mathrm{bd}(\mathbf{S})$. The interior of **S** is the largest open set that is contained in **S**.

Remark 11.5. In $\mathbb{R}$, a set is open if and only if it is the countable union of open intervals. In any space, a set is closed if and only if its complement is open. The middle-third Cantor set is a closed set that is not the union of closed intervals; in fact, the interior of the Cantor set is the empty set.

Definition 11.6. For a map f, a set **V** is a *trapping region*, provided that it is closed, bounded, and the closure of the iterate $f(\mathbf{V})$ is contained in the interior of **V**; that is, provided that

$$\mathrm{cl}(f(\mathbf{V})) \subset \mathrm{int}(\mathbf{V}).$$

This means that the set **V** is mapped well within itself, with no points staying on the boundary. Although the trapping region is closed, we still need to take the closure of the image $f(\mathbf{V})$ because we allow the map to have discontinuities and $f(\mathbf{V})$ is not necessarily closed.

A set **A** is an *attracting set* provided that there exists a trapping region **V** with

$$\mathbf{A} = \bigcap_{j \geq 1} \mathbf{V}_j,$$

where $\mathbf{U}_1 = \mathrm{cl}\,(f(\mathbf{V}))$ and $\mathbf{V}_{j+1} = \mathrm{cl}\,(f(\mathbf{V}_j))$ by induction. We also say that the set **A** has a *trapping region* **V**. Note that

$$\mathbf{V} \supset \mathbf{V}_1 \supset \cdots \supset \mathbf{V}_j \supset \mathbf{V}_{j+1} \supset \cdots$$

and that **A** is a closed nonempty invariant set. Since such an **A** must be invariant and any invariant set contained in **V** must be a subset of **A**, it is the largest invariant set contained in its trapping region **V**.

Definition 11.7. An *attractor* is an attracting set **A** which has no proper subset that is also an attracting set; that is, if $\emptyset \neq \mathbf{B} \subset \mathbf{A}$ is an attracting set, then $\mathbf{B} = \mathbf{A}$.

A *transitive attractor* **A** is an attracting set that is topologically transitive. We usually verify transitivity by showing that there is a point x_0 with $\omega(x_0; f) = \mathbf{A}$. (This last condition automatically shows that there is no proper subset that is an attracting set.)

An invariant set **S** is called *chaotic* provided that f is transitive on **S** and f restricted to **S** has sensitive dependence on initial conditions.

A set **A** is a *chaotic attractor* provided that it is a transitive attractor for which f restricted to **A** has sensitive dependence on initial conditions.

11.2. Chaotic Attractors

Remark 11.8. Since f restricted to a chaotic attractor has sensitive dependence on initial conditions, a chaotic attractor cannot be just a single periodic orbit.

In higher dimensions there is a distinction between sensitive dependence at all points of an attracting set and sensitive dependence when restricted to an attracting set, so we include this as part of our definition. See Examples 13.41 and 7.15.

Remark 11.9. As indicated in the Historical Prologue, T.Y. Li and J. Yorke introduced the word "chaos" in [**Li,75**]. Although their paper contains many precise mathematical properties of the system studied, which ones were necessary for chaos was not precisely stated.

Devaney gave the first precise mathematical definition in [**Dev89**]. His definition concerned the situation where the whole system is chaotic, or the map after it was restricted to an invariant set. The three properties he required were that the system have sensitive dependence on initial conditions, the periodic points be dense in the set, and the map is topologically transitive on the set. We drop the requirement of the density of periodic orbits, because that does not seem essential for the properties desired and is not as observable. See Appendix B for a discussion of the reason why a "generic" system has the periodic orbits dense in any attractor.

There have been several other definitions given for a chaotic attractor. In Section 11.3.1, we mention the one given in the book [**All97**] by Alligood, Sauer, and Yorke, which is based on Lyapunov exponents. Basically they require that points move apart at an exponential rate.

Remark 11.10. The idea of the definition of an attracting set is that all points in a neighborhood of **A** must have their ω-limit set inside **A**. An attractor **A** has no smaller attracting set inside itself. In fact, C. Conley showed that an attractor with no proper attracting subsets has a type of recurrence. Every point in such a set **A** is chain recurrent and there is an ϵ-chain between any two points in **A**. What this means is that, given two points p and q in **A** and any small size of jumps $\epsilon > 0$, there is a sequence of points $x_0 = p$, x_1, ..., $x_n = q$ such that $|f(x_{i-1}) - x_i| < \epsilon$. Thus, there is an orbit with errors from p to q. There is also such an ϵ-chain from p back to p again. See [**Rob99**] for more details.

Example 11.11. For examples of attracting sets and attractors, consider a map f with five fixed points $p_1 < p_2 < p_3 < p_4 < p_5$, where p_1, p_3, and p_5 are attracting and p_2 and p_4 are repelling. Assume that $\delta > 0$ is such that, for $i = 1, 3, 5$,

$$f([p_i - \delta, p_i + \delta]) \subset (p_i - \delta, p_i + \delta) \quad \text{and}$$
$$\bigcap_{j \geq 0} f^j([p_i - \delta, p_i + \delta]) = \{p_i\}.$$

Thus, $\{p_i\}$ for $i = 1, 3, 5$ are attracting sets. Since there can be no proper attracting subsets, these sets are also attractors.

In addition,

$$\bigcap_{j \geq 0} f^j([p_1 - \delta, p_5 + \delta]) = [p_1, p_5],$$

$$\bigcap_{j \geq 0} f^j([p_1 - \delta, p_3 + \delta]) = [p_1, p_3], \quad \text{and}$$

$$\bigcap_{j \geq 0} f^j([p_3 - \delta, p_5 + \delta]) = [p_3, p_5],$$

so $[p_1, p_5]$, $[p_1, p_3]$, and $[p_3, p_5]$ are attracting sets. Since each of them contains two of the attractors, $\{p_1\}$, $\{p_3\}$, and $\{p_5\}$, these intervals are not attractors.

The next theorem interprets earlier results about the tent map and rooftop map in terms of our new terminology.

Theorem 11.4. **a.** *The interval $[0, 1]$ is a chaotic invariant set for the tent map $T = T_2$. The interval $[0, 1]$ does not have a trapping region for the tent map in the whole line $\mathbb{R}$, so it is not an attractor within $\mathbb{R}$.*

b. *The interval $[0, 1]$ is a chaotic attractor for the rooftop map given in Example 11.3.*

J. Milnor introduced the next definition of an attractor, which gives a very different meaning to the term. In order to give this definition, we start with the definition of the basin of attraction of an invariant set.

Definition 11.12. The *basin of attraction* of a closed invariant set $\mathbf{A}$ is the set of points with ω-limit set in $\mathbf{A}$:

$$\mathscr{B}(\mathbf{A}; f) = \{\, x_0 : \omega(x_0; f) \subset \mathbf{A} \,\}.$$

The *Lebesgue measure* of a set $\mathscr{B}$, $\lambda(\mathscr{B})$, is the generalized "length" (for subsets of $\mathbb{R}$); if the basin is a countable union of open intervals, then it is just the sum of the lengths of the individual open intervals. See Section 11.4 for more details.

Definition 11.13. A closed invariant set $\mathbf{A}$ is called a *Milnor attractor* for f provided that the following two conditions hold:

(i) The basin of attraction of $\mathbf{A}$ has positive Lebesgue measure, $\lambda(\mathscr{B}(\mathbf{A}; f)) > 0$.
(ii) There is no smaller closed invariant set $\mathbf{A}' \subset \mathbf{A}$ such that $\mathscr{B}(\mathbf{A}; f) \setminus \mathscr{B}(\mathbf{A}'; f)$ has Lebesgue measure zero.

The origin is a Milnor attractor of the map $f(x) = x + x^2$ since $\mathscr{B}(0; f) = [-1, 0]$. However, the origin is only attracts from one side and repels on the other, so there is no trapping region for 0 and it is not an attractor by our definition.

Remark 11.14. The idea of a Milnor attractor is that it is observable. With a randomly chosen initial condition, there is a positive probability of landing on the basin of attraction of $\mathbf{A}$. His way of saying that the set $\mathbf{A}$ has no proper subset that is an attractor is that there is no smaller closed set that has the same size basin of attraction in terms of measure. This definition makes sense for one-dimensional maps, but leads to calling sets attractors in higher dimensions that do not really attract in the usual sense of the word. Therefore, we do not take Milnor's definition

11.2. Chaotic Attractors

as the principal one for an attractor. When we use his definition, we identify the set as a Milnor attractor.

11.2.1. Expanding Maps with Discontinuities. For a closed bounded interval **J** to be a chaotic attractor for a map f, nearby points in **J** must move apart under iteration. Therefore, the map f must be expanding, at least on the average. However, the interval must be invariant, so there must be either folding or cutting (i.e, the interval must be subdivided into subintervals on which the map is expanding, and the division points must correspond to either a change of direction of the image, or to a discontinuity). Notice that the map in Example 11.3 and the tent map have stretching and folding, while the doubling map has stretching and cutting. In this subsection, we give more examples of chaotic attractors for maps with a discontinuity that have stretching and cutting. We also give explicit conditions that insure these discontinuous maps have a chaotic attractor.

Because we need nearby orbits to separate, we consider expanding maps with derivative $|f'(x)| > 1$. Because we want to allow cutting, we consider maps with discontinuities. In fact, the folding we are considering is a discontinuity of the derivative, so we include them among the discontinuities (e.g., 0.5 is considered a discontinuity of the derivative for the tent map T).

We start by considering maps that have a single discontinuity and that are increasing everywhere, such as the one-dimensional Lorenz map. Section 7.3.3 discusses the manner in which this type of map arises from the Lorenz system of differential equations. Later, we consider other expanding maps with single discontinuities and either sign of the derivative. We end by making some comments on piecewise expanding maps with a finite number of discontinuities.

One-dimensional Lorenz map

In Section 7.3, we discussed how the flow for the geometric model of the Lorenz system of differential equations induces a one-dimensional map of a certain type. Computer simulation of the actual Lorenz system of differential equations yields a similar one-dimensional map. We now discuss why one-dimensional maps of this type are topologically transitive and have sensitive dependence on initial conditions, which is a theorem due to R. Williams. This result is similar to the results for the doubling map given in Sections 10.3 and 10.4 (and also Section 7.3 of Part 1). These maps also have an interval as an attractor, so the maps have a chaotic attractor. Later in this section, we see examples of expanding maps whose attractors are made up of a finite number of subintervals rather than a single interval.

In the ensuing discussion, we often use the right-hand and left-hand limits at a point of discontinuity c. Provided the limits exist, we define

$$f(c-) = \lim_{x<c,\ x\to c} f(x) \quad \text{and}$$
$$f(c+) = \lim_{x>c,\ x\to c} f(x).$$

We often consider the forward iterates of $f(c-)$ and $f(c+)$ to be the forward orbit of the discontinuity c. We also write $f([r,c])$ and $f([c,r])$ for $\mathrm{cl}(f([c,r)))$ and $\mathrm{cl}(f((c,r]))$.

Before stating the theorem, we give a simple example.

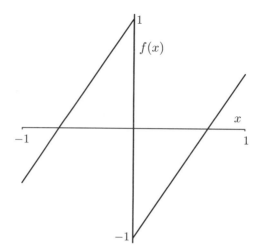

Figure 2. Graph of linear Lorenz one-dimensional map for Example 11.15

Example 11.15. Consider the expanding map in Figure 2 given by

$$f(x) = \begin{cases} \sqrt{3}\,x + 1 & \text{for } x < 0, \\ \sqrt{3}\,x - 1 & \text{for } x > 0. \end{cases}$$

The interval $[f(0+), f(0-)] = [-1, 1]$ is determined by the two images of 0, and is invariant. The two end points are mapped into the interior, $-1 < f(\pm 1) < 1$. In fact $f\left(\pm 2/\sqrt{3}\right) = \pm 1$, so any interval $[-L, L]$ with $1 < L \leq 2/\sqrt{3}$ is a trapping region for the attracting set $[-1, 1]$ since $f([-L, L]) = [-1, 1]$ and $[-1, 1]$ is invariant. It does not have a Markov partition because the forward orbits of $f(0+)$ and $f(0-)$ are not eventually periodic and both contain an infinite number of points. However, the next theorem shows that f has sensitive dependence on initial conditions and is topologically transitive on $[-1, 1]$, and so $[-1, 1]$ is a chaotic attractor. See Figures 3 and 4. In fact the theorem shows that for any small interval $\mathbf{J} \subset [-1, 1]$, there are two successive iterates of $\mathbf{J}$ that contain 0 and so some further iterate covers all of $[-1, 1]$. To see that the hypotheses of the theorem are satisfied, note that $f(0+) = -1$, $f(-1) = 1 - \sqrt{3} < 0$, $f^2(-1) = f(1 - \sqrt{3}) = \sqrt{3} - 2 < 0$, $f(0-) = 1$, $f(1) = \sqrt{3} - 1 > 0$, and $f^2(1) = f(\sqrt{3} - 1) = 2 - \sqrt{3} > 0$.

Theorem 11.5 (Williams). *Assume the following: The map f is a differentiable function from $\mathbb{R} \setminus \{c\}$ to $\mathbb{R}$. The one-sided limits at the single discontinuity at the point c exist and are denoted by $f(c-) = b$ and $f(c+) = a$. The iterates of the points a and b each stay on the same side of the discontinuity c for two iterates; more specifically,*

$$a < f(a) < f^2(a) < c < f^2(b) < f(b) < b.$$

For x is $[a, c) \cup (c, b]$, the derivative satisfies $f'(x) \geq \beta > \sqrt{2}$. In particular, f is increasing on both sides of the discontinuity.

With the above hypotheses, the following three statements follow:

11.2. Chaotic Attractors

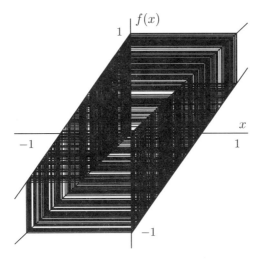

Figure 3. Plot of graphical iteration of a dense orbit for Example 11.15

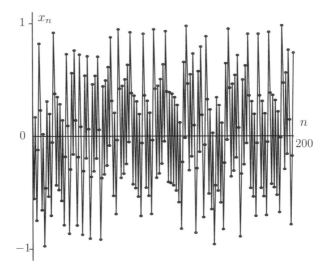

Figure 4. Iteration of a single x_0 for Example 11.15

a. The map f has sensitive dependence on initial conditions on $[a,b]$.

b. For any subinterval $\mathbf{J} \subset [a,b]$ of positive length, there is an iterate n such that $f^n(\mathbf{J}) = (a,b)$.

c. The map f is topologically transitive on $[a,b]$ and so $[a,b]$ is a chaotic attractor.

The proof of the theorem is given in Section 11.6 at the end of this chapter.

Remark 11.16. The same result is true if $f(x)$ is decreasing on both sides of the discontinuity. The changes in the statement of the assumptions are as follows: The one-sided limits at the single discontinuity $f(c-) = a$ and $f(c+) = b$ exist and satisfy $a < c < f(a) < b$, $b < f^2(a) < c$, $a < f(b) < c < b$, and $c < f^2(b) < b$. The derivative of $f(x)$ satisfies $f'(x) \leq -\beta < -\sqrt{2}$ for all x in $[a, c) \cup (c, b]$.

If $f(x)$ is increasing on one side and decreasing on the other side, the result is not true without more assumptions. See Example 11.19.

Expanding maps with one discontinuity

The next theorem states that an expanding map with a single discontinuity has a transitive Milnor attractor, even if it does not satisfy the conditions of Williams' Theorem. It also gives necessary and sufficient conditions for the invariant set to have a trapping region and so be an attractor in our sense.

Theorem 11.6. *Assume that $f : \mathbb{R} \smallsetminus \{c\} \to \mathbb{R}$ is a differentiable function. The point c is the one discontinuity of $f(x)$ and/or $f'(x)$. Assume that the one-sided limits of $f(x)$ exist at c and are denoted by $f(c-) = r_1^-$ and $f(c+) = r_1^+$. Let $r_2^- = f(r_1^-)$, $r_2^+ = f(r_1^+)$, and*

$$a = \min\{r_1^+, r_1^-, r_2^+, r_2^-\} \quad \text{and}$$
$$b = \max\{r_1^+, r_1^-, r_2^+, r_2^-\}.$$

Assume that $a < c < b$, $a \leq f(a) \leq b$, and $a \leq f(b) \leq b$, so $[a, b]$ is positively invariant. Assume that the derivative satisfies $|f'(x)| \geq \beta > 1$ for $x \in [a, c) \cup (c, b]$.

Then the following hold.

a. *There is a topologically transitive invariant subset $\mathbf{A}$ of $[a, b]$ that is the finite union of closed intervals (perhaps the whole interval $[a, b]$). The point c of discontinuity is in the interior of one of the intervals of $\mathbf{A}$. The end points of the intervals of $\mathbf{A}$ are in the forward orbits of the points r_1^- and r_1^+. The forward orbit of an arbitrarily small open interval $(c - \delta, c + \delta)$ about c covers all of $\mathbf{A}$ and determines it.*

b. *The map f has sensitive dependence on initial conditions when restricted to $\mathbf{A}$.*

c. *The basin of attraction of $\mathbf{A}$, $\mathscr{B}(\mathbf{A}; f)$, is dense and open in $[a, b]$, so $\mathbf{A}$ is a Milnor attractor.*

d. *The invariant set $\mathbf{A}$ has a trapping region if and only if there is no periodic orbit that lies entirely on the end points of $\mathbf{A}$. (We allow a periodic point that goes through c and $r_1^\pm$.) If these conditions hold then $\mathbf{A}$ is a chaotic attractor.*

Remark 11.17. T.Y. Li and J. Yorke [**Li,78**] proved results that are essentially parts (a-c) of the previous theorem. Their proof uses an invariant measure, a topic discussed in Section 11.4 later in the chapter, and does not use the terminology given above. Later, Morales and Pujals [**Mor97**] gave a more topological proof of a similar result. Finally, Y. Choi [**Cho04**] gave the condition in (d) that is necessary and sufficient to have a trapping region.

We give two examples of maps whose attractor consists of more than one interval. The first example has positive slope on both sides like the Lorenz map, and the second example has both positive and negative slopes.

11.2. Chaotic Attractors

Example 11.18. Consider the example

$$f(x) = \begin{cases} 1.25\,x + 1 & \text{for } x < 0, \\ 1.25\,x - 1.05 & \text{for } x > 0. \end{cases}$$

Notice that $f'(x) = 1.25 < \sqrt{2}$, so Williams theorem does not apply. The ends of the invariant interval are $a = f(0+) = r_1^+ = -1.05$ and $b = f(0-) = r_1^- = 1$; they have images in the open interval, $f(a) = -0.3125$, $f(b) = 0.2 \in (a,b)$. Since $f(1.6) = 0.95$ and $f(-1.6) = -1$, $\text{cl}(f([-1.6, 1.6])) = \text{cl}((-1.05, 1)) = [-1.05, 1]$ and $[-1.6, 1.6]$ is a trapping region for the attracting set $[-1.05, 1]$.

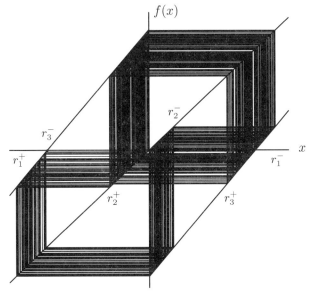

Figure 5. Plot of graphical iteration of a dense orbit in the attractor for Example 11.18

Graphical iteration shown in Figure 5 indicates that the attractor is not all of $[-1.05, 1]$ but is a subset made up of three intervals. We can find this attractor by looking at the iterates of $r_1^\pm$. The following table gives the first four values of $r_j^\pm = f^{j-1}(r_1^\pm)$.

j	r_j^-	r_j^+
1	1.0	-1.050
2	0.2	-0.312
3	-0.8	0.609
4	0.0	-0.288

The order of these points on the line is as follows:

$$a = r_1^+ < r_3^- < r_2^+ < r_4^+ < r_4^- = 0 < r_2^- < r_3^+ < r_1^- = b.$$

Because $f'(x) > 0$ everywhere, for $x > 0$ but near 0, then $f(x) > f(0+) = r_1^+$. By induction, if $x > r_j^+$ but near r_j^+, then $f(x) > f(r_j^+) = r_{j+1}^+$. Thus, any attracting set with 0 in its interior must contain all points just to the right of all the r_j^+ and the r_j^+ can only be left end points of the attractor. Similarly, if $x < 0$ but near 0, then $f(x) < f(0-) = r_1^-$. By induction, the attractor contains all points just to the left of all the r_j^- and they can only be right end points of the attractor. Since $\{r_1^-, r_2^-, r_3^-\}$ are the only nonzero values for the r_j^-, the right end points of the three intervals must be these three points. These intervals must include all the points r_j^+, so the interval containing 0, must extend from r_2^- down to r_2^+. Thus, we are led to consider the union of the three closed intervals

$$\mathbf{A} = [r_1^+, r_3^-] \cup [r_2^+, r_2^-] \cup [r_3^+, r_1^-],$$

where $[r_2^+, r_2^-] = [r_2^+, 0] \cup [0, r_2^-]$. See Figure 5. The set $\mathbf{A}$ is invariant since

$$f([r_1^+, r_3^-]) = [r_2^+, 0], \qquad f([r_2^+, 0]) = [r_3^+, r_1^-],$$
$$f([0, r_2^-]) = [r_1^+, r_3^-], \qquad f([r_3^+, r_1^-]) = [r_4^+, r_2^-] \subset [r_2^+, r_2^-].$$

Since $r_4^+, r_4^- \in (r_2^+, r_2^-)$ are in the interior of $\mathbf{A}$, the end points of $\mathbf{A}$ eventually get mapped into its interior, no periodic orbit is contained entirely on the end points, a trapping region for $\mathbf{A}$ can be found, and $\mathbf{A}$ is an attracting set. In fact by experimenting with points, a trapping region is

$$\mathbf{V} = [-1.07, -0.7] \cup [-0.35, 0.24] \cup [0.562, 1.02].$$

We next show that $\mathbf{A}$ is transitive, and so an attractor. By Theorem 11.6, an attracting set $\mathbf{A}$ that contains the single discontinuity 0 is transitive provided that the forward orbit of an arbitrarily small open interval $(-\delta, \delta)$ about 0 covers all of $\mathbf{A}$. In our example, the one-sided small intervals $(-\delta, 0) \subset [r_2^+, 0]$ have a forward orbits that cover $\mathbf{A}$. First we check the iterates of the whole interval $[r_2^+, 0]$:

$$f([r_2^+, 0]) = [r_3^+, r_1^-],$$
$$f^2([r_2^+, 0]) = f([r_3^+, r_1^-]) = [r_4^+, r_2^-] \supset [0, r_2^-],$$
$$f^3([r_2^+, 0]) \supset f([0, r_2^-]) = [r_1^+, r_3^-],$$
$$f^4([r_2^+, 0]) \supset f([r_1^+, r_3^-]) = [r_2^+, 0].$$

Since $f^4([r_2^+, 0]) \supset [r_2^+, 0]$ and $f^4(0-) = 0$, there is a subinterval $[x_1, 0]$ in $[r_2^+, 0]$ such that $f^4([x_1, 0]) = [r_2^+, 0]$ and f^4 is an expansion from $[x_1, 0]$ onto $[r_2^+, 0]$. Because $f^4|_{[x_1, 0]}$ is an expansion, for any smaller interval $(-\delta, 0] \subset [0, x_1]$, there is an an integer j such that $f^{4j}((-\delta, 0]) \supset [r_2^+, 0]$. Thus, the the forward orbit of any small $(-\delta, 0]$ must contain the entire interval $[r_2^+, 0]$ and its iterates, hence all of $\mathbf{A}$. By the theorem, $\mathbf{A}$ is transitive and a chaotic attractor made up of three intervals.

Outside the attractor, a repelling period-2 orbit is found in $(r_3^-, r_3^+) \cup (r_2^-, r_3^+)$: For $x < 0$,

$$0 = f^2(x) - x = \frac{5}{4}\left(\frac{5}{4}x + 1\right) - 1.05 - x$$
$$= \frac{9}{16}x + 1.25 - 1.05 = \frac{9}{16}x + \frac{1}{5},$$
$$x = -\frac{16}{45} \approx -0.3555,$$

11.2. Chaotic Attractors

and $f(-16/45) = 5/9 \approx 0.5555$. All the other points in $(r_3^-, r_3^+) \cup (r_2^-, r_3^+)$ have ω-limit sets contained in $\mathbf{A}$,

$$\mathscr{B}(\mathbf{A}; f) \supset [-1.6, 1.6] \setminus \{-16/45, 5/9\}.$$

Another way to find a trapping region for the attractor is to cut out two small intervals about this repelling period-2 orbit.

The function in the preceding example has positive slope on both sides, which is similar to the Lorenz map and the assumption of the Williams theorem. The next example has one positive slope and one negative slope. This type of example falls within Theorem 11.6, but not Theorem 11.5 even if the slopes were greater than $\sqrt{2}$ in absolute value.

Example 11.19. For $r_1^- = 116/81 \approx 1.432$, consider the following map with both positive and negative slopes,

$$f(x) = \begin{cases} 1.25\, x + r_1^- & \text{for } x < 0, \\ -1.25\, x + 1 & \text{for } x > 0. \end{cases}$$

Again, let $r_{j+1}^- = f^j(r_1^-)$, $r_1^+ = f(0+) = 1$, and $r_{j+1}^+ = f^j(r_1^+)$. The value of r_1^- is chosen so that r_3^- is the fixed point $p > 0$ that satisfies $p = -(5/4)\,p + 1$, $(9/4)\,p = 1$, or $p = 4/9$. The first few values of $r_j^\pm$ are given in the following table.

j	r_j^+	r_j^-
1	1.000	1.432
2	-0.250	-0.790
3	1.120	0.444
4	-0.399	0.444
5	0.933	0.444

The order of these points on the line is as follows:

$$a = r_2^- < r_4^+ < r_2^+ < 0 < r_3^- = p < r_5^+ < r_1^+ < r_3^+ < r_1^- = b.$$

The invariant interval $[a, b]$ has $a = r_2^- \approx -0.79$ and $b = r_1^- \approx 1.43$. A slightly larger interval $[-0.82, 1.45]$ is a trapping region for $[r_2^-, r_1^-]$: $f(1.45) = -0.8125$, $f(-0.82) \approx -0.407$, so $\mathrm{cl}(f([-0.82, 1.45])) \subset (-0.82,\, , 1.45])$.

Graphical iteration shown in Figure 6 indicates that we should look for an attractor $\mathbf{A}$ composed of three intervals. Because $f'(x) > 0$ for $x < 0$, if $x < 0$ then $f(x) < f(0-) = r_1^-$. The attractor $\mathbf{A}$ must contain points to the left of 0, and so of r_1^-. For the same reason, if $r_j^\pm < 0$, then $\mathbf{A}$ must be on the same side of $r_{j+1}^\pm$ as of $r_j^\pm$. Because $f'(x) < 0$ for $x > 0$, if $r_j^\pm > 0$, then $\mathbf{A}$ set must be on the opposite side of $r_{j+1}^\pm$ than of $r_j^\pm$. The attractor $\mathbf{A}$ must contain points to the right of 0 and so to the left of r_1^+. Also, $\mathbf{A}$ must contain points to the left of r_3^- and to the right of $r_4^- = r_3^-$, so r_3^- cannot be an end point of $\mathbf{A}$. Also, the discussion just given implies that, if all the end points are among $\{p_j^\pm\}_{j=1}^5$, then the left end points of $\mathbf{A}$ must be $\{r_2^-, r_2^+, r_3^+\}$, and right end points must be among the points $\{r_1^+, r_1^-, r_4^+, r_5^+\}$. For the attracting set to contain all the $p_j^\pm$, it must be the three intervals

$$\mathbf{A} = [r_2^-, r_4^+] \cup [r_2^+, r_1^+] \cup [r_3^+, r_1^-].$$

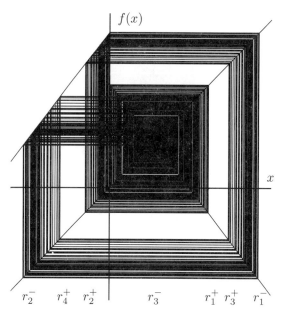

Figure 6. Plot of graphical iteration of a dense orbit in the attractor for Example 11.19

The set **A** is invariant because

$$f([r_3^+, r_1^-]) = [r_2^-, r_4^+], \qquad f([0, r_1^+]) = [r_2^+, r_1^+],$$
$$f([r_2^+, 0]) = [r_3^+, r_1^-], \qquad f([r_2^-, r_4^+]) = [r_3^-, r_5^+] \subset [0, r_2^+].$$

Since $r_5^+, r_3^- \in (r_2^+, r^+)$ are in the interior of **A**, the end points of **A** eventually get mapped into its interior, and no periodic orbit is contained entirely on the end points. By Theorem 11.6, **A** has a trapping region and is an attracting set.

The chaotic attractor guaranteed by Theorem 11.6 must contain an interval about 0. As argued earlier, it must contain to the points on both sides of $p = {}^4\!/\!_5 = r_3^- = r_4^-$, so the attractor contains a small interval $\mathbf{J} \subset (0, r_1^+) \subset \mathbf{A}$ about p. Iterates of **J** keep expanding until one contains 0. Thus, there is a positive integer k such that $f^k(\mathbf{J}) \supset [0, p]$, $f^{k+1}(\mathbf{J}) \supset [p, r_1^+)$, and the iterates of **J** contain all of **A**. This shows that the chaotic attractor must be all of **A**.

We now show that there is a repelling period-2 orbit in the gaps of **A**. Assuming that $x > 0$,

$$0 = f^2(x) - x = \frac{5}{4}\left(-\frac{5}{4}x + 1\right) + r_1^- - x$$
$$= -\frac{-25 - 16}{16}x + \frac{5}{4} + r_1^-,$$
$$q_0 = \frac{16}{41}\left(\frac{5}{4} + \frac{116}{81}\right) \approx 1.047, \qquad \text{and}$$
$$q_1 = f(q_0) = \frac{16}{41}\left(1 - \frac{5 \cdot 116}{4 \cdot 81}\right) \approx -0.308.$$

11.2. Chaotic Attractors

All points in the intervals $(r_4^+, r_2^-) \cup (r_1^+, r_3^-)$ besides the period-2 orbit have ω-limit sets contained in **A**, and $[-0.82, 1.45]$ is a trapping region for $[r_2^-, r_1^-]$, so that

$$\mathscr{B}(\mathbf{A}; f) \supset [-0.82, 1.45] \smallsetminus \{q_0, q_1\}.$$

The next example finds a parameter value for a function for which a periodic orbit is contained entirely in the end points of the invariant set, so it is a Milnor attractor but not an attractor in our sense.

Example 11.20 (Choi). For $0 < \mu < 1$, consider the family of functions

$$f_\mu(x) = \begin{cases} \frac{1}{\mu} x + 1 & \text{for } x < 0, \\ \frac{1}{\mu} x + \left(1 - \mu - \frac{1}{\mu}\right) & \text{for } x \geq 0. \end{cases}$$

The images of the discontinuity are

$$r_1^- = f(0-) = 1 \quad \text{and}$$
$$r_1^+ = f(0+) = 1 - \mu - \frac{1}{\mu},$$

which give the invariant interval $[a, b] = [r_1^+, r_1^-]$. The iterates of $r_1^- = 1$ return to 0:

$$r_2^- = f(r_1^-) = 1 - \mu,$$
$$r_3^- = f(1 - \mu) = \frac{1-\mu}{\mu} + 1 - \mu - \frac{1}{\mu} = -\mu, \quad \text{and}$$
$$r_4^- = f(-\mu) = -1 + 1 = 0.$$

The iterates of $r_1^+ = 1 - \mu - 1/\mu$ are as follows:

$$r_2^+ = f(r_1^+) = \frac{1}{\mu} - \frac{1}{\mu^2},$$
$$r_3^+ = f(r_2^+) = \frac{1}{\mu^2} - \frac{1}{\mu^3} + 1 > 0, \quad \text{and}$$
$$r_4^+ = f(r_3^+) = \frac{1}{\mu} + \frac{1}{\mu^3} - \frac{1}{\mu^4} 1 - \mu - \frac{1}{\mu}$$
$$= 1 - \mu + \frac{1}{\mu^3} - \frac{1}{\mu^4}.$$

We consider the union of intervals

$$\mathbf{A} = [r_1^+, r_3^-] \cup [r_2^+, r_2^-] \cup [r_3^+, r_1^-].$$

For this to be invariant, we need $r_2^+ \leq r_4^+ = f(r_3^+)$:

$$\frac{1}{\mu} - \frac{1}{\mu^2} \leq 1 - \mu + \frac{1}{\mu^3} - \frac{1}{\mu^4},$$
$$\mu^3 - \mu^2 \leq \mu^4 - \mu^5 + \mu - 1,$$
$$0 \leq -\mu^5 + \mu^4 - \mu^3 + \mu^2 + \mu - 1$$
$$\leq (1 - \mu)(\mu^4 + \mu^2 - 1).$$

The right-hand side is zero for $\mu = \mu_0$, where

$$0 = \mu_0^4 + \mu_0^2 - 1,$$

$$\mu_0^2 = \frac{-1 + \sqrt{1+4}}{2} = \frac{\sqrt{5}-1}{2},$$

$$\mu_0 = \sqrt{\frac{\sqrt{5}-1}{2}} \approx 0.7861513.$$

For $\mu_0 \leq \mu < 1$, $(1-\mu)(\mu^4 + \mu^2 - 1) > 0$, and the union of intervals

$$\mathbf{A} = [r_1^+, r_3^-] \cup [r_2^+, r_2^-] \cup [r_3^+, r_1^-]$$

is invariant. For $0 < \mu < \mu_0$, the whole interval $[r_1^+, r_1^-]$ is the attractor.

To determine whether the set $\mathbf{A}$ is an attractor or not, we need to see if there is a periodic orbit that is entirely contained in its boundary. The period-2 orbit, with $p < 0 < q$ is given as follows:

$$p = f(q) = \frac{q}{\mu} + 1 - \mu - \frac{1}{\mu},$$

$$q = f(p) = \frac{q}{\mu^2} + \frac{1}{\mu} - 1 - \frac{1}{\mu^2} + 1$$

$$= \frac{q}{\mu^2} + \frac{1}{\mu} - \frac{1}{\mu^2},$$

$$\mu^2 q = q + \mu - 1,$$

$$(\mu^2 - 1)q = \mu - 1,$$

$$q = \frac{\mu - 1}{\mu^2 - 1} = \frac{1}{\mu + 1},$$

and

$$p = f(q) = \frac{1}{\mu(\mu+1)} + 1 - \mu - \frac{1}{\mu}$$

$$= \frac{1 + \mu + \mu^2 - \mu^2 - \mu^3 - \mu - 1}{\mu(\mu+1)}$$

$$= \frac{-\mu^2}{\mu + 1}.$$

For a given μ, the period-2 orbit lies on the end points of the intervals, if $p = r_2^+$:

$$\frac{-\mu^2}{\mu+1} = \frac{\mu-1}{\mu^2},$$

$$-\mu^4 = (\mu-1)(\mu+1) = \mu^2 - 1, \quad \text{or}$$

$$0 = \mu^4 + \mu^2 - 1.$$

We previously showed this equality occurs for $\mu = \mu_0$. For the parameter value μ_0, the union of intervals

$$\mathbf{A} = [r_1^+, r_3^-] \cup [r_2^+, r_2^-] \cup [r_3^+, r_1^-]$$

is invariant and has the period-2 orbit $\{p, q\}$ on part of its boundary. Since this is a repelling periodic orbit, **A** cannot have a trapping region and is not an attractor by our definition, but it is a Milnor attractor.

For $\mu < \mu_0$, there is only one interval in the attractor, and for $\mu > \mu_0$, there are three intervals, but the invariant set is an attractor with a trapping region.

Expanding maps with several discontinuities

Both Li–Yorke and Choi discussed the case in which there are several discontinuities. In this case, there can be one or several attractors. However, the number of attractors has to be less than or equal to the number of discontinuities.

Theorem 11.7. *Assume that $f : \mathbb{R} \setminus \{c_i\}_{i=1}^k \to \mathbb{R}$ is a differentiable function with k discontinuities that satisfies the following conditions.*

(i) *Assume that the one-sided limits of $f(x)$ exist at each of the c_i and are denoted by $r_{1,i}^- = f(c_i-)$ and $r_{1,i}^+ = f(c_i+)$. Also denote $r_{j+1,i}^{\pm} = f^j(r_{1,i}^{\pm})$.*

(ii) *Let $a = \min\{r_{j,i}^{\pm}\}$ and $b = \max\{r_{j,i}^{\pm}\}$. Assume that $a < f(a) < b$ and $a < f(b) < b$.*

(iii) *Assume that the derivative satisfies $|f'(x)| \geq \beta > 1$ for all x in $[a, b] \setminus \{c_1, \ldots, c_k\}$.*

Then, there is a finite collection of subsets $\mathbf{A}_1, \ldots, \mathbf{A}_n$ of $[a_0, b_0]$ that has the following properties:

a. *Each $\mathbf{A}_j$ is the finite union of closed intervals (perhaps only one interval). The end points of each $\mathbf{A}_j$ are among the $r_{j,i}^{\pm}$.*

b. *Each $\mathbf{A}_j$ contains at least one of the points of discontinuity in its interior, so $1 \leq n \leq k$.*

c. *The interiors of the $\mathbf{A}_j$ are pairwise disjoint, so $\text{int}(\mathbf{A}_i) \cap \text{int}(\mathbf{A}_j) = \emptyset$ if $i \neq j$.*

d. *Each $\mathbf{A}_j$ is a transitive Milnor attractor, and f has sensitive dependence on initial conditions when restricted to $\mathbf{A}_j$.*

e. *The set of points with ω-limit sets in one of the $\mathbf{A}_j$, $\bigcup_j \{x : \omega(x; f) \subset \mathbf{A}_j\} = \bigcup_j \mathscr{B}(\mathbf{A}_j; f)$, is dense and open in $[a, b]$.*

f. *If there is no periodic orbit that lies entirely on the end points of one of the $\mathbf{A}_j$, then $\mathbf{A}_j$ has a trapping region and is a chaotic attractor. (We allow a periodic point that goes through the point of discontinuity c_j.)*

Exercises 11.2

1. Let f be the map
$$f(x) = \frac{4}{\pi} \arctan(x).$$
Determine all the attracting sets for f. Which of these sets are attractors? Why are there no chaotic attractors? Note that $f(0) = 0$, $f(1) = 1$, and $f(-1) = -1$.

2. Consider the map
$$f(x) = \begin{cases} 1 + \frac{3}{2}x & \text{for } x \leq 0, \\ 1 - 2x & \text{for } x \geq 0. \end{cases}$$

  a. Show that $[-1, 1]$ has a trapping region for f. More precisely, show that there is a trapping region $\mathbf{U}$ for f such that $[-1, 1]$ is the largest invariant set in $\mathbf{U}$. Hint: Find $a < -1$ such that $f(a) = -1$. Then, find $b > 1$ such $a < f(b) < -1$. Conclude that $f([a, b]) \subset (a, b)$ and $f^n([a, b]) = [-1, 1]$ for $n \geq 2$.
  b. Show that f has a Markov partition on $[-1, 1]$.
  c. Is f is transitive on $[-1, 1]$? If so, why?
  d. Does f have a chaotic attractor? If so, what is it?

3. Consider the map
$$f(x) = \begin{cases} 3(1-x) + 1 & \text{for } 0 \leq x \leq 1, \\ 3(x-1) + 1 & \text{for } 1 \leq x \leq 2, \\ 3(3-x) + 1 & \text{for } 2 \leq x \leq 3, \\ 3(x-3) + 1 & \text{for } 3 \leq x \leq 4, \\ 3(5-x) + 1 & \text{for } 4 \leq x \leq 5. \end{cases}$$

  a. Show that f has a trapping region for an attracting set.
  b. Show that the attracting set is a chaotic attractor (i.e., (i) show f restricted to the attractor has sensitive dependence on initial condition, and (ii) f has a point whose limit set is the attractor).

4. Consider the map
$$f(x) = \begin{cases} \dfrac{x}{3} & \text{for } x \leq 0 \\ 3x & \text{for } 0 \leq x \leq 1 \\ 3(2-x) & \text{for } 1 \leq x \leq 2 \\ 3(x-2) & \text{for } 2 \leq x \leq 3 \\ 2 + \dfrac{x}{3} & \text{for } 3 \leq x. \end{cases}$$

  a. Show that f has a trapping region for an attracting set.
  b. Show that the attracting set is a chaotic attractor (i.e., (i) show f restricted to the attractor has sensitive dependence on initial condition, and (ii) f has a point whose limit set is the attractor).

5. (Choi) Consider the map
$$f(x) = \begin{cases} -1.3x - 0.25 & \text{for } x < 0, \\ -1.2x + 0.2 & \text{for } x \geq 0. \end{cases}$$

  a. Calculate the first three iterates of $r_1^\pm = f(0\pm)$.
  b. Show that there is a set $\mathbf{A}$ made up of three intervals, and that $\mathbf{A}$ is invariant. The middle interval contains 0 in its interior.
  c. Find the two fixed points, one positive and one negative. Show that the fixed points are not in the set $\mathbf{A}$ found in the preceding part, but are in the gaps between the intervals.

11.3. Lyapunov Exponents

d. Show that the forward iterates of the end points of **A** eventually fall into the interior of **A**. This shows that the set **A** has a trapping region and is a chaotic attractor.

6. Consider the map $f(x) = |4x(1-x)|$ on $\mathbb{R}$. Show that f has a chaotic attractor.
7. Consider the map

$$f(x) = \begin{cases} \frac{4}{3}x + 18 & \text{for } x \leq 0, \\ -\frac{7}{6}x + 18 & \text{for } 0 \leq x \leq 6, \\ -5(x-6) + 11 & \text{for } 6 \leq x \leq 10, \\ -\frac{9}{8}(x-10) - 9 & \text{for } 10 \leq x. \end{cases}$$

The important discontinuity of the derivative is at $x = 0$. At $x = 6$ and 10, the sign of the derivative does not change.

a. Show that the set $[-18, -9] \cup [-6, 6] \cup [10, 18]$ is an invariant set. (This set is actually a chaotic attractor.)

b. Show that the gaps $(-9, -6) \cup (6, 10)$ contains a repelling invariant Cantor set modeled on a subshift of finite type. What is the subshift?

11.3. Lyapunov Exponents

Sensitive dependence on initial conditions is a good mathematical concept and directly specifies the behavior we are seeking to describes. We can verify that some maps satisfy this condition, but it is not amenable to a numerical calculation using a computer or for experimental data. In this section, we define the Lyapunov exponent of an orbit, which is possible to calculate numerically using a computer and also from experimental data. When the Lyapunov exponent is positive, this indicates that the system has sensitive dependence; when the Lyapunov exponent is negative, this indicates that the orbit is going to an attracting periodic orbit (at least in one dimension).

We want a measurement of instability. For a period-n point p, if $|(f^n)'(p)| > 1$, then the orbit is repelling. Because we want the quantity to converge when we take more iterates, we check whether $|(f^n)'(p)|^{1/n}$, rather than $|(f^n)'(p)|$, is greater than one or less than one.

We now consider nonperiodic points. For two points x_0 and $x_0 + \delta$, at the nth iterate, by a Taylor expansion

$$|f^n(x_0 + \delta) - f^n(x_0)| \approx |(f^n)'(x_0)| \cdot |\delta| \quad \text{or}$$

$$\frac{|f^n(x_0 + \delta) - f^n(x_0)|}{|\delta|} \approx |(f^n)'(x_0)|.$$

Therefore after n iterates, lengths are approximately stretched or contracted by a factor of $|(f^n)'(x_0)|$. If $|f'(f^j(x_0))|$ is approximately L for each iterate, then $|(f^n)'(x_0)| \approx L^n$, or

$$L \approx |(f^n)'(x_0)|^{1/n}$$

is the average factor by which distances are expanded for each iterate. Passing to the limit gives the average expansion of the iterates along the whole forward orbit.

Since it is easier to add and divide than to multiply and take an nth root, the usual practice is to take logarithms, or

$$\ell = \ln(L) \approx \frac{\ln(|(f^n)'(x_0)|)}{n},$$

and then to take the limit as n goes to infinity.

With this motivation, we state the definition.

Definition 11.21. Let f be a map from $\mathbb{R}$ to itself that has a derivative. The *Lyapunov multiplier* of an initial condition x_0 for the map f is defined to be

$$L(x_0; f) = \lim_{n \to \infty} |(f^n)'(x_0)|^{1/n},$$

when this limit exists. The *Lyapunov exponent* of an initial condition x_0 for the map f is defined to be the logarithm of this quantity,

$$\ell(x_0; f) = \lim_{n \to \infty} \frac{1}{n} \ln(|(f^n)'(x_0)|),$$

when this limit exists. In the calculation of the multiplier and exponent, we do not want to calculate the composition of the function. By the chain rule,

$$|(f^n)'(x_0)| = |f'(x_{n-1})| \cdot |f'(x_{n-2})| \cdots |f'(x_1)| \cdot |f'(x_0)| \quad \text{and}$$
$$\ln(|(f^n)'(x_0)|) = \ln(|f'(x_{n-1})|) + \cdots + \ln(|f'(x_1)|) + \ln(|f'(x_0)|),$$

where $x_j = f^j(x_0)$. Therefore, these quantities become

$$L(x_0; f) = \lim_{n \to \infty} \left[\prod_{j=0}^{n-1} |f'(x_j)| \right]^{1/n} \quad \text{and}$$

$$\ell(x_0; f) = \lim_{n \to \infty} \frac{1}{n} \sum_{j=0}^{n-1} \ln(|f'(x_j)|).$$

Remark 11.22. Notice that $\ell(x_0; f) = \ln(L(x_0; f))$, so that $L(x_0; f)$ is greater than one exactly when $\ell(x_0; f)$ is positive.

For the Lyapunov exponent, $(1/n)\sum_{j=0}^{n-1} \ln(|f'(x_j)|)$ is an average of the logarithm of the absolute value of the derivative along the first n iterates of the point. Passing to the limit gives the average of the logarithm of the absolute value of the derivative along the whole orbit.

For the Lyapunov multiplier, $|(f^n)'(x_0)|^{1/n} = \left[\prod_{j=0}^{n-1} |f'(x_j)| \right]^{1/n}$ is the geometric average of the expansion for each iterate. Passing to the limit gives the geometric average of the expansion along the whole orbit.

Remark 11.23. Notice that if $f'(x_j) = 0$ along the orbit, then the Lyapunov multiplier $L(x_0; f) = 0$ and the Lyapunov exponent $\ell(x_0; f) = -\infty$.

First, we check the Lyapunov exponent of a point on a periodic orbit and points in the basin of attraction of a periodic orbit.

11.3. Lyapunov Exponents

Theorem 11.8. *Let $x_j = f^j(x_0)$ and $p_j = f^j(p_0)$ be the points on the orbits of x_0 and p_0.*

- **a.** *If $f'(x_j) \neq 0$ for $0 \leq j < k$, then $\ell(x_k; f) = \ell(x_0; f)$ for any other point x_k on the forward orbit of x_0.*
- **b.** *If p_0 is a period-n point, then $\ell(p_0; f)$ exists and*

$$\ell(p_0; f) = \frac{1}{n} \sum_{j=0}^{n-1} \ln(|f'(p_j)|).$$

- **c.** *Assume that $\ell(p_0; f)$ exists, the forward orbit $\mathcal{O}_f^+(p_0)$ is bounded, and that $f'(p_j) \neq 0$ for all $j \geq 0$. Assume that x_0 is a point for which $f'(x_j) \neq 0$ for all $j \geq 0$ and $\lim_{j \to \infty} |x_j - p_j| = 0$. Then, $\ell(x_0; f)$ exists and equals $\ell(p_0; f)$. In particular, if p_0 is a period-n point, and x_0 is in the basin of attraction of the orbit for p_0, with $f'(x_j) \neq 0$, then $\ell(x_0; f) = \ell(p_0; f)$.*

Proof. (a) Dropping off a finite number of terms from the infinite average does not change the limit.

(b) The same terms are repeated over and over, so the limit is just the average over the orbit.

(c) Let $\epsilon > 0$. There is an $N > 0$ such that

$$|\ln(|f'(x_j)|) - \ln(|f'(p_j)|)| < \epsilon \qquad \text{for all } j \geq N.$$

The average after N is the same as the average of all the terms, so

$$|\ell(x_0; f) - \ell(p_0; f)| = \lim_{n \to \infty} \frac{\left| \sum_{j=N+1}^{N+n} [\ln(|f'(x_j)|) - \ln(|f'(p_j)|)] \right|}{n}$$

$$\leq \lim_{n \to \infty} \frac{\sum_{j=N+1}^{N+n} |\ln(|f'(x_j)|) - \ln(|f'(p_j)|)|}{n}$$

$$\leq \lim_{n \to \infty} \frac{\sum_{j=N+1}^{N+n} \epsilon}{n} = \epsilon.$$

Since this is true for any small ϵ, they must be equal, as claimed.

The case of the periodic orbit follows from the general case. $\square$

Example 11.24. Consider the logistic map $g_a(x) = ax(1-x)$, with $g_a'(x) = a - 2ax$.

(I) First, assume that $1 < a < 3$. Then, all points in $(0, 1)$ are in the basin of attraction of the fixed point $p_a = (a-1)/a$. The Lyapunov exponent of p_a is $\ell(p_a; g_a) = \ln(g_a'(p_a)) = \ln|2 - a| < 0$. For all points $0 < x_0 < 1$ such that the orbit $f^j(x_0) \neq 0.5$ for all $j \geq 0$, $\ell(x_0; g_a) = \ell(p_a; g_a) = \ln|2 - a| < 0$. Note that, for $a = 2$, $p_2 = 0.5$ and $\ell(0.5; g_2) = -\infty$.

(II) Next, assume that $3 < a < 1 + \sqrt{6}$. Then, the period-2 orbit is attracting. The points are

$$q_{\pm} = \frac{1}{2} + \frac{1}{2a} \pm \frac{1}{2a}\sqrt{(a-3)(a+1)}.$$

The derivative is

$$g_a'(q_{\pm}) = -1 \mp \sqrt{(a-3)(a+1)},$$

so
$$\ell(q_\pm; g_a) = \frac{1}{2}\left[\ln(|-1+\sqrt{(a-3)(a+1)}|) + \ln(|-1-\sqrt{(a-3)(a+1)}|)\right]$$
$$= \frac{1}{2}\left[\ln(|1-(a-3)(a+1)|)\right]$$
$$= \frac{1}{2}\left[\ln(|-a^2+2a+4|)\right]$$
$$< 0.$$

The point p_a is still a fixed point, but for these parameter values,
$$\ell(p_a; g_a) = \ln|2-a| > 0.$$
For any point x_0 in $(0,1)$ one the following cases holds: (i) It lands exactly on p_a for some iterate, $f^n(x_0) = p_a$, and has $\ell(x_0; g_a) = \ln|2-a| > 0$. (ii) It lands on the critical point 0.5 for some iterate, $f^n(x_0) = 0.5$, and $\ell(x_0; g_a) = -\infty$. (iii) It is in the basin of attraction of the period-2 orbit and does not pass through 0.5, and $\ell(x_0; g_a) = \ell(q_+; g_a)$.

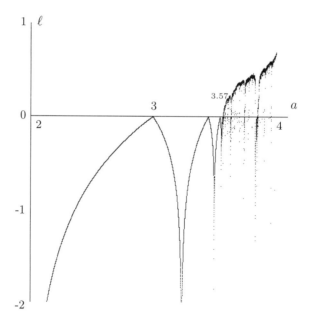

Figure 7. Plot of the Lyapunov exponent for the logistic family as a function of the parameter a, $\ell = \ell(g_a(0.5); g_a)$.

In Figure 7, we plot the value of a numerically calculated Lyapunov exponent of $x_0 = g_a(0.5)$ for the logistic family g_a as a function of the parameter a, $\ell(g_a(0.5); g_a)$. Notice that the value is negative in ranges of the parameter where there is an attracting periodic orbit. At $a = 3$, where the fixed point changes from attracting to repelling and an attracting period-2 orbit is born, the Lyapunov exponent is 0. Just above $a = 3.57$, the Lyapunov exponent is positive. This is the same range for which the (Feigenbaum) bifurcation diagram given in Figure 30 of Chapter 9 shows a whole interval of points in the limit set. For even larger values of a, near 3.83,

11.3. Lyapunov Exponents

there is another short parameter interval in which there is an attracting period-3 orbit and the Lyapunov exponent is negative.

Example 11.25. Let
$$T(x) = \begin{cases} 2x & \text{if } x \leq 0.5, \\ 2(1-x) & \text{if } x \geq 0.5, \end{cases}$$
be the tent map with slope ± 2. The derivative $|T'(x)| = 2$ for $x \neq 0.5$, so $\ell(x_0; T) = \ln(2) > 0$ as long as $x_j = f^j(x_0) \neq 0.5$ for all $j \geq 0$.

Example 11.26. Let $G(x) = 4x(1-x)$ be the logistic map for $a = 4$. The point 0 is a fixed point and $G(1) = 0$, so
$$\ell(0; G) = \ell(1; G) = \ln|G'(0)| = \ln(4).$$
The point 0.5 has Lyapunov exponent equal to minus infinity, as does any point whose orbit passes through the point 0.5.

Now consider an $0 < x_0 < 1$ such that $x_j = G^j(x_0) \neq 0.5$ for all $j \geq 0$. (Since the only point that goes to 1 is 0.5, this insures that the orbit never goes through 1 or 0 as well.) The map G is conjugate to T by $x = C(y) = \sin^2(\ln \pi y 2) = 1/2 - 1/2 \cos(\pi y)$, so $G^n(x) = C \circ T^n \circ C^{-1}(x)$. (We use x's for the variables of G and y's for the variables of T.) Also,
$$0 \leq C'(y) = \frac{\pi}{2} \sin(\pi y) \leq \frac{\pi}{2},$$
with the derivative equal to zero only when $y = 0$ or 1. Letting $x_j = G^j(x_0)$ and $y_j = C^{-1}(x_j) = T^j(y_0)$, we get
$$\ell(x_0; G) = \lim_{n \to \infty} \frac{1}{n} \ln(|(G^n)'(x_0)|)$$
$$= \lim_{n \to \infty} \frac{1}{n} \left[\ln(|C'(y_n)|) + \ln(|(T^n)'(y_0)|) + \ln(|(C^{-1})'(x_0)|) \right]$$
$$= 0 + \ln(2) + \lim_{n \to \infty} \frac{1}{n} \ln(|C'(y_n)|)$$
$$\leq \ln(2).$$

In the preceding calculation, $\ln(|(C^{-1})'(x_0)|) \neq 0$ is a fixed quantity, so dividing by n and taking the limit, we get zero. Similarly,
$$0 < |C'(y_n)| < \frac{\pi}{2} \quad \text{and}$$
$$-\infty < \frac{1}{n} \ln(|C'(y_n)|) < \frac{1}{n} \ln(\pi/2),$$
so the limit is less than or equal to zero, and
$$\ell(x_0; G) \leq \ln(2)$$
for all points for which the limit exists. In fact, if
$$\lim_{n \to \infty} \frac{1}{n} \ln(|C'(y_n)|) = 0,$$
then $\ell(x_0; G) = \ln(2)$. In particular, any periodic orbit has $\ell(x_0; G) = \ln(2)$ and is repelling.

Also, for any $\delta > 0$, if the T-orbit is completely contained inside $[\delta, 1-\delta]$, then

$$\ln(|C'(y_n)|) \geq \frac{\pi \sin(\pi\delta)}{2} > 0$$

and the Lyapunov exponent is $\ln(2)$. For example, if we take an x_0 for which its symbol sequence does not contain L repeated N times in a row (i.e., L^N never appears), then the T-orbit stays in the interval $[1/2^N, 1 - 1/2^{N+1}]$ and $\ell(x_0; G) = \ln(2)$. This shows that there are many orbits with Lyapunov exponent equal to $\ln(2)$.

The point x^*, which we showed has a dense orbit, has a symbol sequence including L^N for arbitrarily large N, so it does not fall into the last case. (Since x^* has a dense orbit, its orbit comes arbitrarily near both 0 and 1.) However, for its symbol sequence $\mathbf{s}^*$, before

$$s_n \ldots s_{n+N-1} = L^N,$$

we must list all the strings of length $1, 2, \ldots, N-1$, so

$$n > \sum_{j=1}^{N-1} 2^j$$

$$= 2\left(\sum_{j=1}^{N-2} 2^j\right)$$

$$= 2\left(\frac{2^{N-1}-1}{2-1}\right)$$

$$= 2^N - 2$$

$$> 2^{N-1}.$$

Therefore, $(\pi/2) \sin(\pi y_n) \geq (\pi/2) \pi 2^{1-N}/2 = \pi^2 2^{-(N+1)}$ (where we divide by 2 to estimate the sine function from below), and

$$0 \geq \lim_{n \to \infty} \frac{1}{n} \ln(|C'(y_n)|) \geq \lim_{N \to \infty} \frac{\ln\left(\pi^2 \, 2^{-(N+1)}\right)}{n} \geq \lim_{N \to \infty} \frac{-(N+1)\ln(2)}{2^{N-1}} = 0.$$

Thus, the point x^*, with a dense orbit, also has $\ell(x_o; G) = \ln(2)$.

Notice that the numerical value calculated for the Lyapunov exponent of g_4 in Figure 7 is approximately $\ln(2) \approx 0.693$.

Theorem 11.9. *Assume that $f : \mathbb{R} \to \mathbb{R}$ is a differentiable expanding map possibly with discontinuities with $|f'(x)| \geq \beta > 1$ at points other than the discontinuities. Assume that $x_j = f^j(x_0)$ is an orbit that does not go through the discontinuities. Then $\ell(x_0; f) \geq \ln(\beta)$.*

The theorem holds because the Lyapunov exponent is the average of numbers greater than or equal to $\ln(\beta)$.

11.3.1. Test for Chaotic Attractors.

In this section, we combine the use of Lyapunov exponents with the idea of a limit set to develop criteria for a chaotic attractor that are more easily verifiable in experiments or numerical work. We do not take these criteria as our definition, because there are various pathologies that are difficult to rule out using this approach, especially in higher dimensions. However, for "most systems", the pathologies do not occur and these criteria provide a good test to see whether the map has a chaotic attractor.

Test for a Chaotic Attractor: An experimental criterion for a set $\mathbf{A}$ to be a *chaotic attractor* for a map f from $\mathbb{R}$ to $\mathbb{R}$ is the following:

(1) There are many points $x_0^1, \ldots, x_0^k$ with distinct orbits that have their ω-limit sets equal to all of $\mathbf{A}$, or

$$\omega(x_0^i; f) = \mathbf{A} \qquad \text{for } 0 \leq i \leq k.$$

(2) The Lyapunov exponents of the x_0^i are positive, $\ell(x_0^i; f) > 0$.

Remark 11.27. What we have called a criterion or test for a chaotic attractor is essentially what Alligood, Sauer, and Yorke presented in their book [**All97**] as the definition of a chaotic attractor. They give the definition in several contexts, but for a map of one variable, they call a set $\mathbf{A}$ a *chaotic attractor* for f provided that (i) $\mathbf{A}$ is a Milnor attractor and (ii) there is a point x_0 in $\mathbf{A}$ with $\omega(x_0; f) = \mathbf{A}$ and $\ell(x_0; f) > 0$.

Remark 11.28. In our definition, the fact that there are many distinct points with the same ω limit set indicates that the measure of the basin of attraction is positive and $\mathbf{A}$ is a Milnor attractor. In many cases there will be a trapping region for such Milnor attractors, but this is not experimentally verifiable for most cases. We do not require that the points x_0^i are in $\mathbf{A}$, so we do not necessarily know that f is topologically transitive on $\mathbf{A}$. However, the existence of many points with $\omega(\mathbf{x}_0^i; \mathbf{F}) = \mathbf{A}$ gives an indication that this is true, so $\mathbf{A}$ should be a chaotic attractor.

In this criterion, a positive Lyapunov exponent replaces the assumption of sensitive dependence on initial conditions used in the more mathematical definition given in Section 11.2.

Most systems satisfying the preceding test also satisfy the earlier definition. Thus, we are tacitly assuming that the system is "generic".

For more discussion of the ways generic properties relate to the definition of a chaotic attractor, see the discussion in Appendix B.

Exercises 11.3

1. The map $g_a(x) = a\,x\,(1-x)$ for $a = 3\frac{1}{6} = {}^{19}\!/_6$ has a period-2 orbit at $q_1 = {}^{10}\!/_{19}$ and $q_2 = {}^{15}\!/_{19}$ and fixed points at 0 and $p_a = {}^{13}\!/_{19}$.

    a. Find the Lyapunov exponents for the two fixed points 0 and p_a. Are the fixed points attracting or repelling?

b. Find the Lyapunov exponent for the point q_1 for $g_{19/6}$. Is the orbit attracting or repelling?
   c. Find the Lyapunov exponent for the point $x_0 = g_{19/6}(0.5)$. Why is this the correct answer? Hint: The map has a negative Schwarzian derivative. Also, we do not start at the critical point 0.5 but at the image of the critical point.
   d. What values are there for Lyapunov exponents for x_0 in $[0,1]$.
2. Numerically calculate the Lyapunov exponent for the point $x_a = g_a(0.5)$ and function $g_a(x) = a\,x\,(1-x)$ using the values $a = 3.81$, 3.83, 3.85, and 3.87.
3. Let f be a piecewise expanding map on $[a,b]$ with $f([a,b]) = [a,b]$. We do not assume that f is continuous, but that there are points of discontinuity or lack of derivative $a = p_0 < p_1 < \cdots < p_k = b$, with $|f'(x)| \geq s > 1$ for all $p_{j-1} < x < p_j$ and $1 \leq j \leq k$. Let x_0 be any point whose forward orbit does not pass through one of the points of discontinuity p_j. Explain why the Lyapunov exponent of x_0 is positive.
4. Consider the function $g_5(x) = 5\,x\,(1-x)$. Assume that x_0 is a point whose whole orbit stays in $[0,1]$ and which has a Lyapunov exponent (the limit exists). Prove that this exponent is positive, $\ell(x_0; g_5) > 0$.
5. Assume that the point x_0 has a Lyapunov exponent for a map f. Show that the Lyapunov exponent of x_0 for a power f^k is $k\ell(x_0; f)$; that is, show that
$$\ell(x_0, f^k) = k\,\ell(x_0; f).$$
6. Let $f(x) = 3x + \sin(x)$.
   a. Find the Lyapunov exponent for $x = 0$.
   b. Show that, for any x_0 for which the Lyapunov exponent exists, it satisfies $\ln(2) \leq \ell(x_0; f) \leq \ln(4)$.
7. Consider the logistic map $g_5(x) = 5\,x\,(1-x)$ which has an invariant Cantor set Λ_{g_5}.
   a. For what constants $1 < A < B$, is
$$A \leq |g'_5(x)| \leq B$$
   for points x in Λ_{g_5}. Try to find optimal values of A and B.
   b. Since the Lyapunov exponent $\ell(x_0; g_5)$ for a point x_0 in Λ_{g_5} is the limit of averages of values $\ln(|g'_5(x)|)$, which is between $\ln(A)$ and $\ln(B)$, what estimate can you give for the Lyapunov exponent for any point x_0 in Λ_{g_5}?
8. Explain why the rooftop map given in Example 11.3 satisfies the "Test for a Chaotic Attractor."

11.4. Invariant Measures

We have used the term "measure" several times as a generalization of the length of an interval. In this section, we make this term more precise and give examples that are connected to the dynamics of a map. When discussing chaotic dynamics, we sometimes want to know how frequently the orbit returns to different parts of a set and not just that the orbit is dense in a set. An invariant measure captures this frequency of return. A natural measure is one for which the frequency of return

11.4. Invariant Measures

of the orbits of many different points agrees with the measure. A natural measure that is spread out over a whole interval implies transitivity in a strong sense: Not only do orbits have ω-limit sets equal to the interval, but also the orbit returns near all points in the interval with positive frequency.

Before defining the particular type of measure determined by the dynamics of a function, we give the general properties of any measure. When developing the general theory of integration in mathematics, the concept of a measure is introduced. In our treatment, we restrict ourselves to examples of measures on the line, so the measures are generalizations of length. However, we give the definitions so they apply in any dimension. In two dimensions, a measure is a type of generalized area; in three dimensions, it is a type of generalized volume, etc.

The measures we consider are defined on all the open and closed sets and sets formed by intersections and unions of these. Although there are pathological sets for which the measure is not defined, this fact need not concern you since we focus on the measure of open and closed sets.

Definition 11.29. Consider subsets of a Euclidean space $\mathbb{R}^m$, or more generally, some metric space $\mathbf{X}$. The collection of all *Borel sets of* $\mathbf{X}$, $\mathscr{B}$, is the smallest collection of sets that has the following properties:

1. The collection $\mathscr{B}$ includes all the open and closed sets of $\mathbf{X}$.
2. If $\mathbf{S}$ is in $\mathscr{B}$, then $\mathbf{X} \setminus \mathbf{S}$ is also in $\mathscr{B}$.
3. A union of a finite or denumerable number of sets in $\mathscr{B}$ yields a set in $\mathscr{B}$.
4. An intersection of a finite or denumerable number of sets in $\mathscr{B}$ yields a set in $\mathscr{B}$.

Property (4) follows from properties (2) and (3). A set in $\mathscr{B}$ is called a *Borel set*.

Definition 11.30. A *Borel measure* μ is function on $\mathscr{B}$ that has the following properties:

1. The measure of any $\mathbf{S}$ in $\mathscr{B}$, is a nonnegative number, $\mu(\mathbf{S}) \geq 0$.
2. The measure is *countably additive*: that is, the measure of the disjoint union of finitely many or a countably infinite number of sets (so $\mathbf{S}_i \cap \mathbf{S}_j = \emptyset$ if $i \neq j$) is equal to the sum of the measures of the individual sets,
$$\mu(\mathbf{S}_1 \cup \cdots \cup \mathbf{S}_k) = \mu(\mathbf{S}_1) + \cdots + \mu(\mathbf{S}_k) \quad \text{and}$$
$$\mu\left(\bigcup_{j=1}^{\infty} \mathbf{S}_j\right) = \sum_{j=1}^{\infty} \mu(\mathbf{S}_j).$$

If the measure of the entire space equals one, then the measure is called a *probability measure*.

Definition 11.31. The *support* of a measure μ, $\mathrm{supp}(\mu)$, is defined to be the smallest closed set $\mathbf{S}_0$ such that all the measure is on $\mathbf{S}_0$; that is, if $\mathbf{S}$ is closed and $\mu(\mathbf{X} \setminus \mathbf{S}) = 0$, then $\mathbf{S} \supset \mathbf{S}_0$. For a probability measure, $\mathbf{S}_0$ is the smallest closed set such that $\mu(\mathbf{S}_0) = 1$,
$$\mathrm{supp}(\mu) = \bigcap \{\, \mathbf{S} : \mu(\mathbf{S}) = 1 \text{ and } \mathbf{S} \text{ is closed} \,\}.$$

Definition 11.32. A measure on an interval $[a, b]$ in the line such that the measure of any subinterval is its usual length is called *Lebesgue measure* and is denoted by λ. In the plane $\mathbb{R}^2$, if the measure of any rectangular region is its area, then the measure is also called *Lebesgue measure* on $\mathbb{R}^2$, and is also denoted by λ. Similarly, the usual volume on any $\mathbb{R}^n$ is called Lebesgue measure.

Lemma 11.10. *Let μ be a Borel measure and let $\mathbf{S}_1$ and $\mathbf{S}_2$ be two Borel sets, with $\mathbf{S}_2 \subset \mathbf{S}_1$. Then, $\mu(\mathbf{S}_2) \leq \mu(\mathbf{S}_1)$.*

Proof. $\mu(\mathbf{S}_1) = \mu(\mathbf{S}_1 \setminus \mathbf{S}_2) + \mu(\mathbf{S}_2) \geq \mu(\mathbf{S}_2)$ since $\mu(\mathbf{S}_1 \setminus \mathbf{S}_2) \geq 0$. $\square$

Example 11.33. If $\mathbf{K}$ is the middle-third Cantor set, then for any $n \geq 1$, $\mathbf{K}$ is contained in a set $\mathbf{K}_n$ of 2^n intervals of length $1/3^n$ each, so of total length $(2/3)^n$. By the lemma, the Lebesgue measure of $\mathbf{K}$ satisfies

$$\lambda(\mathbf{K}) \leq \lambda(\mathbf{K}_n) = \left(\frac{2}{3}\right)^n \qquad \text{for all } n \geq 1.$$

Thus, $\lambda(\mathbf{K}) = 0$, the Lebesgue measure of the Cantor set is zero.

We use the notation $\chi_{\mathbf{S}}(q)$ for the *characteristic function* of the set $\mathbf{S}$, which is equal to 1 if q is in $\mathbf{S}$ and is equal to 0 if q is not in $\mathbf{S}$:

$$\chi_{\mathbf{S}}(q) = \begin{cases} 1 & \text{if } q \in \mathbf{S}, \\ 0 & \text{if } q \notin \mathbf{S}. \end{cases}$$

Here, χ is the lower case Greek letter "chi".

Thus, Lebesgue measure of a closed interval $[a, b]$ in $\mathbb{R}$ is equal to the integral of $\chi_{[a,b]}(x)$:

$$\lambda([a,b]) = b - a$$
$$= \int_a^b 1 \, dx$$
$$= \int_{-\infty}^{\infty} \chi_{[a,b]}(x) \, dx.$$

If the total space is an interval $[a, b]$, then we can "normalize" Lebesgue measure so that the measure of all of $[a, b]$ is one: We define

$$\mu_{[a,b]}(\mathbf{S}) = \frac{1}{b-a} \lambda(\mathbf{S} \cap [a,b])$$
$$= \frac{1}{b-a} \int_a^b \chi_{\mathbf{S}}(x) \, dx$$
$$= \int_{-\infty}^{\infty} \chi_{\mathbf{S}}(x) \frac{\chi_{[a,b]}(x)}{b-a} \, dx.$$

The function

$$\rho(x) = \frac{\chi_{[a,b]}(x)}{b-a}$$

is called the *density function* of this measure.

11.4. Invariant Measures

Definition 11.34. A real-valued integrable function $\rho(x)$ is a *density function* for a measure, provided that $\rho(x) \geq 0$ for all x, and it induces a measure μ_ρ by

$$\mu_\rho(\mathbf{S}) = \int \chi_\mathbf{S}(x)\, \rho(x)\, dx.$$

In particular, the measure of a closed interval $[x_1, x_2]$ is

$$\mu_\rho([x_1, x_2]) = \int_{x_1}^{x_2} \rho(x)\, dx.$$

This measure is a probability measure provided that $\int_{-\infty}^{\infty} \rho(x)\, dx = 1$.

The density function for the *normal distribution* is

$$\rho(x) = \frac{1}{\sqrt{2\pi}}\, e^{-x^2/2}$$

For a measure given by a density function, if a set $\mathbf{S}$ has Lebesgue measure zero then its measure $\mu_\rho(\mathbf{S}) = 0$. A measure with this property is called *absolutely continuous* with respect to Lebesgue measure. This property is also a sufficient condition for a density function. We mainly consider measures that have density functions.

Definition 11.35. Let μ be a Borel measure on a metric space $\mathbf{X}$ and let f be a function that takes $\mathbf{X}$ to itself. The measure μ is called *invariant* for the function f, provided that

$$\mu(f^{-1}(\mathbf{S})) = \mu(\mathbf{S})$$

for all Borel sets $\mathbf{S}$ in $\mathbf{X}$. If μ is an invariant measure for a function f, we also say that f *preserves the measure* μ or f is *measure preserving* for μ.

The inverse image $f^{-1}(\mathbf{S})$ is used to define an invariant measure because we need to check the measure of all the points that are mapped to the set $\mathbf{S}$ by f. Think of the measure as giving a distribution of mass around the space. Take a Borel set $\mathbf{S}$ with $\mathbf{S}_{-1} = f^{-1}(\mathbf{S})$ as its preimage. To be measure preserving, all the mass in $\mathbf{S}_{-1}$ is moved by f to the set $\mathbf{S}$ and must equal the mass that is there before, $\mu(\mathbf{S})$.

Theorem 11.11. *Both the doubling map D and the tent map T preserve Lebesgue measure on $[0, 1]$. Therefore, the density function $\rho(x) \equiv 1$ defines an invariant measure on $[0, 1]$.*

Proof. For the tent map,

$$T^{-1}([x, y]) = \left[\frac{x}{2}, \frac{y}{2}\right] \cup \left[1 - \frac{y}{2}, 1 - \frac{x}{2}\right]$$

and

$$\begin{aligned}\lambda(T^{-1}([x,y])) &= \lambda\left(\left[\frac{x}{2}, \frac{y}{2}\right]\right) + \lambda\left(\left[1 - \frac{y}{2}, 1 - \frac{x}{2}\right]\right) \\ &= \frac{y-x}{2} + \frac{y-x}{2} \\ &= y - x \\ &= \lambda([x, y]).\end{aligned}$$

Thus, T preserves the measure of closed intervals. Notice that the inverse image of an interval contains two subintervals; each interval is half the length, but together they have a total length which is the same. The calculation for open intervals is similar. Since T preserves the measure of both open and closed intervals, it follows that it preserves it for all Borel sets induced by them.

The proof for the doubling map is similar. □

We can use the conjugacy of the logistic map with the tent map to obtain an invariant measure for the logistic map G as stated in the following theorem.

Theorem 11.12. *The logistic map $G(x) = 4x(1-x)$ has an invariant probability measure $\mu_G = \mu_{\rho_G}$ with density function*

$$\rho_G(x) = \frac{1}{\pi\sqrt{x-x^2}}.$$

Remark 11.36. The graph of this density function is depicted in Figure 8. Note that the density function ρ_G is unbounded on $[0,1]$. Since G takes an interval of moderate length near 0.5 to a very short interval near 1, the density function is very large near 1. The points near 1 go to points near 0, so the density function is also very large near 0. In contrast, we shall show that the density functions for piecewise linear maps with the Markov property are bounded.

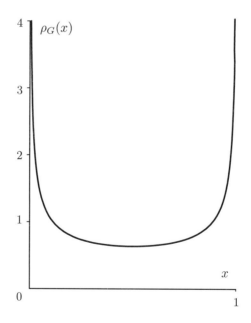

Figure 8. Density function ρ_G for $G(x) = 4x(1-x)$.

Proof. Theorem 9.16 showed that $G \circ C(s) = C \circ T(s)$ for the tent map T (with slope 2) and the conjugacy

$$x = C(s) = \frac{1-\cos(\pi s)}{2} = \sin^2\left(\frac{\pi s}{2}\right).$$

11.4. Invariant Measures

The conjugacy induces a measure from Lebesgue measure λ by
$$\mu_G(\mathbf{S}) = \lambda\left(C^{-1}(\mathbf{S})\right).$$
The total measure is one because $\mu_G([0,1]) = \lambda(C^{-1}[0,1]) = \lambda([0,1]) = 1$.

The conjugacy implies that
$$C^{-1}\left(G^{-1}(\mathbf{S})\right) = (G \circ C)^{-1}(\mathbf{S}) = (C \circ T)^{-1}(\mathbf{S}) = T^{-1}\left(C^{-1}(\mathbf{S})\right).$$
The tent map preserves Lebesgue measure by Theorem 11.11, so for a Borel set $\mathbf{S}$,
$$\begin{aligned}
\mu_G\left(G^{-1}(\mathbf{S})\right) &= \lambda\left(C^{-1}\left(G^{-1}(\mathbf{S})\right)\right) \\
&= \lambda\left(T^{-1}\left(C^{-1}(\mathbf{S})\right)\right) \\
&= \lambda\left(C^{-1}(\mathbf{S})\right) \\
&= \mu_G(\mathbf{S}),
\end{aligned}$$
and μ_G is an invariant measure for G.

The measure of an interval $([0, x_0]$ can be expressed as the integral
$$\mu_G([0, x_0]) = \lambda\left(C^{-1}([0, x_0])\right) = \int_{s \in C^{-1}([0, x_0])} 1\, ds.$$
If we make the substitution $x = C(s) = \sin^2\left(\pi s/2\right)$, then
$$\begin{aligned}
dx &= \pi \sin\left(\pi s/2\right) \cos\left(\pi s/2\right)\, ds \\
&= \pi \sin\left(\pi s/2\right) \sqrt{1 - \sin^2\left(\pi s/2\right)}\, ds \\
&= \pi \sqrt{x}\, \sqrt{1-x}\, ds \\
&= \pi \sqrt{x - x^2}\, ds, \quad \text{and} \\
ds &= \frac{dx}{\pi\sqrt{x-x^2}},
\end{aligned}$$
so
$$\mu_G([0, x_0]) = \int_0^{x_0} \frac{dx}{\pi\sqrt{x-x^2}}.$$
Therefore, the density function for the invariant measure μ_G is given by
$$\rho_G(x) = \frac{1}{\pi\sqrt{x-x^2}}.$$
$\square$

11.4.1. Piecewise Linear Maps with Markov Property. In this section, we consider a case where we can determine the invariant measure by linear algebra. Assume that $f : [q_0, q_k] \to [q_0, q_k]$ is piecewise linear map with the the Markov property for a partition. $q_0 < q_1 < \cdots < q_k$, for which $[q_0, q_k]$ is positively invariant. More precisely, for each $1 \leq i \leq k$, assume that f is linear on the open interval $\mathbf{A}_i = (q_{i-1}, q_i)$ and β_i be the absolute value of the slope of f on $\mathbf{A}_i$, $\beta_i = |f'(x)|$ for $x \in \mathbf{A}_i$. Let $L_i = q_i - q_{i-1}$ be the length of $\mathbf{A}_i$ and (t_{ij}) the transition matrix
$$t_{ij} = \begin{cases} 1 & \text{if } f(\mathbf{A}_i) \supset \mathbf{A}_j, \\ 0 & \text{if } f(\mathbf{A}_i) \cap \text{int}(\mathbf{A}_j) = \emptyset \end{cases}$$
(i.e., t_{ij} is 1 if it is possible to go from the ith interval $\mathbf{A}_i$ to the jth interval $\mathbf{A}_j$).

Assume that there is a density function such that on each $\mathbf{A}_i$ it takes on the constant value of ρ_i. Let $m_i = \mu_\rho(\mathbf{A}_i) = \rho_i L_i$ be the total measure of mass of $\mathbf{A}_i$ and $\lambda(\mathbf{J})$ be the length of an interval $\mathbf{J}$. The length $\lambda(f(\mathbf{A}_i)) = \beta_i L_i$. Because f is linear on $\mathbf{A}_i$, the proportion of the length $\mathbf{A}_i$ going onto $\mathbf{A}_j$ is the length $t_{ij} \lambda(\mathbf{A}_j)$ divided by this total length of $f(\mathbf{A}_i)$, or

$$\frac{\lambda\left(f^{-1}(\mathbf{A}_j) \cap \mathbf{A}_i\right)}{\lambda(\mathbf{A}_i)} = \frac{t_{ij} \lambda(\mathbf{A}_j)}{\lambda(f(\mathbf{A}_i))} = \frac{t_{ij} L_j}{\beta_i L_i}.$$

Note, that if $t_{ij} = 0$ and $f^{-1}(\mathbf{A}_j) \cap \mathbf{A}_i = \emptyset$, then this equation still holds. Since the density is constant on $\mathbf{A}_i$, this is also the ratio of the measures and

$$\mu_\rho\left(f^{-1}(\mathbf{A}_j) \cap \mathbf{A}_i\right) = \frac{t_{ij} L_j}{\beta_i L_i} m_i.$$

Summing over the preimages of all the $\mathbf{A}_j$,

$$\mu_\rho(f^{-1}(\mathbf{A}_j)) = \sum_i \mu_\rho(f^{-1}(\mathbf{A}_j) \cap \mathbf{A}_i) = \sum_i \frac{t_{ij} L_j}{\beta_i L_i} m_i.$$

If $\mathbf{m}^* = (m_1^*, \ldots, m_k^*)$ is row vector for an invariant measure, then

$$m_j^* = \mu_{\rho^*}(\mathbf{A}_j) = \mu_{\rho^*}(f^{-1}(\mathbf{A}_j)) = \sum_i \frac{t_{ij} L_j}{\beta_i L_i} m_i^*.$$

In matrix notation,

$$\mathbf{m}^* = \mathbf{m}^* \mathbf{M}, \quad \text{where}$$

$$\mathbf{M} = \left(\frac{t_{ij} L_j}{\beta_i L_i}\right)$$

and i gives the row and j gives the column of $\mathbf{M}$.

Because the total length of $f(\mathbf{A}_i)$ is the sum of the lengths of the intervals covered by $f(\mathbf{A}_i)$, $\beta_i L_i = \sum_j t_{ij} L_j$ and the sum of any row of $\mathbf{M}$ is one,

$$\sum_j \frac{t_{ij} L_j}{L_i \beta_i} = 1.$$

Letting $\mathbf{e} = (1, \ldots, 1)^\mathsf{T}$ be the column vector with all components equal to one, $\mathbf{M}\mathbf{e} = \mathbf{e}$ since all the row sums are equal to one. Thus, 1 is an eigenvalue of $\mathbf{M}$ and also of $\mathbf{M}^\mathsf{T}$, and there is a row eigenvector $\mathbf{m}^* = (m_1^*, \ldots, m_k^*)$ such that

$$\mathbf{M}^\mathsf{T} \mathbf{m}^{*\mathsf{T}} = \mathbf{m}^{*\mathsf{T}} \quad \text{or} \quad \mathbf{m}^* \mathbf{M} = \mathbf{m}^*;$$

that is, $\mathbf{m}^*$ is invariant by matrix multiplication by $\mathbf{M}$ on the right. Also because all the entries of $\mathbf{M}$ are nonnegative, the row eigenvector $\mathbf{m}^*$ can be chosen with all nonnegative components by the Perron–Frobenius Theorem 12.7. By scaling, we can insure that the sum of the components of $\mathbf{m}^*$ is one,

$$\sum_j m_j^* = 1,$$

yielding a probability measure. To get the invariant density function on each interval, let $\rho_j^* = m_j^*/L_j$ be the measure of the jth interval $\mathbf{A}_j$, divided by its length.

11.4. Invariant Measures

The probability transition matrices that we consider have the additional property that there is a power $n_0 > 0$ for which $\mathbf{M}^{n_0}$ has all positive entries (i.e., $\mathbf{M}$ is aperiodic). This condition implies that the transition graph is irreducible (i.e., it is possible to get from any vertex of the transition graph to any other vertex by an allowable path in the transition graph). For an aperiodic probability transition matrix, the Perron–Frobenius Theorem 12.7 implies that all the other eigenvalues λ_j of $\mathbf{M}$ have $|\lambda_j| < 1$. Also if $\mathbf{m} = (m_1, \ldots, m_k)$ is any initial guess for an invariant distribution of mass or measure on the intervals with all $m_j > 0$ and $\sum_j m_j = 1$, then $\mathbf{mM}^n$ converges to the eigenvector $\mathbf{m}^*$ with all $m_j^* > 0$. See Section 12.5.1 for more details on the Perron–Frobenius theorem. Notice that, if $\sum_j m_j = 1$ and $\mathbf{m}' = \mathbf{mM}$, then

$$\sum_j m_j' = \sum_j \sum_k \frac{m_k t_{kj} L_j}{L_k \beta_k} = \sum_k m_k \sum_j \frac{t_{kj} L_j}{L_k \beta_k} = \sum_k m_k \cdot 1 = 1,$$

so the sum remains one. This insures that the sum of the components of the limiting vector is also equal to one.

Theorem 11.13. *Assume that $f : [q_0, q_k] \subset \mathbb{R} \to [q_0, q_k]$ is a piecewise linear map with the Markov partition for a partition $q_0 < \cdots < q_k$, $L_i = q_i - q_{i-1}$, $|f'(x)| \geq \beta_i > 1$, transition matrix $\mathbf{T} = (t_{ij})$, and $\mathbf{M} = \left(\dfrac{t_{ij} L_j}{L_i \beta_i}\right)$ the transition matrix on the measures of intervals.*

a. Then $\mathbf{M}$ has 1 as an eigenvalue with $1 \geq |\lambda_j|$ for any other eigenvalue. There is a row eigenvector $\mathbf{m}^*$ for the eigenvalue 1 of $\mathbf{M}$ satisfying $\mathbf{m}^* \mathbf{M} = \mathbf{m}^*$, $m_j^* \geq 0$ for all j, and $m_1^* + \cdots + m_k^* = 1$. The corresponding densities of the invariant measure are given by A density function for an invariant probability measure for f takes on the constant values $\rho_j^* = {m_j^*}/{L_j}$ on each $\mathbf{A}_j$.

b. Assume that $\mathbf{T}$ is aperiodic (eventually positive). Then $1 > |\lambda_j|$ for any other eigenvalue of $\mathbf{M}$ and all the m_j^*, $\rho_j^* > 0$.

Example 11.37. We consider a piecewise expanding linear map with the Markov property that has four intervals. Let $\mathbf{A}_1$ have length $L_1 = 2$ and map onto $\mathbf{A}_2$ with slope $\beta_1 = {}^3\!/_2$. Let $\mathbf{A}_2$ have length $L_2 = 2({}^3\!/_2) = 3$ and map onto $\mathbf{A}_3$ with slope $\beta_2 = {}^4\!/_3$. Let $\mathbf{A}_3$ have length $L_3 = 3({}^4\!/_3) = 4$ and map onto $\mathbf{A}_4$ and $\mathbf{A}_3$ with slope $\beta_3 = {}^5\!/_4$. Let $\mathbf{A}_4$ have length $L_4 = 4({}^5\!/_4) - 4 = 1$ and map onto $\mathbf{A}_1$ with slope $\beta_4 = 2$. See Figure 9.

The transition matrix on measures for this example is

$$\mathbf{M} = \begin{pmatrix} 0 & 1 & 0 & 0 \\ 0 & 0 & 1 & 0 \\ 0 & 0 & \frac{4}{5} & \frac{1}{5} \\ 1 & 0 & 0 & 0 \end{pmatrix}.$$

Notice that all the measure from $\mathbf{A}_1$ goes to $\mathbf{A}_2$ (giving the first row of $\mathbf{M}$), all the measure from $\mathbf{A}_2$ goes to $\mathbf{A}_3$, ${}^4\!/_5$ of the measure from $\mathbf{A}_3$ goes to $\mathbf{A}_3$ and ${}^1\!/_5$ of the measure from $\mathbf{A}_3$ goes to $\mathbf{A}_3$, and all the measure from $\mathbf{A}_4$ goes to $\mathbf{A}_1$.

To find the row eigenvector for the eigenvalue 1 of $\mathbf{M}$, we take the transpose so that we can perform row reduction and find a column eigenvector for $\mathbf{M}^\mathsf{T}$. We

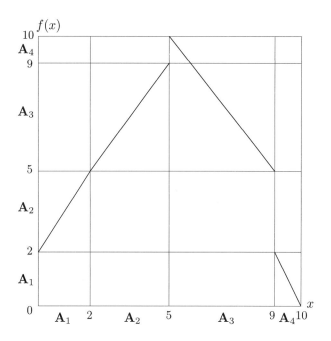

Figure 9. Piecewise expanding map for Example 11.37.

need $\mathbf{M}^\mathsf{T}\mathbf{m}^\mathsf{T} - \mathbf{m}^\mathsf{T} = \mathbf{0}$, so we row reduce $\mathbf{M}^\mathsf{T} - \mathbf{I}$:

$$\mathbf{M}^\mathsf{T} - \mathbf{I} = \begin{pmatrix} 0-1 & 0 & 0 & 1 \\ 1 & 0-1 & 0 & 0 \\ 0 & 1 & \frac{4}{5}-1 & 0 \\ 0 & 0 & \frac{1}{5} & 0-1 \end{pmatrix} \sim \begin{pmatrix} 1 & 0 & 0 & -1 \\ 0 & -1 & 0 & 1 \\ 0 & 1 & -\frac{1}{5} & 0 \\ 0 & 0 & \frac{1}{5} & -1 \end{pmatrix}$$

$$\sim \begin{pmatrix} 1 & 0 & 0 & -1 \\ 0 & 1 & 0 & -1 \\ 0 & 0 & -\frac{1}{5} & 1 \\ 0 & 0 & \frac{1}{5} & -1 \end{pmatrix} \sim \begin{pmatrix} 1 & 0 & 0 & -1 \\ 0 & 1 & 0 & -1 \\ 0 & 0 & 1 & -5 \\ 0 & 0 & 0 & 0 \end{pmatrix}.$$

Since the eigenvector satisfies $\left(\mathbf{M}^\mathsf{T} - \mathbf{I}\right)\mathbf{m}^\mathsf{T} = \mathbf{0}$,

$$m_1 - m_4 = 0,$$
$$m_2 - m_4 = 0,$$
$$m_3 - 5\,m_4 = 0;$$

$m_1 = m_4 = m_2$ and $m_3 = 5\,m_4$; or,

$$\mathbf{m} = (m_4, m_4, 5\,m_4, m_4).$$

The sum of the components is $8\,m_4$, so we need to choose $m_4 = 1/8$ to make the sum one:

$$\mathbf{m}^* = \left(\frac{1}{8}, \frac{1}{8}, \frac{5}{8}, \frac{1}{8}\right).$$

11.4. Invariant Measures

The densities for the invariant measure are found by dividing by the lengths of the intervals,

$$\rho_1^* = \frac{m_1*}{L_1} = \frac{1}{8} \cdot \frac{1}{2} = \frac{1}{16},$$
$$\rho_2^* = \frac{m_2*}{L_2} = \frac{1}{8} \cdot \frac{1}{3} = \frac{1}{24},$$
$$\rho_3^* = \frac{m_3*}{L_3} = \frac{5}{8} \cdot \frac{1}{4} = \frac{5}{32}, \quad \text{and}$$
$$\rho_4^* = \frac{m_4*}{L_4} = \frac{1}{8} \cdot \frac{1}{1} = \frac{1}{8}.$$

The graph of the density function is given in Figure 10.

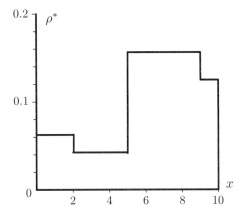

Figure 10. Density function for Example 11.37.

11.4.2. Frequency Measures. After considering invariant measures for piecewise linear maps with the Markov property, we turn in this section to a general way to obtain a dynamically defined invariant measure.

For an initial condition x_0, the orbit $\mathcal{O}_f^+(x_0)$ spends most of its time near to $\omega(x_0, f)$. The frequency measure gives different weights to the points $\omega(x_0, f)$ depending on how frequently the orbit comes near that point. The time average along a whole forward orbit of a continuous function $g(x)$ on the phase space $\mathbb{R}$ for the initial condition x_0 is defined by

$$g^*(x_0) = \lim_{n \to \infty} \frac{1}{n} \sum_{j=0}^{n-1} g(f^j(x_0)).$$

Such a function on our phase space is a single measurement such as temperature, pressure, velocity, or other measurable quantity. In particular, the Lyapunov exponent of a point x_0 involves the continuous function

$$g(x) = \ln|f'(x)|,$$

since

$$\ell(x_0) = \lim_{n\to\infty} \frac{1}{n} \sum_{j=0}^{n-1} \ln|f'(f^j(x_0))| = \lim_{n\to\infty} \frac{1}{n} \sum_{j=0}^{n-1} g(f^j(x_0)) = g^*(x_0).$$

The space average for the frequency measure μ_{x_0} is the integral of the product of $g(x)$ and the density function $\rho_{x_0}(x)$,

$$\int g(x)\,\rho_{x_0}(x)\,dx.$$

The frequency measure is defined so that the time average of a continuous function equals the space average with respect to the frequency density function ρ_{x_0}; that is,

$$g^*(x_0) = \int g(x)\,\rho_{x_0}(x)\,dx.$$

The density function $\rho_{x_0}(x)$ gives different weights to points x in $\omega(x_0;f)$ depending on how frequently the orbit comes near that point.

In the definition of the frequency measure of a closed set, we need to form small open sets surrounding it. For a closed set $\mathbf{S}$, the distance from a point x to the set $\mathbf{S}$ is

$$d(x,\mathbf{S}) = \min\{\,|x - p| : p \in \mathbf{S}\,\}.$$

The *r-neighborhood* of a closed set $\mathbf{S}$ is

$$\mathbf{B}(\mathbf{S},r) = \{\,x : d(x,\mathbf{S}) < r\,\} \quad \text{and}$$

$$\mathbf{S} = \bigcap_{r>0} \mathbf{B}(\mathbf{S},r) = \{\,x : d(x,\mathbf{S}) = 0\,\}.$$

Definition 11.38. Fix an initial condition x_0 for which the forward orbit $\mathcal{O}_f^+(x_0)$ is bounded (i.e., there is a constant $C > 0$ such that $|f^j(x_0)| \leq C$ for all $j \geq 0$). We define the *frequency measure* for x_0 and f first for open sets and then for closed sets.

For an open set $\mathbf{U}$, define

$$\nu(\mathbf{U},n,x_0) = \#\{\,j : f^j(x_0) \in \mathbf{U} \text{ for } 0 \leq j < n\,\},$$

which is the number of times the iterate of x_0 is in the set $\mathbf{U}$ during the first n iterates. Here, ν is the lower case Greek letter "nu". Define the frequency measure of the open set $\mathbf{U}$ by

$$\mu_{x_0,f}(\mathbf{U}) = \lim_{n\to\infty} \frac{\nu(\mathbf{U},n,x_0)}{n}.$$

The limit of this ratio $\nu(\mathbf{U},n,x_0)/n$ as n goes to infinity equals the limit of the proportion of time the orbit spends in the open set, so $0 \leq \mu_{x_0,f}(\mathbf{U}) \leq 1$.

For a closed set $\mathbf{S}$, we define

$$\mu_{x_0,f}(\mathbf{S}) = \lim_{r\to 0+} \mu_{x_0,f}(\mathbf{B}(\mathbf{S},r)),$$

where the limit is taken over $r > 0$ as r approaches 0. Notice that $\mathbf{B}(\mathbf{S},r)$ is an open set, so $\mu_{x_0,f}(\mathbf{B}(\mathbf{S},r))$ is previously defined. The frequency measure of closed sets does not require that the orbit of x_0 hit the set $\mathbf{S}$, but only that it comes near the set; this definition has some similarity to the the definition of the ω-limit set.

11.4. Invariant Measures

The measure on open and closed sets induces the measure on all the other Borel sets by the properties of a measure.

The frequency measure is like a rain gage, showing the relative number of times the orbit hits the various subsets.

The next theorem states that a frequency measure has the properties of a measure and is invariant

Theorem 11.14. *Assume that x_0 is an initial condition with bounded forward orbit for a continuous function f on a Euclidean space $\mathbb{R}^m$. (If it is just a metric space $\mathbf{X}$, then we need to assume that the forward orbit is contained in a compact subset of $\mathbf{X}$.) Let $\mu_{x_0,f}$ be the corresponding frequency measure.*

a. *For any Borel set $\mathbf{S}$, $0 \leq \mu_{x_0,f}(\mathbf{S}) \leq 1$.*

b. *The measure of the total space $\mathbb{R}^m$ is one: $\mu_{x_0,f}(\mathbb{R}^m) = 1$.*

c. *If $\{\,\mathbf{S}_j : j \in \mathscr{I}\,\}$ are a finite or denumerable collection of disjoint sets, $\mathbf{S}_i \cap \mathbf{S}_j = \emptyset$ for $i \neq j$, then*

$$\mu_{x_0,f}\left(\bigcup_{j \in \mathscr{I}} \mathbf{S}_j\right) = \sum_{j \in \mathscr{I}} \mu_{x_0,f}(\mathbf{S}_j),$$

so the measure is countably additive.

d. *If $\mathbf{S}$ is a Borel set, then*

$$\mu_{x_0,f}(f^{-1}(\mathbf{S})) = \mu_{x_0,f}(\mathbf{S}),$$

so $\mu_{x_0,f}$ is an invariant measure for f.

e. *The support of the measure is contained in the ω-limit set, $\mathrm{supp}(\mu_{x_0,f}) \subset \omega(x_0; f)$.*

Idea of proof: Parts **(a)** and **(b)** of the theorem follow fairly directly from the definitions.

(c) We consider only two sets. First, if $\mathbf{U}_1$ and $\mathbf{U}_2$ are disjoint open sets, then

$$\nu(\mathbf{U}_1 \cup \mathbf{U}_2, n, x_0) = \nu(\mathbf{U}_1, n, x_0) + \nu(\mathbf{U}_2, n, x_0)$$

for any n, because the sets are disjoint. Therefore,

$$\begin{aligned}\mu_{x_0,f}(\mathbf{U}_1 \cup \mathbf{U}_2) &= \lim_{n \to \infty} \frac{\nu(\mathbf{U}_1 \cup \mathbf{U}_2, n, x_0)}{n} \\ &= \lim_{n \to \infty} \frac{\nu(\mathbf{U}_1, n, x_0)}{n} + \lim_{n \to \infty} \frac{\nu(\mathbf{U}_2, n, x_0)}{n} \\ &= \mu_{x_0,f}(\mathbf{U}_1) + \mu_{x_0,f}(\mathbf{U}_2).\end{aligned}$$

Next, consider two disjoint closed sets $\mathbf{S}_1$ and $\mathbf{S}_2$. For sufficiently small $r > 0$, $\mathbf{B}(\mathbf{S}_1, r)$ and $\mathbf{B}(\mathbf{S}_2, r)$ are disjoint and

$$\mathbf{B}(\mathbf{S}_1 \cup \mathbf{S}_2, r) = \mathbf{B}(\mathbf{S}_1, r) \cup \mathbf{B}(\mathbf{S}_2, r).$$

Therefore, the limits add, or

$$\mu_{x_0,f}(\mathbf{S}_1 \cup \mathbf{S}_2) = \lim_{r \to 0^+} \mu_{x_0,f}(\mathbf{B}(\mathbf{S}_1 \cup \mathbf{S}_2, r))$$
$$= \lim_{r \to 0^+} \mu_{x_0,f}(\mathbf{B}(\mathbf{S}_1, r)) + \lim_{r \to 0^+} \mu_{x_0,f}(\mathbf{B}(\mathbf{S}_2, r))$$
$$= \mu_{x_0,f}(\mathbf{S}_1) + \mu_{x_0,f}(\mathbf{S}_2).$$

The result for a finite or denumerable number of open or closed sets is similar. Finally, the result for open and closed sets implies the result for all Borel sets.

(d) For open sets, $f^{-1}(\mathbf{U})$ is also an open set because f is continuous, and the orbit must pass through $f^{-1}(\mathbf{U})$ to reach $\mathbf{U}$, $f^j(x_0) \in f^{-1}(\mathbf{U})$ if and only if $f^{j+1}(x_0) \in \mathbf{U}$, so the count

$$\#\{ j : f^j(x_0) \in f^{-1}(\mathbf{U}) \text{ for } 0 \le j < n \,\}$$
$$= \#\{ j : f^j(x_0) \in \mathbf{U} \text{ for } 1 \le j < n+1 \,\},$$

and

$$|\nu(f^{-1}(\mathbf{U}), n, x_0) - \nu(\mathbf{U}, n+1, x_0)| \le 1.$$

Therefore,

$$\mu_{x_0,f}(f^{-1}(\mathbf{U})) = \lim_{n \to \infty} \frac{\nu(f^{-1}(\mathbf{U}), n, x_0)}{n}$$
$$= \lim_{n \to \infty} \frac{\nu(\mathbf{U}, n, x_0)}{n}$$
$$= \mu_{x_0,f}(\mathbf{U})$$

for open sets.

For a closed bounded set $\mathbf{S}$, because f is (uniformly) continuous,

$$\bigcap_{r>0} f^{-1}(\mathbf{B}(\mathbf{S}, r)) = f^{-1}(\mathbf{S}),$$

so

$$\mu_{x_0,f}(f^{-1}(\mathbf{S})) = \mu_{x_0,f}\left(\bigcap_{r>0} f^{-1}(\mathbf{B}(\mathbf{S}, r)) \right)$$
$$= \lim_{r \to 0^+} \mu_{x_0,f}\left(f^{-1}(\mathbf{B}(\mathbf{S}, r)) \right)$$
$$= \lim_{r \to 0^+} \mu_{x_0,f}\left(\mathbf{B}(\mathbf{S}, r) \right)$$
$$= \mu_{x_0,f}(\mathbf{S}).$$

For the property on open and closed sets, the property follows for all Borel sets.

(e) Since an orbit tends toward its ω-limit set, the support of the frequency measure is a subset of the ω-limit set, $\mathrm{supp}(\mu_{x_0,f}) \subset \omega(x_0;f)$. In most cases, this should be an equality, but there can be exceptional cases in which the orbit of x_0 only rarely comes near a point y in $\omega(x_0;f)$, so y is not in $\mathrm{supp}(\mu_{x_0,f})$. □

Example 11.39. In this example, we describe how it is possible for the support of a frequency measure to be smaller than the ω-limit set by modifying the construction of the sequence $\mathbf{s}^*$ that has the full two shift as its ω-limit set. The new sequence $\mathbf{t}^*$ still has its ω-limit set equal to the full two shift but its frequency measure has support at a single point (a single symbol sequence), $\mu_{\mathbf{t}^*} = \delta_{\mathbf{0}}$. Let $\mathbf{t}^*$ have

11.4. Invariant Measures

the same strings as $\mathbf{s}^*$, but after every string of length k in $\mathbf{s}^*$ add a sting of length k^2 of all 0's. Thus, each string of length k is replaced by a string of length $k + k^2$ with 0 occurring at least k^2 times. Thus, the fraction of 0's in the new string is at least $k^2/(k+k^2) = k/(1+k)$. As k goes to infinity the fraction goes to one. Therefore the support of the frequency measure must be in the set $\{\mathbf{s} : s_0 = 0\}$. Similar reasoning shows that the support must be in each of the sets $\{\mathbf{s} : s_i = 0 \text{ for } 0 \leq i < j\}$ and so in $\{\mathbf{0}\}$.

Using the semiconjugacy k from Σ_2^+ to $[0,1]$ for the tent map, the point $y^* = k(\mathbf{t}^*)$ has $\omega(y^*, T) = [0,1]$ and $\mu_{y^*,T} = \delta_0$ with an atomic measure at the origin.

A measure that does not depend in a delicate way on the choice of the initial condition x_0 is called a natural measure, as indicated precisely in the next definition.

Definition 11.40. A measure μ^* is called a *natural measure* for f, provided that the set of points x_0, for which the frequency measure $\mu_{x_0,f}$ for x_0 equals μ^*, has positive Lebesgue measure; that is

$$\lambda(\{x_0 : \mu_{x_0,f} = \mu^*\}) > 0.$$

The fact that the set of points x_0, for which $\mu_{x_0,f} = \mu^*$ has positive Lebesgue measure, means that a choice of an initial condition within a reasonable part of the phase space often yields a point that generates the measure. Thus, there is a positive chance, in terms of Lebesgue measure, of selecting a point for which $\omega(x_0; f) = \mathrm{supp}(\mu^*)$, and $\mathrm{supp}(\mu^*)$ is a Milnor attractor.

The next theorem asserts that a few of the invariant measures we discussed earlier are natural measures.

Theorem 11.15. a. *Lebesgue measure on $[0,1]$ is a natural measure for both the doubling map and the tent map.*

b. *The measure with density function $\rho_G(x) = \dfrac{1}{\pi\sqrt{x-x^2}}$ is a natural measure for the logistic map $G(x) = 4x(1-x)$.*

Since Lebesgue measure is a natural measure for the tent map and the logistic map is conjugate to the logistic map, the resulting measure μ_G is a natural measure for G. The fact that $\rho_G(x) > 0$ for $0 < x < 1$, indicates that an orbit $\mathcal{O}_f^+(x_0)$, which generates the natural measure, is not only dense in $[0,1]$, but also has a positive frequency in every short interval in $[0,1]$. Thus, the map G is transitive and, so, chaotic in a strong sense of the term. Also, the fact that the density function is not bounded on $[0,1]$ but goes to infinity as x approaches both 0 and 1 implies that the typical orbit that generates the natural measure spends more time near these points than in the middle of the interval.

The frequency measure of a specific point can be different than the μ_f natural measure, μ_f. For example, if p is a period-n point for f then $\mu_{p,f}$ is a measure whose support equals to the periodic orbit. Let $\delta_q(\mathbf{S})$ be the measure given by

$$\delta_q(\mathbf{S}) = \begin{cases} 1 & \text{if } q \in \mathbf{S}, \\ 0 & \text{if } q \notin \mathbf{S}. \end{cases}$$

Exercise 3 in this section asks the reader to show that if p is a period-n point for f, then

$$\mu_{p,f}(\mathbf{S}) = \frac{1}{n}\sum_{j=0}^{n-1}\delta_{f^j(p)}(\mathbf{S}).$$

Thus, $\mu_{p,f}(\mathbf{S})$ just counts the proportion of the orbit $\mathcal{O}_f^+(p)$ that lies in a set $\mathbf{S}$. For the tent map, this is not equal to its natural measure.

Numerical estimate of frequency measure

A numerical calculation of the frequency measure for $f(x)$ on an invariant interval $[a,b]$ is make as follows. Let $\mathbf{A}_j = \left[a + \frac{(j-1)(b-a)}{k}, a + \frac{j(b-a)}{k}\right)$ for $1 \leq j \leq k$ be the subdivision of $[a,b]$ into k subintervals of equal length. Fix a large number of iterates N. Let $\nu(j) = \nu(\mathbf{A}_j, N, x_0)$ and $\nu(j)/N$ be the proportion of times $f^i(x_0)$ returns to $\mathbf{A}_j$ for $0 \leq i < N$. The approximate constant density on $\mathbf{A}_j$, ρ_j, is chosen so that $\frac{\nu(j)}{N} = \rho_j \lambda(\mathbf{A}_j) = \frac{\rho_j(b-a)}{k}$, or $\rho_j = \frac{k\,\nu(j)}{N(b-a)}$. The plot of ρ_j over each $\mathbf{A}_j$ is called a *histogram* or *frequency plot*.

Figure 11(a) gives the histogram for the logistic map $G(x) = 4x(1-x)$ with $N = 10,000$ iterates and with $k = 100$ subintervals. Figure 11(b) shows both the histogram and the plot of the actual density function ρ_G. Notice the good agreement between the two plots.

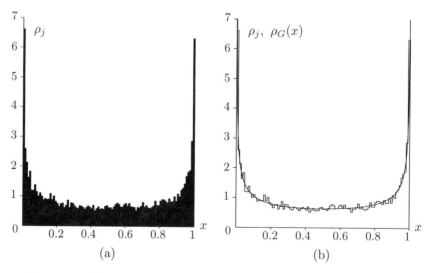

Figure 11. (a) Histogram of $G(x) = 4x(1-x)$ for 10,000 iterates and $k = 100$ subintervals. (b) Graph of both the histogram and density function.

For a function in higher dimensions, we often make a computer plot such as 3 with one black pixel for the location of each iterate. The darkness of an open region $\mathbf{U}$ of the image indicates how often the orbit returns (to different points) in the region; this darkness corresponds to the size of the frequency measure $\mu_{x_0,f}(\mathbf{U})$.

11.4. Invariant Measures

11.4.3. Piecewise Expanding Maps. In this section, we consider frequency measures for piecewise nonlinear expanding maps, without assuming the Markov property and allowing discontinuities. We define a Perron-Frobenius operator on density functions that generalizes the matrix method we presented for piecewise linear maps with a Markov partition.

We start by modifying the earlier method for piecewise linear maps with a Markov partition to see how the densities transform. For an eventually positive (or aperiodic) transition matrix, if $\mathbf{m}_0 = (m_{1,0}, \ldots, m_{k,0})$ is an initial guess and $\mathbf{m}_{\ell+1} = \mathbf{m}_\ell \mathbf{M}$ are defined inductively, then $\mathbf{m}_\ell$ converges to the eigenvector $\mathbf{m}^*$ giving the invariant measures of the intervals. The components of the sequence of measures are given by

$$m_{j,\ell+1} = \sum_i \frac{t_{ij} L_j}{\beta_i L_i} m_{i,\ell}.$$

If we let $\rho_{j,\ell} = m_{j,\ell}/L_j$ be the corresponding densities, then

$$\rho_{j,\ell+1} = \frac{m_{j,\ell+1}}{L_j} = \frac{1}{L_j} \sum_i \frac{t_{ij} L_j}{\beta_i L_i} m_{i,\ell}$$

$$= \sum_i \frac{t_{ij}}{\beta_i L_i} L_i \rho_{i,\ell} = \sum_i t_{ij} \frac{\rho_{i,\ell}}{\beta_i}.$$

Thus to calculate $\rho_{j,\ell+1}$, we find all intervals $\mathbf{A}_i$ that map onto $\mathbf{A}_j$ and divide the density $\rho_{i,\ell}$ on each by its slope β_i and take the sum.

We now turn to a nonlinear map which possibly has a finite number of discontinuities or points where possibly the derivative does not exist. We assume that there are points

$$q_0 < q_1 < \cdots < q_k$$

such that f restricted to each open interval $\mathbf{A}_j = (q_{j-1}, q_j)$ is C^2, with a bound on the first and second derivatives. Moreover, the map is expanding, so we assume that there is an $s > 1$ such that $|f'(x)| \geq s$ for all $q_0 < x < q_k$, with $x \neq q_j$ for any j. Finally, we assume that the interval $[q_0, q_k]$ is positively invariant, so $f(x) \in [q_0, q_k]$ for all x in $[q_0, q_k]$.

For such a nonlinear map, analogous to the piecewise linear case, we want a construction of a sequence of density functions that converge to a density function of an invariant measure. Starting with $\rho_0(x) = \frac{1}{q_k - q_0}$, assume that we have defined densities up to $\rho_\ell(x)$. Near any preimage $y = f^{-1}(x)$, the image is infinitesimally stretched by $|f'(y)|$, so the density $\rho_\ell(y)$ is thinned out by $\frac{1}{|f'(y)|}$ and so is taken by f to $\frac{\rho_\ell(y)}{|f'(y)|}$. Summing the terms for all the preimages of x,

$$\rho_{\ell+1}(x) = P(\rho_\ell)(x) = \sum_{y \in f^{-1}(x)} \frac{\rho_\ell(y)}{|f'(y)|}.$$

Note that we divided by the slope in the linear case and divide by the derivative in the nonlinear case. We sum over all preimages in both cases. This operator P, which takes one density function to another density function, is called the *Perron–Frobenius operator*. Lasota and Yorke [**Las73**] proved that the limit of the average of the first n density functions converges to a density function $\rho^*(x)$,

$$\rho^*(x) = \lim_{n\to\infty} \frac{1}{n} \sum_{\ell=0}^{n-1} \rho_\ell(x).$$

It is necessary to take the average of the sequence of densities because we are not assuming any condition like aperiodicity. The construction guarantees that $\rho^*(x)$ is the density function for an invariant measure μ_{ρ^*}. We summarize these results in the following theorem.

Theorem 11.16 (Lasota, Li, and Yorke). *Let f be a possibly discontinuous piecewise C^2 expanding function with expanding factor $\beta > 1$. We assume that f and f' have only finitely many discontinuities $q_0, \ldots, q_k$, and that $[q_0, q_k]$ is positively invariant. Then the following hold.*

a. *There is an invariant measure μ_{ρ^*} with a bounded density function $\rho^*(x)$, $\rho^*(x) \leq C$, for all x in $[q_0, q_k]$, so $\mu_{\rho^*}([a', b']) \leq C|b' - a'|$ for any closed interval $[a', b']$ in $[q_0, q_k]$. In fact, if a set $\mathbf{S}$ has Lebesgue measure zero, $\lambda(\mathbf{S}) = 0$, then it has zero measure for the invariant measure, $\mu_{\rho^*}(\mathbf{S}) = 0$.*

b. *The measure μ_{ρ^*} can be represented as the sum of a finite number of measures $\mu_{\rho^*} = \mu_{\rho_1} + \cdots + \mu_{\rho_k}$ such that (i) each μ_{ρ_j} is a natural measure with $\omega(x_0^j, f) = \mathrm{supp}(\mu_{\rho_j})$ for a point x_0^j and (ii) the interiors of the supports of the ρ_j are disjoint; that is,*

$$\mathrm{int}\{x : \rho_i(x) > 0\} \cap \mathrm{int}\{x : \rho_j(x) > 0\} = \emptyset \quad \text{for } i \neq j.$$

c. *If there is a point x_0 with $\omega(x_0; f) = [q_0, q_k]$, then the density $\rho^*(x) > 0$ on $[q_0, q_k]$ and μ_{ρ^*} is a natural measure (i.e., $k = 1$ in part (b) and there is no decomposition).*

Idea of the proof. We gave the idea of the proof of part (**a**) before the statement of the theorem.

(**b**) Li and Yorke [**Li,75**] showed that the density function for an invariant probability measure of part (a) can be split up into a finite number of parts, each with its own density function, and the interiors of the supports of the different density functions are disjoint. Each of the separate density functions gives an invariant measure that is a natural measure.

(**c**) If there is a point x_0 for which $\omega(x_0; f)$ equals the whole interval, then all of the limiting measure cannot be split up, and so the limiting measure itself is a natural measure. □

Example 11.41. Let

$$f(x) = \frac{4}{3} x \pmod{1}$$

be the map with one discontinuity at $x = 3/4$. It does not have a Markov partition because $(4/3)^n$ is never an integer, so $f^n(1)$ never returns to 0. Thus, we cannot use the matrix method for piecewise linear maps with a Markov partition. We construct the first few density functions by applying the Perron-Frobenius operator, which indicates the form of the invariant density function.

11.4. Invariant Measures

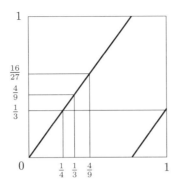

Figure 12. Graph of function for Example 11.41

Take $\rho_0(x) \equiv 1$ on $[0,1]$. Points $0 \leq x \leq 1/3$ have two preimages, while those with $1/3 < x < 1$ have one and $x = 1$ has none, so

$$\rho_1(x) = \begin{cases} \frac{3}{2} & \text{for } 0 \leq x \leq \frac{1}{3}, \\ \frac{3}{4} & \text{for } \frac{1}{3} < x < 1, \\ 0 & \text{for } x = 1. \end{cases}$$

To calculate the next density function $\rho_2(x) = P(\rho_1)(x)$, we use the following preimages: Points $0 \leq x < 1/3$ have one preimage in $[0, 1/4)$ and another in $[3/4, 1)$; points $(1/3, 4/9]$ have only one preimage in $(1/4, 1/3]$; finally, points $(4/9, 1)$ have only one preimage in $(1/3, 3/4)$. Using the values of $\rho_1(x)$ in these respective intervals, we get

$$\rho_2(x) = \begin{cases} \frac{3}{2} \cdot \frac{3}{4} + \frac{3}{4} \cdot \frac{3}{4} = \frac{27}{16} & \text{for } 0 \leq x < \frac{1}{3}, \\ \frac{3}{2} \cdot \frac{3}{4} = \frac{9}{8} & \text{for } \frac{1}{3} < x \leq \frac{4}{9}, \\ \frac{3}{4} \cdot \frac{3}{4} = \frac{9}{16} & \text{for } \frac{4}{9} < x < 1, \end{cases}$$

By similar considerations,

$$\rho_3(x) = \begin{cases} \frac{27}{16} \cdot \frac{3}{4} + \frac{9}{16} \cdot \frac{3}{4} = \frac{27}{16} & \text{for } 0 \leq x \leq \frac{1}{3}, \\ \frac{27}{16} \cdot \frac{3}{4} = \frac{81}{64} & \text{for } \frac{1}{3} < x \leq \frac{4}{9}, \\ \frac{9}{8} \cdot \frac{3}{4} = \frac{27}{32} & \text{for } \frac{4}{9} < x \leq \frac{16}{27}, \\ \frac{9}{16} \cdot \frac{3}{4} = \frac{27}{64} & \text{for } \frac{16}{27} < x < 1. \end{cases}$$

We do not explicitly calculate more of the density functions in the sequence, but Figure 13(a) shows the graph of $\rho_{15}(x)$, which was calculated using Maple. In Figure 13(b), we plot the histogram for 500,000 iterates and 500 subintervals of $[0, 1]$. Notice the relative good agreement between these two plots. The spikes in the histogram are caused by the fact that we did not use an infinite orbit; the true limiting density is monotonically decreasing.

Example 11.42. The piecewise expanding maps with chaotic attractors given in Examples 11.18 and 11.19 also do not have Markov partitions. In the Figures 14

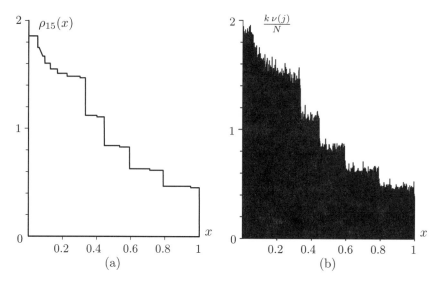

Figure 13. Example 11.41, $f(x) = \frac{4}{3}x \pmod 1$. (a) Graph of the approximate density function $\rho_{15}(x)$ calculated by the Perron–Frobenius operator. (b) Graph of histogram for 500,000 iterates.

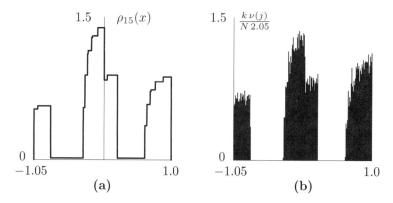

Figure 14. Plots for Example 11.18. (a) Graph of approximate density function $\rho_{15}(x)$. (b) Graph of histogram for 10,000 iterates and $k = 200$ subdivisions.

and 15, we plot (a) the approximate densities $\rho_{15}(x)$ calculated with the Perron-Frobenius operator and (b) the histograms of a frequency for 10,000 iterates. For the histogram, each subinterval is $w = \dfrac{b-a}{k}$ wide, so we plot $\rho_j = \dfrac{k\,\nu(j)}{N\,(b-a)}$. Note that each of the measure is supported on three intervals that correspond to their respective attractors.

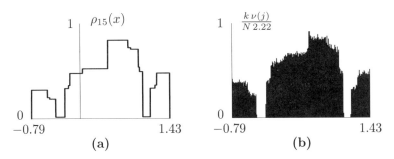

Figure 15. Plots for Example 11.19. (a) Graph of approximate density function $\rho_{15}(x)$. (b) Graph of histogram for 10,000 iterates and $k = 200$ subdivisions.

Exercises 11.4

1. Explain why the rotation
$$R_\alpha(x) = x + \alpha \pmod{1}$$
preserves Lebesgue measure on $[0, 1]$.

2. Show that the tripling map $f(x) = 3x \pmod{1}$ preserves Lebesgue measure on $[0, 1]$.

3. Assume that p is a period-n point for f. Explain why
$$\mu_{p,f}(\mathbf{S}) = \frac{1}{n} \sum_{j=0}^{n-1} \delta_{f^j(p)}(\mathbf{S}).$$

4. Consider the map
$$f(x) = \begin{cases} \frac{3}{2}x & \text{for } 0 \leq x \leq \frac{2}{3}, \\ 2x - \frac{4}{3} & \text{for } \frac{2}{3} < x \leq 1. \end{cases}$$

   a. Draw the graph of f. Also, explain why f is an expanding map that has a Markov partition.
   b. Give the transition matrix $\mathbf{M}$ on measures of the subintervals, and find the invariant measures $\mathbf{m}^*$.
   c. Give the densities ρ_i^* that correspond to the invariant measures $\mathbf{m}^*$. Sketch the graph of the density function ρ^* that takes on the values ρ_i^*.
   d. Show that the Perron–Frobenius operator for this map preserves the density function ρ^* found in part (c).

5. Consider the map
$$f(x) = \begin{cases} 4x & \text{for } 0 \leq x \leq \frac{1}{4}, \\ -\frac{3}{2}x + \frac{11}{8} & \text{for } \frac{1}{4} < x \leq \frac{3}{4}, \\ 2x - \frac{5}{4} & \text{for } \frac{3}{4} < x \leq 1. \end{cases}$$

a. Draw the graph of f. Also, explain why f is an expanding map that has a Markov partition.

b. Give the transition matrix $\mathbf{M}$ on measures of the subintervals, and find the invariant measures $\mathbf{m}^*$.

c. Give the densities ρ_i^* that correspond to the invariant measures $\mathbf{m}^*$. Sketch the graph of the density function ρ^* that takes on the values ρ_i^*.

d. What is the set of all possible periods for f?

11.5. Applications

11.5.1. Capital Accumulation. Business cycles can be modeled by the amount produced and a savings level. A discrete-time version of the Solow growth model is given as follows. We assume a Cobb–Douglas production function $Y_n = A K_n^\alpha L_n^{1-\alpha}$ with $0 < \alpha < 1$ and $A > 0$, where K_n is the capital in period n and L_n is the labor. Let $k_n = K_n/L_n$ be the ratio of capital to labor in period n. Let $s(k_n)$ be the possibly nonlinear saving function, so $K_{n+1} = s(k_n) Y_n$. The labor is assume to grow by $L_{n+1} = (1+\lambda) L_n$. Then

$$k_{n+1} = \frac{K_{n+1}}{L_{n+1}} = s(k_n) \frac{A K^\alpha L^{1-\alpha}}{(1+\lambda) L_n}$$
$$= s(k_n) \frac{A}{1+\lambda} k_n^\alpha.$$

If $\sigma = s(k_n)$ is a constant, then we get the function

$$k_{n+1} = f(k_n) = \frac{\sigma A}{1+\lambda} k_n^\alpha$$

to iterate, as we indicated in Chapter 8. This type of function cannot have any chaotic behavior.

R. Day indicated that it is possible to get irregular or chaotic growth cycles by using a nonlinear saving function. See [**Day82**]. Also, compare with Section 3.12 in [**Sho02**]. Day assumes that the interest rate r_n is given by $r_n = \beta y_n / k_n = \beta A k_n^{\alpha - 1}$ and the per capita savings that satisfy

$$s(k_n) y_n = a \left(1 - \frac{b}{r_n}\right) k_n = a \left(1 - \frac{b k_n^{1-\alpha}}{\beta A}\right) k_n, \quad \text{so}$$

$$k_{n+1} = g(k_n) = \frac{a}{1+\lambda} \left(1 - \frac{b k_n^{1-\alpha}}{\beta A}\right) k_n$$

(11.1) $$= c \left(1 - B k_n^{1-\alpha}\right) k_n,$$

where $c = a/(1+\lambda)$ and $B = b/(\beta A)$. This function g has many of the properties of the logistic map with one maximum and $g(0) = 0 = g\left(B^{1/(1-\alpha)}\right)$. It has dynamic behavior like the logistic map. In particular, Figure 16 shows the graphical iteration of the single point 0.5 for the parameter values $\alpha = 0.5$, $B = 1$, and $c = 6.5$. This orbit appears to be dense in an interval approximately equal to $[0.117, 0.963]$. Thus, this map appears to have a chaotic attractor for these parameter values.

11.5. Applications

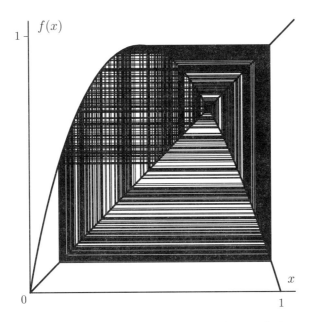

Figure 16. Plot of graphical iteration for the capital accumulation model for $\alpha = 0.5$, $B = 1$, and $c = 6.5$.

11.5.2. Experience dependent utility preferences. In [**Ben81**], Benhabib and Day gave an example where erratic or chaotic behavior can result in a situation where the preferences depend on experience. Assume that an agent maximizes a Cobb-Douglas utility function $U = x^a y^{1-a}$ subject to the budget constraint $px + py = w$. Lagrange multipliers shows that the maximum occurs at

$$x = \frac{w}{p} a \quad \text{and} \quad y = \frac{w}{q}(1-a).$$

They assume that the parameter a at time period $n+1$ depends on x_n and y_n by the relationship

$$a_{n+1} = b\, x_n\, y_n,$$

for a constant b. Using this dependence and the fact that $qy_n = w - px_n$, we get the map

$$\begin{aligned} x_{n+1} &= \frac{w}{p} b\, x_n\, y_n = \frac{w}{p} b\, x_n \left(\frac{w - p\, x_n}{q}\right) \\ &= \left(\frac{wbp}{pq}\right) x_n \left(\frac{w}{p} - x_n\right). \end{aligned}$$

Making the substitution $\dfrac{w}{p} z = x$, we get the map

(11.2) $$z_{n+1} = \left(\frac{w^2 bp}{q}\right) z_n (1 - z_n).$$

The parameter

(11.3) $$A = \frac{w^2 bp}{q}$$

determines the type of behavior. To insure that x_n does not converge to 0, we need $A > 1$. To insure that $w \geq p\,x_n = w\,z_n$ in the wealth constraint, we need $A \leq 4$. As we discussed earlier for the logistic map, the whole sequence of period doubling bifurcations is finished by $A^* \approx 3.57$. Chaotic behavior occurs for many parameter values with $A^* < A < 4$.

11.5.3. Chaotic Blood Cell Population. As discussed in Section 9.7.3, A. Lasota [**Las77**] proposed a discrete model for the population of red blood cells given by the function

$$f_a(x) = (1-a)x + b\,x^r e^{-sx}.$$

On the basis of data, the parameter values taken are $r = 8$, $s = 16$, and $b = 1.1 \times 10^6$.

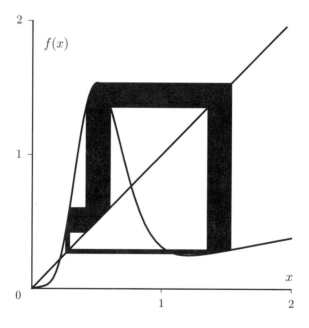

Figure 17. Plot of graphical iteration for the blood cell model with $a = 0.81$

For $a = 0.81$, computer simulation indicates there is a chaotic attractor. Figure 17 plots the graphical iteration of a single orbit, starting at 0.5. This plot gives numerical evidence that there is a point with an orbit dense in the set made up of three subintervals, approximately $[0.262, 0.293] \cup [0.419, 0.64] \cup [1.353, 1.54]$; hence, numerical investigation suggests that $f_{0.81}$ has a chaotic attractor as defined in Section 11.2. Notice that, although the nature of the attractor is similar to those for expanding maps discussed in Section 11.2.1, this map is not expanding since $f_{0.81}$ has a critical point in the attractor.

See [**Mar99**] for further discussion and references for this model of blood cell population.

11.5.4. Mixing of Fluids. J. Ottino and his co-workers have used the ideas of horseshoes to describe mechanisms of mixing of fluids. The mixing mechanism involves stretching and folding of the two-dimensional fluid. To get "chaotic" mixing, he uses a time dependent stirring processes. Through this process, he attains what corresponds to a homoclinic point and a horseshoe map. Further development of his work is given in [**Ott89a**] and [**Ott89b**] by J. Ottino.

11.6. Theory and Proofs

Properties of limit sets

In part (a) of the theorem, we use the definition of a closed set given in Appendix A.2 and in Section 11.2. The details of the proof of this part of the theorem are not important and the result is used mainly to prove part (c).

Theorem 11.17. **a.** *For any map f and for any point x_0, $\omega(x_0; f)$ is a closed set.*
  b. *The set $\omega(x_0; f)$ is positively invariant.*
  c. *If y_0 is a point in $\omega(x_0; f)$, then $\omega(y_0; f) \subset \omega(x_0; f)$.*

Proof. (a) Let p be a point in the closure of $\omega(x_0; f)$, $r > 0$, and N. Since p is in the closure, there is a point q in $(p-r, p+r)$. Take $r' > 0$ such that $(q-r', q+r') \subset (p-r, p+r)$. Because q is in the limit set, there is an $n \geq N$ such that $f^n(x_0) \in (q-r', q+r') \subset (p-r, p+r)$. Because such an n exists for every $r > 0$, and N, p is in the limit set. This shows that the closure of $\omega(x_0; f)$ equals $\omega(x_0; f)$, so $\omega(x_0; f)$ is closed.

 (b) If $f^{k_j}(x_0)$ converges to y_0, then $f^{k_j+1}(x_0)$ converges to $f(y_0)$, so $f(y_0)$ is in the limit set $\omega(x_0; f)$. Thus, $\omega(x_0; f)$ is positively invariant.

 (c) If y_0 is in $\omega(x_0; f)$, then $\mathcal{O}_f^+(y_0) \subset \omega(x_0; f)$ since it is positively invariant by part (b). Since $\omega(x_0; f)$ is closed, the ω-limit points of y_0 must be in $\omega(x_0; f)$. □

One-dimensional Lorenz map

Restatement of Theorem 11.5. *Assume the following: The map f is a differentiable function from $\mathbb{R} \smallsetminus \{c\}$ to $\mathbb{R}$. The one-sided limits at the single discontinuity at the point c exist and are denoted by $f(c-) = b$ and $f(c+) = a$. The iterates of the points a and b each stay on the same side of the discontinuity c for two iterates; more specifically,*

$$a < f(a) < f^2(a) < c < f^2(b) < f(b) < b.$$

For x is $[a, c) \cup (c, b]$, the derivative satisfies $f'(x) \geq \beta > \sqrt{2}$. In particular, f is increasing on both sides of the discontinuity.

  With the above hypotheses, the following three statements follow:

  a. *The map f has sensitive dependence on initial conditions on $[a, b]$.*
  b. *For any subinterval $\mathbf{J} \subset [a, b]$ of positive length, there is an iterate n such that $f^n(\mathbf{J}) = (a, b)$.*
  c. *The map f is topologically transitive on $[a, b]$ and so $[a, b]$ is a chaotic attractor.*

Proof. (a) Take any two nearby points $x_0 < y_0$. If x_0 and y_0 are on the same side of c, then by the mean value theorem, there is a point z_0 between x_0 and y_0 such that
$$|y_1 - x_1| = |f(y_0) - f(x_0)|$$
$$= |f'(z_0)(y_0 - x_0)|$$
$$\geq \beta |y_0 - x_0|.$$
Repeating the argument, as long as $x_j = f^j(x_0)$ and $y_j = f^j(y_0)$ are on the same side of c for $0 \leq j < k$, then
$$|y_k - x_k| \geq \beta |y_{k-1} - x_{k-1}|$$
$$\geq \beta^k |y_0 - x_0|.$$
Since this number grows, eventually x_k and y_k must be on opposite sides of c. Because the map is monotone on each side, $x_k < c < y_k$.

We have $f(c^-) = b > c$ and $f(c^+) = a < c$. Let $r > 0$ be a value such that $f(x) \geq c + (r/2)$ for $c - r \leq x < c$ and $f(y) \leq c - (r/2)$ for $c < y \leq c + r$. We show that this r works in the definition of sensitive dependence. If $y_k - x_k > r$, then we are done. Otherwise, both points are within distance r of c and
$$x_{k+1} - y_{k+1} \geq c + \frac{r}{2} - \left(c - \frac{r}{2}\right) = r.$$
In any case, some iterates are farther apart than r, and we have proved part (a).

(b) Let $\mathbf{J}$ be an interval with positive length in $[a, b]$. We define a sequence of intervals by induction. Let
$$\mathbf{J}_0 = \begin{cases} \mathbf{J} & \text{if } c \notin \mathbf{J}, \\ \text{the longer subinterval of } \mathbf{J} \setminus \{c\} & \text{if } c \in \mathbf{J}. \end{cases}$$
Then by induction, let
$$\mathbf{J}_{k+1} = \begin{cases} f(\mathbf{J}_k) & \text{if } c \notin f(\mathbf{J}_k), \\ \text{the longer subinterval of } f(\mathbf{J}_k) \setminus \{c\} & \text{if } c \in f(\mathbf{J}_k). \end{cases}$$
By the mean value theorem again, if $c \notin f(\mathbf{J}_k)$, $\lambda(\mathbf{J}_{k+1}) \geq \beta \lambda(\mathbf{J}_k)$. If $c \in f(\mathbf{J}_k)$, then $\lambda(\mathbf{J}_{k+1}) \geq (\beta/2) \lambda(\mathbf{J}_k)$, since we cut the interval into two parts and take the longer part. Taking the second iterate, we have
$$\lambda(\mathbf{J}_{k+2}) \geq \begin{cases} \beta^2 \lambda(\mathbf{J}_k) & \text{if } c \notin f(\mathbf{J}_k) \cup f(\mathbf{J}_{k+1}), \\ \dfrac{\beta^2}{2} \lambda(\mathbf{J}_k) & \text{if } c \notin f(\mathbf{J}_k) \cap f(\mathbf{J}_{k+1}), \\ \dfrac{\beta^2}{4} \lambda(\mathbf{J}_k) & \text{if } c \in f(\mathbf{J}_k) \cap f(\mathbf{J}_{k+1}). \end{cases}$$
The first case assumes that c is in neither $f(\mathbf{J}_k)$ nor $f(\mathbf{J}_{k+1})$; the second case assumes that c is not in both $f(\mathbf{J}_k)$ and $f(\mathbf{J}_{k+1})$; the last case assumes that c is in both $f(\mathbf{J}_k)$ and $f(\mathbf{J}_{k+1})$ (i.e., in two successive iterates). Since $\beta > \sqrt{2}$, $\beta^2/2 > 1$, and the length of every second iterate grows until two successive iterates $f(\mathbf{J}_{n-4})$ and $f(\mathbf{J}_{n-3})$ contain c.

11.6. Theory and Proofs

Because c is in $f(\mathbf{J}_{n-4})$, c is one of the end points of $\mathbf{J}_{n-3}$. By assumption (ii) on the function, $f(c^-) = b$ and $f(c^+) = a$, so $f(\mathbf{J}_{n-3})$ limits on either a or b; because $f(\mathbf{J}_{n-3})$ contains c, it contains either the interval $(a, c]$ or $[c, b)$. The two cases are similar, so we consider only the second case in which it contains $[c, b)$. Then,
$$f^2(\mathbf{J}_{n-3}) \supset f((c, b)) \supset (a, c] \cup [c, f(b)).$$
Taking the next iterate, we obtain
$$\begin{aligned} f^3(\mathbf{J}_{n-3}) &\supset f((a, c)) \cup f((c, f(b))) \\ &\supset [c, b) \cup (a, f^2(b)) \\ &\supset [c, b) \cup (a, c] \\ &= (a, b), \end{aligned}$$
where the last inclusion holds by assumption (iii) on the function. Since $f^n(\mathbf{J}) \supset f^3(\mathbf{J}_{n-3})$, this proves the claim.

(c) Part (b) proves that, for any two open intervals $\mathbf{J}_1$ and $\mathbf{J}_2$ in $[a, b]$, there is an n such that $f^n(\mathbf{J}_1) \cap \mathbf{J}_2 \neq \emptyset$. By the theorem that follows, called the Birkhoff transitivity theorem, there is a point with a dense orbit in $[a, b]$. □

The case in which the derivative is positive on one side of the discontinuity and negative on the other side is more complicated. Y. Choi presents some results in [**Cho04**]. Also, see [**Rob00**] for a summary of these results.

The next result gives a precise statement of the Birkhoff transitivity theorem.

Theorem 11.18 (Birkhoff Transitivity Theorem). *Assume that $\mathbf{F}$ is a continuous map from $\mathbb{R}^n$ to itself that takes a closed subset $\mathbf{X}$ to itself. (In general, $\mathbf{X}$ can be a complete metric space with countable basis.) Assume that, for every two open sets $\mathbf{U}$ and $\mathbf{V}$ of $\mathbf{X}$, there is an iterate k such that $\mathbf{F}^k(\mathbf{U})$ intersects $\mathbf{V}$. Then, there is a dense subset $\mathbf{Y}$ of $\mathbf{X}$ such that, for every $\mathbf{p}$ in $\mathbf{Y}$, the forward orbit of $\mathbf{p}$, $\mathcal{O}_\mathbf{F}^+(\mathbf{p})$, is dense in $\mathbf{X}$. Moreover, if $\mathbf{F}$ is not necessarily continuous on all of $\mathbf{X}$ or $\mathbb{R}^n$, but is continuous on a set $\mathbf{X}'$ that is dense and open in $\mathbf{X}$, then the same result is true.*

For more discussion of this theorem, see [**Rob99**].

Chapter 12

Periodic Points of Higher Dimensional Maps

Chapters 9 to 11 have considered the iteration of a function of a single variable. Chapters 12 to 14 consider the iteration of a function of several variables. In this context, the function **F** takes a point in $\mathbb{R}^n$ and gives back another point in the same space $\mathbb{R}^n$. In Section 8.2, we gave a few examples of such functions.

This chapter concentrates on the periodic points of such functions and the behavior near these periodic points. We start with linear maps so we can understand the typical behavior near fixed or periodic points. Next, we discuss the stability types and classification of periodic points. For those periodic points with some contracting directions and some expanding directions (saddles), we introduce the "stable manifold", which consists of those points that tend toward the periodic orbit. We end this chapter by considering a more global situation of certain maps on tori; we can find many different periodic points for these particular maps.

12.1. Dynamics of Linear Maps

For a nonlinear map with a fixed point, the linear terms at the fixed point determine many of the features of the phase portrait of the nonlinear map near the fixed point. Therefore, although we are primarily interested in nonlinear maps, we start by considering the iteration of linear maps.

Let **A** be an $n \times n$ matrix with real entries. For a point **x** in $\mathbb{R}^n$, the matrix product **Ax** gives a new point in $\mathbb{R}^n$. This map is often called the *linear map* or *linear transformation* induced by **A**, but we do not worry about distinguishing between the matrix and the linear map it induces. (The matrix depends on a choice of basis for $\mathbb{R}^n$; a different basis gives a different matrix.)

The origin is always a fixed point for a linear map since $\mathbf{A}\mathbf{0} = \mathbf{0}$ for any $n \times n$ matrix $\mathbf{A}$. Also, matrix multiplication has the linearity property that

$$\mathbf{A}(a\mathbf{x} + b\mathbf{y}) = a\,\mathbf{A}\mathbf{x} + b\,\mathbf{A}\mathbf{y}$$

for any two vectors $\mathbf{x}$ and $\mathbf{y}$ and any two scalars a and b.

Example 12.1. Consider the linear map

$$L\begin{pmatrix} x_1 \\ x_2 \end{pmatrix} = \begin{pmatrix} \frac{4}{5} & 0 \\ \frac{4}{5} & \frac{1}{5} \end{pmatrix} \begin{pmatrix} x_1 \\ x_2 \end{pmatrix}.$$

If we start with a point $\begin{pmatrix} x_1^{(0)} \\ x_2^{(0)} \end{pmatrix}$, then the iterates are

$$\begin{pmatrix} x_1^{(1)} \\ x_2^{(1)} \end{pmatrix} = \begin{pmatrix} \frac{4}{5} & 0 \\ \frac{4}{5} & \frac{1}{5} \end{pmatrix} \begin{pmatrix} x_1^{(0)} \\ x_2^{(0)} \end{pmatrix} = \begin{pmatrix} \frac{4}{5} x_1^{(0)} \\ \frac{4}{5} x_1^{(0)} + \frac{1}{5} x_2^{(0)} \end{pmatrix},$$

$$\begin{pmatrix} x_1^{(2)} \\ x_2^{(2)} \end{pmatrix} = \begin{pmatrix} \frac{4}{5} & 0 \\ \frac{4}{5} & \frac{1}{5} \end{pmatrix} \begin{pmatrix} x_1^{(1)} \\ x_2^{(1)} \end{pmatrix} = \begin{pmatrix} \left(\frac{4}{5}\right)^2 x_1^{(0)} \\ \left(\frac{4^2}{5^2} + \frac{4}{5^2}\right) x_1^{(0)} + \left(\frac{1}{5}\right)^2 x_2^{(0)} \end{pmatrix},$$

$$\begin{pmatrix} x_1^{(3)} \\ x_2^{(3)} \end{pmatrix} = \begin{pmatrix} \frac{4}{5} & 0 \\ \frac{4}{5} & \frac{1}{5} \end{pmatrix} \begin{pmatrix} x_1^{(2)} \\ x_2^{(2)} \end{pmatrix} = \begin{pmatrix} \left(\frac{4}{5}\right)^3 x_1^{(0)} \\ \left(\frac{4^3}{5^3} + \frac{4^2}{5^3} + \frac{4}{5^3}\right) x_1^{(0)} + \left(\frac{1}{5}\right)^3 x_2^{(0)} \end{pmatrix}.$$

Notice that forward iterates involve powers of $\frac{4}{5}$ and $\frac{1}{5}$, which are the entries on the diagonal (the eigenvalues).

Since the determinant of the matrix of the linear map is $\frac{4}{25} \neq 0$, the matrix has an inverse

$$\frac{25}{4}\begin{pmatrix} \frac{1}{5} & 0 \\ -\frac{4}{5} & \frac{4}{5} \end{pmatrix} = \begin{pmatrix} \frac{5}{4} & 0 \\ -5 & 5 \end{pmatrix}.$$

Therefore, it is possible to take backward iterates

$$\begin{pmatrix} x_1^{(-1)} \\ x_2^{(-1)} \end{pmatrix} = L^{-1}\begin{pmatrix} x_1^{(0)} \\ x_2^{(0)} \end{pmatrix} = \begin{pmatrix} \frac{5}{4} & 0 \\ -5 & 5 \end{pmatrix}\begin{pmatrix} x_1^{(0)} \\ x_2^{(0)} \end{pmatrix}.$$

Again, notice that backward iterates involve powers of $\frac{5}{4}$ and 5, which are the inverses of the entries on the diagonal (and the eigenvalues). We return to a more complete analysis of this linear map in Example 12.2.

We now give notation for a general linear map $\mathbf{A}$ that we used in the preceding example. If we start with a point (or vector) $\mathbf{x}^{(0)}$ in $\mathbb{R}^n$, then we can repeatedly apply the matrix $\mathbf{A}$ to obtain new points that we label as $\mathbf{x}^{(1)}, \mathbf{x}^{(2)}, \ldots$:

$$\mathbf{x}^{(1)} = \mathbf{A}\mathbf{x}^{(0)} \quad \text{and}$$
$$\mathbf{x}^{(k)} = \mathbf{A}\mathbf{x}^{(k-1)} = \mathbf{A}^k \mathbf{x}^{(0)} \quad \text{for } k \geq 1.$$

12.1. Linear Maps

If $\mathbf{A}$ has a nonzero determinant, then it has an inverse $\mathbf{A}^{-1}$, and we can form the backward iterates

$$\mathbf{x}^{(-1)} = \mathbf{A}^{-1}\mathbf{x}^{(0)} \quad \text{and}$$
$$\mathbf{x}^{(-k)} = \mathbf{A}^{-1}\mathbf{x}^{(-k+1)} = \mathbf{A}^{-k}\mathbf{x}^{(0)} \quad \text{for } -k \leq -1.$$

The points with the simplest dynamics are those that are taken into a scalar multiple of themselves; that is, points which satisfy

$$\mathbf{A}\mathbf{v} = \lambda\mathbf{v} \quad \text{for } \mathbf{v} \neq \mathbf{0}.$$

A number λ is an *eigenvalue* of a matrix $\mathbf{A}$ if there is a nonzero vector $\mathbf{v}$ such that

$$\mathbf{A}\mathbf{v} = \lambda\mathbf{v}.$$

The vector $\mathbf{v}$ is called an *eigenvector* of $\mathbf{A}$, corresponding to the eigenvalue λ. These vectors and numbers satisfy

$$(\mathbf{A} - \lambda\mathbf{I})\mathbf{v} = \mathbf{0}$$

for a nonzero vector $\mathbf{v}$, so

$$\det(\mathbf{A} - \lambda\mathbf{I}) = 0.$$

This preceding equation is called the *characteristic equation* of the matrix $\mathbf{A}$.

Eigenvalues and Eigenvectors

To find the eigenvalues and eigenvectors, the first step is to solve the characteristic equation

$$\det(\mathbf{A} - \lambda\mathbf{I}) = 0,$$

for the roots $\lambda_1, \ldots, \lambda_n$. The λ_j are the *eigenvalues*, and can be real or complex. For each root λ_j, row reduce $\mathbf{A} - \lambda_j\mathbf{I}$ to find a nonzero solution $\mathbf{v}^j$ of the equation

$$(\mathbf{A} - \lambda_j\mathbf{I})\mathbf{v} = \mathbf{0}.$$

The vector $\mathbf{v}^j$ is called an *eigenvector* for the eigenvalue λ_j.

The characteristic equation is especially easy to write down for a two-by-two matrix.

Theorem 12.1. *For a two-by-two matrix* $\mathbf{A} = \begin{pmatrix} a & b \\ c & d \end{pmatrix}$ *with trace* $\tau = \operatorname{tr}(\mathbf{A}) = a + d$ *and determinant* $\Delta = \det(\mathbf{A}) = ad - bc$, *the the characteristic equation is* $0 = \lambda^2 - \tau\lambda + \Delta$.

Proof. The characteristic equation is

$$0 = \det\begin{pmatrix} a - \lambda & b \\ c & d - \lambda \end{pmatrix} = (a-\lambda)(d-\lambda) - bc = \lambda^2 - (a+d)\lambda + (ad-bc). \quad \square$$

If the eigenvalues are distinct (i.e., $\lambda_i \neq \lambda_j$ when $i \neq j$), then it can be shown that the n eigenvectors $\mathbf{v}^1, \ldots, \mathbf{v}^n$ are *linearly independent*; that is, if

$$c_1 \mathbf{v}^1 + \cdots + c_n \mathbf{v}^n = \mathbf{0}$$

then all the $c_j = 0$. This property can be expressed by saying that the only solution of the matrix equation

$$(\mathbf{v}^1, \ldots, \mathbf{v}^n) \begin{pmatrix} c_1 \\ \vdots \\ c_n \end{pmatrix} = \mathbf{0}$$

is $c_1 = \cdots = c_n = 0$ (i.e., the homogeneous equation has only the trivial solution). The homogeneous equation has only the trivial solution if and only if

$$\det(\mathbf{v}^1, \ldots, \mathbf{v}^n) \neq 0.$$

Also, for any column vector $\mathbf{x} = (x_1, \ldots, x_n)^\mathsf{T}$, it is possible to solve

$$(\mathbf{v}^1, \ldots, \mathbf{v}^n) \begin{pmatrix} y_1 \\ \vdots \\ y_n \end{pmatrix} = \begin{pmatrix} x_1 \\ \vdots \\ x_n \end{pmatrix}$$

for the coefficients y_i, so any $\mathbf{x}$ can be written as a linear combination of the eigenvectors, or

$$\mathbf{x} = y_1 \mathbf{v}^1 + \cdots + y_n \mathbf{v}^n.$$

Because any vector can be written uniquely as a linear combination, any set of n vectors in $\mathbb{R}^n$ for which $\det(\mathbf{v}^1, \ldots, \mathbf{v}^n) \neq 0$ is called a *basis* of $\mathbb{R}^n$.

If $\mathbf{v}^j$ is an eigenvector for the eigenvalue λ_j, then $\mathbf{A} \mathbf{v}^j = \lambda_j \mathbf{v}^j$, and by induction

$$\mathbf{A}^2 \mathbf{v}^j = \mathbf{A} \left(\mathbf{A} \mathbf{v}^j \right) = \mathbf{A} \left(\lambda_j \mathbf{v}^j \right) = \lambda_j \left(\mathbf{A} \mathbf{v}^j \right) = \lambda_j \left(\lambda_j \mathbf{v}^j \right) = \lambda_j^2 \mathbf{v}^j,$$
$$\mathbf{A}^3 \mathbf{v}^j = \mathbf{A} \left(\mathbf{A}^2 \mathbf{v}^j \right) = \mathbf{A} \left(\lambda_j^2 \mathbf{v}^j \right) = \lambda_j^2 \left(\mathbf{A} \mathbf{v}^j \right) = \lambda_j^2 \left(\lambda_j \mathbf{v}^j \right) = \lambda_j^3 \mathbf{v}^j, \quad \text{and}$$
$$\mathbf{A}^k \mathbf{v}^j = \mathbf{A} \left(\mathbf{A}^{k-1} \mathbf{v}^j \right) = \mathbf{A} \left(\lambda_j^{k-1} \mathbf{v}^j \right) = \lambda_j^{k-1} \left(\mathbf{A} \mathbf{v}^j \right) = \lambda_j^{k-1} \left(\lambda_j \mathbf{v}^j \right) = \lambda_j^k \mathbf{v}^j$$
$$\text{for } k \geq 1.$$

If $\mathbf{A}^{-1}$ exists, then all the eigenvalues $\lambda_j \neq 0$ and for negative powers,

$$\mathbf{A}^{-1} \mathbf{v}^j = \lambda_j^{-1} \mathbf{v}^j \quad \text{and}$$
$$\mathbf{A}^{-k} \mathbf{v}^j = \lambda_j^{-k} \mathbf{v}^j \quad \text{for } -k \leq -1.$$

Combining, we have

$$\mathbf{A}^k \mathbf{v}^j = \lambda_j^k \mathbf{v}^j \quad \text{for all integers } k.$$

Note that, if $|\lambda_j| < 1$, then $\|\mathbf{A}^k \mathbf{v}^j\| = |\lambda_j|^k \|\mathbf{v}^j\|$ goes to zero as k goes to infinity and goes to infinity as k goes to minus infinity. If $|\lambda_j| > 1$, then $\|\mathbf{A}^k \mathbf{v}^j\| = |\lambda_j|^k \|\mathbf{v}^j\|$ goes to infinity as k goes to infinity and goes to zero as k goes to minus infinity.

Real distinct eigenvalues

Assume that $n = 2$ and that there are two real eigenvalues $\lambda_1 \neq \lambda_2$, and $|\lambda_1| < |\lambda_2| < 1$. Let $\mathbf{v}^1$ and $\mathbf{v}^2$ be the corresponding eigenvectors. Then, any point $\mathbf{x}$ can be written as a combination of the eigenvectors:

$$\mathbf{x} = y_1 \mathbf{v}^1 + y_2 \mathbf{v}^2.$$

Then,

$$\mathbf{A}^k \mathbf{x} = y_1 \mathbf{A}^k \mathbf{v}^1 + y_2 \mathbf{A}^k \mathbf{v}^2$$
$$= y_1 \lambda_1^k \mathbf{v}^1 + y_2 \lambda_2^k \mathbf{v}^2.$$

Taking absolute values and applying the triangle inequality yields

$$\|\mathbf{A}^k \mathbf{x}\| = \|y_1 \lambda_1^k \mathbf{v}^1 + y_2 \lambda_2^k \mathbf{v}^2\|$$
$$\leq |y_1| \cdot |\lambda_1|^k \cdot \|\mathbf{v}^1\| + |y_2| \cdot |\lambda_2|^k \cdot \|\mathbf{v}^2\|,$$

which goes to zero as k goes to infinity.

For a linear map with distinct eigenvalues that are each real and of absolute value less than one, the origin is called a *linear stable node* (or just *stable node*).

Example 12.2 (Linear Stable Node). Consider the linear map

$$L \begin{pmatrix} x_1 \\ x_2 \end{pmatrix} = \begin{pmatrix} \frac{4}{5} & 0 \\ \frac{4}{5} & \frac{1}{5} \end{pmatrix} \begin{pmatrix} x_1 \\ x_2 \end{pmatrix}.$$

As stated for the general situation, the origin $\mathbf{0}$ is a fixed point of this linear map.

The characteristic equation,

$$0 = \det \begin{pmatrix} \frac{4}{5} - \lambda & 0 \\ \frac{4}{5} & \frac{1}{5} - \lambda \end{pmatrix} = \left(\frac{4}{5} - \lambda\right)\left(\frac{1}{5} - \lambda\right),$$

has roots $\lambda = 1/5, 4/5$, both of which are real and less than one in absolute value.

For $\lambda_2 = 4/5$,

$$\mathbf{A} - \frac{4}{5}\mathbf{I} = \begin{pmatrix} 0 & 0 \\ \frac{4}{5} & -\frac{3}{5} \end{pmatrix} \sim \begin{pmatrix} 4 & -3 \\ 0 & 0 \end{pmatrix},$$

where we write $\sim$ when one matrix can be row reduced to the other. Thus, the components of an eigenvector satisfy $4v_1 - 3v_2 = 0$, $v_1 = \frac{3}{4} v_2$, which has a solution

$$\mathbf{v}^2 = \begin{pmatrix} 3 \\ 4 \end{pmatrix}.$$

The reader can either repeat the preceding process for $\lambda_1 = \frac{1}{5}$ or just check that an eigenvector is

$$\mathbf{v}^1 = \begin{pmatrix} 0 \\ 1 \end{pmatrix}.$$

Since both of these eigenvalues are less than one in absolute value, $\mathbf{A}^k \mathbf{x}$ goes to zero as k goes to infinity for any point $\mathbf{x}$, so $\mathbf{0}$ is an attracting fixed point for the linear map. However,

$$\mathbf{A}\begin{pmatrix} x_1 \\ 0 \end{pmatrix} = \begin{pmatrix} \frac{4}{5}x_1 \\ \frac{4}{5}x_1 \end{pmatrix} \quad \text{and}$$

$$\left\|\mathbf{A}\begin{pmatrix} x_1 \\ 0 \end{pmatrix}\right\| = |x_1|\frac{4\sqrt{2}}{5} > \left\|\begin{pmatrix} x_1 \\ 0 \end{pmatrix}\right\|.$$

Thus, the first iterate of $\begin{pmatrix} x_1 \\ 0 \end{pmatrix}$ is farther from the origin even though higher iterates eventually converge to zero.

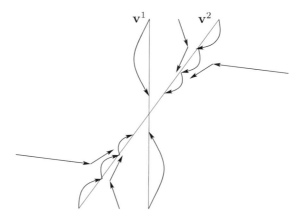

Figure 1. Phase portrait for a linear stable node

If $\mathbf{x} = y_1\mathbf{v}^1 + y_2\mathbf{v}^2$ is written as a linear combination of the eigenvectors, then

$$\mathbf{A}\left(y_1\mathbf{v}^1 + y_2\mathbf{v}^2\right) = y_1\,\mathbf{A}\mathbf{v}^1 + y_2\,\mathbf{A}\mathbf{v}^2$$

$$= y_1\frac{1}{5}\mathbf{v}^1 + y_2\frac{4}{5}\mathbf{v}^2.$$

Taking powers, we get

$$\mathbf{A}^k\left(y_1\mathbf{v}^1 + y_2\mathbf{v}^2\right) = y_1\left(\frac{1}{5}\right)^k\mathbf{v}^1 + y_2\left(\frac{4}{5}\right)^k\mathbf{v}^2.$$

Notice that $\left(1/5\right)^k$ goes to zero faster than $\left(4/5\right)^k$, so if $y_2 \neq 0$, then the iterates approach the origin in a direction that tends to the line generated by the eigenvector for $4/5$:

$$\mathbf{A}^k\left(y_1\mathbf{v}^1 + y_2\mathbf{v}^2\right) = \left(\frac{4}{5}\right)^k\left[y_1\left(\frac{1}{4}\right)^k\mathbf{v}^1 + y_2\,\mathbf{v}^2\right],$$

and the part inside the square brackets approaches the vector $y_2\mathbf{v}^2$. Thus, the orbits approach the origin in a direction asymptotic to the eigendirection of the weaker contraction. Figure 1 shows the phase portrait of this map using several representative orbits.

Table 1 gives the general procedure for drawing the phase portrait for a stable node.

12.1. Linear Maps

> **Phase portrait for a linear stable node**
>
> Consider the case where a 2×2 matrix $\mathbf{A}$ has eigenvalues that are real and distinct with $|\lambda_1| < |\lambda_2| < 1$ and that they have eigenvectors $\mathbf{v}^1$ and $\mathbf{v}^2$ respectively.
>
> (1) Draw two straight lines through the origin in the directions of the two eigenvectors, $\mathbf{v}^1$ and $\mathbf{v}^2$. Mark representative orbits on the four half-lines by arrows between points on various orbits.
>
> (2) In each of the four regions between the eigenvector lines, draw in representative orbits which are linear combinations of orbits that start at $\mathbf{v}^1$ and $\mathbf{v}^2$:
> $$\mathbf{A}^k \left(y_1 \mathbf{v}^1 + y_2 \mathbf{v}^2 \right) = y_1 \lambda_1^k \mathbf{v}^1 + y_2 \lambda_2^k \mathbf{v}^2.$$
> If $y_2 \neq 0$, then as k goes to infinity, the orbit approaches the origin in a direction asymptotic to the line for the eigenvector $\mathbf{v}^2$ corresponding to the eigenvalue whose absolute value is closer to one.
>
> **Table 1**

If $n = 2$, both eigenvalues are real, and $|\lambda_1| > |\lambda_2| > 1$, then the iterates of points go to infinity as k goes to infinity. Backward iterates go to the origin as k goes to minus infinity. The origin for a linear map of this type with all the eigenvalues real, of absolute value greater than one, and distinct, is called a *linear unstable node*. The procedure to draw the phase portrait of an unstable node is similar to that for a stable node, with obvious changes between k going to infinity and minus infinity. Exercise 1 in this section contains an example of a linear unstable node.

Finally, assume that $|\lambda_1| < 1 < |\lambda_2|$ for $n = 2$. The origin for this situation is called a *linear saddle*. Let $\mathbf{v}^1$ and $\mathbf{v}^2$ be the corresponding eigenvectors. Any point can be written as a linear combination of the eigenvectors, $\mathbf{x} = y_1 \mathbf{v}^1 + y_2 \mathbf{v}^2$. Then, the iterates

$$\mathbf{A}^k \mathbf{x} = y_1 \lambda_1^k \mathbf{v}^1 + y_2 \lambda_2^k \mathbf{v}^2$$

have the first term go to zero and the second term go to infinity (if $y_2 \neq 0$), so

$$\mathbf{A}^k \mathbf{x} - y_2 \lambda_2^k \mathbf{v}^2 = y_1 \lambda_1^k \mathbf{v}^1$$

goes to zero. Thus, iterates are asymptotic to the line generated by $\mathbf{v}^2$. Backward iterates are asymptotic to the line generated by $\mathbf{v}^1$:

$$\mathbf{A}^{-k} \mathbf{x} - y_1 \lambda_1^{-k} \mathbf{v}^1 = y_2 \lambda_2^{-k} \mathbf{v}^2$$

goes to zero as $-k$ goes to minus infinity. If both y_1 and y_2 are nonzero, then $\|\mathbf{A}^k \mathbf{x}\|$ goes to infinity as k goes to both plus and minus infinity.

Example 12.3 (Saddle). Consider

$$L \begin{pmatrix} x_1 \\ x_2 \end{pmatrix} = \begin{pmatrix} \frac{7}{2} & -\frac{3}{2} \\ 3 & -1 \end{pmatrix} \begin{pmatrix} x_1 \\ x_2 \end{pmatrix}.$$

The characteristic equation,
$$0 = \lambda^2 - \frac{5}{2}\lambda + 1,$$
has roots $\lambda = 1/2, 2$. A direct calculation shows that the corresponding eigenvectors are
$$\mathbf{v}^1 = \begin{pmatrix} 1 \\ 2 \end{pmatrix} \quad \text{and} \quad \mathbf{v}^2 = \begin{pmatrix} 1 \\ 1 \end{pmatrix}.$$
See Figure 2 for the phase portrait of the iterates.

Notice that $\|\mathbf{A}^k \mathbf{v}^1\|$ goes to zero as k goes to infinity and goes to infinity as k goes to minus infinity. On the other hand, $\|\mathbf{A}^k \mathbf{v}^2\|$ goes to infinity as k goes to infinity and goes to zero as k goes to minus infinity. For any other point $\mathbf{x} = y_1 \mathbf{v}^1 + y_2 \mathbf{v}^2$ with $y_1 \neq 0$ and $y_2 \neq 0$, $\|\mathbf{A}^k \mathbf{x}\|$ goes to infinity as k goes to either infinity or minus infinity.

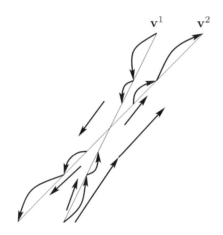

Figure 2. Phase portrait for a linear saddle

Phase portrait for a linear saddle

(1) Draw two straight lines through the origin in the directions of the two eigenvectors. Mark representative orbits on the four half-lines by arrows between points on various orbits, with the orbits along one line approaching the origin as k goes to infinity, and the orbits along the other line going to infinity.

(2) In each of the four regions between the straight line solutions, draw in representative orbits which are linear combinations of orbits that start at $\mathbf{v}^1$ and $\mathbf{v}^2$.

Table 2

The eigenvalues can be negative, as the next example illustrates.

12.1. Linear Maps

Example 12.4 (Negative Eigenvalues). Consider

$$L\begin{pmatrix} x_1 \\ x_2 \end{pmatrix} = \begin{pmatrix} \frac{1}{2} & \frac{3}{2} \\ 1 & 1 \end{pmatrix} \begin{pmatrix} x_1 \\ x_2 \end{pmatrix}.$$

The characteristic equation,

$$0 = \lambda^2 - \frac{3}{2}\lambda - 1,$$

has roots $\lambda = -1/2, 2$. A direct calculation shows that the corresponding eigenvectors are

$$\begin{pmatrix} 3 \\ -2 \end{pmatrix} \quad \text{and} \quad \begin{pmatrix} 1 \\ 1 \end{pmatrix}.$$

The image,

$$L\begin{pmatrix} 3 \\ -2 \end{pmatrix} = -\frac{1}{2}\begin{pmatrix} 3 \\ -2 \end{pmatrix},$$

is on the opposite side of the origin. The second iterate returns to the original side but has been multiplied by $(-1/2)^2 = 1/4$:

$$L^2 \begin{pmatrix} 3 \\ -2 \end{pmatrix} = \frac{1}{4}\begin{pmatrix} 3 \\ -2 \end{pmatrix}.$$

Thus, the linear map is a flip on the line spanned by the eigenvector for the eigenvalue $-1/2$. See Figure 3.

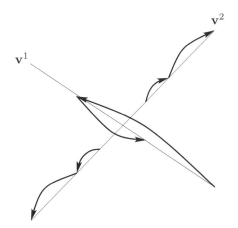

Figure 3. Phase portrait for a linear saddle with a flip

Repeated eigenvalues

If there is a repeated eigenvalue, there are not always as many independent eigenvectors as the multiplicity of the eigenvalue as a root of the characteristic polynomial. For $n = 2$, if there are two independent eigenvectors for a multiple eigenvalue, then the matrix has to be a diagonal matrix; for example,

$$\begin{pmatrix} 0.5 & 0 \\ 0 & 0.5 \end{pmatrix}.$$

However, a matrix such as
$$\begin{pmatrix} 0.5 & 1 \\ 0 & 0.5 \end{pmatrix}$$
has only one independent eigenvector $\mathbf{v} = (1,0)^\mathsf{T}$. To find a second independent solution, the theory of linear algebra shows that the nonhomogeneous equation
$$(\mathbf{A} - \lambda\mathbf{I})\mathbf{w} = \mathbf{v}$$
has a solution for a vector $\mathbf{w}$. The vector $\mathbf{w}$ is called a generalized eigenvector for λ. (See Appendix A.3 for further discussion of generalized eigenvectors.) Then,
$$\begin{aligned}
\mathbf{A}\mathbf{w} &= \lambda\mathbf{w} + \mathbf{v}, \\
\mathbf{A}^2\mathbf{w} &= \lambda\mathbf{A}\mathbf{w} + \mathbf{A}\mathbf{v} \\
&= \lambda(\lambda\mathbf{w} + \mathbf{v}) + \lambda\mathbf{v} \\
&= \lambda^2\mathbf{w} + 2\lambda\mathbf{v}, \\
\mathbf{A}^3\mathbf{w} &= \mathbf{A}^2\mathbf{A}\mathbf{w} \\
&= \lambda\mathbf{A}^2\mathbf{w} + \mathbf{A}^2\mathbf{v} \\
&= \lambda(\lambda^2\mathbf{w} + 2\lambda\mathbf{v}) + \lambda^2\mathbf{v} \\
&= \lambda^3\mathbf{w} + 3\lambda^2\mathbf{v}, \quad \text{and} \\
\mathbf{A}^k\mathbf{w} &= \mathbf{A}^{k-1}\mathbf{A}\mathbf{w} \\
&= \lambda\mathbf{A}^{k-1}\mathbf{w} + \mathbf{A}^{k-1}\mathbf{v} \\
&= \lambda\left(\lambda^{k-1}\mathbf{w} + (k-1)\lambda^{k-2}\mathbf{v}\right) + \lambda^{k-1}\mathbf{v} \\
&= \lambda^k\mathbf{w} + k\lambda^{k-1}\mathbf{v}.
\end{aligned}$$

Notice that, if $|\lambda| < 1$, then these iterates still go to the origin, since $k\lambda^{k-1}$ goes to zero. However, the first iterate can get longer. For example, if
$$\mathbf{A} = \begin{pmatrix} a & 1 \\ 0 & a \end{pmatrix}$$
with $0 < a < 1$, then
$$\left\| \mathbf{A}\begin{pmatrix} 0 \\ 1 \end{pmatrix} \right\| = \left\| \begin{pmatrix} 1 \\ a \end{pmatrix} \right\| = \sqrt{1 + a^2} > 1.$$

For $n = 2$, with a multiple eigenvalue λ where $|\lambda| < 1$ but only one independent eigenvector, the origin is called a *degenerate stable node*. If the multiple eigenvalue has $|\lambda| > 1$ and only one independent eigenvector, then the origin is called a *degenerate unstable node*.

Example 12.5 (Degenerate stable node). Consider
$$\mathbf{A} = \begin{pmatrix} -\frac{1}{2} & 2 \\ -\frac{1}{2} & \frac{3}{2} \end{pmatrix},$$

12.1. Linear Maps

which has the single eigenvalue $1/2$ with multiplicity two. To find an eigenvector, we row reduce $\mathbf{A} - \frac{1}{2}\mathbf{I}$:

$$\mathbf{A} - \frac{1}{2}\mathbf{I} = \begin{pmatrix} -1 & 2 \\ -\frac{1}{2} & 1 \end{pmatrix} \sim \begin{pmatrix} -1 & 2 \\ 0 & 0 \end{pmatrix}.$$

Thus, we need $-v_1 + 2v_2 = 0$ and $v_1 = 2v_2$. So an eigenvector is

$$\mathbf{v} = \begin{pmatrix} 2 \\ 1 \end{pmatrix}.$$

To find the vector $\mathbf{w}$, we row reduce the augmented matrix:

$$\begin{pmatrix} -1 & 2 & | & 2 \\ -\frac{1}{2} & 1 & | & 1 \end{pmatrix} \sim \begin{pmatrix} -1 & 2 & | & 2 \\ 0 & 0 & | & 0 \end{pmatrix}.$$

Thus, we need $-w_1 + 2w_2 = 2$; one solution is $w_1 = 0$ and $w_2 = 1$, or

$$\mathbf{w} = \begin{pmatrix} 0 \\ 1 \end{pmatrix}.$$

Then,

$$\mathbf{A}\mathbf{w} = \frac{1}{2}\mathbf{w} + \mathbf{v},$$
$$\mathbf{A}^2\mathbf{w} = \frac{1}{2}\left(\frac{1}{2}\mathbf{w} + \mathbf{v}\right) + \frac{1}{2}\mathbf{v} = \frac{1}{2^2}\mathbf{w} + \frac{2}{2}\mathbf{v},$$
$$\mathbf{A}^3\mathbf{w} = \frac{1}{2^2}\left(\frac{1}{2}\mathbf{w} + \mathbf{v}\right) + \frac{2}{2^2}\mathbf{v} = \frac{1}{2^3}\mathbf{w} + \frac{3}{2^2}\mathbf{v}, \quad \text{and}$$
$$\mathbf{A}^k\mathbf{w} = \frac{1}{2^{k-1}}\left(\frac{1}{2}\mathbf{w} + \mathbf{v}\right) + \frac{k-1}{2^{k-1}}\mathbf{v} = \frac{1}{2^k}\mathbf{w} + \frac{k}{2^{k-1}}\mathbf{v}.$$

The first few iterates of $\mathbf{w}$ move farther from the origin, but eventually they tend into the origin. Since $\mathbf{A}^k = \frac{k}{2^{k-1}}\left[\mathbf{v} + \frac{1}{2k}\mathbf{w}\right]$ approaches the line generated by $\mathbf{v}$, every orbit approaches the origin in a direction asymptotic to the line generated by the eigenvector. See Figure 4 for the phase portrait.

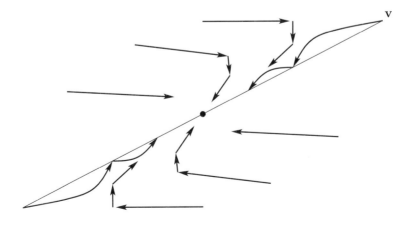

Figure 4. Phase portrait for a degenerate stable node

The general procedure for drawing the phase portrait for two equal real eigenvalues is given in Table 3.

Phase portrait for two equal real eigenvalues

We consider the stable case; the case of an unstable system is similar, with obvious changes between k going to infinity and minus infinity. First, assume that there are two independent eigenvectors (and the matrix is diagonal).

If there are two independent eigenvectors, then all solutions go straight in toward the origin. The origin of this system is called a *stable star*; all points go toward the origin along straight lines.

Next assume that there is only one independent eigenvector $\mathbf{v}$, and a second generalized eigenvector $\mathbf{w}$, where $(\mathbf{A} - \lambda\mathbf{I})\mathbf{w} = \mathbf{v}$.

(1) Draw the two orbits that move in straight lines toward the origin along the line generated by $\mathbf{v}$, connecting points on the orbits by arrows.

(2) Next, draw the orbit that has initial condition $\mathbf{w}$ and then comes in toward the origin, with limiting displacement vector from the origin being a multiple of the vector $\mathbf{v}$ (i.e., the orbit $\lambda^k \mathbf{w} + k\lambda^{k-1}\mathbf{v}$).

(3) Draw the orbit with initial condition $-\mathbf{w}$ which should be just the reflection through the origin of the previous orbit.

Table 3

Complex eigenvalues

For $n = 2$, the matrices that are in simplest form with a pair of complex eigenvalues are given by

$$\mathbf{A} = \begin{pmatrix} a & -b \\ b & a \end{pmatrix} = \begin{pmatrix} |\lambda|\cos\theta & -|\lambda|\sin\theta \\ |\lambda|\sin\theta & |\lambda|\cos\theta \end{pmatrix} = |\lambda| \begin{pmatrix} \cos\theta & -\sin\theta \\ \sin\theta & \cos\theta \end{pmatrix},$$

where $|\lambda| = \sqrt{a^2 + b^2}$, $a = |\lambda|\cos\theta$, and $b = |\lambda|\sin\theta$. A direct check shows that the eigenvalues are

$$a \pm i\,b = |\lambda|\left(\cos\theta \pm i\,\sin\theta\right).$$

The matrix

$$R(\theta) = \begin{pmatrix} \cos\theta & -\sin\theta \\ \sin\theta & \cos\theta \end{pmatrix}$$

corresponds to a rotation through an angle of θ, and the scalar $|\lambda|$ gives the expansion or contraction factor. Taking powers, we see that

$$\mathbf{A}^k = |\lambda|^k R(k\theta)$$

is a rotation through an angle of $k\theta$, and an expansion or contraction by $|\lambda|^k$. If $|\lambda| < 1$, then the map is a contraction, and the origin is called a *stable focus*; if $|\lambda| > 1$, then the map is an expansion, and the origin is called an *unstable focus*; and if $|\lambda| = 1$, then the map is a rotation, and the origin is called a *linear center*.

12.1. Linear Maps

Finally, if every eigenvalue of **A** has $|\lambda| < 1$ (whether they are real or complex), then the origin is called a *linear sink*. If every eigenvalue of **A** has $|\lambda| > 1$ (whether it is real or complex), then the origin is called a *linear source*.

Example 12.6. Let
$$\mathbf{A} = \begin{pmatrix} -1 & 4 \\ -8 & 7 \end{pmatrix}.$$
Then, the characteristic equation is $0 = \lambda^2 - 6\lambda + 25$, which has roots $\lambda = 3 \pm i\,4$. So, $|\lambda| = \sqrt{3^2 + 4^2} = 5 > 1$, and the map is an expansion. See Figure 5.

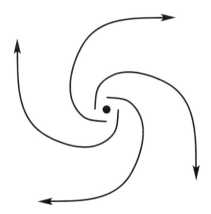

Figure 5. Phase portrait for a linear unstable focus

Phase portrait for a pair of complex eigenvalues

Assume that the eigenvalues are $\lambda = |\lambda|\,e^{\pm i\theta}$ with $\theta \neq 0, \pi$, so the eigenvalues are not real.

(1) If $|\lambda| = 1$, then the origin is a *linear center* with each orbit lying on an ellipse around the origin. The direction of motion can be either clockwise or counterclockwise, depending on the entries in the matrix. To determine the direction, check the image of $(1,0)^\mathsf{T}$ (or any other vector).

(2) If $|\lambda| < 1$, then the origin is a *stable focus*, and the orbits rotate either clockwise or counterclockwise around the origin as they come in toward it.

(3) If $|\lambda| > 1$, then the origin is an *unstable focus*, and the orbits rotate either clockwise or counterclockwise around the origin as the iterates get larger and larger in absolute value.

Table 4

Norm of a matrix

We often want to know the largest amount by which a linear map stretches a vector: The ratio $\|\mathbf{Av}\|/\|\mathbf{v}\|$ is the amount the vector $\mathbf{v}$ is stretched, and the maximum of this ratio over all nonzero vectors $\mathbf{v}$ gives the maximum amount of stretching. Because $\mathbf{v}/\|\mathbf{v}\|$ is a unit vector, and

$$\frac{\|\mathbf{Av}\|}{\|\mathbf{v}\|} = \left\|\mathbf{A}\frac{\mathbf{v}}{\|\mathbf{v}\|}\right\|,$$

it follows that

$$\max_{\mathbf{x}\neq 0}\left\{\frac{\|\mathbf{Av}\|}{\|\mathbf{v}\|}\right\} = \max_{\|\mathbf{u}\|=1}\|\mathbf{Au}\|.$$

We define the *norm* $\|\mathbf{A}\|$ of the matrix $\mathbf{A}$ by

$$\|\mathbf{A}\| = \max_{\|\mathbf{u}\|=1}\|\mathbf{Au}\|.$$

Notice that $\|\mathbf{A}\| \geq 0$. In fact, the norm $\|\mathbf{A}\|$ is the square root of the largest eigenvalue of the symmetric matrix $\mathbf{A}^\mathsf{T}\mathbf{A}$. See Appendix A.3.

Usually, we do not need to know the exact value of the norm but only that such a number exists. We use the norm, at least implicitly, to see that the linear part of a map determines many of the properties of the phase portrait near a fixed point.

Example 12.7. Consider the matrix

$$\mathbf{A} = \begin{pmatrix} a & 1 \\ 0 & a \end{pmatrix}$$

with multiple eigenvalue a, where $0 < a < 1$. The norm of powers of $\mathbf{A}$ goes to zero, as the following argument shows. The power

$$\mathbf{A}^k = \begin{pmatrix} a^k & k\,a^{k-1} \\ 0 & a^k \end{pmatrix} \quad \text{and}$$

$$\left(\mathbf{A}^k\right)^\mathsf{T}\mathbf{A}^k = \begin{pmatrix} a^{2k} & ka^{2k-1} \\ ka^{2k-1} & a^{2k} + k^2 a^{2k-2} \end{pmatrix}.$$

The eigenvalues of this matrix are

$$\frac{2a^{2k} + k^2 a^{2k-2} \pm \sqrt{4k^2 a^{4k-2} + k^4 a^{4k-4}}}{2}$$

$$= a^{2k} + a^{2k-2}\frac{k^2}{2} \pm a^{2k-2}\frac{k^2}{2}\sqrt{4a^2/k^2 + 1}.$$

The largest eigenvalue is found by taking the positive square root, which is close to $a^{2k} + a^{2k-2}k^2$ for large k. The norm $\|\mathbf{A}^k\|$, which is the square root of the largest eigenvalue, is close to $\sqrt{a^{2k} + a^{2k-2}k^2}$. Both a^{2k} and $a^{2k-2}k^2$ go to zero as k goes to infinity, so the norm of $\mathbf{A}^k$ goes to zero, and the iterates of any point converge to the origin.

Exercises 12.1

1. Find the eigenvalues and draw the phase portraits for the linear maps that the following matrices represent:

   a. $\begin{pmatrix} 0.5 & 0.125 \\ 0.5 & 0.5 \end{pmatrix}$
   b. $\begin{pmatrix} 3 & 1 \\ 1 & 3 \end{pmatrix}$
   c. $\begin{pmatrix} 0.25 & 0.25 \\ -0.5 & 1 \end{pmatrix}$
   d. $\begin{pmatrix} 1 & 1 \\ 1 & 0 \end{pmatrix}$
   e. $\begin{pmatrix} 0.4 & 0.2 \\ -0.2 & 0.4 \end{pmatrix}$
   f. $\begin{pmatrix} 1 & -3 \\ 3 & 1 \end{pmatrix}$

12.2. Classification of Periodic Points

The preceding section presented the phase portraits of various types of linear maps. In this section, we turn to the different concepts of stability for periodic points. We gave them earlier for one-dimensional maps, but we repeat them with the notation for higher dimensional maps. When reading the definitions, the reader should think of the linear examples given in the preceding section, but the definitions apply also to nonlinear maps. After giving the definitions, we indicate conditions on the eigenvalues of a linear map that imply the different types of stability.

For a point $\mathbf{p}_0$ in $\mathbb{R}^n$, the *open ball of radius* r (without the boundary) is the set
$$\mathbf{B}(\mathbf{p}_0, r) = \{\mathbf{x} \in \mathbb{R}^n : \|\mathbf{x} - \mathbf{p}_0\| < r\}.$$
Sometimes, people call this the *open disk*, especially when $n = 2$.

Definition 12.8. A period-k point $\mathbf{p}_0$ for a map $\mathbf{F}$ is called *Lyapunov stable* provided that, for each $r > 0$, there exists a $\delta > 0$ such that, if $\mathbf{x}$ is in $\mathbf{B}(\mathbf{p}_0, \delta)$, then $\mathbf{F}^j(\mathbf{x})$ is in $\mathbf{B}(\mathbf{F}^j(\mathbf{p}_0), r)$ for all $j \geq 0$. We often write *L-stable* for Lyapunov stable.

A period-k point $\mathbf{p}_0$ for a map $\mathbf{F}$ is called *unstable* provided that it is not Lyapunov stable. Thus, $\mathbf{p}_0$ is unstable provided that there exists an $r_0 > 0$ such that, for each $\delta > 0$, there exists a point $\mathbf{x}_\delta$ in $\mathbf{B}(\mathbf{p}_0, \delta)$ and an $j_\delta \geq 0$ such that $\|\mathbf{F}^{j_\delta}(\mathbf{x}_\delta) - \mathbf{F}^{j_\delta}(\mathbf{p}_0)\| \geq r_0$.

The linear stable node in Example 12.2 shows that it is often necessary to take $\delta < r$. In two dimensions, linear maps for which the origin is L-stable include linear stable nodes, degenerate stable nodes, linear stable foci, and linear centers. Theorem 12.2 gives precise conditions for a linear map to be L-stable in any dimension. Linear maps for which the origin is unstable include linear saddles, unstable nodes, and unstable foci.

Definition 12.9. A period-k point $\mathbf{p}_0$ for a map $\mathbf{F}$ is called *attracting* (or *asymptotically stable*, or a *periodic sink*) provided that

(i) it is Lyapunov stable and

(ii) there is a $\delta_1 > 0$ such that, if $\mathbf{x}$ is in $\mathbf{B}(\mathbf{p}_0, \delta_1)$, then $\|\mathbf{F}^j(\mathbf{x}) - \mathbf{F}^j(\mathbf{p}_0)\|$ goes to zero as j goes to infinity.

It follows that, if $\mathbf{p}_0$ is a periodic sink and $\mathbf{x}_0$ is a point in the set $\mathbf{B}(\mathbf{p}_0, \delta_1)$ with δ_1 as in the definition, then $\omega(\mathbf{x}_0, \mathbf{F}) = \mathcal{O}_{\mathbf{F}}^+(\mathbf{p}_0)$ is the orbit of $\mathbf{p}_0$.

The origin is an attracting fixed point for all linear maps for which all the eigenvalues have absolute value less than one, which includes linear stable nodes, degenerate stable nodes, and linear stable foci. For a linear map on $\mathbb{R}^2$ with a pair of complex eigenvalues of absolute value equal to one (the origin is a linear center), the origin is L-stable but not attracting.

Definition 12.10. A period-k point $\mathbf{p}_0$ for a map $\mathbf{F}$ is called *repelling* (or a *periodic source*) provided that there is an $r_1 > 0$ such that, if $\mathbf{x} \neq \mathbf{p}_0$ is in $\mathbf{B}(\mathbf{p}_0, r_1)$, then there exists a $j = j_{\mathbf{x}}$ such that

$$\|\mathbf{F}^{j_{\mathbf{x}}}(\mathbf{x}) - \mathbf{F}^{j_{\mathbf{x}}}(\mathbf{p}_0)\| \geq r_1.$$

We really should add the condition to the definition of a repelling periodic point that $\mathbf{p}_0$ is L-stable for $\mathbf{F}^{-1}$, but we wait until the end of the section when we discuss the inverse of a map.

A periodic point is unstable if some points move away, and it is repelling if all points move away. The origin is repelling for all linear maps for which all the eigenvalues have absolute value greater than one; this includes linear unstable nodes, degenerate unstable nodes, and unstable foci. Note that a saddle fixed point is unstable but not repelling.

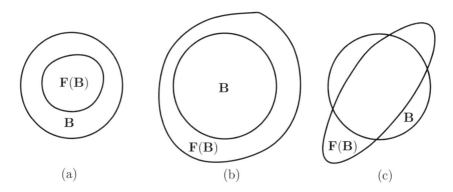

Figure 6. (a) Attracting. (b) Repelling. (c) Unstable but not repelling (saddle).

Theorem 12.2. *The stability type of a linear map* $\mathbf{Ax}$ *is classification as follows.*

  a. *If all the eigenvalues λ_j of $\mathbf{A}$ have $|\lambda_j| < 1$, then the origin is attracting.*

  b. *If all the eigenvalues λ_j of $\mathbf{A}$ have $|\lambda_j| \leq 1$, and each eigenvalue with $|\lambda_j| = 1$ has multiplicity one, then the origin is L-stable.*

  c. *If one eigenvalue λ_{j_0} of $\mathbf{A}$ has $|\lambda_{j_0}| > 1$, then the origin is unstable.*

  d. *If all the eigenvalues λ_j of $\mathbf{A}$ have $|\lambda_j| > 1$, then the origin is repelling.*

12.2. Classification of Periodic Points

Example 12.11. Consider the map in polar coordinates given by

$$\mathbf{F}\begin{pmatrix}r\\ \theta\end{pmatrix} = \begin{pmatrix} \dfrac{r+1}{2} \\ \theta^2\,(2\pi - \theta)^2 \ (\mathrm{mod}\ 2\pi)\end{pmatrix}$$

with fixed point

$$\mathbf{F}\begin{pmatrix}1\\ 0\end{pmatrix} = \begin{pmatrix}1\\ 0\end{pmatrix}.$$

We show that this fixed point is not Lyapunov stable but satisfies condition (ii) of Definition 12.9 for an attracting fixed point. Because $\theta^2\,(2\pi - \theta)^2 > 0$ for $0 < \theta < 2\pi$, the orbit of every point on $r = 1$ goes around the circle and has $\omega((1, \theta_0)) = (0, 2\pi) = (1, 0)\ (\mathrm{mod}\ 2\pi)$. Similarly, for iterates of (r_0, θ_0) with $r_0 \neq 0$, r_j goes to 1 and θ_j goes to 2π, so $\omega(r_0, \theta_0) = (1, 0)$. Thus, the fixed point satisfies condition (ii) for asymptotic stability. However, the fixed point is not L-stable, since points with small $\theta_0 > 0$ go around near $\theta = \pi$ before returning to near $\theta_0 = 0$. See Figure 7.

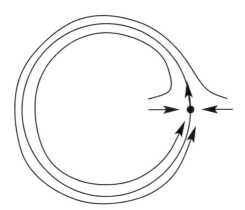

Figure 7. Unstable fixed point for Example 12.11

For nonlinear maps, we consider the matrix of partial derivatives to determine the *linearized stability* of a periodic point. Then, Theorem 12.3 asserts the extent to which the periodic point has the same stability type for the nonlinear map as the linearized map.

Definition 12.12. Let $\mathbf{F}$ be a nonlinear map from $\mathbb{R}^n$ to $\mathbb{R}^n$ with coordinate functions F_i. The *matrix of partial derivatives* at a point $\mathbf{p}$, or the *derivative*, is the $n \times n$ matrix

$$D\mathbf{F}_{(\mathbf{p})} = \left(\frac{\partial F_i}{\partial x_j}(\mathbf{p})\right) = \begin{pmatrix} \dfrac{\partial F_1}{\partial x_1}(\mathbf{p}) & \cdots & \dfrac{\partial F_1}{\partial x_n}(\mathbf{p}) \\ \vdots & & \vdots \\ \dfrac{\partial F_n}{\partial x_1}(\mathbf{p}) & \cdots & \dfrac{\partial F_n}{\partial x_n}(\mathbf{p}) \end{pmatrix}.$$

The row is determined by the coordinate function and the column is determined by the variable used to calculate the partial derivative.

Appendix A.1 indicates the manner in which the Taylor expansion extends to multiple variables to show that

$$\mathbf{F}(\mathbf{x}) = \mathbf{F}(\mathbf{p}) + D\mathbf{F}_{(\mathbf{p})}(\mathbf{x} - \mathbf{p}) + O(\|\mathbf{x} - \mathbf{p}\|^2),$$

where $|O(\|\mathbf{x} - \mathbf{p}\|^2)| \leq C \|\mathbf{x} - \mathbf{p}\|^2$ for some constant $C > 0$, so it is a higher order term. Since the term $O(\|\mathbf{x} - \mathbf{p}\|^2)$ is smaller than $D\mathbf{F}_{(\mathbf{p})}(\mathbf{x} - \mathbf{p})$ for small displacements, this expansion can be used to show that the linear terms dominate near a fixed point.

For an iterate of $\mathbf{p}_0$, if we let $\mathbf{p}_j = \mathbf{F}^j(\mathbf{p}_0)$, then by the chain rule for partial derivatives applied to the matrices, we obtain

$$D(\mathbf{F}^k)_{(\mathbf{p}_0)} = D\mathbf{F}_{(\mathbf{p}_{k-1})} D\mathbf{F}_{(\mathbf{p}_{k-2})} \cdots D\mathbf{F}_{(\mathbf{p}_1)} D\mathbf{F}_{(\mathbf{p}_0)},$$

which is a product of the matrices. In one variable, the order of multiplication of the derivatives does not matter, but here in higher dimensions, the order of matrix multiplication does matter: the matrix to the extreme right is evaluated at the starting point and is the first to act on the vector; the matrix to the extreme left is evaluated at the last point $\mathbf{p}_{k-1}$ and is the last to act on the vector.

Example 12.13. Consider the map

$$\mathbf{F}\begin{pmatrix} x \\ y \end{pmatrix} = \begin{pmatrix} x + y^2 \\ x \end{pmatrix}.$$

Its derivative is

$$D\mathbf{F}_{(x,y)} = \begin{pmatrix} 1 & 2y \\ 1 & 0 \end{pmatrix}.$$

If $\mathbf{p}_0 = \begin{pmatrix} 2 \\ 1 \end{pmatrix}$, then $\mathbf{p}_1 = \begin{pmatrix} 3 \\ 2 \end{pmatrix}$, and

$$D(\mathbf{F}^2)_{(2,1)} = D\mathbf{F}_{(3,2)} \, D\mathbf{F}_{(2,1)}$$
$$= \begin{pmatrix} 1 & 4 \\ 1 & 0 \end{pmatrix} \begin{pmatrix} 1 & 2 \\ 1 & 0 \end{pmatrix} = \begin{pmatrix} 5 & 2 \\ 1 & 2 \end{pmatrix}.$$

We often want to specify that a function has a certain number of partial derivatives, we introduce the following terminology.

Definition 12.14. For an integer $r \geq 1$, a map $\mathbf{F}(\mathbf{x})$ from $\mathbb{R}^n$ to $\mathbb{R}^n$ is said to be C^r, provided that $\mathbf{F}$ is continuous and all partial derivatives up to order r exist and are continuous; that is,

$$\frac{\partial^k \mathbf{F}}{\partial^{i_1} x_1 \cdots \partial^{i_n} x_n}(\mathbf{x})$$

exists for all $(i_1, \ldots, i_n)$ with all $i_j \geq 0$ and $i_1 + \cdots + i_n \leq r$, and these partial derivatives are continuous as a function of $\mathbf{x}$. If partial derivatives of all orders exist (i.e., $\mathbf{F}$ is C^r for all r), $\mathbf{F}$ is said to be C^∞.

The next theorem gives the stability of periodic points for nonlinear maps.

Theorem 12.3. Let $\mathbf{F}$ from $\mathbb{R}^n$ to $\mathbb{R}^n$ be C^2. Assume $\mathbf{p}_0$ is a period-k point. Let $\lambda_1, \ldots, \lambda_n$ be the eigenvalues of $D(\mathbf{F}^k)_{(\mathbf{p}_0)}$.

a. If all the eigenvalues λ_j of $D(\mathbf{F}^k)_{(\mathbf{p}_0)}$ have $|\lambda_j| < 1$, then the periodic orbit $\mathcal{O}_\mathbf{F}^+(\mathbf{p}_0)$ is attracting.

12.2. Classification of Periodic Points

b. If one eigenvalue λ_{j_0} of $D(\mathbf{F}^k)_{(\mathbf{p}_0)}$ has $|\lambda_{j_0}| > 1$, then the periodic orbit $\mathcal{O}_\mathbf{F}^+(\mathbf{p}_0)$ is unstable.

c. If all the eigenvalues λ_j of $D(\mathbf{F}^k)_{(\mathbf{p}_0)}$ have $|\lambda_j| > 1$, then the periodic orbit $\mathcal{O}_\mathbf{F}^+(\mathbf{p}_0)$ is repelling.

Remark 12.15. If all the eigenvalues λ_j of $D(\mathbf{F}^k)_{(\mathbf{p}_0)}$ have $|\lambda_j| \leq 1$, at least one eigenvalue has $|\lambda_{j_0}| = 1$, and each eigenvalue with $|\lambda_j| = 1$ has multiplicity one, then the periodic point could be attracting, L-stable, or unstable: The linear terms are not sufficient to determine the stability type.

Definition 12.16. A period-k point $\mathbf{p}_0$ is called *hyperbolic* provided that $|\lambda_j| \neq 1$ for each eigenvalue λ_j of $D(\mathbf{F}^k)_{(\mathbf{p}_0)}$.

A period-k point $\mathbf{p}_0$ is called a *saddle* provided that it is hyperbolic, $|\lambda_{j_1}| < 1$ for some j_1, and $|\lambda_{j_2}| > 1$ for some other j_2 (i.e., there are both an attracting direction and a repelling direction).

Example 12.17. We consider the stability types of the fixed points for the nonlinear map given by

$$\mathbf{F}(x, y) = (y, a \sin(x) - y).$$

The matrix of partial derivatives is

$$D\mathbf{F}_{(x,y)} = \begin{pmatrix} 0 & 1 \\ a\cos(x) & -1 \end{pmatrix}.$$

The fixed points satisfy

$$x = y \quad \text{and}$$
$$y = a \sin(x) - y.$$

Substituting x for y, we find that the second equation becomes

$$2x = a \sin(x) \quad \text{or}$$
$$\sin(x) = \frac{2x}{a}.$$

There is always a fixed point with $x = y = 0$. For $0 < a < 2$, this is the only fixed point, since the slope of $2x/a$ is then greater than one. For $a > 2$, there is at least a second fixed point (x_a, x_a) with $0 < x_a < \pi$, since the slope of $2x/a$ is less than one.

At the origin,

$$D\mathbf{F}_{(0,0)} = \begin{pmatrix} 0 & 1 \\ a & -1 \end{pmatrix},$$

which has characteristic equation $\lambda^2 + \lambda - a = 0$ and eigenvalues

$$\lambda_\pm = \frac{-1 \pm \sqrt{1 + 4a}}{2}.$$

We consider various ranges of the parameter a where the eigenvalues are real or complex.

For $a < -1/4$, the eigenvalues are not real, and

$$|\lambda_\pm| = \sqrt{\frac{1}{4} + \frac{-4a-1}{4}} = \sqrt{-a} = \sqrt{|a|}.$$

Thus, for $a < -1$, the origin is a repelling fixed point with complex eigenvalues; for $-1 < a < -1/4$, the origin is an attracting fixed point with complex eigenvalues.

For $a \geq -1/4$, the eigenvalues are real. The eigenvalue $\lambda_- = -1$ for $a = 0$; therefore, for $-1/4 \leq a < 0$, we have $-1 < \lambda_- \leq -1/2$ and $-1/2 \leq \lambda_+ < 0$, and the origin is attracting. The eigenvalue $\lambda_+ = 1$ for $a = 2$; therefore, for $0 < a < 2$, we have $\lambda_- < -1$ and $0 < \lambda_+ < 1$ and the origin is a saddle. Finally, for $2 < a$, we have that $\lambda_- < -2$ and $1 < \lambda_+$ and the origin is repelling.

Summarizing the stability of the origin, we have the following properties:

$$
\begin{array}{ll}
a < -1 & \text{source with complex eigenvalues,} \\
-1 < a < -\dfrac{1}{4} & \text{sink with complex eigenvalues,} \\
-\dfrac{1}{4} < a < 0 & \text{sink with real eigenvalues,} \\
0 < a < 2 & \text{saddle with real eigenvalues,} \\
2 < a & \text{source with real eigenvalues.}
\end{array}
$$

Turning to the fixed point (x_a, x_a) for $a > 2$, the matrix of partial derivatives is

$$D\mathbf{F}_{(x,y)} = \begin{pmatrix} 0 & 1 \\ a\cos(x_a) & -1 \end{pmatrix}.$$

It has characteristic equation $\lambda^2 + \lambda - a\cos(x_a) = 0$ and has eigenvalues

$$\lambda_\pm = \frac{-1 \pm \sqrt{1 + 4a\cos(x_a)}}{2}.$$

The eigenvalue $\lambda_- = -1$ when $\cos(x_a) = 0$, $x_a = \pi/2$, $a\sin(\pi/2) = \pi$, or $a = \pi$. For this parameter value of $a = \pi$, $\lambda_- = -1$ and $\lambda_+ = 0$. For $2 < a < \pi$, $0 < x_a < \pi/2$. Since $\cos(x_a) > 0$, $\lambda_- < -1$ and $\lambda_+ > 0$. For this range of parameters, we want to see that $\lambda_+ < 1$, or $\sqrt{1 + 4a\cos(x_a)} < 3$. We would need

$$3 > \sqrt{1 + 4a\cos(x_a)},$$
$$\cos(x_a) < \frac{2}{a} = \frac{\sin(x_a)}{x_a}, \quad \text{or}$$
$$x_a < \tan(x_a).$$

This last inequality is true when $0 < x_a < \pi/2$, so the earlier inequalities are true. Therefore, for $2 < a < \pi$, $0 < \lambda_+ < 1$, and (x_a, x_a) is a saddle point. In particular, for $a = 3$, $x_3 \approx 1.50$, $\lambda_- \approx -1.189$, and $\lambda_+ \approx 0.189$.

For $a > \pi$, the eigenvalues can be real or complex. Since x_a is an increasing function of a, there is a bifurcation value a_2 such that the eigenvalues are real for $\pi < a \leq a_2$ and complex for $a_2 < a$. For $\pi < a \leq a_2$, the eigenvalues are real and $\cos(x_a) < 0$, so $0 \leq \sqrt{1 + 4a\cos(x_a)} < 1$, $-1 < \lambda_- \leq \lambda_+ < 0$, and (x_a, x_a) is a sink.

12.2. Classification of Periodic Points

For $a_2 < a$,

$$|\lambda_\pm| = \frac{\sqrt{1 - (1 + 4a\cos(x_a))}}{2}$$
$$= \sqrt{a|\cos(x_a)|}$$
$$= \sqrt{a}\left(1 - \sin^2(x_a)\right)^{1/2}$$
$$= \sqrt{a}\left(1 - \frac{4x_a^2}{a^2}\right)^{1/2},$$

which goes to infinity as a goes to infinity. Thus, this fixed point becomes a source for large a, and there is a bifurcation value a_3 such that $|\lambda_\pm| < 1$ for $a_2 < a < a_3$, and $|\lambda_\pm| > 1$ for $a_3 < a$.

The following list summarizes the stability of the fixed points (x_a, x_a):

$2 < a < \pi$	saddle with real eigenvalues,
$\pi < a \leq a_2$	sink with real eigenvalues,
$a_2 < a < a_3$	sink with complex eigenvalues,
$a_3 < a$	source with complex eigenvalues.

We do not consider the fixed points with $x > \pi$, which occur for sufficiently large parameter values.

The Hénon map

M. Hénon introduced the map that bears his name as an example with a specific quadratic formula that could be iterated on a computer. This map also illustrates the ideas of stability for fixed points and period-2 points. Note its similarity to the logistic map.

The Hénon map is given by

$$\mathbf{F}(x, y) = (a - x^2 - b\,y, x).$$

The two constants a and b are parameters. Some authors, including Hénon, write the map differently and the parameter b has the opposite sign.

Fixed Points of the Hénon Map

The fixed points satisfy

$$(a - x^2 - b\,y, x) = (x, y)$$

or

$$y = x \quad \text{and}$$
$$x = a - x^2 - b\,y.$$

Substituting x for y, we obtain the equation

$$0 = x^2 + (b+1)\,x - a$$

with roots

$$x_\pm = \frac{-(b+1) \pm \sqrt{(1+b)^2 + 4a}}{2}.$$

Thus, the two fixed points are
$$(x_+, x_+) \quad \text{and} \quad (x_-, x_-).$$
The fixed points are real if
$$(1+b)^2 + 4a \geq 0.$$
For general parameter values, the matrix of partial derivatives is
$$D\mathbf{F}_{(x,y)} = \begin{pmatrix} -2x & -b \\ 1 & 0 \end{pmatrix}.$$

The determinant, $\det(D\mathbf{F}_{(x,y)}) = b$ is the factor by which area changes. If b is negative, then the map reflects in one direction and is orientation reversing. (Again, in the original paper, $b < 0$ corresponds to orientation preserving and $b > 0$ corresponds to orientation reversing.)

For example, if $a = 0$ and $b = -0.3$, then
$$x_{\pm} = \frac{-0.7 \pm \sqrt{(0.7)^2}}{2} = 0, \, -0.7,$$
and the two fixed points are
$$(0, 0) \quad \text{and} \quad (-0.7, -0.7).$$

At the fixed point $(0, 0)$, for $a = 0$ and $b = -0.3$,
$$D\mathbf{F}_{(0,0)} = \begin{pmatrix} 0 & 0.3 \\ 1 & 0 \end{pmatrix},$$
which has characteristic equation
$$0 = \lambda^2 - 0.3,$$
and eigenvalues $\lambda = \pm\sqrt{0.3} \approx \pm 0.5477$. Since $|\lambda| = \sqrt{0.3} < 1$ for both eigenvalues, the fixed point $(0, 0)$ is attracting.

At the other fixed point $(-0.7, -0.7)$,
$$D\mathbf{F}_{(-0.7,-0.7)} = \begin{pmatrix} 1.4 & 0.3 \\ 1 & 0 \end{pmatrix},$$
which has characteristic equation
$$0 = \lambda^2 - 1.4\lambda - 0.3,$$
and eigenvalues
$$\lambda = \frac{1.4 \pm \sqrt{(1.4)^2 + 1.2}}{2}$$
$$= \frac{1.4 \pm \sqrt{3.16}}{2}$$
$$\approx 1.5888, \, -0.1888.$$

Since $|-0.1888| < 1$ and $|1.5888| > 1$, the fixed point $(-0.7, -0.7)$ is a saddle and unstable, but not repelling.

12.2. Classification of Periodic Points

Period-2 Points of the Hénon Map

A period-2 point satisfies

$$(x_1, y_1) = \mathbf{F}(x_0, y_0) = (a - x_0^2 - b\,y_0, x_0) \quad \text{and}$$
$$(x_0, y_0) = \mathbf{F}(x_1, y_1) = (a - x_1^2 - b\,y_1, x_1).$$

It follows that $y_1 = x_0$, $y_0 = x_1$,

$$x_0 = a - x_1^2 - b\,y_1$$
$$= a - y_0^2 - b\,x_0,$$
$$0 = y_0^2 + (1+b)x_0 - a,$$

and

$$x_1 = a - x_0^2 - b\,y_0$$
$$= a - y_1^2 - b\,x_1,$$
$$0 = y_1^2 + (1+b)x_1 - a.$$

We leave it as an exercise to show that the points on a period-2 orbit satisfy

$$x_j + y_j = 1 + b.$$

Substituting $x_0 = 1 + b - y_0$ into the equation $0 = y_0^2 + (1+b)x_0 - a$, we obtain

$$0 = y_0^2 - (1+b)y_0 + (1+b)^2 - a.$$

We now take parameter values for which this equation has roots with simple expressions, $b = -0.3$ and $a = 0.57$. We leave it to the reader to check that both fixed points are saddles for these parameter values. The value y_0 for the period-2 point satisfies

$$0 = y_0^2 - (0.7)y_0 + (0.7)^2 - 0.57$$
$$= y_0^2 - 0.7\,y_0 - 0.08,$$

with roots $y_0 = 0.8$, -0.1. Thus, the period-2 orbit is

$$\left\{ \begin{pmatrix} 0.8 \\ -0.1 \end{pmatrix}, \begin{pmatrix} -0.1 \\ 0.8 \end{pmatrix} \right\}, \quad \text{with}$$
$$\mathbf{F}\begin{pmatrix} 0.8 \\ -0.1 \end{pmatrix} = \begin{pmatrix} -0.1 \\ 0.8 \end{pmatrix} \quad \text{and}$$
$$\mathbf{F}\begin{pmatrix} -0.1 \\ 0.8 \end{pmatrix} = \begin{pmatrix} 0.8 \\ -0.1 \end{pmatrix}.$$

The matrix of partial derivatives is

$$D\mathbf{F}^2_{(0.8,-0.1)} = D\mathbf{F}_{(-0.1,0.8)}\,D\mathbf{F}_{(0.8,-0.1)}$$
$$= \begin{pmatrix} 0.2 & 0.3 \\ 1 & 0 \end{pmatrix} \begin{pmatrix} -1.6 & 0.3 \\ 1 & 0 \end{pmatrix} = \begin{pmatrix} -0.02 & 0.06 \\ -1.6 & 0.3 \end{pmatrix}.$$

The characteristic equation is

$$0 = \lambda^2 - 0.28\,\lambda + 0.09,$$

with eigenvalues
$$\lambda = \frac{0.28 \pm \sqrt{(0.28)^2 - 0.36}}{2}$$
$$= 0.14 \pm i\sqrt{0.0704}$$
$$\approx 0.14 \pm i\, 0.2653.$$

The absolute value of the eigenvalues satisfies
$$|\lambda| = \sqrt{(0.14)^2 + 0.0704}$$
$$= \sqrt{0.09}$$
$$= 0.3 < 1.$$

Therefore, the period-2 orbit is attracting.

Inverse and α-limit set

We defined the ω-limit set in Section 11.1. We now proceed to define the α-limit set if the map has an inverse and it is possible to follow the orbit for backward iterates. We have used inverses before when defining conjugacy for one-dimensional maps and for linear maps. However, we want to solidify the idea of an inverse for a nonlinear map with more variables. Therefore, we start by defining an inverse of a function and calculating the inverse of the Hénon map.

A map is *one to one* provided that if $\mathbf{x}_1 \neq \mathbf{x}_2$ then $\mathbf{F}(\mathbf{x}_1) \neq \mathbf{F}(\mathbf{x}_2)$. For a continuous function on the line, this is the same as saying that the function is monotone, either increasing everywhere or decreasing everywhere.

If a function is one to one, it is possible to define an inverse function defined on the image of the function. The *image* of a function $\mathbf{F}$ is the set of all $\mathbf{y}$ for which there is some $\mathbf{x}$ such that $\mathbf{F}(\mathbf{x}) = \mathbf{y}$,

$$\text{image}(\mathbf{F}) = \{\, \mathbf{y} : \mathbf{F}(\mathbf{x}) = \mathbf{y} \text{ for some } \mathbf{x} \text{ in the domain of } \mathbf{F}\,\}.$$

The map $\mathbf{F}$ from $\mathbf{X}$ to $\mathbf{Y}$ is called *onto* provided image$(\mathbf{F}) = \mathbf{Y}$. Sometimes we say the map $\mathbf{F}$ is onto $\mathbf{Y}$ to emphasize the image.

Assume that $\mathbf{F}$ is a one-to-one function from a space $\mathbf{X}$ to a space $\mathbf{Y}$ with image $\mathbf{J}$. Then, the *inverse* $\mathbf{F}^{-1}$ is defined from $\mathbf{J}$ onto $\mathbf{X}$ by $\mathbf{F}^{-1}(\mathbf{y}_0) = \mathbf{x}_0$ if and only if $\mathbf{F}(\mathbf{x}_0) = \mathbf{y}_0$. The compositions $\mathbf{F}^{-1} \circ \mathbf{F}$ is the identity on $\mathbf{X}$ and $\mathbf{F} \circ \mathbf{F}^{-1}$ is the identity on $\mathbf{J}$.

Definition 12.18. Let $\mathbf{U}$ and $\mathbf{V}$ be two open sets in $\mathbb{R}^n$ (or metric spaces). A map $\mathbf{h}$ from $\mathbf{U}$ to $\mathbf{V}$ is called a *homeomorphism* provided that (i) $\mathbf{h}$ is continuous, (ii) $\mathbf{h}$ is one to one on $\mathbf{U}$, (iii) $\mathbf{h}$ is onto $\mathbf{V}$, that is $\mathbf{h}(\mathbf{U}) = \mathbf{V}$, and (iv) the inverse $\mathbf{h}^{-1}$ is a continuous map from $\mathbf{V}$ to $\mathbf{U}$.

Definition 12.19. Let $\mathbf{U}$ and $\mathbf{V}$ be two open sets in $\mathbb{R}^n$. For an integer $r \geq 1$, a C^r *diffeomorphism* $\mathbf{F}$ from $\mathbf{U}$ to $\mathbf{V}$ is a homeomorphism from $\mathbf{U}$ to $\mathbf{V}$ such that $\mathbf{F}$ and its inverse $\mathbf{F}^{-1}$ are C^r.

For a diffeomorphism, it follows that $\det(D\mathbf{F}_{(\mathbf{x})}) \neq 0$ at all points $\mathbf{x}$ and that
$$D(\mathbf{F}^{-1})_{(\mathbf{y})} = \left(D\mathbf{F}_{(\mathbf{x})}\right)^{-1},$$
where $\mathbf{y} = \mathbf{F}(\mathbf{x})$.

12.2. Classification of Periodic Points

Example 12.20. We calculate the inverse of the Hénon map. If
$$(x_1, y_1) = \mathbf{F}(x_0, y_0) = (a - x_0^2 - b\,y_0, x_0),$$
then
$$x_1 = a - x_0^2 - b\,y_0,$$
$$y_1 = x_0.$$
Substituting y_1 for x_0 and solving for y_0, we get
$$x_1 = a - y_1^2 - b\,y_0,$$
$$b\,y_0 = a - y_1^2 - x_1,$$
$$y_0 = \frac{a - y_1^2 - x_1}{b}.$$
Therefore, we have expressions for x_0 and y_0 in terms of x_1 and y_1, and the inverse is given by
$$(x_0, y_0) = \mathbf{F}^{-1}(x_1, y_1) = \left(y_1, \frac{a - y_1^2 - x_1}{b}\right).$$
Since both the map and its inverse have continuous partial derivatives of all orders, the Hénon map is a C^∞ diffeomorphism from $\mathbb{R}^2$ onto $\mathbb{R}^2$.

Definition 12.21. If the map $\mathbf{F}$ has an inverse, then the *orbit of a point* $\mathbf{x}_0$ is the set of all both forward and backward iterates of $\mathbf{x}_0$ and denote it by
$$\mathcal{O}_{\mathbf{F}}(\mathbf{x}_0) = \{\,\mathbf{F}^j(\mathbf{x}_0) : -\infty < j < \infty\,\}.$$

Definition 12.22. Let $\mathbf{F}$ be a map on a space $\mathbf{X}$, and let $\mathbf{x}_0$ be an initial condition in $\mathbf{X}$. Assume that $\mathbf{F}$ has an inverse $\mathbf{F}^{-1}$. A point $\mathbf{q}$ is an α-*limit point of* $\mathbf{x}_0$ *for the map* $\mathbf{F}$ provided that there is a sequence of iterates $-k_j$ going to minus infinity such that $\mathbf{F}^{-k_j}(\mathbf{x}_0)$ converges to $\mathbf{q}$. This means that the orbit of $\mathbf{x}_0$ keeps coming back close to $\mathbf{q}$ under backward iteration. A more precise way of saying this is that, for any $\epsilon > 0$ and for any $N > 0$, there exists a $k \geq N$ such that
$$\|\mathbf{F}^{-k}(\mathbf{x}_0) - \mathbf{q}\| < \epsilon.$$

The α-*limit set of* $\mathbf{x}_0$ *for* $\mathbf{F}$ is the set of all α-limit points of $\mathbf{x}_0$ for $\mathbf{F}$:
$$\alpha(\mathbf{x}_0; \mathbf{F}) = \{\,\mathbf{q} : \mathbf{q} \text{ is an } \alpha\text{-limit point of } \mathbf{x}_0 \text{ for } \mathbf{F}\,\}.$$

The property of a periodic point being repelling can now be expressed in terms of the inverse of a map. A period-k point $\mathbf{p}_0$ for a homeomorphism $\mathbf{F}$ on $\mathbb{R}^n$ is *repelling*, provided that it is attracting for $\mathbf{F}^{-1}$; that is, the following conditions are satisfied:

(i) For every $r > 0$, there is a $\delta > 0$ such that
$$\mathbf{F}^{-j}(\mathbf{B}(\mathbf{p}_0, \delta)) \subset \mathbf{B}(\mathbf{F}^{-j}(\mathbf{p}_0), r)$$
for all $j \geq 0$ (so $-j \leq 0$).

(ii) There is an $r_1 > 0$ such that, if $\mathbf{x}_0$ is in $\mathbf{B}(\mathbf{p}_0, r_1)$, then $\alpha(\mathbf{x}_0; \mathbf{F}) = \mathcal{O}_{\mathbf{F}}(\mathbf{p}_0)$.

Exercises 12.2

1. Determine the stability type of the fixed point at the origin for each of the linear maps in Exercise 1 of Section 12.1.

2. Show that the linear map with matrix
$$\begin{pmatrix} 1 & 1 \\ 0 & 1 \end{pmatrix}$$
is unstable.

3. Let
$$\mathbf{F}\begin{pmatrix} x \\ y \end{pmatrix} = \begin{pmatrix} x + y + x^2 \\ 2x + 3y \end{pmatrix}.$$
Find the fixed points and classify them as source, saddle, sink, or none of these.

4. Let
$$\mathbf{F}\begin{pmatrix} x \\ y \end{pmatrix} = \begin{pmatrix} 2xy + y \\ 3y - x \end{pmatrix}.$$
Find the fixed points and classify them as source, saddle, sink, or none of these.

5. Let
$$\mathbf{F}\begin{pmatrix} x \\ y \end{pmatrix} = \begin{pmatrix} \frac{1}{2}x + y^2 \\ \frac{1}{4}x + \frac{3}{2}y \end{pmatrix}.$$
Find the fixed points and classify them as source, saddle, sink, or none of these.

6. Consider the Hénon map.
   a. Show that, if (x_+, x_+) and (x_-, x_-) are the two fixed points, then
   $$x_+ + x_- = -1 - b.$$
   b. Show that, if $\{(x_0, y_0), (x_1, y_1)\}$ is a period-2 orbit, then
   $$1 + b = x_0 + y_0 = x_1 + y_1.$$

7. Consider the Hénon map with $b = -0.2$.
   a. Show that, for $a \geq -0.16 = a_0$, there are fixed points. Find the eigenvalues of the single fixed point for $a = -0.16$.
   b. Show that, for $a > -0.16$, $x_- < -0.4$, $\lambda_+ > 1$, and $\lambda_- = -0.2/\lambda_+$ satisfies $-1 < \lambda_- < 0$, so (x_-, x_-) is a saddle point.
   c. Show that, for the fixed point (x_+, x_+), the eigenvalue $\lambda_- = -1$ for $a = 0.48 = a_1$. Using the continuity of the eigenvalues, conclude that the fixed point is attracting for $-0.16 < a < 0.48$.
   d. Show that there is a period-2 orbit for $a > 0.48$, and that the product of the two values of x on the orbit is $0.8^2 - a$, $x_0 x_1 = 0.8^2 - a$. Hint: Use the results of the previous exercise about period-2 orbits for the Hénon map.
   e. Show that the characteristic equation for this period-2 orbit is
   $$\lambda^2 - (4x_0 x_1 + 0.4)\lambda + 0.04 = 0.$$

Letting $-\mu = x_0 x_1$, show that one of the eigenvalues is -1 when

$$0 = 3\mu + 0.4\mu - 0.84 \quad \text{or}$$
$$\mu = \frac{2.8}{6}.$$

Show that this occurs for $a = 0.64 + {}^{2.8}\!/_6 = a_2$. Also, for $a_1 < a < a_2$, the period-2 orbit is attracting.

8. Let $a > 0$, and define $\mathbf{F}_a$ from $\mathbb{R}^2$ to itself by

$$\mathbf{F}_a(x, y) = (1 - ax^2 + y, x).$$

  a. Find all period-2 points for $\mathbf{F}_a$.
  b. Find all values of a for which the period-2 cycle is of saddle type.

12.3. Stable Manifolds

A linear map for which the origin is a saddle has contracting and expanding directions. If a nonlinear map has a fixed point for which the matrix of partial derivatives is a saddle, then the nonlinear map has curves (or surfaces), for which the points on these curves have a similar type of behavior with respect to the fixed point.

Considering a nonlinear map in two dimensions, with a saddle fixed, there is a curve of points, called the *stable manifold*, that converge to the fixed point under forward iteration. There is another curve of points, called the *unstable manifold*, that converge to the fixed point under backward iteration. These curves are important in separating the points that pass the fixed point on one side from those that pass the fixed point on the other side, so in two dimensions they are sometimes called *separatrices*.

These stable and unstable manifolds are important for a variety of reasons. A stable manifold of one saddle fixed point can form the boundary of the basin of attraction of an attracting fixed point as illustrated in Section 12.3.2. In Section 13.3, we explain how an intersection of stable and unstable manifolds of a fixed point implies the existence of an invariant set on which the map has complicated dynamics.

In higher dimensions, these curves are replaced by surfaces or higher dimensional "manifolds." In mathematics, the term *manifold* refers to curves, surfaces, and higher dimensional objects. (In common usage, manifold means many times or an object with many openings.) Appendix A.2 provides a more thorough discussion of the term.

If the origin is a saddle fixed point for a linear map on the plane, then there is a line on which the map is a contraction and another line on which the map is an expansion. For example, for

$$\mathbf{A} = \begin{pmatrix} \frac{1}{2} & 0 \\ 0 & 2 \end{pmatrix}$$

and $k \geq 0$,

$$\mathbf{A}^k \begin{pmatrix} x_1 \\ 0 \end{pmatrix} = \tfrac{1}{2^k} \begin{pmatrix} x_1 \\ 0 \end{pmatrix} \quad \text{and}$$

$$\mathbf{A}^k \begin{pmatrix} 0 \\ x_2 \end{pmatrix} = 2^k \begin{pmatrix} 0 \\ x_2 \end{pmatrix} \quad \text{or}$$

$$\mathbf{A}^{-k} \begin{pmatrix} 0 \\ x_2 \end{pmatrix} = 2^{-k} \begin{pmatrix} 0 \\ x_2 \end{pmatrix}.$$

The points on the eigenspace for $\lambda = 1/2$ are the set of all points $\mathbf{x}$ such that $\mathbf{A}^k \mathbf{x}$ goes to $\mathbf{0}$ as k goes to infinity. This set of these points is called the *stable manifold of the origin*, denoted by

$$W^s(\mathbf{0}, \mathbf{A}) = \left\{ \mathbf{x} : \mathbf{A}^k \mathbf{x} \text{ goes to } \mathbf{0} \text{ as } k \text{ goes to } \infty \right\} = \left\{ \begin{pmatrix} x_1 \\ 0 \end{pmatrix} \right\}.$$

The points on the eigenspace for $\lambda = 2$ are the set of all points $\mathbf{x}$ such that $\mathbf{A}^k \mathbf{x}$ goes to $\mathbf{0}$ as k goes to minus infinity. This collection of point is called the *unstable manifold of the origin*, denoted by

$$W^u(\mathbf{0}, \mathbf{A}) = \left\{ \mathbf{x} : \mathbf{A}^k \mathbf{x} \text{ goes to } \mathbf{0} \text{ as } k \text{ goes to } -\infty \right\} = \left\{ \begin{pmatrix} 0 \\ x_2 \end{pmatrix} \right\}.$$

Now, consider a nonlinear map $\mathbf{F}$ in $\mathbb{R}^2$ with a fixed point $\mathbf{p}$, such that the linearization $\mathbf{A} = D\mathbf{F}_{(\mathbf{p})}$ has a saddle at the origin. The main theorem of this section says that there are invariant curves tangent to the lines $\mathbf{p} + W^s(\mathbf{0}, \mathbf{A})$ and $\mathbf{p} + W^u(\mathbf{0}, \mathbf{A})$ that consist of points whose iterates $\mathbf{F}^k(\mathbf{x})$ tend to $\mathbf{p}$ as k goes to infinity or minus infinity, respectively:

$$W^s(\mathbf{p}, \mathbf{F}) = \left\{ \mathbf{x} : \|\mathbf{F}^k(\mathbf{x}) - \mathbf{p}\| \text{ goes to 0 as } k \text{ goes to } \infty \right\} \quad \text{and}$$
$$W^u(\mathbf{p}, \mathbf{F}) = \left\{ \mathbf{x} : \|\mathbf{F}^k(\mathbf{x}) - \mathbf{p}\| \text{ goes to 0 as } k \text{ goes to } -\infty \right\}.$$

We start with an example in which we can calculate these curves explicitly.

Example 12.23. Consider the map

$$\mathbf{F} \begin{pmatrix} x \\ y \end{pmatrix} = \begin{pmatrix} \tfrac{1}{2} x \\ 2y + x^2 \end{pmatrix},$$

which has a fixed point at the origin. The derivative is

$$D\mathbf{F}_{(\mathbf{0})} = \begin{pmatrix} \tfrac{1}{2} & 0 \\ 0 & 2 \end{pmatrix}.$$

The map preserves the y-axis on which it is an expansion, so that

$$\mathbf{F} \begin{pmatrix} 0 \\ y \end{pmatrix} = \begin{pmatrix} 0 \\ 2y \end{pmatrix}.$$

Thus, the iterates of the points by the inverse map,

$$\mathbf{F}^{-k} \begin{pmatrix} 0 \\ y \end{pmatrix} = \begin{pmatrix} 0 \\ 2^{-k} y \end{pmatrix},$$

12.3. Stable Manifolds

converge to the fixed point at the origin. Therefore, the unstable manifold of the fixed point is the y-axis:

$$W^u(\mathbf{0}, \mathbf{F}) = \{ (0, y)^\mathsf{T} \}.$$

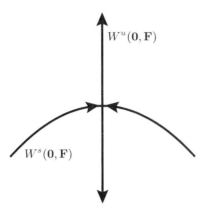

Figure 8. Stable and unstable manifolds for Example 12.23

The stable manifold is not as easy to find. Assume that it is the graph over the x-axis of a function $y = \psi(x)$. Since the curve goes through the origin, $\psi(0) = 0$; since the curve is tangent to the eigendirection $(1, 0)^\mathsf{T}$, $\psi'(0) = 0$. Express ψ in terms of a power series,

$$y = \psi(x) = a_2 x^2 + a_3 x^3 + \cdots,$$

with no constant or linear term. Then, the iterate is expressed as

$$\begin{pmatrix} x_1 \\ \sum_{j=2}^{\infty} a_j x_1^j \end{pmatrix} = \mathbf{F}\begin{pmatrix} x \\ \sum_{j=2}^{\infty} a_j x^j \end{pmatrix}$$

$$= \begin{pmatrix} \frac{1}{2}x \\ x^2 + \sum_{j=2}^{\infty} 2\, a_j x^j \end{pmatrix}$$

$$= \begin{pmatrix} \frac{1}{2}x \\ (2\,a_2 + 1) x^2 + \sum_{j=3}^{\infty} 2\, a_j x^j \end{pmatrix}$$

$$= \begin{pmatrix} x_1 \\ 4\,(2\,a_2 + 1) x_1^2 + \sum_{j=3}^{\infty} 2^{j+1}\, a_j x_1^j \end{pmatrix},$$

where we have used $2\,x_1 = x$. Equating the coefficient of x_1^2, $a_2 = 4(2\,a_2 + 1)$. So, $-4 = 7a_2$, and $a_2 = {-4}/{7}$. Equating the coefficient of x_1^j for $j > 2$, $a_j = 2^{j+1} a_j$. So, $(2^{j+1} - 1)a_j = 0$ and $a_j = 0$. Thus, the invariant curve is given by

$$y = \psi(x) = -\frac{4}{7}x^2.$$

Since this curve is invariant by $\mathbf{F}$, the iterates of points on the curve satisfy

$$\begin{pmatrix} x_1 \\ \psi(x_1) \end{pmatrix} = \mathbf{F}\begin{pmatrix} x_0 \\ \psi(x_0) \end{pmatrix} = \begin{pmatrix} \frac{1}{2}x_0 \\ \psi(x_1) \end{pmatrix} \quad \text{and}$$

$$\begin{pmatrix} x_k \\ \psi(x_k) \end{pmatrix} = \mathbf{F}\begin{pmatrix} x_{k-1} \\ \psi(x_{k-1}) \end{pmatrix} = \begin{pmatrix} \frac{1}{2}x_{k-1} \\ \psi(x_k) \end{pmatrix} = \begin{pmatrix} \frac{1}{2^k}x_0 \\ \psi(x_k) \end{pmatrix}.$$

Thus, $x_k = x_0 2^{-k}$ goes to zero as k goes to infinity. Also, since ψ is continuous, $y_k = \psi(x_k)$ must also go to zero. Thus, the pair of points (x_k, y_k) goes to the fixed point at the origin, and

$$W^s(\mathbf{0}, \mathbf{F}) = \left\{ \left(x, -\frac{4}{7}x^2\right)^\mathsf{T} \right\}.$$

See Figure 8.

We next give the general definition of the stable and unstable manifold of a periodic point.

Definition 12.24. Let $\mathbf{F}$ be a diffeomorphism and let $\mathbf{p}$ be a period-q point of $\mathbf{F}$. The *stable manifold of a period-q point* $\mathbf{p}$ is the set of points whose iterates are asymptotic to the iterates of $\mathbf{p}$. More specifically,

$$W^s(\mathbf{p}, \mathbf{F}) = \{\mathbf{x} : \|\mathbf{F}^k(\mathbf{x}) - \mathbf{F}^k(\mathbf{p})\| \text{ goes to } 0 \text{ as } k \text{ goes to } \infty \}.$$

The *stable manifold of the orbit of a period-q point* $\mathbf{p}$ is the set of points whose iterates are asymptotic to the iterates of $\mathbf{p}$, or

$$W^s(\mathcal{O}_\mathbf{F}(\mathbf{p}), \mathbf{F}) = \bigcup_{j=0}^{q-1} W^s(\mathbf{F}^j(\mathbf{p}), \mathbf{F}) = \{\mathbf{x} : \omega(\mathbf{x}) = \mathcal{O}_\mathbf{F}(\mathbf{p}) \}.$$

The *unstable manifold of a period-q point* $\mathbf{p}$ is the set of points whose iterates are backwardly asymptotic to the iterates of $\mathbf{p}$:

$$W^u(\mathbf{p}, \mathbf{F}) = \{\mathbf{x} : \|\mathbf{F}^k(\mathbf{x}) - \mathbf{F}^k(\mathbf{p})\| \text{ goes to } 0 \text{ as } k \text{ goes to } -\infty \}.$$

The *unstable manifold of the orbit of a period-q point* $\mathbf{p}$ is the set of points that are backwardly asymptotic to the orbit of $\mathbf{p}$:

$$W^u(\mathcal{O}_\mathbf{F}(\mathbf{p}), \mathbf{F}) = \bigcup_{j=0}^{q-1} W^u(\mathbf{F}^j(\mathbf{p}), \mathbf{F}) = \{\mathbf{x} : \alpha(\mathbf{x}) = \mathcal{O}_\mathbf{F}(\mathbf{p}) \}.$$

We also want to define the local stable and unstable manifold. For $\delta > 0$, the *local stable manifold of* $\mathbf{p}$ *of size* δ is the set

$$W^s_\delta(\mathbf{p}, \mathbf{F}) = \{\mathbf{x} \in W^s(\mathbf{p}, \mathbf{F}) : \|\mathbf{F}^k(\mathbf{x}) - \mathbf{F}^k(\mathbf{p})\| \leq \delta \text{ for all } k \geq 0 \}.$$

Similarly, for $\delta > 0$, the *local unstable manifold of* $\mathbf{p}$ *of size* δ is the set

$$W^u_\delta(\mathbf{p}, \mathbf{F}) = \{\mathbf{x} \in W^u(\mathbf{p}, \mathbf{F}) : \|\mathbf{F}^k(\mathbf{x}) - \mathbf{F}^k(\mathbf{p})\| \leq \delta \text{ for all } k \leq 0 \}$$
$$= \{\mathbf{x} : \|\mathbf{F}^k(\mathbf{x}) - \mathbf{F}^k(\mathbf{p})\| \leq \delta \text{ for all } k \leq 0 \}.$$

12.3. Stable Manifolds

Remark 12.25. If **p** is an attracting fixed point, then the stable manifold of **p** is what we earlier called the *basin of attraction*:
$$W^s(\mathbf{p}, \mathbf{F}) = \mathscr{B}(\mathbf{p}; \mathbf{F}).$$
From this point on, we use the notation for the stable manifold for this basin of attraction.

The next theorem characterizes the local stable and unstable manifolds of a periodic point.

Theorem 12.4 (Stable Manifold Theorem). *Let* $\mathbf{F}$ *be a* C^r *diffeomorphism on* $\mathbb{R}^2$ *for* $r \geq 1$, *and let* $\mathbf{p}$ *be a saddle period-q point with eigenvalues* λ_s *and* λ_u, *with* $|\lambda_s| < 1$ *and* $|\lambda_u| > 1$. *For sufficiently small* $\delta > 0$, *the local stable and unstable manifolds of* $\mathbf{p}$ *of size* δ, $W^s_\delta(\mathbf{p}, \mathbf{F})$ *and* $W^u_\delta(\mathbf{p}, \mathbf{F})$, *are* C^r *curves tangent to the directions given by the eigenvectors for the eigenvalues* λ_s *and* λ_u, *respectively. Moreover, they are characterized by*

$$W^s_\delta(\mathbf{p}, \mathbf{F}) = \{\mathbf{x} : \|\mathbf{F}^k(\mathbf{x}) - \mathbf{F}^k(\mathbf{p})\| \leq \delta \text{ for all } k \geq 0\} \quad \text{and}$$
$$W^u_\delta(\mathbf{p}, \mathbf{F}) = \{\mathbf{x} : \|\mathbf{F}^k(\mathbf{x}) - \mathbf{F}^k(\mathbf{p})\| \leq \delta \text{ for all } k \leq 0\}.$$

These curves are locally invariant in the sense that

$$\mathbf{F}(W^s_\delta(\mathbf{p}, \mathbf{F})) \subset W^s_\delta(\mathbf{F}(\mathbf{p}), \mathbf{F}) \quad \text{and}$$
$$\mathbf{F}(W^u_\delta(\mathbf{p}, \mathbf{F})) \supset W^u_\delta(\mathbf{F}(\mathbf{p}), \mathbf{F}).$$

The (global) stable and unstable manifolds are equal to the union of the iterates of the local stable and unstable manifolds, respectively:

$$W^s(\mathbf{p}, \mathbf{F}) = \bigcup_{k=1}^{\infty} \mathbf{F}^{-k}(W^s_\delta(\mathbf{F}^k(\mathbf{p}), \mathbf{F})),$$
$$W^u(\mathbf{p}, \mathbf{F}) = \bigcup_{k=1}^{\infty} \mathbf{F}^k(W^u_\delta(\mathbf{F}^{-k}(\mathbf{p}), \mathbf{F})).$$

The global stable manifold cannot cross itself, but can wind around in a complicated fashion. See Figure 9 for an example of this phenomenon for the Hénon map with $a = 1.6$ and $b = -0.3$. The stable manifold of the fixed saddle point $\mathbf{p} = (x_-, x_-)$ contains the curves in the figure that are more vertical than horizontal and also points outside the frame of the figure; the unstable manifold contains the curves that are more horizontal than vertical, which stay completely in the window. (The unstable manifold looks like it has sharp corners, but it really has just very sharp smooth bends.) The figure indicates the second fixed point $\mathbf{q} = (x_+, x_+)$, but not its stable and unstable manifolds.

Remark 12.26. Poincaré discovered the importance of the stable and unstable manifolds in the last quarter of the nineteenth century. In particular, if the unstable manifold of a periodic orbit crosses the stable manifold of the same periodic orbit, then each manifold is embedded in a complicated manner and the map has an invariant set with complicated dynamics modeled by symbolic dynamics. He used these ideas to explain why an N-body problem governed by Newtonian laws of attraction, such as our solar system, can be unstable. His ideas relate to the construction of a horseshoe for a transverse homoclinic point given in Section 13.3.

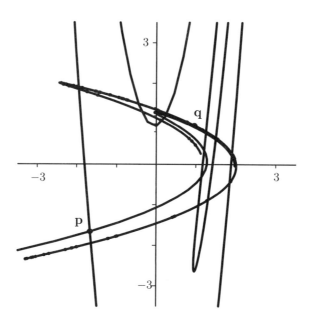

Figure 9. Stable and unstable manifolds of the fixed point **p** for the Hénon map with $a = 1.6$ and $b = -0.3$. Note that the stable manifold run out of the window shown.

12.3.1. Numerical Calculation of the Stable Manifold. Theorem 12.4 asserts the existence of stable and unstable manifolds. Using computer simulation, there are various ways to estimate stable and unstable manifolds for diffeomorphisms in two dimensions. The simplest technique is to pretend that the local unstable manifold is a short line segment through the fixed point **p** in the direction of the unstable eigenvector $\mathbf{v}^u$:

$$W^u_\delta(\mathbf{p}) = \{\,\mathbf{p} + t\,\mathbf{v}^u : -\delta \leq t \leq \delta\,\}.$$

Iterating points on $W^u_\delta(\mathbf{p})$ determines the global unstable manifold. Assume that we place N evenly spaced points on one side of this local unstable manifold; then,

$$\mathbf{x}^j = \mathbf{p} + \left(\frac{j}{N}\right)\delta\mathbf{v}^u,$$

for $1 \leq j \leq N$. Taking a few iterates gives a fairly long unstable manifold, namely,

$$\{\,\mathbf{F}^k(\mathbf{x}^j) : 1 \leq j \leq N\,\},$$

for a fixed $k > 0$. For example, we could take $N = 1000$ and $k = 5$.

The stable manifold could be found by a similar method using backward iterates. Let

$$\mathbf{y}^j = \mathbf{p} + \left(\frac{j}{N}\right)\delta\mathbf{v}^s,$$

for $1 \leq j \leq N$. Then,

$$\{\,\mathbf{F}^{-k}(\mathbf{y}^j) : 1 \leq j \leq N\,\},$$

for a fixed $k > 0$, approximates the stable manifold.

This primitive method is often used to plot the stable and unstable manifolds of a fixed point in the plane. There are a couple of problems that cause the locations of these numerically calculated curves to be imprecise, especially far out on the curves. First, the local stable and unstable manifolds are approximated with line segments. Second, the numerical iteration of points has round-off errors. Finally, the points on the curve can spread apart as higher iterates are taken. However, the contracting and expanding nature of the map itself often causes the general shape of the manifolds to be correct even if one cannot verify their numerical accuracy.

The book [**Par89**] by Parker and Chua gives a better algorithm that varies the number of points along the unstable manifold to keep them evenly spaced farther out on the manifold. Their algorithm also increases the number of points used when the curve bends more sharply.

To obtain greater accuracy, a shooting method can be used. In this approach, one checks to see on which side of $\mathbf{p}$ the iterates of points $\mathbf{x}$ pass. If one point $\mathbf{x}_1$ passes on one side and another point $\mathbf{x}_2$ passes on the other side, then $\mathbf{x}_3 = (^1/_2)(\mathbf{x}_1 + \mathbf{x}_2)$ is a better guess. This new point $\mathbf{x}_3$ passes on the opposite side of $\mathbf{p}$ than does one of the original points. Using these two points we can continue to refine which point is on the stable manifold.

12.3.2. Basin Boundaries. As we mentioned at the beginning of Section 12.3, the stable manifold of a saddle fixed point can form the boundary of the basin of attraction of an attracting fixed point. We use the Hénon family of maps with $b = 0.3$ to illustrate this principle:
$$\mathbf{F}_{a,0.3}(x,y) = (a - x^2 - 0.3\,y, x).$$
This family has fixed points for $a \geq {-1.3^2}/4 = -0.4225$. When the fixed points first appear ($a \approx -0.4225$), one of the fixed points $\mathbf{p} = (x_-, x_-)$ is a saddle point and the other $\mathbf{q} = (x_+, x_+)$ is an attracting fixed point. The stable manifold of $\mathbf{q}$ is an open set in the plane, previously called the basin of attraction of the fixed point. For values of a near -0.4225, computer studies show that the basin of attraction $W^s(\mathbf{q}; F_{a,0.3})$ has the stable manifold $W^s(\mathbf{p}; F_{a,0.3})$ as its boundary. See Figure 10. This is true not only for the Hénon family but also for any family of maps of the plane that creates a pair of fixed points as a parameter is varied in the same manner as the Hénon family. (See [**Pat87**].)

12.3.3. Stable Manifolds in Higher Dimension. There is a version of the stable manifold theorem for a saddle periodic point in $\mathbb{R}^n$. Assume that $\mathbf{p}$ is a period-q point for a diffeomorphism $\mathbf{F}$ in $\mathbb{R}^n$. Let $\mathbf{A} = D(\mathbf{F}^q)_{(\mathbf{p})}$, with eigenvalues λ_j for $1 \leq j \leq n$. Assume $|\lambda_j| < 1$ for $1 \leq j \leq n_s < n$ and $|\lambda_j| > 1$ for $n_s + 1 \leq j \leq n$. Let $\mathbf{v}^j$ be a corresponding eigenvector or generalized eigenvector for λ_j. Then the *stable eigenspace* at $\mathbf{p}$ is the set of vectors
$$\mathbb{E}^s = \mathrm{span}\{\,\mathbf{v}^j : 1 \leq j \leq n_s\,\} = \{\,y_1 \mathbf{v}^1 + \cdots + y_{n_s} \mathbf{v}^{n_s} : y_1, \ldots, y_{n_s} \in \mathbb{R}\,\}.$$
Translating this subspace to $\mathbf{p}$ gives the hyperplane $\mathbf{p} + \mathbb{E}^s$. Similarly, the *unstable eigenspace* at $\mathbf{p}$ is the set of vectors
$$\mathbb{E}^u = \mathrm{span}\{\,\mathbf{v}^j : n_s + 1 \leq j \leq n\,\}$$
$$= \{\,y_{n_s+1} \mathbf{v}^{n_s+1} + \cdots + y_n \mathbf{v}^n : y_{n_s+1}, \ldots, y_n \in \mathbb{R}\,\},$$

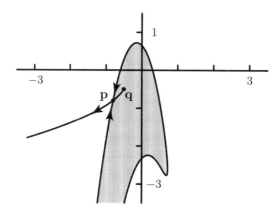

Figure 10. Stable manifold of the saddle fixed point **p** as the boundary of the basin of attraction of a second fixed point **q** for the Hénon map with $a = -0.4$ and $b = 0.3$

and the corresponding hyperplane through **p** is $\mathbf{p} + \mathbb{E}^u$.

For a saddle periodic point, both $\mathbb{E}^s$ and $\mathbb{E}^u$ contain nonzero vectors. In $\mathbb{R}^3$, one is a line and the other is a plane.

Theorem 12.5. *Let* **F** *be a* C^r *diffeomorphism on* $\mathbb{R}^n$ *for* $r \geq 1$, *and let* **p** *be a saddle period-q point. For sufficiently small* $\delta > 0$, *the local stable and unstable manifolds of* **p** *of size* δ, $W_\delta^s(\mathbf{p}, \mathbf{F})$ *and* $W_\delta^u(\mathbf{p}, \mathbf{F})$, *are* C^r *manifolds given as graphs over the hyperplanes* $\mathbf{p} + \mathbb{E}^s$ *and* $\mathbf{p} + \mathbb{E}^u$, *respectively. The manifold* $W_\delta^s(\mathbf{p}, \mathbf{F})$ *is tangent to the hyperplane* $\mathbf{p} + \mathbb{E}^s$ *at* **p**, *and* $W_\delta^u(\mathbf{p}, \mathbf{F})$ *is tangent to* $\mathbf{p} + \mathbb{E}^u$ *at* **p**. *These manifolds are locally invariant; that is,*

$$\mathbf{F}(W_\delta^s(\mathbf{p}, \mathbf{F})) \subset W_\delta^s(\mathbf{F}(\mathbf{p}), \mathbf{F})$$
$$\mathbf{F}(W_\delta^u(\mathbf{p}, \mathbf{F})) \supset W_\delta^u(\mathbf{F}(\mathbf{p}), \mathbf{F}).$$

The global stable and unstable manifolds are the unions of the iterates of the local stable and unstable manifolds, respectively:

$$W^s(\mathbf{p}, \mathbf{F}) = \bigcup_{k=1}^\infty \mathbf{F}^{-k}(W_\delta^s(\mathbf{F}^k(\mathbf{p}), \mathbf{F})) \quad \text{and}$$
$$W^u(\mathbf{p}, \mathbf{F}) = \bigcup_{k=1}^\infty \mathbf{F}^k(W_\delta^u(\mathbf{F}^{-k}(\mathbf{p}), \mathbf{F})).$$

Example 12.27. Consider the map

$$\mathbf{F}(x, y, z) = \left(\frac{1}{2}x, \frac{1}{2}y, 2z + x^2 + y^2\right),$$

which has the origin as a fixed point. The matrix of partial derivatives at the origin is

$$D\mathbf{F}_{(\mathbf{0})} = \begin{pmatrix} \frac{1}{2} & 0 & 0 \\ 0 & \frac{1}{2} & 0 \\ 0 & 0 & 2 \end{pmatrix}.$$

The map preserves the z-axis, and
$$\mathbf{F}\begin{pmatrix}0\\0\\z\end{pmatrix}=\begin{pmatrix}0\\0\\2z\end{pmatrix}.$$
So, $\mathbf{F}$ expands the z-axis, and
$$W^u(\mathbf{0},\mathbf{F})=\{\,(0,0,z)^\mathsf{T}:z\in\mathbb{R}\,\}.$$
Just as for the map in $\mathbb{R}^2$ given in Example 12.23, the stable manifold is given by
$$W^s(\mathbf{0},\mathbf{F})=\left\{\,(x,y,z):z=-\frac{4}{7}(x^2+y^2)\,\right\}.$$

Exercises 12.3

1. Consider the map
$$\begin{pmatrix}x_1\\y_1\end{pmatrix}=\mathbf{F}\begin{pmatrix}x\\y\end{pmatrix}=\begin{pmatrix}0.5\,x-4\,y^3\\2\,y\end{pmatrix}.$$
   a. Find the inverse of $\mathbf{F}$.
   b. Find the stable and unstable manifolds of the fixed point at the origin.

2. Consider the map
$$\begin{pmatrix}x_1\\y_1\\z_1\end{pmatrix}=\mathbf{F}\begin{pmatrix}x\\y\\z\end{pmatrix}=\begin{pmatrix}0.5\,x-4\,y^3+8\,z^2\\2\,y\\4\,z\end{pmatrix}.$$
   a. Find the inverse of $\mathbf{F}$.
   b. Find the stable and unstable manifolds of the fixed point at the origin.

12.4. Hyperbolic Toral Automorphisms

This section considers a particular type of map on a torus. These maps have infinitely many periodic points that are dense in the total space. They can be thought of as higher dimensional maps related to the doubling map in one dimension, but they are now have an inverse.

S. Smale popularized these maps as examples with infinitely many periodic points. These are concrete examples of systems studied extensively by D.V. Anosov, hence they are often called *Anosov diffeomorphisms*. They are defined on a torus by means of a hyperbolic linear map on a Euclidean space, so they are also called *hyperbolic toral automorphisms*.

The *torus* or *two torus* $\mathbb{T}^2$ is the set of points (x,y) for which each coordinate is taken modulo one. You can think of each variable as being an angular variable modulo 1. This space can be considered as the surface of a donut or bagel by the following process. Any number x taken modulo 1 can be represented by a number x' with $0\leq x'<1$. Thus, the set of points in $\mathbb{T}^2$ is related to the unit square:
$$\mathbf{S}=\{\,(x,y):0\leq x\leq 1,\ 0\leq y\leq 1\,\}.$$

If **S** is rolled in the x-direction and the line segment $\{0\} \times [0,1]$ is glued to the line segment $\{1\} \times [0,1]$, with each point $(0, y)$ glued to the point $(1, y)$, then the result is a cylinder. See Figure 11(b). Subsequently, the cylinder is bent in the y-direction so that the "circle" $[0,1] \times \{0\}$ is glued to the "circle" $[0,1] \times \{1\}$, then the ends of the cylinder are brought together to form a surface like the surface of a donut. See Figure 11(c). Because $x = 0$ is glued to $x = 1$, a point whose x-coordinate is near 0 is close to a point whose x-coordinate is near 1. The same thing is true for points with y near 0 and 1.

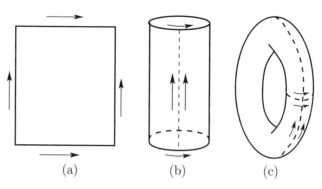

(a) (b) (c)

Figure 11. Construction of a torus: (a) Unit square. (b) Cylinder formed by attaching vertical edges together. (c) Torus formed from cylinder by attaching horizontal edges together.

The hyperbolic toral automorphisms are defined by means of a linear map on $\mathbb{R}^2$. Therefore, we introduce a notation for the "quotient map" π that takes a point in $\mathbb{R}^2$ and gives a point in the torus. This quotient map is essential the same as taking points modulo 1 in each coordinate. If

$$x' = x \ (\bmod \ 1) \quad \text{and} \quad y' = y \ (\bmod \ 1), \quad \text{then}$$
$$x' = x + m \quad \text{and} \quad y' = y + n,$$

for two integers m and n (positive, zero, or negative). If such integers exist, then the points (x, y) and (x', y') in $\mathbb{R}^2$ represent the same point in the torus. Let π be the map from $\mathbb{R}^2$ to $\mathbb{T}^2$ such that $\pi(x, y) = \pi(x', y')$ if and only if there are two integers m and n such that $x' = x + m$ and $y' = y + n$. This projection accomplishes the gluing we described previously.

We introduce hyperbolic toral automorphisms by means of a specific example,

(12.1)
$$\mathbf{A} = \begin{pmatrix} 2 & 1 \\ 1 & 1 \end{pmatrix}.$$

This matrix has integer entries and a determinant equal to 1. (Examples with a determinant equal to -1 also work.) Because the determinant is 1, the inverse also has integer entries:

$$\mathbf{A}^{-1} = \begin{pmatrix} 1 & -1 \\ -1 & 2 \end{pmatrix}.$$

The matrix **A** induces a linear map from $\mathbb{R}^2$ to itself:

$$\mathbf{A} : \mathbb{R}^2 \to \mathbb{R}^2.$$

12.4. Hyperbolic Toral Automorphisms

If $\pi(x, y) = \pi(x', y')$, with $x' = x + m$ and $y' = y + n$, then

$$\mathbf{A}\begin{pmatrix} x' \\ y' \end{pmatrix} = \mathbf{A}\begin{pmatrix} x+m \\ y+n \end{pmatrix} = \mathbf{A}\begin{pmatrix} x \\ y \end{pmatrix} + \mathbf{A}\begin{pmatrix} m \\ n \end{pmatrix} = \mathbf{A}\begin{pmatrix} x \\ y \end{pmatrix} + \begin{pmatrix} 2m+n \\ m+n \end{pmatrix},$$

and therefore,

$$\pi\left(\mathbf{A}\begin{pmatrix} x' \\ y' \end{pmatrix}\right) = \pi\left(\mathbf{A}\begin{pmatrix} x \\ y \end{pmatrix}\right).$$

Thus, two points that are identified by π are taken to two points identified by π, so $\mathbf{A}$ induces a map $\mathbf{F_A}$ from $\mathbb{T}^2$ to $\mathbb{T}^2$. In fact, if $\mathbf{p}$ is a point in $\mathbb{T}^2$, and (x, y) is any point with $\pi(x, y) = \mathbf{p}$, then

$$\mathbf{F_A}(\mathbf{p}) = \pi\left(\mathbf{A}\begin{pmatrix} x \\ y \end{pmatrix}\right)$$

makes the map well defined. Thus, if x and y are taken as modulo one variables, then to find the image by $\mathbf{F_A}$, we find $\mathbf{A}\begin{pmatrix} x \\ y \end{pmatrix}$ and take each new coordinate modulo one.

Because $\mathbf{A}^{-1}$ also has integer entries, it also induces a map

$$\mathbf{F_A^{-1}} : \mathbb{T}^2 \to \mathbb{T}^2,$$

which is the inverse of $\mathbf{F_A}$; so, $\mathbf{F_A}$ is one to one and onto $\mathbb{T}^2$.

Figure 12 shows the image of the unit square $\mathbf{S}$ by the matrix $\mathbf{A}$. The regions that are identified are labeled with the same letters, so it can be seen that, with the identifications, the image $\mathbf{A}(\mathbf{S})$ covers the unit square exactly once.

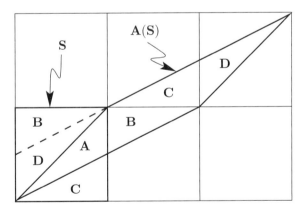

Figure 12. The image of the unit square by the toral automorphism $\mathbf{F_A}$

The eigenvalues of $\mathbf{A}$ are

$$\lambda_1 = \frac{3+\sqrt{5}}{2} > 1 \quad \text{and} \quad 0 < \lambda_2 = \frac{3-\sqrt{5}}{2} < 1,$$

so the matrix is hyperbolic.

For any 2×2 matrix $\mathbf{A}$ which (i) has integer entries, (ii) has a determinant equal to ± 1, and (iii) has eigenvalues with $0 < |\lambda_2| < 1 < |\lambda_1|$, the induced map

$\mathbf{F_A}$ is called a *hyperbolic toral automorphism* or an *Anosov diffeomorphism*. For an $n \times n$ matrix $\mathbf{A}$, the assumptions are basically the same, excepts for (iii′) all the eigenvalues have $|\lambda_i| \neq 1$, at least one eigenvalue has $|\lambda_j| > 1$, and at least one eigenvalue has $|\lambda_k| < 1$.

We next discuss the periodic points of the map $\mathbf{F_A}$. The following theorem concerns the periodic points of a hyperbolic toral automorphism.

Theorem 12.6. *Assume that $\mathbf{F_A}$ is a hyperbolic toral automorphism on $\mathbb{T}^2$. Then the periodic points of $\mathbf{F_A}$ are exactly the points $\pi(x, y)$ where both x and y are rational numbers. Therefore, there are infinitely many periodic points, and they are dense in $\mathbb{T}^2$.*

Furthermore, all the periodic points are hyperbolic: if $\mathbf{p}$ is a period-k point, then the eigenvalues of $D(\mathbf{F_A^k})_{(\mathbf{p})}$ are λ_1^k and λ_2^k, where

$$\lambda_1 = \frac{3+\sqrt{5}}{2} > 1 \quad \text{and} \quad 0 < \lambda_2 = \frac{3-\sqrt{5}}{2} < 1.$$

Proof. Consider the set of rational points that all have the same denominator q:

$$\mathbf{Q}_q = \left\{ \left(\frac{m}{q}, \frac{n}{q} \right) : 0 \leq m < q,\ 0 \leq n < q \right\}.$$

These numbers are not necessarily written in lowest terms, so there can be a common factor of m and q or n and q. Since

$$\mathbf{A} \begin{pmatrix} \frac{m}{q} \\ \frac{n}{q} \end{pmatrix} = \begin{pmatrix} \frac{2m+n}{q} \\ \frac{m+n}{q} \end{pmatrix},$$

the map $\mathbf{F_A}$ takes $\pi(\mathbf{Q}_q)$ to itself. Since $\mathbf{F_A^{-1}}$ also preserves $\pi(\mathbf{Q}_q)$, $\mathbf{F_A}$ restricted to $\pi(\mathbf{Q}_q)$ must be one to one and onto (i.e., a permutation of the finite set $\mathbf{Q}_q$ with q^2 elements). Since this set is finite, each point must be periodic, with period less than or equal to q^2.

The union of these sets,

$$\bigcup_{q=1}^{\infty} \mathbf{Q}_q = \left\{ \begin{pmatrix} x \\ y \end{pmatrix} : 0 \leq x < 1,\ 0 \leq y < 1,\ x \text{ and } y \text{ are rational} \right\},$$

is dense in the unit square. Therefore, the projection of this union by π is dense in $\mathbb{T}^2$. This proves that the periodic points are dense in $\mathbb{T}^2$.

The eigenvalues of $\mathbf{A}$ are λ_1 and λ_2, as given in the statement of the theorem. The derivative of the map $\mathbf{F_A}$ at all points is $\mathbf{A}$. So, if $\mathbf{p}$ is a period-k point, then

$$D\left(\mathbf{F_A^k}\right)_{(\mathbf{p})} = \mathbf{A}^k.$$

The eigenvalues of $\mathbf{A}^k$ are the kth power of those of $\mathbf{A}$, and so they are λ_1^k and λ_2^k as stated in the theorem.

12.4. Hyperbolic Toral Automorphisms

Now, assume that $\mathbf{p}$ is periodic, so $\mathbf{F}_{\mathbf{A}}^{k}(\mathbf{p}) = \mathbf{p}$. Then, if $\pi(x,y) = \mathbf{p}$,

$$\mathbf{A}^k \begin{pmatrix} x \\ y \end{pmatrix} = \begin{pmatrix} x \\ y \end{pmatrix} + \begin{pmatrix} m \\ n \end{pmatrix} \quad \text{and}$$

$$(\mathbf{A}^k - \mathbf{I}) \begin{pmatrix} x \\ y \end{pmatrix} = \begin{pmatrix} m \\ n \end{pmatrix}.$$

Since the eigenvalues of $\mathbf{A}^k - \mathbf{I}$ are $\lambda_1^k - 1 \neq 0$ and $\lambda_2^k - 1 \neq 0$, $\mathbf{A}^k - \mathbf{I}$ is invertible with integer entries, so the inverse has rational entries with denominator equal to the determinant of $\mathbf{A}^k - \mathbf{I}$. This shows that the vector

$$\begin{pmatrix} x \\ y \end{pmatrix} = (\mathbf{A}^k - \mathbf{I})^{-1} \begin{pmatrix} m \\ n \end{pmatrix}$$

in $\mathbb{R}^2$ has rational coordinates, and $\mathbf{p} = \pi(x,y)$ is the projection of a vector with rational coordinates. This is what we wanted to prove. $\square$

Example 12.28. For $(1/3, 2/3)^\top$ and the matrix $\mathbf{A}$ given in equation (12.1),

$$\mathbf{A} \begin{pmatrix} \frac{1}{3} \\ \frac{2}{3} \end{pmatrix} = \begin{pmatrix} \frac{2 \cdot 1 + 1 \cdot 2}{3} \\ \frac{2+1}{3} \end{pmatrix} = \begin{pmatrix} \frac{1}{3} \\ \frac{0}{3} \end{pmatrix} \begin{pmatrix} \text{mod } 1 \\ \text{mod } 1 \end{pmatrix},$$

$$\mathbf{A} \begin{pmatrix} \frac{1}{3} \\ \frac{0}{3} \end{pmatrix} = \begin{pmatrix} \frac{2}{3} \\ \frac{1}{3} \end{pmatrix},$$

$$\mathbf{A} \begin{pmatrix} \frac{2}{3} \\ \frac{1}{3} \end{pmatrix} = \begin{pmatrix} \frac{6}{3} \\ \frac{3}{3} \end{pmatrix} = \begin{pmatrix} \frac{2}{3} \\ \frac{0}{3} \end{pmatrix} \begin{pmatrix} \text{mod } 1 \\ \text{mod } 1 \end{pmatrix}, \quad \text{and}$$

$$\mathbf{A} \begin{pmatrix} \frac{2}{3} \\ \frac{0}{3} \end{pmatrix} = \begin{pmatrix} \frac{4}{3} \\ \frac{2}{3} \end{pmatrix} = \begin{pmatrix} \frac{1}{3} \\ \frac{2}{3} \end{pmatrix} \begin{pmatrix} \text{mod } 1 \\ \text{mod } 1 \end{pmatrix}$$

gives an orbit of period four.

Stable manifolds

If two points $\overline{\mathbf{p}}$ and $\overline{\mathbf{q}}$ in $\mathbb{R}^2$ differ by a scalar multiple of the stable eigenvector, or

$$\overline{\mathbf{q}} - \overline{\mathbf{p}} = t\,\mathbf{v}^s,$$

then the distance between their iterates goes to zero under forward iteration:

$$\|\mathbf{A}^k \overline{\mathbf{q}} - \mathbf{A}^k \overline{\mathbf{p}}\| = |t|\,\lambda_s^k\,\|\mathbf{v}^s\|.$$

Therefore, in $\mathbb{R}^2$, $\overline{\mathbf{q}} = \overline{\mathbf{p}} + t\,\mathbf{v}^s$ is on the stable manifold of $\overline{\mathbf{p}}$:

$$W^s(\overline{\mathbf{p}}, \mathbf{A}) = \{\,\overline{\mathbf{p}} + t\,\mathbf{v}^s : t \in \mathbb{R}\,\}.$$

In $\mathbb{R}^2$, this stable manifold is a line. Because the stable eigenvector has irrational slope, the stable manifold in $\mathbb{T}^2$ is dense; that is,

$$W^s(\mathbf{p}, \mathbf{F}_{\mathbf{A}}) = \pi W^s(\overline{\mathbf{p}}, \mathbf{A}),$$

where $\pi \overline{\mathbf{p}} = \mathbf{p}$.

In the same way,
$$W^u(\overline{\mathbf{p}}, \mathbf{A}) = \{\overline{\mathbf{p}} + t\mathbf{v}^u : t \in \mathbb{R}\}$$
and
$$W^u(\mathbf{p}, \mathbf{F_A}) = \pi W^u(\overline{\mathbf{p}}, \mathbf{A})$$
is dense in $\mathbb{T}^2$.

Exercises 12.4

1. Find the eigenvectors for
$$\begin{pmatrix} 2 & 1 \\ 1 & 1 \end{pmatrix}.$$
Use these to draw the local stable and unstable manifolds of the origin in the plane.

2. The slopes of the eigenvectors of the matrix
$$\begin{pmatrix} 2 & 1 \\ 1 & 1 \end{pmatrix}$$
are irrational. Explain why this forces the stable and unstable manifolds of the origin to be dense in the torus. Hint: Consider the intersection of these manifolds with $x = 0 \pmod 1$, which are multiples of an irrational number.

12.5. Applications

12.5.1. Markov Chains. In discussing expanding maps with Markov partitions in Section 11.4.1, we encountered matrices that have row sum equal to one. These matrices lead to *Markov chains*, which we discuss in more detail in this section.

Assume that some material is spread out among the n sites, with $x_i^{(0)} \geq 0$ the amount of material at the ith site at time 0. Assume that $m_{ij} \geq 0$ is the probability of going from the ith site to the jth site, so that $x_i^{(0)} m_{ij}$ is the amount of material from the ith site that is returned to the jth site at time 1. The total amount at the jth site at time 1 is the sum of the material from all the sites, or
$$x_j^{(1)} = \sum_i x_i^{(0)} m_{ij}.$$
Let $\mathbf{M} = (m_{ij})$ be the corresponding $n \times n$ matrix (where i gives the row and j gives the column of the entry), and let
$$\mathbf{x}^{(q)} = \left(x_1^{(q)}, \ldots, x_n^{(q)}\right)$$
be the row vector of the amount of material at time q at all the sites. Using this matrix notation,
$$\mathbf{x}^{(1)} = \mathbf{x}^{(0)} \mathbf{M},$$
and, more generally,
$$\mathbf{x}^{(q)} = \mathbf{x}^{(q-1)} \mathbf{M} = \mathbf{x}^{(0)} \mathbf{M}^q,$$
for the transition from the distribution at time $q-1$ to time q.

12.5. Applications

The sum of the probabilities of going from the ith site to some other site is one, so $\sum_j m_{ij} = 1$ and each of the rows sums of $\mathbf{M}$ is one. We assume that is possible to make a transition to each of the jth sites from some other site, so each column of $\mathbf{M}$ has some nonzero entry. Finally, we assume that from at least one site there are more than one possible following site, so some row has more than one positive entry.

Notice that the total amount of material at time q is the same as at time $q-1$ and so the same as at time 0:

$$\sum_j x_j^{(q)} = \sum_j \left(\sum_i x_i^{(q-1)} m_{ij} \right)$$
$$= \sum_i \left(\sum_j m_{ij} \right) x_i^{(q-1)}$$
$$= \sum_i x_i^{(q-1)}$$

(we use the fact that the row sums are 1). Call the total amount $X = \sum_j x_j^{(0)} = \sum_j x_j^{(q)}$. Then,

$$p_i^{(q)} = \frac{x_i^{(q)}}{X}$$

is the proportion of the material at the ith site at time q. Letting

$$\mathbf{p}^{(q)} = (p_1^{(q)}, \ldots, p_n^{(q)}) = \frac{1}{X} \mathbf{x}^{(q)}$$

be the row vector of these proportions, we see that

$$\mathbf{p}^{(q-1)} \mathbf{M} = \frac{\mathbf{x}^{(q-1)}}{X} \mathbf{M} = \frac{1}{X} \left(\mathbf{x}^{(q-1)} \mathbf{M} \right) = \frac{1}{X} \mathbf{x}^{(q)} = \mathbf{p}^{(q)},$$

also transforms through multiplication by the matrix $\mathbf{M}$.

We summarize the assumptions on the matrix $\mathbf{M}$ in the next definition.

Definition 12.29. An $n \times n$ matrix $\mathbf{M}$ with real entries m_{ij} is called a *stochastic matrix* or *probability transition matrix* provided that the following conditions are satisfied:

(i) All the entries m_{ij} satisfy $0 \leq m_{ij} \leq 1$.

(ii) Each row sums to one, $\sum_j m_{ij} = 1$ for each i.

(iii) Each column has some nonzero entry.

(iv) Some row has more than one nonzero entry.

Definition 12.30. A stochastic matrix $\mathbf{M}$ is called *aperiodic* (or *eventually positive*) provided that there is a $q_0 > 0$ such that $\mathbf{M}^{q_0}$ has all positive entries (i.e., for this iterate, it is possible to make a transition from any site to any other site). It then follows that $\mathbf{M}^q$ has all positive entries for $q \geq q_0$. An aperiodic stochastic matrix automatically satisfies conditions (iii) and (iv) in the definition of a stochastic matrix.

Definition 12.31. As in the case for a transition matrix, a stochastic matrix is called *irreducible* provided that it is possible to get from each site i_1 to each other site i_2 by making a finite number of transitions. In other words, for any pair of sites (i_1, i_2), there are indices j_k, with $1 \leq j_k \leq n$ for $k = 0, \ldots, q$ (where q can depend on (i_1, i_2)), such that $j_0 = i_1$, $j_q = i_2$, and $m_{j_{k-1} j_k} > 0$ for $k = 1, \ldots, q$ (i.e., the (i_1, i_2)-entry of $\mathbf{M}^q$ is nonzero).

One way to insure that a stochastic matrix is aperiodic is for it to be irreducible and also to have one positive entry on the diagonal. We concentrate on aperiodic matrices, which are therefore irreducible.

For a stochastic matrix $\mathbf{M}$, 1 is always an eigenvalue with column eigenvector $(1, \ldots, 1)^\mathsf{T}$:

$$\mathbf{M} \begin{pmatrix} 1 \\ \vdots \\ 1 \end{pmatrix} = \begin{pmatrix} \sum_j m_{1j} \\ \vdots \\ \sum_j m_{nj} \end{pmatrix} = \begin{pmatrix} 1 \\ \vdots \\ 1 \end{pmatrix}.$$

Since $\mathbf{M}^\mathsf{T}$ and $\mathbf{M}$ have the same eigenvalues, $\mathbf{M}^\mathsf{T}$ always has 1 as an eigenvalue, with eigenvector $(p_1^*, \ldots, p_n^*)^\mathsf{T}$. This is equivalent to saying that the row vector $(p_1^*, \ldots, p_n^*)$ satisfies

$$(p_1^*, \ldots, p_n^*) = (p_1^*, \ldots, p_n^*) \mathbf{M}.$$

Theorem 12.7 says that this vector can be chosen with $p_j^* > 0$ and $\sum_j p_j^* = 1$. When $\mathbf{M}$ is aperiodic, Theorem 12.7 further states that (i) all other eigenvalues satisfy $|\lambda_j| < 1$ and (ii) if $(p_1, \ldots, p_n)$ is any initial probability distribution, then $(p_1, \ldots, p_n) \mathbf{M}^q$ converges to $(p_1^*, \ldots, p_n^*)$ as q goes to infinity.

Before stating the general result, we give some examples.

Example 12.32. Let

$$\mathbf{M} = \begin{pmatrix} 0.5 & 0.3 & 0.2 \\ 0.2 & 0.8 & 0 \\ 0.3 & 0.3 & 0.4 \end{pmatrix}.$$

This matrix has eigenvalues 1, 0.5, and 0.2. (We do not give the characteristic polynomial, but do derive an eigenvector $\mathbf{v} = (v_1, v_2, v_3)$ for each of these values.)

For $\lambda = 1$,

$$\mathbf{M}^\mathsf{T} - \mathbf{I} = \begin{pmatrix} -0.5 & 0.2 & 0.3 \\ 0.3 & -0.2 & 0.3 \\ 0.2 & 0 & -0.6 \end{pmatrix} \sim \begin{pmatrix} 1 & -0.4 & -0.6 \\ 0 & -0.08 & 0.48 \\ 0 & 0.08 & -0.48 \end{pmatrix} \sim \begin{pmatrix} 1 & 0 & -3 \\ 0 & 1 & -6 \\ 0 & 0 & 0 \end{pmatrix}.$$

Thus, $v_1 = 3v_3$ and $v_2 = 6v_3$. Since we want $1 = v_1 + v_2 + v_3 = (3 + 6 + 1)v_3 = 10v_3$, $v_3 = 0.1$, and $\mathbf{p}^* = \mathbf{v}^1 = (0.3, 0.6, 0.1)$.

For $\lambda_2 = 0.5$,

$$\mathbf{M}^\mathsf{T} - 0.5\,\mathbf{I} = \begin{pmatrix} 0 & 0.2 & 0.3 \\ 0.3 & 0.3 & 0.3 \\ 0.2 & 0 & -0.1 \end{pmatrix} \sim \begin{pmatrix} 2 & 0 & -1 \\ 0 & 2 & 3 \\ 1 & 1 & 1 \end{pmatrix}$$

$$\sim \begin{pmatrix} 2 & 0 & -1 \\ 0 & 2 & 3 \\ 0 & 1 & 1.5 \end{pmatrix} \sim \begin{pmatrix} 2 & 0 & -1 \\ 0 & 2 & 3 \\ 0 & 0 & 0 \end{pmatrix}.$$

Thus, $2v_1 = v_3$, $2v_2 = -3v_3$, and $\mathbf{v}^2 = (1, -3, 2)$. Notice that $v_1 + v_2 + v_3 = 1 - 3 + 2 = 0$. This is always the case for the eigenvectors of the other eigenvalues.

12.5. Applications

For $\lambda_3 = 0.2$,

$$\mathbf{M}^\mathsf{T} - 0.2\,\mathbf{I} = \begin{pmatrix} 0.3 & 0.2 & 0.3 \\ 0.3 & 0.6 & 0.3 \\ 0.2 & 0 & 0.2 \end{pmatrix} \sim \begin{pmatrix} 1 & 0 & 1 \\ 3 & 6 & 3 \\ 3 & 2 & 3 \end{pmatrix}$$

$$\sim \begin{pmatrix} 1 & 0 & 1 \\ 0 & 6 & 0 \\ 0 & 2 & 0 \end{pmatrix} \sim \begin{pmatrix} 1 & 0 & 1 \\ 0 & 1 & 0 \\ 0 & 0 & 0 \end{pmatrix}.$$

Thus, $v_1 = -v_3$, $v_2 = 0$, and $\mathbf{v}^3 = (1, 0, -1)$. Again, $v_1 + v_2 + v_3 = 1 + 0 - 1 = 0$.

If the original distribution is given by

$$\mathbf{p}^{(0)} = (0.45, 0.45, 0.1) = (0.3, 0.6, 0.1) + \frac{1}{20}(1, -3, 2) + \frac{1}{10}(1, 0, -1),$$

then

$$\mathbf{p}^{(0)}\mathbf{M}^q = (0.3, 0.6, 0.1) + \frac{1}{20}\left(\frac{1}{2}\right)^q (1, -3, 2) + \frac{1}{10}\left(\frac{1}{5}\right)^q (1, 0, -1),$$

which converges to the distribution $\mathbf{v}^1 = (0.3, 0.6, 0.1)$ as q goes to infinity. This convergence of the iterates holds for any initial distribution $\mathbf{p}^{(0)}$. See Theorem 12.7.

Example 12.33 (Complex Eigenvalues). The following stochastic matrix illustrates the fact that an aperiodic stochastic matrix can have complex eigenvalues. Let

$$\mathbf{M} = \begin{pmatrix} 0.6 & 0.1 & 0.3 \\ 0.3 & 0.6 & 0.1 \\ 0.1 & 0.3 & 0.6 \end{pmatrix}.$$

The eigenvalues are $\lambda = 1$ and $0.4 \pm i\,0.1\sqrt{3}$. Notice that $|0.4 \pm i\,0.1\sqrt{3}| = \sqrt{0.16 + 0.03} = \sqrt{0.19} < 1$.

Example 12.34 (Not Aperiodic). An example of a stochastic matrix that is not aperiodic (nor irreducible) is given by

$$\mathbf{M} = \begin{pmatrix} 0.8 & 0.2 & 0 & 0 \\ 0.3 & 0.7 & 0 & 0 \\ 0 & 0 & 0.6 & 0.4 \\ 0 & 0 & 0.3 & 0.7 \end{pmatrix},$$

which has eigenvalues $\lambda = 1, 1, 0.5$, and 0.3. Notice that it is possible to go between sites 1 and 2, and it is possible to go between and sites 3 and 4, but it is not possible to go from the sites 1 and 2 to the sites 3 and 4.

An example of a stochastic matrix that is irreducible, but not aperiodic, is given by

$$\mathbf{M} = \begin{pmatrix} 0 & 0 & 0.8 & 0.2 \\ 0 & 0 & 0.3 & 0.7 \\ 1 & 0 & 0 & 0 \\ 0 & 1 & 0 & 0 \end{pmatrix},$$

which has eigenvalues $\lambda = 1, -1$, and $\pm\sqrt{0.5}$. Here, it is possible to get from any site to any other site, but starting at site one, the odd iterates are always at either sites 3 or 4 and the even iterates are always at either sites 1 or 2. Thus, there is no one power for which all the transition probabilities are positive. Therefore, $\mathbf{M}$

is not aperiodic. Also, this matrix has another eigenvalue -1 with absolute value equal to one.

Theorem 12.7 (Perron–Frobenius). *Let $\mathbf{M}$ be an aperiodic stochastic matrix.*

a. *The matrix $\mathbf{M}$ has 1 as an eigenvalue of multiplicity one (i.e., 1 is a simple root of the characteristic equation). A row eigenvector $\mathbf{p}^*$ for eigenvalue 1 can be chosen with all positive entries and $\sum_j p_j^* = 1$.*

b. *All the other eigenvalues λ_j have $|\lambda_j| < 1$. If $\mathbf{v}^k$ is a row eigenvector for λ_k, then $\sum_j v_j^k = 0$.*

c. *If $\mathbf{p}$ is any probability distribution with all $p_j > 0$ and $\sum_j p_j = 1$, then*

$$\mathbf{p} = \mathbf{p}^* + \sum_{j=2}^{n} y_k \mathbf{v}^k$$

for some $y_2, \ldots, y_n$. Also, $\mathbf{p}\mathbf{M}^q$ converges to $\mathbf{p}^*$ as q goes to infinity.

Sketch of the proof. We give a sketch of the proof based on the proof using ideas from dynamical systems given in [**Rob99**].

We assume in what follows that all the $m_{ij} > 0$, which can be done by taking a power of $\mathbf{M}$ if necessary.

(**a**) As noted previously, $\mathbf{M}$ always has 1 as an eigenvalues, so it is always an eigenvalue of $\mathbf{M}^\mathsf{T}$ (i.e., $\mathbf{M}$ has a row eigenvector for the eigenvalue 1). To discuss the multiplicity, we assume that there is another column eigenvector $\mathbf{v}$, with $\mathbf{M}\mathbf{v} = \mathbf{v}$ and not all the v_j are equal. Assume that k is the index for which $|v_k|$ is the largest component. By scalar multiplication by -1, if necessary, we can take v_k positive. Thus, $v_k = |v_k| \geq |v_j|$ for all j and $v_k > |v_\ell|$ for some ℓ. Then,

$$v_k = \sum_j m_{kj} v_j < \sum_j m_{kj} v_k = v_k.$$

The strict inequality uses the fact that all the $m_{ij} > 0$ (i.e., that $\mathbf{M}$ is aperiodic). Since this shows $v_k > v_k$, the contradiction implies that there are no such other vectors, and so that, there can be only one eigenvector for the eigenvalue 1.

To complete the proof, we would have to consider the case in which 1 is a multiple eigenvalue with only one eigenvector. We leave this detail to the reference.

(**b**) Case (i): Assume that $\lambda \neq 1$ is a real eigenvalue. Again, assume that $\mathbf{M}\mathbf{v} = \lambda\mathbf{v}$. Let k be such that $v_k = |v_k| \geq |v_j|$ for all j and $v_k > |v_\ell|$ for some ℓ. Then,

$$\lambda v_k = \sum_j m_{kj} v_j < \sum_j m_{kj} v_k = v_k.$$

This shows that $\lambda v_k < v_k$, so $\lambda < 1$.

We now show that $\lambda > -1$. Notice that $v_j \geq -v_k$ for all j and $v_\ell > -v_k$ for some ℓ. Therefore,

$$\lambda v_k = \sum_j m_{kj} v_j > \sum_j m_{kj} (-v_k) = -v_k.$$

This shows that $\lambda v_k > -v_k$, so $\lambda > -1$. Combining, for a real eigenvalue that is not equal to 1, $-1 < \lambda < 1$.

12.5. Applications

Case (ii): Assume that $\lambda = re^{2\pi\omega i}$ is a complex eigenvalue with complex eigenvector $\mathbf{v}$. Here, $r = |\lambda|$ and $e^{2\pi\omega i}$ is complex. Assume that the v_j are chosen with v_k real and $v_k \geq \text{Re}(v_j)$ for all j. Since $\mathbf{M}^q \mathbf{v} = \lambda^q \mathbf{v}$,

$$r^q \text{Re}(e^{2\pi q \omega i}) v_k = \text{Re}(\lambda^q) v_k = \text{Re}((\mathbf{M}^q \mathbf{v})_k) = (\mathbf{M}^q \text{Re}(\mathbf{v}))_k$$
$$= \sum_j \left(m_{kj}^{(q)} \text{Re}(v_j) \right) < \sum_j \left(m_{kj}^{(q)} v_k \right) = v_k.$$

Therefore, $r^q \text{Re}(e^{2\pi q \omega i}) < 1$ for all q. Since we can find a q_1 for which $\text{Re}(e^{2\pi q_1 \omega i})$ is very close to 1, we have $r^{q_1} < 1$ so $r = |\lambda| < 1$.

(c) Let $\mathbf{p}$ be a probability distribution with $\sum_j p_j = 1$. The eigenvectors are a basis, so there exist $y_1, \ldots, y_n$ such that

$$\mathbf{p} = \sum_{i=1}^n y_i \mathbf{v}^i.$$

Here, all the vectors are row vectors, and $\mathbf{v}^1 = \mathbf{p}^*$. Then,

$$1 = \sum_j p_j = y_1 \sum_j v_j^1 + \sum_{i=2}^n y_i \sum_j v_j^i = y_1 + \sum_{i=2}^n y_i \cdot 0 = y_1.$$

Thus,

$$\mathbf{p} = \mathbf{v}^1 + \sum_{i=2}^n y_i \mathbf{v}^i,$$

as claimed.

Writing the iteration as if all the eigenvalues are real, we have

$$\mathbf{p}\mathbf{M}^q = \mathbf{v}^1 \mathbf{M}^q + \sum_{i=2}^n y_i \mathbf{v}^i \mathbf{M}^q = \mathbf{v}^1 + \sum_{i=2}^n y_i \lambda_i^q \mathbf{v}^i,$$

which tends to $\mathbf{v}^1$ because all the $|\lambda_i^q| < 1$ for $i \geq 2$. $\square$

12.5.2. Newton Map in $\mathbb{R}^n$. We discussed the Newton map of one scalar variable in Section 9.3.1. A similar method applies to finding zeroes of a function from $\mathbb{R}^n$ to $\mathbb{R}^n$.

Let $\mathbf{F}$ be a map from $\mathbb{R}^n$ to $\mathbb{R}^n$. Assume that $\mathbf{x}_j$ approximates a zero of $\mathbf{F}$. The linear approximation of $\mathbf{F}$ at $\mathbf{x}_j$ is given by

$$\mathbf{y} = \mathbf{F}(\mathbf{x}_j) + D\mathbf{F}_{(\mathbf{x}_j)}(\mathbf{x} - \mathbf{x}_j).$$

Setting $\mathbf{y} = \mathbf{0}$ and solving for the variable $\mathbf{x}$ defines $\mathbf{x}_{j+1}$:

$$\mathbf{0} = \mathbf{F}(\mathbf{x}_j) + D\mathbf{F}_{(\mathbf{x}_j)}(\mathbf{x}_{j+1} - \mathbf{x}_j)$$
$$-\mathbf{F}(\mathbf{x}_j) = D\mathbf{F}_{(\mathbf{x}_j)}(\mathbf{x}_{j+1} - \mathbf{x}_j)$$
$$-\left(D\mathbf{F}_{(\mathbf{x}_j)}\right)^{-1} \mathbf{F}(\mathbf{x}_j) = \mathbf{x}_{j+1} - \mathbf{x}_j$$
$$\mathbf{x}_{j+1} = \mathbf{x}_j - \left(D\mathbf{F}_{(\mathbf{x}_j)}\right)^{-1} \mathbf{F}(\mathbf{x}_j).$$

This equation defines the associated Newton map by

$$\mathbf{N_F}(\mathbf{x}) = \mathbf{x} - \left(D\mathbf{F}_{(\mathbf{x})}\right)^{-1} \mathbf{F}(\mathbf{x}).$$

Note that $\mathbf{N_F}(\mathbf{x}) = \mathbf{x}$ if and only if $\mathbf{F}(\mathbf{x}) = \mathbf{0}$.

Just as in the case in one variable, the derivative of $\mathbf{N_F}$ at a zero of $\mathbf{F}$ is zero. Assume that $\mathbf{F(p) = 0}$. Let $\mathbf{I}$ be the identity matrix, which is the diagonal matrix with ones down the diagonal. Then,

$$D\left(\mathbf{N_F}\right)_\mathbf{p} = \mathbf{I} - \left(D\mathbf{F_{(p)}}\right)^{-1}\left(D\mathbf{F_{(p)}}\right) - D\left[\left(D\mathbf{F_{(\cdot)}}\right)^{-1}\right]_{(\mathbf{p})}\mathbf{F(p)}$$

$$= \mathbf{I} - \mathbf{I} - D\left[\left(D\mathbf{F_{(\cdot)}}\right)^{-1}\right]_{(\mathbf{p})}\mathbf{0}$$

$$= \mathbf{0}.$$

The mysterious term $D\left[\left(D\mathbf{F_{(\cdot)}}\right)^{-1}\right]_{(\mathbf{p})}$ is the derivative of the derivative (second derivative), where the dot represents the variable with which it is differentiated. Because this term acts on the zero vector, we do not need to know what it is. Using the Taylor expansion to degree two,

$$\|\mathbf{N_F(x) - p}\| \leq C \|\mathbf{x - p}\|^2$$

for $\mathbf{x}$ near $\mathbf{p}$, where $C > 0$ is some constant depending on the second partial derivatives. Thus, starting near the zero leads to orbits that rapidly converge to $\mathbf{p}$ (i.e., $\mathbf{p}$ is *superattracting*).

12.5.3. Beetle Population Model. In Example 8.5, we introduced a model of the flour beetle with three stages of population: larva, pupa, and adult. The function for the model given there is

$$L_{n+1} = b\,A_n e^{-C_{LA}A_n} e^{-C_{LL}L_n},$$
$$P_{n+1} = (1 - \mu_L)\,L_n,$$
$$A_{n+1} = P_n\,e^{-C_{PA}A_n} + (1 - \mu_A)\,A_n,$$

where $b > 0$, $0 \leq \mu_L \leq 1$, $0 \leq \mu_A \leq 1$, $C_{LA} \geq 0$, $C_{LL} \geq 0$, and $C_{PA} \geq 0$. The quantities $\exp(-C_{LA}A_n)$ and $\exp(-C_{LL}L_n)$ are the probabilities that an egg is not eaten by the adult population A_n and the larvae population L_n, respectively; the quantity $\exp(-C_{PA}A_n)$ is the survival probability of the pupae into adulthood; b is the birth rate of larva in terms of the adult population, μ_L is the death rate of larva (that do not transform into pupa), and μ_A is the death rate of adults. The analysis of the fixed points for this system is given in the book [**Bra01**] by F. Brauer and C. Castillo–Chávez. We follow their treatment for the simplified case presented.

To analyze this discrete system, we make the simplifying assumption that $C_{LL} = 0$ (i.e., we neglect the cannibalism of the eggs by the larva). With this assumption, the function becomes

$$\begin{pmatrix} L_{n+1} \\ P_{n+1} \\ A_{n+1} \end{pmatrix} = \mathbf{F} \begin{pmatrix} L_n \\ P_n \\ A_n \end{pmatrix} = \begin{pmatrix} b\,A_n e^{-C_{LA}A_n} \\ (1 - \mu_L)\,L_n \\ P_n\,e^{-C_{PA}A_n} + (1 - \mu_A)\,A_n \end{pmatrix}.$$

The fixed points of $\mathbf{F}$ satisfy

$$L = b\,A e^{-C_{LA}A},$$
$$P = (1 - \mu_L)\,L, \quad \text{and}$$
$$P = A\,\mu_A\,e^{C_{PA}A}.$$

12.5. Applications

Eliminating P gives
$$L = b\,A e^{-C_{LA} A} \quad \text{and}$$
$$(1-\mu_L)\,L = A\,\mu_A\,e^{C_{PA} A}.$$

Certainly, $L = A = P = 0$ is one solution, called the *extinction fixed point*. If these variables are not zero, then we can take the ratio of these two equations, which yields
$$(1-\mu_L) = \frac{\mu_A}{b}\,e^{(C_{LA}+C_{PA})A^*},$$
$$e^{(C_{LA}+C_{PA})A^*} = \frac{b\,(1-\mu_L)}{\mu_A}, \quad \text{or}$$
$$A^* = \frac{\ln(\theta)}{C_{LA}+C_{PA}},$$
where
$$\theta = \frac{b\,(1-\mu_L)}{\mu_A}.$$

Once we have solved for A^*, it follows that
$$L^* = b\,A^* e^{-C_{LA} A^*} \quad \text{and}$$
$$P^* = (1-\mu_L)\,L^* = (1-\mu_L)\,b\,A e^{-C_{LA} A^*}.$$

When $\theta > 1$, this gives a second fixed point (L^*, P^*, A^*), with all the populations positive, which is called a *survival fixed point*.

The matrix of partial derivatives of $\mathbf{F}$ is
$$\begin{pmatrix} 0 & 0 & be^{-C_{LA}A}(1-C_{LA}A) \\ (1-\mu_L) & 0 & 0 \\ 0 & e^{-C_{PA}A} & 1-\mu_A - PC_{PA}e^{-C_{PA}A} \end{pmatrix}.$$

At the extinction fixed point $(0,0,0)$, it reduces to
$$\begin{pmatrix} 0 & 0 & b \\ (1-\mu_L) & 0 & 0 \\ 0 & 1 & 1-\mu_A \end{pmatrix},$$

which has characteristic equation
$$0 = \lambda^3 + a_1\lambda^2 + a_2\lambda + a_3 \quad \text{with}$$
$$a_1 = -(1-\mu_A),$$
$$a_2 = 0, \quad \text{and}$$
$$a_3 = -b(1-\mu_L).$$

The next lemma, which appeared in the paper [**Sam41**], by P. Samuelson, gives a criterion for stability of a fixed point for a function of three variables. (Also see [**Bra01**].)

Lemma 12.8. *Assume that* $p(\lambda) = \lambda^3 + a_1\lambda^2 + a_2\lambda + a_3$.

   a. *If either* $1 + a_1 + a_2 + a_3 < 0$ *or* $1 - a_1 + a_2 - a_3 < 0$, *then the absolute value of at least one of the roots is greater than one.*

b. *If*

$$1 + a_1 + a_2 + a_3 > 0, \qquad 1 - a_1 + a_2 - a_3 > 0,$$
$$3 + a_1 - a_2 - 3a_3 > 0, \qquad 1 + a_1 a_3 - a_2 - a_3^2 > 0,$$

then the absolute value of each of the roots (real or complex) is less than one.

Note that $p(1) = 1 + a_1 + a_2 + a_3$ and the coefficient of λ^3 is positive, so if $p(1) < 0$, then there is a real root greater than one. Similarly, $p(-1) = -1 + a_1 - a_2 + a_3$, so if $p(-1) > 0$, then there is a real root less than minus one that has absolute value greater than one.

For the characteristic equation for the extinction fixed point, if $\theta > 1$, then

$$\begin{aligned} 1 + a_1 + a_2 + a_3 &= 1 - (1 - \mu_A) - b(1 - \mu_L) \\ &= \mu_A - b(1 - \mu_L) \\ &< 0, \end{aligned}$$

and the fixed point is unstable. If $\theta < 1$, then the first quantity is greater than zero. The second and third quantities are positive as follows:

$$1 - a_1 + a_2 - a_3 = 1 + (1 - \mu_A) + b(1 - \mu_L) > 1 > 0 \quad \text{and}$$
$$3 + a_1 - a_2 - 3a_3 = 3 - (1 - \mu_A) + 3b(1 - \mu_L) > 3 - 1 > 0.$$

For the last quantity,

$$\begin{aligned} 1 + a_1 a_3 - a_2 - a_3^2 &= 1 + (1 - \mu_A) b (1 - \mu_L) - b^2 (1 - \mu_L)^2 \\ &> 1 - b^2 (1 - \mu_L)^2 \\ &> 1 - \mu_A^2 \\ &> 0, \end{aligned}$$

since $\theta < 1$. Thus, the extinction fixed point $(0,0,0)$ is attracting for $\theta < 1$ and is unstable for $\theta > 1$.

When $\theta > 1$, the extinction fixed point is unstable and the survival fixed point has all positive populations. Using the equations of the fixed point, we can write the matrix of partial derivatives at the survival fixed point as

$$\begin{pmatrix} 0 & 0 & \frac{L^*}{A^*} - C_{LA} L^* \\ (1 - \mu_L) & 0 & 0 \\ 0 & e^{-C_{PA} A^*} & 1 - \mu_A - \mu_A C_{PA} A^* \end{pmatrix}.$$

This matrix has characteristic equation

$$0 = \lambda^3 + a_1 \lambda^2 + a_2 \lambda + a_3 \qquad \text{with}$$
$$a_1 = -1 + \mu_A + \mu_A C_{PA} A^*,$$
$$a_2 = 0, \qquad \text{and}$$
$$a_3 = -(1 - \mu_L) \left(\frac{L^*}{A^*} - C_{LA} L^* \right) e^{-C_{PA} A^*}.$$

12.5. Applications

So the quantities from Lemma 12.8 for the characteristic polynomial of the survival fixed point are

$$1 + a_1 + a_2 + a_3$$
$$= \mu_A + \mu_A C_{PA} A^* - (1 - \mu_L)\left(\frac{L^*}{A^*} - C_{LA}L^*\right)e^{-C_{PA}A^*},$$

$$1 - a_1 + a_2 - a_3$$
$$= 2 - \mu_A - \mu_A C_{PA} A^* + (1 - \mu_L)\left(\frac{L^*}{A^*} - C_{LA}L^*\right)e^{-C_{PA}A^*},$$

$$3 + a_1 - a_2 - 3a_3$$
$$= 2 + \mu_A + \mu_A C_{PA} A^* + +3(1 - \mu_L)\left(\frac{L^*}{A^*} + 3C_{LA}L^*\right)e^{-C_{PA}A^*},$$

$$1 + a_1 a_3 - a_2 - a_3^2$$
$$= 1 + (1 - \mu_A - \mu_A C_{PA} A^*))(1 - \mu_L)\left(\frac{L^*}{A^*} + C_{LA}L^*\right)e^{-C_{PA}A^*}$$
$$- (1 - \mu_L)^2 \left(\frac{L^*}{A^*} + C_{LA}L^*\right)^2 e^{-2C_{PA}A^*}.$$

The values of these four quantities for $b = 4.88$, $\mu_L = 0.2$, $\mu_A = 0.01$, $C_{LA} = 0.01$, and $C_{PA} = 0.005$ are all positive, $0.5967\ldots$, $1.9403\ldots$, $1.9405\ldots$, and $0.9702\ldots$, so the fixed point is attracting. If two of the parameters are changed to $\mu_A = 0.96$ and $C_{PA} = 0.5$ and the other parameter values are left the same, then $1 + a_1 a_3 - a_2 - a_3^2 = -1.0669\ldots$, and the survival fixed point becomes unstable. Figure 13 shows the plot of a single orbit for the last set of parameter values; notice that this orbit winds around in a complicated manner (i.e., it appears chaotic).

12.5.4. A Discrete Epidemic Model. Some populations have fixed time intervals between generations, so a map is a better model for the population than a differential equation. In this subsection, we consider a single population that can become infected, and then recover to become susceptible again. In this setting, it would be a matter of distinct periods when the infection takes place. This is based on Section 2.9 of [**Bra01**] and on the work of C. Castillo–Chávez and A.A. Yakubu referenced there. Compare with Section 4.7.2, where we considered an epidemic model based on a system of differential equations.

At the nth stage, S_n is the size of the susceptible population and I_n is the size of the infected population. The model takes the form

$$S_{n+1} = \Lambda + S_n e^{-\mu}e^{-\alpha I_n} + I_n e^{-\mu}[1 - e^{-\sigma}],$$
$$I_{n+1} = S_n e^{-\mu}[1 - e^{-\alpha I_n}] + I_n e^{-\mu}e^{-\sigma},$$

where Λ, μ, α, and σ are all positive parameters. The total population $T_n = S_n + I_n$ satisfies the iteration $T_{n+1} = \Lambda + e^{-\mu}T_n \equiv g(T_n)$, which has a unique fixed point at

$$T^* = \frac{\Lambda}{1 - e^{-\mu}},$$

and this fixed point is attracting.

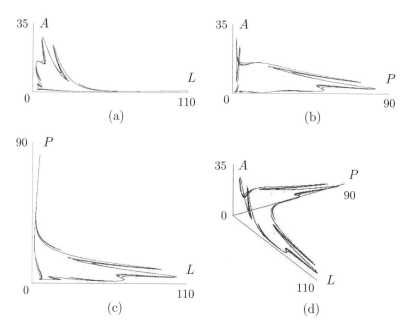

Figure 13. Plot of an orbit for the beetle population for $b = 4.88$, $\mu_L = 0.2$, $\mu_A = 0.96$, $C_{LA} = 0.01$, and $C_{PA} = 0.5$. (a) A as a function of L, (b) A as a function of P, (c) P as a function of L, and (d) three-dimensional plot of L, P, and A

One fixed point of the two-dimensional system is $S = T^*$ and $I = 0$. The matrix of partial derivatives of the two-dimensional map at this fixed point is

$$\begin{pmatrix} e^{-\mu} & e^{-\mu}[1 - e^{-\sigma}] \\ 0 & e^{-\mu-\sigma} + \alpha T^* e^{-\mu} \end{pmatrix},$$

which has eigenvalues $e^{-\mu}$ and $e^{-\mu-\sigma} + \alpha T^* e^{-\mu}$. The first eigenvalue is always less than one, $0 < e^{-\mu} < 1$, and corresponds to the fact that T^* is an attracting fixed point for $g(T_n)$. The second eigenvalue is always positive and is less than one for some parameter values:

$$1 \overset{?}{>} e^{-\mu-\sigma} + \alpha T^* e^{-\mu},$$

$$1 - e^{-\mu-\sigma} \overset{?}{>} \alpha T^* e^{-\mu},$$

$$1 \overset{?}{>} \frac{\alpha T^* e^{-\mu}}{1 - e^{-\mu-\sigma}} \equiv R_0.$$

This last quantity R_0 is called the reproductive number. Thus, for $R_0 < 1$, this fixed point is attracting. For $R_0 > 1$, this fixed point is a saddle and is unstable.

If the total population is at equilibrium with $T_n = T^*$, then $S_n = T^* - I_n$, and we can get a function of a single variable,

$$I_{n+1} = (T^* - I_n) e^{-\mu}[1 - e^{-\alpha I_n}] + I_n e^{-\mu-\sigma} \equiv f(I_n).$$

The derivative at $I = 0$ is $f'(0) = e^{-\mu-\sigma} + \alpha T^* e^{-\mu}$. For $R_0 < 1$, $0 < f'(0) < 1$ and $I = 0$ is an attracting fixed point for f. In fact for $0 < I \leq T^*$, $0 < f(I) < I$ and $I = 0$ is a globally attracting fixed point for f, attracting all points on $(0, T^*]$.

12.5. Applications

For $R_0 > 1$, $1 < f'(0)$ and $I = 0$ is unstable. Since $f(T^*) = T^* e^{-\mu-\sigma} < T^*$, there is a second fixed point I^* with $0 < I^* < T^*$. We do not discuss the stability of this fixed point.

12.5.5. One-Locus Genetic Model. A one-locus genetic model from mathematical biology provides a relatively simple situation in which a nonlinear system has a globally attracting fixed point. Our treatment is based on that in J. Hofbauer and K. Sigmund, [**Hof88**].

The characteristics of an organism are determined by a pair of strands of chromosomes. In humans, each strand contains 46 chromosomes. A single trait can be determined by several pairs of genes occurring at one or several sites of chromosomes, called the *locus of the trait*. We consider the simplest case in which the locus is a single site or single locus. Usually, there can be several types of genes that may occupy a locus. These genes are called the *alleles*. We label them $A_1, \ldots, A_n$. We assume that the allele A_i occurs with frequency $p_i \geq 0$, with $p_1 + \cdots + p_n = 1$, so the pair $A_i A_j$ occurs with frequency $p_i p_j$ in the original generation. The frequency vector is given by
$$\mathbf{p} = (p_1, \ldots, p_n)^\mathsf{T}.$$

By random mating, we assume that the gene pair (A_i, A_j) also occurs with probability $p_i p_j$ in the next generation. If N is the total number of alleles in this next generation, then there are $p_i p_j N$ of the gene pair (A_i, A_j). Let $w_{ij} \geq 0$ be the probability that the gene pair (A_i, A_j) survives to adulthood. The order in which the genes appear on the two strands is not assumed to affect their viability, so $w_{ij} = w_{ji}$. The selection matrix
$$\mathbf{W} = (w_{ij})$$
is therefore symmetric. Let $\mathbf{T}(\mathbf{p}) = \mathbf{p}'$ be the frequency of the alleles in the next generation at adulthood, and p'_{ij} be the frequency of the gene pair (A_i, A_j). Then,
$$p'_{ij} = \frac{w_{ij} p_i p_j N}{\sum_{k,\ell} w_{k\ell} p_k p_\ell N} = \frac{w_{ij} p_i p_j}{\sum_{k,\ell} w_{k\ell} p_k p_\ell} = \frac{w_{ij} p_i p_j}{\mathbf{p}^\mathsf{T} \mathbf{W} \mathbf{p}}$$
and
$$p'_i = \sum_j p'_{ij} = p_i \frac{\sum_j w_{ij} p_j}{\sum_{k,\ell} w_{k\ell} p_k p_\ell} = p_i \frac{(\mathbf{W}\mathbf{p})_i}{\mathbf{p}^\mathsf{T} \mathbf{W} \mathbf{p}},$$
where v_i denotes the ith component of the vector $\mathbf{v}$.

The state space of the possible $\mathbf{p}$ is the simplex
$$\mathbf{S} = \{\mathbf{p} = (p_1, \ldots, p_n)^\mathsf{T} : \sum_i p_i = 1,\ 0 \leq p_i \leq 1 \text{ for } i = 1, \ldots, n\,\}.$$

We are not going to check stability by means of the linearization, but by means of a real-valued function. This approach is similar to the use of Lyapunov functions, which we present for systems of ordinary differential equations in Section 5.3.

Theorem 12.9 (Fundamental theorem of natural selection)**.** *Let $\mathbf{T}$ be the map from $\mathbf{S}$ to itself given by*
$$(\mathbf{T}(\mathbf{p}))_i = p_i \frac{(\mathbf{W}\mathbf{p})_i}{\mathbf{p}^\mathsf{T} \mathbf{W} \mathbf{p}}.$$

Let $\overline{w}(\mathbf{p}) = \mathbf{p}^\mathsf{T} \mathbf{W} \mathbf{p}$ be the average fitness of the population. Then, with each iteration, the average fitness is nondecreasing, or

$$\overline{w}(\mathbf{T}(\mathbf{p})) \geq \overline{w}(\mathbf{p}),$$

and there is equality if and only if $\mathbf{T}(\mathbf{p}) = \mathbf{p}$ is a fixed point.

Therefore, if there is a fixed point $\mathbf{p}^*$ with all $p_i^* > 0$, then any $\mathbf{p}$ with $p_i > 0$ is in the basin of attraction of $\mathbf{p}^*$ and the ω-limit of $\mathbf{p}$ is $\mathbf{p}^*$, $\omega(\mathbf{p}; \mathbf{T}) = \mathbf{p}^*$.

Proof. We need to use some facts about inequalities. First, the geometric mean is no larger than the arithmetic mean, or

$$\sqrt{ab} \leq \frac{a+b}{2},$$

for $a, b > 0$.

The function $f(x) = x^\alpha$, with $\alpha > 1$, has $f''(x) > 0$ and is convex: so, for p_1, $p_2 > 0$, $p_1 + p_2 = 1$, and $0 < x_1 < x_2$,

$$(p_1 x_1 + p_2 x_2)^\alpha = f(p_1 x_1 + p_2 x_2) < p_1 f(x_1) + p_2 f(x_2) = p_1 (x_1)^\alpha + p_2 (x_2)^\alpha.$$

Similarly, for n points $x_i > 0$ and $p_i > 0$, with $p_1 + \cdots + p_n = 1$,

$$(12.2) \qquad \left(\sum_i p_i x_i\right)^\alpha = f\left(\sum_i p_i x_i\right) \leq \sum_i p_i f(x_i) = \sum_i p_i (x_i)^\alpha.$$

Equality occurs if and only if all the x_i are equal.

Now we can start the calculation of the change in the average fitness:

$$\overline{w}(\mathbf{T}(\mathbf{p})) = \frac{\sum_{i,j} p_i (\mathbf{W}\mathbf{p})_i w_{ij} p_j (\mathbf{W}\mathbf{p})_j}{\overline{w}(\mathbf{p})^2}.$$

We can multiply across by the denominator and get

$$\overline{w}(\mathbf{p})^2 \overline{w}(\mathbf{T}(\mathbf{p})) = \sum_{i,j,k} p_i w_{ik} p_k w_{ij} p_j (\mathbf{W}\mathbf{p})_j.$$

Interchanging the roles of j and k, we get

$$\overline{w}(\mathbf{p})^2 \overline{w}(\mathbf{T}(\mathbf{p})) = \sum_{i,j,k} p_i w_{ij} p_j w_{ik} p_k (\mathbf{W}\mathbf{p})_k.$$

12.5. Applications

Taking the average and then applying the inequality for the arithmetic and geometric means yields

$$\overline{w}(\mathbf{p})^2 \overline{w}(\mathbf{T}(\mathbf{p})) = \sum_{i,j,k} p_i p_j p_k w_{ij} w_{ik} \left[\frac{(\mathbf{W}\mathbf{p})_j + (\mathbf{W}\mathbf{p})_k}{2} \right]$$

$$\geq \sum_{i,j,k} p_i p_j p_k w_{ij} w_{ik} (\mathbf{W}\mathbf{p})_j^{\frac{1}{2}} (\mathbf{W}\mathbf{p})_k^{\frac{1}{2}}$$

$$= \sum_i p_i \left[\sum_j w_{ij} p_j (\mathbf{W}\mathbf{p})_j^{\frac{1}{2}} \sum_k w_{ik} p_k (\mathbf{W}\mathbf{p})_k^{\frac{1}{2}} \right]$$

$$= \sum_i p_i \left[\sum_j w_{ij} p_j (\mathbf{W}\mathbf{p})_j^{\frac{1}{2}} \right]^2.$$

Applying inequality (12.2) for $\alpha = 2$, we get

$$\overline{w}(\mathbf{p})^2 \overline{w}(\mathbf{T}(\mathbf{p})) \geq \left[\sum_i p_i \sum_j w_{ij} p_j (\mathbf{W}\mathbf{p})_j^{\frac{1}{2}} \right]^2$$

$$= \left[\sum_j p_j (\mathbf{W}\mathbf{p})_j^{\frac{1}{2}} \sum_i p_i w_{ij} \right]^2$$

$$= \left[\sum_j p_j (\mathbf{W}\mathbf{p})_j^{\frac{1}{2}} (\mathbf{W}\mathbf{p})_j \right]^2$$

$$= \left[\sum_j p_j (\mathbf{W}\mathbf{p})_j^{\frac{3}{2}} \right]^2.$$

Applying inequality (12.2) again for $\alpha = 3/2$, we obtain

$$\overline{w}(\mathbf{p})^2 \overline{w}(\mathbf{T}(\mathbf{p})) \geq \left[\sum_j p_j (\mathbf{W}\mathbf{p})_j \right]^{\frac{3}{2} \cdot 2} = \overline{w}(\mathbf{p})^3.$$

Dividing by $\overline{w}(\mathbf{p})^2$, we get the desired result.

In the last inequality, there is equality if and only if all the $(\mathbf{W}\mathbf{p})_j$ are equal. However, if $(\mathbf{W}\mathbf{p})_j = c$, then $\overline{w}(\mathbf{p}) = \sum_i p_i c = c$ and

$$(\mathbf{T}(\mathbf{p}))_i = p_i \frac{c}{c} = p_i$$

for all i, so $\mathbf{p}$ is a fixed point. □

Exercises 12.5

1. Consider the stochastic matrix

$$\mathbf{M} = \begin{pmatrix} 0.8 & 0.1 & 0.1 \\ 0.2 & 0.7 & 0.1 \\ 0.1 & 0.3 & 0.6 \end{pmatrix}.$$

   Find the steady-state probability distribution $\mathbf{p}^*$.

2. A simpler SIS epidemic model, which has been proposed by L.J.S. Allen for certain situations, [All94], is

$$S_{n+1} = S_n \left(1 - \frac{\alpha}{N} I_n\right) + \gamma I_n,$$
$$I_{n+1} = I_n \left(1 - \gamma + \frac{\alpha}{N} S_n\right),$$

   where $N = I_0 + S_0$, and the parameters are $\alpha > 0$, $\gamma > 0$, and $\alpha < 1 + \gamma$.
   a. Show that $S_n + I_n = N$ for all $n \geq 0$.
   b. Show that making the substitution $S_n = N - I_n$ and setting $p = 1 - \gamma + \alpha$ results in the following function of a single variable:

$$I_{n+1} = f(I_n) = p I_n \left(1 - \frac{\alpha}{pN} I_n\right).$$

   c. Show that the fixed point $I = 0$ is attracting if and only if $R = \alpha/\gamma < 1$.
   d. For $R > 1$, show that there is a second fixed point at $I^* = N(p-1)/\alpha$. Show that this fixed point is attracting if $\alpha < 2 + \gamma$.

3. A model for inflation and unemployment, proposed by Ahemed, El-Misiery, and Agiza [Ahm99], is given by the function

$$\begin{pmatrix} U_{n+1} \\ I_{n+1} \end{pmatrix} = \begin{pmatrix} U_n - b(I_n - m) \\ I_n + (c-1)f(U_n) + f(U_n - b(I_n - m)) \end{pmatrix},$$

   where $f(x) = -\alpha + \beta e^x$, where α, β, m, and b are all positive parameters, and $0 < c < 1$.
   a. Show that the unique fixed point is given by

$$\begin{pmatrix} U^* \\ I^* \end{pmatrix} = \begin{pmatrix} \ln(\alpha/\beta) \\ m \end{pmatrix}.$$

   b. Let Δ be the determinant of the matrix of partial derivatives at the fixed point. For what restriction on the parameter values is $0 < \Delta < 1$?
   c. Show that the eigenvalues of the fixed point are

$$\lambda_\pm = 1 - \frac{\alpha b}{2} \pm \frac{1}{2}\sqrt{\alpha^2 b^2 - 4\alpha bc}.$$

   d. For what restriction on the parameter values are the eigenvalues complex (i.e., not real)? For these parameter values, is the fixed point attracting or repelling? Hint: Remember that the product of the eigenvalues is equal to the determinant.

e. Assume that the parameters are chosen so that the eigenvalues at the fixed point are real. (i) Show that, if $ab < 2$, then $0 < \lambda_+ < 1$. (ii) Show that, if $ab < 2$ and $\Delta > 0$, then $\lambda_- > 0$. In this last case, is the fixed point attracting or repelling?

12.6. Theory and Proofs

Linearization near periodic points

To a large extent the linearized equations at a fixed point determine the features of the phase portrait near the fixed point. In particular, two maps are conjugate provided that there is a change of coordinates that takes the orbits of one system into the orbits of the other. In this section, we make this idea more precise; in particular, we need to be more specific about how many derivatives the change of coordinates has. The results in this section have proofs beyond the scope of this text. We merely state the theorems and provide references for the proofs.

The statements of the theorems do not involve a conjugacy on the whole space $\mathbb{R}^n$, but from neighborhoods of the fixed point to a neighborhood of the origin. Therefore, we repeat the definition of conjugacy to introduce this idea of taking a subset.

Definition 12.35. We say that **F** on **U** is *topologically conjugate* to **G** on **V** provided that there is a homeomorphism **h** from **U** to **V** such that

$$\mathbf{h} \circ \mathbf{F}(\mathbf{x}) = \mathbf{G} \circ \mathbf{h}(\mathbf{x})$$

for all **x** in **U**. Such a map **h** is a continuous change of coordinates.

For an integer $r \geq 1$, these maps are called C^r *conjugate* on open sets **U** and **V** provided that there is a C^r diffeomorphism **h** from **U** to **V** such that

$$\mathbf{h} \circ \mathbf{F}(\mathbf{x}) = \mathbf{G} \circ \mathbf{h}(\mathbf{x});$$

thus, both **h** and $\mathbf{h}^{-1}$ are C^r. Such a map **h** is a differentiable change of coordinates.

In the theorems involving the linearization of a map **F**, we assume that the derivative is invertible at the fixed point $\mathbf{x}^*$, $\det(D\mathbf{F}_{(\mathbf{x}^*)}) \neq 0$. By the inverse function theorem, for such maps, there are two neighborhoods $\mathbf{U}_1$ and $\mathbf{U}_2$ about $\mathbf{x}^*$ such that **F** is a diffeomorphism from $\mathbf{U}_1$ to $\mathbf{U}_2$. We call **F** a *local diffeomorphism* at $\mathbf{x}^*$.

As we state below, there is a general theorem called the Grobman–Hartman theorem which states that a nonlinear map with a hyperbolic fixed point is topologically conjugate to the linear map induced by the derivative at the fixed point. Unfortunately, this topological conjugacy does not tell us much about the features of the phase portrait. Any two linear maps in the same dimension for which the origin is attracting are topologically conjugate on all of $\mathbb{R}^n$. See [**Rob99**]. In particular, a linear system for which the origin is a stable focus is topologically conjugate to one for which the origin is a stable node. Thus, a continuous change of coordinates does not preserve the property that trajectories spiral in toward a fixed point or do not spiral. A differentiable change of coordinates does preserves such features. Unfortunately, the theorems that imply the existence of a differentiable change of coordinates require more assumptions.

We now state these results. They also apply to periodic points, but we state them only for fixed points, where the statements are simpler. In the theorems,
$$\mathbf{A} = D\mathbf{F}_{(\mathbf{x}^*)}$$
is the matrix of partial derivatives, and
$$D\mathbf{F}_{(\mathbf{x}^*)}\,\mathbf{y}$$
is the associated linear map.

Theorem 12.10 (Grobman–Hartman). *Let $\mathbf{F}$ be a C^1 map with a hyperbolic fixed point $\mathbf{x}^*$ such that $\det(D\mathbf{h}_{(\mathbf{x}^*)}) \neq 0$.*

Then, there are open sets $\mathbf{U}$ containing $\mathbf{x}^*$ and $\mathbf{V}$ containing the origin, such that $\mathbf{F}(\mathbf{x})$ on $\mathbf{U}$ is topologically conjugate to the associated linear map $D\mathbf{F}_{(\mathbf{x}^*)}\,\mathbf{y}$ on $\mathbf{V}$.

See [**Rob99**] for a proof.

To get a differentiable change of coordinates we need to treat special cases or add more assumptions. The result with more derivatives is due to Sternberg, but requires a "nonresonance" requirement on the eigenvalues.

Theorem 12.11 (Sternberg). *Let $\mathbf{F}$ be a C^∞ map with a hyperbolic fixed point $\mathbf{x}^*$ such that $\det(D\mathbf{h}_{(\mathbf{x}^*)}) \neq 0$.*

Let $\lambda_1, \ldots, \lambda_n$ be the eigenvalues of $D\mathbf{F}_{(\mathbf{x}^*)}$, and assume that each λ_k has algebraic multiplicity one. Moreover, assume that
$$\lambda_k \neq \lambda_1^{m_1} \lambda_2^{m_2} \cdots \lambda_n^{m_n},$$
for each k and any nonnegative integers m_j with $\sum_j m_j \geq 2$. (This condition is called multiplicative nonresonance of the eigenvalues.)

Then there are open sets $\mathbf{U}$ containing $\mathbf{x}^*$ and $\mathbf{V}$ containing the origin such that $\mathbf{F}$ on $\mathbf{U}$ is C^∞ conjugate to the linear map $D\mathbf{F}_{(\mathbf{x}^*)}\,\mathbf{y}$ on $\mathbf{V}$.

The proof of this theorem is much harder than Grobman–Hartman theorem proof. It also requires the nonresonance of the eigenvalues. The book [**Har82**] contains a proof. The next two theorems state results about a C^1 conjugacy. Hartman's theorem has no assumption of the nonresonance of the eigenvalues.

Theorem 12.12 (Hartman). *Let $\mathbf{F}$ be a C^1 map with a hyperbolic attracting fixed point $\mathbf{x}^*$ such that $\det(D\mathbf{h}_{(\mathbf{x}^*)}) \neq 0$. Then, there exist open sets $\mathbf{U}$ containing $\mathbf{x}^*$ and $\mathbf{V}$ containing the origin such that $\mathbf{F}$ on $\mathbf{U}$ is C^1 conjugate to the linear map $D\mathbf{F}_{(\mathbf{x}^*)}\,\mathbf{y}$ on $\mathbf{V}$.*

Theorem 12.13 (Belickii [**Bel72**]). *Let $\mathbf{F}$ be a C^2 map with a hyperbolic attracting fixed point $\mathbf{x}^*$ such that $\det(D\mathbf{h}_{(\mathbf{x}^*)}) \neq 0$. Assume that the eigenvalues of $D\mathbf{F}_{(\mathbf{x}^*)}$ satisfy a multiplicative nonresonance assumption that says that $\lambda_k \neq \lambda_i \lambda_j$ for any three eigenvalues (including $\lambda_k \neq \lambda_i^2$).*

Then, there exist open sets $\mathbf{U}$ containing $\mathbf{x}^*$ and $\mathbf{V}$ containing the origin such that the map $\mathbf{F}$ on $\mathbf{U}$ is C^1 conjugate to the linear map $D\mathbf{F}_{(\mathbf{x}^*)}\,\mathbf{y}$ on $\mathbf{V}$.

Chapter 13

Invariant Sets for Higher Dimensional Maps

In this chapter, we consider invariant sets for maps with two or more variables. By a treatment analogous to the one used for the tent map and the logistic map G in one dimension, we use symbolic dynamics to show that a nonlinear map has sensitive dependence on initial conditions and a dense orbit in an invariant set. We start with the geometric horseshoes of Smale, which has an invariant set that is a two-dimensional analog of the Cantor sets arising from tent maps considered in Chapter 10. We follow by defining symbolic dynamics for a nonlinear map in higher dimensions, using correctly aligned boxes. Just as the geometric horseshoe has certain rectangles whose images cross each other, a map is said to have a Markov partition if there are nonlinear boxes whose images are correctly aligned with the set of boxes. The toral automorphisms were among the first maps for which Markov partitions were constructed; we illustrate ideas by giving a Markov partition for some simple examples of toral automorphisms in two dimensions. We also make the connection between correctly aligned boxes and shadowing (i.e., finding a true orbit near an orbit with errors). Going back to ideas of Poincaré, we show how an intersection of stable and unstable manifolds leads to a horseshoe and, so, complicated dynamics. These topics all relate to carrying over symbolic dynamics to maps in two or more dimensions.

We discuss chaotic attractors for maps in multidimensions: the definitions are essentially the same as those in one dimension, but the examples give new insights into the concepts and the manner in which they arise. We introduce Lyapunov exponents for maps in more dimensions, which is more complicated than in one dimension. Just as for one-dimensional maps, Lyapunov exponents give a more computable quantity than the definition of sensitive dependence on initial conditions. We relate Lyapunov exponents and computed orbits to a test for a chaotic attractor. This treatment carries over the concepts presented in Chapter 11 in one dimension to maps in two or more dimensions. This discussion broadens the arena where these ideas can be applied.

13.1. Geometric Horseshoe

To start our consideration of multidimensional maps with invariant sets more complicated than a periodic orbit, we give an example that is piecewise affine (linear plus a constant) in the regions that contain the invariant set. In this regard, this example is like the tent map with slope 3, which produced the standard middle-third Cantor set. S. Smale introduced this example as a model for the type of invariant sets that arise from homoclinic intersections of stable and unstable manifolds as discussed in Section 13.3. The map is specified as affine on two horizontal strips, is given by a geometric description on the gap between the strips, and is extended to the whole plane so other points eventually enter the region where the description is given. Since the map is not given by a formula in all of $\mathbb{R}^2$, but its properties are described geometrically, the invariant set is called a *geometric horseshoe*, or sometimes, the *Smale horseshoe*.

Description of the Horseshoe Map

The *horseshoe diffeomorphism* $\mathbf{F}$ is first prescribed on a stadium shaped region

$$\mathbf{U} = \mathbf{S}' \cup \mathbf{E}_0 \cup \mathbf{E}_1, \quad \text{where}$$

$$\mathbf{S}' = \left[-\frac{1}{16}, \frac{17}{16}\right] \times \left[-\frac{1}{16}, \frac{17}{16}\right]$$

is a square region and $\mathbf{E}_0$ and $\mathbf{E}_1$ are semicircular regions added at the top and bottom of $\mathbf{S}'$ as indicated in Figure 1(a). The map is described geometrically on $\mathbf{U}$ as the composition of two maps, $\mathbf{F}(x, y) = \mathbf{F}_0(\mathbf{L}(x, y))$: (i) The linear map $\mathbf{L}$ from $\mathbf{U}$ to $\mathbb{R}^2$ contracts the first coordinate by $1/3$ and expands the second coordinate by 4. (ii) The second map $\mathbf{F}_0$ bends the region $\mathbf{L}(\mathbf{U})$ into a horseshoe shaped object and places the image inside $\mathbf{U}$. See Figure 1(b). The name of the map derives from the fact that $\mathbf{F}(\mathbf{U})$ has the shape of a horseshoe. By advanced mathematical reasons (methods from differential topology), the map can be extended to the whole plane in a manner so that other points eventually enter the region $\mathbf{U}$.

To give a more precise description, we divide the region $\mathbf{S}'$ into three subregions,

$$\mathbf{H}'_0 = \left\{ (x, y) \in \mathbf{S}' : -1/16 \leq y \leq 1/4 \right\},$$
$$\mathbf{G}' = \left\{ (x, y) \in \mathbf{S}' : 1/4 \leq y \leq 3/4 \right\}, \quad \text{and}$$
$$\mathbf{H}'_1 = \left\{ (x, y) \in \mathbf{S}' : 3/4 \leq y \leq 17/16 \right\}.$$

The horseshoe diffeomorphism $\mathbf{F}$ is given by

$$\mathbf{F}(x, y) = \begin{cases} \left(\tfrac{1}{3}x, 4y\right) & \text{for } (x, y) \in \mathbf{H}'_0, \\ \left(1 - \tfrac{1}{3}x, 4 - 4y\right) & \text{for } (x, y) \in \mathbf{H}'_1, \end{cases}$$

and $\mathbf{F}$ bends in a nonlinear fashion in $\mathbf{G}'$. The region $\mathbf{E}_0$ is mapped into itself by a contraction, so we can specify the map to have a single attracting fixed point $\mathbf{p}_0$ in this region whose basin of attraction includes all of $\mathbf{E}_0$. The region $\mathbf{E}_1$ is mapped into $\mathbf{E}_0$ so it is also in the basin of attraction of $\mathbf{p}_0$. In the analysis, it is very

13.1. Geometric Horseshoe

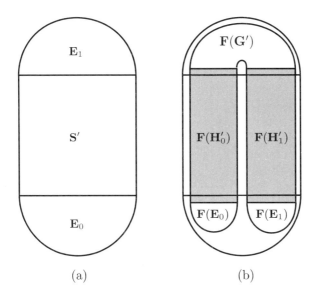

Figure 1. Region $\mathbf{U} = \mathbf{S}' \cup \mathbf{E}_0 \cup \mathbf{E}_1$ and the image of $\mathbf{F}(\mathbf{U})$ for the Geometric Horseshoe

important that inside $\mathbf{H}_0' \cup \mathbf{H}_1'$, vertical directions are taken to vertical directions and horizontal directions are taken to horizontal directions.

Because points (x, y) with $-1/16 \leq y < 0$ or $1 < y \leq 17/16$ leave $\mathbf{S}'$ under forward iteration, and points (x, y) with $-1/16 \leq x < 0$ or $1 < x \leq 17/16$ leave under backward iteration, the only points which stay in $\mathbf{S}'$ for all iterates also stay in the unit square

$$\mathbf{S} = [0, 1] \times [0, 1];$$

that is,

$$\bigcap_{j=-\infty}^{\infty} \mathbf{F}^j(\mathbf{S}') = \bigcap_{j=-\infty}^{\infty} \mathbf{F}^j(\mathbf{S}).$$

Therefore, we switch over to using the unit square.

There are two horizontal strips in $\mathbf{S}$ whose images also lie in $\mathbf{S}$:

$$\mathbf{H}_0 = \mathbf{H}_0' \cap \mathbf{S} = \left\{ (x, y) : 0 \leq x \leq 1 \text{ and } 0 \leq y \leq 1/4 \right\} \quad \text{and}$$
$$\mathbf{H}_1 = \mathbf{H}_1' \cap \mathbf{S} = \left\{ (x, y) : 0 \leq x \leq 1 \text{ and } 3/4 \leq y \leq 1 \right\}.$$

The images by $\mathbf{F}$ of these two horizontal strips are the two vertical strips in $\mathbf{S}$ given by

$$\mathbf{V}_0 = \left\{ (x, y) : 0 \leq x \leq 1/3 \text{ and } 0 \leq y \leq 1 \right\} \quad \text{and}$$
$$\mathbf{V}_1 = \left\{ (x, y) : 2/3 \leq x \leq 1 \text{ and } 0 \leq y \leq 1 \right\}.$$

The map $\mathbf{F}$ takes $\mathbf{H}_0$ to $\mathbf{V}_0$ by means of a contraction of $1/3$ in the first coordinate and an expansion of 4 in the second coordinate. It takes $\mathbf{H}_1$ to $\mathbf{V}_1$ by means of a rotation by 180 degrees and a contraction of $1/3$ in the first coordinate and an expansion of 4 in the second coordinate. See Figure 2.

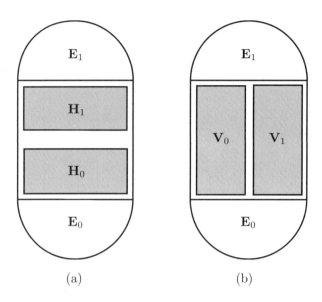

Figure 2. Horizontal and Vertical Strips in **S**

The inverse map is given by

$$\mathbf{F}^{-1}(x,y) = \begin{cases} \left(3x, \tfrac{1}{4}y\right) & \text{for } (x,y) \in \mathbf{V}_0, \\ \left(3 - 3x, 1 - \tfrac{1}{4}y\right) & \text{for } (x,y) \in \mathbf{V}_1, \end{cases}$$

and $\mathbf{F}^{-1}$ maps $\mathbf{V}_0$ onto $\mathbf{H}_0$ and $\mathbf{V}_1$ onto $\mathbf{H}_1$.

Invariant Set for the Horseshoe Map

Before discussing the symbolic dynamics, we indicate which points stay in **S** for all iterates. We start with notation for the finite intersections of iterates by

$$\mathbf{S}_m^n = \bigcap_{j=m}^{n} \mathbf{F}^j(\mathbf{S}) = \{\, \mathbf{x} : \mathbf{F}^{-j}(\mathbf{x}) \in \mathbf{S} \text{ for } m \leq j \leq n \,\}.$$

The next lemma describes the sets $\mathbf{S}_m^n$ for various choices of m and n.

Lemma 13.1. a. *For forward iteration with $n \geq 0$, $\mathbf{S}_0^n$ is the union of 2^n vertical strips, each of width $\left(\tfrac{1}{3}\right)^n$, and $\mathbf{S}_0^\infty$ is a Cantor set of vertical line segments, $\mathbf{C}_1 \times [0,1]$.*

b. *For backward iteration with $-n \leq 0$, $\mathbf{S}_{-n}^0$ is the union of 2^n horizontal strips, each of height $\left(\tfrac{1}{4}\right)^n$, and $\mathbf{S}_{-\infty}^0$ is a Cantor set of horizontal line segments, $[0,1] \times \mathbf{C}_2$.*

c. *The points that stay in **S** for all iterates are in the invariant set*

$$\Lambda = \mathbf{S}_{-\infty}^\infty = \mathbf{C}_1 \times \mathbf{C}_2,$$

which is the Cartesian product of two Cantor sets.

Proof. As can be seen from the definitions,

$$\mathbf{S}_0^1 = \mathbf{S} \cap \mathbf{F}(\mathbf{S}) = \mathbf{V}_0 \cup \mathbf{V}_1.$$

13.1. Geometric Horseshoe

This set is the union of 2 vertical strips each of width $1/3$.

Putting the set $\mathbf{F}(\mathbf{S})$ in the intersection twice, we have

$$\begin{aligned}\mathbf{S}_0^2 &= [\mathbf{S} \cap \mathbf{F}(\mathbf{S})] \cap [\mathbf{F}(\mathbf{S}) \cap \mathbf{F}^2(\mathbf{S})] \\ &= [\mathbf{S} \cap \mathbf{F}(\mathbf{S})] \cap \mathbf{F}(\mathbf{S} \cap \mathbf{F}(\mathbf{S})) \\ &= \mathbf{S}_0^1 \cap \mathbf{F}(\mathbf{S}_0^1).\end{aligned}$$

Using $\mathbf{S}_0^1 = \mathbf{V}_0 \cup \mathbf{V}_1$ and $\mathbf{F}(\mathbf{H}_j) = \mathbf{V}_j$, we get

$$\begin{aligned}\mathbf{S}_0^2 &= \left[\mathbf{V}_0 \cap \mathbf{F}(\mathbf{S}_0^1)\right] \cup \left[\mathbf{V}_1 \cap \mathbf{F}(\mathbf{S}_0^1)\right] \\ &= \mathbf{F}\left(\mathbf{H}_0 \cap \mathbf{S}_0^1\right) \cup \mathbf{F}\left(\mathbf{H}_1 \cap \mathbf{S}_0^1\right).\end{aligned}$$

On $\mathbf{H}_0$, the map is a contraction by $1/3$ in the horizontal direction and an expansion by 4 in the vertical direction; therefore, the set $\mathbf{F}\left(\mathbf{H}_0 \cap \mathbf{S}_0^1\right)$ has 2 vertical strips in $\mathbf{V}_0$, each of width $1/3$ of the width of the strips in $\mathbf{S}_0^1$, so of width $\left(1/3\right)^2$. These two strips each reach from $y = 0$ to $y = 1$. Similarly, on $\mathbf{H}_1$, the map is a contraction by $1/3$ in the horizontal direction and an expansion by 4 in the vertical direction, followed by a rotation of 180 degrees; therefore, the set $\mathbf{F}\left(\mathbf{H}_1 \cap \mathbf{S}_0^1\right)$ has 2 vertical strips in $\mathbf{V}_1$, each of width $\left(1/3\right)^2$. Combining, we see that $\mathbf{S}_0^2$ has $2+2 = 2^2$ vertical strips each of width $\left(1/3\right)^2$, going all the way from $y = 0$ to $y = 1$. See Figure 3.

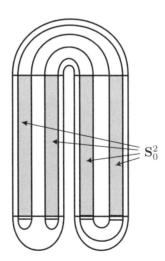

Figure 3. The set $\mathbf{S}_0^2 = \mathbf{F}^2(\mathbf{S}) \cap \mathbf{F}(\mathbf{S}) \cap \mathbf{S}$

Continuing by induction, at the next stage, $\mathbf{S}_0^n$ is formed by intersecting the previous set of vertical strips $\mathbf{S}_0^{n-1}$ with $\mathbf{H}_0$ and taking their image together with intersecting $\mathbf{S}_0^{n-1}$ with $\mathbf{H}_1$ and taking their image. Each of these sets gives 2^{n-1} vertical strips for a total of $2^{n-1} + 2^{n-1} = 2 \cdot 2^{n-1} = 2^n$. The width of the strips in $\mathbf{S}_0^{n-1}$ is $\left(1/3\right)^{n-1}$, by induction, so their images are of width $1/3 \left(1/3\right)^{n-1} = \left(1/3\right)^n$. The strips $\mathbf{H}_j \cap \mathbf{S}_0^{n-1}$ go all the way from the top to the bottom of $\mathbf{H}_j$, so their images go all the way from $y = 0$ to $y = 1$. The more formal derivation of the images is as follows:

$$\begin{aligned}
\mathbf{S}_0^n &= [\mathbf{S} \cap \mathbf{F}(\mathbf{S})] \cap [\mathbf{F}(\mathbf{S}) \cap \cdots \cap \mathbf{F}^n(\mathbf{S})] \\
&= \mathbf{S}_0^1 \cap \mathbf{F}(\mathbf{S}_0^{n-1}) \\
&= [\mathbf{V}_0 \cap \mathbf{F}(\mathbf{S}_0^{n-1})] \cup [\mathbf{V}_1 \cap \mathbf{F}(\mathbf{S}_0^{n-1})] \\
&= \mathbf{F}\left(\mathbf{H}_0 \cap \mathbf{S}_0^{n-1}\right) \cup \mathbf{F}\left(\mathbf{H}_1 \cap \mathbf{S}_0^{n-1}\right).
\end{aligned}$$

Taking the limit as n goes to infinity, we see that $\mathbf{S}_0^\infty$ is a Cantor set of vertical line segments,

$$\mathbf{S}_0^\infty = \mathbf{C}_1 \times [0, 1].$$

Turning to the negative iterates, since $\mathbf{F}^{-1}$ maps $\mathbf{V}_0$ to $\mathbf{H}_0$ and $\mathbf{V}_1$ to $\mathbf{H}_1$,

$$\begin{aligned}
\mathbf{S}_{-1}^0 &= \mathbf{S} \cap \mathbf{F}^{-1}(\mathbf{S}) \\
&= \mathbf{F}^{-1}(\mathbf{F}(\mathbf{S})) \cap \mathbf{F}^{-1}(\mathbf{S}) \\
&= \mathbf{F}^{-1}\left(\mathbf{S}_0^1\right) \\
&= \mathbf{F}^{-1}\left(\mathbf{V}_0 \cup \mathbf{V}_1\right) \\
&= \mathbf{H}_0 \cup \mathbf{H}_1.
\end{aligned}$$

This set is the union of two horizontal strips, each of height $1/4$. By induction, we see that

$$\begin{aligned}
\mathbf{S}_{-n}^0 &= \left[\mathbf{F}^{-n}(\mathbf{S}) \cap \cdots \cap \mathbf{F}^{-1}(\mathbf{S})\right] \cap \left[\mathbf{F}^{-1}(\mathbf{S}) \cap \mathbf{S}\right] \\
&= \mathbf{F}^{-1}\left(\mathbf{S}_{-n+1}^0\right) \cap \mathbf{S}_{-1}^0 \\
&= \left[\mathbf{F}^{-1}\left(\mathbf{S}_{-n+1}^0\right) \cap \mathbf{H}_0\right] \cup \left[\mathbf{F}^{-1}\left(\mathbf{S}_{-n+1}^0\right) \cap \mathbf{H}_1\right] \\
&= \mathbf{F}^{-1}\left(\mathbf{S}_{-n+1}^0 \cap \mathbf{V}_0\right) \cup \mathbf{F}^{-1}\left(\mathbf{S}_{-n+1}^0 \cap \mathbf{V}_1\right)
\end{aligned}$$

is the union of $2 \cdot 2^{n-1} = 2^n$ horizontal strips. Since $\mathbf{F}^{-1}$ is a vertical contraction by $1/4$ and a horizontal expansion by 3, each strip in $\mathbf{S}_{-n}^0$ has a height of $1/4 \left(1/4\right)^{n-1} = \left(1/4\right)^n$, and extends all the way from $x = 0$ to $x = 1$. Taking the limit as n goes to infinity, $\mathbf{S}_{-\infty}^0$ is a Cantor set of horizontal line segments:

$$\mathbf{S}_{-\infty}^0 = [0, 1] \times \mathbf{C}_2.$$

Next, we take the intersections over both forward and backward iterates:

$$\begin{aligned}
\Lambda = \mathbf{S}_{-\infty}^\infty &= \mathbf{S}_0^\infty \cap \mathbf{S}_{-\infty}^0 \\
&= \mathbf{C}_1 \times [0, 1] \cap [0, 1] \times \mathbf{C}_2 \\
&= \mathbf{C}_1 \times \mathbf{C}_2.
\end{aligned}$$

The set Λ is the Cartesian product of two Cantor sets and is invariant by both $\mathbf{F}$ and $\mathbf{F}^{-1}$. It is also the set of all points which stay in $\mathbf{S}$ for all iterates. □

Itinerary map

We want to describe the dynamics of $\mathbf{F}$ on its invariant set Λ by means of symbolic dynamics. This description is possible because $\mathbf{F}$ is an expansion in the y-coordinate and a contraction in the x-coordinate. For a point that stays in $\mathbf{S}$ for all iterates, its y-coordinate is determined by the rough location of its forward iterates, just as in the case for the tent map with slope 3. The rough location of an iterate is specified by a 0 or a 1, which specifies whether it is in $\mathbf{H}_0$ or $\mathbf{H}_1$. The map is a contraction in the x-coordinate, but its inverse is an expansion in this direction. Therefore, for a point that stays in $\mathbf{S}$ for all iterates, its x-coordinate is determined by the rough location of its backward iterates. The location of a backward iterate is given by specifying a 0 or a 1 on the negative indices. This bi-infinite sequence of 0's and 1's specifies the point by means of symbolic dynamics. These symbol sequences are used to show the existence of periodic points and a point with a dense orbit in Λ.

A point in $\mathbf{S} \setminus [\mathbf{H}_0 \cup \mathbf{H}_1]$ leaves $\mathbf{S}$ under the first iteration. Therefore, $\Lambda \subset \mathbf{H}_0 \cup \mathbf{H}_1$. Similarly, for a point $\mathbf{x}$ to be in Λ, $\mathbf{F}^j(\mathbf{x})$ must be in $\mathbf{H}_0 \cup \mathbf{H}_1$ for all j. For each positive or negative j, let s_j be either 0 or 1, depending on which horizontal strip contains $\mathbf{F}^j(\mathbf{x})$; that is,

$$s_j = \begin{cases} 0 & \text{if } \mathbf{F}^j(\mathbf{x}) \in \mathbf{H}_0, \\ 1 & \text{if } \mathbf{F}^j(\mathbf{x}) \in \mathbf{H}_1, \end{cases} \quad \text{or}$$

$$\mathbf{F}^j(\mathbf{x}) \in \mathbf{H}_{s_j} \qquad \text{for all } j.$$

We put the sequences of 0's and 1's into a single bi-infinite sequence of symbols:

$$\mathbf{s} = \cdots s_{-2} s_{-1} . s_0 s_1 s_2 \cdots.$$

The "decimal point" separates the s_j with zero or positive j from those with negative j. It would be better to put a box around s_0 to indicate that it gives the current location, but we stick with the notation that is easier to write.

The *itinerary map*

$$h(\mathbf{x}) = \mathbf{s} = \cdots s_{-2} s_{-1} . s_0 s_1 s_2 \cdots$$

takes a point $\mathbf{x}$ in the invariant set Λ and assigns the bi-infinite sequence of 0's and 1's by the rule

$$\mathbf{F}^j(\mathbf{x}) \in \mathbf{H}_{s_j} \qquad \text{for all } j.$$

The set of all such bi-infinite sequences is called the *two-sided full shift on two symbols* and is denoted by Σ_2, where the 2 indicates that there are two symbols. (Remember that Σ_2^+ is used to denote one-sided sequences of two symbols.) Σ_2 is also called the *symbol space* for the geometric horseshoe. The *shift map* σ on Σ_2 is defined by

$$\mathbf{t} = \sigma(\mathbf{s}) = \sigma(\cdots s_{-2} s_{-1} . s_0 s_1 s_2 \cdots) = \cdots s_{-2} s_{-1} s_0 . s_1 s_2 \cdots$$

or

$$t_j = s_{j+1} \qquad \text{for all } j.$$

The distance on Σ_2 is defined as in Section 10.3 for Σ_2^+,

$$d(\mathbf{s}, \mathbf{t}) = \sum_j \frac{\delta(s_j, t_j)}{3^j}.$$

Theorem 13.2. *The shift map σ is continuous on Σ_2 and has the following properties:*

    a. *There is a point $\mathbf{s}^*$ in Σ_2 such that both the forward orbit $\mathcal{O}_\sigma^+(\mathbf{x}^*)$ and the backward orbit $\mathcal{O}_\sigma^-(\mathbf{x}^*)$ are dense in Σ_2.*

    b. *The shift map σ has sensitive dependence on initial conditions.*

Proof. (a) Let $\mathbf{s}^*$ be the bi-infinite sequence with $s_{-j}^* = s_j^*$ and let the positive symbols be the same as those used in Theorem 10.10 for the one-sided shift space. We leave the details of the proof that the forward and backward orbits are dense to the reader.

(b) The proof is essentially the same as for Theorem 10.15. $\square$

Orbits from symbol sequences

Starting with a sequence $\mathbf{s}$ in Σ_2, we form the intersection

$$\{\, \mathbf{x} : \mathbf{F}^j(\mathbf{x}) \in \mathbf{H}_{s_j} \text{ for } -\infty < j < \infty \,\} = \bigcap_{j=-\infty}^{\infty} \mathbf{F}^{-j}(\mathbf{H}_{s_j}).$$

The two preceding sets are equal because $\mathbf{F}^j(\mathbf{x})$ is in $\mathbf{H}_{s_j}$ if and only if $\mathbf{x}$ is in $\mathbf{F}^{-j}(\mathbf{H}_{s_j})$. For any symbol sequence, we want to see that this intersection is nonempty and that it is a single point. For any choice of $s_0, s_1, \ldots s_{n-1}$, let

$$\mathbf{B}_{.s_0 s_1 \cdots s_{n-1}} = \bigcap_{j=0}^{n-1} \mathbf{F}^{-j}(\mathbf{H}_{s_j})$$
$$= \{\, \mathbf{x} : \mathbf{F}^j(\mathbf{x}) \in \mathbf{H}_{s_j} \text{ for } 0 \leq j \leq n-1 \,\}.$$

Notice that s_0 just to the right of the decimal point is the symbol corresponding to the present location. From the definition of $\mathbf{B}_{.s_0}$, it follows that

$$\mathbf{B}_{.0} = \mathbf{H}_0 \quad \text{and} \quad \mathbf{B}_{.1} = \mathbf{H}_1.$$

Since $\mathbf{F}^{-1}$ maps $\mathbf{V}_0$ to $\mathbf{B}_{.0} = \mathbf{H}_0$ by a contraction in the second coordinate by $1/4$, we have

$$\mathbf{B}_{.00} = \mathbf{H}_0 \cap \mathbf{F}^{-1}(\mathbf{H}_0) = \mathbf{F}^{-1}(\mathbf{V}_0 \cap \mathbf{H}_0) \quad \text{and}$$
$$\mathbf{B}_{.01} = \mathbf{H}_0 \cap \mathbf{F}^{-1}(\mathbf{H}_1) = \mathbf{F}^{-1}(\mathbf{V}_0 \cap \mathbf{H}_1).$$

Each horizontal strip $\mathbf{H}_i$ is intersected with $\mathbf{V}_0$ and its image is a strip of height equal to $1/4$ of the height of $\mathbf{H}_i$), $(1/4)^2$. Similarly,

$$\mathbf{B}_{.10} = \mathbf{H}_1 \cap \mathbf{F}^{-1}(\mathbf{H}_0) = \mathbf{F}^{-1}(\mathbf{V}_1 \cap \mathbf{H}_0) \quad \text{and}$$
$$\mathbf{B}_{.11} = \mathbf{H}_1 \cap \mathbf{F}^{-1}(\mathbf{H}_1) = \mathbf{F}^{-1}(\mathbf{V}_1 \cap \mathbf{H}_1).$$

Each of these strips is also of height $(1/4)^2$. Altogether,

$$\mathbf{S}_{-2}^0 = \mathbf{B}_{.00} \cup \mathbf{B}_{.01} \cup \mathbf{B}_{.10} \cup \mathbf{B}_{.11},$$

13.1. Geometric Horseshoe

where each horizontal strip is labeled with the two symbols that give its present location and the location of the first iterate. See Figure 4(b).

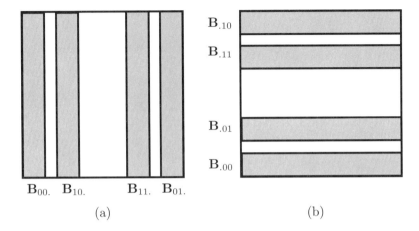

Figure 4. (a) Boxes $\mathbf{B}_{.s_0 s_1}$ and (b) boxes $\mathbf{B}_{s_{-2} s_{-1}}$.

Continuing by induction, we see that

$$\mathbf{B}_{.s_0 s_1 \cdots s_{n-1}} = \mathbf{H}_{s_0} \cap \mathbf{F}^{-1}(\mathbf{B}_{.s_1 \cdots s_{n-1}}) = \mathbf{F}^{-1}(\mathbf{V}_{s_0} \cap \mathbf{B}_{.s_1 \cdots s_{n-1}})$$

is a strip of height

$$\tfrac{1}{4}(\text{height of } \mathbf{B}_{.s_1 \cdots s_{n-1}}) = \tfrac{1}{4}\left(\tfrac{1}{4}\right)^{n-1} = \left(\tfrac{1}{4}\right)^n.$$

The union of all these strips gives all of $\mathbf{S}^0_{-n}$,

$$\mathbf{S}^0_{-n} = \bigcup_{s_0, s_1, \ldots, s_{n-1} = 0, 1} \mathbf{B}_{.s_0 \cdots s_{n-1}},$$

and each horizontal strip is labeled with the n symbols that give its present location and the location of the $n-1$ iterates.

We now repeat the process for negative indexes. For any choice of s_{-1}, s_{-2}, $\ldots$, s_{-n}, let

$$\mathbf{B}_{s_{-n} \cdots s_{-1}.} = \bigcap_{j=-n}^{-1} \mathbf{F}^{-j}(\mathbf{H}_{s_j}) = \bigcap_{j=1}^{n} \mathbf{F}^{j}(\mathbf{H}_{s_{-j}})$$
$$= \{\, \mathbf{x} : \mathbf{F}^j(\mathbf{x}) \in \mathbf{H}_{s_j} \text{ for } -n \leq j \leq -1 \,\}.$$

Notice that

$$\mathbf{B}_{s_{-1}.} = \mathbf{F}(\mathbf{H}_{s_{-1}}) = \mathbf{V}_{s_{-1}}$$

is a vertical strip of width $1/3$. Continuing by induction, since $\mathbf{F}$ maps $\mathbf{H}_i$ to $\mathbf{V}_i$ by a contraction in the first coordinate by $1/3$, it follows that

$$\begin{aligned}\mathbf{B}_{s_{-n}s_{-n+1}\cdots s_{-1}.} &= \mathbf{F}\left(\mathbf{H}_{s_{-1}}\right)\cap \mathbf{F}^2\left(\mathbf{H}_{s_{-2}}\right)\cap\cdots\cap \mathbf{F}^n\left(\mathbf{H}_{s_{-n}}\right)\\ &= \mathbf{F}\left(\mathbf{H}_{s_{-1}}\right)\cap \mathbf{F}\left(\mathbf{F}\left(\mathbf{H}_{s_{-2}}\right)\cap\cdots\cap \mathbf{F}^{n-1}(\mathbf{H}_{s_{-n}})\right)\\ &= \mathbf{F}\left(\mathbf{H}_{s_{-1}}\right)\cap \mathbf{F}\left(\mathbf{B}_{s_{-n}\cdots s_{-2}.}\right)\\ &= \mathbf{F}\left(\mathbf{H}_{s_{-1}}\cap \mathbf{B}_{s_{-n}\cdots s_{-2}.}\right)\end{aligned}$$

is a strip of width

$$\tfrac{1}{3}\left(\text{width of } \mathbf{B}_{s_{-n}\cdots s_{-2}.}\right) = \tfrac{1}{3}\left(\tfrac{1}{3}\right)^{n-1} = \left(\tfrac{1}{3}\right)^n.$$

Again, the union of all these vertical strips gives all of $\mathbf{S}_0^n$; that is,

$$\mathbf{S}_n^0 = \bigcup_{s_{-1},\ldots,s_{-n}=0,1} \mathbf{B}_{s_{-n}\cdots s_{-1}.},$$

and each horizontal strip is labeled with the n symbols that give its present location and the location of the first $n-1$ backward iterates. See Figure 4(a).

Combining the forward and backward iterates shows that

$$\mathbf{B}_{s_{-n}s_{-n+1}\cdots s_{-1}.s_0\cdots s_{n-1}} = \bigcap_{j=-n}^{n-1} \mathbf{F}^{-j}(\mathbf{H}_{s_j}) = \mathbf{B}_{s_{-n}s_1\cdots s_{-1}.}\cap \mathbf{B}_{.s_0 s_1\cdots s_{n-1}}$$

is a "box" of size $(1/3)^n$ by $(1/4)^n$. This box includes its boundary, so it is closed in the usual mathematical terminology. See Figure 5.

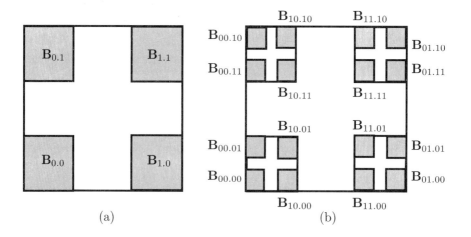

Figure 5. Boxes $\mathbf{B}_{s_{-1}.s_0}$ and Boxes $\mathbf{B}_{s_{-2}s_{-1}.s_0 s_1}$

Taking the limit as n goes to infinity yields

$$\bigcap_{j=-\infty}^{\infty} \mathbf{F}^{-j}(\mathbf{H}_{s_j}) = \bigcap_{n=1}^{\infty} \mathbf{B}_{s_{-n}\cdots s_{-1}.s_0\cdots s_{n-1}}$$

13.1. Geometric Horseshoe

is nonempty because it is the nested intersection of closed bounded sets (compact sets). The sets are nested because

$$\mathbf{B}_{s_{-n-1} \cdots s_{-1}.s_0 \cdots s_n} \subset \mathbf{B}_{s_{-n} \cdots s_{-1}.s_0 \cdots s_{n-1}}.$$

The size goes to zero in both directions as n goes to infinity, so this is a single point. We denote this point by $k(\mathbf{s})$:

$$\{k(\mathbf{s})\} = \bigcap_{j=-\infty}^{\infty} \mathbf{F}^{-j}(\mathbf{H}_{s_j}) = \bigcap_{n=1}^{\infty} \mathbf{B}_{s_{-n} \cdots s_{-1}.s_0 \cdots s_{n-1}}.$$

This map k goes from the symbol space Σ_2 into the invariant set Λ in the plane; it is the inverse of the itinerary map h defined earlier. If $\mathbf{x}$ is a point in Λ and $\mathbf{s} = h(\mathbf{x})$, then the point $\mathbf{x}$ is in $\bigcap_{j=-\infty}^{\infty} \mathbf{F}^{-j}(\mathbf{H}_{s_j})$, so, by uniqueness, $\mathbf{x} = k(\mathbf{s})$, or

$$k \circ h(\mathbf{x}) = \mathbf{x} \quad \text{and}$$
$$h \circ k(\mathbf{s}) = \mathbf{s}.$$

This shows that both maps are one to one and onto.

The next theorem summarizes the properties of the maps h and k.

Theorem 13.3. *Let $\mathbf{F}$ be the map for the geometric horseshoe with invariant set Λ described previously. Then the following properties are true:*

a. *For any bi-infinite sequence $\mathbf{s}$ in Σ_2, the intersection $\bigcap_{j=-\infty}^{\infty} \mathbf{F}^{-j}(\mathbf{H}_{s_j})$ is a unique point that we denote by $k(\mathbf{s})$. With the distance on Σ_2 defined defined in this section, k is a continuous map from the symbol space Σ_2 into the invariant set Λ in the plane. Also, the itinerary h is continuous from Λ to Σ_2.*

b. *The maps h and k are inverses of each other, $k \circ h = \text{id}$ and $h \circ k = \text{id}$. Moreover, (i) both k and h are one to one, (ii) the map h is onto Σ_2 and k is onto Λ, and (iii) the maps h and k are both continuous, so they are homeomorphisms.*

c. *The map k is a conjugacy, $\mathbf{F} \circ k = k \circ \sigma$. So also, $h \circ \mathbf{F} = \sigma \circ h$.*

d. *A symbol $\mathbf{s}$ is periodic if and only if $k(\mathbf{s})$ is periodic for $\mathbf{F}$.*

e. *There is a point $\mathbf{x}^*$ in Λ such that both the forward orbit $\mathcal{O}_{\mathbf{F}}^+(\mathbf{x}^*)$ and the backward orbit $\mathcal{O}_{\mathbf{F}}^-(\mathbf{x}^*)$ are dense in Λ.*

f. *The map $\mathbf{F}$ has sensitive dependence on initial conditions when restricted to Λ.*

Proof. Parts (a) – (d) and (f) follow by the construction of the symbolic dynamics and the treatment in one dimension. (Compare with Theorems 10.11 and 10.18.) Parts (e) and (f) follow from Theorem 13.2. □

Stable and Unstable Manifolds

Before leaving this example, we want to discuss the stable and unstable manifolds of points. For points in $\Lambda \subset \mathbf{S}$, the derivative is of the form

$$D\mathbf{F}_{(x,y)} = \begin{pmatrix} \pm\frac{1}{3} & 0 \\ 0 & \pm 4 \end{pmatrix}.$$

Since these matrices all have the same form and are diagonal, the derivative of an iterate is also of a simple form:

$$D\mathbf{F}^k_{(x,y)} = \begin{pmatrix} \pm\frac{1}{3^k} & 0 \\ 0 & \pm 4^k \end{pmatrix}.$$

Therefore, all points in a horizontal line segment through a point $\mathbf{p} = (p_1, p_2)$ in Λ are contracted under forward iterates, and

$$W^s(\mathbf{p}, \mathbf{F}) \supset \left[-\frac{1}{16}, \frac{17}{16}\right] \times \{p_2\}.$$

This line segment is a piece of the stable manifold even though it is not of a specified radius; so, we just call it a *local stable manifold* and use the notation

$$W^s_{\text{loc}}(\mathbf{p}, \mathbf{F}) = \left[-\frac{1}{16}, \frac{17}{16}\right] \times \{p_2\}.$$

Using backward iterates, the vertical line segments are contracted, so a *local unstable manifold* through a point $\mathbf{p}$ in Λ is a vertical line segment,

$$W^u_{\text{loc}}(\mathbf{p}, \mathbf{F}) = \{p_1\} \times \left[-\frac{1}{16}, \frac{17}{16}\right].$$

We denote the unstable manifold of the invariant set by the union of the unstable manifolds of points in Λ as

$$W^u(\Lambda, \mathbf{F}) = \bigcup_{\mathbf{x} \in \Lambda} W^u(\mathbf{x}, \mathbf{F}).$$

If we use the trapping region $\mathbf{U} = \mathbf{S}' \cup \mathbf{E}_0 \cup \mathbf{E}_1$, then

$$\bigcap_{j=0}^{\infty} \mathbf{F}^j(\mathbf{U}) = W^u(\Lambda, \mathbf{F}) \cup \{\mathbf{p}_0\},$$

where $\mathbf{p}_0$ is the attracting fixed point in $\mathbf{E}_0$. (See Sections 11.2 and 13.4 for a discussion of trapping regions.) Clearly, the local unstable manifolds of points in Λ are in the neighborhood $\mathbf{U}$. Since this trapping region is positively invariant, all of the unstable manifolds are contained in all forward iterates of $\mathbf{U}$ and hence are also contained in the intersection. Other points in $\mathbf{U}$ are in the stable manifold of either a point in Λ or of $\mathbf{p}_0$. This set is an attracting set, but is not indecomposable, since both Λ and $\mathbf{p}_0$ are proper invariant subsets with isolating neighborhoods of the type specified in the definition of indecomposable given in Section 13.4.

13.1.1. Basin Boundaries. Section 12.3.2 discusses a simple example in which the boundary of the basin of an attracting fixed point is the stable manifold of a saddle fixed point. In this section, we give an example in which the boundary between the basin of two attracting fixed points is made up of the stable manifold of a horseshoe, so the basin boundary is much more complicated.

The map considered is like the geometric horseshoe in which a box is mapped across itself three times. Let $\mathbf{F}$ be a map that takes the box $\mathbf{S}$ across itself three times as given in Figure 6. We extend the map to regions $\mathbf{E}_1$ below $\mathbf{S}$ and $\mathbf{E}_2$ above $\mathbf{S}$ so that $\mathbf{F}$ contracts each $\mathbf{E}_j$ into itself. The map can be specified so that each $\mathbf{E}_j$

13.1. Geometric Horseshoe

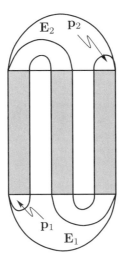

Figure 6. Horseshoe with three strips

contains a fixed point sink $\mathbf{p}_j$ and $W^s(\mathbf{p}_j; \mathbf{F}) \supset \mathbf{E}_j$. The set of points in $\mathbf{S}$ whose first iterates are also in $\mathbf{S}$ consists of three horizontal substrips, so that

$$\mathbf{S} \cap \mathbf{F}^{-1}(\mathbf{S}) = \mathbf{H}_1 \cup \mathbf{H}_2 \cup \mathbf{H}_3.$$

These horizontal strips are separated from each other by two gaps $\mathbf{G}_1$ and $\mathbf{G}_2$ which are labeled so that $\mathbf{F}(\mathbf{G}_j) \subset \mathbf{E}_j$. See Figure 7.

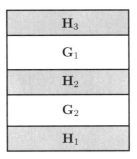

Figure 7. Horizontal strips and gaps

The set of points in $\mathbf{S}$ whose first two iterates are in $\mathbf{S}$ is made of 3^2 horizontal strips separated by

$$\mathbf{G}_1 \cup \mathbf{G}_2 \cup \left(\mathbf{H}_1 \cap \mathbf{F}^{-1}(\mathbf{G}_1)\right) \cup \left(\mathbf{H}_1 \cap \mathbf{F}^{-1}(\mathbf{G}_2)\right) \cup \left(\mathbf{H}_2 \cap \mathbf{F}^{-1}(\mathbf{G}_1)\right)$$
$$\cup \left(\mathbf{H}_2 \cap \mathbf{F}^{-1}(\mathbf{G}_2)\right) \cup \left(\mathbf{H}_3 \cap \mathbf{F}^{-1}(\mathbf{G}_1)\right) \cup \left(\mathbf{H}_3 \cap \mathbf{F}^{-1}(\mathbf{G}_2)\right).$$

The intersection of all the backward iterates gives the set of points, all of whose forward iterates stay in $\mathbf{S}$; that is,

$$\bigcap_{j=-\infty}^{0} \mathbf{F}^j(\mathbf{S}) = [0,1] \times \mathbf{C}_2,$$

where $\mathbf{C}_2$ is a Cantor set created using three subintervals at each stage. The forward iterates of $\mathbf{S}$ give vertical strips of points whose backward iterates stay in $\mathbf{S}$, or

$$\bigcap_{j=0}^{\infty} \mathbf{F}^j(\mathbf{S}) = \mathbf{C}_1 \times [0,1],$$

where $\mathbf{C}_1$ is a Cantor set created using three subintervals at each stage. Then,

$$\Lambda = \bigcap_{j=-\infty}^{\infty} \mathbf{F}^j(\mathbf{S}) = \mathbf{C}_1 \times \mathbf{C}_2$$

is the set of points that stay in $\mathbf{S}$ for all iterates.

The local stable manifolds of points in Λ are horizontal intervals $[0,1] \times \{y\}$. If we denote the stable manifold of the invariant set by

$$W^s(\Lambda, \mathbf{F}) = \bigcup_{\mathbf{x} \in \Lambda} W^s(\mathbf{x}, \mathbf{F}),$$

then

$$W^s(\Lambda) \cap \mathbf{S} = [0,1] \times \mathbf{C}_2 = \bigcap_{j=-\infty}^{0} \mathbf{F}^j(\mathbf{S}).$$

The basin of $\mathbf{p}_i$ contains all the points in $\mathbf{E}_i$. Since $\mathbf{G}_i$ maps into $\mathbf{E}_i$, it follows that

$$\mathbf{E}_i \cup \mathbf{G}_i \subset W^s(\mathbf{p}_i).$$

This basin also includes all the points that map into $\mathbf{E}_i \cup \mathbf{G}_i$, so it includes points in

$$\mathbf{F}^{-k}\left(\mathbf{E}_i \cup \mathbf{G}_i\right) \cap \bigcap_{j=0}^{k-1} \mathbf{F}^{-j}\left(\mathbf{H}_{s_j}\right),$$

where each s_j is a 1, 2, or 3. Since the sets

$$\mathbf{F}^{-k}\left(\mathbf{E}_1 \cup \mathbf{G}_1\right) \cap \bigcap_{j=0}^{k-1} \mathbf{F}^{-j}(\mathbf{H}_{s_j}) \quad \text{and}$$

$$\mathbf{F}^{-k}\left(\mathbf{E}_2 \cup \mathbf{G}_2\right) \cap \bigcap_{j=0}^{k-1} \mathbf{F}^{-j}(\mathbf{H}_{s_j})$$

are on either side of

$$\mathbf{F}^{-k}\left(\mathbf{H}_2\right) \cap \bigcap_{j=0}^{k-1} \mathbf{F}^{-j}(\mathbf{H}_{s_j}),$$

each of the basins $W^s(\mathbf{p}_i)$ has $W^s(\Lambda) \cap \mathbf{S}$ in its boundary:

$$\mathrm{bd}(W^s(\mathbf{p}_1)) \cap \mathrm{bd}(W^s(\mathbf{p}_2)) \supset W^s(\Lambda) \cap \mathbf{S} = [0,1] \times C_2.$$

Thus, the stable manifold of Λ, which contains a Cantor set of curves, forms the boundary between the basins of the two sinks and separates points that tend toward one sink from points that tend toward the other sink. Therefore, it may be difficult to actually land on the horseshoe invariant set or its stable manifold, but it still is observable as the boundary between the basins of one sink and another.

Exercises 13.1

1. Consider the geometric horseshoe map given in the text.
   a. Find the coordinates in the plane of the two fixed points. Hint: Consider the two different definitions of $\mathbf{F}$ in $\mathbf{H}_0$ and $\mathbf{H}_1$:
   $$\mathbf{F}_0(x,y) = \left(\frac{1}{3}x, 4y\right) \quad \text{for } (x,y) \in \mathbf{H}_0,$$
   $$\mathbf{F}_1(x,y) = \left(1 - \frac{1}{3}x, 4 - 4y\right) \quad \text{for } (x,y) \in \mathbf{H}_1.$$
   So, one fixed point satisfies $\mathbf{F}_0(x,y) = (x,y)$, and the other fixed point satisfies $\mathbf{F}_1(x,y) = (x,y)$.
   b. Find the points on the period-2 orbit. Hint: One point $\mathbf{q}$ on the orbit satisfies $\mathbf{F}_1 \circ \mathbf{F}_0(x,y) = (x,y)$, and the other point is $\mathbf{F}_0(\mathbf{q})$.

2. Consider the geometric horseshoe map given in Section 13.1. Describe the part of the stable and unstable manifolds $W^s(\mathbf{0}, \mathbf{F}) \cap \mathbf{S}'$ and $W^u(\mathbf{0}, \mathbf{F}) \cap \mathbf{S}'$, that are in $\mathbf{S}'$. Note that this is not just the local stable and unstable manifold, but the whole manifold.

3. Consider the map $\mathbf{G}$ from $\mathbb{R}^2$ to itself given by the formulas
$$\mathbf{G}(x,y) = \begin{cases} \left(\frac{1}{3}x, 3y\right) & \text{for } y < \frac{1}{2}, \\ \left(\frac{2}{3} + \frac{1}{3}x, 3y - 2\right) & \text{for } y \geq \frac{1}{2}. \end{cases}$$
Notice that $\mathbf{G}$ is discontinuous along $y = 1/2$.
   a. Let $\mathbf{S} = [0,1] \times [0,1]$ and $\mathbf{S}_m^n = \bigcap_{j=m}^n \mathbf{G}^j(\mathbf{S})$. Describe the sets $\mathbf{S}_0^1$, $\mathbf{S}_0^2$, $\mathbf{S}_0^n$, $\mathbf{S}_0^\infty = \bigcap_{j=0}^\infty \mathbf{G}^j(\mathbf{S})$, $\mathbf{S}_{-1}^0$, $\mathbf{S}_{-2}^0$, $\mathbf{S}_{-n}^0$, $\mathbf{S}_{-\infty}^0 = \bigcap_{j=-\infty}^0 \mathbf{G}^j(\mathbf{S})$, and $\mathbf{S}_{-\infty}^\infty = \bigcap_{j=-\infty}^\infty \mathbf{G}^j(\mathbf{S})$.
   b. How does this example differ from the geometric horseshoe?

4. Let $\mathbf{S} = [0,1] \times [0,1]$ be the unit square. Consider the map $\mathbf{G}$ from $\mathbf{S}$ into $\mathbb{R}^2$ given by the formulas
$$\mathbf{G}(x,y) = \begin{cases} \left(\frac{1}{5}x, 5y\right) & \text{for } 0 \leq y < \frac{1}{3}, \\ \left(\frac{2}{5} + \frac{1}{5}x, 5y - 2\right) & \text{for } \frac{1}{3} \leq y < \frac{2}{3}, \\ \left(\frac{4}{5} + \frac{1}{5}x, 5y - 4\right) & \text{for } \frac{2}{3} \leq y \leq 1. \end{cases}$$
Notice that $\mathbf{G}$ is discontinuous along $y = 1/3$ and $2/3$.
   a. Describe the sets $\mathbf{S}_0^1 = \bigcap_{j=0}^1 \mathbf{G}^j(\mathbf{S})$, $\mathbf{S}_0^2 = \bigcap_{j=0}^2 \mathbf{G}^j(\mathbf{S})$, $\mathbf{S}_0^n = \bigcap_{j=0}^n \mathbf{G}^j(\mathbf{S})$, $\mathbf{S}_0^\infty = \bigcap_{j=0}^\infty \mathbf{G}^j(\mathbf{S})$, $\mathbf{S}_{-1}^0 = \bigcap_{j=-1}^0 \mathbf{G}^j(\mathbf{S})$, $\mathbf{S}_{-2}^0 = \bigcap_{j=-2}^0 \mathbf{G}^j(\mathbf{S})$, $\mathbf{S}_{-n}^0 = \bigcap_{j=-n}^0 \mathbf{G}^j(\mathbf{S})$, $\mathbf{S}_{-\infty}^0 = \bigcap_{j=-\infty}^0 \mathbf{G}^j(\mathbf{S})$, and $\mathbf{S}_{-\infty}^\infty = \bigcap_{j=-\infty}^\infty \mathbf{G}^j(\mathbf{S})$.
   b. How does this example differ from the geometric horseshoe?

5. Explain more fully why the geometric horseshoe has a point $\mathbf{x}^*$ with both $\mathcal{O}_\mathbf{F}^+(\mathbf{x}^*)$ and $\mathcal{O}_\mathbf{F}^-(\mathbf{x}^*)$ being dense in Λ.

6. Consider the map $\mathbf{F}$ for the geometric horseshoe. Let $\mathbf{L}^s = W_{\text{loc}}^s(\mathbf{0})$ and $\mathbf{L}^u = W_{\text{loc}}^u(\mathbf{0})$ be the local stable and unstable manifolds of the fixed point at the origin. Draw the images of $\mathbf{L}^u$ by $\mathbf{F}^3$ and $\mathbf{L}^s$ by $\mathbf{F}^{-3}$.

13.2. Symbolic Dynamics

The goal of this section is to identify the essential properties of the manner in which the rectangles $\mathbf{H}_i$ are mapped across each other for the geometric horseshoe, so that we can get symbolic dynamics for any nonlinear maps satisfying them. A collection of rectangles with these essential properties is called a Markov partition.

A Markov partition in one dimension is a collection of closed intervals $\{\mathbf{I}_j\}_{j=1}^k$ such that, if $f(\text{int}(\mathbf{I}_i)) \cap \text{int}(\mathbf{I}_j) = \emptyset$, then $f(\mathbf{I}_i) \supset \mathbf{I}_j$. A Markov partition in two dimensions is a collection of closed sets $\{\mathbf{R}_j\}_{j=1}^k$ such that, if $f(\text{int}(\mathbf{R}_i)) \cap \text{int}(\mathbf{R}_j) = \emptyset$, then there is an interval in one direction in $\mathbf{R}_i$ that maps across the interval in the corresponding direction in $\mathbf{R}_j$, and there is an interval in a second direction in $\mathbf{R}_i$ that maps inside the interval in the second direction in $\mathbf{R}_j$. Often these conditions are given in terms of the derivative of the nonlinear map and a condition called *hyperbolicity* that generalizes a saddle fixed point. We give the condition in terms of just the features of the function itself.

When there is one rectangle, we want to find at least one point that stays in $\mathbf{R}$ for all iterates, and we want to know that there is a fixed point in $\mathbf{R}$. For a sequence of rectangles, we want to find at least one point that goes through the sequence of rectangles $\mathbf{R}_j$; if the sequence of rectangles $\mathbf{R}_j$ is periodic, we want to find at least one periodic point that goes through the sequence of rectangles.

In the first subsection, we use model rectangles for which the first coordinate is the inflowing direction and the second coordinate is the overflowing coordinate. In the second subsection, we consider images of these model rectangles by homeomorphisms. This latter context is the one that is needed when applying the concepts to nonlinear maps. Also, the emphasis of the first subsection is on finding fixed points, while in the second subsection we use the methods to induce symbolic dynamics for an invariant set.

13.2.1. Correctly Aligned Rectangle. We start by giving the definitions in two dimensions and then consider higher dimensions briefly.

Two dimensions

We use model rectangles for which the first coordinate is the inflowing direction and the second coordinate is the overflowing coordinate. This first definition is long and involves the compatibility of the map and properties of the rectangle chosen. We write the conditions in the case in which the first coordinate is inflowing and the second coordinate is overflowing. The role of these coordinates can be reversed as follows from the treatment of Markov partitions.

Definition 13.1. Assume that $\mathbf{F}$ is a continuous map from $\mathbb{R}^2$ onto itself, with a continuous inverse (i.e., $\mathbf{F}$ is a homeomorphism of $\mathbb{R}^2$). The image by $\mathbf{F}$ of a rectangle

$$\mathbf{R} = [a_1, b_1] \times [a_2, b_2]$$

is said to be *correctly aligned* with itself provided that there is a choice of overflowing and inflowing directions such that the four conditions that follow are satisfied. See Figure 8. To indicate that choices of overflowing and inflowing directions have been made, we call $\mathbf{R}$ an *M-rectangle*.

13.2. Symbolic Dynamics

We separate the boundary $\mathrm{bd}(\mathbf{R})$ of $\mathbf{R}$ into the sides that are assumed to map "inside" $\mathbf{R}$,
$$\partial^{\mathrm{in}}(\mathbf{R}) = (\{a_1\} \times [a_2, b_2]) \cup (\{b_1\} \times [a_2, b_2])$$
and the top and bottom edges that are assumed to map "outside" $\mathbf{R}$,
$$\partial^{\mathrm{out}}(\mathbf{R}) = ([a_1, b_1] \times \{a_2\}) \cup ([a_1, b_1] \times \{b_2\}).$$
The four conditions are as follows:

a. The intersection
$$\mathbf{F}(\mathrm{int}(\mathbf{R})) \cap \mathrm{int}(\mathbf{R})$$
is one connected piece.

b. The image by $\mathbf{F}$ of the interior of $\mathbf{R}$ does not intersect incoming sides $\partial^{\mathrm{in}}(\mathbf{R})$,
$$\mathbf{F}(\mathrm{int}(\mathbf{R})) \cap \partial^{\mathrm{in}}(\mathbf{R}) = \emptyset.$$

c. The image $\mathbf{F}(\partial^{\mathrm{out}}(\mathbf{R}))$ of the outgoing top and bottom does not intersect the interior of $\mathbf{R}$,
$$\mathbf{F}(\partial^{\mathrm{out}}(\mathbf{R})) \cap \mathrm{int}(\mathbf{R}) = \emptyset.$$

d. For any x in $[a_1, b_1]$, the image of each vertical line segment $\{x\} \times [a_2, b_2]$ stretches from the bottom to the top of $\mathbf{R}$, separating the two sides; in particular, there is a subinterval $[a'_x, b'_x] \subset [a_2, b_2]$ depending on x such that
$$\mathbf{F}(\{x\} \times [a'_x, b'_x]) \subset \mathbf{R},$$
$$\mathbf{F}(x, a'_x) \in \partial^{\mathrm{out}}(\mathbf{R}), \quad \text{and}$$
$$\mathbf{F}(x, b'_x) \in \partial^{\mathrm{out}}(\mathbf{R}),$$
where the two image points $\mathbf{F}(x, a'_x)$ and $\mathbf{F}(x, b'_x)$ are in opposite ends of $\mathbf{R}$. Thus, we assume that $\mathbf{F}$ takes the overflowing direction of $\mathbf{R}$ across $\mathbf{R}$ in the overflowing direction.

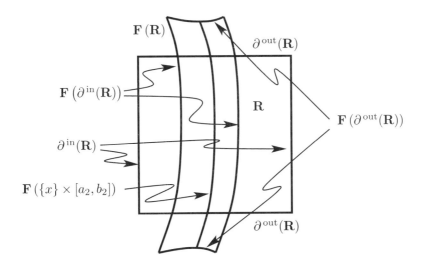

Figure 8. Aligned image of an M-rectangle $\mathbf{R}$ with itself

Remark 13.2. In the definition, we use the interiors of the rectangles, just as we used the open intervals in the definition in one dimension. A more straightforward assumption to replace (a) would be the stronger assumption that

a1. the intersection $\mathbf{F}(\mathbf{R}) \cap \mathbf{R}$ is one connected piece.

However, in our consideration of hyperbolic toral automorphisms and the solenoid, we can use fewer M-rectangles if we use the weaker assumption given in the definition.

Remark 13.3. If the image of $\mathbf{R}$ by $\mathbf{F}$ is correctly aligned with itself, it follows that the image of $\mathbf{R}$ by $\mathbf{F}^{-1}$ is correctly aligned with itself.

a'. The intersection
$$\mathbf{F}^{-1}\left(\operatorname{int}(\mathbf{R})\right) \cap \operatorname{int}(\mathbf{R})$$
is one connected piece.

b'. The interior of the inverse image $\mathbf{F}^{-1}(\mathbf{R})$ does not intersect the outgoing top or bottom of $\mathbf{R}$,
$$\operatorname{int}\left(\mathbf{F}^{-1}(\mathbf{R})\right) \cap \partial^{\operatorname{out}}(\mathbf{R}) = \emptyset.$$

c'. The inverse image $\mathbf{F}^{-1}(\partial^{\operatorname{in}}(\mathbf{R}))$ of the incoming sides does not intersect the interior of the previous M-rectangle $\mathbf{R}$,
$$\mathbf{F}^{-1}\left(\partial^{\operatorname{in}}(\mathbf{R})\right) \cap \operatorname{int}(\mathbf{R}) = \emptyset.$$

d'. For any y in $[a_2, b_2]$, the image by $\mathbf{F}^{-1}$ of the horizontal line segment $[a_1, b_1] \times \{y\}$ stretches from the right side to the left side of $\mathbf{R}$, separating the top from the bottom; in particular, there is a subinterval $[a'_y, b'_y]$ of $[a_1, b_1]$ depending on y such that
$$\mathbf{F}^{-1}\left([a'_y, b'_y] \times \{y\}\right) \subset \mathbf{R},$$
$$\mathbf{F}^{-1}\left(a'_y, y\right) \in \partial^{\operatorname{in}}(\mathbf{R}), \qquad \text{and}$$
$$\mathbf{F}^{-1}\left(b'_y, y\right) \in \partial^{\operatorname{in}}(\mathbf{R}),$$

where the two inverse image points are in opposite sides of $\mathbf{R}$.

Now, we assume that the image of an M-rectangle $\mathbf{R}$ is correctly aligned with itself. In this discussion, we make assumption (a1) rather than (a), so we can avoid using interiors and closures. In the next subsection on Markov partitions, we return to the more general assumption. The second iterate of $\mathbf{R}$ is a vertical strip that still separates the two sides:

$$\bigcap_{j=0}^{2} \mathbf{F}^j(\mathbf{R}) = \mathbf{R} \cap \mathbf{F}(\mathbf{R}) \cap \mathbf{F}^2(\mathbf{R}) = \mathbf{R} \cap \mathbf{F}\left(\mathbf{R} \cap \mathbf{F}(\mathbf{R})\right).$$

Continuing by induction shows that

$$\bigcap_{j=0}^{n} \mathbf{F}^j(\mathbf{R}) = \mathbf{R} \cap \mathbf{F}\left(\bigcap_{j=0}^{n-1} \mathbf{F}^j(\mathbf{R})\right)$$

13.2. Symbolic Dynamics

is a vertical strip that separates the two sides, and so does

$$\bigcap_{j=0}^{\infty} \mathbf{F}^j(\mathbf{R}) = \bigcap_{n=0}^{\infty} \bigcap_{j=0}^{n} \mathbf{F}^j(\mathbf{R}).$$

This infinite intersection is often just a curve connecting the top of **R** to the bottom.

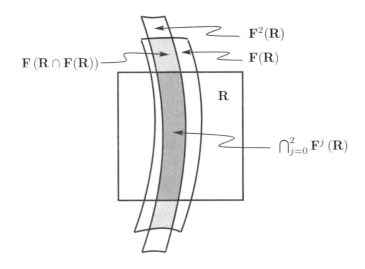

Figure 9. The intersection $\bigcap_{j=0}^{2} \mathbf{F}^j(\mathbf{R})$

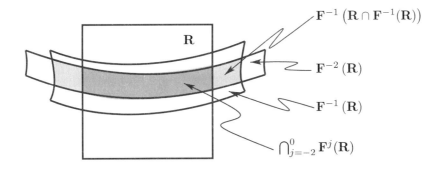

Figure 10. The intersection $\bigcap_{j=-2}^{0} \mathbf{F}^j(\mathbf{R})$

Taking backward iterates, $\mathbf{R} \cap \mathbf{F}^{-1}(\mathbf{R})$ is a horizontal strip that separates the top from the bottom. The second backward iterate is a horizontal strip that still separates the top from the bottom, or

$$\bigcap_{j=-2}^{0} \mathbf{F}^j(\mathbf{R}) = \mathbf{R} \cap \mathbf{F}^{-1}(\mathbf{R}) \cap \mathbf{F}^{-2}(\mathbf{R}) = \mathbf{R} \cap \mathbf{F}^{-1}(\mathbf{R} \cap \mathbf{F}^{-1}(\mathbf{R})).$$

Continuing by induction, we see that

$$\bigcap_{j=-n}^{0} \mathbf{F}^j(\mathbf{R}) = \mathbf{R} \cap \mathbf{F}^{-1}\left(\bigcap_{j=-n+1}^{0} \mathbf{F}^j(\mathbf{R})\right)$$

is a horizontal strip that separates the top from the bottom. The infinite intersection,

$$\bigcap_{j=-\infty}^{0} \mathbf{F}^j(\mathbf{R}) = \bigcap_{n=0}^{\infty} \bigcap_{j=-n}^{0} \mathbf{F}^j(\mathbf{R}),$$

also separates the top of $\mathbf{R}$ from the bottom and, in the simplest cases, is a curve connecting the two sides. See Figure 11.

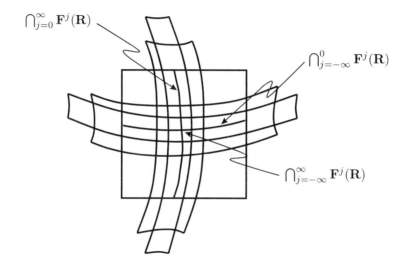

Figure 11. The intersection $\bigcap_{j=-\infty}^{\infty} \mathbf{F}^{(}\mathbf{R})$

Because $\bigcap_{j=-\infty}^{0} \mathbf{F}^j(\mathbf{R})$ separates the top from the bottom and $\bigcap_{j=0}^{\infty} \mathbf{F}^j(\mathbf{R})$ separates the two sides, they must intersect and $\bigcap_{j=-\infty}^{\infty} \mathbf{F}^j(\mathbf{R})$ is nonempty. See Figure 11. We summarize this result in the next theorem.

Theorem 13.4. *Assume that* $\mathbf{F}$ *is a homeomorphism of* $\mathbb{R}^2$ *and the image of an M-rectangle* $\mathbf{R}$ *is correctly aligned with itself. Then, the intersection*

$$\bigcap_{j=-\infty}^{\infty} \mathbf{F}^j(\mathbf{R})$$

is nonempty. Thus, there is at least one point $\mathbf{x}_0$ *such that* $\mathbf{F}^j(\mathbf{x}_0)$ *is in* $\mathbf{R}$ *for all* j.

Fixed points in two dimensions

We would like to know that there is a fixed point in the nonempty intersection guaranteed by Theorem 13.4. There are some general results in topology which insure that this is true. (These sets satisfy the Lefschetz property.) There are also some arguments using conditions on the derivatives of **F**, such as those given in the proof of the stable manifold theorem, which prove that the intersection is a single point, and therefore a fixed point. These assumptions involve the fact that the map **F** is a contraction in the first coordinate and an expansion in the second coordinate. See [**Rob99**] for a detailed presentation of these ideas. We give a topological argument that we only describe in two dimensions. We consider the manner in which the displacement vector $\mathbf{x} - \mathbf{F}(\mathbf{x})$ winds around as the point $\mathbf{x}$ varies around the boundary of an M-rectangle.

Definition 13.4. Assume that **F** is a map on $\mathbb{R}^2$ and that **R** is a region in the plane such that the following properties hold: (i) The boundary of **F**, bd(**R**), is made up of a single closed curve, and (ii) the map **F** has no fixed points on the boundary of **R**, $\mathbf{F}(\mathbf{x}) \neq \mathbf{x}$ for **x** in bd(**R**). Define an induced map **G** from the boundary bd(**R**) to the unit circle by

$$\mathbf{G}(\mathbf{x}) = \frac{\mathbf{x} - \mathbf{F}(\mathbf{x})}{\|\mathbf{x} - \mathbf{F}(\mathbf{x})\|}.$$

As the point **x** goes around the boundary in a counterclockwise direction, measure the increase in the angle of $\mathbf{G}(\mathbf{x})$. It has to return to the original angle plus some integer i multiple of 2π. Call this integer i the *index of* **F** *for the region* **R**. This integer can be positive, zero, or negative, depending on whether the displacement vector winds around in the same direction as the boundary is traversed, has no net rotation, or winds around in the opposite direction as the boundary is traversed.

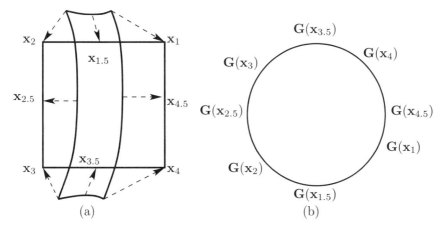

Figure 12. Index of an M-rectangle correctly aligned with itself. (a) The vectors $\mathbf{x} - \mathbf{F}(\mathbf{x})$. (b) The vectors $\mathbf{G}(\mathbf{x})$.

Consider the image of an M-rectangle that is correctly aligned with itself as given in Figure 12(a). In the figure, the corners of the rectangle are labeled $\mathbf{x}_1$, $\mathbf{x}_2$, $\mathbf{x}_3$, and $\mathbf{x}_4$. A point on the edge between $\mathbf{x}_1$ and $\mathbf{x}_2$ is labeled with a number

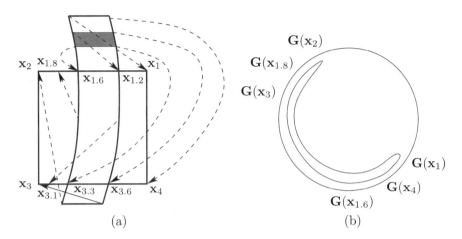

Figure 13. Index of an M-rectangle not-correctly aligned with itself. (a) The vectors $\mathbf{x} - \mathbf{F}(\mathbf{x})$. (b) The vectors $\mathbf{G}(\mathbf{x})$.

between 1 and 2 (e.g., $\mathbf{x}_{1.5}$). Part (b) of the figure shows the images of these points by the index map $\mathbf{G}$. Since the boundary of the rectangle is transversed in a counterclockwise direction and the image of the boundary is transversed once in a clockwise direction, the index is -1. The index is still nonzero even if there is a flip in some of the coordinates. For a map with a source having eigenvalues $\lambda_1 > \lambda_2 > 1$, the index is 1.

It is possible for the image of a rectangle to have the same shape as one that is correctly aligned, but have index zero and not be correctly aligned. For the map indicated in Figure 13, the index is zero, since the image of $\mathbf{G}$ does not go all the way around the circle. The map need not have a fixed point in $\mathbf{R}$. The map for Figure 13(a) is taken so that the image of $\mathbf{R} \cap \mathbf{F}(\mathbf{R})$ is the shaded area. Since the image of $\mathbf{R} \cap \mathbf{F}(\mathbf{R})$ is disjoint from itself, there can be no fixed points. Also,

$$\mathbf{R} \cap \mathbf{F}(\mathbf{R}) \cap \mathbf{F}^2(\mathbf{R}) = \emptyset,$$

so no orbits stay in the rectangle for all iterates. Such a map is

$$\mathbf{F}(x, y) = \left(-\frac{1}{2} y, 8x + 6\right),$$

with M-rectangle $[-1, 1] \times [-1, 1]$. Its only fixed point is $(-3/5, 6/5)$, which is outside $[-1, 1] \times [-1, 1]$.

The next result holds for any M-rectangle whose image is correctly aligned with itself.

Theorem 13.5. *Let $\mathbf{F}$ by a homeomorphism of $\mathbb{R}^2$ that has an M-rectangle $\mathbf{R}$ whose image is correctly aligned with itself. Then, either (i) $\mathbf{F}$ has a fixed point on the boundary of $\mathbf{R}$ (and the index is not defined), or (ii) the index of $\mathbf{R}$ for $\mathbf{F}$ is nonzero.*

The next theorem relates the index and a fixed point in $\mathbf{R}$.

Theorem 13.6. *Assume that $\mathbf{F}$ is a map of $\mathbb{R}^2$ with a nonzero index for a region $\mathbf{R}$ whose boundary is a single closed curve. Then, $\mathbf{F}$ has a fixed point in $\mathbf{R}$.*

13.2. Symbolic Dynamics

In particular, if the image of an M-rectangle $\mathbf{R}$ is correctly aligned with itself by $\mathbf{F}$, then $\mathbf{F}$ has a fixed point in $\mathbf{R}$.

Proof. We assume that there is no fixed point in $\mathbf{R}$ and get a contradiction.

For an M-rectangle, we can shrink the boundary of $\mathbf{R}$ to a point. Assume that $\gamma_s(t)$ is a one-parameter family of closed curves in $\mathbf{R}$ such that $\gamma_0(t)$ goes around the boundary of $\mathbf{R}$ in a counterclockwise direction as t varies from 0 to 2π, $\gamma_s(0) = \gamma_s(2\pi)$ for all $0 \leq s \leq 1$, and $\gamma_1(t)$ is a single point in $\mathbf{R}$, independent of t. If there are no fixed points in $\mathbf{R}$, then for each value of s with $0 \leq s \leq 1$, we can define the map

$$\mathbf{G}_s(t) = \frac{\gamma_s(t) - \mathbf{F}(\gamma_s(t))}{\|\gamma_s(t) - \mathbf{F}(\gamma_s(t))\|}$$

and

$$\theta_s(t) = \mathrm{angle}(\mathbf{G}_s(t)).$$

By the definition of the index, $\theta_0(2\pi) - \theta_0(0) = k\,2\pi$, where k is the index. As s varies from 0 to 1, $\dfrac{\theta_s(2\pi)}{2\pi}$ must remain an integer, so it must remain equal to $k \neq 0$. However, $\gamma_1(t)$ does not move, so $\mathbf{G}_1(t)$ is a constant and $\theta_1(2\pi) - \theta_1(0) = 0$. This contradiction proves that $\mathbf{F}$ must have a fixed point in $\mathbf{R}$. $\square$

Higher dimensions

We now turn to stating the definition of a correctly aligned M-rectangle in higher dimensions. Given a homeomorphism, the choice of an M-rectangle depends on the number of overflowing and inflowing directions. We also use round balls about the origin in the overflowing and inflowing subspaces, but these could be replaced by rectangular regions. (Also, in their use in Markov partitions, we use homeomorphic images of this standard M-rectangle.) For an integer k and $r > 0$, let

$$\bar{\mathbf{B}}^k(\mathbf{0}, r) = \{\, \mathbf{x} \in \mathbb{R}^k : \|\mathbf{x}\| \leq r \,\} \quad \text{and}$$
$$\mathrm{bd}\left(\bar{\mathbf{B}}^k(\mathbf{0}, r)\right) = \{\, \mathbf{x} \in \mathbb{R}^k : \|\mathbf{x}\| = r \,\}$$

be the closed ball and its boundary in $\mathbb{R}^k$.

Definition 13.5. We assume that $\mathbf{F}$ is a homeomorphism map from $\mathbb{R}^n$ onto itself. The image by $\mathbf{F}$ of an M-rectangle, $\mathbf{R} = \bar{\mathbf{B}}^{n_1}(\mathbf{0}, r_1) \times \bar{\mathbf{B}}^{n_2}(\mathbf{0}, r_2)$, where $n_1 + n_2 = n$, is *correctly aligned* with itself, provided that the four conditions listed momentarily are satisfied. We separate the boundary $\mathrm{bd}(\mathbf{R})$ of $\mathbf{R}$ into the part that maps "inside" $\mathbf{R}$,

$$\partial^{\mathrm{in}}(\mathbf{R}) = \mathrm{bd}\left(\bar{\mathbf{B}}^{n_1}(\mathbf{0}, r_1)\right) \times \bar{\mathbf{B}}^{n_2}(\mathbf{0}, r_2),$$

and the part that maps "outside" $\mathbf{R}$,

$$\partial^{\mathrm{out}}(\mathbf{R}) = \bar{\mathbf{B}}^{n_1}(\mathbf{0}, r_1) \times \mathrm{bd}\left(\bar{\mathbf{B}}^{n_2}(\mathbf{0}, r_2)\right).$$

The four conditions are the following:

a. The intersection
$$\mathbf{F}\left(\mathrm{int}(\mathbf{R})\right) \cap \mathrm{int}(\mathbf{R})$$
is one connected piece.

b. The image by $\mathbf{F}$ of the interior of $\mathbf{R}$ does not intersect the incoming part of the boundary of $\mathbf{R}$,
$$\mathbf{F}\left(\operatorname{int}(\mathbf{R})\right) \cap \partial^{\operatorname{in}}(\mathbf{R}) = \emptyset.$$

c. The image of the outgoing part of the boundary, $\mathbf{F}\left(\partial^{\operatorname{out}}(\mathbf{R})\right)$, does not intersect the interior of the next M-rectangle $\mathbf{R}$,
$$\mathbf{F}\left(\partial^{\operatorname{out}}(\mathbf{R})\right) \cap \operatorname{int}(\mathbf{R}) = \emptyset.$$

d. For any $\mathbf{x}$ in $\bar{\mathbf{B}}^{n_1}(\mathbf{0}, r_1)$, the image of the vertical disk $\{\mathbf{x}\} \times \bar{\mathbf{B}}^{n_2}(\mathbf{0}, r_2)$ stretches across $\mathbf{R}$ in all the overflowing directions.

See Figure 8 for the two-dimensional figure.

There is a variety of ways to make more precise the condition that the image of a vertical disk is stretched across the next M-rectangle in the overflowing directions. One way is in terms of assumptions on the derivatives. Another way is in terms of "homology theory" from algebraic topology. The homology generalizes the idea of index that we used in two dimensions. A third way was recently developed by M. Gidea and P. Zgliczyński. See [Gid02] or [Gid03]. They assume that the nonlinear map can be deformed or changed into one in a standard form.

d1. There is a continuous map $\mathbf{G}$ from $[0,1] \times \mathbf{R}$ into $\mathbb{R}^n$ that satisfies the following three conditions:

(i) $\mathbf{G}(0, \mathbf{z}) = \mathbf{F}(\mathbf{z})$.

(ii) For every $0 \leq s \leq 1$, the image of $\mathbf{R}$ by $\mathbf{G}(s, \cdot)$ satisfies the preceding conditions (a) to (c), namely,
$$\mathbf{G}\left(s, \operatorname{int}(\mathbf{R})\right) \cap \operatorname{int}(\mathbf{R}) \quad \text{is connected},$$
$$\operatorname{int}\left(\mathbf{G}(s, \mathbf{R})\right) \cap \partial^{\operatorname{in}}(\mathbf{R}) = \emptyset, \quad \text{and}$$
$$\mathbf{G}(s, \partial^{\operatorname{out}}(\mathbf{R})) \cap \operatorname{int}(\mathbf{R}) = \emptyset.$$

(iii) The map for $s = 1$ is an affine map in the overflowing direction; that is,
$$\mathbf{G}(1, (\mathbf{x}_1, \mathbf{x}_2)) = \mathbf{a} + (\mathbf{0}, \mathbf{A}\mathbf{x}_2),$$
with $\mathbf{A}$ an $n_2 \times n_2$ matrix and $\operatorname{sign}(\det(\mathbf{A})) \neq 0$.

Because the deformation does not pull the image off the $\mathbf{R}$ or through the sides, the map $\mathbf{G}(0, \cdot)$ crosses the M-rectangle in the same way that $\mathbf{G}(1, \cdot)$ does.

From these assumptions, it follows that the inverse $\mathbf{F}^{-1}$ satisfies the conditions, with the roles of the overflowing and inflowing directions reversed.

a'. The intersection
$$\mathbf{F}^{-1}\left(\operatorname{int}(\mathbf{R})\right) \cap \operatorname{int}(\mathbf{R})$$
is one connected piece.

b'. The interior of the image $\mathbf{F}^{-1}(\mathbf{R})$ does not intersect the outgoing part of the boundary of $\mathbf{R}$,
$$\operatorname{int}\left(\mathbf{F}^{-1}(\mathbf{R})\right) \cap \partial^{\operatorname{out}}(\mathbf{R}) = \emptyset.$$

c'. The image of the incoming part of the boundary by $\mathbf{F}^{-1}$ does not intersect the interior of the M-rectangle $\mathbf{R}$,
$$\mathbf{F}^{-1}\left(\partial^{\operatorname{in}}(\mathbf{R})\right) \cap \operatorname{int}(\mathbf{R}) = \emptyset.$$

d'. For any **y** in $\bar{\mathbf{B}}^{n_2}(\mathbf{0}, r_2)$, the image by $\mathbf{F}^{-1}$ of a horizontal disk $\bar{\mathbf{B}}^{n_1}(\mathbf{0}, r_1) \times \{\mathbf{y}\}$ stretches across **R** in all the inflowing directions.

With these changes in the definitions, Theorems 13.4 and 13.6 are still true. To prove these results in higher dimensions, we need to either use more advanced ideas from algebraic topology, or assume the map is a contraction and expansion in the different directions. (cf. [**Rob99**].)

13.2.2. Markov Partition. For a nonlinear map, the boxes that are correctly aligned are usually not be as simple as those given in the preceding section but are images of actual rectangles.

Definition 13.6. Assume that **F** is a homeomorphism of $\mathbb{R}^n$. A finite collection of closed sets $\{\mathbf{R}_j\}_{j=1}^J$ in $\mathbb{R}^n$ is called a *Markov partition for* **F**, or is said to have the *Markov property*, provided that the following conditions are satisfied:

(1) Assume that n_1 and n_2 are positive integers with $n_1 + n_2 = n$, $\bar{\mathbf{B}}^{n_1}(\mathbf{0}, 1)$ is a closed ball in $\mathbb{R}^{n_1}$, and $\bar{\mathbf{B}}^{n_2}(\mathbf{0}, 1)$ is a closed ball in $\mathbb{R}^{n_2}$. The M-rectangle

$$\mathbf{B} = \bar{\mathbf{B}}^{n_1}(\mathbf{0}, 1) \times \bar{\mathbf{B}}^{n_2}(\mathbf{0}, 1),$$

is a model for the Markov rectangles. In two dimensions with $n_1 = n_2 = 1$, **B** is a square, $\mathbf{B} = [-1, 1] \times [-1, 1]$.

(2) For each $\mathbf{R}_j$, there is a homeomorphism ϕ_j from the standard box **B** onto $\mathbf{R}_j$). The images by ϕ_j of the different parts of the boundary of **B** are given similar labels for $\mathbf{R}_j$:

$$\partial^{\text{in}}(\mathbf{R}_j) = \phi_j(\partial^{\text{in}}(\mathbf{B})) \quad \text{and} \quad \partial^{\text{out}}(\mathbf{R}_j) = \phi_j(\partial^{\text{out}}(\mathbf{B})).$$

(3) The interiors of the $\mathbf{R}_i$ are disjoint, $\text{int}(\mathbf{R}_i) \cap \text{int}(\mathbf{R}_j) = \emptyset$ for $i \neq j$.

(4) If the image by of the interior of $\mathbf{R}_i$ intersects the interior of $\mathbf{R}_j$, $\mathbf{F}(\text{int}(\mathbf{R}_i)) \cap \text{int}(\mathbf{R}_j) \neq \emptyset$, then the image of $\mathbf{R}_i$ by **F** is correctly aligned with $\mathbf{R}_j$ in the sense that the image of **B** by $\phi_j^{-1} \circ \mathbf{F} \circ \phi_i$ is correctly aligned with **B**, and the following four conditions are satisfied:
   a. The intersection $\mathbf{F}(\text{int}(\mathbf{R}_i)) \cap \text{int}(\mathbf{R}_j)$ is one connected piece.
   b. The image of the interior of $\mathbf{R}_i$ does not intersect the incoming part of the boundary $\partial^{\text{in}}(\mathbf{R}_j)$,

   $$\mathbf{F}(\text{int}(\mathbf{R}_i)) \cap \partial^{\text{in}}(\mathbf{R}_j) = \emptyset.$$

   c. The image $\mathbf{F}(\partial^{\text{out}}(\mathbf{R}_i))$ of the outgoing boundary does not intersect the interior of the next M-rectangle $\mathbf{R}_j$,

   $$\mathbf{F}(\partial^{\text{out}}(\mathbf{R}_i)) \cap \text{int}(\mathbf{R}_j) = \emptyset.$$

   d. For any **x** in $\bar{\mathbf{B}}^{n_1}(\mathbf{0}, 1)$, the image of $\phi_i(\{\mathbf{x} \times \bar{\mathbf{B}}^{n_2}(\mathbf{0}, 1)\})$ by **F** stretches across $\mathbf{R}_j$ in all the overflowing directions.

The sets $\mathbf{R}_i$ in the Markov partition are called *Markov rectangles*.

Remark 13.7. We allow $\mathbf{F}(\text{int}(\mathbf{R}_i)) \cap \text{int}(\mathbf{R}_j)$ to be empty for some i and j. We allow the image of $\mathbf{R}_i$ by **F** to cross $\mathbf{R}_j$ at most once, because in the definition of being correctly aligned we assume that $\mathbf{F}(\text{int}(\mathbf{R}_i)) \cap \text{int}(\mathbf{R}_j)$ is connected.

Remark 13.8. Usually, each $\mathbf{R}_j$ is a diffeomorphic image and frequently an affine image of $\mathbf{B}$. In the latter case, there is a matrix $\mathbf{A}_j$ and a constant point $\mathbf{p}_j$ such that $\mathbf{R}_j = \mathbf{p}_j + \mathbf{A}_j(\mathbf{B})$.

Remark 13.9. The definition of the Markov property is compatible with the one given in one dimension. However, a one-dimensional map has only an overflowing direction and no inflowing direction.

Remark 13.10. In choosing a Markov partition for a map, the rectangles must be chosen with their overflowing and inflowing directions so that the map takes the overflowing direction in one rectangle across the overflowing direction of any rectangle it intersects.

Remark 13.11. The Markov property has been used in a probabilistic sense for many years. Ya. Sinai and R. Bowen carried these ideas over to dynamical systems in the late 1960s and early 1970s.

Definition 13.12. For a collection of boxes with the Markov property for $\mathbf{F}$, we define an *associated transition graph* $\mathscr{G}$ by letting the vertices be the labels of the boxes in the partition, $\{1, \ldots, J\}$, and putting a directed edge from i to j if and only if
$$\mathbf{F}(\text{int}(\mathbf{R}_i)) \cap \text{int}(\mathbf{R}_j) \neq \emptyset$$
(i.e., the image of $\mathbf{R}_i$ by $\mathbf{F}$ has a nonempty crossing with $\mathbf{R}_j$ that is correctly aligned).

We also form the *transition matrix* $\mathbf{T} = (t_{ij})$ by
$$t_{ij} = \begin{cases} 0 & \text{if } \mathbf{F}(\text{int}(\mathbf{R}_i)) \cap \text{int}(\mathbf{R}_j) = \emptyset, \\ 1 & \text{if } \mathbf{F}(\text{int}(\mathbf{R}_i)) \cap \text{int}(\mathbf{R}_j) \neq \emptyset. \end{cases}$$
The nonzero entries give the allowable transitions between symbols.

Definition 13.13. Fix a transition graph $\mathscr{G}$ with J vertices or a $J \times J$ transition matrix $\mathbf{T}$. The *full two-sided shift space on J symbols* is the set Σ_J of all bi-infinite sequences of the symbols $1, \ldots, J$. A symbol sequence $\mathbf{s}$ in Σ_J is *allowable* for $\mathscr{G}$ or $\mathbf{T}$ provided that there is an edge in $\mathscr{G}$ from s_i to s_{i+1} for every i (i.e., $t_{s_i s_{i+1}} = 1$ in the transition matrix). Let $\Sigma_{\mathscr{G}}$ be the set of all allowable symbol sequences. (It is also written as $\Sigma_{\mathbf{T}}$ if it is specified by the transition matrix $\mathbf{T}$.) The space $\Sigma_{\mathscr{G}}$ is called a *subshift of finite type* because there is a finite set of rules given by the transition graph that indicate which symbols are allowed to follow which other symbols.

Notice, that for the horseshoe, all the entries of the transition matrix are 1 and there are edges connecting all the vertices of the transition graph.

Theorem 13.7. *Assume that $\mathscr{G}$ is an irreducible transition graph and some vertex has more than one edge going out. Let $\Sigma_{\mathscr{G}}$ be the associated two-sided subshift of finite type. Then the shift map is topologically transitive on $\Sigma_{\mathscr{G}}$ and has sensitive dependence on initial conditions when restricted to $\Sigma_{\mathscr{G}}$.*

The proof is essentially the same as that of Theorem 10.21 for the one-sided subshift.

13.2. Symbolic Dynamics

Theorem 13.8. *Assume that* $\mathbf{F}$ *is a homeomorphism on* $\mathbb{R}^n$. *Assume that* $\{\mathbf{R}_i\}_{i=1}^{J}$ *is a Markov partition for* $\mathbf{F}$ *with transition graph* $\mathscr{G}$ *and associated subshift of finite type* $\Sigma_{\mathscr{G}}$.

a. *Write* $\mathbf{S}^\circ$ *for* $\bigcup_{i=1}^{J} \text{int}(\mathbf{R}_i)$, *and*

$$\Lambda = \bigcap_{k=0}^{\infty} \text{cl}\left(\bigcap_{j=-k}^{k} \mathbf{F}^{-j}(\mathbf{S}^\circ)\right).$$

Then, there is an itinerary function h *from* Λ *to* $\Sigma_{\mathscr{G}}$ *(i.e., if* $\mathbf{x}$ *is in* Λ, *then the itinerary* $h(\mathbf{x})$ *is an allowable sequence). Note that, if the point* $\mathbf{F}^j(\mathbf{x})$ *is on the boundary of two or more rectangles* $\mathbf{R}_i$, *then it is necessary to make a choice for* s_j.

b. *Assume that* $\mathbf{s}$ *in* $\Sigma_{\mathscr{G}}$ *is an allowable bi-infinite symbol sequence. Then,*

$$\bigcap_{k=0}^{\infty} \text{cl}\left(\bigcap_{j=-k}^{k} \mathbf{F}^{-j}(\text{int}(\mathbf{R}_{s_j}))\right) \neq \emptyset$$

and there exists a point $\mathbf{x_s}$ *such that* $h(\mathbf{x_s}) = \mathbf{s}$ *(i.e.,* $\mathbf{F}^j(\mathbf{x_s})$ *is in* $\mathbf{R}_{s_j}$ *for all* j). *Thus, the itinerary function* h *is onto all of* $\Sigma_{\mathscr{G}}$.

c. *Assume that* $\mathbf{s}$ *in* $\Sigma_{\mathscr{G}}$ *is an allowable bi-infinite symbol sequence that has period-p (i.e.,* $s_{j+p} = s_j$ *for all* j *and there is no shorter period than* p). *Then, there exists a point* $\mathbf{x_s}$ *such that* $h(\mathbf{x_s}) = \mathbf{s}$ *and* $\mathbf{F}^p(\mathbf{x_s}) = \mathbf{x_s}$. *Consequently, the period of* $\mathbf{x_s}$ *divides* p. *If the point* $\mathbf{x_s}$ *is not on the boundary of* $\mathbf{R}_{s_0}$, *then the period is exactly* p.

d. *Assume that the transition graph is irreducible. Further, assume that*

$$\bigcap_{k=0}^{\infty} \text{cl}\left(\bigcap_{j=-k}^{k} \mathbf{F}^{-j}(\text{int}(\mathbf{R}_{s_j}))\right)$$

is a single point for each allowable sequence $\mathbf{s}$. *Then, the map* $\mathbf{F}$ *is topologically transitive on* Λ, *and* $\mathbf{F}$ *restricted to* Λ *has sensitive dependence on initial conditions.*

Remark 13.14. The intersections in the theorem can be simplified if we add the following additional assumption to the list in Definition 13.6:

5. If
$$\mathbf{F}(\text{int}(\mathbf{R}_i)) \cap \text{int}(\mathbf{R}_j) \neq \emptyset,$$
then
$$\mathbf{F}(\mathbf{R}_i) \cap \mathbf{R}_j = \text{cl}(\mathbf{F}(\text{int}(\mathbf{R}_i)) \cap \text{int}(\mathbf{R}_j)).$$

This condition says that there are no extra intersections on the ends that are not related to the images crossing. With assumption (5) added, we can replace the intersections in the theorem as follows: we replace

$$\bigcap_{k=0}^{\infty} \text{cl}\left(\bigcap_{j=-k}^{k} \mathbf{F}^{-j}(\mathbf{S}^\circ)\right) \quad \text{with} \quad \bigcap_{j=-\infty}^{\infty} \mathbf{F}^{-j}(\mathbf{S})$$

and
$$\bigcap_{k=0}^{\infty} \text{cl}\left(\bigcap_{j=-k}^{k} \mathbf{F}^{-j}\left(\text{int}\left(\mathbf{R}_{s_j}\right)\right)\right) \quad \text{with} \quad \bigcap_{j=-\infty}^{\infty} \mathbf{F}^{-j}\left(\mathbf{R}_{s_j}\right).$$

where $\mathbf{S} = \bigcup_{i=1}^{J} \mathbf{R}_i$.

Remark 13.15. In part (d) of the theorem, we need to assume that all the intersections
$$\bigcap_{k=0}^{\infty} \text{cl}\left(\bigcap_{j=-k}^{k} \mathbf{F}^{-j}\left(\text{int}\left(\mathbf{R}_{s_j}\right)\right)\right)$$
are points in order to get sensitive dependence, because the points of this set stay close together for all iterates.

We apply the preceding theorem to the Hénon map with large value of a.

Example 13.16 (Hénon map). We apply these ideas to the Hénon map
$$\mathbf{F}\begin{pmatrix} x \\ y \end{pmatrix} = \begin{pmatrix} a - by - x^2 \\ x \end{pmatrix},$$
for $b = 0.3$ and $a = 5$. We consider the square
$$\mathbf{S} = [-3, 3] \times [-3, 3].$$

The images of the corners are
$$\mathbf{F}\begin{pmatrix} \pm 3 \\ 3 \end{pmatrix} = \begin{pmatrix} 5 - 0.9 - 9 \\ \pm 3 \end{pmatrix} = \begin{pmatrix} -4.9 \\ \pm 3 \end{pmatrix} \quad \text{and}$$
$$\mathbf{F}\begin{pmatrix} \pm 3 \\ -3 \end{pmatrix} = \begin{pmatrix} 5 + 0.9 - 9 \\ \pm 3 \end{pmatrix} = \begin{pmatrix} -3.1 \\ \pm 3 \end{pmatrix}.$$

The image of a vertical line segment in $\mathbf{S}$ is a horizontal line segment, and the

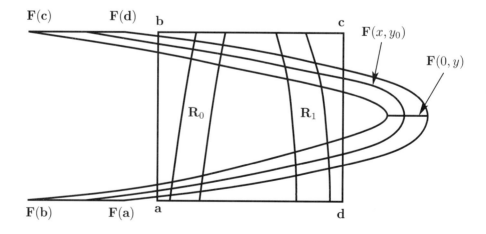

Figure 14. Image of the square $\mathbf{S}$ by the Hénon map

13.2. Symbolic Dynamics

image of a horizontal line segment in $\mathbf{S}$ is a parabola that cuts across $\mathbf{S}$ twice:

$$\mathbf{F}\begin{pmatrix} x_0 \\ y \end{pmatrix} = \begin{pmatrix} 5 - 0.3\,y - x_0^2 \\ x_0 \end{pmatrix},$$

$$\mathbf{F}\begin{pmatrix} x \\ y_0 \end{pmatrix} = \begin{pmatrix} 5 - 0.3\,y_0 - x \\ x \end{pmatrix}.$$

See Figure 14.

The intersection $\mathbf{S} \cap \mathbf{F}^{-1}(\mathbf{S})$ is the union of two strips $\mathbf{R}_0$ and $\mathbf{R}_1$ that reach from the bottom to the top of $\mathbf{S}$. The image of $\mathbf{R}_0 \cup \mathbf{R}_1$ equals $\mathbf{S} \cap \mathbf{F}(\mathbf{S})$, which is the union of two strips that reach between the two sides of $\mathbf{S}$. These later strips play the roles of $\mathbf{V}_0$ and $\mathbf{V}_1$ for the horseshoe, while $\mathbf{R}_0$ and $\mathbf{R}_1$ play the roles of $\mathbf{H}_0$ and $\mathbf{H}_1$. These strips $\mathbf{R}_0$ and $\mathbf{R}_1$ are the M-rectangles in the Markov partition. These M-rectangles can be determined by the inverse map

$$\mathbf{F}^{-1}\begin{pmatrix} x \\ y \end{pmatrix} = \begin{pmatrix} y \\ \dfrac{5 - x - y^2}{0.3} \end{pmatrix}.$$

The sides of the strips that are mapped to $x = \pm 3$ are the parabolas

$$\mathbf{F}^{-1}\begin{pmatrix} 3 \\ y \end{pmatrix} = \begin{pmatrix} y \\ \dfrac{2 - y^2}{0.3} \end{pmatrix} \quad \text{and}$$

$$\mathbf{F}^{-1}\begin{pmatrix} -3 \\ y \end{pmatrix} = \begin{pmatrix} y \\ \dfrac{8 - y^2}{0.3} \end{pmatrix}.$$

These sides are the "outgoing" part of the boundary. The part of the boundary on $y = \pm 3$ is the "incoming" part of the boundary, which gets mapped to the horizontal part of the boundaries of $\mathbf{F}(\mathbf{S}) \cap \mathbf{S}$.

The pair $\{\mathbf{R}_0, \mathbf{R}_1\}$ forms a Markov partition. The transition matrix has all 1's. For each bi-infinite string of 0's and 1's, $\mathbf{s}$, the intersection

$$\bigcap_{j=-\infty}^{\infty} \mathbf{F}^{-j}\left(\mathbf{R}_{s_j}\right)$$

is nonempty. It is not obvious that it is a single point, but this can be shown by further detailed analysis, which is beyond the scope of this book. See [**Rob99**]. However, even without knowing that the intersection is a point, the preceding results show that periodic symbol sequences correspond to periodic points. We can also determine nonperiodic points by using symbol sequences that are not periodic.

13.2.3. Markov Partitions for Toral Automorphisms.
Toral automorphisms were among the first maps for which Markov partitions were constructed. The Markov partition connects the dynamics of the hyperbolic toral automorphism $\mathbf{F}_\mathbf{A}$ on $\mathbb{T}^2$ with a subshift of finite type for a transition matrix $\mathbf{T}$. R. Adler and B. Weiss (1970) showed that, if the original matrix $\mathbf{A}$ inducing the hyperbolic toral automorphism has all positive entries, then it is possible to find a Markov partition with two M-rectangles having a transition matrix $\mathbf{T}$ that is the same as $\mathbf{A}$. (See Remark 13.20 for comments about transition matrices with entries bigger than one.)

Several other people, including M. Snavely [**Sna91**] and E. Rykken [**Ryk98**], have given further details on constructing such a Markov partition. In this section, we consider only the case, in which the construction is easiest to give.

The Markov partition can be constructed in a more specific manner than using the general definitions we have given. Some of the properties are closer to Bowen's original definition, but the use of the covering map from the plane to the torus is particular to toral automorphisms.

For the toral automorphisms, we take a Markov partition $\mathscr{R} = \{\mathbf{R}_i\}_{i=1}^m$ to cover the whole torus:

$$\mathbb{T}^2 = \bigcup_{i=1}^{m} \mathbf{R}_i.$$

Each M-rectangle can be given as $\mathbf{R}_i = \pi(\bar{\mathbf{R}}_i)$, where $\bar{\mathbf{R}}_i$ is a parallelogram in $\mathbb{R}^2$ and the projection π is a homeomorphism on the interior of $\bar{\mathbf{R}}_i$. We represent the projection of the interior of $\bar{\mathbf{R}}_i$ by $\mathbf{R}_i^\circ$ and call it the interior; that is,

$$\mathbf{R}_i^\circ = \pi\left(\operatorname{int}(\bar{\mathbf{R}}_i)\right).$$

We require that

$$\mathbf{R}_j^\circ \cap \mathbf{R}_i^\circ = \emptyset \qquad \text{if } j \neq i.$$

We call

$$\partial(\mathbf{R}_i) = \mathbf{R}_i \setminus \mathbf{R}_i^\circ$$

the boundary of the rectangle in $\mathbf{T}^2$ even though it might be larger than the boundary in $\mathbf{T}^2$ in the usual sense of the term. In the Example 13.17, $\mathbf{R}_1^\circ$ is larger than the interior in the torus $\operatorname{int}(\mathbf{R}_1)$ since it touches itself along $\partial(\mathbf{R}_1)$; also, $\partial(\mathbf{R}_1)$ is larger than the usual topological boundary of $\mathbf{R}_1$ in the torus for the same reason.

The boundary of this parallelogram is made up of pieces of stable and unstable manifolds of the induced map on $\mathbb{R}^2$. In fact, for our examples, the boundaries can be made up of line segments $\mathbf{L}$ such that $\pi(\mathbf{L})$ is contained in $W^s(\mathbf{0}) \cup W^u(\mathbf{0})$, and

$$\partial^{\text{in}}(\mathbf{R}_i) \subset W^u(\pi(\mathbf{0})) \quad \text{and}$$
$$\partial^{\text{out}}(\mathbf{R}_i) \subset W^s(\pi(\mathbf{0})).$$

We allow two points on the boundary of $\bar{\mathbf{R}}_i$ to be equivalent modulo 1 (i.e., the map π can take two points on the boundary to the same point in the torus). The map π can be many to one on the boundary of $\bar{\mathbf{R}}_i$. As a consequence, an M-rectangle $\mathbf{R}_i$ can touch itself on the boundary.

For the covering property, we merely require that $\mathbf{F}(\mathbf{R}_i^\circ) \cap \mathbf{R}_j^\circ$ is one connected piece if this intersection is not empty. (Thus, we allow other contacts on the image of the boundary.)

In the plane, the stable and unstable manifolds are lines that do not return to a rectangle, so the intersection $W^\sigma(\bar{\mathbf{z}}) \cap \bar{\mathbf{R}}_j$ is a line segment for $\sigma = u$ or s. Therefore, for $\bar{\mathbf{z}}$ in $\bar{\mathbf{R}}_j$, $\mathbf{z} = \pi(\bar{\mathbf{z}})$ in $\mathbf{R}_j$, and $\sigma = u$ or s, we define

$$W^\sigma(\mathbf{z}, \mathbf{R}_j) = \pi\left(W^\sigma(\bar{\mathbf{z}}) \cap \bar{\mathbf{R}}_j\right).$$

13.2. Symbolic Dynamics

These stable and unstable manifolds in the rectangles then satisfy the following property: If $\mathbf{z} \in \mathbf{R}_i^\circ$ and $\mathbf{F}(\mathbf{z}) \in \mathbf{R}_j^\circ$, then

$$\mathbf{F}(W^u(\mathbf{z}, \mathbf{R}_i)) \supset W^u(\mathbf{F}(\mathbf{z}), \mathbf{R}_j) \quad \text{and}$$
$$\mathbf{F}(W^s(\mathbf{z}, \mathbf{R}_i)) \subset W^s(\mathbf{F}(\mathbf{z}), \mathbf{R}_j).$$

(Bowen had essentially this property as part of his definition of an M-rectangle.)

Examples

Example 13.17. We find a Markov partition for the toral automorphism $\mathbf{F} = \mathbf{F_A}$ induced by the matrix $\mathbf{A} = \begin{pmatrix} 1 & 1 \\ 1 & 0 \end{pmatrix}$. The eigenvalues are $\lambda_u = \frac{1}{2}(1+\sqrt{5}) > 1$ with eigenvector $\mathbf{v}^u = (2, -1+\sqrt{5})^\mathsf{T}$ and $-1 < \lambda_s = \frac{1}{2}(1-\sqrt{5}) < 0$ with eigenvector $\mathbf{v}^s = (2, -1-\sqrt{5})^\mathsf{T}$. The eigenvector $\mathbf{v}^u$ has positive slope and $\mathbf{v}^s$ has negative slope.

The *lattice points* in $\mathbb{R}^2$ are all the points with both coordinates equal an integer; these are the points $\mathbf{p}$ with $\pi(\mathbf{p}) = \pi(\mathbf{0})$.

To form the rectangles $\bar{\mathbf{R}}_j$ in $\mathbb{R}^2$, take the part of the unstable manifold that goes above and to the right of the lattice point as shown in Figure 15. Take the part of the stable manifold from the lattice point downward to the point $\bar{\mathbf{a}}$, where it hits the unstable line segment drawn above. Let $[\mathbf{0}, \bar{\mathbf{a}}]_s$ be the line segment in the stable manifold from $\mathbf{0}$ to $\bar{\mathbf{a}}$, and in general, let $[\bar{\mathbf{x}}, \bar{\mathbf{y}}]_s$ be the line segment in the stable manifold from $\bar{\mathbf{x}}$ to $\bar{\mathbf{y}}$. Also, extend the stable manifold upward from a lattice point to the point $\bar{\mathbf{b}}$, where it hits the unstable line segment drawn above. See Figure 15. These can be chosen so that $\bar{\mathbf{F}}(\bar{\mathbf{a}}) = \bar{\mathbf{b}}$. Let $\bar{\mathbf{c}} = \bar{\mathbf{F}}(\bar{\mathbf{b}})$ and $\bar{\mathbf{c}}' = \bar{\mathbf{c}} + (1,1)$ so $\pi(\bar{\mathbf{c}}) = \pi(\bar{\mathbf{c}}')$. Finally, extend the unstable manifold to the point $\bar{\mathbf{c}}'$, where it hits the line segment $[\bar{\mathbf{a}}, \bar{\mathbf{b}}]_s$ in the stable manifold. These line segments, $[\bar{\mathbf{a}}, \bar{\mathbf{b}}]_s$ in $W^s(\mathbf{0})$ and $[\mathbf{0}, \bar{\mathbf{c}}']_u$ in $W^u(\mathbf{0})$ (and their translates in $\mathbb{R}^2$), define two rectangles $\bar{\mathbf{R}}_1$ and $\bar{\mathbf{R}}_2$ in $\mathbb{R}^2$, and hence, $\mathbf{R}_1$ and $\mathbf{R}_2$ in $\mathbb{T}^2$. See Figure 15.

To determine the images of the rectangles by $\mathbf{F}$, we first consider the images of the points $\bar{\mathbf{a}}$, $\bar{\mathbf{b}}$, and $\bar{\mathbf{c}}$: we have $\bar{\mathbf{F}}(\bar{\mathbf{a}}) = \bar{\mathbf{b}}$, $\bar{\mathbf{F}}(\bar{\mathbf{b}}) = \bar{\mathbf{c}}$, and $\mathbf{e} = \mathbf{F}(\bar{\mathbf{c}}) \in [\mathbf{0}, \bar{\mathbf{b}}]_s$. See Figure 15. We have labeled multiple points that project down on same point in the torus with the same letter, distinguishing them with primes. With these images, it follows that

$$\mathbf{F}(\mathbf{R}_1) \quad \text{crosses } \mathbf{R}_1 \text{ and } \mathbf{R}_2 \text{ in } \mathbb{T}^2 \text{ and}$$
$$\mathbf{F}(\mathbf{R}_2) \quad \text{crosses } \mathbf{R}_1 \text{ in } \mathbb{T}^2.$$

See Figure 16. The incoming part of the boundary of each $\mathbf{R}_j$, $\partial^{\text{in}}(\mathbf{R}_j)$, is made up of pieces of unstable manifolds, and the outgoing part of the boundary of each $\mathbf{R}_j$, $\partial^{\text{out}}(\mathbf{R}_j)$, is made up of pieces of stable manifolds. The pair of rectangles $\{\mathbf{R}_1, \mathbf{R}_2\}$ have the properties of a *Markov partition* for $\mathbf{F}$: (1) each rectangle is a parallelogram, (2) the collection of rectangles covers $\mathbb{T}^2$ and the interiors of $\mathbf{R}_1$ and $\mathbf{R}_2$ are disjoint, and (3) if $\mathbf{F}(\mathbf{R}_i^\circ) \cap \mathbf{R}_j^\circ \neq \emptyset$, then the following hold. (a) $\mathbf{F}(\mathbf{R}_i^\circ) \cap \mathbf{R}_j^\circ$ is one connected piece. (b) $\mathbf{F}(\mathbf{R}_i^\circ)$ does not intersect $\partial^{\text{in}}(\mathbf{R}_i)$. (c) $\mathbf{F}(\mathbf{R}_i)$ reaches all the way across $\mathbf{R}_j$ in the unstable direction so $\mathbf{F}(\partial^{\text{out}}(\mathbf{R}_i))$ does not intersect $\mathbf{R}_i^\circ$. (d) For any $\mathbf{z}$ in $\mathbf{R}_i$, $\mathbf{F}(W^u(\mathbf{z}, \mathbf{R}_i))$ stretches across $\mathbf{R}_j$ in the unstable direction.

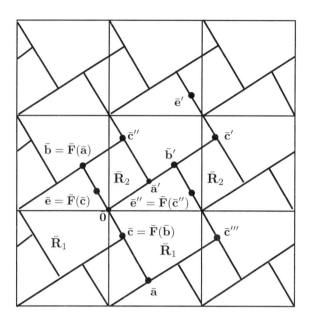

Figure 15. Rectangles for Example 13.17

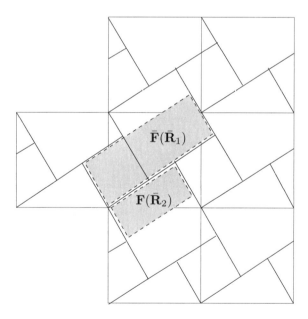

Figure 16. Images of rectangles for Example 13.17

The transition matrix is

$$\mathbf{T} = \begin{pmatrix} 1 & 1 \\ 1 & 0 \end{pmatrix}.$$

13.2. Symbolic Dynamics

The symbol space $\Sigma_\mathbf{T}$ consists of the allowable sequences, and $\sigma_\mathbf{T} = \sigma|\Sigma_\mathbf{B}$ is the shift map restricted to the subshift $\Sigma_\mathbf{B}$. The semiconjugacy map k is defined from $\Sigma_\mathbf{B}$ onto $\mathbb{T}^2$ by

$$k(\mathbf{s}) = \bigcap_{n=0}^{\infty} \mathrm{cl}\left(\bigcap_{j=-n}^{n} \mathbf{F}^{-j}(\mathbf{R}_{s_j}^\circ)\right).$$

We take the images of the interiors because $\mathbf{R}_{s_1} \cap \mathbf{F}^{-1}(\mathbf{R}_{s_2})$ does not always equal

$$\mathrm{cl}\left(\mathbf{R}_{s_1}^\circ \cap \mathbf{F}^{-1}(\mathbf{R}_{s_2}^\circ)\right),$$

but can have extra points whose images are on the boundary of $\mathbf{R}_{s_2}$. (We must put up with this annoyance in order to be able to use fewer rectangles.) Theorem 13.9 shows that k is a semiconjugacy: continuous, onto, and $\mathbf{F} \circ k = k \circ \sigma_\mathbf{T}$. In fact, it proves that k is, at most, four to one since there are only two rectangles.

Remark 13.18. In the preceding example, for a point $\mathbf{p} \in \mathrm{bd}(\mathbf{R}_i)$, there are at least two choices of rectangles to which $\mathbf{p}$ belongs. Therefore, there is no way to assign a unique symbol sequence to points on the boundary of a rectangle (i.e., the itinerary map h is not always uniquely defined, or is discontinuous). In fact, since the torus is connected and the symbol space $\Sigma_\mathbf{T}$ is not, there cannot be a continuous map from $\mathbb{T}^2$ onto $\Sigma_\mathbf{T}$.

Example 13.19. As a second example of a hyperbolic toral automorphism, let $\mathbf{B} = \begin{pmatrix} 2 & 1 \\ 1 & 1 \end{pmatrix}$. As we previously noted, if $\mathbf{A} = \begin{pmatrix} 1 & 1 \\ 1 & 0 \end{pmatrix}$, then $\mathbf{A}^2 = \mathbf{B}$. Let $\mathbf{G} = \mathbf{F_B}$ and $\mathbf{F} = \mathbf{F_A}$, so $\mathbf{G} = \mathbf{F}^2$.

The rectangles $\mathbf{R}_1$ and $\mathbf{R}_2$ from Example 13.17 are still rectangles for this matrix. This partition satisfies most of the conditions of a Markov partition, but the image of $\mathbf{R}_1$ by $\mathbf{G}$ crosses $\mathbf{R}_1$ twice, so $\mathbf{G}(\mathbf{R}_1^\circ) \cap \mathbf{R}_1^\circ$ is not connected. The transition matrix, counting geometric crossings (adjacency matrix), is $\mathbf{B}$.

If we want a transition matrix with only 0's and 1's, we must subdivide the rectangles (split symbols) by taking components of $\mathbf{R}_1 \cap \mathbf{F_B}(\mathbf{R}_1)$: let the rectangle

$$\mathbf{R}_{1a} = \pi\left(\bar{\mathbf{R}}_1 \cap L_\mathbf{B}(\bar{\mathbf{R}}_1)\right),$$

where $L_\mathbf{B}$ is the map on $\mathbb{R}^2$, and

$$\mathbf{R}_{1b} = \mathrm{cl}(\mathbf{R}_1 \setminus \mathbf{R}_{1a}).$$

These rectangles can also be formed by extending the unstable manifold of the origin until it intersects the stable line segment $[\mathbf{0}, \mathbf{b}]_s$ translated by $(2, 1)$ at the point $\mathbf{e}''' = \mathbf{F}(\mathbf{c}) + (2, 1)$. See Figures 17 and 15. A direct check shows that

$$\begin{aligned}\mathbf{F_B}(\mathbf{R}_{1a}) &\quad \text{crosses} \quad \mathbf{R}_{1a}, \mathbf{R}_{1b} \text{ and } \mathbf{R}_2, \\ \mathbf{F_B}(\mathbf{R}_{1b}) &\quad \text{crosses} \quad \mathbf{R}_{1a}, \mathbf{R}_{1b} \text{ and } \mathbf{R}_2, \\ \mathbf{F_B}(\mathbf{R}_2) &\quad \text{crosses} \quad \mathbf{R}_{1b} \text{ and } \mathbf{R}_2.\end{aligned}$$

Thus, the transition matrix is

$$\mathbf{T} = \begin{pmatrix} 1 & 1 & 1 \\ 1 & 1 & 1 \\ 0 & 1 & 1 \end{pmatrix}.$$

This transition matrix has characteristic polynomial $p(\lambda) = -\lambda(\lambda^2 - 3\lambda + 1)$, and has eigenvalues equal to 0, λ_u^2, and λ_s^2, where λ_u and λ_s are the eigenvalues of $\mathbf{A}$. Thus, the eigenvalues of $\mathbf{T}$ are those of $\mathbf{B}$, together with the extra eigenvalue of 0. In general, the eigenvalues of the transition matrix are always equal to plus or minus the eigenvalues of the original matrix inducing the hyperbolic toral automorphism, together with possibly 0 and/or roots of unity. See [**Sna91**].

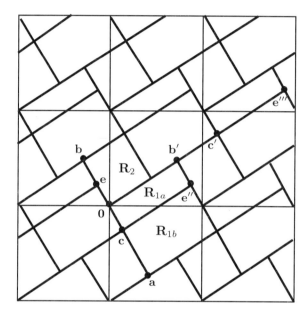

Figure 17. Markov partition for Example 13.19

Remark 13.20. In this preceding example, with $\mathbf{B} = \begin{pmatrix} 2 & 1 \\ 1 & 1 \end{pmatrix}$, if we allowed the image of a rectangle to cross more than one time, then we could use the Markov partition with only two rectangles, $\mathbf{R}_1$ and $\mathbf{R}_2$. If we allow multiple crossings, then we must allow the transition matrix to have integer entries that are larger than 1 (i.e., we obtain an adjacency matrix with nonnegative integer entries). This situation is easier to handle using the transition graphs, where multiple edges connecting vertices are permitted. We do not pursue this connection. See Franks (1982).

We can now state the result about the semiconjugacy of the subshift of finite type and the hyperbolic toral automorphism.

Theorem 13.9. *Let* $\mathscr{R} = \{\mathbf{R}_i\}_{i=1}^m$ *be a Markov partition for a hyperbolic toral automorphism on* $\mathbb{T}^2$ *with transition matrix* $\mathbf{T}$ *and*

$$\bigcup_{j=1}^m \mathbf{R}_j = \mathbb{T}^2.$$

13.2. Symbolic Dynamics

Let $(\Sigma_{\mathbf{T}}, \sigma_{\mathbf{T}})$ be the shift space and define $k : \Sigma_{\mathbf{T}} \to \mathbb{T}^2$ by

$$k(\mathbf{s}) = \bigcap_{n=0}^{\infty} \mathrm{cl}\left(\bigcap_{j=-n}^{n} \mathbf{F}^{-j}(\mathbf{R}^{\circ}_{s_j}) \right).$$

Then, k is a finite-to-one semiconjugacy from $\sigma_{\mathbf{T}}$ to $\mathbf{F}$. In fact, k is at most m^2 to one, where m is the number of rectangles in the partition.

A sketch of the proof is given in Section 13.7.

Remark 13.21. Because we assume that only $\mathbf{F}(\mathbf{R}^{\circ}_i) \cap \mathbf{R}^{\circ}_j$ is connected and not $\mathbf{F}(\mathbf{R}_i) \cap \mathbf{R}_j$, it is necessary to take the interiors and then closure in the definition of k. See $\mathbf{F}(\mathbf{R}_1)$ and $\mathbf{R}_2$ in Figure 16 for an example for which these are different. If we assume that $\mathbf{F}(\mathbf{R}_i) \cap \mathbf{R}_j$ is connected for every intersection, then we could just use the simpler definition of k given by

$$k(\mathbf{s}) = \bigcap_{j=-\infty}^{\infty} \mathbf{F}^{-j}(\mathbf{R}_{s_j}).$$

The problem is that $\mathbf{F}(\mathbf{R}^{\circ}_i) \cap \mathbf{R}^{\circ}_j$ can be nonempty and $\mathbf{F}(\mathbf{R}_i)$ can abut the boundary of $\mathbf{R}_j$ at points for which there are no nearby interior points, so

$$\mathrm{cl}\left(\mathbf{F}(\mathbf{R}^{\circ}_i) \cap \mathbf{R}^{\circ}_j \right) \neq \mathbf{F}(\mathbf{R}_i) \cap \mathbf{R}_j.$$

See Example 13.17. We allow such intersections on the boundary in order to find Markov partitions with fewer rectangles. This forces us to use this slightly more complicated definition of k.

13.2.4. Shadowing. When a map is iterated using a computer, there can be round-off errors at each iteration. If the map has some expanding directions and sensitive dependence on initial conditions, then small round-off errors can be amplified by further iteration. E. Lorenz discovered this phenomenon in his famous study of a model for convection rolls in the atmosphere. With this uncertainty in the future iterates, what validity does a computed orbit have as a model for an actual orbit? The idea of shadowing and its validity for many maps indicates that an approximate orbit calculated by the computer is often near to a real orbit with slightly different initial conditions. In this section, we sketch these ideas.

Definition 13.22. A sequence of points $\{\mathbf{x}_j\}_{j=-\infty}^{\infty}$, or a finite string $\{\mathbf{x}_j\}_{j=k_1}^{k_2}$, is called an ϵ-*chain* or ϵ-*pseudo-orbit* for a map $\mathbf{F}$, provided that

$$\|\mathbf{F}(\mathbf{x}_{j-1}) - \mathbf{x}_j\| < \epsilon$$

for all j. Thus, an ϵ-chain is an orbit with errors or small jumps.

An ϵ-chain $\{\mathbf{x}_j\}_{j=-\infty}^{\infty}$ can be δ-*shadowed*, provided that there is an initial condition $\mathbf{y}_0$ with

$$\|\mathbf{F}^j(\mathbf{y}_0) - \mathbf{x}_j\| < \delta$$

for all j.

Definition 13.23. To get shadowing on an invariant set, we need to assume that the diffeomorphism $\mathbf{F}$ has expanding and contracting directions near each point of $\mathbf{A}$. An invariant set with such a property is called *hyperbolic*. Examples of maps with this feature are the geometric horseshoe and hyperbolic toral automorphisms. Precisely, we assume that there is a $\delta_0 > 0$ such that, for every δ with $0 < \delta \leq \delta_0$, there is an $\epsilon > 0$, that depends of δ, for which the following holds: near each point $\mathbf{x}$ in $\mathbf{A}$, there is a small box $\mathbf{R_x}$ of small size δ such that, if $\mathbf{x}_1$ in $\mathbf{A}$ is within ϵ of $\mathbf{F}(\mathbf{x}_0)$, then the image of $\mathbf{R}_{\mathbf{x}_0}$ by $\mathbf{F}$ is correctly aligned with $\mathbf{R}_{\mathbf{x}_1}$.

Theorem 13.10. *Assume that a diffeomorphism is hyperbolic on an invariant set $\mathbf{A}$. Then, for all $\delta > 0$, there is an $\epsilon > 0$, depending on δ, such that any ϵ-chain for $\mathbf{F}$ can be δ-shadowed by a real orbit; that is, if $\{\mathbf{x}_j\}_{j=-\infty}^{\infty}$ is an ϵ-chain for $\mathbf{F}$, then there is an initial condition $\mathbf{y}_0$ with $\|\mathbf{F}^j(\mathbf{y}_0) - \mathbf{x}_j\| < \delta$ for all j.*

This result follows from the proof of Theorem 13.8(d) applied to a sequence of correctly aligned M-rectangles. For a more precise treatment using "hyperbolicity"; see [**Rob99**].

Further extensions of this idea have been developed by J. Yorke and coauthors, who show that, even without the assumption of the correctly aligned boxes (hyperbolicity), an ϵ-chain for the map can be δ-shadowed for a long period of time.

Exercises 13.2

1. Check the index of the square $[-1, 1] \times [-1, 1]$ for the following maps:
   a. Let $\mathbf{F}(x, y) = \left(-2x, \frac{1}{2}y\right)$.
   b. Let $\mathbf{F}(x, y) = (-2x, 2y)$.
   c. Let $\mathbf{F}(x, y) = (-2x, -2y)$.
   d. Let $\mathbf{F}(x, y) = \left(\frac{1}{2}x, \frac{1}{2}y\right)$.
   e. Let $\mathbf{F}(x, y) = \left(x + \frac{1}{2}x^2, \frac{1}{2}y\right)$.
   f. Let $\mathbf{F}(x, y) = \left(8y - 5, \frac{1}{2}x\right)$.

2. Consider the map defined by
$$\mathbf{F}(x, y) = \begin{cases} \left(-3x + 1, \frac{1}{3}y\right) & \text{if } 0 \leq x \leq \frac{1}{3}, \\ \left(x - \frac{2}{3}, 1 - \frac{1}{3}y\right) & \text{if } \frac{2}{3} \leq x \leq 1. \end{cases}$$

In the middle gap, $1/3 < x < 2/3$, the map bends around the end. See Figure 18.

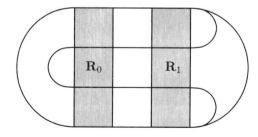

Figure 18. Exercise 2

a. Let
$$\mathbf{R}_0 = \left\{ (x,y) : 0 \leq x \leq \frac{1}{3},\ 0 \leq y \leq 1 \right\},$$
$$\mathbf{R}_1 = \left\{ (x,y) : \frac{2}{3} \leq x \leq 1,\ 0 \leq y \leq 1 \right\}.$$

Explain why $\{\mathbf{R}_0, \mathbf{R}_1\}$ satisfies the conditions of a Markov partition for $\mathbf{F}$. What are $\partial^{\text{in}}(\mathbf{R}_0)$, $\partial^{\text{out}}(\mathbf{R}_0)$, $\partial^{\text{in}}(\mathbf{R}_1)$, and $\partial^{\text{out}}(\mathbf{R}_1)$? What is the transition graph?

b. How many fixed points are there? How many points of period 2? Hint: Use the transition graph.

c. Why is there a dense orbit in the invariant set?

3. What is the transition graph for the map in Exercise 3 of Section 13.1?

4. Consider the map given by
$$\mathbf{F}\begin{pmatrix} x \\ y \end{pmatrix} = \begin{cases} \begin{pmatrix} \frac{1}{4}x + \frac{1}{8}\sin(4\pi y) \\ 4y \end{pmatrix} & \text{for } y < \frac{1}{2}, \\ \begin{pmatrix} \frac{1}{4}x + \frac{3}{4} - \frac{1}{8}\sin(4\pi(y-0.75)) \\ 4y - 3 \end{pmatrix} & \text{for } \frac{1}{2} \leq y. \end{cases}$$

a. Define the rectangles $\mathbf{R}_0 = [0,1] \times [0, 0.25]$ and $\mathbf{R}_1 = [0,1] \times [0.75, 1]$. Show that $\{\mathbf{R}_0, \mathbf{R}_1\}$ is a Markov partition (i.e., show that the images of $\mathbf{R}_0$ and $\mathbf{R}_1$ by $\mathbf{F}$ are correctly aligned with $\mathbf{R}_0$ and $\mathbf{R}_1$). In particular, what are $\partial^{\text{in}}(\mathbf{R}_j)$ and $\partial^{\text{out}}(\mathbf{R}_j)$ for $j = 1, 2$? What are the images of these sets?

b. What is the index of the map from $\mathbf{R}_0$ to itself? What is the index of the map from $\mathbf{R}_1$ to itself? Explain the answer in terms of the images of several points by the map
$$\mathbf{G}(\mathbf{x}) = \frac{\mathbf{x} - \mathbf{F}(\mathbf{x})}{\|\mathbf{x} - \mathbf{F}(\mathbf{x})\|}.$$

5. Consider the map given in polar coordinates by
$$\mathbf{F}\begin{pmatrix} \theta \\ r \end{pmatrix} = \begin{pmatrix} 8\theta - \pi/8 \\ 1 + r/16 + (2\theta)/\pi \end{pmatrix}$$

for $0 \leq \theta \leq \pi/2$ and $1 \leq r \leq 2$. (Notice this definition is for only part of the plane.) Let
$$\mathbf{R}_0 = \{(\theta, r) : 0 \leq \theta \leq 3\pi/32,\ 1 \leq r \leq 2\},$$
$$\mathbf{R}_1 = \{(\theta, r) : \pi/4 \leq \theta \leq 11\pi/32,\ 1 \leq r \leq 2\}.$$

Show that $\{\mathbf{R}_0, \mathbf{R}_1\}$ satisfies the conditions of a Markov partition for the $\mathbf{G}$.

6. Let $\mathbf{S} = [0,1] \times [0,1]$ be the unit square that contains four horizontal strips $\mathbf{H}_0$, $\mathbf{H}_1$, $\mathbf{H}_2$, and $\mathbf{H}_3$ of height $1/10$, and four vertical strips $\mathbf{V}_0$, $\mathbf{V}_1$, $\mathbf{V}_2$, and $\mathbf{V}_3$ of width $1/12$. See Figure 19. Consider a map $\mathbf{F}$ from $\mathbb{R}^2$ to itself such that it maps $\mathbf{H}_j$ onto $\mathbf{V}_j$ by stretching by 10 in the vertical direction and contracting by $1/12$ in the horizontal direction, and with the appropriate translation. The

strips $\mathbf{H}_1$ and $\mathbf{H}_3$ are also flipped over. Notice that
$$\mathbf{S} \cap \mathbf{F}(\mathbf{S}) = \mathbf{V}_0 \cup \mathbf{V}_1 \cup \mathbf{V}_2 \cup \mathbf{V}_3 \quad \text{and}$$
$$\mathbf{S} \cap \mathbf{F}^{-1}(\mathbf{S}) = \mathbf{H}_0 \cup \mathbf{H}_1 \cup \mathbf{H}_2 \cup \mathbf{H}_3.$$

a. How many vertical strips does the intersection
$$\mathbf{S} \cap \mathbf{F}(\mathbf{S}) \cap \mathbf{F}^2(\mathbf{S})$$
contain and how wide are they?

b. How many horizontal strips does the intersection
$$\mathbf{S} \cap \mathbf{F}^{-1}(\mathbf{S}) \cap \mathbf{F}^{-2}(\mathbf{S})$$
contain and how high are they?

c. Notice that
$$\mathbf{V}_j = \{\mathbf{x} : \mathbf{F}^{-1}(\mathbf{x}) \in \mathbf{H}_j\} \equiv \mathbf{S}_{j\cdot}.$$
Let
$$\mathbf{S}_{s_{-2}s_{-1}\cdot} = \{\mathbf{x} : \mathbf{F}^{-1}(\mathbf{x}) \in H_{s_{-1}} \text{ and } \mathbf{F}^{-2}(\mathbf{x}) \in H_{s_{-2}}\}.$$
State the order of the vertical strips
$$\mathbf{S}_{00\cdot}, \mathbf{S}_{10\cdot}, \ldots, \mathbf{S}_{33\cdot}.$$

d. How many fixed points does the map $\mathbf{F}$ have?

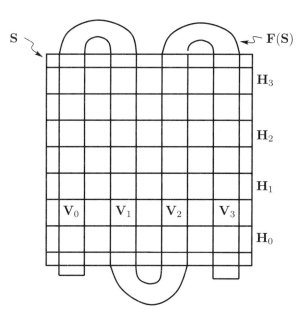

Figure 19. Exercise 6

7. Let $f : \mathbb{T}^2 \to \mathbb{T}^2$ be the diffeomorphism induced by the matrix
$$A = \begin{pmatrix} 1 & 1 \\ 1 & 0 \end{pmatrix}.$$

Form a Markov partition with three rectangles by using the line segment $[\mathbf{a}, \mathbf{b}]_s$ as in the text, and an unstable line segment $[\mathbf{g}, \mathbf{c}]_u$, where $\mathbf{g}$ is determined by extending the unstable manifold of the origin through the origin so that it terminates at a point $\mathbf{g} \in [\mathbf{a}, \mathbf{0}]_s$. Thus, $\mathbf{0}$ is within the unstable segment $[\mathbf{g}, \mathbf{c}]_u$. Determine the transition matrix B for this partition. Determine the three eigenvalues for the transition matrix. How do the eigenvalues compare with the eigenvalues for A?

8. (A horseshoe as a subsystem of a hyperbolic toral automorphism.) Let $f_{A_2} : \mathbb{T}^2 \to \mathbb{T}^2$ be the diffeomorphism induced by the matrix

$$A_2 = \begin{pmatrix} 2 & 1 \\ 1 & 1 \end{pmatrix}$$

discussed in Example 13.19. Let R_{1a} be the rectangle used in the Markov partition for this diffeomorphism. Let $g = f_{A_2}^2$ and

$$\Lambda = \bigcap_{j=-\infty}^{\infty} g^j(R_{1a}).$$

Prove that $g : \Lambda \to \Lambda$ is topologically conjugate to the two-sided full two-shift $\sigma : \Sigma_2 \to \Sigma_2$. Hint: R_{1a} plays the role that S played in the construction of the geometric horseshoe. Prove that $g(R_{1a}) \cap R_{1a}$ is made up of two disjoint rectangles. (These rectangles are similar to V_1 and V_2 in the geometric horseshoe.) Looking at the transition matrix for the Markov partition for f may help.

9. What is the tranition graph for the map given in Figure 20.

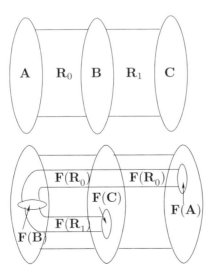

Figure 20. For Exercise 13.2.9

13.3. Homoclinic Points and Horseshoes

So far, we have described the geometric horseshoe and showed that a horseshoe occurs for the Hénon map for certain parameter values. In this section, using the machinery of Markov partitions, we show that a horseshoe occurs whenever stable and unstable manifolds for the same periodic orbit intersect. Thus, horseshoes are very prevalent. Since a horseshoe implies sensitive dependence on initial conditions, sensitive dependence is very prevalent. Henri Poincaré discovered this connection in the last quarter century of the 1800s. See Section 13.6.1 for further discussion.

Definition 13.24. Assume that **p** is a periodic saddle point. A *homoclinic point* for the orbit of **p** is a point **q** in both the stable and unstable manifold of the orbit of **p**, other than the orbit of **p** itself; that is,

$$\mathbf{q} \in [W^s(\mathcal{O}_\mathbf{F}(\mathbf{p})) \cap W^u(\mathcal{O}_\mathbf{F}(\mathbf{p}))] \setminus \mathcal{O}_\mathbf{F}(\mathbf{p}).$$

It follows that $\omega(\mathbf{q}) = \alpha(\mathbf{q}) = \mathcal{O}_\mathbf{F}(\mathbf{p})$.

Figure 21 gives a simplified picture of a homoclinic point for a saddle fixed point. Figure 9 shows homoclinic intersections for stable and unstable manifolds of a fixed point for the Hénon map. Notice that there are many such homoclinic intersections; once there is one, then the whole orbit of that point gives other homoclinic intersections.

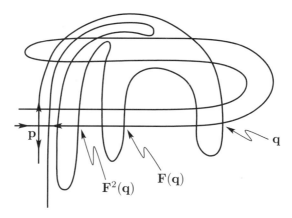

Figure 21. A transverse homoclinic point **q**

Definition 13.25. Two stable and unstable manifolds $W^s(\mathbf{p}_1)$ and $W^u(\mathbf{p}_2)$ in $\mathbb{R}^2$ are *transverse at* **q** provided that either (i) they do not intersect at **q** or (ii) a vector tangent to $W^s(\mathbf{p}_1)$ at **q** is not parallel to the vector tangent to $W^u(\mathbf{p}_2)$ at **q**.

In higher dimensions of $\mathbb{R}^n$, the requirement is that the set of vectors tangent to $W^s(\mathbf{p}_1)$ at **q** together with the set of vectors tangent to $W^u(\mathbf{p}_2)$ at **q** span all of $\mathbb{R}^n$.

Figure 21 shows a transverse homoclinic point at **q** for $W^s(\mathbf{p})$ and $W^u(\mathbf{p})$. The homoclinic point **q** in Figure 22 is nontransverse.

13.3. Homoclinic Points

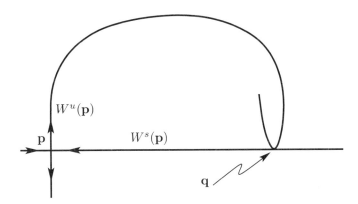

Figure 22. A nontransverse homoclinic point **q**

With this definition, we can state the main result of this section. We state the result both for an invariant set for a power of **F** conjugate to the full two shift and for an invariant set for **F** itself conjugate to a subshift of finite type.

Theorem 13.11 (Transverse homoclinic point). *Assume that a diffeomorphism* **F** *on* $\mathbb{R}^n$ *has a saddle periodic point* **p** *whose orbit has a transverse homoclinic point* **q** *with*

$$\mathbf{q} \in [W^s(\mathcal{O}_\mathbf{F}(\mathbf{p})) \cap W^u(\mathcal{O}_\mathbf{F}(\mathbf{p}))] \setminus \mathcal{O}_\mathbf{F}(\mathbf{p}),$$

and $W^s(\mathcal{O}_\mathbf{F}(\mathbf{p}))$ *is transverse to* $W^u(\mathcal{O}_\mathbf{F}(\mathbf{p}))$ *at* **q**.

a. *Then, there is a power of the map,* $\mathbf{F}^k$ *with* $k > 0$, *that has an invariant set* Λ_k *containing* **p** *and* **q** *such that* $\mathbf{F}^k$ *restricted to* Λ_k *is conjugate to the shift map on the full shift space on two symbols* Σ_2.

b. *The map* **F** *has an invariant set* Λ *containing* **p** *and* **q** *such that* **F** *restricted to* Λ *is conjugate to the shift map on an irreducible subshift of finite type* $\Sigma_{\mathcal{G}}$. *See Figure 23 for the transition graph in the case in which* **p** *is a fixed point. In particular,* **F** *has infinitely many periodic orbits and sensitive dependence on initial conditions when restricted to* Λ.

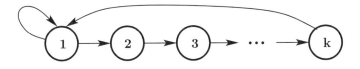

Figure 23. Transition graph for subshift from homoclinic point

Remark 13.26. In the case in which **p** is a fixed point, the set $\mathbf{A} = \{\mathbf{p}\} \cup \mathcal{O}(\mathbf{q})$ is invariant and has the properties of Theorem 13.10. Many different ϵ-chains can be formed by following the orbit of **q** until it is within ϵ of **p**, jumping to **p**, staying at **p** for an arbitrary number of iterates, and then jumping off onto the orbit of **q** again which returns to **q** and back to near **p**. The set of orbits that shadow such

ϵ-chains forms the invariant set of part (b) of the theorem, with transition graph given in Figure 23.

These ϵ-chains give a way of visualizing the transition graph. The symbol 1 corresponds to the fixed point $\mathbf{p}$; the symbol 2 corresponds to a point $\mathbf{F}^{-k_-}(\mathbf{q})$ on the backward orbit through $\mathbf{q}$ that is very close to $\mathbf{p}$; the symbols j for $2 \leq j \leq k$ corresponds to the points $\mathbf{F}^{-k_-+j-2}(\mathbf{q})$; k is chosen so that $\mathbf{F}^{-k_-+k-2}(\mathbf{q})$ is again very close to $\mathbf{p}$. The jumps from $\mathbf{F}^{-k_-+k-2}(\mathbf{q})$ to $\mathbf{p}$ and from $\mathbf{p}$ to $\mathbf{F}^{-k_-}(\mathbf{q})$ are small, with size "ϵ". Any ϵ-chain in this set of orbits corresponds to a real orbit of $\mathbf{F}$ that shadows this ϵ-chain.

Idea of the construction. In the following discussion, we consider the case of a saddle fixed point $\mathbf{p}$ with transverse homoclinic point $\mathbf{q}$ in $\mathbb{R}^2$, although we state the theorem for the general case of a saddle periodic orbit in $\mathbb{R}^n$.

(a) For a carefully chosen rectangle $\mathbf{B}$ near $\mathbf{p}$ and a negative power $-k_- < 0$,

$$\mathbf{S} = \mathbf{F}^{-k_-}(\mathbf{B})$$

is a thin strip containing the stable manifold of $\mathbf{p}$ out to the homoclinic point $\mathbf{q}$. Then, for $k_+ > 0$ and $k = k_+ + k_-$ chosen correctly, the strip $\mathbf{S}$ maps across itself twice in a correctly aligned manner by the map $\mathbf{F}^k$:

$$\mathbf{F}^{k_+}(\mathbf{B}) = \mathbf{F}^k(\mathbf{S}).$$

See Figure 24. The choices of $\mathbf{B}$, k_-, and k_+ are interrelated and must be made at the same time. See [**Rob99**] for more careful choices. Once we have the image of $\mathbf{S}$ by $\mathbf{F}^k$ mapping across itself twice, we obtain an invariant set that can be modeled on two symbols, i.e., we have a horseshoe. In fact, using the transversality, with more work, we can show that there is only one point for every symbol sequence. The uniqueness of the point gives a conjugacy to the shift space and implies sensitive dependence on initial conditions.

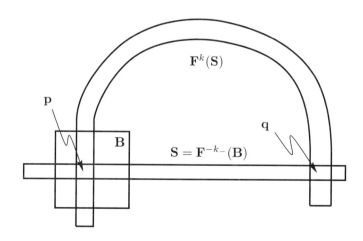

Figure 24. Boxes for a homoclinic point

(b) Let **B** be as in the proof of (a), and let $\mathbf{R}_1$ be the M-rectangle from part (a) containing **q** (i.e., $\mathbf{R}_1$ is the component of $\mathbf{F}^n(\mathbf{S}) \cap \mathbf{S}$ containing **q**). Then,

$$\mathbf{F}^j(\mathbf{R}_1) \cap \mathbf{B} \begin{cases} = \emptyset & \text{for } -k_+ < j < k_-, \\ \neq \emptyset & \text{for } j = -k_+, k_-. \end{cases}$$

In fact, $\mathbf{F}^{k_-}(\mathbf{R}_1)$ is correctly aligned with **B**, and $\mathbf{F}(\mathbf{B})$ is correctly aligned with $\mathbf{F}^{-k_+ +1}(\mathbf{R}_1)$. Thus, we can use the M-rectangles

$$\mathbf{R}'_1 = \mathbf{B},$$
$$\mathbf{R}'_2 = \mathbf{F}^{-k_+ +1}(\mathbf{R}_1), \quad \text{and}$$
$$\mathbf{R}'_j = \mathbf{F}^{-k_+ +j-1}(\mathbf{R}_1) \quad \text{for } 2 \leq j \leq k_+ + k_- = k.$$

These M-rectangles form a Markov partition and have the transition graph given in Figure 23. The subshift can be understood in terms of Remark 13.26. □

Exercises 13.3

1. Assume that a diffeomorphism **F** of $\mathbb{R}^2$ has two saddle fixed points $\mathbf{p}_1$ and $\mathbf{p}_2$ with heteroclinic points

$$\mathbf{q}_1 \in W^u(\mathbf{p}_1) \cap W^s(\mathbf{p}_2) \quad \text{and}$$
$$\mathbf{q}_2 \in W^u(\mathbf{p}_2) \cap W^s(\mathbf{p}_1).$$

   Discuss why the set

$$\mathbf{A} = \{\mathbf{p}_1, \mathbf{p}_2\} \cup \mathcal{O}(\mathbf{q}_1) \cup \mathcal{O}(\mathbf{q}_2)$$

   can be included in an invariant set Λ that is conjugate to a subshift of finite type $\Sigma_{\mathcal{G}}$. What should be the transition graph $\mathcal{G}$ for the subshift of finite type?

2. Consider the geometric horseshoe presented in Section 13.1. It has fixed points at **0** and $\mathbf{p} = \left(^3\!/_4, ^4\!/_5\right)$. By determining the stable and unstable manifolds of these two points more than once across the square, show that there is a transverse homoclinic point at $\left(^1\!/_3, ^3\!/_4\right)$ for **0**, and a transverse homoclinic point at $\left(^1\!/_4, ^4\!/_5\right)$ for **p**.

13.4. Attractors

The definitions related to a chaotic attractor in higher dimensions are the same as those given earlier for one dimension, but we repeat them because it is helpful to rethink their meaning in terms of the different geometry of the situation. We also give an example of an attractor in $\mathbb{R}^3$ called the solenoid. This invariant set has the characteristic form of locally being the Cartesian product of a Cantor set and a curve. Finally, we discuss the Hénon attractor.

In these definitions, let **F** be a continuous map from $\mathbb{R}^n$ to itself.

Definition 13.27. A *trapping region for* **F** is a closed, bounded set **U** such that
$$\mathbf{F}(\mathbf{U}) \subset \operatorname{int}(\mathbf{U}).$$
Thus, the image of the closed set maps inside the interior of the set. If **F** is not continuous, then it is necessary to take the closure and assume that $\operatorname{cl}(\mathbf{F}(\mathbf{U})) \subset \operatorname{int}(\mathbf{U})$. See Section 11.2.

A set **A** is an *attracting set*, provided that there is a trapping region **U** such that
$$\mathbf{A} = \bigcap_{j=0}^{\infty} \mathbf{F}^j(\mathbf{U}).$$
The set **U** is called a *trapping region for* **A**.

The *basin of attraction* of the attracting set is
$$\mathscr{B}(\mathbf{A}; \mathbf{F}) = \{\, \mathbf{x} : \omega(\mathbf{x}; \mathbf{F}) \subset \mathbf{A} \,\}.$$

Since **A** is invariant and any invariant set contained in **U** must be a subset of **A**, it is the largest invariant set in **U**. Note that every point **x** in **U** has its ω-limit set inside **A**, $\omega(\mathbf{x}) \subset \mathbf{A}$.

Definition 13.28. An *attractor* is an attracting set **A** which has no proper subset that is an attracting set, i.e., if **A**′ is an attracting set with $\emptyset \neq \mathbf{A}' \subset \mathbf{A}$, then $\mathbf{A}' = \mathbf{A}$. In particular, an attracting set **A** for which there is a point $\mathbf{x}_0$ (in the ambient space) such that $\omega(\mathbf{x}; \mathbf{F}) = \mathbf{A}$ is an attractor.

A *transitive attractor* is an attracting set **A** for which there is a point $\mathbf{x}_0$ in **A** such that $\omega(\mathbf{x}; \mathbf{F}) = \mathbf{A}$.

The invariant set for the geometric horseshoe is not an attractor, but it is isolated according to the following definition.

Definition 13.29. An invariant set **S** for a homeomorphism **F** is called *isolated* if there is a closed, bounded region **U** that contains **S** in its interior such that
$$\mathbf{S} = \bigcap_{j=-\infty}^{\infty} \mathbf{F}^j(\mathbf{U}).$$
Note that this intersection is allowed to go from minus infinity to plus infinity. The set **U** is called an *isolating neighborhood* for the invariant set **S**.

See Section 13.7 for further discussion of isolated invariant sets that are not attractors.

The next theorem implies that an attracting set contains all the unstable manifolds of points in the attracting set. For this theorem, we extend the definition of an unstable manifold to points that are not necessarily periodic by defining
$$W^u(\mathbf{p}, \mathbf{F}) = \{\, \mathbf{x} : |\mathbf{F}^j(\mathbf{x}) - \mathbf{F}^j(\mathbf{p})| \text{ goes to zero as } j \text{ goes to } -\infty \,\}.$$

Theorem 13.12. *Assume that* **A** *is an attracting set for* **F** *and* **p** *is a point in* **A**. *Then, the unstable manifold of* **p** *is contained inside* **A**, $W^u(\mathbf{p}, \mathbf{F}) \subset \mathbf{A}$.

13.4. Attractors

Proof. Let $\mathbf{U}$ be a trapping region for $\mathbf{A}$. Because $\mathbf{U}$ is a neighborhood of $\mathbf{A}$, there is an $r > 0$ such that the local unstable manifolds $W_r^u(\mathbf{x}, \mathbf{F}) \subset \mathbf{U}$ for any $\mathbf{x}$ in $\mathbf{A}$. Then, for $k > 0$ and $j > 0$,

$$W_r^u(\mathbf{F}^{-k}(\mathbf{p}), \mathbf{F}) \subset \mathbf{F}^j\left(W_r^u(\mathbf{F}^{-k-j}(\mathbf{p}), \mathbf{F})\right) \subset \mathbf{F}^j(\mathbf{U}),$$
$$\mathbf{F}^k\left(W_r^u(\mathbf{F}^{-k}(\mathbf{p}), \mathbf{F})\right) \subset \mathbf{F}^{k+j}\left(W_r^u(\mathbf{F}^{-k-j}(\mathbf{p}), \mathbf{F})\right) \subset \mathbf{F}^{k+j}(\mathbf{U}),$$

$$\mathbf{F}^k\left(W_r^u(\mathbf{F}^{-k}(\mathbf{p}), \mathbf{F})\right) \subset \bigcap_{j=0}^{\infty} \mathbf{F}^{k+j}(\mathbf{U}) = \mathbf{A}, \qquad \text{and}$$

$$W^u(\mathbf{p}, \mathbf{F}) = \bigcup_{k=0}^{\infty} \mathbf{F}^k\left(W_r^u(\mathbf{F}^{-k}(\mathbf{p}), \mathbf{F})\right) \subset \mathbf{A}.$$

This proves the theorem. $\square$

Just as in the one-dimensional case, we define a chaotic attractor in terms of sensitive dependence on initial conditions.

Definition 13.30. A map $\mathbf{F}$ has *sensitive dependence on initial conditions when restricted to an invariant set* $\mathbf{S}$ provided that there is an $r > 0$ such that, for any point $\mathbf{q}$ in $\mathbf{S}$ and any $\delta > 0$, there is a point $\mathbf{x}$ in $\mathbf{S}$ with $\|\mathbf{x} - \mathbf{q}\| < \delta$ and a $k > 0$ such that

$$\|\mathbf{F}^k(\mathbf{q}) - \mathbf{F}^k(\mathbf{x})\| \geq r.$$

Definition 13.31. An isolated invariant set $\mathbf{S}$ is called *chaotic* provided that it satisfies the following conditions:

1. The map $\mathbf{F}$ is transitive on $\mathbf{S}$.
2. The map $\mathbf{F}$ restricted to $\mathbf{S}$ has sensitive dependence on initial conditions.

Thus, a *chaotic attractor* $\mathbf{A}$ is a transitive attractor such that the map $\mathbf{F}$ restricted to $\mathbf{A}$ has sensitive dependence on initial conditions.

Remark 13.32. In the second condition, we require that the system has sensitive dependence when restricted to the attractor and not just sensitive dependence on initial conditions at all points in the attractor within the ambient space. Example 13.41 at the end of the section (and Example 7.15) shows that these two conditions are not the same: Our stronger assumption eliminates the type of nonchaotic dynamic behavior that these examples exhibit.

Example 13.33 (Chaotic Attractor for Noninvertible Map). Consider the noninvertible map in polar coordinates given by

$$\mathbf{G}(\theta, r) = (2\theta \ (\bmod\ 2\pi), \sqrt{r}).$$

The set $r = 1$ is invariant. Since $G_2(r) = \sqrt{r}$ has $G_2'(r) = 1/(2\sqrt{r})$ and $G_2'(1) = 1/2 < 1$, the invariant set $\{r = 1\}$ is attracting. The map G_1 is transitive, so the circle $\{r = 1\}$ is a transitive attractor. Also, the map G_1 has sensitive dependence on initial conditions, so $\mathbf{G}$ restricted to the attractor has sensitive dependence. Thus, $\{r = 1\}$ is a chaotic attractor for $\mathbf{G}$.

Example 13.34 (Solenoid). The solenoid is an attractor in $\mathbb{R}^3$. We consider coordinates for which two of the dimensions are given by a single complex variable z in $\mathbb{C}$. We let
$$\mathbf{D} = \{\, z \in \mathbb{C} : |z| \leq 1 \,\}$$
be the unit disk in the complex plane. We use a real variable t modulo 1 to give the coordinate on a circle, S^1. Then, a solid torus can be given by
$$\mathbf{U} = S^1 \times \mathbf{D} = \{\, (t \ (\text{mod } 1), z) \in S^1 \times \mathbf{D} \,\}.$$
Define a map $\mathbf{F}$ from $\mathbf{U}$ into itself by stretching $\mathbf{U}$ in the t direction to be twice as long, shrinking in the z coordinate, and then wrapping the image around inside $\mathbf{U}$ twice in the S^1 direction, so that the first coordinate function is given by $F_1(t) = 2t \ (\text{mod } 1)$ and
$$\mathbf{F}(t,z) = \left(2t \ (\text{mod } 1),\ \beta z + \frac{1}{2} e^{2\pi t i} \right),$$
with $0 < \beta < 1/2$. The region $\mathbf{U}$ is a trapping region for $\mathbf{F}$ with attracting set
$$\Lambda = \bigcap_{k \geq 0} \mathbf{F}^k(\mathbf{U}).$$
We see later that Λ is actually an attractor. See Figure 25.

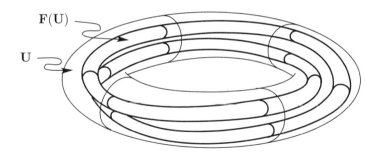

Figure 25. Trapping neighborhood $\mathbf{U}$ for the solenoid and its image $\mathbf{F}(\mathbf{U})$

Let
$$\mathbf{D}(t_0) = \{t_0\} \times \mathbf{D}$$
be one of the disks with a fixed "angle" t_0. The absolute value
$$\left| \frac{1}{2} e^{2\pi t i} \right| = \frac{1}{2} \left| e^{2\pi t i} \right| = \frac{1}{2},$$
so the image of one of the disks $\mathbf{D}(t_0)$ is a disk of radius β with center at the point $(1/2)e^{2\pi t i}$, which is on the circle of radius $1/2$. The two values t and $t + 1/2$ map to the same point, $F_1(t) = F_1\left(t + 1/2\right)$. In fact, if $F_1(t) = t_0$, then $t = {}^{t_0}\!/_2$ or ${}^{t_0}\!/_2 + 1/2$. Thus,
$$\mathbf{F}(\mathbf{U}) \cap \mathbf{D}(t_0) = \mathbf{F}\left(\mathbf{D}\left(\frac{t_0}{2}\right)\right) \cup \mathbf{F}\left(\mathbf{D}\left(\frac{t_0 + 1}{2}\right)\right)$$
consists of two subdisks with centers at
$$\frac{1}{2} e^{\pi t_0 i} \quad \text{and} \quad \frac{1}{2} e^{\pi i (t_0+1) i} = \frac{1}{2} e^{\pi t_0 i} e^{\pi i} = -\frac{1}{2} e^{\pi t_0 i}.$$

13.4. Attractors

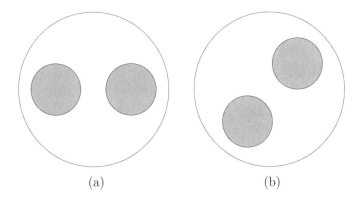

(a) (b)

Figure 26. $\mathbf{F}(\mathbf{U}) \cap \mathbf{D}(t_0)$ for $t_0 = 0$ and $\frac{1}{6}$

Thus, these centers are directly opposite each other at a distance 1 apart. The disks are disjoint since $2\beta < 1$. See Figure 26.

The second image,

$$\mathbf{F}^2(\mathbf{U}) \cap \mathbf{D}(t_0) = \mathbf{F}\left(\mathbf{F}(\mathbf{U}) \cap \mathbf{D}\left(\frac{t_0}{2}\right)\right) \cup \mathbf{F}\left(\mathbf{F}(\mathbf{U}) \cap \mathbf{D}\left(\frac{t_0+1}{2}\right)\right)$$
$$= \mathbf{F}^2\left(\mathbf{D}\left(\frac{t_0}{4}\right)\right) \cup \mathbf{F}^2\left(\mathbf{D}\left(\frac{t_0+2}{4}\right)\right)$$
$$\cup \mathbf{F}^2\left(\mathbf{D}\left(\frac{t_0+1}{4}\right)\right) \cup \mathbf{F}^2\left(\mathbf{D}\left(\frac{t_0+3}{4}\right)\right),$$

is the union of $2^2 = 4$ disjoint disks of radius β^2. Continuing by induction, $\mathbf{F}^k(\mathbf{U}) \cap \mathbf{D}(t_0)$ is the union of 2^k disjoint disks of radius β^k. The intersection

$$\mathbf{D}(t_0) \cap \bigcap_{k \geq 0} \mathbf{F}^k(\mathbf{U})$$
$$= \mathbf{D}(t_0) \cap \Lambda$$

is a Cantor set of points, sometimes called Cantor dust. The set $\mathbf{D}(t_0) \cap \Lambda$ is closed and perfect, individual points are connected components, and it has uncountably many points.

A disk $\mathbf{D}(t_0)$ is contracted by a factor of β and taken into another disk $\mathbf{D}(t_1)$. This continues, so the local stable manifolds of points are just one of these two-dimensional disks, or

$$W^s_{\text{loc}}(t_0, z_0) = \mathbf{D}(t_0).$$

A short line segment $\mathbf{L} = [t_0, t_1] \times \{z\}$ is stretched in the t direction and taken around the solid torus $\mathbf{U}$. Continuing to iterate this line segment, makes it grow in length and wrap more and more around the solid torus $\mathbf{U}$. Therefore, the unstable manifolds are curves that wind infinitely around inside the attractor in the t direction. In fact, each of the $W^u(t_0, z_0)$ winds densely in Λ. Locally, each point in the Cantor set $\mathbf{D}(t_0) \cap \Lambda$ has a curve in the unstable manifold running through it. Therefore, locally the attractor is a Cantor set of curves.

The trapping region $\mathbf{U}$ and each of its iterates $\mathbf{F}^k(\mathbf{U})$ are connected and nested inside one another. It follows that the intersection Λ is connected. However, there are two different points (t_1, z_1) and (t_2, z_2) that cannot be connected by a curve contained entirely inside Λ (i.e., Λ is not path connected).

Notice that, for this example, in which the map is invertible, the images $\mathbf{F}^k(\mathbf{U})$ get smaller as k increases: There is no finite k for which $\mathbf{F}^k(\mathbf{U}) = \Lambda$. This feature is different from the examples we gave of attractors for one-dimensional maps, and is a characteristic of attractors for invertible maps (diffeomorphisms) in higher dimensions.

Markov partition: Let
$$\mathbf{R}_0 = \left[0, \frac{1}{2}\right] \times \mathbf{D} \quad \text{and}$$
$$\mathbf{R}_1 = \left[\frac{1}{2}, 1\right] \times \mathbf{D}.$$

Then, $\mathbf{F}(\text{int}(\mathbf{R}_i)) \cap \text{int}(\mathbf{R}_j)$ is connected for each pair of i and j, but $\mathbf{F}(\mathbf{R}_i) \cap \mathbf{R}_j$ is not (e.g., $\mathbf{F}(\mathbf{R}_0) \cap \mathbf{R}_0$ has a second disk in $\{0\} \times \mathbf{D}$). The other properties for a Markov partition are direct. The different parts of the boundary are given by

$$\partial^{\text{out}}(\mathbf{R}_0) = \mathbf{D}(0) \cup \mathbf{D}\left(\frac{1}{2}\right),$$
$$\partial^{\text{out}}(\mathbf{R}_1) = \mathbf{D}(\frac{1}{2}) \cup \mathbf{D}(1),$$
$$\partial^{\text{in}}(\mathbf{R}_0) = \left[0, \frac{1}{2}\right] \times \{z \in \mathbb{C} : |z| = 1\}, \quad \text{and}$$
$$\partial^{\text{in}}(\mathbf{R}_1) = \left[\frac{1}{2}, 1\right] \times \{z \in \mathbb{C} : |z| = 1\}.$$

Because all transitions are allowed, the symbol space is the full two shift, Σ_2. Because the shift map is topologically transitive on Σ_2, $\mathbf{F}$ is topologically transitive on Λ. This implies that Λ is an attractor, and not just an attracting set. Also, because the periodic points are dense for the shift map in Σ_2, the periodic points of $\mathbf{F}$ are dense in Λ.

The solenoid has sensitive dependence on initial conditions when restricted to the attractor because the first coordinate map (the doubling map) is expanding. Therefore, this set for the solenoid is a chaotic attractor much like the one for the noninvertible map in Example 13.33, except that the solenoid is invertible.

Example 13.35. This example is related to one Martelli gives in [**Mar99**] of a chaotic invariant set. Let
$$\mathbf{U} = S^1 \times \mathbf{D} \times S^1 = \{(t \ (\text{mod } 1), z, \tau \ (\text{mod } 1)) \in S^1 \times \mathbf{D}\}$$
be the Cartesian product of the solid torus of the solenoid and the circle. Consider the map on $\mathbf{U}$ given by
$$\mathbf{F}(t, z, \tau) = \left(2t \ (\text{mod } 1), \beta z + \frac{1}{2} e^{2\pi t i}, \tau + \omega\right),$$
where $0 < \beta < 1/2$ and ω irrational. Let Λ be the attracting set for this trapping region. The first two coordinates are topologically transitive because of the discussion of the solenoid. The third coordinate function is an irrational rotation, and so it is topologically transitive. Therefore, there is a point (t_0, z_0, τ_0) whose

13.4. Attractors

ω-limit set equals the whole set Λ. The first coordinate insures the map has sensitive dependence on initial conditions. Therefore this is a chaotic attractor by our definition.

Example 13.36 (Nonchaotic Attractor with Sensitive Dependence). Consider the map

$$\mathbf{F}(\theta, \phi, x) = \left(\theta + \tfrac{1}{2}\sin(\phi) \ (\mathrm{mod}\ 2\pi\,), \ \phi \ (\mathrm{mod}\ 2\pi\,), \ \tfrac{1}{2}x\right).$$

The set $\mathbf{A} = \{\, x = 0 \,\}$ is an attracting set. The coordinate ϕ is a constant along an orbit, so $\mathbf{F}$ is not transitive on $\mathbf{A}$. If $\sin(\phi_0)$ is irrational, then $\mathbf{F}$ is an irrational rotation on the circle $\{\, \phi = \phi_0,\ x = 0\,\}$. Any attracting set must contain complete copies of these circles, and they are dense in $\mathbf{A}$. Therefore, there no proper subset of $\mathbf{A}$ that are attracting sets, and $\mathbf{A}$ is an attractor. If $\phi \neq \phi'$, then the difference of the first coordinates of the iterates

$$\mathbf{F}^n(\theta, \phi, 0) = \left(\theta + \frac{n}{2}\sin(\phi) \ (\mathrm{mod}\ 2\pi\,),\ \phi \ (\mathrm{mod}\ 2\pi\,),\ 0\right) \quad \text{and}$$

$$\mathbf{F}^n(\theta', \phi', 0) = \left(\theta' + \frac{n}{2}\sin(\phi') \ (\mathrm{mod}\ 2\pi\,),\ \phi' \ (\mathrm{mod}\ 2\pi\,),\ 0\right) \quad \text{is}$$

$$|\theta_n - \theta'_n| = \frac{n}{2}|\sin(\phi) - \sin(\phi')| \ (\mathrm{mod}\ 2\pi\,).$$

This grows until it is approximately $1/2$. Therefore, this map has sensitive dependence on initial conditions when restricted to $\mathbf{A}$. Since $\mathbf{F}$ restricted to $\mathbf{A}$ is not transitive, it is a nonchaotic attractor for which $\mathbf{F}$ restricted to $\mathbf{A}$ has sensitive dependence on initial conditions. Since the dynamics is merely a shear on the attractor, is seems right to not call it chaotic.

Example 13.37. Unfortunately, the map in this example does not seem chaotic, but it does meet our definition of a chaotic attractor.

For *beta* an irrational number, consider that map

$$\mathbf{F}\begin{pmatrix} x \\ y \end{pmatrix} = \begin{pmatrix} x + \beta \ (\mathrm{mod}\ 1) \\ x + y \ (\mathrm{mod}\ 1) \end{pmatrix}.$$

Since both variables are taken modulo one, this is a map on the two torus.

The map can be seem to have sensitive dependence on initial conditions as follows. For any point (x_0, y_0), let $(x'_0, y'_0) = (x_0 = \delta, y_0)$, $\mathbf{F}^n(x_0, y_0) = (x_n, y_n)$, and $\mathbf{F}^n(x'_0, y'_0) = (x'_n, y'_n)$. Then, $|y'_n - y)n| = n\beta (\mathrm{mod}\ 1)$, which can be made greater than $1/4$ for correctly chosen large n.

To see that $\mathbf{F}$ is topologically transitive, we use the Birkhoff Transitivity Theorem 11.18. Any two open sets $\mathbf{U}$ and $\mathbf{V}$ contain squares,

$$\left[x_0, x_0 + \frac{1}{k}\right] \times \left[y_0, y_0 + \frac{1}{k}\right] \subset \mathbf{U} \quad \text{and}$$

$$\left[x'_0, x'_0 + \frac{2}{k}\right] \times \left[y'_0, y'_0 + \frac{2}{k}\right] \subset \mathbf{V}.$$

The edge $\mathbf{F}^n\left(\left[x_0, x_0 + \tfrac{1}{k}\right] \times \{y_0\}\right)$ looked at in $\mathbb{R}^2$ is a line segment from (x_n, y_n) to $(x_n + 1/k, y_n + n/k)$. For $n > k$, this line segment wraps at least once around the circle in the y-direction. For correctly chosen $n > k$, $\left[x_n, x_n + 1/k\right] \subset \left[x'_0, x'_0 + \tfrac{2}{k}\right]$, so the line segment intersects **V**. Since this is true for any pair of open sets, **F** is topologically transitive.

Thus, we have shown that **F** is chaotic by our definition on all of the two torus. Since the sensitive dependence comes form a shear, this map does not seem very "chaotic". In the next section, we define Lyapunov exponents. The Lyapunov exponents for this map are both zero, which is a statement that the points do not move apart at an exponential rate but only linearly.

Example 13.38 (Hénon). The traditional values of the parameters for the Hénon map that give an attractor are $a = 1.4$ and $b = -0.3$. There is a saddle fixed point **p** part of whose stable and unstable manifolds look like those in Figure 27. Take the positively invariant closed set Ω whose boundary consists of pieces of stable and unstable manifolds as in Figure 27. Let

$$\Lambda = \bigcap_{k \geq 0} \mathbf{F}^k(\Omega).$$

The area contracts by $|\det(D\mathbf{F}_{(x,y)})| = |b| = 0.3$ for each iterate. Thus, the area of the iterates $\mathbf{F}^k(\Omega)$ goes to zero, and Λ must have zero area (zero measure).

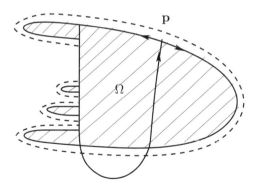

Figure 27. Positively invariant region for Hénon map

To construct a trapping region for Λ, it is necessary to enlarge Ω slightly on the part of the boundary made up of $W^u(\mathbf{p})$. Since the map contracts in toward $W^u(\mathbf{p})$ near **p**, this curve can be chosen so it maps closer to Ω. See dashed curve in Figure 27. This enlarged region $\mathbf{U} \supset \Omega$ still is a trapping region for Λ:

$$\bigcap_{k \geq 0} \mathbf{F}^k(\mathbf{U}) = \bigcap_{k \geq 0} \mathbf{F}^k(\Omega) = \Lambda.$$

This argument shows that Λ is an attracting set.

Numerical evidence indicates that all points of Λ are chain recurrent. Figure 28 shows the iterate of one point in the attractor. This result has been proven rigorously by M. Benedicks and L. Carleson for values of b close to 0, but not for the value $b = -0.3$ itself. See [**Ben91**] and [**Ben93**].

By Theorem 13.12, the unstable manifold of **p** must be entirely contained inside Λ. Because the lengths of the segments along the stable manifold go to zero under iteration, it can be shown that $W^u(\mathbf{p})$ is dense in Λ. See [**Rob99**].

13.4. Attractors

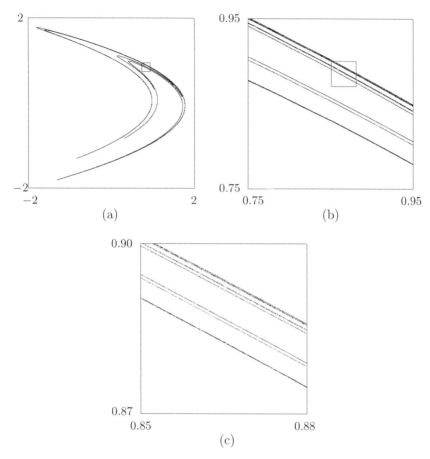

Figure 28. One orbit in the Hénon attractor for $a = 1.4$ and $b = -0.3$. (a) The window shown is $[-2, 2] \times [-2, 2]$. (b) The window shown is $[0.75, 0.95] \times [0.75, 0.95]$. (c) The window shown is $[0.85, 0.88] \times [0.97, 0.90]$.

Locally, the attractor has the general form of a Cantor set of curves. See Figure 28. When part of the attractor is enlarged, as from Figure 28(b) to Figure 28(c), the form of the attractor does not change much.

Example 13.39 (Beetle population model). In Section 12.5.3, we gave certain parameters for the model of beetle population where a chaotic attractor is seen numerically. See Figure 13 in Chapter 12.

Example 13.40. The geometric horseshoe is an example of a chaotic invariant set that is not an attractor. Theorem 13.3 proves that there is a point whose ω-limit set is the invariant set, and that the map restricted to the invariant set has sensitive dependence on initial conditions. Remark 13.53 gives the pair of trapping regions which determine the isolating neighborhood that implies that it is what we call a basic invariant set.

Example 13.41. This example illustrates the distinction between sensitive dependence at all points of an attracting set and sensitive dependence when restricted to an attracting set. It is easier to express in terms of solutions of differential equations, so we start that way. Consider the system of differential equations in polar coordinates given by

$$\frac{dr}{dt} = -\frac{1}{2}(r-1)^3,$$
$$\frac{d\theta}{dt} = \alpha 2\pi + \frac{1}{2}(r-1).$$

We write $r(t; r_0)$ and $\theta(t; r_0, \theta_0)$ for the solution with $r(0; r_0) = r_0$ and $\theta(0; r_0, \theta_0) = \theta_0$. By separation of variables, for $r_0 \neq 1$, we find that

$$\frac{-2 \frac{dr}{dt}}{(r-1)^3} = 1,$$
$$(r(t; r_0) - 1)^{-2} = t + C,$$
$$r(t; r_0) - 1 = \frac{S}{\sqrt{t+C}},$$

where

$$S = \text{sign}(r_0 - 1) = \begin{cases} 1 & \text{if } r_0 - 1 \geq 0, \\ -1 & \text{if } r_0 - 1 < 0. \end{cases}$$

By evaluating at $t = 0$, we see that

$$r_0 - 1 = \frac{S}{\sqrt{C}} \quad \text{or}$$
$$C = (r_0 - 1)^{-2}.$$

Thus,

$$r(t; r_0) = \begin{cases} 1 & \text{for } r_0 = 1, \\ 1 + \dfrac{\text{sign}(r_0 - 1)}{\sqrt{t + (r_0 - 1)^{-2}}} & \text{for } r_0 \neq 1. \end{cases}$$

Substituting the expression for $r(t; r_0)$ into the differential equation for θ, we get

$$\frac{d\theta}{dt} = \alpha 2\pi + \frac{S}{2\sqrt{t+C}},$$
$$\theta(t; r_0, \theta_0) = \theta_0 + \alpha 2\pi t + S\left(\sqrt{t+C} - \sqrt{C}\right) \pmod{2\pi},$$
$$\theta(1; r_0, \theta_0) = \theta_0 + \alpha 2\pi + S\left(\sqrt{1+C} - \sqrt{C}\right) \pmod{2\pi}.$$

Combining the solutions for $r(1; r_0)$ and $\theta(1; r_0, \theta_0)$, we can define the time one map,

$$\mathbf{F}\begin{pmatrix} r_0 \\ \theta_0 \end{pmatrix} = \begin{pmatrix} 1 + \dfrac{\text{sign}(r_0 - 1)}{\sqrt{1 + (r_0 - 1)^{-2}}} \\ \theta_0 + \alpha 2\pi + S\left(\sqrt{1+C} - \sqrt{C}\right) \pmod{2\pi} \end{pmatrix},$$

where $C = (r_0 - 1)^{-2}$ and $S = \text{sign}(r_0 - 1)$.

Let

$$\mathbf{U} = \{(r, \theta) : 0.5 \leq r \leq 1.5,\ 0 \leq \theta \leq 2\pi\}.$$

Then,

$$\mathbf{F}(\mathbf{U}) \subset \text{int}(\mathbf{U}) = \{\,(r, \theta) : 0.5 < r < 1.5,\ 0 \leq \theta \leq 2\pi\,\} \quad \text{and}$$

$$\bigcap_{j \geq 0} \mathbf{F}^j(\mathbf{U}) = \{\,(1, \theta) : 0 \leq \theta \leq 2\pi\,\}.$$

Thus, $\mathbf{A} = \{\,(1, \theta) : 0 \leq \theta \leq 2\pi\,\}$ is an attracting set with trapping region $\mathbf{U}$.

For $r_0 = 1$, $r(n; 1) = 1$, $\theta(n; 1, \theta_0) = \theta_0 + \alpha 2\pi n$, and the map is like the rotation R_α. (The α is multiplied by the factor 2π because the variable is taken modulo 2π rather than modulo 1.) If α is irrational, then $\omega((1, \theta_0); \mathbf{F}) = \mathbf{A}$ (i.e., $\mathbf{F}$ is topologically transitive on the circle $r = 1$). Thus, for α irrational, the set $\mathbf{A}$ is an attractor.

The restriction of $\mathbf{F}$ to $\mathbf{A}$ does not have sensitive dependence on initial conditions because it preserves lengths, so $\mathbf{A}$ is not a chaotic attractor.

However, if $r_0 > 1$, then

$$\theta(n; r_0, \theta_0) - [\theta_0 + \alpha 2\pi n] = \sqrt{n + (r_0 - 1)^{-2}} - (r_0 - 1)^{-1}$$

goes to infinity as n goes to infinity. Thus, the iterates of points off the circle run ahead of the iterates on the circle, causing the map $\mathbf{F}$ to have sensitive dependence on initial conditions at all points of $\mathbf{A}$ when considered as a map on the two-dimensional ambient space.

Exercises 13.4

1. Consider the Hénon map with parameters $a = 1.4$ and $b = -0.3$. Let $\mathbf{R}$ be the quadrilateral with the four vertices $(1.857, 0.63)$, $(-1.875, 1.99)$, $(-1.49, -2.2)$, and $(1.752, -0.60)$. Show that $\mathbf{R}$ is a trapping region.

2. The Lozi map is a piecewise affine map that is related to the Hénon map. Let

$$\mathbf{F}\begin{pmatrix} x \\ y \end{pmatrix} = \begin{pmatrix} 1 + y - A|x| \\ Bx \end{pmatrix},$$

where A and B are parameters. Take $B = 1/2$ and $A = 3/2$.
   a. Prove that $\mathbf{F}$ has two fixed points, one of which $\mathbf{p}_1$ lies in the first quadrant.
   b. Prove that the unstable manifold of $\mathbf{p}_1$, $W^u(\mathbf{p}_1; \mathbf{F})$, contains a line segment $\mathbf{L}$ that intersects the x-axis at a point $\mathbf{q}$ and the y-axis at $\mathbf{F}^{-1}(\mathbf{q})$.
   c. Draw the line segments $\mathbf{L}$, $\mathbf{F}(\mathbf{L})$, and $\mathbf{F}^2(\mathbf{L})$.
   d. Let $\mathbf{U}$ be the triangular region with corners at $\mathbf{q}$, $\mathbf{F}(\mathbf{q})$ and $\mathbf{F}^2(\mathbf{q})$. Prove that $\mathbf{U}$ is a trapping region.

3. Consider the solenoid map $\mathbf{F}$ given in Example 13.34. Consider the three sets
$$\mathbf{R}_0 = \{\,(t, z) \in \mathbf{N} : 0 \leq t \leq 1/3\,\},$$
$$\mathbf{R}_1 = \{\,(t, z) \in \mathbf{N} : 1/3 \leq t \leq 2/3\,\}, \quad \text{and}$$
$$\mathbf{R}_2 = \{\,(t, z) \in \mathbf{N} : 2/3 \leq t \leq 1\,\}.$$

Show that these form a Markov partition. What is the transition graph?

4. Consider the map
$$\mathbf{F}\begin{pmatrix} t \\ z \end{pmatrix} = \begin{pmatrix} 4t \pmod{1} \\ \beta z + 0.5\, e^{2\pi t i} \end{pmatrix},$$
which is a modification of the solenoid.
   a. Show that, for $0 < \beta < 1/(2\sqrt{2})$, the map is one to one on the trapping neighborhood $\mathbf{N}$ used for the solenoid.
   b. Explain why $\mathbf{F}$ has sensitive dependence on initial conditions when restricted to the attracting set determined by the trapping region $\mathbf{N}$?
   c. What is a Markov partition for this map? What is the transition graph for this partition? Explain why the attracting set is an attractor. Hint: It is necessary to take more than two M-rectangles to get the intersections to be connected.

13.5. Lyapunov Exponents

Just as in our treatment in one dimension, Lyapunov exponents are quantities that are numerically computable and are related to sensitive dependence on initial conditions.

For a differentiable map f from $\mathbb{R}$ to itself, we defined the Lyapunov exponent of a point p to be the following limit:
$$\ell(p, f) = \lim_{k \to \infty} \frac{1}{k} \ln \left| (f^k)'(p) \right|.$$
For maps $\mathbf{F}$ from $\mathbb{R}^n$ to itself, there are several different directions, which can have different rates of expansion or contraction. Thus, we need to define a number that depends on a direction or vector $\mathbf{v}$, as well as, on the point $\mathbf{p}$. By carefully picking n different vectors, we get n principal rates of expansion and contraction for the point $\mathbf{p}$.

Definition 13.42. Let $\mathbf{F}$ from $\mathbb{R}^n$ to itself be differentiable, and let $\mathbf{p}$ be a point with a bounded forward orbit, $\mathcal{O}_\mathbf{F}^+(\mathbf{p})$. Let $\mathbf{v}$ be a vector in $\mathbb{R}^n$, usually a unit vector (i.e., of length one). Define the *Lyapunov exponent of* $\mathbf{F}$ *at* $\mathbf{p}$ *in the direction of* $\mathbf{v}$ to be the following limit,
$$\ell(\mathbf{p}, \mathbf{v}; \mathbf{F}) = \lim_{k \to \infty} \frac{1}{k} \ln \left\| D(\mathbf{F}^k)_{(\mathbf{p})} \mathbf{v} \right\|.$$
Just as in the one-dimensional case, we calculate the Lyapunov exponent by using the chain rule to write
$$D(\mathbf{F}^k)_{(\mathbf{p})} = D\mathbf{F}_{(\mathbf{p}_{k-1})} \cdots D\mathbf{F}_{(\mathbf{p}_0)},$$
where $\mathbf{p}_j = \mathbf{F}^j(\mathbf{p})$. Because the order of multiplication of matrices matters, it is important to multiply the matrices in the order given. We can also let $\mathbf{v}_0 = \mathbf{v}$ and define $\mathbf{v}_k = D\mathbf{F}_{\mathbf{p}_{k-1}} \mathbf{v}_{k-1}$. Then, $\ell(\mathbf{p}, \mathbf{v}; \mathbf{F}) = \lim_{k \to \infty} (1/k) \ln(\|\mathbf{v}_{k-1}\|)$.

Consider a diffeomorphism $\mathbf{F}$ in n dimensions. For most initial conditions $\mathbf{p}$ for which the forward orbit is bounded, the Lyapunov exponents exist for all vectors $\mathbf{v}$. There are at most n distinct values for $\ell(\mathbf{p}, \mathbf{v}; \mathbf{F})$ as $\mathbf{v}$ varies. If we count multiplicities, then there are exactly n values, $\ell_1(\mathbf{p}; \mathbf{F}) = \ell(\mathbf{p}, \mathbf{v}_1; \mathbf{F})$, $\ell_2(\mathbf{p}; \mathbf{F}) = \ell(\mathbf{p}, \mathbf{v}_2; \mathbf{F})$, ..., $\ell_n(\mathbf{p}; \mathbf{F}) = \ell(\mathbf{p}, \mathbf{v}_n; \mathbf{F})$. We can order these so that
$$\ell_1(\mathbf{p}; \mathbf{F}) \geq \ell_2(\mathbf{p}; \mathbf{F}) \geq \cdots \geq \ell_n(\mathbf{p}; \mathbf{F}).$$

13.5. Lyapunov Exponents

In Subsection 13.5.1, we explain the fact that there are n rates in terms of eigenvalues of matrices. Because the growth rate of a combination of quantities is determined by the largest growth rate, most initial vectors will grow at the rate $\ell_1(\mathbf{p}; \mathbf{F})$. In fact, if $\mathbf{v}$ can be written as a linear combination of $\mathbf{v}_j, \ldots, \mathbf{v}_n$, namely

$$\mathbf{v} = a_j\,\mathbf{v}_j + a_{j+1}\,\mathbf{v}_{j+1} + \cdots + a_n\,\mathbf{v}_n$$

with $a_j \neq 0$, then

$$\ell(\mathbf{p}, \mathbf{v}; \mathbf{F}) = \ell(\mathbf{p}, \mathbf{v}_j; \mathbf{F}) = \ell_j(\mathbf{p}; \mathbf{F}).$$

For the case in which a diffeomorphism has a derivative whose determinant is a constant, such as the Hénon map or the solenoid map, the next theorem gives a condition of the sum of all the Lyapunov exponents.

Theorem 13.13. *Assume that $\mathbf{F}$ is a diffeomorphism in n dimensions with $\Delta = |\det(D\mathbf{F_p})|$, where Δ is a constant.*

Then, for any point $\mathbf{p}$ for which the Lyapunov exponents exist,

$$\ell_1(\mathbf{p}; \mathbf{F}) + \cdots + \ell_n(\mathbf{p}; \mathbf{F}) = \ln(\Delta).$$

In Section 13.5.1, we show why this theorem is true after we have discussed why Lyapunov exponents are the limit of eigenvalues of a sequence of certain matrices.

We come back later to discuss practical algorithms to calculate the Lyapunov exponents numerically. At the moment, we just remark that the Lyapunov exponents are similar to the logarithms of the eigenvalues of a periodic point. In fact, we have the next theorem.

Theorem 13.14. *Assume that $\mathbf{p}$ is a period-m point for a differentiable map $\mathbf{F}$. Let λ_j be the eigenvalues of $D(\mathbf{F}^m)_{(\mathbf{p})}$ with generalized eigenvectors $\mathbf{v}^j$. (If an eigenvalue $\lambda_j = \alpha_j + i\,\beta_j$ is complex, then we have to use the complex eigenvector, with real and imaginary parts, for the vectors $\mathbf{v}^j$ for $\alpha_j \pm i\,\beta_j$.) Then, the Lyapunov exponents are*

$$\ell(\mathbf{p}, \mathbf{v}^j; \mathbf{F}) = \frac{1}{m}\ln|\lambda_j|.$$

We need to divide by the period, because the eigenvalues are for a power of the map and not for $\mathbf{F}$ itself.

The next theorem is like Theorem 11.8 for one-dimensional maps, except that a diffeomorphism cannot have a point for which $\det(D\mathbf{F_p}) = 0$, so we do not have to worry about the zeros of the derivative.

Theorem 13.15. *Assume that $\mathbf{F}$ is a diffeomorphism in $\mathbb{R}^n$. Assume further that $\ell_i(\mathbf{p}_0; \mathbf{F})$ exist for $1 \leq i \leq n$, and that $\mathbf{x}_0$ is a point for which*

$$\lim_{j \to \infty} \|\mathbf{F}^j(\mathbf{x}_0) - \mathbf{F}^j(\mathbf{p}_0)\| = 0.$$

Then, $\ell_i(\mathbf{x}_0; \mathbf{F})$ exists and $\ell_i(\mathbf{x}_0; \mathbf{F}) = \ell_i(\mathbf{p}_0; \mathbf{F})$ for $1 \leq i \leq n$.

In particular, if $\mathbf{p}_0$ is a periodic point, and $\mathbf{x}_0$ is on the stable manifold of the orbit for $\mathbf{p}_0$, then $\ell_i(\mathbf{x}_0; \mathbf{F}) = \ell_i(\mathbf{p}_0; \mathbf{F})$ for $1 \leq i \leq n$.

We end the section with two simple examples in which the expanding and contracting directions are constant, and for which the Lyapunov exponents can be computed by hand. In Section 13.5.2, we describe the method for computing the exponents numerically for examples in which the expanding directions are not all the same at all points.

Example 13.43. For the geometric horseshoe, the derivative at all points has the form

$$D\mathbf{F}_{(x,y)} = \begin{pmatrix} \pm\frac{1}{3} & 0 \\ 0 & \pm 4 \end{pmatrix}.$$

Since this derivative preserves the horizontal and vertical directions, if $\mathbf{p}$ is a point in the invariant set Λ, then

$$D(\mathbf{F}^k)_\mathbf{p} \begin{pmatrix} 1 \\ 0 \end{pmatrix} = \pm \left(\frac{1}{3}\right)^k \begin{pmatrix} 1 \\ 0 \end{pmatrix},$$

$$D(\mathbf{F}^k)_\mathbf{p} \begin{pmatrix} 0 \\ 1 \end{pmatrix} = \pm 4^k \begin{pmatrix} 0 \\ 1 \end{pmatrix},$$

$$\ell(\mathbf{p}, (1,0)^\mathsf{T}; \mathbf{F}) = \ln\left(\frac{1}{3}\right) = -\ln(3), \quad \text{and}$$

$$\ell(\mathbf{p}, (0,1)^\mathsf{T}; \mathbf{F}) = \ln(4).$$

Therefore, this map has two nonzero Lyapunov exponents: One is positive and one is negative. Also,

$$\Delta = \left|\det\left(D\mathbf{F}_{(x,y)}\right)\right| = \frac{4}{3}$$

is a constant and $\ln(\Delta)$ is the sum of the Lyapunov exponents.

Example 13.44. For a second example, consider the solenoid. The derivative of the map is

$$D\mathbf{F}_{(t_0,z_0)} = \begin{pmatrix} 2 & 0 \\ \pi i & \beta \mathbf{I}_2 \end{pmatrix},$$

where $\mathbf{I}_2$ is the two-by-two identity matrix, because β acts on a two-dimensional space. Because the diagonal entries of the derivative are 2, β, and β, the determinant of the derivative is

$$\Delta = \left|\det\left(D\mathbf{F}_{(x,y)}\right)\right| = 2\beta^2.$$

Vectors $\mathbf{v} = (0, \mathbf{v}_2)^\mathsf{T}$, tangent to one of the disks $\mathbf{D}(t_0)$, are contracted by a factor of $\beta < \frac{1}{2}$ and taken to a vector of the same type. Therefore,

$$D\mathbf{F}^k_{(t_0,z_0)} \begin{pmatrix} 0 \\ \mathbf{v}_2 \end{pmatrix} = \begin{pmatrix} 0 \\ \beta^k \mathbf{v}_2 \end{pmatrix}$$

and

$$\ell((t_0, z_0), (0, \mathbf{v}_2)^\mathsf{T}; \mathbf{F}) = \ln(\beta) < 0.$$

Since these disks are two-dimensional (z is in $\mathbb{C}$), the Lyapunov exponent $\ln(\beta)$ has multiplicity two.

By Theorem 13.13, the third Lyapunov exponent is

$$\ln(2\beta^2) - 2\ln(\beta) = \ln(2).$$

This can also be seen from the fact that the derivative stretches vectors of the form $(v_1, \mathbf{v}_2)^\mathsf{T}$ with $v_1 \neq 0$ by a factor of two in the first component. The second component is contracted. Therefore, the growth rate is dominated by the first component, and

$$\ell((t_0, z_0), (v_1, \mathbf{v}_2)^\mathsf{T}); \mathbf{F}) = \ln(2) > 0.$$

13.5. Lyapunov Exponents

This example has two negative Lyapunov exponents (two contracting directions) and one positive Lyapunov exponent (one expanding direction).

Example 13.45. For the Hénon map with $a = 1.4$ and $b = -0.3$, the first Lyapunov exponent can be calculated to be $\ell_1(\mathbf{p}; \mathbf{F}) = 0.42\ldots$. Because, the determinant of the derivative is a constant, it is not necessary to numerically calculate the second Lyapunov exponent. We can just derive it from the first exponent and the logarithm of the determinant. The sum of the Lyapunov exponents equals

$$\ln(\Delta) = \ln\left(|\det(D\mathbf{F}_\mathbf{x})|\right) = \ln(0.3).$$

Therefore, the second Lyapunov exponent is

$$\begin{aligned}\ell_2(\mathbf{p}; \mathbf{F}) &= \ln(0.3) - \ell_1(\mathbf{p}; \mathbf{F}) \\ &= -1.20\cdots - 0.42\ldots \\ &= -1.62\ldots.\end{aligned}$$

13.5.1. Lyapunov Exponents from Axes of Ellipsoids.

In this subsection, we show the relationship between the Lyapunov exponents and the lengths of the axes of the image of a sphere by the derivative map. This discussion helps explain why there are n Lyapunov exponents in n dimensions. Also, we use it to relate the sum of the Lyapunov exponents and the determinant of the derivative of the diffeomorphism.

Image of all unit vectors by a matrix

We start by considering the image of all unit vectors by a linear map. We use the notation

$$\mathscr{U} = \{\mathbf{x} \in \mathbb{R}^n : \|\mathbf{x}\| = 1\}$$

for the set of all unit vectors in $\mathbb{R}^n$. In $\mathbb{R}^2$, $\mathscr{U}$ is the unit circle; in three dimensions, $\mathscr{U}$ is the usual sphere. In higher dimensions, this set is called the $n-1$ sphere because there are $n-1$ directions to move while staying on this set.

In dimension two, the image of the unit circle $\mathscr{U}$ by a linear map is an ellipse; in dimensions three and above, the image of the unit sphere $\mathscr{U}$ by a linear map is an ellipsoid. The lengths of the axes of the ellipse or ellipsoid give the different fundamental amounts by which vectors are stretched or contracted by the linear map. For the sequence of linear maps given by the derivatives of powers of a map, these expansion rates determine the Lyapunov exponents of the orbit.

Theorem 13.16. *Let $\mathbf{A}$ be a real $n \times n$ matrix.*

a. *Then the eigenvalues of $\mathbf{A}^\mathsf{T}\mathbf{A}$ and $\mathbf{A}\mathbf{A}^\mathsf{T}$ are the same, $\{\lambda_1, \ldots, \lambda_n\}$. Moreover, all the λ_j are real and $\lambda_j \geq 0$. Let $s_j = \sqrt{\lambda_j}$. Let $\{\mathbf{v}^1, \ldots, \mathbf{v}^n\}$ be corresponding eigenvectors of $\mathbf{A}^\mathsf{T}\mathbf{A}$ and $s_j\mathbf{u}^j = \mathbf{A}\mathbf{v}^j$. Then $\{\mathbf{u}^1, \ldots, \mathbf{u}^n\}$ are eigenvectors for $\mathbf{A}\mathbf{A}^\mathsf{T}$, and these two sets of eigenvectors can be chosen to be orthonormal, $\mathbf{v}^k \cdot \mathbf{v}^j = 0 = \mathbf{u}^k \cdot \mathbf{u}^j$ for $k \neq j$ and $\|\mathbf{v}^j\| = 1 = \|\mathbf{u}^j\|$. Finally, the image of $\mathscr{U}$ by $\mathbf{A}$ is an ellipsoid with semiaxes given by the vectors $s_j\mathbf{u}^j = \mathbf{A}\mathbf{v}^j$.*

b. *If we define the matrix $\mathbf{P}$ by $\mathbf{P}\mathbf{v}^j = s_j\mathbf{v}^j$, then $\mathbf{P}$ is a positive semidefinite symmetric matrix. If we let $\mathbf{Q} = \mathbf{A}\mathbf{P}^{-1}$, then $\mathbf{Q}$ is an orthogonal matrix. Finally, $\mathbf{A} = \mathbf{Q}\mathbf{P}$. Thus, any matrix $\mathbf{A}$ can be written as the product of an orthogonal matrix and a positive semidefinite symmetric matrix.*

Further discussion and a proof of this result are presented in Section 13.7. The axes of the ellipsoid $\mathbf{A}(\mathscr{U})$ and their lengths are expressed in terms of eigenvalues and eigenvectors of the matrix $\mathbf{A}^\mathsf{T}\mathbf{A}$. In this section, we merely give an example in which we calculate these quantities before returning to the discussion of Lyapunov exponents.

Example 13.46. Let
$$\mathbf{A} = \begin{pmatrix} \frac{2}{3} & 1 \\ 0 & \frac{2}{3} \end{pmatrix}.$$

Then,
$$\mathbf{A}^\mathsf{T}\mathbf{A} = \begin{pmatrix} \frac{4}{9} & \frac{2}{3} \\ \frac{2}{3} & 1 + \frac{4}{9} \end{pmatrix},$$

which has characteristic equation
$$\lambda^2 - \frac{17}{9}\lambda + \frac{16}{81} = 0,$$

and eigenvalues
$$\lambda = \frac{1}{9}, \frac{16}{9}.$$

So, even though both eigenvalues of $\mathbf{A}$ are less than one, the largest eigenvalue of $\mathbf{A}^\mathsf{T}\mathbf{A}$ is $^{16}/_9$ which is greater than one, and the norm of $\mathbf{A}$ is $\sqrt{^{16}/_9} = {}^4/_3$, which is greater than one.

The Lyapunov exponents are related to the logarithms of the eigenvalues of the product of matrices, so we continue this example by considering powers of the same matrix. As we have seen before,

$$\mathbf{A}^k = \begin{pmatrix} \left(\frac{2}{3}\right)^k & k\left(\frac{2}{3}\right)^{k-1} \\ 0 & \left(\frac{2}{3}\right)^k \end{pmatrix},$$

$$\left(\mathbf{A}^k\right)^\mathsf{T}\mathbf{A}^k = \left(\frac{2}{3}\right)^{2k}\mathbf{B}_k, \qquad \text{where}$$

$$\mathbf{B}_k = \begin{pmatrix} 1 & k\left(\frac{3}{2}\right) \\ k\left(\frac{3}{2}\right) & 1 + k^2\left(\frac{9}{4}\right) \end{pmatrix}.$$

The eigenvalues of $\mathbf{B}_k$ are roots of
$$\lambda^2 - \left(2 + \frac{9k^2}{4}\right)\lambda + 1 = 0,$$

which are
$$\lambda = \frac{2 + \frac{9k^2}{4} \pm \sqrt{\left(2 + \frac{9k^2}{4}\right)^2 - 4}}{2}$$
$$= \frac{8 + 9k^2 \pm \sqrt{81k^4 + 144k^2}}{8}.$$

Thus, the eigenvalues of $\left(\mathbf{A}^k\right)^\mathsf{T}\mathbf{A}^k$ are
$$\lambda_j^{(k)} = \left(\frac{2}{3}\right)^{2k}\left[\frac{8 + 9k^2 \pm 9k^2\sqrt{1 + {}^{16}/(9k^2)}}{8}\right].$$

13.5. Lyapunov Exponents

Table 1 shows that the larger of these two eigenvalues, $\lambda_1^{(k)}$, first grows and then decreases as k increases. The table also includes the square root, $s_1^{(k)} = \sqrt{\lambda_1^{(k)}}$, which arises in the discussion of the Lyapunov exponents. The rate that these eigenvalues go to zero is the same as the rate of $(2/3)^{2k}$, where $2/3$ is the multiple eigenvalue of the original matrix $\mathbf{A}$.

k	$\lambda_1^{(k)}$	$s_1^{(k)}$
1	1.7777	1.3333
2	2.1540	1.4677
3	1.9400	1.3928
4	1.4816	1.2172
5	1.0098	1.0049
6	0.6396	0.7997

Table 1. The larger eigenvalue $\lambda_1^{(k)}$ of $(\mathbf{A}^k)^\mathsf{T} \mathbf{A}^k$ and its square root $s_1^{(k)}$ for Example 13.46

In the calculations of the Lyapunov exponents, $s_j^{(k)} = \sqrt{\lambda_j^{(k)}}$, and

$$\begin{aligned}
\ell_j &= \lim_{k \to \infty} \frac{1}{k} \ln\left(s_j^{(k)}\right) \\
&= \lim_{k \to \infty} \left[\frac{1}{k} \ln\left(\frac{2}{3}\right)^k + \frac{1}{2k} \ln\left(\frac{8 + 9k^2 \pm 9k^2 \sqrt{1 + 16/(9k^2)}}{8} \right) \right] \\
&= \ln\left(\frac{2}{3}\right) + \lim_{k \to \infty} \frac{\ln\left(8 + 9k^2 \pm 9k^2 \sqrt{1 + 16/(9k^2)}\right) - \ln(8)}{2k} \\
&= \ln\left(\frac{2}{3}\right),
\end{aligned}$$

since $\ln(k^2) = 2\ln(k)$ grows more slowly than k. Thus, both Lyapunov exponents equal the logarithm of the eigenvalue of $\mathbf{A}$, $\ln(2/3)$.

Lyapunov exponents as limits of eigenvalues

Returning to the general situation, assume that $\mathbf{F}$ is a differentiable map from $\mathbb{R}^n$ to itself and that $\mathbf{p}_0$ is a point with a bounded orbit. Let

$$\mathbf{A}_k = D\left(\mathbf{F}^k\right)_{(\mathbf{p}_0)},$$

and let $\left(s_j^{(k)}\right)^2$ be the eigenvalues of $(\mathbf{A}_k)^\mathsf{T} \mathbf{A}_k$ for $1 \leq j \leq n$. Order them so that

$$s_1^{(k)} \geq s_2^{(k)} \geq \cdots \geq s_n^{(k)}.$$

Therefore, the image $\mathbf{A}_k(\mathscr{U})$ is an ellipsoid with axes of length $s_1^{(k)}, \ldots, s_n^{(k)}$. The Oseledec multiplicative ergodic theorem stated momentarily shows that the jth Lyapunov exponent is the limit of the logarithm of the kth root of $s_j^{(k)}$, or

$$\ell_j(\mathbf{p}_0; \mathbf{F}) = \lim_{k \to \infty} \frac{1}{k} \ln(s_j^{(k)}).$$

In particular, this shows that there are n Lyapunov exponents (counting multiplicities) in dimension n.

We state the Oseledec theorem in terms of a trapping region with an attracting set. Its statement involves any invariant measure, including the frequency measures $\mu_{\mathbf{x}_0;\mathbf{F}}$. After stating this theorem, we reinterpret it for the case in which the frequency measures all give the same invariant measure (i.e., there is a natural or Sinai–Ruelle–Bowen measure). We leave further details to the references (e.g., [**Rob99**] or [**Kat95**]).

Theorem 13.17 (Oseledec multiplicative ergodic theorem). *Let $\mathbf{F}$ be a C^2 diffeomorphism on $\mathbb{R}^n$ (or other n-dimensional space). Let $\mathbf{U}$ be an isolating neighborhood for an isolated invariant set $\mathbf{S}$. Then, there is a subset $\mathbf{S}' \subset \mathbf{S}$ such that the following results hold:*

a. *The set $\mathbf{S}'$ is invariant, $\mathbf{F}(\mathbf{S}') = \mathbf{S}'$.*

b. *If μ is any invariant probability measure with support in $\mathbf{U}$, and hence in $\mathbf{S}$, then $\mathbf{S}'$ is of full μ-measure, or $\mu(\mathbf{S}') = 1$.*

c. *Take a point $\mathbf{x}_0$ in $\mathbf{S}'$, and let $\left(s_j^{(k)}\right)^2$ be the eigenvalues of*
$$\left(D\left(\mathbf{F}^k\right)_{(\mathbf{x}_0)}\right)^\mathsf{T} D\left(\mathbf{F}^k\right)_{(\mathbf{x}_0)}.$$
Order them so that $s_1^{(k)} \geq s_2^{(k)} \geq \cdots \geq s_n^{(k)}$. Then, the limits of $(1/k)\ln\left(s_j^{(k)}\right)$ exist for $1 \leq j \leq n$ and are the Lyapunov exponents for the point $\mathbf{x}_0$; that is,
$$\ell_j(\mathbf{x}_0) = \lim_{k \to \infty} \frac{1}{k} \ln\left(s_j^{(k)}\right) \quad \text{and}$$
$$\ell_1(\mathbf{x}_0; \mathbf{F}) \geq \ell_2(\mathbf{x}_0; \mathbf{F}) \geq \cdots \geq \ell_n(\mathbf{x}_0; \mathbf{F}).$$

d. *For each point $\mathbf{x}_0$ in $\mathbf{S}'$, there is a basis of vectors $\mathbf{v}_{\mathbf{x}_0}^1, \ldots, \mathbf{v}_{\mathbf{x}_0}^n$ such that $\ell(\mathbf{x}_0, \mathbf{v}_{\mathbf{x}_0}^j; \mathbf{F})$ exists and equals $\ell_j(\mathbf{x}_0; \mathbf{F})$ for $1 \leq j \leq n$.*

e. *If $\mathbf{x}_0$ is in $\mathbf{S}'$ and $\mathbf{v}$ is a vector that can be written as $\mathbf{v} = a_j\,\mathbf{v}_{\mathbf{x}_0}^j + \cdots + a_n\,\mathbf{v}_{\mathbf{x}_0}^n$, with $a_j \neq 0$, then*
$$\ell(\mathbf{x}_0, \mathbf{v}; \mathbf{F}) = \ell(\mathbf{x}_0, \mathbf{v}_{\mathbf{x}_0}^j; \mathbf{F}).$$

We now interpret this theorem for the case in which an attractor has a natural or Sinai–Ruelle–Bowen measure μ. In this situation, there is a trapping region $\mathbf{U}$ and a subset $\mathbf{U}' \subset \mathbf{U}$ of full measure in terms of Lebesgue measure λ, $\lambda(\mathbf{U}') = \lambda(\mathbf{U})$ and $\lambda(\mathbf{U} \setminus \mathbf{U}') = 0$, such that $\mu_{\mathbf{x}_0;\mathbf{F}} = \mu$ for $\mathbf{x}_0$ in $\mathbf{U}'$. Because the frequency measure for a point $\mathbf{x}_0$ in $\mathbf{U}'$ equals the natural measure μ, the set $\mathbf{U}'$ can be used in place of $\mathbf{A}'$ in parts (c) and (d) of the theorem. Because the frequency measure is the same for all the points in $\mathbf{U}'$, the natural measure μ is "ergodic" and the Lyapunov exponents are independent of the point $\mathbf{x}_0$ in $\mathbf{U}'$. Thus, there are numbers $\ell_1, \ldots, \ell_n$ such that $\ell_j(\mathbf{x}_0; \mathbf{F}) = \ell_j$ for all $\mathbf{x}_0$ in $\mathbf{U}'$ and $1 \leq j \leq n$. The significance of getting the result for a set of full Lebesgue measure is that this means that most initial conditions in the basin (in terms of Lebesgue measure) have Lyapunov exponents. Thus, if the initial conditions are chosen at random, the Lyapunov exponents exist and are independent of the choice of the initial condition. Thus, the Lyapunov

13.5. Lyapunov Exponents

exponents are a feature of the attractor for the diffeomorphism and not of the particular initial condition. We summarize these ideas in the next corollary.

Corollary 13.18. *Let $\mathbf{F}$ be a C^2 diffeomorphism on $\mathbb{R}^n$ (or other n-dimensional space) that has a bounded attractor $\mathbf{A}$ with a natural or Sinai–Ruelle–Bowen measure μ. Then, there is a trapping region $\mathbf{U}$ for the attractor $\mathbf{A}$, and a subset $\mathbf{U}' \subset \mathbf{U}$ such that (i) the frequency measures for any point $\mathbf{x}_0$ in $\mathbf{U}'$ equals μ, $\mu_{\mathbf{x}_0;\mathbf{F}} = \mu$, and (ii) the following results hold:*

a. *The set $\mathbf{U}'$ is invariant relative to $\mathbf{U}$; that is,*
$$\mathbf{F}(\mathbf{U}') = \mathbf{U}' \cap \mathbf{F}(\mathbf{U}).$$

b. *The subset $\mathbf{U}' \subset \mathbf{U}$ has full measure in terms of Lebesgue measure λ, $\lambda(\mathbf{U}') = \lambda(\mathbf{U})$ or $\lambda(\mathbf{U} \setminus \mathbf{U}') = 0$.*

c. *There are numbers*
$$\ell_1 \geq \ell_2 \geq \cdots \geq \ell_n,$$
independent of the points, that give the Lyapunov exponents at each point of $\mathbf{U}'$. More precisely, the following holds: For any point $\mathbf{x}_0$ in $\mathbf{U}'$, let $\left(s_j^{(k)}\right)^2$ be the eigenvalues of
$$\left(D\left(\mathbf{F}^k\right)_{(\mathbf{x}_0)}\right)^{\mathsf{T}} D\left(\mathbf{F}^k\right)_{(\mathbf{p}_0)}.$$
Order them so that $s_1^{(k)} \geq s_2^{(k)} \geq \cdots \geq s_n^{(k)}$. Then, the limits of $(^1/_k) \ln\left(s_j^{(k)}\right)$ exist for $1 \leq j \leq n$ and are the Lyapunov exponents for the point $\mathbf{x}_0$; that is,
$$\lim_{k \to \infty} \frac{1}{k} \ln\left(s_j^{(k)}\right) = \ell_j(\mathbf{x}_0) = \ell_j,$$
for $1 \leq j \leq n$,

d. *For each point $\mathbf{x}_0$ in $\mathbf{U}'$, there is a basis of vectors $\mathbf{v}_{\mathbf{x}_0}^1$, ..., $\mathbf{v}_{\mathbf{x}_0}^n$ such that $\ell(\mathbf{x}_0, \mathbf{v}_{\mathbf{x}_0}^j; \mathbf{F})$ exists and equals ℓ_j for $1 \leq j \leq n$.*

e. *If $\mathbf{x}_0$ is in $\mathbf{U}'$ and $\mathbf{v}$ is a vector that can be written as a linear combination of $\mathbf{v}_{\mathbf{x}_0}^j$ through $\mathbf{v}_{\mathbf{x}_0}^n$, $\mathbf{v} = a_j \mathbf{v}_{\mathbf{x}_0}^j + \cdots + a_n \mathbf{v}_{\mathbf{x}_0}^n$ with $a_j \neq 0$, then*
$$\ell(\mathbf{x}_0, \mathbf{v}; \mathbf{F}) = \ell_j.$$

Next, we show that Theorem 13.13 is also a corollary of the Oseledec multiplicative ergodic theorem.

Proof of Theorem 13.13. The product of the eigenvalues equals the determinant of a matrix, so
$$\left(s_1^{(k)}\right)^2 \cdots \left(s_n^{(k)}\right)^2 = \det\left(\left(D\left(\mathbf{F}^k\right)_{(\mathbf{p}_0)}\right)^{\mathsf{T}} D\left(\mathbf{F}^k\right)_{(\mathbf{p}_0)}\right).$$

But the determinant of a product of matrices is the product of the determinants, and the determinant of the transpose of a matrix equals the determinant of a matrix, so
$$\det\left(\left(D\left(\mathbf{F}^k\right)_{(\mathbf{p}_0)}\right)^{\mathsf{T}} D\left(\mathbf{F}^k\right)_{(\mathbf{p}_0)}\right) = \left(\det\left(D\left(\mathbf{F}^k\right)_{(\mathbf{p}_0)}\right)\right)^2.$$

Since
$$D\left(\mathbf{F}^k\right)_{(\mathbf{p}_0)} = D\mathbf{F}_{(\mathbf{p}_{k-1})} \cdots D\mathbf{F}_{(\mathbf{p}_0)},$$
we get
$$\begin{aligned}\left(s_1^{(k)}\right)^2 \cdots \left(s_n^{(k)}\right)^2 &= \left(\det\left(D\left(\mathbf{F}^k\right)_{(\mathbf{p}_0)}\right)\right)^2 \\ &= \left(\det\left(D\mathbf{F}_{(\mathbf{p}_{k-1})}\right) \cdots \det\left(D\mathbf{F}_{(\mathbf{p}_0)}\right)\right)^2 \\ &= \left(\Delta^k\right)^2 = \Delta^{2k}.\end{aligned}$$
Therefore,
$$\begin{aligned}\ln(\Delta) &= \frac{1}{2k}\ln\left(\Delta^{2k}\right) = \lim_{k\to\infty}\frac{1}{2k}\ln\left(\left(s_1^{(k)}\right)^2 \cdots \left(s_n^{(k)}\right)^2\right) \\ &= \lim_{k\to\infty}\frac{1}{k}\ln\left(s_1^{(k)}\right) + \cdots + \frac{1}{k}\ln\left(s_n^{(k)}\right) \\ &= \ell_1(\mathbf{p}_0) + \cdots + \ell_n(\mathbf{p}_0).\end{aligned}$$
$\square$

13.5.2. Numerical Calculation of Lyapunov Exponents. So far, we have defined Lyapunov exponents and related them to the limit of eigenvalues for a sequence of matrices. In this section, we describe the process used to calculate them numerically.

For a matrix $\mathbf{A}$ that has a saddle point at the origin in $\mathbb{R}^2$, for most initial vectors $\mathbf{v}$, the iterates $\mathbf{A}^k\mathbf{v}$ tend toward the direction of the expanding eigenvector. In the same way in $\mathbb{R}^n$, for most initial vectors $\mathbf{v}$, the action of the derivatives of the kth power, $D\left(\mathbf{F}^k\right)_{(\mathbf{p}_0)}\mathbf{v}$, tends toward the direction of greatest increase as k goes to infinity. Therefore, for most of these vectors
$$\ell(\mathbf{p}_0, \mathbf{v}; \mathbf{F}) = \ell_1(\mathbf{p}_0; \mathbf{F}).$$
It is difficult to pick a vector that will give the second largest rate of growth, except in rare cases in which the derivative is a constant matrix. Therefore, we do not try to determine this vector, but instead we use the Gram–Schmidt process of taking a set of independent vectors and making them into a set of orthogonal vectors.

We write down the process in $\mathbb{R}^3$, but it can easily be extended to any dimension. Let $\mathbf{p}_j = \mathbf{F}^j(\mathbf{p}_0)$. Start with a set of unit vectors $\mathbf{v}_1^0, \mathbf{v}_2^0, \mathbf{v}_3^0$ that are mutually orthogonal (perpendicular). Take the image of these vectors by the derivative at $\mathbf{p}_0$:
$$\mathbf{w}_j^1 = D\mathbf{F}_{(\mathbf{p}_0)}\mathbf{v}_j^0 \qquad \text{for } j = 1, 2, 3.$$
These vectors should be thought of as having $\mathbf{p}_1$ as base point. Change these vectors into a set of orthogonal vectors by applying the Gram–Schmidt process:
$$\begin{aligned}\mathbf{v}_1^1 &= \mathbf{w}_1^1, \\ \mathbf{v}_2^1 &= \mathbf{w}_2^1 - \left(\frac{\mathbf{w}_2^1 \cdot \mathbf{v}_1^1}{\|\mathbf{v}_1^1\|^2}\right)\mathbf{v}_1^1, \\ \mathbf{v}_3^1 &= \mathbf{w}_3^1 - \left(\frac{\mathbf{w}_3^1 \cdot \mathbf{v}_1^1}{\|\mathbf{v}_1^1\|^2}\right)\mathbf{v}_1^1 - \left(\frac{\mathbf{w}_3^1 \cdot \mathbf{v}_2^1}{\|\mathbf{v}_2^1\|^2}\right)\mathbf{v}_2^1.\end{aligned}$$

Notice that $\mathbf{v}_2^1$ is perpendicular to $\mathbf{v}_1^1$, and that the area of the parallelogram spanned by $\mathbf{v}_1^1$ and $\mathbf{v}_2^1$ is the same as that spanned by $\mathbf{w}_1^1$ and $\mathbf{w}_2^1$. Also, note that $\mathbf{v}_3^1$ is perpendicular to $\mathbf{v}_1^1$ and $\mathbf{v}_2^1$, and that the volume of the parallelepiped spanned by $\mathbf{v}_1^1$, $\mathbf{v}_2^1$, and $\mathbf{v}_3^1$ is the same as that spanned by $\mathbf{w}_1^1$, $\mathbf{w}_2^1$, and $\mathbf{w}_3^1$.

Continue by induction, assuming that we have defined $\mathbf{v}_1^{k-1}$, $\mathbf{v}_2^{k-1}$, and $\mathbf{v}_3^{k-1}$. Take the image of these vectors by the derivative at $\mathbf{p}_{k-1}$:

$$\mathbf{w}_j^k = D\mathbf{F}_{(\mathbf{p}_{k-1})} \mathbf{v}_j^{k-1} \qquad \text{for } j = 1, 2, 3.$$

These vectors should be thought of as having $\mathbf{p}_k$ as base point. Change these vectors into a set of orthogonal vectors by applying the Gram–Schmidt process:

$$\mathbf{v}_1^k = \mathbf{w}_1^k,$$
$$\mathbf{v}_2^k = \mathbf{w}_2^k - \left(\frac{\mathbf{w}_2^k \cdot \mathbf{v}_1^k}{\|\mathbf{v}_1^k\|^2} \right) \mathbf{v}_1^k,$$
$$\mathbf{v}_3^k = \mathbf{w}_3^k - \left(\frac{\mathbf{w}_3^k \cdot \mathbf{v}_1^k}{\|\mathbf{v}_1^k\|^2} \right) \mathbf{v}_1^k - \left(\frac{\mathbf{w}_3^k \cdot \mathbf{v}_2^k}{\|\mathbf{v}_2^k\|^2} \right) \mathbf{v}_2^k.$$

Notice that the area of the parallelogram spanned by $\mathbf{v}_1^k$ and $\mathbf{v}_2^k$ is the same as that spanned by $\mathbf{w}_1^k$ and $\mathbf{w}_2^k$. Also, note that the volume of the parallelepiped spanned by $\mathbf{v}_1^k$, $\mathbf{v}_2^k$, and $\mathbf{v}_3^k$ is the same as that spanned by $\mathbf{w}_1^k$, $\mathbf{w}_2^k$, and $\mathbf{w}_3^k$.

It is not obvious, but $s_j^{(k)}$ grows at the same rate as $\|\mathbf{v}_j^k\|$, where $\left(s_j^{(k)}\right)^2$ is the jth eigenvalue of $D\left(\mathbf{F}^k\right)^{\mathsf{T}}_{(\mathbf{p}_0)} D\left(\mathbf{F}^k\right)_{(\mathbf{p}_0)}$. Thus,

$$\ell_j(\mathbf{p}_0; \mathbf{F}) = \lim_{k \to \infty} \frac{1}{k} \ln \|\mathbf{v}_j^k\|.$$

By making the vectors perpendicular at each step, the growth rate of the length of the second vector gives the second Lyapunov exponent and the growth rate of the length of the third vector gives the third Lyapunov exponent.

13.5.3. Test for Chaotic Attractors.
Just as we did for the one-dimensional case and for systems of differential equations, we use Lyapunov exponents to give a criterion for a chaotic attractor. This criterion is more easily verifiable in experiments or numerical work than the definition we have presented. We do not take this criterion as our definition, because there are various pathologies that are difficult to rule out using this approach. However, for "most systems," the pathologies do not occur and this criterion is a good test to see whether a chaotic attractor is present in the system.

We assume that all the Lyapunov exponents are nonzero, so nearby orbits do not proceed along parallel paths. This criterion also rules out quasiperiodic motion. (See Example 13.50 below and Example 13.41.) If one of the exponents is positive, then small displacements in that direction tend to grow in length (at least infinitesimally).

Test for a Chaotic Attractor. Experimental criteria for a set **A** to be a *chaotic attractor* for a map **F** are the following:

1. There are many points $\mathbf{x}_0^1, \ldots, \mathbf{x}_0^k$ with distinct orbits that have **A** as their ω-limit set, $\omega(\mathbf{x}_0^i; \mathbf{F}) = \mathbf{A}$ for $1 \leq i \leq k$.
2. At least one of the Lyapunov exponents of the $\mathbf{x}_0^i$ is positive, $\ell_1(\mathbf{x}_0^i; \mathbf{F}) > 0$.
3. All the Lyapunov exponents are nonzero, $\ell_j(\mathbf{x}_0^i; \mathbf{F}) \neq 0$ for $1 \leq j \leq n$.

Remark 13.47. In this criterion, a positive Lyapunov exponent replaces the assumption of sensitive dependence on initial conditions used in the more mathematical definition given in Section 13.4. The fact that there are many distinct points with the same ω-limit set indicates that the measure of the basin of attraction is positive and **A** is a Milnor attractor. In many cases, there will be a trapping region for such Milnor attractors, but this is not easily verifiable experimentally. We do not require that the points $\mathbf{x}_0^i$ be in **A**, so we do not necessarily have that **F** is topologically transitive on **A**. However, the existence of many points with $\omega(\mathbf{x}_0^i; \mathbf{F}) = \mathbf{A}$ gives an indication that this is true, so **A** should be a chaotic attractor.

Remark 13.48. Most systems satisfying the preceding test also satisfy the earlier definition. Thus, we are tacitly assuming that the system is "generic." This assumption is not verifiable, but is used to eliminate situations such as Example 7.15 which is related to Example 13.41). This condition implies that the periodic points are dense in **A**, and that the stable and unstable manifolds of fixed points and periodic orbits cross each other. For more discussion of the way in which generic properties relate to the definition of a chaotic attractor see the discussion in Appendix B.

Remark 13.49. What we have called a test for a chaotic attractor is essentially what Alligood, Sauer, and Yorke gave in their book [**All97**] as the definition for a chaotic attractor. They say that a set **A** is a *chaotic attractor* for a map **F** provided that (i) **A** is a Milnor attractor for **F** and (ii) there is a point $\mathbf{x}_0$ in **A** with $\omega(\mathbf{x}_0; \mathbf{F}) = \mathbf{A}$, $\ell_1(\mathbf{x}_0; \mathbf{F}) > 0$, and $\ell_j(\mathbf{x}_0; \mathbf{F}) \neq 0$ for $1 \leq j \leq n$.

Example 13.50 (Quasiperiodic attractor). Consider the map
$$\mathbf{F}\begin{pmatrix} x \\ y \end{pmatrix} = \begin{pmatrix} \cos(2\pi\,\omega) & -\sin(2\pi\,\omega) \\ \sin(2\pi\,\omega) & \cos(2\pi\,\omega) \end{pmatrix} \begin{pmatrix} 3\,x\,(2+x^2+y^2)^{-1} \\ 3\,y\,(2+x^2+y^2)^{-1} \end{pmatrix},$$
where ω is irrational. If we introduce polar coordinates $r^2 = x^2 + y^2$ and $\tan(\theta) = y/x$, then the map becomes
$$\mathbf{G}\begin{pmatrix} r \\ \theta \end{pmatrix} = \begin{pmatrix} 3\,r\,(2+r^2)^{-1} \\ \theta + 2\pi\,\omega \pmod{2\pi} \end{pmatrix}.$$
Letting $f(r) = 3\,r(2+r^2)^{-1}$, $f(1) = 1$, $f'(r) = (6 - 3r^2)(2+r^2)^{-2}$, and $f'(1) = 1/3$, so the set $r = 1$ is an attracting set. Since the motion on this set is an irrational rotation, there is a dense orbit and the set is an attractor. The two variables decouple (the matrix of partial derivatives is diagonal); the Lyapunov exponent in the r variable is $\ln(1/3) = -\ln(3)$, and the Lyapunov exponent in the θ variable is 0. Therefore, the total system has Lyapunov exponents $-\ln(3)$ and 0. Since 0 appears as a Lyapunov exponent, the attractor does not meet the test for a chaotic attractor. Notice that the map restricted to the chaotic attractor does not

have sensitive dependence on initial conditions, so the attractor does not meet the conditions of our definition of a chaotic attractor either. Since an irrational rotation is not "chaotic" in appearance, this classification makes sense.

Exercises 13.5

1. Define the toral automorphism
$$\mathbf{F}_A \begin{pmatrix} x \\ y \end{pmatrix} = \begin{pmatrix} 2x + y \pmod{1} \\ x + y \pmod{1} \end{pmatrix}.$$

    What are the Lyapunov exponents? Hint: The expanding and contracting directions are the same at all points.

2. For each of the following matrices, calculate the length and direction of the axes of the image of the unit circle:

    (a) $\begin{pmatrix} 2 & 0.5 \\ 2 & -0.5 \end{pmatrix}$, (b) $\begin{pmatrix} 2 & 1 \\ -2 & 2 \end{pmatrix}$.

3. Define the map $\mathbf{F}$ by
$$\mathbf{F}\begin{pmatrix} x \\ y \end{pmatrix} = \begin{cases} \begin{pmatrix} \frac{1}{4}x + \frac{1}{8}\sin(2\pi y) \\ 2y \end{pmatrix} & \text{for } y < \frac{1}{2}, \\ \begin{pmatrix} \frac{1}{4}x + \frac{3}{4} + \frac{1}{8}\sin(2\pi y) \\ 2y - 1 \end{pmatrix} & \text{for } \frac{1}{2} \leq y \leq 1. \end{cases}$$

    a. For a point $\mathbf{p}_0 = (x_0, y_0)$, with $0 \leq x_0 \leq 1$ and $0 \leq y_0 \leq 1$, and the vector $\mathbf{v} = (1, 0)$, what is the Lyapunov exponent, $\ell(\mathbf{p}_0, \mathbf{v}; \mathbf{F})$?
    b. What is the other Lyapunov exponent?

4. Consider the modification of the map for the solenoid given by
$$\mathbf{F}\begin{pmatrix} t \\ z \end{pmatrix} = \begin{pmatrix} 4t \pmod{1} \\ \beta z + \frac{1}{2} e^{2\pi t i} \end{pmatrix},$$

    for $0 < \beta < 1/(2\sqrt{2})$ and z a complex number. What are the three Lyapunov exponents for this map?

5. Calculate the Lyapunov exponents for the Hénon map for $a = 1.4$ and $b = -0.3$ using a computer program.

6. Consider the map
$$\mathbf{F}\begin{pmatrix} x \\ y \\ z \end{pmatrix} = \begin{pmatrix} x + \alpha \pmod{1} \\ x + y \pmod{1} \\ 0.5 z \end{pmatrix}.$$

    a. Explain why the map has sensitive dependence on initial conditions when restricted to the set $\mathbb{T}^2 = \{z = 0\}$. So $\mathbb{T}^2$ satisfies our definition of a chaotic attractor.

b. Explain why the Lyapunov exponents are zero, so the invariant set $\mathbb{T}^2$ does not meet the "Test for a Chaotic Attractor." *Remark:* This example shows why some type of exponential separation of points might be appropriate in the definition of a chaotic attractor. (The more technical mathematical concept of positive topological entropy is perhaps another way of avoiding these pathological examples.)

13.6. Applications

13.6.1. Stability of the Solar System. Poincaré

In the historical prologue, we mentioned the work of Henri Poincaré in connection with the motion of the solar system. In fact, in 1885 King Oscar of Sweden, at the urging of G. Mittag-Leffler, offered a prize for anyone who could show whether the solar system was stable or unstable. The prize was awarded to Poincaré in 1889 for a paper in which he introduced the ideas of instability through homoclinic points. He discussed why such homoclinic intersections of stable and unstable manifolds lead to sensitive dependence on initial conditions in systems such as the solar system. A quote from Poincaré's later book *Les Méthodes Nouvelles* describes his insight into the complexity that results from a transverse homoclinic point:

> When we try to represent the figure formed by the [stable and unstable manifolds] and their infinitely many intersections, each corresponding to a doubly asymptotic solution, these intersections form a type of trellis, tissue or grid with infinitely fine mesh. Neither of the two curves must ever cut across itself again, but it must bend back upon itself in a very complex manner in order to cut across all the meshes in the grid an infinite number of times.

Even though Poincaré did not completely prove the instability of the solar system, he was awarded the prize for the originality of his contribution. For a more complete discussion of Poincaré's work, including his original mistake, see the books [**Dia96**] by F. Diacu and P. Holmes and [**Bar97**] by J. Barrow-Green.

Horseshoes in for Newtonian mechanics

Since the time of Poincaré's original paper, much research has been done to further this understanding of stability and instability. The work of S. Smale in the 1960s on the geometric horseshoe and its connection with homoclinic points (or heteroclinic cycles) was a major step forward in this endeavor. His survey paper [**Sma67**] summarized much of his results and further stimulated mathematical research on stability and instability in dynamical systems, as well as the n-body problem and other applications.

The book by J. Moser [**Mos73**] applied these ideas to the problem of several bodies moving under the influence of Newtonian attraction, the so called *n-body problem*. In particular, he gave a very clear treatment of K. Sitnikov's example of oscillation of three bodies given in [**Sit60**]. In its simplest form, two heavy masses are assumed to move on an elliptical orbit in the plane. The third body is assumed so light that it does not influence the motion of the first two bodies, but is influenced by them. This is what is called the *restricted three-body problem*.

13.6. Applications

In Sitnikov's example, the third body moves on the z-axis, which is perpendicular to the plane of motion of the first two masses and through the center of mass of these two masses. When the third mass is at infinity, the first two bodies move on a periodic orbit. This periodic orbit has stable and unstable manifolds in the phase space with the third body of finite distance along the z-axis. It is shown that these stable and unstable manifolds have a transverse homoclinic point. This intersection of the stable and unstable manifolds implies the existence of complicated orbits of the third body, which can be specified in terms of symbolic dynamics. The third body moves up and down the z-axis repeatedly crossing the plane where $z = 0$. The symbols can be given by specifying the number of rotations that the first two bodies make between successive times when the third body crosses $z = 0$. There is an integer N such that, for any sequence of integers n_i with $n_i \geq N$, there is an orbit such that the first two bodies make n_i rotations between the ith and $(i+1)^{\text{st}}$ crossing of the third body with the $z = 0$ plane. Thus, there can be orbits that oscillate wildly, and other orbits that escape to infinity after making a number of oscillations. Although this motion is not very close to that for the solar system, it does indicate that bodies moving under the influence of Newtonian attraction can behave in a very complicated manner.

There have been many other people who have worked on showing the existence of horseshoes for the n-body problem. Z. Xia has a number of very interesting results. Sitnikov's problem for the spatial situation is of lower dimension than that of three bodies moving in the plane. (The third body is constrained to a single line in Sitnikov's example.) For the planar three-body problem, Xia showed the existence of horseshoes and chaotic motion, [**Xia92a**].

In another work [**Xia92b**], he showed the existence of motion for a case in which five bodies, all with positive masses, fly apart in finite time. This answered an old question of Painlevé as to whether it was possible to have a motion that was not defined for all time without a collision. This work answered a question going back to P. Painlevé in the 1890s. Earlier, J. Mather and R. McGehee [**Mat75**] had given a solution on the line that included binary collisions. Since the goal was to have no collisions, Xia's result finished off the problem.

Many other people have contributed to understanding the complexity of possible motion for the n-body problem. D. Saari made several contributions to the mathematical theory, including a classification of the configurations that bodies attain for a motion in which the bodies spread apart. He has a forthcoming introduction to his approach to the study of the n-body problem, [**Saa04**]. The book [**Mey92**] by K. Meyer and G.R. Hall also gives a mathematical introduction to the n-body problem. The book [**Dia96**] by F. Diacu and P. Holmes describes much of the history of this subject. It is presented for nonspecialists but discusses many of the mathematical concepts involved.

Computational studies of the solar system

The work of Poincaré, Smale, and Xia shows that instability is possible for n bodies moving under the influence of mutual Newtonian attraction. These results do not apply very directly to the solar system. Numerical simulations of the solar system by J. Laskar [**Las89**] and G.J. Sussman and J. Wisdom [**Sus92**] indicated that it takes about 5 million years for a separation of initial conditions to grow by a factor of $e \approx 2.718$. Since the age of the solar system is approximately 5 billion years, this is shorter than might be expected, but still very long in terms of a human lifetime.

Thus, the solar system appears very stable in terms of human observations, but there are both mathematical and computational results that indicate that the long-term behavior cannot be predicted with any certainty for all future time on the basis of measurements with even small uncertainty.

13.7. Theory and Proofs

Markov partitions for hyperbolic toral automorphisms

Restatement of Theorem 13.9. *Let $\mathscr{R} = \{\mathbf{R}_i\}_{i=1}^{m}$ be a Markov partition for a hyperbolic toral automorphism on $\mathbb{T}^2$, with transition matrix $\mathbf{T}$ and $\bigcup_{j=1}^{m} \mathbf{R}_j = \mathbb{T}^2$. Let $(\Sigma_{\mathbf{T}}, \sigma_{\mathbf{T}})$ be the shift space and let $k: \Sigma_{\mathbf{T}} \to \mathbb{T}^2$ be defined by*

$$k(\mathbf{s}) = \bigcap_{n=0}^{\infty} \operatorname{cl}\left(\bigcap_{j=-n}^{n} \mathbf{F}^{-j}(\mathbf{R}_{s_j}^{\circ})\right).$$

Then, k is a finite-to-one semiconjugacy from $\sigma_{\mathbf{T}}$ to $\mathbf{F}$. In fact, k is, at most, m^2 to one, where m is the number of rectangles in the partition.

Proof. By the assumptions, $\operatorname{cl}\left(\mathbf{R}_{s_j}^{\circ} \cap \mathbf{F}^{-1}\left(\mathbf{R}_{s_{j+1}}^{\circ}\right)\right)$ is a nonempty subrectangle that reaches all the way across in the stable direction. By induction,

$$\operatorname{cl}\left(\bigcap_{j=n}^{n+i} \mathbf{F}^{-j}\left(\mathbf{R}_{s_j}^{\circ}\right)\right)$$

is a nonempty subrectangle that reaches all the way across in the stable direction for any $n, j \geq 0$. The width of this set in the unstable direction decreases exponentially at the rate given by one over the expansion constant. Thus,

$$\bigcap_{n=0}^{\infty} \operatorname{cl}\left(\bigcap_{j=0}^{n} \mathbf{F}^{-j}(\mathbf{R}_{s_j}^{\circ})\right) = W^s(\mathbf{p}_1, \mathbf{R}_{s_0}),$$

for some $\mathbf{p}_1 \in \mathbf{R}_{s_0}$. Similarly,

$$\bigcap_{n=0}^{-\infty} \operatorname{cl}\left(\bigcap_{j=n}^{0} \mathbf{F}^{-j}(\mathbf{R}_{s_j}^{\circ})\right) = W^u(\mathbf{p}_2, \mathbf{R}_{s_0}),$$

for some $\mathbf{p}_2 \in \mathbf{R}_{s_0}$. Therefore,

$$\bigcap_{n=\infty}^{\infty} \operatorname{cl}\left(\bigcap_{j=0}^{n} \mathbf{F}^{-j}(\mathbf{R}_{s_j}^{\circ})\right) = W^s(\mathbf{p}_1, \mathbf{R}_{s_0}) \cap W^u(\mathbf{p}_2, \mathbf{R}_{s_0})$$

is a unique point $\mathbf{p} = k(\mathbf{s})$. This shows that k is a well-defined map. By arguments such as those used for the horseshoe, k is continuous, onto, and a semiconjugacy.

Next, we show that k is, at most, m^2 to one, where m is the number of partitions. Let $\mathbf{p} = k(\mathbf{s})$. If $\mathbf{F}^j(\mathbf{p})$ is in $\mathbf{R}_{s_j}^{\circ}$ and not just $\mathbf{R}_{s_j}$ for all j, then $k^{-1}(\mathbf{p})$ is a unique symbol sequence $\mathbf{s}$, because $\mathbf{F}^j(\mathbf{p}) \notin \mathbf{R}_i$ for $i \neq s_j$. Therefore, we have to worry only if $\mathbf{F}^j(\mathbf{p})$ is on the boundary of some rectangle $\mathbf{R}_i$. The boundary of any M-rectangle is made up of $\partial^{\text{in}}(\mathbf{R}_i)$ and $\partial^{\text{out}}(\mathbf{R}_i)$, where $\partial^{\text{in}}(\mathbf{R}_i)$ is the union of unstable manifolds $W^u(\mathbf{z}, \mathbf{R}_i)$, and $\partial^{\text{out}}(\mathbf{R}_i)$ is the union of such stable manifolds. If $\mathbf{F}^j(\mathbf{p}) \in \partial^{\text{out}}(\mathbf{R}_{s_j})$, then $\mathbf{F}^i(\mathbf{p}) \in \partial^{\text{out}}(\mathbf{R}_{s_i})$ for $i \geq j$. There are, at most, m choices for s_j. Since the transitions of interiors are unique, a choice for s_j determines the choices of s_i for $i \geq j$. Similarly, if $\mathbf{F}^{j'}(\mathbf{p}) \in \partial^{\text{in}}(\mathbf{R}_{s_{j'}})$, then a choice for $s_{j'}$ determines the choices of s_i for $i \leq j'$. Combining shows that there are, at most, m^2 choices as claimed. $\square$

Basic invariant sets

We want to find the invariant sets that determine the main features of the dynamics, the *basic invariant sets*. The invariant set for the geometric horseshoe is also transitive, so it should be a basic invariant set. However, any hyperbolic saddle periodic orbit in this invariant set is also an isolated invariant set, so this invariant set does have proper subsets that are isolated invariant sets. Therefore, we cannot just modify the criterion used to define an attractor: an attracting set with no proper subsets that are attracting sets. The theory developed by C. Conley gives rise to the following alternative definition.

Definition 13.51. An isolated invariant set $\mathbf{S}$ is called *basic*, provided that the following two conditions are satisfied:

(i) There are two trapping regions $\mathbf{U}_1$ and $\mathbf{U}_2$ such that $\mathbf{U} = \mathbf{U}_1 \setminus \mathbf{U}_2$ is an isolating neighborhood for $\mathbf{S}$. Notice that $\mathbf{U}_1$ is a trapping region that contains $\mathbf{S}$ and $\mathbf{U}_2 \cap \mathbf{S} = \emptyset$.

(ii) There is no isolated invariant set $\mathbf{S}' \subset \mathbf{S}$ with $\emptyset \neq \mathbf{S}' \neq \mathbf{S}$ such that $\mathbf{S}'$ has an isolating neighborhood of the type given in condition (i), $\mathbf{U}' = \mathbf{U}'_1 \setminus \mathbf{U}'_2$ where both $\mathbf{U}'_1$ and $\mathbf{U}'_2$ are trapping regions.

Remark 13.52. Assume that $\mathbf{S}$ is an isolated invariant set with an isolating neighborhood of the type of condition (i) of the definition. Then $\mathbf{S}$ is basic provided that either it is transitive or there exists a point $\mathbf{x}_0$ such that the ω-limit set of $\mathbf{x}_0$ equals all of $\mathbf{S}$, $\omega(\mathbf{x}_0) = \mathbf{S}$.

Remark 13.53. The geometric horseshoe Λ of Section 13.1 is a basic isolated invariant set. The sets

$$\mathbf{U}_1 = \mathbf{S}' \cup \mathbf{E}_0 \cup \mathbf{E}_1 \quad \text{and}$$
$$\mathbf{U}_2 = \mathbf{E}_0 \cup \mathbf{E}_1$$

are trapping regions that satisfy condition (i) of the definition, with $\mathbf{S}' = \mathbf{U}_1 \setminus \mathbf{U}_2$ being the isolating neighborhood of Λ. The invariant set is transitive, so it is basic.

Conley proved a very important characterization of basic invariant sets, and so for the sets we call attractors. The conditions are given in terms of ϵ-chains (or ϵ-pseudo-orbits). The following definition is the same as the one given for one dimension in Chapter 11, but we repeat it here.

Definition 13.54. A point $\mathbf{p}$ is called *chain recurrent* provided that, for every $\epsilon > 0$, there is an ϵ-chain from $\mathbf{p}$ back to $\mathbf{p}$: $\mathbf{p} = \mathbf{x}_0, \mathbf{x}_2, \ldots, \mathbf{p} = \mathbf{x}_k$ with $|\mathbf{F}(\mathbf{x}_{j-1}) - \mathbf{x}_j| < \epsilon$ for $1 \leq j \leq k$. The number of points k can depend on ϵ.

Theorem 13.19 (Conley). *Assume that* **S** *is an isolated invariant set with a trapping region of the type given in condition (i) of the definition. Then,* **S** *is basic if and only if every point in* **S** *is chain recurrent.*

In particular, if **A** *is an attracting set, then the following are equivalent: (i)* **A** *is an attractor. (ii) Every point in* **A** *is chain recurrent.*

See [**Rob99**] for the proof of the equivalent result for differential equations.

The linear image of a sphere is an ellipsoid

Restatement of Theorem 13.16. *Let* **A** *be a real* $n \times n$ *matrix. Let* $\mathscr{U}$ *be the set of vectors of length one in* $\mathbb{R}^n$.

a. *Then, the eigenvalues of* $\mathbf{A}^\mathsf{T}\mathbf{A}$ *and* $\mathbf{A}\mathbf{A}^\mathsf{T}$ *are the same,* $\{\lambda_1, \ldots, \lambda_n\}$. *Moreover, all the* λ_j *are real and* $\lambda_j \geq 0$. *Let* $s_j = \sqrt{\lambda_j}$. *Let* $\{\mathbf{v}^1, \ldots, \mathbf{v}^n\}$ *be the corresponding eigenvectors of* $\mathbf{A}^\mathsf{T}\mathbf{A}$ *and* $s_j \mathbf{u}^j = \mathbf{A}\mathbf{v}^j$. *Then,* $\{\mathbf{u}^1, \ldots, \mathbf{u}^n\}$ *are the eigenvectors for* $\mathbf{A}\mathbf{A}^\mathsf{T}$, *and these two sets of eigenvectors can be chosen to be orthonormal,* $\mathbf{v}^k \cdot \mathbf{v}^j = 0 = \mathbf{u}^k \cdot \mathbf{u}^j$ *for* $k \neq j$ *and* $\|\mathbf{v}^j\| = 1 = \|\mathbf{u}^j\|$. *Finally, the image of the n-sphere by* **A**, $\mathbf{A}(\mathscr{U})$, *is an ellipsoid with semiaxes given by the vectors* $s_j \mathbf{u}^j = \mathbf{A}\mathbf{v}^j$.

b. *Also, if we define the matrix* **P** *by* $\mathbf{P}\mathbf{v}^j = s_j \mathbf{v}^j$, *then* **P** *is a positive semidefinite symmetric matrix. The matrix* $\mathbf{Q} = \mathbf{A}\mathbf{P}^{-1}$ *is an orthogonal matrix. Finally,* $\mathbf{A} = \mathbf{Q}\mathbf{P}$. *Thus, any matrix* **A** *can be written as the product of an orthogonal matrix and a positive semidefinite symmetric matrix.*

Proof. The ratio $\|\mathbf{A}\mathbf{v}\|/\|\mathbf{v}\|$ is the amount the vector **v** is stretched when acted on by **A**. The square of the length of $\mathbf{A}\mathbf{v}$,

$$\|\mathbf{A}\mathbf{v}\|^2 = (\mathbf{A}\mathbf{v})^\mathsf{T}\mathbf{A}\mathbf{v} = \mathbf{v}^\mathsf{T}\mathbf{A}^\mathsf{T}\mathbf{A}\mathbf{v},$$

relates to the matrix $\mathbf{A}^\mathsf{T}\mathbf{A}$. The eigenvalues λ_j of this matrix are real since it is symmetric: $(\mathbf{A}^\mathsf{T}\mathbf{A})^\mathsf{T} = \mathbf{A}^\mathsf{T}(\mathbf{A}^\mathsf{T})^\mathsf{T} = \mathbf{A}^\mathsf{T}\mathbf{A}$. The corresponding eigenvectors $\mathbf{v}^j$ can be taken as an orthonormal basis (i.e., length one and mutually perpendicular: $\mathbf{v}^k \cdot \mathbf{v}^j = 0$ for $k \neq j$ and $\mathbf{v}^j \cdot \mathbf{v}^j = \|\mathbf{v}^j\|^2 = 1$). Also,

$$\begin{aligned} 0 \leq \|\mathbf{A}\mathbf{v}^j\|^2 &= (\mathbf{A}\mathbf{v}^j)^\mathsf{T}(\mathbf{A}\mathbf{v}^j) \\ &= (\mathbf{v}^j)^\mathsf{T}\mathbf{A}^\mathsf{T}\mathbf{A}\mathbf{v}^j = (\mathbf{v}^j)^\mathsf{T}\lambda_j \mathbf{v}^j \\ &= \lambda_j \|\mathbf{v}^j\|^2 = \lambda_j. \end{aligned}$$

Therefore, the eigenvalues λ_j of $\mathbf{A}^\mathsf{T}\mathbf{A}$ are all nonnegative. (Such a symmetric matrix $\mathbf{A}^\mathsf{T}\mathbf{A}$ is called *positive semidefinite*.) This **first step** has shown that the eigenvalues λ_j of $\mathbf{A}^\mathsf{T}\mathbf{A}$ are real and nonnegative. Their corresponding eigenvectors $\{\mathbf{v}^j\}$ form an orthonormal basis. Let $s_j = \sqrt{\lambda_j}$.

If we define **P** to be the matrix such that $\mathbf{P}\mathbf{v}^j = s_j \mathbf{v}^j$, then $\mathbf{P}^2 \mathbf{v}^j = s_j^2 \mathbf{v}^j = \mathbf{A}^\mathsf{T}\mathbf{A}\mathbf{v}^j$ and **P** is a positive semidefinite symmetric matrix. Since $\mathbf{P}^2 = \mathbf{A}^\mathsf{T}\mathbf{A}$, **P** can be interpreted as the square root of $\mathbf{A}^\mathsf{T}\mathbf{A}$, $\mathbf{P} = \sqrt{\mathbf{A}^\mathsf{T}\mathbf{A}}$.

The eigenvectors $\mathbf{v}^j$ of $\mathbf{A}^\mathsf{T}\mathbf{A}$ turn out to be the directions that determine the various amounts of stretching of vectors by **A**; their images $\mathbf{A}\mathbf{v}^j = s_j \mathbf{u}^j$ give the axes of the ellipsoid.

13.7. Theory and Proofs

The **second step** is to show that the $\mathbf{u}^j = (1/s_j)\mathbf{A}\mathbf{v}^j$ are eigenvectors of the matrix $\mathbf{A}\mathbf{A}^\mathsf{T}$ with the same eigenvalues λ_j. For simplicity, we have assumed that all the $\lambda_j > 0$; if $s_j = 0$ for $1 \leq j \leq k$, then we pick these $\mathbf{u}^j$ as unit vectors that are perpendicular to $\mathbf{u}^i$ for $k < i \leq n$. We omit the details for this possibility. The proof of the second step is a straightforward calculation that follows from the definitions and regrouping the matrix product:

$$(\mathbf{A}\mathbf{A}^\mathsf{T})\mathbf{u}^j = (\mathbf{A}\mathbf{A}^\mathsf{T})\frac{1}{s_j}\mathbf{A}\mathbf{v}^j$$
$$= \frac{1}{s_j}\mathbf{A}(\mathbf{A}^\mathsf{T}\mathbf{A})\mathbf{v}^j$$
$$= \frac{1}{s_j}\mathbf{A}\lambda_j\mathbf{v}^j$$
$$= \lambda_j \mathbf{u}^j.$$

The **third step** is to show that the vectors $\{\mathbf{u}^j\}$ form an orthonormal basis (i.e., $\mathbf{u}^k \cdot \mathbf{u}^j = 0$ for $k \neq j$ and $\mathbf{u}^j \cdot \mathbf{u}^j = \|\mathbf{u}^j\|^2 = 1$). Again, this is a straightforward calculation:

$$\mathbf{u}^k \cdot \mathbf{u}^j = (\mathbf{u}^k)^\mathsf{T} \mathbf{u}^j = \left(\frac{1}{s_k}\mathbf{A}\mathbf{v}^k\right)^\mathsf{T}\left(\frac{1}{s_j}\mathbf{A}\mathbf{v}^j\right)$$
$$= \frac{1}{s_k s_j}(\mathbf{v}^k)^\mathsf{T}(\mathbf{A}^\mathsf{T}\mathbf{A})\mathbf{v}^j = \frac{1}{s_k s_j}(\mathbf{v}^k)^\mathsf{T}\lambda_j\mathbf{v}^j$$
$$= \frac{\lambda_j}{s_k s_j}\mathbf{v}^k \cdot \mathbf{v}^j = \begin{cases} 0 & \text{for } k \neq j, \\ 1 & \text{for } k = j. \end{cases}$$

The **fourth step** is to show that the image $\mathbf{A}(\mathscr{U})$ is an ellipsoid. To determine the image, we express any vector $\mathbf{x} \in \mathscr{U}$ in terms of the orthonormal basis $\{\mathbf{v}^j\}$, rather than the standard basis: $\mathbf{x} = x_1\mathbf{v}^1 + \cdots + x_n\mathbf{v}^n \in \mathscr{U}$ and $\|\mathbf{x}\|^2 = x_1^2 + \cdots x_n^2 = 1$. The image point is

$$\mathbf{y} = \mathbf{A}\mathbf{x} = x_1\mathbf{A}\mathbf{v}^1 + \cdots + x_n\mathbf{A}\mathbf{v}^n$$
$$= x_1 s_1 \mathbf{u}^1 + \cdots + x_n s_n \mathbf{u}^n.$$

Thus, the component of the image point $\mathbf{y} = \mathbf{A}\mathbf{x}$ has components $y_j = x_j s_j$ in terms of the basis $\{\mathbf{u}^j\}$. These components satisfy

$$\frac{y_1^2}{s_1^2} + \cdots + \frac{y_n^2}{s_n^2} = x_1^2 + \cdots + x_n^2 = 1,$$

so $\mathbf{y} = \mathbf{A}\mathbf{x}$ is on the ellipsoid

$$\left\{ y_1\mathbf{u}^1 + \cdots + y_n\mathbf{u}^n \; : \; \frac{y_1^2}{s_1^2} + \cdots + \frac{y_n^2}{s_n^2} = 1 \right\},$$

with axes along the vectors $\mathbf{u}^j$, and with the length of the axes $s_j = \sqrt{\lambda_j}$; that is,

$$\mathbf{A}(\mathscr{U}) = \left\{ y_1\mathbf{u}^1 + \cdots + y_n\mathbf{u}^n \; : \; \frac{y_1^2}{s_1^2} + \cdots + \frac{y_n^2}{s_n^2} = 1 \right\}.$$

The **fifth and final step** is to write the matrix $\mathbf{A}$ as the product of an orthogonal matrix and a positive semidefinite symmetric matrix. As we said at the end of step one, there is a positive semidefinite symmetric matrix $\mathbf{P}$ such that $\mathbf{P}^2 = \mathbf{A}^\mathsf{T}\mathbf{A}$. If we let $\mathbf{Q} = \mathbf{A}\mathbf{P}^{-1}$, then

$$\begin{aligned}\mathbf{Q}^\mathsf{T}\mathbf{Q} &= (\mathbf{P}^{-1})^\mathsf{T}\mathbf{A}^\mathsf{T}\mathbf{A}\mathbf{P}^{-1} \\ &= \mathbf{P}^{-1}\mathbf{P}^2\mathbf{P}^{-1} \\ &= \mathbf{I},\end{aligned}$$

so $\mathbf{Q}$ is an orthogonal matrix. (In terms of the bases given above, $\mathbf{Q}\mathbf{v}^j = \mathbf{u}^j$.) Thus, any square matrix $\mathbf{A}$ can be written as $\mathbf{A} = \mathbf{Q}\mathbf{P}$, where $\mathbf{Q}$ is orthogonal and $\mathbf{P}$ is positive semidefinite symmetric. By a similar argument, it can also be written as $\mathbf{A} = \mathbf{P}'\mathbf{Q}'$, where $\mathbf{P}'$ is positive semidefinite symmetric and $\mathbf{Q}'$ is orthogonal.

We can further decompose the matrix $\mathbf{A}$. Write $\mathbf{A} = \mathbf{Q}\mathbf{P}$ as in the preceding. Then, the positive semidefinite symmetric matrix $\mathbf{P}$ can be conjugated to a nonnegative diagonal matrix Λ by an orthogonal matrix $\mathbf{Q}_1$. Then $\mathbf{A} = \mathbf{Q}\mathbf{Q}_1^{-1}\Lambda\mathbf{Q}_1 = \mathbf{Q}_2\Lambda\mathbf{Q}_1$, where $\mathbf{Q}_1$ and $\mathbf{Q}_2$ are orthogonal, and Λ is a nonnegative diagonal matrix. $\square$

Chapter 14

Fractals

B. Mandelbrot coined the term *fractal* in the 1960s for a geometric object that has complicated or irregular geometry and is roughly self-similar.

A set is called *self-similar* if arbitrarily small pieces of a set can be magnified to reproduce the whole set, (i.e., it looks the same at smaller scales). Cantor sets are examples of self-similar sets. Many attractors for nonlinear systems and other sets Mandelbrot considered are not exactly self-similar but reproduce the important irregular nature of the object at arbitrarily small scales. Therefore the requirement for a fractal is that is it roughly self-similar.

The other property of being irregular is specified in terms of some type of "fractal dimension." The box dimension is such a dimension, and it can have noninteger values, as well as integer values. This dimension measures the extent to which a set is not a curve or a surface (i.e., it is "strange"). We also give two other fractal dimensions, Lyapunov dimension and correlation dimension, that are defined in terms of the dynamics of the map, rather than the complexity of an invariant set. These dimensions measure the extent to which a map behaves in an irregular or chaotic manner. A fractal is required to have a fractal dimension that is not an integer in order to insure that it has irregular or complicated geometry.

The union of all the unstable manifolds of the horseshoe form a set that is a fractal, so the basin boundary discussed in Section 13.1.1 is also a fractal. Chaotic attractors are usually fractal sets. For example, the Hénon attractor at small scales is approximately the Cartesian product of a Cantor set and a line segment; thus, it is roughly self-similar and has a box dimension that is not an integer.

Besides introducing fractal dimension, this chapter also considers iterated-function systems that are another way to generate fractals. These sets are easier to produce than chaotic attractors for nonlinear maps and can have some interesting appearance and geometry.

14.1. Box Dimension

Because the set of points in a chaotic attractor is often a complicated set, it is sometimes called a "strange attractor." In this section, we introduce a measurement of a set called *box dimension*. Because this quantity is not always an integer, it is sometimes called a *fractal dimension*. These ideas go back at least to Hausdorff in the early 1900s, who introduced a different fractal dimension, called the Hausdorff dimension, which has certain advantages mathematically, but is not as easy to define or compute. The box dimension was introduced by A.N. Kolmogorov, who originally called it the *capacity* of a set, but this term has other meanings, so the name has been replaced by the box dimension. A more thorough treatment of different dimensions at an undergraduate level is given in the book by J. Edgar [**Edg90**].

Our treatment of box dimension deals only with subsets of $\mathbb{R}^n$ but makes sense in sets on which a distance is defined, a metric space. The box dimension of the closure of a set is the same as the box dimension of the original set, so it is sufficient to consider closed sets.

Definition 14.1. In $\mathbb{R}^n$, consider a fixed grid of boxes of size $r > 0$:

$$\bar{\mathbf{B}}(j_1,\ldots,j_n;r) = \{ (x_1,\ldots,x_n) \in \mathbb{R}^n : j_i r \leq x_i < (j_i+1)r \text{ for } 1 \leq i \leq n \}$$

for integers $j_1, \ldots, j_n$. For a set $\mathbf{S}$ in $\mathbb{R}^n$, let

$$N(r, \mathbf{S})$$

be the number of boxes of the preceding type that $\mathbf{S}$ intersects. In Figure 1, $N(r, \mathbf{S}) = 65$.

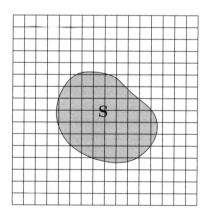

Figure 1. Grid lines for box dimension

If $\mathbf{S}$ is a line segment of length L, the length is approximately equal to the product of the size of the box r times the number $N(r, \mathbf{S})$ of boxes that intersect $\mathbf{S}$; that is,

$$L \sim r\, N(r, \mathbf{S}), \quad \text{so}$$

$$N(r, \mathbf{S}) \sim \frac{L}{r}.$$

14.1. Box Dimension

If **S** is a disk in a plane of area A, the area is approximately equal to $r^2 N(r, \mathbf{S})$, or

$$A \sim r^2 N(r, \mathbf{S}), \quad \text{so}$$
$$N(r, \mathbf{S}) \sim \frac{A}{r^2}.$$

If **S** is an object with d-dimensional volume V, the volume is approximately equal to $r^d N(r, \mathbf{S})$, or

$$V \sim r^d N(r, \mathbf{S}), \quad \text{so}$$
$$N(r, \mathbf{S}) \sim \frac{V}{r^d}.$$

To solve for the exponent d, we take the logarithm and then the limit:

$$V \sim r^d N(r, \mathbf{S}),$$
$$\ln(V) \sim d \ln(r) + \ln(N(r, \mathbf{S})),$$
$$d \sim \frac{\ln(V)}{\ln(r)} - \frac{\ln(N(r, \mathbf{S}))}{\ln(r)} = \frac{\ln(V)}{\ln(r)} + \frac{\ln(N(r, \mathbf{S}))}{\ln(1/r)}.$$

As r goes to zero, the first term on the right of the last line goes to zero, because $\ln(r)$ goes to minus infinity. So, it makes sense to define the dimension as the limit of the second term.

Definition 14.2. Let **S** be a set in $\mathbb{R}^n$ and let $N(r, \mathbf{S})$ be the number of boxes of size r for a fixed grid, $\bar{\mathbf{B}}(j_1, \ldots, j_n; r)$, that **S** intersects. The *box dimension* or *box-counting dimension* of **S** is defined to be the following limit, provided that it exists:

$$\dim{}_b(\mathbf{S}) = \lim_{r \to 0} \frac{\ln(N(r, \mathbf{S}))}{\ln(1/r)}.$$

The next two examples give sets with noninteger values for the box dimensions.

Example 14.3. Consider the set

$$\mathbf{S} = \{0\} \cup \left\{ \frac{1}{i} : i \geq 1 \right\}.$$

Let us choose

$$r_k = \frac{1}{k} - \frac{1}{k+1} = \frac{1}{k(k+1)}.$$

Divide $[0, 1]$ into equal fixed intervals of length r_k. For $i \geq k$, the pairs of points $1/(i+1)$ and $1/i$ are closer together than r_k, so each of the intervals $[(j-1)r_k, jr_k)$ with $jr_k \leq 1/k$ is hit by **S**. There are

$$\frac{\frac{1}{k}}{\frac{1}{k(k+1)}} = k+1$$

of these intervals. This leaves the k points $1, \ldots, 1/k$ that are in separate intervals of length r_k. See Figure 2. Therefore,

$$N(r_k, \mathbf{S}) = (k+1) + k = 2k+1.$$

Figure 2. Eight intervals for Example 14.3 with $k = 4$

We do not use all values of r, but the choices of the numbers r_k satisfy Theorem 14.2, so we can use them to calculate the dimension:

$$\begin{aligned}\dim{}_\mathrm{b}(\mathbf{S}) &= \lim_{k \to \infty} \frac{\ln\left(N(r_k, \mathbf{S})\right)}{\ln\left(r_k^{-1}\right)} \\ &= \lim_{k \to \infty} \frac{\ln(2k+1)}{\ln(k(k+1))} \\ &= \lim_{k \to \infty} \frac{\ln(k) + \ln\left(2 + \frac{1}{k}\right)}{2\ln(k) + \ln\left(1 + \frac{1}{k}\right)} \\ &= \frac{1}{2}.\end{aligned}$$

Thus, even though this set is only a countable number of points, the box dimension is positive and not an integer.

Example 14.4. In Section 10.5, we introduced the middle-third Cantor set. We now consider a more general Cantor set. Let $0 < b < 0.5$ and $0 < a < 1$ be numbers such that $a + 2b = 1$. Let $\mathbf{K}_0(a) = [0,1]$ be the unit interval. Assume that, at the first stage, an open interval of length a is removed from the middle of $\mathbf{K}_0(a)$, leaving two closed intervals of length b. Let $\mathbf{K}_1(a)$ be the union of these two intervals. Now, from each of the two intervals of $\mathbf{K}_1(a)$, remove the middle a portion, leaving a total of four intervals, each of length b^2. Let the union of these 2^2 closed intervals, each of length b^2, be denoted by $\mathbf{K}_2(a)$. Continuing, at each stage remove the middle a of each of the closed intervals. Thus, at the jth stage, $\mathbf{K}_j(a)$ is the union of 2^j intervals, each of length b^j. The set

$$\mathbf{K}(a) = \bigcap_{j \geq 0} \mathbf{K}_j(a),$$

is called the *middle-a Cantor set*.

If we take $r_j = b^j$ and do not use a fixed grid to determine the boxes, then we can cover $\mathbf{K}(a)$ with 2^j closed intervals of length b^j, so $N(b^j, \mathbf{K}(a)) = 2^j$. Because these choices of r_j are powers of a fixed number, they can be used to calculate the dimension:

$$\begin{aligned}\dim{}_\mathrm{b}(\mathbf{K}(a)) &= \lim_{j \to \infty} \frac{\ln(2^j)}{\ln(b^{-j})} \\ &= \lim_{j \to \infty} \frac{j \ln(2)}{j \ln\left(1/b\right)} \\ &= \frac{\ln(2)}{\ln(1/b)} < 1.\end{aligned}$$

14.1. Box Dimension

Notice that, for the middle-third Cantor set, the dimension is $\ln(2)/\ln(3) \approx 0.6309$. The box dimensions for all these middle-a Cantor sets is positive, even though the sets have empty interiors and are totally disconnected.

We already used two simplifications in the preceding example. First, we did not use a fixed grid, but placed the boxes (intervals) to cover the set most efficiently. Second, we did not use all values of r, but just a sequence of values. As third simplification in higher dimensions, it is possible to use different shaped "boxes" (e.g. disks or balls in two or three dimensions). After giving some definitions, we summarize these simplifications in two theorems.

Definition 14.5. Sometimes, we do not want to keep the grid of boxes fixed. Let
$$N'(r, \mathbf{S})$$
be the minimum number of closed boxes (squares or cubes) of size r needed to cover the set $\mathbf{S}$ without fixing the grid.

Sometimes, we want to use round balls rather than boxes to cover the set. Let
$$N''(r, \mathbf{S})$$
be the minimum number of closed (round) balls $\overline{\mathbf{B}}(\mathbf{x}, r)$ of size r that are needed to cover $\mathbf{S}$.

Theorem 14.1. *Let $\mathbf{S}$ be a set in $\mathbb{R}^n$ for which one of the limits,*
$$\lim_{r \to 0} \frac{\ln(N'(r, \mathbf{S}))}{\ln(1/r)} \quad or$$
$$\lim_{r \to 0} \frac{\ln(N''(r, \mathbf{S}))}{\ln(1/r)},$$
exists and is equal to d. Then the box dimension $\dim_b(\mathbf{S})$ exists and is equal to d.

The proof is given in Section 14.4.

Theorem 14.2. *Let $\mathbf{S}$ be a set in $\mathbb{R}^n$. Assume that there is a sequence of sizes r_j, with*
$$0 = \lim_{j \to \infty} r_j \quad and \quad 1 = \lim_{j \to \infty} \frac{\ln(r_{j+1})}{\ln(r_j)}$$
such that the following limit exists:
$$d = \lim_{j \to \infty} \frac{\ln(N'(r_j, \mathbf{S}))}{\ln(1/r_j)}.$$
Then, $\dim_b(\mathbf{S})$ exists and is equal to d.

The proof is in Section 14.4.

Remark 14.6. If we take $r_j = b^j$ with $b < 1$, then these values satisfy Theorem 14.2:
$$\lim_{j \to \infty} \frac{\ln(r_{j+1})}{\ln(r_j)} = \lim_{j \to \infty} \frac{\ln(b^{j+1})}{\ln(b^j)} = \lim_{j \to \infty} \frac{(j+1)\ln(b)}{j \ln(b)}$$
$$= \lim_{j \to \infty} \frac{j+1}{j} = 1.$$

The reader can check that the sequence $r_j = 1/(j(j+1))$, as in Example 14.3, also satisfies the theorem.

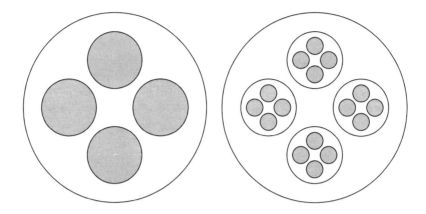

Figure 3. The sets $\mathbf{K}_1$ and $\mathbf{K}_2$ for Example 14.7.

Example 14.7. The set in this example is related to the Cantor set in the line, but is a subset of $\mathbb{R}^2$. It is like the type of set obtained by intersecting the solenoid with one of the disks, but we use four disks rather than just two. See Figure 3. Let $\mathbf{K}_0$ be the disk of radius 1 in the plane. Let $\mathbf{K}_1$ be 4 disks of radius a with centers at $(\pm 0.5, 0)$ and $(0, \pm 0.5)$. A direct check shows that these new disks are disjoint if $a < \sqrt{2}/4$. At the next stage, let $\mathbf{K}_2$ be 4^2 disks of radius a^2, with 4 disks in each of the disks of $\mathbf{K}_1$. It is possible to place a scaled copy of $\mathbf{K}_1$ in each of the disks of $\mathbf{K}_1$ to accomplish this. See Figure 3. Continuing by induction, we see that $\mathbf{K}_j$ is 4^j disks of radius a^j. Let $\mathbf{K} = \bigcap_{j \geq 0} \mathbf{K}_j$. Then,

$$\dim_{\mathrm{b}}(\mathbf{K}) = \lim_{j \to \infty} \frac{\ln(N''(a^j, \mathbf{K}))}{\ln(a^{-j})} = \lim_{j \to \infty} \frac{j \ln(4)}{j \ln(1/a)} = \frac{\ln(4)}{\ln(1/a)}.$$

If we take

$$\frac{1}{4} < a < \frac{\sqrt{2}}{4},$$

then $\ln(1/a) < \ln(4)$ and $\dim_{\mathrm{b}}(\mathbf{K}) > 1$. Thus, the box dimension of the set can be greater than one, even though it is totally disconnected and the set contains no curves.

For a specific chaotic attractor in $\mathbb{R}^n$, it is possible to form a grid in $\mathbb{R}^n$ and then calculate how many of the boxes are hit by the iterate of a point whose orbit is dense in the attractor. This count gives an estimate of the box dimension of the attractor. A set such as that in Figure 18 for the forced Duffing equation should have box dimension slightly larger than one. This is a very memory-intensive calculation for a set in $\mathbb{R}^2$, because it is necessary to form an array with the number of entries on the order of r^{-2}: for $r = 0.01$, the array has on the order of 1000 entries, and for smaller r, there are even more entries. In the next example, we discuss the box dimension of the Hénon attractor more carefully.

14.1. Box Dimension

Example 14.8 (Hénon Attractor). Consider the Hénon attractor for $a = 1.4$ and $b = -0.3$. The attractor is approximately self-similar, as we discussed in the introduction to this chapter. Figure 4 shows the attractor at several scales; at small scales the attractor looks like the product of a Cantor set and a line interval.

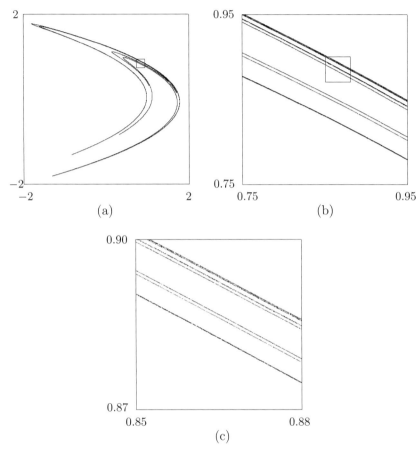

Figure 4. The Hénon attractor for $a = 1.4$ and $b = -0.3$. (a) The window shown is $[-2, 2] \times [-2, 2]$. (b) The window shown is $[0.75, 0.95] \times [0.75, 0.95]$. (c) The window shown is $[0.85, 0.88] \times [0.87, 0.90]$.

Our numerical calculation of the box dimension is based on the number of boxes hit by the attractor for $r = 2^{-3}, 2^{-4}, \ldots, 2^{-7}$. Figure 5 shows the attractor and a grid of size $r = 2^{-3} = 1/8$. The attractor hits 160 of these boxes, $N(1/8) = 160$. Table 1 gives the number $N(2^{-k})$ of boxes hit by an orbit that winds through the attractor for 1,000,000 iterates. Figure 6 plots the values of $\log_2(N(2^{-k}))$ versus $k = \log_2(1/2^{-k})$ for $k = 3, \ldots, 7$. The logarithm base 2 is used, since the logarithm of the values $1/r = 2^k$ are integers, and ratios of the logarithms are the same as using the natural logarithm. Because there might be a constant in the expression $N(r) \sim C\,r^{-d}$, a better estimate is obtained by taking the slope of the line best fitted to the data points. The slope of the line in Figure 6 is 1.24. In [**Nus98**], H. Nusse and J. Yorke give the box dimensions for these parameter values as 1.245 ± 0.093.

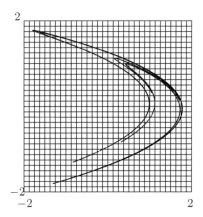

Figure 5. The Hénon attractor for $a = 1.4$ and $b = -0.3$ with the grid of size $r = \frac{1}{8}$.

k	$N(2^{-k})$
3	160
4	388
5	908
6	2152
7	4947

Table 1. Values of $N(2^{-k})$ for the Hénon map.

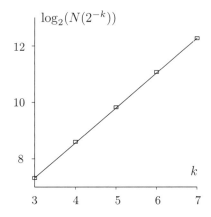

Figure 6. Plot of $\log_2(N(2^{-k}))$ as a function of $k = \log_2(1/2^{-k})$ for $k = 2, \ldots, 7$.

Before turning to two different types of Cantor sets, we state a theorem that relates box dimension and Lebesgue measure.

Theorem 14.3. *Let* **S** *be a bounded subset of* $\mathbb{R}^n$, *with box dimension less than* n. *Then, the n-dimensional Lebesgue measure of* **S** *is zero.*

The proof is given in Section 14.4.

14.1. Box Dimension

Example 14.9 (A Non-symmetric Cantor Set). We have considered Cantor sets in which the two sides left at each stage are the same length. Now, consider the case in which the left interval has length $\beta > 0$ and the right interval has length $\gamma > 0$, with $\gamma \neq \beta$. If the gap removed has length α, then $\beta + \alpha + \gamma = 1$. Repeat this process at each step. This Cantor set can be defined using the two functions

$$f_1(x) = \beta\, x \quad \text{and}$$
$$f_2(x) = \gamma\, x + 1 - \gamma.$$

Let $\mathbf{K}_0 = [0, 1]$ and define $\mathbf{K}_j$ by induction as

$$\mathbf{K}_j = f_1(\mathbf{K}_{j-1}) \cup f_2(\mathbf{K}_{j-1}).$$

Let

$$\mathbf{K} = \bigcap_{j=0}^{\infty} \mathbf{K}_j$$

be the resulting nonuniform Cantor set. Then,

$$f_1(\mathbf{K}_0) = [0, \beta],$$
$$f_2(\mathbf{K}_0) = [1 - \gamma, 1], \quad \text{and}$$
$$\mathbf{K}_1 = f_1(\mathbf{K}_0) \cup f_2(\mathbf{K}_0) = [0, \beta] \cup [1 - \gamma, 1].$$

Next,

$$f_1(\mathbf{K}_1) = \left[0, \beta^2\right] \cup [\beta(1 - \gamma), \beta] \subset f_1(\mathbf{K}_0) \subset \mathbf{K}_1,$$
$$f_2(\mathbf{K}_1) = [1 - \gamma, 1 - \gamma + \beta\,\gamma)] \cup \left[1 - \gamma^2, 1\right] \subset f_2(\mathbf{K}_0) \subset \mathbf{K}_1, \quad \text{and}$$
$$\mathbf{K}_2 = f_1(\mathbf{K}_1) \cup f_2(\mathbf{K}_1) \subset \mathbf{K}_1 \subset \mathbf{K}_0.$$

The lengths of the intervals in $\mathbf{K}_2$ are β^2, $\beta\gamma$, $\beta\gamma$, and γ^2. By induction, the set $\mathbf{K}_j$ contains 2^j intervals whose lengths vary among

$$\beta^j,\ \beta^{j-1}\gamma,\ \ldots,\ \beta\gamma^{j-1},\ \gamma^j.$$

For small $r > 0$, it is not easy to determine how many intervals of length r are needed to cover this Cantor set.

We give an intuitive argument for the box dimension that actually gives the right answer. We point out the step in which the derivation is not rigorous.

Let $N(r)$ be the number of intervals of length r needed to cover $\mathbf{K}$. Let $\mathbf{I}_\beta = [0, \beta]$ and $\mathbf{I}_\gamma = [1 - \gamma, 1]$, so $\mathbf{K}_1 = \mathbf{I}_\beta \cup \mathbf{I}_\gamma$ is the set after one interval is removed. Let $N(r, \mathbf{I}_\beta)$ be the number of intervals of length r needed to cover $\mathbf{I}_\beta \cap \mathbf{K}$, and let $N(r, \mathbf{I}_\gamma)$ be the number of intervals of length r needed to cover $\mathbf{I}_\gamma \cap \mathbf{K}$. The sum of the number for the two halves must equal the number of intervals needed to cover the whole Cantor set $\mathbf{K}$; that is,

$$N(r) = N(r, \mathbf{I}_\beta) + N(r, \mathbf{I}_\gamma).$$

The subset $\mathbf{K} \cap \mathbf{I}_\beta$, scaled by the factor $1/\beta$, equals all of $\mathbf{K}$, or $\mathbf{K} = \tfrac{1}{\beta}(\mathbf{K} \cap \mathbf{I}_\beta)$, so

$$N(r, \mathbf{I}_\beta) = N(r/\beta).$$

The subset $\mathbf{K} \cap \mathbf{I}_\gamma$, scaled by the factor $1/\gamma$ and translated, equals all of $\mathbf{K}$, so

$$N(r, \mathbf{I}_\gamma) = N(r/\gamma).$$

Combining, we have rigorously shown that
$$N(r) = N(r, \mathbf{I}_\beta) + N(r, \mathbf{I}_\gamma)$$
$$= N\left(r/\beta\right) + N\left(r/\gamma\right).$$

However, now we assume that
$$N(r) = C\, r^{-d}$$
gives an exact count of the intervals needed of size r for some C, rather than just being an asymptotic limit. If this equality were true, then substituting in the above count in $N(r) = N\left(r/\beta\right) + N\left(r/\gamma\right)$ would yield
$$C\, r^{-d} = C\left(\frac{r}{\beta}\right)^{-d} + C\left(\frac{r}{\gamma}\right)^{-d} \quad \text{and}$$
$$1 = \beta^d + \gamma^d.$$

To show that this equation defines a unique $0 < d < 1$, let $h(d) = \beta^d + \gamma^d$. Then, $h(0) = 2$, $h(1) = \beta + \gamma < 1$, and $h'(d) = \ln(\beta)\beta^{d-1} + \ln(\gamma)\gamma^{d-1} < 0$. Therefore, there is a unique d with $0 < d < 1$ such that $h(d) = 1$, as claimed. See Figure 7.

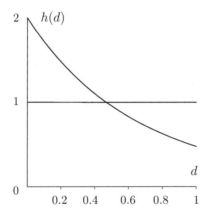

Figure 7. Graph of $h(d)$.

The value d satisfying $h(d) = 1$ is called the *scaling dimension* or *similarity dimension*. A theorem due to Moran proves that the scaling dimension equals the box dimension. See [**Edg90**] for its precise statement and proof.

Notice that when $\gamma = \beta$, $d = \ln(2)/\ln(\beta^{-1})$, as before:
$$1 = 2\,\beta^d,$$
$$0 = \ln(2) - d\,\ln(1/\beta),$$
$$d = \ln(2)/\ln(1/\beta).$$

We can also consider a Cantor set formed by leaving n intervals of length β_1, ..., β_n with gaps between them at the first stage, and then repeating at every stage by scaling the process at the first stage. For such a construction, the scaling dimension is the value of d such that
$$1 = \beta_1^d + \cdots + \beta_n^d.$$

Again, by Moran's theorem, the value of d that solves this equation also gives the box dimension.

Example 14.10 (A Cantor set with positive measure). All of the Cantor sets in the line that we have considered so far have a box dimension less than 1 and Lebesgue measure of zero (by Theorem 14.3). A Cantor set formed by removing a smaller percentage of the intervals at each step can have positive Lebesgue measure and a box dimension equal to 1.

Let $\mathbf{K}_0 = [0,1]$ be the unit interval. At the first stage, remove the middle $1/4$ of the interval, to obtain

$$\mathbf{K}_1 = \left[0, \frac{3}{8}\right] \cup \left[\frac{5}{8}, 1\right].$$

To form $\mathbf{K}_2$, remove two intervals of length $1/4^2$, for a total length removed of $2/4^2 = 1/4 \, (1/2)$. At the kth stage, remove 2^{k-1} intervals of length $1/4^k$ each for a total length removed of $2^{k-1}/4^k = 1/4 \, (1/2)^{k-1}$. The total length removed at all stages is

$$\frac{1}{4} + \frac{1}{4}\left(\frac{1}{2}\right) + \cdots + \frac{1}{4}\left(\frac{1}{2}\right)^{k-1} + \cdots$$

$$= \frac{1}{4}\left(1 + \frac{1}{2} + \cdots + \left(\frac{1}{2}\right)^{k-1} + \cdots\right)$$

$$= \frac{1}{4}\left(\frac{1}{1-\frac{1}{2}}\right)$$

$$= \frac{1}{2}.$$

The Lebesgue measure of the part left, $\mathbf{K} = \bigcap_{k=0}^{\infty} \mathbf{K}_k$, is $1 - 1/2 = 1/2$. Thus, this Cantor set has positive Lebesgue measure, and so must have box dimension equal to one.

The difference between this Cantor set and a middle-α Cantor set is that there is not a fixed proportion removed from each interval. In fact, the proportion removed goes to zero.

Exercises 14.1

1. Calculate the box dimension of the set $\mathbf{S}$ in $\mathbb{R}$ made up of the points

$$\mathbf{S} = \{0\} \cup \left\{\frac{1}{i^2} : i \geq 1\right\}.$$

2. Calculate the box dimension of the set $\mathbf{S}$ in $\mathbb{R}$ made up of the points

$$\mathbf{S} = \{0\} \cup \left\{\frac{1}{2^i} : i \geq 1\right\}.$$

3. For $p \geq 1$, calculate the box dimension of the set $\mathbf{S}$ in $\mathbb{R}$ made up of the points

$$\mathbf{S} = \{0\} \cup \left\{\frac{1}{i^p} : i \geq 1\right\}.$$

4. For $p > 1$ be fixed, calculate the box dimension of the set **S** in $\mathbb{R}$ made up of the points
$$\mathbf{S} = \{0\} \cup \left\{\frac{1}{p^k} : k \geq 1\right\}.$$
Consider the Cantor set formed by removing two intervals of length $1/5$ and leaving three intervals of length $1/5$. Continue this process by induction. At each stage, for each interval **J** from the previous stage, remove two intervals of $1/5^{\text{th}}$ the length of **J** and leave three intervals of length $1/5^{\text{th}}$ the length of **J**. Calculate the box dimension of the resulting Cantor set.

5. What is the scaling dimension of a set formed by removing an open interval of length $1/4$ and leaving two intervals, one of length $1/4$ and one of length $1/2$? Hint: Let $z = (1/2)^d$ in the equation $1 = (1/2)^d + (1/4)^d$.

6. Let **C** be the middle-third set in the line. Calculate the box dimension of
$$\mathbf{S} = \mathbf{C} \times \mathbf{C} = \{(x, y) \in \mathbb{R}^2 : x, y \in \mathbf{C}\},$$
the Cartesian product of the Cantor set with itself.

14.2. Dimension of Orbits

The box dimension presents certain difficulties in terms of computation. The primary problem is that the box dimension has a large storage requirement for a computer calculation. For this reason, P. Grassberger and I. Procaccia proposed the correlation dimension. Another difference between the correlation dimension and the box dimension is that the box dimension is a measurement applied to a set, such as an attractor, while by contrast, the correlation dimension is applied directly to the dynamics of an orbit.

J. Kaplan and J. Yorke introduced another type of dimension that also applies to an orbit, which is calculated from the Lyapunov exponents and is called the Lyapunov dimension. It is also easier to calculate than the box dimension. The main disadvantage of this measurement is the lack of a good intuitive explanation of why it should be called a dimension.

14.2.1. Correlation Dimension. The correlation dimension measures how often an orbit returns close to another point on the orbit. If the points are contained in a curve and roughly fill up the curve, then the number of points within a distance of r is approximately a linear function of r. If the points fill up a surface, then the number of points within a distance of r is approximately a quadratic function of r. That explanation motivates the next definition.

Definition 14.11. Assume that **F** is a map from $\mathbb{R}^n$ to itself with a bounded forward orbit of a point, $\mathcal{O}_\mathbf{F}^+(\mathbf{p}_0)$. Let $\mathbf{p}_j = \mathbf{F}^j(\mathbf{p}_0)$, and consider the set of m points
$$\mathbf{S}_m = \{\mathbf{p}_j : 0 \leq j < m\}.$$
There are $m(m-1)/2$ pairs (i, j) with $0 \leq i < j < m$. Consider the fraction of the pair of indices for which the pair of points $\mathbf{p}_i$ and $\mathbf{p}_j$ are within a distance $r > 0$

from each other:
$$C(r,m) = \frac{\#\{(i,j) : 0 \leq i < j < m \text{ and } \|\mathbf{p}_i - \mathbf{p}_j\| < r\}}{\#\{(i,j) : 0 \leq i < j < m\}}.$$

This number $C(r,m)$ is between 0 and 1. The limit
$$C(r) = \lim_{m \to \infty} C(r,m)$$
is the proportion of pairs of points on the whole orbit that are within a distance r of each other. Finally, the *correlation dimension of the orbit* is defined as the limit
$$\dim{}_c(\mathcal{O}_{\mathbf{F}}^+(\mathbf{p}_0)) = \lim_{r \to 0} \frac{\ln(C(r))}{\ln(r)}.$$

If $d = \dim{}_c(\mathcal{O}_{\mathbf{F}}^+(\mathbf{p}_0))$, then $C(r)$ scales like r^d, $C(r) \sim r^d$. Notice that $C(r) \leq 1$ and $r < 1$, so both logarithms are negative and we do not need to take the negative of either to get a nonnegative limit.

Example 14.12. Consider the doubling map $D(x) = 2x \pmod 1$. It has some points x_0 whose frequency measure is a uniform density on $[0,1]$. For small $r > 0$, $C(r) \sim r$ and $\dim{}_c(\mathcal{O}_D^+(x_0)) = 1$.

Example 14.13. Consider the map on $[0,1] \times [0,1]$ given by
$$\mathbf{F}(x,y) = (2x \pmod 1, 3x \pmod 1).$$

This map has some points $\mathbf{p}_0 = (x_0, y_0)$ whose frequency measure is a uniform density in the square. For a small $r > 0$, approximately r^2 of the pairs of points are within a distance of r, or
$$C(r) \sim r^2 \quad \text{and}$$
$$\dim{}_c(\mathcal{O}_F^+(\mathbf{p}_0)) = 2.$$

Example 14.14. Consider the tent map T_b with slope $b > 2$ that forms the middle $\alpha = 1 - 2/b$ Cantor set C_α. Then, a typical orbit that stays in $[0,1]$ for all iterates is uniformly distributed among the 2^k intervals of length $1/b^k$, so
$$C(b^{-k}) \sim \frac{1}{2^k} \quad \text{and}$$
$$\dim{}_c(\mathcal{O}_{T_b}^+(\mathbf{p}_0)) = \frac{\ln(2)}{\ln(b)} < 1.$$

Note that this number is the same as the box dimension for this Cantor set.

Numerical data for the Hénon map for $a = 1.4$ and $b = -0.3$ indicate that the correlation dimension for a typical point is 1.23, which is slightly less than the box dimension of 1.245.

For more details about the correlation dimension, as well as other practical issues in the use of fractal dimensions, see the book by E. Ott, T. Sauer, and J. Yorke [**Ott94**]

14.2.2. Lyapunov Dimension. Another fractal dimension, called the *Lyapunov dimension*, was introduced by J. Kaplan and J. Yorke based on the Lyapunov exponents of an orbit. The Lyapunov exponents of an orbit can be calculated numerically without problems in the amount of the data stored, so the Lyapunov dimension is also more easily computable than the box dimension. When the map has uniform expansion and contraction, it often gives the same answer as the box dimension. For nonuniformly expanding maps, the answer will probably be close but different. The Lyapunov dimension, like the correlation dimension, is also based on an orbit rather than on a set of points. The idea is that the orbit should be asymptotic to an attractor, and the Lyapunov dimension gives information about the attractor that the orbit approaches.

Definition 14.15. Let $\mathbf{F}$ be a map on $\mathbb{R}^n$. Let $\mathcal{O}_\mathbf{F}^+(\mathbf{p}_0)$ be a bounded forward orbit having Lyapunov exponents $\ell_j = \ell_j(\mathbf{p}_0; \mathbf{F})$, with $\ell_1 \geq \ell_2 \geq \cdots \geq \ell_n$. Let k be the integer for which

$$\ell_1 + \cdots + \ell_k \geq 0 \quad \text{and}$$
$$\ell_1 + \cdots + \ell_{k+1} < 0.$$

Then, the *Lyapunov dimension* of the orbit is

$$\dim_\mathrm{L}(\mathcal{O}_\mathbf{F}^+(\mathbf{p}_0)) = k + \frac{\ell_1 + \cdots + \ell_k}{|\ell_{k+1}|}.$$

Notice that $\ell_1 + \cdots + \ell_k < |\ell_{k+1}|$, so $\dim_\mathrm{L}(\mathcal{O}_\mathbf{F}^+(\mathbf{p}_0)) < k+1$.

A diffeomorphism with an attractor decreases volume near the attractor, so $\ell_1 + \cdots + \ell_n < 0$ and $\dim_\mathrm{L}(\mathcal{O}_\mathbf{F}^+(\mathbf{p}_0)) < n$. If the attractor has a positive Lyapunov exponent, then $k \geq 1$. Thus, an attractor with a positive Lyapunov exponent has a Lyapunov dimension greater than one and less than n.

Example 14.16. The Hénon map is not uniformly expanding and contracting. For $a = 1.4$ and $b = -0.3$, $\ell_1 \approx 0.42$ and $\ell_2 \approx -1.62$. Thus, its Lyapunov dimension is approximately

$$1 + \frac{0.42}{1.62} \approx 1.26.$$

On the other hand, the box dimension is about 1.245.

The next theorem states that, when the attractor is roughly the same at all points, then the Lyapunov dimension equals the box dimension. A map $\mathbf{F}$ is said to be *uniformly hyperbolic on an invariant set* provided that there is a constant $\lambda > 1$ such that the rate of expansion is greater than λ and the rate of contraction is at least $1/\lambda$ at all points of the invariant set.

Theorem 14.4. *Let $\mathbf{F}$ be a diffeomorphism that is uniformly hyperbolic on a chaotic attractor $\mathbf{A}$. Assume that $\mathbf{p}_0$ is a point with $\omega(\mathbf{p}_0, \mathbf{F}) = \mathbf{A}$. Then, the Lyapunov dimension for this orbit equals the box dimension of the attractor $\mathbf{A}$.*

Idea of the proof: The following heuristic argument can be made rigorous in the uniformly expanding and contracting case. Let $\ell_j = \ell_j(\mathbf{p}_0; \mathbf{F})$.

First, consider the case in two dimensions. Because the attractor is chaotic, all directions cannot be contracting and $\ell_1 > 0$. The area of the trapping region

14.2. Dimension of Orbits

is decreased by iteration so $\ell_1 + \ell_2 < 0$ and $\ell_2 < 0$. Let $N(r)$ be the number of boxes of size r needed to cover the attractor $\mathbf{A}$. After k iterates, the image of each box by $\mathbf{F}^k$ is a new box with dimensions of approximately $r\,e^{\ell_1 k}$ by $r\,e^{\ell_2 k}$. To cover each of these boxes with smaller boxes of size $r\,e^{\ell_2 k}$ will take approximately $e^{\ell_1 k}/e^{\ell_2 k} = e^{(\ell_1-\ell_2)k}$ boxes. Since there are $N(r)$ of these images, the total number of boxes of size $r\,e^{\ell_2 k}$ needed to cover the attractor is approximately

$$N(r\,e^{\ell_2 k}) \approx N(r)\,e^{(\ell_1-\ell_2)k}.$$

Taking the limit as k goes to infinity, we have

$$\begin{aligned}
\dim{}_{\mathrm{b}}(\mathbf{A}) &= \lim_{k\to\infty} \frac{\ln\left(N(r\,e^{\ell_2 k})\right)}{\ln(r^{-1}\,e^{-\ell_2 k})} \\
&= \lim_{k\to\infty} \frac{\ln(N(r)) + k(\ell_1 - \ell_2)}{\ln(r^{-1}) - k\,\ell_2} \\
&= \frac{\ell_1 - \ell_2}{-\ell_2} = 1 + \frac{\ell_1}{|\ell_2|} \\
&= \dim{}_{\mathrm{L}}(\mathcal{O}_{\mathbf{F}}^{+}(\mathbf{p}_0)).
\end{aligned}$$

Next, consider the case of a uniform expansion and contraction in $\mathbb{R}^3$, where $\ell_1 + \ell_2 > 0$ but $\ell_1 + \ell_2 + \ell_3 < 0$. The solenoid that wraps around four times with a contraction of $\beta > 1/4$ is an example in which this is true, even though $\ell_2 < 0$. Let $N(r)$ be the number of boxes of size r needed to cover the attractor. After k iterates, the image of each box by $\mathbf{F}^k$ is a new box with dimensions of $r\,e^{\ell_1 k}$ by $r\,e^{\ell_2 k}$ by $r\,e^{\ell_3 k}$. To cover each of these boxes with smaller boxes of size $r\,e^{\ell_3 k}$ takes approximately $e^{\ell_1 k}/e^{\ell_3 k} = e^{(\ell_1-\ell_3)k}$ boxes in the first direction and $e^{\ell_2 k}/e^{\ell_3 k} = e^{(\ell_2-\ell_3)k}$ boxes in the second direction. Combining, we see that it takes a total of about

$$e^{(\ell_1-\ell_3)k}\,e^{(\ell_2-\ell_3)k} = e^{(\ell_1+\ell_2-2\ell_3)k}$$

boxes to cover each image box. Since there are $N(r)$ of these image boxes, the total number of boxes of size $r\,e^{\ell_3 k}$ needed to cover the attractor is approximately given by

$$N(r\,e^{\ell_3 k}) \approx N(r)\,e^{(\ell_1+\ell_2-2\ell_3)k}.$$

Taking the limit as k goes to infinity, we obtain

$$\begin{aligned}
\dim{}_{\mathrm{b}}(\mathbf{A}) &= \lim_{k\to\infty} \frac{\ln\left(N(r\,e^{\ell_2 k})\right)}{\ln(r^{-1}\,e^{-\ell_3 k})} \\
&= \lim_{k\to\infty} \frac{\ln\left(N(r)\,e^{(\ell_1+\ell_2-2\ell_3)k}\right)}{\ln(r^{-1}\,e^{-\ell_3 k})} \\
&= \lim_{k\to\infty} \frac{\ln(N(r)) + k(\ell_1+\ell_2-2\ell_3)}{\ln(r^{-1}) - k\,\ell_3} \\
&= \frac{\ell_1+\ell_2-2\ell_3}{-\ell_3} = 2 + \frac{\ell_1+\ell_2}{|\ell_3|} \\
&= \dim{}_{\mathrm{L}}(\mathcal{O}_{\mathbf{F}}^{+}(\mathbf{p}_0)).
\end{aligned}$$

Thus, for this case, the Lyapunov dimension equals the box dimension as well. □

Exercises 14.2

1. A skinny baker's map is given by

$$\mathbf{B}\begin{pmatrix} x \\ y \end{pmatrix} = \begin{cases} \begin{pmatrix} \frac{1}{3}x \\ 2y \end{pmatrix} & \text{for } 0 \leq y < \frac{1}{2}, \\ \begin{pmatrix} \frac{1}{3}x + \frac{2}{3} \\ 2y - 1 \end{pmatrix} & \text{for } \frac{1}{2} \leq y \leq 1. \end{cases}$$

   In the calculation of the correlation dimension, for an orbit with a uniformly dense orbit in the attractor, the quantity $C(3^{-j}) \approx 2^{-j} \cdot 3^{-j}$. (The fraction 2^{-j} of the orbit come within a distance 3^{-j} in the x-direction, and 3^{-j} of the orbit come within a distance 3^{-j} in the y-direction.) Using this as an equality, what is the correlation dimension.

2. Let $\mathbf{F}$ be the map determining the solenoid Λ defined in Example 13.34. What is the Lyapunov dimension of the solenoid?

3. A skinny baker's map is given by

$$\mathbf{B}\begin{pmatrix} x \\ y \end{pmatrix} = \begin{cases} \begin{pmatrix} \frac{1}{3}x \\ 2y \end{pmatrix} & \text{for } 0 \leq y < \frac{1}{2}, \\ \begin{pmatrix} \frac{1}{3}x + \frac{2}{3} \\ 2y - 1 \end{pmatrix} & \text{for } \frac{1}{2} \leq y \leq 1. \end{cases}$$

   a. If $\mathbf{S} = [0,1] \times [0,1]$, explain why $\Lambda = \bigcap_{j=0}^{\infty} \mathbf{B}^j(\mathbf{S})$ is the Cartesian product of the middle-third Cantor set and the unit interval ($\Lambda = \mathbf{C} \times [0,1]$).
   b. What are the Lyapunov exponents of any point in $\mathbf{S}$?
   c. What is the Lyapunov dimension of Λ?

14.3. Iterated-Function Systems

This section presents a different and easy way to generate complicated sets using what are called iterated-function systems. M. Barnsley popularized these systems, [Bar88]. The appearance of the sets constructed can be very diverse and interesting. Some of these sets relate to invariant sets generated by a single map and some do not.

In the introduction to this chapter, we gave the definition of self-similar. We repeat it here, because this is the first section that uses the definition and many iterated-function systems are naturally self-similar.

Definition 14.17. A set is called *self-similar* if arbitrarily small pieces of it can be magnified to give the whole set.

We start with a different construction of Cantor sets. Then, we apply these ideas and constructions of this example to sets in the plane.

Example 14.18 (Cantor set). We return to Example 14.9, which was expressed in terms of two functions,

$$f_1(x) = \beta x \quad \text{and}$$
$$f_2(x) = \gamma x + 1 - \gamma,$$

with $\beta, \gamma > 0$ and $\beta + \gamma < 1$, so $\beta < 1 - \gamma$. The unit interval is invariant by these two functions, so let $\mathbf{K}_0 = [0,1]$. Then,

$$f_1(\mathbf{K}_0) = [0, \beta] \subset \mathbf{K}_0 \quad \text{and}$$
$$f_2(\mathbf{K}_0) = [1 - \gamma, 1] \subset \mathbf{K}_0.$$

Therefore, defining $\mathbf{K}_1$ as the union, we get

$$\mathbf{K}_1 = f_1(\mathbf{K}_0) \cup f_2(\mathbf{K}_0) = [0, \beta] \cup [1 - \gamma, 1] \subset \mathbf{K}_0.$$

Then,

$$f_1(\mathbf{K}_1) = [0, \beta^2] \cup [\beta(1-\gamma), \beta]$$
$$\subset f_1(\mathbf{K}_0) \subset \mathbf{K}_1 \subset \mathbf{K}_0 \quad \text{and}$$
$$f_2(\mathbf{K}_1) = [1 - \gamma, 1 + \beta\gamma - \gamma] \cup [1 - \gamma^2, 1]$$
$$\subset f_2(\mathbf{K}_0) \subset \mathbf{K}_1 \subset \mathbf{K}_0.$$

Defining $\mathbf{K}_2$ as the union yields

$$\mathbf{K}_2 = f_1(\mathbf{K}_1) \cup f_2(\mathbf{K}_1),$$

which has $2^2 = 4$ intervals of lengths β^2, $\beta\gamma$, or γ^2. Letting $\delta = \max\{\beta, \gamma\}$, all the lengths are less than δ^2. By induction,

$$\mathbf{K}_j = f_1(\mathbf{K}_{j-1}) \cup f_2(\mathbf{K}_{j-1})$$

has 2^j intervals whose lengths are all less than δ^j. The intersection

$$\mathbf{A} = \bigcap_{j=0}^{\infty} \mathbf{K}_j$$

is the nonsymmetric Cantor set discussed in Example 14.9.

The important feature of this construction is that f_1 and f_2 are two contractions, each of which takes the unit interval into itself, and the images are disjoint.

Notice that f_1 takes the set $\mathbf{A}$ onto the subset $\mathbf{A} \cap [0, \beta]$ by a similarity contraction; so, its inverse f_1^{-1} takes the subset $\mathbf{A} \cap [0, \beta]$ onto $\mathbf{A}$ by a similarity expansion. In the same way, f_2^{-1} takes the subset $\mathbf{A} \cap [1-\gamma, 1]$ onto $\mathbf{A}$ by a similarity expansion. Taking further iterates, for a choice of $s_0, \ldots, s_j$, with each s_i equal to either 0 or 1, the composition $f_{s_j} \circ \cdots \circ f_{s_0}$ takes $\mathbf{A}$ onto the subset $\mathbf{A} \cap f_{s_j} \circ \cdots \circ f_{s_0}[0,1]$ by a similarity contraction, and its inverse takes the subset onto the whole attractor $\mathbf{A}$ by a similarity expansion. Thus, the set $\mathbf{A}$ is self-similar.

Definition 14.19. An *iterated-function system on* $\mathbb{R}^n$ is a finite collection of maps $\mathscr{F} = \{\mathbf{F}_1, \ldots, \mathbf{F}_k\}$ with $k \geq 1$ defined from $\mathbb{R}^n$ to $\mathbb{R}^n$. We refer to such a system as an *IFS*. The action of the IFS on compact sets (closed and bounded sets) is defined by

$$\mathscr{F}(\mathbf{S}) = \mathbf{F}_1(\mathbf{S}) \cup \cdots \cup \mathbf{F}_k(\mathbf{S}).$$

Definition 14.20. A map $\mathbf{F}_i$ is a contraction if there is an $0 < r_i < 1$ such that
$$\|\mathbf{F}_i(\mathbf{x}) - \mathbf{F}_i(\mathbf{y})\| \le r_i \|\mathbf{x} - \mathbf{y}\|$$
for all points $\mathbf{x}$ and $\mathbf{y}$. The smallest value r_i that works for all x and y is called the *contraction constant*. If each map $\mathbf{F}_i$ in an IFS is a contraction, with contraction constant r_i, then it is called an IFS of contractions, and the value $r = \max\{r_i\}$ is called a contraction constant for the IFS.

The collection of maps is an iterated-function system of *affine maps*, provided that each map is given as
$$\mathbf{F}_i(\mathbf{x}) = \mathbf{L}_i \mathbf{x} + \mathbf{c}_i,$$
where each $\mathbf{L}_i$ is a linear map and each $\mathbf{c}_i$ is a constant point in $\mathbb{R}^n$. One of these affine maps is a contraction exactly when the norm of the linear map is less than one, $\|\mathbf{L}_i\| < 1$.

If the map $\mathbf{F}_i$ satisfies
$$\|\mathbf{F}_i(\mathbf{x}) - \mathbf{F}_i(\mathbf{y})\| = r_i \|\mathbf{x} - \mathbf{y}\|$$
for all points $\mathbf{x}$ and $\mathbf{y}$, then it takes an object $\mathbf{S}$ into an object that is similar (i.e., the same shape, but possibly different scale). Such a map can be written as
$$\mathbf{F}_i(\mathbf{x}) = r_i\, \mathbf{Q}_i \mathbf{x} + \mathbf{c}_i,$$
where $\mathbf{Q}_i$ is an orthogonal matrix. Such a map is called a *similarity*. In almost all of our examples, all the maps are similarity contractions by a uniform factor $0 < r < 1$ that works for all the maps, $r_i = r$. Such a system is called an *iterated-function system of similarity contractions by a uniform factor r*.

Example 14.21 (Sierpinski gasket). Start with a filled-in equilateral triangle $\mathbf{T}_0$ whose sides are each of length 1 and that has vertices at $(0,0)$, $(1,0)$, and $\left(1/2, \sqrt{3}/2\right)$. Let $\mathbf{F}_1$ be a contraction by $1/2$ toward the vertex at $(0,0)$, let $\mathbf{F}_2$ be a contraction by $1/2$ toward the vertex at $(1,0)$, and let $\mathbf{F}_3$ be a contraction by $1/2$ toward the vertex at $(1/2, \sqrt{3}/2)$:
$$\mathbf{F}_1(\mathbf{x}) = \frac{1}{2}\mathbf{x},$$
$$\mathbf{F}_2(\mathbf{x}) = \frac{1}{2}\left(\mathbf{x} - (1,0)\right) + (1,0) = \frac{1}{2}\mathbf{x} + \left(\frac{1}{2}, 0\right), \quad \text{and}$$
$$\mathbf{F}_3(\mathbf{x}) = \frac{1}{2}\left(\mathbf{x} - \left(\frac{1}{2}, \frac{\sqrt{3}}{2}\right)\right) + \left(\frac{1}{2}, \frac{\sqrt{3}}{2}\right) = \frac{1}{2}\mathbf{x} + \left(\frac{1}{4}, \frac{\sqrt{3}}{4}\right).$$
By the description, it follows that $\mathbf{F}_j(\mathbf{T}_0) \subset \mathbf{T}_0$ for $1 \le j \le 3$.

Inductively define
$$\mathbf{T}_j = \mathscr{F}(\mathbf{T}_{j-1}) = \mathbf{F}_1(\mathbf{T}_{j-1}) \cup \mathbf{F}_2(\mathbf{T}_{j-1}) \cup \mathbf{F}_3(\mathbf{T}_{j-1}).$$
Then, $\mathbf{T}_1$ consists of three filled-in equilateral triangles with sides of length $1/2$. See Figure 8(a). By induction, we see that $\mathbf{T}_j$ has 3^j filled-in equilateral triangles with sides of length $1/2^j$. See Figure 8(a–d) for $\mathbf{T}_1$ to $\mathbf{T}_4$. The intersection
$$\mathbf{G} = \bigcap_{j=0}^{\infty} \mathbf{T}_j$$
is called the *Sierpinski gasket*.

14.3. Iterated-Function Systems

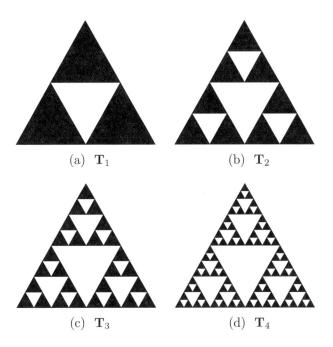

Figure 8. The sets T_1, T_2, T_3, and T_4 for the Sierpinski gasket

The Sierpinski gasket is self-similar just as is a Cantor set.

Since there are 3^j triangles of size 2^{-j} at the jth stage, and they intersect only on the corners, the box dimension of the gasket satisfies

$$\dim_{\mathrm{b}}(\mathbf{G}) = \lim_{j\to\infty} \frac{\ln(3^j)}{\ln(2^j)} = \frac{\ln(3)}{\ln(2)} > 1.$$

Example 14.22 (Koch curve). The final example is described in a slightly different way, since we do not have a set that is mapped into itself. Let $\mathbf{S}_0$ be a line segment of length one. Let $\mathbf{S}_1$ be the curve made up of four line segments, each of length $1/3$ as shown in Figure 9(a). Each of the four maps $\mathbf{F}_i$ is a contraction by a factor of $1/3$. The maps are taken so that each image $\mathbf{F}_i(\mathbf{S}_0)$ equals the ith subinterval of $\mathbf{S}_1$. Thus, $\mathbf{F}_1$ is merely a contraction with no rotation or translation; $\mathbf{F}_2$ rotates by an angle of $\pi/3$ and translates by $\mathbf{c}_2 = (1/3, 0)$; $\mathbf{F}_3$ rotates by an angle of $-\pi/3$ and translates by $\mathbf{c}_3 = (1/2, \sqrt{3}/6)$; and $\mathbf{F}_4$ is only a translation by $\mathbf{c}_4 = (2/3, 0)$. With these choices, $\mathscr{F}(\mathbf{S}_0) = \mathbf{S}_1$.

Applying the iterated-function system again, the image $\mathbf{F}_i(\mathbf{S}_1)$ has end points at the end points of the ith subinterval of $\mathbf{S}_1$. The union of these gives $\mathbf{S}_2$:

$$\mathbf{S}_2 = \mathscr{F}(\mathbf{S}_1) = \mathbf{F}_1(\mathbf{S}_1) \cup \mathbf{F}_2(\mathbf{S}_1) \cup \mathbf{F}_3(\mathbf{S}_1) \cup \mathbf{F}_4(\mathbf{S}_1).$$

See Figure 9(b). This curve is made up of 4^2 line segments of length $1/3^2$ each, for a total length of $(4/3)^2$.

By induction, define

$$\mathbf{S}_j = \mathscr{F}(\mathbf{S}_{j-1}) = \mathbf{F}_1(\mathbf{S}_{j-1}) \cup \mathbf{F}_2(\mathbf{S}_{j-1}) \cup \mathbf{F}_3(\mathbf{S}_{j-1}) \cup \mathbf{F}_4(\mathbf{S}_{j-1}),$$

which is made up of 4^j line segments of length $1/3^j$ each, for a total length of $(4/3)^j$. Let **K** be the limiting curve for the sets $\mathbf{S}_j$. Because $(4/3)^j$ goes to infinity as j goes to infinity, the limiting set **K** is a continuous curve with infinite length. Saying that the curve has infinite length means that it does not really have a length, it is *nonrectifiable*.

Since there are 4^j line segments of size 3^{-j} at the jth stage and they intersect only on the end points, the box dimension of the attractor is

$$\dim{}_b(\mathbf{K}) = \lim_{j \to \infty} \frac{\ln(4^j)}{\ln(3^j)} = \frac{\ln(4)}{\ln(3)} \quad \text{and} \quad 1 < \dim{}_b(\mathbf{K}) < 2.$$

Thus, **K** is a continuous curve with box dimension greater than one.

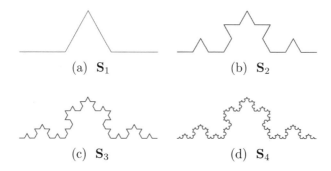

Figure 9. The sets $\mathbf{S}_1$, $\mathbf{S}_2$, $\mathbf{S}_3$, and $\mathbf{S}_4$ for the Koch curve

The next theorem states that an iterated-function system made up of a collection of contraction mappings has a unique compact invariant set.

Theorem 14.5. *Let $\mathscr{F} = \{\mathbf{F}_1, \ldots, \mathbf{F}_k\}$ be an IFS of contractions on $\mathbb{R}^n$ with a contraction constant r_i for each function. Let $r = \max\{r_i\}$ be the contraction constant of the IFS.*

a. *Let $R = \max_i \left\{ \dfrac{\|\mathbf{F}_i(\mathbf{0})\|}{1 - r_i} \right\}$ and let $\overline{\mathbf{B}}(R) = \{\mathbf{x} \in \mathbb{R}^n : \|\mathbf{x}\| \leq R\}$. Then, this ball is positively invariant by the IFS,*

$$\mathscr{F}(\overline{\mathbf{B}}(R)) \subset \overline{\mathbf{B}}(R).$$

b. *Let $\mathbf{S}_0$ be any compact set such that $\mathscr{F}(\mathbf{S}_0) \subset \mathbf{S}_0$. Inductively define $\mathbf{S}_m = \mathscr{F}(\mathbf{S}_{m-1})$, and $\mathbf{A} = \bigcap_{m=0}^{\infty} \mathbf{S}_m$. Then, the system maps $\mathbf{A}$ onto $\mathbf{A}$, $\mathscr{F}(\mathbf{A}) = \mathbf{A}$. Also, $\mathbf{A}$ is the unique compact invariant set for the iterated-function system. More precisely, if $\mathbf{A}'$ is any other nonempty compact set such that $\mathbf{A}' = \mathscr{F}(\mathbf{A}')$, then $\mathbf{A}' = \mathbf{A}$. In particular, if $\mathbf{S}_0$ is any compact set such that $\mathscr{F}(\mathbf{S}_0) \subset \mathbf{S}_0$, then $\mathbf{A} \subset \mathbf{S}_0$.*

c. *The box dimension of the attractor $\mathbf{A}$ satisfies $\dim{}_b(\mathbf{A}) \leq \dfrac{\ln(k)}{\ln(1/r)}$. If $\mathscr{F}$ is an IFS of similarity contractions by a uniform factor r, and there is a compact positively invariant set $\mathbf{S}_0$ for $\mathscr{F}$, with $\mathbf{F}_i(\text{int}(\mathbf{S}_0)) \cap \mathbf{F}_j(\text{int}(\mathbf{S}_0)) = \emptyset$ for all $i \neq j$, then the box dimension of the attractor $\mathbf{A}$ is given by $\dim{}_b(\mathbf{A}) = \dfrac{\ln(k)}{\ln(1/r)}$.*

The proof is in Section 14.4.

Remark 14.23. In the preceding examples, rather than picking a large ball $\overline{\mathbf{B}}(R)$ and looking at its image by the iterated-function system, we use a set more closely related to the system. Thus, for the Cantor set, we use an interval; for the Sierpinski gasket, we use a triangle; for the Koch curve, we do not use a set mapped into itself, but use successive iterates of a line segment.

14.3.1. Iterated-Function Systems Acting on Sets. We have seen already that a single contracting map on $\mathbb{R}^n$ has a fixed point that attracts all the points. In this subsection, we consider the map $\mathscr{F}$ induced by an iterated-function system on compact subsets of $\mathbb{R}^n$. For a iterated-function system of contracting maps, we see that $\mathscr{F}$ has a unique fixed compact set that is the invariant sets seen in the preceding subsection.

To make this precise, we have to start by defining a distance between compact subsets of $\mathbb{R}^n$. In order to insure that there is a fixed point, we need to know that the space $\mathscr{K}(\mathbb{R}^n)$ of compact subsets of $\mathbb{R}^n$ is complete in terms of the distance defined. For example, the irrational numbers are not complete: the map $f(x) = x/2$ is a contraction on the irrational numbers but does not have a fixed point in the irrational numbers. (The only fixed point is the rational number 0.)

We show that the induced action of an iterated-function system of contracting maps is a contraction in terms of the distance between these sets. It follows that the action has a fixed compact set, which is the attractor for the iterated-function system.

Distance between compact sets

Definition 14.24. Given a set $\mathbf{X}$, a function that takes two points $\mathbf{x}$ and $\mathbf{y}$ in $\mathbf{X}$ and yields a number $d(\mathbf{x}, \mathbf{y})$ is called a *metric* or *distance* d on $\mathbf{X}$, provided that it has the following four properties for all points $\mathbf{x}$, $\mathbf{y}$ and $\mathbf{z}$ in $\mathbf{X}$:

(i) $d(\mathbf{x}, \mathbf{y}) = d(\mathbf{y}, \mathbf{x})$.
(ii) $0 \leq d(\mathbf{x}, \mathbf{y}) < \infty$.
(iii) $d(\mathbf{x}, \mathbf{y}) = 0$ if and only if $\mathbf{x} = \mathbf{y}$.
(iv) $d(\mathbf{x}, \mathbf{z}) \leq d(\mathbf{x}, \mathbf{y}) + d(\mathbf{y}, \mathbf{z})$.

The set $\mathbf{X}$, together with a metric d, is called a *metric space*.

Let $\mathscr{K}(\mathbb{R}^n)$ denote the collection of all nonempty compact subsets of $\mathbb{R}^n$:

$$\mathscr{K}(\mathbb{R}^n) = \{\, \mathbf{A} \subset \mathbb{R}^n : \mathbf{A} \text{ is nonempty and compact }\,\}.$$

Remember that a subset of $\mathbb{R}^n$ is compact if and only if it is closed and bounded.

The distance between compact sets in $\mathscr{K}(\mathbb{R}^n)$ is given in terms of the distance from a point $\mathbf{x}$ to a nonempty closed set $\mathbf{S}$, which is given by

$$d(\mathbf{x}, \mathbf{S}) = \min\{\, \|\mathbf{x} - \mathbf{y}\| : \mathbf{y} \in \mathbf{S}\,\}.$$

Because the set is closed, the minimum is attained. See Figure 10. Also, because the set is closed, $d(\mathbf{x}, \mathbf{S}) = 0$ if and only if $\mathbf{x}$ is in the set $\mathbf{S}$. Using the distance between a point and a compact set $\mathbf{A}$, we define a closed neighborhood by

$$\mathscr{N}_r(\mathbf{A}) = \{\, \mathbf{x} \in \mathbb{R}^n : d(\mathbf{x}, \mathbf{A}) \leq r\,\}.$$

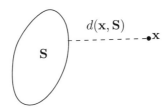

Figure 10. Distance from a point to a compact set

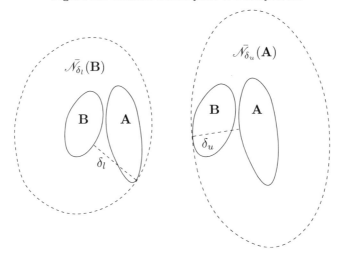

Figure 11. The lower distance $\delta_l = \delta_l(\mathbf{A}, \mathbf{B})$ and the upper distance $\delta_u = \delta_u(\mathbf{A}, \mathbf{B})$ between two sets $\mathbf{A}$ and $\mathbf{B}$

The distance between compact sets is defined in terms of two "hemimetrics". (A hemimetric between two compact sets can be zero without them being equal.) Let $\mathbf{A}$ and $\mathbf{B}$ be sets in $\mathscr{K}(\mathbb{R}^n)$. The *lower hemimetric* on $\mathscr{K}(\mathbb{R}^n)$ is defined by

$$\delta_l(\mathbf{A}, \mathbf{B}) = \min\{r \geq 0 : \mathbf{A} \subset \bar{\mathscr{N}}_r(\mathbf{B})\}$$
$$= \max\{d(\mathbf{a}, \mathbf{B}) : \mathbf{a} \in \mathbf{A}\}.$$

Similarly, the *upper hemimetric* is defined by

$$\delta_u(\mathbf{A}, \mathbf{B}) = \min\{r \geq 0 : \mathbf{B} \subset \bar{\mathscr{N}}_r(\mathbf{A})\}$$
$$= \max\{d(\mathbf{b}, \mathbf{A}) : \mathbf{b} \in \mathbf{B}\}$$
$$= \delta_l(\mathbf{B}, \mathbf{A}).$$

The *Hausdorff metric* is defined by

$$\delta(\mathbf{A}, \mathbf{B}) = \max\{\delta_l(\mathbf{A}, \mathbf{B}), \delta_u(\mathbf{A}, \mathbf{B})\}$$
$$= \min\{r > 0 : \mathbf{A} \subset \bar{\mathscr{N}}_r(\mathbf{B}) \text{ and } \mathbf{B} \subset \bar{\mathscr{N}}_r(\mathbf{A})\}$$
$$= \max\{d(\mathbf{a}, \mathbf{B}), d(\mathbf{b}, \mathbf{A}) : \mathbf{a} \in \mathbf{A},\ \mathbf{b} \in \mathbf{B}\},$$

for $\mathbf{A}, \mathbf{B} \in \mathscr{K}(\mathbb{R}^n)$.

The next theorem states that δ is a metric on $\mathscr{K}(\mathbb{R}^n)$.

14.3. Iterated-Function Systems

Theorem 14.6. *Let* **A**, **B**, *and* **C** *be sets in* $\mathscr{K}(\mathbb{R}^n)$.

  a. *The hemimetric δ_l satisfies the following conditions:*
  (i′) $\delta_l(\mathbf{A}, \mathbf{B})$ *is not necessarily equal to* $\delta_l(\mathbf{B}, \mathbf{A})$.
  (ii′) $0 \leq \delta_l(\mathbf{A}, \mathbf{B}) < \infty$.
  (iii′) $\delta_l(\mathbf{A}, \mathbf{B}) = 0$ *if and only if* $\mathbf{A} \subset \mathbf{B}$, *and*
  (iv′) $\delta_l(\mathbf{A}, \mathbf{C}) \leq \delta_l(\mathbf{A}, \mathbf{B}) + \delta_l(\mathbf{B}, \mathbf{C})$.

  b. *The hemimetric δ_u satisfies the same conditions stated for δ_l, but with (iii′) replaced by $\delta_u(\mathbf{A}, \mathbf{B}) = 0$ if and only if $\mathbf{B} \subset \mathbf{A}$.*

  c. *The Hausdorff metric δ is a metric on the space $\mathscr{K}(\mathbb{R}^n)$.*

For the proof, see Section 14.4.

As we mentioned in the opening paragraphs of this subsection, we need to know that the space is complete in order to show that a contraction on it has a fixed point. For the real line, we can define completeness in terms of limits of increasing sequences of points. In Euclidean spaces and general metric spaces, this is not possible. We give some definitions to express this concept.

Definition 14.25. Let **X** be a metric space with a distance d that satisfies the conditions of Definition 14.24.

A sequence of points $\{\mathbf{x}^i\}_{i=0}^{\infty}$ is a *Cauchy sequence* provided that, for any $\epsilon > 0$, there is an I such that $\|\mathbf{x}^i - \mathbf{x}^j\| < \epsilon$ for all $i, j \geq I$.

A metric space **X** with a distance d is *complete* provided that, for any Cauchy sequence $\{\mathbf{x}^i\}_{i=0}^{\infty}$ in **X**, there is a limit $\mathbf{x}^\infty$ in **X** such that $\|\mathbf{x}^i - \mathbf{x}^\infty\|$ goes to zero as i goes to infinity. Thus, any sequence of points that get "close" to each other (as in a Cauchy sequence) must converge to some point $\mathbf{x}^\infty$ in the space.

Theorem 14.7. *The space $\mathscr{K}(\mathbb{R}^n)$ is complete with respect to the Hausdorff metric δ.*

The proof of the theorem needs some detailed constructions and is not trivial. See Section 14.4.

Contraction maps applied to sets

If **S** is a compact set and **F** is a continuous function, then $\mathbf{F}(\mathbf{S})$ is also compact; thus, **F** takes a set **S** in $\mathscr{K}(\mathbb{R}^n)$ and gives back a set $\mathbf{F}(\mathbf{S})$ in $\mathscr{K}(\mathbb{R}^n)$. If $\mathscr{F} = \{\mathbf{F}_1, \ldots, \mathbf{F}_k\}$ is an IFS on $\mathbb{R}^n$, then $\mathscr{F}(\mathbf{S}) = \mathbf{F}_1(\mathbf{S}) \cup \cdots \cup \mathbf{F}_k(\mathbf{S})$ induces a map from $\mathscr{K}(\mathbb{R}^n)$ to $\mathscr{K}(\mathbb{R}^n)$.

The next theorem says that, if **F** is a contraction when acting on points in $\mathbb{R}^n$, then it induces a contraction on $\mathscr{K}(\mathbb{R}^n)$.

Theorem 14.8. *Assume that **F** is a single contraction on $\mathbb{R}^n$ with contraction constant $r < 1$. Then, **F** induces a contraction on $\mathscr{K}(\mathbb{R}^n)$ with the same contraction constant r.*

The proof is in Section 14.4.

Theorem 14.9. *Assume that $\mathscr{F} = \{\mathbf{F}_1, \ldots, \mathbf{F}_k\}$ is an IFS of contractions on $\mathbb{R}^n$, with contraction constants $r_1, \ldots, r_k$, and $r = \max\{r_i : 1 \leq i \leq k\} < 1$. Then, the induced action of $\mathscr{F}$ on $\mathscr{K}(\mathbb{R}^n)$ is a contraction with contraction constant r.*

The proof is in Section 14.4.

Just as a single contraction on $\mathbb{R}^n$ has a fixed point, the induced action of an IFS of contractions on $\mathscr{K}(\mathbb{R}^n)$ has a set which is fixed. This uses the fact that $\mathscr{K}(\mathbb{R}^n)$ is complete.

Theorem 14.10. *Assume that $\mathscr{F} = \{\mathbf{F}_1, \ldots, \mathbf{F}_k\}$ is an iterated-function system of contractions on $\mathbb{R}^n$, with a contraction constant $0 < r < 1$. Then, $\mathscr{F}$ has a unique fixed set $\mathbf{A}$ in $\mathscr{K}(\mathbb{R}^n)$ that is the attractor for the iterated-function system.*

The proof is in Section 14.4.

14.3.2. Probabilistic Action of Iterated-Function Systems.

We have seen how a iterated-function system of contractions has an attractor for the system by means of the action of $\mathscr{F}$ on $\mathscr{K}(\mathbb{R}^n)$. Another way to generate the attractor is to plot an orbit obtained by probabilistically choosing a map from the IFS.

Theorem 14.11. *Let $\mathscr{F} = \{\mathbf{F}_1, \ldots, \mathbf{F}_k\}$ be an IFS on $\mathbb{R}^n$, with contraction constants $r_1, \ldots, r_k$, $r = \max\{r_i : 1 \le i \le k\} < 1$, and an invariant compact set $\mathbf{A}$. Then, for any initial condition $\mathbf{x}_0$ in $\mathbb{R}^n$, if $\mathbf{x}_j = \mathbf{F}_{s_j}(\mathbf{x}_{j-1})$ is an orbit obtained by choosing a sequence of maps from the IFS, then the distance of $\mathbf{x}_j$ to $\mathbf{A}$ goes to zero at least as fast as r^j times the original distance of $\mathbf{x}_0$ to $\mathbf{A}$, $d(\mathbf{x}_j, \mathbf{A}) \le r^j \, d(\mathbf{x}_0, \mathbf{A})$.*

The proof is in Section 14.4.

Besides acting on sets, an iterated-function system can also act on points by picking one of the functions from the system at random and applying it to the point. As the system is repeatedly applied in this probabilistic manner, the orbit fills up the attractor for the system.

Definition 14.26. An *iterated-function system with probabilities* is a finite collection of maps $\mathscr{F} = \{\mathbf{F}_1, \ldots, \mathbf{F}_k\}$ with $k \ge 1$ defined from some $\mathbb{R}^n$ to $\mathbb{R}^n$, together with some positive weights $p_1, p_2, \ldots, p_k$, such that $0 < p_i < 1$ for $1 \le i \le k$ and $p_1 + \cdots + p_k = 1$. Each p_i is the probability of choosing $\mathbf{F}_i$.

Example 14.27 (Cantor set). Take a point x_0 in $[0, 1]$. Flip a coin. If it shows heads, then map the point to $x_1 = f_1(x_0) = \frac{1}{3}x_0$; if the coin shows tails, then map the point to $x_1 = f_2(x_0) = \frac{1}{3}x_0 + \frac{2}{3}$. Continue to flip the coin and take f_1 or f_2, depending on whether the coin shows heads or tails, $x_n = f_1(x_{n-1})$ or $f_2(x_{n-1})$. For a typical random sequence of heads and tails, the sequence of points x_n is dense in the middle-third Cantor set, $\mathbf{K}$. Even if x_0 is outside of the interval $[0, 1]$, the iterates converge to the set $\mathbf{K}$. Thus, this set is an attractor for this pair of functions.

Example 14.28 (Sierpinski gasket). Let

$$\mathscr{F} = \{\mathbf{F}_1, \mathbf{F}_2, \mathbf{F}_3\}$$

be the three functions used to define the Sierpinski gasket. Given an initial point $\mathbf{x}_0$ in the equilateral triangle $\mathbf{T}_0$, pick one of the maps with equal probability of one third each. (This could be implemented by rolling a single die and using $\mathbf{F}_1$ if the outcome is 1 or 2, using $\mathbf{F}_2$ if the outcome is 3 or 4, and using $\mathbf{F}_3$ if the outcome is 5 or 6.) For a typical random sequence of these maps $\mathbf{F}_{s_j}$, the iterates of the initial point $\mathbf{x}_i = \mathbf{F}_{s_j}(\mathbf{x}_{i-1})$ are dense in the set $\mathbf{G}$ given in Example 14.21.

14.3. Iterated-Function Systems

Part (b) of the next theorem gives the assumptions that imply than an orbit of point generated by a random sequence of functions from the IFS is dense in the attractor.

Theorem 14.12. *Let Σ_k^+ with shift map σ be the one-sided full shift on k symbols. Let $\{p_1, \ldots, p_k\}$ be a system of weights with $0 < p_i < 1$ for all $1 \leq i \leq k$ and $p_1 + \cdots + p_k = 1$.*

a. *Then, there is a probability measure $\mu_\mathbf{p}$ defined on Σ_k^+ that is invariant by σ and that satisfies the following: given $a_0, \ldots, a_n \in \{1, \ldots k\}$, then*

$$\mu_\mathbf{p}(\{\mathbf{s} : s_i = a_0\}) = p_{a_0} \quad \text{and}$$

$$\mu_\mathbf{p}(\{\mathbf{s} : s_i = a_i \text{ for } 0 \leq i \leq n\}) = \prod_{i=0}^{n} p_{a_i}.$$

b. *There is a set $\Sigma^* \subset \Sigma_k^+$ of symbol sequences of full measure, $\mu_\mathbf{p}(\Sigma^*) = 1$, such that any $\mathbf{s}$ in Σ^* contains all finite strings (i.e., for any finite string $a_0 \ldots a_{m-1}$, there is a j such that $a_i = s_{j+i}$ for $0 \leq i \leq m-1$). It follows that the ω-limit by the map σ of any symbol sequence $\mathbf{s}$ in Σ^* is all of Σ_k^+, $\omega(\mathbf{s}; \sigma) = \Sigma_k^+$.*

The proof is in Section 14.4.

The proof of the next corollary is an easy consequence of this theorem.

Corollary 14.13. *Let $\mathscr{F} = \{\mathbf{F}_1, \ldots, \mathbf{F}_k\}$ with weights $p_1, \ldots, p_k$ be a contracting iterated-function system with probabilities. Let $\Sigma^* \subset \Sigma_k^+$ be the set of the preceding theorem and let $\mathbf{A}$ be the attractor given in Theorem 14.5 for the iterated-function system.*

a. *If $\mathbf{s}$ is in Σ^* and $\mathbf{x}_0$ is a point in $\mathbf{A}$, then the set*

$$\{\mathbf{F}_{s_i} \circ \cdots \circ \mathbf{F}_{s_0}(\mathbf{x}_0)\}_{i=0}^{\infty}$$

is dense in $\mathbf{A}$.

b. *If $\mathbf{s}$ is in Σ^* and $\mathbf{x}_0$ is a point in $\mathbb{R}^n$, then the closure of the set*

$$\{\mathbf{F}_{s_i} \circ \cdots \circ \mathbf{F}_{s_0}(\mathbf{x}_0)\}_{i=0}^{\infty}$$

contains $\mathbf{A}$.

Proof. (a) Assume that $\mathbf{x}_0$ is in $\mathbf{A}$, and $\mathbf{S}_0$ is a closed and bounded set that is positively invariant by the iterated-function system. An element $\mathbf{s}$ in Σ^* contains all finite strings. Pick a finite string, $a_0 \ldots a_{m-1}$. Then, there is a j such that $a_i = s_{j+i}$ for $0 \leq i \leq m-1$. Thus,

$$\mathbf{F}_{s_{j+m-1}} \circ \cdots \circ \mathbf{F}_{s_0}(\mathbf{x}_0) \in \mathbf{F}_{a_{m-1}} \circ \cdots \circ \mathbf{F}_{a_0}(\mathbf{S}_0).$$

But, the sets $\mathbf{F}_{a_{m-1}} \circ \cdots \circ \mathbf{F}_{a_0}(\mathbf{S}_0)$ are arbitrarily small subsets of $\mathbf{A}$. Therefore, by this sequence of maps, the orbit of $\mathbf{x}_0$ is dense in $\mathbf{A}$.

(b) If $\mathbf{x}_0$ is an arbitrary point in $\mathbb{R}^n$, after a few iterates, the point $\mathbf{F}_{s_j} \circ \cdots \circ \mathbf{F}_{s_0}(\mathbf{x}_0)$ is very near $\mathbf{A}$ by Theorem 14.11. Thus, part (b) follows by the argument of part (a). $\square$

14.3.3. Determining the Iterated-Function System.
Starting with a fractal **A**, we would like to determine an iterated-function system that generates it. For all the examples considered, the functions are affine and so are of the form

$$\mathbf{F}_i \begin{pmatrix} x \\ y \end{pmatrix} = \begin{pmatrix} a_i & b_i \\ c_i & d_i \end{pmatrix} \begin{pmatrix} x \\ y \end{pmatrix} + \begin{pmatrix} e_i \\ f_i \end{pmatrix}.$$

It can be tricky to find the IFS, but we give some examples in which this is possible.

Example 14.29. The fractal in Figure 12 clearly has some subsets that are similar to the whole fractal. Each corner contains a copy of the fractal that is $(1/3)^{\text{rd}}$ the size of the whole. There is a fifth copy in the center, which is also $(1/3)^{\text{rd}}$ the size of the whole. Since these copies do not involve any rotation, we can take the matrix to be multiplication by $1/3$; that is,

$$\begin{pmatrix} a_i & b_i \\ c_i & d_i \end{pmatrix} = \begin{pmatrix} \frac{1}{3} & 0 \\ 0 & \frac{1}{3} \end{pmatrix},$$

for $1 \leq i \leq 5$. For each function, the point $(e_i, f_i)^\mathsf{T}$ is the bottom left corner of the subset. The bottom left copy contains the origin, so it is not displaced and we can take $e_1 = f_1 = 0$. The bottom right copy is displaced to $(2/3, 0)^\mathsf{T}$. Continuing in that fashion, we get the list of values for the affine maps given in Table 2.

i	a_i	b_i	c_i	d_i	e_i	f_i
1	$\frac{1}{3}$	0	0	$\frac{1}{3}$	0	0
2	$\frac{1}{3}$	0	0	$\frac{1}{3}$	$\frac{2}{3}$	0
3	$\frac{1}{3}$	0	0	$\frac{1}{3}$	0	$\frac{2}{3}$
4	$\frac{1}{3}$	0	0	$\frac{1}{3}$	$\frac{2}{3}$	$\frac{2}{3}$
5	$\frac{1}{3}$	0	0	$\frac{1}{3}$	$\frac{1}{3}$	$\frac{1}{3}$

Table 2. Parameter values for Example 14.29

Figure 12. Fractal for Example 14.29

14.3. Iterated-Function Systems

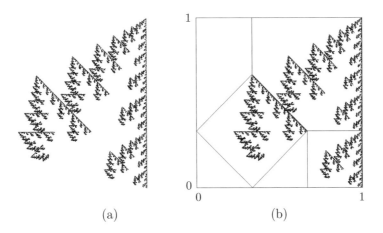

Figure 13. Fractal for Example 14.30

Example 14.30. This next example, illustrated in Figure 13, is more complicated than the preceding one and involves some rotation. A close inspection of the figure indicates that there is a $\frac{2}{3}$-sized copy of the whole in the upper right, a $\frac{1}{3}$-sized copy of the whole in the lower right. and a rotated intermediate sized copy of the whole in the upper left. The first two maps are easy to write down:

$$\mathbf{F}_1\begin{pmatrix}x\\y\end{pmatrix} = \begin{pmatrix}\frac{2}{3} & 0\\ 0 & \frac{2}{3}\end{pmatrix}\begin{pmatrix}x\\y\end{pmatrix} + \begin{pmatrix}\frac{1}{3}\\\frac{1}{3}\end{pmatrix} \quad \text{and}$$

$$\mathbf{F}_2\begin{pmatrix}x\\y\end{pmatrix} = \begin{pmatrix}\frac{1}{3} & 0\\ 0 & \frac{1}{3}\end{pmatrix}\begin{pmatrix}x\\y\end{pmatrix} + \begin{pmatrix}\frac{2}{3}\\0\end{pmatrix}.$$

For the third affine map, we see that

$$\mathbf{F}_3\begin{pmatrix}0\\0\end{pmatrix} = \begin{pmatrix}\frac{1}{3}\\0\end{pmatrix}, \quad \mathbf{F}_3\begin{pmatrix}1\\0\end{pmatrix} = \begin{pmatrix}\frac{2}{3}\\\frac{1}{3}\end{pmatrix}, \quad \text{and} \quad \mathbf{F}_3\begin{pmatrix}0\\1\end{pmatrix} = \begin{pmatrix}0\\\frac{1}{3}\end{pmatrix}.$$

From the image of $(0,0)$, we see that $e_3 = 1/3$ and $f_3 = 0$. Using these values and the images of the other two points, we can determine that

$$a_3 = \frac{1}{3}, \quad b_3 = -\frac{1}{3}, \quad c_c = \frac{1}{3}, \quad \text{and} \quad d_3 = \frac{1}{3}.$$

Thus, Table 3 gives all the values for the three functions.

i	a_i	b_i	c_i	d_i	e_i	f_i
1	$\frac{2}{3}$	0	0	$\frac{2}{3}$	$\frac{1}{3}$	$\frac{1}{3}$
2	$\frac{1}{3}$	0	0	$\frac{1}{3}$	$\frac{2}{3}$	0
3	$\frac{1}{3}$	$-\frac{1}{3}$	$\frac{1}{3}$	$\frac{1}{3}$	$\frac{1}{3}$	0

Table 3. Parameter values for Example 14.30

Exercises 14.3

1. (Sierpinski carpet) Consider the unit square $\mathbf{S}_0 = [0,1] \times [0,1]$ and the four maps $\mathbf{F}_i$ which contract by a factor of one third toward each of the four corners of the unit square. Let $\mathscr{F}$ be the IFS $\{\mathbf{F}_1, \mathbf{F}_2, \mathbf{F}_3, \mathbf{F}_4\}$.
   a. Write an expression for each of the four functions.
   b. Sketch the first two images of $\mathbf{S}_0$ by the iterated-function system, $\mathbf{S}_1 = \mathscr{F}(\mathbf{S}_0)$ and $\mathbf{S}_2 = \mathscr{F}(\mathbf{S}_1)$.
   c. What is the box dimension of the attractor $\mathbf{A} = \bigcap_{k=0}^{\infty} \mathbf{S}_k$?

2. Consider the iterated-function system $\mathscr{F}$ with the following three maps on $\mathbb{R}^2$:

$$\mathbf{F}_1 \begin{pmatrix} x \\ y \end{pmatrix} = 0.5 \begin{pmatrix} x \\ y \end{pmatrix},$$

$$\mathbf{F}_2 \begin{pmatrix} x \\ y \end{pmatrix} = 0.5 \begin{pmatrix} x \\ y \end{pmatrix} + \begin{pmatrix} 0.5 \\ 0 \end{pmatrix}, \quad \text{and}$$

$$\mathbf{F}_3 \begin{pmatrix} x \\ y \end{pmatrix} = 0.5 \begin{pmatrix} x \\ y \end{pmatrix} + \begin{pmatrix} 0 \\ 0.5 \end{pmatrix}.$$

   a. Let $\mathbf{S}$ be the unit square, $[0,1] \times [0,1]$. Describe and sketch the images $\mathscr{F}(\mathbf{S})$, $\mathscr{F}^2(\mathbf{S})$.
   b. How many boxes of size $(0.5)^3$ are needed to cover the attractor?
   c. What is the box dimension of the attractor for the iterated-function system? (Derive the answer, don't just write down a number.)

3. Which of the listed functions are contraction mappings on the real line? For those which are, find the smallest possible contraction constant r such that $|f(x) - f(y)| \leq r|x - y|$ for all x and y in $\mathbb{R}$.
   a. $f(x) = \frac{1}{2} - x$,
   b. $f(x) = -\frac{1}{3}x + 6$,
   c. $f(x) = \sin(x)$,
   d. $f(x) = \frac{1}{2}\sin(x)$.

4. For each of the following pairs of sets $\mathbf{A}$ and $\mathbf{B}$, what are the distances between the sets, $\delta_l(\mathbf{A}, \mathbf{B})$, $\delta_u(\mathbf{A}, \mathbf{B})$, and $\delta(\mathbf{A}, \mathbf{B})$?
   a. Let $\mathbf{A}$ be the unit square and let $\mathbf{B}$ be the unit disk; that is,
   $$\mathbf{A} = \{(x, y) : 0 \leq x, y \leq 1\},$$
   $$\mathbf{B} = \{(x, y) : x^2 + y^2 \leq 1\}.$$
   b. Let $\mathbf{A}$ be the square $[-1, 1] \times [-1, 1]$ and let $\mathbf{B}$ be the unit disk as in part (a); that is,
   $$\mathbf{A} = \{(x, y) : -1 \leq x, y \leq 1\}.$$
   c. Let $\mathbf{A} = [-1, 1] \times [-1, 1]$ as in part (b), and
   $$\mathbf{B} = \{(x, y) : x^2 + y^2 \leq 2\}.$$

5. Let $\mathbf{X}$ be the set of rational numbers between 1 and 2; that is,
$$\mathbf{X} = \left\{ \frac{p}{q} : 1 \leq \frac{p}{q} \leq 2 \right\}.$$

Let d be the usual distance on **X** inherited from the real line, $d(x,y) = |x-y|$. (The space **X** is not complete with this metric.) Define the function (Newton map) N from **X** to itself by

$$N(x) = x - \frac{x^2 - 2}{2x}.$$

a. Check that, if x is in **X**, then $N(x)$ is in **X**.
b. Show that N is a contraction mapping on **X**. Hint: The map N takes the real interval $[1,2]$ into itself, and $|N'(x)| < 1$.
c. Show that N does not have a fixed point in **X**. Explain why that is possible.

6. Let $f(x) = x/2$, $g(x) = (x+1)/2$, and $\mathscr{F} = \{f, g\}$. What is the unique fixed point of $\mathscr{F}$ on the collection $\mathscr{K}(\mathbb{R})$ of compact sets of $\mathbb{R}$?

7. Consider the fractals in Figure 14. What IFS of contractions has each of these sets as its attractor? Hint: For each IFS, there are three functions, one with a rotation.

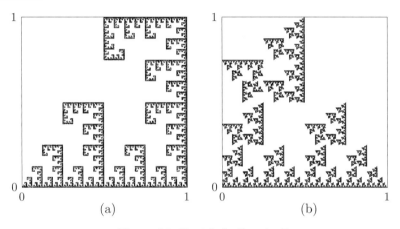

Figure 14. Fractals for Exercise 7

14.4. Theory and Proofs

Box dimension

Restatement of Theorem 14.1. *Let* **S** *be a set in* $\mathbb{R}^n$ *for which one of the limits,*

$$\lim_{r \to 0} \frac{\ln(N'(r, \mathbf{S}))}{\ln(1/r)} \quad or$$

$$\lim_{r \to 0} \frac{\ln(N''(r, \mathbf{S}))}{\ln(1/r)},$$

exists and equals d. *Then the box dimension* $\dim_b(\mathbf{S})$ *exists and equals* d.

Proof. We consider the case of boxes without a fixed grid, $N'(r, \mathbf{S})$, but the other case is similar. For a given $r > 0$, clearly $N'(r, \mathbf{S}) \leq N(r, \mathbf{S})$, because a cover with boxes for a fixed grid are certainly boxes that do not necessarily have a fixed grid. Also, consider a cover of **S** by boxes of size r, without fixed grid, $\{\mathbf{B}_1, \ldots, \mathbf{B}_{N'(r,\mathbf{S})}\}$. Each box $\mathbf{B}_i$ intersects at most 2 boxes in each direction for the fixed grid, or 2^n boxes of size r with the fixed grid. Therefore, $N(r, \mathbf{S}) \leq 2^n N'(r, \mathbf{S})$. Taking logarithms, dividing by $\ln(1/r)$, we get

$$N'(r, \mathbf{S}) \leq N(r, \mathbf{S}) \leq 2^n N'(r, \mathbf{S}),$$

$$\frac{\ln(N'(r, \mathbf{S}))}{\ln(1/r)} \leq \frac{\ln(N(r, \mathbf{S}))}{\ln(1/r)} \leq \frac{\ln(N'(r, \mathbf{S})) + n\ln(2)}{\ln(1/r)}.$$

Taking the limit,

$$d = \lim_{r \to 0} \frac{\ln(N'(r, \mathbf{S}))}{\ln(1/r)} \leq \lim_{r \to 0} \frac{\ln(N(r, \mathbf{S}))}{\ln(1/r)} \leq \lim_{r \to 0} \frac{\ln(N'(r, \mathbf{S})) + n\ln(2)}{\ln(1/r)} = d.$$

Since the first and last term are equal, the middle limit must exist and must be equal to d:

$$\dim_b(\mathbf{S}) = \lim_{r \to 0} \frac{\ln(N(r, \mathbf{S}))}{\ln(1/r)} = d.$$

This completes the proof. □

Restatement of Theorem 14.2. *Let* **S** *be a set in* $\mathbb{R}^n$. *Assume that there is a sequence of sizes* r_j, *with*

$$0 = \lim_{j \to \infty} r_j \quad \text{and} \quad 1 = \lim_{j \to \infty} \frac{\ln(r_{j+1})}{\ln(r_j)}$$

such that the following limit exists:

$$d = \lim_{j \to \infty} \frac{\ln(N'(r_j, \mathbf{S}))}{\ln(1/r_j)}.$$

Then, $\dim_b(\mathbf{S})$ *exists and equals* d.

Proof. We could use $N(r_j, \mathbf{S})$ for a fixed grid set of boxes, but instead we argue with arbitrary boxes used to define $N'(r, \mathbf{S})$. For any $r > 0$, there is a j such that $r_j \geq r > r_{j+1}$. Because more smaller boxes are needed than larger boxes,

$$N'(r_j, \mathbf{S}) \leq N'(r, \mathbf{S}) \leq N'(r_{j+1}, \mathbf{S}).$$

Using the fact that $\ln(1/r_{j+1}) \geq \ln(1/r) \geq \ln(1/r_j)$, we obtain

$$\frac{\ln(1/r_j)}{\ln(1/r_{j+1})} \frac{N'(r_j, \mathbf{S})}{\ln(1/r_j)} \leq \frac{N'(r, \mathbf{S})}{\ln(1/r)} \leq \frac{\ln(1/r_{j+1})}{\ln(1/r_j)} \frac{N'(r_{j+1}, \mathbf{S})}{\ln(1/r_{j+1})}.$$

From the assumption of the theorem,

$$\lim_{j \to \infty} \frac{\ln(1/r_j)}{\ln(1/r_{j+1})} = \lim_{j \to \infty} \frac{\ln(1/r_{j+1})}{\ln(1/r_j)} = 1,$$

so the limit of the first and last terms is d, and it follows that the limit of the middle term exists and equals d,

$$\dim_b(\mathbf{S}) = \lim_{r \to 0} \frac{N'(r, \mathbf{S})}{\ln(1/r)} = d.$$

□

14.4. Theory and Proofs

Restatement of Theorem 14.3. *Let $\mathbf{S}$ be a bounded subset of $\mathbb{R}^n$, with box dimension less than n. Then, the n-dimensional Lebesgue measure of $\mathbf{S}$ is zero; that is, $\lambda(\mathbf{S}) = 0$.*

Proof. The set $\mathbf{S}$ is contained in $N(r)$ boxes whose length is r on each side. Therefore, the Lebesgue measure of $\mathbf{S}$ is less than $N(r)\, r^n$. Let $d = \lim_{r \to 0} \frac{\ln(N(r))}{\ln(r^{-1})} < n$ and let $d_1 = (d+n)/2 < n$. There is an $r_0 > 0$ such that, for any r with $0 < r \leq r_0$,
$$\frac{\ln(N(r))}{\ln(r^{-1})} \leq d_1,$$
$$\ln(N(r)) \leq \ln(r^{-d_1}),$$
$$N(r) \leq r^{-d_1}, \quad \text{and}$$
$$\lambda(\mathbf{S}) \leq N(r)\, r^n \leq r^{n-d_1}.$$

Since $n - d_1 > 0$, the right-hand side of the last inequality goes to zero as r goes to 0, so $\lambda(\mathbf{S})$ must be zero. $\square$

Iterated-function systems

Restatement of Theorem 14.5. *Let $\mathscr{F} = \{\mathbf{F}_1, \ldots, \mathbf{F}_k\}$ be an IFS of contractions on $\mathbb{R}^n$ with a contraction constant r_i for each function. Let $r = \max\{r_i\}$ be the contraction constant of the IFS.*

a. *Let $R = \max_i \left\{ \frac{\|\mathbf{F}_i(\mathbf{0})\|}{1 - r_i} \right\}$ and let $\overline{\mathbf{B}}(R) = \{\mathbf{x} \in \mathbb{R}^n : \|\mathbf{x}\| \leq R\}$. Then, this ball is positively invariant by the IFS,*
$$\mathscr{F}(\overline{\mathbf{B}}(R)) \subset \overline{\mathbf{B}}(R).$$

b. *Let $\mathbf{S}_0$ be any compact set such that $\mathscr{F}(\mathbf{S}_0) \subset \mathbf{S}_0$. Inductively define $\mathbf{S}_m = \mathscr{F}(\mathbf{S}_{m-1})$ and $\mathbf{A} = \bigcap_{m=0}^{\infty} \mathbf{S}_m$. Then, the system maps $\mathbf{A}$ onto $\mathbf{A}$, $\mathscr{F}(\mathbf{A}) = \mathbf{A}$. Also, $\mathbf{A}$ is the unique compact invariant set for the iterated-function system. More precisely, if $\mathbf{A}'$ is any other nonempty compact set such that $\mathbf{A}' = \mathscr{F}(\mathbf{A}')$, then $\mathbf{A}' = \mathbf{A}$. In particular, if $\mathbf{S}_0$ is any compact set such that $\mathbf{F}_i(\mathbf{S}_0) \subset \mathbf{S}_0$ for all $1 \leq i \leq k$, then $\mathbf{A} \subset \mathbf{S}_0$.*

c. *The box dimension of the attractor $\mathbf{A}$ satisfies $\dim_b(\mathbf{A}) \leq \frac{\ln(k)}{\ln(1/r)}$. If $\mathscr{F}$ is an IFS of similarity contractions by a uniform factor r, and there is a compact positively invariant set $\mathbf{S}_0$ for $\mathscr{F}$ with $\mathbf{F}_i(\mathrm{int}(\mathbf{S}_0)) \cap \mathbf{F}_j(\mathrm{int}(\mathbf{S}_0)) = \emptyset$ for all $i \neq j$, then the box dimension of the attractor $\mathbf{A}$ is given by $\dim_b(\mathbf{A}) = \frac{\ln(k)}{\ln(1/r)}$.*

Proof (a): By the choice of R, $\|\mathbf{F}_i(\mathbf{0})\| \leq R(1 - r_i)$. For $\|\mathbf{x}\| \leq R$,
$$\|\mathbf{F}_i(\mathbf{x})\| \leq \|\mathbf{F}_i(\mathbf{x}) - \mathbf{F}_i(\mathbf{0})\| + \|\mathbf{F}_i(\mathbf{0})\|$$
$$\leq r_i \|\mathbf{x} - \mathbf{0}\| + \|\mathbf{F}_i(\mathbf{0})\|$$
$$\leq r_i R + R(1 - r_i)$$
$$= R.$$

This shows that the ball $\overline{\mathbf{B}}(R)$ is positively invariant by each $\mathbf{F}_i$ and so by $\mathscr{F}$. $\square$

Lemma 14.14. *Let $\mathbf{A}$ be the set defined in Theorem 14.5. Then, the iterated-function system maps $\mathbf{A}$ onto $\mathbf{A}$.*

Proof. Take $\mathbf{x}^*$ in $\mathbf{A}$. Since $\mathbf{A} = \bigcap_{m=0}^{\infty} \mathscr{F}(\mathbf{S}_m)$, there exist $\mathbf{x}^m$ in $\mathbf{S}_m$ and i_m such that $\mathbf{F}_{i_m}(\mathbf{x}^m) = \mathbf{x}^*$. There must by an infinite number of the i_m that are the same, so we take a subsequence m_j such that $i_{m_j} = i^*$. Since $\mathbf{x}^{m_j}$ is in $\mathbf{S}_{m_j} \subset \mathbf{S}_{m_{j_0}} \subset \mathbf{S}_0$ for $j \geq j_0$, and $\mathbf{S}_{m_{j_0}}$ is closed and bounded (compact), a subsequence of the $\mathbf{x}^{m_j}$ must converge to a point $\mathbf{x}^{\infty}$ in $\mathbf{S}_{m_{j_0}}$. Letting j_0 go to infinity, $\mathbf{x}^{\infty}$ is in $\mathbf{A}$. By continuity, since $\mathbf{F}_{i^*}(\mathbf{x}^{m_j}) = \mathbf{x}^*$ for all j, the limit $\mathbf{F}_{i^*}(\mathbf{x}^{\infty}) = \mathbf{x}^*$. This shows that the function system is onto. □

Let $d(\mathbf{x}, \mathbf{S}) = \min\{\,\|\mathbf{x} - \mathbf{y}\| : \mathbf{y} \in \mathbf{S}\,\}$ be the distance from the point $\mathbf{x}$ to the nonempty closed set as defined in Section 14.3.1.

Lemma 14.15. *Let $\mathbf{F}$ have contraction constant $0 < r < 1$. If $\mathbf{S}$ is a positively invariant set for $\mathbf{F}$, $\mathbf{F}(\mathbf{S}) \subset \mathbf{S}$, then $d\,(\mathbf{F}(\mathbf{x}), \mathbf{S}) \leq r\,d(\mathbf{x}, \mathbf{S})$ for any point $\mathbf{x}$.*

Proof. For any pair of points $\mathbf{x}$ and $\mathbf{y}$, $\|\mathbf{F}(\mathbf{x}) - \mathbf{F}(\mathbf{y})\| \leq r\|\mathbf{x} - \mathbf{y}\|$, so

$$\begin{aligned} d\,(\mathbf{F}(\mathbf{x}), \mathbf{F}(\mathbf{S})) &= \min\{\,\|\mathbf{F}(\mathbf{x}) - \mathbf{y}'\| : \mathbf{y}' \in \mathbf{F}(\mathbf{S})\,\} \\ &= \min\{\,\|\mathbf{F}(\mathbf{x}) - \mathbf{F}(\mathbf{y})\| : \mathbf{y} \in \mathbf{S}\,\} \\ &\leq r\,\min\{\,\|\mathbf{x} - \mathbf{y}\| : \mathbf{y} \in \mathbf{S}\,\} \\ &= r\,d(\mathbf{x}, \mathbf{S}). \end{aligned}$$

If $\mathbf{S}$ is a positively invariant set, then

$$\begin{aligned} d(\mathbf{F}(\mathbf{x}), \mathbf{S}) &\leq d\,(\mathbf{F}(\mathbf{x}), \mathbf{F}(\mathbf{S})) \\ &\leq r\,d(\mathbf{x}, \mathbf{S}). \end{aligned}$$

□

Proof of Theorem 14.5(b): Now, assume that there is a nonempty invariant set $\mathbf{A}'$ that is not equal to $\mathbf{A}$. Note that both $\mathbf{A}'$ and $\mathbf{A}$ are positively invariant by all the functions in the iterated-function system, so Lemma 14.15 applies to both sets.

Assume that there is a point $\mathbf{x}$ in $\mathbf{A}'$, but not in $\mathbf{A}$, so $d(\mathbf{x}, \mathbf{A}) > 0$. Let $\mathbf{x}^*$ be the point in $\mathbf{A}'$ such that

$$d(\mathbf{x}^*, \mathbf{A}) = \max\{d(\mathbf{x}, \mathbf{A}) : \mathbf{x} \in \mathbf{A}'\,\}$$

(i.e., $\mathbf{x}^*$ is the point in $\mathbf{A}'$ that is furthest away from $\mathbf{A}$). But, for all $\mathbf{x}$ in $\mathbf{A}'$,

$$\begin{aligned} d(\mathbf{F}_i(\mathbf{x}), \mathbf{A}) &\leq r\,d(\mathbf{x}, \mathbf{A}) \\ &\leq r\,d(\mathbf{x}^*, \mathbf{A}) \\ &< d(\mathbf{x}^*, \mathbf{A}). \end{aligned}$$

Since, this is true for all i, $\mathbf{A}'$ cannot be equal to $\mathscr{F}(\mathbf{A}')$, contradicting the assumption that $\mathbf{A}'$ is invariant. Therefore, $\mathbf{A}' \subset \mathbf{A}$.

Reversing the roles of $\mathbf{A}$ and $\mathbf{A}'$, $\mathbf{A} \subset \mathbf{A}'$. Combining, we have $\mathbf{A} = \mathbf{A}'$. □

Distance between compact sets

Restatement of Theorem 14.6. *Let* $\mathbf{A}$, $\mathbf{B}$, *and* $\mathbf{C}$ *be sets in* $\mathscr{K}(\mathbb{R}^n)$.

- **a.** *The hemimetric* δ_l *satisfies the following conditions:*
  - (i') $\delta_l(\mathbf{A}, \mathbf{B})$ *is not necessarily equal to* $\delta_l(\mathbf{B}, \mathbf{A})$.
  - (ii') $0 \le \delta_l(\mathbf{A}, \mathbf{B}) < \infty$.
  - (iii') $\delta_l(\mathbf{A}, \mathbf{B}) = 0$ *if and only if* $\mathbf{A} \subset \mathbf{B}$, *and*
  - (iv') $\delta_l(\mathbf{A}, \mathbf{C}) \le \delta_l(\mathbf{A}, \mathbf{B}) + \delta_l(\mathbf{B}, \mathbf{C})$.
- **b.** *The hemimetric* δ_u *satisfies the following conditions:*
  - (i') $\delta_u(\mathbf{A}, \mathbf{B})$ *is not necessarily equal to* $\delta_u(\mathbf{B}, \mathbf{A})$.
  - (ii') $0 \le \delta_u(\mathbf{A}, \mathbf{B}) < \infty$.
  - (iii') $\delta_u(\mathbf{A}, \mathbf{B}) = 0$ *if and only if* $\mathbf{B} \subset \mathbf{A}$.
  - (iv') $\delta_u(\mathbf{A}, \mathbf{C}) \le \delta_u(\mathbf{A}, \mathbf{B}) + \delta_u(\mathbf{B}, \mathbf{C})$.
- **c.** *The Hausdorff metric* δ *is a metric on the space* $\mathscr{K}(\mathbb{R}^n)$.

Proof. (a.i') This is clear from the definitions.

(a.ii') It is clear from the definitions that $0 \le d_l(\mathbf{A}, \mathbf{B})$, for each two sets $\mathbf{A}, \mathbf{B} \in \mathscr{K}(\mathbb{R}^n)$. The metric is finite because the sets are compact.

(a.iii') If $\mathbf{A} \subset \mathbf{B}$, then it is clear that $\delta_l(\mathbf{A}, \mathbf{B}) = 0$. Next, if $\delta_l(\mathbf{A}, \mathbf{B}) = 0$, then $\mathbf{A} \subset \bar{\mathscr{N}}_r(\mathbf{B})$ for all $r > 0$, so $\mathbf{A} \subset \bigcap_{r>0} \bar{\mathscr{N}}_r(\mathbf{B}) = \mathbf{B}$.

(a.iv') Note that, if $\mathbf{A} \subset \bar{\mathscr{N}}_{r'}(\mathbf{B})$ and $\mathbf{B} \subset \bar{\mathscr{N}}_r(\mathbf{C})$, then $\mathbf{A} \subset \bar{\mathscr{N}}_{r'}(\mathbf{B}) \subset \bar{\mathscr{N}}_{r'+r}(\mathbf{C})$. Taking the minimum, we get

$$\begin{aligned}
\delta_l(\mathbf{A}, \mathbf{C}) &= \min\{s \ge 0 : \mathbf{A} \subset \bar{\mathscr{N}}_s(\mathbf{C})\} \\
&= \min\{r + r' : \mathbf{A} \subset \bar{\mathscr{N}}_{r+r'}(\mathbf{C}),\ r \ge 0,\ r' \ge 0\} \\
&\le \min\{r + r' : \mathbf{A} \subset \bar{\mathscr{N}}_{r'}(\mathbf{B}) \text{ and } \mathbf{B} \subset \bar{\mathscr{N}}_r(\mathbf{C}),\ r \ge 0,\ r' \ge 0\} \\
&\le \min\{r' \ge 0 : \mathbf{A} \subset \bar{\mathscr{N}}_{r'}(\mathbf{B})\} + \min\{r \ge 0 : \mathbf{B} \subset \bar{\mathscr{N}}_r(\mathbf{C})\} \\
&= \delta_l(\mathbf{A}, \mathbf{B}) + \delta_l(\mathbf{B}, \mathbf{C}).
\end{aligned}$$

The proof of (b) is similar to (a).

(c) The conditions for δ to be a metric are satisfied as follows.

(i) Clearly, $\delta(\mathbf{A}, \mathbf{B}) = \delta(\mathbf{B}, \mathbf{A})$ from the definition.

(ii) For a pair of sets $\mathbf{A}, \mathbf{B} \in \mathscr{K}(\mathbb{R}^n)$, by (a.ii') and (b.ii'), $0 \le \delta(\mathbf{A}, \mathbf{B}) < \infty$.

(iii) $\delta(\mathbf{A}, \mathbf{A}) = 0$ is clear. Also if $\delta(\mathbf{A}, \mathbf{B}) = 0$, then $\delta_l(\mathbf{A}, \mathbf{B}) = \delta_u(\mathbf{A}, \mathbf{B}) = 0$. By (a.iii') and (b.iii'), $\mathbf{A} \subset \mathbf{B} \subset \mathbf{A}$, so $\mathbf{A} = \mathbf{B}$. Thus, $\delta(\mathbf{A}, \mathbf{B}) = 0$ if and only if $\mathbf{A} = \mathbf{B}$.

(iv) Combining (a.iv') and (b.iv'), $\delta(\mathbf{A}, \mathbf{C}) \le \delta(\mathbf{A}, \mathbf{B}) + \delta(\mathbf{B}, \mathbf{C})$. $\square$

The next theorem says that the space $\mathscr{K}(\mathbb{R}^n)$ is complete, and so a contraction on it has a fixed point.

Restatement of Theorem 14.7. *The space* $\mathscr{K}(\mathbb{R}^n)$ *is complete with respect to the Hausdorff metric* δ.

Proof. The proof of the theorem is not trivial, and uses some constructions.

To show that it is complete, we must take a Cauchy sequence of sets $\mathbf{A}_k \in \mathcal{K}(\mathbb{R}^n)$. Let $\mathbf{A}_\infty$ be the set of cluster points of sequences $\{\mathbf{x}_k\}$, where $\mathbf{x}_k \in \mathbf{A}_k$. Thus, for any point $\mathbf{x}$ in $\mathbf{A}_\infty$, there is a subsequence of $\mathbf{x}_{k_j}$ of the preceding type that converges to $\mathbf{x}$. Now, the sequence of sets $\mathbf{A}_k$ is Cauchy, so given $\epsilon > 0$, there exists an N_ϵ such that $\delta(\mathbf{A}_k, \mathbf{A}_m) \leq \epsilon$ for $k, m \geq N_\epsilon$. Because $\mathbf{A}_\infty$ is the set of cluster points of sequences of points in the $\mathbf{A}_k$, the set $\mathbf{A}_\infty \subset \bar{\mathcal{N}}_\epsilon(\mathbf{A}_k)$ for $k \geq N_\epsilon$. It easily follows that $\mathbf{A}_\infty$ is closed and bounded, so $\mathbf{A}_\infty \in \mathcal{K}(\mathbb{R}^n)$.

Next, we show that $\mathbf{A}_k \subset \bar{\mathcal{N}}_\epsilon(\mathbf{A}_\infty)$ for $k \geq N_\epsilon$. Fix $k \geq N_\epsilon$ and take a point $\mathbf{x}_k^0$ in $\mathbf{A}_k$. For $m \geq N_\epsilon$, $\mathbf{A}_k \subset \bar{\mathcal{N}}_\epsilon(\mathbf{A}_m)$, so there exists $\mathbf{x}_m$ in $\mathbf{A}_m$ with $d(\mathbf{x}_k^0, \mathbf{x}_m) \leq \epsilon$. In fact, since the sequence $\mathbf{A}_m$ is Cauchy, we can take the sequence of points $\mathbf{x}_m$ with the preceding property to be a Cauchy sequence in $\mathbb{R}^n$. Therefore, they converge to a point $\mathbf{x}_\infty$ in $\mathbb{R}^n$. By the definition of $\mathbf{A}_\infty$, $\mathbf{x}_\infty$ is a point in $\mathbf{A}_\infty$. Therefore, $d(\mathbf{x}_k^0, \mathbf{x}_\infty) = \lim_{m \to \infty} d(\mathbf{x}_k^0, \mathbf{x}_m) \leq \epsilon$, and $d(\mathbf{x}_k^0, \mathbf{A}_\infty) \leq \epsilon$. Since this is true for any $\mathbf{x}_k^0$ in $\mathbf{A}_k$, $\mathbf{A}_k \subset \bar{\mathcal{N}}_\epsilon(\mathbf{A}_\infty)$.

Combining, we see that, for any $k \geq N_\epsilon$, $\delta(\mathbf{A}_k, \mathbf{A}_\infty) \leq \epsilon$. This shows that the sequence of sets $\mathbf{A}_k$ converges to a set $\mathbf{A}_\infty$ in $\mathcal{K}(\mathbb{R}^n)$. □

Contraction maps applied to sets

Restatement of Theorem 14.8. *Assume that $\mathbf{F}$ is a single contraction on $\mathbb{R}^n$ with contraction constant $r < 1$. Then, $\mathbf{F}$ induces a contraction on $\mathcal{K}(\mathbb{R}^n)$ with the same contraction constant r.*

Proof. Let $\mathbf{A}$ and $\mathbf{B}$ be two sets in $\mathcal{K}(\mathbb{R}^n)$. Then,

$$\delta_l(\mathbf{F}(\mathbf{A}), \mathbf{F}(\mathbf{B})) = \max\{\, d(\mathbf{F}(\mathbf{a}), \mathbf{F}(\mathbf{B})) : \mathbf{a} \in \mathbf{A} \,\}$$
$$\leq \max\{\, r\, d(\mathbf{a}, \mathbf{B}) : \mathbf{a} \in \mathbf{A} \,\}$$
$$= r\, \delta_l(\mathbf{A}, \mathbf{B}).$$

In the same way,

$$\delta_u(\mathbf{F}(\mathbf{A}), \mathbf{F}(\mathbf{B})) \leq r\, \delta_u(\mathbf{A}, \mathbf{B}).$$

Combining, we have

$$d(\mathbf{F}(\mathbf{A}), \mathbf{F}(\mathbf{B})) = \max\{\, \delta_l(\mathbf{F}(\mathbf{A}), \mathbf{F}(\mathbf{B})),\ \delta_u(\mathbf{F}(\mathbf{A}), \mathbf{F}(\mathbf{B})) \,\}$$
$$\leq \max\{\, r\, \delta_l(\mathbf{A}, \mathbf{B}),\ r\, \delta_u(\mathbf{A}, \mathbf{B}) \,\}$$
$$= r\, d(\mathbf{A}, \mathbf{B}).$$

□

Restatement of Theorem 14.9. *Assume that $\mathscr{F} = \{\mathbf{F}_1, \ldots, \mathbf{F}_k\}$ is an IFS of contractions on $\mathbb{R}^n$, with contraction constants $r_1, \ldots, r_k$, and $r = \max\{r_i : 1 \leq i \leq k\} < 1$. Then, the induced action of $\mathscr{F}$ on $\mathcal{K}(\mathbb{R}^n)$ is a contraction with contraction constant r.*

Proof. For the hemimetric δ_l,

$$\delta_l(\mathscr{F}(\mathbf{A}), \mathscr{F}(\mathbf{B})) = \delta_l\left(\bigcup_i \mathbf{F}_i(\mathbf{A}), \bigcup_j \mathbf{F}_j(\mathbf{B})\right)$$

$$= \max_i \left\{ \delta_l\left(\mathbf{F}_i(\mathbf{A}), \bigcup_j \mathbf{F}_j(\mathbf{B})\right) \right\}$$

$$\leq \max_i \left\{ \delta_l(\mathbf{F}_i(\mathbf{A}), \mathbf{F}_i(\mathbf{B})) \right\}$$

$$\leq \max_i \left\{ r_i \, \delta_l(\mathbf{A}, \mathbf{B}) \right\}$$

$$\leq r \, d(\mathbf{A}, \mathbf{B}).$$

By a similar argument,

$$\delta_u(\mathscr{F}(\mathbf{A}), \mathscr{F}(\mathbf{B})) \leq r \, d(\mathbf{A}, \mathbf{B}).$$

Combining, we get

$$d(\mathscr{F}(\mathbf{A}), \mathscr{F}(\mathbf{B})) \leq r \, d(\mathbf{A}, \mathbf{B}).$$

□

Restatement of Theorem 14.10. *Assume that $\mathscr{F} = \{\mathbf{F}_1, \ldots, \mathbf{F}_k\}$ is an iterated-function system of contractions on $\mathbb{R}^n$, with a contraction constant $0 < r < 1$. Then, $\mathscr{F}$ has a unique fixed set $\mathbf{A}$ in $\mathscr{K}(\mathbb{R}^n)$ that is the attractor for the iterated-function system.*

Proof. The IFS of contractions induces a contraction on $\mathscr{K}(\mathbb{R}^n)$. The space $\mathscr{K}(\mathbb{R}^n)$ is complete, so starting with any set $\mathbf{S}_0$, the sequence of sets $\mathbf{S}_j = \mathscr{F}(\mathbf{S}_{j-1})$ converges to the unique "fixed point" $\mathbf{A}$ in $\mathscr{K}(\mathbb{R}^n)$, $\mathscr{F}(\mathbf{A}) = \mathbf{A}$. □

Probabilistic action

Restatement of Theorem 14.11. *Let $\mathscr{F} = \{\mathbf{F}_1, \ldots, \mathbf{F}_k\}$ be an IFS on $\mathbb{R}^n$, with contraction constants $r_1, \ldots, r_k$, $r = \max\{r_i : 1 \leq i \leq k\} < 1$, and an invariant compact set $\mathbf{A}$. Then, for any initial condition $\mathbf{x}_0$ in $\mathbb{R}^n$, if $\mathbf{x}_j = \mathbf{F}_{s_j}(\mathbf{x}_{j-1})$ is an orbit obtained by choosing a sequence of maps from the IFS, then the distance of $\mathbf{x}_j$ to $\mathbf{A}$ goes to zero at least as fast as r^j times the original distance of $\mathbf{x}_0$ to $\mathbf{A}$, $d(\mathbf{x}_j, \mathbf{A}) \leq r^j \, d(\mathbf{x}_0, \mathbf{A})$.*

Proof. Take any $\mathbf{x}_0$ in $\mathbb{R}^n$. By Lemma 14.15, for any i,

$$d(\mathbf{F}_i(\mathbf{x}_0), \mathbf{A}) \leq r \, d(\mathbf{x}_0, \mathbf{A}).$$

By induction, taking any sequence of these maps, we get

$$d(\mathbf{F}_{s_j} \circ \cdots \circ \mathbf{F}_{s_0}(\mathbf{x}_0), \mathbf{A}) \leq r \, d(\mathbf{F}_{s_{j-1}} \circ \cdots \circ \mathbf{F}_{s_0}(\mathbf{x}_0), \mathbf{A}) \leq \cdots \leq r^{j+1} \, d(\mathbf{x}_0, \mathbf{A}).$$

□

Restatement of Theorem 14.12. *Let Σ_k^+ with shift map σ be the one-sided full shift on k symbols. Let $\{p_1, \ldots, p_k\}$ be a system of weights, with $0 < p_i < 1$ for all $1 \leq i \leq k$ and $p_1 + \cdots + p_k = 1$.*

a. Then, there is a probability measure $\mu_\mathbf{p}$ defined on Σ_k^+ that is invariant by σ and that satisfies the following: given $a_0, \ldots, a_n \in \{1, \ldots k\}$, then

$$\mu_\mathbf{p}(\{\mathbf{s} : s_i = a_0\}) = p_{a_0} \quad \text{and}$$

$$\mu_\mathbf{p}(\{\mathbf{s} : s_i = a_i \text{ for } 0 \leq i \leq n\}) = \prod_{i=0}^{n} p_{a_i}.$$

b. There is a set of symbol sequences $\Sigma^* \subset \Sigma_k^+$ of full measure, $\mu_\mathbf{p}(\Sigma^*) = 1$, such that any $\mathbf{s}$ in Σ^* contains all finite strings (i.e., for any finite string $a_0 \ldots a_{m-1}$, there is a j such that $a_i = s_{j+i}$ for $0 \leq i \leq m-1$). It follows that the ω-limit by the map σ of any symbol sequence $\mathbf{s}$ in Σ^* is all of Σ_k^+, $\omega(\mathbf{s}; \sigma) = \Sigma_k^+$.

Proof. (a) The statement of the theorem defines the measure on the cylinder sets, where the first few symbols are specified. Since these sets are both open and closed in Σ_k^+, the properties of measures can be used to determine the measure on any Borel set.

(b) Fix a finite string $\mathbf{w} = a_0 \ldots a_{m-1}$. Its probability is

$$p_\mathbf{w} = \prod_{i=0}^{m-1} p_{a_i}.$$

Let $\Sigma(\mathbf{w})$ be the symbol sequences in Σ_k^+ that contain $\mathbf{w}$ somewhere in their sequences. The probability that $\mathbf{w}$ is not the first m symbols is $1 - p_\mathbf{w}$. The probability that $\mathbf{w}$ is not the first m symbols nor symbols from m to $2m-1$ is $(1 - p_\mathbf{w})^2$. Continuing, the probability that $\mathbf{w}$ is not one of the first j blocks of symbols of length m is $(1 - p_\mathbf{w})^j$. Since this number goes to zero, the probability that $\mathbf{w}$ is not one of the blocks $s_{mi}, \ldots, s_{m(i+1)-1}$ for any i is 0, and $\Sigma(\mathbf{w})$ has measure 1. (There are other ways for $\mathbf{w}$ to appear in the symbol sequence, but that just makes the set larger and the measure cannot get larger than one.)

Let

$$\Sigma^* = \bigcap_\mathbf{w} \Sigma(\mathbf{w}).$$

Since there are only countably many words,

$$\mu_\mathbf{p}\left(\bigcup_\mathbf{w} (\Sigma_k^+ \setminus \Sigma(\mathbf{w}))\right) \leq \sum_\mathbf{w} \mu_\mathbf{p}(\Sigma_k^+ \setminus \Sigma(\mathbf{w})) = \sum_\mathbf{w} 0 = 0$$

and

$$\mu_\mathbf{p}(\Sigma^*) = 1 - \mu_\mathbf{p}(\Sigma_k^+ \setminus \Sigma^*) = 1 - \mu_\mathbf{p}\left(\bigcup_\mathbf{w} \Sigma_k^+ \setminus \Sigma(\mathbf{w})\right) = 1.$$

$\square$

Appendix A

Background and Terminology

A.1. Calculus Background and Notation

Continuity is defined in Appendix A.2, Definition A.18.

A.1.1. Functions of one variable.

Definition A.1. A function f from $\mathbb{R}$ to $\mathbb{R}$ is *continuously differentiable* or C^1 provided that f is continuous and $f'(x)$ is a continuous function of x. It is *continuously differentiable of order r* or C^r for some integer $r \geq 1$, provided that f is continuous and the first r derivatives of f are continuous functions of x. If the function is C^r for all positive integers r, then it is called C^∞.

Theorem A.1 (Mean Value Theorem). *Let f from $\mathbb{R}$ to $\mathbb{R}$ be a C^1 function. Then, given any two points x_1 and x_2, there is a point z between x_1 and x_2 such that*

$$f(x_2) - f(x_1) = f'(z)(x_2 - x_1).$$

Theorem A.2 (First-order Taylor expansion). *If $f : \mathbb{R} \to \mathbb{R}$ is a C^2 function, then*

$$f(t) = f(0) + f'(0)\,t + \frac{1}{2}f''(t_1)t^2,$$

where t_1 is between 0 and t. Since $|f''(t_1)|$ is bounded by a constant C for $|t_1| \leq 1$,

$$|f''(t_1)t^2| \leq C\,|t|^2.$$

We write

$$f''(t_1)t^2 = O(|t|^2)$$

to mean that the absolute value of $f''(t_1)t^2$ is less than $C\,|t|^2$ for some constant C.

A.1.2. Functions of several variables.

Definition A.2. Let $\mathbf{F}$ be a function from $\mathbb{R}^m$ to $\mathbb{R}^n$. We write the coordinate functions as $F_i(\mathbf{x})$, so

$$\mathbf{F}(\mathbf{x}) = \begin{pmatrix} F_1(\mathbf{x}) \\ \vdots \\ F_n(\mathbf{x}) \end{pmatrix}.$$

For an integer $r \geq 1$, the function $\mathbf{F}$ is said to be *continuously differentiable or order r* or C^r provided that $\mathbf{F}$ is continuous and all partial derivatives up to order r exist and are continuous; that is,

$$\frac{\partial^k F_i}{\partial^{j_1} x_1 \partial^{j_2} x_2 \cdots \partial^{j_m} x_m}(\mathbf{x})$$

exists and is continuous for all $j + \cdots + j_m = k \leq r$, with $j_1 \geq 0$, ..., $j_m \geq 0$. If partial derivatives of all orders exist (i.e., $\mathbf{F}$ is C^r for all r), then it is called C^∞.

To simplify the notation, we use the notation

$$D\mathbf{F}_{(\mathbf{p})} = \left(\frac{\partial F_i}{\partial x_j}(\mathbf{p}) \right)$$

for the *matrix of partial derivatives* of the coordinate functions of the function. We also call this the *derivative* of $\mathbf{F}$ at $\mathbf{p}$. Many people put the point of evaluating the derivative on the same line, $D\mathbf{F}(\mathbf{p})$; we put it as a subscript so that, when several of these are multiplied (as in the next theorem), the line is easier to read. The reader should use care to distinguish this notation from that which uses the subscript to indicate the variable with which a derivative is taken.

Using this notation, we can write the chain rule for this function as a matrix product as given in the next theorem.

Theorem A.3 (Chain Rule). *Let $\mathbf{F}$ be a C^1 function from $\mathbb{R}^k$ to $\mathbb{R}^m$ and let $\mathbf{G}$ be a C^1 function from $\mathbb{R}^m$ to $\mathbb{R}^n$. Then, the matrix of partial derivatives of the composition $\mathbf{G} \circ \mathbf{F}$ is the product of the matrix of partial derivatives of $\mathbf{G}$ times the matrix of partial derivatives of $\mathbf{F}$, or*

$$D(\mathbf{G} \circ \mathbf{F})_{(\mathbf{x})} = D\mathbf{G}_{(\mathbf{F}(\mathbf{x}))} \cdot D\mathbf{F}_{(\mathbf{x})}.$$

Theorem A.4 (First-order Taylor expansion). *Assume that $\mathbf{F}$ is a C^2 function from $\mathbb{R}^m$ to $\mathbb{R}^n$ with coordinate functions F_i. Then*

$$\mathbf{F}(\mathbf{x}) = \mathbf{F}(\mathbf{p}) + D\mathbf{F}_{(\mathbf{p})}(\mathbf{x} - \mathbf{p}) + O(|\mathbf{x} - \mathbf{p}|^2).$$

Idea of proof: Fix an i and let $g(t) = F_i(\mathbf{p}_t)$, where $\mathbf{p}_t = \mathbf{p} + t(\mathbf{x} - \mathbf{p})$. Then $g(0) = F_i(\mathbf{x})$, $g(1) = F_i(\mathbf{x})$, and

$$g'(t) = \sum_j \frac{\partial F_i}{\partial x_j}(\mathbf{p}_t)(x_j - p_j),$$

$$g''(t_{i,1}) = \sum_{j,k} \frac{\partial^2 F_i}{\partial x_j \partial x_k}(\mathbf{p}_{t_1})(x_j - p_j)(x_k - p_k),$$

$$F_i(\mathbf{x}) = F_i(\mathbf{p}) + \left(\frac{\partial F_i}{\partial x_1}(\mathbf{p}), \ldots, \frac{\partial F_i}{\partial x_n}(\mathbf{p})\right) \begin{pmatrix} x_1 - p_1 \\ \vdots \\ x_n - p_n \end{pmatrix} + O(|\mathbf{x} - \mathbf{p}|^2),$$

$$\mathbf{F}(\mathbf{x}) = \mathbf{F}(\mathbf{p}) + \begin{pmatrix} \frac{\partial F_1}{\partial x_1}(\mathbf{p}) & \cdots & \frac{\partial F_1}{\partial x_n}(\mathbf{p}) \\ \vdots & & \vdots \\ \frac{\partial F_n}{\partial x_1}(\mathbf{p}) & \cdots & \frac{\partial F_n}{\partial x_n}(\mathbf{p}) \end{pmatrix} \begin{pmatrix} x_1 - p_1 \\ \vdots \\ x_n - p_n \end{pmatrix} + O(|\mathbf{x} - \mathbf{p}|^2), \text{ and}$$

$$\mathbf{F}(\mathbf{x}) = \mathbf{F}(\mathbf{p}) + D\mathbf{F}_{(\mathbf{p})}(\mathbf{x} - \mathbf{p}) + O(|\mathbf{x} - \mathbf{p}|^2).$$

Since the value of $t_{1,i}$ depends on the coordinate function, it is not possible to find one point at which to evaluate all of the second partial derivatives, as is the case for a function of one variable. It is possible to calculate the derivative $D\mathbf{F}_{(\mathbf{x})}$ when $\mathbf{F}$ is only C^1, but the remainder in the form given requires that the function be C^2. □

A.2. Analysis and Topology Terminology

There are various concepts and terminology from analysis and topology that we use in this book. These concepts are usually covered in an undergraduate course in real analysis or advanced calculus. We have not assumed that the reader has taken such a course, so we collect together some of this terminology so that we can use it in the book. For a more complete treatment see [**Bar00a**], [**Tho01**], [**Lew93**], [**Mar93**], or [**Wad00**].

Real line

Definition A.3. It is possible to have an increasing sequence of rational numbers that converges to an irrational number and not a rational number. Thus, the set of rational numbers is not complete. However, the real numbers $\mathbb{R}$ are *complete*. This property can be defined in several ways. For the real numbers, it can be defined as follows: Let x_n be an increasing sequence of real numbers that is bounded above by a number b; that is,

$$x_0 < x_1 < \cdots < x_{n-1} < x_n < \cdots < b.$$

Then there is a real number x_∞ such that x_n converges to x_∞ (i.e., for any $\epsilon > 0$, there is an N such that $|x_n - x_\infty| < \epsilon$ for all $n \geq N$).

We use the continuous version of completeness of the real numbers that follows in the proof of Lemma 4.11.

Theorem A.5. *Assume that $g(t)$ is a function from $\mathbb{R}$ to $\mathbb{R}$ that is a bounded and monotonically increasing (i.e., if $t_1 < t_2$, then $g(t_1) \leq g(t_2)$). Then,*
$$\lim_{t \to \infty} g(t)$$
exists and is finite.

Theorem A.6 (Intermediate Value Theorem in $\mathbb{R}$). *Assume that f is a continuous function from $\mathbb{R}$ to $\mathbb{R}$ and $f(x_1) < a < f(x_2)$. Then, there is some point x_3 between x_1 and x_2 such that $f(x_3) = a$.*

Theorem A.7. *If*
$$[a_0, b_0] \supset [a_1, b_1] \supset \cdots \supset [a_k, b_k] \supset \cdots$$
is an infinite sequence of nonempty closed bounded intervals, then
$$\bigcap_{k \geq 0} [a_k, b_k] \neq \emptyset.$$

This theorem follows from the completeness of $\mathbb{R}$ stated above: The left end points are increasing and are bounded above by any of the a_k, so converge to a point p in all the intervals. This property is a particular case of the intersection of nested compact sets discussed in what follows.

Properties of sets

Assume that $\mathbf{S}_1$ and $\mathbf{S}_2$ are two sets. Then the *set difference* of $\mathbf{S}_1$ and $\mathbf{S}_2$ is
$$\mathbf{S}_1 \setminus \mathbf{S}_2 = \{\, \mathbf{x} \in \mathbf{S}_1 : \mathbf{x} \notin \mathbf{S}_2 \,\},$$
where $\mathbf{x} \in \mathbf{S}_1$ means that $\mathbf{x}$ is in $\mathbf{S}_1$ and $\mathbf{x} \notin \mathbf{S}_2$ means that $\mathbf{x}$ is not in $\mathbf{S}_2$. If $\mathbf{X}$ is the total ambient space and $\mathbf{S}$ is a subset, then $\mathbf{S}^c = \mathbf{X} \setminus \mathbf{S}$.

At times in the discussion, we talk about open sets, closed sets, interior of a set, boundary of a set, and neighborhood of a point. All of these definitions make sense in $\mathbb{R}$, $\mathbb{R}^n$ or a metric space $\mathbf{X}$ (i.e., there is a distance d defined between points of $\mathbf{X}$). We write the definitions using $\mathbb{R}^n$.

For $r > 0$ and a point $\mathbf{x}_0$ in $\mathbb{R}^n$, the *open ball of radius r about the point $\mathbf{p}$* is the set
$$\mathbf{B}(\mathbf{p}, r) = \{\, \mathbf{x} \in \mathbb{R}^n : \|\mathbf{x} - \mathbf{p}\| < r \,\}.$$

If we change the inequality to a less than or equal to sign, then we get the *closed ball of radius r*:
$$\overline{\mathbf{B}}(\mathbf{p}, r) = \{\, \mathbf{x} \in \mathbb{R}^n : \|\mathbf{x} - \mathbf{p}\| \leq r \,\}.$$
In R, $\mathbf{B}(p, r) = (p - r, p + r)$ and $\overline{\mathbf{B}}(p, r) = [p - r, p + r]$.

Definition A.4. For a general set $\mathbf{S} \subset \mathbb{R}^n$, the boundary of $\mathbf{S}$, denoted by $\mathrm{bd}(\mathbf{S})$, is the set of all the points $\mathbf{p}$ such that $\mathbf{B}(\mathbf{p}, r)$ contains points in both $\mathbf{S}$ and $\mathbf{S}^c = \mathbb{R}^n \setminus \mathbf{S}$ for all $r > 0$,
$$\mathrm{bd}(\mathbf{S}) = \{\, \mathbf{p} : \text{for all } r > 0, \mathbf{B}(\mathbf{p}, r) \cap \mathbf{S} \neq \emptyset \,\&\, \mathbf{B}(\mathbf{p}, r) \cap \mathbf{S}^c \neq \emptyset \,\}.$$
Note that $\mathrm{bd}(\mathbf{S}) = \mathrm{bd}(\mathbf{S}^c)$

A.2. Analysis

If **S** is the finite union of intervals in $\mathbb{R}$, then the boundary of **S** is the set of all the end points of **S**.

The boundary of a ball is the sphere; that is,
$$\mathrm{bd}(\mathbf{B}(\mathbf{x_0}, r)) = \mathrm{bd}(\overline{\mathbf{B}}(\mathbf{x_0}, r)) = \{\mathbf{x} \in \mathbb{R}^n : \|\mathbf{x} - \mathbf{x_0}\| = r\}.$$

Definition A.5. A set **U** is *open* provided that it contains no boundary points, $\mathbf{U} \cap \mathrm{bd}(\mathbf{U}) = \emptyset$.

This is equivalent to saying that a set **U** in $\mathbb{R}^n$ is open provided that, for each point **p** in **U**, there is an $r > 0$ such that the open ball of radius r about **p** is contained in **U**:
$$\mathbf{B}(\mathbf{p}, r) \subset \mathbf{U}.$$

In $\mathbb{R}$, an open set is the countable union of open intervals.

Definition A.6. A set **S** in $\mathbb{R}$ is *closed* provided that it contains all its boundary points, $\mathrm{bd}(\mathbf{S}) \subset \mathbf{S}$. This is equivalent to saying that $\mathbf{S}^c$ is open.

For a closed set $\mathbf{C} \subset \mathbb{R}^n$, if $\mathbf{x}_j$ is a sequence of points in **C** that converge to a point $\mathbf{x}_\infty$ in the ambient space $\mathbb{R}^n$, then $\mathbf{x}_\infty$ is in **C**.

An arbitrary union of open sets is open and an arbitrary intersection of closed sets is closed.

Definition A.7. The *interior* of a set **S** in $\mathbb{R}$ is formed by removing the boundary from **S**, and is denoted by $\mathrm{int}(\mathbf{S}) = \mathbf{S} \setminus \mathrm{bd}(\mathbf{S})$. The interior of **S** is the largest open set that is contained in **S**. The interior of a set is also the set of all points that have a small ball about them that is contained entirely inside **S**:
$$\mathrm{int}(\mathbf{S}) = \{\mathbf{x} \in \mathbf{S} : \text{there is an } r > 0 \text{ such that } \mathbf{B}(\mathbf{x}_0, r) \subset \mathbf{S}\}.$$

Definition A.8. The *closure* of a set **S** is the union of **S** and its boundary, and is denoted by $\mathrm{cl}(\mathbf{S}) = \mathbf{S} \cup \mathrm{bd}(\mathbf{S})$. The closure of **S** is also the smallest closed set that contains **S**. This is equivalent to saying that
$$\mathrm{cl}(\mathbf{S}) = \{\mathbf{p} : \mathbf{B}(\mathbf{p}, r) \cap \mathbf{S} \neq \emptyset \text{ for all } r > 0\}.$$

For balls in $\mathbb{R}^n$,
$$\mathrm{cl}(\mathbf{B}(\mathbf{x}_0, r)) = \overline{\mathbf{B}}(\mathbf{x}_0, r) \quad \text{and} \quad \mathrm{int}(\overline{\mathbf{B}}(\mathbf{x}_0, r)) = \mathbf{B}(\mathbf{x}_0, r).$$

Other relationships between the boundary, closure, and interior are
$$\mathrm{bd}(\mathbf{S}) = \mathrm{cl}(\mathbf{S}) \setminus \mathrm{int}(\mathbf{S})$$
$$= \mathrm{cl}(\mathbf{S}) \cap \mathrm{cl}(\mathbf{S}^c), \quad \text{and}$$
$$\mathrm{cl}(\mathbf{S}) = (\mathrm{int}(\mathbf{S}^c))^c.$$

Definition A.9. Given a point **a** in $\mathbb{R}^n$, a *neighborhood of* **a** is any set that contains **a** in its interior. In particular, an open set is a neighborhood of any point in the set.

Definition A.10. A set **S** in $\mathbb{R}^n$ is *bounded* provided that there is a constant $C > 0$ such that $\mathbf{S} \subset \mathbf{B}(\mathbf{0}, C)$.

Definition A.11. For sets $\mathbf{A} \subset \mathbf{S} \subset \mathbb{R}^n$, the set $\mathbf{A}$ is *dense* in $\mathbf{S}$ provided that arbitrarily close to each point $\mathbf{p}$ in $\mathbf{S}$, there is a point in $\mathbf{A}$. More precisely, for each $\mathbf{p}$ in $\mathbf{S}$ and each $\epsilon > 0$, there is a point $\mathbf{a}$ in $\mathbf{A}$ such that $\|\mathbf{a} - \mathbf{p}\| < \epsilon$. This is equivalent to saying that $\operatorname{cl}(\mathbf{A}) \supset \mathbf{S}$.

Definition A.12. A set $\mathbf{S}$ in $\mathbb{R}^n$ is *compact* provided that it is closed and bounded. (This is not the usual definition, but is one of the standard criteria for a subset of $\mathbb{R}^n$ to be compact.)

Definition A.13. An interval $[a, b]$ in $\mathbb{R}$ has a property called connected. A set $\mathbf{S}$ in $\mathbb{R}^n$ is *connected* provided that, if $\mathbf{S}$ is contained in the union of two open sets $\mathbf{U}$ and $\mathbf{V}$, $\mathbf{S} \subset \mathbf{U} \cup \mathbf{V}$, and $\mathbf{U}$ and $\mathbf{V}$ do not intersect on $\mathbf{S}$, $\mathbf{S} \cap \mathbf{U} \cap \mathbf{V} = \emptyset$, then either (i) $\mathbf{S} \subset \mathbf{U}$ and $\mathbf{S} \cap \mathbf{V} = \emptyset$, or (ii) $\mathbf{S} \subset \mathbf{V}$ and $\mathbf{S} \cap \mathbf{U} = \emptyset$.

One application of compact sets relates to the intersection of nested compact sets. If $\mathbf{S}_0 \supset \mathbf{S}_1 \supset \cdots \supset \mathbf{S}_k \supset \cdots$ is an infinite sequence of nested nonempty compact subsets of $\mathbb{R}^n$, then

$$\bigcap_{k \geq 0} \mathbf{S}_k \neq \emptyset.$$

If the nested compact sets are also connected, then $\bigcap_{j=1}^{\infty} \mathbf{S}_j$ is connected,

Definition A.14. The *distance* from a point $\mathbf{p}$ to a closed set $\mathbf{S}$ in $\mathbb{R}^n$ is the minimum of the distance of $\mathbf{p}$ to a point $\mathbf{x}$ in $\mathbf{S}$,

$$d(\mathbf{p}, \mathbf{S}) = \min\{\, \|\mathbf{p} - \mathbf{x}\| : \mathbf{x} \in \mathbf{S}\,\}.$$

See Figure 1.

It is possible to define the distance to a set that is not closed, but the wording is not as simple. The distance from a point $\mathbf{p}$ to an arbitrary set $\mathbf{S}$ is the largest number $r \geq 0$ such that

$$r \leq \|\mathbf{x} - \mathbf{p}\|,$$

where $\mathbf{x}$ is a point in $\mathbf{S}$. This largest number r is called the *greatest lower bound* (or *infimum*) of the numbers $\|\mathbf{x} - \mathbf{p}\|$. Thus, we write

$$d(\mathbf{p}, \mathbf{S}) = \operatorname{glb}\{\, \|\mathbf{x} - \mathbf{p}\| : \mathbf{x} \in \mathbf{S}\,\}.$$

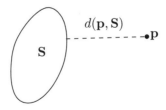

Figure 1. Distance from a point to a set

Characterization of a Cantor set

Definition A.15. A closed set with empty interior is called *nowhere dense*. Thus, a closed set $\mathbf{S}$ is nowhere dense provided that for any point $\mathbf{p}$ in $\mathbf{S}$, there are points in $\mathbf{S}^c$ arbitrarily close to $\mathbf{p}$; that is, for any point $\mathbf{p}$ in $\mathbf{S}$ and any $r > 0$, the open ball $\mathbf{B}(\mathbf{p}, r)$ contains points that are not in $\mathbf{S}$.

A.2. Analysis

Definition A.16. A set **S** in $\mathbb{R}^n$ is called *totally disconnected* provided that, for any two points **x** and **y** in **S**, there are two disjoint open sets **U** and **V**, with **x** in **U** and **y** in **V** and $\mathbf{S} \subset \mathbf{U} \cup \mathbf{V}$.

Definition A.17. A set **S** is called *perfect* provided that, for every point p in **S** and for every $\delta > 0$, the ball $\mathbf{B}(\mathbf{p}, \delta)$ contains points of **S** distinct from p).

A subset **S** of $\mathbb{R}$ is a *Cantor set* provided that it is closed, perfect, and nowhere dense. A subset **S** of $\mathbb{R}^n$ is a *Cantor set* provided that it is closed, perfect, and totally disconnected.

Function on $\mathbb{R}^n$

Definition A.18. A function **F** from a subset **U** of $\mathbb{R}^m$ to $\mathbb{R}^n$ is *continuous* at a point $\mathbf{x}_0$ in **U** provided that a small change in the point $\mathbf{x}_0$ to **x** leads to a small change in the value $\mathbf{F}(\mathbf{x})$. It is necessary to be able to specify the error $\epsilon > 0$ in the outcome values and then find a bound on the uncertainty $\delta > 0$ of the input point that has at most that error; more precisely, for every $\epsilon > 0$, there is a $\delta > 0$ such that $\|\mathbf{F}(\mathbf{x}) - \mathbf{F}(\mathbf{x}_0)\| < \epsilon$ whenever $\|\mathbf{x} - \mathbf{x}_0\| < \delta$. If the function **F** is continuous at all points of the set **U**, then we say that **F** is *continuous* or **F** is *continuous on* **U**. For a function from $\mathbb{R}$ to $\mathbb{R}$, the graph of a continuous function f has no jumps or discontinuities (i.e., it is possible to draw the graph without picking up the pen drawing it).

The next theorem generalizes the intermediate value theorem to the case of a function from a connected subset of $\mathbb{R}^n$ into $\mathbb{R}$.

Theorem A.8 (Intermediate Value Theorem on $\mathbb{R}^n$). *Assume that F is a continuous real-valued function on $\mathbb{R}^n$, **U** is a connected subset of $\mathbb{R}^n$, $\mathbf{x}_1$ and $\mathbf{x}_2$ are in **U**, and $F(\mathbf{x}_1) < a < F(\mathbf{x}_2)$. Then, there is a point $\mathbf{x}_3$ in **U** such that $F(\mathbf{x}_3) = a$.*

One of the important ways we get open and closed sets is by means of a continuous function. Assume that f from $\mathbb{R}^n$ to $\mathbb{R}$ is continuous, and look at the sets

$$f^{-1}(C) \equiv \{\, \mathbf{x} \in \mathbb{R}^n : f(\mathbf{x}) = C \,\},$$
$$\mathbf{U} = f^{-1}((-\infty, C)) \equiv \{\, \mathbf{x} \in \mathbb{R}^n : f(\mathbf{x}) < C \,\}, \quad \text{and}$$
$$\mathbf{K} = f^{-1}((-\infty, C]) \equiv \{\, \mathbf{x} \in \mathbb{R}^n : f(\mathbf{x}) \leq C \,\}.$$

Then, the sets $f^{-1}(C)$ and **K** are closed and **U** is open. The closure of **U** is contained in **K**, but there might be points with $f(\mathbf{x}) = C$ and no nearby points with $f(\mathbf{x}) < C$. The boundary of both **U** and **K** is contained in $f^{-1}(C)$. However, to be a boundary point there must be nearby points whose values are larger than C and less than C. The interior of **K** contains **U** (i.e., any point in **U** is in the interior of **K**). However, there can be points on $f^{-1}(C)$ in the interior of **C**. For example, if $f(x) = x^3 - 3x$, then -1 is in the interior of $f^{-1}((-\infty, 2])$, since all nearby points have values less than or equal to 2.

More generally, let f from $\mathbb{R}^m$ to $\mathbb{R}^n$ be continuous, let **U** be an open set in $\mathbb{R}^n$, and let **C** be a closed set in $\mathbb{R}^n$. Then, $f^{-1}(\mathbf{U})$ is open in $\mathbb{R}^m$ and $f^{-1}(\mathbf{C})$ is closed in $\mathbb{R}^m$.

If a set $\mathbf{S}$ in $\mathbb{R}^m$ is compact and f from $\mathbb{R}^m$ to $\mathbb{R}$ is continuous, then there are points $\mathbf{x}_{\min}$ and $\mathbf{x}_{\max}$ in $\mathbf{S}$ such that
$$f(\mathbf{x}_{\min}) \leq f(\mathbf{x}) \leq f(\mathbf{x}_{\max})$$
for all points $\mathbf{x}$ in $\mathbf{S}$. Thus, f attains its maximum and minimum values on a compact set. More generally, if $\mathbf{S}$ in $\mathbb{R}^m$ is compact and f from $\mathbb{R}^m$ to $\mathbb{R}^n$ is continuous, then $f(\mathbf{S})$ is compact in $\mathbb{R}^n$.

When we discuss limit sets, we use the fact that a bounded sequence of points in $\mathbb{R}^n$ has a limit point. This argument uses compactness. Let $C > 0$ and assume $\mathbf{a}_j$ is a sequence of points in $\overline{\mathbf{B}}(\mathbf{0}, C)$. Then, there is a subsequence $\mathbf{a}_{j_i}$ and a point $\mathbf{a}^*$ such that $\mathbf{a}_{j_i}$ converges to $\mathbf{a}^*$ (i.e., given any $\epsilon > 0$, there is an N such that $\|\mathbf{a}_{j_i} - \mathbf{a}^*\| < \epsilon$ whenever $i \geq N$).

There is a generalization of this last property to metric spaces.

Definition A.19. Given a set $\mathbf{X}$, a function d that takes two points $\mathbf{x}$ and $\mathbf{y}$ in $\mathbf{X}$ and yields a number $d(\mathbf{x}, \mathbf{y})$ is called a *metric* or *distance* on $\mathbf{X}$, provided that it has the following four properties for all points $\mathbf{x}$, $\mathbf{y}$ and $\mathbf{z}$ in $\mathbf{X}$:

(i) $\delta(\mathbf{x}, \mathbf{y}) = \delta(\mathbf{y}, \mathbf{x})$.

(ii) $0 \leq \delta(\mathbf{x}, \mathbf{y}) < \infty$.

(iii) $\delta(\mathbf{x}, \mathbf{y}) = 0$ if and only if $\mathbf{x} = \mathbf{y}$.

(iv) $\delta(\mathbf{x}, \mathbf{z}) \leq \delta(\mathbf{x}, \mathbf{y}) + \delta(\mathbf{y}, \mathbf{z})$.

The set $\mathbf{X}$, together with a metric d, is called a *metric space*.

Let $\mathbf{X}$ be a metric space with metric d. A subset $\mathbf{A}$ of $\mathbf{X}$ is *compact* provided that, for any sequence of points $\{\mathbf{x}_j\}$ in $\mathbf{A}$, there is a subsequence $\{\mathbf{x}_{j_i}\}$ that converges to a point $\mathbf{x}_\infty$ in $\mathbf{A}$.

A sequence of points $\{\mathbf{x}_j\}$ in a metric space $\mathbf{X}$ with metric d is called a *Cauchy sequence* provided that, for all $\epsilon > 0$, there is an N_ϵ such that
$$d(\mathbf{x}_j, \mathbf{x}_k) < \epsilon$$
for all $j, k \geq N_\epsilon$.

A metric space $\mathbf{X}$ with metric d is called *complete* provided that for any Cauchy sequence $\{\mathbf{x}_j\}$ in $\mathbf{X}$ converges to a point $\mathbf{x}_\infty$ in $\mathbf{X}$.

A.2.1. Manifold. The concept of a *manifold* is not usually covered in a calculus course: it is a general term in mathematics that includes both curves and surfaces, as well as higher dimensional objects. The word "manifold" in nontechnical English usage means many, or a pipe with many lateral outlets (as an exhaust pipe connected to several cylinders in an engine). Thus, the technical mathematical usage emphasizes that the set of points has (possibly) many different directions of motion within the set.

Curves and surfaces are treated in multidimensional calculus. A curve is a smooth image of a line with no corners, and a surface is the smooth image of a piece of a plane with no folds or corners. Curves are often described as the differentiable image of a function $\mathbf{g}$ from (an open interval in) the real line into three-dimensional space $\mathbb{R}^3$: Thus, a curve is given parametrically, with the parameter being the value t in the real line. Surfaces are usually given as graphs of functions from the plane into the reals, with the graph being in three-dimensional space. Some calculus books discuss parametrically defined surfaces, which allows the use of spherical coordinates to define (most of) the surface of a sphere.

We use the term manifold in "stable manifold" and "unstable manifold" because the dimension varies depending on the stability type of the fixed point, periodic orbit, or other invariant set. These manifolds are defined locally as graphs over the eigenspaces near the fixed points. A more thorough treatment is given in [**Gui74**]. Also, see [**Rob99**].

A.3. Matrix Algebra

In this appendix, we review some important concepts in matrix algebra. Most of these should be covered in a first course in linear algebra. A few of them are not usually covered but can be expressed in terms of the material of such a course.

A.3.1. Eigenvalues and eigenvectors. Assume that $\mathbf{A}$ is an $n \times n$ matrix with real entries. The determinant $\det(\mathbf{A})$ is defined in most books on linear algebra. The trace $\mathrm{tr}(\mathbf{A})$ is the sum of the entries down the diagonal:

$$\mathrm{tr}(\mathbf{A}) = a_{11} + \cdots + a_{nn}.$$

The transpose of $\mathbf{A}$, denoted by $\mathbf{A}^\mathsf{T}$, is the matrix with the rows and columns interchanged: if $\mathbf{A} = (a_{ij})$ and $(b_{ij}) = \mathbf{B} = \mathbf{A}^\mathsf{T}$, then $b_{ij} = a_{ji}$ for all pairs i and j.

The *characteristic equation* of an $n \times n$ matrix $\mathbf{A}$ is the equation

$$0 = \det(\mathbf{A} - \lambda \mathbf{I}).$$

The roots of the characteristic equation are called the *eigenvalues*. If $\lambda_1, \ldots, \lambda_n$ are the eigenvalues, with possible multiplicity, then the determinant is the product of the eigenvalues and the trace is the sum of the eigenvalues:

$$\det(\mathbf{A}) = \lambda_1 \lambda_2 \cdots \lambda_n \quad \text{and}$$
$$\mathrm{tr}(\mathbf{A}) = \lambda_1 + \lambda_2 + \cdots + \lambda_n.$$

We use a couple other facts about eigenvalues. Assume that the eigenvalues of $\mathbf{A}$ are $\lambda_1, \ldots, \lambda_n$. Then, the eigenvalues of $\mathbf{A}^k$ for an integer $k > 1$ are $\lambda_1^k, \ldots, \lambda_n^k$. If $\mathbf{A}$ has an inverse, then this relationship also holds for negative values of k. In particular, the eigenvalues of the inverse $\mathbf{A}^{-1}$ are $\lambda_1^{-1}, \ldots, \lambda_n^{-1}$.

If $\mathbf{B}$ is linearly conjugate to $\mathbf{A}$, $\mathbf{B} = \mathbf{C}^{-1} \mathbf{A} \mathbf{C}$ for an invertible matrix $\mathbf{C}$, then the eigenvalues of $\mathbf{A}$ and $\mathbf{B}$ are the same.

Finally, the eigenvalues of $\mathbf{A}$ and $\mathbf{A}^\mathsf{T}$ are the same.

A *diagonal matrix* is a matrix with nonzero entries only on the diagonal:

$$\mathrm{diag}(a_1, \ldots, a_n) = \begin{pmatrix} a_1 & 0 & \cdots & 0 \\ 0 & a_2 & \cdots & 0 \\ \vdots & \vdots & \ddots & \vdots \\ 0 & 0 & \cdots & a_n \end{pmatrix}.$$

A matrix $\mathbf{A}$ is *diagonalizable*, provided that there is a linear conjugacy to a diagonal matrix, or
$$\mathbf{C}^{-1}\mathbf{A}\mathbf{C} = \text{diag}(a_1, \ldots, a_n).$$

Vectors $\{\mathbf{w}^1, \ldots, \mathbf{w}^k\}$ are said to be *linearly independent* provided that, if
$$c_1\mathbf{w}^1 + \cdots + c_k\mathbf{w}^k = \mathbf{0},$$
then all the $c_j = 0$. In particular, a set of n vectors $\{\mathbf{w}^1, \ldots, \mathbf{w}^n\}$ in $\mathbb{R}^n$ are linearly independent if and only if
$$\det\left(\mathbf{w}^1, \ldots, \mathbf{w}^n\right) \neq 0.$$

If $\{\mathbf{w}^1, \ldots, \mathbf{w}^n\}$ are n linearly independent vectors in $\mathbb{R}^n$, then any other vector $\mathbf{x}$ in $\mathbb{R}^n$ can be uniquely written as a *linear combination* of the $\mathbf{w}^j$; that is,
$$\mathbf{x} = y_1\mathbf{w}^1 + \cdots + y_n\mathbf{w}^n.$$

This last statement follows, because it is possible to solve
$$\left(\mathbf{w}^1, \ldots, \mathbf{w}^n\right) \begin{pmatrix} y_1 \\ \vdots \\ y_n \end{pmatrix} = \mathbf{x}$$
for $y_1, \ldots, y_n$. For this reason, if $\{\mathbf{w}^1, \ldots, \mathbf{w}^n\}$ is a set of n linearly independent vectors in $\mathbb{R}^n$, it is called a *basis of* $\mathbb{R}^n$.

Combining the preceding definitions and statements, we can assert that, if $\lambda_1, \ldots, \lambda_n$ are n distinct eigenvalues with eigenvectors $\mathbf{v}^1, \ldots, \mathbf{v}^n$, then the set of vectors $\{\mathbf{v}^1, \ldots, \mathbf{v}^n\}$ forms a basis of $\mathbb{R}^n$.

A.3.2. Generalized eigenvectors. The Cayley–Hamilton theorem states that, if
$$p(x) = (-1)^n x^n + a_{n-1} x^{n-1} + \cdots + a_0$$
is the characteristic polynomial for $\mathbf{A}$, then
$$\mathbf{0} = (-1)^n \mathbf{A}^n + a_{n-1}\mathbf{A}^{n-1} + \cdots + a_0 \mathbf{I}.$$
In particular, if $\lambda_1, \ldots, \lambda_q$ are the distinct eigenvalues of $\mathbf{A}$ with algebraic multiplicities $m_1, \ldots, m_q$,
$$p(x) = (x - \lambda_1)^{m_1} \cdots (x - \lambda_q)^{m_q},$$
then
$$S_k = \{\mathbf{v} : (\mathbf{A} - \lambda_k \mathbf{I})^{m_k} \mathbf{v} = \mathbf{0}\}$$
is a vector subspace of dimension m_k. Thus, the geometric multiplicity, which is the dimension of S_k, is the same as the algebraic multiplicities, the multiplicity in the characteristic equation. Vectors in S_k are called *generalized eigenvectors*. Thus, even when there is no basis of eigenvectors, there is always a basis of generalized eigenvectors.

Take a real eigenvalue λ_k of multiplicity $m_k > 1$. Assume that $\mathbf{v}^{(r)}$ is a vector with
$$(\mathbf{A} - \lambda_k \mathbf{I})^r \mathbf{v}^{(r)} = \mathbf{0} \quad \text{but} \quad (\mathbf{A} - \lambda_k \mathbf{I})^{r-1} \mathbf{v}^{(r)} \neq \mathbf{0},$$

A.3. Matrix Algebra

with $1 < r \leq m_k$. Setting $\mathbf{v}^{(r-j)} = (\mathbf{A} - \lambda_k \mathbf{I})^j \mathbf{v}^{(r)}$, we get

$$(\mathbf{A} - \lambda_k \mathbf{I})\mathbf{v}^{(r)} = \mathbf{v}^{(r-1)},$$
$$(\mathbf{A} - \lambda_k \mathbf{I})\mathbf{v}^{(r-1)} = \mathbf{v}^{(r-2)},$$
$$\vdots \qquad \vdots$$
$$(\mathbf{A} - \lambda_k \mathbf{I})\mathbf{v}^{(2)} = \mathbf{v}^{(1)},$$
$$(\mathbf{A} - \lambda_k \mathbf{I})\mathbf{v}^{(1)} = \mathbf{0},$$

or

$$\mathbf{A}\mathbf{v}^{(r)} = \lambda_k \mathbf{v}^{(r)} + \mathbf{v}^{(r-1)},$$
$$\mathbf{A}\mathbf{v}^{(r-1)} = \lambda_k \mathbf{v}^{(r-1)} + \mathbf{v}^{(r-2)},$$
$$\vdots \qquad \vdots$$
$$\mathbf{A}\mathbf{v}^{(2)} = \lambda_k \mathbf{v}^{(2)} + \mathbf{v}^{(1)},$$
$$\mathbf{A}\mathbf{v}^{(1)} = \lambda_k \mathbf{v}^{(1)}.$$

In terms of this partial basis, there is an $r \times r$ subblock of the form

$$\mathbf{C}_\ell = \begin{pmatrix} \lambda_k & 1 & 0 & \cdots & 0 & 0 \\ 0 & \lambda_k & 1 & \cdots & 0 & 0 \\ 0 & 0 & \lambda_k & \cdots & 0 & 0 \\ \vdots & \vdots & \vdots & \ddots & \vdots & \vdots \\ 0 & 0 & 0 & \cdots & \lambda_k & 1 \\ 0 & 0 & 0 & \cdots & 0 & \lambda_k \end{pmatrix}.$$

The Jordan canonical form theorem says that there are enough blocks of this form, together with eigenvectors, to span the total subspace S_k. Therefore, in the case of a repeated real eigenvector, the matrix $\mathbf{A}$ on S_k can be represented by blocks of the form $\mathbf{C}_\ell$ plus a diagonal matrix.

Next, we assume that $\lambda_k = \alpha_k + i\beta_k$ is a complex eigenvalue of multiplicity $m_k > 1$. If there are not as many eigenvectors as the dimension of the multiplicity, then there are blocks of the form

$$\mathbf{D}_\ell = \begin{pmatrix} \mathbf{B}_k & \mathbf{I} & \cdots & 0 & 0 \\ 0 & \mathbf{B}_k & \cdots & 0 & 0 \\ \vdots & \vdots & \ddots & \vdots & \vdots \\ 0 & 0 & \cdots & \mathbf{B}_k & \mathbf{I} \\ 0 & 0 & \cdots & 0 & \mathbf{B}_k \end{pmatrix},$$

where $\mathbf{B}_k$ is the 2×2 block with entries α_k and $\pm \beta_k$ previously defined.

A.3.3. Norm of a matrix. We often want to know the largest amount by which a linear map stretches a vector, which is $\|\mathbf{A}\mathbf{v}\|/\|\mathbf{v}\|$. Because

$$\frac{\|\mathbf{A}\mathbf{v}\|}{\|\mathbf{v}\|} = \left\|\mathbf{A}\frac{\mathbf{v}}{\|\mathbf{v}\|}\right\|,$$

it follows that
$$\max_{\mathbf{x} \neq \mathbf{0}} \left\{ \frac{\|\mathbf{A}\mathbf{v}\|}{\|\mathbf{v}\|} \right\} = \max_{\|\mathbf{u}\|=1} \|\mathbf{A}\mathbf{u}\|.$$
We define the *norm* of the matrix $\mathbf{A}$, $\|\mathbf{A}\|$, to be this maximum:
$$\|\mathbf{A}\| = \max_{\|\mathbf{u}\|=1} \|\mathbf{A}\mathbf{u}\|.$$
Notice that $\|\mathbf{A}\| \geq 0$, and $\|\mathbf{A}\| = 0$ if and only if $\mathbf{A} = \mathbf{0}$, the matrix with all entries equal to 0.

We want to be able to find the norm in terms of the eigenvalues of a matrix that is derived from $\mathbf{A}$. In fact, the norm $\|\mathbf{A}\|$ is the square root of the largest eigenvalue of the symmetric matrix $\mathbf{A}^\mathsf{T}\mathbf{A}$, as the next proposition shows.

Proposition A.9. *Let $\mathbf{A}$ be a square $n \times n$ matrix.*

a. *The matrix $\mathbf{A}^\mathsf{T}\mathbf{A}$ is symmetric, so it has real eigenvalues, all of which are nonnegative.*

b. *If λ_1 is the largest eigenvalue of $\mathbf{A}^\mathsf{T}\mathbf{A}$, then the norm of $\mathbf{A}$ is the square root of λ_1,*
$$\|\mathbf{A}\| = \sqrt{\lambda_1}.$$

Proof. (a) The matrix $\mathbf{A}^\mathsf{T}\mathbf{A}$ is symmetric, so it has real eigenvalues and the eigenvectors can be chosen to be perpendicular and of length one (orthonormal). If λ is an eigenvalue for $\mathbf{v}$, then
$$0 \leq \|\mathbf{A}\mathbf{v}\|^2 = \mathbf{v}^\mathsf{T}\mathbf{A}^\mathsf{T}\mathbf{A}\mathbf{v} = \mathbf{v}^\mathsf{T}\lambda\mathbf{v} = \lambda\|\mathbf{v}\|^2,$$
so the eigenvalues must be nonnegative.

(b) For the matrix $\mathbf{A}^\mathsf{T}\mathbf{A}$, let $0 \leq \lambda_n \leq \lambda_{n-1} \leq \cdots \leq \lambda_1$ be the eigenvalues, with the corresponding orthonormal basis of eigenvectors $\mathbf{v}^1, \ldots, \mathbf{v}^n$. If
$$\mathbf{x} = y_1\mathbf{v}^1 + \cdots + y_n\mathbf{v}^n,$$
with $\|\mathbf{x}\| = 1$, then $\|\mathbf{x}\| = \sqrt{y_1^2 + \cdots + y_n^2} = 1$, and
$$\begin{aligned}
\|\mathbf{A}\mathbf{x}\|^2 &= \mathbf{x}^\mathsf{T}(\mathbf{A}^\mathsf{T}\mathbf{A})\mathbf{x} \\
&= (y_1\mathbf{v}^1 + \cdots + y_n\mathbf{v}^n)^\mathsf{T}(y_1\lambda_1\mathbf{v}^1 + \cdots + y_n\lambda_n\mathbf{v}^n) \\
&= y_1^2\lambda_1\|\mathbf{v}^1\| + \cdots + y_n^2\lambda_n\|\mathbf{v}^n\| \\
&= y_1^2\lambda_1 + \cdots + y_n^2\lambda_n.
\end{aligned}$$
The maximum of this quantity for $y_1^2 + \cdots + y_n^2 = 1$ occurs for $y_1 = 1$ and $y_j = 0$, for $j \geq 2$. Therefore, the norm is given as the square root of this largest eigenvalue of $\mathbf{A}^\mathsf{T}\mathbf{A}$, or
$$\|\mathbf{A}\| = \sqrt{\lambda_1}.$$
$\square$

Usually, we do not need to know the exact value of the norm, just that such a number exists. We use this at least implicitly to see that the linear part of a map determines many of the properties near a fixed point.

Appendix B

Generic Properties

We remarked in our discussion of chaotic attractors, for both systems of differential equations and iteration of functions, that it is not enough to assume that the attractor has sensitive dependence with respect to initial conditions in the ambient space; it must have sensitive dependence on initial conditions when restricted to the attractor. Examples 7.15 and 13.41 deal with attractors that have sensitive dependence in the ambient space, but not when restricted to the attractor; neither of these examples have motion that should be called chaotic. For Example 7.15, if the small term $0.01\cos(\tau)$ is added to the equation for $\dot{y}$ for this system, then the attractor appears to become chaotic. See Figure 19 in Chapter 7 for the plot of the Poincaré map for this perturbation. The attractor changes dramatically under a small perturbation, so the original system is not generic.

The idea of a *generic* condition is that any system should be able to be approximated by a new system that has the desired property. In this section, we discuss several generic properties. For more details on this topic, see Chapter XI of [**Rob99**].

The search for generic properties was inaugurated by S. Smale in the 1960s. See [**Sma67**]. This search was part of the program to find which systems were structurally stable (i.e., systems whose essential dynamics did not change much with small perturbations). The hope was that most systems were structurally stable. This turned out not to be the case; there are much richer possibilities for dynamics than just those for structurally stable systems. However, generic properties still show us what characteristic are likely to be robust under small perturbations of the system. Since most systems are vulnerable to small outside disturbances, it seems reasonable that the ones observed are generic.

When discussing generic properties, it is important to specify how many derivatives of the function are only slightly changed. A C^1-approximation is one in which both the values of the two functions and of their first derivatives are almost equal, but nothing is said about the difference of their second or higher derivatives. A C^r-approximation is one in which both the values of the two functions and any of their partial derivatives up to order r are close. Thus, if g is near f in the C^2 sense, this is more stringent than just being close in the C^1 sense. A C^1-*generic property* is one that is true for most functions in the C^1 sense (i.e., any function can be C^1-approximated by one with the property). Similarly, a C^r-*generic property* is one that is true for most functions in the C^r sense (i.e., any function can be C^r-approximated by one with the property). Thus, it is harder to show a property is C^2-generic than to show it is C^1-generic. The first few properties we discuss have only been proved to be C^1-generic and not C^2-generic or any other C^r-generic for $r \geq 2$.

We start with some C^1-generic properties that are related to our definition of a chaotic attractor. Charles Pugh proved what is called the *general density theorem*: For a C^1-generic set of systems, the set $\mathscr{P}(\phi)$ of periodic orbits and fixed points is dense in any ω-limit set for ϕ. Thus, for a generic property (such as the density of the set $\mathscr{P}$ in any ω-limit set), any system can be approximated by one for which the property is valid. This theorem is based on what is called the C^1-*closing lemma* of Pugh: This theorem states that any system ϕ with a point **p** in an ω-limit set of ϕ can be C^1-approximated by a new system ψ for which **p** is a periodic point.

A more recent theorem of C. Bonatti and S. Crovisier shows that, for a C^1-generic system, the periodic orbits are dense in any attractor. More generally, the periodic orbits are dense in the chain recurrent set. See [**Bon03**]. This work is based on a very significant extension of the C^1-closing lemma by S. Hayashi, called the connecting lemma. Bonatti and Crovisier also show that, C^1-generically, any attractor (with a trapping region) is topologically transitive. Thus, for a C^1-generic system with an attractor **A**, the system is topologically transitive on **A**, and the periodic points are dense in **A**. This means that, for a generic system, the restriction to a chaotic attractor in our sense is chaotic in the sense that Devaney defined.

Now we turn to properties that are not directly connected to our definition of a chaotic attractor, but do show why some of our examples are not generic. These features are C^r-generic for any $r \geq 1$.

One nongeneric feature from Example 7.15 is that the stable and unstable manifolds of the periodic orbit γ coincide:

$$\gamma = \{\,(0,0,\tau) : 0 \leq \tau \leq 2\pi\,\} \quad \text{and}$$
$$W^u(\gamma) = W^s(\gamma).$$

For most systems, the stable and unstable manifolds of periodic orbits and fixed points must cross "transversely," and not coincide as is the case for this example.

To introduce the idea of transversality, we first consider linear subspaces S_1 and S_2 of $\mathbb{R}^n$. The origin is in both subspaces. Consider the sum of the subspaces formed by adding together all possible vectors in each subspace:

$$S_1 + S_2 = \{\,\mathbf{v}_1 + \mathbf{v}_2 : \mathbf{v}_1 \in S_1 \ \mathbf{v}_2 \in S_2\,\}.$$

If $S_1 + S_2$ is all of $\mathbb{R}^n$ (i.e., vectors from the two subspaces span all of $\mathbb{R}^n$), then we say that they are transverse at the origin. In $\mathbb{R}^3$, a line and a plane both through the origin are transverse provided that the line is not contained in the plane. Two lines cannot be transverse in $\mathbb{R}^3$ since their vectors give only a plane. Two planes through the origin that do not coincide are transverse in $\mathbb{R}^3$, even though they intersect along a line.

Two parametrized curves $\gamma_1(t)$ and $\gamma_2(t)$ in $\mathbb{R}^2$ are transverse provided that the vectors $\gamma_1'(t_1)$ and $\gamma_2'(t_2)$ are not parallel whenever $\gamma_1(t_1) = \gamma_2(t_2)$.

The objects we want to be transverse are the stable and unstable manifolds of fixed points or periodic orbits (or just periodic orbits for iteration of a function). These manifolds can be parametrized near any point by functions $\mathbf{G}$ from some open set $\mathbf{U}$ in a Euclidean space $\mathbb{R}^k$ into the ambient space $\mathbb{R}^n$:

$$\mathbf{G}(\mathbf{t}) = \begin{pmatrix} F_1(t_1, \ldots, t_k) \\ \vdots \\ F_n(t_1, \ldots, t_k) \end{pmatrix}.$$

For this to be a good parametrization, the rank of the matrix of partial derivatives of $\mathbf{G}$ at all points of $\mathbf{U}$ must be k, the dimension of the parameter space. In most cases considered in this book, $\mathbf{G}$ is actually a graph in which certain variables are given in term of the other variables, such as

$$\mathbf{G}(x_1, \ldots, x_k) = \begin{pmatrix} x_1 \\ \vdots \\ x_k \\ F_{k+1}(x_1, \ldots, x_k) \\ \cdots \\ F_n(x_1, \ldots, x_k) \end{pmatrix}.$$

For a manifold $\mathbf{S}$, we denote the set of all the tangent vectors a point $\mathbf{p}$ by

$$T_{\mathbf{p}} \mathbf{S}$$

and call this the *tangent space to the manifold at* $\mathbf{p}$. If $\mathbf{G}$ is a parametrization of a manifold $\mathbf{S}$ near $\mathbf{G}(\mathbf{a}) = \mathbf{p}$, then the image of the matrix of partial derivatives of $\mathbf{G}$ at $\mathbf{a}$ equals to the tangent space to $\mathbf{S}$ at $\mathbf{p}$; that is,

$$D\mathbf{G}_{(\mathbf{a})} \mathbb{R}^k = \{ \mathbf{v} = D\mathbf{G}_{(\mathbf{a})} \mathbf{w} : \mathbf{w} \in \mathbb{R}^k \} = T_{\mathbf{p}} \mathbf{S}.$$

For more details on manifolds and tangent spaces see [**Gui74**].

We can now give the definition of transversality.

Definition B.1. Let $\mathbf{S}_1$ and $\mathbf{S}_2$ be two manifolds in $\mathbb{R}^n$. If $\mathbf{p}$ is a point in both $\mathbf{S}_1$ and $\mathbf{S}_2$, we say that $\mathbf{S}_1$ and $\mathbf{S}_2$ are *transverse at* $\mathbf{p}$, provided that

$$T_{\mathbf{p}} \mathbf{S}_1 + T_{\mathbf{p}} \mathbf{S}_2 = \mathbb{R}^n.$$

We say that $\mathbf{S}_1$ and $\mathbf{S}_2$ are *transverse* provided that $\mathbf{S}_1$ and $\mathbf{S}_2$ are transverse at any point $\mathbf{p}$ that is in both $\mathbf{S}_1$ and $\mathbf{S}_2$. Thus, if $\mathbf{S}_1$ and $\mathbf{S}_2$ do not intersect, they are transverse.

In the next definition and theorem, $\mathfrak{X}^1(\mathbb{R}^n)$ is the set of C^1 vector fields on $\mathbb{R}^n$, and $\text{Diff}^1(\mathbb{R}^n)$ is the set of C^1 diffeomorphisms on $\mathbb{R}^n$. A vector field $\mathbf{F}$ in $\mathfrak{X}^1(\mathbb{R}^n)$ induces a system of differential equations on $\mathbb{R}^n$ (or other manifold) by $\dot{\mathbf{x}} = \mathbf{F}(\mathbf{x})$.

We let $\mathscr{F}^1(\mathbb{R}^n)$ be either $\mathfrak{X}^1(\mathbb{R}^n)$ or $\text{Diff}^1(\mathbb{R}^n)$. The C^1 topology on $\mathscr{F}^1(\mathbb{R}^n)$ says that two elements of $\mathscr{F}^1(\mathbb{R}^n)$ are close if their values and partial derivatives are close at every point. (Because $\mathbb{R}^n$ is not compact, to say that two elements of

$\mathscr{F}^1(\mathbb{R}^n)$ are close, we need to specify the closeness at each point using a continuous real-valued function. We skip this detail.)

Definition B.2. We first give this definition for vector fields and then for diffeomorphisms.

Let $\mathbf{F}(\mathbf{x})$ be a vector field in $\mathfrak{X}^1(\mathbb{R}^n)$ (or a vector field on another manifold) that induces the system of differential equations. We say that $\mathbf{F}$ satisfies the *Kupka–Smale property* provided that the following two properties are satisfied:

1. All fixed points and periodic orbits are hyperbolic. Thus, all the eigenvalues of fixed points have nonzero real parts, and all the characteristic multipliers of periodic orbits have absolute value not equal to one.

2. Let $\mathbf{S}_1$ be the unstable manifold of any fixed point or periodic orbit. Let $\mathbf{S}_2$ be the stable manifold of any fixed point or periodic orbit. Then, $\mathbf{S}_1$ and $\mathbf{S}_2$ are transverse.

Let $\mathbf{F}(\mathbf{x})$ be a diffeomorphism on $\mathbb{R}^n$ (or other manifold). We say that $\mathbf{F}$ satisfies the Kupka–Smale property provided that the following two properties are satisfied:

1. All the periodic orbits are hyperbolic. Thus, all the eigenvalues for periodic orbits have absolute value not equal to one.

2. Let $\mathbf{S}_1$ be the unstable manifold of any orbit point. Let $\mathbf{S}_2$ be the stable manifold of any periodic orbit. Then, $\mathbf{S}_1$ and $\mathbf{S}_2$ are transverse.

The next theorem states that the Kupka–Smale property is generic.

Theorem B.1. *Let $r \geq 1$ and $\mathscr{F}^r(\mathbb{R}^n)$ be either $\mathfrak{X}^r(\mathbb{R}^n)$ or $\mathrm{Diff}^r(\mathbb{R}^n)$ with the C^r topology. Let $\mathscr{K}^r(\mathbb{R}^n)$ be the systems in $\mathscr{F}^r(\mathbb{R}^n)$ that satisfy the Kupka–Smale property. Then, $\mathscr{K}^r(\mathbb{R}^n)$ is dense in $\mathscr{F}^r(\mathbb{R}^n)$.*

For more details, see [**Rob99**] or [**Sma67**].

Generic systems that satisfy the general density theorem and the Kupka–Smale property cannot exhibit the special features of Examples 7.15 and 13.41, and have more of the dynamic behavior that intuitively we think a chaotic system should possess.

Bibliography

[Ack69] Ackerman, E., L. Gatewood, J. Rosevear, and G. Molnar, *Blood glucose regulation and diabetes*, Concepts and Models of Biomathematics (F. Heinmets, ed.), Marcel Dekker, 1969, pp. 131–156.

[Ahm99] Ahmed, A., El-Misiery, and H.N. Agiza, *On controlling chaos in an inflation-unemployment dynamical system*, Chaos Solitons Fractals **10** (1999), 1567–1570.

[Ale68a] Alekseev, V.M., *Quasirandom dynamical systems, I*, Math. USSR-Sb. **5** (1968), 73–128.

[Ale68b] _____, *Quasirandom dynamical systems, II*, Math. USSR-Sb. **6** (1968), 505–560.

[Ale69] _____, *Quasirandom dynamical systems, III*, Math. USSR-Sb. **7** (1969), 1–43.

[All94] Allen, L.J.S., *Some discrete-time SI, SIS, and SIR epidemic models*, Math. Biosciences **124** (1994), 83–105.

[All97] Alligood, K., T. Sauer, and J. Yorke, *Chaos: An Introduction to Dynamical Systems*, Springer–Verlag, New York–Berlin–Heidelberg, 1997.

[And37] Andronov, A.A. and L. Pontryagin, *Systèmes grossier*, Dokl. Akad. Nauk. SSSR **14** (1937), 247–251.

[Bar88] Barnsley, M., *Fractals Everywhere*, Academic Press, Inc., New York, 1988.

[Bar97] Barrow-Green, J., *Poincaré and the three body problem*, Amer. Math. Soc., Prividence RI, 1997.

[Bar00a] Bartle, R. and D. Shubert, *Introduction to Real Analysis*, John Wiley & Sons, Inc., New York, 2000.

[Bar00b] Barton, R. and K. Burns, *A simple special case of Sharkovskii's Theorem*, Amer. Math. Monthly **107** (2000), 932–933.

[Bel72] Belickii, G. R., *Functional equations and conjugacy of local diffeomorphisms of finite smooth class*, Dokl. Akad. Nauk. SSSR **13** (1972), 56–59.

[Ben81] Benhabib, J. and R.H. Day, *Rational choice and erratic behaviour*, Review of Economic Studies **48** (1981), 459–471.

[Ben91] Benedicks, M. and L. Carleson, *The dynamics of the Hénon map*, Annals of Math. **133** (1991), 73–169.

[Ben93] Benedicks, M. and L.S. Young, *Sinai–Bowen–Ruelle measures for certain Hénon maps*, Invent. Math. **112** (1993), 541–576.

[Bir27] Birkhoff, G. D., *Dynamical Systems*, American Mathematical Society, Providence RI, 1927.

[Blo80] Block, L., J. Guckenheimer, M. Misiurewicz, and L.S. Young, *Periodic points and topological entropy of one dimensional maps*, Lect. Notes in Math. **819** (New York–Berlin–Heidelberg), Springer–Verlag, 1980, pp. 18–34.

[Bon03] Bonatti, C. and S. Crovisier, *Recurrence et Genericity*, C.R. Math. Acad. Sci. Paris **336** (2003), 839–844.

[Bra69] Brauer, F. and J. Nohel, *Qualitative Theory of Ordinary DIfferential Equations*, Benjamin, Inc., New York and Amsterdam, 1969.

[Bra73] Braun, M., *Differential Equations and Their Applications*, Springer–Verlag, New York–Berlin–Heidelberg, 1973.

[Bra78] Braun, M., C. Coleman, and D. Drew, *Differential Equation Models*, Springer–Verlag, New York–Berlin–Heidelberg, 1978.

[Bra01] Brauer, F. and C. Castillo-Chávez, *Mathematical Models in Population Biology and Epidemiology*, Springer–Verlag, New York–Berlin–Heidelberg, 2001.

[Bur78] Burden, L. and J.D. Faires,, *Numerical Analysis*, PWS-Kent Publ. Co., Boston, 1978.

[Bur11] Burns, K. and B. Hasselblatt, *The Sharkovsky Theorem: A Natural Direct Proof*, Amer. Math. Monthly **118** (2011), 229–244.

[Car81] Carr, J., *Applications of Center Manifold Theory*, Springer–Verlag, New York–Berlin–Heidelberg, 1981.

[Chi84] Chiang, A., *Fundamental Methods of Mathematical Economics*, McGraw Hill, New York, 1984.

[Cho77] Chow, S.N., J. Mallet-Paret, *Integral averaging and bifurcation*, J. Differential Equations **26** (1977), 112–159.

[Cho82] Chow, S.N. and J. Hale, *Methods of Bifurcation Theory*, Springer–Verlag, New York–Berlin–Heidelberg, 1982.

[Cho04] Choi, Y., *Attractors from one dimensional Lorenz-like maps*, Discrete Contin. Dyn. Syst. **11** (2004), 715–730.

[Cos95] Costantino, R.F., R.A. Desharnais, J.M. Cushing, and B. Dennis, *Chaotic dynamics in an insect population*, Science **275** (1995), 389–391.

[Cou78] Coullet, P. and C. Tresser, *Itération d'endomorphismes et groupe de renormalisation*, J. de Physique Colloque **39** (1978), C5–C25.

[Day82] Day, R., *Irregular growth cycles*, Amer. Economic Review **72** (1982), 406–414.

[Dev89] Devaney, R., *An introduction to chaotic dynamical systems*, Addison–Wesley Publ. Co., New York & Reading, MA, 1989.

[Dev92] _____, *A first course in chaotic dynamical systems*, Addison–Wesley Publ. Co., New York & Reading, MA, 1992.

[Dia96] Diacu, F. and P. Holmes, *Celestial Encounters*, Princeton University Press, Princeton New Jersey, 1996.

[Du,02] Du, B.-S. , *A simple proof of sharkovkii's theorem*, Institute of Mathematics, Academia Sinica, Taipei Taiwan, 2002.

[Edg90] Edgar, G., *Measure, Topology, and Fractal Geometry*, Springer–Verlag, New York–Berlin–Heidelberg, 1990.

Bibliography

[Enn97] Enns, R. and G. McGuire, *Nonlinear Physics with Maple for Scientists and Engineers*, Birkhäuser, Boston–Basel–Berlin, 1997.

[Fei78] Feigenbaum, M., *Quantitative universality for a class f non-linear transformations*, J. Stat. Phys. **21** (1978), 25–52.

[Gid02] Gidea, M. and P. Zgliczyński, *Covering relations for multidimensional dynamical systems*, http://www.im.uj.edu.pl/~zgliczyn, 2002.

[Gid03] Gidea, M. and C. Robinson, *Topologically crossing heteroclinic connnections to invariant tori*, J. Diff. Equat. **193** (2003), 49–74.

[Gin00] Gintis, H., *Game Theory Evolving*, Princeton University Press, Princeton, 2000.

[Gle87] Gleick, J., *Chaos: Making a New Science*, Penguin Books, New York, London, 1987.

[Guc76] Guckenheimer, J., *A strange, strange attractor*, Hopf Bifurcation and Its Applications (New York–Berlin–Heidelberg) (Marsden and McCracken, eds.), Springer–Verlag, 1976, pp. 368–381.

[Guc80] Guckenheimer, J. and R. Williams, *Structural stability of the Lorenz attractor*, Publ. Math. I.E.H.S. **50** (1980), 73–100.

[Guc83] Guckenheimer, J. and P. Holmes, *Nonlinear Oscillations, Dynamical Systems and Bifurcations of Vector Fields*, Springer–Verlag, New York–Berlin–Heidelberg, 1983.

[Gui74] Guillemin, V. and A. Pollack, *Differential Topology*, Prentice Hall, Englewood Cliffs, NJ, 1974.

[Gul92] Gulick, D., *Encounters with chaos*, McGraw Hill, New York, et al., 1992.

[Hah67] Hahn, W., *Stability of Motion*, Springer–Verlag, New York–Berlin–Heidelberg, 1967.

[Hal69] Hale, J., *Ordinary differential equations*, Wiley-Interscience, New York–London–Sydney–Toronto, 1969.

[Hal91] Hale, J. and H. Koçak, *Dynamics and bifurcations*, Springer–Verlag, New York–Berlin–Heidelberg, 1991.

[Har82] Hartman, P., *Ordinary differential equations*, 2nd ed., Birkäuser, Boston, Basel, and Stuttgart, 1982.

[Has74] Hassell, M.P., *Density-dependence in single-species populations*, J. Animal Ecology **44** (1974), 283–296.

[Hir74] Hirsch, M. and S. Smale, *Differential Equations, Dynamical Systems, and Linear Algebra*, Academic Press, New York and London, 1974.

[Hof88] Hofbauer, J. and K. Sigmund, *The Theory of Evolution and Dynamical Systems*, Cambridge University Press, New York and Cambridge, UK, 1988.

[Hop82] Hopfield, J.J., *Neural networks and physical systems with emergent collective computational abilitites*, Proc. Natl. Acad. Sci. USA **79** (1982), 2554–2558.

[Hur84] Hurley, M. and C. Martin, *Newton's algorithm and chaotic dynamical systems*, SIAM J. Math. Anal. **15** (1984), 238–252.

[Jor87] Jordan, D.W. and P. Smith, *Nonlinear Ordinary Differential Equations, second edition*, Oxford University Press, Oxford, 1987.

[Kap95] Kaplan, D. and L. Glass, *Understanding Nonlinear Dynamics*, Springer–Verlag, New York–Berlin–Heidelberg, 1995.

[Kat95] Katok, H. and B. Hasselblatt, *Introduction to the Modern Theory of Dynamical Systems*, Cambridge University Press, Cambridge, UK and New York, 1995.

[Kul02] Kulenović, M., *Discrete Dynamical Systems and Difference Equations with Mathematica*, Chapman & Hall/CRC, Boca Raton FL, 2002.

[Lan84] Lanford, O.E., *A shorter proof of the existence of Feigenbaum fixed point*, Commun. Math. Phys. **96** (1984), 521–538.

[LaS61] LaSalle, J. P. and S. Lefschetz, *Stability by Liapunov's Direct Method*, Academic Press, New York, 1961.

[Las73] Lasota, A. and J. Yorke, *On the existence of invariant measures for piecewise monotonic transformations*, Transactions Amer. Math. Soc. **186** (1973), 481–488.

[Las77] Lasota, A., *Ergodic problems in biology*, Soc. Math. France, Astérisque **50** (1977), 239–250.

[Las89] Laskar, J., *A numerical experiment on the chaotic behavior of the solar system*, Nature **338** (1989), 237–238.

[Lay01] Lay, S., *Analysis with an Introduction to Proof*, 3rd ed., Prentice Hall, Englewood Cliffs, NJ, 2001.

[Lew93] Lewin, J. and M. Lewin, *An Introduction to Mathematical Analysis*, McGraw-Hill, New York, 1993.

[Li,75] Li, T. and J. Yorke, *Period three implies chaos*, Amer. Math. Monthly **82** (1975), 985–992.

[Li,78] ———, *Ergodic transformations from an interval into itself*, Transactions of Amer. Math. Soc. **335** (1978), 183–192.

[Lor63] Lorenz, E.N., *Deterministic nonperiodic flow*, J. Atmos. Sci. **20** (1963), 130–141.

[Lot25] Lotka, A.J., *Elements of Phsical Biology*, Williams & Wilkins, Baltimore, 1925.

[Lyn01] Lynch, S., *Dynamical Systems with Applications using Maple*, Birkhäuser, Boston–Basel–Berlin, 2001.

[Mar76] Marsden, J. and M. McCracken, *Hopf bifurcation and its applications*, Springer–Verlag, New York–Berlin–Heidelberg, 1976.

[Mar93] Marsden, J. and M. Hoffman, *Elementary classical analysis*, 2nd ed., W. H. Freeman and Co., New York, 1993.

[Mar99] Martelli, M., *Introduction to discrete dynamical systems and chaos*, Wiley–Interscience Publ., New York, 1999.

[Mat75] Mather, J. and R. McGehee, *Solutions of the collinear four-body problem which become unbounded in finite time*, Dynamical Systems Theory and Applications, Lecture Notes in Physics Vol. 38 (New York–Berlin–Heidelberg) (J. Moser, ed.), Springer–Verlag, 1975, pp. 573–587.

[May75] May, R., *Stability and complexity in model ecosystems*, 2nd ed., Princeton Univ. Press, Princeton, NJ, 1975.

[Mey92] Meyer, K.R. and G.R. Hall, *Introduction to hamiltonian dynamical systems and the n-body problem*, Springer–Verlag, New York–Berlin–Heidelberg, 1992.

[Mor97] Morales, C.A. and E.R. Pujals, *Singular strange attractors on the boundary of Morse–Smale systems*, Annales Econle Norm. Sup. **30** (1997), 693–717.

[Mos73] Moser, J., *Stable and Random Motions in Dynamical Systems*, Princeton University Press, Princeton, NJ, 1973.

[Mur89] Murray, J.D., *Mathematical biology*, Springer–Verlag, New York–Berlin–Heidelberg, 1989.

[Nus98] Nusse, H. and J. Yorke, *Dynamcis: Numerical exploration*, Springer–Verlag, New York–Berlin–Heidelberg, 1998.

[Ott89a] Ottino, J., *The Kinematics of Mixing: Stretching, Chaos, and Transport*, Cambridge University Press, Cambridge, UK and New York NY, 1989.

[Ott89b] ———, *The Mixing of Fluids*, Scientific American **260** (1989), 56–67.

[Ott94] Ott, E., T. Sauer, and J.A. Yorke, *Coping with Chaos: Analysis of chaotic data and the exploitation of chaotic systems*, J. Wiley, New York, 1994.

[Par89] Parker, T.S. and L.O. Chua, *Practical Numerical Algorithms for Chaotic Systems*, Springer–Verlag, New York–Berlin–Heidelberg, 1989.

[Pat87] Patterson, S. and C. Robinson, *Basins of sinks near homoclinic tangencies*, Dynamical Systems and bifurcation theory, Pitman Research Notes in Math. (New York) (M. I. Camacho, M. J. Pacifico and F. Takens, ed.), John Wiley & Sons, Inc., 1987, pp. 347–376.

[Per82] Percival, I. and D. Richards, *Introduction to Dynamics*, Cambridge University Press, Cambridge–New York–New Rochelle, 1982.

[Pol04] Polking, J. and D. Arnold, *Ordinary Differential Equations using Matlab*, Prentice Hall, Englewood Cliffs, NJ, 2004.

[Ric54] Ricker, W.E., *Stock and recruitment*, J. Fisheries Research Board of Canada **11** (1954), 559–623.

[Rob99] Robinson, C., *Dynamical Systems: Stability, Symbolic Dynamics, and Chaos*, CRC Press, Boca Raton, London, New York, Washington, D.C., 1999.

[Rob00] ———, *Nonsymmetric Lorenz attractors from a homoclinic bifurcation*, SIAM J. Math. Analysis **32** (2000), 119–141.

[Ros00] Ross, C. and J. Sorensen, *Will the real bifurcation diagram please stand up!*, College Math. J. **31** (2000), 2–14.

[Rue71] Ruelle, D. and F. Takens, *On the nature of turbulence*, Commun. Math. Phys. **20** (1971), 167–192.

[Ryk98] Rykken, E., *Markov partitions for hyperbolic toral automorphisms of $\mathbf{t}^2$*, Rocky Mountain J. Math. **28** (1998), 1103–1124.

[Saa84] Saari, D. and J. Urenko, *Newton's method, circle maps, and chaotic motion*, Amer. Math. Monthly **91** (1984), 3–17.

[Saa04] Saari, D., *Celestial Mechanics*, Regional Conference Series in Mathematics, Amer. Math. Soc. for College Board of the Mathematical Sciences, Providence RI, 2004.

[Sam41] Samuelson, P., *Conditions that a root of a polynomial be less than unity in absolute value*, Ann. Math. Stat. **12** (1941), 360–364.

[Sch06] Schecter, S., *Notes on Game Theory*, Preprint North Carolina State University, 2006.

[Sha64] Sharkovskii, A.N., *Coexistence of cycles of a contiuous map of a line into itself*, Ukrainian Math. J. **16** (1964), 61–71.

[Sho02] Shone, R., *Economic Dynamics: Phase Diagrams and their Economic Application*, Cambridge University Press, New York and Cambridge, UK, 2002.

[Sin78] Singer, D., *Stable orbits and bifurcation of maps of the interval*, SIAM J. Appl. Math. **35** (1978), 260–267.

[Sit60] Sitnikov, K., *Existence of oscillating motions for the three-body problem*, Dokl. Akad. Nauk. USSR **133** (1960), 303–306.

[Sma67] Smale, S., *Differentiable dynamical systems*, Bull. Amer. Math. Soc **73** (1967), 747–817.

[Smi95] Smith, H., *Monotone Dynamical Systems*, American Math. Soc., Providence RI, 1995.

[Sna91] Snavely, M., *Markov partitions for the two-dimensional torus*, Proc. Amer. Math. Soc **113** (1991), 517–527.

[Spa82] Sparrow, C., *The Lorenz Equations: Bifurcations, Chaos, and Strange Attractors*, Springer–Verlag, New York–Berlin–Heidelberg, 1982.

[Ste89] Stewart, I., *Does God Play Dice?, The Mathematics of Chaos*, Blackwell, Cambridge, MA, 1989.

[Str78] Straffin, P. D., *Periodic orbits of continuous functions*, Math. Mag. **51** (1978), 99–105.

[Str94] Strogatz, S., *Nonlinear Dynamics and Chaos*, Addison–Wesley Publ. Co., Reading MA, 1994.

[Sus92] Sussman, J. and J. Wisdom, *Chaotic evolution of the solar system*, Science **257** (1992), 56–62.

[Tau01] Taubes, C., *Modeling Differential Equtions in Biology*, Prentice Hall, Englewood Cliffs, NJ, 2001.

[Tho01] Thomson, B., J. Bruckner, and A. Bruckner, *Elementary Real Analysis*, Prentice Hall, Englewood Cliffs, NJ, 2001.

[Thu01] H. Thunberg, *Periodicity versus chaos in one-dimensional dynamics*, SIAM Review **43** (2001), 3–30.

[Tuc99] Tucker, W., *The Lorenz system exists*, C. R. Acad. Sci. Paris Sér. I Math. **328** (1999), 1197–1202.

[Ued92] Ueda, Y., *Strange attractors and the origin of chaos*, Nonlinear Science Today **2** (1992), 1–16.

[Ued73] Ueda, Y. et al., *Computer simulation of nonlinear ordinary differential equations and nonperiodic oscillations*, Trans. IECE Japan **56-A** (1973), 218–225, (English translation) Scripta, pp. 27-34.

[Vol31] Volterra, V., *A Mathematical Theory of the Struggle for Life*, Gauthier–Villars, Paris, 1931.

[Wad00] Wade, W., *An Introduction of Analysis, Second Edition*, Prentice Hall, Englewood Cliffs, NJ, 2000.

[Wal83] Waltman, P., *Competition Models in Population Biology*, Soc. for Indust. and Applied Math., Philadelphia, 1983.

[Wil02] Williams, A., *Asymptotic Stability of Nonsymmetirc Neural Networks by Sink Symmetrization*, Ph.D. thesis, Northwestern University, Evanston, Illinois, 2002.

[Xia92a] Xia, Z., *Melnikov method and transversl homoclinic points in the restricted three-body problem*, J. Differential Equations **96** (1992), 170–184.

[Xia92b] ———, *The existence of noncollision singularities in Newtonian systems*, Annals of Mathematics **135** (1992), 411–468.

Index

C^1, 367, 705
C^∞, 167, 367, 558, 705, 706
C^r, 167, 367, 558, 705, 706
$D\mathbf{F}_{(\mathbf{x})}$, 137
$L^{-1}(C)$, 174
$O(x^k)$, 89, 394
$W^s(\mathbf{0})$, 128
$W^s(\mathbf{x}^*)$, 115
$W^u(\mathbf{x}^*)$, 115
Fix(f), 356
Per(n, f), 356
Σ_2, 603, 604
Σ_N^+, 442
$\Sigma_{\mathcal{G}}^+$, 468
$\chi_\mathbf{S}(q)$, 516
$\in$, 708
$\notin$, 708
$\overline{\mathbf{B}}(\mathbf{p}, r)$, 708
$\phi(t; \mathbf{x}_0)$, 80
$\mathbf{B}(\mathbf{p}, r)$, 708
$d(\mathbf{p}, \mathbf{S})$, 710
f^n, 354
$\mathbf{B}(\mathbf{p}, r)$, 286
$\mathbf{u}^j$, 82
$\mathcal{O}_f^+(x_0)$, 355
$\mathcal{O}_\mathbf{F}(\mathbf{x}_0)$, 565

affinely conjugate, 407
Alekseev, V. M., xix
Alligood, K., 513, 660
allowable string, 471
allowable symbol sequence, 468
α-limit set, 110, 565

Andronov–Hopf bifurcation, 213, 235
Anosov diffeomorphism, 575
Anosov, D.V., 575
aperiodic, 7, 521, 581
    stochastic matrix, 581
    transition matrix, 521
Arnold, V., xix
asymptotically stable, 115, 368, 556
    globally, 115, 128
    orbitally, 214
atmospheric convection, 335
attracting, 6, 556
    fixed point, 115, 119, 123
    period-n point, 368
    periodic orbit, 214, 256
attracting set, 287, 311, 640
attractor, 287, 311, 492, 640
    chaotic, 293, 311, 492, 513, 641, 660
    Hénon, 646
    Lorenz, 304, 306
    Milnor, 289, 494
    Rössler, 313
    solenoid, 642
    test for chaotic, 329, 513, 660
    transitive, 287, 492, 640

ball, 286
    closed, 708
    open, 286, 708
basic invariant set, 647, 665
basic invariant sets, 665
basin of attraction, 115, 128, 136, 139, 184, 366, 494, 571, 640

727

basis, 15, 544, 714
beetle, flour, 349, 586
Belickii Theorem, 596
Belousov–Zhabotinsky chemical
    reaction, xix, 263
Bendixson criterion, 248
Benedicks, M., 646
bifurcation, 232, 391
    Andronov–Hopf, 235
    homoclinic, 245
    period doubling, 397
    saddle-node, 392
    subcritical, 233
    supercritical, 233
    tangential, 392
    value, 232
bifurcation diagram, 392
bifurcation value, 391
binary expansion, 437
Birkhoff Transitivity Theorem, 539
Birkhoff, G. D., xviii, xix, 539
blood cells, 415, 536
Bonatti, C., 718
Borel measure, 515
Borel set, 515
boundary, 286, 481, 492, 708
bounded, 286, 709
bounding function, 191, 219, 222, 242
Bowen, R., 622
box dimension, 671
Brusselator, 224
butterfly effect, xx, 291, 453
BZ chemical reaction, 263

Cantor set, 316, 457, 482, 711
    non-symmetric, 676, 685
    positive measure, 679
capacity, 671
capital accumulation, 412, 534
Carleson, L., 646
Castillo–Chávez, C., 589
Cauchy sequence, 712
celestial mechanics, xix
center
    linear, 552
    nonlinear, 170, 180
chain recurrent, 288, 493, 665
chaotic, xvii, 7, 470
chaotic attractor, 293, 492, 641
    test, 329, 513, 660
chaotic invariant set, 492, 641
characteristic equation, 4, 543, 713

characteristic function, 516
characteristic multipliers, 256
chemostat, 152, 154
closed, 286, 481, 492, 709
closed ball, 708
closed orbit, 214
closure, 286, 492, 709
Cobb–Douglas production function,
    346, 412, 534
cobweb method of iteration, 365
compact, 286, 710, 712
compartmental model, 53
competitive market, 134
competitive populations, 145
complete, 707, 712
complex eigenvalues, 29
conjugacy, 407
conjugate, 407, 595
    C^r, 167, 595
Conley, C., 288, 665
connected, 710
conservative systems, 173
constant coefficients, 12
continuous, 705, 711
continuously differentiable, 367, 705
contraction, 385
    constant, 686
correctly aligned M-rectangles, 613
correlation dimension, 681
Coullet, P., xix, 400
countable, 457
countably additive, 515
Cournot duopoly model, 58
critical point, 196, 386
cross section, 217
cutting, 454

damped harmonic oscillator, 3
damping, 183
degenerate stable node, 38
    linear map, 550
degenerate unstable node
    linear map, 550
dense, 710
dense orbit, 439
dense periodic points, 447
density function, 517
derivative, 82, 137, 557, 706
determinant, 713
diabetes, 54
diagonal, 358
diagonal matrix, 714

diagonalizable, 714
diffeomorphism, 167, 564
differentiably conjugate, 407
dimension
  box, 671
  correlation, 681
  fractal, 670
  Hausdorff, 670
  Lyapunov, 682
  similarity, 678
direction fields, 129
dissipative, 183
distance, 712
distance from a point to a set, 710
divergence, 243
double-well potential, 174
doubling map, 337, 356, 439
doubling the periods, 433, 436
Duffing equation, 200
  forced, 317
Dulac criterion, 248
dynamical systems, xviii

eigenvalue, 543, 713
eigenvalues of a fixed point, 137
eigenvector, 543
electric circuit, 264
elliptic center, 30
energy, 173
epidemic model, 156, 158, 589
ϵ-chain, 493, 631, 638, 665
equilibrium point, 4, 81, 120
equivalent flows, 167
Euler method, 85
  higher dimensions, 91
eventually positive stochastic matrix, 581
evolutionary game theory, 202
expanding factor, 467
expanding, piecewise, 467

Feigenbaum constant, 402
Feigenbaum, M., xix, 400
Fibonacci recurrence relation, 472, 474
first return map, 216
first variation equation, 82, 102
fixed point, 4, 81, 119, 120, 354
fixed point for iteration, 356
flour beetle, 349, 586
flow, 77, 80
folding, 454
food chain, 190

forced Duffing equation, 317
forced oscillator, 317
forward orbit, 355
fractal, 669
fractal dimension, 670
frequency measure, 524
frequency plot, 528
friction, 183
full shift on N-symbols, 442
full shift on two symbols, 603
full shift space on two symbols, 637
full two-sided shift on finite number of symbols, 622
fundamental domain, 421
fundamental matrix solution, 15
fundamental set of solutions, 15

game theory, evolutionary, 202
generalized eigenvector, 714
generalized eigenvectors, 68
generic, 717
geometric horseshoe, 598
geometric series, 437
Gleick, J., xx
globally asymptotically stable, 115, 128
gradient, 196
gradient system of differential equations, 196
graph of a function, 358
graphical method of iteration, 365
Grassberger, P., 680
Grobman–Hartman Theorem, 596
Gronwall's inequality, 101

Hamiltonian differential equations, xix
hard spring, 200
harmonic oscillator
  coupled, 45
  damped, 49
  uncoupled, 44
Hartman Theorem, 596
harvesting, 125
Hausdorff dimension, 670
Hayashi, S., 718
Hénon attractor, 350
Hénon map, 350, 561, 566, 571, 624, 636, 646, 649, 653, 675, 681, 682
Hénon, M., 350
Herman, M., xix
heteroclinic orbit, 178, 179
Heun method, 86
  higher dimensions, 92

Hirsch, M., 264
histogram, 528
homeomorphism, 167, 406, 564
homoclinic bifurcation, 213, 245
homoclinic orbit, 177
homoclinic point, 636
Hopf bifurcation, 235
hyperbolic, 559
hyperbolic fixed point, 137
hyperbolic periodic orbit, 256
hyperbolic toral automorphism, 575
hyperbolicity, 612, 632

IFS, 685
image, 564
improved Euler method, 86
   higher dimensions, 92
inflation, 54
initial condition, 13
input-output, 55
integral of motion, 4
interior, 286, 424, 492, 709
invariant, 112
   negatively, 112
   positively, 112, 379, 439
invariant measure, 517
invariant set, 439
   basic, 647, 665
inverse, 564
irrational rotation, 660
irreducible, 581
irreducible transition graph, 468
irreducible transition matrix, 472
irreducible word, 429
isocline, 126
isolated invariant set, 640
isolating neighborhood, 640
iterate, 256
iterated-function system, 685
   with probabilities, 692
iterates of a function, 354
itinerary, 426, 430
   map, 444, 603, 629

Jordan canonical form, 67, 715

Kaplan, J., 680, 682
Keynesian IS-LM model, 58
kinetic energy, 173
King Oscar's prize, 662
Koch curve, 687
Kolmogorov, A.N., xix, 671

Kupka–Smale property, 720

L-stable, 115, 368, 555
   orbitally, 214
Lanford, O.E., 402
Lasota, A., 530
least period, 356
Lebesgue measure, 516
length of an interval, 443
Leontief, 55
level set, 174
Li, T.Y., xviii, 293, 424, 493, 498, 530
Lienard equation, 229, 265
limit cycle, 215
limit set
   α-limit set, 110, 565
   ω-limit set, 110, 488
linear combination, 14, 714
linear transformation, 541
linearized stability, 557
linearized system, 137
linearly conjugate, 407
linearly independent, 544, 714
linearly independent set of solutions, 15
Liouville formula, 15, 63, 248, 282, 325
Lipschitz, 97
local diffeomorphism, 595
local stable manifold, 138, 570, 608
local unstable manifold, 139
logistic differential equation, 120
logistic equation, 79
logistic function, 345, 363
Lorenz differential equations, 297
Lorenz, E., xix, 291, 331
Lotka, A.J., xix
Lotka–Volterra equations, 145, 169, 201, 202, 269
Lyapunov dimension, 682
Lyapunov exponent, 321
Lyapunov function, 186, 210
   weak, 186
Lyapunov stable, 115, 368, 555

M-rectangle, 613
Mandelbrot, B., 669
manifold, 139, 567, 712
   local stable, 138
   local unstable, 139
   stable, 138, 570
   unstable, 138
Maple, 25, 89, 129, 177
Markov chain, 351, 580

Markov partition, 466, 621
Markov rectangles, 621
Mathematica, 25, 89, 129, 177
Mather, J., 663
Matlab, 25, 89, 129, 177
matrix of partial derivatives, 82, 137, 557, 706
May, R., xix, 345
McGehee, R., 663
measure, 515
  Borel, 515
  frequency, 524
  Lebesgue, 516
  natural, 527, 656
measure preserving, 517
metastasis of tumor cells, 52
metric, 442
metric space, 689, 712
middle-third Cantor set, 457
Milnor attractor, 289, 494
Milnor, J., 289, 494
mod, 70, 254, 356
modulo, 70, 254, 356
Moser, J., xix, 662

natural measure, 527, 656
negatively invariant, 112
neighborhood, 524, 709
neural network, 208
Newton map, 348, 373, 585
Newton method for roots, 347
Newton, I., xvii
node, 26
nonhomogeneous linear system, 49
nonlinear center, 170, 180
nonrectifiable, 688
nonresonance, 596
norm of a matrix, 61, 554, 716
nowhere dense, 482, 710
nullclines, 126
numerical methods, 84

ω-limit set, 110, 488
one degree of freedom, 173
one to one, 406, 564
onto, 406
open, 286, 492, 709
open ball, 286, 555, 708
orbit, 80, 355, 565
  forward, 355
  periodic, 356
orbitally ω-attracting, 214

orbitally asymptotically stable, 214, 262
orbitally L-stable, 214, 262
Oregonator system, 263
oscillator, 173
  coupled harmonic, 45
  forced, 317
  uncoupled harmonic, 44
  undamped nonlinear, 200
Oseledec Multiplicative Ergodic Theorem, 656
Ottino, J., 537

partition, 428
pendulum, 177
  with damping, 184
perfect, 482, 711
period, 5, 81, 214, 356
period doubling bifurcation, 397
period doubling cascade, 399
period-n point for iteration, 356
periodic, 5, 214
periodic orbit, 5, 81, 214
  stable, 256
periodic point, 356
periodic sink, 214, 556
periodic source, 215
permutation matrix, 472
Perron–Frobenius operator, 529
Perron–Frobenius theorem, 520
phase plane, 23
phase portrait, 23, 126, 129
phase space, 23
Picard iteration scheme, 77, 99
piecewise expanding, 467
Poincaré map, 216, 217, 231, 251, 255, 273
  Lorenz equations, 307
Poincaré, H., xvii, 636, 662
Poincaré–Bendixson theorem, 219
populations, 150, 412, 477
  competitive, 145
  epidemic, 156, 158, 589
  food chain, 190
  Hassel model, 413
  predator–prey, 169, 241, 265
  Ricker model, 413, 416, 477
  SIR model, 156, 158
  SIS model, 589
  Verhulst model, 412
positively invariant, 112, 379, 439
potential energy, 173
predator–prey system, 169, 241, 265

principal Lyapunov exponents, 329
probability transition matrix, 581
Procaccia, I., 680
pseudo-orbit, 631, 665
Pugh, C., 718

quasiperiodic, 45, 52, 295, 330, 660
quasiperiodic function, 48

rationally independent, 47
reducible word, 429
repeated eigenvalues, 35
repelling
    fixed point, 119, 123
    periodic point, 368, 556, 565
repelling fixed point, 116
repelling periodic point, 556, 565
replicator system of differential
        equations, 203
Ricker model, 416
rooftop map, 490
Rössler attractor, 313
Rössler, O., xx
rotary solutions, 179
Ruelle, D., xviii, 316
Runge–Kutta method, 87
    higher dimensions, 93

saddle, 24
    linear map, 547
saddle fixed point, 137
saddle periodic point, 559
Sauer, T., 513, 660
scaling dimension, 678
Schwarzian derivative, 388
second-order scalar equations, 41
self-excited oscillator, 229
self-similar, 669, 684
semiconjugacy, 407
semistable, 123, 215, 368
sensitive dependence on initial
        conditions, xx, 8, 291, 452, 641
    at points in a set, 291, 452
    when restricted to a set, 291, 452
separation of variables, 77
separatrix, 567
set difference, 708
shadowed, 631
Sharkovskii ordering, 431
Sharkovskii, A. N., 424
shed map, 465
shift map, 442, 604

shift space, 468, 604
    full two-sided, 622
Siegel, C. L., xix
Sierpinski gasket, 686
similarity, 686
similarity dimension, 678
Sinai, Ya., 622
Sinai–Ruelle–Bowen measure, 656
Singer, D., 386
sink, 115, 368
    linear map, 553
SIR model, 156, 158
SIS model, 589
Sitnikov, K., xix, 662
six-twelve potential, 200
Smale horseshoe, 598
Smale, S., xviii, 575, 598, 662, 717
soft spring, 200
solution of a linear equation, 13
source, 116, 368, 556
    linear map, 553
sphere, 709
stable
    periodic orbit, 256
stable eigenspace, 573
stable fixed point, 115
stable focus, 32, 552
stable manifold, 115, 135, 138, 570
stable node, 26
    linear map, 545
stable subspace, 139
stair step method of iteration, 365
Sternberg Theorem, 596
Stewart, I., xx
stochastic matrix, 581
strange attractor, xviii, 316
stretching, 454
stretching factor, 467
string, 430
strongly attracting, 187
subcritical bifurcation, 233
subshift of finite type, 465, 468, 471,
        622, 637
superattracting, 369, 586
supercritical bifurcation, 233
support of a measure, 515
symbol space, 442, 603, 604
symbolic dynamics, 423
symbols, 428

Takens, F., xviii, 316
tangent space to a manifold, 719

tangent vectors to a manifold, 719
tent map, 316, 359, 443
tent map of slope r, 384, 455
ternary expansion, 438
test function, 191, 219
time plot of the solution, 22
time-dependent
    differential equation, 85, 91, 257, 317, 352
    linear differential equation, 49, 59, 83, 103, 165, 248, 282, 324
topological conjugacy, 407
topological Markov chain, 471
topologically conjugate, 167, 407, 595
topologically equivalent, 167
topologically transitive, 287, 439, 492
totally disconnected, 711
trace, 63, 713
trajectory, 80
transition graph, 428
    irreducible, 468
transition matrix, 471
transitive, 287, 439, 492
transpose, 12, 713
transversal, 217, 231, 254
transverse, 636
transverse manifolds, 719
trapping region, 286, 304, 492, 640
Tresser, C., xix, 400
Tucker, W., xx, 297

Ueda, Y., xx
unemployment, 54
uniformly hyperbolic, 682
unstable, 115
    periodic orbit, 256
unstable eigenspace, 573
unstable focus, 552
unstable manifold, 115, 135, 138, 570
unstable node, 27
    linear map, 547
unstable periodic point, 368, 555
unstable subspace, 139

Van der Pol equation, 229, 264
Variation of parameters, 50
vector field for the system of equations, 95, 126
Volterra, V., xix
volume change, 247, 281

waterwheel model of Lorenz equations, 332
weakly attracting, 123, 187
weakly repelling fixed point, 124
Williams, R., 311, 495
word, 430
Wronskian, 15

Xia, Z., 663

Yakubu, A.A., 589
Yorke, J., xviii, 293, 424, 493, 498, 513, 530, 660, 680, 682

Published Titles in This Series

19 R. Clark Robinson, An Introduction to Dynamical Systems: Continuous and Discrete, Second Edition, 2012
18 Joseph L. Taylor, Foundations of Analysis, 2012
17 Peter Duren, Invitation to Classical Analysis, 2012
16 Joseph L. Taylor, Complex Variables, 2011
15 Mark A. Pinsky, Partial Differential Equations and Boundary-Value Problems with Applications, Third Edition, 1998
14 Michael E. Taylor, Introduction to Differential Equations, 2011
13 Randall Pruim, Foundations and Applications of Statistics, 2011
12 John P. D'Angelo, An Introduction to Complex Analysis and Geometry, 2010
11 Mark R. Sepanski, Algebra, 2010
10 Sue E. Goodman, Beginning Topology, 2005
9 Ronald Solomon, Abstract Algebra, 2003
8 I. Martin Isaacs, Geometry for College Students, 2001
7 Victor Goodman and Joseph Stampfli, The Mathematics of Finance, 2001
6 Michael A. Bean, Probability: The Science of Uncertainty, 2001
5 Patrick M. Fitzpatrick, Advanced Calculus, Second Edition, 2006
4 Gerald B. Folland, Fourier Analysis and Its Applications, 1992
3 Bettina Richmond and Thomas Richmond, A Discrete Transition to Advanced Mathematics, 2004
2 David Kincaid and Ward Cheney, Numerical Analysis: Mathematics of Scientific Computing, Third Edition, 2002
1 Edward D. Gaughan, Introduction to Analysis, Fifth Edition, 1998